The Caledonide Orogen—
Scandinavia and Related Areas

INTERNATIONAL GEOLOGICAL CORRELATION PROGRAMME

A contribution to Project No. 27
THE CALEDONIDE OROGEN (IGCP-CO)

Asymmetric folds (F$_2$) in migmatites, Galten Peninsula, Sørøy, north Norway

The Caledonide Orogen— Scandinavia and Related Areas

Part 2

Edited by

D. G. GEE

Geological Survey of Sweden, Uppsala, Sweden

and

B. A. STURT

Geological Institute, University of Bergen, Norway

A Wiley–Interscience Publication

JOHN WILEY & SONS

Chichester · New York · Brisbane · Toronto · Singapore

Library of Congress Cataloging in Publication Data:

Main entry under title:
The Caledonide orogen
  'A Wiley–Interscience publication.'
  Includes index.
  1. Geology, Stratigraphic—Paleozoic—Addresses,
essays, lectures. 2. Orogeny—Scandinavia—Addresses,
essays, lectures. I. Gee, D. G. II. Sturt, B. A.
QE654.C27 1985   551.7′2   84–17275

ISBN 0 471 10504 X
ISBN 0 471 90821 5 (Pt. 1)
ISBN 0 471 90822 3 (Pt. 2)

British Library Cataloguing in Publication Data:

The Caledonide orogen—Scandinavia and
  related areas.
  1. Geology, Stratigraphic—Paleozoic
  2. Geology—Scandinavia
  I. Gee, D. G.   II. Sturt, B. A.
  551   QE654

ISBN 0 471 10504 X
ISBN 0 471 90821 5 (Pt. 1)
ISBN 0 471 90822 3 (Pt. 2)

Typeset by Spire Print Services Ltd., Salisbury, Wilts.
Printed and bound in Great Britain

Contents

Part 2

IGNEOUS ACTIVITY

METAMORPHISM

TECTONIC EVOLUTION

RELATED CALEDONIAN AREAS

CONTENTS

POCKET containing Maps 1-3 and
Fig.1(Key) plus Figs.1-3.
(7 items)

Preface

The need for a comprehensive integrated review of the Scandinavian Caledonides has been apparent for many years and has become particularly pressing during the last decade. Since the late 1960s the impact of the plate tectonic's paradigm has radically changed much of our thinking about the orogen and hence the character of Caledonide research. For over a hundred years it has been known that the structure of the Scandinavian mountains is dominated by easterly transported nappes. The more recent recognition that the higher thrust sheets were emplaced many hundreds of kilometres onto the Baltoscandian Platform from terranes that lay outboard of continent Baltica during the earliest Phanerozoic has significantly influenced our understanding of orogenic processes; it has provided a framework for many interdisciplinary studies devoted to interpreting the origin of the displaced terranes. The identification during the early-mid 1970s of a variety of allochthonous ophiolites and ensimatic arc assemblages has provided further evidence in support of the faunal and palaeomagnetic basis for inferring the existence of a major ocean basin, Iapetus, during the Early Palaeozoic. Baltoscandian miogeoclinal associations have been recognized and shown to pass oceanwards through 'transitional' environments into oceanic terranes. Polycyclic deformation and metamorphism of these eugeoclinal and miogeoclinal terranes commenced in the Early Ordovician, perhaps even in the Late Cambrian, and continued into the Silurian, culminating in the Mid Palaeozoic with the final elimination of the Iapetus Ocean during collision of Baltica and Laurentia.

This review of the Scandinavian Caledonides, set in its North Atlantic context, was planned within the auspices of the International Geological Correlation Programme Project No. 27 The Caledonide Orogen (IGCP-CO). A framework of solicited contributions provides the regional and stratigraphic reviews both of Scandinavia and of the related Caledonide areas and ensures comprehensive cover. In addition, we have included a large number of papers covering the range of ongoing research in the Scandinavian mountains. Thus this book provides both an overview of the regional geology and an insight into current research. A meeting devoted to the presentation of these papers was organized in Uppsala, Sweden, in August 1981 (Abstracts in Terra Cognita, Vol. 1, pp. 27–87). This, the Uppsala Caledonide Symposium (UCS), was attended by over 300 earth scientists; it provided the participants with both a forum for presenting and discussing their papers, and an opportunity to examine aspects of the orogen directly by participating in the variety of excursions (15) arranged before and after the meeting (Conference report in Terra Cognita, Vol. 2, pp. 89–96). In thanking all those who participated in UCS we want to emphasize that this volume would have been impossible without their presence, their contributions and their continued interest.

The Uppsala Caledonide Symposium was sponsored by the Swedish Ministry of Education, the Swedish Natural Science Research Council (NFR), the Swedish National Committee for IGCP, the Swedish

Geodynamics Project and the Geological Survey of Sweden (SGU). Excursions in Norway were supported by the Norwegian Research Council for Science and the Humanities (NAVF), the Norwegian Committee for IGCP, the Geological Survey of Norway (NGU), Bergen and Oslo Universities and the following oil and mining companies: Statoil, Norsk Hydro, Esso Norway, British Petroleum Norway, Amoco Norway, Els Acquitaine, A/S Sulfidmalm (Falconbridge Ltd.).

The Uppsala Caledonide Symposium was held simultaneously and in collaboration with the 8th annual meeting of the European Geophysical Society (EGS). We thank Dr C.-E. Lund and his EGS organizing committee for excellent collaboration and the integration of many of the UCS-EGS activities. These two meetings together were sponsored by Uppsala University, Uppsala Town Council, the Boliden Mining Company, Selection Trust Ltd. (subsidiary of BP Ltd.) and OK Union.

Although the sixty-two contributions concerning the geology of the Scandinavian Caledonides dominate this book, the regional context of this part of the Caledonian-Appalachian foldbelt is well represented in the twenty-two other contributions, mainly from the North Atlantic area, but also from the USSR and Sardinia. The Scandinavian contributions have been organized along the lines of the IGCP-CO sub-projects; regional reviews are followed by groups of papers concerning stratigraphy, structure, igneous rocks and metamorphism, all with introductory chapters. A final section includes various interpretations of the general tectonic evolution of the mountain belt. A variety of other contributions presented at UCS which were considered to be outside the scope of this volume were published in 1983 in the *Geological Journal* and *Lithos*.

We, the editors, are most indebted to all who have contributed papers to the book, allowing it to be the comprehensive volume originally planned. Some authors were quick to present their manuscripts after UCS; others were tardy. To the former we offer the apologies of the latter. We are indebted to our referees, mostly anonymous, selected from all over the Caledonian–Appalachian orogen and elsewhere for their painstaking and constructive criticism of the papers.

The IGCP-CO sub project representatives in Norway and Sweden helped us with the initial assessments within their particular fields of interest. In addition, we have had support from many colleagues at the Geological Survey of Sweden and the Department of Geology in Bergen.

The late Kirsten Spear was of particular help in Uppsala with the organization of the volume and communication with our authors. Much of the final preparation of the book was done during a visit by one of us (DGG) to the Department of Geological Sciences at VPISU in Blacksburg (Virginia, USA) and we thank Prof. L. Glover III and Prof. D. R. Wones for their support. We received invaluable help with the handling of the manuscripts, particularly the proof correcting, from Dr E. R. Gee and Mrs V. Heintz Gee. The subject index was compiled at the Department of Geology in Bergen by Dr H. Brekke and Ø. Nordgulen.

A particular feature of the book is Map 1, the 1 : 2 M Tectonostratigraphic map of the Scandinavian Caledonides. The map arose out of the needs of both IGCP-CO and its sister project IGCP-CCSS (Correlation of Caledonian Stratabound Sulphides) for a common Norwegian–Swedish geological base map, the latter being a prerequisite for the many sub-project compilations of data concerning stratigraphy, sedimentation, plutonism, volcanism, metamorphism, mineralization, time of deformation and basement character being prepared by the projects. The map was initially drawn in Uppsala and presented at a scale of 1 : 1 M for the UCS meeting, based on a great variety of sources both published and unpublished, the latter referred to in the legend of Map 1. Norwegian data were compiled and edited largely by D. Roberts and A. Thon and Swedish by M. B. Stephens and D. G. Gee. Subsequently, an improved topographic base was obtained from the Swedish Land Survey and the map was redrawn by E. Zachrisson at SGU. During this redrafting many minor errors were eliminated; the positions of some of the rock boundaries were modified after consulting an early edition of the new 1 : 1 M Geological Map of Norway. We are greatly indebted to Ebbe Zachrisson for this very time consuming improvement of the UCS edition. The IGCP-CO and CCSS editions of Map 1 are very similar, but not identical; we are indebted to R. Kumpulainen for help with the preparation of the final IGCP-CO edition for this book. Map 1 is accompanied by 1 : 2 M maps showing Bouguer anomalies (Map 2) and Magnetic anomalies (Map 3) over the entire mountain belt. During drafting of these geophysical maps we received advice from L. Eriksson (SGU), H. Henkel (SGU) and D. Dyrelius (Dept. of Solid Earth Physics, Uppsala).

Throughout this marathon, we have enjoyed excellent collaboration with Dr S. Hemmings and her staff at John Wiley & Sons in Chichester. We thank them for making possible this extensive documentation of the Scandinavian Caledonides and Related Areas.

D. G. GEE
B. A. STURT
Editors

Igneous Activity

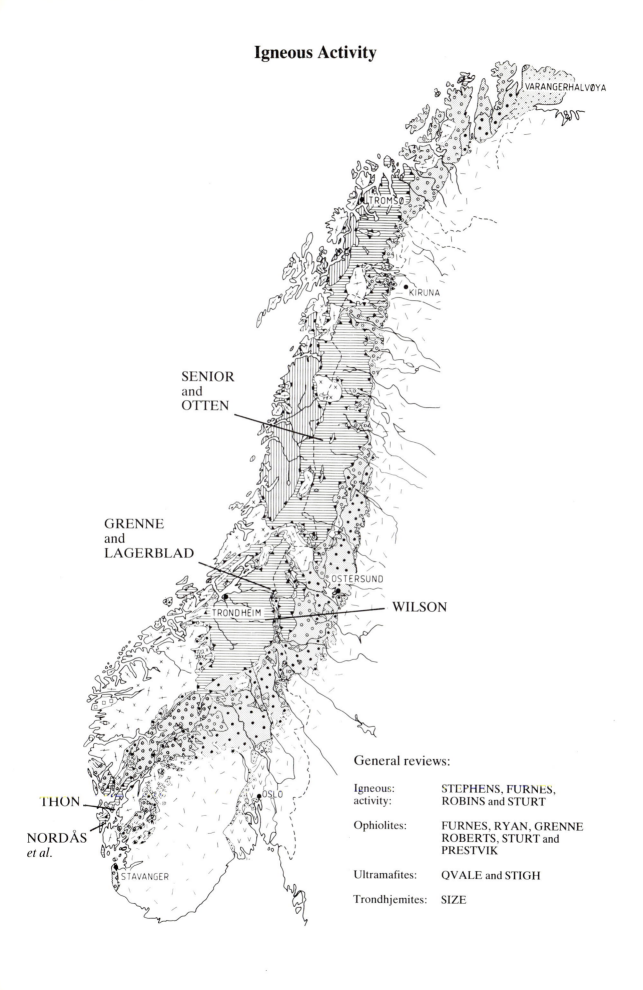

SENIOR
and
OTTEN

GRENNE
and
LAGERBLAD

WILSON

THON

NORDÅS
et al.

VARANGERHALVØYA

TROMSØ

KIRUNA

ØSTERSUND

TRONDHEIM

OSLO

STAVANGER

General reviews:

Igneous: activity:	STEPHENS, FURNES, ROBINS and STURT
Ophiolites:	FURNES, RYAN, GRENNE ROBERTS, STURT and PRESTVIK
Ultramafites:	QVALE and STIGH
Trondhjemites:	SIZE

Igneous activity within the Scandinavian Caledonides

M. B. Stephens[*], **H. Furnes**[†], **B. Robins**[†] and **B. A. Sturt**[*]

[*]*Sveriges geologiska undersökning, Box 670, S–75128 Uppsala, Sweden*

[†]*Geologisk Institutt Avd. A, Universitetet i Bergen, Allégt. 41, N–5014 Bergen, Norway*

ABSTRACT

The igneous rocks of Caledonian age in the Scandinavian Caledonides are concentrated in the upper part of the Middle Allochthon and in higher nappes. An overview is presented of these rocks including remarks on their geochemistry and, in the case of the volcanic and high-level intrusive rocks, an interpretation of their tectonic environments of formation. The volcanic and high-level intrusive rocks are described according to their position in the tectonostratigraphy, different environments being characteristic of successively higher groups of nappes. The synorogenic plutonic rocks are subdivided according to composition and situation in the tectonostratigraphy.

There appears to be evidence in the igneous rocks for the embryonic and youthful stages of ocean (Iapetus) development. Rifting in an ensialic environment related to the opening of Iapetus appears to have occurred after deposition of late Precambrian glaciogenic sediments, while most, if not all, of the ophiolite fragments are pre-late Ordovician in age and one of them (Vassfjellet) is mid-Arenig or older. However, there is some uncertainty as to what extent the ophiolite fragments represent remnants of Iapetus or younger, smaller, marginal basins which opened during the protracted phase of plate convergence and closure of Iapetus. Although there is evidence for igneous activity in the Cambrian and Ordovician related to subduction processes, it is apparent that closure of Iapetus was complex and occurred possibly over a period as long as c. 100 Ma. During the period of plate convergence, different phases of magmatic activity related subduction processes were followed by (Lower and Middle Köli Nappes) or occurred possibly in close temporal relationship with (Støren Nappe) magmatism related to an extensional tectonic régime.

Traces of magmatic evolutionary trends are only present within either a single magmatic province or within limited groups of related nappes. These include: (1) A change upwards from low-K tholeiitic basalts to alkaline basalts in two (Karmøy and Leka) of the ophiolite slices. (2) A development from early tholeiitic magmas through localized high-K calc-alkaline magmas to predominantly alkaline magmas in the Seiland province (Cambrian–early Ordovician). (3) An evolution in both the Lower and Middle Köli Nappes from transitional basalts through bimodal sequences of low-K dacite–rhyodacite and arc-related tholeiitic basalt and basaltic andesite to rifting-related, predominantly tholeiitic basaltic rocks (Ordovician–early Silurian). (4) A change from low-K, arc-related, tholeiitic basalt and basaltic andesite through trondhjemite to granitic rocks in the Upper Köli Nappes and equivalents (Ordovician–Silurian); the granites are spatially associated with gabbro intrusions and show higher K_2O contents and more variable initial $^{87}Sr/^{86}Sr$ ratios than the trondhjemites. (5) An evolution in the Støren Nappe and equivalents from predominantly low-K tholeiitic basalts in ophiolite slices to medium-K, calc-alkaline basalt–rhyolite sequences and either contemporaneous or later plutonic complexes of predominantly quartz diorite to granite extruded/intruded in a convergent plate margin setting (pre-late Arenig–Silurian); the latter is complicated by the eruption of rifting-related, tholeiitic and alkaline basalts in the Ordovician and early Silurian, several of which have been interpreted as forming parts of ophiolites. The concentration of synorogenic granitic rocks of mid to late Ordovician and Silurian age in the highest nappes, in spatial association with synorogenic gabbros and showing variable (0.704–0.721) initial $^{87}Sr/^{86}Sr$ ratios, is thought to reflect increased heat flow during the Ordovician and Silurian in westernmost continental environments.

The data presented here indicate a complex magmatic evolution for the Ordovician and Silurian igneous rocks in the Upper Allochthon. It is suggested that this may be related to the presence of one or more relict oceanic segments and/or the opening of new basins during this period, thus allowing different areas to develop separate magmatic evolutionary trends. Elimination of oceanic segments during Silurian–Devonian continent–continent collision subsequently telescoped together these different sequences.

Introduction

The variably metamorphosed igneous rocks of Caledonian age in the Scandinavian Caledonides occur within a series of flat-lying thrust nappes. Except for bentonites and occasional dykes, igneous activity of Caledonian age is absent in the Autochthon/Parautochthon, Lower and most of the Middle Allochthons (cf. Tectonostratigraphic Map, this volume and Table 1). It is in the upper part of the Middle Allochthon and in higher nappes that Caledonian igneous rocks are present. The nature and geochemical signature of particularly the volcanic and high-level intrusive rocks have been interpreted, in several studies, in terms of different tectonic environments related to the embryonic, youthful and declining stages of Iapetus Ocean development. Silurian–Devonian thrusting related to continent–continent collision of Baltica and Laurentia (Gee 1975) telescoped these environments and emplaced a variety of ensialic and ensimatic igneous suites onto the Baltoscandian platform. In several sequences, this Scandian phase of orogeny (Gee 1975) overprints an earlier, probably Ordovician phase of deformation possibly correlatable with the Finnmarkian orogeny (Sturt et al. 1978) in northern Norway. This earlier phase included obduction of ophiolite fragments and deformation of early arc sequences (Furnes et al. 1980a).

This overview of Caledonian igneous activity is divided into six sections according to the volcanic/high-level intrusive (sections 1–4) or plutonic (sections 5–6) character of the igneous rocks. The volcanic and high-level intrusive rocks are described according to their position in the tectonostratigraphy (Table 1) and the interpretation of their tectonic environments of formation, different environments being represented in successively higher groups of nappes. Those plutonic rocks which are synorogenic and not related to volcanites are subdivided according to composition and position in

the tectonostratigraphy. Using these classification principles, the six sections are concerned respectively with:

1. The rifting-related dolerites and amphibolites in the Middle (Särv Nappe and equivalents) and Upper (Seve Nappes and equivalents) Allochthons, respectively, intruded/extruded in an ensialic environment.
2. The volcanic and subvolcanic rocks related to rifted arcs in the Upper Allochthon (Köli Nappes and equivalents).
3. The fragmentary ophiolite slices and associated within-plate basalts in the Upper Allochthon (including the Støren Nappe) and, possibly, even the Uppermost Allochthon.
4. The mature-arc/active continental margin and continental within-plate volcanic and subvolcanic complexes occurring in the same tectonic unit(s) as the ophiolite slices.
5. Synorogenic basic, ultrabasic, and alkaline intrusive complexes in the Middle Allochthon (Seiland province), the higher nappes of the Upper Allochthon and the Uppermost Allochthon.
6. Synorogenic granitoids concentrated in the higher nappes of the Upper Allochthon and Uppermost Allochthon.

The distribution of Caledonian igneous rocks and place-names referred to in the text are shown in Fig. 1. Certain geochemical characteristics of the volcanic/high-level intrusive complexes are shown in diagrams employing less-mobile elements (Pearce and Cann 1973; Pearce 1975; Miyashiro 1975; Winchester and Floyd 1976). In particular, these help to distinguish alkaline or subalkaline character (TiO_2–Zr/P_2O_5), tholeiitic or calc-alkaline character (TiO_2–FeO^t/MgO) and tectonic environmental affinities (Ti–Zr–Y and Ti–Cr). REE data for the ophiolitic rocks in Norway are reported in a separate paper (Furnes et al., this volume). All Rb–Sr isochron ages quoted in this overview are recalculated to

Table 1 Tectonostratigraphic scheme for the Scandinavian Caledonides based on that proposed for the geological map accompanying this volume. In the scheme adopted here, the Støren Nappe is placed above the Gula and Meråker Nappes (see Gee and Zachrisson 1974; compare Roberts *et al.* 1970 and Wolff and Roberts 1980)

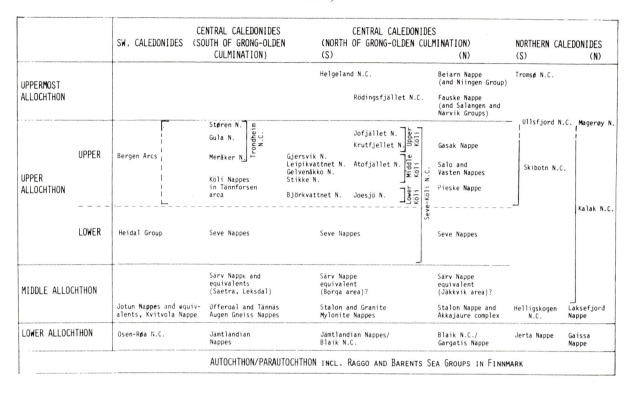

$\lambda^{87}Rb = 1.42 \times 10^{-11}a^{-1}$. The abbreviation MORB used here refers to mid-ocean ridge basalt showing no enrichment in LIL and Ti-group elements, i.e. N-MORB of Sun and Nesbitt (1978) and Wood *et al.* (1979).

This contribution excludes discussion of both the pre-Caledonian igneous rocks occurring within the spatial confines of the orogen and alpine-type ultramafites, the latter being described in a separate paper (Qvale and Stigh this volume). Development of a plate-tectonic model or models is also outside the scope of this paper, although the data discussed here impose constraints on such models. The aim of this contribution is to provide an assessment of Caledonian igneous activity in a tectonostratigraphic context as a basis for a discussion of the magmatic evolution in time and space.

Description of Caledonian Igneous Activity

The Autochthon/Parautochthon, Lower Allochthon, and most of the Middle Allochthon are conspicuous for the virtual absence of Caledonian igneous rocks. A boudinaged granitic dyke (Ekrhovda granite) on the island of Sotra near Bergen (Parautochthon or

Lower Allochthon) has yielded a Rb–Sr whole-rock isochron indicating an age of 463 ± 31 Ma (Sturt *et al.* 1975). Cross-cutting, muscovite-bearing pegmatites are reported in the Kristiansund area of the Western Gneiss Region (Parautochthon or Lower Allochthon) and have been dated by a Rb–Sr whole-rock isochron to 387 ± 7 Ma (Pidgeon and Råheim 1972). In northernmost Norway, basaltic dykes of probable Caledonian age (*c.* 640 Ma, K–Ar whole-rock, Beckinsale *et al.* 1975) occur in the Raggo and Barents Sea Groups within the Autochthon (Siedlecka and Siedlecki 1967; Gayer and Roberts 1973; Roberts 1975a) and the Laksefjord Nappe in the Middle Allochthon; very occasional dolerites have also been reported from the Gaissa Nappe of the Lower Allochthon. The low-grade sediments in these units consist of sandstone-shale (phyllite) sequences derived from a variety of alluvial, coastal and deeper-marine environments as well as glaciogenic sediments in the case of the Gaissa and Laksefjord Nappes. Dyke geochemistries (Roberts 1975a; Gayer and Humphreys 1981) indicate a tholeiitic character and MORB affinity. The only other indication of Caledonian igneous activity in the lower tectonic units is the occurrence of Caradoc (L. Karis personal communication 1979) and Llandovery (Thorslund 1948) bentonites in the

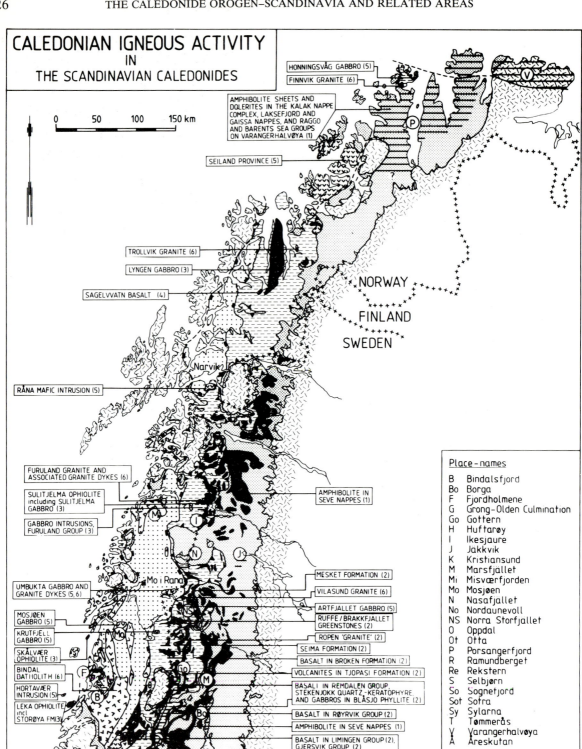

Fig. 1   Distribution of Caledonian igneous rocks in the Scandinavian Caledonides and location of place-names referred to in the text. Tectonostratigraphic base taken from that proposed for the geological map accompanying this volume

Lower Allochthon of central Jämtland (west of Östersund), Sweden. The remainder of this section describes the Caledonian igneous rocks in the upper part of the Middle Allochthon and higher tectonic units according to the scheme outlined in the introduction.

reduce

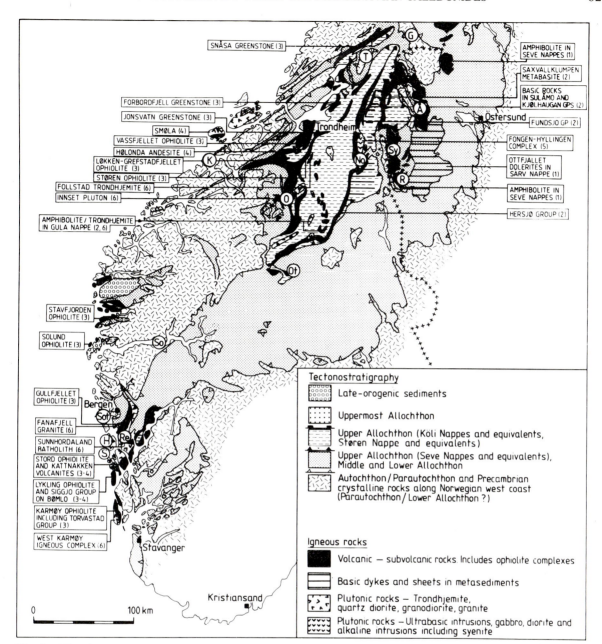

Fig. 1   (Cont.)

1. Rifting-related Ensialic Igneous Activity in the Middle and Upper Allochthons

Ottfjället Dolerites and equivalents (Middle Allochthon)

Basaltic dykes varying in frequency from isolated occurrences to swarms and intruding successions dominated by feldspathic sandstone are a significant feature of the higher tectonic levels in the group of nappes referred to as the Middle Allochthon. The most extensively studied unit is the Särv Nappe (Strömberg 1955, 1961) from the Jämtland area of the Swedish Caledonides.

In the Särv Nappe, basaltic dykes (Ottfjället Dolerites, Holmqvist 1894) cut an unfossiliferous sequence of alluvial and shallow-marine feldspathic sandstones, dolomites and glaciogenic sediments (Röshoff 1975; Kumpulainen 1980). In high-strain zones within and near the base of the nappe, dyke–sediment angles diminish and concordant geometries dominate; this feature is associated with the development of a penetrative fabric and retrogression of original assemblages to paragenesis in the greenschist or lower amphibolite facies. A Rb–Sr whole-rock isochron indicates intrusion at 720 ± 265 Ma (Claesson 1976). K–Ar and $^{40}Ar/^{39}Ar$ investigations indicate the presence of

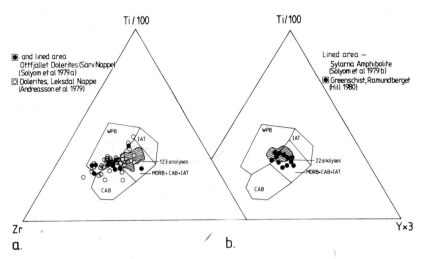

Fig. 2 Ti–Zr–Y plot (after Pearce and Cann 1973) for dolerites in the Särv Nappe and equivalents (a) and for metabasic rocks in the Seve Nappes (b). MORB = Mid-ocean ridge basalt. CAB = Calc-alkaline basalt. IAT = Island-arc tholeiite. WPB = Within-plate basalt

excess Ar and an intrusion age of 665 ± 10 Ma (Claesson and Roddick 1983).

According to Solyom *et al.* (1979a), the predominant dyke lithology is a plagioclase–phyric dolerite; some dykes contain pyroxene and olivine phenocrysts and aphyric varieties also exist. Retrogressed dolerites, occurring particularly near the base of the nappe, contain amphibole, serpentine, epidote, chlorite and biotite. The bulk of the dolerites have a restricted chemical range between quartz- and olivine-bearing tholeiite in the normative ne–ol–di–hy–qz diagram. Fractionation trends and Ti–Zr–Y distribution patterns (Fig. 2) are similar to MORB. A subordinate group of samples contains primary biotite and shows anomalously high TiO_2, K_2O, P_2O_5, Rb, Zr and Nb values; these dolerites show an alkali basaltic character and within-plate basalt (WPB) affinity. Intrusion of the Ottfjället Dolerites is thought to have been associated with continental break-up and the initial stages of Iapetus Ocean development (Gee 1975; Solyom *et al.* 1979a).

Metamorphosed dolerites intruding sandstones within tectonic units which may be correlated with the Särv Nappe have been recognized in the Oppdal area of western Norway (Krill 1980) and the Tømmerås area, northeast of Trondheim (Gee 1977; Kautsky 1978). A Rb–Sr whole-rock isochron age of 745 ± 37 Ma from unfoliated amphibolite dykes preserving primary mineralogies and textures in the Saetra Nappe, in the Oppdal area (Krill 1980), is in good agreement with Claesson's (1976) radiometric age determinations for the Ottfjället Dolerites. The dykes within the upper part of the Leksdal Nappe (Wolff 1979), in the Tømmerås area (Fig. 2), can be divided into tholeiitic and

alkaline types and show mixed MORB/WPB geochemical affinities (Andreasson *et al.* 1979) Further north in Sweden, isolated tectonic lenses of possible Särv Nappe equivalents containing metadolerites and metasandstone sequences occur in the Borga area near the Jämtland–Västerbotten border (Kulling in Strand and Kulling 1972) and near Jäkkvik in southern Norrbotten (Kulling 1982; P.-G. Andreasson personal communication 1981).

Metadolerites in the Kalak Nappe Complex (Middle Allochthon)

In the Kalak Nappe Complex, northern Norway, metadolerite dykes and sheets, deformed and variably altered to hornblende and biotite schists, intrude quartzofeldspathic and pelitic schists (Gayer and Roberts 1971). These rocks have been included within the Middle Allochthon although their metamorphic grade and complex deformational history suggest possible correlation with the rocks in the overlying Seve Nappes occurring further south in the orogen (see below). According to Gayer and Humphreys (1981), the metadolerites are geochemically similar to the basaltic dykes in the underlying Laksefjord Nappe, i.e. tholeiitic and with MORB affinity. The synorogenic plutonic rocks of the Seiland province in the upper and westernmost part of the Kalak Nappe Complex are described in section 5.

Amphibolites in the Seve Nappes (Upper Allochthon)

Amphibolites of variable thickness and often with irregular lens-like geometry are a characteristic feature of the Seve Nappes (Zachrisson 1973) forming the lower part of the Upper Allochthon. The amphibolites usually contain hornblende and plagioclase with varying amounts of epidote-group minerals and garnet. Due to recrystallization under epidote–amphibolite and amphibolite facies and locally under granulite facies conditions (Williams and Zwart 1977), as well as intense deformation, it is difficult to decide, on the basis of field relations and textures, the relative importance of extrusive and intrusive varieties. Trouw (1973), however, reported dykes from the Marsfjället area of Sweden. The amphibolites are associated with quartzofeldspathic, garnet–mica and calcareous schists and locally migmatitic gneiss. In northern Jämtland (van Roermund, this volume) and southern Norrbotten (Andreasson et al., this volume) in Sweden, lenses of variably amphibolitized eclogite are present; in the former area, these occur at two separate tectono-stratigraphic levels.

A limited amount of geochemical work (note particularly Solyom et al. 1979b; Hill 1980) has established the tholeiitic, low-K character of the metabasites and their affinity to MORB (see Fig. 2). Solyom et al. (1979b) emphasized the spilitic character of the amphibolites from Mt. Sylarna, Sweden, and suggested a comagmatic relationship to the Ottfjället Dolerites in the underlying Särv Nappe (see also Strömberg 1969). According to van Roermund (this volume), the eclogites from northern Jämtland show quartz- to olivine-normative tholeiitic compositions.

The age of the amphibolites is controversial and constrained by only a limited number of radiometric age determinations on Seve gneisses (Reymer 1979; Claesson 1982) with which immediately adjacent amphibolites share a common metamorphic–deformational history. Both of the studies cited above establish the presence of pre-Caledonian crustal elements in the Seve Nappes. Thus, one possibility is that the amphibolites are pre-Caledonian in age. However, the petrographic and geochemical affinity of the Ottfjället Dolerites and the Seve amphibolites (Strömberg 1969; Solymon et al. 1979b), taken in relation to the age determination data for the former, suggest an alternative late Proterozoic age for the Seve amphibolites.

2. Igneous Activity Related to Rifted Arcs in the Köli Nappes and Equivalents of the Upper Allochthon

Lower Palaeozoic, fossil-bearing successions, occurring in a number of thrust nappes and generally metamorphosed in the greenschist facies, overlie the Seve Nappes and related units of the Upper Alloch-thon. Between the Grong–Olden Culmination and Nasafjället, these units, referred to there as the Köli Nappes, have been investigated in detail and have been divided by Stephens (1980a) into three groups: the Lower, Middle and Upper Köli. Each of these nappe complexes contains several volcanic–subvolcanic associations at different tectonostratigraphic and even lithostratigraphic levels (see especially Fig. 2 in Stephens 1980a). A description of these associations and volcanic–subvolcanic complexes in equivalent tectonic units, both south of the Grong–Olden Culmination and to the north of Nasafjället (Table 1), is presented below.

Pre-Ashgill and Llandovery volcanic–subvolcanic complexes in the Lower Köli

Volcanites in the Lower Köli (Kulling 1933; Stephens 1977a) occur in both pre-Ashgill and Llandovery sequences, being separated by an extensive, shallow-water succession of quartzite conglomerate, quartzite and fossiliferous limestone of Ashgill age (Vojtja and Slätdal Formations). The pre-Ashgill rocks (Tjopasi Group) consist of a heterogeneous sequence of phyllite, protrusive and detrital serpentinite (Stigh 1979), basic or mixed acid–basic volcanites, and an overlying sequence of turbidites and matrix-rich conglomerates. Silurian rocks (Broken, Lövfjäll, Vesken and Viris Formations) comprise fossiliferous graphitic phyllites with intercalations of basic volcanites of Llandovery age passing upwards into various calcareous and quartz-rich turbidites with local conglomerates.

Pre-Ashgill basic rocks in the Tjopasi Group (Seima Formation), directly overlying schists and amphibolites of the Seve Nappes, are composed of plagioclase–phyric basalts, coarsely-fragmental volcanites including an agglomerate unit, and both fine- and coarse-grained, aphyric basaltic rocks of uncertain volcanic–subvolcanic status. Geochemically, these rocks show a steep trend on the TiO_2–Zr/P_2O_5 diagram, individual points lying close to the line separating alkaline and tholeiitic types, a tholeiite-type trend on the TiO_2–FeO^t/MgO diagram, a Ti–Cr distribution suggesting $TiO_2 > 1.5$ per cent even in the most primitive (c. 400 ppm Cr) basalts and a concentration in the WPB field on the Ti–Zr–Y diagram (Fig. 3). The overlying phyllites, turbidites and conglomeratic rocks (Gilliks Formation) contain minor acid and basic tuffaceous horizons. A fossiliferous unit in the Ikesjaure area of southern Norrbotten, in rocks which can be correlated with the Gilliks Formation, suggests a mid Ordovician age (Kulling in Strand and Kulling 1972) for at least some of the rocks lying stratigraphically above the Seima Formation.

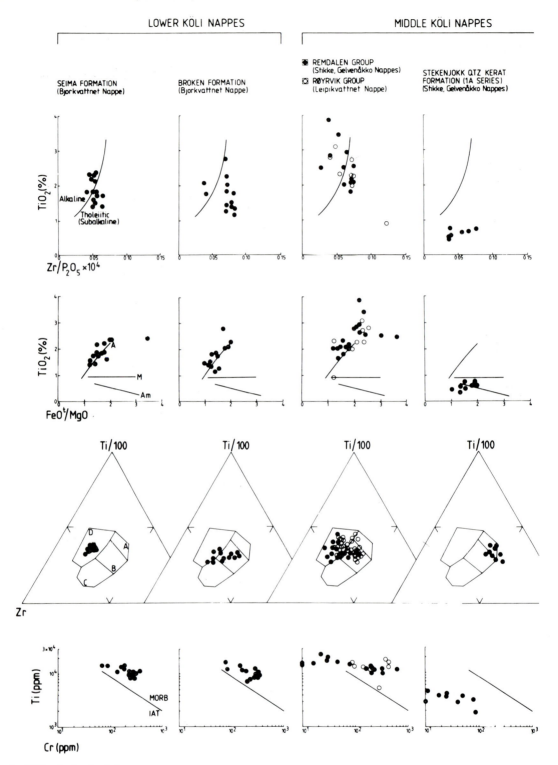

Fig. 3 TiO_2–Zr/P_2O_5 (after Winchester and Floyd 1976), TiO_2–FeO^t/MgO (after Miyashiro, 1975), Ti–Zr–Y (after Pearce and Cann 1973) and Ti–Cr (Pearce 1975) plots for basaltic rocks and basaltic andesites in the Lower and Middle Köli Nappes (Upper Allochthon). A, M and Am on the TiO_2–FeO^t/MgO diagram refer to trend lines for abyssal tholeiite, Macauley Island tholeiitic island-arc series and the Amagi calc-alkaline series. A, B, C and D on the Ti–Zr–Y diagram refer to the fields of IAT, MORB + IAT + CAB, CAB and WPB, respectively (abbreviations as in Fig. 2). Data sources are from Olsen (1980), A. Reinsbakken (unpublished data) and Stephens (1980a, 1982 and unpublished material)

Volcanites belonging to the pre-Ashgill Tjopasi Group outcrop further west and can be traced southwards towards the Grong–Olden Culmination over a strike length of over 100 km (Zachrisson 1969; Sjöstrand 1978; Kollung 1979). These rocks are dominated by well-layered, fine-grained, spilitic basic rocks and quartz- and albite–phyric quartz keratophyre, interpreted as original vitric and vitric–crystal tuffs. Homogeneous and granular-textured albite trondhjemite and coarser basic lenses and sheets with sharp contacts to surrounding rocks probably represent high-level intrusions. Preliminary geochemical data show that some of the basic rocks are apparently both strongly fractionated (low Cr, Ni, Mg) and depleted in the incompatible elements (e.g. Zr, Ti) relative to MORB and display a mildly-tholeiitic trend on the TiO_2–FeO^t/MgO diagram, i.e. show an affinity to island-arc tholeiite. Basaltic rocks showing higher contents of TiO_2 and FeO^t are, however, also present. U–Pb dating of zircons from an albite trondhjemite intrusion indicates an age of 488 ± 5 Ma (Claesson et al. 1983), i.e. early Ordovician. Comparison of pre-Ashgill sequences suggests that the mixed acid–basic volcanites may be stratigraphically younger than the Seima Formation, correlating more probably with the tuffaceous horizons in the Gilliks Formation in the east.

Llandovery volcanites comprise massive and locally pillowed basalts interlayered with fossiliferous graphitic phyllites and calcareous wackes belonging to the Broken Formation. These basaltic rocks are variably spilitic and tholeiitic in character and show MORB(–WPB) affinity (Fig. 3). They are thought to be related to rifting after development of the pre-Ashgill arc complex.

As pointed out by Stephens (1980a), the stratigraphic distribution of turbidite and shallow-water-sequences in the Lower Köli is remarkably similar to the Ordovician and Silurian lithostratigraphy of the Jämtland Supergroup (Gee 1975) in the Lower Allochthon. The presence of Caradoc (L. Karis personal communication 1979) and Llandovery (Thorslund 1948) bentonites in the Jämtland Supergroup strengthens this comparison.

Ordovician(?) volcanic–subvolcanic complexes in the Middle Köli and equivalents

The Middle Köli Nappes, comprising in tectonostratigraphic order from the base upwards, the Stikke, Gelvenåkko, Leipikvattnet and Gjersvik Nappes (Zachrisson 1969; Halls et al. 1977; Stephens 1982) immediately north of the Grong–Olden Culmination, consist of several volcanic–subvolcanic (±phyllite) complexes and calcareous flysch-like sediments often containing basic

extrusions or high-level intrusions (Zachrisson 1964, 1969; Halls et al. 1977; Sjöstrand 1978; Kollung 1979; Lutro 1979).

In the Stikke and Gelvenåkko Nappes, the stratigraphically lowest unit (Sjöstrand 1978; Stephens 1982; but see Zachrisson 1964, 1969) consists of phyllites, often graphitic, basic volcanic and subvolcanic rocks, a minor microcline–quartz porphyry and, in the upper part, crinoidal limestone and quartzite conglomerate (Remdalen Group). The Røyrvik Group, which dominates the overlying Leipikvattnet Nappe, also contains basic volcanites, including pillow lava (Zachrisson 1966; Olsen 1980), pillow breccia and hyaloclastite tuff, in association with phyllites, often graphitic, and recrystallized ribbon chert; lithostratigraphic correlation with the Remdalen Group is likely (see also Sjöstrand 1978; Kollung 1979). The geochemistry of the basic volcanites within the Remdalen and Røyrvik Groups supports such a correlation (Fig. 3). The basic rocks in both groups show mixed tholeiitic and alkaline tendencies, mixed MORB–WPB affinities and nearly flat Ti–Cr distributions with high TiO_2 values ($>$c. 2.0 per cent) even in the most primitive (Cr = 400–600 ppm) basalts.

The Remdalen Group is stratigraphically overlain by a bimodal, acid-dominated, volcanic–subvolcanic complex up to c. 1 km in thickness (Stekenjokk Quartz-Keratophyre) containing, in its upper part, graphitic phyllites. Pyroclastic and high-level intrusive rocks dominate this formation, the former interpreted as recrystallized and deformed vitric and vitric–crystal tuffs of predominantly submarine ash-flow origin (Stephens 1982). A suite of basic, high-level intrusions marks the latest phase of magmatic activity. The Stekenjokk Quartz-Keratophyre is overlain by calcareous turbidites (the Blåsjö Phyllite) containing high-level gabbro intrusions, a basic volcanic–subvolcanic complex (the Lasterfjäll Greenschist) locally separating these formations.

The volcanic and subvolcanic rocks of the Stekenjokk Quartz-Keratophyre are altered to spilite and quartz-keratophyre (Stephens 1980b). Alteration occurred on a regional scale but varies in intensity, apparently being less severe in the high-level intrusions. The variation in the degree of alteration allows some estimation of primary magmatic compositions; basalt, basaltic andesite and low-K dacite–rhyodacite are believed to be represented (Stephens 1982). Three separate fractionation series involving basalts and basaltic andesites have been suggested (Stephens 1982 and Fig. 3):

1. A mildly-tholeiitic series of basalt and basaltic andesite which is fractionated (Cr < 100 ppm) yet highly depleted in Ti (<1.0 per cent TiO_2),

Zr, Y and P relative to MORB (1A series);
2. A tholeiitic series showing MORB affinity and composed of both more primitive and fractionated basalts (1B + 2A series);
3. Fractionated alkaline basalts (2B series).

The late-stage basic intrusions dominate the 1B + 2A and 2B series. The variation between these series is thought to be dictated by partial melting of mantle material under varying P_{load} and P_{H2O} conditions and to a lesser extent by varying mantle compositions. The bulk of the acid rocks in the Stekenjokk Quartz-Keratophyre form a separate series thought to be derived by partial melting of low-K basalt (Stephens 1982). The largely intrusive 1B + 2A and 2B series show geochemical characteristics similar to the basic rocks in the overlying formations, i.e. the Lasterfjäll Greenschist and the gabbros in the Blåsjö Phyllite (Stephens 1981 and Fig. 3). Evolution from a probably ensimatic island-arc (1A series and acid rocks) to a distensional environment (1B + 2A and 2B series in the Stekenjokk Quartz-Keratophyre, Lasterfjäll Greenschist, gabbros in Blåsjö Phyllite) related to incipient arc rifting has been suggested by Stephens (1977b, 1982).

The Gjersvik Nappe (Halls et al. 1977) forms the highest Köli tectonic unit immediately north of the Grong–Olden Culmination. The Gjersvik Group, comprising the upper and western part of the Gjersvik Nappe, consists of a subvolcanic infrastructure of gabbro, diorite, trondhjemite and granodiorite forming the root to a consanguineous submarine volcanic sequence (Halls et al. 1977). The latter is dominated by basaltic and andesitic flows, both massive and pillowed, and high-level intrusions. There are also subordinate but extensive rhyodacite horizons, occurring both as lavas and pyroclastic rocks, as well as acid high-level intrusions. Pre-tectonic alteration of the volcanites to spilite and quartz-keratophyre has occurred (e.g. Halls et al. 1977; Lutro 1979). The majority of the basic lavas in the Gjersvik Group (Bjørkvatnet Formation of Lutro 1979) are tholeiitic and consist of either paler, primitive to fractionated basalts or more highly evolved, darker basalts and basaltic andesites. These rocks contain (see Fig. 3) low or moderate TiO_2 (<2.0 per cent) and incompatible element concentrations even in the most fractionated samples (Cr = 10–100 ppm). Locally, a volcanostratigraphy passing upwards from the paler basalts into the acid volcanites and then the darker basalts and basaltic andesites can be discerned (A. Reinsbakken personal communication 1981). These lavas have been interpreted in terms of an ensimatic island-arc complex (Gale and Roberts 1974; Halls et al. 1977; Lutro 1979), the Ti–Zr–Y distributions (Fig. 3) providing

some support for this interpretation. Grenne and Reinsbakken (1981) have suggested a more complex situation with both subduction-related and rifting related basalts being recognized. More deformed and higher grade basaltic rocks interlayered with acid rocks occur in the western part of the Gjersvik Group (Kleiva and Bjørkvassklumpen Formations of Lutro 1979); they are tholeiitic and show MORB affinity (Fig. 3). The Gjersvik Group is stratigraphically overlain by a flysch-like sequence containing volcanogenic conglomerates, calcareous sandstones and phyllites (the Limingen Group), locally with greenstones interpreted as lavas (Lutro 1979). A four-point Rb–Sr whole-rock isochron composed of samples from separate trondhjemite and gabbro intrusions in the Gjersvik Nappe gives an age of 433 ± 10 Ma (Råheim et al. 1979), interpreted by these authors as an age of deformation/metamorphism.

South of the Grong–Olden Culmination, volcanic and high-level intrusive rocks occur extensively in the Meråker Nappe (Wolff and Roberts 1980). On the basis of geochemistry, lithostratigraphy and the associated volcanogenic ore deposits, Grenne and Reinsbakken (1981) correlated the Gjersvik Group with the volcanic–subvolcanic complexes comprising the Fundsjø Group (Wolff 1967), in the Meråker Nappe, and the Hersjø Group (Rui 1972) lying either in the Meråker or overlying Gula Nappes. Grenne and Lagerblad (this volume) have indicated the presence of lavas, pyroclastic rocks, tuffaceous sediments and high-level intrusions in the Fundsjø Group and stressed the bimodal character of the magmatic activity, basalt (with basaltic andesite) and acid rocks dominating. The basic rocks in the Fundsjø Group show variable chemical composition, volcanites of island-arc tholeiite character occurring together with basic rocks more similar to MORB. Formation of the Fundsjø Group in a rifted-arc situation has been inferred (Grenne and Lagerblad this volume). The correlation of the Gjersvik and Fundsjø Groups is strengthened when it is considered that both groups stratigraphically underlie volcanogenic conglomerates, the pebbles of which can be matched with lithologies within the volcanic–subvolcanic complexes (Oftedahl 1956; Chaloupsky and Fediuk 1967). The stratigraphically overlying Sulåmo (Wolff 1967) and Kjølhaugan (Hardenby 1980) Groups in the Meråker Nappe, possibly equivalent to the Limingen Group north of the Grong–Olden Culmination, consist of calcareous sandstones and phyllites with local conglomerate horizons, basic volcanites and gabbro intrusions Since these calcareous flysch-like sequences associated with basic extrusions and intrusions are overlain by graphitic phyllites of Llandovery age (Getz

1980), an age no younger than Llandovery may be inferred for both the volcanic–subvolcanic complexes of the Fundsjø and Hersjø Groups as well as the Sulåmo and Kjølhaugan Groups. A lower age limit for these units may be obtained from the relationships at Nordaunevoll. At this locality, radioactive, graphitic phyllites containing Tremadoc graptolites are intimately associated with pillow lava belonging to the Hersjø Group and are intruded by dolerite intrusions (Størmer 1941; Nilsen 1971; Gee 1981). The calcareous sediments metamorphosed to a maximum of garnet–hornblende grade in the underlying tectonic unit(s) to the Meråker Nappe (Köli Nappes of Tännforsen area, see Beckholmen 1982) contain tholeiitic basalts showing MORB affinity (Beckholmen and Noro 1981); these rocks lie at tectonostratigraphic levels possibly equivalent to the lower part of the Middle Köli complex.

North of the Grong–Olden Culmination and the type area of the Middle Köli Nappes, volcanic–subvolcanic complexes of dominantly basic or mixed acid–basic character structurally overlying calcareous sediments intruded by gabbros occur in a number of isolated, megatectonic lenses. Conspicuous amongst these are the so-called Ropen 'granite' (Beskow 1929) and equivalents (Lower Laxfjället Unit of Stephens 1977a) in the Atofjället Nappe of Häggbom (1978) in Västerbotten. On the basis of lithostratigraphy, correlation of these rocks with sequences in the Middle Köli Nappes is suggested. In particular, correlation of the volcanic–subvolcanic complexes with the Hersjø–Fundsjø–Gjersvik–(?)Stekenjokk volcanites is favoured.

Volcanic–subvolcanic complexes in the Upper Köli

The Upper Köli Nappes are more or less synonymous with the Storfjäll nappe complex of Kulling (in Strand and Kulling 1972) and equivalent to the Storfjället Nappe of Häggbom (1978). The graphite-bearing mica schists, conglomerates and volcanites within the lower part (the Krutfjellet Nappe) occur in a series of megatectonic lenses connected by high-strain phyllonite zones (Häggbom 1978). In both the Norra Storfjället and Krutfjellet Lenses, these rocks were deformed during regional amphibolite-facies metamorphism, intruded by gabbro and various acid intrusions after the establishment of amphibolite-facies conditions, and subsequently suffered later deformation (Gee and Wilson 1974; Senior 1978; Mørk 1979; Senior and Otten, this volume). Radiometric age-determinations (see section on granitoids) and the local occurrence of crinoids (Kulling in Gavelin and Kulling 1955) suggest that the earlier deformation and amphibolite-facies metamorphism occurred in the Ordovician.

The upper part of the Storfjället Nappe (Jofjället Nappe) consists of phyllites, volcanites, marble, and calcareous sediments metamorphosed in the greenschist facies and repeated in a complex, imbricate structural pattern (Sandwall 1981a; Ramberg 1981).

Within the Norra Storfjället Lens (Krutfjellet Nappe), mainly basic volcanic and subvolcanic rocks occur at apparently three separate tectonostratigraphic levels (Stephens and Senior 1981). The lower level is dominated by volcanic breccia/agglomerate. Aphyric to pyroxene–phyric amphibolite, tentatively interpreted as lavas, volcanic breccia or agglomerate, finely-layered ash-fall tuff, and high-level intrusions constitute the middle unit (Mesket Formation, see Kulling 1938). Pillow lava and fragmental units containing both acid and basic material comprise the upper unit. Geochemical data (Stephens and Senior 1981) suggest that the lower unit and the Mesket Formation are similar and quite distinct from the lavas in the upper unit. The former are composed of mildly-tholeiitic basalt and basaltic andesite showing a wide range in the degree of fractionation, including high-Mg basalts with up to 18 per cent MgO, 120 ppm Cr and 400 ppm Ni. The most fractionated rocks are strongly depleted in the more incompatible elements Ti, P, Zr and Y relative to MORB; affinity to island-arc tholeiite is suggested and a fore-arc basin environment inferred (Stephens and Senior 1981). By contrast, the lavas of the upper volcanic unit are alkaline, mostly strongly fractionated and show WPB affinity.

Both acid and basic tuffaceous units are present at different tectonostratigraphic levels in the lower part of the Jofjället Nappe, an interpretation as a distal pyroclastic facies being suggested for these rocks (Sandwall 1981a). In the middle part of the nappe, two conspicuous units of basic volcanites (the Ruffe and Brackfjället Greenstones), each up to c. 500 m in thickness and separated by highly-deformed, fine-grained, pelitic and quartz-rich phyllites, rest with tectonic contact on phyllites. They have been interpreted by Sandwall (1981a) to be composed of both lavas and high-level intrusions. Geochemically (Sandwall 1981b), both units show evidence of post-eruptive alteration. The Ruffe Greenstone appears to be tholeiitic and shows mixed MORB–WPB affinities, while the tectonically overlying Brackfjället Greenstone appears to be alkaline and shows a clear WPB affinity.

Arguments for the tectonostratigraphic correlation of the Krutfjellet Nappe with the amphibolite-facies rocks of the Gula Nappe (Wolff and Roberts 1980) have been presented by Sandwall (1981a). The Gula Nappe contains minor amphibolite of uncertain status, its most characteristic igneous component, however, being numerous synorogenic trondhjemite intrusions (Size 1979 and see section on granitoids). Based on the correlation of the Krutfjellet and Gula Nappes, the Ruffe and Brackfjället Greenstones would appear to lie at approximately the same tectonic level as the basaltic rocks (ophiolitic) at the base of the Støren Nappe, which overlies the Gula Nappe. Lithostratigraphic correlation of these basaltic sequences is supported by a comparison of the respective geochemical data (Sandwall 1981b). North of Nasafjället, amphibolites within the Gasak Nappe (Kautsky 1953; Nicholson and Rutland 1969) and Skibotn Nappe Complex (Padget 1955; Binns 1978; Gayer and Humphreys 1981) are possible equivalents (Binns 1978; Gee and Zachrisson 1979) to the basic volcanic and subvolcanic rocks within the higher-grade part of the Storfjället Nappe.

Stephens and Gee (this volume) have suggested that the volcanosedimentary associations preserved in the Upper and Middle Köli Nappes have been derived from the fore-arc (Storfjället Nappe), main magmatic arc (Gjersvik Nappe) and back-arc (Stikke, Gelvenåkko and Leipikvattnet Nappes) facies of a low-K, mildly-tholeiitic, rifted-arc complex. The basement to this arc, dominated by fine-grained phyllites and basaltic rocks showing both tholeiitic and alkaline characteristics and mixed MORB–WPB affinities, is preserved in the back-arc and fore-arc situations. Spatial relationships prior to Scandian thrusting suggest eastwards-directed subduction. On the basis of the fossil occurrences south of the Grong–Olden Culmination, it would appear that the rifted-arc complex is early Ordovician (in part Tremadoc) in age while the later calcareous flysch-like sediments with basic extrusions and intrusions are post-Tremadoc and pre-Llandovery in age.

3. Ophiolite Fragments and Associated Within-plate Basalts in the Upper and Uppermost Allochthons

In the area between Karmøy and Lyngen in the Norwegian Caledonides (Fig. 1), several rock associations show geological features and internal relationships that qualify each of them as being part of an ophiolite. The presence of an ophiolite pseudostratigraphy, including, in several cases, a conformable cap of pelagic sediments, has been used to assess ophiolite affinity. Emphasis has also been placed on a geochemistry consistent with the formation of the basic rocks at a spreading centre. The ophiolite fragments occur in the upper part of the Upper Allochthon and possibly even in the Uppermost Allochthon (e.g. Skålvaer). The ophiolites recognized in the Støren Nappe, in the Trondheim region,

ion, are thought to lie either at the same tectonostratigraphic level (Roberts et al. 1970; Wolff and Roberts 1980) or above (Gee and Zachrisson 1974) the successions related to rifted arcs described in section 2. The latter interpretation is illustrated in Table 1 and followed here.

A review of the ophiolite fragments in the Norwegian Caledonides was presented by Furnes et al. (1980a) and has been complemented by Furnes et al. (this volume). Detailed descriptions are available for the Karmøy (Sturt et al. 1980a), Lykling (Nordås et al., this volume), Gullfjellet (Thon, this volume), Løkken-Grefstadfjellet (Grenne et al. 1980; Ryan et al. 1980), Vassfjellet (Grenne et al. 1980; Grenne in Wolff et al. 1980), Leka Prestvik 1974, 1980) and Sulitjelma (Boyle 1980; Boyle et al., this volume) ophiolites, and the predominantly basic lavas of the Forbordfjell and Jonsvatn Greenstones (Grenne and Roberts 1980) which, together with the Snåsa Greenstones, have been interpreted by Furnes et al. (this volume) as forming parts of ophiolite sequences.

A well-developed ultrabasic complex of layered and tectonized dunites, pyroxenites and peridotites, all variably serpentinized, is represented on Leka. The other ophiolite fragments contain only minor ultrabasic rocks, usually represented by serpentinites. Mostly non-vesicular, sometimes variolitic pillow lavas are present in all complexes, and sheeted dyke and/or sill complexes as well as layered/high-level gabbros are usually well developed. In the Støren and Stavfjorden ophiolites as well as the Forbordfjell–Jonsvatn and Snåsa Greenstones, however, only small gabbro bodies and occasional dykes are associated with the pillow lavas. Plagiogranites, usually confined to the upper part of the gabbroic zone or the sheeted intrusive complex, are usually present. Minor to substantial developments of pelagic sediments, conformably overlying or intercalated with pillow lavas in the upper part of the volcanic zone, occur in most of the sequences. In the case of Stord and Skålvaer, the interpretation of the very fragmented and incomplete sequences as representing ophiolites is based solely on pillow-lava geochemistry.

At Sulitjelma, recent results (Boyle et al. 1979) have demonstrated that the succession, calcareous sediments containing high-level gabbro intrusions (Furulund Group) overlain by sediments interpreted as pelagic, pillow lavas and sheeted basic intrusions (Sulitjelma Amphibolite Group), is inverted. Boyle (1980) later interpreted the field relationships between the Sulitjelma Amphibolite Group and the overlying Sulitjelma Gabbro as indicating their inclusion in the same magmatic complex and inferred that they formed the upper part of an inverted ophiolite sequence. Reservations concerning this comagmatic relationship and inclusion of the Sulitjelma Gabbro in the ophiolite complex focus around:

1. Earlier work (Kautsky 1953; Nicholson and Rutland 1969; Kulling 1982) indicating a tectonic break between the Sulitjelma Amphibolite Group and the Sulitjelma Gabbro, i.e. the base of the Gasak Nappe.
2. Intrusion of the Sulitjelma Gabbro under high-pressure (6 kbar) conditions (Mason 1980) into already deformed and metamorphosed sediments (the Sulitjelma Schist Sequence), subsequent deformation of the gabbro demonstrating its synorogenic character.
3. Regional tectonostratigraphic considerations (Stephens et al., this volume) suggesting correlation of Kautsky's (1953) Gasak Nappe with the Krutfjellet Nappe and the underlying Sulitjelma Amphibolite and Furulund Groups with sequences in the Middle Köli Nappes; as indicated in section 2, the Krutfjellet Nappe contains synorogenic gabbros and granitoids occurring as isolated intrusions within schists, while the sequences in the Middle Köli Nappes have been interpreted as being formed in a probably ensimatic rifted-arc environment.

The ophiolite fragments along the southwest coast of Norway (Karmøy, Lykling, Stord, Gullfjellet, Solund, Stavfjorden) as well as the Leka and Lyngen ophiolites are unconformably overlain either by a transgressive sedimentary cover or, in the case of the Lykling and Stord ophiolites, predominantly subaerial acid volcanites. Where faunal or radiometric age determination data are available in the cover sequences to these ophiolites, a pre-late Ordovician age both for the fragments themselves and for their emplacement may be inferred.

Earlier interpretations of the stratigraphy in the Støren Nappe (Vogt 1945; Carstens 1960) correlated all the greenstones now recognized as more or less complete ophiolite sequences. The Støren and Vassfjellet ophiolites (Støren Group), occurring in separate thrust slices with movement zones along their lower contacts (Gale and Roberts 1974; Grenne et al. 1980), are each overlain by volcaniclastic sediments belonging to the Lower Hovin Group. Fossil evidence in these separate slices (Vogt 1945; Bruton and Bockelie 1980) suggests pre-Caradoc and pre-late Arenig ages, respectively, for these ophiolites. In the case of the Løkken–Grefstadfjellet ophiolite, Ryan et al. (1980) described graptolitic shales of mid–late Arenig age to both underlie and overlie thin pillow basalts thought to be lateral equivalents of the ophiolite. This interpretation of the field relationships led Ryan et al. (1980) to suggest that the Løkken–Grefstadfjellet ophiolite occurs within the Lower Hovin Group and is of late Arenig age. Although fossil evidence is lacking in the enclosing sediments to the Forbordfjell and Jonsvatn Greenstones, they are also interpreted to lie stratigraphically within the Lower Hovin Group and assumed to be post–mid Arenig to mid Ordovician in age (Roberts 1975b; Wolff 1979; Grenne and Roberts 1980). A mid Ordovician or older age has been suggested (Spjeldnaes, this volume) for a fossiliferous limestone intercalated with basalt (Roberts 1980a) in the Snåsa sequence. Thus, present interpretations of the field relationships combined with the limited biostratigraphic data are favouring ophiolites of different ages in the Støren Nappe. Further discussion of the stratigraphic relationships within the Støren Nappe may be found in Furnes et al. (this volume) and Stephens and Gee (this volume). Based on the matching of pebble lithologies in conglomerates of the Lower Hovin Group with lithologies in the ophiolite slices, Furnes et al. (1980a) proposed an early Ordovician or earlier age for emplacement of the ophiolites belonging to the Støren Group. They equated this tectonic event with the Finnmarkian phase of orogenesis (Sturt et al. 1978).

Certain geochemical characteristics of the basalts in the different ophiolites are presented in Fig. 4. In the TiO_2–Zr/P_2O_5 diagram, the data for most of the ophiolites plot in the subalkaline field. By contrast, the data from Karmøy, Lykling and Gullfjellet show a pronounced spread. In the TiO_2–FeO^t/MgO diagram, the data from Stord, Solund, Stavfjorden, Løkken, Jonsvatn and Skålvaer define the same trend as MORB, whereas those from Vassfjellet, Støren, Leka and Sulitjelma show trends which are partly similar to that of MORB, and partly intermediate between MORB and IAT (island-arc tholeiite). The Karmøy, Lykling, Gullfjellet, and Forbordfjell basalts show a characteristically large spread in this diagram. In the Ti–Zr–Y diagram, the data from Stord, Solund, Stavfjorden, Vassfjellet, Løkken, Jonsvatn, Leka and Skålvær plot almost without exception within the field which includes MORB, whereas those from Støren plot along the boundary between the WPB field and the field including MORB, and those from Sulitjelma in both the CAB field and the field including MORB. The data from Forbordfjell show a clear division into two groups, one plotting in the field including MORB, the other in the WPB field. Again, the basalts from Karmøy, Lykling and Gullfjellet show the most prominent spread of data, plotting mostly in the fields of IAT and WPB and only a few in the field including MORB. In the Ti–Cr diagram, the data from most of the sequences plot well within the MORB field, whereas those from Karmøy, Lykling and Gullfjellet mostly plot in the field of IAT.

In conclusion, it appears that some of the ophiolites show the influence of island-arc activity (Karmøy, Lykling, Gullfjellet and Sulitjelma), whereas the Støren ophiolite and Forbordfjell Greenstone show mixed WPB–MORB characteristics; the other ophiolite fragments (Stord, Solund, Stavfjorden, Vassfjellet, Løkken, Leka and Skålvaer) as well as the Jonsvatn Greenstone show typical MORB characteristics. Based on the thickness (700 m minimum) of the pelagic sediments in the Torvastad Group of the Karmøy ophiolite, Sturt *et al*. (1980a) suggested generation of this ophiolite at the spreading ridge of a former major ocean (Iapetus). Arguments have been presented for generation of the ophiolites in the Støren Nappe both at the Iapetus spreading ridge (Støren, Vassfjellet and Løkken ophiolites; see Roberts and Gale 1978; Grenne *et al*. 1980) and in a younger, smaller, marginal basin east of and broadly contemporaneous with the mature Smøla–Hølonda arc (see section 4; Løkken–Grefstadfjellet ophiolite as well as Forbordfjell, Jonsvatn and Snåsa Greenstones, see Gale and Roberts 1974; Roberts 1980b; Grenne and Roberts 1980; Ryan *et al*. 1980).

Basic volcanites intercalated within pelagic sediments are associated with two of the ophiolite fragments, i.e. Karmøy and Leka. On Karmøy, this sedimentary/volcanic succession forms the cap to the ophiolite, whereas on Leka, it is in tectonic contact with ultrabasic rocks. On Karmøy, the cap sequence, assigned to the Torvastad Group (Solli 1981), has a minimum thickness of 5 km. The stratigraphically lowest part (the Velle Formation), approximately 2 km in thickness, consists principally of coarse greywackes with associated ribbon cherts, metalliferous sediments, and phyllites; it is thought to represent sedimentation in an oceanic fracture zone (Solli 1981). Higher in the sequence, cherts and phyllites are the dominant sediments. Intercalated within these sediments are two major horizons of pillow lava, approximately 2 km thick. The sedimentary/volcanic sequence associated with the Leka ophiolite, the Storøya Formation (Prestvik 1974), has a less well-developed sedimentary sequence of cherts and phyllites but excellently preserved pillow lavas and pillow breccias. The geochemistry of these pillow basalts from Karmøy and Leka are shown in Fig. 4. In the TiO_2–Zr/P_2O_5 diagram, the basalts from the Torvastad Group on Karmøy plot well within the alkaline field and, in the Ti–Zr–Y diagram, they plot firmly within the WPB field, distinctly different from the strictly ophiolitic basalts with which they are associated.

4. Mature-arc/Active Continental Margin and Within-plate Igneous Activity Associated with Ophiolites in the Upper Allochthon

The volcanites described here lie in the same tectonic unit(s) as the ophiolite fragments described in the previous section. They occur in successions observed or inferred to lie stratigraphically above the obducted ophiolite complexes. They have been subdivided into two types, according to their affinity to mature-arc/active continental margin and continental within-plate complexes.

Ordovician mature-arc/active continental margin volcanites

The island of Smøla, which lies approximately 120 km west of Trondheim (Fig. 1) and is interpreted as lying within the Støren Nappe (see Tectonostratigraphic Map, this volume), contains volcanites showing a complete range in composition from basalt to rhyolite. The volcanites interdigitate with late Arenig–early Llanvirn limestones (Bruton and Bockelie 1979), conglomerates and other sediments. They are intruded by granite, granodiorite, quartz diorite and gabbro, the quartz diorite yielding a Rb–Sr whole-rock isochron indicating an age of 436 ± 7 Ma (Sundvoll and Roberts 1977). The basalts and basaltic andesites are dominantly brecciated, several metre-thick flows and occasional pillow lavas, whereas the dacitic to rhyolitic lavas are flow banded. In addition to the lavas, various tuffaceous rocks occur. Some geochemical trends for the Smøla volcanites are shown in Fig. 5. They show medium-K, calc-alkaline characteristics, consistent with magmatism in a mature island arc (Roberts 1980b) or in an active continental margin setting. The precise nature of the substrate to the Smøla island arc is uncertain but, based on regional considerations, Roberts (1980b) suggested that the arc was built upon a basement of obducted ophiolite.

Andesitic intrusions, fissure flows and explosive breccias associated with late Arenig–early Llanvirn limestones, shales and conglomerates (Lower Hovin Group) have also been described from the Hølonda area in the Støren Nappe of the Trondheim region (Vogt 1945; Bruton and Bockelie 1980). They stratigraphically overlie the Vassfjellet and Løkken–Grefstadfjellet ophiolites and have been suggested to belong to an island-arc system (Loeschke 1976). Both Bruton and Bockelie (1980) and Roberts (1980b) have compared these rocks to the Smøla volcanites, further west. The andesites in the Hølonda area pass stratigraphically upwards into rhyolites and volcaniclastic sediments of Caradoc and possibly Ashgill age (Bruton and Bockelie

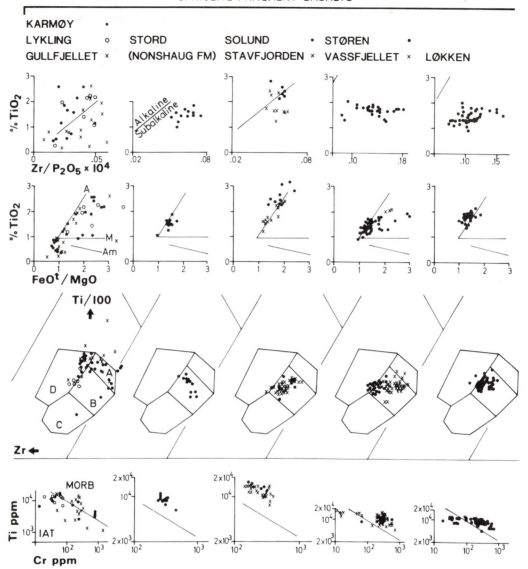

Fig. 4   TiO$_2$–Zr/P$_2$O$_5$ (after Winchester and Floyd 1976), TiO$_2$–FeO^t/MgO (after Miyashiro 1975), Ti–Zr–Y (after Pearce and Cann 1973) and Ti–Cr (Pearce 1975) plots for basaltic rocks from ophiolite fragments (including associated within-plate basalts) in the Upper Allochthon (including Støren Nappe) and possible Uppermost Allochthon. Abbrevia-

1980). Gale and Roberts (1974) and Roberts (1980b) have suggested that the late Arenig to Ashgill volcanosedimentary succession in the Hølonda area was deposited in a marginal basin east of the main Smøla arc. Easterly-directed subduction was inferred on the basis of increasing K$_2$O contents in contemporaneous volcanites from Smøla in the west to Hølonda further east. Roberts (1980b) and Grenne and Roberts (1980) have further proposed that the Smøla arc was active at about the same time as the generation of the ophiolitic rocks interpreted to be interlayered within the Lower Hovin Group. In detail, however, the Løkken–Grefstadfjellet ophiolite is older than the Hølonda andesites while the age

of the Forbordfjell, Jonsvatn and Snøsa Greenstones is not known.

Well-developed, post-ophiolite volcanic sequences also occur on the neighbouring islands of Bømlo and Stord in southwest Norway (Fig. 1). These sequences, attaining thicknesses of 3 km and 7 km, respectively, are predominantly composed of subaerial basaltic to rhyolitic lavas and pyroclastics, ignimbrites dominating the acid volcanites (Lippard 1976; Nordås et al., this volume).On Bømlo, the contact between the Lykling ophiolite and the later subaerial volcanites (assigned to the Siggjo Group) is well exposed; it can be demonstrated that the ophiolite had been deformed and eroded prior to

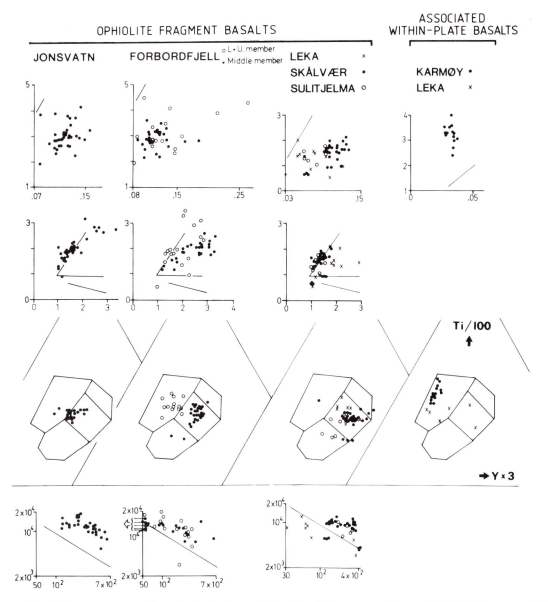

tions as in Fig. 3. Data sources are from A. Boyle (unpublished data), Furnes and Faerseth (1975), Furnes *et al*. (1976, 1980b, 1982), Gale (1974), Grenne and Roberts (1980), Grenne *et al*. (1980), Gustavson (1978, Prestvik (1974, 1980), Prestvik and Roaldset (1978), Sturt *et al*. (1980a)

deposition of the volcanites of the Siggjo Group (Nordås *et al*., this volume). A Rb–Sr whole-rock isochron from the rhyolites on Stord (the Kattnakken volcanites) indicates an age of 445 ± 5 Ma and an initial $^{87}Sr/^{86}Sr$ ratio of 0.7071 (Priem and Torske 1973). A suite of undeformed dolerite dykes and sills of tholeiitic composition cut the Kattnakken volcanites; $^{40}Ar/^{39}Ar$ analyses of three samples have given maximum plateau ages of around 430 Ma (Lippard and Mitchell 1980).

The chemical compositions of the Bømlo and Stord basic volcanites (the Siggjo Group and Kattnakken volcanites, respectively) are shown on Fig. 5; they characteristically show pronounced variations,

particularly with regard to TiO_2 content, both within and between sequences. In the TiO_2–Zr/P_2O_5 diagram, the volcanites from both sequences plot in both the alkaline and subalkaline fields, the basalts with higher contents of TiO_2 on Bømlo plotting in two distinct groups. Both the basalts showing higher contents of TiO_2 on Bømlo and the Stord data show enrichment in TiO_2 with increasing FeO^t/MgO ratios and, in general, the Bømlo volcanites have higher TiO_2 contents than the Stord volcanites. The volcanites from the two sequences even plot differently in the Ti–Zr–Y diagram, the Stord data mainly in the CAB field and the Bømlo data in the WPB field. The data from Stord suggest a calc-alkaline arc or active

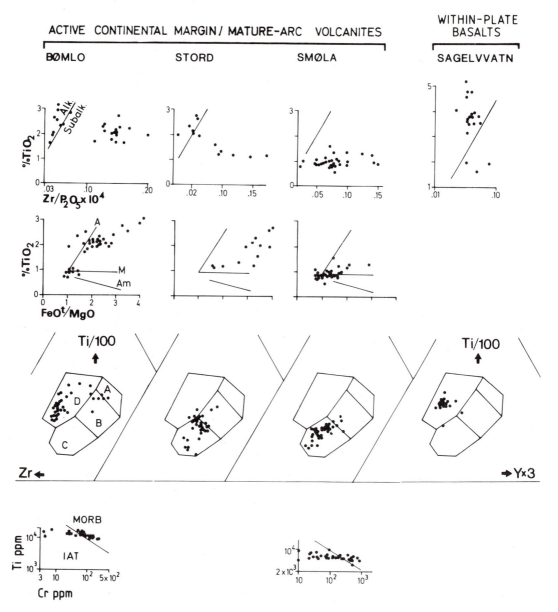

Fig. 5 TiO_2–Zr/P_2O_5 (after Winchester and Floyd 1976), TiO_2–FeO^t/MgO (after Miyashiro 1975), Ti–Zr–Y (after Pearce and Cann, 1973) and Ti–Cr (Pearce 1975) plots for basaltic rocks from active continental margin, mature-arc and continental within-plate complexes associated with ophiolites in the Upper Allochthon (including Støren Nappe). Abbreviations as in Fig. 3. Data sources are from Bjørlykke and Olaussen (1981), Furnes *et al.* (1978), Lippard (1976), Nordås *et al.* (this volume), Roberts (1980b)

continental margin environment similar to the volcanites on Smøla, although basalts with higher contents of TiO_2 appear to be absent on Smøla. The data from Bømlo presented here are more difficult to interpret and may represent different suites of basalts. Based on more complete geochemical data, Nordås *et al.* (this volume) have shown that the basic rocks of the Siggjo Group resemble the basalts from the Snake River and Yellowstone areas bordering the Basin and Range Province in western USA, suggesting an origin in a continental setting. Even though it is thought that the post-ophiolite volca-

nites on Bømlo and Stord represent the same major episode of magmatism, their different geochemical features may perhaps reflect an evolution as two separate central volcanic complexes.

Silurian continental within-plate volcanites

The Sagelvvatn shallow-marine to subaerial basalts within the Ullsfjord Nappe Complex (Binns 1978) in northern Norway are associated with supratidal platform carbonates of late Llandovery age (Bjørlykke and Olaussen 1981). These rocks occur in the trans-

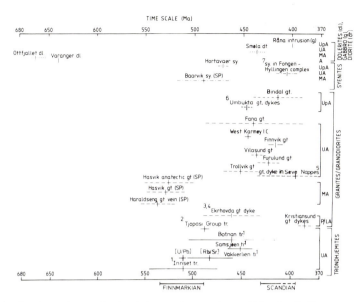

Fig. 6   Ages of Caledonian intrusive rocks (mainly granites and trondhjemites). Data sources for intrusions without a number have been taken from the compilation work by Ryan and Sturt (this volume). Other data sources are (1) I. Klingspor and D. G. Gee personal communication 1982, (2) Claesson *et al.* (1983), (3) Sturt *et al.* (1974), (4) Brueckner (1972), (5) Claesson (1982), (6) Claesson (1979) and (7) Wilson and Pedersen (1981), A = Autochthon, P = Para-autochthon, LA = Lower Allochthon, MA = Middle Allochthon, UA = Upper Allochthon, UpA = Uppermost Allochthon, SP = Seiland province

gressive sedimentary sequence lying stratigraphically above the Lyngen Gabbro, interpreted (Furnes *et al.*, this volume) as part of an ophiolite complex. The geochemical patterns of the basalts are shown in Fig. 5. In the TiO_2–Zr/P_2O_5 diagram, most of the data plot in the alkaline field and a few in the subalkaline field. In the Ti–Zr–Y diagram, the data generally plot well within the field of WPB.

5. Synorogenic Basic, Ultrabasic and Alkaline Intrusive Complexes

The Scandinavian Caledonides contain several isolated basic, ultrabasic and alkaline complexes of widely variable petrographic character as well as a single large magmatic province, the Seiland province, in which numerous intrusions were emplaced at different times in the early Caledonian structural–metamorphic evolution of this area of northern Norway. These complexes were emplaced into a variety of host rocks and possess contact-metamorphic aureoles which overprint pre-existing metamorphic fabrics and parageneses. Their depth of emplacement was such that the aureoles and the complexes themselves have been reworked during subsequent events in the continuing Caledonian tectonothermal evolution; they are, thus, synorogenic in character. In contrast to the intrusive rocks mentioned in earlier sections, no related volcanites are

known to be associated with these complexes. Apart from the Seiland province in the upper part of the Kalak Nappe Complex, the latter thought to belong to the group of nappes referred to as the Middle Allochthon, these complexes are restricted to the higher nappes of the Upper Allochthon and the Uppermost Allochthon. They are described according to their position in the tectonostratigraphy. Age-determinations are summarized in Fig. 6.

Seiland province in the Kalak Nappe Complex (Middle Allochthon)

Basic complexes

In the Seiland province in northern Norway, gabbroic complexes were emplaced during inter-kinematic episodes within the two phases of Finnmarkian deformation as well as during the intervening main phase of porphyroblastesis (Sturt *et al.* 1978). The gabbro complexes intrude a metamorphosed sedimentary sequence of probable late Precambrian to mid Cambrian age (Sturt *et al.* 1978) as well as the Precambrian gneissic basement to this sequence (Ramsay and Sturt 1977; Akselsen 1980; Ramsay *et al.* 1979). The Seiland province is associated with a 110 mgal positive gravity anomaly suggesting a downward extension into the lower crust (Brooks 1970). According to Robins and Gardner (1975), the magmatic activity is related to upwelling of a mantle diapir above a progressively-steepening, eastwards-dipping Benioff Zone. The basic magmatic activity appears to have been protracted but episodic. Rb–Sr whole-rock isochron ages suggest repeated emplacement of magmas from 540 ± 17 Ma to 490 ± 27 Ma (Sturt *et al.* 1978).

Concordant contacts appear to be the rule. Individual intrusions form extensive, thick sheets, steep-sided chonolithic bodies roughly ovate in plan (Robins and Gardner 1974) or lopoliths, in one case with a broad dyke-like keel (Hooper 1971). In all the major bodies, cumulate structures and textures are variably preserved and some exhibit cryptic variations (Robins and Gardner 1974; Gardner 1980; Robins 1982). The largest of the intrusions contain cumulate sequences 1–8 km in thickness.

Plutons emplaced prior to the second phase of deformation are uniformly tholeiitic in character whilst the later intrusions in the eastern part of the province generally appear to be related to alkaline olivine basalt magmas (Robins and Gardner 1975; Robins 1982). An exception to this pattern is the tholeiitic Lille Kufjord gabbro, the lower part of which contains repeated cyclic units in which peridotites grade upwards into leucogabbros (Oosterom 1955; Robins and Gardner 1975). During an early magmatic event in the second phase of deformation, calc-alkaline intrusions including high-K diorites, monzonites and monzogranites were emplaced in a restricted part of the Seiland province (Speedyman 1967; McClenaghan 1972; Robins and Gardner 1975).

The intrusions are associated with extensive aureoles in which Barrovian amphibolite-facies parageneses in metasediments are overprinted by high-grade cordierite- and almandine-bearing contact-metamorphic parageneses (Robins 1971; Sturt and Taylor 1972; Gardener 1980). The inner parts of aureoles are commonly composed of rheomorphic breccias with granitoid neosomes (Robins 1982). Inclusions of metasediments on the scale of cm-size xenoliths to large rafts are common both near the margins of plutons and within their layered sequences (Hooper 1971; Robins and Gardner 1974; Gardner 1980; Robins 1982). They show clear evidence of extensive reaction with the enclosing magmas (Gardner 1980; Robins 1982) and assimilation would appear to have played a significant role in the fractionation of at least some of the individual magma pulses.

Ultrabasic complexes

The high-temperature intrusive peridotites present in the Seiland province appear to be unique in the Scandinavian Caledonides. They bear certain similarities, however, with the synorogenic ultramafic intrusions of Alaska and the Ural mountains (Taylor 1967; Irvine 1974). The peridotites of the Seiland province were emplaced late in the second major phase of Finnmarkian deformation. They are, however, locally deformed and cut by alkaline dykes probably intruded at 490 ± 27 Ma (Sturt *et al.* 1978).

The ultrabasic intrusions contain relatively iron-rich picrites, variably amphibole- and plagioclase-bearing olivine clinopyroxenites, wehrlite and dunite (Oosterom 1963; Robins 1971; Bennett 1974; Sturt *et al.* 1980b). Orthopyroxene appears to be restricted to contacts with metasediments and noritic gabbros. All of the major rock types have clear intrusive relationships with the host layered gabbros. Dyke complexes in which ultrabasic rocks cut gabbro are spectacularly developed along the margins of the Bumannsfjord and Melkvann intrusions (Robins 1971; Sturt *et al.* 1980b). In the Reinfjord complex, pyroxenites, lherzolites and wehrlites exhibit gently-dipping rhythmic layering; contacts with metasediments and layered gabbro vary from concordant to strongly discordant (Bennett 1974). Pyroxene-rich ultrabasic rocks are developed along contacts with gabbro and around gabbro xenoliths and rafts. More than half of the Reinfjord complex consists of a dunite core. Dunite plays a lesser role in the Stjernøy, Melkvann and Bumannsfjord plutons. The latter intrusions do not contain cumulate structures and their petrographic variation appears to be partly due to assimilation of gabbro by the ultramafites. This mechanism is ascribed a major role in the evolution of the Bummannsfjord pluton (Sturt *et al.* 1980b). In the Melkvann body, some wehrlites and dunites appear to be secondary and produced by the replacement of olivine pyroxenites (cf. Hess 1960; Jackson 1961; Irvine 1974), particularly in association with transgressive wehrlitic dykes. Field evidence suggests the co-existence of peridotitic and gabbroic magmas, possibly due to anatexis (Sturt *et al.* 1980b) or remobilization of unconsolidated gabbroic cumulates.

The ultrabasic rocks are cut in all of the complexes by ultrabasic dykes, swarms of veins of highly variable composition, as well as regionally-developed picritic to olivine basaltic dykes (Robins and Takla 1979). The latter and some of the olivine pyroxenite dykes occasionally contain albite-bearing lherzolite and dunite nodules of probable mantle derivation (Robins 1975). The lherzolite nodules show textural evidence of re-equilibration during decreasing pressures (Robins 1975).

Both Bennett (1974) and Sturt *et al.* (1980b) assign mobile ultrabasic magmas a major role in the evolution of the complexes.

Alkaline intrusions and carbonatites

In the Seiland province, alkaline rocks, carbonatites and fenites form three major complexes emplaced into synorogenic gabbros or their metasedimentary host rocks during the closing stages of the Finnmarkian tectonothermal event. Dykes of carbonatite, alkaline pyroxenite and both feldspathoidal and nepheline-free syenite pegmatites are also widely developed in the province. Several rock types represented in the province are not otherwise known from orogenic terrains.

In the Breivikbotn complex on Sørøy, metasediments and metagabbros are intruded and fenitized by melanocratic to leucocratic syenites and later carbonatites and carbonatite breccias (Sturt and Ramsay 1965). The syenites are characterized by K-feldspar, albite–oligoclase aegerine augite, hastingsitic amphibole and calcite. Melanite appears to be a metamorphic mineral. Basic dykes were emplaced into the syenites prior to strong shearing and contain upper greenschist-facies parageneses. A syenite aplite which intrudes the syenites and is cut by a carbonatite has been dated by a Rb–Sr whole-rock isochron to 490 ± 27 Ma (Sturt *et al.* 1978). The banded carbonatite breccias contain highly-deformed inclusions of the adjacent rocks and appear to have been emplaced into their present position during a late expression of the second regional phase of deformation. The carbonatites are calcite-bearing and generally contain aegerine augite, amphibole, sphene and apatite. Some massive carbonatite sheets contain, in addition, significant amounts of biotite, and the carbonatite breccias frequently contain feldspars and other xenocrysts derived from inclusions, possibly during deformation.

The calcite silicocarbonatites within the Lillebukt alkaline complex (Heier 1961) are associated with laminated nepheline-free amphibole syenites, potassium-rich nepheline syenites and alkaline pyroxenites, all associated with characteristic fenites developed from their gabbroic hosts (Robins 1980). The apatite-rich alkaline pyroxenites occur in a dense swarm of parallel dykes which forms an envelope around much of the carbonatite. The swarm suggests that the complex had a ring form prior to the intense late deformation (Skogen 1980). The alkaline pyroxenites may be cumulates related to nephelinitic magmas (Kjøsnes 1981). Basic and ultrabasic dykes emplaced and deformed at various times during the magmatic evolution of the complex are metamorphosed in the upper greenschist facies. A late phase of zeolite-facies metamorphism has been recorded from the nepheline syenite intrusions (Bruland 1980). The mineralogical and geochemical variations in the highly xenolithic carbonatites can largely be explained by extensive assimilation of their host rocks (Strand 1981).

The Pollen carbonatite intrudes strongly fenitized gabbroic cumulates and was deformed during over-thrusting of a thrust slice containing metagabbros, syenodiorites and perthositic syenites. Fenitization transformed the gabbros into silicocarbonatites (Robins and Tysseland 1980, 1983).

Synorogenic alkaline pyroxenite dykes were emplaced by dilation into fenitized gabbro cut by basic and ultrabasic dykes on the island of Seiland (Robins 1974; Robins and Tysseland 1979). They were subsequently intruded by basic dykes and syenite pegmatites (Robins 1974).

Syenite pegmatites and sodic nepheline syenite pegmatites are widespread in the Seiland province They may be up to 100 m wide and several km in length and intrude metasediments, gabbros and ultrabasic rocks (Barth 1927; Robins 1972; Sturt and Ramsay 1965; Appleyard 1965). Narrow fenitic aureoles are developed along the contacts of the pegmatites (Robins and Tysseland 1979). The nepheline syenite pegmatites form zoned intrusions with carbonatite pegmatite (Robins 1972). Structural relationships between the syenite and carbonatite pegmatites suggest the simultaneous emplacement of immiscible liquids (Robins 1972). The pegmatites are locally deformed with the production of gneissic and mylonitic fabrics. On Sørøy, the pegmatites are closely associated with sodic nepheline syenite gneisses which appear to have developed through alkali metasomatism of both metasediments and gabbros (Sturt and Ramsay 1965; Appleyard 1965, 1974).

Basic and ultrabasic intrusions in the Upper Allochthon

Several large synorogenic basic intrusions are present in the higher nappes of the Upper Allochthon. These include: the Fongen–Hyllingen complex lying either in the Meråker or overlying Gula Nappe of the Trøndelag region, the Artfjället and Krutfjell gabbros in the Krutfjellet Nappe (Upper Köli Nappes) of Västerbotten, Sweden and southern Nordland, Norway, and the Honningsvåg complex including both basic and ultrabasic intrusions in the Magerøy Nappe of the Finnmark region. A brief description of these intrusions is provided below. The synorogenic Sulitjelma Gabbro, lying at a similar tectonostratigraphic level as the Artfjället and Krutfjell gabbros, has been referred to earlier in the context of the Sulitjelma ophiolite (see section 3).

The Fongen–Hyllingen complex

This complex is a major, strongly-differentiated pluton emplaced into probable Ordovician metavolcanites and metapelites (Hersjø Group). A Rb–Sr whole-rock isochron from syenitic differentiates has yielded an age of 405 ± 9 Ma (Wilson and Pedersen 1981). Emplacement followed deformation and biotite-grade metamorphism and pre-dated the main regional porphyroblastesis (Olesen et al. 1973). The contact metamorphic paragenesis suggests intrusion at 5–6 kb and the inner part of the aureole is characterized by partial melting (Wilson and Olesen 1975).

The intrusion probably has the form of a bowl or sheet underlain by a feeder-dyke (Wilson and Olesen 1975). Rhythmic layering and a wide range of cumulate structures are developed within the intrusion. The extensive cryptic variation gives evidence of the emplacement of several batches of a potassium-poor olivine tholeiite parent (Esbensen et al. 1978) which fractionated under moderate and increasing P_{H_2O} (Wilson et al. 1981). The fractionation exhibited by the complex is intermediate between typical tholeiitic and calc-alkaline trends and resulted in small amounts of syenitic differentiates.

The Artfjället gabbro

This gabbro is one of several intrusions ranging in composition from gabbroic to granitic within the Norra Storfjället Lens of the Krutfjellet Nappe (Upper Köli Nappes). It was emplaced into metasediments during the second phase of deformation, after the establishment of amphibolite-facies metamorphic conditions and prior to nappe emplacement (Senior and Otten, this volume). The gabbro is associated with a contact aureole of calcsilicate and garnet–kyanite hornfelses (Senior and Otten, this volume). Magmatic layering is locally well developed. The cumulate sequence of olivine gabbros overlain by picrites, olivine gabbros and gabbroic norites, and the associated dyke rocks appear to have sub-alkaline affinities intermediate between tholeiitic and calc-alkaline magma types (Senior and Otten, this volume).

The Krutfjell gabbro complex

This gabbro complex resembles in many respects the Artfjället gabbro. It lies within the Krutfjellet Nappe but in a separate megatectonic lens along strike from the lens in which the Artfjället gabbro occurs. The gabbro intruded immediately after the peak of medium-grade regional metamorphism of the metasedimentary host rocks (Mørk 1979). Contact

metamorphism was accompanied by anatexis in a narrow zone close to the gabbro margin.

The Honningsvåg complex

The subalkaline Honningsvåg complex was emplaced into metasediments of Silurian (early Llandovery) age within the Magerøy Nappe and produced a high-grade contact-metamorphic aureole (Holtedahl et al. 1960; Curry 1975). The complex covers about 45 km^2 and is roughly funnel shaped, extending to a depth of more than 6 km (Lønne and Sellevoll 1975). The evolution of the complex was protracted and took place during the first of two major phases of deformation and the intervening interkinematic episode. Sheets and sills up to 1 km thick of olivine-free two-pyroxene metagabbro were emplaced during the first phase of deformation; a layered sequence up to 2 km thick exhibiting at least five cyclic units of peridotite, allivalite and eucrite crystallized interkinematically; homogeneous olivine-bearing two-pyroxene gabbros and quartz gabbros were subsequently emplaced along the margins of the layered intrusion; xenolithic dykes and sheets ranging in composition from dunite to picrite represent the latest magmatic event in the complex and appear to have been closely associated with remobilization of part of the layered sequence (Curry 1975).

Basic and alkaline intrusions in the Uppermost Allochthon

Synorogenic basic and alkaline complexes intruding rocks of uncertain age are conspicuous in the Uppermost Allochthon. Basic intrusions include the Umbukta complex in the Rödingsfjället Nappe Complex of southern Nordland, Norway and Västerbotten, Sweden, the Råna mafic intrusion in the Narvik Group of northern Nordland, lying at a tectonostratigraphical level possibly equivalent to the Rödingsfjället Nappe Complex (Stephens et al., this volume), and the Mosjøen gabbro in the Helgeland Nappe Complex of southern Nordland. Alkaline intrusions occur at Misvaerfjorden in Nordland, near the contact between the Beiarn and Fauske Nappes, and on Hortavaer in Nord-Trøndelag, possibly within the Helgeland Nappe Complex. A brief description of these intrusions, except the Mosjøen gabbro for which data are at present lacking, follows below.

The Umbukta complex

This complex is situated within the Rödingsfjället Nappe Complex and was emplaced earlier than granitic dykes dated to 447 ± 7 Ma (Claesson 1979;

Flodberg and Stigh 1981). The complex contains ultrabasic and gabbroic cumulates cut by a dense swarm of basic dykes. The latter are subalkaline and have geochemical similarities with within-plate tholeiitic basalts (Flodberg and Stigh 1981).

The Råna mafic intrusion

This intrusion was emplaced into two-mica and calc-silicate gneisses belonging to the Narvik Group prior to the third phase of deformation and during amphibolite-facies metamorphism of the country rock (Boyd and Mathiesen 1979). On the basis of Rb–Sr whole-rock and mineral isochrons, the intrusion has been dated to 400 Ma; this is the same as the age of metamorphism inferred from the country rocks (Boyd and Mathiesen 1979).

The intrusion may have the form of an inverted truncated cone. Rhythmic layering is present locally but is generally uncommon. The various rock units in the pluton have a concentric distribution, marginal peridotites and pyroxenites surrounding norites and gabbros which become quartz-bearing towards the core of the body (Foslie 1920, 1921; Boyd and Mathiesen 1979). The textural evidence of an olivine-liquid reaction relationship which resulted in the crystallization of orthopyroxene, and the interstitial quartz, orthoclase, biotite and amphibole present in the central quartz norites and gabbros (Foslie 1920, 1921; Boyd and Mathiesen 1979) can leave little doubt as to the subalkaline character of the Råna parental magma.

The Skar intrusion

This intrusion, exposed beside Misvaerfjorden in Nordland, is emplaced discordantly into mica schists within the Venset Group close to the contact between the Beiarn and Fauske Nappes (Farrow 1974; Cooper and Bradshaw 1980). It consists of lensoid bodies and cross-cutting dykes of deformed and metamorphosed biotite–apatite pyroxenite and mesotype hornblende syenite. The alkaline rocks are cut by granitic dykes and also occur as xenoliths in the adjacent, in part gneissose Brekfjell granite (Farrow 1974). Small amounts of quartz are present in the syenites and locally in the pyroxenites.

Hortavaer

On Hortavaer, there occur variable monzogabbros and syenites. The contact with nearby mica gneisses, possibly belonging to the Helgeland Nappe Complex (Gustavson 1975), is not exposed. The monzogabbros contain marble and calc-silicate inclusions and are cut by xenolithic syenite and granite dykes (Gus-

tavson and Prestvik 1979). The main body of syenite outcrops around the basic rocks and also encloses marble rafts. Most of the rocks contain accessory calcite, and the most leucocratic syenites are quartz-bearing. A Rb–Sr whole-rock isochron from the complex has given an age of 471 ± 5 Ma (B. Sundvoll personal communication in Gustavson and Prestvik 1979).

6. Synorogenic Granitoids

Granitoids in the Scandinavian Caledonides show a wide range of compositions. Indeed, the limited number of detailed studies of individual 'granites' which are available demonstrate variation from diorite or quartz diorite to granite, even within the bodies themselves. Granitoids forming a subvolcanic infrastructure to island-arc volcanites, plagiogranite associated with ophiolites and subordinate acid intrusions within the Seiland province have already been referred to and will not be discussed further here. The granitoids treated here are not directly associated with volcanites. They are either discordant to pre-existing metamorphic fabrics or intrude obducted ophiolite fragments and are themselves affected by later deformation; they are, thus, synorogenic in character. The granitoids, age-determination data for which are summarized in Fig. 6, are treated according to their position in the tectonostratigraphy. They are concentrated in the higher nappes of the Upper Allochthon and in the Uppermost Allochthon. Exceptions, not discussed in any further detail, include granitic dykes on the island of Sotra (Bergen) and in the Kristiansund region belonging to the Parautochthon or Lower Allochthon (see earlier discussion) as well as migmatitic rocks and synorogenic pegmatite on Åreskutan in the Seve Nappes (Upper Allochthon) dated by Rb–Sr whole-rock techniques to 414 ± 27 Ma and 395 ± 40 Ma, respectively (Claesson 1982).

Trondhjemite and granite in the Trondheim Nappe Complex and equivalents (Upper Allochthon)

Intrusive rocks of trondhjemitic affinity have a fairly widespread distribution in the Trondheim Nappe Complex (Wolff and Roberts 1980). They show a range in composition from trondhjemite *per se* through tonalite to occasional granodiorite and form relatively small bodies, the largest being the Innset Pluton. They are mainly concentrated in the Gula Nappe, though they are also present in the Støren and Meråker Nappes where they cut both the earliest Ordovician and/or pre-Ordovician volcanic units (the Støren, Fundsjø and Hersjø Groups), and overlying sediments. Higher in the Ordovician succession, clasts of trondhjemitic rocks are found in conglomerates. The petrogenetic model that best fits the data from trondhjemites at Follstad, near Støren, appeals to equilibrium melting of a low-K_2O tholeiitic basalt (Size 1979). The low initial $^{87}Sr/^{86}Sr$ ratios (c. 0.7050, Klingspor and Gee 1981) are compatible with this model. Rb–Sr whole-rock isochron ages from various trondhjemites (Fig. 6) range from 509 to 451 Ma and a zircon U–Pb age from the Vakkerlien trondhjemite gave $509 \pm \frac{5}{4}$ Ma (I. Klingspor and D. G. Gee personal communication 1982).

The trondhjemitic rocks generally cut deformational structures (foliation and folds) in the host rocks though, in the Gula Nappe, some are apparently affected by the earliest deformational episodes. The occurrence of a *Dictyonema* fauna at Nordaunevoll (Størmer 1941), in sediments probably influenced by this early deformation (Klingspor and Gee 1981), combined with the age-dating results, implies that at least some deformation occurred in the early Ordovician. The effects of later deformation and at least greenschist facies metamorphism is reflected in the Silurian dates obtained from Rb–Sr mineral isochrons and K–Ar dating results (Wilson et al. 1973).

Further north in the Scandinavian Caledonides, trondhjemite and granite occur within the Krutfjellet and Gasak Nappes and the Skibotn Nappe Complex, i.e. in tectonic units thought to be equivalent to the Gula Nappe in the Trondheim Nappe Complex (Binns 1978; Sandwall 1981a).

Within the Norra Storfjället Lens of the Krutfjellet Nappe (Häggbom 1978), dykes and sills of trondhjemite and granite (Senior 1978) as well as larger plutons of microcline–biotite granite, tonalite and quartz-bearing hornblende diorite (Mulder 1951) intruded country rock schists and amphibolites after two phases of deformation and amphibolite-facies metamorphism (Senior and Otten, this volume) but prior to later deformation including thrusting. Rb–Sr whole-rock isochrons for the Vilasund granite (Gee and Wilson 1974) and migmatites within this lens (Reymer 1979) indicate ages of 438 ± 6 Ma and 442 ± 30 Ma, respectively, and quite distinct initial $^{87}Sr/^{86}Sr$ ratios (0.7055 and 0.72, respectively).

The granitic rocks within the Gasak Nappe of the Sulitjelma area (Wilson 1981) intruded synorogenically the non-fossiliferous Sulitjelma Schist Sequence and yield Rb–Sr whole-rock isochron ages of 424 ± 11 Ma (Furulund granite), 422 ± 8 Ma (Baldoaivve dykes) and 418 ± 14 Ma (Duoldagop dyke). Initial $^{87}Sr/^{86}Sr$ ratios are variable (0.709, 0.704, and 0.715, respectively). According to Wilson (1981) the K-rich Furulund granite intruded

between the first and second phases of deformation, after thrusting of the schists over lower-grade fossiliferous units (Ordovician or Silurian).

According to K. B. Zwaan (personal communication 1981), the Trollvik granite in the Skibotn Nappe Complex of North Troms cuts deformational structures interpreted to be of Finnmarkian age. A Rb–Sr whole-rock isochron indicates an intrusion age of 452 ± 13 Ma and a high (0.7105) initial $^{87}Sr/^{86}Sr$ ratio (Dangla et al. 1978).

Granite in the undifferentiated Upper Allochthon of the southwest and north Norwegian Caledonides

The West Karmøy Igneous Complex

This complex consists of a succession of intrusive phases ranging from quartz diorite through adamellite and granodiorite to biotite-granite (Ledru 1980). Along its eastern contact, the complex intrudes the plutonic zone of the Karmøy ophiolite. In the southern part of Karmøy, the complex is overlain by a Middle(?)–Upper Ordovician/Silurian transgressive sedimentary sequence, dominated by coarse clastics (Skudeneset Group). Several of the intrusions are highly xenolithic, containing fragments derived both from the Karmøy ophiolite and also a gneissic basement complex. The geochronology (Sturt et al. 1979; Ledru 1980) of late granodioritic dykes in the complex, indicating an age of approximately 445 Ma, suggests that emplacement occurred only shortly prior to major uplift and unroofing of the complex and deposition of the overlying transgressive clastic sequence. The dating of the granodioritic dykes fits closely with the age obtained from the Kattnakken rhyolites on Stord (Priem and Torske 1973) which were erupted in a continental environment. Based on the presence of the gneissic xenoliths and the high (0.7210) initial $^{87}Sr/^{86}Sr$ ratio, Ledru (1980) suggested generation of the West Karmøy Igneous Complex by deep-seated crustal anatexis in a thickened crust of Precambrian basement gneisses beneath the Karmøy ophiolite. He considered that partial melting began as the result of crustal thickening related to ophiolite obduction. Alternatively, a model involving subduction of an oceanic plate beneath a continental margin of Andean type may be realistic.

Sunnhordaland Batholith

The term Sunnhordaland Batholith was suggested by Kolderup and Kolderup (1940) to refer to the impressive development of granitic rocks between Bergen and Bømlo/Stord. Here, major plutons are clearly intrusive into the country rocks. The plutons are of at least two generations, intruding prior to and after deposition of the Moberg Conglomerate. Reconnaissance work by one of us (B. A. Sturt with T. B. Andersen) on the island of Selbjørn indicates that pre-Moberg granites are both syn- to post-orogenic with respect to a schist/psammite sequence metamorphosed in amphibolite facies and unconformably overlain by the Moberg Conglomerate. The unconformity beneath the Moberg Conglomerate is cut by later granites and the rocks both above and beneath the unconformity are affected by contact metamorphism. In general, the age-relations between the individual plutons is little understood and only limited work on their petrography has been carried out. Reconnaissance work suggests that the granite plutons on Rekstern and Huftarøy are normally zoned bodies with porphyritic monzonite in their marginal portions and granites occupying the central portions of the massifs. Satellite dykes of the Rekstern pluton cut the rocks of the Holdhus Group (late Ordovician/early Silurian in age) and are also affected by the deformation of this sequence; similar observations were made in relation to the later Selbjørn granites. Unfortunately, no geochronology is yet available for the granitic and associated rocks of the Sunnhordaland Batholith.

The Fanafjell/Krosnes granites

These granites lie above a late thrust plane. They contain several intrusive phases dominated by a medium- to coarse-grained microcline granite, cut by finer-grained porphyritic varieties (Kolderup and Kolderup 1940). The Fana granite has yielded a Rb–Sr whole-rock isochron age of 441 ± 50 Ma and an initial $^{87}Sr/^{86}Sr$ ratio of 0.7063 (± 0.0037) (Brueckner 1972).

The Finnvik Granite

The Finnvik Granite on Magerøy intruded a metasedimentary sequence, in part younger than early Llandovery in age, and was emplaced at the metamorphic maximum between two principal phases of folding (Andersen 1981). It is, thus, synorogenic with respect to the Scandian tectonothermal activity. A Rb–Sr whole-rock isochron from the granite indicates an age of 411 ± 11 Ma (Andersen et al. 1982).

Granitoids in the Uppermost Allochthon

Granitic dykes near Umbukta in the Rödingsfjället Nappe Complex

Granitic dykes, up to a few metres in thickness and generally medium-grained, intrude partly migmatitic schists and the basic/ultrabasic Umbukta complex in the Rödingsfjället Nappe Complex of northwestern Västerbotten, Sweden and central Nordland, Norway. The dykes cross-cut the complex foliation in the schists and are only locally affected by later deformation. A Rb–Sr whole-rock isochron for these dykes gives an age of 447 ± 7 Ma (Claesson 1979) in close agreement with the dates for the Vilasund granite and the migmatites in the underlying Krutfjellet Nappe. The relatively low initial $^{87}Sr/^{86}Sr$ ratio (0.7060) of the dykes has been interpreted (Claesson 1979) in terms of anatexis of relatively young sediments.

Bindal Batholith in the Helgeland Nappe Complex

The Bindal Batholith, lying in the Helgeland Nappe Complex, is a general term that can be applied to the large intrusive massif extending from south of Bindalsfjord to north of Mosjøen (Rekstad 1910, 1915, 1917; Gustavson 1976; Myrland 1972, 1974). There are a large number of individual plutons in this batholithic terrain and the lithologies vary considerably in composition. Myrland (1972) divided the southern part of the terrain into two major massifs. The southern one he termed the Bindal Massif consisting of even-grained granite/granodiorite, porphyritic granite/granodiorite, quartz diorite, monzonite/monzodiorite and gabbroid rocks. Reconnaissance work by one of us (B. A. Sturt 1980) suggests that the rocks of the Bindal Massif intruded through both a metasedimentary sequence and ophiolitic rocks. Dating by Priem et al. (1975) provided an age of 415 ± 26 Ma for late-stage phases of the granitic massif and an initial $^{87}Sr/^{86}Sr$ ratio of 0.7077. The northern massif, termed the Velfjord Massif, consists of hornblende gabbro, hypersthene monzodiorite, diorite with transition to monzodiorite, quartz monzonite and granitic/quartz dioritic rocks (Myrland 1972). There is no reliable control on the age of the country rocks, though Myrland (1972) cites a whole-rock Rb–Sr age of 512 ± 36 Ma for gneisses from Fjordholmene. There is similarly no constraint that can be placed on the timing of individual plutons. Råheim (personal communication 1981) has indications of a late Precambrian age for some of the granitic rocks in the northwest of the batholithic terrain but, as yet, it is not possible to show how these rocks fit into the general picture.

Concluding Remarks on Magma Evolution in Time and Space

In any model for the tectonic evolution of the Scan-

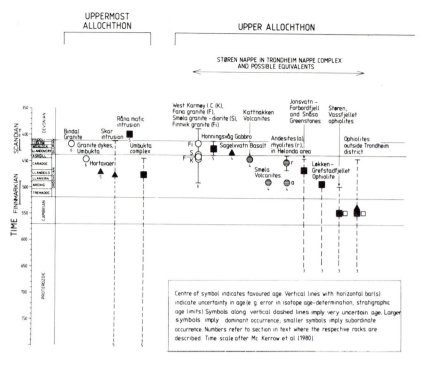

Fig. 7   Magmatic rocks in time and space in the Scandinavian Caledonides

dinavian Caledonides, assessment needs to be made of the changes in magma type both in space—along and across the orogen—and in time. Such assessment is hampered by uncertainties in along-strike correlation, limited knowledge of the individual displacements of nappes containing the magmatic sequences and a paucity of age-determination data on the magmatic rocks. It is perhaps the two latter factors which pose the most serious problems in assessing the magmatic evolution of the orogen, virtually all conclusions being highly tentative. However, with these limitations in mind, an attempt has been made to assess at least certain aspects of this evolution (see Fig. 7). In this figure, different magma types have been plotted against time (vertical axis) and position in the tectonostratigraphy (horizontal axis), successively higher tectonic units being derived from environments progressively further west of the present Norwegian west coast (Gee 1978).

The earliest magmatism, preserved mainly in the Middle Allochthon and possibly also in the Seve Nappes of the Upper Allochthon, is basic and predominantly tholeiitic in composition; a subordinate group of basic dykes in the Middle Allochthon is alkaline and shows higher Ti, K, P. Rb and Zr contents. The geochemistry of these rocks and the nature of the enclosing metasediments have been interpreted in terms of rifting and initial ocean-basin (Iapetus) opening in an ensialic environment (Sol-

yom et al. 1979a, b; Hill 1980). The tectonostratigraphic position suggests that this ensialic environment formed the outer part of the Baltoscandian miogeocline (Gee 1975).

The ophiolite fragments lie in distinctly higher tectonic units. Although the ophiolites (Furnes et al., this volume) interpreted to lie within the Lower Hovin Group of the Støren Nappe are thought to have been generated in an Ordovician marginal basin (Gale and Roberts 1974; Roberts 1980b; Grenne and Roberts 1980; Ryan et al. 1980), other ophiolites, in particular the Karmøy ophiolite (Sturt et al., 1980a), are thought to have been derived from the Iapetus Ocean itself (Furnes et al. 1980a). However, since the lower age limit for the possible Iapetus-derived ophiolites is not known, the temporal relationship betweeen these fragments of oceanic crust and the initial rifting episode cannot, as yet, be resolved. The oldest minimum age is provided by the Vassfjellet ophiolite in the Støren Nappe where a mid Arenig or earlier age for the ophiolite may be inferred. The locally substantial development of pelagic sediments within most of the ophiolite complexes suggests derivation from truly ensimatic environments. In the ophiolites, low-K tholeiitic basic rocks dominate. However, in the Karmøy and Leka ophiolites, an evolution to alkaline basalt intercalated with pelagic sediments is apparent.

The presumed later early Palaeozoic magmatic

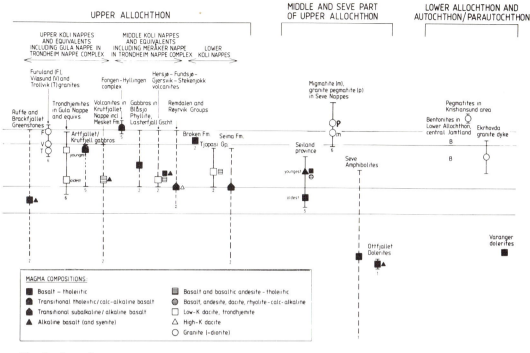

Fig. 7   (cont.)

history is complex with no simple evolutionary trend being evident (Fig. 7). A distinctive feature of this period is the occurrence of both mildly-tholeiitic basalts and basaltic andesites with low K, Ti and Cr contents in intimate association with low-K dacite–rhyodacite, and medium-K calc-alkaline basalts to rhyolites. Such rocks as well as the plutonic complex in the Seiland province have been interpreted to indicate the establishment of one or more convergent plate margins (Gale and Roberts 1974; Robins and Gardner 1975; Stephens 1977b, 1982; Roberts 1980b; Nordås et al., this volume). However, such magmatism was not restricted to a single time interval but occurred at different times probably throughout the Cambrian and Ordovician; such a picture suggests a complex closure history for Iapetus. Eastwards-directed subduction has been suggested for the Cambrian-early Ordovician Seiland plutonic province, the low-K, mildly-tholeiitic arc complex of early Ordovician (in part Tremadoc) age in the Middle and Upper Köli Nappes, and the calc-alkaline sequences of late Arenig–Ashgill (?) age in the Støren Nappe. Relics of a low-K, mildly-tholeiitic arc of early and possibly mid Ordovician age are also present in the Lower Köli Nappes. Igneous activity related to subduction processes was followed by (Lower and Middle Köli Nappes) or occurred possibly in close temporal relationship with (Støren Nappe) purely basic magmatism of predominantly tholeiitic character and of Ordovician to

early Silurian age. Compared with the basic rocks thought to be generated during subduction-related processes, these tholeiitic rocks show higher Ti (and Cr) contents and higher degrees of Ti (and Fe) enrichment. Such magmas show closer affinity to an extensional rather than subduction-related tectonic régime.

Probably the most complete documentation of the evolution of particularly basic magmas emerges from the various intrusions in the Seiland province (Robins and Gardner 1975). Here, a temporal sequence from tholeiitic basalt magma through localized high-K calc-alkaline magmas to predominantly alkaline magmas has been recognized (Fig. 7). This magmatic evolution, thought to be related to subduction processes along an active continental margin (Robins and Gardner 1975), occurred during Finnmarkian deformation over a period of c. 50 Ma (Cambrian–early Ordovician). The tectonic significance of other synorogenic basic intrusions occurring in higher allochthonous units is less well understood. The question of whether any of these basic intrusions were related to the magmatism suggesting extension and incipient basin opening in the Ordovician and early Silurian remains open.

Limited groups of related nappes within the Upper Allochthon show distinct magmatic evolutionary trends during the Ordovician and Silurian. Three such trends are described and their significance inferred (see Fig. 7):

1. In both the Lower and Middle Köli Nappes, the magmatism shows a distinct evolution in time. Transitional basalts thought to lie at the stratigraphic base pass upwards into bimodal sequences consisting of low-K dacite–rhyodacite and subordinate, arc-related (low K, Ti and Cr), tholeiitic basalt and basaltic andesite. These, in turn, are followed upwards by purely basaltic rocks of tholeiitic (and subordinately alkaline) character showing higher Ti and Cr contents and more pronounced Ti (and Fe) enrichment. Such cycles in the Lower and Middle Köli are inferred to be of different ages but are thought to reflect probably ensimatic arc basements and a similar later evolution involving incipient arc rifting (Stephens 1977b, 1982).

2. The limited amount of age-determination data on synorogenic acid intrusions in the Upper Köli Nappes and equivalents tentatively suggest that the granitic rocks (higher K_2O contents) show younger ages and more variable initial $^{87}Sr/^{86}Sr$ ratios than the trondhjemites (lower K_2O contents). The timing of intrusion of granitic melts to the mid to late Ordovician and Silurian and the range in their initial $^{87}Sr/^{86}Sr$ ratios are in good agreement with similar data for such higher-K acid rocks in the Støren and possibly equivalent nappes, and the Uppermost Allochthon. The granitic rocks in the Upper Köli Nappes and equivalents, as in higher tectonic units, are spatially associated with synorogenic gabbro intrusions.

3. In the Støren Nappe and equivalents, medium-K, calc-alkaline basalts to rhyolites either lie or are inferred to lie on top of ophiolite complexes dominated by low-K tholeiitic basalts. Plutonic complexes of quartz diorite to granite are of similar age to or younger than the calc-alkaline volcanites. In the Støren Nappe itself, the Arenig–Llanvirn calc-alkaline basalts to rhyolites and younger acid volcanites are thought to be broadly contemporaneous with rifting-related tholeiitic basalts interpreted as parts of ophiolites (Furnes et al., this volume) and showing higher Ti contents and more pronounced degrees of Ti (and Fe) enrichment compared with the calc-alkaline basalts; a mature arc passing eastwards into a marginal oceanic basin has been inferred (Roberts 1980b; Grenne and Roberts 1980). It is significant that the calc-alkaline volcanites are not developed in the same tectonic unit(s) as the possibly older (see Fig. 7) tholeiitic, arc-related rocks in the underlying Upper and Middle Köli Nappes and equivalents. This implies both a spatial as well as temporal differentiation of the magmas related to plate convergence. It would

appear that the inferred marginal basin in the Støren Nappe was an inter-arc basin with an active calc-alkaline arc to the west and a remnant tholeiitic arc to the east.

In summary, there seems to be evidence in the magmatic rocks of the Scandinavian Caledonides for the embryonic and youthful stages of ocean (Iapetus) development. Although there is also evidence from the magmatic rocks for closure of such an ocean, it is apparent that convergence was complex and occurred possibly over a period as long as c. 100 Ma. During the period of plate convergence, different phases of magmatic activity related to subduction processes were followed by (Lower and Middle Köli Nappes) or occurred possibly in close temporal relationship with (Støren Nappe) magmatism related to rifting and basin opening. For these reasons, no simple cycle of increasing K_2O contents at equivalent SiO_2 contents or diminishing degree of iron enrichment in basaltic rocks has been noted. However, within four more limited groups of related nappes (Lower, Middle, and Upper Köli Nappes and Støren Nappe, and their respective equivalents), certain magmatic evolutionary trends have been detected related to varying Ti and Cr contents and degrees of Ti (and Fe) enrichment in basic rocks, and varying K_2O contents at equivalent SiO_2 values. The former are well illustrated in the Lower and Middle Köli Nappes (and equivalents) and the latter in the Upper Köli Nappes (and equivalents) and, to a lesser extent, in the Støren Nappe. Concentration of truly granitic rocks with variable (0.704–0.721) initial $^{87}Sr/^{86}Sr$ ratios both spatially in the highest nappes (Upper Köli and equivalents, Støren Nappe and equivalents, and Uppermost Allochthon), where they intrude after some deformation and metamorphism, and temporally in the mid to late Ordovician and Silurian is thought to reflect increased heat flow during the Ordovician and Silurian in areas lying furthest out to the west and composed of originally thick or later tectonically-thickened continental crust. The spatial association of these generally poorly known granitic rocks with the large gabbro intrusions may be significant from the point of view of heat input. It is suggested that the spatial differentiation of magmatic evolutionary trends throughout the Ordovician and Silurian may be related to the presence of one or more relict oceanic segments and/or the opening of new basins during this period, thus allowing different areas to develop separate magmatic evolutionary trends. Complete elimination of oceanic segments during Silurian–Devonian continent–continent collision subsequently telescoped together these different sequences.

Acknowledgements

M.B.S. wishes to acknowledge special funding by the Geological Survey of Sweden under the auspices of the IGCP Project No. 60, *Correlation of Caledonian Stratabound Sulphides*. B.A.S., H.F. and B.R. acknowledge the financial support of the Norwegian Research Council for the Sciences and Humanities (N.A.V.F.) during the course of this study. A. Boyle and A. Reinsbakken are thanked for providing unpublished analytical data. The authors wish to thank D. G. Gee for helpful comments on an earlier draft of the manuscript.

References

Akselsen, J. 1980. *Tectonic and metamorphic development of the Pre-Cambrian and Caledonian rocks of Northeastern Seiland, Finnmark.* Unpub. Cand Real thesis, Univ. of Bergen.

Andersen, T. B. 1981. The structure of the Magerøy Nappe, Finnmark, North Norway. *Norges geol. Unders.*, 363, 1–23.

Andersen, T. B., Austrheim, H., Sturt, B. A., Pedersen, S. and Kjaersrud, K. 1982. Rb–Sr whole rock ages from Magerøy, North Norwegian Caledonides. *Norsk geol. Tidsskr.*, 62, 79–85.

Andreasson, P. G., Solyom, Z. and Roberts, D. 1979. Petrochemistry and tectonic significance of basic and alkaline–ultrabasic dykes in the Leksdal Nappe, northern Trondheim Region, Norway. *Norges geol. Unders.*, 348, 47–72.

Andreasson, P. G., Gee, D. G. and Sukotjo, S. this volume. Seve eclogites in the Norrbotten Caledonides, Sweden.

Appleyard, E. C. 1965. Preliminary description of the geology of the Dønnesfjord area, Sørøy. *Norges geol. Unders.*, 231, 144–164.

Appleyard, E. C. 1974. Syn-orogenic igneous alkaline rocks of eastern Ontario and northern Norway. *Lithos*, 7, 147–169.

Barth, T. F. W. 1927. Die Pegmatitgänge des kaledonischen Intrusivgesteine im Seiland-Gebiete. *Morske Vid. Akad. Skr.* I, No. 8.

Beckholmen, M. 1982. Mylonites and pseudo-tachylites associated with thrusting of the Köli Nappes, Tännforsfältet, central Swedish Caledonides. *Geol. Fören. Stockh. Förh.*, 104, 23–32.

Beckholmen, M. and Noro, A. 1981. The Saxvallklumpen metabasite in the garbenschiefer of the Tännforsfältet Köli in Jämtland, Central Swedish Caledonides. *Terra Cognita*, 1, 34.

Beckinsale, R. D., Reading, H. G. and Rex, D. C. 1975. Potassium–argon ages for basic dykes from East Finnmark: stratigraphical and structural implications. *Scott. J. geol.*, 12, 51–65.

Bennett, M. C. 1974. The emplacement of a high temperature peridotite in the Seiland province of the Norwegian Caledonides. *J. geol. Soc. London*, 130, 205–228.

Beskow, G. 1929. Södra Storfjället im südlichen Lappland. *Sver. geol. Unders.*, C 350, 335 pp.

Binns, R. E. 1978. Caledonian nappe correlation and orogenic history in Scandinavia north of lat. 67° N. *Geol. Soc. Am. Bull.*, 89, 1475–1490.

Bjørlykke, A. and Olaussen, S. T. 1981. Silurian sediments, volcanics and mineral deposits in the Sagelvvatn area, Troms, North Norway. *Norges geol. Unders.*, 365, 1–38.

Boyd, R. and Mathiesen, C. O. 1979. The nickel mineralization of the Råna mafic intrusion, Nordland, Norway. *Can. Mineral.*, 17, 287–298.

Boyle, A. P. 1980. The Sulitjelma amphibolites, Norway: Part of a Lower Palaeozoic ophiolite complex? In Panayiotou, A. (Ed.), *Ophiolites. Proc. Int. Ophiolite Symp. Cyprus 1979*, 567–575.

Boyle, A. P., Griffiths, A. J. and Mason, R. 1979. Stratigraphical inversion in the Sulitjelma Area, Central Scandinavian Caledonides. *Geol. Mag.*, 116, 393–402.

Boyle, A. P., Hansen, T. S. and Mason, R. this volume. A new tectonic perspective of the Sulitjelma Region.

Brooks, M. 1970. A gravity survey of coastal areas of West Finnmark, northern Norway. *Q. J. geol. Soc. London*, 125, 171–192.

Brueckner, H. K. 1972. Interpretation of Rb–Sr ages from the Precambrian and Palaeozoic rocks of southern Norway. *Am. J. Sci.*, 272, 334–358.

Bruland, T. 1980. *Lillebukt alkaline complex: Mafic dykes and alteration zones in Nabberen nepheline syenite.* Unpub. Cand. Real. thesis. Univ. of Bergen.

Bruton, D. and Bockelie, J. F. 1979. The Ordovician sedimentary sequence on Smøla, west central Norway. *Norges geol. Unders.*, 348, 21–31.

Bruton, D. L. and Bockelie, J. F. 1980. Geology and palaeontology of the Hølonda area, western Norway—a fragment of North America? In Wones, D. R. (Ed.), *The Caledonides in the USA. Virginia Polytechnic Inst. and State Univ., Dept. of Geol. Sci., Mem.*, 2, 41–47.

Carstens, H. 1960. Stratigraphy and volcanism of the Trondheimsfjord area, Norway. *Norges geol. Unders.*, 212b, 23 pp.

Chaloupsky, J. and Fediuk, F. 1967. Geology of the western and north-eastern part of the Meråker area. *Norges geol. Unders.*, 245, 9–21.

Claesson, S. 1976. The age of the Ottfjället dolerites of the Särv Nappe, Swedish Caledonides. *Geol. Fören. Stockh. Förh.*, 98, 370–374.

Claesson, S. 1979. Pre-Silurian orogenic deformation in the north–central Scandinavian Caledonides. *Geol. Fören. Stockh. Förh.*, 101, 353–356.

Claesson, S. 1982. Caledonian metamorphism of Proterozoic Seve rocks on Mt. Åreskutan, southern Swedish Caledonides. *Geol. Fören Stockh. Förh.*, 103, 291–304.

Claesson, S. and Roddick, J. C. 1983. $^{40}Ar/^{39}Ar$ data on the age and metamorphism of the Ottfjället dolerites, Särv Nappe, Swedish Caledonides. *Lithos*, 16, 61–73.

Claesson, S., Klingspor, I. and Stephens, M. B. 1983. U–Pb and Rb–Sr isotopic data on the Ordovician volcanic/subvolcanic complex from the Tjopasi Group, Köli Nappes, Swedish Caledonides. *Geol. Fören. Stockh. Förh.*, 105, 9–15.

Cooper, M. A. and Bradshaw, R. 1980. The significance of basement gneiss domes in the tectonic evolution of the Salta region, Norway. *J. geol. Soc. London*, 137, 231–240.

Curry, C. J. 1975. *A regional study of the geology of the Magerøy basic igneous complex and its envelope.* Unpub. Ph.D. thesis, Univ. of Dundee.

Dangla, P., Demange, J. C., Ploquin, A., Quernadel, J. M. and Sonet, J. 1978. Données géochronologiques sur les Calédonides Scandinaves septentrionales (Troms, Norvège du Nord). *C. R. Acad. Sci. Paris,* **286,** 1653–1656.

Esbensen, K. H., Thy, P. and Wilson, J. R. 1978. A note on the cumulate stratigraphy of the Fongen–Hyllingen gabbro complex. Trondheim Region, Norway. *Norsk geol. Tidsskr.,* **58,** 103–107.

Farrow, C. M. 1974. *The geology of the Skjerstad area, Nordland, North Norway.* Unpub. Ph.D. thesis, Univ. of Bristol.

Flodberg, K. and Stigh, J. 1981. The Umbukta ultramafic complex in the Rödingsfjället nappe, northern Swedish Caledonides. *Terra Cognita,* **1,** 42.

Foslie, S. 1920. Raana noritfelt. Differentiation ved 'squeezing'. *Norges geol. Unders., Årbok 1920 III,* 1–52.

Foslie, S. 1921. Field observations in Northern Norway bearing on magmatic differentiation. *J. geol.,* **29,** 701–719.

Furnes, H. and Faerseth, R. B. 1975. Interpretation of the preliminary trace element data from the Lower Palaeozoic greenstone sequences on Stord, west Norway. *Norsk geol. Tidsskr.,* **55,** 157–169.

Furnes, H., Skjerlie, F. J. and Tysseland, M. 1976. Plate tectonic model based on greenstone geochemistry in the Late Precambrian–Lower Palaeozoic sequence in the Solund–Stavfjorden areas, west Norway. *Norsk geol. Tidsskr.,* **56,** 161–186.

Furnes, H., Faerseth, R. B. and Tysseland, M. 1978. Petrogenesis of continental metabasalts from the S. W. Norwegian Caledonides. *N. Jb. Miner. Abh.,* **132,** 34–51.

Furnes, H., Roberts, D., Sturt, B. A., Thon, A. and Gale, G. H. 1980a. Ophiolite fragments in the Scandinavian Caledonides. In Panayiotou, A. (Ed.), *Ophiolites. Proc. Int. Ophiolite Symp. Cyprus 1979,* 582–600.

Furnes, H., Sturt, B. A. and Griffin, W. L. 1980b. Trace element geochemistry of metabasalts from the Karmøy ophiolite, southwest Norwegian Caledonides. *Earth Planet. Sci. Lett.,* **50,** 75–91.

Furnes, H., Thon, A., Nordås, J. and Garman, L. B. 1982. Geochemistry of Caledonian metabasalts from some Norwegian ophiolite fragments. *Contrib. Mineral. Petrol.,* **79,** 295–307.

Furnes, H., Ryan, P. D., Grenne, T., Roberts, D., Sturt, B. A. and Prestvik. T. this volume. Geological and geochemical classification of the ophiolite fragments in the Scandinavian Caledonides.

Gale, G. H. 1974. Geokjemiske undersøkelser av vulkanitter og intrusiver i Mitt- og Syd-Norge. *NGU rapport nr. 1228 B.*

Gale, G. H. and Roberts, D. 1974. Trace element geochemistry of Norwegian Lower Palaeozoic basic volcanics and its tectonic implications. *Earth Planet. Sci. Lett.,* **22,** 380–390.

Gardner, P. M. 1980. *The geology and petrology of the Hasvik gabbro. Sørøy, Northern Norway.* Unpub. Ph.D. thesis, Univ. of London.

Gavelin, S. and Kulling, O. 1955. Beskrivning till berggrundskarta över Västerbottens län. *Sver. geol. Unders.,* **Ca 37,** 296 pp.

Gayer, R. A. and Humphreys, R. J. 1981. Tectonic modelling of the Finnmark and Troms Caledonides based on high level igneous rock geochemistry. *Terra Cognita,* **1,** 44.

Gayer, R. A. and Roberts, J. D. 1971. The structural relationships of the Caledonian nappe of Porsangerfjord, West Finnmark, N. Norway. *Norges geol. Unders.,* **269,** 21–67.

Gayer, R. A. and Roberts, J. D. 1973. Stratigraphic review of the Finnmark Caledonides, with possible tectonic implications. *Proc. Geol. Assoc.,* **84,** 405–428.

Gee, D. G. 1975. A tectonic model for the central part of the Scandinavian Caledonides. *Am. J. Sci.,* **275-A,** 468–515.

Gee, D. G. 1977. Extension of the Offerdal and Särv Nappes and the Seve Supergroup into northern Trøndelag. *Norsk geol. Tidsskr.,* **57,** 163–170.

Gee, D. G. 1978. Nappe displacement in the Scandinavian Caledonides. *Tectonophysics,* **47,** 393–419.

Gee, D. G. 1981. The Dictyonema-bearing phyllites at Nordaunevoll, eastern Trøndelag. *Norsk geol. Tidsskr.,* **61,** 93–95.

Gee, D. G. and Wilson, M. R. 1974. The age of orogenic deformation in the Swedish Caledonides. *Am. J. Sci.,* **274,** 1–9.

Gee, D. G. and Zachrisson, E. 1974. Comments on stratigraphy, faunal provinces and structure of the metamorphic allochthon, central Scandinavian Caledonides. *Geol. Fören. Stockh. Förh.,* **96,** 61–66.

Gee, D. G. and Zachrisson, E. 1979. The Caledonides in Sweden. *Sver. geol. Unders.,* **C 769,** 48 pp.

Getz, A. 1890. Graptolitførende skiferzoner i det trondhjemske. *Nyt. Mag. Naturv.,* **3,** 31–42.

Grenne, T. and Roberts, D. 1980. Geochemistry and volcanic setting of the Ordovician Forbordfjell and Jonsvatn greenstones, Trondheim Region, central Norwegian Caledonides. *Contrib. Mineral. Petrol.,* **74,** 374–386.

Grenne, T., and Lagerblad, B. this volume. The Fundsjø Group, Central Norway—a Lower Palaeozoic island arc sequence: geochemistry and regional implications.

Grenne, T. and Reinsbakken, A. 1981. Possible correlations of island arc greenstone belts and related sulphide deposits from the Grong and eastern Trondheim districts of the central Norwegian Caledonides. *Trans. Inst. Min. Metall. (sect. B: abstract Appl. Earth Sci.),* **90,** B 59.

Grenne, T., Grammeltvedt, G. and Vokes, F. M. 1980. Cyprus-type sulphide deposits in the western Trondheim district, central Norwegian Caledonides. In Panayiotou, A. (Ed.), *Ophiolites. Proc. Int. Ophiolite Symp. Cyprus 1979,* 727–743.

Gustavson, M. 1975. The low-grade rocks of the Skålvaer area, S. Helgeland and their relationship to high-grade rocks of the Helgeland Nappe Complex. *Norges geol. Unders.,* **322,** 13–33.

Gustavson, M. 1976. Helgelandsflesa. Beskrivelse til det bergrunnsgeologiske kart H 19–1:100 000. *Norges geol. Unders.,* **328,** 1–23.

Gustavson, M. 1978. Geochemistry of the Skålvaer greenstone, and a geotectonic model for the Caledonides of Helgeland, north Norway. *Norsk geol. Tidsskr.,* **58,** 161–174.

Gustavson, M. and Prestvik, T. 1979. The igneous complex of Hortavaer, Nord-Trøndelag, Central Norway. *Norges geol. Unders.,* **348,** 73–92.

Halls, C., Reinsbakken, A., Ferriday, I., Haugen, A. and Rankin, A. 1977. Geological setting of the Skorovas orebody within the allochthonous volcanic stratigraphy of the Gjersvik Nappe, central Norway. In *Volcanic Processes in Ore Genesis. Spec. Pap. No. 7, Inst. Min. Metall.—Geol. Soc. London,* 128–151.

Hardenby, C. 1980. Geology of the Kjølhaugen area, eastern Trøndelag, central Scandinavian Caledonides. *Geol. Fören. Stockh. Förh.*, 102, 475–492.

Heier, K. S. 1961. Layered gabbro, hornblendite, carbonatite and nepheline syenite on Stjernøy, North Norway. *Norsk geol. Tidsskr.*, 41, 109–155.

Hess, H. H. 1960. Stillwater igneous complex, Montana: a quantitative mineralogical study. *Geol. Soc. America, Mem.*, 80, 230 pp.

Hill, T. 1980. Geochemistry of the greenschists in relation to the Cu–Fe deposit in the Ramundberget area, central Swedish Caledonides. *Norges geol. Unders.*, 360, 195–210.

Holmqvist, P. J. 1894. Om diabasen på Ottfjället i Jemtland. *Geol. Fören. Stockh. Förh.*, 16, 175–192.

Holtedahl, O., Føyn, S. and Reitan, P. H. 1960. Aspects of the geology of Northern Norway. *Norges geol. Unders.*, 212a, 1–66.

Hooper, P. R. 1971. The mafic and ultramafic intrusions of S. W. Finnmark and North Troms. *Norges geol. Unders.*, 269, 147–158.

Häggbom, O. 1978. Polyphase deformation of a discontinuous nappe in the central Scandinavian Caledonides. *Geol. Fören. Stockh. Förh.*, 100, 349–354.

Irvine, T. N. 1974. Petrology of the Duke Island ultramafic complex, Southeastern Alaska. *Geol. Soc. Am. Mem.*, 138, 240 pp.

Jackson, E. D. 1961. Primary textures and mineral associations in the ultramafic zone of the Stillwater complex, Montana. *U.S. Geol. Survey Prof. Paper*, 358, 1–106.

Kautsky, F. E. 1978. New occurrences of megalenses of the Särv Nappe in northern Trøndelag, Norway. *Norsk geol. Tidsskr.*, 58, 237–240.

Kautsky, G. 1953. Der geologische Bau des Sulitjelma–Salojauregebietes in den nordskandinavischen Kaledoniden. *Sver. geol. Unders.*, C 528, 232 pp.

Kjøsnes, K. 1981. *Lillebukt alkaline complex: Fenitization of layered mafic rocks.* Unpub. Cand. Real. thesis, Univ. of Bergen.

Klingspor, I. and Gee, D. G. 1981. Isotopic age-determination studies of the Trøndelag trondhjemites. *Terra Cognita*, 1, 55.

Kolderup, C. F. and Kolderup, N. H. 1940. Geology of the Bergen Arc System. *Bergens Mus. Skrift.*, 20, 1–137.

Kollung, S. 1979. Stratigraphy and major structures of the Grong District, Nord-Trøndelag. *Norges geol. Unders.*, 354, 1–51.

Krill, A. G. 1980. Tectonics of the Oppdal area, central Norway. *Geol. Fören. Stockh. Förh.*, 102, 523–530.

Kulling, O. 1933. Bergbyggnaden inom Björkvattnet–Virisen-området i Våasterbottensfjällens centrala del. *Geol. Fören. Stockh. Förh.*, 55, 167–422.

Kulling, O. 1938. Grönstenarnas placering inom Västerbottensfjällens kambrosilurstratigrafi. *Geol. Fören. Stockh. Förh.*, 60, 153–176.

Kulling, O. 1982. Översikt över södra Norrbottensfjällens kaledonberggrund. *Sver. geol. Unders.*, Ba 26, 295 pp.

Kumpulainen, R. 1980. Upper Proterozoic stratigraphy and depositional environments of the Tossåsfjället Group, Särv Nappe, southern Swedish Caledonides. *Geol. Fören. Stockh. Förh.*, 102, 531–550.

Ledru, P. 1980. Evolution structurale et magmatique du complexe plutonique de Karmøy. *Bull. Soc. géol. minéral. Bretagne*, 12, 1–106.

Lippard, S. J. 1976. Preliminary investigations of some Ordovician volcanics from Stord, West Norway. *Norges geol. Unders.*, 327, 41–66.

Lippard, S. J. and Mitchell, J. G. 1980. Late Caledonian dolerites from the Kattnakken area, Stord, S. W. Norway, their ages and tectonic significance. *Norges geol. Unders.*, 358, 47–62.

Loeschke, J. 1976. Petrochemistry of eugeosynclinal magmatic rocks of the area around Trondheim (Central Norwegian Caledonides). *N. Jb. Miner. Abh.*, 128, 41–72.

Lutro, O. 1979. The geology of the Gjersvik area, Nord-Trøndelag. *Norges geol. Unders.*, 354, 53–100.

Lønne, W. and Sellevoll, M. A. 1975. A reconnaissance gravity survey of Magerøy, Finnmark, Northern Norway. *Norges geol. Unders.*, 319, 1–15.

McClenaghan, J. 1972. *Igneous and metamorphic geology of the Børfjord area, Sørøy, Northern Norway.* Unpub. Ph.D. thesis, Univ. of London.

McKerrow, W. S., Lambert, R. St. J. and Chamberlain, V. E. 1980. The Ordovician, Silurian and Devonian time scale. *Earth Planet. Sci. Lett.*, 51, 1–8.

Mason, R. 1980. Temperature and pressure estimates in the contact aureole of the Sulitjelma gabbro, Norway: Implications for an ophiolite origin. In Panayiotou, A. (Ed.), *Ophiolites. Proc. Int. Ophiolite Symp. Cyprus 1979*, 576–581.

Minsaas, O. 1981. *Lyngenhalvøyas geologi, med spesiell vekt på den sedimentologiske utvikling av de ordovisisk siluriske klastiske sekvenser som overligger Lyngen gabbro-kompleks.* Unpub. Cand. Real. thesis, Univ. of Bergen, 295 pp.

Miyashiro, A. 1975. Classification, characteristics, and origin of ophiolites. *J. geol.*, 83, 249–281.

Mulder, C. J. 1951. *Geology and petrology of the region between Lake Överuman and Tärnasjön (southern Swedish Lapland).* Diss., Univ. of Amsterdam.

Myrland, R. 1972. Velfjord. Beskrivelse til det bergrunns-geologiske gradteigskart I 18–1:100 000. *Norges geol. Unders.*, 274, T–30.

Myrland, R. 1974. Bergrunnsgeologiske kart Bindal 1:100 000. *Norges geol. Unders.*

Mørk, M. B. E. 1979. *Metamorf utvikling og gabbrointrusjon på Krutfjell.* Thesis, Univ. of Oslo, 307 pp.

Nicholson, R. and Rutland, R. W. R. 1969. A section across the Norwegian Caledonides; Bodø to Sulitjelma. *Norges geol. Unders.*, 260, 86 pp.

Nilsen, O. 1971. Sulphide mineralization and wall rock alteration at Rødhammaren Mine, Sør-Trøndelag, Norway. *Norsk geol. Tidsskr.*, 51, 329–354.

Nordås, J., Amaliksen, K. G., Brekke, H., Suthren, R. J., Furnes, H., Sturt, B. A. and Robins, B. this volume. Lithostratigraphy and petrochemistry of Caledonian rocks on Bømlo, Southwest Norway.

Oftedahl, Chr. 1956. Om Grongkulminasjonen og Grongfeltets skyvedekker. *Norges geol. Unders.*, 195, 57–64.

Olesen, N. Ø., Hansen, E. S., Kristensen, L. H. and Thyrsted, T. 1973. A preliminary account on the geology of the Selbu–Tydal area, the Trondheim region, Central Norwegian Caledonides. *Leidse geol. Mededel.*, 49, 259–276.

Olsen, J. 1980. Genesis of the Joma stratiform sulphide deposit, central Norwegian Caledonides. *Proc. 5th IAGOD symposium, Alta, Utah 1978*, 1, 745–757.

Oosterom, M. G. 1955. Some notes on the Lille Kufjord layered gabbro, Seiland, Finnmark, Northern Norway. *Norges geol. Unders.*, 195, 73–87.

Oosterom, M. G. 1963. The ultramafites and layered gabbro sequences in the granulite facies rocks on Stjernøy, Finnmark, Norway. *Leidse geol. Mededel.*, 28, 179–296.

Padget, P. 1955. The geology of the Caledonides of the Birtavarre region, Troms, northern Norway. *Norges geol. Unders.,* **192**, 1–107.

Pearce, J. A. 1975. Basalt geochemistry used to investigate past tectonic environments on Cyprus. *Tectonophysics,* **25**, 41–68.

Pearce, J. A. and Cann, J. R. 1973. Tectonic setting of basic volcanic rocks determined using trace element analyses. *Earth Planet. Sci. Lett.,* **19**, 290–300.

Pidgeon, R. T. and Råheim, A. 1972. Geochronological investigation of the minor intrusive rocks from Kristiansund, west Norway. *Norsk geol. Tidsskr.,* **52**, 241–256.

Prestvik, T. 1974. Supracrustal rocks of Leka, Nord-Trøndelag. *Norges geol. Unders.,* **311**, 65–87.

Prestvik, T. 1980. The Caledonian ophiolite complex of Leka, north central Norway. In Panayiotou, A. (Ed.), *Ophiolites. Proc. Int. Ophiolite Symp. Cyprus 1979,* 555–566.

Prestvik, T. and Roaldset, E. 1978. Rare earth element abundances in Caledonian metavolcanics from the island of Leka, Norway. *Geochem. J.,* **12**, 89–100.

Priem, H. N. A. and Torske, T. 1973. Rb–Sr isochron age of Caledonian acid volcanics from Stord, W. Norway. *Norges geol. Unders.,* **300**, 83–85.

Priem, H. N. A., Boelrijk, N. A. I M., Hebeda, E. H., Verdurmen, E. A. T. and Verschure, R. H. 1975. Isotopic dating of the Caledonian Bindal and Svenningdal granitic massifs, central Norway. *Norges geol. Unders.,* **319**, 29–36.

Qvale, H. and Stigh, J. this volume. Ultramafic rocks in the Scandinavian Caledonides.

Ramberg, I. B. 1981. The Brakfjellet tectonic lens; evidence of pinch-and-swell in the Caledonides of Nordland, north central Norway. *Norsk geol. Tidsskr.,* **61**, 87–91.

Ramsay, D. M. and Sturt, B. A. 1977. A sub-Caledonian unconformity within the Finnmarkian nappe sequence and its regional significance. *Norges geol. Unders.,* **334**, 107–116.

Ramsay, D. M., Sturt, B. A. and Andersen, T. B. 1979. The sub-Caledonian unconformity on Hjelmsøy—New evidence of primary basement/cover relations in the Finnmarkian nappe sequence. *Norges geol. Unders.,* **351**, 1–12.

Rekstad, J. 1910. Beskrivelse til det geologiske kart over Bindal og Leka. *Norges geol. Unders.,* **53**, 1–37.

Rekstad, J. 1915. Helgelands ytre kystrand. *Norges geol. Unders.,* **75**, 1–53.

Rekstad, J. 1917. Vega. Beskrivelse til det geologiske generalkart. *Norges geol. Unders.,* **80**, 1–85.

Reymer, A. P. S. 1979. *Investigations into the metamorphic nappes of the central Scandinavian Caledonides on the basis of Rb–Sr and K–Ar age determinations.* Thesis, Univ. of Leiden, 123 pp.

Roberts, D. 1974. Hammerfest. Beskrivelse til de 1:250 000 bergrunnsgeologiske kart. *Norges geol. Unders.,* **301**, 1–66.

Roberts, D. 1975a. Geochemistry of dolerite and metadolerite dykes from Varanger Peninsula, Finnmark, North Norway. *Norges geol. Unders.,* **322**, 55–72.

Roberts, D. 1975b. The Stokkvola conglomerate—a revised stratigraphical position. *Norsk geol. Tidsskr.,* **55**, 361–371.

Roberts, D. 1980a Geokjemien av Snåsagrønnsteinene—en framgangsrapport. *Unpub. report, Norges geol. Unders.,* 18 pp.

Roberts, D. 1980b. Petrochemistry and palaeogeographic setting of Ordovician volcanic rocks of Smøla, Central Norway. *Norges geol. Unders.,* **359**, 43–60.

Roberts, D. and Gale, G. 1978. The Caledonian–Appalachian Iapetus Ocean. In Tarling, D. H. (Ed.), *Evolution of the Earth's Crust,* Academic Press, London, 255–342.

Roberts, D., Springer, J. and Wolff, F. Chr. 1970. Evolution of the Caledonides in the northern Trondheim region, Central Norway: a review. *Geol. Mag.,* **107**, 133–145.

Robins, B. 1971. *The plutonic geology of the Caledonian complex of Southern Seiland, Northern Norway.* Unpub. Ph.D. thesis, Univ. of Leeds.

Robins, B. 1972. Syenite–carbonatite relationships in the Seiland gabbro province, Finnmark, Northern Norway. *Norges geol. Unders.,* **272**, 43–58.

Robins, B. 1974. Synorogenic alkaline pyroxenite dykes on Seiland, northern Norway. *Norsk geol. Tidsskr.,* **54**, 247–268.

Robins, B. 1975. Ultramafic nodules from Seiland, northern Norway. *Lithos,* **8**, 15–27.

Robins, B. 1980. The evolution of the Lillebukt alkaline complex, Stjernøy, Norway (abs.). *Lithos,* **13**, 219–220.

Robins, B. 1982. The geology and petrology of the Rognsund intrusion, West Finnmark, Northern Norway. *Norges geol. Unders.,* **371**, 1–55.

Robins, B. and Gardner, P. M. 1974. Synorogenic layered basic intrusions in the Seiland petrographic province, Finnmark. *Norges geol. Unders.,* **312**, 91–130.

Robins, B. and Gardner, P. M. 1975. The magmatic evolution of the Seiland province, and Caledonian plate boundaries in Northern Norway. *Earth Planet. Sci. Lett.,* **26**, 167–177.

Robins, B. and Takla, M. A. 1979. Geology and geochemistry of a metamorphosed picrite–ankaramite dyke suite from the Seiland province, northern Norway. *Norsk geol. Tidsskr.,* **59**, 67–95.

Robins, B. and Tysseland, M. 1979. Fenitization of some mafic igneous rocks in the Seiland province, northern Norway. *Norsk geol. Tidsskr.,* **59**, 1–23.

Robins, B. and Tysseland, M. 1980. Fenitization of mafic cumulates by the Pollen carbonatite, Finnmark, Norway (abs.). *Lithos,* **13**, 220.

Robins, B. and Tysseland, M. 1983. The geology, geochemistry and origin of ultrabasic fenites associated with the Pollen carbonatite (Finnmark, Norway). *Chem. Geol.,* **40**, 65–95.

Roermund, H. L. M. van. this volume. Eclogites of the Seve Nappe, central Scandinavian Caledonides.

Rui, I. 1972. Geology of the Røros district, south-eastern Trondheim region with a special study of the Kjøliskarvene–Holtsjøen area. *Norsk geol. Tidsskr.,* **52**, 1–22.

Ryan, P. D. and Sturt, B. A. this volume. Early Caledonian orogenesis in northwestern Europe.

Ryan, P. D., Williams, D. M. and Skevington, D. 1980. A revised interpretation of the Ordovician stratigraphy of Sør Trøndelag, and its implications for the evolution of the Scandinavian Caledonides. In Wones, D. R. (Ed.), *The Caledonides in the USA. Virginia Polytechnic Inst. and State Univ., Dept. Geol. Sci., Mem.,* **2**, 99–105.

Råheim, A., Gale, G. H. and Roberts, D. 1979. Rb, Sr ages of basement gneisses and supracrustal rocks of the Grong area, Nord-Trøndelag, Norway. *Norges geol. Unders.,* **354**, 131–142.

Röshoff, K. 1975. A probable glaciogenic sediment in the Särv Nappe, central Swedish Caledonides. *Geol. Fören. Stockh. Förh.,* **97**, 192–195.

Sandwall, J. 1981a. Caledonian geology of the Jofjället area, Västerbotten county, Sweden. *Sver.geol. Unders.,* **C 778**, 105 pp.

Sandwall, J. 1981b. Greenstones related to rifting and ocean basin opening in the Jofjället area, central Swedish Caledonides. *Geol. Fören. Stockh. Förh.,* **103**, 421–428.

Senior, A. 1978. The Artfjället gabbro, Västerbottens län. *Abst. XIII Nord. geol. Vinter. København 1978,* 64.

Senior, A. and Otten, M. T. this volume. The Artfjället gabbro and its bearing on the evolution of the Storjället Nappe, central Swedish Caledonides.

Siedlecka, A. and Siedlecki, S. 1967. Some new aspects of the geology of Varanger Peninsula (Northern Norway). *Norges geol. Unders.,* **247**, 288–306.

Size, W. B. 1979. Petrology, geochemistry, and genesis of the type area trondhjemite in the Trondheim region, central Norwegian Caledonides. *Norges geol. Unders.,* **351**, 51–76.

Sjögren, H. 1900. Öfersigt af Sulitjelma-områdets geologi. *Geol. Fören. Stockh. Förh.,* **22**, 437–462.

Sjöstrand, T. 1978. Caledonian geology of the Kvarnbergsvattnet area, northern Jämtland, central Sweden. *Sver. geol. Unders.,* **C 735**, 107 pp.

Skogen, J. H. 1980. *Lillebukt alkaline kompleks: Karbonatittens indre strukturer og dens metamorfe og tektoniske Otvikling.* Unpub. Cand. Real. thesis, Univ. of Bergen.

Solli, T. 1981. *The geology of the Torvastad Group, the cap rocks to the Karmøy ophiolite.* Unpub. Cand. Real. thesis, Univ. of Bergen.

Solyom, Z., Gorbatschev, R. and Johansson, I. 1979a. The Ottfjället Dolerites. Geochemistry of the dyke swarm in relation to the geodynamics of the Caledonide orogen in central Scandinavia. *Sver. geol. Unders.,* **C 756**, 38 pp.

Solyom, Z., Andreasson, P. G. and Johansson, I. 1979b. Geochemistry of amphibolites from Mt. Sylarna, Central Scandinavian Caledonides. *Geol. Fören. Stockh. Förh.,* **101**, 17–27.

Speedyman, D. L. 1967. *The plutonic geology of the Husfjord igneous complex, Sørøy, Northern Norway.* Unpub. Ph.D. thesis, Univ. of London.

Speedyman, D. L. 1972. Mechanism of emplacement of plutonic igneous rocks in the Husfjord area of Sørøy, west Finnmark. *Norges geol. Unders.,* **272**, 35–42.

Spjeldnaes, N. this volume. Biostratigraphy of the Scandinavian Caledonides.

Stephens, M. B. 1977a. Stratigraphy and relationship between folding, metamorphism and thrusting in the Tärna–Björkvattnet area, northern Swedish Caledonides. *Sver. geol. Unders.,* **C 726**, 146 pp.

Stephens, M. B. 1977b. The Stekenjokk volcanites—segment of a Lower Palaeozoic island arc complex. In Bjørklykke, A., Lindahl, I. and Vokes, F. M. (Eds), *Kaledonske Malmforekomster. BVLI'S Tekniske Virksomhet, Trondheim,* 24–36.

Stephens, M. B. 1980a. Occurrence, nature and tectonic significance of volcanic and high-level intrusive rocks within the Swedish Caledonides. In Wones, D. R. (Ed.), *The Caledonides in the USA. Virginia Polytechnic Inst. and State Univ., Dept. Geol. Sci., Mem.,* **2**, 289–298.

Stephens, M. B. 1980b. Spilitization, element release and formation of massive sulphides in the Stekenjokk volcanites, central Swedish Caledonides. *Norges geol. Unders.,* **360**, 159–193.

Stephens, M. B. 1981. Evidence for Ordovician arc build-up and arc splitting in the Upper Allochthon of central Scandinavia. *Terra Cognita,* **1**, 75.

Stephens, M. B. 1982. Field relationships, petrochemistry and petrogenesis of the Stekenjokk volcanites, central Swedish Caledonides. *Sver. geol. Unders.,* **C 786**, 111 pp.

Stephens, M. B. and Senior, A. 1981. The Norra Storfjället lens—an example of fore-arc basin sedimentation and volcanism in the Scandinavian Caledonides. *Terra Cognita,* **1**, 76–77.

Stephens, M. B. and Gee, D. G. this volume. A tectonic model for the evolution of the eugeoclinai terranes in the central Scandinavian Caledonides.

Stephens, M. B., Gustavson, M. Ramberg, I. and Zachrisson, E. this volume. The Caledonides of central–north Scandinavia—a tectonostratigraphic overview.

Stigh, J. 1979. Ultramafites and detrital serpentinites in the central and southern parts of the Caledonian Allochthon in Scandinavia. *Geol. Inst., Chalmers Tekniska Høgskola och Göteborgs Univ. Publ.,* **A 27**, 222 pp.

Strand, T. 1981. *Lillebukt alkaline kompleks: karbonatittens mineralogi og petrokjemi.* Unpub. Cand. Real. thesis, Univ. of Bergen.

Strand, T. and Kulling, O. 1972. *Scandinavian Caledonides,* John Wiley and Sons Ltd., London, 302 pp.

Strømberg, A. 1955. Zum Gebirgsbau der Skanden in mittleren Härjedalen. *Uppsala Univ. Geol. Inst. Bull.,* **35**, 199–243.

Strömberg, A. 1961. On the tectonics of the Caledonides in the south-western part of the county of Jämtland, Sweden. *Uppsala Univ. Geol. Inst. Bull.,* **39**, 92 pp.

Strömberg, A. 1969. Initial Caledonian magmatism in Jämtland area, Sweden. *Am Assoc. Petroleum Geologists, Mem.,* **12**, 375–387.

Sturt, B. A. and Ramsay, D. M. 1965. The alkaline complex of the Breivikbotn area, Sørøy, Northern Norway. *Norges geol. Unders.,* **231**, 1–142.

Sturt, B. A. and Taylor, J. 1972. The timing and environment of emplacement of the Storelv gabbro, Sørøy. *Norges geol. Unders.,* **272**, 1–34.

Sturt, B. A. and Thon, A. 1978. A major early Caledonian igneous complex and a profound unconformity in the Lower Palaeozoic sequence of Karmøy, southwest Norway. *Norsk geol. Tidsskr.,* **58**, 221–228.

Sturt, B. A., Skarpenes, O., Ohanian, A. T. and Pringle, I. R. 1975. Reconnaissance Rb/Sr isochron study in the Bergen Arc System and regional implications. *Nature,* **253**, 595–599.

Sturt, B. A. Pringle, I. R. and Ramsay, D. M. 1978. The Finnmarkian phase of the Caledonian Orogeny. *J. geol. Soc. London,* **135**, 547–610.

Sturt, B. A. Thon, A. and Furnes, H. 1979. The Karmøy ophiolite, southwest Norway. *Geology,* **7**, 316–320.

Sturt, B. A., Thon, A. and Furnes, H. 1980a. The geology and preliminary geochemistry of the Karmøy ophiolite, S. W. Norway. In Panayiotou, A. (Ed.), *Ophiolites. Proc. Int. Ophiolite Symp. Cyprus 1979,* 538–554.

Sturt, B. A., Speedyman, D. L. and Griffin, W. L. 1980b. The Nordre Bumandsfjord ultramafic pluton, Seiland, North Norway, Part 1: Field relations. *Norges geol. Unders.,* **358**, 1–30.

Størmer, L. 1941. Dictyonema shales outside the Oslo region. *Norsk geol. Tidsskr.,* **20**, 161–170.

Sun, S. and Nesbitt, R. W. 1978. Geochemical regularities and genetic significance of ophiolitic basalts. *Geology,* **6**, 689–693.

Sundvoll, B. and Roberts, D. 1977. Framgangsrapport på datering og geokjemi av eruptivbergarter på Smøla og Hitra. Unpub. report, *Norges geol. Unders*.

Taylor, H. P. 1967. The zoned ultramafic complexes of southeastern Alaska. In Wyllie, P. J. (Ed.), *Ultramafic and Related Rocks*, Wiley & Sons Inc., New York, 96–118.

Thon, A. this volume. The Gullfjellet ophiolite complex and the structural evolution of the major Bergen arc, west Norwegian Caledonides.

Thorslund, P. 1948. De siluriska lagren ovan Pentameruskalkstenen i Jämtland. *Sver. geol. Unders.*, **C 494**, 39 pp.

Trouw, R. A. J. 1973. Structural geology of the Marsfjällen area, Caledonides of Våsterbotten, Sweden, *Sver. geol. Unders.*, **C 689**, 115 pp.

Vogt, T. 1927. Sulitelmafeltets geologi og petrografi. *Norges geol. Unders.*, **121**, 560 pp.

Vogt, T. 1945. The geology of part of the Hølonda–Horg district, a type area in the Trondheim Region. *Norsk geol. Tidsskr.*, **25**, 449–528.

Williams, P. F. and Zwart, H. J. 1977. A model for the development of the Seve–Köli Caledonian Nappe Complex. In Saxena, S. K. and Bhattacharji, S. (Eds), *Energetics of Geological Processes*, Springer, Berlin, 169–187.

Wilson, J. R. and Olesen, N. Ø. 1975. The form of the Fongen–Hyllingen gabbro complex, Trondheim Region, Norway. *Norsk geol. Tidsskr.*, **55**, 423–439.

Wilson, J. R. and Pedersen, S. 1981. The age of the synorogenic Fongen–Hyllingen complex, Trondheim region, Norway. *Geol. Fören. Stockh. Förh.*, **103**, 429–435.

Wilson, J. R., Esbensen, K. H. and Thy, P. 1981. Igneous petrology of the synorogenic Fongen–Hyllingen layered basic complex, South–Central Scandinavian Caledonides. *J. Petrol.*, **22**, 584–627.

Wilson, M. R. 1981. Geochronological results from Sulitjelma, Norway. *Terra Cognita*, **1**, 82.

Wilson, M. R. Roberts, D. and Wolff, F. C. 1973. Age determinations from the Trondheim region Caledonides, Norway: a preliminary report. *Norges geol. Unders.*, **288**, 53–63.

Winchester, J. A. and Floyd, P. A. 1976. Geochemical magma type discrimination: application to altered and metamorphosed basic igneous rocks. *Earth Planet. Sci. Lett.*, **28**, 459–469.

Wolff, F. Chr. 1967. Geology of the Meråker area as a key to the eastern part of the Trondheim region. *Norges geol. Unders.*, **245**, 123–142.

Wolff, F. Chr. 1979. Beskrivelse til de bergrunnsgeologiske kartbladene Trondheim og Östersund 1:250 000. *Norges geol. Unders.*, **353**, 1–76.

Wolff, F. Chr. and Roberts, D. 1980. Geology of the Trondheim region. *Norges geol. Unders.*, **356**, 117–128.

Wolff, F. Chr., Roberts, D., Siedlecka, A. Oftedahl, Chr. and Grenne, T. 1980. Guide to excursions across part of the Trondheim Region, Central Norwegian Caledonides. *Norges geol. Unders.*, **356**, 129–167.

Wood, D. A., Joron, J.-L. and Treuil, M. 1979. A reappraisal of the use of trace elements to classify and discriminate between magma series in different tectonic settings. *Earth Planet. Sci. Lett.*, **45**, 326–336.

Zachrisson, E. 1964. The Remdalen Syncline. *Sver. geol. Unders.*, **C 596**, 53 pp.

Zachrisson, E. 1966. A pillow lava locality in the Grong District, Norway. *Norsk geol. Tidsskr.*, **46**, 375–378.

Zachrisson, E. 1969. Caledonian geology of Northern Jämtland–Southern Västerbotten, *Sver. geol. Unders.*, **C 644**, 33 pp.

Zachrisson, E. 1973. The westerly extension of Seve rocks within the Seve–Köli Nappe Complex in the Scandinavian Caledonides. *Geol. Fören. Stockh. Förh.*, **95**, 243–251.

The Caledonide Orogen—Scandinavia and Related Areas
Edited by D. G. Gee and B. A. Sturt
© 1985 John Wiley & Sons Ltd

Geological and geochemical classification of the ophiolitic fragments in the Scandinavian Caledonides

H. Furnes,* **P. D. Ryan,**† **T. Grenne,**‡ **D. Roberts,**§ **B. A. Sturt*** and **T. Prestvik**‡

*Geologisk Institutt, Avd. A, Allegt. 41,5014 Bergen, Norway
†Dept. of Geology, University College, Galway, Ireland
‡Geologisk Institutt, Universitetet i Trondheim, N–7034 Trondheim–N.T.H., Norway
§Norges geologiske Undersøkelse, P.O. Box 3006, N–7001 Trondheim, Norway

ABSTRACT

Two groups of ophiolites are recognized in the Scandinavian Caledonides, group I representing major ocean or alternatively mature marginal basin, and group II small, younger marginal basins. The main field characteristics of group I include a well-developed ophiolite pseudostratigraphy, and a variable development of oceanic sediments. They were apparently obducted eastwards on to a Baltoscandian Precambrian basement craton or its marginal sediments during the Finnmarkian orogenic event. Group II sequences have a less well-developed pseudostratigraphy; they are intercalated with volcaniclastic and siliclastic sediments, and are post-Finnmarkian (mostly Middle Ordovician) in age.

The most characteristic REE patterns of the metabasalts are either flat, LREE-depleted, or define convex-upward curves. For most of the assemblages the REE patterns are internally similar, but each one may show significant differences relative to the others. However, all the above-mentioned features may occur within any one sequence. Considering total trace element patterns, group I defines two groups characterized by pronounced differences in several incompatible trace element ratios. Some of the metabasalts of the group II sequences show trace element patterns that reflect a measure of island arc involvement whereas others are more similar to MORB.

Introduction

This paper discusses the classification of metabasite complexes within the Scandinavian Caledonides that have been described as segments of ophiolites (Fig. 1) either on the basis of their geochemical patterns or on the preservation of ophiolite pseudo-stratigraphy (Furnes et al. 1980a), or both. On the basis of recent field studies and geochemical dating (Furnes et al. 1983) it is now possible to recognize at least two fundamental groups of ophiolitic complexes: group I complexes are believed to have originated in a major ocean basin (Iapetus) or a marginal basin, whose geological evolution was such that the ophiolite was removed from sources of arc type volcaniclastic or siliclastic sedimentation for a considerable time after its formation; group II complexes are thought to have developed in small marginal basins near to a supply of arc type volcaniclastic and/or siliclastic sediments. This paper will discuss the evidence on which this classification is

based and the geochemical patterns of the two types of ophiolitic complex recognized to date.

The characteristics of the group I complexes are:
1. They usually exhibit a well-developed ophiolite pseudostratigraphy;
2. They show MORB geochemical trends consistent with their formation at a spreading centre;
3. In most cases they have a relatively thick, conformable cover (cap) sequence of oceanic sediments (this includes chert–lutite associations, metalliferous sediments, and clastic sediments derived exclusively from the ophiolite);
4. Higher levels of the pseudostratigraphy are cut by oceanic plagiogranites (sensu Coleman and Peterman 1975);
5. Later volcanic units intercalated with the oceanic sediments are of ocean island type;
6. Available evidence points to their obduction on to the Baltic shield or a microcontinent with mariginal sediments during the early Ordovician Finnmarkian event.

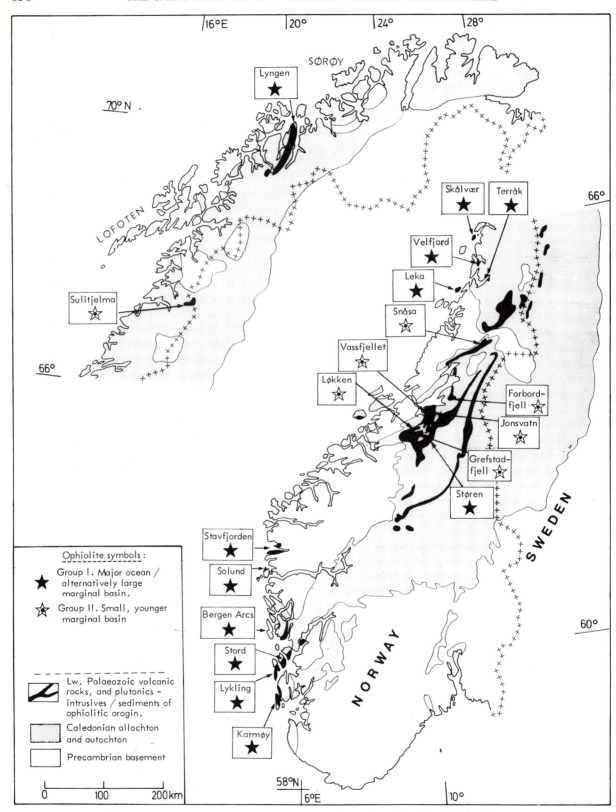

Fig. 1   Distribution of ophiolite fragments in the Scandinavian Caledonides

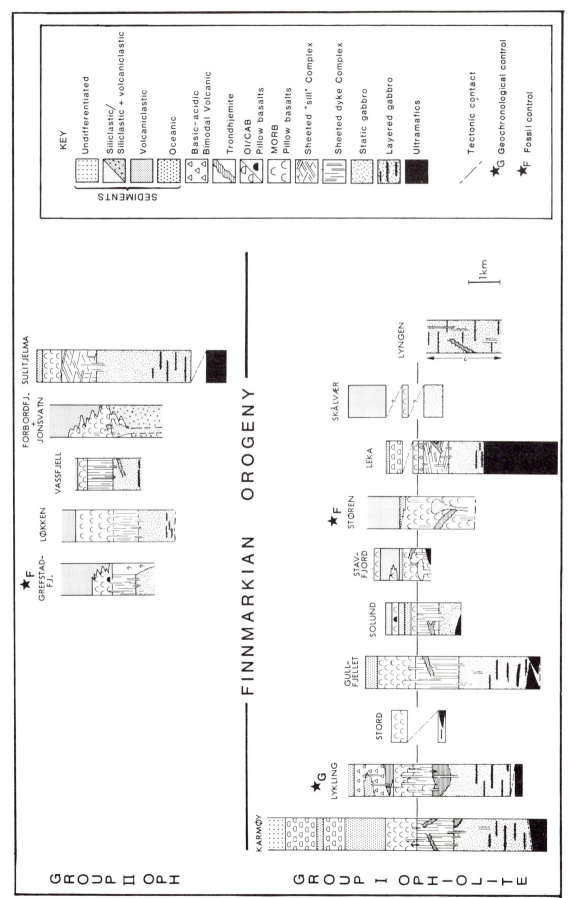

Fig. 2 Generalized composite profiles through the pseudostratigraphy of the group I and II ophiolite sequences, and their associated sediments. Sources for these compilations are given in the text

Columnar sections through the proposed group I complexes are given in Fig. 2.

The characteristic features of the group II (small marginal basin) complexes are:

1. They locally contain a partial ophiolite pseudo-stratigraphy, but well-developed layered gabbro or ultramafic zones have not been recorded to date;
2. They exhibit MORB and intermediate MORB/CAB or MORB/WPB geochemical trends;
3. They pass laterally and vertically into siliclastic or arc-derived volcaniclastic sediments;
4. They are broadly coeval with arc volcanism;
5. All candidates presently recognized are of early to mid Ordovician age, are post-Finnmarkian, and were thrust into their present location during the main Middle to Late Silurian phase of the Caledonian orogeny (the Scandinavian, or Scandian event).

Columnar sections through the group II complexes are given in Fig. 2.

As many of the complexes discussed here have been described in detail elsewhere (see references), the following account of their geology will concentrate upon new information and those aspects relevant to the proposed classification. Their geochemistry will be discussed in a later section.

Field Characteristics

The group I ophiolite fragments

These are characterized by having been subjected to two distinctive tectonometamorphic cycles during Caledonian orogenic evolution, i.e. both Finnmarkian and Scandian (e.g. Sturt et al. 1984). They represent material derived from an oceanic crust which was essentially pre-Ordovician in age, thus providing a record from Iapetus in its early development stage. The evidence for a pre-Ordovician age for this oceanic crust is varied and ranges from geochronology and faunal/stratigraphic control to event correlation (Furnes et al. 1980a; Sturt et al. (1984). There is also evidence for the existence of an early primitive ensimatic island arc, probably representing an initial stage of subduction. These island arc products were eventually obducted together with the subjacent oceanic crust during the Finnmarkian orogenic phase.

It is particularly the Karmøy, Lykling, Gullfjell, and Støren ophiolites which exhibit sufficient diagnostic features for being classified as group I ophiolites. The remaining ophiolite fragments which, in this paper, are also proposed as belonging to this group, do not yet have such well defined stratigraphic or radiometric constraints as those mentioned above.

Karmøy

The Karmøy ophiolite with its almost complete pseudostratigraphy (Fig. 2) was first described by Sturt and Thon (1978). This ophiolite exhibits all the characteristics of a group I complex (see Sturt et al. 1979, 1980; Furnes et al. 1980b). Recent studies have concentrated upon the sediments of the Torvastad Group, the uppermost part of the ophiolite, which contains a chert–lutite sequence and metalliferous sediments typical of oceanic deposits (Solli 1981). Intercalated with the sediments of the Torvastad Group are pillow lavas of oceanic island geochemistry and coarse clastic sediments derived exclusively from the underlying ophiolite. The rocks of the Torvastad Group are considered to have developed in an oceanic fracture zone or transform fault environment (Solli 1981). The age of the Karmøy ophiolite is unknown, but the thickness of oceanic pelagic sediments suggests a long interval before obduction during the Finnmarkian event which must pre-date the emplacement of the West Karmøy igneous complex (ca. 450 Ma) and the subsequent unconformable deposition of the late Ordovician Skudeneset Group.

Lykling

The plutonic zone of the Lykling ophiolite (Amaliksen 1983), attaining a total thickness of more than 4 km, is mainly layered and non-layered gabbros, gabbro pegmatites, and microgabbros at the highest levels. The microgabbro is succeeded by a sheeted dyke complex and pillow lavas towards the top. A bimodal sequence of extrusive greenstones and quartz-keratophyres, interpreted to represent ensimatic island arc volcanism (Nordås et al. this volume; Amaliksen 1983), rests on the ophiolite. Quartz-keratophyres from this sequence have been dated by the Rb–Sr whole rock method, yielding an age of 535 ± 46 Ma (Furnes et al. 1983). A remarkable breccia, consisting exclusively of coarse ophiolite detritus interbedded with sandstones, lutites, and cherts and resting upon both the ophiolite and the associated ensimatic island arc volcanics, may represent transform fault activity (Amaliksen 1983). The ophiolite and above-mentioned associated rocks are unconformably overlain by the Middle Ordovician Siggjo Complex of subaerial rhyolitic to basaltic lavas and pyroclastics (Nordås et al. this volume).

Stord

The Stord complex comprises a thin sequence of pillow lavas with some cherts and black shales cut by dykes that are in fault contact with gabbros and serpentinites transected by rare plagiogranites (Thon *et al.* 1980).

Gulfjell

The Gulfjell ophiolite (Major Bergen Arc) shows many similarities with the Lykling and Karmøy ophiolites. Plutonic rocks, mainly represented by layered and non-layered gabbros, sheeted dykes, and volcanic rocks, are all well exposed, but ultramafic tectonites are absent. The ophiolite had been deformed and deeply eroded prior to the deposition of the unconformably overlying Holdhus Group (Mid?–Upper Ordovician).

Solund and Stavfjorden

They exhibit a partial pseudostratigraphy (including plagiogranites and tectonized serpentinites) and are overlain by thin deposits of oceanic affinity; however, these pass conformably upwards into volcaniclastic sequences containing calc-alkaline, tholeiitic, and possibly alkaline volcanic products (Furnes *et al.* 1976). Both complexes are unconformably overlain by Upper Ordovician deposits.

Støren

Although the Støren Group was the first of the metabasite suites in which an OFB geochemistry was reported (Gale and Roberts 1972, 1974), a complete ophiolite pseudostratigraphy does not occur in the type area. The Støren Group is composed mainly of greenschist facies massive or pillow basalts, pillow breccias, and aquagene tuffs, intercalated with cherts, jaspers, and dark phyllites in the upper parts of the succession. Dolerite dykes occur in one small area near the tectonized base of the basaltic greenstones (Fig. 2). The Støren unit is overlain unconformably by the mid–late Arenig and younger sediments of the Lower Hovin Group (Blake 1962, Skevington 1963; Ryan *et al.* 1980).

Leka

The Leka ophiolite exhibits an almost complete pseudostratigraphy (Fig. 2) (Prestvik 1972, 1974, 1980; Prestvik and Roaldset 1978). It has the best developed ultramafic complex presently recorded, containing more than 2 km of strongly tectonized, banded ultramafics. The sheeted complex is of inter-est in that it is apparently dominated by sheets and sills rather than dykes. Pillow basalts of ocean island affinity (Prestvik and Roaldset 1978) and overlying oceanic sediments occur on Storøya. This sequence is in tectonic contact with the ophiolite. The Leka ophiolite was deformed, metamorphosed, weathered, and eroded, and then unconformably overlain by continental sediments presumably of late Ordovician age (Sturt *et al.* this volume), although fossils are lacking.

Skålvær, Velfjord, and Terråk

The Skålvær greenstone is exposed on offshore islands, in Nordland and its geological relationships with surrounding rock units are uncertain. However, the spatial association of greenstones with volcaniclastic sediments together with geochemical data, have led to the suggestion that the volcanite assemblage developed in a back-arc basin (Gustavson 1978).

In the Terråk area of Tosenfjord, serpentinites, cumulate gabbros, diabase dykes (in part sheeted), and pillow lavas are present and appear to show a reasonably well-developed ophiolite pseudostratigraphy. No data on the geochemistry of these rocks are yet available. They are cut and disrupted by part of the Bindal batholith, dated at 414 ± 26 Ma (Priem *et al.* 1975), though no other age limits can yet be set. In Velfjord a massif of serpentinized periodotites with prominent chromite layering is present at Heggfjord (Myrland 1972).

Lyngen

The Lyngen mafic (Randall 1971; Munday 1974) comprises a strongly tectonized layered gabbro with subsidiary greenstones, dykes of MORB geochemistry, and lensoid ultramafic bodies. The complex was deformed prior to the deposition of Middle to Upper Ordovician continental sediments (Minsaas 1981) in all probability during the Finnmarkian orogenic event.

The group II ophiolite fragments

The group II complexes are recognized mainly in the Trondheim region. In this region it had commonly been assumed, until quite recently, that all major greenstone complexes were correlatives of the Støren Group, which Vogt (1945) noted was separated from the early–mid Ordovician sediments of the Lower Hovin Group by a major tectonic disturbance. Mapping over the past decade has shown, however, that the greenstone–gabbro complexes from the Grefstadfjell, Vassfjell, Løkken, Jonsvatn, Forbordfjell, and Snåsavatn areas form part of the Lower Hovin sequence, and are not correlative with The Støren Group (Roberts *et al*. in press). As this represents an important modification of the accepted stratigraphy for the area, the evidence for making this claim will be briefly reviewed below.

Stratigraphic relationships between the Grefstad-fjell, Løkken, and Vassfjell ophiolites are not known in detail, but all three complexes show geochemical similarities and share the same tectonometamorphic history as the enveloping sediments of the Lower Hovin Group. None of these ophiolite units carries any pre-Lower Hovin Group higher grade fabric as would be expected had they been affected by the earliest Ordovician upper greenschist to amphibolite facies event recorded in the Trondheim area (Klingspor and Gee 1981); this tectonothermal event is that of the post-Støren Group (Trondheim) disturbance. Although the bases of the Løkken and Vass-fjell complexes are unknown, these units, like the Grefstadfjell complex, are directly overlain by sediments of the middle part of the Lower Hovin Group. Accepting that these complexes are broadly correlative, representing portions of an originally extensive pre-latest Arenig (Ya 2) ophiolite, then they must now occupy a major SE-facing F_1 isoclinal fold-nappe with the Løkken complex on the inverted limb and the Grefstadfjell–Vassfjell units on the normal limb (Roberts *et al*. in press).

Løkken

The Løkken ophiolite occurs approximately along strike from the Vassfjell complex and some 5 km north of the Grefstadfjell greenstones. The complex, which is structurally inverted, contains gabbros, sheeted dykes and basic volcanites (often pillowed) intercalated with black cherts, shales, and jaspers at higher stratigraphic levels (Fig. 2) (Grenne *et al*. 1980).

Vassfjell

This complex, which is now considered to be a correlative of the Løkken, exhibits a partial ophiolite pseudostratigraphy (Fig. 2) of gabbro, sheeted dykes, and basaltic greenstones (Grenne *et al*. 1980). It is unconformably overlain by sediments of the Lower Hovin Group, derived from the ophiolite. Associated with the Vassfjell unit, across a fault contact, is a thick breccia of ophiolite and exotic detritus with fragments several tens of metres, or even hundreds of metres, across (Grenne 1980), interpreted as an olistostrome.

Grefstadfjell

Pillow lavas near the top of the Grefstadfjell lava pile are intercalated with graptolitic black shales, manganiferous sediments, and cherts. Graptolite fragments recovered from these shales indicate a possible age range of late Arenig to Llanvirn, and are generally similar to those recovered from the Bogo Formation of the Lower Hovin Group. This fauna is presently being reinvestigated by Dr. B. Erdtmann and a more precise age determination may become available in the near future. The manganiferous sediments are in sedimentological contact with and locally channelled by volcanic breccias that intercalate with fossiliferous sediments of the Bogo Formation. At the eastern extremity of the complex, vesicular pillow lavas representing the highest pseudostratigraphic levels are intercalated with breccias and sandstones that are assigned to the Lower Hovin Group (Ryan *et al*. 1980).

Jonsvatn and Forbordfjell

The Jonsvatn and Forbordfjell greenstones consist of lava piles up to 2 km thick within the Lower Hovin Group. Based on regional correlations of adjacent sediments, an age of Middle Llanvirn to Lower Llandeilo has been suggested for the volcanites (Roberts 1975). Although dominantly composed of massive and pillowed greenstones, the sequence also contains cherts, quartz-keratophyres, and tuffs. Subconcordant gabbro and cross-cutting dolerite dykes locally intrude the volcanites (Grenne and Roberts 1980).

Snåsavatn

Greenstones occur in the Snåsavatn region and towards their base interdigitate with limestone yielding a fauna of probable late Arenig–Llanvirn age (Roberts 1980). The greenstone complex consists of massive and tuffitic schistose greenstone with metagabbro, and in places intermediate and silicic volcanites (Roberts 1967). Geochemically, the lowermost lavas are of calc-alkaline affinity, whereas the bulk of the greenstones are tholeiites of essentially ocean floor character (see the section Geochemistry) considered to have accumulated in a marginal basin setting.

Sulitjelma

The Sulitjelma magmatic complex shows a well-developed pseudostratigraphy and is overlain by graphitic and ferromagnesian schists containing thin keratophyres (Boyle 1980). The gabbro complex metamorphoses shelf sediments, and although it is not clear whether this is a contact or obduction metamorphism (Mason 1980), this would suggest evolution in a back-arc basin close to or at the continental margin. The low angle between the intrusions of the sheeted dyke complex and flow layering in the overlying lavas (Fig. 2) is probably not a primary feature of this highly deformed complex.

Geochemistry

In this paper we deal only with the trace elements of the metabasalts, summing up all available data. More comprehensive descriptions and presentation of geochemical data can be found in previously published papers, or papers in preparation (see references in Figs. 3–5). Within each of the ophiolite fragments, the metabasalts may show profound variations in the total abundances and ratios of incompatible trace elements. Where ratios between incompatible trace elements show only minor variations, we present average values; variations in concentrations we regard as being due to fractional crystallization and/or minor differences in the degree of partial melting. In the geochemical description we follow the classification of the ophiolites as shown in Figs. 1 and 2.

The REE patterns of the metabasalts from group I ophiolites are shown in Fig. 3. Those from Karmøy characteristically show a progressive depletion from Sm to La; from Sm to Lu they define a rather flat pattern and a small negative Eu anomaly. The amphibolites from the Minor Bergen Arc show a REE pattern which is nearly identical with that of the Karmøy metabasalts. The REE pattern of the metabasalts from Lykling (between Karmøy and Bergen), however, differs from the two above-mentioned in being much more depleted in La and Ce, and by showing a convex-downward pattern from Nd through Yb. The metabasalts from Gullfjell (the Major Bergen Arc) define a variety of REE patterns, varying from LREE-depleted, slightly LREE-enriched, to variously shaped convex-upward patterns. A convex-upward REE pattern is also characteristic of the Solund and Stavfjorden metabasalts, the former also having a pronounced negative Eu anomaly. The REE patterns of the metabasalts from Støren are slightly LREE enriched, while the Leka metabasalts are slightly depleted in LREE. In contrast to the REE patterns of the above-mentioned sequences in the Trøndelag region are those from the Skålvær area further to the north. In this region the greenstones, like those from the Major Bergen Arc, define a variety of patterns, from LREE-enriched, LREE-depleted, to convex-upward patterns.

The REE patterns of the metabasalts from group II ophiolite assemblages are shown in Fig. 4, and those from Forbordfjell and Jonsvatn vary from slightly depleted to slightly enriched in LREE. The Løkken, Vassfjell, and Grefstadfjell metabasalts define patterns depicting flat REE to LREE depletion.

In Fig. 5a the average values for the trace elements Sr, Rb, Ba, Ta, Nb, Ce, Zr, Hf, Sm, Ti, Y, and Yb, which are (more or less) incompatible elements with respect to basaltic melts, together with Sc and Cr, for each of the ophiolite fragments have been normalized against an average MORB (Pearce 1980). In this diagram, the data show distinct groupings. The metabasalts from Stord, Solund, Stavfjord, Støren, Leka, Skålvær, and Sulitjelma define nearly flat trends and are indeed similar to an average MORB with respect to these element ratios. The metabasalts from Karmøy, Lykling, and the Bergen Arcs, on the other hand, are significantly different from those mentioned above in showing certain negative anomalies, especially with respect to Y, Zr, Ta, and Nb.

Fig. 5b shows the MORB-normalized trace element patterns of the Middle Ordovician metabasalts from Snåsavatn, Forbordfjell, Jonsvatn, Løkken, Vassfjell, and Grefstadfjell which, on field and geochemical evidence, have been interpreted as being of marginal basin origin. The metabasalts from Snåsavatn define two distinct groupings, the main one showing a tholeiitic trend and the other a calc-alkaline trend (Roberts 1982). This is also clearly shown on Fig. 5b, where the average calc-alkaline metabasalts (group B) are much more enriched in all trace elements than the tholeiitic group A, except for Ti, Y, and Cr. Characteristic features of both groups of the Snåsavatn greenstones, though most pronounced for the calc-alkaline group, are the high concentrations of K and Ba, low Nb concentrations, and the progressive decrease in concentrations from Ba through Y. Some features of this trend are also reflected in the lower and upper members of the Forbordfjell metabasalts (Fig. 5b). The Vassfjell, Løkken, Grefstadfjell, and Jonsvatn metabasalts show rather flat trace element patterns (Fig. 5). The lower and upper greenstones from Forbordfjell, however, show a progressive enrichment from Yb through Sr (Fig. 5), which is also the case for basalts from some marginal basins (Fig. 6).

Above, we have demonstrated that the REE patterns and the total trace element concentrations of the metabasalts may show similarities as well as clear differences. In the following we will therefore compare their trace element patterns with those of basalts from various tectonic environments.

In Fig. 6 the MORB-normalized trace element patterns of basalts from some island arcs, marginal

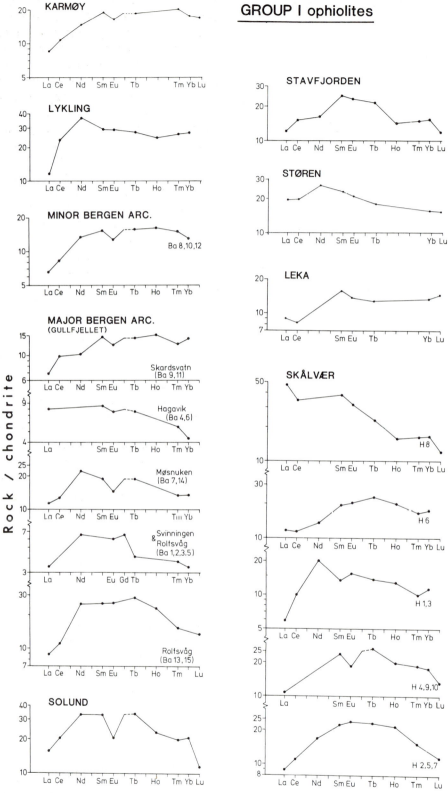

Fig. 3   REE patterns of metabasalts from group I ophiolite fragments. Chondrite data from Frey *et al*. (1968). REE data from the metabasalts are from the following sources: Karmøy: Furnes *et al*. (1982); Lykling, Minor and Major Bergen Arcs, Solund, Stavfjorden, and Skålvær: Furnes *et al*. (1982); Støren: Loeschke (1976a,b), Loeschke and Schock (1980), Barker and Millard (1979); Leka: Prestvik (1980)

GROUP II ophiolites

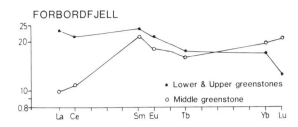

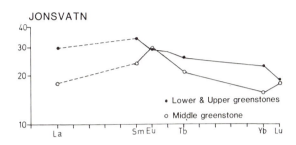

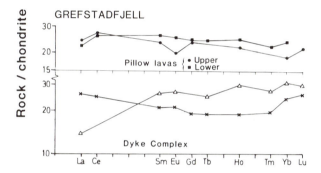

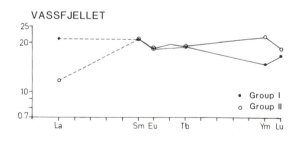

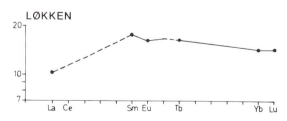

Fig. 4   REE patterns of metabasalts from group II ophiolite fragments. Chondrite data as above. REE data from the metabasalts are from the following sources. Forbordfjell and Jonsvatn: Grenne and Roberts (1980); Grefstadfjell: P. D. Ryan and H. Furnes (unpubl. data); Vassfjell and Løkken: Grenne et al. (1980)

basins and ocean islands are shown, and despite large variations in the absolute concentrations, element ratios stay nearly constant for basalts from the same tectonic environment. An important feature to emphasize is the similarity, particularly the pronounced negative Nb anomalies, between the island arc basalts and the young marginal basin basalts.

On comparing the trace element patterns of the metabasalts from Karmøy, Lykling, and the Bergen Arcs with those from modern island arcs and marginal basins and the average MORB (Fig. 6), no obvious similarites exist although it should be borne in mind that the highest MORB-normalized element values for the Lykling ophiolites are K and Rb, and for Karmøy it is Ba. Although we have reservations regarding the discriminative use of these particular elements in metamorphic rocks, we will nonetheless keep the possibility open that these may be features inherited from island arc activity.

The calc-alkaline basalts from Snåsavatn define a similar trend to that of the island arc basalts, whereas the tholeiitic basalts from Snåsavatn and the Lower and Upper greenstone members from Forbordfjell show comparable patterns to those of young marginal basin basalts except for the negative Nb anomalies (Fig. 5b). The Middle greenstone member from Forbordfjell, and those from Jonsvatn and parts of Grefstadfjell, on the other hand, are essentially indistinguishable from MORB. This geochemical trend of the marginal basin basalts, from the inclusion of an island-arc pattern at Snåsavatn to a progressively more MORB-like character further south, is an interesting feature which is difficult to explain at present. If the various volcanic sequences from Snåsavatn (north) to Grefstadfjell (south) developed in the same marginal basin at approximately the same time, it probably means that from north to south one progressively moved away from the island arc. This appears to fit with the picture of the regional geology, and the location of the mature arc, as described by Roberts (1980). Alternatively, this feature may also partly be a reflection of time, as discussed by Grenne and Roberts (1980), although constraints on the precise age of many of the volcanite complexes are minimal. Here, the Snåsavatn and Grefstadfjell metabasalts may represent an earlier stage of marginal basin spreading than those of Forbordfjell and Jonsvatn.

In summary it can be stated that the REE patterns of the metabasalts from Gullfjell and Skålvær show significant within-sequence variation, whereas those from all the others show little internal but large between-sequence variations. In general they can be matched with those found in both major oceans and marginal basins. In most cases the Rock/MORB characteristics of metabasalts from groups I and II

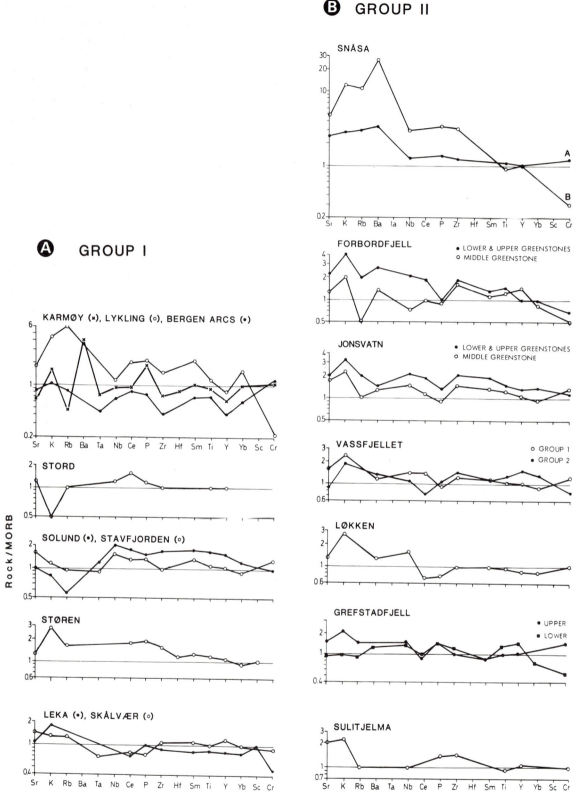

Fig. 5 MORB-normalized trace element patterns of the group I and II ophiolitic metabasalts of the Scandinavian Caledonides, MORB values from Pearce (1980). Trace element data of the metabasalts; Karmøy: Furnes *et al*. (1980b); Lykling: Furnes *et al*. (1982); Stord: Furnes and Færseth (1975); Bergen Arcs: Furnes *et al*. (1982); Støren: Loeschke (1976a,b), Loeschke and Schock (1980), Barker and Millard (1979); Vassfjellet and Løkken: Grenne *et al*. (1980); Forbordfjell and Jonsvatn: Grenne and Roberts (1980); Grefstadfjell: P. D. Ryan (unpubl. data); Snåsa (Groups A and B represent the structurally highest and lowest levels, respectively): D. Roberts (unpubl. data); Leka: Prestvik (1980); Skålvær: Gustavson (1978), Furnes *et al*. (1982); Sulitjelma: A. Boyle (unpubl. data)

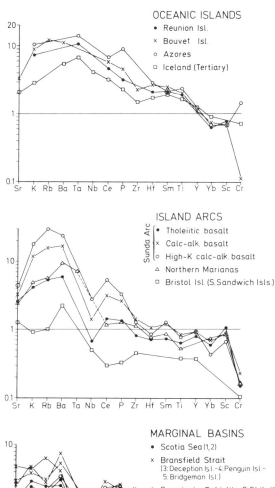

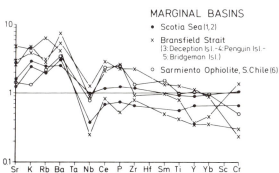

Fig. 6 MORB-normalized trace element patterns of some island arcs, marginal basins, and oceanic islands. Data sources: see Furnes *et al.* (1982)

show flat patterns. However, a clear island arc influence is recorded in some of the group II sequences, particularly Snåsavatn, and also in the group I ophiolites such as Karmøy, Lykling, and Gullfjell.

Summary

This classification of the ophiolite fragments in the Scandinavian Caledonides relies primarily upon the nature and age of the sedimentary component and its relationship to the magmatic members of the ophiolitic complexes, their structural history, age relationships (geochronological and palaeontological), and also partly on the geochemical patterns

of the latter. Recent studies have tended to concentrate upon the magmatic rocks within the ophiolitic fragments, and only in a few cases have the sediments been studied in sufficient detail to allow the complexes to be classified. The group I ophiolites in the area between Leka and Solund show normal MORB geochemistry, whereas those from Bergen, Lykling, and Karmøy also reflect a component of island arc influence. In the case of the group II complexes more information on their sedimentary components are available, and the complexes of Grefstadfjell, Løkken, Vassfjell, Jonsvatn, and Forbordfjell are considered as representative examples.

The different ages of the group I and II complexes, either suspected or known at the present time, may be substantiated by further research, and if this is the case it will prove critically important to the understanding of the geotectonic evolution of the Scandinavian Caledonides. However, it must be emphasized that in only two complexes have the associated or adjacent sediments so far yeilded fossils (Grefstadfjell and Snåsavatn), and that the ages of all the others are based on lithostratigraphic or tectonostratigraphic arguments, with the exception of Lykling which, on the basis of Rb–Sr whole rock dating, has been shown to have a minimum age of 535 ± 46 Ma. The Støren complex is pre-Middle Arenig, and the Karmøy ophiolite must have developed at some considerable time prior to its obduction and subsequent intrusion by the 450 Ma West Karmøy Igneous Complex. Minimum ages for many other complexes are provided only by the mid to late Ordovician unconformable cover. This, however, is not diagnostic of either group as such relationships are recorded in both group I (Karmøy) and group II (Grefstadfjell) complexes. Only those complexes that contain a penetrative pre-mid to late Ordovician tectonometamorphic fabric are here assigned a pre-Finnmarkain age. The lack of a full pseudostratigraphy in group II complexes may not be important for their classification as such. Indeed, some of the best developed ophiolites, as for example the Osman ophiolite, have been shown to have developed in a back-arc basin (Pearce *et al.* 1981). It is therefore interesting to note that it has recently been suggested that an ultramafic complex in southeastern Trøndelag could represent the lower levels of an ophiolite complex formed in a back-arc environment (Moore and Hultin 1980).

Acknowledgements

Dr. A. Boyle kindly let us use some of his unpublished geochemical data from the Sulitjelma ophiolite, and E. Irgens and J. Lien prepared the illustrations. We express our thanks to these persons. International Geological Correlation Programme. Norwegian contribution no. 62 to Project 27–Caledonide Orogen.

References

Amaliksen, K. G. 1983. *The geology of the Lykling ophiolite complex, Bømlo, SW Norway.* Unpubl. Cand. Real. thesis, Univ. of Bergen, 417 pp.

Barker, F. and Millard, H. T. 1979. Geochemistry of the type trondhjemite and three associated rocks, Norway. In Barker, F. (Ed.), *Trondhjemite, Dacites and Related Rocks.* Elsevier Sci. Publ. Com., 517–529.

Blake, D. H. 1962. A new Lower Ordovician graptolite fauna from the Trondheim region. *Norsk geol. Tidsskr.*, **42**, 223–238.

Boyle, A. P. 1980. The Sulitjelma amphibolities, Norway: part of a Lower Palaeozoic ophiolite complex? *Proc. Int. Ophiolite Symp., Cyprus 1979*, 567–575.

Coleman, R. G. and Peterman, Z. E. 1975. Oceanic plagiogranites. *J. Geophys. Res.*, **80**, 1099–1108.

Frey, F. A., Haskin, M. A., Poetz, J. A. and Haskin, L. A. 1968. Rare earth abundances in some basaltic rocks. *J. Geophys. Res.*, **73**, 6085–6098.

Furnes, H., Skjerlie, F. J. and Tysseland, M. 1976. Plate liminary trace element data from the Lower Palaeozoic greenstone sequences on Stord, west Norway. *Norsk geol. Tidsskr.*, **55**, 157–169.

Furnes, H., Skjerlie, F. J. and Tysseland, M. 1976. Plate tectonic model based on greenstone geochemistry in the Late Precambrian–Lower Palaeozoic sequence in the Solund–Stavfjorden areas, west Norway. *Norsk geol. Tidsskr.*, **56**, 161–186.

Furnes, H., Roberts, D., Sturt, B. A., Thon, A. and Gale, G. H. 1980a. Ophiolite fragments in the Scandinavian Caledonides. *Proc. Int. Ophiolite Symp., Cyprus 1979*, 582–600.

Furnes, H., Sturt, B. A. and Griffin, W. L. 1980b. Trace element geochemistry of metabasalts from the Karmøy ophiolite, southwest Norwegian Caledonides. *Earth Planet. Sci. Lett.*, **50**, 75–92.

Furnes, H., Thon, A., Nordås, J. and Garmann, L. B. 1982. Trace element geochemistry of metabasalts from some Norwegian ophiolite fragments. *Contrib. Mineral. Petrol.*, **79**, 295–307.

Furnes, H., Austrheim, H., Amaliksen, K. G. and Nordås, J. 1983. Evidence for an incipient early Caledonian (Cambrian) orogenic phase in southwestern Norway. *Geol. Mag.*, **120**, 607–612.

Gale, G. H. and Roberts, D. 1972. Palaeogeographic implications of greenstone petrochemistry in the southern Norwegian Caledonides. *Nature*, London, **238**, 60–61.

Gale, G. H. and Roberts, D. 1974. Trace element geochemistry of Norwegian Lower Palaeozoic basic volcanics and its tectonic implications. *Earth Planet. Sci. Lett.*, **22**, 380–390.

Grenne, T. 1980 Excursion across parts of the Trondheim Region, Central Norwegian Caledonides. *Norges geol. Unders.*, **356**, 159–164.

Grenne, T., Grammeltvedt, G. and Vokes, F. M. 1980. Cyprus-type sulphide deposits in the western Trondheim district, central Norwegian Caledonides. *Proc. Int. Ophiolite Symp., Cyprus 1979*, 727–743.

Grenne, T. and Roberts, D. 1980. Geochemistry and volcanic setting of the Ordovician Forbordfjell and Jonsvatn greenstones, Trondheim Region, central Norwegian Caledonides. *Contrib. Mineral. Petrol.*, **74**, 374–386.

Gustavson, M. 1978. Geochemistry of the Skålvær greenstone, and a geotectonic model for the Caledonides of Helgeland, north Norway. *Norsk geol. Tidsskr.*, **58**, 161–174.

Klingspor, I. and Gee, D. G. 1981. Isotopic age-determination studies of the Trøndelag trondhjemites. *Terra cognita*, **1**, 55.

Loeschke, J. 1976a. Major element variations in Ordovician pillow lavas of the Støren Group, Trondheim Region, Norway. *Norsk geol. Tidsskr.*, **56**, 141–159.

Loeschke, J. 1976b. Petrochemistry of eugeosynclinal magmatic rocks of the area around Trondheim (Central Norwegian Caledonides). *Neues Jb. Miner. Abh.*, **128**, 41–72.

Loeschke, J. and Schock, H. H. 1980. Rare earth element contents of Norwegian greenstones and their geotectonic implications. *Norsk geol. Tidsskr.*, **60**, 29–37.

Mason, R. 1980. Temperature and pressure estimates in the contact aureole of the Sulitjelma gabbro, Norway: implications for an ophiolite origin. *Proc. Int. Ophiolite Symp., Cyprus 1979*, 576–581.

Minsaas, O. 1981. *Lyngenhalvøyas geologi, med spesiell vekt på den sedimentologiske utvikling av de ordovisisk–siluriske klastiske sekvenser som overligger Lyngen gabbro-kompleks.* Unpbl. Cand. Real. thesis, Univ. of Bergen.

Moore, A. C. and Hultin, I. 1980. Petrology, mineralogy, and origin of the Feragen ultra-mafic body, Sør–Trøndelag. Norway. *Norsk geol. Tidsskr.*, **60**, 235–254.

Munday, R. J. C. 1974. The geology of the northern half of the Lyngen Peninsula, Troms, Norway. *Norsk geol. Tidsskr.*, **54**, 49–62.

Myrland, R. 1972. Veltjord. Beskrivelse til det berggrunnsgeologiske gradteigskart I 18—1:100000. *Norges geol. Unders.*, **274**, *Skrifter,1*.

Nordås, J., Amaliksen, K. G., Brekke, H., Suthren, R., Furnes, H., Sturt, B. A. and Robins, B. this volume. Lithostratigraphy and petrochemistry of Caledonian rocks on Bømlo, S.W. Norway.

Pearce, J. A. 1980. Geochemical evidence for the genesis and eruptive setting of lavas from Tethyan ophiolites. *Proc. Int. Ophiolite Symp., Cyprus 1979*, 261–272.

Pearce, J. A., Alabaster, T., Shelton, A. W. and Searle, M. P. 1981. The Oman ophiolite as a Cretaceous arc-basin complex: evidence and implications. *Phil. Trans. R. Soc. London* **A300**, 299–317.

Prestvik, T. 1972. Alpine-type mafic and ultramafic rocks of Leka, Nord-Trøndelag. *Norges geol. Unders.*, **273**, 23–34.

Prestvik, T. 1974. Supracrustal rocks of Leka, Nord-Trøndelag. *Norges geol. Unders.*, **311**, 65–87.

Prestvik, T. 1980. The Caledonian ophiolite complex of Leka, north–central Norway. *Proc. Int. Ophiolite Symp., Cyprus 1979*, 555–566.

Prestvik, T. and Roaldset, E. 1978. Rare earth element abundances in Caledonian metavolcanics from the island of Leka, Norway. *Geochem. Jour.*, **12**, 89–100.

Priem, H. N. A., Boelrijk, N. A. I. M., Hebeda, E. H., Verdurman, E. A. Th. and Verschure, R. H. 1975. Isotopic dating of the Caledonian Bindal and Svenningdal granitic massif, central Norway. *Norges geol. Unders.*, **319**, 29–36.

Randall, B. A. O. 1971. An outline of the geology of the Lyngen Peninsula, Troms, Norway. *Norges geol. Unders.*, **269**, 68–71.

Roberts, D. 1967. Geological investigations in the Snåsa–Lurudal area. Nord-Trøndelag. *Norges geol. Unders.*, **257**, 18–38.

Roberts, D. 1975. The Stokvolla conglomerate—a revised stratigraphical position. *Norsk geol. Tidsskr.*, **55**, 361–371.

Roberts, D. 1980. Petrochemistry and palaeogeographic setting of Ordovician volcanic rocks of Smøla, Central Norway. *Norges geol. Unders.*, **359**, 43–60.

Roberts, D. 1982. Disparate geochemical patterns from the Snåsavatn greenstone, Nord–Trøndelag, Central Norway. *Norges geol. Unders.*, **373**, 63–73.

Roberts, D., Grenne, T. and Ryan, P. D. in press. Ordovician marginal basin development in the central Norwegian Caledonides. *J. geol. Soc. London.*

Ryan, P., Williams, D. M. and Skevington, D. 1980. A revised interpretation of the Ordovician stratigraphy of Sør Trøndelag, and its implications for the evolution of the Scandinavian Caledonides. *Proc. 'The Caledonides in the USA' I.G.C.P. project 27: Caledonide Orogen.* 1979 meeting, Blacksburg, Virginia.

Skevington, D. 1963. On the age of the Boge shale. *Norsk geol. Tidsskr.*, **43**, 257–260.

Solli, T. 1981. *The Geology of the Torvastad Group, the Cap Rocks to the Karmøy Ophiolite.* Unpubl. Cand. Real. thesis, Univ. of Bergen.

Sturt, B. A., Andersen, T. B. and Furnes, H. this volume. The Skei Group, Leka: an unconformable clastic sequence overlying the Leka Ophiolite.

Sturt, B. A. and Thon, A. 1978. An ophiolite complex of probable early Caledonian age discovered on Karmøy. *Nature*, London, **275**, 538–539.

Sturt, B. A., Thon, A. and Furnes, H. 1979. The field relationships of the Karmøy Ophiolite. *Geology*, 7, 316–320.

Sturt, B. A., Thon, A. and Furnes, H. 1980. The geology and preliminary geochemistry of the Karmøy Ophiolite, SW. Norway. *Proc. Int. Ophiolite Symp., Cyprus 1979*, 538–554.

Sturt, B. A., Roberts, D. and Furnes, H. 1984. A conspectus of Scandinavian Caledonian ophiolites. In Gass, I. G., Lippard, S. T. and Shelton, A. W. (Eds), Ophiolites and Oceanic Lithosphere. *Spec. Publ. Geol. Soc. Lond.*, 381–391.

Thon, A. this volume. The Gullfjellet ophiolite and the structural evolution of the Major Bergen Arc, west Norwegian Caledonides.

Thon, A., Magnus, C. and Breivik, H. 19870. The stratigraphy of the Dyvikvågen Group, Stord: a revision. *Norges geol. Unders.*, **359**, 31–42.

Vogt, Th. 1945. The geology of the Hølonda–Horg district, a type of area in the Trondheim Region. *Norsk geol. Tidsskr.*, **25**, 449–528.

The Caledonide Orogen—Scandinavia and Related Areas
Edited by D. G. Gee and B. A. Sturt

The Gullfjellet ophiolite complex and the structural evolution of the major Bergen arc, west Norwegian Caledonides

A. Thon

Statoil, Postboks 300, 4033 Forus, Norway

ABSTRACT

Structurally the Gullfjellet Ophiolite Complex forms part of the Major Bergen Arc. In the west the Bergen Arcs are underlain by Precambrian gneisses and in the east they overlie the Precambrian and Lower Palaeozoic rocks of the Bergsdalen Nappes. The Gullfjellet Ophiolite Complex has a minimum thickness of 5 km. Several intrusive phases can be recognized in the top of the plutonic zone, which contains homogenous gabbros with minor trondhjemites. The gabbros grade into a well-developed sheeted dyke complex with gabbro screens in the lowest part, whilst in the upper part of the dyke complex steeply-dipping pillow lava screens occur. The dyke complex is overlain by a gently eastward-dipping pillow lava sequence. Beneath the Gullfjellet massif occurs a multicomponent tectonostratigraphic unit—the Samnanger Complex. In the western part of the latter, slices of ophiolitic rocks predominate, and in the eastern part sheets of gneisses and a sequence of continental prism metasediments predominate. The Samnanger Complex contains a series of tectonic lenses of serpentinites which possibly represent dunites thrust up from the base of the ophiolite. The obduction of the Gullfjellet Ophiolite occurred in pre-Middle/Upper Ordovician times. The ophiolite was deeply eroded and unconformably overlain by an Upper Ordovician/Lower Silurian cover sequence, the Holdhus Group. During the end-Silurian orogenesis, the cover sediments were folded into pinched-in synclines and the entire complex was translated further eastwards. The Samnanger Complex may thus represent a gigantic imbricate structure involving both ophiolitic rocks and possible continental prism sediments and their gneissic substrate.

Introduction

Structurally the Gullfjellet Ophiolite Complex forms part of the Major Bergen Arc, which is one of a series of thrust sheets comprising the Bergen Arcs. The Bergen Arcs (Fig. 1) consists of rocks both of Lower Palaeozoic and Precambrian age (Kolderup and Kolderup 1940; Sturt *et al.* 1975). A recent summary of the tectonostratigraphy of the Bergen Arcs is presented in Sturt and Thon (1978a). To the west, the Bergen Arcs are underlain by the Øygarden Gneiss Complex consisting of Precambrian gneisses which show varying degrees of Caledonian reworking. It is not possible to say whether this unit is true basement or if it is allochthonous. To the east the Bergen Arcs overlie the Bergsdalen Nappes

which contain rocks of both Precambrian and Lower Palaeozoic age (Kvale 1960; Pringle *et al.* 1975). The Major and Minor Bergen Arcs consist of rocks of Lower Palaeozoic age and a number of imbricated thrust sheets of highly retrograded and mylonitized Precambrian gneisses. It is possible to distinguish between two major rock units of Lower Palaeozoic age separated by a regional stratigraphic unconformity (Sturt and Thon 1976; Naterstad 1976; Færseth *et al.* 1977; Henriksen 1981). These were termed the Samnanger Complex (oldest) and the Holdhus Group (youngest) respectively by Færseth *et al.* (1977). The Holdhus Group includes metasediments of low metamorphic grade which contain an Upper Ordovician (Ashgillian) shallow marine fauna (Resch 1882; Kolderup 1914; Kiær 1929;

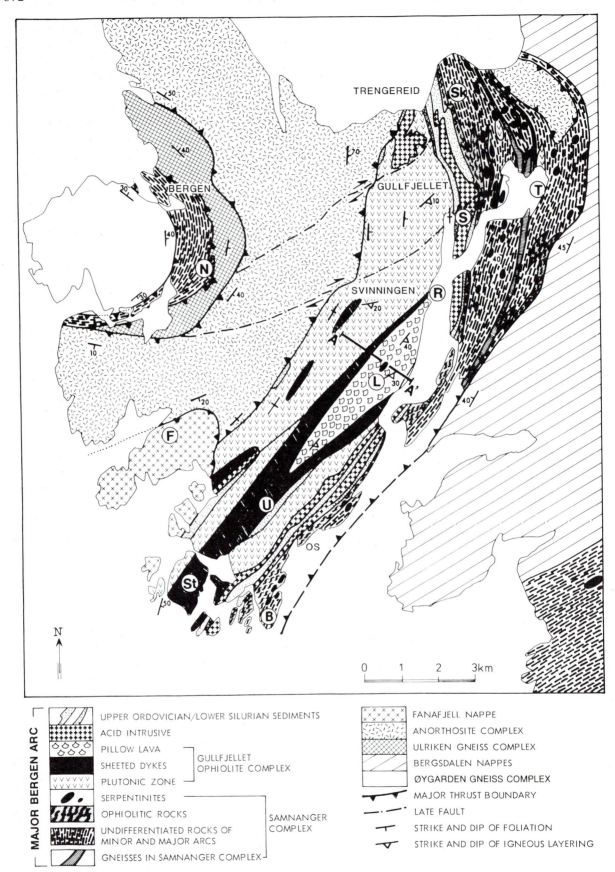

Fig. 1   Geological map of the central part of the Bergen Arcs. Partly from Kolderup and Kolderup (1940), Færseth *et al.* (1977), and also unpublished information from Ingdahl (1981). Abbreviations on map: B—Bjørnatrynet, F—Fanafjell, L—Lønningdal, N—Nattlandsfjell, R—Rolvsvåg, S—Skardsvåg, Sk—Skulstad, St—Strøno, T—Tysse, U—Ulven

Kolderup and Kolderup 1940). These schist arcs are overlain by a series of tectonic units which are, from the tectonically lowest to the highest unit, the Ulriken Gneiss Nappe Complex, the Anorthosite Complex, and the Fanafjell Nappe (Sturt and Thon 1978a).

Geology of the Major Bergen Arc

The Gullfjellet Ophiolite Complex was previously described as saussurite gabbros on the Geological Map of the Bergen District (Kolderup and Kolderup 1940). Torske (1972) described pillow lava structures within this complex northeast of Strøno (St, Fig. 1) and Inderhaug (1975 and unpublished) interpreted the basic rocks northwest of Ulven (U, Fig. 1) as submarine volcanics displaying ocean-floor geochemical characteristics. Inderhaug (1975) claimed that the metavolcanics were intruded by dolerites, gabbros, diorites, and quartz diorites associated with ensimatic island arc magmatism. Furnes *et al.* (1980), based on Inderhaug's data, interpreted the basic rocks of the Gullfjellet–Strøno tract as a dismembered ophiolite fragment, and they proposed that this ophiolite fragment is part of a much larger slab thrust onto crystalline rocks of the Baltic Shield in a late Cambrian/early Ordovician orogenic event.

In the Samnanger area east of Gullfjellet (Fig. 1), Færseth *et al.* (1977) recognized two fundamental rock units of the Major Bergen Arc, i.e. the Samnanger Complex and the unconformably overlying Holdhus Group of Upper Ordovician/Lower Silurian age. The Samnanger Complex is mainly composed of metasedimentary schists, however, in its original definition (Færseth *et al.* 1977) included a series of metaperidotites (serpentinites), gabbros, quartz diorites, and greenstones. Within the greenstones, both pillow lava structures and dyke complexes are recorded and geochemically these greenstones are similar to the ones from the Gullfjellet Ophiolite (Fig. 4; Furnes *et al.* 1982). In Fig. 1 these greenstone lenses are shown as ophiolitic rocks in the Samnanger Complex.

Within the Samnanger Complex several linear belts of serpentinite lenses can be traced from Os up to the northernmost extremity of the Arcs (Kolderup and Kolderup 1940). They are typical Alpine type serpentinites. Chromite banding is locally preserved. They show evidence of polyphasel structural and metamorphic evolution (Bøe 1978). They were interpreted to have originated from dunites by Bøe (1978) who suggested that they represent fractional crystallization products of a picritic magma.

A rather different rock association dominates in the eastern part of the Samnanger Complex, including garnet–mica schists, graphite-bearing mica schists, quartzites, quartz-conglomerates, local marbles, and minor garnet-bearing amphibolites. These lithologies are interleaved with sheets of strongly retrograded and mylonitized Precambrian gneisses still locally preserving relics of granulite facies metamorphism. Naterstad (1976) suggested that an original cover to one of these gneiss sheets could be recognized in the area south of Tysse (T, Fig. 1). Henriksen (1981) working on Osterøy in the northern part of the Major Bergen Arc suggested that the metapelites and psammites represented a sedimentary prism accumulated on a continental margin to the Iapetus ocean floor.

The rocks of the Samnanger Complex show evidence of at least two major tectonometamorphic events (Færseth *et al.* 1977). The earliest phase reached upper greenschist/low amphibolite facies conditions and was associated with a polyphase structural development resulting in recumbent folding and imbricate structures. Inderhaug (1975) described inversion of the cumulate layering in the Liafjell gabbro southeast of Ulven (U, Fig. 1) prior to deposition of Upper Ordovician/Lower Silurian sediments. At Moldvika north of Os (Fig. 1) pillow lavas were inverted prior to deposition of the Moberg Conglomerate of Upper Ordovician age (Sturt and Thon 1976). A series of quartz diorites intruded previously-deformed greenstones and gabbros (Færseth *et al.* 1977).

The Upper Ordovician/Lower Silurian Holdhus Group sediments were deposited unconformably on the rocks of the Gullfjellet Ophiolite Complex and the Samnanger Complex in fault-bounded troughs controlled by the structural grain in the substrate (Thon this volume).

During the final *mise en place* the whole sequence of the Major Bergen Arc was translated further eastwards accentuating the older structural trends. During this tectonometamorphic episode the rocks of the Samnanger Complex underwent retrograde metamorphism whilst in the sediments of the Holdhus Group the metamorphism was prograde reaching middle greenschist facies. The Holdhus Group sediments are preserved in variable squeezed out synclines between blocks of the substrate forming the typical linear outcrop pattern (Fig. 1).

Pseudostratigraphy of the Gullfjellet Ophiolite Complex

The distribution of the main units of the Gullfjellet Ophiolite Complex and its relation to the surround-

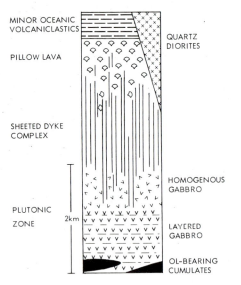

Fig. 2   Reconstructed idealized pseudostratigraphy of the Gullfjellet Opiolite Complex

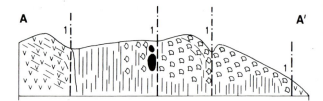

Fig. 3   Structural section through the Gullfjellet Ophiolite. Location shown on Fig. 1 (not to scale)

ing units is shown in Fig. 1. In the west the Gullfjellet Ophiolite Complex is in fault contact with rocks of the Anorthosite Complex and the Fanafjell Nappe. Along this contact isolated lenses of limestones and quartzose sediments occur, which are considered to represent remnants of the younger fossilbearing Holdhus Group. The eastern contact of the Gullfjellet Ophiolite is a fault with rocks of the Samnanger Complex and partially with rocks of the Holdhus Group.

The internal stratigraphy of the Gullfjellet Ophiolite Complex is complicated due to the frequently faulted nature of the main contacts and also due to strong shearing and faulting at various levels of the complex. Despite this it is possible to work out a pseudostratigraphy for the complex, bearing in mind that the thicknesses presented are minimum values (Fig. 2).

Plutonic zone

The rocks of the lowest exposed stratigraphic level consist of serpentinized olivine-bearing cumulates with a well-developed igneous layering. A lower level of the plutonic zone may be represented by the serpentinite lenses in the Samnanger Complex. Further evidence for a dunitic lower part of the ophiolite is found north of Lønningdal (L, Fig. 1) where a serpentinite lens has been emplaced along steep faults close to the sheeted dyke pillow lava contact (Fig. 3). It is suggested that the lenses were torn off the base of the ophiolite during eastward obduction and emplaced along thrust planes into the Samnanger Complex.

The bulk of the rocks of the plutonic zone, however, are saussurite gabbros, and the original

mineralogy is rarely preserved. In the best studied area, the Liafjell gabbro southeast of Ulven (U, Fig. 1), Inderhaug (1975 and unpublished) described three cumulate sequences each showing internal cryptic layering. In the lowest part of cumulate sequence two Inderhaug (1975) reports a variation in the plagioclase content from An_{86-64} and in the olivine content from Fo_{87-77}. Towards the top of the first cumulate sequence he reports An_{57} + cpx + ilmenite-magnetite and ore. The main part of the plutonic zone is composed of saussurite gabbros containing clinozoisite + actinolite + talc or albite + clinozoisite + epidote + hornblende + chlorite + quartz.

Layering is best preserved in the interior parts of the plutonic zone such as at Svinningen and the southeastern part of Gullfjellet (Fig. 1). Towards the margins of the plutonic zone new flaser-gabbro textures develop. Generally, towards the top of the plutonic zone, the layering becomes less obvious and the gabbro is more homogeneous developing an isotropic fabric. The homogeneous gabbros are locally intruded by sheets and veins of trondhjemite. The sheets and veins are cut by the dykes of the sheeted complex. In the valley west of Lønningdal (L, Fig. 1) there is a transition from homogeneous gabbro to distinct dykes. The thickness of the transition zone is in the order of 200 m. Unfortunately continuous exposures are not available. The exposed thickness of the plutonic zone is in the order of 3 km.

Sheeted dyke complex

The sheeted dyke complex forms a zone of subvertical dykes that can be traced continuously along strike from the southwestern shore of Strøno (St, Fig. 1) to Rolvsvåg (R, Fig. 1). At Strøno the zone is about 1.5 km wide. Zones of high dyke density in the gabbro are also found west of the main zone (Fig. 1). Within the plutonic zone at Gullfjellet there are areas where dykes occupy more than 50 per cent of the outcrop. The dykes which trend in a northeast direction are doleritic, but also gabbroic dykes, feldsparphyric dykes, rare pyroxene (now amphibole) phyric dykes, and trondhjemitic dykes have been observed. They are metamorphosed in typical greenschist facies with epidote + actinolite + albite +

quartz and ore. Chilled margins have been observed. The main contact with the plutonic zone and the sheeted complex is faulted, although a transition zone is present west of Lønningdal.

Pillow lava

Overlying the dyke complex and apparently fed from them is an eastward younging pillow lava sequence more than 1.5 km thick. The bedding planes dip about 40° to the southeast. The best exposed sections of the lavas are seen in the mountains north and south of Lønningdal (L, Fig. 1). Lønningdal is a NW–SE trending glacial valley and in the road sections on the southwestern shore of Lønningdalsvatn sheeted dykes are exposed while pillow lavas outcrop on the hillside.

The pillow lavas are often variolitic indicating eruptions at depths in the order of 1500–2000 m (Furnes 1973). The lavas are metamorphosed and show typical greenschist facies mineral assemblages. North of Lønningdal (L, Fig. 1) a serpentinite lens with fault contacts on either side is found within the pillow lava sequence.

One structural feature of the lava sequence requires special comment. The bedding planes of pillow lavas in the upper part of the dyke complex are steep, i.e. parallel to the dykes, whereas in the thicker units a flatter attitude is prevalent. In the thicker units the dykes are still vertical. At present the author has no explanation for this phenomenon, but would suggest that the steep dips of bedding in the pillow lava screens in the dyke complex are related to faulting during emplacement of the dykes.

Oceanic sediments

Minor occurrences of jasper and Mn-rich sediments have been found associated with pillow lava and dykes within the western part of the Samnanger Complex northwest of Bjørnatrynet (B, Fig. 1) (Ingdahl personal communication 1981).

Quartz diorites

Quartz diorites intrude the rocks of the Gullfjellet Ophiolite Complex and the Samnanger Complex. They intrude deformed greenstone and were subsequently deformed prior to their unconformable truncation by the Upper Ordovician conglomerates (Sturt and Thon 1976; Færseth et al. 1977).

Ophiolitic rocks in the Samnanger Complex

The western part of the Samnanger Complex is composed mainly of greenstones and greenschists. The degree of deformation is generally higher than in the Gullfjellet massif, but locally it is still possible to recognize primary pillow shapes and internal dyke contacts. Sturt and Thon (1976) described inverted pillow lavas from Moldvika just north of Os (Fig. 1). In the area between Os and Bjørnatrynet (Fig. 1), Ingdahl (personal communication) has mapped several pillow lava and sheeted dyke horizons with ocean-floor geochemical characteristics. At Skardsvåg (S, Fig. 1) a dyke complex about 200 m wide is found again displaying ocean-floor geochemical characteristics. At Skulstad (Sk, Fig. 1) both dykes and pillow lavas are present (Færseth et al. 1977). On Osterøy, in the northern part of the Major Bergen Arc, Henriksen (1981) mapped possible ophiolite fragments within the Samnanger Complex. In the Minor Bergen Arc sheets of flaser-gabbro, containing numerous basic dykes, are common and at Nattlandsfjellet (N, Fig. 1) local dyke contacts in a greenstone can still be recognized. Geochemically the greenstones at Nattlandsfjellet show similarities with those from the Gullfjellet Ophiolite Complex (Furnes et al. 1982).

Comments on the structure of the Gullfjellet Ophiolite Complex

A section through the Gullfjellet Ophiolite Complex is shown in Fig. 3 (see Fig. 1 for location). The main units are in fault contact and within each unit numerous shear zones and faults occur. As can be seen from the profile (Fig. 3), the northwestern block has moved up relative to the eastern. In these major shear zones primary igneous layering in the gabbros and bedding of the pillow lavas become vertical. The resultant rock types are flaser-gabbros and chlorite schists. The emplacement of the serpentinite lens north of Lønningdal has taken place along such faults.

Geochemistry

Fig. 4 shows the FeO^t–FeO^t/MgO and TiO_2/MgO diagrams from rocks both from the Gullfjellet Ophiolite Complex and from the ophiolitic rocks within the Samnanger Complex. No significant differences can be determined between the analyses from the two groups. Furnes et al. (1982) presented geochemical data from the Gullfjellet Ophiolite Complex and several other Norwegian ophiolite fragments. They concluded that geochemically the Gullfjellet Ophiolite reflects partial melts from a mid-oceanic mantle, or probably a back-arc basin mantle which differed from that of the northwestern Norwegian ophiolites. For further discussion see paper by Furnes et al. (1982) and this volume).

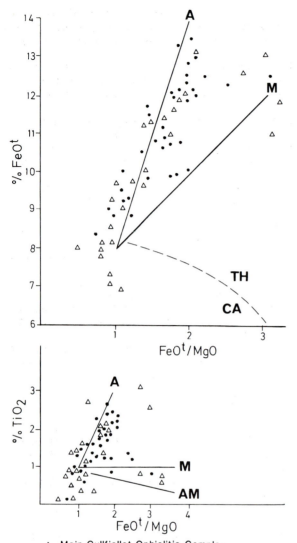

△ Main Gullfjellet Ophiolitic Complex

• Ophiolitic rocks, Western Samnanger Complex

Fig. 4 Plot of FeO^t–FeO^t/MgO and TiO_2–FeO^t/MgO diagrams for rocks of the main Gullfjellet Ophiolite Complex and ophiolitic rocks in the Samnanger Complex. The field boundaries separating the tholeiitic (TH) field boundaries and the calc-alkaline (CA) series, and the trend lines for abyssal tholeiites (A), Macauley island arc tholeiitic series (M), and the Amagi calc-alkaline series are after Miyashiro (1975)

Summary and Regional Correlations

A number of recent geochemical studies on the Caledonian greenstone sequences in western Norway have given independent evidence that they represent ocean-floor basalts of the Iapetus Ocean (see summary by Furnes et al. 1980 and this volume). The Karmøy Ophiolite (Sturt and Thon 1978b; Sturt et al. 1980) gave the first well-documented geological evidence in favour of such an interpretation. Furnes et al. (this volume) grouped the inferred

ophiolites into two main categories, type I having developed in relation to a major ocean or a large mature marginal basin, and type II in a small marginal basin of later age. The Gullfjellet Ophiolite Complex was grouped as 'unclassified' together with the Lykling Ophiolite some 50 km further south (Nordås et al. this volume).

The original age of the Gullfjellet Ophiolite Complex is not clear, but is generally believed to be of Cambrian age. A recent Rb–Sr whole rock isochron age from trondhjemitic extrusives above the Lykling Ophiolite of 535 ± 46 Ma strengthens this idea (Furnes et al. 1983).

The obduction of the Gullfjellet Ophiolite occurred during a major orogenic phase prior to Upper Ordovician times, given by the age of the unconformably overlying Holdhus Group. Correlation with nearby areas dates this event more precisely. A minimum age for the emplacement of the Karmøy Ophiolite is 450 Ma (Sturt and Thon 1978b; Sturt et al. 1980). On the island of Bømlo the Lykling Ophiolite is unconformably overlain by the Siggjo volcanics (Nordås et al. this volume) which consist mainly of subaerial volcanics. A Rb–Sr whole rock isochron date of 464 ± 16 Ma (Furnes et al. 1983) for the Siggjo volcanics places the minimum age for the obduction of the Lykling Ophiolite in the Middle Ordovician.

Based on the data presented in this paper a schematic model for the evolution of the Gullfjellet Ophiolite Complex and the Major Bergen Arc is outlined:

1. Development of ocean floor in a major ocean (Iapetus) or a large marginal basin during Cambrian times.

2. Closing of the ocean/marginal basin in Lower/Middle Ordovician times. The Gullfjellet Ophiolite Complex was obducted onto the continental prism of Baltic Shield crystallines, represented by the eastern Samnanger sequence. This thrusting caused imbrication and recumbent folding. Chromite-bearing dunites were sheared off from the base of the ophiolite and occur now as serpentinite lenses in the Samnanger Complex.

3. Extensive erosion of the ophiolite and Samnanger Complex and deposition of shallow marine coarse clastics overlain by carbonates during Upper Ordovician/Lower Silurian times.

4. Final closure of the Iapetus Ocean resulting in a continent–continent collision in Upper Silurian times. The entire nappe sequence of the Bergen Arcs moved further eastward on a major sole thrust. Older structural trends in the Gullfjellet Ophiolite Complex and in the Samnanger Complex were reactivated and the rocks under-

went retrograde metamorphism. The Upper Ordovician/Lower Silurian cover was squeezed in between the rigid blocks of the substrate forming typical pinched-in synclines.

Acknowledgements

Prof. B. A. Sturt read an early draft of this manuscript and made several important improvements. For this and good companionship during the fieldwork in the western Caledonides he is thanked. I must also thank an unknown referee for a thorough hammering of the first draft. I also thank Drs H. Furnes (University of Bergen) and P. D. Ryan (Galway, Ireland) for many stimulating discussions on west Norwegian ophiolites. C. m. S. Ingdahl is thanked for help during some of the fieldwork and also for access to unpublished material. S. Trovik is thanked for the FeO determinations. Miss E. Irgens and Mr M. Aadachi kindly prepared the illustrations. Financial support has been given by the Norwegian Council of Science and the Humanities (NAVF grant No D.48.22–17 and D.48.22–26). Statoil A/S Stavanger, Norway, are thanked for financial support for attending the UCS meeting in Uppsala (August 1981).

References

Bøe, R. 1978. Unpublished thesis, University of Bergen.

Furnes, H. 1973. Variolitic structure in Ordovician pillow lavas and its possible significance as an environmental indicator. *Geology*, **1**, 27–30.

Furnes, H., Roberts, D., Sturt, B. A., Thon, A. and Gale, G. 1980. Ophiolite fragments in the Scandinavian Caledonides. *Proc. Int. Ophiolite Symp. Nicosia, Cyprus 1979*, 582–600.

Furnes, H., Thon, A., Nordås, J. and Garmann, L. B. 1982. Geochemistry of Caledonian metabasalts from some Norwegian ophiolite fragments. *Contrib. mineral. Petrol.*, **79**, 295–307.

Furnes, H., Austrheim, H., Amaliksen, K. G. and Nordås, J. 1983. Evidence for an incipient early Caledonian (Cambrian) orogenic phase in southwestern Norway. *Geol. Mag.*, **120**, 607–612.

Furnes, H., Ryan, P. D., Grenne, T., Roberts, D., Sturt, B. A. and Prestvik, T. this volume. Geological and geochemical classification of the ophiolitic fragments in the Scandinavian Caledonides.

Færseth, R. B., Thon, A., Larsen, S. G. L., Sivertsen, A.

and Elvestad, L. 1977. Geology of the Lower Palaeozoic rocks in the Samnanger–Osterøy area, Major Bergen Arc, Western Norway. *Norges geol. Unders.*, **334**, 19–58.

Henriksen, H. 1981. A major unconformity within the metamorphic Lower Palaeozoic sequence of the Major Bergen Arc, Osterøy area, Western Norway. *Norges geol. Unders.*, **367**, 65–75.

Kiær, J. 1929. Den fossilførende ordovisiske–siluriske lagrekke på Stord. *Bergens Mus. Årbok*, **11**, 145–159.

Kolderup, C. F. 1914. Fjellbygningen i strøket mellom Sørfjorden og Sammnangerfjorden i Bergensfeltet. *Bergens Mus. Årbok*, **8**, 257 pp.

Kolderup, C. F. and Kolderup, N. H. 1940. Geology of the Bergen Arcs. *Bergens Mus. Skr.*, **20**, 137 pp.

Kvale, A. 1960. The nappe area in the Caledonides on Western Norway. *Excursion Guide. Norges geol. Unders.*, **212** e, 43 pp.

Inderhaug, J. 1975. Unpublished thesis, University of Bergen.

Miyashiro, A. 1975. Classifications, characteristics and origin of ophiolites. *J. Geol.*, **83**, 249–281.

Naterstad, J. 1976. Comment on the Lower Palaeozoic unconformity in West Norway. *Am. J. Sci.*, **276**, 394–397.

Nordås, J., Amaliksen, K. G., Brekke, H., Suthren, R., Furnes, H., Sturt, B. A. and Robins, B. this volume. Lithostratigraphy and petrochemistry of Caledonian rocks on Bømlo, southwest Norway.

Pringle, I. R., Kvale, A. and Anonsen, L. B. 1975. The age of the Hernes granite, Lower Bergsdalen Nappe, western Norway. *Norsk geol. Tidsskr.*, **55**, 191–195.

Reusch, H. 1882. Silurfossiler og pressede konglomerater i Bergensskifrene. *Univ. program 1. halvår. Christiania*, 152 pp.

Sturt, B. A., Skarpnes, O., Ohanian, A. T. and Pringle, I. R. 1975. Reconnaissance Rb/Sr isochron study in the Bergen Arc System and regional implications. *Nature, London*, **253**, 595–599.

Sturt, B. A. and Thon, A. 1976. The age of orogenic deformation in the Swedish Caledonides, discussion. *Am. J. Sci.*, **276**, 385–390.

Sturt, B. A. and Thon, A. 1978a. Caledonides of southern Norway. In *IGCP Project 27. Caledonian–Appalachian Orogen of the North Atlantic Region. Geol. Surv. Can. Pap.*, **78–13**, 39–47.

Sturt, B. A. and Thon, A. 1978b. An ophiolite complex of probable early Caledonian age discovered on Karmøy. *Nature, London*, **275**, 538–539.

Sturt, B. A., Thon, A. and Furnes, H. 1980. The geology and preliminary geochemistry of the Karmøy Ophiolite, SW Norway. *Proc. Int. Ophiolite Symp. Cyprus 1979*, 538–554.

Thon, A. this volume. Late Ordovician and Early Silurian cover sequences to the west Norwegian ophiolite fragments: stratigraphy and structural evolution.

Torske, T. 1972. Intercalated pillow structures in greenstone in the Os area, Western Norway, with remarks on pillow terminology. *Norges geol. Unders.*, **273**, 37–42.

The Caledonide Orogen—Scandinavia and Related Areas
Edited by D. G. Gee and B. A. Sturt
© 1985 John Wiley & Sons Ltd

Lithostratigraphy and petrochemistry of Caledonian rocks on Bømlo, southwest Norway

J. Nordås, K. G. Amaliksen, H. Brekke,[*] **R. J. Suthren**[†] and **H. Furnes, B. A. Sturt, B. Robins**[*]

[*]*Geologisk Institutt, Avd. A, Allégt. 41, 5014 Bergen, Norway*
[†]*Oxford Polytechnic, Department of Geology and Physical Sciences, Headington, Oxford OX3 0BP, England*

ABSTRACT

The central and southern parts of the island of Bømlo comprise at least four major rock complexes. The oldest, the Lykling Ophiolitic Complex (L.O.C.), may be sub-divided into a lower igneous zone, a mixed extrusive/sedimentary zone (Cambrian), both of which have been intruded by voluminous plagiogranites, and an upper sedimentary sequence. The lower igneous zone, containing serpentinite, layered and homogeneous gabbro, trondhjemite, massive greenstone, a basic dyke complex, and pillow lava, probably formed either at a major ocean or a marginal basin spreading centre. The unconformably overlying mixed extrusive/sedimentary zone, containing basaltic and trondhjemitic extrusives interbedded with volcaniclastics, chert–lutite and jasper, is both lithologically and geochemically comparable with rock sequences representing primitive island arcs. The upper sedimentary sequence, unconformably overlying both the lower igneous and the mixed extrusive/sedimentary zones, contains a series of coarse breccias, which are thought to represent sedimentation in an oceanic fracture zone.

The Siggjo Complex (Middle to Late Ordovician) unconformably overlies the L.O.C. It comprises predominantly subaerial lavas, pyroclastic falls and flows, ranging in composition from basalt to rhyolite, as well as minor continental sediments.

The Vikafjord Group unconformably overlies both the L.O.C. and the Siggjo Complex. The group consists of alluvial conglomerates, Ashgillian marine limestones and clastics, capped by a thick sequence of subaerial lavas and tuffs with a similar chemistry to that of the Siggjo Complex.

The Langevåg Group, of supposed Silurian age, is divided into six formations. The two lowermost formations consist of subaerial, calc-alkaline volcanics overlain by submarine volcaniclastic breccias, tuffs, and cherts. The upper formation comprise greywackes, bedded cherts, and submarine greenstones of tholeiitic to transitional tholeiitic/alkaline character, and suggest the development of an ensialic marginal basin.

Introduction

The island of Bømlo, located between Bergen and Haugesund on the western coast of Norway (Fig. 1), is part of the Sunnhordland magmatic province of the Caledonides of southwest Norway (Strand 1972). The rocks on Bømlo belong to Strand's 'western facies Cambro-Silurian sequence', and form part of the upper allochthon in the Hardanger-vidda–Ryfylke area (Solli *et al.* 1978). Apart from the description of the volcano-sedimentary succession between Vika and Lykling (Fig. 1) by Songstad (1971), there have been few attempts to reconstruct the geological history of Bømlo since the pioneering work of Reusch (1888).

According to Kolderup (1931) a NE–SW trending high-angle fault passes through Bømlo and separates a Cambro-Silurian metasedimentary succession in the southeast from an igneous complex, the Sunnhordland Massif, to the northwest. This fault, or fault system, certainly exists, but the geological subdivision is by no means so simple (Fig. 1). Results of recent mapping by the authors show that the southern part of the island of Bømlo can be divided into four major lithostratigraphic units: the Lykling Ophiolitic Complex (L.O.C.), the Siggjo Complex,

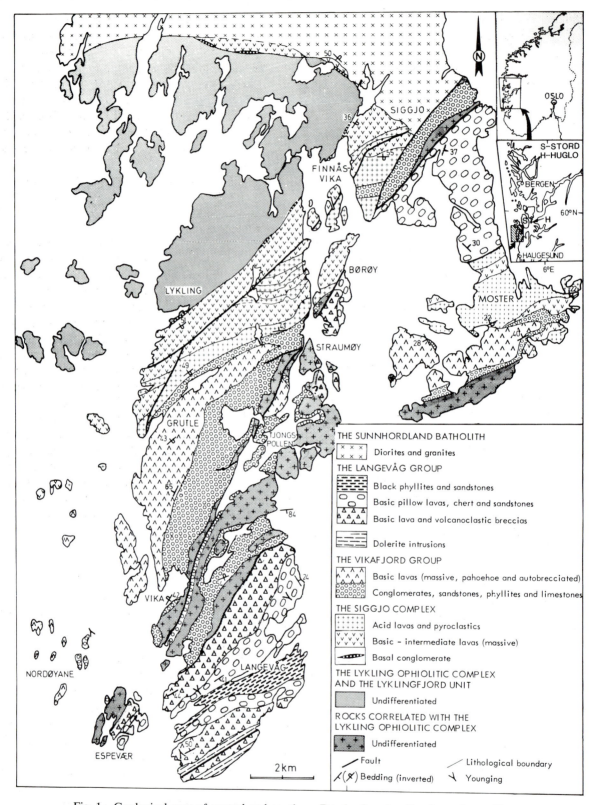

Fig. 1   Geological map of central and southern Bømlo, Sunnhordland, southwest Norway

the Vikafjord Group, and the Langevåg Group. The L.O.C. and the Siggjo Complex are limited to the north by later dioritic to granitic intrusions belonging to the later Sunnhordland batholith.

This subdivision of the area supersedes that of Færseth (1982) who postulated the existence of four lithostructural units: the Holsnøy Complex, the Hardangerfjord Group, the Dyvikvågen Group, and the Sunnhordland Igneous Complex, believed to be separated by tectonic contacts.

The present account deals with stratigraphic relationships between and within these major units as newly defined and, on the basis of the lithological associations represented and the geochemistry of their contained extrusive and intrusive igneous rocks, attempts to outline the types of geotectonic environment in which they may have formed.

The Lykling Ophiolitic Complex

The Lykling Ophiolitic Complex (L.O.C.) crops out in an area to the north of Lykling and presents an almost complete ophiolite pseudostratigraphy despite having been invaded by voluminous plagiogranites.

Rock assemblages which are correlated with this complex crop out in two other areas:

1. Within a NE–SW trending zone which extends across the southern part of the island from Espevær to Straumøy (Fig. 1). This zone contains gabbro and trondhjemite, as well as greenschist and greenstone derived from both dykes and pillow lavas.
2. Within an area on the southernmost part of Moster (Fig. 1). This area comprises homogeneous and rhythmically-layered gabbro, gabbro pegmatite, and subordinate greenstone.

The following description deals with the different parts of the L.O.C., and a simplified geological map is shown in Fig. 2. The generalized stratigraphic relationships within the complex are shown in Fig. 3. The L.O.C. is in thrust contact with the structurally underlying Lyklingfjord Unit, a metasedimentary sequence whose stratigraphic position is poorly understood, and has been folded into a major, NE–SW trending synform (Fig. 2). The complex can be subdivided into a lower igneous zone, a mixed extrusive/sedimentary zone and an upper sedimentary sequence. The two former have been invaded by voluminous plagiogranites.

The lower igneous zone

The structurally lowest part of the L.O.C. is characterized by lenses of chromite-bearing serpentinite. These are strongly brecciated and contain serpentinite fragments ranging in diameter from less than 1 cm to almost 1.5 m. The serpentinites are mylonitized towards their margins.

The overlying strongly saussuritized gabbro is over 4 km thick. Locally it contains small-scale rhythmic layering. Modally graded layers and uniform leucocratic layers are particularly common and are separated by homogeneous cumulates. Along certain horizons the rhythmic layering is highly disturbed by slumping and the layering is transected by swarms of minor faults with cm-scale displacements which formed during the magmatic evolution of the complex. Much of the gabbro lacks rhythmic layering and is composed of homogeneous gabbro, gabbro pegmatite, and at the highest levels, microgabbro.

The gabbro is separated by a trondhjemite intrusion from a greenstone sequence which is more than 1 km thick. This sequence has microgabbro at its base and contains massive greenstone, a dyke swarm in places reaching a density of 100%, and pillow lava towards the top. The vesicularity of the pillow lavas is around 10%. Vesicles are up to 2 mm in diameter, the majority, however, are less than 1 mm across. According to Moore (1965) a vesicularity of this magnitude suggests extrusion at non-abyssal depths. The apparent hybridization within members of the dyke swarm together with the curved and irregular margins of the greenstone fragments within intrusive quartz-keratophyre and greenstone breccias suggests the co-existence of acidic and basic magmas.

The dyke rocks, gabbro, and plagiogranite have been locally affected by penetrative in situ brecciation and consist of sub-angular to sub-rounded fragments generally less than 1 cm across. These breccias resemble those present in the Bay of Islands Complex described by Williams and Malpas (1972). Their fabric is thought to be due to hydraulic fracturing or seismic events.

The mixed extrusive/sedimentary zone

The mixed extrusive/sedimentary zone of the complex, resting on both gabbro and pillow lava of the lower igneous zone (Fig. 2), is a bimodal, mainly extrusive sequence of metabasalt and quartz-keratophyre interbedded with sedimentary horizons. Locally it is unconformably overlain by breccias belonging to the upper sedimentary sequence.

The basic volcanics include massive greenstone, vesicular and non-vesicular pillow lava, pillow breccia, hyaloclastite, and pahoehoe lava. The latter is confined to the Finnåsvika area (Fig. 2) where it occurs as a pile of flattened, highly-vesicular, and elongated toes. The vesicles are concentrated in zones parallel to the margins in the upper part of the toes. The distribution of vesicles together with gas cavities with arched roofs and flat floors, indicate that the lava flow faces northwards (Furnes and Lippard 1979).

Extrusive quartz-keratophyres, geochemically similar to the trondhjemites which intrude the complex

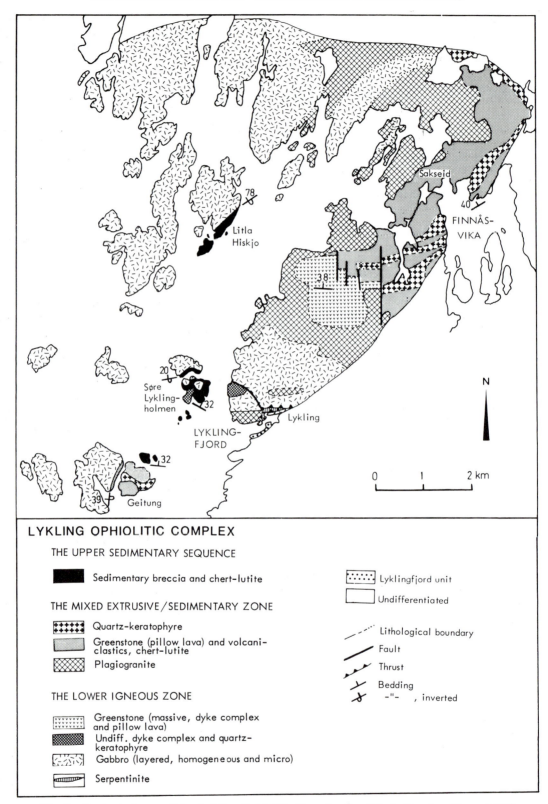

Fig. 2   Geological map of the Lykling Ophiolitic Complex (modified after Amaliksen 1980)

are interbedded with greenstones. These are well developed on the island of Geitung and east of Sakseid (Fig. 2) where they occur as flows, as tuffaceous, welded horizons and as hyaloclastites. The quartz-keratophyre lavas consist of both porphyritic and aphanitic, autobrecciated, flow-foliated and flow-folded flows, locally highly vesicular. Some of the blocks included in the flows have shapes suggesting

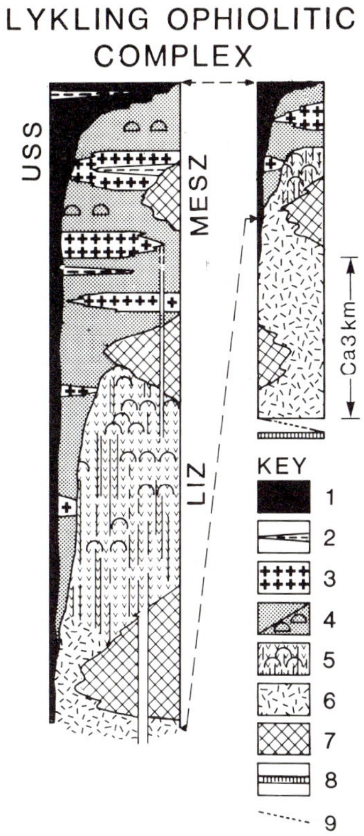

LYKLING OPHIOLITIC COMPLEX

Fig. 3 Generalized pseudostratigraphy of the Lykling Ophiolitic Complex. 1. Coarse sedimentary breccia; 2. Chert–lutite; 3. Quartz-keratophyre; 4. Extrusive greenstone pillow lava; 5. Dyke complex, massive greenstone and pillow lava; 6. Gabbro; 7. Plagiogranite; Serpentinite; 9. Tectonic contact; USS The upper sedimentary sequence; MESZ The mixed extrusive/sedimentary zone; LIZ The lower igneous zone. (Modified after Amaliksen 1980)

plastic deformation, whereas in the hyaloclastites angular obsidian fragments have been derived by brittle fragmentation. The latter deposits show many similarities with the subglacial hyaloclastites described from Iceland (Furnes *et al.* 1980).

Volcaniclastic breccias, containing both acid and basic fragments, occur together with the acid and basic extrusives. These breccias may superficially resemble ordinary clastic sediments, but their matrices are quartz-keratophyre or greenstone. They are particularly abundant in the higher levels of the extrusive sequence.

Several chert–lutite horizons are interbedded with the extrusives (Fig. 3). These contain relics of recrystallized undetermined radiolaria and are often enriched in iron.

Plagiogranites

North of Lykling rocks of both the lower igneous and the mixed extrusive/sedimentary zones of the complex are intruded by tonalite and large bodies of

trondhjemite and quartz-keratophyre dykes (Fig. 2). Trondhjemite and quartz-keratophyre constitute more than 6 per cent of the intrusive rocks within the L.O.C.

The upper sedimentary sequence

On the islets north of Geitung (Fig. 2) sedimentary breccias occur above volcaniclastic breccias belonging to the mixed extrusive/sedimentary zone of the L.O.C. Similar breccias crop out on Søre Lyklingholmen where they unconformably overlie both a sheeted dyke complex and gabbro, and on the island north of Søre Lyklingholmen and on Litla Hiskjo where they rest on rhythmically-layered gabbro alone. On Søre Lyklingholmen the breccias are interbedded with conglomerates, sandstones, lutites, and chert–lutites. Certain of the chert–lutite horizons contain up to 4.5 wt% MnO. Thin, tuffaceous layers are sporadically present. The chert–lutites sometimes contain abundant deformed and recrystallized objects tentatively interpreted as representing radiolaria and acritarchs.

The breccias form a series of both coarsening-upwards and fining-upwards cycles. They appear to represent submarine talus and sediment gravity-flow deposits, some of the latter units having strongly erosional lower contacts which cut down into the underlying units. The breccias on Søre Lyklingholmen contain clasts derived exclusively from the ophiolitic complex ranging from millimetre size grains to blocks more than 100 m across. Many of the blocks are angular, though others are sub-rounded. This rounding is apparently primary and due to the presence of curved fractures in the source area; fractures of this type are abundant in the large gabbro blocks. A section through the thickest part of the sedimentary sequence, whose base is not seen, contains black slate in its lower levels overlain by chert–lutite, lutite, and sandstone which in turn is succeeded by clast-supported breccias with clasts of greenstone, microgabbro, gabbro pegmatite, trondhjemite, quartz-keratophyre, and jasper. Higher in the sequence the clasts are almost exclusively of gabbro with some microgabbro, and the breccias are clast-supported often with extremely little matrix. The matrix is mainly epidote and may, in part, be of hydrothermal origin. The total thickness of the sequence is 190 m. It would appear that the clasts in the breccias show an inverted ophiolite pseudostratigraphy. Ultramafic clasts have, however, not been recorded. The clasts show variable imprints of pre-erosional deformation and metamorphic textures such as brecciation and gneissic textures presumably developed during ocean-floor metamorphism.

The Siggjo Complex

The Siggjo Complex unconformably overlies the L.O.C. (Fig. 1) and consists predominantly of volcanic rocks together with minor intercalations of sedimentary rocks (Fig. 4).

The volcanics, comprising basic, intermediate, and acid lavas, and pyroclastics are largely subaerial. The basic and intermediate volcanics are generally lava flows, the acidic volcanics are mainly pyroclastics.

The basic lavas are of both aphyric and plagioclase-phyric types. They are strongly altered and typically occur in 0.5–5 m thick flow units with scoriaceous bases and tops. The central parts of flows are massive in the lower parts of the stratigraphic sequence and autobrecciated higher up. The massive flow units are uniformly highly vesicular, and the large lateral extent of each flow unit suggests eruption of relatively low-viscosity lavas.

Thin air-fall tuffs are associated with many of the basic lava flows and give evidence of intermittent explosive activity.

The intermediate flows are both thicker and more massive than the basic flow units. Both brecciated and slaggy layers occur locally and probably represent tops and bases of flow units.

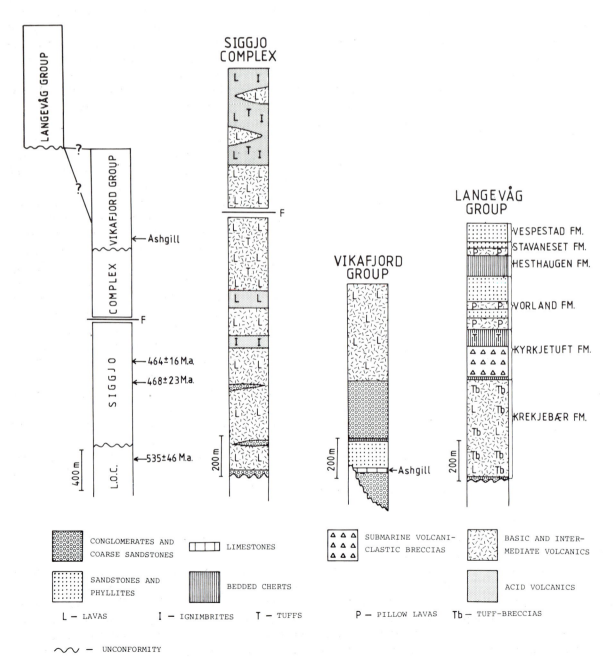

Fig. 4   The stratigraphy of the post-ophiolitic rock units of Bømlo (modified after Brekke 1983)

The acid pyroclastic rocks are dominated by ash-flows carrying lithic fragments at their bases and showing eutaxitic textures and fiamme, representing collapsed pumice fragments, higher up. Occasionally, the size of the fiamme increases towards the tops of individual flow units. Concentration of lithic fragments at the base of a flow unit and reverse grading of pumice inclusions upwards is typical of ash-flow deposits (Sparks et al. 1973). The centres of some flow units have been so strongly welded, however, that all trace of a vitroclastic texture has been lost; all that remains is a homogeneous devitrified glass.

Welded breccias, some tens of metres thick and traceable laterally over hundreds of metres, locally seem to occupy depressions in an irregular pre-existing topography. These breccias are normally-graded and contain rhyolite clasts up to 1 m across at their bases and welded lapilli tuffs at their tops. The eutaxitic matrix suggests emplacement at high temperature, and they are believed to represent pyroclastic flows. In addition, welded air-fall tuffs, produced by rapid emplacement of hot pyroclastic material have been described from the area (Suthren and Furnes 1980).

There is evidence of subaerial reworking of the acidic pyroclastic material in the form of discontinuous conglomerates, and tuffs containing sedimentary structures.

Acid lavas, flow-brecciated at their upper and lower margins and with folded flow-banding in their centres, locally dominate over the pyroclastics.

The Vikafjord Group

The Vikafjord Group rests unconformably on the L.O.C. and its correlatives as well as the Siggjo Complex (Fig. 1). It consists of alluvial and shallow-marine sediments succeeded by subaerial volcanics (Fig. 4). Due to an accident of erosion, the group is preserved in the cores of synclines (Fig. 1).

At the base of the group is an impersistent, matrix-supported, unsorted conglomerate, containing clasts derived from the local substrate. The thickness of the conglomerate varies from 0 m to 250 m laterally, indicating a rough palaeotopography. The conglomerates are overlain by limestones and calcareous phyllites containing a fauna of Ashgillian/Llandoverian age (Reusch 1888; Kiær 1929; R.B. Neuman personal communication; N. Aarhus personal communication; B. Neuman personal communication) including rugose and tabulate corals, gastropods, brachiopods, and bryozoa. The limestones are succeeded by a coarsening-upwards turbidite sequence which in turn is followed by sub-arkosic, plane-parallel laminated sandstones sug-gesting a coastline progradation from the west. These rocks are capped by fine-grained phyllites and chert, indicating a rapid transgression.

The lower part of the group is overlain by resedimented conglomerates and coarse sandstones. The conglomerates are clast-supported, laterally persistent, normal or inverse-to-normal graded, or ungraded. The conglomerates alternate with laterally persistent beds of coarse to medium, massive or plane-parallel laminated sandstone, which often show gradational contacts to the conglomerate bed below. In the southeast the coarse clastics interfinger with fine-grained phyllites and coarsening-upwards cycles from phyllite to conglomerate are developed.

The conglomerate–sandstone facies is interpreted as submarine, resedimented conglomerates of the types described by Davies and Walker (1974) and Walker (1975), and are taken to represent an ancient fan-delta that built out from a tectonically active coastline to the west. This facies is correlated with the post-Lower Llandoverian Utslettefjell Conglomerate on the adjacent island of Stord (Fig. 1, inset map) (Færseth and Steel 1978).

On top of the resedimented conglomerates there is a thick pile of subaerial, mafic lavas (Fig. 4). In the Grutle area (Fig. 1) ovoid bodies suggestive of pahoehoe toes occur in association with more massive flow units. These flows are succeeded by plagioclase-phyric, strongly brecciated lavas alternating with subordinate massive lavas. The brecciated lavas contain blocks, up to 1 m across, which are generally angular, but some blocks are highly rounded, flattened, or amoeboid in shape. The auto-brecciated lavas are thought to represent viscous block-lava flows. Many of the brecciated tops of flow units show evidence of rounding and transport to form thin but laterally persistent conglomerates. Thin air-fall tuffs occurring in association with the conglomerates are rich in sedimentary structures attributed to subaerial reworking. The degree of reworking increases upwards within the sequence of the plagioclase–phyric lavas, and laterally the lavas pass into a thinner sequence of entirely reworked material consisting of clasts derived from the lavas. This indicates that the original lava pile was of limited extent.

The Langevåg Group

The Langevåg Group crops out in a syncline on the southernmost part of Bømlo and along strike to the northeast on Moster and Stord (Fig. 1). The rocks of this group, which are probably the youngest supracrustals in the area, have a distinctive lithostratigraphy which has been informally subdivided at the formational level (Fig. 4).

The *Krekjebær Formation* comprises subaerial, mainly mafic, volcaniclastic breccia and minor lava. The formation is dominated by poorly sorted tuff breccias with non-vesicular, angular to subrounded, basic lava fragments up to 50 cm across containing phenocrysts of clinopyroxene and altered plagioclase. The matrix is composed of fragmented crystals (occasionally euhedral) and lava fragments ranging in size from ash to lapilli. Bed thicknesses vary from 0.30 to 7–8 m and beds typically wedge out laterally within 10–100 m. A number of beds display a crude internal stratification defined by strings of fragments parallel to bedding, and the upper part of beds show normal grading. These tuff-breccias are interbedded with aa-flows of porphyritic lava similar to the lava-fragments in the breccias. Impact structures made by blocks of lava on occasional bedded tuffs are sometimes observed.

The tuff-breccias are interpreted as the result of phreatic explosions caused by degassed aa-lava flowing over wet ground. Similar deposits have been described by Fisher (1968) and Macdonald (1972, p. 192).

The *Kyrkjetuft Formation* comprises submarine volcaniclastic debris flows and turbidites, waterlain tuff, chert, and a possible submarine pyroclastic flow.

The formation is dominated by green, sandy, epiclastic, laterally-extensive beds with abundant rounded feldspar grains and fragments of basic volcanics. Bed thicknesses vary from less than 20 cm up to 6 m, but are usually between 1–2 m. Beds thicker than 1 m are characteristically poorly sorted and often contain feldspar-phyric, matrix-supported clasts up to 60 cm across. A few clasts of intermediate rocks with eutaxitic textures are also present. Crude internal planar stratification of orientated fragments is seen and some of the beds have erosional bases.

Interbedded with the volcaniclastic debris flows are thinly-bedded waterlain tuffs and graded, tuffaceous turbidites, mudstones and laminated, green, radiolarian chert–lutites.

A single example of a possible pyroclastic flow has been recorded. This 2 m thick unit is composed entirely of lava and crystal fragments and appears to have been derived from a single eruption.

Towards the top of the Kyrkjetuft Formation there is a pronounced upward-fining; thinly bedded waterlain tuffs and tuffaceous sandstones, mudstones, and chert–lutites dominate.

The *Vorland Formation* consists of green phyllite and sandstone interbedded with pillowed and massive greenstone. The sandstones contain clastic quartz and plagioclase. They are poorly sorted and occur in beds up to one metre thick, usually 15–30 cm, and show normal grading. They are interpreted as turbidites. The interbedded greenstones are laterally impersistent and their maximum thicknesses vary from 10–50 m. Much of the pillow lavas are quite vesicular, suggesting limited water depths, but non-vesicular flows are also found. Also interbedded with the turbidites are red, bedded, radiolarian cherts of varying lateral extent and thickness.

The *Hesthaugen Formation* is the thickest chert-unit in the Langevåg Group and may be traced continuously for 3 km along strike and has a maximum thickness of 100 m.

The *Stavaneset Formation* consists of lithologies similar to those of the Vorland Formation.

The *Vespestad Formation* is the uppermost unit in the Langevåg Group and contains quartz-rich turbidites and black phyllites.

Age Relationships

Stratigraphic relationships between the L.O.C., Siggjo Complex, Vikafjord Group and Langevåg Group are summarized on Fig. 4. The L.O.C. is the oldest rock unit within the area studied. An extrusive quartz-keratophyre from the upper part of the L.O.C. has yielded a Rb/Sr whole-rock isochron age ($\lambda^{87}Rb = 1.42. \times 10^{-11}$ yr^{-1}) of 535 ± 46 Ma (Furnes *et al.* 1983).

The ophiolitic complex underwent folding and deep erosion prior to the deposition of the Siggjo Complex. This is evident from the local inversion of the L.O.C. prior to the deposition of the Siggjo Complex and from the way in which the basal conglomerate in the Siggjo Complex rests on various levels of the L.O.C.

At Grutle and on Moster, the Vikafjord Group crops out in the cores of major synclines (Fig. 1). Along the northern limbs of these structures the Vikafjord Group rests unconformably on the Siggjo Complex, while along the southern limbs it overlies rocks correlated with the L.O.C. Hence the Siggjo Complex is overstepped by the Vikafjord Group, probably due to a period of block faulting and tilting and/or weak folding prior to erosion of the Siggjo Complex.

Rhyolites from the Siggjo Complex and its lithostratigraphic equivalents on the island of Stord (Lippard 1976) have yielded Rb/Sr whole-rock isochron ages ($\lambda^{87}Rb = 1.42 \times 10^{-11}$ yr^{-1}) of 464 ± 16 Ma (Furnes *et al.* 1983) and 445 ± 5 Ma (Priem and Torske 1973) respectively. According to the timescale of Gale *et al.* (1980) these ages span the period from the Llanvirnian to the Upper Caradocian. As the fossiliferous limestones in the Vikafjord Group are of Ashgillian age, deformation and erosion of the Siggjo Complex probably took place in Upper Caradocian/Lower Ashgillian times.

On Espevær (Fig. 1) the Langevåg Group is in unconformable contact with the correlatives of the L.O.C. Primary contacts between the Langevåg Group and the Vikafjord Group or Siggjo Complex have not, however, been observed on Bømlo. On the island of Huglo, northeast of Bømlo (Fig. 1), rocks correlated with the Langevåg Group stratigraphically overlie fossiliferous limestones of probable Ashgillian age (Monsen 1937; Færseth 1982; C. Magnus personal communication). Hence the Langevåg Group is inferred to have a Lower Silurian age.

The fan-delta deposits of the Vikafjord Group and the correlated Utslettefjell Conglomerate on Stord, show a slight angular discordance to the underlying fossiliferous strata. If this unconformity has a regional significance, it may be correlated with the inferred post-Ashgillian unconformity below the Langevåg Group, implying that the lower part of the Langevåg Group and the upper part of the Vikafjord Group are coeval.

Geochemistry

Selected immobile major and trace elements have been employed to illustrate the geochemical characteristics of volcanic rocks from the four complexes and groups described. Apart from variations which can be portrayed in TiO_2 vs. Zr (Fig. 5c) and Th–Hf–Ta diagrams (Fig. 7), in which all the major volcanic compositions are represented, the discussion is restricted to the geochemistry of the basic metavolcanics.

The Lykling Ophiolitic Complex

The metabasalts from the lower igneous and the mixed extrusive/sedimentary zones of the complex are geochemically indistinguishable in the variation diagrams presented in Fig. 5. In the TiO_2 vs. FeO^t/MgO diagram (Fig. 5a) no definite trend can be discerned. However, the metabasalts clearly are enriched in TiO_2 relative to typical calc-alkaline basalts of similar FeO^t/MgO ratios. An unambiguous subalkaline feature is the consistently low Nb/Y-ratio (Fig. 5b). In the TiO_2 vs. Zr diagram (Fig. 5c), which discriminates between mid-ocean ridge basalts (MORB), within-plate lavas (WPL), and island arc lavas (IAL), the ophiolitic greenstones mainly fall within the MORB field which overlaps the fields for WPL and IAL. A within-plate origin can, however, be rejected as the greenstones plot in the IAT or MORB fields in the Zr/Y vs. Zr diagram (Fig. 5d) which discriminates WPB from the two others. Although in this diagram some of the greenstones show distinct MORB or IAT charac-

teristics, many of the analyses plot in the field where MORB and IAT overlap. To overcome this problem a Ti vs. Cr plot has been used where the proposed fields for low-potassium tholeiites (LKT) and ocean floor basalts (OFB) do not overlap (Fig. 6). The majority of the greenstones from the L.O.C. fall on the LKT-side of the dividing line.

A primitive island arc environment is also suggested by the Th–Hf–Ta ratios of four metabasalts from the mixed extrusive/sedimentary zone of the complex (Fig. 7).

The Siggjo Complex

In Fig. 5a analyses of metabasalts from the Siggjo Complex are scattered, but show an indistinct trend of increasing TiO_2 with increasing FeO^t/MgO, similar to some tholeiitic series. The relatively low Nb/Y ratios and the horizontal distribution of analyses in Fig. 5b is also typical of subalkaline basalts.

Due to their relatively high TiO_2 and Zr concentrations the basic volcanics from the Siggjo Complex plot in the fields for WPB in both Fig. 5c and 5d. However, in the Th–Hf–Ta diagram (Fig. 7), the Siggjo volcanics plot within the field for calc-alkaline volcanics of destructive plate margins. The analyses define a trend of constant Hf/Ta ratio which may be due to crustal contamination of magmas with compositions plotting originally in the tholeiitic WPB field (Wood 1980).

The chemical characteristics of metabasalts from the Siggjo Complex, especially the relatively high TiO_2 (1.37–2.92 wt%), Zr (93–253 ppm), FeO^t (7.91–12.44 wt%), and K_2O (0.48–3.12 wt%) contents resemble the tholeiitic basalts from the Snake River and Yellowstone areas bordering the Basin and Range Province in western U.S.A. (Leeman and Rogers 1970). This suggests an origin in a continental setting. An initial $^{87}Sr/^{86}Sr$ ratio of 0.70740 (Furnes et al. 1983) for a rhyolite in the complex supports this inference.

The Vikafjord Group

Metabasalts from the Vikafjord Group are geochemically very similar to those of the Siggjo Complex (Fig. 5), though K_2O contents are slightly lower. An origin in an intracontinental setting is hence also suggested for the Vikafjord Group.

The Langevåg Group

The geochemistry of metabasalts from the Krekjebær Formation forming the lowest part of the Langevåg Group is distinctly different from that of the submarine lavas of the younger Vorland and

LYKLING OPHIOLITIC COMPLEX   SIGGJO COMPLEX

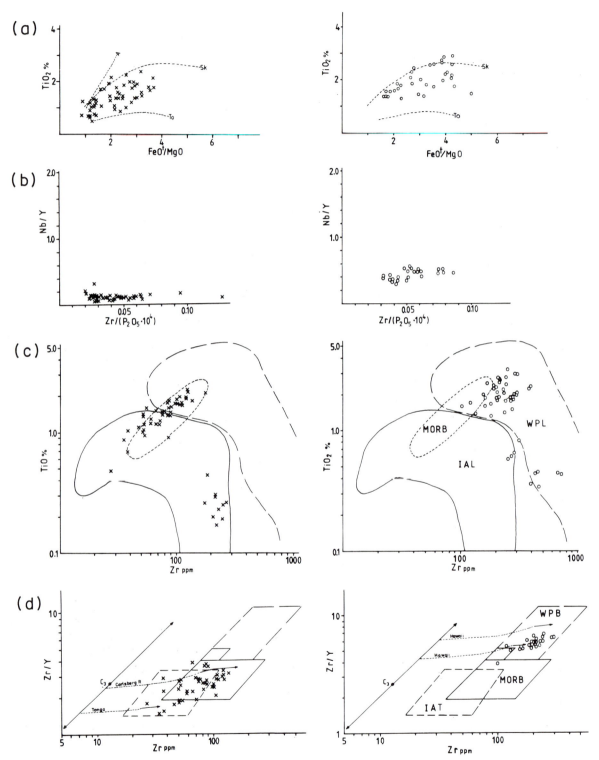

Fig. 5 Geochemical plots of basaltic volcanics from the Lykling Ophiolitic Complex, the Siggjo Complex, the Vikafjord Group, and the Langevåg Group. (a) TiO$_2$ vs. FeO^t/MgO. K—Kilauea (tholeiitic); A—abyssal tholeiite; Sk—Skærgaard (tholeiitic); To—Tofua (island arc tholeiite); Am—Amagi (calc–Alkaline). (After Miyashiro 1975). (b) Nb/Y vs. Zr/(P$_2$O$_5$ × 10^4). (After Floyd and Winchester 1975). (c) TiO$_2$ vs. Zr, discriminating between mid-ocean ridge basalts (MORB), volcanic arc lavas (IAL), and within-plate lavas (WPL). (After Pearce 1980). (Figures include points for all major volcanic types.) (d) Zr/Y vs. Zr, discriminating between mid-ocean ridge basalts (MORB), island arc tholeiites (IAT), and within plate basalts (WPB). Genetic pathways for basalts of known settings are included. These are modelled by the combination of partial melting (dashed lines) of a garnet-free source which is enriched or depleted

VIKAFJORD GROUP

LANGEVÅG GROUP

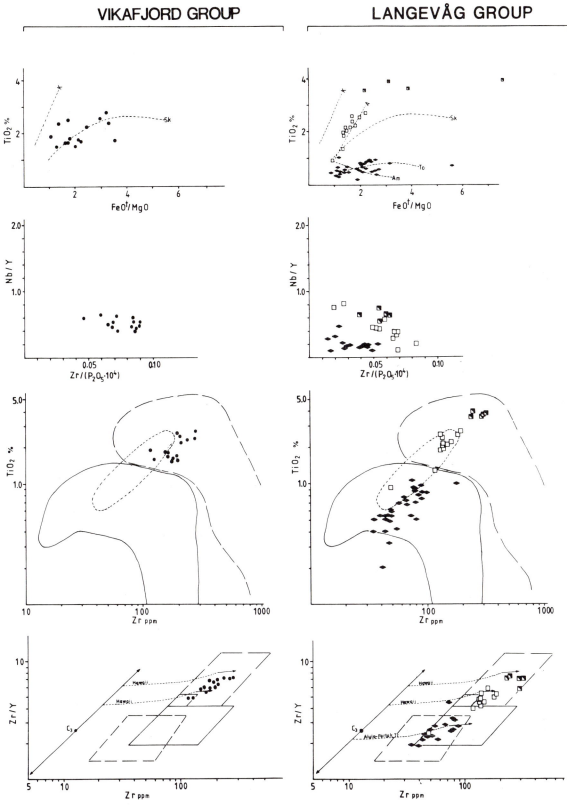

relative to C_3 chondrite composition, and subsequent fractional crystallization (solid lines). See Pearce and Norry (1979) for further details

Symbols: ✕ —Lykling Ophiolitic Complex
          ○ —Siggjo Complex
          ● —Vikafjord Group
          ◨ —Stavaneset Fm. ⎫
          □ —Vorland Fm.   ⎬ Langevåg Group
          ◆ —Krekjebær Fm.   ⎭

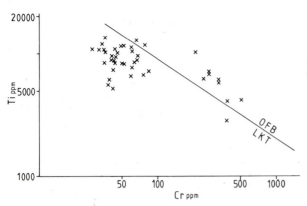

Fig. 6 Ti vs. Cr plot discriminating between low-K tholeiites (LKT) and ocean-floor basalts (OFB). (After Pearce 1975). Symbols as in Fig. 5

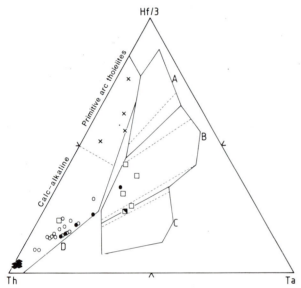

Fig. 7 Th–Hf–Ta discrimination diagram. Field A: N-type MORB. Field B: E-type MORB and tholeiitic within-plate basalts and differentiates. Field C: Alkaline within-plate basalts and differentiates. Field D: Destructive plate-margin basalts and differentiates. (After Wood 1980. Analyses from Brekke et al. in preparation). Symbols as in Fig. 5

Stavaneset Formations. The subaerial, porphyritic volcanics of the Krekjebær Formation are consistently low in TiO_2 (Fig. 5a) and Zr (Fig. 5c) and enriched in Th relative to Hf and Ta (Fig. 7), suggesting that they belong to a calc-alkaline series. This is in accordance with high Al_2O_3 contents (up to 19 per cent in aphyric samples).

Discrimination diagrams (Fig. 5c and d) suggest a volcanic arc environment for these volcanics. The low K/Rb-ratios (195–240), high La/Nb (5.0–8.5), Th/Nb (2.8–5.7) and La/Ta (129–178) ratios and high K_2O (0.5–2.6 wt%) strongly suggest a continental setting for this volcanic arc (see Jakeš and White 1972; Saunders et al. 1980).

The submarine, basic lavas of the Vorland and Stavaneset Formations (Fig. 5) show distinctly different geochemical signatures. They generally have high concentrations of TiO_2 (up to 3.99 wt% in the Stavaneset Formation). The negative correlation between the Nb/Y and Zr/P_2O_5 ratios (Fig. 5b) indicates a transitional tholeiitic/alkaline character. In the discrimination diagrams they generally plot in the WPB fields (Fig. 5c and d). However, analyses from the Vorland Formation show certain MORB-like features (Fig. 5a and c).

Such a geochemical diversity has been described from basalts interbedded with greywackes and cherts in ensialic marginal basins in the incipient stages of opening behind ensialic magmatic arcs (Dalziel et al. 1974; Saunders et al. 1979; Weaver et al. 1979).

Discussion

The lower igneous zone of the L.O.C. is comparable with the lower parts of the nearby Karmøy Ophiolite (Sturt et al. 1980) and the Gullfjellet Ophiolite Complex (Thon this volume), whilst the lithological associations present in the mixed extrusive/sedimentary zone and the upper sedimentary sequence have not hitherto been recorded from any other Scandinavian ophiolite fragment (Furnes et al. this volume). As shown in the preceding description greenstones from the L.O.C. have geochemical similarities with both island arc tholeiites and ocean-ridge basalts although approach the first more closely. It is, however, uncertain whether the lower igneous zone of the L.O.C. was formed at a major ocean or a marginal basin spreading centre.

The large volumes of acidic rocks within the L.O.C. are unlikely to have been generated by differentiation of basaltic magmas. A derivation by partial melting at or above a subducted slab of oceanic crust is preferred.

The mixed extrusive/sedimentary zone of the L.O.C. shows many similarities with the association of keratophyre, spilite, and radiolarian tuff in the Water Island Formation in the Caribbean island arcs, as described by Donnelly (1966). The Water Island Formation is interpreted as a primitive island arc association (Donnelly and Rogers 1980).

The sedimentary breccias of the L.O.C. with reverse ophiolite pseudostratigraphy in their clast population are consistent with the progressive erosion of evolving fault-scarps. The recorded section shows many similarities with the sedimentary breccias described from the Arakapas fault belt in the southeast part of the Troodos Complex, Cyprus (Simonian and Gass 1978) which is regarded as representing a fossil transform fault. The similarities are expressed by lithologies present, stratigraphy, grading, roundness, bed thickness, bed contacts, derivation of clasts from the underlying ophiolites, and a lack of serpentinite clasts. An origin related to the development of an oceanic fracture zone is suggested for the breccias of the L.O.C. As these breccias rest locally on primitive island arc volcanics it may be postulated that they accumulated at a submarine primitive island arc/fracture zone intersection rather than a ridge/fracture zone intersection (Amaliksen 1983).

The Siggjo Complex unconformably overlies the previously deformed (=obducted?) ophiolite and it seems likely that it developed in association with an active plate margin. Its geochemistry strongly suggests magma generation in an intracontinental setting. The occurrence of the bimodal basaltic andesite–rhyolite suite and the geochemistry of the metabasalts is very similar to the tholeiitic suites described from the High Lava Plains in the Basin and Range Province (Leeman and Rogers 1970; Armstrong 1978; Christiansen and McKee 1978). An active continental margin setting has already been proposed for the Kattnakken volcanics on Stord (Lippard 1976) and is also suggested for the Siggjo Complex (Nordås in preparation).

The lower part of the Vikafjord Group represents a hiatus in the volcanic activity. The basal coarse clastics and unconformable relationship to the underlying Siggjo Complex gives evidence for a phase of deformation after the cessation of the volcanism represented by the Siggjo Complex and prior to the deposition of the Vikafjord Group. A renewal of tectonic activity is indicated by the coarse clastics of the fan-delta in the upper parts of the group, and was followed by a revival of Siggjo-type volcanism in Llandoverian times.

During Llandoverian times or somewhat later, calc-alkaline volcanism represented by the Krekjebær Formation at the base of the Langevåg Group commenced. The chemistry of these rocks suggests the existence of an ensialic magmatic arc related to a subduction zone. The contrasting chemical characteristics between the Siggjo Complex and Vikafjord Group and the Krekjebær Formation of the Langevåg Group is comparable to the contrast between the High Lava Plains and the Cascades of the Western U.S.A. (see Armstrong 1978).

The greywacke, chert, and pillow lava in the upper part of the Langevåg Group may be traced for 100 km along strike NE-wards east of the subaerial volcanics and indicate that a marine basin developed here in Lower Silurian times. This basin was probably floored by ophiolites obducted over continental crust and transgressed the volcanics to the west as it widened. The transitional tholeiitic/alkaline character of the pillow lavas suggests a tensional regime during basin development (Brekke 1983).

Conclusion

The Lykling Ophiolitic Complex probably represents a fragment of oceanic crust on which a volcanic arc was built. Evidence strongly suggestive of active fracture zone tectonics has been presented. The ophiolitic rocks were folded and locally inverted, either during or subsequent to emplacement. This deformation was followed by uplift and erosion prior to the development of a volcanosedimentary sequence representing an active continental margin, possibly comparable to the Western Cordillera of the U.S.A. A marginal basin subsequently developed on continental crust behind the volcanic terrain. The authors consider that the continental margin tectonic and magmatic activity developed in Mid-Ordovician to Early Silurian times, predating the major nappe translations of the region.

Acknowledgements

The authors would like to thank Dr. R. E. Bevins for access to unpublished data. An early draft of this paper has benefited from comments by Drs. P. D. Ryan, D. Roberts and R. A. Jamieson. We are grateful to M. Tysseland for assistance during the analytical work and to E. Irgens and J. Ellingsen for drawing the figures. The fieldwork was supported by Norges Geologiske Undersøkelse, Orkla Industrier A/S, and Norges Almenvitenskapelige Forskningsråd (Norwegian Research Council for the Science and Humanities).

International Geological Correlation Programme, Norwegian contribution no. 58 to Project 27—Caledonide Orogen.

References

Amaliksen, K. G. 1980. Gullforekomster på Bømlo. *Norges geol. Unders.*, Rapp. nr. **1750/35A**.

Amaliksen, K. G. 1983. *The geology of the Lykling Ophiolitic Complex, Bømlo, SW Norway*. Unpublished Cand. Real. thesis, Univ, of Bergen.

Armstrong, R. L. 1978. Cenozoic igneous history of the U.S. Cordillera from lat 42° to 49°N. In Smith, R. B. and Eaton, G. P. (Eds), *Cenozoic Tectonics and Regional Geophysics of the Western Cordillera. Geol. Soc. Am. Mem.*, **152**, 265–282.

Brekke, H. 1983. *The Caledonian Geological Patterns of Moster and Southern Bømlo. Evidence for Lower Palaeozoic Magmatic Arc Development*. Unpublished Cand. Real. thesis, Univ. of Bergen.

Brekke, H., Furnes, H., Nordås, J. and Hertogen, J. 1984. Lower Palaeozoic convergent plate margin volcanism on Bømlo, SW. Norway, and its bearing on the tectonic environments of the Norwegian Caledonides. *J. geol. Soc.*, London.

Christiansen, R. L. and McKee, E. H. 1978. Late Cenozoic volcanic and tectonic evolution of the Great Basin and Columbia intermontane regions. In Smith,

R. B. and Eaton, G. P. (Eds), *Cenozoic Tectonics and Regional Geophysics of the Western Cordillera*. Geol. Soc. Am. Mem., **152**, 283–311.

Dalziel, I. W. D., de Wit, M. J. and Palmer, K. F. 1974. Fossil marginal basin in the southern Andes. *Nature, London*, **250**, 291–294.

Davies, I. C. and Walker, R. G. 1974. Transport and deposition of resedimented conglomerates: the Cap Enrage Formation, Cambro-Ordovician, Gaspé, Quebec. *J. Sed. Petrol.*, **44**, 1200–1216.

Donnelly, T. W. 1966. Geology of St. Thomas and St. John, U.S. Virgin Islands. In Hess, H. H. (Ed.), *Caribbean Geological Investigations*. Geol. Soc. Am. Mem., **98**, 85–176.

Donnelly, T. W. and Rogers, J. J. W. 1980. Igneous series in island arcs: the northeastern Caribbean compared with worldwide island-arc assemblages. *Bull. Volcanol.*, **43–2**, 347–382.

Fisher, R. V. 1968. Puu Hou littoral cones, Hawaii. *Geol. Rundschau*, **57**, 837–864.

Floyd, P. A. and Winchester, J. A. 1975. Magma type and tectonic setting discrimination using immobile elements. *Earth Planet. Sci. Lett.*, **27**, 211–218.

Furnes, H., Austrheim, H., Amaliksen, K. G. and Nordås, J. 1983. Evidence for an incipient early Caledonian (Cambrian) orogenic phase in southwestern Norway. **120**, 607–612.

Furnes, H., Fridleifsson, I. B. and Atkins, F. B. 1980. Subglacial volcanics—on the formation of acid hyaloclastites. *J. Volcanol. Geotherm. Res.*, **8**, 95–110.

Furnes, H. and Lippard, S. J. 1979. On the significance of Caledonian pahoehoe, aa, and pillow lava from Bømlo, SW Norway. *Norsk geol. Tidsskr.*, **59**, 107–114.

Furnes, H., Ryan, P. D., Grenne, T., Roberts, D., Sturt, B. A. and Prestvik, T. this volume. Geological and geochemical classification of the ophiolitic fragments in the Scandinavian Caledonides.

Fæseth, R. B. 1982. Geology of southern Stord and adjacent islands, southwest Norwegian Caledonides. *Norges geol. Unders.*, **371**, 57–112.

Færseth, R. B. and Steel, R. J. 1978. Silurian conglomerate sedimentation and tectonics within the Fennoscandian continental margin, Sunnhordland, western Norway. *Norsk geol. Tidsskr.*, **58**, 145–159.

Gale, N. H., Beckingsale, R. D. and Wadge, A. J. 1980. Discussion of a paper by McKerrow, Lambert and Chamberlain on the Ordovician, Silurian and Devonian time scales. *Earth Planet. Sci. Lett.*, **51**, 9–17.

Jakeš, P. and White, A. J. R. 1972. Major and trace element abundances in volcanic rocks of orogenic areas. *Bull. geol. Soc. Am.*, **83**, 29–40.

Kiær, J. 1929. Den fossilførende ordovicisk–silurske lagrekke på Stord. *Bergens Mus. Årbok 1937*, **11**, 92 pp.

Kolderup, N. H. 1931. Oversikt over den Kaledonske fjellkjede på Vestlandet. *Bergens Mus. Årbok*, **13**, 1–43.

Leeman, W. P. and Rogers, J. J. 1970. Late Cenozoic alkali-olivine basalts of the Basin-Range Province, USA. *Contrib. mineral. Petrol.*, **25**, 1–24.

Lippard, S. J. 1976. Preliminary investigations of some Ordovician volcanics from Stord, West Norway. *Norges geol. Unders.*, **327**, 41–66.

Macdonald, G. A. 1972. Volcanoes, Prentice-Hall, New Jersey, 510 pp.

Miyashiro, A. 1975. Classification, characteristics and origin of ophiolites. *J. Geology*, **83**, 249–281.

Monsen, A. 1937. Unpublished field notes, University of Bergen.

Moore, J. G. 1965. Petrology of deep-sea basalt near Hawaii. *Am. J. Sci.*, **263**, 40–52.

Nordås, J. *The geology of Caledonian volcanic rocks in central parts of Bømlo, SW Norway*. Cand. Real. thesis, Univ. of Bergen (in preparation).

Pearce, J. A. 1975. Basalt geochemistry used to investigate past tectonic environments on Cyprus. *Tectonophysics*, **25**, 41–67.

Pearce, J. A. 1980. Geochemical evidence for the genesis and eruptive setting of lavas from Tethyan ophiolites. *Proc. Int. Ophiolite Symp., Cyprus 1979*, 261–272.

Pearce, J. A. and Norry, M. J. 1979. Petrogenetic implications of Ti, Zr, Y and Nb variations in volcanic rocks *Contrib. mineral. Petrol.*, **69**, 33–47.

Priem, H. N. A. and Torske, T. 1973. Rb–Sr isochron age of Caledonian acid volcanics from Stord, W. Norway. *Norges geol. Unders.*, **300**, 83–85.

Reusch, H. 1888. Bømmeløen og Karmøen med omgivelser. *Norges geol. Unders.*, 422 pp.

Saunders, A. D., Tarney, J., Stern, C. R. and Dalziel, I. W. D. 1979. Geochemistry of Mesozoic marginal basin floor igneous rocks from southern Chile. *Bull. geol. Soc. Am.*, **90**, 237–258.

Saunders, A. D., Tarney, J. and Weaver, S. D. 1980. Transverse geochemical variations across the Antarctic Peninsula: implications for the genesis of calc-alkaline magmas. *Earth Planet. Sci. Lett.*, **46**, 344–360.

Simonian, K. O. and Gass, I. G. 1978. Arakpas fault belt, Cyprus: a fossil transform fault. *Bull. geol. Soc. Am.*, **89**, 1220–1230.

Solli, A., Naterstad, J. and Andresen, A. 1978. Structural succession in a part of the outer Hardangerfjord area, West Norway. *Norges geol. Unders.*, **343**, 39–51.

Songstad, P. 1971. *Geologiske Undersøkelser av den Ordovisiske lagrekken Mellom Løkling og Vikafjord, Bømlo, Sunnhordland*, Unpubl. Cand. Real. thesis, Univ. of Bergen

Sparks, R. J. S., Self, S. and Walker, G. P. L. 1973. Products of ignimbrite eruptions. *Geology*, **1**, 115–118.

Strand, T. 1972. The Norwegian Caledonides. In Strand T. and Kulling, O. (Eds), *Scandinavian Caledonides*, Wiley–Interscience, London, 3–145.

Sturt, B. A., Thon, A. and Furnes, H. 1980. The geology and preliminary geochemistry of the Karmøy Ophiolite, SW. Norway. *Proc. Int. Ophiolite Symp., Cyprus 1979*, 538–554.

Suthren, R. J. and Furnes, H. 1980. Origin of some bedded welded tuffs. *Bull. Volcanol.*, **43–1**, 61–71.

Thon, A. this volume. The Gullfjellet ophiolite complex and the structural evolution of the major Bergen arc, west Norwegian Caledonides.

Walker, R. G. 1975. Generalized facies models for resedimented conglomerates of turbidite associations. *Bull. geol. Soc. Am.*, **86**, 737–748.

Weaver, S. D., Saunders, A. D., Pankhurst, R. J. and Tarney, J. 1979. A geochemical study of magmatism associated with the initial stages of back-arc spreading. *Contrib. mineral. Petrol.*, **68**, 151–169.

Williams, H. and Malpas, J. 1972. Sheeted dikes and brecciated dike rocks within transported igneous complexes, Bay of Islands, Western Newfoundland. *Canadian J. Earth Sci.*, **9**, 1216–1229.

Wood, D. A. 1980. The application of a Th–Hf–Ta diagram to problems of tectonomagmaatic classification and to establishing the nature of crustal contamination of basaltic lavas of the British Tertiary Volcanic Province. *Earth Planet. Sci. Lett.*, **50**, 11–30.

The Caledonide Orogen—Scandinavia and Related Areas
Edited by D. G. Gee and B. A. Sturt

Ultramafic rocks in the Scandinavian Caledonides

H. Qvale* and **J. Stigh†**

Norges Geologiske Undersøkelse, c/o Mineralogisk–Geologisk Museum, Sars gt 1, N–Oslo 5, Norway[1]
†*Department of Geology, Chalmers University of Technology and University of Gøteborg, S–412 96 Göteborg, Sweden*

ABSTRACT

Ultramafic rocks of the Scandinavian Caledonides consist mainly of peridotites that are more or less serpentinized. Field evidence, textures, and geochemistry indicate five different categories: (1) Layers and lenses within basic intrusions of Caledonian or Precambrian age. (2) High-T Caledonian ultramafic intrusions. (3) Ultramafic xenoliths in basic intrusions. (4) Widespread 'alpine-type' ultramafites occurring either as solitary bodies or as lower parts of ophiolitic assemblages, with a mineralogy reflecting in part the regional Caledonian metamorphic conditions. This category is subdivided into groups (a) Generally massive, low-Al (<3.8 per cent Al_2O_3) with relict metamorphic (or tectonite) textures and predominantly of (meta-) harzburgitic or dunitic composition. They are associated with supracrustal rocks of Precambrian to Caledonian age, and are characterized by low alkalis and CaO, high Ni and Cr contents, stable MnO (0.13 per cent), and high MgO/FeO. Although now associated with supracrustal rocks they probably represent suboceanic upper mantle material. (b) Often compositionally layered with relict cumulate textures, high to low-Al ultramafites associated with ortho- and paragneisses, metabasites (eclogites, amphibolites, gabbros, etc), and, in the Basal Gneiss Region, also anorthosites. They also have relatively higher contents of CaO, TiO_2, and alkalis and lower MgO/FeO. Ultramafites of this group are interpreted as cumulates, closely related either to associated gabbros and formed at lower oceanic crustal level, or to anorthosites of disputed origin. Dismembering of the original associations has mostly taken place during orogenesis which also caused intermixing with the supracrustal rocks. (5) Detrital serpentinites are widespread in the low/medium grade metamorphic Caledonian allochthon and are generally associated with solitary ultramafites, the latter being regarded as the source rock, having been exposed to erosion on the sea floor.

Introduction

Ultramafic rocks in the Scandinavian Caledonides occur in two main associations: in the Precambrian Basal Gneiss Region of southwestern Norway and in the metamorphosed allochthon of the Seve–Köli Nappe Complex and equivalents, and higher tectonic units (Fig. 1). There are no ultramafites in the autochthonous and parautochthonous upper Proterozoic and lower Palaeozoic sequences. Within the nappes, ultramafites are generally absent in the late Proterozoic sediments of the Lower and Middle Allochthons (*cf.* the Tectonostratigraphic map, this volume).

With respect to characteristic features such as petrography, chemistry, associated rock assemblages, etc., the ultramafic rocks show a considerable variation. Many previous papers have described their occurrences, and presented individual genetic models. We find it appropriate to review the data available and to pay special attention to features which may be of importance in a more general genetic discussion.

For the purpose of the review, a classification system is needed, and an attempt has therefore been made to define generally acceptable categories. This task has proven to be rather difficult, partly due to the overlap between rocks of two or more categories

---

[1] Present address: *Institute for Energy Technology, P.O. Box 40, N-2007 Kjeller, Norway.*

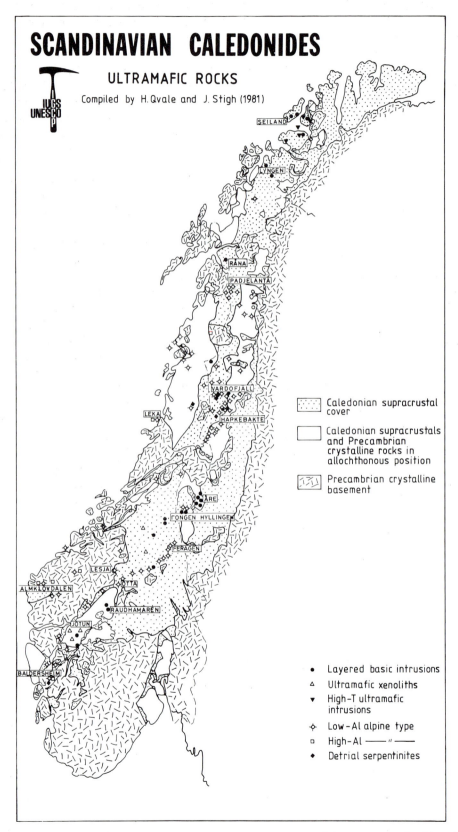

Fig. 1   Ultramafic rocks in the Scandinavian Caledonides. 'Major' localities are named with references to text. Geology simplified after UCS–81 preliminary compilation

and between the categories themselves, and to the fact that the 'classification' to a large extent has to be based on interpretation made by different workers.

Despite these problems we find it possible to treat the occurrences of ultramafic rocks in the following five categories:

1. Layers and lenses within basic intrusions,
2. High-temperature ultramafic intrusions,
3. Ultramafic xenoliths,
4. So-called 'alpine-type' ultramafites, and
5. Detrital serpentinites.

The chemical and petrographical diversity within categories 3, 4, and 5 has led us to subdivide these further into 'a' and 'b' characterized respectively by low Al contents and tectonite textures (a), and (relatively) high Al contents and cumulate textures (b). This subdivision refers to all three categories, but will be discussed individually.

Layers and Lenses within Basic Intrusions

Layers of peridotitic composition appear as integral parts of cumulate, compositional, and often rhythmic layering within layered mafic complexes in the Scandinavian Caledonides. The best known of these are the occurrences within the Seiland Igneous Province in Finnmark, North Norway, the Jotun Nappe Complex, Western Norway, and the Fongen–Hyllingen Complex within the Trondheim Supergroup (Fig. 1).

Lensoid and podiform ultramafic bodies are also commonly associated with layered mafic complexes. However, they exhibit a more independent relationship to the compositional layering of the complex thereby obscuring these features' genetic relationship. Well known examples of such ultramafites occur within the igneous provinces of Seiland, Råna, and the Jotun Nappe Complex which will be discussed individually below.

The ages of the enclosing mafic complexes vary from Precambrian (probably Svecofennian) for the Jotun rocks (Bhanumathi and Bryhni in preparation; Milnes and Koestler, this volume) to synorogenic (around 500 Ma) for the Seiland Complex (e.g. Sturt et al. 1978). Despite this wide range, the complexes preferentially appear at high allochthonous levels (Middle to Uppermost Allochthon) and mostly in Norway. The only known occurrences from the Swedish sector are those in the Åre area in Jämtland (Fig. 1).

In the *Seiland Igneous Complex* lherzolitic horizons occur within the rhythmically layered Hasvik gabbro (Robins and Gardner 1974) or as disconnected peridotitic lenses (Stumpfl and Sturt 1965)

either concentrated near the base of the intrusion, as in the Storelv gabbro, or evenly distributed as in the Breivikbotn intrusion. These bodies are regarded as synorogenic having intruded members of the Kalak Nappe Complex during the Finnmarkian phase of the Caledonian orogeny (around 500 Ma). The gabbros have tholeiitic compositions and are thought to represent the first subalkaline magmatic generation within the province (Robins and Gardner 1974, 1975; Sturt and Roberts 1978).

Around the *Fongen–Hyllingen Complex*, a well-developed contact aureole with cordierite, sillimanite, and andalusite in the enclosing Lower Palaeozoic metasediments and metavolcanics of the Trondheim Nappe Complex documents the *in situ* position of the intrusion (Nilsen 1973; Wilson and Olesen 1975; Wilson *et al.* 1981). The ultramafic rocks are developed either as weakly layered peridotites at the base of the intrusion (Esbensen *et al.* 1978), or as distinct layers, up to 30 m thick, at higher levels. The basal ultramafites contain olivine with a Fo-content below 87 per cent and subordinate pyroxene, chromite, and bytownitic plagioclase. At higher peridotitic levels the Fo-content is below 74 per cent and it coexists with labradoritic plagioclase and orthopyroxene as intercumulus phases (Esbensen *et al.* 1978). The chemistry of the mafic members of the complex indicates a tholeiitic trend of fractionation (Wilson *et al.* 1981).

The *Råna mafic complex* (Fig. 1) is situated east of Narvik in Northern Norway where it has intruded gneisses of the Narvik Group (Gustavson 1972) pre-dating the third of four fold phases. The main body shows a concentric structure with a lateral mostly noritic zone enclosing irregular bands and lenses of peridotitic and pyroxenitic rocks (Boyd and Mathiesen 1979). No conclusive observations have been made on the relative age of the ultramafic *vs.* noritic rocks, although the authors report local intrusions of norite in ultramafites suggesting multiple magma pulses.

The peridotites are mostly harzburgitic with 40–60 per cent cumulus olivine of composition Fo_{86-87} with intercumulus hypersthene, augite, phlogopite, and plagioclase (rare), sulphides, picotite, and Fe–Ti-oxides (Boyd and Mathiesen 1979). Lherzolitic and dunitic rocks are also present. The pyroxenites contain up to 80 per cent hypersthene with intercumulus augite, phlogopite, plagioclase, and opaques.

The Råna peridotites have recently received special attention as exploration target for Ni (Boyd and Mathiesen 1979), and are at present being exploited for the mining of olivine for metallurgical use.

The *Jotun Nappe Complex* represents the uppermost allochthonous unit in southern Norway.

Radiometric age-determinations (Rb/Sr) indicate a Svecofennian age for magmatic crystallization and granulite facies metamorphism (Bhanumathi and Bryhni in preparation; Milnes and Koestler, this volume), and Caledonian age for nappe emplacement. The complex comprises feldspathic igneous rocks (anorthosites, gabbros, mangerites, norites, etc.) and to a minor extent metasediments (e.g. quartzites) enclosing layers and lenses of ultramafic rocks.

Only one of the large bodies, the Raudhammaren body, has been studied in any detail (Dietrichson 1958; Murthy 1973) but is taken as representative. It is composed of olivine (Fo_{80-92}), clinopyroxene and orthopyroxene (both with Al_2O_3 contents between 2 and 3 per cent), and accessory spinel. The variation in composition defines an extensive banding commonly with a thickness of 2–3 m. The contacts with the surrounding foliated gabbroic rocks are sharp and often tectonized. Coronas of two pyroxenes + spinel or two pyroxenes + garnet have been reported by Griffin (1971a) and Bryhni et al. (1983) respectively. No chilled margins or contact metamorphic effects are reported. Ultramafic layers are known from gabbroic intrusions within the Jotun Nappe Complex (e.g. Qvale 1982), but no petrographical or chemical data are available.

The appearance of peridotitic to dunitic inclusions with olivine of composition For_{72-81} (Griffin 1971b; Qvale 1982) in a late generation of anorthositic dykes (Qvale 1982) gives rise to the well known corona textures (Griffin 1971b, 1972; Griffin and Heier 1973).

The Fa-rich olivine and high Al contents of all ultramafites in this province indicate that these rocks were generated as deep-seated crustal cumulates (Dietrichson 1958; Battey 1965). These have subsequently been emplaced either tectonically (larger lenses) or as xenoliths into an environment of higher P and/or lower T (i.e. the granulite facies metamorphism) leading to reaction between olivine and plagioclase (Griffin 1971b).

In the Åre area of Sweden (Fig. 1) some ultramafic bodies of this category occur (Stigh 1979). Here differentiated hornblende-bearing peridotites of layered intrusions occur closely associated with gabbroic rocks.

It should be emphasized that the internal characteristics of layered intrusions may not be significantly different from those of cumulate parts of ophiolite sequences. The interpretation of their origin may therefore change as new data are acquired. One of several examples in the Scandinavian Caledonides it the Lyngen Gabbro Complex, which will be discussed below as an ophiolite complex.

High-temperature Ultramafic Intrusions

High-T ultramafic plutons are well documented from the Seiland Igneous Province in Finnmark, North Norway (Fig. 1). They represent the last stage of magmatic activity during the Finnmarkian phase of the Caledonian orogeny (Sturt et al. 1980a; Bennett 1974). The plutons consist of dunites, wehrlites, and lherzolites. The dunites are regarded as equivalents of the primary magma, with c. 40 per cent MgO and Fo content in the olivine of 72–81 per cent. The intrusion of high-temperature dunite magma caused assimilation of country rock materials producing variation in magma composition from dunite to wehrlites and lherzolites. It also resulted in partial melting of the surrounding gabbro, producing mafic and anorthositic melts that back-vein into ultramafites.

Ultramafic Xenoliths

Ultramafic nodules occur in basic and ultrabasic dykes intruded into the Seiland syenogabbro, at a late stage of the evolution of the Seiland Igneous Province (Fig. 1) (Robins 1975). Both differentiated and simple dykes occur, although the latter are the more common. The composition of these dykes shows large variation; picrite, hornblende–olivine gabbro, hornblende gabbro, bojite, and hornblendite are reported.

The nodules are of two types: (a) a lherzolite-dominated tectonite, and (b) a wehrlitic, poikilitic cumulate. The former is characteristically low in Al and Ca and high in Cr, and resembles the low-Al alpine-type ultramafites discussed later. They are interpreted as relics of a high-T ultramafic body that rose diapirically through the upper mantle and thereby caused the synorogenic magmatism of the area. Analyses of the wehrlitic cumulates are not available.

Ultramafic and mafic nodules have been described from metabasites within the Gula group supracrustal rocks of the Trondheim Supergroup (Nilsen 1974). Based on this description we feel that these rocks should be included in our category 4, alpine-type ultramafites, and they will be discussed later.

Alpine-type Ultramafites

The 'alpine-type' ultramafites are the most widespread category within the Scandinavian Caledonides, but present a problem of definition. The concept of alpine-type ultramafic rocks originated with Benson (1926) and included all occurrences of ultramafites in orogenic belts. Later the concept has been changed to cover lenses and layers of gabbroic and ultramafic rocks occurring in an orogenic environment. According to Thayer (1960) such rocks are '. . . characterized by irregularity of form and internal structure, distribution along eugeosynclinal belts which have been subjected to an alpine type of deformation (hence the name) and features which can be explained best by strong deformation and mixing of previously semi-solid rocks during emplacement.' Or more exactly, Thayer 1960, 1967 specified:

1. Close spatial and structural relations are observed between ultramafic, gabbroic, dioritic, and granophyric members.
2. Contacts to enclosing rocks are usually tectonic.
3. Effects of contact metamorphism are rarely observable.
4. Primary concentric compositional variation is rarely observable.
5. Compositional layering is common, but not traceable over longer distances and is often strongly deformed.
6. The ultramafites are Mg-rich as are the olivines, if present.
7. Chromite, if present, is low in Fe^{2+}

These points separate the alpine-type ultramafites from zoned complexes ('Alaskan-type') and members of layered intrusions.

Several attempts have been made to further subdivide the alpine-type ultramafites, e.g., den Tex (1969) into ophiolitic (or truly alpine-type) and 'root-zone' peridotites; Jackson and Thayer (1972) into lherzolitic and harzburgitic subtypes; Chidester and Cady (1972) into allochthonous sheet-like, intrusive or diapiric and exotic bodies; Naldrett and Cabri (1976) into large obducted sheets, ophiolite complexes, deformed ophiolitic complexes, clastic blocks in melange terrains, and possible diapirs. Den Tex's (1969) term 'root-zone peridotite' and Jackson and Thayer's (1972) 'lherzolitic subtype' probably also include the high-temperature peridotitic intrusions already discussed as our category 2. These contributions, although useful, are not easily applied to the Scandinavian Caledonides, and are therefore left for the discussion of origin in a later chapter.

Ophiolite complexes (Steinmann 1926; Penrose Field Conference 1972; Colemann 1977) characteristically consist of a basal ultramafic complex overlain successively by a gabbroic complex, a basic sheeted dyke complex, and a mafic volcanic complex, and are associated with ribbon cherts and marine sediments, podiform chromitite bodies, and intrusive and extrusive sodic rocks.

In most ophiolite complexes the basal ultramafic zone can be divided into a non-cumulate (tectonized) and a cumulate (non-tectonized) unit. The non-cumulate unit is usually composed of dunites and peridotites of harzburgitic affinities. The rocks are made up of olivine, pyroxene, and chromite with a noteworthy absence of plagioclase. This unit is overlain by finer-grained peridotites of lherzolitic affinity showing typical cumulate textures. England and Davies (1973) pointed out a significant break in chemical composition between the non-cumulate and the cumulate units, and to a difference in chemistry between the olivines and pyroxenes of the two units.

Recently the record of such complexes in the Scandinavian Caledonides has grown dramatically (Furnes *et al.* 1980), often at the expense of the alpine-type ultramafites, e.g. Karmøy (Sturt and Thon 1978; Sturt *et al.* 1980b), Major Bergen Arc (Inderhaug 1975; Færseth *et al.* 1977), Solund–Stavfjorden (Skjerlie 1974; Gale 1975; Furnes *et al.* 1976), Trondheim–Løkken (Gale and Roberts 1972, 1974; Loeschke 1976), Leka (Prestvik 1972, 1980), and Lyngen (Munday 1974; Minsaas and Sturt, this volume).

Consequently we regard the 'alpine-type' ultramafites (in the wide sense) of the Scandinavian Caledonides as belonging to two major types:

1. Ultramafites associated with ophiolitic sequences.
2. Solitary ultramafites, where the ophiolite association is not easily recognized.

The solitary ultramafites, i.e. the alpine type of Thayer (1960) and Moore and Qvale (1977), appear as lens-shaped, elongated bodies lying concordantly within mostly metavolcanic and metasedimentary sequences. Well exposed contacts are strongly deformed, testifying to movements between the ultramafites and the surrounding rocks (Stigh and Ronge 1978). Convincing evidence of contact metamorphism in the surrounding rocks is lacking.

Moore and Qvale (1977) demonstrated the existence of three varieties of alpine-type ultramafites in the Norwegian Caledonides. Type 1 bodies have primary (i.e. pre-orogenic with reference to the Caledonides) olivine, clinopyroxene, orthopyroxene, and chromite, and are partly or completely serpentinized. They occur exclusively in rocks of Cambro-Silurian age and are exemplified by the occurrences at Baldersheim and Feragen (Fig. 1). Type 2 are polymetamorphic metaperidotites or sagvandites dominated by olivine, enstatite, and carbonates, with talc and amphibole commonly present. They occur in medium- to high-grade metamorphic rocks of varying age. Examples are the type-locality for sagvandites at Sagelvvatn in Troms (Pettersen 1883), and the bodies at Hjelmkona (Moore 1977) and Lesja (Fig. 1) in Central Norway. Type 3 also shows a metamorphic mineral association of olivine, orthopyroxene, and minor chromite, with garnet, clinopyroxene, amphibole, and Cr-chlorite as possible extra phases. The rocks show subsequent hydration to varying degrees and occur in high grade metamorphic gneisses of Svecofennian age, associated with anorthosites and eclogites.

This grouping coincides with Jackson and Thayer's (1972) division of the alpine-type ultramafites into harzburgitic and lherzolitic subtypes. However, in many complexes (e.g. Baldersheim and Feragen) the two subtypes coexist to such a degree as to limit the diagnostic values suggested.

The alpine-type ultramafites in the western Norwegian *Basal Gneiss Region* (Fig. 1) are enclosed exclusively in rocks defined as the Fjordane Complex (Bryhni 1966) or its lithological equivalents. These are acid to intermediate ortho/paragneisses, marbles, quartzites, amphibolites, anorthosites, and eclogites, which have been subjected to high-pressure metamorphism (Griffin *et al.*, this volume).

North of Nordfjord the relic high-pressure metamorphic assemblages are represented by the garnet-bearing ultramafites. In the well known localities along Almklovdalen garnet-peridotites (lherzolite, websterite, wehrlite) occur interlayered with garnet clinopyroxenites (Eskola 1921; Carswell 1981) with all gradations into surrounding retrograded chlorite peridotites with olivine + orthopyroxene + chlorite/amphibole (Lappin 1966; Medaris 1980b). Larger bodies of predominantly dunitic composition ($+/-$ chlorite, $+/-$ chromite) with only faintly developed layering are also common in the area. Internal deformation of the ultramafites has been extensive and isoclinal folding of the layering is a general feature, as is the penetrative foliation defined by the preferred orientation of chromium chlorite (Lappin 1966, 1967; Moore and Qvale 1977; Medaris 1980b).

Olivine and orthopyroxenes are highly magnesian (Fo_{88-94}, En_{91-94}) with the most magnesian members within the chlorite-peridotites (Moore and Qvale 1977; Medaris 1980b). The chemical variations suggest a pre-metamorphic layering defined by variation in the amounts of olivine, pyroxenes, and spinel comparable with the layering at Feragen (see Fig. 4).

The origin of these ultramafites is still highly controversial, partly because of disagreement regarding the physical conditions during crystallization of the garnetiferous assemblages. Medaris (1980b) summarizes the types of suggested origin, all of which involve tectonic emplacement of subcontinental ultramafic upper mantle material.

1. Both eclogites and ultramafites equilibrated in the upper mantle before emplacement (O'Hara and Mercy 1963; Lappin 1974; Lappin and Smith 1978).
2. The ultramafites equilibrated in the mantle before emplacement, whereas the eclogites simultaneously equilibrated in the crust (Carswell 1974; Bryhni *et al.* 1977; probably also Cordellier *et al.* 1981).
3. The ultramafites were emplaced prior to crustal 'eclogite facies' metamorphism and subsequently reequilibrated (Brueckner 1977; Carswell and Gibb 1980; Medaris 1980a, 1980b; Carswell 1981).

Recently Brastad (this volume) has pointed out the close relationship, chemically and in the field, between anorthosites and high-Al ultramafites similar to those in Almklovdalen (Fig. 1). He suggests a common low-P origin for the two rock types. This is also supported by Griffin and Qvale (this volume) who show that some ultramafites probably had a spinel–lherzolite or serpentinite mineralogy prior to the high-P metamorphism. Thus, the protoliths of the garnet-peridotites may have formed as cumulates by differentiation of a basic magma, and it is tempting to suggest that the differentiation could have taken place in an oceanic environment. Chemical data on the ultramafites tend to support this model, as will be discussed later, but it is still controversial as it involves the anorthosites, which although reported (e.g. Bezzi and Piccardo 1971; Coleman 1977), are not very common in this environment.

South of Nordfjord in the basal gneisses the ultramafites (Fig. 1) are largely hydrated (Bryhni 1966) thereby obscuring possible 'primary' features. They do, however, show a generally homogenous metadunitic appearance and low-Al composition (Bryhni 1966; Jacobsen 1977).

Along the eastern boundary of the Basal Gneiss Region a large number of ultramafites (Fig. 1) show partly hydrated amphibolite facies assemblages as at Hjelmkona, Nordmøre (Moore 1977) and Lesja (Qvale 1981), representing type 2 of Moore and Qvale (1977). Characteristic features are faint layering defined by varying proportions of primary chromite/magnetite or recrystallized olivine and orthopyroxene, with secondary talc, serpentine, and carbonates occurring interstitially or in veins. Primary structures have been obliterated except those defined by the oxides which have survived during deformation. This type of recrystallization is also found locally throughout the northern part of the Gneiss Region, reflecting late Caledonian metamorphism.

In *Nordland* and *Troms* similar low-Al metaharzburgitic rocks (Fig. 1) dominate the ultramafites (Pettersen 1883; Ohnmacht 1974; Moore and Qvale 1977; Qvale and Ramberg 1981). Some of these localitites show the unusual assemblages orthopyroxene + carbonate (sagvandites). This paragenesis is generated at the expense of forsterite in the presence of a CO_2-rich fluid phase, although most of the carbonate is probably secondary relative to the pyroxene (*cf.* Moore 1977; Schreyer *et al.* 1971; Ohnmacht 1974). These occurrences are closely associated with metasupracrustals, such as pelitic schists and gneisses, carbonates, amphibolites, quartzites, and Al-rich rocks which may represent lateritic deposits (Qvale and Ramberg 1981). Sørensen (1967) reports a gradual transition from carbonate rocks to peridotites in Glomfjord. No chemical analyses which could confirm their sedimentary/metasomatic origin, are available from these localitites. However, analyses from other bodies in this area and further south are typical low-Al, high-Ni and high-Cr alpine-type ultramafites (Qvale unpublished).

Occasional bodies of more homogenous orthopyroxenite in southern Nortdland (Hestmannen; Korneliussen 1977), granular dunite (Rana; Qvale and Ramberg 1981), and garnet-pyroxenite (Dunderlandsdalen; Gjelle 1974) are reported from tectonic units which have been metamorphosed at a higher grade and/or altered by the activity of fluids. Extremely low f_{O_2} is documented by secondary, complex hydrothermal veins containing graphite as the only C-bearing phase (Qvale and Ramberg 1981).

The low-grade metamorphic Lower Palaeozoic sections of the Scandinavian Caledonides contain several alpine-type complexes, though only a few of these have been studied in any detail. The ultramafic/mafic complex at Baldersheim, southwestern Norway (Fig. 1) (Qvale 1978) will be used as an example and a brief summary given of some of the others.

The alpine-type rocks at Baldersheim are enclosed in intensely deformed Ordovician units of metagreywackes and greenschists stratigraphically below the Moberg conglomerate (Sturt and Thon 1976; Naterstad 1976) within the eugeosynclinal series of the 'Faltungsgraben' (Goldschmidt 1913) of Hardanger. The complex comprises more than 100 lensoid bodies up to 500 m long (Fig. 2) (Qvale 1978). Different ultramafic rock-types are usually not in contact with each other. The rock-types may be divided into three groups:

1. Tectonitized low-Al metadunites and metaharzburgites (subordinate),

2. Cumulate ultramafites (harzburgites, lherzolites, websterites), some of which are only found as boulders in high-Al detrital deposits,
3. Mafic cumulates (norites, gabbros and diorites).

The metadunites exhibit a well-developed rhythmic layering defined by variation in the abundances of olivine and chromite (Fig. 3). The present olivine (Fo_{95-98}) overprints an earlier resorbed texture interpreted as being a result of partial melting.

An estimate of the physical conditions for the recrystallization of olivine is based on olivine–chromite equilibria (Evans and Frost 1975) and indicates temperatures around 700°C. These are significantly higher than expected from the greenschist facies metamorphism of the surrounding rocks (c. 450°C at 300–500 MPa), but would be in accordance with ocean floor metamorphism. The Baldersheim Complex is interpreted as a tectonically dismembered series of tectonites and cumulates that originated in an oceanic environment, and thus represents parts of a highly dismembered ophiolite.

A belt of similar ultramafic bodies occurs along the southeastern margin of *Trondheim Region*. Of these the largest lies just east of lake Feragen about 20 km east of Røros, at the tectonic contact between Eocambrian sediments and overlying Røros Schist of Middle to Upper Ordovician age (Rui 1972). This body covers an area of 14–17 km^2 (Grønlie and Rui 1976; Moore and Qvale 1977; Stigh 1979; Moore and Hultin 1980). Primary compositional layering (Fig. 4) is preserved (Moore and Qvale 1977) although modified by later tectonism and recrystallization. Preservation of compositional layering has also been recorded from many other ultramafic bodies within the Scandinavian Caledonides (Eskola 1921; Trouw 1973; Moore and Qvale 1977; Stigh and Ronge 1978; Prestvik 1980; Carswell 1981).

Solitary ultramafites enclosed in supracrustal rocks of the Gula Group are described by Nilsen (1974). They occur included in or close to the Gula amphibolites, and have been deformed and metamorphosed with them. Two types of solitary bodies are recognized here: a metagabbro and a metaperidotite, of which the former type occurs more frequently. Most of the metaperidotites appear as small discs and sheets, but soapstone bodies up to 70×5 m are recorded. The rocks are usually hydrated to a large extent. Relict olivine and diopside occur enclosed in orthopyroxenes and in amphiboles, which are the dominant metamorphic minerals. A primary lherzolitic composition is thereby indicated. The chemistry of the inclusions is characteristically high in Al and Ni (Ni contents are

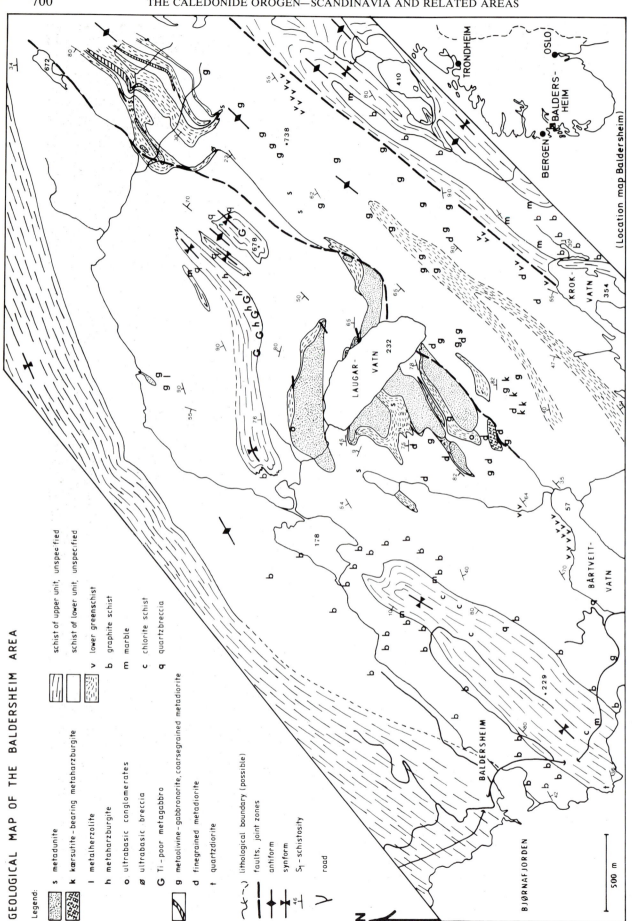

GEOLOGICAL MAP OF THE BALDERSHEIM AREA

Legend:

s	metadunite	
k	kærsutite-bearing metaharzburgite	
l	metalherzolite	
h	metaharzburgite	
o	ultrabasic conglomerates	
ø	ultrabasic breccia	
G	Ti-poor metagabbro	
g	metaolivine-gabbronorite, coarsegrained metadiorite	
d	finegrained metadiorite	
t	quartzdiorite	

schist of upper unit, unspecified
schist of lower unit, unspecified

v	lower greenschist
b	graphite schist
m	marble
c	chlorite schist
q	quartzbreccia

lithological boundary (possible)
faults, joint zones
antiform
synform
S_1 – schistosity
road

500 m

BJØRNAFJORDEN
BALDERSHEIM
BÅRTVEIT-VATN
KROK-VATN
LAUGAR-VATN

TRONDHEIM
OSLO
BALDERS-HEIM
BERGEN

(Location map Baldersheim)

Fig. 2. The Baldersheim Complex. Geological map of the Baldersheim area (Qvale 1978)

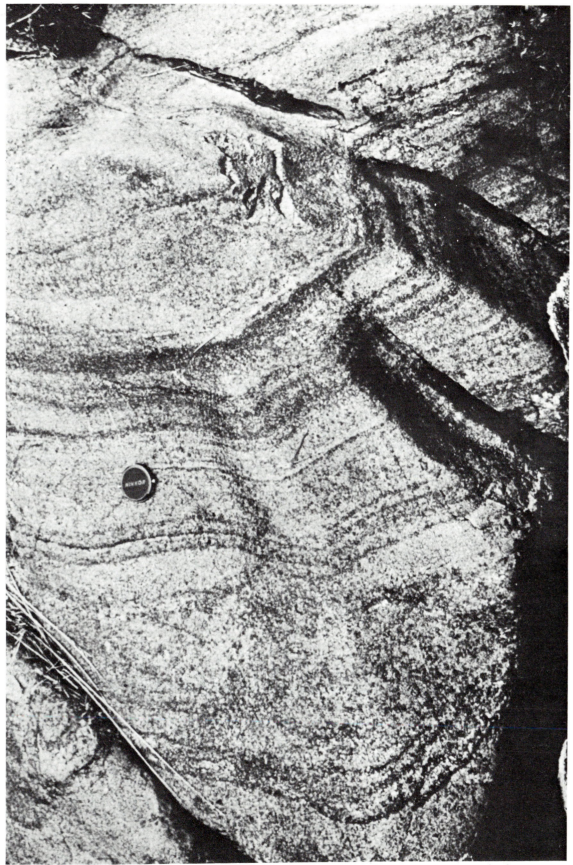

Fig. 3   Compositional layering in tectonitic alpine-type ultramafites, Baldersheim. Weakly deformed layering in chromite-bearing dunite (Fig. 4.37, Qvale 1978. Photo: J. Naterstad)

Fig. 4   Compositional layering in cumulate alpine-type ultramafites. Alternating harzburgite (with rough surface), dunite (smooth surface) and isoclinally folded pyroxenite, Feragen

indicated but not listed) and shows a nearly constant Cr/Ni ratio of about 3.

The similarity with ordinary high-Al lherzolititc alpine-type ultramafites is acknowledged by the author (Nilsen 1974) who suggests that they were inclusions generated as cumulates by differentiation of a basaltic magma, now represented by the Gula metabasites, in an oceanic environment.

At *Leka* a sequence of ultramafic rocks, characterized by variations in both mineralogy and petrography, has been described (Prestvik 1980). The ultramafic complex, appearing as a part of an ophiolite complex, includes both non-layered and layered sequences. The main rock types are serpentinite, clinopyroxenite, and dunite. Mineralogical and chemical data (Prestvik 1980) indicate that the non-layered and layered sequences represent depleted tectonite peridotites and ultramafic cumulates respectively.

The *Lyngen Gabbro* (Fig. 1) in Troms, North Norway, has earlier been described as a layered gabbro enclosing zoned ultramafic rocks of Alaskan type (Munday 1974), although no contact aureoles are preserved. Recent work in the area (R. Boyd, personal communication) however, has, revealed that the Lyngen ultramafites show interlayering of harzburgite, dunite, and pyroxenite, as well as cross-cutting pyroxenite dykes. These rocks are boudinaged and tectonically emplaced into the enclosing gabbro, and the whole zone apparently represents the plutonic zone of an ophiolite complex (Minsaas and Sturt, this volume). The Lyngen Gabbro is overlain unconformably by Ordovician/ Silurian metasediments (Minsaas and Sturt, this volume).

Alpine-type ultramafites are described by several early workers in the Swedish Caledonides (Törnebohm 1877: Kittelfjäll, Västerbotten; Svenonius 1883; Eichstädt 1884; and Du Rietz 1935: several localities in Jämtland, Västerbotten, and Norrbotten). More recently, Zachrisson (1969), Trouw (1973), and Stigh (1979) have discussed the ultramafic bodies and their abundant occurrence within certain stratigraphical units. These ultramafic bodies are dominated by solitary, low-Al alpine-type ultramafites (Stigh 1979). They normally occur as elongated bodies with their longest axes parallel to the regional schistosity in the *Seve–Köli Nappe Complex* (Zachrisson 1969; Stigh 1979) (Fig. 1). No such ultramafites occur in the autochthonous/parautochthonous late Precambrian and Lower Palaeozoic sequences or in the underlying basement. This concentration of ultramafic rocks in the Seve–Köli Nappe Complex, in a broad northeast trending belt, led previous authors (Wilson 1966; Dewey 1969) to suggest that

the zone defines an important tectonic discontinuity. This is apparently not the case, as has been discussed by Nicholson (1971), Gee and Zachrisson (1974), and Gee (1975).

Ultramafic bodies occur in the Eastern Belt in staurolite–kyanite–garnet-bearing rocks of the lower part of the Seve Nappe. The Central Belt of the Seve is dominantly composed of gneisses and amphibolites, partly in granulite facies, and contains very few ultramafites (Fig. 5). Eclogites are locally associated with these rocks (van Roermund and Bakker in press). In the overlying Western Belt, composed of staurolite–kyanite–garnet-bearing schists and amphibolites (e.g. Svartsjöbäcken Group, Trouw 1973) the ultramafites appear frequently. The relative frequency of ultramafic bodies occurring in the different units of the Seve is estimated in Fig. 5.

In the Köli unit of the Seve–Köli Nappe Complex the ultramafites are most numerous in the lower part i.e. the Tjopasi Group (Fig. 5). They occur frequently in the homogeneous phyllites of the Fatmomakk Formation and in the metavolcanics of the Seima Formation. Ultramafites also occur very frequently in the phyllites, greywackes, and conglomerates of the Gilliks formation, whereas they are much less frequent in the overlying Lasterfjäll and Remdalen Groups.

There is an important difference between the ultramafites in the Seve and in the Köli regarding the degree of serpentinization. In the Seve (amphibolite and granulite facies) they are dominantly olivine-rich rocks, and the degree of serpentinization is low compared to the olivine-poor ultramafites of the Köli unit (greenschist facies).

Three textural types of olivine are thought to represent a sequence of crystallization from a primary phase to a late metamorphic crystallization in the low grade metamorphic environments of the Köli (Calon 1979; Stigh 1979): Primary, probably magmatic olivines (Fo_{95-97}) are overprinted by a second generation of coarse-grained ('foam-texture') and a third generation of fine-grained ('mortar-texture') granular olivines.

These case studies suggest that the alpine-type ultramafites of the Scandinavian Caledonides may be separated into two major groups on chemical, petrographical, and structural grounds.

1. The generally (meta-)dunitic to harzburgitic rocks, massive or with only weakly developed layering. If present, relict early textures are tectonitic/metamorphic (i.e. metamorphic peridotites of Coleman 1977).
2. The massive or strongly-layered rocks of dominantly lherzolitic compositions, usually

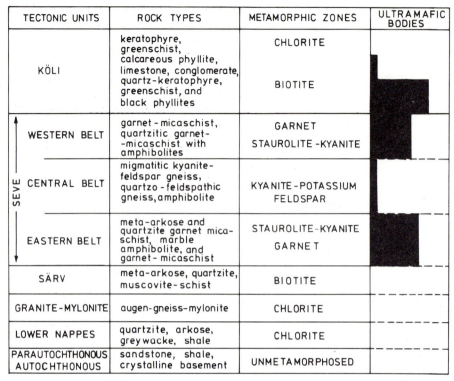

TECTONIC UNITS	ROCK TYPES	METAMORPHIC ZONES	ULTRAMAFIC BODIES
KÖLI	keratophyre, greenschist, calcareous phyllite, limestone, conglomerate, quartz-keratophyre, greenschist, and black phyllites	CHLORITE / BIOTITE	
SEVE WESTERN BELT	garnet-micaschist, quartzitic garnet-micaschist with amphibolites	GARNET / STAUROLITE-KYANITE	
SEVE CENTRAL BELT	migmatitic kyanite-feldspar gneiss, quartzo-feldspathic gneiss, amphibolite	KYANITE-POTASSIUM FELDSPAR	
SEVE EASTERN BELT	meta-arkose and quartzite garnet mica-schist, marble amphibolite, and garnet-micaschist	STAUROLITE-KYANITE / GARNET	
SÄRV	meta-arkose, quartzite, muscovite-schist	BIOTITE	
GRANITE-MYLONITE	augen-gneiss-mylonite	CHLORITE	
LOWER NAPPES	quartzite, arkose, greywacke, shale	CHLORITE	
PARAUTOCHTHONOUS AUTOCHTHONOUS	sandstone, shale, crystalline basement	UNMETAMORPHOSED	

Fig. 5   Alpine-type ultramafites in the Caledonides in Västerbotten. Relative frequency of solitary ultramafites in Västerbotten Caledonides related to their metamorphic zones (after Zwart 1974)

interlayered with rocks of harzburgitic and even dunitic composition. Relict cumulate textures may be present.

The geographical and tectonic distributions are important in the discussion of the origin of these rocks. Solitary, tectonite alpine-type ultramafites (1 above) are totally dominant in the eastern part (Sweden) of the mountains in the allochthonous Seve–Köli Nappe Complex. Further to the west cumulate alpine-type ultramafites (2 above) appear and 'coexist' with tectonites even in well-preserved ophiolite sequences (Furnes et al. 1980; Prestvik 1980).

Detrital Serpentinites

The term 'detrital serpentinite' is preferred for clastic sedimentary rocks composed largely of serpentinite. They form a part of the broader category of sedimentary serpentinites (Lockwood 1971), a term reserved for discussion of all occurrences of serpentinites where sedimentary processes are inferred. 'Psephitic serpentinites' are used here to refer to all those coarse 'clastic' serpentinites whose origin is uncertain (Stigh 1979), and where both tectonic and sedimentary origins are possible.

Detrital serpentinites occur frequently along con-

tinental margins of the world, and are mainly associated with alpine-type geosynclinal environments (Lockwood 1971). They occur as detrital accumulations of bedded serpentinite conglomerates, sandstones, and shales, chaotic breccias (olistostromes), and gravity slide blocks. In the Scandinavian Caledonides the detrital serpentinites are most abundant in low-grade metamorphic rocks of the Köli type (Fig. 6) (Stigh 1979, 1980). They often occur around, and in contact with solitary bodies indicating the nature of the source rock. Monomict serpentinite conglomerates dominate and finer-grained metasediments are very subordinate. With one exception all recorded detrital serpentinites, as well as their source if observable, are characteristically of low-Al type (Stigh 1980). The exception is represented by the Baldersheim complex where high-Al lherzolites appear both as boulders in polymict ultramafic conglomerates and as a solitary body (Qvale 1978).

Various authors (Kulling 1933; Kautsky 1953; Zachrisson 1969) have pointed out the occurrence of the serpentinite conglomerates at a well defined stratigraphic level. The detrital serpentinites are concentrated in units of pre-Ashgillian age, at least in the Västerbotten mountains and in Gudbrandsdalen, but are not bound to a certain stratigraphic level in the Tjopasi Group (Stigh 1979).

The sedimentary origin of the detrital serpenti-

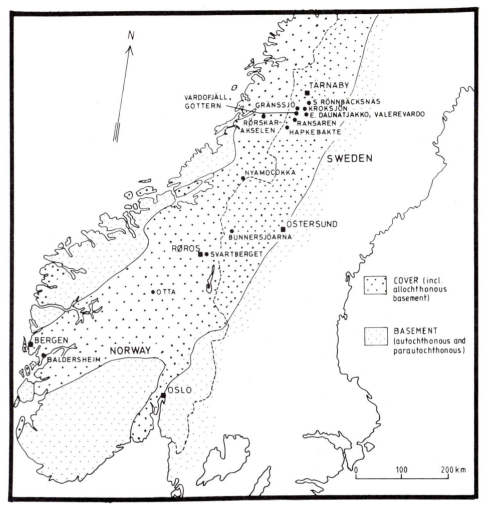

Fig. 6   Detrital serpentinites. Distribution of investigated detrital serpentinites in the Scandinavian Caledonides

nites is evident from the occurrence of stratification, the variation in grain-size (from conglomerates to sandstones and shales), the presence of sedimentary structures such as grading and cross-bedding (Fig. 7a), the heavy mineral concentrations and, locally, the presence of fossils (Fig. 7b). The best known fossiliferous serpentinites in the Scandinavian Caledonides occur at Otta in Gudbrandsdalen, Norway (Hedström 1930; Yochelson 1963; Strand 1951; Jaanusson 1973; Iversen 1981; Bruton and Harper 1981) where a Lower Ordovician age is suggested. Recently, a new Ordovician gastropod locality has been discovered by Holmqvist (1980) at Vardofjäll in Lappland, Sweden. It is not possible to draw any conclusions about the depositional environment from the fauna at Otta because this is almost certainly deposited in gravity-slide accumulations. (V. Jaanusson personal communication 1980).

In the Scandinavian Caledonides the ambiguous psephites and pseudoconglomerates occur as hori-zons irregularly distributed throughout the massive alpine-type ultramafic bodies. Michel (1950) strongly favoured a tectonic origin for the 'conglomerates' by Gunnel Majas Tjärn in the Aunere area, Västerbotten, and Trouw (1973) supported this interpretation to explain the serpentinite conglomerate at Ransarluspen.

Oftedahl (1969) suggested a pyroclastic origin for the conglomerates and Strand (1970) replied to this by restating the old ideas (Törnebohm 1896) that the conglomeratic serpentinites were truly of sedimentary origin.

The rapid accumulation of the sedimentary serpentinites is indicated by poorly sorted, ultramafic, sedimentary material and the lack of extensive lateral gradation.

The most probable transport agents were turbidity currents, underwater landslides or even mudflows, as suggested by Lockwood (1971), causing rapid suspension-type deposition.

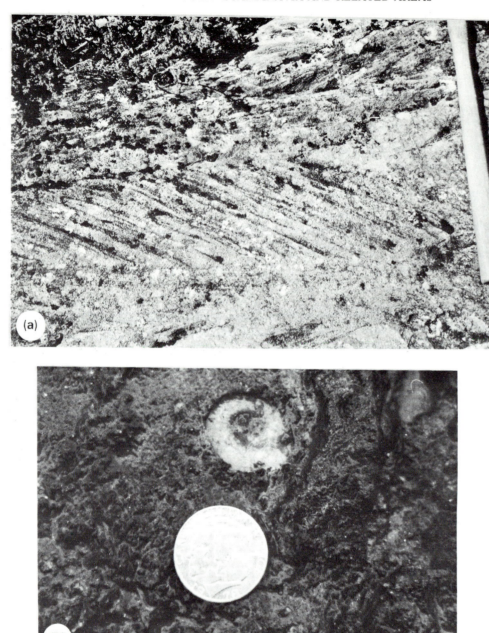

Fig. 7 Detrital serpentinite. (a) Cross-bedding in fine-grained detrital serpentinite. Southernmost outcrop east of Gaisartjåkko, Västerbotten (photo: J. Stigh). (b) Gastropod in calcareous, fine-grained detrital serpentinite, south of Nieritjåkko, Västerbotten (photo: D. Gee)

Chemistry

A comparison of the chemistry of the different categories of ultramafic rocks raises serious problems of systematics, especially as to how representative the analyses in the literature really are. This does not cause any problems for those categories represented by large numbers of samples per occurrence and small relative variations, such as categories 2 (high-T intrusions), 3a (tectonic xenoliths), 4a (tectonitic alpine-type), and 5 (detrital serpentinites). For some of these the number of analyses is large enough to allow the use of average values for those bodies represented by more than one analysis.

However, the cumulates are not so easily treated, as they may be part of compositional layering defined by variation in abundance of major phases, mainly olivine, pyroxenes, and plagioclase. As we are discussing ultramafic rocks the content of plagioclase should not exceed 10 per cent. However, sericitization, uralitization, and saussuritization reduce the plagioclase content, and the rock may appear to be ultramafic although the normative plagioclase exceeds 10 per cent. The available analytical material does not allow us to avoid this inconsistency, which is in part built into the categories concerned (mainly 1, ultramafic members of layered intrusions, but possibly also 3b, cumulate xenoliths, and 4b cumulate alpine-type).

On this basis the condensed chemical data for all categories are presented in Fig. 8 and Table 1.

Category 1. There is a large range in contents of major elements including Al_2O_3 (<1–16 per cent), MgO (10–44 per cent), TiO_2 (0–1.5 per cent), Na_2O (0–2 per cent), and K_2O (0–1 per cent). On the average the rocks are characterized by high Si, Ti, Al, Ca, and alkalis, due to the dominance of clinopyroxene and presence of plagioclase and Ti-ores, and they are characteristically low in Mg, Cr, and Ni.

Category 2. Chemical data on the dunites by Bennett (1974), Sturt *et al.* (1980a), and Griffin (personal communication) show little variation and are generally high in Fe (Av. 18.6 per cent FeO^{tot}) and MgO (Av. 39.1 per cent) but with low Mg/Fe ratio (Table 1) reflecting the low Fo contents in the olivine of these rocks. This also explains the low Ni content compared to the other olivine-dominated categories (3a and 4a).

Category 3. Of the two subdivisions of this group data are available only on the tectonites (3a). They are characterized by high Mg, Cr, and Ni and low Ti, Al, Ca, Na, and K contents. This is similar to the alpine-type ultramafites, which will be discussed below: highly magnesian olivine, pyroxenes, and Cr-rich spinel dominate the tectonites, whereas the primary presence of plagioclase probably affects the chemistry of the cumulates.

Category 4. The tectonite members (4a) of this category are characterized by high Mg, Cr, and Ni contents and MgO/MgO + FeO ratios, and by low Al, Ca, Na, and K contents relative to the cumulate members.

The bimodal distribution is best defined by the Al_2O_3 contents (Fig. 8). The tectonites show only small deviations from the average value 1.1 per cent Al_2O_3, whereas the analyses of the cumulates cover the whole range up to more than 15 per cent Al_2O_3 and an average of 4.8 per cent on a recalculated basis. This has led to the recognition of low-Al and high-Al alpine-type ultramafites. It should be again emphasized that the chemical distribution pattern is based on average values of several analyses from each body.

Summary of categories 1 to 4. The bimodality among the alpine-type ultramafites suggests a general bimodal grouping of all occurrences of rocks of categories 1 to 4. One group consists of the layered intrusions, cumulate xenoliths, and high-Al alpine-type ultramafics. This group is characterized by relatively high Ti, Al, Ca, and Na, and low Mg, Cr, and Ni. The Ti-anomalies are related to the ilmenite usually found in these rocks. Correspondingly, Al and Ca may reflect the clinopyroxene content and Al, Ca, and Na the plagioclase content. K is less depleted in the high-Al alpine-type ultramafites, where K-bearing phases such as amphiboles and phlogopite are more common.

The other group includes categories 3a (tectonite xenoliths) and 4a (low-Al alpine-type ultramafites); it is characterized by high Mg, Cr, and Ni, and low Ti, Al, Ca, Na, and K. The high-Mg content reflects the highly magnesian olivine (Fo_{90-98}) and pyroxenes, whereas Cr reflects the presence of chrome-spinels. Ni is enriched in olivine and magnetite as well as occurring in separate sulphides and/or occasionally in the native state (Hultin 1968).

Mineralogically the high-T intrusions differ from the low-Al ultramafites in one important respect; the Fo content of the olivine is significantly lower, thereby reducing the Mg content and the Mg/Fe ratios of the bulk rock as well as the Ni content which is usually proportional to the Fo content.

Category 5. The detrital serpentinites are with the exception of the lherzolitic material from Baldersheim, chemically similar to the low-Al ultramafites. Some exceptions are, however, evident, e.g. the extremely high-Ca content and the generally higher Ti, Al, Cr, Ni, and Mn and lower Mg and Fe contents. The Ca-enrichment is the result of secondary carbonate precipitation during serpentinization and steatitization. The enrichment of some minor elements may be due in part to incorporated clay minerals (for Al) and possibly also to the enrichment of phases resistant to weathering and erosion. The most significant of these phases are the chrome-spinels, which are enriched in Ni and associated elements in the magnetite rim (Qvale 1977) leading to enrichment of Cr, Al, Mn, Ni, and Co in the sediments. For example, the average Ni content in the detrital serpentinites is 3000 ppm *vs.* 2600 ppm in the low-Al alpine-type ultramafites. The corresponding Cr-contents are 4700 *vs.* 2450 ppm (Stigh 1981). The detrital serpentinites are also enriched in Zr and Sr relative to low-Al alpine-type ultramafites (Stigh 1981). The silica content indicates a depositional environment far from the continent without deposition of quartz in the form of sand-size particles (Stigh 1979).

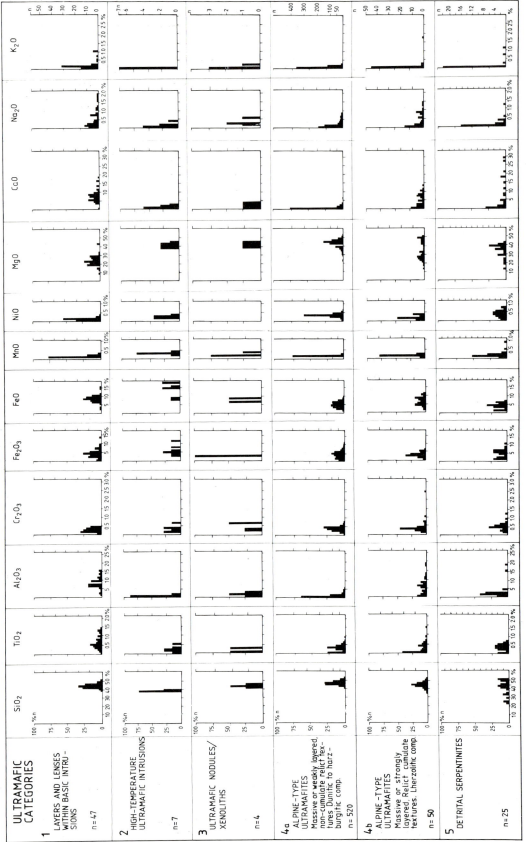

Fig. 8 Chemistry of ultramafic rocks. Histographical presentation of ultramafic rocks in the Scandinavian Caledonides. Generally all individual analyses are used as they appear. However, category 4a are represented by one average sample for each body, because of the high number of available analyses of those rocks. All analyses are recalculated to 100 per cent on volatile-free basis

Table 1 Chemisty of ultramafic rocks. Chemical average values for each category, using data as for Fig. 8. Selected values from Coleman (1977) from ophiolite complexes are included for comparison. All analyses are recalculated to 100 per cent on volatile-free basis

Category	1. Layers and lenses within basic intrusions		2. High temperature ultramafic intrusions[7]		3. Xenoliths a. Tectonites		4. Alpine-type ultramafites a. Tectonites		b. Cumulates[7]	
	n	wt%	n	wt%	n	wt%	n	wt%	n	wt%
SiO_2	47	45.01	16	38.07	4	43.14	520	44.66	64	43.88
TiO_2	47	0.67	16	0.22	4	0.10	454	0.30	64	0.21
Al_2O_3	47	8.39	16	1.20	4	1.58	520	1.08	64	4.75
Cr_2O_3	45	0.25	11	0.29	4	0.45	426	0.38	62	0.40
Fe_2O_3	47	4.47	16	4.85	4	2.42	520	3.98	64	2.96
FeO	47	8.68	16	14.20	4	7.50	520	4.98	64	6.29
MnO	47	0.18	14	0.24	4	0.21	520	0.13	64	0.15
NiO	43	0.11	14	0.18		—	402	0.36	44	0.18
MgO	47	23.61	16	39.13	4	42.27	520	42.77	64	37.05
CaO	47	7.89	16	1.41	4	1.96	519	1.07	64	3.70
Na_2O	47	0.55	11	0.17	4	0.29	517	0.24	64	0.34
K_2O	47	0.19	12	0.04	4	0.08	518	0.07	64	0.07
$\dfrac{MgO}{MgO + FeO}$[1]		0.65		0.68		0.81		0.83		0.81
'loss of ignition'[2]	47	3.29	15	1.83	4	2.56	519	9.29	64	6.82

Category	5. Detrital ultramafites a. 'Low-Al'		b. 'High-Al'		'Average dunites'[3]		'Average harzburgites'[4]		'Average lherzolites'[5]		'Average ultramafic cumulates'[6]	
	n	wt%	n	wt%	n	wt%	n	wt%	n	wt%	n	wt%
SiO_2	23	41.62	2	36.37	9	40.3	11	40.3	4	44.9	4	44.4
TiO_2	23	0.38	2	0.25	8	<0.03	8	0.04	4	0.23	3	0.06
Al_2O_3	23	1.79	2	16.92	9	0.47	11	1.14	4	6.8	4	3.2
Cr_2O_3	23	0.52	2	0.09	7	0.58	9	0.47	4	0.26	1	0.38
Fe_2O_3	23	4.84	2	4.04	9	3.0	11	2.6	4	1.2	4	4.1
FeO	23	4.18	2	4.61	9	5.4	11	5.8	4	7.4	4	5.4
MnO	23	0.23	2	0.17	8	0.12	10	0.11	4	0.1	2	0.11
NiO	23	0.42	2	0.05	7	0.31	9	0.39	1	0.27	1	0.12
MgO	23	33.20	2	37.45	9	49.3	11	45.2	4	32.7	4	35.1
CaO	23	12.52	2	5.96	7	0.40	11	0.9	4	5.6	4	6.6
Na_2O	23	0.18	2	0.10	7	0.07	11	0.13	4	0.5	4	0.15
K_2O	23	0.12	2	0.06	7	0.05	8	0.04	3	0.04	3	0.10
$\dfrac{MgO}{MgO + FeO}$[1]		0.80		0.79		0.86		0.85		0.79		0.79
'loss of ignition'[2]	23	16.57	2	13.20		—		—		—	4	5.85

Notes:
1. Total Fe as FeO.
2. 'Loss of ignition'. Either $H_2O + CO_2 + SO_3$, or loss of ignition.
3. Average dunite. Average of nine worldwide 'typical dunites from ophiolite metamorphic peridotites'. Table 1, Coleman (1977).
4. Average harzburgite. Average of 11 worldwide 'typical harzburgites from ophiolite metamorphic peridotites'. Table 2, Coleman (1977).
5. Average lherzolites. Average of four 'lherzolite subtypes' of ophiolite peridotites. Analyses 1, 2, 3 and 5, Table 3, Coleman (1977).
6. Average ultramafic cumulates from ophiolites. Average of analysis no. 4, 5, 6 and 7 in Table 4, Coleman (1977).
7. Some of the analyses listed by Griffin personal communication (Category 2), and Moore and Hultin (1980) (Category 4b), are averages of more than one individual analysis. These are given full weight in the calculations here, whereas only the averages are plotted in Fig. 8.

References:
Category 1: Dietrichson 1958; Flodberg and Stigh 1981; Heier 1961; Nilsen 1973; Oosterom 1963; Stigh 1979; Stumpfl and Sturt 1965; 2: Bennett 1974; Sturt et al. 1980a; Griffin personal communication; 3a: Robins 1975; 4a: Lappin 1974; Moore 1977; Moore and Qvale 1977; Ohnmacht 1974; Qvale 1978; Qvale unpublished; Stigh 1979; Stigh et al. 1981; Zachrisson and Stigh 1981; 4b: Carswell 1968a,b; Eskola 1921; Mercy and O'Hara 1965; Moore and Hultin 1980; Nilsen 1974; Qvale 1978; Qvale unpublished; 5a: Stigh 1979; 5b: Qvale 1978.

Origin of Ultramafic Rocks

The ultramafites occurring as layers and lenses within basic intrusions are mostly considered to be of a cumulative character as the result of the differentiation of basic magma. The layered intrusions are either intruded *in situ* with development of contact aureoles (e.g. the Fongen–Hyllingen Complex) or tectonically emplaced during orogenesis (e.g. the Jotun Nappe Complex).

High-temperature ultramafic intrusions of the Seiland Igneous Province in Finnmark, North Norway, represent the latest stage of the magmatic activity. The ultramafic magma itself is generated by partial melting of already depleted upper mantle material (Sturt *et al.* 1980a). The intrusion of high-temperature dunite magma caused partial melting of the surrounding gabbro.

Ultramafic nodules and/or xenoliths reported from the Scandinavian Caledonides are very variable in composition. They are of two types, i.e. tectonites regarded as representing relic upper mantle material and cumulates having been generated by cumulation of ultramafic material in the host mafic to ultramafic magma. Both types are emplaced as xenoliths within their magmatic host rocks.

The origin of alpine-type ultramafites is widely debated. We will not review this serial here, but wish to emphasize the close relationship between the ultramafic members of ophiolite sequences and the solitary ultramafic bodies. Chidester and Cady (1972) drew attention to the characteristic 'predominance of peridotites over pyroxenites and gabbros, the absence of cryptic layering and the irregularity of visible layering, the predominance of cataclastic and allotriomorphic textures, the absence of chilled margins and the general absence of high-temperature contact metamorphic effects'.

Naldrett and Cabri (1976) stated that alpine-type bodies were mostly emplaced during the compressive phase of orogenesis. They emphasized that this contrasts with the origin of the layered complexes, thought to be emplaced contemporaneously with eugeosynclinal volcanism, and the Alaskan-type complexes which are interpreted to have been emplaced at a late stage of orogenesis, largely during uplift. Thayer (1960) drew attention to the geochemical distinction between alpine-type ultramafites and ultramafites of layered peridotite-gabbro intrusions.

In many ophiolite complexes the ultramafic rocks are divided into lower non-cumulate ('tectonitic') and an upper cumulate unit with compositions more or less equivalent to the harzburgitic and lherzolitic subtypes of alpine-type ultramafites of Jackson and Thayer (1972), respectively. England and Davies (1973) and Coleman (1977) have pointed out the major chemical differences between ultramafites from the two levels of the ophiolites. These differences correspond well to those recorded between the two subcategories of alpine-type ultramafites in the Scandinavian Caledonides. The tectonite low-Al ultramafites are characterized by high Mg, Cr, and Ni contents, and low Ti, Al, Ca, Na, and K contents, thought to reflect depleted suboceanic mantle. The cumulate high-Al ultramafites, on the other hand, are higher in Ti, Al, Ca, Na, and K and lower in Mg, Cr, and Ni, and this composition indicates a cumulate origin in the lower part of the oceanic crust.

Emplacement of the Alpine-type Ultramafites and Associated Detrital Serpentinites

The emplacement of the ultramafic members of more or less complete ophiolite sequences can be explained as the result of obduction of the oceanic crust (e.g. Karmøy: Sturt *et al.* 1980b). However, many of the Caledonian ophiolite complexes are extensively dismembered as the result of deformation and erosion. Likewise most of the solitary bodies are tectonized to such an extent that they show, at best, only equivocal relations to other components of the ophiolites, and their emplacement cannot be explained by a simple obduction mechanism.

The density contrasts between the rock types in combination with tectonic movements, may explain the observed separation. The density of dunites and peridotites in general normally varies between 3.25 and 3.40 g/cm^3. A combination of serpentinization, dehydration, and protrusion (Lyell 1871; Lockwood 1972) may lead to upward migration of heavy mantle material into crustal positions.

The alpine-type ultramafites in the Scandinavian Caledonides, as described, do not represent a single orogenic episode. The ages of ultramafic protoliths probably span from Svecofennian/Sveconorwegian in the West Norwegian Basal Gneiss Region and the Seve units, to Ordovician in the Köli units and equivalents. Despite this major diversity the rocks have so many features in common that we will suggest the following generalized model.

Early hydration of peridotites to serpentinite took place in an oceanic ridge or trench environment, probably closely connected with pre-orogenic faulting. It could have resulted in protrusion from the non-hydrous upper mantle into an oceanic crustal environment. In a subsequent situation of plate collision accompanied by subduction the partially serpentinized ultramafites would be dragged down and further serpentinized until the decrease in density reversed the process and they started to protrude along shear zones into the overlying geosynclinal sediments. Occasionally these ultramafites and perhaps even the associated sedimentary serpentinites deposited on the oceanic crust were then emplaced by obduction and subsequent gravity-sliding into the Caledonian sedimentary basins. This would produce a melange of intermixed sedimentary and volcanic rocks and ultramafic bodies as exemplified by the Baldersheim Complex (Qvale 1978).

This process requires conversion of dunitic and peridotitic upper mantle material into serpentinite prior to the main orogenic stage. Many authors (Milovanovic and Karamata 1960; Knipper 1965; Dickinson 1966; Oakeshott 1968; Moiseyev 1970) have appealed to this serpentinization and protrusion mechanism for emplacement of ultramafites at high levels in the crust. In orogenic zones, the serpentinites may protrude along thrusts or normal faults (Lockwood 1972). Movement of such a serpentinitic mass occurs by interblock and intergranular displacements that result in a plastic behaviour of the body as a whole (Lockwood 1972).

By contrast, obduction as a mechanism of emplacement does not require serpentinization. In view of the subordinate role of ultramafic rocks in oceanic crust it is likely that this mechanism would have resulted in a larger proportion of the mafic and sedimentary components of ophiolites within the orogenic belt.

The serpentinites in the subduction zone are in a low temperature environment when they protrude into the geosynclinal sediments. The temperatures are higher above the subduction zone (Minear and Toksöz 1970; Oxburgh and Turcotte 1970). When the serpentinites protrude into these areas of higher temperature a dehydration will occur (Lockwood 1972; Raleigh and Paterson 1965; Evans *et al.* 1976) leading to the formation of forsterite from antigorite + brucite at temperatures between 300 and 400°C and pressure below 300 MPa. Or, if no excess Mg is present, antigorite breaks down at T > 500°C and P = 200 MPa, thereby stabilizing forsterite + talc. These reactions have probably given rise to most of the recrystallized olivine observed. If the textures observed have been correctly interpreted, then the serpentinites that protruded into the Seve and Köli Supergroups were probably more serpentinized than they are at present. The serpentinites that protruded into the Seve and Helgeland Nappes Complexes and the Basal Gneiss Complex were subjected to amphibolite or granulite facies metamorphism (*cf.* Griffin *et al.*, this volume; van Roermund and Bakker in press) which led to further and often complete dehydration. The Seve and Basal Gneiss ultramafites are thought never to have risen to the surface; they stopped within the dry complexes, probably owing to the increase in density resulting from dehydration. The high-pressure metamorphism was succeeded by retrogradation under amphibolite facies conditions and later on also by *in situ* serpentinization. By contrast, the ultramafites in the Köli and equivalent low-grade units rose into a sedimentary, water-rich environment and in some cases they reached the surface, where the detrital serpentinites accumulated.

Finally the eugeosynclinal metasediments, gneisses, etc. with their ultramafites were emplaced onto Precambrian craton(s) in a regional phase of nappe formation. Internal tectonic mobilization on a smaller scale within the nappe units took place at this stage causing shearing and faulting along contacts.

Acknowledgments

We wish to thank Drs R. Boyd, D. A. Carswell, W. L. Griffin, L. G. Medaris, and B. A. Sturt for their constructive criticism leading to major improvements of this chapter in the development of the Scandinavian Caledonides.

References

Battey, M. H. 1965. Layered structure in rocks of the Jotunheimen complex. *Mineral. Mag.*, **34**, 35–51.

Bennett, M. C. 1974. Emplacement of high temperature periodotite in Seiland Province of the Norwegian Caledonides. *J. Geol. Soc. London*, **130**, 205–228.

Benson, W. N. 1926. The tectonic conditions accompanying the intrusion of basic and ultrabasic igneous rocks. *Nat. Acad. Sci. Mem.*, **19**, 1, 90 pp.

Bezzi, A. and Piccardo, G. B. 1971. Structural features of the Ligurian ophiolites: petrologic evidence for the 'oceanic' floor of the Northern Apennines geosyncline; a contribution to the problem of alpine type gabbro–peridotite associations. *Mem. Soc. Geol. Ital.*, **10**, 53–63.

Bhanumathi, L. and Bryhni, I. in prep. Petrology of anorthositic rocks between Undredal and Stalheim, Sognefjord, Norway.

Boyd, R. and Mathiesen, C. O. 1979. The nickel mineralization of the Råna mafic intrusion, Nordland, Norway. *Can. Min.*, **17**, 287–298.

Brastad, K. this volume. Relationship between peridotites, anorthosites and eclogites in Bjørkedalen, Western Norway.

Brueckner, H. K. 1977. A crustal origin for eclogites and a mantle origin for garnet peridotites: Sr isotopic evidence from clinopyroxenes. *Contrib. mineral. Petrol.*, **60**, 1–15.

Bruton, D. L. and Harper, D. A. T. 1981. Brachiopods and trilobites of the early Ordovician serpentinite Otta conglomerate, south central Norway. *Norsk geol. Tidsskr.*, **61**, 153–181.

Bryhni, I. 1966. Reconnaissance studies of gneisses, ultrabasites, eclogites and anorthosites in Outer Nordfjord, western Norway. *Norges geol. Unders.*, **241**, 1–68.

Bryhni, I., Krogh, E. J. and Griffin, W. L. 1977. Crustal derivation of Norwegian eclogites: a review. *N. Jb. Miner. Abh.*, **130**, 59–68.

Bryhni, I., Brastad, K. and Jacobsen, V. W. 1983. Subdivision of the Jotun Nappe Complex between Aurlandsfjorden and Nærøyfjorden. *Norges geol. Unders.*, **380**, 23–33.

Calon, T. J. 1979. *A Study of the Alpine-type Peridotites in the Seve–Köli Nappe Complex Central Swedish Caledonides with Special Reference to the Kittelfjäll Peridotite*, thesis, University of Leiden, 236 pp.

Carswell, D. A. 1968. Possible primary upper mantle peridotite in Norwegian basal gneiss. *Lithos*, **1**, 322–355.

Carswell, D. A. 1968b. Picritic magma residual dunite relationships in garnet peridotites at Kalskaret near Tafjord, south Norway. *Contrib. mineral. Petrol.*, **19**, 97–124.

Carswell, D. A. 1974. Comparative equilibration temperatures and pressures of garnet lherzolites in Norwegian gneisses and in kimberlite. *Lithos*, **7**, 113–121.

Carswell, D. A. 1981. Clarification of the petrology and occurrence of garnet lherzolites and eclogite in the vicinity of Rødhaugen, Almklovdalen, West Norway. *Norsk geol. Tidsskr.*, **61**, 249–260.

Carswell, D. A. and Gibb, F. G. F. 1980. The equilibrium conditions and petrogenesis of European crustal garnet lherzolites. *Lithos*, **13**, 19–29.

Chidester, A. H. and Cady, W. M. 1972. Origin and emplacement of alpine-type ultramafic rocks. *Nature, London, Phys. Sci.*, **240**, 27–31.

Coleman, R. G. 1977. *Ophiolites. Ancient Oceanic Lithosphere?*, Springer, New York, 200 pp.

Cordellier, F., Boudier, F. and Boullier, A. M. 1981. Structural study of the Almklovdalen massif (Southern Norway). *Tectonophysics*, **77**, 257–281.

Den Tex, E. 1969. Origin of ultramafic rocks, their tectonic setting and history: a contribution to the discussion of the paper 'The origin of ultramafic and ultrabasic rocks' by P. J. Wyllie. *Tectonophysics*, **7**, 457–488.

Dewey, J. F. 1969. Evolution of the Appalachian–Caledonian orogen. *Nature, London*, **222**, 124–129.

Dickinson, W. R. 1966. Table Mountain serpentinite extrusion in the California Coast Ranges. *Bull. geol. Soc. Am.*, **77**, 415–472.

Dietrichson, B. 1958. Variation diagrams supporting the stratiform, magmatic origin of the Jotun Eruptive Nappes. *Norges geol. Unders.*, **203**, 5–34.

Du Rietz, 1935. Peridotites, serpentinites and soapstones of northern Sweden. *Geol. Fören. Stockholm Förh.*, **57**, 133–206.

Eichstädt, F. 1884. Mikroskopisk undersökning av olivinstenar och serpentiniter från Norrland. *Geol. Fören. Stockholm Förh.*, **7**, 333–368.

England, R. N. and Davies, H. L. 1973. Mineralogy of ultramafic cumulates and tectonites from eastern Papua. *Earth Planet. Sci. Lett.*, **17**, 416–425.

Eskola, P. 1921. On the eclogites of Norway. *Skr. Vidensk. Selsk. Christiania, Mat.–nat. Kl. I*, **8**, 1–118.

Esbensen, K. H., Thy, P. and Wilson, J. R. 1978. A note on the cumulate stratigraphy of the Fongen–Hyllingen gabbro complex. Trondheim Region, Norway. *Norsk geol. Tidsskr.*, **58**, 103–107.

Evans, B. W. and Frost, B. R. 1975. Chrome-spinel in progressive metamorphism—a preliminary analysis. *Geochim. Cosmochim. Acta*, **39**, 959–972.

Evans, B. W., Johannes, W., Oterdoom, H. and Trommsdorff, V. 1976. Stability of chrysolite and antigorite in the serpentinite multisystem. *Schweiz. Min. Petr. Mitt.*, **56**, 79–93.

Flodberg, K. and Stigh, J. 1981. The Umbukta ultramafic and mafic complex in the Rödingfjället Nappe, Northern Swedish Caledonides (Abstract). *Terra cognita*, **1**, 42.

Furnes, H., Skjerlie, F. J. and Tysseland, M. 1976. Plate tectonic model based on greenstone geochemistry in the Late Precambrian–Lower Palaeozoic sequence in the Solund Stavfjorden areas, west Norway. *Norsk geol. Tidsskr.*, **56**, 161–186.

Furnes, H., Roberts, D., Sturt, B. A., Thon, A. and Gale, G. H. 1980. Ophiolite fragments in the Scandinavian Caledonides. In Panayiotou, A. (Ed.), *Ophiolites. Proceedings Int. Ophiolite Symp. Cyprus 1979*, Geol. Surv. Dept., Nicosia, 582–600.

Færseth, R. B., Thon, A., Larsen, S., Sivertsen, A. and Elvestad, L. 1977. Geology of the Lower Palaeozoic rocks in the Samnanger–Osterøy area. Major Bergen Arc, western Norway. *Norges geol. Unders.*, **334**, 19–58.

Gale, G. H. 1975. Ocean floor type basalts from the Grimeli Formation, Stavenes Group, Sunnfjord. *Norges geol. Unders.*, **319**, 47–58.

Gale, G. H. and Roberts, D. 1972. Paleogeographical implication of greenstone petrochemistry in the southern Norwegian Caledonides. *Nature, London, Phys. Sci.*, **238**, 60–61.

Gale, G. H. and Roberts, D. 1974. Trace element geochemistry of Norwegian Lower Paleozoic basic volcanites and its tectonic implications. *Earth Planet. Sci. Lett.*, **22**, 380–390.

Gee, D. G. 1975. A tectonic model for the central part of the Scandinavian Caledonides. *Am. J. Sci.*, **275–A**, 468–515.

Gee, D. G. and Zachrisson, E. 1974. Comments on stratigraphy, faunal provinces, and structure of the metamorphic allochthon, central Scandinavian Caledonides. *Geol. Fören. Stockholm Förh.*, **96**, 61–66.

Gjelle, S. 1974. *En petrografisk/strukturgeologisk undersøkelse i Bjøllanes- området, Rana*, unpubl. Cand. Real. thesis, Univ. of Oslo, 283 pp.

Goldschmidt, V. M. 1913. Geologisch–petrographische Studien im Hochgebirge des südlichen Norwegens. II. Die kaledonische Deformation der südnorwegischen Urgebirgstafel. *Skr. Vidensk. Selsk. Christiana Mat. nat. kl. I*, **19**, 11 pp.

Griffin, W. L. 1971a. Mineral reactions at a peridotite–gneiss contact, Jotunheimen, Norway. *Mineral. Mag.*, **38**, 435–445.

Griffin, W. L. 1971b. Genesis of coronas in anorthosites of the Upper Jotun Nappe, Indre Sogn, Norway. *J. Petrol.*, **12**, 219–243.

Griffin, W. L. 1972. Formation of eclogites and the coronas in anorthosite, Bergen Arcs, Norway. *Geol. Soc. Am. Memoir*, **135**, 37–63.

Griffin, W. L. and Heier, K. S. 1973. Petrological implications of some corona structures. *Lithos*, **6**, 315–335.

Griffin, W. L. and Qvale, H. this volume. Superferrian eclogites and the crustal origin of garnet peridotites, Almklovdalen, Norway.

Griffin, W. L., Austrheim, H., Brastad, K., Bryhni, I., Krill, A., Krogh, E. J., Mørk, M. B. E., Qvale, H. and Tørudbakken, B. this volume. High-pressure metamorphism in the Scandinavian Caledonides.

Grønlie, G. and Rui, I. J. 1976. Gravimetric indications of

basement undulations. *Norsk geol. Tidsskr.*, **56**, 195–202.

Gustavson, M. 1972. The Caledonian mountains chain of the southern Troms and Ofoten areas. III. Structures and structural history. *Norges geol. Unders.*, **283**, 1–56.

Hedström, H. 1930. Om Ordoviciska fossil från Ottadalen i det centrala Norge. *Nor. Vidensk.—Akad., Avh. 1930, I Mat.—nat. kl.*, **10**, 1–10.

Heier, K. S. 1961. Layered gabbro, hornblendite, carbonatite and nepheline syenite on Stjernøy, North Norway. *Norsk geol. Tidsskr.*, **41**, 109–155.

Holmqvist, A. 1980. Ordovician gastropods from Vardofjäll, Swedish Lapland, and the dating of Caledonian serpentinite conglomerates. *Geol. Fören. Stockholm Förh.*, **102**, 493–497.

Hultin, I. 1968. Awaruite (josephinite), a new mineral for Norway. *Norsk geol. Tidsskr.*, **48**, 179–185.

Inderhaug, I. E. 1975. *Liafjellområdets Geologi*, unpubl. cand. real. thesis Univ. of Bergen.

Iversen, E. 1981. Day 3, B. The Vågå area. In Bryhni, I. (Ed.), *The Southern Norwegian Caledonides—Oslo to Sognefjord and Ålesund, Guide to UCS—excursion Al, Preliminary issue*, 53–60.

Jaanusson, V. 1973. Aspects of carbonate sedimentation in the Ordovician of Baltoscandia. *Lethaia*, **6**, 11–34.

Jackson, E. D. and Thayer, T. P. 1972. Some criteria for distinguishing between stratiform, concentric and alpine peridotite–gabbro complexes. *24th Int. Geol. Congr. Montreal*, Sect. 2, 289–296.

Jacobsen, V. W. 1977. *Serpentinisert olivinstein i Nordfjord.*, unpubl. cand. real thesis. Univ. of Oslo, 185 pp.

Kautsky, G. 1953. Der geologische Bau des Sulitelma–Salojauregebiets in den nordscandinavischen Kaledoniden, *Sver. geol. Unders.*, **C. 528**, 228 pp.

Knipper, A. L. 1965. Osobennosti obrazovaniya antiklinaley s serpentinitovymi yadrami (Sevano–Akerianskaya zona Maloga Kavkaza). (The development of anticlines with serpentinite cores (Sevan–Akerin zone of the lesser Caucasus)). *Muskov. Obshch. Ispyt. Prir. By ull. Otdel. geol.*, **15**, 46–58.

Korneliussen, A. 1977. Alpintype peridotitter i Hestmannøyområdet. *Geolognytt*, **12**, 27 (abstract).

Kulling, O. 1933. Bergbyggnaden inom Björkvattnet–Virisen området i Västerbottens-fjällens centrala del. *Geol. Fören. Stockholm Förh.*, **55**, 167–422.

Lappin, M. A. 1966. The field relationships of basic and ultrabasic masses in the Basal Gneiss Complex of Stadlandet and Almklovdalen, Nordfjord, southwest Norway. *Norsk geol. Tidsskr.*, **46**, 439–496.

Lappin, M. A. 1967. Structural and petrofabric studies of the dunites of Almklovdalen, Nordfjord, Norway. In Wyllie, P. J. (Ed.), *Ultramafic and Related Rocks*, Wiley, New York, 183–190.

Lappin, M. A. 1974. Eclogites from the Sunndal–Grubse ultramafic mass, Almklovdalen, Norway and T–P history of the Almklovdalen masses. *J. Petrol.*, **15**, 567–601.

Lappin, M. A. and Smith, D. C. 1978. Mantle-equilibrated orthopyroxene eclogite pods from the basal gneisses in the Selje district, western Norway. *J. Petrol.*, **19**, 530–584.

Lockwood, J. P. 1971. Sedimentary and gravity slide emplacement of alpine-type serpentinite. *Bull. Geol. Soc. Am.*, **82**, 919–936.

Lockwood, J. P. 1972. Possible mechanisms for emplacement of alpine-type serpentinites. *Geol. Soc. Am. Mem.*, **132**, 273–287.

Loeschke, J. 1976. Petrochemistry of eugeosynclinal magmatic rocks of the area around Trondheim (Central Norwegian Caledonides). *N. Jb. miner. Abh.*, **128**, 41–72.

Lyell, C. 1871. The Student's Elements of geology, Harper & Bros., New York, 640 pp.

Medaris, L. G. 1980a. Convergent metamorphism of eclogite and garnet-bearing ultramafic rocks at Lien, Almklovdalen, west Norway. *Nature, London*, **283**, 470–472.

Medaris, L. G. 1980b. Petrogenesis of the Lien peridote and associated eclogites, Almklovdalen, western Norway. *Lithos*, **13**, 339–353.

Mercy, E. L. P. and O'Hara, M. J. 1965. Chemistry of garnet-bearing rocks from South Norwegian peridotites. *Norsk geol. Tidsskr.*, **45**, 323–332.

Michel, H. 1950. Geology and Petrology of the Borkafjäll Region, Broese & Peereboom, Bredo, Netherlands, 138 pp.

Milnes, A. G. and Koestler, A. G. this volume. Geological structure of Jotunheimen, southern Norway.

Milovanovic, B. and Karamata, S. 1960. Ueber den Diapirismus serpentinischer Massen. *Proceedings 21st Int. Geol. Congr. Copenhagen*, **18**, 409–417.

Minear, J. H. and Toksöz, M. N. 1970. Thermal regime of a down-going slab and new global tectonics. *J. Geophys. Res.*, **75**, 1397–1419.

Minsaas, O. and Sturt, B. A. this volume. The Ordovician-Silurian clastic sequence overlying the Lyngen Gabbro Complex and its environmental significance.

Moiseyev, A. N. 1970. Late serpentinite movement in California Coast Ranges—new evidence and its implications. *Bull. geol. Soc. Am.*, **81**, 1721–1732.

Moore, A. C. 1977. The petrography of the Hjelmkona ultramafic (sagvandite), Nordmøre, and its possible regional significance. *Norsk geol. Tidsskr.*, **57**, 55–64.

Moore, A. C. and Hultin, I. 1980. Petrology, mineralogy, and origin of the Feragen ultramafic body, Sør–Trøndelag, Norway. *Norsk geol. Tidsskr.*, **60**, 235–254.

Moore, A. C. and Qvale, H. 1977. Three varieties of alpine type ultramafic rocks in the Norwegian Caledonides and Basal Gneiss Complex. *Lithos*, **10**, 149–161.

Munday, R. J. C. 1974. The geology of the northern half of the Lyngen peninsula, Troms, Norway. *Norsk geol. Tidsskr.*, **54**, 49–62.

Murthy, S. R. 1973. Petrochemistry and origin of the Raudhmaren ultramafites, Jotunheimen. *Norges geol. Unders.*, **300**, 41–52.

Naldrett, A. J. and Cabri, L. J. 1976. Ultramafic and related mafic rocks: their classification and genesis with special references to the concentrations of nickel sulfides and platinum—group elements. *Econ. geol.*, **71**, 1131–1158.

Naterstad, J. 1976. Comment on Lower Palaeozoic inconformity in West Norway. *Am. J. Sci.*, **276**, 394–397.

Nicholson, R. 1971. Faunal provinces and ancient continents in the Scandinavian Caledonides. *Bull. Geol. Soc. Am.*, **82**, 2349–2356.

Nilsen, O. 1973. Petrology of the Hyllingen gabbro complex, Sør–Trøndelag, Norway. *Norsk geol. Tiddskr.*, **53**, 213–231.

Nilsen, O. 1974. Mafic and ultramafic inclusions from the initial (Cambrian?) volcanism in the central Trondheim Region, Norway. *Norsk geol. Tidsskr.*, **54**, 337–359.

Oakeshott, G. B. 1968. Diapiric structures in Diablo Range, California. *Bull. Am. Assoc. Petrol. Geol. Mem.*, **8**, 228–243.

Oftedahl, C. 1969. Caledonian pyroclastic (?) serpentinite in central Norway. *Geol. Soc. Am. Mem.*, **115**, 305–315.

O'Hara, M. J. and Mercy, E. L. P. 1963. Petrology and petrogenesis of some garnetiferous peridotites. *Trans. Roy. Soc. Edinburgh*, **65**, 251–314.

Ohnmacht, W. 1974. Petrogenesis of carbonate-orthopyroxenites (sagvandites) and related rocks from Troms, northern Norway. *J. Petrol.*, **15**, 303–323.

Oosterom, M. G. 1963. The ultramafites and layered gabbro sequences. *Leidse geol. Meded.*, **28**, 177–296.

Oxburgh, E. R. and Turcotte, D. L. 1970. Thermal structure of island arcs. *Bull. Geol. Soc. Am.*, **81**, 1665–1688.

Penrose Field Conference 1972. Ophiolites. *Geotimes*, **17** (12), 24–25.

Pettersen, K. 1883. Sagvanditt, en ny bergart. *Tromsø Mus. Aarshefter*, **6**, 72–80.

Prestvik, T. 1972 Alpine-type mafic and ultramafic rocks of Leka, Nord–Trøndelag. *Norges geol. Unders.*, **272**, 23–34.

Prestvik, T. 1980. The Caledonian ophiolite complex of Leka, north central Norway. In Panayiotou, A. (Ed.), *Ophiolites. Proceedings Int. Ophiolite Symp. Cyprus 1979, Geol. Surv. Dept., Nicosia*, 555–566.

Qvale, H. 1977. Kjemiske variasjoner i chrom-spinell i alpinotype metadunitter fra Baldersheim i Hordaland (in Norwegian with English summary). In Bjørlykke, A. (Ed.), *Kaledonske malmforekomster*, Malmgeologisk symposium, Trondheim 1977, 85–94.

Qvale, H. 1978. *Geologisk undersøkelse av et kaledonsk serpentinittfelt ved baldersheim, Hordaland*, unpubl. cand. real. thesis, Univ. of Oslo, 252. pp.

Qvale, H. 1981. Introduction to Day 5. Røros–Alvdal–Folldal–Lesja. In Qvale, H. and Stigh, J. (Eds), *Ultramafites and Detrital Serpentinites—Caledonian allochthon in Västerbotten, Sweden, and Røros, Norway. Guide to UCS—excursion B13, Preliminary issue*, 76–79.

Qvale, H. 1982. Jotundekkets anothositter: geologi mineralogi og geokjemi. *Norges geol. Unders. Report*, 1560/32, 162 pp. + appendix.

Qvale, H. and Ramberg, I. B. 1981. Day 10. The Sjona Area. In Ramberg, I. B. and Stephens, M. B. (Eds), *The Central Scandinavian Caledonides, Storuman to Mo i Rana. Guide to UCS—excursion A3, Preliminary issue*, 82–89.

Raleigh, C. B. and Paterson, M. S. 1965. Experimental deformation of serpentinite and its tectonic implications. *J. Geophys. Res.*, **70**, 3965–3985.

Ringwood, A. E. 1975. Composition and Petrology of the Earth's Mantle, MacGraw-Hill, New York.

Robins, B. 1975. Ultramafic nodules from Seiland, northern Norway. *Lithos*, **8**, 15–27.

Robins, B. and Gardner, P. N. 1974. Synorogenic layered basic intrusions in the Seiland petrographic province, Finnmark. *Norges geol. Unders.*, **312**, 91–130.

Robins, B. and Gardner, P. N. 1975. The magmatic evolution of the Seiland Province, and Caledonian plate boundaries in northern Norway. *Earth Planet. Sci. Lett.*, **26**, 167–178.

Rui, I. 1972. Geology of the Røros district, south-eastern Trondheim Region, with a special study of the Kjøliskarvene–Holtsjøen area. *Norsk geol. Tidsskr.*, **52**, 1–21.

Schreyer, W., Ohnmacht, W. and Mannchen, J. 1972.

Carbonate-orthopyroxenites (sagvandites) from Troms, northern Norway. *Lithos*, **5**, 345–363.

Skjerlie, F. J. 1974. The Lower Palaeozoic sequence of the Stavfjord district, Sunnfjord. *Norges geol. Unders.*, **302**, 1–32.

Steinmann, G. 1926. Die ophiolitischen Zonen in dem Mediterranean Kettengebirgen. *Comptes Rendus, 14th Int. Geol. Congr. Madrid*, **2**, 638–667.

Stigh, J. 1979. Ultramafites and Detrital Serpentinites in the Central and Southern Parts of the Caledonian Allochthon in Scandinavia, Dept. of Geol., Chalmers Univ. of Technology, Göteborg, Publ. A27, 222 pp.

Stigh, J. 1980. Detrital serpentinites of the Caledonian Allochthon in Scandinavia. In Wones, D. R. (Ed.), *The Caledonides in the USA. Virginia Polytechnic Inst. and State Univ., Dept. Geol. Sci. Mem.*, **2**, 149–156.

Stigh, J. 1981. Content of Ni and Cr in ultramafites of the Caledonian allochthon in Scandinavia and their enrichment in detrital serpentinites, In *An International Symposium on Metallogeny of Mafic and Ultramafic Complexes: The Eastern Mediterranean–Western Asia Area, and its Comparison with Similar Metallogenic Environments in the World. Proceedings*, **2**, 374–389, National Technical University of Athens, Greece.

Stigh, J. and Ronge, B. 1978. Origin and emplacement of two compositional layered ultramafic bodies in the Caledonides of Västerbotten County, Sweden. *Geol. Fören. Stockholm Förh.*, **100**, 317–334.

Stigh, J., Zachrisson, E. and Larkin, S. 1981. Ultramafiter i fjällen; Prospekteringsrapport-analyser. *Sver. geol. Unders., B.R.A.P.*, 124 pp.

Strand, T. 1951. The Sel and Vågå map areas—geology and petrology of a part of the Caledonides of central southern Norway. *Norges geol. Unders.*, **178**, 1–117.

Strand, T. 1970. On the mode of formation of the Otta Serpentinite Conglomerate. *Norsk geol Tidsskr.*, **50**, 393–395.

Stumpfl, E. F. and Sturt, B. A. 1965. A preliminary account of the geochemistry and ore mineral parageneses of some Caledonian basic igneous rocks from Sørøy, northern Norway. *Norges geol. Unders.*, **234**, 196–230.

Sturt, B. A. and Roberts, D. 1978. Caledonides of northernmost Norway (Finnmark). In *IGCP Project 22, Caledonian–Appalachian Orogen in the North Atlantic Region. Geol. Surv. Canada, Pap.*, **78–13**, 17–24.

Sturt, B. A. and Thon, A. 1976. The age of the orogenic deformation in the Swedish Caledonides. Discussion. *Am. J. Sci.*, **276**, 385–390.

Sturt, B. A. and Thon, A. 1978. An ophiolite complex of probable early Caledonian age discovered on Karmøy. *Nature, London*, **275**, 538–539.

Sturt, B. A., Pringle, I. R. and Ramsay, D. M. 1978. The Finnmarkian phase of the Caledonian Orogeny. *J. Geol. Soc. London*, **135**, 597–610.

Sturt, B. A., Speedyman, D. L. and Griffin, W. L. 1980a The Nordre Bumandsfjord Ultramafic Pluton, Seiland, North Norway. Part 1: field relations. *Norges geol. Unders.*, **358**, 1–30.

Sturt, B. A., Thon, A. and Furnes, F. 1980b. The geology and preliminary geochemistry of the Karmøy ophiolite, S.W. Norway. In Panayiotou, A. (Ed.), *Ophiolites. Proceedings Int. Ophiolite Symp. Cyprus 1979, Geol. Surv. Dept. Nicosia*, 538–564.

Svenonius, F. 1883. Om olivinstens- och serpentinitförekomster i Norrland. *Geol. Fören. Stockholm Förh.*, **6**, 342–369.

Sørensen, H. 1967. Metamorphic and metasomatic processes in the formation of ultramafic rocks. In Wyllie, P. J. (Ed.), *Ultramafic and Related Rocks*, Wiley, New York, 204–212.

Thayer, T. P. 1960. Some critical differences between alpine-type and stratiform peridotite–gabbro complexes. *21st Int. Geol. Congr. Copenhagen*, part 13, 247–249.

Thayer, T. P. 1967. Chemical and structural relations of ultramafic and feldspathic rocks in alpine intrusive complexes. In Wyllie, P. J. (Ed.), *Ultramafic and Related Rocks*, Wiley, New York, 222–239.

Trouw, R. A. J. 1973. Structural geology of the Marsfjällen area, Caledonides of Västerbotten, Sweden. *Sver. geol. Unders.*, **C689**, 1–115.

Törnebohm, A. E. 1877. Olivinsten från Kittelsfjäll. *Geol. Fören. Stockholm Förh.*, **3**, 250–252.

Törnebohm, A. E. 1896. Grunddragen af det centrala Skandinaviens bergbyggnad. *Kgl. Svenska Vetensk. akad. Handl.*, **28**(5), 1–212.

van Roermund, H. L. M. and Bakker, E. in press. Structure and metamorphism of the Tangen–Innviken Area, Seve Nappes, Central Scandinavian Caledonides. *Geol. Fören. Stockh. Förh.*

Wilson, J. R., Esbensen, K. H. and Thy, P. 1981. Igneous petrology of the synorogenic Fongen–Hyllingen Layered Basic Complex, south–central Scandinavian Caledonides, *J. Petrol.*, **22**, 584–627.

Wilson, J. R. and Olesen, N. Ø. 1975. The form of the Fongen–Hyllingen gabbro complex, Trondheim region, Norway, *Norsk geol. Tidsskf.*, **55**, 423–439.

Wilson, J. T. 1966. Did the Atlantic close and re-open? *Nature, London*, **211**, 676–681.

Yochelson, E. L. 1963. Gastropoda from the Otta conglomerate. *Norsk geol. Tidsskr.*, **43**, 75–82.

Zachrisson, E. 1969. Caledonian geology of Northern Jämtland–Southern Västerbotten. *Sver. geol. Unders.*, **C644**, 33 pp.

Zachrisson, E. and Stigh, J. 1981. Ultramafiter i fjällen; Prospekteringsrapport. *Sver. geol. Unders., B.R.A.P.*, 55 pp.

Zwart, H. J. 1974. Structure and metamorphism in the Seve–Köli Nappe Complex (Scandinavian Caledonides) and its implications concerning the formation of metamorphic nappes. In *Centenaire de la Société Geologique de Belgique, Géologie des Domaines Cristallins*, Liège, 129–144.

The Caledonide Orogen—Scandinavia and Related Areas
Edited by D. G. Gee and B. A. Sturt

The synorogenic Fongen–Hyllingen layered basic complex, Trondheim region, Norway

J. R. Wilson

Geologisk Institut, Universitetsparken, 8000 Århus C, Denmark

ABSTRACT

The 160 km^2 Fongen–Hyllingen massif, located 60 km southeast of Trondheim, was intruded synorogenically at 5–6 Kb into Fundsjø Group rocks of the Trondheim Nappe Complex during the Scandian orogenic event. Almost 10,000 m of rhythmically layered rocks vary from chrome-spinel bearing dunits to quartz-bearing ferrosyenites. The latter provide a zircon U–Pb age of $426 \pm {8 \atop 2}$ Ma and a Rb–Sr whole rock age of 405 ± 9 Ma, dates which probably record the time of crystallization of the late differentiates and cooling after amphibolite facies regional metamorphism respectively. Extreme fractionation of a tholeiitic basalt under fairly high and increasing pH$_2$O conditions produced mineral fractionation trends intermediate between those of calc-alkaline and tholeiitic affinities. Periodic magma addition is indicated by reversals in mineral fractionation trends with stratigraphic height. Strongly discordant relations between rhythmic and phase/cryptic layering indicate *in situ* crystallization for the origin of the complex.

Introduction

Synorogenic layered basic intrusions in the Scandinavian Caledonides occur extensively in the Finnmark Caledonides, but are less widely developed in the younger parts of the fold belt. The 160 km^2 Fongen–Hyllingen complex is the largest layered intrusion in the central Scandinavian Caledonides and provides a valuable marker for the timing of regional structural and metamorphic events during orogeny.

Regional Setting and Age

The complex lies about 60 km southeast of Trondheim, Norway, and is emplaced in rocks of the allochthonous Trondheim Nappe Complex (Fig. 1) which has been divided into three major subsidiary units; from west to east the Støren, Gula, and Meråker Nappes (Wolff 1979; Wolff and Roberts 1980). The Fongen–Hyllingen complex is situated entirely within the Fundsjø Group, stratigraphically the lowest unit of the Meråker Nappe, which consists of metabasic rocks, including pillow lavas and dyke complexes, together with quartz keratophyres and metapelites/semipelites.

The Støren Group, stratigraphically the lowest unit of the Støren Nappe and dominated by basaltic greenstones of ocean floor tholeiite character (Gale and Roberts 1974), and the Fundsjø Group, which is mainly of island arc affinity (Furnes *et al.* 1980), are proposed to have been emplaced above the Gula Group during the main obduction of ocean ophiolites in pre-Middle Ordovician (probably pre-Middle Arenig) times (Furnes *et al.* 1980). For the Fundsjø Group, obduction must have been in post-Tremadocian times since black shales belonging to this unit contain *Dictyonema flabelliforme* at the Nordaunevoll locality (Gee 1981) within the contact aureole of the Fongen–Hyllingen complex.

The Fundsjø Group is overlain to the east by a conglomerate (the Lille Fundsjø Conglomerate with locally derived clasts including greenstones, metagabbros, and quartz keratophyres) marking the base of the Sulåmo Group which in turn passes up into the Kjølhaug Group. The succeeding Slågan Group contains graptolites of Llandoverian age.

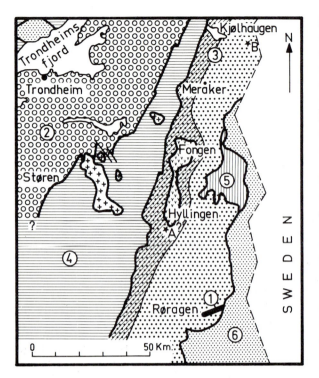

Fig. 1. Map of the Trondheim region, modified after Wolff and Roberts (1980). (1) Røragen Old Red Sandstone deposits; (2) Støren Nappe; (3) Meråker Nappe—with diagonal lines = Fundsjø Group (Hersjø Formation in south (Rui 1972)); (4) Gula Nappe; (5) Øyfjell and Essandsjø Nappes; (6) Autochthonous units. Crosses—trondhjemite/diorite; V ornament—Fongen –Hyllingen complex; A—Dictyonema locality; B—Kjølhaugen (Slågan Group) Lower Silurian graptolite locality

Olesen *et al*. (1973) found that the Gula and Fundsjø Groups and the Sulåmo Group had been through comparable deformation. However, Hardenby *et al*. (1981) and Furnes *et al*. (1980) consider that the Gula and Fundsjø Groups were deformed and metamorphosed prior to deposition of the Sulåmo and younger groups.

The Fongen–Hyllingen complex has a well-developed contact metamorphic aureole in metapelites of the Fundsjø Group and extending into the Sulåmo Group, with cordiere, sillimanite, and andalusite zones, each with almandine garnet, indicating a pressure of 5–6 Kb during intrusion (Olesen *et al*. 1973). The complex contains inclusions of hornfelsed, folded country rocks and cross-cuts at least an early metamorphic foliation, intrusion taking place during an early stage of the main penetrative deformation (D_2 of Olesen *et al*. 1973) (Fig. 2). The complex is itself penetratively deformed within well-defined shear zones, especially in its narrowest part (Figs 3A and 3B).

The peak of regional metamorphism in amphibo-lite facies followed intrusion and evidence from overprinting of the contact metamorphic aureole (e.g. andalusite → kyanite) indicates pressures and temperatures up to about 7 Kb and 600°C respectively. Regional metamorphism of the basic complex was largely dependent on the availability of a hydrous phase and rocks retaining the primary igneous mineralogy are patchily, but widely, distributed throughout most of the complex.

The main penetrative deformation was followed by several phases of folding, the most important of which was the Selbu–Tydal fold phase of Olesen *et al*. (1973). The Fongen–Hyllingen complex is situated partly in the core and partly on the western limb of the Tydal synform and its present form is to a large extent due to this location (Wilson and Olesen 1975).

Syenitic fractionation products in the eastern Hyllingen area grading down into layered ferrogabbro, have yielded a Rb–Sr whole rock age of 405 ± 9 Ma (decay constant $\lambda^{87}Rb = 1.42 \times 10^{-11}a^{-1}$; 9 point isochron; initial ratio 0.7047 ± 0.0002; MSWD 0.28) which Wilson and Pedersen (1981) interpreted as recording the time of magmatic crystallization of these final differentiates. However, rocks from the same locality as the most Rb-rich sample on the isochron provide a zircon U–Pb age of 426 ± $^8_2$ Ma (Wilson *et al*. 1983). While this confirms that intrusion took place in the Silurian, there is an interval of some 20 Ma between the ages. It seems likely that the zircon age records the time of magmatic crystallization, the whole rock system remaining open to Rb and Sr until cooling after regional metamorphism. The explanation for this probably lies in the low temperature of crystallization of these near H_2O-saturate extreme differentiates (Wilson *et al*. 1981a, 1983).

Palaeomagnetic studies of rhythmically layered gabbroic rocks from the Fongen area (Abrahamsen *et al*. 1979) have revealed two stable magnetic components which indicate either a major, late Caledonian geomagnetic excursion, or *c*. 90° rotation of a crustal block containing the Fongen–Hyllingen intrusion, possibly during nappe emplacement.

Field Relations

The field relations of the Fongen–Hyllingen complex have been described by Wilson *et al*. (1981a) in which references to earlier work can be found. A new map of the complex is presented in Figs. 3A and B and only a brief outline of the geology is given here.

The complex is discordant to a major lithological boundary within the Fundsjø Group (Olesen 1974; Fig. 3) and on an outcrop scale the contact is discor-

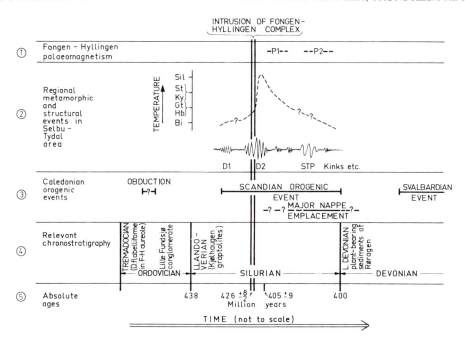

Fig. 2.   Events in the east Trondheim region related to the timing of intrusion of the Fongen–Hyllingen complex. (1) from Abrahamsen *et al.* (1979); (2) from Olesen *et al.* (1973); STP = Selbu–Tydal phase; according to Furnes *et al.* (1980) and Hardenby *et al.* (1981), D_1 should be moved to before deposition of the Lille Fundsjø Conglomerate; (3) from Furnes *et al.* (1980) and Roberts and Sturt (1980). The term 'Scandian' (Gee, 1975) is to be preferred to 'Scandinavian' for the main Caledonian orogenic event (Häggborn, 1982); (4) from Wolff and Roberts (1980); (5) Ordovician–Silurian boundary from McKerrow *et al.* (1980); Silurian–Devonian boundary is 'compromise' age from Gale *et al.* (1980); $426 \pm \frac{8}{2}$ Ma is zircon age (Wilson *et al.* 1983); 405 ± 9 Ma is Rb–Sr whole rock isochron age (Wilson and Pedersen 1981)

dant to early folds. Between the hornfelsed country rocks and the rhythmically layered basic rocks is usually developed a zone of unlayered, broadly gabbroic rocks which are frequently amphibole-bearing and pegmatitic. In the extreme northern part of the complex, rocks dominated by medium-grained olivine gabbronorite and hornblende–plagioclase pegmatite extend some 3 km northwest of Litlefongen, forming a protrusion which is markedly discordant to country rock lithological boundaries. Numerous country rock inclusions are orientated with their long axes NW–SE, parallel to a locally developed steep igneous banding that is strongly discordant to the rhythmic layering in the rocks to the north and east of Litlefongen. It is suggested that these rocks represent a feeder conduit to at least part of the complex. To the east of this presumed feeder occur the stratigraphically lowest layered rocks in the complex, consisting of dunites and troctolites. The rhythmic layering in the northern part of the complex (*ca.* 6200 m thick) is folded by the Selbu–Tydal phase, the fold hinges plunging roughly to the south at 15–50° (profiles 1 and 2 in Fig. 3) and giving marginal overturning and broad open to tight folds (profiles 3 and 4 in Fig. 3). The folds with roughly E–W striking axial surfaces in the Ruten area (profile 2) do not appear to be related to any

regional event. The trace of the 'Transitional Series' in the northern part of the complex is based on phase layering and taken from Wilson *et al.* (1981a).

Individual layers cannot usually be traced along the strike from outcrop to outcrop but in the Ruten area some olivine-rich units, up to 30 m thick, can be followed for up to 3 km. Evidence of way-up is provided by erosional features, sedimentary deformational structures and mineral (density) grading. Magmatic load structures from the Ruten area have been described by Thy and Wilson (1980).

Stratigraphic relationships between the northern and southern parts of the complex have not been established because of extensive penetrative deformation and poor exposures in the intermediate River Nea area. In the Hyllingen area the strike of the layering swings gently from NNE in the north to SE in the south, and youngs consistently to the east. A maximum thickness of *c.* 3600 m is developed and there is a complete upward gradation from layered gabbroic rocks to quartz-bearing ferrosyenites which vein country rock amphibolites in the Jensfjellet area. There are no marginal rocks developed along the eastern border in the Hyllingen area, and there is no evidence of crystallization down from the roof.

Work in progress in the Hyllingen area has revealed that inclusions of large fine to medium

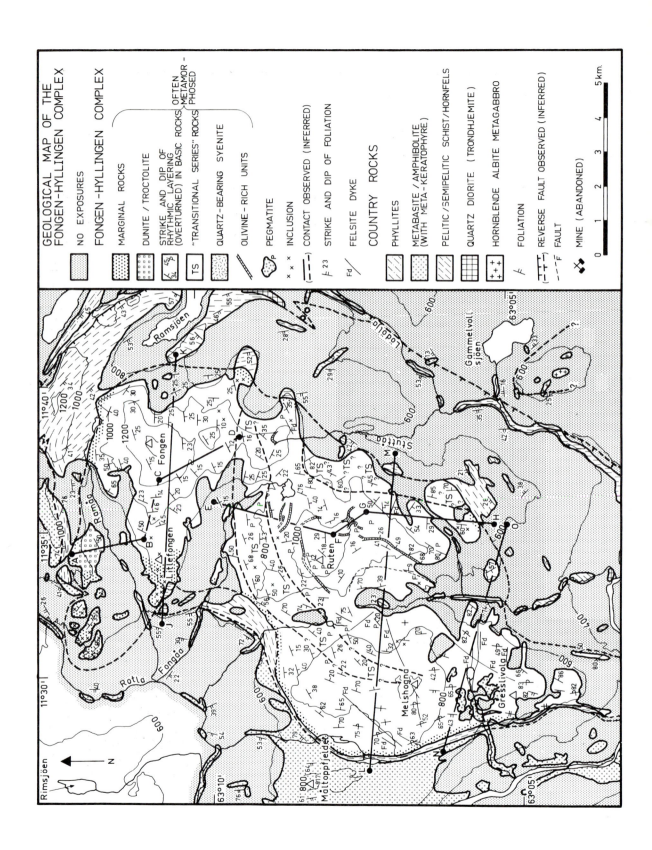

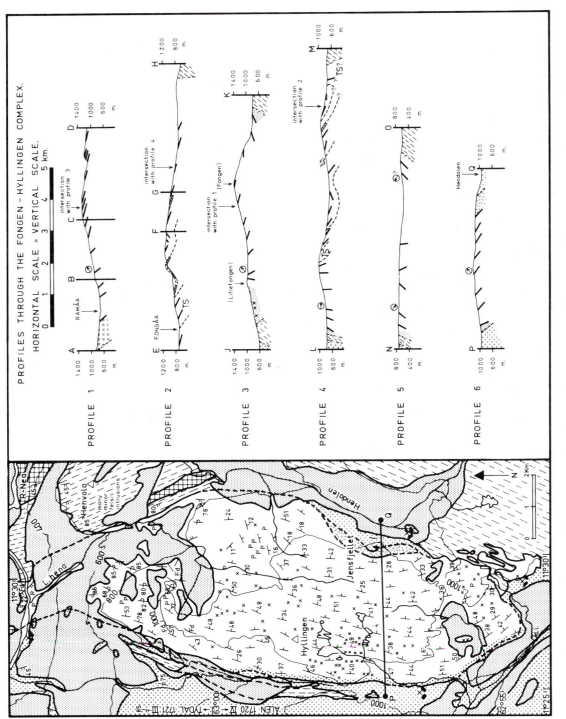

Fig. 3   Geological map of the northern (3A) and southern (3B) parts of the Fongen–Hyllingen complex. Country rock geology in part after Olesen (1974), Kisch (1962), and Nilsen (1973). Fongen–Hyllingen geology based on unpublished work by P. Thy, K. H. Esbensen, S. B. Larsen, J. O. Svane, and J. R. Wilson. Directions of younging are shown by arrows inside circles in the profiles in Fig. 3B

grained basic volcanic hornfels with occasional plagioclase phyric metabasic dykes, clearly similar to the country rocks to the west, are particularly abundant. These inclusions are mostly plate-like bodies which can be over 1 km long and 100 m wide and have their long axes parallel to the rhythmic layering in the gabbroic host. The absence of impact structures related to these large inclusions, which may be more or less in place, is taken to imply that magma emplacement was, at least locally, permissive.

Broadly granitic pegmatites usually occur in or near shear zones and were generated in relation to penetrative deformation. Fine-grained trondhjemitic dykes are chilled against foliated metagabbroic rocks and form distinct features in the massif and in the country rocks.

Mineralogy and Petrology

Wilson *et al.* (1981a) have established a series of cumulate stratigraphic zones based on phase layering. The sequence of entry(+) and exit(−) of cumulus minerals is fairly consistent: olivine(+); Cr-spinel(+); plagioclase(+); Ca-rich pyroxene(+); Ca-poor pyroxene(+); Cr-spinel(−); olivine(−); olivine(+); Fe–Ti oxides(+); calcic amphibole(+); apatite(+); biotite(+); zircon(+); quartz(+); Ca-

poor pyroxene(−); K-feldspar(+); olivine(−); allanite(+); apatite(−).

The major compositional variations of the main solid solution phases (Fig. 4) indicate extreme fractionation. Taking the complex as a whole, olivine varies from Fo_{86} to Fo_0 and there is a hiatus between about Fo_{71} and Fo_{61} in the Fongen layered sequence, although olivines within this range occur elsewhere in the massif (Wilson *et al.* 1981a). Plagioclase ranges from An_{80} to An_1; Ca-rich pyroxenes from $Wo_{45}En_{44}Fs_{11}$ to $Wo_{47}En_0Fs_{53}$; Ca-poor pyroxenes from $Wo_2En_{67}Fs_{31}$ to $Wo_2En_{17}Fs_{81}$; calcic amphiboles range from titanian–pargasite to Mg-free ferroedenite. Crystallization of zircon and allanite in the uppermost c. 400 m and c. 180 m respectively of the Hyllingen sequence (profile IIIC in Fig. 3 of Wilson *et al.* 1981a) indicates exteme concentration of Zr and REE in the fractionating melt, elements which are normally incompatible in basaltic liquids. Enrichment of phosphorus in the liquid led to the crystallization of apatite in the more evolved rocks; this crystallization eventually depleted the liquid in P to such an extent that apatite is absent in the most extreme differentiates.

The mineralogy and fractionation trend (Fig. 4) clearly imply crystallization under fairly high and increasing pH_2O conditions. These features include the crystallization of calcic amphibole over a wide fractionation range; the presence of biotite in the

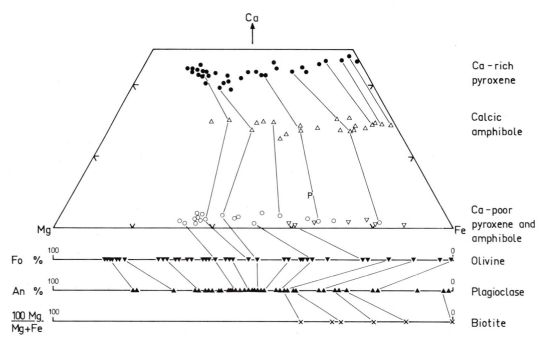

Fig. 4 Compositional variation of the main solid solution magmatic phases from the Fongen–Hyllingen complex based on bulk microprobe analyses. P = first occurrence of inverted pigeonite. The most Fe-rich Ca-poor pyroxenes are rimmed by apparently primary magmatic cummingtonite/grunerite. Based in part on Wilson *et al.* (1981a)

late differentiates; the wide miscibility gap of the coexisting pyroxenes; the late appearance of inverted pigeonite; the crystallization of apparently magmatic cummingtonite/grunerite in the later stages of fractionation; and the coexistence of albite and orthoclase in the final differentiates. Water vapour pressure was not high enough to produce a calc-alkaline fractionation trend since anhydrous mafic phases reach Mg-free compositions unlike, for example, the Guadalupe complex (Best 1963; Best and Mercy 1967). However, pH_2O was considerably higher than in a typical tholeiitic layered intrusion and the fractionation trend can be considered as being, in many respects, intermediate between tholeiitic and calc-alkaline (Wilson et al. 1981b). Crystallization of Fe–Ti oxides probably commenced at a maximum temperature of 1070°C and indicates oxygen fugacity conditions of 10^{-9} bar, close to the NNO buffer curve (Thy 1982).

A tholeiitic parent magma is implied by the precipitation of both Ca-rich and Ca-poor pyroxenes, the quartz-bearing nature of the final differentiates, and the occurrence of a hiatus in olivine crystallization. The absence of large volumes of granitic differentiates indicates that the initial magma was not a calc-alkaline basalt. The compositions of rocks from the presumed feeder (Wilson et al. 1981a) together with the nature of the fractionation products suggest that the parental magma was a high-alumina basalt.

While the highest temperature rocks in the complex occur at the base of the Fongen area and the lowest temperature rocks at the top of the Hyllingen sequence, the intervening layered cumulates reveal many reversals with stratigraphic height (Fig. 5) (Esbensen et al. 1978) which is taken to indicate the repeated influx of fresh magma. Assuming that the successive batches of magma were of constant composition, reversals to irregular compositions which never approach that of the lowest cumulates (these have Fo_{86}; the most Mg-rich olivine at the base of a reversal is Fo_{73} in the Ruten area) cannot be explained by a model involving the expulsion of residual magma as proposed, for example, for the Muskox intrusion (Irvine 1980). The extreme composition of the final differentiates in the Hyllingen area indicates that the magma chamber was closed to the exit of material when the final magma batch crystallized.

A chamber open to the influx of fresh magma and closed to the exit of material was envisaged by Wilson et al. (1981a), successive magma batches mixing with the residual, fractionated liquid from the previous batch; the magnitude of the reversal depends on the relative volumes of the primitive and the evolved batches. The magma chamber probably expanded both vertically upwards and laterally towards the

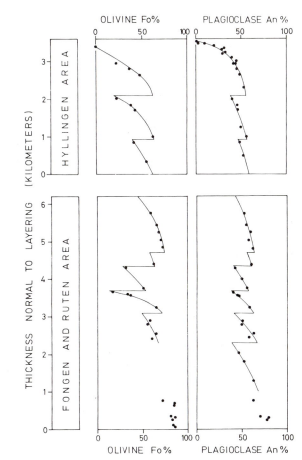

Fig. 5 Olivine and plagioclase compositional variation with stratigraphic height. Olivine Fo_{86}–Fo_0; plagioclase An_{80}–$An_{1.5}$. Reversals reflect the periodic influx of new magma

present south as a result of this repeated influx of large volumes of basaltic magma. While a feeder is preserved which probably provided magma for the early batch(es), the source of the later batches, and three-dimensional shape of the intrusion, are unknown.

It is generally assumed that cryptic and phase layering are concordant with rhythmic layering in layered intrusions. However, in the Hyllingen area, phase and cryptic layering are markedly discordant to rhythmic layering and appear to be more closely related to distance from the walls and/or roof of the intrusion (Wilson and Larsen 1982). For example, using the strike of the rhythmic layering as a reference plane, zircon occurs at a stratigraphic level about 1200 m higher in profile 6 in Fig. 3B than some 2500 m further south. Likewise, the compositions of cumulus plagioclase and olivine become increasingly evolved as the contact is approached, varying along the strike of the rhythmic layering from about An_{50}:Fo_{50} near the centre of profile 6 (Fig. 3B) to about An_{30}:Fo_{10} near the southern margin. Similar features, though much less marked, have been noted in the Skaergaard complex (McBirney and Noyes 1979) and favour an *in situ* crystallization model for the origin of the Fongen–Hyllingen complex. Crystallization of a stratified magma column along an inclined floor is favoured for the Hyllingen sequence by Wilson and Larsen (in press).

Acknowledgements

Financial support for research on the Fongen–Hyllingen complex has been provided by the Carlsberg Foundation and the Danish Natural Science Research Council. I am grateful to Niels Ø. Olesen and other colleagues from the Geology Department, University of Aarhus, for their advice and criticism.

References

Abrahamsen, N., Wilson, J. R., Thy, P., Olesen, N. Ø. and Esbensen, K. H. 1979. Palaeomagnetism of the Fongen–Hyllingen gabbro complex, southern Scandinavian Caledonides; plate rotation or polar shift? *Geophys. J. R. Astr. Soc*., **59**, 231–248.

Best, M. G. 1963. Petrology of the Guadalupe igneous complex, southwestern Sierra Nevada foothills, California. *J. Petrol*., **4**, 223–259.

Best, M. G. and Mercy, E. L. P. 1967. Composition and crystallisation of mafic minerals in the Guadalupe igneous complex, California. *Am. Miner*., **52**, 436–474.

Esbensen, K. H., Thy, P. and Wilson, J. R. 1978. A note on the cumulate stratigraphy of the Fongen–Hyllingen gabbro complex, Trondheim Region, Norway. *Norsk geol. Tidsskr*., **58**, 103–107.

Furnes, H., Roberts, D., Sturt, B. A., Thon, A. and Gale, G. H. 1980. Ophiolite fragments in the Scandinavian Caledonides. *Proc. Internat. Ophiolite Symposium, Cyprus 1979*, 582–600.

Gale, G. H. and Roberts, D. 1974. Trace element geochemistry of Norwegian Lower Paleozoic basic volcanics and its tectonic implications. *Earth Planet. Sci. Lett*., **22**, 380–390.

Gale, N. H., Beckinsale, R. D. and Wadge, A. J. 1980. Discussion of a paper by McKerrow, Lambert and Chamberlain on the Ordovician, Silurian and Devonian time scales. *Earth Planet. Sci. Lett*., **51**, 9–17.

Gee, D. G. 1975. A tectonic model for the central part of the Scandinavian Caledonides. *Am. J. Sci*., **275A**, 468–515.

Gee, D. G. 1981. The Dictyonema-bearing phyllites at Nordaunevoll, eastern Trøndelag, Norway. *Norsk geol. Tidsskr*., **61**, 93–95.

Häggbom, O. 1982. The Scandian episode of the Caledonian diastrophism in Scandinavia: nomenclature. *Geol. Fören. Stockholm Förh*., **104**, 304.

Hardenby, C., Lagerblad, B. and Andreasson, P. G. 1981. Structural development of the northern Trondheim Nappe Complex, Central Scandinavian Caledonides. *Uppsala Caledonides Symposium Abstracts in Terra Cognita*, **1**, 50.

Irvine, T. N. 1980. Magmatic infiltration metasomatism, double-diffusive fractional crystallisation, and adcumulus growth in the Muskox intrusion and other layered intrusions. In Hargraves, R. B. (Ed.), *Physics of Magmatic Processes*, Princeton University Press.

Kisch, H. J. 1962. *Petrographical and geological investigations in the south-western Tydal Region, Sør-Trøndelag, Norway*. Acad. Proefschrift University of Amsterdam.

McBirney, A. R. and Noyes, R. M. 1979. Crystallization and layering of the Skægaard Intrusion. *J. Petrol*., **20**, 487–554.

McKerrow, W. S., Lambert, R. St. J. and Chamberlain, V. E. 1980. The Ordovician, Silurian and Devonian time scales. *Earth Planet. Sci. Lett*., **51**, 1–8.

Nilsen, O. 1973. Petrology of the Hyllingen gabbro complex, Sør-Trøndelag, Norway, *Norsk geol. Tidsskr*., **53**, 213–231.

Olesen, N. Ø. 1974. Geological Map of the Heggest Area, the Trondheim region, University of Leiden.

Olesen, N.Ø., Hansen, E. S., Kristensen, L. H. and Thyrsted, T. 1973. A preliminary account on the geology of the Selbu–Tydal area, the Trondheim region, Central Norwegian Caledonides. *Leidse geol. Meded*., **49**, 259–276.

Roberts, D. and Sturt, B. A. 1980. Caledonian deformation in Norway. *J. Geol. Soc. London*, **137**, 241–250.

Rui, I. 1972. Geology of the Røros district, south-eastern Trondheim region, with a special study of the Kjøliskarvene–Holtsjøen area. *Norsk geol. Tidsskr*., **52**, 1–21.

Thy, P. 1982. Titanomagnetite and ilmenite in the Fongen–Hyllingen basic complex, Norway. *Lithos*, **15**, 1–16.

Thy, P. and Wilson, J. R. 1980. Primary igneous load-cast deformation structures in the Fongen–Hyllingen layered basic intrusion, Trondheim Region, Norway. *Geol. Mag*., **117**, 363–371.

Wilson, J. R., Esbensen, K. H. and Thy, P. 1981a. Igneous petrology of the synorogenic Fongen–Hyllingen layered basic complex, South-Central Scandinavian Caledonides. *J. Petrol*., **22**, 584–627.

Wilson, J. R., Esbensen, K. H. and Thy, P. 1981b. A new pyroxene fractionation trend from a layered basic intrusion. *Nature, London*, **290**, 325–326.

Wilson, J. R., Hansen, B. T. and Pedersen, S. 1983. Zircon U–Pb evidence for the age of the Fongen–Hyllingen complex, Trondheim region, Norway. *Geol. Fören. Stockholm Förh*. **105**, 68–70.

Wilson, J. R. and Larsen, S. B. 1982. Discordant layering relations in the Fongen–Hyllingen basic intrusion. *Nature, London* **299**, 625–626.

Wilson, J. R. and Larsen, S. B. in press. Evolution of a magma chamber: evidence from the Fongen–Hyllingen layered mafic intrusion, Norway. *Geol. Mag*.

Wilson, J. R. and Olesen, N. Ø. 1975. The form of the Fongen–Hyllingen gabbro complex, Trondheim Region, Norway. *Norsk geol. Tidsskr*., **55**, 423–439.

Wilson, J. R. and Pedersen, S. 1981. The age of the synorogenic Fongen–Hyllingen complex, Trondheim region, Norway. *Geol. Fören. i Stockholm Förh*., **103**, 429–435.

Wolff, F. C. 1979. Beskrivelse til de berggrunnsgeologiske kartbladene Trondheim og Østersund 1:250 000. *Norges geol. Unders*., **353**, 77 pp.

Wolff, F. C. and Roberts, D. 1980. Geology of the Trondheim region. In *Excursions Across Part of the Trondheim Region, Central Norwegian Caledonides. Norges geol. Unders*., **356**, 117–167.

The Caledonide Orogen—Scandinavia and Related Areas
Edited by D. G. Gee and B. A. Sturt
© 1985 John Wiley & Sons Ltd

The Artfjället gabbro and its bearing on the evolution of the Storfjället Nappe, central Swedish Caledonides

A. Senior and **M. T. Otten**

Vakgroep PMKGB, Instituut voor Aardwetenschappen, Rijksuniversiteit Utrecht, Postbus 80021, 3508 TA Utrecht, The Netherlands

ABSTRACT

The Artfjället gabbro is one of a series of synorogenic plutons intruded into the Storfjället nappe. The Artfjället gabbro is the most basic of these, containing olivine gabbro and norite. A contact aureole of calc-silicate and garnet–kyanite hornfels was developed around the gabbro. The contact metamorphism overprinted the beginning of regional amphibolite-facies metamorphism, which occurred after D_1. Folds in calc-silicate xenoliths within the gabbro indicate that the gabbro intruded between D_{2a} (folding event) and D_{2b} (flattening event).

Later metamorphism, in part contemporaneous with D_3 and mainly of greenschist facies, with concomitant intrusion of trondhjemitic and granitic dykes and sills, affected the margins and eastern part of the gabbro.

The Artfjället gabbro is suggested to have been intruded in a convergent plate boundary setting.

Introduction

In the Storfjället nappe (Central Swedish Caledonides) a series of synorogenic intrusions occurs, comprising olivine gabbro, norite, diorite and granite. Of these the Artfjället is the most basic and contains the largest range of differentiation products in a single intrusion.

A study of the gabbro enables the time relationships between igneous, metamorphic, and tectonic processes to be established. Its geochemistry indicates the geotectonic setting of the Krutfjellet part of the Storfjället nappe in the developing Caledonian orogeny.

Geological Setting

The Artfjället gabbro occurs in northeastern Artfjället (66°N, 15°E). Structurally it is situated in the Storfjället nappe, the highest unit in the Köli nappe complex, according to Stephens (1980). The Storfjället nappe has been subdivided by Häggbom (1980) into the high-grade Krutfjellet nappe, in which the Artfjället gabbro occurs, and the overlying low-grade Jofjället nappe. The high-grade Krutfjellet nappe consists of a string of lenses (Häggbom 1980), of which the Norra Storfjället lens comprises the area studied.

The lens consists of metasedimentary and metavolcanic rocks, which vary in metamorphic grade from lower- to upper-amphibolite facies. Stephens and Senior (1981) postulate an island-arc origin for these rocks. Locally they have been intruded by gabbroic and granitic plutons. Trondhjemite and granite sills, dykes, and veins intrude the gabbro and country rocks.

At least four deformation phases have affected the nappe, the first two (D_1) and (D_2) comprising isoclinal folds and the regional foliation (S_R). D_2 may be divided into a folding event (D_{2a}) and a flattening event (D_{2b}). D_3 and D_4 comprise tight and open structures, which fold the S_R. D_{2b} and D_3 are probably contemporaneous with thrusting.

Regional structure

Only major D_3 and D_4 structures have been recognized. Earlier isoclinal structures are however suggested by the repetition of rock units, especially in the area near the Telemast (Fig. 1). In the northern and eastern part of the area studied a major D_4 synform is present (Fig. 1), whose axial trace follows the lake Överuman, striking inland near Smilanäset. Towards the south it trends north–south and becomes more open until it disappears near the gabbro sill.

The area around the Telemast is dominated by a large D_3 synform and antiform (Fig. 1). These are fairly open, trend northwest–southeast and plunge gently northwest.

In the southern and western part of the area no major structures have been recognized. Generally the regional foliation defines a dome shape around the gabbro.

A shear zone occurs along the eastern margin of the gabbro. Both gabbro and country rocks have been deformed strongly here, producing an east-plunging stretching lineation, steepening of the regional foliation towards the gabbro, thinning and disappearing of rock units and irregular folds. The gabbro sill was disconnected from the gabbro with a displacement of a few hundred metres. Metamorphism accompanied the deformation, as shown by the development of flasered and foliated metagabbros and coating of shear planes by chlorite and talc. From the occurrence of calc-silicate re-entrants it is inferred that shearing took place mainly along the margin of the gabbro. The shearing is probably of D_3 origin.

Regional metamorphic rocks

In the area studied the rocks of the Krutfjellet unit comprise predominantly garnet–mica schist, graphite schist, layered cal-silicate, amphibolite and migmatitic gneiss. In the metapelitic rocks garnet, kyanite and staurolite are abundant and in the migmatitic gneiss sillimanite also occurs. The regional metamorphism has a polyphase character, but due to the strong deformation around the gabbro, retrograde metamorphism obscures the earlier parageneses. It is not yet possible to establish the relative timing of the deformation and metamorphism in the area studied.

The Artfjället Gabbro

This gabbro (described as norite, type Artfjäll by Backlund and Quensel (1929) and as gabbro–norite complex by Budding (1951)) forms an elliptical body, outcropping over some 20 km². A gabbro sill, which crosses lake Överuman, is found at its southeastern side. Except for the sheared eastern margin, all contacts with country rocks are primary, as shown by the presence of a contact aureole.

Stratigraphy

Four major zones have been recognized in the gabbro, designated Basal Zone (BZ), Lower Zone (LZ), Middle Zone (MZ), and Upper Zone (UZ).

The *Basal Zone* is confined to the eastern part of the gabbro (Fig. 1) and consists of olivine gabbro and norite, which have been extensively affected by later metamorphism. The gabbro sill is considered to form an extension of the BZ, although other zones may be present as well, but are not recognizable because of strong tectonization and concomitant metamorphism. The grain size in the BZ is highly variable, ranging from fine to very coarse (pegmatoid). The reversal in differentiation at the boundary between the BZ and LZ, as brought out by the mineral chemistry, indicates that the BZ probably represents an early pulse of more differentiated magma. Similar zones have been found in the Seiland Province (Robins and Gardner 1974).

The *Lower Zone* is also confined to the eastern part of the gabbro. It consists mainly of picrite and olivine-rich gabbro (olivine more than 30 vol per cent). The picrite consists of medium-grained granular olivine embedded in coarse poikilitic plagioclase and augite. Chrome-spinel is a common accessory. The LZ rocks display a pronounced brown weathering colour.

The *Middle Zone* occupies more than half of the exposed part of the gabbro. It is similar to the BZ and consists of olivine gabbro (olivine less than 20 vol per cent) with minor norite. Grain-size variations are in general irregular, but locally define a layering. Pegmatoid gabbro is relatively abundant.

The *Upper Zone* consists mainly of norite. Two separate units are recognized, Upper Zone A (UZA) and B (UZB) (Fig. 1), which may represent parts of the same zone, now dissected by erosion, although a strong difference in texture exists between the two parts. In the UZA the grain size varies from medium to very coarse (pegmatoid) and the norite is subophitic. The rocks in the UZB are generally fine grained and granular and are texturally similar to the dolerites found in the gabbro.

The boundaries between the zones are usually transitional and rather difficult to define.

Layering

The layering in the gabbro occurs mainly in the UZA and in the MZ, in the vicinity of the UZA (Fig. 1). The layering is defined by grain size and/or modal variations (Fig. 2). The layers occur isolated or in units and disappear rapidly along strike. Cross-cutting relationships and other sedimentary-looking structures are sometimes found. Igneous lamination is extremely rare in the gabbro.

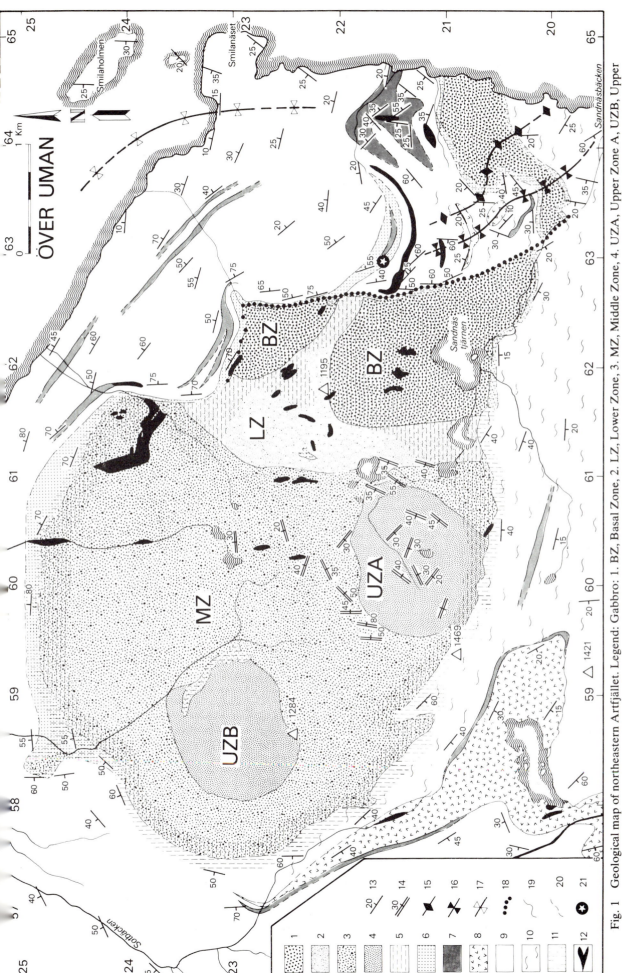

Fig. 1 Geological map of northeastern Artfjället. Legend: Gabbro: 1. BZ, Basal Zone, 2. LZ, Lower Zone, 3. MZ, Middle Zone, 4. UZA, Upper Zone A, UZB, Upper Zone B, 5. Metagabbro. Regional metamorphic rocks: 6. Calc-silicate, 7. Graphite schist, 8. Metavolcanite, 9. Garnet–mica schist and gneiss, 10. Garnet–mica schist and gneiss, migmatitic, 11. Contact hornfels, 12. Leucocratic dykes and sills, 13. Dip and strike of regional foliation, 14. Dip and strike of magmatic layering, 15. D₃ antiform, 16. D₃ synform, 17. D₄ synform, 18. Shear zone, 19. Geological boundary, observed, 20. Geological boundary, inferred, 21. Telemast

Fig. 2. Typical magmatic layering in the Artfjället gabbro, defined by strong grain-size variations. Note irregular appearance of the layers

The layering was not necessarily horizontal when it formed, especially in scour structures, where it may intersect at angles up to 30°. Nevertheless the presence of a synform with moderate plunge to the south is inferred in the UZA and surroundings (Fig. 1). This synform may be a D_3 feature.

Internal structure

The internal structure of the gabbro was deduced from the distribution of the zones and the attitude of the magmatic layering. The structure is interpreted to consist of N–S trending open folds, plunging to both north and south.

Pegmatoid gabbro

Pegmatoid gabbro, i.e. gabbro with a grain size between very coarse (1 cm) and extremely coarse (up to 50 cm), is fairly common in the BZ, MZ and UZA. The mineralogy is similar to that of the enveloping gabbro. The pegmatoid gabbro occurs in patches or zones, the latter often cross-cutting layering. Sometimes multiple zones are present (Fig. 3).

Dolerites

Basic dykes varying in thickness from a few centimetres to several metres, are found throughout the gabbro, but rarely outside. They may be classified as olivine-dolerite, orthopyroxene-dolerite or quartz-dolerite with a few transitional types. In shape they vary from irregular, often branching, to tabular. Concomitant with this variation a decrease in grain size is present. Tabular dolerites sometimes display chilled margins. Cross-cutting relationships between dolerites always show the tabular types to be younger. Often dolerites intrude along zones of pegmatoid gabbro (Fig. 3).

Mineralogically the dolerites are akin to the gabbro. Primary magmatic hornblende and biotite are usually abundant in the dolerites.

The regular decrease in grain size, the shape of the dolerites, the rare occurrence of chilled margins, the relationship with pegmatoid gabbro, the confinement to the gabbro and the mineralogy all indicate a genetic link between the gabbro and the dolerites.

Mineralogy of the gabbro and dolerites

Plagioclase is a ubiquitous constituent of the gabbro. Generally it is rather abundant. In the picrite it is very coarse grained and poikilitic, in all other rocks it is fine to coarse grained and hypidiomorphic to xenomorphic.

The compositional range is An_{75}–An_{55} (mol per cent) (Fig. 4), most rocks having An_{65}–An_{60}. Zoning is usually present, but the change in composition is

Fig. 3. Dolerite intruded along a zone of pegmatoid gabbro. In this area the magmatic layering is approximately horizontal

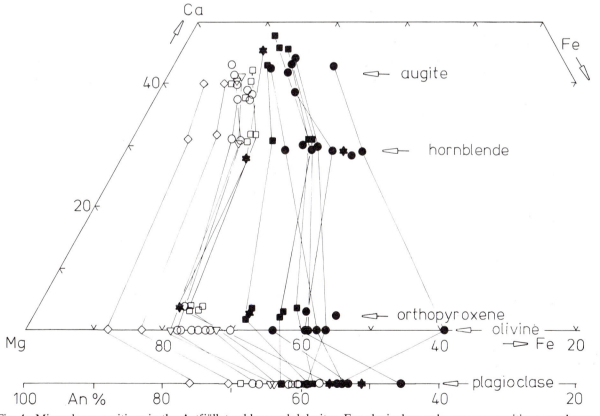

Fig. 4. Mineral compositions in the Artfjället gabbro and dolerites. For plagioclase only core compositions are shown. The hornblende in the gabbro is of subsolidus origin. Legend: open symbols denote gabbro: inverted triangle BZ, diamond LZ, circle MZ, and square UZ; solid symbols indicate dolerites: circle olivine-dolerite, square orthopyroxene-dolerite, and star quartz-dolerite

generally less than 5 per cent An. The zoning may take the form of simple rims or of complex oscillatory or resorption patterns and may be normal or reversed. Plagioclase in the dolerites varies from An_{75}–An_{18} and is usually more strongly zoned, especially when phenocrystal.

Augite is also ubiquitous, except in a few rare norites in the UZB, which contain abundant orthopyroxene and virtually no augite. Generally augite is poikilitic or interstitial, its grain size varying from fine to extremely coarse, independent of the grain size of the other minerals. Its composition varies from $Mg_{55}Fe_5Ca_{40}$ to $Mg_{45}Fe_{15}Ca_{40}$ (Fig. 4), with fairly high Al_2O_3 (4–6 wt per cent). Compositionally the augite is similar to that in calc-alkaline plutonic mafic rocks (Best and Mercy 1967; Arculus and Wills 1980) and to the augite of the early stages of differentiation of the Fongen–Hyllingen intrusion (Thy 1977; Wilson *et al.* 1981a, 1981b). In the dol-

erites the composition of the augite is extended to $Mg_{34}Fe_{23}Ca_{43}$, with Al_2O_3 significantly lower in matrix pyroxene (1.5 wt per cent) than in phenocrysts (4–5 wt per cent).

Olivine occurs abundantly (30–70 vol per cent) in the LZ, in smaller amounts in the BZ and MZ and is rare in the UZA. It is rarely found together with orthopyroxene in the gabbro. In olivine-rich rocks it is granular to hypidiomorphic, with decreasing modal percentages it becomes irregular and amoeboid or even poikilitic.

Its composition varies from Fo_{88} to Fo_{70} (Fig. 4). Olivine in the dolerites is more fayalitic (Fo_{65}–Fo_{40}) than in the gabbro. It is noteworthy that the olivine compositions in the olivine-dolerites are extended to higher Fe/(Mg + Fe) ratios than the orthopyroxene in the orthopyroxene-dolerites, indicating that the latter are not simply related by fractional crystallization to the olivine-dolerites.

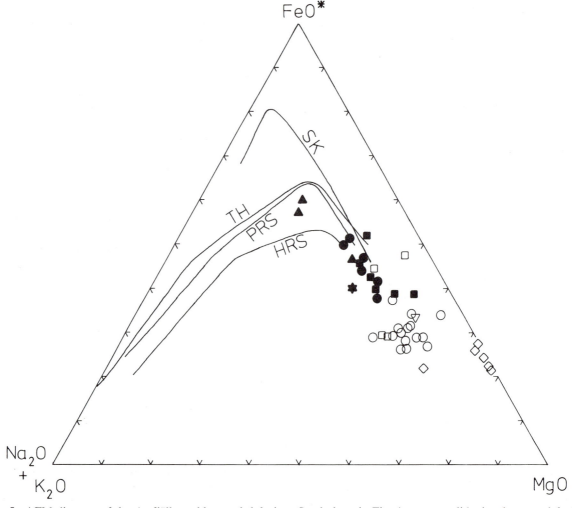

Fig. 5. AFM diagram of the Artfjället gabbro and dolerites. Symbols as in Fig. 4, except solid triangle: metadolerite. Trends shown are for Skaergaard liquid (SK; Wager and Brown 1968), Thingmuli (TH; Carmichael 1964) and Kuno's pigeonitic rock series (PRS) and hypersthenic rock series (HRS; after Gill 1981)

Often olivine is enveloped by a double-rim corona consisting of orthopyroxene and hornblende ± spinel.

Orthopyroxene occurs in the BZ, MZ, and UZ, in the latter often abundantly. It forms fine to very coarse grains, usually hypidiomorphic, but sometimes interstitial or poikilitic. Its composition is $Mg_{75}Fe_{20}Ca_5$ (Fig. 4). Like the augite it is aluminous (2–3 wt per cent Al_2O_3). Orthopyroxene in the dolerites varies from $Mg_{68}Fe_{29}Ca_3$ to $Mg_{45}Fe_{52}Ca_3$ and is also lower in Al_2O_3 (1.5 wt per cent in matrix pyroxene). The phenocrysts in one quartz-dolerite are compositionally identical to the orthopyroxene in the gabbro.

Chrome-spinel is an accessory (1–2 vol per cent) in the picrite, where it forms very fine hypidiomorphic grains, rarely enclosed in olivine. It has $Cr/(Cr + Al)$ 0.2 and $Mg/(Mg + Fe)$ 0.65–0.55.

Ilmenite and *sulphides* (aggregates of pyrrhotite, pentlandite, and chalcopyrite) are common accessories in gabbro and dolerites.

Hornblende is a common constituent of the gabbro, occurring in variable quantities and types. In the dolerites it is usually more abundant, often replacing pyroxenes. The hornblende forms rims around and blebs of lamellae in pyroxenes, constitutes the outer rim of coronas between olivine and plagioclase, and occurs as interstitial and poikilitic grains. In some dolerites coarse-grained poikilitic brown hornblende encloses strongly corroded relics of pyroxenes. Usually the hornblende is a brown variety (kaersutite to titanian pargasite) with Al_2O_3 of 10–15 wt per cent and TiO_2 of 3–5 wt per cent, but especially in the coronas it may grade into or consist wholly of a colourless (low Ti)) amphibole, the latter often symplectitic with spinel. Except for the brown hornblende in the dolerites these amphiboles are of subsolidus origin.

Biotite occurs sparsely in the gabbro and more abundantly in the dolerites, usually originating through alteration of ilmenite or orthopyroxene.

Chemistry of the gabbro and dolerites

The major and trace element geochemistry of the gabbro and the dolerites is discussed by Otten (doctoral thesis, in preparation 1982). The relation between the gabbro and dolerites is supported by their chemistry, although complexities arise with regard to the olivine-rich nature of the LZ picrites and the relatively magnesian UZ orthopyroxenes, which may be the result of magma mixing.

The dolerites define a strongly tholeiitic trend (Fig. 5) and are generally very similar to mid-ocean ridge basalts (MORB), having e.g. TiO_2 contents between 1 and 3 wt per cent. The general increase of SiO_2, K_2O and some related trace elements such as Rb, Ba, Sr, and the light rare earth elements from the olivine-dolerites through the orthopyroxene-dolerites to the quartz-dolerites shows that the dolerites were not simply derived from a common parent by fractional crystallization and the abundance of these elements exceeds those in MORB.

Furthermore, the occurrence of magmatic hornblende and biotite in the dolerites shows that the magma must have had a higher water content than MORB.

Consequently it appears that the dolerites combine some characteristics of MORB and some from volcanic rocks at convergent plate boundaries, pointing to a possible origin in an environment similar to present-day marginal basins (Hart *et al.* 1972; Hawkins 1977; Tarney *et al.* 1977, 1981).

Xenoliths

Many xenoliths are found in the gabbro, with dimensions ranging from a few metres to one kilometre. They consist predominantly of calc-silicate hornfels with only two small gneiss xenoliths which occur near the highest part of the intrusion (UZB). The calc-silicate xenoliths occur mainly in the BZ and LZ, except for a large slab near the UZB, and are randomly orientated. The nature and abundance of these xenoliths suggest that roof subsidence has taken place during solidification of the lower part of the gabbro.

Two generations of folds are found in the xenoliths, the first isoclinal and the second open to tight. These might be correlated with D_1/D_2 and D_3 respectively, but this is unlikely because the gabbro sill lies in a D_3 antiform and synform and the lineation and foliation in the metagabbro are of pre-D_4 origin, as shown by overprinting relationships. Furthermore, leucocratic veins and dykes, in part demonstrably syn-D_3, intrude the gabbro and xenoliths and the accompanying fluids caused low-grade metamorphism in the gabbro. Therefore a correlation with D_1 and D_{2a} is preferred. The latter is believed to have been more open before the flattening event (D_{2b}) took place.

Contact aureole

Except at the sheared eastern margin the gabbro is surrounded by a contact aureole approximately 100 metres thick. At the northern and southeastern contact this consists of calc-silicate hornfels. Garnet–kyanite hornfels is present at the southern and western contacts, where the contact zone is agmatitic and commonly contains calc-silicate blocks. This indicates that also here calc-silicate initially formed the contact, but was disrupted at a later stage.

Fig. 6. Coarse patches and veins of plagioclase and augite developed during contact metamorphism in finely layered calc-silicate. Visible part of hammer is 30 cm long

The calc-silicate hornfels is a conspicuously layered rock, with alternating mafic and felsic layers, which vary from less than 1 mm up to a few centimetres. With increasing grade of contact metamorphism, coarse patches of augite and plagioclase develop, destroying the layering (Fig. 6). Gabbroic veins are often seen near the contact with the gabbro, but it is not clear whether these represent intruded gabbroic material or partial melting of calc-silicate.

The calc-silicate hornfels is strongly affected by retrograde metamorphism, which obscures the original hornfels mineral assemblage. Plagioclase (An_{50}), quartz, augite, orthopyroxene, scapolite and wollastonite have been found.

The garnet–kyanite hornfels is a massive pink rock, which is well exposed along the ridge at the southern contact of the gabbro. Garnet forms large, strongly poikilitic grains, frequently enclosing quartz, biotite, sillimanite, rutile and ilmenite. Both sillimanite and kyanite occur in the hornfels, but sillimanite is either enclosed in garnet or originated by later breakdown of kyanite. The latter forms large porphyroblasts. Plagioclase is almost wholly lacking, indicating that the hornfels has been depleted of a partial melt fraction (K-feldspar is not present in the hornfels or the migmatitic gneiss at northeastern Artfjället). It is not clear what has happened to this melt fraction.

Above the agmatitic zone the contact metamorphism decreases in intensity. The grain size of the garnet and kyanite diminishes and the rocks lose their massive character as biotite increases in abundance and they pass gradually into migmatitic gneisses and schists.

Although the exact relationship between sillimanite and kyanite is not clear, the abundance of randomly orientated kyanite and its coarser grain size as compared to the regional metamorphic kyanite, suggest that kyanite was the stable aluminosilicate mineral during at least part of the contact metamorphism. The pressure implied by contact-metamorphic kyanite (more than 10 Kb) is, however, incompatible with the occurrence of plagioclase in the gabbro and it is probable that sillimanite formed initially (as demonstrated by the fibrous inclusions in garnet) and kyanite was developed at a later, somewhat cooler stage. A second generation of sillimanite replaces the kyanite.

Metamorphism

This is well developed around the margins of the gabbro and in its eastern part. Minor effects are discernible throughout the gabbro, often along cracks or contacts with leucocratic veins and dykes (trondhjemite and granite). Generally greenschist assemblages were formed, comprising green amphibole,

Table 1 History of the Storfjället nappe

Tectonic activity		Metamorphic activity	Igneous activity
D₄	open folding		
D₃	open-tight folding, thrusting	⌐regional retrogradation	trondhjemite and granite sills,
		⌐ gabbro metamorphism	dykes and veins
D₂ᵦ	flattening, thrusting	? -	
D₂ₐ	folding	⌐ regional amphibolite facies	Vilasund granite
		⌐(442 ± 30 Ma, Rb–Sr)	(438 ± 6 Ma, Rb–Sr)
			Artfjället gabbro
			(447 Ma, K–Ar)
D₁	isoclinal folding		

clinozoisite, and albite-rich plagioclase, but locally (higher-grade?) assemblages of brownish-green hornblende, garnet and more calcia plagioclase may precede the greenschist metamorphism.

The greenschist metamorphism of the gabbro is obviously related to addition of fluid phases. As there is a relation between deformation, metamorphism, and intrusion of the leucocratic veins and dykes, it is probable that the fluids accompanied these intrusive rocks.

Summary and Discussion

The time sequence of events as deduced from the area studied, although still not very well constrained, is presented in Table 1. The isotopic ages shown are from Reymer (1979; his 511 Ma K–Ar age of calc-silicate xenolith hornblende has been discarded, as it very probably is due to excess Ar) and from Gee and Wilson (1974).

The history of the Norra Storfjället lens of the Krutfjellet nappe is tentatively interpreted as follows. An island arc was developed by accumulation of sediments and volcanics (Stephens and Senior 1981). During the Ordovician these rocks were transported to a deep crustal level (some 30 km?), possibly by a subduction-related mechanism (Stephens personal communication 1981) and metamorphosed. Subsequent intrusion of the Artfjället gabbro may have occurred at a convergent plate boundary (marginal basin?). Intrusion of granites of somewhat younger age suggests a transition towards a calc-alkaline magma type (island arc or continental margin?).

Acknowledgements

Study of the Artfjället gabbro was financially supported by the Netherlands Organization for the Advancement of Pure Research (Z.W.O.) to A. Senior, and the Molengraaff-Fonds to M. T. Otten.

Electron microprobe analyses were performed at the EPMA laboratory of the Instituut voor Aardwetenschappen, Vrije Universiteit Amsterdam, with financial and personnel support from Z.W.O.–W.A.C.O.M. (Research Group for Analytical Chemistry of Rocks and Minerals). Assistance by Drs. C. Kieft, P. Maaskant and by W. J. Lustenhouwer is gratefully acknowledged. We would like to thank Dr. J. R. Wilson and P. Thy for making available unpublished material on the Fongen–Hyllingen intrusion and Dr. M. Barton for reading the manuscript.

References

Arculus, R. J. and Wills, K. J. A. 1980. The petrology of plutonic blocks and inclusions from the Lesser Antilles island arc. *J. Petrol.*, **21**, 743–799.

Backlund, H. G. and Quensel, P. 1929. Karta över berggrunden inom Västerbottens fjällområde. *Sver. geol. Unders., Ser. Ca*, **21**.

Best, M. G. and Mercy, E. L. P. 1967. Composition and crystallization of mafic minerals in the Guadelupe igneous complex, California. *Am. Mineral.*, **52**, 436–474.

Budding, A. J. 1951. *Geology and petrology of the northeastern Artfjäll, Swedish Lapland.* Academisch Proefschrift, Amsterdam.

Carmichael, I. S. E. 1964. The petrology of Thingmuli, a Tertiary volcano in eastern Iceland. *J. Petrol.*, **5**, 435–460.

Gee, D. G. and Wilson, M. R. 1974. The age of orogenic deformation in the Swedish Caledonides. *Am. J. Sci.*, **274**, 1–9.

Gill, J. B. 1981. Orogenic Andesites and Plate Tectonics, Minerals and Rocks, **16**, Springer, Berlin.

Hart, S. R., Glassley, W. E. and Karig, D. E. 1972. Basalts and sea floor spreading behind the Mariana island arc. *Earth Planet. Sci. Lett.*, **15**, 12–18.

Hawkins, J. W. Jr. 1977. Petrologic and geochemical characteristics of marginal basin basalts. In Talwani, M. and Pitman, W. C. III (eds), *Island Arcs, Deep Sea Trenches and Back-arc Basins*, American Geophysical Union, 355–365.

Häggbom, O. 1980. Polyphase deformation of a discontinuous nappe in the central Scandinavian Caledonides. *Geol. Fören. Stockh. Förhandl.*, **100**, 349–354.

Reymer, A. P. S. 1979. *Investigations into the metamorphic nappes of the central Scandinavian Caledonides.* Academisch Proefschrift, Leiden.

Robins, B. and Gardner, P. M. 1974. Synorogenic layered intrusions in the Seiland Petrographical Province, Finnmark. *Norges geol. Unders.*, **312**, 91–130.

Stephens, M. B. 1980. Occurrence, nature and tectonic significance of volcanic and high-level intrusive rocks within the Swedish Caledonides. In Wones, D. R. (ed.), *The Caledonides in the USA. Virginia Polytechnic Inst. State Univ. Dept. Geol. Sci. Mem.*, **2**, 289–298.

Stephens, M. B. and Senior, A. 1981. The Norra Storfjället lens—an example of fore-arc basin sedimentation and volcanism in the Scandinavian Caledonides. *Terra cognita*, **1**, 76–77.

Tarney, J., Saunders, A. D. and Weaver, S. D. 1977. Geochemistry of volcanic rocks from the island arcs and marginal basins of the Scotia arc region. In Talwani, M. and Pitman, W. C. III (eds), *Island Arcs, Deep Sea Trenches and Back-arc Basins*, American Geophysical Union, 367–377.

Tarney, J., Saunders, A. D., Matthews, D. P., Wood, D. A. and Marsh, N. G. 1981. Geochemical aspects of back-arc spreading in the Scotia Sea and western Pacific. *Phil. Trans. R. Soc. London*, **A300**, 263–285.

Thy, P. 1977. *En petrografisk og geokemisk undersøgelse af den centrale del (Ruten) af Fongen–Hyllingen gabbro Komplekset, Sør–Trøndelag, Norge.* Specialeafhandling Geol. Inst. Århus Univ. (Denmark) (unp published).

Wager, L. R. and Brown, G. M. 1968. Layered Igneous Rocks, Oliver & Boyd, Edinburgh.

Wilson, J. R., Esbensen, K. H. and Thy, P. 1981a. A new pyroxene fractionation trend from a layered basic intrusion. *Nature, London*, **290**, 325–326.

Wilson, J. R., Esbensen, K. H. and Thy, P. 1981b. Igneous petrology of the synorogenic Fongen–Hyllingen layered basic complex, South–Central Scandinavian Caledonides. *J. Petrol.*, **22**, 584–627.

The Caledonide Orogen—Scandinavia and Related Areas
Edited by D. G. Gee and B. A. Sturt

Origin of trondhjemite in relation to Appalachian–Caledonide palaeotectonic settings

W. B. Size

Department of Geology, Emory University, Atlanta, Georgia 30322, U.S.A.

ABSTRACT

Trondhjemite magmatism is frequently related to continental margin tectonics. However, melts of such low K_2O content can originate from polygenic paths as evidenced by their geochemistry, associated rocks, and structural setting in the Appalachian–Caledonide orogen. Trondhjemites commonly have a spatial relationship with basalts (low-potassium tholeiites, greenstones, or amphibolites), copper sulphide mineralization, and major structural features, such as nappes and overthrusts, indicative of subduction/obduction zones. Such settings place constraints on trondhjemite petrogenesis and include: (1) incipient anatexis (metatexis) of basaltic rocks in orogenic zones (Trondheim region, Norway), (2) migmatization of high-grade (K-depleted) gneiss and granulite crustal rocks (Whiteside Complex, North Carolina), (3) partial remelting of altered ophiolites, possibly depleted in potassium due to sea water exchange (Bay of Isands Complex, Newfoundland), (4) partial melting of ensimatic sediments (greywacke) from island arc and back-arc sequences (Shelbourne Pluton, Nova Scotia), (5) remobilization and anatexis of tonalitic rocks due to renewed tectonomagmatic activity (Elkahatchee, Almond, and Blakes Ferry Plutons, Alabama). In addition, trondhjemites may originate as late-stage immiscible liquids from mid-ocean ridge basalts or from fractional crystallization in a gabbro–tonalite–trondhjemite suite (Uusikaupunki–Kalanti Complex, Finland). Trondhjemitic rocks can apparently form in a series of stages starting with plagiogranite in ophiolite sequences. Such rocks are characterized by very low K_2O content, about 0.5%, relatively low Al_2O_3 values, usually less than 15%, and LREE depletion with a negative Eu-anomaly. Second stage, anatectic trondhjemitic rocks in orogenic zones generally have a higher K_2O content, about 1.5%, slightly higher Al_2O_3 values, ranging from 14–16%, and a LREE-enriched fractionated pattern with little or no Eu-anomaly. Third stage, trondhjemite gneisses can have higher K_2O content, usually greater than 2%, but having more variation, higher Al_2O_3 values, greater than 15% and a highly fractionated LREE enrichment pattrn with a positive Eu-anomaly.

Introduction and Trondhjemitic Petrogenetic Models

Rocks of trondhjemitic composition were important components during early stages of crustal margin accretion and irreversible differentiation. However the environment which generated abundant Archean trondhjemite grey-gneiss may not have analogues in the Appalachian–Caledonide orogen due to increased thickness of continental crust (Barker *et al.* 1976). In younger orogens trondhje-mite occurs as small intrusive bodies contained within allochthons and as localized differentiates in ophiolite sequences. Here low isotopic ratios indicate that their immediate precursors were predominantly juvenile additions to the crust and intermediate stages, if any, had short residence times.

The radiometric age of ophiolite-bound trondhjemite serves as a lower age limit for obduction of allochthons onto the continental margin. Relatively younger and undeformed trondhjemite contained within allochthons date the upper age limit of major deformation.

General characteristics of trondhjemite are given by Barker (1979) and a discussion of the Trondheim type-area is given by Size (1979). Trondhjemite although following the general calc-alkaline trend shows a strong Na-enrichment. Trondhjemite can originate from polygenic paths with contributions coming from both chemical and mechanical processes. During partial melting a significant portion of the residue may accompany the anatectic melt as autoliths which may disaggregate or melt adiabatically as the melt rises, thus tending to erase evidence of crystal–magma mixing. Therefore even the most reliable group of elements used in the study of trondhjemite petrogenesis, such as REE and trace elements that are relatively immobile, are of questionable usefulness if autocrysts were mechanically added or subtracted randomly in the melt (i.e. Eu and Sr in plagioclase or Rb and Ba in biotite).

The model that best fits the data for the type area trondhjemite is partial melting of a low-K_2O tholeiitic basalt during anatexis in an orogenic zone (Size 1979). This is based on: (1) the common association of trondhjemite with basalt, gabbro and amphibolite of ocean floor affinities, (2) the restriction of trondhjemite to allochthonous units, (3) low initial $^{87}Sr/^{87}Sr$ ratios for trondhjemite, i.e. close to that of a basaltic parent, (4) oxygen isotope values similar to basalt, (5) results of partial melting experiments on low-K_2O basalts in temperature and pressure ranges associated with subduction zones and regional metamorphism, and (6) trace element and REE-analyses partition coefficients that support a partial melting model for trondhjemite.

Results of melting experiments on low-K_2O basalts (Helz 1976) show that corundum-normative, trondhjemite melt is produced at about 825°C (P_{H_2O} = 5 Kb). In the K_2O–SiO_2 diagram (Fig. 1) the composition melting curve for Picture Gorge tholeiite (solid triangle) used by Helz is plotted between 700–930°C. In Fig. 1 the first melt is richer in K_2O (i.e. granitic) which decreases at higher temperatures. The pattern for natural trondhjemite compositions is concentrated about the melting curve and centred in the 825°C temperature range. Complete basalt melting compositions are graphed in Fig. 2 and show that primary melt at 700°C is essentially a granite (9.8% melt) while at 825°C the melt is trondhjemitic in composition. For comparison the average composition of the Follstad type-area trondhjemite and the average Støren gabbro composition are shown.

Also of interest are liquid immiscibility experiments of Dixon and Rutherford (1979) using a mid-ocean ridge basalt (solid square in Fig. 1). At

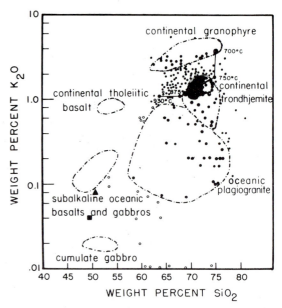

Fig. 1   Semilog plot of SiO_2 vs. K_2O for granophyre, trondhjemite, plagiogranite, and related rocks (Size 1979; Barker 1979). Compositional fields after Coleman (1977). Dots = trondhjemite; open circles = plagiogranite; solid triangle = Picture Gorge tholeiite starting material in partial melting experiment (Helz 1976) from 700–930°C as shown by line connecting larger dots. Solid square = MORB used by Dixon and Rutherford (1979) as starting material for liquid immiscibility experiment. Shaded area represents the compositional field for the Follstad, Norway type-area trondhjemite.

1010°C, after 95% crystallization the remaining magma splits into a ferrobasalt and a trondhjemite melt. Composition of the latter falls along the melting curve of Helz between 825–875°C (Fig. 1).

Trondhjemite–plagiogranite Association in Ophiolite–greenstone Assemblages

Occurring within, and genetically related to, ophiolite slices and greenstones are small masses of both pre-tectonic oceanic plagiogranite and syn- to late-tectonic trondhjemite (Saunders *et al.* 1979; Brown *et al.* 1979). Discovery of ophiolite fragments intercalated with previously identified allochthonous ocean floor tholeiites (greenstones) in the Norwegian Caledonides (Furnes *et al.* 1980) has strengthened the earlier view of Gale and Roberts (1974) that these rocks represent oceanic lithosphere obducted eastward onto the Baltic continental margin during early Palaeozoic time. Examples include the Karmøy ophiolite (Sturt *et al.* 1979), Trondheim–Løkken ophiolite (Furnes *et al.* 1980) and Leka ophiolite (Prestvik 1972).

Timing and petrogenesis of these leucocratic plutonic rocks and their volcanic correlatives (quartz

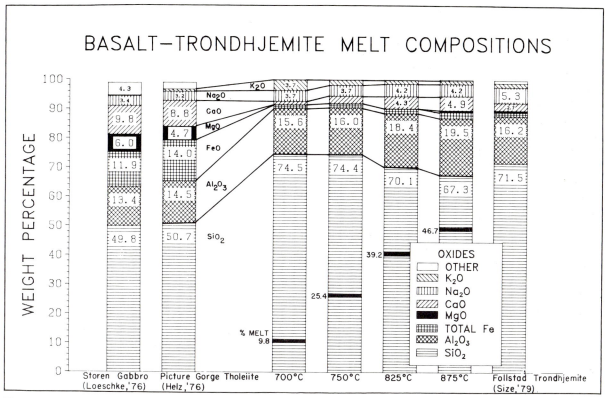

Fig. 2   Bar graph showing partial melt compositions for Picture Gorge tholeiite between 700–875°C (after Helz 1976).
Follstad, Norway type-area trondhjemite and Støren gabbro shown for comparison

keratophyres and dacites) is related to the formation of oceanic crust at spreading centres, plate convergence, and ocean floor subduction and island arc magmatism; but not apparently to any ensuing ensialic, within plate, magmatism. Plagiogranites and trondhjemites can be used to correlate magmatic sequences to timing of tectonic sequences such as in the Appalachian–Caledonide orogen. However, since they commonly occur together and outwardly resemble one another as leucocratic, mainly quartz–plagioclase plutonic rocks, differentiating them requires closer inspection (Table 1).

Oceanic plagiogranites are directly related to primary crystallization processes in ophiolites and are most commonly concentrated in a zone above the cumulate gabbro and below the plated-downward roof and sheeted dykes. For example the Sarmiento ophiolite in southern Chile (Saunders et al. 1979) includes both trondhjemite and plagiogranite, which together with granophyre, make up approximately 10% of the complex. Plagiogranite forms an almost continuous layer just below the sheeted dyke complex. It contains only about 0.01 wt% K_2O which is extremely low, even for plagiogranites, and may indicate that hydrothermal activity at the spreading centre leached K (plus Rb and Ba) shortly after, or during their forma-

tion. The Sarmiento plagiogranite is also associated with ferrogabbros and ferrobasalts. This association coincides with the immiscibility experiments on primitive oceanic tholeiite from the Galapagos spreading centre (Dixon and Rutherford 1979). In the Sarmiento complex later stage trondhjemite dykes cut plagiogranite and invade the upper zone of the ophiolite. Analogues in the Appalachian–Caledonides are generally more deformed and metamorphosed thus masking the original features. The ophiolite fragment in the Elmtree inlier of northern New Brunswick (Rast and Stringer 1980) may contain two generations of trondhjemitic rocks concentrated in the gabbro–diabasic dykes sequence (Rast personal communication). The earlier phase may actually be plagiogranite and the later phase, trondhjemite.

In Norway, the position of plagiogranite in the Karmøy ophiolite is above the plutonic zone, into the sheeted dyke complex. It is of interest to note that plagiogranite comprises an estimated 2–3% of the Karmøy ophiolite (Furnes et al. 1980) which is similar to the amount of plagiogranite produced in silicate liquid immiscibility experiments on oceanic lithosphere (Dixon and Rutherford 1979).

In the central Norwegian Caledonides early work by Gale and Roberts (1974) showed that

Table 1 Characteristic properties distinguishing oceanic plagiogranite from anatectic trondhjemite

Characteristics of Oceanic Plagiogranites	Characteristics of Anatectic Trondhjemites
1. Restricted to ophiolites; commonly situated in the 'sandwich zone' above the cumulate gabbros as a layer or as small intrusions into the overlying sheeted dykes.	1. Commonly associated, but not restricted to, the upper zones of ophiolites. Also occur as small intrusions in ocean floor basalts (greenstone, amphibolites, etc.) and as members of gabbro–trondhjemite complexes.
2. Developed by a *primary* process of fractional crystallization or liquid immiscibility during ophiolite crystallization.	2. Developed by a *secondary* process of partial remelting of low-K tholeiitic basalt or perhaps remobilization of other low-K protoliths in orogenic zones (i.e. greywacke or quartz-diorite).
3. Compositionally heterogeneous.	3. Compositionally homogeneous, unless mechanically disturbed (i.e. crystal–melt mixing).
4. Mafic minerals commonly are hornblende and pyroxene.	4. Mafic mineral commonly is biotite.
5. Wide range of An-content in highly-zoned plagioclase. An_{10-60}.	5. Narrow range of An-content in zoned plagioclase. An_{10-30}.
6. Usually less than 15 wt.% Al_2O_3.	6. Usually greater than 15 wt.% Al_2O_3.
7. Extremely low K_2O content (0.01–1.0 wt.%)	7. Low K_2O content (1.0–3.0 wt.%).
8. Plot near the end of the tholeiitic trend on AFM diagram.	8. Plot near the end of the calc-alkaline trend on AFM diagram.
9. Low Rb content (≈ 5 ppm); low Rb/Sr ratio (<0.015).	9. Higher Rb content (20–50 ppm); higher Rb/Sr ratio (>0.015).
10. Slightly fractionated HREE enrichment pattern; negative Eu-anomaly.	10. Highly fractionated LREE enrichment pattern; possible small negative or positive Eu-anomaly.

greenstones represent ocean floor and island arc type tholeiites. More recently, large bodies of gabbro have been identified in the Løkken and Vassfjell areas (Støren Group) representative of ophiolite slices (Grenne *et al*. 1980). Trondhjemites (and plagiogranites?) are also present in these rocks, in the lower Støren Group greenstones, and in the tectonically underlying Gula Group further to the east at the type trondhjemite locality at Follstad (Size 1979). Trondhjemites and/or plagiogranites are also present in the cumulate gabbro and sheeted dyke sequence of the Leka ophiolite (Furnes *et al*. 1980).

Ophiolite fragments within the Løkken and Vassfjell greenstones contain massive sulphide deposits concentrated above the gabbro unit (Grenne *et al*. 1980). Coleman (1977) showed that Cu is depleted in both the gabbro (1–14 ppm Cu) and the plagiogranite (1 ppm) members of ophiolites compared to that in average tholeiitic basalts (127 ppm). This suggests a post-igneous hydrothermal removal of Cu (also Fe, Mn, and Ni), possibly as a result of low-grade thermal (200°C) alteration. A stockwerk zone commonly underlies massive sulphide bodies suggesting channelways for circulation of ore-forming brines. Such stringers or feeder channels have been identified at the Løkken deposit (Grenne *et al*. 1980). A similar interpretation of metal-rich feeder channels above the gabbro–plagiogranite 'sandwich' zone in oceanic lithosphere has been made for the

spreading zone at the Galapagos Rift (EPRSG 1981).

Plagiogranite and trondhjemite are apparently totally lacking in the later stage ensialic, within plate magmatism in the Scandinavian Caledonides. A crustal origin for rocks produced in this setting is evidenced by a higher $^{87}Sr/^{86}Sr$ initial ratio, such as granodiorite from the West Karmøy igneous complex (0.7210), which has xenoliths of crustal rocks, and a much higher K_2O content (Furnes *et al*. 1980).

Disassociation between Continental Crust and Trondhjemite Petrogenesis

Trondhjemites have characteristics that collectively indicate they had a more primitive source, such as oceanic lithosphere, rather than continental crustal rocks. Of importance is the low initial $^{87}Sr/^{86}Sr$ ratios for trondhjemites in the range of 0.7015–0.7039 for various early Palaeozoic trondhjemites (Peterman 1979) and, more specifically, 0.7049–0.7053 for trondhjemites from the central Norwegian Caledonides (Klingspor and Gee 1981). The Rb/Sr ratios for trondhjemites of Precambrian to Cretaceous age define a tight incremental growth curve indicating a source region with a fairly uniform Rb/Sr ratio through time (i.e. mantle derived) (Peterman 1979). In addition the K_2O content is anomalously low (1–3 wt. per cent) in trondhjemites compared to ensialic, crustally derived rocks. Field evidence also shows that trondhjemites are commonly associated with greenstones, basalts, and gabbros of more oceanic lithosphere derivation. When such rock sequences are translated onto the continental margin the trondhjemites are restricted to the allochthon and are not found invading the tectonically-underlying crustal rocks, which indicates that trondhjemites were spatially and petrogenetically removed from continental crustal interaction.

It seems reasonable to conclude that trondhjemite activity is related to the period of plate convergence and subsequent subduction of oceanic crust and lithosphere. For example, in the Trondheim region the trondhjemites were emplaced into the Støren (ocean floor basalts) and Gula (back-arc sediments?) Groups some time after obduction of the ophiolite fragments in pre-Middle Arenig times (Furnes et al. 1980); but prior to final overthrusting of these allochthonous units onto the continental margin in mid-Silurian times (Roberts 1978).

In the Norwegian Caledonides (see general map this volume) trondhjemite petrogenesis was episodic with three or four periods of activity identified (Size 1979). This may directly relate to the model proposed for this region as having developed in phases of spreading followed by minor crustal thickening events from Ordovician to at least early Silurian time (Grenne and Roberts 1980).

Recent radiometric age dates for trondhjemites from the Trøndelag region (Klingspor and Gee 1981) have yielded early and middle Ordovician ages. These ages place trondhjemite activity after the pre-Middle Arenig obduction of the ophiolite slices in the Støren greenstone onto the continental margin (Furnes et al. 1980) but prior to the post-Llandovery age for final nappe translation onto western Norway (Gee 1978).

A question not yet resolved is that, according to the above ages for the trondhjemites, they would pre-date or be synchronous with the main thermotectonic event in the central Caledonides, given as a minimum age of 438 ± 12 Ma for the last metamorphic phase (Wilson et al. 1973). However many of the trondhjemites in this region, including the type locality of Follstad (Size 1979) show only faint traces of metamorphic and tectonic overprinting compared to the degree of deformation in the enclosing Støren and Gula Group rocks which indicates that the trondhjemites formed late during an earlier thermotectonic event. A possible explanation is that only those trondhjemites nearest to active margins of allochthonous units were deformed while those more in the interior were insulated or buttressed against deformation by the enclosing rocks.

Tectonomagmatic Relationship of Trondhjemite to the Southern Appalachian Orogen

The southern Appalachian orogen consists of Palaeozoic sedimentary rocks contained within imbricate thrust-bounded blocks or nappes. General northwestern movement along the southeastward-dipping thrust planes juxtaposed rocks of varying lithologies, deformation styles, and degrees of metamorphism (Fig. 3) (Thomas et al. 1980). The Brevard Fault zone (Fig. 3) separates high-grade schists, gneisses, and migmatites to the southeast in the Inner Piedmont from lower grade, folded, and faulted metasediments to the northwest. The Brevard fault may have served as a root zone for early to mid-Palaeozoic thrusting in the Blue Ridge (Hatcher and Odom 1980). Major thrusting in this region appears to have been episodic with perhaps three periods of movement. Nappe transgression was initiated during a Taconic metamorphism (Hatcher and Odom 1980). Renewed movement in the Acadian is evidenced from the K/Ar age dates from blastomylonites along fault zones. Final emplacement of the nappe sequence appears to be a post-metamorphic event (Alleghanian?) as evidenced by truncation and juxtaposition of metamorphic isograds (Hatcher 1981).

The major orogenic event in the Southern Appalachians is dated at post-early Pennsylvanian, but was probably active episodically through much of Palaeozoic time (Thomas et al. 1980). The thermal peak of regional metamorphism in the southwestern Blue Ridge is put at pre-Middle to Upper Ordovician while in the Alabama Piedmont peak metamorphism occurred later during Lower to Middle Devonian time (Tull 1978). Absolute age determinations of magmatic rocks provide not only evidence for the polyphasal activity but also constraints for bracketing episodic deformational events.

Trondhjemites and associated rocks are concentrated along a narrow, northeasterly trending tectonic unit just north of the Brevard Zone named the Tallapoosa Block in Alabama (Fig. 4). These rocks also occur northeastward in the Blue Ridge (i.e. Villa Rica trondhjemite, Georgia; Whiteside trondhjemite, North Carolina). The Tallapoosa Block is a unit of the Northern Alabama Piedmont bounded by the Brevard Fault zone to the southeast and the Valley and Ridge province to the northwest. The North Alabama Piedmont contains additional lithotectonic units; the Coosa Block and the Talladega Block towards the northwest. These three units are bounded by thrust faults that have imbricated and cut out lithologies (Tull 1978). The Talladega and Piedmont zones have recently been defined as suspect terranes (Williams and Hatcher 1982).

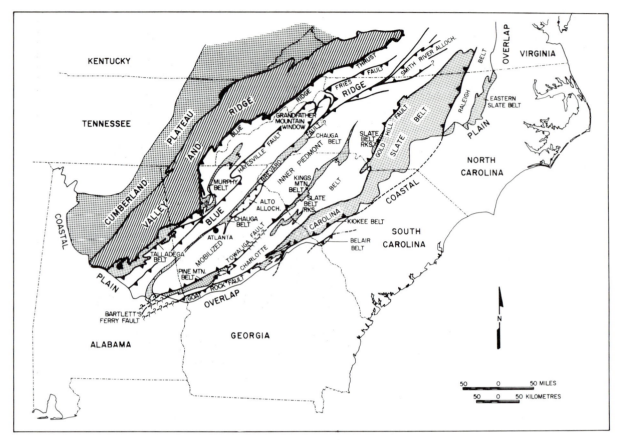

Fig. 3   Map of tectonic subdivisions of Southern Appalachians (Hatcher and Zietz 1980). Trondhjemite occurrences are restricted to the Inner Piedmont and Blue Ridge

Trondhjemite and associated rocks in the Tallapoosa Block include over 20 intrusions ranging in area from over 1000 km^2 for the Elkahatchee Quartz Diorite to less than one kilometre square. Most intrusions within the Tallapoosa Block intrude Wedowee Group rocks, mainly graphitic metapelites (Fig. 4). The age of the Wedowee Group is uncertain but may be late Precambrian.

There is little evidence of Precambrian crust occurring in the region north of the Brevard Zone, except for gneissic clasts of possible Grenville age in the Devonian age Lay Dam Formation in the Talladega terrane (Tull 1978).

The apparent lack of crustal rocks in this region could be accounted for by a present shallow level of erosion. However, it is of interest that initial $^{87}Sr/^{86}Sr$ ratios for the granitic and trondhjemitic rocks in the Southern Appalachian Piedmont are very low. The 325–265 Ma granitic rocks have initial Sr ratios between 0.7024 to 0.7052, which indicate an upper mantle to lower crustal source (Fullagar and Butler 1979).

Trondhjemitic rocks of the Northern Alabama Piedmont are restricted to one structural block or allochthon, the Tallapoosa (Fig. 4). In the southwestern portion of the Tallapoosa Block, the

Eklahatchee Quartz Diorite outcrops over an area greater than 1000 km^2 (65 km long by 20 km wide), and is one of the largest plutons north of the Brevard Zone. It is also one of the oldest plutons in the Northern Alabama Piedmont, and has yielded a U–Pb date from zircons at 516 Ma and 490 ± 26 Ma from whole-rock Rb/Sr. The Elkahatchee Quartz Diorite has a low initial $^{87}Sr/^{86}Sr$ ratio of 0.7036 ± 0.0004 (Russell 1978). The modal mineral averages of the Elkahatchee Quartz Diorite show a high quartz content coupled with an extremely low K-spar content and intermediate plagioclase composition, classifying the rock more as a tonalite or biotite-rich trondhjemite.

	Elkahatchee 'Tonalite' average modal composition	Blakes Ferry trondhjemite average modal composition
quartz	27.5%	26.8%
plagioclase	45.2%	63.0%
	(An$_{35-40}$)	(An$_{5-18}$)
K-feldspar	0.2%	2.5%
biotite	16.0%	3.5%
muscovite	5.0%	1.9%
epidote	5.3%	1.9%
accessories	0.8%	0.9%

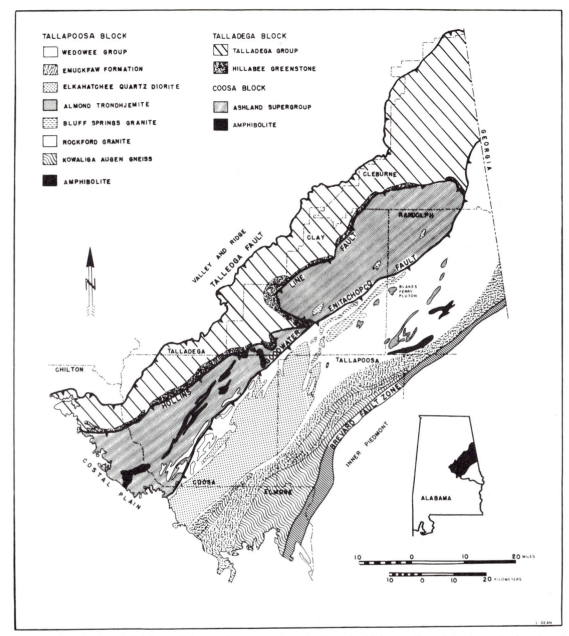

Fig. 4   Map of geologic subdivisions of the southernmost Appalachians in Alabama. Trondhjemite and related rocks are confined to the Tallapoosa block (allochthon)

Further northeastward along structural trend of the Tallapoosa Block, the Elkahatchee Quartz Diorite contains an increasing proportion of trondhjemitic dykes and isolated pods, some with biotite rims (Fig. 5). This grades into a series of over 20 intrusions collectively called the Almond trondhjemites (Fig. 4). They show varying degrees of deformation, with the Blakes Ferry trondhjemite showing the least amount of deformation, and only minor metamorphic overprinting.

Field and petrologic data indicate that the Almond trondhjemite originated by partial anatexis of a tonalitic protolith similar to the Elkahatchee

'tonalite'. Tonalitic xenoliths are abundant in the Blakes Ferry trondhjemite. They vary from angular with little evidence of resorption, to rounded with biotite rims, and finally to ghost outlines with faintly dispersed mafic schlieren (Fig. 5). The composition and texture of the angular xenoliths is comparable to the Elkahatchee Quartz Diorite except for a lower anorthite content in plagioclase and less biotite, which is in accord with a partial anatexis process. The AFM diagram (Fig. 5) shows that the Elkahatchee Quartz Diorite and tonalitic xenoliths in Blakes Ferry trondhjemite plot together in the middle of the gabbro–trondhjemite trend, while

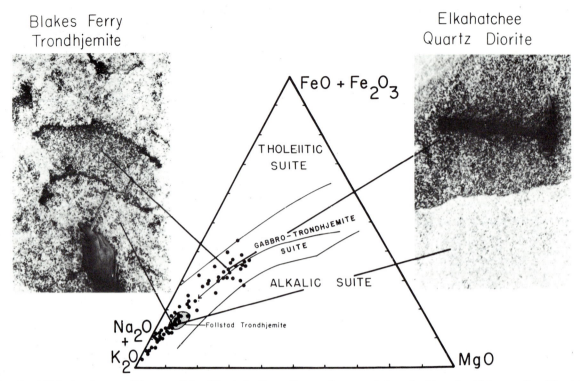

Fig. 5   AFM-plot for tonalitic–trondhjemitic rocks from the southernmost Appalachians in Alabama. See Fig. 4 for location of samples. Gabbro–trondhjemite trend after Barker (1979). Xenolith in Blakes Ferry trondhjemite plots together with suspected Elkahatchee Quartz Diorite protolith. Anatectic melt composition in Blakes Ferry trondhjemite plots together with leucocratic remelted zones in Elkahatchee Quartz Diorite. Follstad, Norway type-area trondhjemite shown for comparison

both Blakes Ferry trondhjemite and leucocratic zones from the Elkahatchee Quartz Diorite plot together at the lower end of the trend.

Additional evidence that a tonalitic rock similar to the Elkahatchee Quartz Diorite was the protolith for later trondhjemite magmatism is shown in the rare earth elements (Fig. 6). The Elkahatchee Quartz Diorite (EQD–1) has the greatest sum of REE (160.58 ppm) and shows a LREE-enriched fractionated pattern with an almost negligible + Eu-anomaly. A tonalite xenolith in Blakes Ferry trondhjemite (BF–19) has the same LREE-enrichment pattern but shifted down due to lower sum of REE (61.34 ppm), and has a slight negative Eu-anomaly. These differences can be accounted for by removal of plagioclase during anatexis (EQD–1) = 56.8% plagioclase; BF–19 has 26.3% plagioclase).

Both plagioclase and biotite in dacite have partition coefficients less than 1.0 for all REE (except Eu in plagioclase) (McCarthy and Hasty 1976; Arth 1979). However in biotite tonalite they would be the main REE-bearing minerals. Therefore partial removal of both plagioclase and biotite during anatexis will affect REE distribution to the point of reducing the sum of rare earths and producing a

negative Eu-anomaly in the restite (i.e. partially resorbed xenoliths). Samples of the Blakes Ferry trondhjemite (BF–24, BF–28) taken away from xenolith borders show that the LREE-enrichment pattern is still present but is lower than both protolith samples (EQD–1 and BF–19) shown by the lower sum of rare earths (BF–24 = 37.34 ppm, BF–28 = 26.24 ppm). The positive Eu-anomaly in these samples indicates a mechanical addition of plagioclase restite in the anatectic melt as evidenced by the high percentage of broken xenocrystic clusters (average 63.0%) in the Blakes Ferry trondhjemite. Disaggregated xenoliths in the Blakes Ferry trondhjemite show that only a limited amount of plagioclase and biotite remelting took place. Rather, these two minerals were mixed in with the anatectic melt as the xenocrysts were liberated and mechanically carried with the magma. This is evidenced in the field where trondhjemite is mineralogically extremely heterogeneous, even over a few centimetres or near ghost outlines of former xenoliths. Therefore essential differences between Elkahatchee Quartz Diorite (tonalitic protolith) and (anatectically derived) Blakes Ferry trondhjemite can be accounted for by partial remelting coupled with a mechanical addition of solid crystal phases.

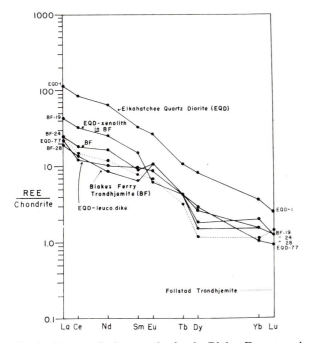

Fig. 6 Rare earth element plot for the Blakes Ferry trondhjemite, xenolith within the trondhjemite, and the Elkahatchee Quartz Diorite suspected protolith. Follstad, Norway type-area trondhjemite trend shown for comparison. REE were analysed by INAA using 0.5 g samples (with standards) irradiated for 1 hour at a thermal flux of 5×10^{12} n cm^{-2} s^{-1}. Samples were cooled for 8 days and counted between 2000–10 000 seconds using Ge(Li) detectors with a resolution of 1.7 kev at 1332 kev

Concluding Remarks

1. Age dates for late stage trondhjemites restricted within allochthonous oceanic crust provide a maximum age estimate for nappe emplacement or obduction onto the continental margin. If present, the ages of ensialic, crustal derived, magmatic rocks that might cut the allochthon and underlying basement would, together with the trondhjemite age, bracket the time for nappe translation onto the continental margin.

2. During partial melting in a subduction zone, where a constant supply of new material is brought in, the melt derived is from the similar, least refractory, portions of the heterogeneous parent. Petrogenetic models based on variation diagrams from such rocks that show evidence of anatexis with mechanical additions and/or subtractions through crystal–magma mixing require special attention to the chemical vs. mechanical controlled trends.

References

Albuquerque, G. A. R. de 1977. Geochemistry of the tonalitic and granitic rocks of the Nova Scotia southern plutons. *Geochim. Cosmochim. Acta*, **41**, 1–13.

Arth, J. G. 1979. Some trace elements in trondhjemites—their implications to magma genesis and paleotectonic setting. In Baker, F. (Ed.), *Trondhjemites, Dacites and Related Rocks*, Elsevier Scientific Publishing Company, Amsterdam, 123–132.

Arth, J. G. and Barker, F. 1976. Rare-earth partitioning between hornblende and dacitic liquids and implications for the genesis of trondhjemitic–tonalitic magmas. *Geology*, **4**, 534–536.

Arth, J. G. and Barker, F. 1978. Geochemistry of the gabbro–tonalitic–trondhjemite suite of southwest Finland and its implications for the origin of tonalitic and trondhjemitic magmas. *J. Petrol.*, **19**, 289–316.

Barker, F. (Ed.), 1979. Trondhjemites, Dacites and Related Rocks, Elsevier Scientific Publishing Company, Amsterdam, 659 p.

Barker, F. and Arth, J. G. 1976. Generation of trondhjemitic–tonalitic liquids and Archean bimodal trondhjemite–basalt suites. *Geology*, **4**, 596–600.

Barker, F., Arth, J. G., Peterson, Z. G. and Friedman, I. 1976. The 1.7–1.8 b.y.-old trondhjemites of southwestern Colorado and northern New Mexico: geochemistry and depth of genesis. *Bull. geol. Soc. Am.*, **87**, 189–198.

Barker, F. and Millard, H. T. Jr. 1979. Geochemistry of the type area trondhjemite and three associated rocks, Norway. In Barker, F. (ed.), *Trondhjemites, Dacites and Related Rocks*, Elsevier Scientific Publishing Company, Amsterdam, 517–530.

Bouseily, A. M. and El Sakkary, A. A. 1975. The relation between Rb, Ba and Sr in granitic rocks. *Chem. Geol.*, **16**, 207–219.

Brown, E. H., Bradshaw, J. Y. and Mustoe, G. E. 1979. Plagiogranite and keratophyre in ophiolite on Fidalgo Island, Washington. *Bull. geol. Soc. Am.*, **90**, 493–507.

Coleman, R. G. 1977. Ophiolites—Ancient Oceanic Lithosphere?, Springer-Verlag, Berlin, 229 p.

Coleman, R. G. and Peterman, Z. G. 1975. Oceanic plagiogranite. *J. Geophys. Res.*, **80**, 1099–1108.

Coleman, R. G. and Donato, M. M. 1979. Oceanic plagiogranite revisited. In Barker, F. (Ed.), *Trondhjemites, Dacites and Related Rocks*, Elsevier Scientific Publishing Company, Amsterdam, 149–168.

DeFant, M. C. 1980. *A Geochemical and Petrogenic Analysis of the Almond and Blakes Ferry Plutons, Randolph County, Alabama*, unpublished M.S. thesis, Univ. of Alabama.

Deininger, R. W. 1975. Granitic rocks in the Northern Alabama Piedmont. In Neathery, T. L. and Tull, J. F. (Eds), *Geologic Profiles of the Northern Alabama Piedmont. Alabama geol. Soc. Guidebook, 13th Ann. Field Trip*, 49–62.

Dixon, S. and Rutherford, M. J. 1979. Plagiogranites as late-stage immiscible liquids in ophiolite and mid-ocean ridge suites: an experimental study. *Earth Planet. Sci. Lett.*, **45**, 45–60.

East Pacific Rise Study Group 1981. Crustal processes of the mid-ocean ridge. *Science*, **213**, 31–40.

Fullagar, P. D. and Butler, J. R. 1979. 325 to 265 M.Y.-old granitic plutons in the Piedmont of the southern Appalachians. *Am. J. Sci.*, **279**, 161–185.

Furnes, H., Roberts, D., Sturt, B. A., Thon, A. and Gale, G. H. 1980. Ophiolite fragments in the Scandinavian Caledonides. In *Ophiolites. Proceedings International Ophiolite Symposium, Cyprus Geol. Surv.*, 582–599.

Gale, G. H. and Roberts, D. 1974. Trace element geochemistry of Norwegian Lower Paleozoic basic volcanics and its tectonic implications. *Earth Planet. Sci. Lett.*, **22**, 380–390.

Gault, H. R. 1945. Petrography, structures, and petrolfabrics of the Pickneyville Quartz Diorite, Alabama. *Bull. Geol. Soc. Am.*, **56**, 181–246.

Gee, D. G. 1978. Nappe displacement in the Scandinavian Caledonides. *Tectonophysics*, **47**, 393–419.

Gee, D. G. 1980. Basement–cover relationship in the central Scandinavian Caledonides. *Geol. Fören. Stockholm Förh.*, **102**, 4, 455–474.

Grenne, T., Grammeltvedt, G. and Vokes, F. M. 1980. Cyprus-type sulphide deposits in the western Trondheim district, central Norwegian Caledonides. In *Ophiolites. Proceedings International Ophiolite Symposium. Cyprus Geol. Surv.*, 727–743.

Grenne, T. and Roberts, D. 1980. Geochemistry and volcanic setting of the Ordovician Forbordfjell and Jonsvatn Greenstones, Trondheim region, central Norwegian Caledonides. *Contrib. mineral. Petrol.*, **74**, 375–386.

Hatcher, R. D. Jr. 1981. Thrusts and nappes in the North American Appalachian Orogen. *J. Geol. Soc. London*, **138**, 491–499.

Hatcher, R. D. Jr. and Odom, A. L. 1980. Timing of thrusting in the southern Appalachians, U.S.A.: model for orogeny. *J. Geol. Soc. London*, **137**, 321–327.

Hatcher, R. D. Jr. and Zietz, I. 1980. Tectonic implications of regional aeromagnetic and gravity data from the Southern Appalachians. In Wones, D. R. (ed.), *The Caledonides in the U.S.A. Dep. of Geol. Sciences, Virginia Polytechnic Institute and State University Mem.*, **2**, 235–244.

Helz, R. T. 1976. Phase relations of basalt in their melting ranges at P_{H_2O} = 5Kb. Part II, Melt compositions. *J. Petrol.*, **17**, 139–193.

Klingspor, I. and Gee, D. G. 1981. Isotopic age-determination studies of the Trøndelag trondhjemites. *Abstract. Terra cognita*, **1**, 55.

Loeschke, J. 1976. Petrochemistry of eugeosynclinal magmatic rocks of the area around Trondheim (central Norwegian Caledonides). *N. Jb. miner. Abh.*, **128**, pt. 1, 41–72.

McCarthy, T. S. and Hasty, R. A. 1976. Trace element distribution patterns and their relationship to the crystallization of granitic melts. *Geochim. Cosmoschim. Acta*, **40**, 1351–1358.

Neathery, T. L. and Tull, J. F. (Eds), 1975. Geologic profiles of the Northern Alabama Piedmont. *Alabama Geol. Soc. Guidebook, 13th Ann. Field Trip*, 173 pp.

Payne, J. G. and D. F. Strong, 1979. Origin of the Twillingate trondhjemite; North–Central Newfoundland: partial melting in the roots of an island arc. In Barker, F. (Ed.), *Trondhjemites, Dacites and Related Rocks*, Elsevier Scientific Publishing Company, Amsterdam, 489–516.

Peterman, Z. E. 1979. Strontium isotope geochemistry of Late Archean to Late Cretaceous tonalites and trondhjemites. In Barker, F. (Ed.), *Trondhjemites, Dacites, and Related Rocks*, Elsevier Scientific Publishing Company, Amsterdam, 133–148.

Piwinski, A. J. 1968. Experimental studies on igneous rock series, Central Sierra Nevada Batholith, California. *J. Geol.*, **76**, 548–570.

Prestvik, T. 1972. Alpine-type mafic and ultramafic rocks of Leka, Nord-Trøndelag. *Norges geol. Unders.*, **273**, 23–34.

Rast, N. and Stringer, P. 1980. A geotraverse across a deformed Ordovician ophiolite and its Silurian cover, northern New Brunswick, Canada. *Tectonophysics*, **69**, 221–245.

Roberts, D. 1978. Caledonides of south central Norway. *Can. Geol. Surv. Pap.*, **78–13**, 31–37.

Russell, G. S. 1978. *U–Pb, Rb–Sr, and K–Ar Isotopic Studies Bearing on the Tectonic Development of the Southernmost Appalachian Orogen, Alabama*, unpubl Ph.D. thesis, Florida State University, 197 pp.

Saunders, A. D., Tarney, S., Stern, D. R. and Dalziel, I. W. D. 1979. Geochemistry of Mesozoic marginal basin floor igneous rocks from southern Chile. *Bull. Geol. Soc. Am.*, **90**, 237–258.

Size, Wm. B. 1979. Petrology, geochemistry and genesis of the type area trondhjemite in the Trondheim region, Central Norwegian Caledonides. *Norges geol. Unders.*, **351**, 51–76.

Streickeisen, A. 1967. Classification and nomenclature of igneous rocks. *N. Jb. miner. Abh.*, **102**, 144–214.

Sturt, A. A., Thon, A. and Furnes, H. 1979. The Karmoy ophiolite, southwest Norway. *Geology*, **7**, 316–320.

Thomas W. A., Tull, J. F., Bearce, D. N., Russell, G. and Odom, A. L. 1980. Geologic synthesis of the southernmost Appalachians, Alabama and Georgia. In Wones, D. R. (Ed.), *The Caledonides in the U.S.A. Department of Geological Sciences, Virginia Polytechnic Institute and State University, Mem.*, **2**, 91–98.

Tull, J. F. 1978. Structural development of the Alabama Piedmont northwest of the Brevard Zone. *Am. J. Sci.*, **278**, 4, 442–460.

Tull, J. F. 1980. Overview of the sequence and timing of deformational events in the southern Appalachians: evidence from the crystalline rocks, North Carolina to Alabama. In Wones, D. R. (Ed.), *The Caledonides in the U.S.A. Dep. Geol. Sci., Virginia Polytechnic Institute and State University, Mem.*, **2**, 161–178.

Tull, J. F. and Stow, S. H. 1980. The Hillabee Greenstone: a mafic volcanic complex in the Appalachian Piedmont of Alabama. *Geol. Soc. Am.*, **91**, 27–36.

Williams, H. and Hatcher, R. D. Jr. 1982. Suspect terranes and accretionary history of the Appalachian orogen. *Geology*, **10**, 530–536.

Wilson, M. R., Roberts, D. and Wolff, F. C. 1973. Age determinations from the Trondheim region, Caledonides, Norway: A preliminary report. *Norges geol. Unders.*, **288**, 53–63.

Wolff, F. C. 1976. Geologisk kart over Norge, berggrunnskart Trondheim 1:250 000. *Norges geol. Unders.*

Wyllie, P. J. 1977. Crustal anatexis: an experimental review. *Tectonophysics*, **43**, 41–71.

The Caledonide Orogen—Scandinavia and Related Areas
Edited by D. G. Gee and B. A. Sturt
© 1985 John Wiley & Sons Ltd

The Fundsjø Group, central Norway—a Lower Palaeozoic island arc sequence: geochemistry and regional implications

T. Grenne[*] and **B. Lagerblad**[†]

Geologisk Institutt, Universitetet i Trondheim, N–7034 Trondheim–NTH, Norway (Present address: Orkla Industrier A/S, N–7332 Løkken Verk, Norway)
†*Geologiska Institutionen, Avd. för mineralogi and petrologi, Lunds Universitet, Sölvegatan 13, S–223 62 Lund, Sweden*

ABSTRACT

The Caledonian Trondheim Nappe Complex contains two major volcanic sequences, the Fundsjø and the Støren Groups. Although geographically separated, the two have usually been correlated and interpreted as occupying the same tectono-stratigraphic position. The pre-Middle Arenig Støren Group in the western Trondheim Region comprises a thick succession of pillowed and massive lavas with intercalated cherty sediments. The metavolcanites are almost all basaltic and their geochemical patterns range from those typical of modern mid-ocean ridge basalts (MORB) to within-plate, possibly oceanic island basalts.

The Fundsjø Group, in the eastern Trondheim Region, is lithologically and geochemically more complex, with a high proportion of acidic metavolcanites intimately mixed with metabasites, pyroclastic rocks, and tuffaceous sediments. In addition to the felsic rocks, the group also contains amphibolites. These show a wider silica range, and a higher mean SiO_2 content, than the Støren greenstones, their compositions ranging from basaltic through basaltic andesitic, to andesitic. A subdivision of the rocks, based on major and trace element contents and trends, indicates:

1. A distinctive group of metabasalts to andesites with the geochemical signature of island arc tholeiites and characterized by very low TiO_2 and Zr abundances,
2. Amphibolites with element trends and abundances transitional between MORB and island arc tholeiites, and finally
3. A group of metabasites showing similarities to MORB.

The Fundsjø volcanites are interpreted as having formed in a Tremadoc to late Cambrian immature, ensimatic island arc setting. Probably, incipient rifting of this volcanic arc caused the formation of the transitional and MORB-type magmas now occurring as intrusives (or extrusives) that are intimately mixed with the original arc-type tholeiites. Subsequent compression, leading to a lower Ordovician deformation and metamorphism of this remnant arc, was followed by a new period of tension. This is indicated by intrusion of a swarm of porphyritic dolerites, and extrusion of geochemically comparable MORB-type lavas in the younger Sulåmo Group. It is tentatively suggested that this later period of crustal distension is coupled to the formation of the Sulåmo sedimentary basin.

Though the eastern and western sequences of the Trondheim Nappe Complex show many similarities, the significant differences described in this paper suggest that they represent two separate tectonic segments, rather than a continuous nappe sheet as indicated in previous models.

Introduction

In recent years, geochemical studies of basic metavolcanic rocks have been widely applied in determining the original tectonic environment of eruption of deformed and metamorphosed volcanite sequences. In the Norwegian Caledonides, the first investigations of this type by Gale and Roberts (1972, 1974) were followed by several detailed and regional contributions on metavolcanite petrochemistry. The ideas developed during these chemical studies, some of them summarized in

Furnes *et al.* (1980), have contributed much to the understanding of the palaeotectonic evolution of the Caledonian orogen. In addition, geochemistry has provided new information about the palaeogeographic setting and geographical distribution of the important group of volcanite-hosted massive sulphide deposits in the Caledonides.

In the Caledonian Trondheim Nappe Complex of the Trondheim Region (Fig. 1) previous work has been concentrated on the greenstones in its western part, where pre-Middle Arenig ophiolite fragments of the Støren Group and its equivalents have been recognized, together with Ordovician island arc, and

continental margin metavolcanites (Grenne *et al.* 1980; Roberts 1980; Grenne and Roberts 1980; Roberts *et al.* in press).

The metavolcanic Fundsjø Group in the eastern part of the Trondheim Region has previously been correlated with the geographically separate Støren Group (Wolff 1979). In the absence of chemical investigations of the Fundsjø Group, the correlations have been based mainly on lithological comparisons and regional structural interpretations (see below). In this paper we present 38 major and trace element analyses of basic metavolcanites sampled from traverses across the northern part of the

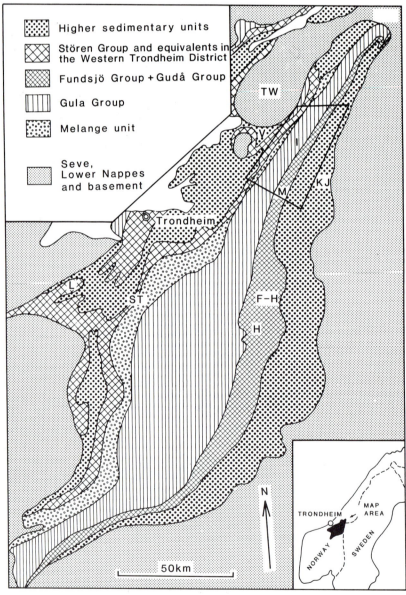

Fig. 1 Simplified geology of the Trondheim Region, showing the major volcanic complexes and tectonic units of the Trondheim Nappe Complex. TW: Tømmerås window; G–O: Grong–Olden basement culmination; V: Verdal; I: Inndal; M: Meråker; KJ: Kjølhaugan; L: Løkken; ST: Støren; F–H: Fongen–Hyllingen gabbro complex; H: Holtsjøen; D: Dombås

Fundsjø Group in the Meråker–Verdal area. These analyses are compared with published results from the well-investigated parts of the Støren Group and its presumed equivalents south and southwest of Trondheim, and with 15 new analyses of Støren Group greenstones from the Inndal–Verdal area. In the eastern part of the investigated area of Fundsjø Group rocks, a swarm of porphyritic metadolerites cut an older deformational fabric. Twelve major and trace element analyses of these intrusions are also considered, together with the chemistry of greenstones in the stratigraphically higher sedimentary units.

Regional Setting

The Trondheim Nappe (Kulling 1961; Wolff 1967) or Nappe Complex (Guezou 1978; Wolff 1979) is bounded by tectonic contacts and is contained in a major depression of the Precambrian basement. Geographically, the Gula Group or complex (Wolff and Roberts 1980) forms a central axis subdividing the nappe complex into eastern and western districts (Fig. 1). In each of these districts, the lowermost part of the stratigraphy is occupied by thick metavolcanic sequences. In the western district, greenstones of the Støren Group are overlain by sediments and volcanics of the Lower Hovin, Upper Hovin, and Horg Groups. In the eastern district, greenstones and amphibolites of the Fundsjø Group are overlain (see below) by the mainly sedimentary Sulåmo, Kjølhaug, Slågån, and Liafjellet Groups. The apparent symmetry of the two districts has been the basis of several stratigraphic and structural interpretations: either viewing the Trondheim Nappe as deformed in a megasynform (Bugge 1912) or interpreting the Gula Complex as a high-grade metamorphic core of a mushroom-shaped antiform where the eastern and western sedimentary sequences occupied inverted limbs (Wolff 1964; 1979; Roberts et al. 1970; Olesen et al. 1973; Roberts 1978). This straightforward near-symmetry of units in the eastern and western districts is now questionable with the recognition of a separate low-grade sequence between the Gula Group proper and the Støren greenstones to the west. This contains a rock assemblage resembling a mélange (Horne 1979), and is thus in this paper referred to as the 'mélange unit'.

The Støren Group

The greenstones of the Støren Group occur in the western part of the Trondheim district immediately to the west of either the Gula or the mélange unit. They are tectonically inverted and young north-

westwards from the contact with the mélange unit at the boundary towards the Gula Group (Horne 1979). This Støren Group sensu stricto has a thickness of ca 4 km in the area southeast of Trondheim. Along the strike, to the NNE, these greenstones wedge out tectonically but reappear in the same tectonostratigraphic position at Inndalen and Verdalen, which lie about 7 km WNW of the part of the Fundsjø Group investigated in this study. The greenstones of this area are homogeneous and usually lack well-developed primary structures, although pillows can be recognized in rocks that have escaped strong deformation. Locally, minor amounts of acidic rocks of uncertain origin occur. The Støren Group consists essentially of pillowed and massive metabasalts, subvolcanic dykes, intercalated recrystallized jaspers, quartzites, and banded quartz schists (Torske 1965). Sedimentary pyrite and pyrrhotite deposits ('vasskis') and iron ores are often associated with the quartzitic sediments.

Thick greenstone sequences in the Trondheim–Løkken area south and southwest of Trondheim have previously been correlated with the Støren Group s.s. (Vogt 1945; Chaloupsky 1970). Recent mapping of these greenstones has revealed relatively well-preserved ophiolite fragments containing a sheeted doleritic dyke-complex, with a thickness of up to 1 km, and which caps large bodies of metagabbro. The sheeted dykes are overlain by a thick unit of metabasalts (Grenne et al. 1980). The dykes represent the feeders to the sequence of pillowed and massive lavas, that contain frequent intercalated volcanic breccias, cherts, jaspers, and stratiform massive sulphide ores.

The geochemistry of the Støren Group metavolcanics and their supposed equivalents in the western Trondheim district is well known from previous studies (Gale and Roberts 1974; Gale 1974; Grenne et al. 1980). The rocks are chemically homogeneous and almost wholly basaltic. They exhibit close similarities to the tholeiitic basalts of oceanic spreading ridges. The minor acidic rocks occurring within the greenstones are mostly trondhjemitic intrusives (Vogt 1945; Torske 1965) which are probably unrelated to the basaltic volcanism. A few small bodies of plagiogranite occur within, and are comagmatic, with the ophiolitic gabbros (Grenne et al. 1980). Although acid extrusive rocks are apparently lacking in the little deformed 'classical' parts of the Støren Group, the general lithology is somewhat more complex farther to the northwest, in the vicinity of Trondheim, in the unit known as the Bymark greenstones. In that area, the greenstones contain a few continuous, thin, quartz keratophyres that may represent acid effusives (Oftedahl 1968).

Overlying, fossiliferous sediments indicate that

the Støren Group *sensu stricto* (*s.s.*) is pre-Middle Arenig in age. It has been interpreted as a fragment of a possibly Cambrian oceanic crust that was obducted on to the Baltic plate prior to the mid-Arenig (Furnes *et al.* 1980). It must be emphasized, however, that the exact age and palaeotectonic environment of the Støren Group *s.s.* is still somewhat unclear. The pre-Upper Arenig ophiolites in the Løkken and Trondheim–Vessfjell areas (see Fig. 1), which have been regarded as equivalents to the Støren *s.s.*, are now interpreted in terms of crustal distension in an early Ordovician marginal basin, rather than representing fragments of true Iapetus crust (Roberts *et al.* in press).

The Støren Group has undergone three main deformation episodes (Roberts 1978), however, it differs from the Fundsjø Group in that the greenstones usually exhibit fairly well-preserved primary volcanic structures. Investigations by one of us (B.L.) in the Meråker–Verdalen part of the Støren Group indicate two groups of folds separated by an episode of shearing, although the homogeneous character of the rocks and the lack of banding make it difficult to carry out a detailed examination of the structures.

The Fundsjø Group

Volcanic rocks of the Fundsjø Group can be traced from the Grong–Olden culmination in the north to Dombås in the south, i.e. along the entire length of the Trøndelag depression. The results presented in this paper are derived from the northern Meråker–Verdalen area, though similar rock types and geological patterns are also found further south.

The volcanic rocks consist of interlayered basic and acidic units. The rocks have been intensely deformed and are commonly finely banded with individual layers varying from mm to dm thickness. Particularly in the eastern part, where deformation was less intense, but also elsewhere, thicker units are sometimes present.

Thick lenses of albite–granite (Wolff 1973, 1979) occur in the western part of the area. These intensely deformed rocks are of uncertain origin though they appear to intrude the greenstones and the banded sequences. The albite-granites may possibly represent products of the metamorphism of thick units of acidic volcanics. The chemistry (Wolff 1973) and mineral assemblages resemble both the acidic members of the banded sequences and the trondhjemites. A rough estimate, in the Meråker–Faeren area, suggests that acid volcanics make up about 15–20 % of the volcanic complex. However, if the albite-granite belongs to the volcanics, almost a half of that com-

plex is of acidic composition. The acid volcanic rocks are also extensively developed to the south (Rui 1972; Olesen *et al.* 1973; Guezou 1978). The volcanic rocks appear to be bimodal in composition, with a few intermediate rock types.

The thick units of greenstones are commonly homogeneous, but tectonically elongated pillows can be recognized locally. Further south, pillows are common (*cf.* Rui 1972; Olesen *et al.* 1973; Guezou 1978; Nilsen 1978). In the east, where deformation and metamorphism have been less intense, some other relics of primary igneous textures are found. The original grain shapes of pyroxene and plagioclase can be recognized in the greenstones as can relic ophitic and porphyritic textures. The acidic volcanics have porphyritic textures featuring phenocrysts of euhedral sodic plagioclase in a fine-grained matrix of quartz and albite, with minor amounts of secondary mafic minerals. In the Trondheim region, the acid volcanic rocks are known as quartz keratophyres. The high quartz contents, the occurrence of sodium-rich plagioclase as the only feldspar, and their chemical composition (Wolff 1973) classify them as sodic rhyolites.

The eastern occurrence of the Fundsjø Group in the Verdalen–Meråker area, contains subordinate layers of coarse tuffs (Wolff 1973). In its western part, tuffite and tuff horizons are common in the vicinity of the Gudå Group. Further south, pyroclastic and volcanoclastic rocks become more prominent and are often mixed with pelitic debris (Kisch 1962; Olesen *et al.* 1973). In the southernmost part of the Trondheim region, volcanic greenstone units are separated by rhythmic (Musa-) sequences, which consist of marble, graphite schists, and pelitic schist of tuffaceous origin (Guezou 1978). Agglomerates have been reported by Olesen *et al.* (1973) and Nilsen (1978).

West of the Fundsjø Group, occurs a thin veneer of sediments of the Gudå Group between the Fundsjø greenstones and rocks typical of the Gula Group (Lagerblad 1983). The eastern part of the Gudå Group is transitional to the Fundsjø Group and consists of volcanoclastic and pyroclastic units interlayered with more massive greenstones. In the west, in the vicinity of the Gula Group, increasing amounts of pelitic, quartzose, and carbonaceous materials occur together with the volcanics. There are also horizons of marble and graphite schist. Sedimentary breccias and agglomerates can be recognized. A distinct alumina-rich staurolite-bearing pelitic schist occurs close to the more semipelitic rocks of the Gula Group. Quartzitic pebbles are found both in volcanoclastic and more pelitic horizons. A similar unit, Tverråi Formation, occurs in the southernmost (Guezou 1978) and the central part of the Trondheim region (Rui 1972) where it occupies a tectonostratigraphic level similar to that of the Gudå.

The lithology of the Gudå Group demonstrates that at least part of it is composed of, or associated with, volcanics. The Gudå was formed in a transitional environment where volcanic, erosional, and marine processes were interacting. The volcanoclastic and sedimentary units occurring within the southern part of the Fundsjø Group show similarities with Gudå Group lithologies and also represent comparatively shallow-water environments.

A dolerite dyke swarm has been recognized in the eastern part of the Fundsjø Group in the Faeren–Meråker area. These dolerites can be distinguished from other basic rocks by their well-developed porphyritic textures and different relationships to the tectonic structures. The dykes have chilled, fine-grained margins and coarser, markedly porphyritic interiors. They have been metamorphosed, but original grain shapes of pyroxene can sometimes still be recognized. The phenocrysts consist of epidotized plagioclase.

In the Inndalen valley, one porphyritic dolerite dyke has been found in the Gudå Group. Several other porphyritic dolerites occur around the Hermansnasa Gabbro Complex, but none has been found in the central part of the Fundsjø Group in the area investigated. The dolerite in Inndalen has a significantly higher Al_2O_3 content (20–23%, Wolff 1973; Lagerblad in press) than the dykes described in the present paper. Porphyritic dolerites have also been described from around the Hyllingen Gabbro Complex, and a genetic relationship has been suggested (Nilsen 1971). The analysed dolerites are all from the eastern side of the Fundsjø Group, and the conclusions given here are drawn from these rocks.

East of the outcrops of the Fundsjø Group, younger sediments form a sequence extending to the thrust boundary of the Trondheim Nappe Complex. Part of the debris in these rocks has been derived from the Fundsjø Group. The pebble material in a conglomerate horizon at the contact between the Fundsjø and Sulåmo Groups (the Lille Fundsjø Conglomerate) is composed mainly of material from extrusive and intrusive rocks found in the Fundsjø Group; however, it does not contain pebbles of the porphyritic dolerites. The field relationships of the conglomerate horizon are somewhat unclear, but one narrow greenstone dyke that is similar in chemistry to the porphyritic dolerites cuts the bedding of the conglomerate (Hardenby 1980) and indicates that at least some sedimentation occurred before the intrusion of the dolerites. Dolerites are also found immediately to the east of the conglomerate, but they do not occur higher up in the Sulåmo Group, or

in the other units of the eastern sediments. With the exception of the Turifoss greenstones (Wolff 1973, 1979) and late gabbro bodies, the higher sedimentary units contain no metabasaltic rocks. An early deformation event has affected the Gula, Gudå, and Fundsjø Groups but not the eastern sediments (Hardenby et al. 1981). Therefore, there is probably a major unconformity between the Fundsjø Group and the overlying sediments (Wolff 1967; Furnes et al. 1980; Hardenby et al. 1981)

Deformation

The Fundsjø and Gula Groups have suffered at least four phases of deformation (Lagerblad 1983). Prior to the late Silurian thrusting of the Trondheim Nappe Complex, they had been affected by two phases of folding and now exhibit isoclinal, steeply dipping, mutually interfering folds. After an intense deformation that produced the regional schistosity, the rocks were refolded into folds with an open style and with horizontal fold axes. The shearing giving the regional schistosity can probably be related to the thrusting of the Trondheim Nappe Complex onto the Baltic Shield (Lagerblad 1983).

In contrast, higher sedimentary units to the east have not been affected by the first phases of folding, but the post-thrusting phases are better developed (Hardenby et al. 1981). During the first phase of folding the Fundsjø Group was intruded by diorites. The albite-granites also possibly belong to this period (see above). The porphyritic dolerites intruded after the second phase of folding, before the development of the regional foliation (Fig. 2 and Lagerblad 1983). All units except the Gula Complex have been intruded by gabbros, probably equivalents of the Fongen–Hyllingen Gabbro that has yielded a late Silurian Rb/Sr age (see below).

Age

A Dictyonema-bearing black schist of Tremadocian age (Størmer 1941) has been found in the Holtsjøen area in the same tectonostratigraphic position as the Gudå Group (Rui 1972). In the Gula, trondhjemite dykes of Lower Ordovician age (Klingspor and Gee, 1981) cut previously deformed rocks. This deformation is probably the one that gave the now steeply dipping folds in the Gula and Fundsjø Groups in the investigated area. The sediments of the Kjølhaug Group contain graptolites of Llandovery age (Getz 1890). The last event dated is the intrusion of the Fongen–Hyllingen Gabbro Complex into the

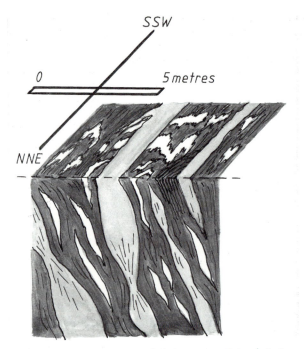

Fig. 2   Sketch section showing the cross-cutting relation-ship between the early deformation (D_1/D_2) of the Fundsjø amphibolites (dark grey), quartz-keratophyres (white), and the later porphyritic metadolerites (light grey). The later, regional schistosity and the associated boudinaging (D_4) of the porphyrites is indicated. Locality: eastern part of the Fundsjø Group, Inndalen valley

Fundsjø Group. The age of late differentiates in the gabbro is 405 ± 9 Ma (Wilson 1981). The peak of regional metamorphism in the Gula and Fundsjø Groups occurred after the intrusion of the gabbro (Birkeland and Nilsen 1972) but before the thrust-ing of the Trondheim Nappe Complex (Roberts *et al.* 1970; Olesen *et al.* 1973; Lagerblad 1983).

Thus the Fundsjø Group must have been depo-sited, at least in part, during the Tremadocian and underwent deformation during the early Ordovician. Sedimentation then continued up into the Silurian. All rocks of the Trondheim Nappe Complex were thrust onto the Baltic Shield at the end of the Silurian.

Metamorphism

The greenstones contain Ca-amphibole, plagioclase, epidote, and chlorite as major phases, and quartz, ilmenite, and magnetite as accessory minerals. Cal-cite is sometimes present in rocks of low metamorphic grades and garnet occurs in higher grade rocks where the chemistry was suitable. The maximum metamorphic grade of high amphibolite facies was reached in the western part of the Fundsjø Group (Lagerblad 1983). The metamorphic grade

decreases gradually and epidote–amphibolite and greenschist-facies rocks are found at the eastern con-tact.

Metamorphic textures have been studied from fif-teen of the analysed greenstones from the Fundsjø Group in the Inndalen valley that traverses the Faeren area. In the east the amphiboles are often zoned with actinolitic cores and hornblendic rims. This zoning is probably a result of continuous reac-tions and incomplete re-equilibrium during prograde metamorphic growth. At higher metamorphic grades, the Ca-amphiboles are more hornblendic. In the central part of the Fundsjø Group in Inndalen, there is a break in the anorthite content of plagio-clase, across the peristerite gap, from albite (An 5 per cent) to oligoclase (An 15 per cent). The modal contents of plagioclase and Ca-amphibole increase with rising metamorphic grade, while the amounts of epidote and chlorite decrease.

The rocks from the Støren Group have mineral assemblages resembling those of the lowest grade rocks from the Fundsjø Group. The Støren green-stones from the southern areas have been affected by metamorphism ranging between low and middle greenschist facies. If the chemical differences be-tween the Støren and Fundsjø Groups are due to the higher metamorphic grade in the latter, there must have been an open chemical system. The gap in the anorthite contents of the plagioclase and the calcite breakdown are discontinuous reactions, in which H_2O and CO_2 leave the system and the Ca contents of the rocks may change. Elements such as Al, and to some extent Fe and Mg enter other mineral phases in reactions coupled to the discontinuous reactions, but most elements participate in continu-ous reactions and thus remain within small, closed, chemical systems.

Very few veins or other features related to chemi-cal transport and partial melting have been observed in the greenstones. No significant chemical differ-ences can be observed between Fundsjø Group greenstones of high and low metamorphic grade. The chemical differences between the greenstones of the Fundsjø and Støren Groups must therefore be a primary feature.

Geochemistry

Thirty-five samples of amphibolites from the Fundsjø Group have been analysed for major and trace elements by a combination of X-ray fluores-cence and wet chemical methods (Na_2O, K_2O, and P_2O_5). The analyses were performed at the Geologi-cal Survey of Norway and the Dept. of Geology in Lund. For comparison, 15 new analyses of green-

Table 1   Mean major and trace element composition of the Fundsjø Group amphibolites, the later porphyritic metadolerites, the Turifoss greenstones in the Sulåmo Group (with standard deviations in brackets) and the Støren Group greenstones. Data for average ocean-floor and island arc tholeiites from Pearce (1975). FeO$^{(tot)}$: total iron as FeO. Numbers of analysed samples in parentheses. Major elements normalized to 100 per cent.

	Fundsjø Group amphibolites			Porphyritic dykes (13)	Turifoss greenstones (19)		Støren Group greenstones (15)	Average ocean-floor basalt	Average island arc tholeiite
	I (12)	II (14)	III (9)						
SiO$_2$	53.7	53.2	50.7	50.0	50.3	[1.2]	50.0	49.9	52.9
TiO$_2$	0.59	1.56	1.85	1.69	1.91	[0.28]	1.88	1.43	0.83
Al$_2$O$_3$	15.8	15.6	14.8	16.6	13.8	[0.6]	15.3	16.2	16.8
FeO$^{(tot)}$	8.7	11.1	11.4	9.1	12.3	[1.3]	10.8	10.2	10.4
MnO	0.20	0.24	0.21	0.18	0.21	[0.02]	0.20		
MgO	7.9	5.7	7.0	7.2	7.2	[0.5]	7.7	7.7	6.1
CaO	9.0	7.2	9.9	11.3	11.5	[1.0]	9.9	11.4	10.5
Na$_2$O	3.8	5.0	3.9	3.6	2.6	[0.6]	3.7	2.8	2.1
K$_2$O	0.38	0.20	0.22	0.24	0.23	[0.31]	0.25	0.24	0.44
P$_2$O$_5$	0.06	0.12	0.09	0.09	0.08	[0.03]	0.18		
ppm									
Zr	32	76	115	114	128	[28]	124	92	52
Y	19	33	40	33	42	[6]	35	30	19
Sr	183	167	213	232	144	[43]	265	131	207
Cr	234	62	183	226	180	[86]	263	310	160

stones from the Støren Group are also presented below. The chemistry of the late porphyritic dykes that intrude amphibolites of the Fundsjø Group will be discussed briefly on the basis of 13 analyses. The means and standard deviations of 19 samples of Turifoss greenstones are included in Table 1. All samples from the Fundsjø and Støren Groups are from the Verdalen–Meråker area.

In the following discussion, the Fundsjø Group amphibolites are divided into separate series, the subdivision being based solely on the geochemistry of the rocks. Major and trace element contents, and a visual discrimination of plots on variation diagrams has made it possible to distinguish between three geochemically distinctive volcanite types. Although there is a significant difference between the geochemical patterns of 'type I' on the one hand, and 'type II' and 'type III' on the other, the difference between the latter two types is not always obvious (cf. Figs. 3, 4, 5, where the fields occupied by the various volcanite types are indicated on the diagrams). It appears that the samples of types II and III amphibolites represent a gamut of several magma types, which range from compositions of type II to those of type III. The subdivisions into types II and III is therefore somewhat artificial. The chemistry of the Fundsjø greenstones is summarized in Table 1, which also shows the mean compositions of the late porphyritic dykes and the Støren Group greenstones.

The conspicuous diversity of the Fundsjø Group

amphibolites is illustrated in Fig. 3, which shows the apparently different fractionation trends of the three types. The most obvious difference is the much lower contents of total iron and TiO$_2$ in type I as compared to types II and III (see Table 1). In addition, types II and, in particular, type III show marked enrichment trends of iron and TiO$_2$. Type III is similar to the abyssal tholeiite trend of Miyashiro (1975). However, the low-TiO$_2$ type I amphibolites have flatter trends that are comparable to those of tholeiitic volcanites from the Tonga–Kermadec and Izu–Bonin island arcs of the southwestern Pacific (Fig. 3).

The late porphyritic dykes and the greenstones of the Støren Group show uniform chemistries with well-defined steep FeO- and TiO$_2$-enrichment trends (Fig. 3) that are similar to those of abyssal tholeiites. Some of the Støren samples appear to be highly fractionated. They approach ferrobasaltic compositions with high total iron and TiO$_2$, and relatively low MgO contents. The porphyrites comprise a very narrow compositional range with total FeO/MgO varying only between 1 and 1.5.

Apart from the above-mentioned differences, the lower SiO$_2$ contents of the Støren greenstones (and also of the porphyritic dykes that intrude the Fundsjø amphibolites), distinguish them from the amphibolites in the eastern district (Table 1 and Fig. 3). Whereas the former two rock suites are entirely basaltic, nearly all analysed Fundsjø Group amphibolites have SiO$_2$ contents exceeding 50 per

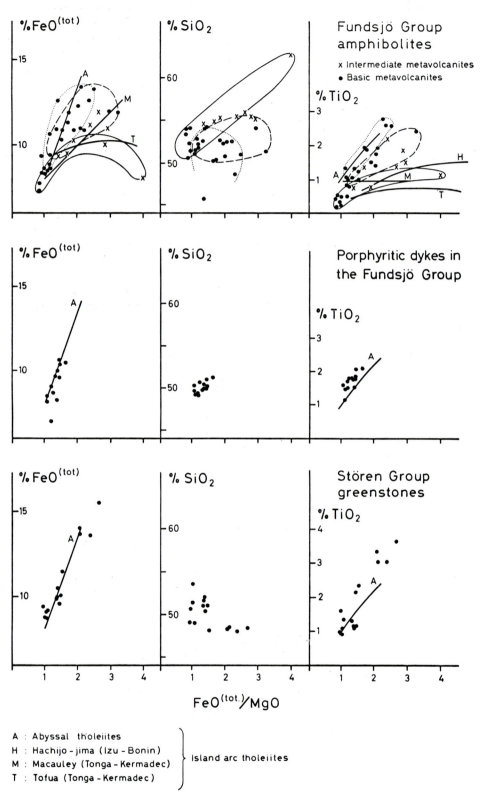

Fig. 3   FeO^{tot}, SiO$_2$, and TiO$_2$ versus FeO/MgO plots of the Fundsjø Group amphibolites, the later porphyritic dykes in the Fundsjø Group and the Støren greenstones (FeO$^{(tot)}$ = total Fe as FeO). Trend lines from Miyashiro (1975). In this diagram and in Figs. 4, 5, and 6, the symbols and boundary lines used are as follows. Filled circles: basic rocks (< 54 per cent SiO$_2$ on a volatile-free basis); crosses: intermediate rocks (> 54 per cent SiO$_2$); full line: type I amphibolites; broken line: type II amphibolites; stippled line: type III amphibolites

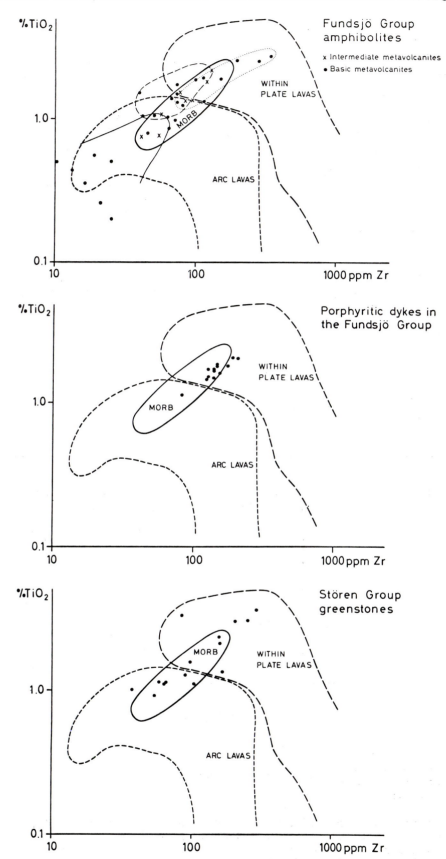

Fig. 4   TiO$_2$ versus Zr variation diagram for the Fundsjø amphibolites, the porphyritic dykes and the Støren green-stones. Boundary lines for arc lavas, within place lavas, and mid-ocean ridge basalts are from Pearch (1980). Symbols and boundary lines for the Fundsjø metavolcanites as in Fig. 3

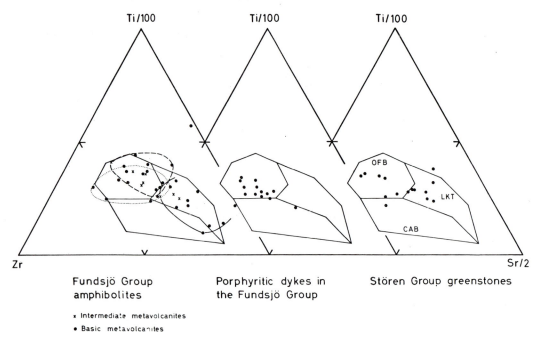

Fig. 5   Ti–Zr–Sr plot (Pearce and Cann 1973) of the Fundsjø Group amphibolites, the porphyritic metadolerites and the Støren greenstones. OFB: ocean-floor basalts; LKT: island arc tholeiites; CAB: calc-alkaline basalts. Symbols and boundary lines as in Fig. 3

cent (calculated on a volatile-free basis). This difference might be due to alteration, but probably it reflects a slightly more intermediate primary composition of the Fundsjø amphibolites. In fact, eight samples of the 35 analysed Fundsjø amphibolites can be classified as 'intermediate' (more than 54 per cent SiO_2), even though non-basic rocks were consistently avoided during the sampling.

The ocean-floor basaltic characteristics of the porphyritic dykes and the metabasalts of the Støren Group are clearly seen in Fig. 4. The porphyrites plot reasonably well within the MORB field; the Støren volcanites fall partly in the field of within-plate basalts. The Fundsjø amphibolites on the other hand, show a greater scatter and plot partly in the field of island arc lavas. This illustrates their heterogeneous chemistry; the type II and III analyses plot both in the MORB and the within-plate fields.

The Ti–Zr–Sr diagram of Pearce and Cann (1973) utilizes a relatively mobile element (Sr) to discriminate between island arc lavas and mid-ocean ridge basalts. Plots of our analyses in this diagram (Fig. 5) are quite consistent with that in the other diagrams (Figs. 3 and 4) indicating a lack of large scale alteration. The diversity of the amphibolites of the Fundsjø Group is seen also in Fig. 5 where the analyses cover both the ocean-floor basalt field and the field of island arc tholeiites. However, by treating types I, II, and III separately the low-TiO$_2$ type (I) can be classified almost exclusively as island arc

tholeiites. Types II and III overlap in this diagram (Fig. 5), showing a Ti/Zr/Sr-chemistry broadly similar to that of ocean-floor volcanites. With the exception of one sample, the analyses of porphyritic dykes plot in the field of ocean-floor basalts. The comparatively dense clustering indicates that there was little relative mobility of these elements.

Most of the Støren Group metabasalts plot well inside or close to the boundary of the OFB field (Fig. 5). As seen on Fig. 4, these greenstones apparently include within-plate type basalts (possibly off-axis or ocean-island type), with TiO$_2$ values as high as 3.4 per cent. A range of compositions from OFB through to WPB is suggested also by the Ti–Zr–Y diagram of Pearce and Cann (1973) (not shown here). The minor spread of plots outside the OFB field into the 'arc tholeiite' field on Fig. 5 is caused by this composite OFB–WPB character of the Støren Group, as the Ti–Zr–Sr discrimination diagram (Fig. 5) is designed to separate only normal oceanic spreading ridge basalts from island arc basalts.

The Ti–Cr diagram of Pearce (1975) clearly illustrates the different geochemistry of the Fundsjø amphibolites relative to the later porphyrite intrusions, and the Støren Group metavolcanics (Fig. 6). The latter two suites have an ocean-floor basaltic character with relatively high contents of Cr and Ti. They therefore plot exclusively on the upper, OFB-side of the division line.

The wide scatter of Ti–Cr values in the Fundsjø

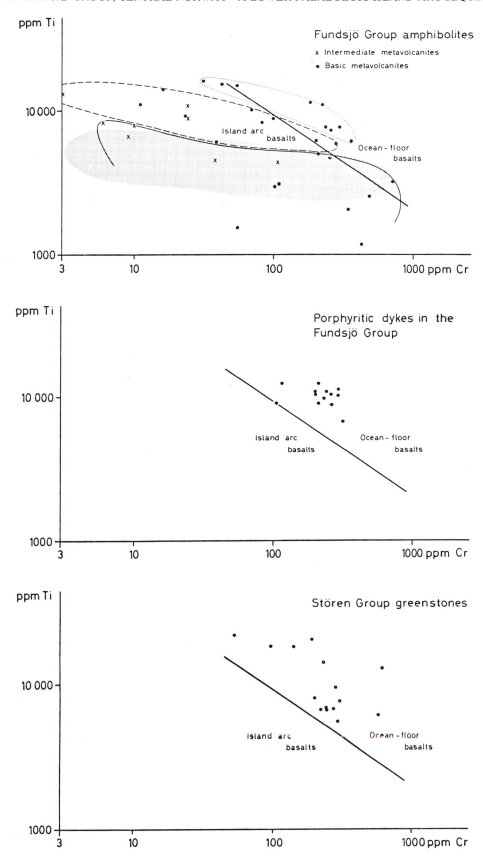

Fig. 6   Ti versus Cr variation diagram for the Fundsjø amphibolites, the porphyritic dykes, and the Støren Group greenstones. Discrimination line between island arc and ocean-floor basalts from Pearce (1975). Shaded area: island arc tholeiites from Izu–Bonin (from Shiraki *et al*. 1978). Symbols and boundary lines as in Fig. 3

amphibolites (Fig. 6) bears no resemblance to the distribution patterns for the porphyrites or the Støren greenstones. Characteristically, they show large variations in Cr content, from about 700 ppm in the least fractionated basalt, to c. 10 ppm in some metabasites and even lower in some of the basaltic andesites and andesites. More than 70 per cent of the amphibolites analysed plot within the island arc tholeiite field below the division line. Furthermore, by separating the samples into the types I, II, and III, the rocks of the low-TiO_2 type (I), which show the most definite island-arc tholeiite trends and compositions in the preceding diagrams, are almost exclusively classed as arc volcanites (Fig. 6). However, some of the metabasalts have much higher Cr contents than the average value of 160 ppm given by Pearce (1975). The mean Cr of type I amphibolites is 234 ppm (Table 1), which is close to the average of 250 ppm given for ocean-floor basalts (Pearce 1980). Nevertheless, some island arcs exist showing the same range of Cr at similar Ti levels. For instance, island arc tholeiites from Izu–Bonin (the shaded area in Fig. 6) can have as much as 652 ppm Cr in their least fractionated basaltic lavas (Shiraki *et al.* 1978).

The amphibolites of type II define a trend parallel to that of type I (Fig. 6), but have higher Ti values. This group of amphibolites which shows differentiation trends intermediate between those typical of ocean-floor and island arc tholeiitic rocks (Fig. 3), is classified as arc volcanites in Fig. 6, although many of the samples are obviously much richer in Ti (up to about 2.3 per cent TiO_2 in intermediate rocks) than normal subduction-related igneous rocks (see Figs. 3 and 4).

The Fundsjø Group amphibolites of type III bear no similarities to island arc volcanites (Fig. 6). They have generally higher Zr and TiO_2 values than type II (see Table 1 and Fig. 4) but appear to be slightly more primitive with respect to the FeO/MgO ratio (Fig. 3). They have a considerably higher Cr content than type II (Table 1), and the majority of these metabasites are thus similar to ocean-floor basalts (Fig. 6).

The trace and minor element characteristics of the three types of Fundsjø Group amphibolites are summarized in Fig. 7. The mean element contents are normalized to those of an average mid-ocean ridge basalt (Pearce 1980). A typical island arc tholeiite composition from Pearce (1980) is shown for comparison. This diagram also shows data for the Støren Group greenstones and the porphyritic dykes intruding the Fundsjø amphibolites. Although only 8 of the 16 elements constituting the diagram have been analysed, the similarity of the Støren greenstones and the prophyrites to normal spreading-ridge

basalts is striking; since the normalized patterns are nearly flat.

In contrast, the Fundsjø amphibolites exhibit markedly different trace element patterns for the three separate types (Fig. 7). For all analysed elements there is a very close similarity between type I and the Tongan tholeiite shown in the diagram. The depletion in stable elements and the enrichment in alkali elements that characterize island arc tholeiites relative to MORB is obvious in the pattern for the Fundsjø amphibolites, the only significant difference being their somewhat higher Cr contents.

Types II and III, on the other hand, have relatively flat, MORB-like patterns (Fig. 7). Type II is strongly depleted in Cr in comparison to MORB and shows a progressive increase in Zr to Ti, that is typical of trends found in island arc tholeiites (Pearce 1980). The slight enrichment from Sr to Ba may be a result of element mobility and low abundances of these alkali elements in our samples.

Discussion

The analyses of Støren Group greenstones from the Verdalen area confirm previous studies from areas farther south in the Trondheim district, indicating that these rocks represent fragments of oceanic crust (see above). The present study suggests that there may also be a component of ocean island volcanism present. The Støren rocks range from typical, primitive, mid-ocean ridge basalts to TiO_2-rich ferrobasalts and there is nothing to suggest that these lavas were related to subduction zone generated magmas.

Parts of the Fundsjø Group (the type I amphibolites) exhibit a definite island arc tholeiite character in the trends, element abundances and ratios of our diagrams (Figs. 3 to 7). This suite of rocks range from primitive, low-TiO_2, basaltic rocks with high Cr values, similar to the least differentiated Izu–Mariana island arc basalts (Shiraki *et al.* 1978), to differentiated low-TiO_2–low-Cr basalts and basaltic andesites, and to andesites.

Since low-TiO_2, basic volcanites are relatable to subduction zones, the type I Fundsjø amphibolites were probably formed in an island arc setting. The complete lack of calc-alkaline metavolcanics and the primitive nature of the basaltic rocks suggests that they originated in an immature ensimatic arc on oceanic crust.

A relatively high proportion of the Fundsjø amphibolites (types II and III) have a chemistry akin to that of mid-ocean ridge basalts. Such chemical compositions are not found in recent arcs; their chemistry rather suggests a tensional regime for the

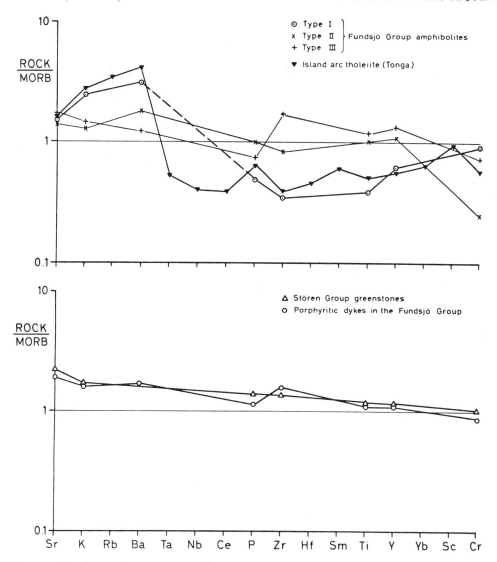

Fig. 7   MORB-normalized trace element patterns (average) of the Fundsjø amphibolites, the porphyritic metadolerites, and the Støren greenstones. MORB values and values for the Tongan arc tholeiite from Pearce (1980)

formation of these relatively TiO_2-rich magmas. Two possible settings are considered:

1. The MORB-like amphibolites (types II and III) may represent the ocean-floor basement to the island arc (type I amphibolites); and the two groups have been interleaved as a consequence of the intense deformation.
2. The MORB-like amphibolites may represent igneous activity following a shift from a subduction to a tensional environment, i.e. incipient rifting of the arc and, eventually, the formation of an inter-arc marginal basin; a model comparable to the one proposed by Stephens (1981) for the Stekenjokk igneous rocks some 200 km to the NNE.

Of these two possibilities the latter is perhaps the most reasonable one in view of the composition of the type II amphibolites, that have a chemistry intermediate between normal MORB and island arc tholeiites. Similar transitional compositions have been described from the Lau marginal basin that separates the active Tonga–Kermadec arc from a remnant arc, the Lau–Colville ridge (Reay et al. 1974; Gill 1976). Whether these amphibolites represent dykes intruding the island arc volcanics or are rifting/tensional lavas is, however, open to debate. Both possibilities seem reasonable, but the strong deformation suffered by the Fundsjø Group has destroyed most primary structures and thus there is little stratigraphic control.

An island arc character of at least part of the Fundsjø Group is also supported by the intercalated pyroclastic, volcanoclastic, and sedimentary materials, which are in part shallow water deposits. The volcanic part of the Gudå Group that is associated with the Fundsjø Group shows that at least some parts of the volcanic complex were eroded while volcanic activity was still in progress (Lagerblad 1983). In contrast, the frequent occurrence of variolitic and non-vesicular lavas in the Støren Group indicates a deep water environment (Furnes *et al.* 1980; Grenne *et al.* 1980).

By analogy to the Støren greenstones, the swarm of late porphyritic dykes, that intrude the previously deformed Fundsjø Group amphibolites shows no sign of being subduction-related. The major and trace element contents, ratios, and trends are broadly similar to those of ocean-floor basalts, which suggests mantle derivation and relatively shallow fractional crystallization.

Conclusions and Regional Implications

The Støren and Fundsjø Groups show significant geochemical differences (although there are also some similarities). The Støren rocks show geochemical signatures, substantiated by lithology, that clearly relate the metabasalts to tensional processes and the formation of a new oceanic crust in pre-Middle Arenig time; this may have occurred in a large marginal basin or a major ocean. The Fundsjø Group, on the other hand, formed at least partly in a Tremadocian immature island arc setting. Before burial under a cover of sediments, the Fundsjø rocks had undergone deformation, metamorphism and uplift followed by erosion (Furnes *et al.* 1980; Hardenby *et al.* 1981). Although the sediments covering the two volcanic groups are lithologically comparable, fossils are rare and previous attempts at correlation for this higher part of the stratigraphy contain substantial elements of speculation. Furthermore, although it is still possible that the Fundsjø and Støren Groups are broadly time-correlatives, they originated in completely different palaeotectonic environments. This rules out the previously accepted, simple correlation between the two volcanic complexes.

The Fundsjø Group has been affected by a significantly higher grade of metamorphism than the Støren Group. At least the Fundsjø Group reached its peak metamorphism just prior to the culmination of the late Silurian thrusting (see above). This implies that the Fundsjø Group was in a different and deeper position than the Støren Group during this orogenic event. The tectonostratigraphic position of the Gula Group is critical in all interpretations. The Gula, Gudå, and Fundsjø Groups to the east of the central axis of the Trondheim Nappe Complex have all been affected by the same early phases of folding and metamorphism (Lagerblad 1983). However, the previously accepted interpreta-

tion of the Støren–Gula contact has been complicated by the recent recognition of an intervening mélange unit (Horne 1979). The Gula Complex is bounded in the west by tectonic contacts that are characterized by partly recrystallized mylonites, and a regional metamorphic break occurs between the Gula and and Mélange unit (Lagerblad 1983). This suggests that the western and eastern rock sequences of the Trondheim Nappe Complex may not be strictly continuous and it appears feasible to treat its two halves as segments separated by a major thrust plane. In the area of this investigation the tectonic break is situated between the Gula and the mélange unit to the east of the Støren or, locally, the Støren Group proper. This interpretation conflicts with the assumption of a continuous nappe stratigraphy as required by current interpretations of the Trondheim Nappe Complex as a continuous major synformal or antiformal structure (see above). Evidently the apparent resemblance in general lithostratigraphy and the high tectonostratigraphic positions of the two halves of the Trondheim Nappe Complex within the Caledonides point to somewhat comparable depositional histories. However, the data presented above suggest that the two halves of the nappe complex may represent separate tectonic slices of a composite, volcanosedimentary assemblage that had been broken up during or prior to the late Silurian thrusting on to the Baltic Shield. The inferences drawn here from the differences in metamorphic grade of the Fundsjø and the Støren Groups are supported by similar differences in the degree of deformation, which is more intense in the east. This interpretation can be extended to all of the rock sequences to the west and east of the Gula Group.

The Tectonic Significance of the Porphyritic Dolerite Dykes in the Fundsjø Group

The porphyritic dolerite dykes in the Fundsjø Group cut the early structures of the host rocks and must have been intruded after the first phases of folding, and thus cannot be related to the primary volcanics of the Fundsjø. Their occurrence is restricted to the Fundsjø and, possibly, the lowermost part of the Sulåmo Group. No similar dykes are found in the Gula, but the structures cut by the dykes in the Fundsjø are similar to some of the structures in the Gula. This indicates that at the time of the dyke intrusion, the Fundsjø and the Gula Groups had been sufficiently close to each other to undergo common deformation. The absence of porphyritic dykes in the Gula may be considered an indirect argument for suggesting that the presently exposed parts of the Fundsjø and Gula were not immediately juxtaposed, but occupied neighbouring geographical positions.

The intrusion of mantle-derived dykes must have been due to a tensional event after the early Ordovician tectonic superposition of the Fundsjø upon the Gula but before the late Silurian thrusting. Chemical and field evidence suggest that the dyke swarm shows no genetic relationship with the gabbros occurring in this region (see above). Our dykes therefore appear to be different from the dykes associated with the Fongen–Hyllingen gabbro. However, there is a considerable chemical similarity to the effusive Turifoss greenstones in the lowermost sedimentary unit of the Sulåmo Group (Table 1). As with the dolerites, the Turifoss greenstones have patterns of immobile trace elements that resemble those of ocean-floor basalts (Figs. 5, 6, and 7). The small differences in major and trace element contents can be ascribed to a somewhat higher degree of fractionation for the effusives, that in contrast to the porphyritic dykes, may also have suffered severe submarine weathering and consequent chemical alteration. Hence, on a geochemical basis, both could be related to the same episode of igneous activity. Because the Sulåmo Group was deposited after two phases of folding in the Fundsjø, the chemical similarity between the dolerites and the Turifoss greenstones may suggest that the tensional episode responsible for the intrusion of the dolerite dykes was associated with the formation of the sedimentary basin in which the Sulåmo Group and, subsequently, the still higher sedimentary rock units (Slågån, Kjølhaugan, and Liafjell Groups) were deposited.

If the proposed correlation between the porphyritic dolerites and the Turifoss effusives is correct, the time interval for the intrusion of the dolerites can be narrowed to between the common deformation of the Gula and Fundsjø Groups in the early Ordovician, and the erosion of the Fundsjø, coupled to the deposition of most of the Sulåmo sediments.

Finally, the geological setting of both the Turifoss greenstones and the porphyrite dykes in the Fundsjø does not accord with their formation as part of a juvenile ocean floor. The formal 'ocean-floor' chemical affinities of these rock therefore cannot be interpreted literally, but only as an indication of the general geotectonic regime and the mantle sources of magma that was mobilized during the generation of the Sulåmo sedimentary basin.

Acknowledgements

The work carried out by T. Grenne was financed through grant No D.48.22–12 'Greenstone geochemistry and sulphide formation in the Caledonian mountain chain' from the Norwegian Research Council for Science and Humanities (NAVF). The work carried out by B. Lagerblad was financed by the Swedish Natural Science Research Council (NFR) via the Swedish Geodynamic Project. The authors are grateful to Drs. D. Roberts, F. M. Vokes, R. Gorbatschev, and A. Reinsbakken for comments on the first draft of the manuscript, and for correcting the English text. Chemical analyses were carried out by the Geological Survey of Norway and by Ingrid Johansson at the Dept. of Geology, University of Lund.

References

Birkeland, T. and Nilsen, O. 1972. Contact metamorphism associated with gabbro in the Trondheim region. *Norges geol. Unders.*, **273**, 13–22.

Bugge, C. 1912. Lagfølgen i Trondhjemsfeltet. (Aarbok for 1912, Nr. 2). *Norges geol. Unders.*, **61**, 27 pp.

Chaloupsky, J. 1970. 'Geology of the Hølonda–Hulsjøen area, Trondheim region. *Norges Geol. Unders.*, **266**, 277–304.

Furnes, H., Roberts, D., Sturt, B. A., Thon, A. and Gale, G. H. 1980. Ophiolite fragments in the Scandinavian Caledonides. *Proc. Intern. Ophiolite Symp.*, Geol. Survey, Cyprus, 582–600.

Gale, G. H. 1974. Reconnaissance geochemical survey of Caledonian volcanics and intrusives in central and south Norway. *Norges geol. Unders.*, Rpt. 1228.

Gale, G. H. and Roberts, D. 1972. Paleogeographical implications of greenstone petrochemistry in the southern Norwegian Caledonides. *Nature*, London, *Phys. Sci.*, **238**, 82, 60–61.

Gale, G. H. and Roberts, D. 1974. Trace element geochemistry of Norwegian Lower Palaeozoic basic volcanics and its tectonic implications. *Earth Planet. Sci. Lett.*, **22**, 380–390.

Getz, A. 1890. Graptolitførende skiferzoner i det trondhjemske. *Nyt Mag. Naturv.*, **31**, 31–42.

Gill, J. B. 1976. Composition and age of Lau Basin and Ridge volcanic rocks: implications for evolution of an interarc basin and remnant arc. *Bull. Geol. Soc. Am.*, **87**, 1384–1395.

Grenne, T., Grammeltvedt, G. and Vokes, F. M. 1980. Cyprus-type sulphide deposits in the western Trondheim District, central Norwegian Caledonides. *Proc. Intern. Ophiolite Symp.*, Geol. Survey Cyprus, 582–600.

Grenne, T. and Roberts, D. 1980. Geochemistry and volcanic setting of the Ordovician Forbordfjell and Jonsvatn greenstones, Trondheim Region, central Norwegian Caledonides. *Contrib. Mineral. Petrol.*, **74**, 374–386.

Guezou, J.-C. 1978. Geology and structure of the Dombås–Lesja Area, southern Trondheim region, south–central Norway. *Norges geol. Unders.*, **340**, 1–34.

Hardenby, C. 1980. Geology of the Kjølhaugan area, eastern Trøndelag, central Scandinavian Caledonides. *Geol. Fören. Stockh. Förh.*, **102**, 475–493.

Hardenby, C., Lagerblad, B. and Andréasson, P. G. 1981. Structural development of the northern Trondheim Nappe Complex, central Scandinavian Caledonides.

(Abstract, Uppsala Caledonide Symposium.) *Terra Cognita*, **1**, 50.

Horne, G. S. 1979. Mélange in the Trondheim Nappe suggests a new tectonic model for the central Norwegian Caledonides. *Nature*, London, **281**, 267–270.

Kisch, H. J. 1962. *Petrographical and Geological Investigations in the South-western Tydal Region*. Thesis, Amsterdam, 136 pp.

Klingspor, I. and Gee, D. G. 1981. Isotopic age-determination studies of the Trøndelag trondhjemites. (Abstract, Uppsala Caledonide symposium.) *Terra Cognita*, **1**, 55.

Kulling, O. 1961. On age and tectonic position of the Valdres Sparagmite. *Geol. Fören. Stockh. Förh.*, **83**, 210–214.

Lagerblad, B. 1983. Tectono-metamorphic relationship of the Gula Group–Fundsjø Group contact in the Inndalen–Faeren area, Trøndelag, central Norwegian Caledonides. *Geol. Fören Stockh. Förh.*, **105**, 131–155.

Miyashiro, A. 1975. Classification, characteristics and origin of ophiolites. *J. Geol.*, **83**, 249–281.

Nilsen, O. 1971. Sulphide mineralization and wall rock alteration at Rødhammeren mine, Sør–Trøndelag, Norway. *Norsk geol. Tidsskr.*, **51**, 329–354.

Nilsen, O. 1978. Caledonian sulphide deposits and minor iron-formations from the southern Trondheim region, Norway. *Norges geol. Unders.*, **340**, 35–85.

Olesen, N. Ø., Hansen, E. S., Kristensen, L. H. and Thyrsted, T. 1973. A preliminary account on the geology of the Selbu–Tydal area, the Trondheim region, central Norwegian Caledonides. *Leidse Geol. Meded.*, **49**, 259–276.

Oftedahl, Chr. 1968. Greenstone volcanoes in the central Norwegian Caledonides. *Geol. Rdsch.*, **57**, 920–930.

Pearce, J. A. 1975. Basalt geochemistry used to investigate past tectonic environments on Cyprus. *Tectonophysics*, **25**, 41–67.

Pearce, J. A. 1980. Geochemical evidence for the genesis and eruptive setting of lavas from Tethyan ophiolites. *Proc. Intern. Ophiolite Symp.*, *Geol. Survey Cyprus*, 582–600.

Pearce, J. A. and Cann, J. R. 1973. Tectonic setting of basic volcanic rocks determined using trace element analyses. *Earth Planet. Sci. Lett.*, **19**, 290–300.

Reay, A., Rooke, J. M., Wallace, R. G. and Whelan, P. 1974. Lavas from Niuafo'ou Island, Tonga, resemble ocean-floor basalts. *Geology*, **2**, 605–606.

Roberts, D. 1978. Caledonides of south central Norway. In McLaren, D. J. (Ed.), *IGCP Project 27, Caledonian-Appalachian Orogen of the North Atlantic Region*. *Geol. Pap. Surv. Can.*, 78–13, 31–37.

Roberts, D., Springer, J. and Wolff, F. Chr. 1970. Evolution of the Caledonides in northern Trondheim region, central Norway: a review. *Geol. Mag.*, **107**, 133–145.

Roberts, D. 1980. Petrochemistry and palaeogeographic setting of Ordovician volcanic rocks on Smøla, central Norway. *Norges geol. Unders.*, **359**, 43–60.

Roberts, D., Grenne, T. and Ryan, P. D. Ordovician marginal basin development in the central Norwegian Caledonides. *Volcanic processes in marginal basins, 1982, Symposium, Proceedings* in press.

Rui, I. J. 1972. Geology of the Røros district, south eastern Trondheim region with a special study of the Kjøli-skarvene–Holtsjøen area. *Norsk geol. Tidsskr.*, **52**, 1–21.

Rui, I. J., and Bakke, I. 1975. Stratabound sulphide mineralization in the Kjøli area, Røros district, Norwegian Caledonides. *Norsk Geol. Tidsskr.*, **55**, 51–75.

Ryan, P. D., Williams, D. M. and Skevington, D. 1980. A revised interpretation of the Ordovician stratigraphy of Sør–Trøndelag and its implications for the evolution of the Scandinavian Caledonides. *Mem. Virginia Poly. Inst. and State Univ.*, **2**, 99–103.

Shiraki, K., Kuroda, N., Maruyama, S. and Urano, H. 1978. Evolution of the Tertiary volcanic rocks in the Izu–Mariana arc. *Bull. Volc.*, **41**, 548–562.

Stephens, M. B. 1981. Evidence for Ordovician arc build-up and arc splitting in the Upper Allochthon of central Scandinavia (Abstract, Uppsala Caledonide symposium.) *Terra Cognita*, **1**, 75–76.

Størmer, L. 1941. Dictyonema Shales outside the Oslo region. *Norsk geol. Tidsskr.*, **20**, 161–167.

Torske, T. 1965. Geology of the Mostadmarka and Selbustrand area, Trøndelag. *Norges geol. Unders.*, **232**, 1–83.

Vogt, Th. 1945. The geology of part of the Hølonda–Horg district, a type area in the Trondheim region. *Norsk geol. Tidsskr.*, **25**, 449–528.

Wilson, J. R. 1981. The synorogenic Fongen–Hyllingen layered basic intrusion, Trondheim region, Norway—a review. (Abstract, Uppsala Caledonide Symposium.) *Terra Cognita*, **1**, 82.

Wolff, F. Chr. 1964. Stratigraphical position of the Gudå conglomerate zone. *Norges geol. Unders.*, **227**, 85–91

Wolff, F. Chr. 1967. Geology of the Meråker area as a key to the eastern part of the Trondheim region. *Norges geol. Unders.*, **245**, 123–142.

Wolff, F. Chr. 1973. Meråker og Faeren. Beskrivelse til de berggrunnsgeologiske kart (AMS-M711) 1721 I og 1722 II—1:50 000. *Norges geol. Unders.*, **295**, 1–42.

Wolff, F. Chr. 1979. Beskrivelse til de berggrunns-geologiske kart Trondheim og Østersund 1:250 000. *Norges geol. Unders.*, **353**, 76 pp.

Wolff, F. Chr. and Roberts, D. 1980. Geology of the Trondheim region. In Wolff, F. Chr. (Ed.), *Excursions Across Part of the Trondheim Region, Central Norwegian Caledonides*. *Norges geol. Unders.*, **356**, 117–128.

Metamorphism

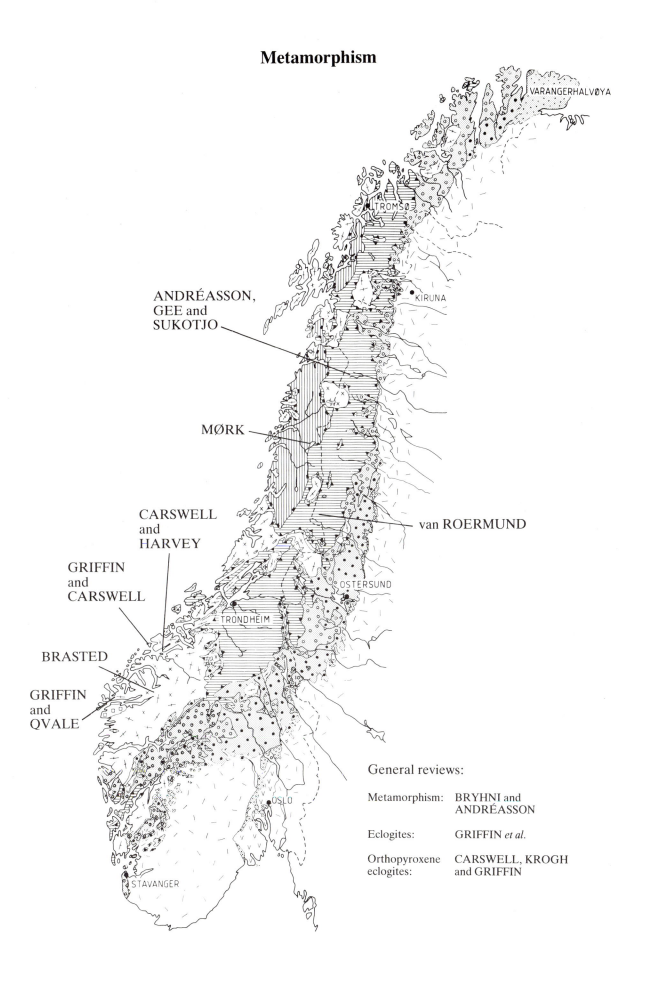

VARANGERHALVØYA

TROMSØ

KIRUNA

ANDRÉASSON,
GEE and
SUKOTJO

MØRK

van ROERMUND

CARSWELL
and
HARVEY

OSTERSUND

GRIFFIN
and
CARSWELL

TRONDHEIM

BRASTED

GRIFFIN
and
QVALE

General reviews:

Metamorphism: BRYHNI and
                  ANDRÉASSON

Eclogites: GRIFFIN *et al.*

Orthopyroxene   CARSWELL, KROGH
eclogites: and GRIFFIN

OSLO

STAVANGER

The Caledonide Orogen—Scandinavia and Related Areas
Edited by D. G. Gee and B. A. Sturt

Metamorphism in the Scandinavian Caledonides

I. Bryhni[*] and **P.-G. Andréasson**[†]

[*]*Mineralogisk–Geologisk Museum, Sars gate 1, N–OSLO 5, Norway*
[†]*Dept. Mineralogy and Petrology, University of Lund, Sölvegatan 13, S–223 62 LUND, Sweden*

ABSTRACT

Various erosional levels in the Scandinavian Caledonides display rocks from a wide range of shallow to very deep-seated metamorphic environments. Extensive nappe displacement transported and often telescoped original isograd patterns. In the central and eastern belts of the orogen, the present-day pattern of regional metamorphism therefore largely coincides with the nappe pattern, and multiply metamorphosed high-grade units are locally juxtaposed with low-grade units which bear witness to only a single metamorphic event.

The tectonometamorphic development was episodic with major events in the Late Precambrian, Middle Cambrian to Middle Ordovician, Late Silurian, and Devonian.

Precambrian elements constitute large parts of the orogen. In the lower and middle allochthonous units in the east, they were only slightly affected by Caledonian metamorphism. The upper and uppermost allochthons and some of the basement regions in the west were, on the other hand, strongly overprinted by Caledonian metamorphism, probably culminating with eclogites formed at progressively higher pressures and temperatures westwards.

Such variations can be understood in terms of westwards subduction of the margin of the Baltic Shield to depths exceeding fifty kilometres along the present coast. This probably Late Silurian collisional phase was followed by tectonic stripping of basement, final thrusting onto the foreland, rapid upheaval, and erosion.

Introduction

General introductions to metamorphism in the Scandinavian Caledonides have been given previously by M. Gustavson, A. G. B. Strömberg, and H. J. Zwart in *Metamorphic Map of Europe* (Zwart 1978). More recent reviews were provided by Bryhni and Brastad (1980), Andréasson and Gorbatschev (1980), and Bryhni (1983). All these studies emphasize problems in discriminating between Caledonian and earlier metamorphic events. Difficulties are also related to incomplete knowledge about the effects of various Caledonian events (Fig. 1) e.g. the 'Svalbardian', 'Horg', 'Ekne', and 'Trondheim disturbances' of Vogt (1929, 1945) or the 'Finnmarkian' and Late Silurian ('Scandinavian', 'Scandian') events of Ramsay and Sturt (1976) and Gee (1975). Only a few fossil finds and scattered and often controversial radiometric data are as yet available to assess age relations. To avoid this problem, we will consider all metamorphic rocks within the Scandinavian Caledonides irrespective of age; in most cases only peak metamorphism is referred to. Facies conventions (Table 1) are those agreed upon by the working group for the IGCP–CO Metamorphic Map (Fisher 1980) although other schemes (e.g. Zwart *et al.* 1967) have also been of much guidance.

On the basis of the tectonostratigraphic subdivision of the Scandinavian Caledonides presented in this volume (accompaning map 1), we first review metamorphism in the various tectonostratigraphic units. Following a discussion of the timing of metamorphism, a model for the metamorphic evolution of the Scandinavian Caledonides is briefly outlined.

Metamorphism in the Tectonostratigraphic Sequence

Materials subjected to Caledonian metamorphism were of two essential types:

1. Precambrian crystalline continental basement, and nappes, derived from such basement.
2. Upper Riphean and Vendian to Lower Palaeozoic sediments deposited on the Baltoscandian margin together with sediments, volcanics, and plutonites from oceanic, island arc, and trench environments.

The IGCP Tectonostratigraphic Map of the Scandinavian Caledonides (this volume) recognizes the following principal units: Autochthon, Parautochthon, Lower Allochthon, Middle Allochthon, Upper and Uppermost Allochthons, and Old Red Sandstone basins. Schematic illustrations of the vertical variation of metamorphism throughout the tectonostratigraphic succession are given in Fig. 2 and Table 2. Reference is also made to the simplified metamorphic map in Fig. 3.

Autochthon and Parautochthon

A thin or only locally developed Late Precambrian–Lower Palaeozoic cover overlies crystalline basement of Svecokarelian to Sveconorvegian (approx. Grenvillian) age (Gorbatschev, this volume). Basement was only weakly affected by Caledonian metamorphism in thc east but is increasingly 'Caledonized' westwards, toward the presumed hinterland, especially in upper parts and along thrust zones. In the cover, metamorphism increases from subgreenschist facies in the east to greenschist facies in the west or possibly even higher grade at the coast.

It should be noticed that some of the basement windows and culminations treated here may not be autochthonous but rather belong to the Lower Allochthon.

Basement

A weak Caledonian influence on the basement can be observed far to the east of the present-day Caledonian Front. Thus, prehnite–pumpellyite retrogressed rocks from Dalarna (Sweden) yield K–Ar whole-rock ages up to some 300 Ma lower than the about 900 Ma ages obtained from non-retrogressed rocks (Verschure 1981). Prehnite and pumpellyite are locally developed also in the basement east of Stavanger in southern Norway (Verschure *et al.* 1980). In the latter area, basement rocks within a ten kilometres wide zone along the nappe front underwent prograde metamorphism up to greenschist facies. Stilpnomelane and fine-grained green biotite become gradually more and more common towards the Caledonian Front, while brown biotite, pyroxenes, and plagioclase break down. K–Ar and Rb–Sr ages of the green biotite indicate Caledonian origin while 'fresh' brown biotite retains its Sveconorvegian age. Orthopyroxene is altered first into serpentine minerals and then into amphibole as the front is approached.

In the southwest, basal conglomerates uncon-

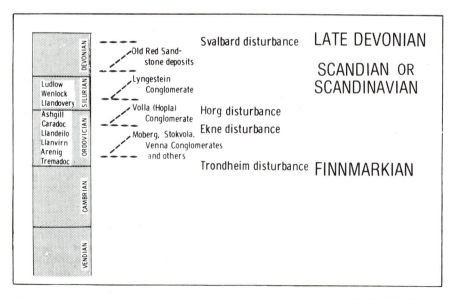

Fig. 1   Timing of diastrophism in the Scandinavian Caledonides, mainly after Vogt (1929, 1945), Gee (1975), Ramsay and Sturt (1976), and Sturt *et al.* (1978)

Table 1   Facies conventions used in this paper are those agreed upon at the Symposium in Blacksburgh, Va., September 1979 (Fisher 1980): cf. also *Metamorphic Map of Europe, Explanatory Text* (Zwart 1978). Blueschist facies is excluded

Symbol	Name	Diagnostic minerals	Common minerals	Forbidden minerals
W	Weakly metamorphosed	Kaolinite montmorillonite	Chlorite white mica dolomite + dolomite + quartz ankerite	
GS	Greenschist facies	Chlorite + actinolite biotite in pelites	Epidote chloritoid albite muscovite calcite dolomite stilpnomelane Mn-garnet	Staurolite kyanite andalusite cordierite plagioclase An_{10} prehnite + pumpellyite
E	Epidote amphibolite facies	Al-garnet in pelites albite of oligoclase + epidote + hornblende	Biotite chlorite muscovite	Staurolite kyanite andalusite cordierite
A_{AA}	Lower amphibolite facies	Staurolite or kyanite in pelites diopside in calc-silicates	Garnet biotite muscovite hornblende + plagioclase cordierite	Sillimanite ortho- + clinopyroxene actinolite + chlorite cordierite + garnet
A_{AB}	Medium amphibolite facies	Sillimanite	Same as A_{AA}	Sillimanite + K-feldspar
A_{AC}	High amphibolite facies	Sillimanite + K-feldspar	Same as A_{AA}	
GN	Granulite facies	Ortho- + clinopyroxene	Garnet cordierite K-feldspar sillimanite hornblende calcic plagioclase	Staurolite muscovite epidote zoisite

formably overlying basement are preserved at places. However, thrusting often produced a tectonic contact zone in which basement received a pervasive foliation parallel to that in the cover. At Hardangervidda, at least two sets of folds are found in the topmost basement and the rocks carry low greenschist facies parageneses with stilpnomelane and muscovite (Naterstad and Jorde, 1981). In areas of intense deformation, metamorphism in the contact zones reached lower amphibolite facies whereas nearby contacts were unaffected. The same development is described from the Tømmerås Window (Andréasson and Lagerblad 1980). Selective mass transport resulted in local enrichment of kyanite in several deep thrust zones. Fist size nodules of kyanite occur in the basement–cover thrust zones of the Tømmerås and Nasafjäll Windows. Kinetic

effects obviously have to be called upon to explain local variation in the Caledonian metamorphic rejuvenation of basement underlying basal thrusts.

A section across the Caledonides north of Trondheim may serve to illustrate the progressive 'Caledonization' of the basement–cover contact westwards. In the east (Jämtland), metamorphism is only epizonal, but increase via greenschist facies and lower amphibolite facies towards the Norwegian coast, with garnet, kyanite, and rare staurolite stable in the basal cover of the Tømmerås Window. To the west of the window, there is a dramatic increase in ductility of deformation of basement and cover, and basement–cover contacts are strictly concordant.

In Nordland, there are rows of basement massifs which are relatively unaffected by Caledonian metamorphism in the east but strongly affected in

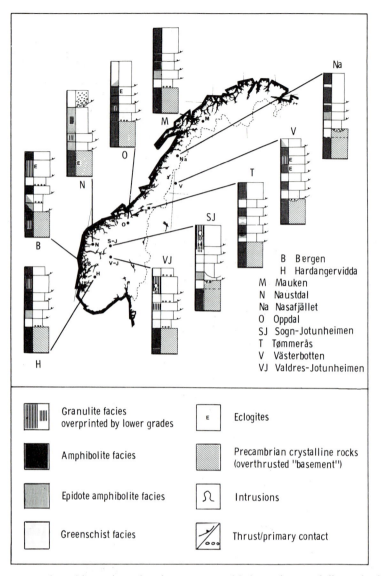

Fig. 2   Simplified tectonostratigraphic sections showing metamorphic inversions and discontinuities related to thrusting in the Scandinavian Caledonides

the west. In the Nasafjäll Window, autochthonous cover metasediments in the east carry quartz + albite + muscovite + biotite/chlorite (Gjelle 1978). In the west, the foliated uppermost basement contains garnetiferous amphibolites (J. E. Lindqvist personal communication 1982).

Near Narvik, the eugeosynclinal sequence was thrust upon crystalline basement of the Lofoten–Vesterålen area. Tull (1977) recognized a Caledonian influence on the basement 15–20 km away from the thrust. However, despite the amphibolite facies metamorphism of the overriding cover, metamorphism in the basement was of low grade. Farther north, at Vanna, basement is relatively unaffected; however, here metamorphism in the overriding cover did not exceed lower greenschist facies (Binns et al. 1980). At Lofoten, thrusting at amphibolite facies conditions heated and

deformed the immediately underlying basement (Bartley 1981).

Clearly, several factors may have contributed to the Caledonian metamorphism of basement. While heat from below is considered unimportant because the available time was far too short for the geotherm to recover (Andréasson 1982), heat flux from the overriding allochthon, as indicated by the examples from Vanna and Lofoten respectively, must be taken into account. Locally, strain induced prograde metamorphism, including strain enhanced mobility of fluids, mass transfer, and strain heating, was effective. Survival of Precambrian parageneses and microstructures in the basement below strongly metamorphosed nappes, can be explained by the absence, in particular in granulite facies terrains, of H_2O to catalyse reactions and promote diffusional transport.

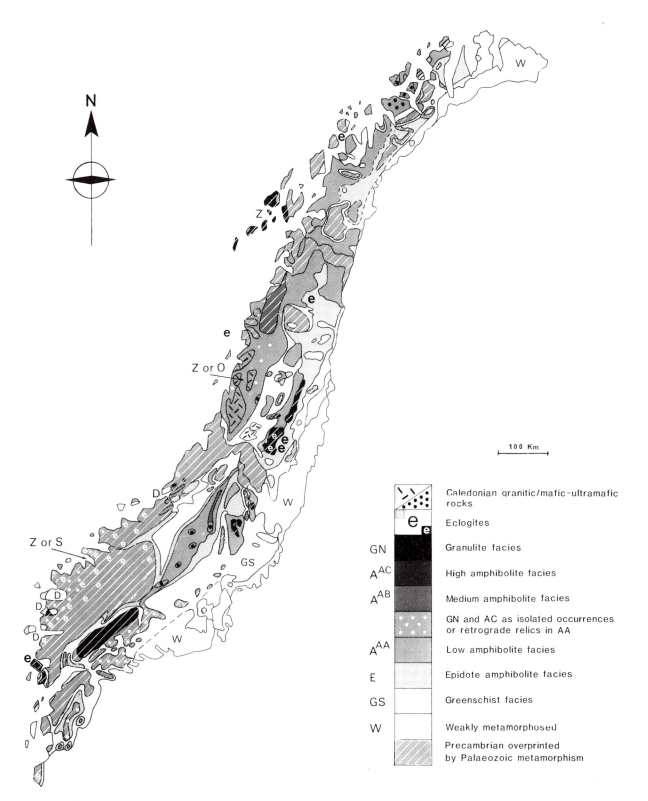

Fig. 3 Metamorphism in the Scandinavian Caledonides, irrespective of age. Much of the indicated grade of metamorphism is related to Precambrian rocks which became transported as nappes and only in part overprinted during the Caledonian orogeny. Facies conventions are those agreed upon by the IGCP working group for metamorphism (Fisher 1980); cf. Table 1

Table 2 Tectonostratigraphic succession, with grade of metamorphism indicated in a very simplified way

Unit	Age	Metamorphism	Nappe complexes
Uppermost Allochthon	Precambrian crystalline rocks ('basement') and possibly Lower Palaeozoic cover strongly metamorphosed during Caledonian orogeny. Multiple tectonometamorphic events and basement–cover relations.	Lower to medium amphibolite facies. At places, greenschist facies rocks, with either tectonic of Barrovian relations to amphibolite facies terrains. Eclogites occur in the Tromsø area and on Træna. Contact metamorphism around syntectonic intrusions.	Beiarn Helgeland Rödingsfjället Tromsø
Upper Allochthon	Precambrian crystalline rocks ('basement') overlain by Upper Proterozoic and Lower Palaeozoic supracrustals strongly metamorphosed during Caledonian orogeny. Multiple tectonometamorphic events in some lower elements.	Variation from greenschist facies to upper amphibolite facies, in part related to the presence of Precambrian elements, in part to (transported) Barrovian zonation. Syntectonic intrusions during peak regional metamorphism. Part of the Seve carries granulites, retro-eclogites and rare glaucophane (retrograde product). Local cover of Devonian rocks in subgreenschist to greenschist facies.	Mageröy, Lyngen Gasak Gjersvik Meråker, Gula, Støren Köli, Storfjället Seve Sunnhordaland
Middle Allochthon	Precambrian crystalline rocks ('basement') overlain by Upper Proterozoic sediments. Basement–cover relations. In N. Norway, extensive igneous and metamorphic development during the early Caledonian Finnmarkian event.	Precambrian elements in variable grade from low (Raipas) to granulite facies (Jotun, Bergen Anorthosite Complex). Caledonian metamorphism of Precambrian elements ('basement') is mainly dynamic and of low grade in the east but may increase westwards in south Norway where it may produce eclogites in zones of high strain. Sedimentary cover in greenschist facies (N. Norway: higher).	Laksefjord, Kalak/Reisa Akkajaure Särv, Sætra, Leksdal Offerdal, Risberget Kvitvola, Valdres Jotun, Bergen Anorthosite Hardangervidda/Ryfylke
Lower Allochthon	Locally preserved Precambrian crystalline rocks ('basement') below dominant Upper Proterozoic and Lower Palaeozoic sediments.	Precambrian crystalline rocks are mainly dynamically metamorphosed in Caledonian time and have retained Precambrian assemblages up to amphibolite facies. Cover is dominantly in greenschist facies or lower; in south Norway possibly reaching amphibolite facies.	Gaissa/Tierta Rautas Jämtlandian Osen/Røa Bergsdalen
Western Gneiss Region	Precambrian crystalline 'basement', possibly also Lower Palaeozoic rocks in a Caledonian nappe complex.	According to one hypothesis, a Precambrian block became remetamorphosed during Caledonian orogeny up to amphibolite, granulite and eclogite conditions. Later retrogression brought part of the block back in epidote amphibolite of greenschist facies. An alternative hypothesis claims that the region suffered only mild Caledonian influence and that high grade assemblages survived from the Precambrian.	Jostedalen Fjordane (may contain units correlative to the lower, middle and upper allochthons)
Autochthon Parautochthon	Precambrian crystalline rocks ('basement') overlain by Upper Proterozoic and Lower Palaeozoic cover.	Caledonian metamorphism varies from subgreenschist facies (local prehnite–pumpellyite) to greenschist facies and even higher in shear (thrust) zones in the east, or in western areas.	

Cover

The cover consists essentially of Vendian–Cambrian psammitic rocks ('sparagmites') and sandstones, shales, and limestones of Lower Palaeozoic age.

Near Oslo, the Lower Palaeozoic cover sequence has a distinct schistosity in pelitic beds and locally also strain-distorted fossils, but the grade of Caledonian metamorphism probably never reached greenschist facies. Sericitic mica is recrystallized 1M muscovite, suggesting near absence of metamorphism (Bjørlykke 1965). Primary organic material in the shales has, however, been transformed into graphitoidal carbon, and conodont colouration indicates temperatures of 300°C or more (T. G. Bockelie, personal communication 1978; Bergström 1980). At Hardangervidda, similar rocks have undergone polyphase deformation and metamorphism up to chlorite and biotite grades and further westwards (in the Hardangerfjord area) even to garnet grade.

In Jämtland, the lowermost autochthonous cover carries newly formed, minute flakes of chlorite and slightly aligned white mica. Illite crystallinity and vitrinite reflectance of higher autochthonous–parautochthonous units testify to a progressive zonal increase of temperature northwestwards, from diagenetic conditions (<200°C) around Östersund in the east to epizone metamorphism >350°C) sixty kilometres farther west (Kisch 1980). Temperatures indicated by conodont colouration were slightly higher than estimated from illite and vitrinite (Löfgren 1978; Bergström 1980). The general increase in temperature westwards through Jämtland probably reflects depth of burial rather than proximity to a front of 'hot' nappes, since in all likelihood, 'hot' nappes once covered the entire belt.

The thin zone of cover along the Caledonian Front from Jämtland to Finnmark is mostly developed as shales and sandstones/quartzites. Biotite is as a rule absent, but there are flattened detrital grains with incipient recrystallization at their margins, undulatory grains and formation of chlorite in cracks (Olesen 1971). Gustavson (1966) found it difficult to decide whether the chlorite and mica of the Hyolithus Zone in southern Troms were of clastic or metamorphic origin; however, the frequency of quartz–chlorite–sphene veins increases towards the overriding sole thrust. Also the clastic character of sandstones decreases towards the thrust.

A large area of unmetamorphosed or subgreenschist facies rocks make up the marginal part of the Caledonides of Finnmark. In the extreme north (Barents Sea Group), metamorphism involved recrystallization of matrix in greywackes, formation of cleavage in pelites and growth of tiny micas (seri-

cite with some chlorite). Grade did not exceed lowest greenschist facies (Roberts 1972). Recrystallization textures have only rarely been observed (Jøsang 1971).

In summary, Caledonian metamorphism in the autochthonous–parautochthonous cover is of very low grade and appears to be related to depth of burial or concentration of strain.

Western Gneiss Region

In southwestern Norway, phyllites and schists of the Lower Allochthon override gneisses of the Western Gneiss Region. The latter has previously been regarded as autochthonous basement. However, recognition of the sole thrust in the Grong–Olden Culmination and the Oppdal area (Gee 1975, 1980) and seismic evidence of a low-velocity zone beneath the Western Gneiss Region (Mykkeltveit et al., 1980) have recently indicated that it may be allochthonous. This review will therefore treat the Western Gneiss Region in a separate section.

Basement–cover relations along the southeastern margin (e.g. at Skjolden) display progressive 'Caledonization' of basement upwards towards the overriding metasediments of low metamorphic grade. There are various categories of basement–cover contacts (Bryhni and Brastad 1980) characterized by increasing reorientation of earlier unconformable relations into quasiconcordant contacts and superimposed Caledonian metamorphism up to lower amphibolite facies. The usually unconformable eastern margin of the gneiss region was in places (e.g. inner Sognefjord) sheared into flaggy white mica gneisses within a zone 10–70 m thick, where two different phases of greenschist facies metamorphism can be recognized (Banham 1968). Irregular structures in the basement are transposed into mylonitic banding in the vicinity of the thrust. This banding resulted from extreme attenuation (Roberts, J. L. 1977). The contact can be traced northwards to the Oppdal area, where basement and cover appear to be folded together into regional folds under amphibolite facies conditions. Since such basement–cover contacts are easily erased with increasing ductile deformation and metamorphism westwards, it is quite possible that nappes occur in the Western Gneiss Region, by analogy with the situation in the Lower and Middle Allochthons of the eastern margin (Gee 1980).

Most of the Western Gneiss Region comprises rocks in amphibolite or lower facies. In the west, this metamorphism overprinted granulite and eclogite parageneses. The age of the eclogitization and subsequent retrogression is at present uncertain but is likely to be late Caledonian (Griffin and Brueckner

1980; Krill and Griffin 1981), while some of the high grade metamorphism is considered also Precambrian in age.

Granulites occur in various environments: *discrete bodies* of orthogneiss mappable on a regional scale (Måløy, Flatraket, Kolåstind) though often gradational into surrounding amphibolite facies gneisses via retrogression; *dolerites* carrying granulite facies assemblages mainly developed as coronas; *veins and layers in eclogite bodies* and as *reaction zones* related to bodies of anorthosite and ultramafite. Other granulite facies rocks occur in *thrust sheets of strongly retrogressed anorthosite-bearing orthogneiss* (Middle Allochthon?) interfolded with feldspathic quartzite and other metasediments.

There are several indications to suggest that granulite facies relicts in the coastal amphibolite-facies terrain were once cofacial with the eclogites of the same area, as first suggested by Kolderup (1955). Clinopyroxene-plagioclase symplectite pseudomorphs after omphacite and even fresh omphacite have been preserved as inclusions within plagioclase-bearing gneisses (Mysen and Heier 1972), and Krogh (1980) concluded that the maximum metamorphic conditions for eclogites and adjacent garnet granulites from the Kristiansund area were comparable ($750° \pm 50°C$, 18.5 ± 3.0 kb). Having different rheological properties and a higher f_{H_2O} as compared to the eclogites; the gneisses are likely to suffer retrogression more readily.

The dolerites of the Western Gneiss Region contain high-pressure granulite facies assemblages developed in coronas (Gjelsvik 1952), which can be found as integral constituents of some dolerite intrusions. Griffin and Råheim (1973) grouped the reactions in different stages; where the product of the last stage is indistinguishable from an eclogite. Interestingly, it appears that shear stress may greatly enhance the final production of eclogite; doleritic bodies are often eclogitized along margins or internal shear zones.

Eclogites are very characteristic constituents of the western part of the gneiss region, where they appear to have been produced at progressively higher pressures and temperatures northwestwards during a Caledonian (?) continent–continent collision. The estimated conditions of formation increase from $500°C$, 12 kb in Sunnfjord in the southeast to about $800°C$, 18 kb along the coast of Møre in the northwest (Krogh 1977; Griffin *et al.*, this volume). The suite of rocks found together with the eclogites is composed essentially of any of the minerals *clinopyroxene, garnet, phengite–quartz–kyanite, quartz–kyanite–zoisite* together with *garnet–omphacite* gneiss and *garnet–kyanite–mica* gneiss. All of these rocks appear to be gradational into true eclo-

gites composed of garnet and omphacite. Calc-silicate rocks associated with the eclogites carry almandine–grossular garnet solid solutions.

Ultramafic bodies occurring in the eclogite terrain contain rocks composed essentially of (*olivine–*) *garnet–diopside* (i.e. garnet clinopyroxenite), (*olivine–*) *garnet–orthopyroxene–diopside* (i.e. garnet websterite), and garnet peridotites (i.e. garnet wehrlite, garnet lherzolite, garnet harzburgite). Some of these rocks (e.g. garnet clinopyroxenites and websterite) are visually very similar to true eclogites and were termed 'eclogites' by Eskola (1921) and others.

The age and origin of eclogites and ultramafic rocks in the Western Gneiss Region is still in dispute (see Griffin *et al.* this volume for a review). The classical two rival petrogenetic theories: *in situ metamorphism* or *solid emplacement from the mantle* each has its adherents. A clue to the metamorphic history is given by the inclusions in garnets of some of the eclogites. Barroisitic–carinthinic amphibole in the cores gives way to omphacite at the rims (Bryhni and Griffin 1971; Krogh 1982). Already Backlund (1936) interpreted the amphibole inclusions in eclogite garnets as 'Gepanzerte Relikte'. Thus, at least some of the eclogites may have evolved from amphibolitic protoliths, and in general, the wide compositional range can be taken as an indication that the eclogites formed from a variety of rocks (basalts, dolerites, gabbros, and even marly sediments) by *in situ* metamorphism under extreme pressure.

Lower Allochthon

In the Lower Allochthon, sheets of Precambrian crystalline rocks are relatively subordinate in the foreland; they dominate in the hinterland. They are present as tectonic wedges or sheets in which primary Precambrian mineral assemblages have been replaced by products of lower grade and dynamic metamorphism. Caledonian metamorphism is typically of greenschist facies. However, internal parts of nappes may be almost unaffected while sections close to the thrusts or situated far to the west towards the hinterland appear to have epidote amphibolite facies or even higher grade.

Precambrian crystalline rocks ('Basement')

Crystalline Precambrian rocks occur as short-travelled tectonic wedges with greenschist facies assemblages superposed upon Precambrian igneous or metamorphic assemblages. Such locally retrogressed 'basement' sheets can be found in allochthonous phyllites and schists in, for instance, the

Stavanger–Sognefjorden area. Other vast areas of allochthonous Precambrian crystalline rocks are found in the Oppdal, Tømmerås, and Grong–Olden districts of the central Caledonides. There, superposed Caledonian metamorphism locally reached epidote amphibolite to amphibolite facies. In the Nasafjäll Window, Precambrian rhyolitic porphyries of the Gargatis Nappe may be excellently preserved only some tens of metres away from thrust zones, as are ripple marks in the overlying sandstones of the Laisvall Group.

Cover

The cover of the Lower Allochthon consists of Late Riphean to Early Cambrian arkoses ('sparagmites') and quartzites and Early Palaeozoic shales, phyllites, and schists. There is an increase in metamorphic grade westwards, but greenschist facies phyllites are the most widely distributed rocks, carrying biotite or chlorite. Intercalations of higher-grade basement sheets sometimes make it difficult to assess grade of regional metamorphism. Thus, in the schists near Stavanger, lenses of muscovite–chloritoid–kyanite schist, garnet–chloritoid schist, and relics of sillimanite and andalusite parageneses occur (Sigmond 1978). This succession may actually represent a pre- or early Caledonian tectonic interlayering within a Caledonian cover sequence.

Metamorphism sometimes increases step-wise upwards from one tectonic sheet to another. Thus, in the Nedstrand area, near Hugesund, there are successive steps upsection from the chlorite zone at the base to the garnet zone at the top (Riis 1977). Such variation is clearly related to telescoping of the nappe pile.

In other places, there are gradual transitions across the subgreenschist–greenschist facies boundary. Sparagmites in southeastern Norway (Kumpulainen and Nystuen, this volume) crystallized muscovite of both 1M and 2M type in the southeast but only 2M muscovite and locally stilpnomelane in the northwest (Englund 1973). In Sweden, 'sparagmites' made up a considerable part of the Lower Allochthon. Here, increasing metamorphism can be observed as a colour change from red/violet to grey/green (Kulling 1942). The process includes sericitization of feldspars, formation of chlorite and epidote, and aggregation of red iron oxide dust into larger grains of magnetite and hematite. Kulling (1942) recognized three stages of metamorphism on the basis of phyllosilicate mineralogy: (1) 'lowest grade', with only sericitization; (2) 'medium-grade', with sericite + chlorite; and (3) 'highest grade', with sericite, chlorite, and green biotite. This metamorphism is essentially strain-induced and of local significance.

Middle Allochthon

This allochthon was derived from crystalline Precambrian basement of the Baltoscandian margin and the clastic wedge deposited on the margin in Late Riphean to Cambrian time. Thick masses of crystalline rocks with often well-preserved Precambrian metamorphic assemblages cover vast areas of southern Norway. The Caledonian metamorphic overprint is generally cataclastic and of low grade. In the west and north, the situation is partly different. Along the west coast, Caledonian overprint is strong and units presumed to relate to the Middle Allochthon have been deeply folded into the basement with obliteration of original contacts and with metamorphism to amphibolite facies. Even eclogites appear to have formed in these western areas, in zones of high Caledonian (?) shear (near Stavanger and Bergen). In northernmost Norway, the Middle Allochthon includes also Caledonian intrusions and interleaved basement and cover, with metamorphism varying from greenschist to granulite facies.

Cover units in the east bear witness to mainly greenschist facies metamorphism. The interior of thicker (up to six kilometres) units was only weakly if at all, affected by Caledonian metamorphism. On the other hand, cover units of the Middle Allochthon possibly make up part of higher grade terrains along the west coast (cf. the section on the Western Gneiss Region).

Precambrian crystalline rocks ('Basement')

The overthrust complex of Stavanger–Hardangervidda carries a distinctive high-pressure granulite unit sandwiched between sheets of lower metamorphic grade. The crystalline units of the Middle Allochthon are often such Precambrian granulites which largely escaped Caledonian overprint. Typical examples in the Jotun Nappe Complex are syenitic to dioritic orthogneisses with microperthite, two pyroxenes and locally garnet, and gabbroid–anorthositic varieties with plagioclase, two pyroxenes, and locally abundant garnet. Mineral reactions include: olivine + plagioclase = orthopyroxene + clinopyroxene + spinel and orthopyroxene + clinopyroxene + spinel = garnet + pyroxene. In gabbro–anorthosites, olivine is rimmed successively by orthopyroxene, clinopyroxene, and garnet, although olivine may often be replaced by orthopyroxene, and clinopyroxene/garnet may be replaced by amphibole. Griffin (1971) recognized the following history of formation of coronites in

Sogn:

1. Crystallization from a mafic magma, at pressures less than 8 kb, and precipitation of olivine + plagioclase + spinel ± aluminous pyroxene.
2. Slow cooling accompanied by an increase in pressure to about 10–12 kb. This resulted in a reaction between olivine and plagioclase to form orthopyroxene–Tschermakitic clinopyroxene + spinel coronas, followed by a reaction between olivine, pyroxenes, spinel, and plagioclase to form garnet and less Tschermakitic pyroxene.
3. Formation of pargasitic amphibole in most coronas, at the expense of garnet and clinopyroxene.
4. Rapid decompression, resulting in symplectitic decomposition of garnet to pyroxenes, plagioclase, and spinel.

Austrheim (1978) suggested a more complex history for the Bergen area, where events of granulite metamorphism alternated with magmatic and deformational events. Thus, Precambrian mangerites with granulite facies assemblages appear to have intruded into an even older granulite terrain. Caledonian (?) deformation resulted locally in fine-grained shear zones with eclogite parageneses (Austrheim and Griffin 1982), but essentially produced mylonites, blastomylonites, and greenschist facies assemblages in the thrust zones. Laminated, schistose or flaggy gneisses at the base of the crystalline complex of the Stavanger–Jotunheimen area have been interpreted (Müller and Wurm 1969; Banham *et al.* 1979) as low-grade metasediments but are rather dynamically metamorphosed rocks (Bryhni 1981).

In Sweden, augen gneisses of greenschist to epidote amphibolite facies grade make up a considerable part of the Middle Allochthon (Tännäs Augen Gneiss Nappe; Granite–Mylonite Nappe). The vast Akkajaure Complex in the north is dominated by sheets of Precambrian syenites; however, gabbroic–anorthositic rocks similar to those in the Jotun Nappe Complex in southern Norway are also known in this part of northern Sweden.

Cover

Shales, quartzites, phyllites, 'Valdres sparagmites' and other metapsammites are often strongly deformed with pronounced schistosity and locally containing strongly flattened or elongate pebbles. In eastern terrains, parageneses seldom indicate metamorphism above greenschist facies. Metapsammites of the Kvitvola, Offerdal, and Särv Nappes are typically of greenschist facies. However, the Särv

Nappe has probably not been above 300°C since the intrusion of the prominent Ottfjäll dolerite dyke swarm about 665 Ma ago (Claesson and Roddick 1983), despite Caledonian metamorphism and transport at least 200 km. Contact metamorphic parageneses are locally preserved in the wall-rock of the dykes (Andréasson and Gorbatschev 1980). Caledonian effects are as a rule restricted to nappe margins, where dolerite dykes have been rotated into parallelism with the margin and foliated. Postkinematic garnets may overgrow flattened plagioclase of such deformed dykes.

In northernmost Norway, nappes referred to as Middle Allochthon differ from those described above by containing an important element of Caledonian intrusions. Sturt and Taylor (1972) related rocks of low-pressure granulite facies (biotite–cordierite–almandine subfacies) in the Seiland Igneous Province to contact metamorphism during the thermal peak of kyanite–sillimanite grade regional metamorphism. Pre-Karelian granulite facies rocks are also present in between the igneous rocks. The Kalak–Reisa Nappe Complex consists of metasediments of assumed Vendian to Cambrian (locally even Ordovician) age together with minor Precambrian elements. Metamorphism varies from greenschist to amphibolite facies with several minor occurrences of granulites. Thus, felsic rocks of Aursfjordbotten carry relics with kyanite, sillimanite, garnet, and clinopyroxene while metabasites contain two pyroxenes, plagioclase, and garnet (Bergh 1980).

In the Middle Allochthon, we meet the first signs of metamorphism related to Caledonian orogenic events which occurred far (200–400 km) to the west of the present-day coast, including contact metamorphism related to riftogenic dyke swarms or syntectonic intrusions. If Caledonian in age, eclogites of the Middle Allochthon require deep orogenic burial. There are, nevertheless, parts of both 'basement' and cover which appear to have escaped significant Caledonian metamorphism.

Upper Allochthon

In this allochthon, a succession of nappes of eugeosynclinal derivation and of generally low grade occurs together with high grade crystalline and supracrustal rocks. A typical example is the Seve–Köli Nappe Complex where Seve rocks were metamorphosed in amphibolite to granulite facies while the Köli attained essentially greenschist facies grade. However, Köli terrains also contain tectonic lenses of higher grade. Along the Norwegian coast, fragments of several ophiolite complexes are typically developed in greenschist or epidote amphibolite facies; some are of lowest greenschist facies and excellently preserved. Trondhjemitic and gabbroic intrusions within higher parts of the Upper Allochthon caused local andalusite–cordierite contact metamorphism. The presence of fossils in eugeosynclinal rocks proved a Early Palaeozoic (up to Lower Silurian) age for this part of the nappe complex. An internal unconformity of pre-Middle Arenig age separates at least two main stages of the metamorphic development. Higher grade members of the Upper Allochthon probably contain both Precambrian and Lower Palaeozoic rocks, metamorphosed during the Caledonian orogeny. Devonian Old Red Sandstone basins were possibly affected by the last movements of the Upper Allochthon.

In the Seve Nappes, upper amphibolite–granulite facies rocks occur as a distinct tectonic unit. The high-grade Central Seve Belt of Jämtland and southern Västerbotten is sandwiched between amphibolite facies belts. Retroeclogites occurring in the Central Belt formed at higher temperatures than those found in the underlying (Eastern) belt (van Roermud 1982). Low temperature eclogites occur also in the Seve of northern Sweden (Andréasson et al. this volume), where, moreover, glaucophane has recently been found (Stephens and van Roermund personal communication 1982).

Mt Åreskutan, a classical locality for pyroxene granulites and other metamorphics occurring on top of greenschist facies, fossiliferous Silurian rocks, is built up by three units. The uppermost consists of gneisses with sillimanite, kyanite, cordierite, and pyrope-rich garnet, and banded basic rocks carrying two pyroxenes, hornblende, and spinel. A late amphibolite facies foliation defined by garnet, biotite, and kyanite wraps around a relict granulite facies foliation (Törnebohm 1872; Arnbom 1980).

The age of the high-grade mineral assemblages in the Seve is still disputed, but the weight of radiometric evidence is presently in favour of a Caledonian metamorphism of Precambrian protoliths (Claesson 1981).

The Köli Nappes rest tectonically on the Seve Nappes, although the contact is transitional or tectonically complex in places. Basic metavolcanic rocks and associated rocks of ophiolite affinity in the Köli Nappes were metamorphosed in greenschist or epidote amphibolite facies (Furnes et al. 1980). Evidence of ocean floor metamorphism has been recorded from the Karmøy and Grong ophiolite fragments. In the former, the evidence includes folded hornblende fabrics cut by sheeted dykes, and clastic grains of amphibole in sediments overlying pillow lavas. While the ultramafites of amphibolite facies and higher grade terrains (i.e. Seve and corresponding units) are mainly serpentinized dunites and garnet peridotites, those of the Köli greenschist facies terrains occur mainly as antigoritic serpentinites. Zones of actinolite, talc, and chlorite can be found along the margins of the bodies (Moore and Qvale 1977; Stigh 1979).

High grade rocks of higher tectonic levels (e.g. the Trondheim Nappe Complex) of the Upper Allochthon yield examples of very complex polyphase metamorphism (Andréasson 1981, 1982). Guezou (1979) recognized the following successive stages of metamorphism in the central Trondheim region:

M_0: Relicts of andalusite, staurolite, and garnet found in the Gula Nappe only.

M_1: Andesine/labradorite + diopside + orthoamphibole + garnet + biotite in calc schists of the Gula Nappe.

M_2: Biotite + sodic plagioclase + tremolite/actinolite + epidote + quartz passing into biotite + garnet + hornblende + plagioclase. This stage is widespread throughout the Trondheim Nappe Complex and was related to large-scale recumbent F_2 folding.

M_3: Albite/oligoclase + actinolitic amphibole + phengite + garnet + epidote + quartz in greenschist facies terrains and kyanite + almandine + oligoclase + hornblende + paragonite in amphibolite facies terrains. These parageneses syn- to postdated tight F_3 folding.

M_4: Postkinematic (-thrusting) growth of biotite, garnet, staurolite, hornblende, possibly related to a thermal event postdating nappe emplacement.

M_5: General retrogression with growth of chlorite, sericite, epidote, albite, and quartz, related to kinking, faulting and cataclastic deformation.

All these metamorphic stages were probably Caledonian. At least M_0–M_1 predate the intrusion of trondhjemite dykes dated older than the Middle Ordovician (Klingspor and Gee 1981), and may be related to the Late Cambrian–Early Ordovician event recognized from various parts of the orogen (Kvale 1960; Sturt and Thon 1975; Ramsay and Sturt 1976). M_2–M_4 may be related to the Late Silurian nappe displacement with its complex structural evolution. Guezou (1979) related the M_0 relicts to early contact metamorphic events.

Several units of the Upper Allochthon were intruded by gabbroid and trondhjemitic rocks. In some cases, contact metamorphic parageneses survived later regional metamorphism. At Frøya–Bremangerland, cordierite–andalusite hornfelses occur adjacent to granodiorite and gabbros (Bryhni and Lyse in preparation). Subsequent regional metamorphism changed andalusite porphyroblasts into micaceous aggregates; locally into margarite or kyanite. However, since relict porphyroblasts are found up to three kilometres from the nearest exposed igneous contacts, it cannot be excluded that this metamorphism was related to an early regional thermal even rather than to local igneous intrusions. Gabbros in the Trondheim region intruded prior to the peak of regional metamorphism (Birkeland and Nilsen 1972; Olesen *et al.* 1973; Bøe 1974). Contact metamorphic minerals largely vanished or were pseudomorphosed into kyanite, paragonite, etc. during successive later events. However, also here, relicts of andalusite, staurolite, and garnet are found far away from any exposed intrusive, and may be interpreted in terms of an early regional thermal metamorphism (Dudek *et al.* 1973; Guezou 1979), provided intrusions do not occur at depth or occured above the present erosion surface.

There are several published descriptions of progressive Barrovian zonation of metamorphism in the Upper Allochthon. In the Trondheim region, Goldschmidt (1915) mapped the regional distribution of chlorite, biotite, garnet, and limesilicate/staurolite–kyanite zones. Although this distribution has been confirmed and extended or only slightly modified in later studies, the interpretation of this pattern is fundamentally different. In general, grade of metamorphism increases down the tectonostratigraphic sections of the Trondheim Nappe (upper) and Seve–Köli Nappe complexes respectively. After a break at the amphibolite facies base of the Gula Nappe, metamorphism again increases downwards from greenschist facies in the Köli Nappes to amphibolite facies in the Seve Nappes.

The metamorphic pattern of the Trondheim region accordingly represents transported metamorphism; the energy sources for the different events occurred at different orogenic levels at least 500 km to the west of the Trondheim region. However, while the pseudo-isozones are in most cases cut by thrusts, the pattern is not entirely controlled by the tectonostratigraphy, because the garnet zone locally transects the margin of the Gula Nappe. This could be taken to indicate that the nappes were hot and buried at the time of juxtaposition.

In the Köli terrain of Jämtland, there is a Barrovian type zonation of metamorphism from chlorite zone, through biotite zone to garnet–hornblende zone at the base in the thrust contact with the Seve nappes. However, since garnet and hornblende developed largely as postkinematic porphyroblasts (garben growth), the observed zonation is not necessarily cofacial (Sjöstrand 1978).

Vogt's (1927) detailed study of the Sulitjelma area showed that metamorphism increases steadily westwards and that the retrogression of primary

gabbroic assemblages can be represented by four metamorphic zones. The zonation is of Barrovian type although oligoclase or a more calcic plagioclase is the typical index mineral instead of staurolite or kyanite due to the somewhat calcareous and feldspathic character of the pelites in the area. The basal thrust of the Gasak Nappe corresponds in the east to the junction between greenschist and amphibolite facies (Mason 1967, 1971; Henley 1971). Below the thrust, plagioclase compositions are well above An_{20} and there appears to be no sharp break across the thrust, as judged also from garnet chemistry. The combined hornblende–plagioclase zone is flat or dipping gently westwards, with higher-grade rocks overlying lower-grade. This inversion may be due either to heat flux from the overlying nappe or long term flux from adjacent gabbro intrusions (Henley 1971). Recent work has shown that the Sulitjelma Gabbro intruded beneath the upper part of a possible ophiolite sequence (Boyle 1980) at a temperature of 650°C and a pressure of 6 kb (Mason 1980). The emplacement occurred nearly simultaneous with the attainment of peak temperatures of regional metamorphism and at higher pressures that can be expected for an ocean floor intrusion.

In northernmost Norway, the Magerøy Nappe features a sequence of Upper Ordovician to Lower Silurian sedimentary rocks metamorphosed during the Late Silurian phase (Andersen 1979). A Barrovian zonal sequence (biotite–almandine–staurolite—kyanite–sillimanite) occurs. Peak metamorphism coincided with mafic and granitic intrusions (411 Ma; Andersen *et al.* 1982). The event affected also underlying rocks of the Kalak Nappe Complex, as indicated by resetting of Rb–Sr isotope systems at 410 Ma.

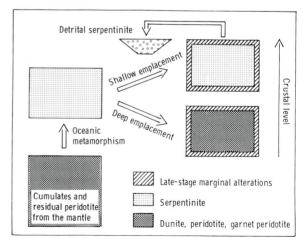

Fig. 4 Model for production of the variety of ultramafites within the orogenic belt by different depth of emplacement or different subsequent metamorphism. It is presumed that serpentinites originally formed by oceanic metamorphism of mantle rocks

In summary, the Upper Allochthon bears witness to a complex Caledonian metamorphic development, in 'basement' rocks overprinting Precambrian metamorphism. The nappes represent very deep, intermediate and very shallow orogenic levels, as evidenced by eclogites, Barrovian zonation and excellently preserved, almost unmetamorphosed pillow lavas respectively. Such a range of P–T environments rather than an overall Barrovian facies type is only to be expected in plate collisional orogenic belts (Andréasson and Gorbatschev 1980, p. 348). While ultramafites of other tectonic units (e.g. Seve and Western Gneiss Region) may be products of granulite or amphibolite facies dehydration, those of the greenschist facies terrains (Köli) may have been emplaced in a cold, water-rich or even surface environment (Fig. 4).

Uppermost Allochthon

Migmatites, gneisses, amphibolites, schists, calc-silicate rocks, and marbles cover large areas of western Nordland and southern Troms. Metamorphism is mostly of amphibolite facies, with garnet, staurolite, kyanite, and sillimanite as frequent minerals. Caledonian metamorphism and structures dominate but reworked Precambrian elements also occur. Rare eclogites and the presence of low grade units indicate juxtaposition of units with different metamorphic histories. Thus, Riis and Ramberg (1981) concluded that units identified as Precambrian crystalline rocks ('basement') retained evidence of a more complicated metamorphic history (migmatization and two amphibolite facies events) than that found in the cover (only a single amphibolite facies event). According to these authors, a large amount of the Uppermost Allochthon is of Precambrian derivation.

The Helgeland Nappe Complex rests partly on medium grade rocks (Rödingsfjället Nappe) and partly on low grade of the Upper Allochthon (Köli). The basal thrust is developed as a thick zone of ultramylonites, mylonites, and blastomylonites, locally passing upwards into gneisses. Variations from 'almandine–staurolite' to 'almandine–kyanite–muscovite subfacies' were described by Nissen (1974). Although a distinct metamorphic zonation on a regional scale has not been detected, staurolite, kyanite, and sillimanite occur in the west.

Low grade greenstones and probably also ophiolites occur on the coast of Nordland, either within the Helgeland Nappe Complex or in underlying offshore units. Gustavson (1978) mapped zones with chlorite, biotite, almandine +chloritoid, and staurolite and calculated a geothermal gradient of c. 33°C km^{-1}. Retro-eclogites occurring on the island of Træna

(Gustavson 1979) may belong to the Helgeland Nappe Complex or an underlying unit.

Large volumes of gabbroid and granitoid intrusions were emplaced at various stages during the Late Precambrian–Caledonian evolution of the Helgeland Nappe Complex. Some of these intrusions were syntectonic and generated contact metamorphism closely following the major amphibolite facies regional metamorphism (Mørk 1981). Evidence of high temperature metamorphism also includes a rare garnet–perthitic microcline assemblage in the Rödingsfjället Nappe and extensive distribution of sillimanite (e.g. Beiarn Nappe).

The Tromsø Nappe Complex is an important element of the Uppermost Allochthon in northernmost Norway. It was metamorphosed mainly in amphibolite facies. The uppermost of the three lithological divisions, the Tromsdalstind Sequence, consists of garnet–kyanite or clinopyroxene-bearing rocks which contain retro-eclogites. Krogh *et al.* (1982) estimated maximum P–T conditions for these rocks to 700–750°C and 10–15 kb. Retrogression with introduction of water locally caused partial melting of the eclogites, leaving garnet as a residual phase. Radiometric dating indicates that this high grade event is of Caledonian age (433 ± 11 Ma).

Metamorphism in the Uppermost Allochthon resembles that in the Upper Allochthon, with a wide P–T range and polyphase development. Syntectonic magmatism contributed to the energy budget of regional metamorphism. Where present, zonal patterns represent transported metamorphism.

Old Red Sandstone

Devonian sandstones and conglomerates of western Norway rest with unconformable contacts on the Upper and Middle(?) Allochthon. Along their eastern borders, they have tectonic boundaries to underlying units. This indicates, that Devonian sediments were tectonically transported together with their underlying rocks at a late stage of the Caledonian orogeny. The Middle Devonian sediments are folded and slightly metamorphosed, mainly of subgreenschist grade, but strongly dynamic and of greenschist grade in the vicinity of the thrust boundaries of the basin. Phyllonites have been produced in the thrust zones together with typical mylonites and blastomylonites.

Timing of Polymetamorphism

As described in previous sections, many of the tectonic elements in the Scandinavian Caledonides are of Precambrian age and have been involved in metamorphic events prior to Caledonian polymetamorphism. Parts of the crystalline basement, and Precambrian crystalline rocks intercalated in the nappes suffered Sveconorwegian (c. Grenvillian) metamorphism overprinted upon older Precambrian events. In some areas, 'basement' includes virtually unmetamorphosed sediments, which, locally, became involved in Caledonian thrusting and metamorphism. The effects of the Caledonian imprint on these ancient constituents of the orogenic belt varied from only slight in the marginal parts (Lower and Middle Allochthon) to quite considerable in areas closer to the orogenic interior or in far-travelled nappes (Western Gneiss Region, Upper, and Uppermost Allochthons).

Caledonian polymetamorphism increases in complexity from lower to higher allochthonous units, i.e. with the travel distance of the nappes. However, excellently preserved and almost unmetamorphosed parts of far-travelled nappes nevertheless exist. Very complex relations are encountered in the essentially eugeosynclinal Upper Allochthon (e.g. the Trondheim Nappe Complex), where five successive stages of metamorphic recrystallization superposed on an early contact or regional thermal event have been recognized. Similar evidence has been recorded elsewhere and interpreted in terms of two major orogenic episodes, one in the Late Ordovician–Early Cambrian (Grampian, Finnmarkian) and the other in the Late Silurian. In the following, some of the recorded features will be broadly grouped according to presumed events and age.

Late Precambrian

Some of the Upper Proterozoic sedimentary successions and volcanics retain vestiges of metamorphism dating from the opening and early stage of the Iapetus Ocean. Psammites of the Särv Nappe appear to have been metamorphosed prior to or during intrusion of riftogenic dyke swarms 665 ± 10 Ma ago (Claesson and Roddick 1983). Ophiolites formed at a later stage carry folded fabrics cut by sheeted dykes and clastic grains of amphibole in pelagic sediments. This presumed early ocean floor metamorphism must have occurred prior to obduction of the ophiolites onto the continental margin.

Middle Cambrian to Middle Ordovician

The obducted ophiolites as well as older rocks in various tectonic units underwent greenschist to amphibolite facies metamorphism prior to deposition of transgressive sequences dated to Late Ordovician–Early Silurian. In the southwestern Caledonides, this early Caledonian metamorphism increases northwards from greenschist facies with stilpnomelane near Stavanger to low amphibolite facies with staurolite near Bergen. The age of the event is pre-Ashgillian, as evident from fossil dating of the overlying metasediments, and possibly bracketed within the interval 465–535 Ma (Rb–Sr ages; Furnes et al. 1983).

In the Trondheim Nappe Complex, this early metamorphism affected Cambrian rocks but predated fossiliferous Arenig beds and trondhjemite intrusions dated as 480–450 Ma old (Klingspor and Gee 1981). It represented the peak of Barrovian metamorphism and occurred prior to the final emplacement of the nappe complex.

The Finnmarkian event in North Norway produced rocks of up to middle amphibolite facies grade and also migmatites in Middle Cambrian–Early Ordovician time, with a climax of regional metamorphism probably at 535 ± 17 Ma and uplift/cooling at 486 ± 27 Ma (Sturt et al. 1978). Metamorphism in the central belt (Sørøy) probably commenced prior to that of the marginal zone and certainly reached a more advanced stage than the apparently simultaneous events occurring at the southwestern edge of the Scandinavian Caledonides.

Middle Silurian to Early Devonian

This phase (Scandian, Scandinavian) involved renewed metamorphism and folding in the ophiolite-bearing older succession as well as metamorphism of the overlying Upper Ordovician–Lower Silurian cover sequence. The event was accompanied or followed by southeastwards translation of major nappes onto the foreland. Vogt (1929) showed that this major orogenic phase took place at the Ludlovian–Downtonian boundary; for marginal parts of the orogenic belt, he suggested a slightly later, Erian movement.

Middle Silurian to Early Devonian tectonometamorphism affected sedimentary successions dated by fossils up to Late Llandovery in western Norway, up to Wenlock (?) in Sweden and up to Downtonian in southeastern Norway. Nappe emplacement must postdate igneous events dated at c. 425 Ma (U–Pb; Wilson et al. 1983). An upper age bracket for the phase is provided by the ORS deposition in Trøndelag which extended into the Downton (Pridoli); possibly even into the Ludlow.

The magnitude of Scandian or Scandinavian metamorphism varied considerably. A most important aspect is the potential production of eclogites and granulites during this phase (Griffin and Brueckner 1980). If the mainly Sm–Nd ages obtained so far do in fact give the age of formation of these high-pressure metamorphic rocks, then we will have to assign also an impressive amount of post-eclogite or post-granulite migmatitization and retrograde metamorphism to the Late Silurian event. A large body of mineral ages and K–Ar whole rock ages around 380–400 Ma testify to the waning stages of the phase or subsequent cooling during uplift.

Devonian

The sediments of Old Red Sandstone basins were folded and thrust during an event (Svalbardian phase of Vogt 1929) which probably took place at the Middle/Upper Devonian boundary. Metamorphism was mainly dynamic and related to faults and thrusts, but reached lower greenschist facies in places. In the Solund district, shearing of conglomerates can be recognized up to two and a half kilometres away from the basal thrust (Nilsen 1968). Some of the dynamic metamorphism related to faults (such as the prominent ENE–WSW lineaments of the Trondheim region) may well be related to the 'Svalbardian' phase.

A Model for the Metamorphic Evolution of the Scandinavian Caledonides

Our present knowledge about the complex metamorphic evolution of the Scandinavian Caledonides allows application of only a very simplified model for a Wilson cycle (Fig. 5):

Weak *Late Precambrian metamorphism* may have been related to continental rifting (Fig. 5A, B), accompanied by deposition of thick successions of psammitic sediments and intrusion of an extensive dolerite dyke swarm. At this stage, metamorphism was probably caused by burial and riftogenic volcanism c. 650–750 Ma ago. Subsequent oceanic contraction was accompanied by low grade metamorphism now recorded in some of the ophiolite fragments.

The *Middle Cambrian to Middle Ordovician metamorphism* occurred during multistage contraction and closure of the Iapetus Ocean (Fig. 5C, D). This took place along sutures located several (at least 500) kilometres to the west of the the present orogen. Polarity of subduction is disputed but essentially unknown, and may have changed through time.

The model permits obduction of oceanic crust, igneous activity, and high temperature metamorphism followed by intermediate pressure metamorphism up to lower amphibolite facies. Immature island arc activity (c. 535 Ma), obduction of ophiolites upon continental margin and subsequent Andean type

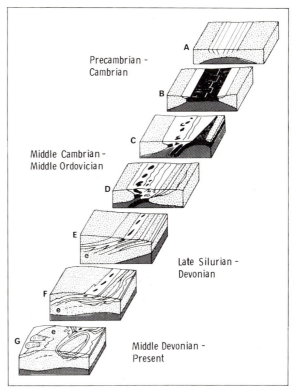

Fig. 5   A model for the tectonometamorphic development of the Scandinavian Caledonides with rifting/drifting in the Late Precambrian–Cambrian (A,B), oceanic contraction with obduction in the Middle Cambrian–Middle Ordovician (C,D), and continental collision with extensive formation of thrust nappes during the Late Silurian–Devonian. 'e' stands for production of eclogites within the crust

volcanism (465 Ma; Furnes *et al.* 1983) are important stages of this event. In north Norway, the Finnmarkian event had probably reached an advanced stage in its tectonometamorphic development 535 Ma ago, with intermediate pressure metamorphism in amphibolite facies and formation of migmatites.

Middle Silurian to Early Devonian metamorphism reflects the collisional stage, whereby the Baltoscandian margin became overridden by a thick succession of imbricated basement slices and far-travelled thrust nappes (Gee 1975; Roberts and Gale 1978; Cuthbert *et al.* 1983). This was the major stage of regional metamorphism, dynamic metamorphism and inversion of the metamorphic gradients along the whole 1800 km length of the orogenic belt in Scandinavia (Fig. 5E, F). Many of the nappes derived from the basement or its cover were transported as sheets with virtually no internal Caledonian metamorphism (e.g. Jotun Nappe Complex, Särv Nappe) while others underwent prethrusting amphibolite and possibly even granulite and eclogite facies metamorphism (e.g. Seve Nappe).

The extensive formation of eclogites and granulites in the Western Gneiss Region can, if really of

Late Silurian age, be related to the pressure increase and thermal blanketing effects of the overriding nappes. The westwards increasing temperatures and pressures of eclogite formation indicate that the nappe overburden increased in that direction, possibly as a result of westwards (WNW) subduction of the Baltoscandian margin beneath the North American plate (Griffin *et al* this volume). The crustal thickening caused by the imbrication and nappe stacking facilitated partial melting and ductile deformation at depth and made it easier for Caledonian structures to develop and in places completely obliterate pre-existing structures. Rapid uplift and denudation must have followed and high-pressure assemblages were subjected to extensive retrogression.

The simple picture presented above relies heavily on the interpretation of the radiometric ages of 407–447 Ma (Griffin and Brueckner 1980) as the real age of eclogite formation. However, it must be admitted, that it is difficult to understand the close juxtaposition of eclogite-bearing gneisses and low grade Lower Palaeozoic rocks along the coast, if the eclogites really are of Late Silurian age.

Devonian metamorphism can be related to extensional and compressional regimes producing faults, folds and thrusts. Metamorphism was evidently mainly dynamic and of greenschist facies or lower grade.

The regional distribution of metamorphic facies can readily be mapped throughout the Scandinavian Caledonides, but the age of metamorphism is often uncertain. Over vast areas, we see mainly pre-Caledonian elements or the effects of Late Silurian thrusting and metamorphism upon older rocks. The main stage of Caledonian development is in some areas superposed upon rocks formed during Middle Cambrian to Middle Ordovician events, which in their turn may be superposed upon products of Precambrian metamorphism. Disentanglement of these age relations forms a major challenge of future research. In addition, the Scandinavian Caledonides provide unique opportunities, for the study of dynamic metamorphism, the interdependence of deformation and metamorphism and the evolution of geotherms in extensive overthrust belts.

Acknowledgements

We wish to thank a number of colleagues who contributed to this review by informal information and stimulating interest. In particular, we acknowledge assistance from T. B. Andersen, W. L. Griffin, and B. A. Sturt for critical reading of an early draft of the manuscript and constructive comments. The authors received financial support from the Norwegian Council for Science and Humanities (NAVF) Grant No. D.46.22–0.25 (I.B.) and the Swedish Natural Science Research Council Grant No. G3559–115 (P.G.A).

References

Andersen, T. B. 1979. *The Geology of South-western Magerøy, with Special Reference to the Tectono-metamorphic Development.* Unpublished Cand. Real. Thesis, Univ. of Bergen, 338pp.

Andersen, T. B., Austrheim, H., Sturt, B. A., Pedersen, S. and Kjærsrud, K. 1982. Rb–Sr whole rock ages from Magerøy, North Norwegian Caledonides. *Norsk geol. Tidsskr.*, **62**, 79–85.

Andréasson, P.-G. 1980. Metamorphism in the Tömmerås area, western Scandinavian Caledonides. *Geol. Fören. Stockh. Förh.* **101**, 273–290.

Andréasson, P.-G. 1981. P–T path of the Gula Nappe, central Scandes. A case history of Caledonian polymetamorphism. *Terra Cognita* **1**, 31 (Abstract).

Andréasson, P.-G. 1982. Structural, metamorphic and petrochemical aspects of nappe displacement in the central Scandies. *Covering article for a dissertation, Department of Mineralogy and Petrology. Institute of Geology, Univ. of Lund,* 44 pp.

Andréasson, P.-G. and Gorbatschev, R. 1980. Metamorphism in extensive nappe terrains: a study of the Central Scandinavian Caledonides. *Geol. Fören. Stockh. Förh.*, **102**, 335–357.

Andréasson, P.-G. and Lagerblad, B. 1980. Occurrence and significance of inverted metamorphic gradients in the western Scandinavian Caledonides. *J. Geol. Soc. London*, **137**, 219–230.

Andreasson, P.-G., Gee, D. G., and Sukotjo, S. this volume. Seve eclogites in the Norrbotten Caledonides, Sweden.

Arnbom, J.-O. 1980. The metamorphism of the Seve Nappes at Åreskutan, Swedish Caledonides. *Geol. Fören. Stockh. Förh.*, **102**, 359–371.

Austrheim, H. 1978. *The Metamorphic Evolution of the Granulite Facies Rocks of Radøy, with Special Emphasis on the Rocks of the Mangerite Complex,* Unpublished Cand. Real. Thesis, Univ. of Bergen, 265 pp.

Austrheim, H. and Griffin, W. L. 1982. Shear deformation and eclogite formation within granulite-facies anorthosites of the Bergen arcs, western Norway. *Terra Cognita*, **2**, 315.

Backlund, H. G. 1936. Zur genetischen Deutung der Eklogite. *Geol. Rundsch.* **27**, 47–61.

Banham, P. H. 1968. The basal gneisses and basement contact of the Hestbrepiggan area, North Jotunheimen, Norway. *Norges geol. Unders.*, **252**, 1–77.

Banham, P. H., Gibbs, A. D. and Hopper, F. W. M. 1979. Geological evidence in favour of a Jotunheimen suture. *Nature*, **277**, 289–291.

Bartley, J. M. 1981. Lithostratigraphy of the Storvann Group. East Hinnøy, North Norway and its regional implications. *Norges geol. Unders.*, **370**, 11–24.

Bergh, S. G. 1980. *Stratigrafiske, Struktur geologiske og Metamorfe Undersøkelser av Kaledonske Bergarter Vest for balsfjord i Troms.* Unpublished Cand. Real. Thesis, Universitetet i Tromsø, 168 pp.

Bergström, S. 1980. Conodonts as paleotemperature tools in Ordovician rocks of the Caledonides and adjacent areas in Scandinavia and the British Isles. *Geol. Fören. Stockh. Förh.,* **102**, 377–391.

Binns, R. E., Chroston, P. N. and Matthews, D. W. 1980. Low-grade sediments on Precambrian gneiss on Vanna, Northern Norway. *Norges geol. Unders.,* **359**, 61–70.

Birkeland, T. and Nilsen, O. 1972. Contact metamorphism associated with gabbros in the Trondheim region. *Norges geol. Unders.,* **273**, 13–22.

Bjørlykke, K. 1965. The geochemistry and mineralogy of some shales from the Oslo region. *Norsk geol. Tidsskr.,* **45**, 435–456.

Boyle, P. A. 1980. The Sulitjelma amphibolites: Part of a Lower Paleozoic ophiolite complex?. In Panayiotou, A. (Ed.), *Ophiolites. Proc. Int. Ophiolite Symp., Cryprus 1979*, Nicosia, Cyprus, 567–581.

Bryhni, I. 1979. Bunnen av Joktundekket (Abstract). 14. Nord. geol. vintermøte 1980. *Geolognytt,* **13**, 12.

Bryhni, I. 1983. Regional overview of metamorphism in the Scandinavian Caledonides. In Schenck, P. A. (Ed.). *Regional Trends in the Geology of the Appalachian— Caledonian–Hercynian–Mauritanide Orogen*, D. Reidel, Dordtrecht, 193–204.

Bryhni, I. and Griffin, W. L. 1971. Zoning in eclogite garnets from Nordfjord, West Norway. *Contr. Mineral. Petrol.,* **36**, 112–125.

Bryhni, I. and Brastad, K. 1980. Caledonian regional metamorphism in Norway. *J. geol. Soc. London,* **137**, 251–259.

Bøe, P. 1974. Petrography of the Gula Group in Hessdalen, southeastern Trondheim region, with special reference to the paragonitization of andalusite pseudomorphs. *Norges geol. Unders.,* **304**, 33–46.

Claesson, S. 1981. *Ages, origins and Metamorphic Histories of Some Major Nappe Units in the Southern Swedish Caledonides*. Doctoral Dissertation, Univ. of Stockholm, 21 pp.

Claesson, S. and Roddick, J. C. 1983. $^{40}Ar/^{39}Ar$ data on the age and metamorphism of the Ottfjället dolerites Särv Nappe, Swedish Caledonides. *Lithos,* **16**, 61–73.

Cuthbert, S. J., Harvey, M. A. and Carswell, D. A. 1983. A tectonic model for the metamorphic evolution of the Basal Gneiss Complex, western south Norway. *J. metamorphic Geol.* 63–90.

Dudek, A., Fediuk, F., Suk, M. and Wolff, F. C. 1973. Metamorphism of the Færen area, central Norwegian Caledonides. *Norges geol. Unders.,* **289**, 1–14.

Englund, J.-O. 1973. Geochemistry and mineralogy of pelitic rocks from the Hedmark Group and the Cambro-Ordovician sequence, southern Norway. *Norges geol. Unders.,* **286**, 1–60.

Eskola, P. 1921. On the eclogites of Norway. *Skr. norske Vidensk.-Akad. Oslo. Kl.* **I, 8**, 118 pp.

Fisher, G. W. 1980. Metamorphism. *Caledonide Orogen Project, U.S. Working Group and Specialist Study Groups, Newsletter,* **2**, 6–9.

Furnes, H., Roberts, D., Sturt, B. A., Thon, A. and Gale, G. H. 1980. Ophiolite fragments in the Scandinavian Caledonides. In Panayiotou, A. (Ed.), *Ophiolites. Proc. Int. Ophiolite Symp., Cyprus 1979*, Nicosia, Cyprus, 582–600.

Furnes, H., Austrheim, H., Amalkisen, K. G. and Nordås, J. 1983. Evidence for an incipient early Caledonian (Cambrian) orogenic phase in southwestern Norway. *Geol. Mag.,* **120**, 607–612.

Gee, D. G. 1975. A Geotraverse through the Scandinavian Caledonides Östersund to Trondheim. *Sveriges geol. Unders.,* **C 717**, 66 pp.

Gee, D. G. 1980. Basement–cover relationships in the central Scandinavian Caledonides. *Geol. Fören. Stockh. Förh.,* **102**, 455–474.

Gjelle, S. 1978. Geology and structure of the Bjøllånes area, Rana, Nordland. *Norges geol. Unders.,* **343**, 1–37.

Gjelsvik, T. 1952. Metamorphosed dolerites in the gneiss area of Sunnmøre on the west coast of southern Norway. *Norsk geol. Tidsskr.,* **30**, 31–134.

Goldschmidt, V. M. 1915. Die Kalksilikatgneise und Kalksilikat-Glimmerschiefer des Trondheim-Gebietes. *Vitensk. Selsk. Skr. I, Mat. Nat. K.,* **10**, 1–37.

Gorbatschev, R. this volume. The Precambrian basement of the Scandinavian Caledonides.

Griffin, W. L. 1971. Genesis of coronas in anorthosites of the Upper Jotun Nappe, Indre Sogn, Norway. *J. Petrology,* **12**, 219–243.

Griffin, W. L. and Råheim, A. 1973. Convergent metamorphism of eclogites and dolerites, Kristiansund area, Norway. *Lithos,* **6**, 21–40.

Griffin, W. L. and Brueckner, H. K. 1980. Caledonian Sm–Nd ages and a crustal origin for Norwegian eclogites. *Nature,* **285**, 319–321.

Griffin, W. L., Austrheim, H., Brastad, K., Bryhni, I., Krill, A. G., Krogh, E. J., Mørk, M. B. E., Qvale, H. and Tørudbakken, B. this volume. High pressure metamorphism in the Scandinavian Caledonides.

Guezou, J.-C. 1979. Geology and structure of the Dombås–Lesja area, southern Trondheim region, south central Norway. *Norges geol. Unders.,* **340**, 1–34.

Gustavson, M. 1966. The Caledonian mountain chain of the southern Troms and Ofoten areas; Part 1, basement rocks and Caledonian metasediments. *Norges geol. Unders.,* **283**, 1–56.

Gustavson, M. 1978. The low-grade rocks of the Skålvær area, S. Helgeland, and their relation to high-grade rocks of the Helgeland Nappe Complex. *Norges geol. Unders.,* **332**, 13–33.

Gustavson, M. 1979. Et meta-eklogitførende gneiskompleks på ytre Helgeland. *Geolognytt,* **12**, 12.

Henley, K. J. 1971. The structural and metamorphic history of the Sulitjelma region, with special reference to the nappe hypothesis. *Norges geol. Unders.,* **269**, 77–82.

Jøsang, O. 1971. Petrografiske undersøkelser ved Vardø. *Norges geol. Unders.,* **270**, 109–128.

Kisch, H. J. 1980. Incipient metamorphism of Cambro-Silurian clastic rocks from the Jämtland Supergroup, central Scandinavian Caledonides, western Sweden: illite crystallinity and 'vitrinite reflectance'. *J. geol. Soc. London,* **137**, 271–288.

Klingspor, I. and Gee, D. G. 1981. Isotopic age-determination studies in the Trøndelag trondhjemites. *Terra Cognita,* **1**, 55 (Abstract).

Kolderup, N.-H. 1955. Sammenhengen mellom, skifer, gneiser og eklogitter i Nordvesttavlen. *Geol. Fören. Stockh. Förh.,* **77**, 257–264.

Krill, A. G. and Griffin, W. L. 1981. Interpretation of Rb–Sr dates from the Western Gneiss Region; a cautionary note. *Norsk geol. Tidsskr.,* **61**, 83–86.

Krogh, E. J. 1977. Evidence for a Precambrian continent–continent collision in western Norway. *Nature, London,* **267**, 17–19.

Krogh, E. J. 1980. Compatible P–T conditions for eclogites and surrounding gneisses in the Kristiansund Area, western Norway. *Contrib. Mineral. Petrol.,* **75**, 378–393.

Krogh, E. J. 1982. Metamorphic evolution of Norwegian country-rock eclogites, as deduced from mineral inclusions and compositional zoning in garnets. *Lithos,* **15,** 305–321.

Krogh, E. J., Andresen, A., Bryhni, I. and Kristensen, S.-E. 1982. Tectonic setting, age and petrology of ecolgites within the uppermost tectonic unit of the Scandinavian Caledonides, Tromsø area, northern Norway. *Terra Cognita,* **2,** 316.

Kulling, O. 1942. Grunddragen av fjällkedjerandens bergbyggnad inom Västerbottens Län. *Sveriges geol. Unders.,* **C 445,** 319 pp.

Kumpulainen, R. and Nystuen, J.-P. this volume. Late Proterozoic basin evolution and sedimentation in the westernmost part of Baltoscandia.

Kvale, A. 1960. The nappe area of the Caledonides in western Norway. Excursion guide. *Norges geol. Unders.,* **212,** 1–43.

Löfgren, A. 1978. Arenigian and Llanvirnian conodonts from Jämtland, northern Sweden. *Sveriges geol. Unders.,* **Ca 37,** 101–296.

Mason, R. 1967. The field relations of the Sulitjelma gabbro, Nordland. *Norsk geol. Tidsskr.,* **47,** 237–248.

Mason, R. 1971. The chemistry and structure of the Sulitjelma gabbro. *Norges geol. Unders.,* **269,** 108–141.

Mason, R. 1980. Temperature and pressure estimate in the contact aureole of the Sulitjelma gabbro, Norway: implications for an ophiolite origin. In Panayiotou, A. (ed.), *Ophiolites. Proc. Int. Ophiolite Symp., Cyprus 1979,* Nicosia, Cyprus, 576–581.

Moore, A. C. and Qvale, H. 1977. Three varieties of alpine-type ultramafic rocks in the Norwegian Caledonides and Basal Gneiss Complex. *Lithos,* **10,** 149–161.

Müller, G. and Wurm, F. 1969. Die Gesteine der Inselgruppe Randøy–Fogn. Beiträge zur Metamorphose und zum Aufbau der Kambro-silurischen Gesteine des Stavanger–Gebietes. I. *Norsk geol. Tidsskr.,* **49,** 97–114.

Mykkeltvedt, S., Husebye, E. S. and Oftedahl, C. 1980. Subduction of the Iapetus ocean crust beneath the Møre Gneiss region, southern Norway. *Nature, London,* **288,** 473–475.

Mysen, B. O. and Heier, K. S. 1972. Petrogenesis of eclogites in high grade metamorphic gneisses, exemplified by the Hareidlandet eclogite, western Norway. *Contr. Miner. Petrol.,* **36,** 73–94.

Mørk, M. B. E. 1981. Geology and metamorphism of the Krutfjell area, Nordland, Norway. *Terra Cognita,* **1,** 60.

Naterstad, J. and Jorde, K. 1981. Basement–cover relationships in southern Norway–Hardangervidda to Karmøy. *UCS Excursion No. B6 Preliminary guide,* 37 pp.

Nilsen, T. H. 1968. The relationship of sedimentation to tectonics in the Solund Devonian district of southwestern Norway. *Norges geol. Unders.,* **259,** 1–108.

Nissen, A. L. 1974. Mosjøen, beskrivelse til det berggrunnsgeologiske kart I 17–1:100 000. *Norges geol. Unders.,* **307,** 29 pp.

Olesen, N. Ø. 1971. The relative chronology of fold phases, metamorphism and thrust movements in the Caledonides of Troms, North Norway. *Norsk geol. Tidsskr.,* **51,** 355–377.

Olesen, N. Ø., Hansen, E. S., Kristensen, L. H. and Thyrsted, T. 1973. A preliminary account of the geology of the Selbu–Tydal area, the Trondheim region, central Norwegian Caledonides. *Leidse Geol. Meddel.,* **49,** 259–278.

Ramsay, D. M. and Sturt, B. A. 1976. The syn-

metamorphic emplacement of the Magerøy Nappe. *Norsk geol. Tidsskr.,* **56,** 291–307.

Riis, F. 1977. *En petrografisk–strukturgeologisk undersøkelse av Nedstrand-området, Ryfylke.* Univ. i Oslo. Hovedoppgave.

Riis, F. and Ramberg, I. B. 1981. The Uppermost Allochton—the rødingsfjället and the Helgeland Complex in a segment south of Ranafjorden. *Terra Cognita,* **1,** 69.

Roberts, D. 1972. Tectonic deformation in the Barents Sea region of Varanger Peninsula. *Norges geol. Unders.,* **282,** 11–39.

Roberts, D. and Gale, G. H. 1978. The Caledonian–Appalachian Iapetus Ocean. In Tarling, D. H. (Ed.), *Evolution of the Earth's crust,* Academic Press, London, 255–343.

Roberts, J. L. 1977. Allochthonous origin of the Jotunheim Massif in southern Norway: a reconnaissance study along its northwestern margin. *J. geol. Soc. London,* **334,** 351–362.

Roermund, H. L. M. van. 1982. *On Eclogites from the Seve Nappe, Jämtland, Central Scandinavian Caledonides,* Unpublished Thesis, Rijksuniversiteit te Utrecht, 99 pp.

Sigmond, E. M. O. 1978. Beskrivelse til det berggrunnsgeologiske kartblad Sauda 1:250 000. *Norges geol. Unders.,* **341,** 1–94.

Sjöstrand, T. 1978. Caledonian geology of the Kvarnbergsvattnet area, northern Jämtland, central Sweden, Stratigraphy, metamorphism, deformation. *Sveriges geol. Unders.,* **C 735,** 1–107.

Sneltvedt, H. S. 1982. En strukturgeologisk og petrografisk undersøkelse av området vest for Svartisen, mellom Glomfjord og Holandsfjord, Nordland. *Geolognytt,* **17,** 47–48.

Stigh, J. 1979. Ultramafites and detrital serpentinites in the central and southern parts of the Caledonian allochthon in Scandinavia. *Chalmers Tekn. Högskola och Göteborgs Univ. Dissertation Publ.,* **A 27,** 222 pp.

Sturt, B. A. and Taylor, J. 1972. The timing and environment of emplacement of the Storelv gabbro, Sørøy. *Norges geol. Unders.,* **272,** 1–34.

Sturt, B. A. and Thon, A. 1975. The age of orogenic deformation in the Swedish Caleodnides. *Am. J. Sci.* **276,** 385–390.

Sturt, B. A. and Roberts, D. 1978. Caledonides of northernmost Norway. In Tozer, E. T. and Schenk, P. E. (Eds), *Caledonian–Appalachian orogen of the North Atlantic Region. Geol. Surv. Can. Paper* **78–13,** 17–24.

Sturt, B. A., Pringle, I. R. and Ramsay, D. M. 1978. The Finnmarkian phase of the Caledonian orogeny. *J. geol. Soc. London,* **135,** 597–610.

Tull, J. F. 1977. Geology and structure of Vestvågøy, Lofoten, north Norway. *Norges geol. Unders.,* **333,** 1–59.

Törnebohm, A. E. 1872. En geognostisk profil öfver den skandinaviska fjällryggen mellan Östersund och Levanger. *Sveriges geol. Unders.,* **C 6,** 24 pp.

Verschure, R. H. 1981. The extent of Sveconorwegian and Caledonian imprints in Norway and Sweden. (Abstract) *Seventh Eur. Colloq. Geochron. Cosmochron. Isotope Geology,* Jerusalem, Israel.

Verschure, R. H., Andriessen, P. A. M., Boelrijk, N. A. I. M. Hebeda, E. H., Maijer, C., Priem, H. N. A. and Verdumen, E. A. Th. 1980. On the thermal stability of Rb–Sr and K–Ar biotite systems: evidence from coexisting Sveconorwegian (ca. 870 mA) and Caledonian (ca. 400 Ma) biotites in SW Norway. *Contr. Miner. Petrol.*, **74**, 245–252.

Vogt, T. 1927. Sulitjelmafeltets geologi og petrografi. *Norges geol. Unders.*, **121**, 1–556.

Vogt, T. 1929. Den norske fjellkjedes revolusjonshistorie. *Norsk geol. Tidsskr.*, **10**, 97–115.

Vogt, T. 1945. The geology of part of the Hølonda-Horg district, a type area in the Trondheim region. *Norsk geol. Tidsskr.*, **25**, 449–528.

Wilson, J. R., Hansen, B. T. and Pedersen, S. 1983. Zircon U–Pb evidence for the age of the Fongen–Hyllingen complex, Trondheim region, Norway. *Geol. Fören. Stockh. Förh.*, **105**, 68–70.

Zwart, H. J. 1978. Some additional remarks on metamorphism in the Swedish Caledonides. In Zwart, H. J. (ed.), *Metamorphic Map of Europe 1:25 00. Explanatory Test.* 73–74.

Zwart, H. J., Corvalan, J., James, H. L., Miyashiro, A., Saggerson, E. P., Sobolev, U. S. Subramaniam, A. P. and Vallace, T. G. 1967. A scheme of metamorphic facies for the cartographic representation of regional metamorphic belts. *Geol. Newsletter*, **2**, 57–72.

The Caledonide Orogen—Scandinavia and Related Areas
Edited by D. G. Gee and B. A. Sturt
© 1985 John Wiley & Sons Ltd

High-pressure metamorphism in the Scandinavian Caledonides

W. L. Griffin, H. Austrheim, K. Brastad, I. Bryhni, A. G. Krill, E. J. Krogh, M. B. E. Mørk, H. Qvale and **B. Tørudbakken**

Mineralogisk—Geologisk Museum, Sars gate 1, Oslo 5, Norway

ABSTRACT

The Western Gneiss Region of Norway includes (in addition to paragneisses and orthogneisses) peridotites, anorthosites, gabbros, and coarse-grained intermediate–acid rocks ('rapakivi granites', 'mangerites'). All of these rock types enclose eclogites. Igneous mineral assemblages in gabbros, rapakivi granites, and anorthosites are overprinted by high-pressure metamorphic mineral assemblages, producing eclogites in mafic rocks or equivalent high-pressure granulites in felsic rocks. Despite pervasive later amphibolitization, high-P assemblages are locally preserved in the gneisses along the coast of western Norway.

Low-P protoliths can be demonstrated for many eclogites. Prograde metamorphism to eclogite facies is demonstrated by inclusion suites within garnet grains and by zoning of eclogite minerals. Sr, Nd, Ar, and O isotopes suggest extensive interaction of some eclogites, especially the opx-bearing varieties, with their country rocks before or during metamorphism. The regional pattern of K_D(gnt/cpx) and X_{jd}^{cpx} shows that T and P increased toward the coastline, reaching maximum values near 800°C and 18–20 Kb on the islands north of Alesund. Lowest values are found in Sunnfjord (500°C, 10 Kb). This regional pattern is interpreted as the result of continental subduction during collision of the Baltic and American plates.

Radiometric data indicate that while most of the protoliths were of Precambrian age, the eclogite facies metamorphism may be Caledonian. Eclogites are found in Caledonian nappes in Tromsø and in Sweden, implying that subduction preceded the (late) thrusting. However, data from the Oppdal area suggest that early-thrusted nappes were themselves involved in the subduction.

Introduction

Eclogites and related high-pressure metamorphic rocks occur in many places in the Scandinavian Caledonides (Fig. 1). The origin of these rocks has been debated for many years, in part because the model adopted for their age and origin will constrain tectonic models for the Caledonide orogen as a whole. Studies carried out largely during the IGCP program strongly suggest that most, and perhaps all, of the high-pressure rocks originated by Caledonian metamorphism of low-pressure protoliths, ranging in age from Svecofennian to Palaeozoic. This paper summarizes the results of these studies and discusses the tectonic consequences.

In this paper we will use the term 'eclogite facies' to refer to the physical conditions (P, T, f_{H_2O}, etc.) under which basic rocks may be converted to eclogite (omphacite + garnet) assemblages. Numerous experimental studies (see summary by Green and Ringwood 1972) demonstrate that the pressure of the granulite–eclogite transition is strongly dependent on bulk composition. Over a wide range of pressures below that of the albite = jadeite + quartz transition, eclogite mineral assemblages in some basic rocks will coexist with high-pressure granulite (clinopyroxene + garnet + plagioclase + quartz) assemblages in more felsic rocks. Smith (1982) has suggested that 'eclogite facies' be restricted to conditions where plagioclase is unstable in all compositions, i.e. to pressures above the albite–jadeite transition. However, this definition is far too restrictive. A usable definition of eclogite facies must include the true eclogites (garnet + omphacite ± qtz) that may form from a wide range of basic compositions at pressures below the albite breakdown. If the definitions of the eclogite- and high-pressure granulite facies are both based on *basaltic* compositions (*s.l.*), as suggested originally by DeWaard (1965) then there is no ambiguity involved in our usage of these terms.

Eclogites have been described from several places in the Caledonian nappes (Fig. 1). In the high-level Tromsø Nappe Complex (Krogh and Andresen 1979; Andresen *et al.* 1981), they occur together with marbles, calc-silicate gneisses, garnet–mica schists, quartzo-feldspathic gneisses, and ultramafic rocks. The true eclogites (omphacite + garnet ± quartz) resemble eclogites elsewhere in the Caledonides, while the eclogite-like calc-silicate rocks consist mainly of grossular-rich garnets and Na-poor pyroxenes. No overlap in mineral composition has yet been found between the two rock types. Similar associations, apparently in a similar tectonic

position, occur along the north Norwegian coast (Glomfjord, Træna). In the Seve Nappe of the Nasafjäll area of northern Sweden, heavily retrograded eclogites occur as boudined, doubly-folded mafic layers in micaceous quartzites (Andréasson *et al.* this volume). Better-preserved examples are found in the Seve rocks of the area farther south (van Roermund this volume).

In the overthrust rocks of the Bergen Arcs, Griffin (1972) described eclogites associated with granulite-facies anorthosites and mangerites, and interpreted them as the result of a single cooling history from magmatic conditions through granulite facies to eclogite facies. However, new structural and radiometric evidence (Austrheim 1981; Austrheim and Griffin 1982) indicates that the lower-T eclogite metamorphism has been superposed on Precambrian granulite-facies assemblages. Similar rocks of the Jotun Nappe in the Sognefjord area contain no eclogites, only Precambrian granulite-facies assemblages overprinted by later retrograde assemblages (Griffin 1971).

Eclogites have been reported from Svalbard, both as lenses in gneisses (Gee 1966), as in western Norway, and associated with glaucophane schists (Ohta 1979). Both types are as yet poorly known.

The Western Gneiss Region (WGR) of Norway, extending from the Sognefjord to the Trondheimsfjord (Fig. 1), is a classic area for high-pressure rocks. This area includes, in addition to variegated paragneisses and orthogneisses, numerous peridotites, anorthosites, gabbros, and rapakivi granites, all of which enclose eclogites. The WGR is generally regarded as a basement complex, although it locally has been shown to include infolded, polymetamorphic nappe units of younger age (Krill 1980). Mykkeltveit *et al.* (1980) have proposed that the entire WGR is allochthonous, as suggested by recent models for the Piedmont belt of the southern Appalachians (Harris and Bayer 1979).

Because of its large dimensions (*ca.* 350 × 100 km; Fig. 1) the WGR offers the best possibility to study the relations of the high-pressure rocks to the surrounding gneisses on a regional scale. The following discussion will therefore concentrate on the evidence from this area, and return to the problems of the nappe eclogites. Descriptions of many localities mentioned here may be found in Griffin and Mørk (1981).

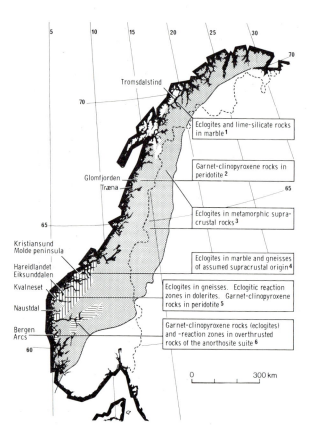

Fig. 1 Regional distribution of eclogitic rocks in the Scandinavian Caledonide belt. Western Gneiss Region (WGR) and Bergen Arcs/Jotun nappes indicated by vertical and horizontal ruling respectively. Selected references: 1. Landmark 1973; 2. Sørensen 1955; 3. Gee and Sjöström personal communication 1976; Andréasson this volume; van Roermund this volume; 4. Hernes 1954; Råheim 1972; 5. Bryhni *et al.* 1970; Griffin and Råheim 1973; O'Hara and Mercy 1963; Carswell 1974; Moore and Qvale 1977; 6. Griffin 1971, 1972

Geological Setting: Western Gneiss Region (WGR)

Gneisses

Numerous attempts have been made to divide the rocks of the WGR into mappable units, on a local scale. Bryhni (1966) suggested that these units could be grouped in two major complexes: a heterogeneous association of supracrustal rocks, anorthosites, and augen gneisses (Fjordane complex) in the west and northwest, and a more homogeneous, migmatitic orthogneiss group (Jostedal complex) in the eastern part. Eclogites were originally believed to occur only in the Fjordane complex, but they also have been found in areas mapped as Jostedal complex (Krogh 1980a).

The association of rock types indicates that the gneisses of the Fjordane complex (*s.l.*) are largely supracrustal in origin. They include pelitic and 'granitic' migmatites, marbles, quartzites, and amphibolites, often intimately interlayered. The supracrustal rocks enclose gabbro, anorthosites, and peridotites, and are intruded by several types of intermediate to acidic igneous rocks, now variably foliated. The most distinctive of these is a suite of 'rapakivi granites' (also described as 'augen gneiss' or 'mangerite'). The least deformed of these show plagioclase-rimmed orthoclase megacrysts from 1 to 20 cm across (Bryhni 1966; Lappin *et al.* 1979; Carswell and Harvey 1981, 1984).

The gneisses are dominated by amphibolite-facies mineralogy, but enclose numerous relics of high-pressure granulite assemblages. In the less foliated parts of the rapakivi granites, the assemblage cpx- + gnt + qtz + ksp + kyan occurs interstitially to the megacrysts and as coronas on them. In the medium-grained, foliated parts of some (Måløy gneiss: Bryhni 1966; Malinconico personal communication; loc. 12, 13, Fig. 3A) the same assemblage forms an equilibrium mosaic texture. Mysen and Heier (1972) described omphacite-bearing garnet–biotite gneisses from the Hareidland area (loc. 31, Fig. 3A), and many small occurrences of high-P granulites have been registered from Gurskøy and the islands outside Hareidland (loc. 29, 30, Fig. 3A) (Rosenqvist 1956; Griffin and Mørk 1981). Krogh (1980b) described high-P granulites from the Kristiansund (loc. 46–49, Fig. 3A) area. Felsic high-P granulites also occur interlayered with eclogites on the islands Fjørtoft (loc. 36, Fig. 3A) and Harøy, north of Ålesund (Mørk and Griffin in preparation), and together with eclogites and marbles on the Molde Peninsula (loc. 43, 44, Fig. 3A) (Griffin and Mørk 1981; Harvey in preparation). In all of these occurrences the granulite-facies assemblages show various degrees of retrogression to amphibolite-facies assemblages.

In the Adula Nappe of the Lepontine Alps, felsic gneisses enclosing eclogites locally contain relics of a high-pressure garnet + phengite assemblage (Heinrich 1982). In most of these gneisses this assemblage has broken down by the reaction gnt + pheng →

biotite + feldspars ± muscovite + H_2O; the released H_2O has contributed to hydration of the eclogites. The breakdown reaction commonly produces a distinctive poikiloblastic intergrowth of micas and feldspars. This microstructure has now been observed in numerous samples of gneiss from the WGR, and may be evidence that high-P metamorphism of the felsic rocks was much more widespread than previously recognized.

Eclogites in the gneisses

Eclogites enclosed in the gneisses ('external eclogites', 'country-rock eclogites') characteristically occur as pods or lenses from decimetres to tens of metres in length, apparently representing boudined layers (Lappin 1966). These eclogites have conformable contacts with the gneisses, though internal layering is commonly necked or truncated by the contact. Some boudins are the detached hinges of large isoclinal folds. Eclogites occur as thin (10–30 cm) concordant layers in supracrustal sequences of the Kristiansund area (Griffin and Råheim 1973), while on the Molde Peninsula thick sequences of marble are interlayered with eclogite and show gradational contacts (Hernes 1954; Harvey in preparation; Griffin and Mørk 1981). Eclogites appear as groups of tabular bodies (dyke swarms) in massive orthogneisses of the Molde–Otrøy area (loc. 37–39) (Carswell and Harvey 1981) and in Sunnfjord (loc. 2–10) (Krogh 1980a; Cuthbert 1981). Internally complex eclogite bodies with dimensions of kilometres occur on Hareidlandet (Mysen and Heier 1972; Schmidt 1963) and in Sunnfjord (Krogh 1980a; Griffin and Mørk 1981).

The country-rock eclogites are dominantly bimineralic (omphacite + gnt ± qtz ± rutile). The Sunnfjord eclogites commonly have a chloromelanitic omphacite (Binns 1967; Krogh 1980a), and impure jadeites have been reported from the Selje area (Lappin and Smith 1978; Smith *et al.* 1980). The garnets show a wide range in Fe/Mg, but limited variation in Ca/(Ca + Fe + Mg). Kyanite-bearing eclogites, often with clinozoisite, are common. Orthopyroxene is usually restricted to Mg-rich, Ca, Na-poor compositions. Glaucophane occurs in Sunnfjord eclogites (Krogh 1980a), but tremolitic and barroisitic amphiboles are more common. Phengite is relatively common in aluminous compositions, while phlogopite appears in some magnesian rocks. Paragonite is abundant in some of the Sunnfjord eclogites. In many cases, textural evidence suggests that these hydrous phases are in equilibrium with pyroxene and garnet, and this is confirmed by studies of element distribution (Krogh and Råheim 1978).

Several stages of retrograde metamorphism can be recognized in some country-rock eclogites. The first involves recrystallization to a cpx + gnt + plag + qtz assemblage (high-P granulite); this may be either replaced or overstepped by opx + cpx + plag + qtz (relict gnt) assemblages (intermediate-P granulite; Mysen and Heier 1972; Griffin and Råheim 1973; Carswell *et al.* 1982; Krogh 1982). Development of hydrous amphibolite-facies assemblages (amph + plag ± bio ± epidote) affects both the granulite-facies assemblages and the primary eclogite assemblages. The breakdown of omphacite to diopside + plagioclase symplectites, often associated with oxidation (Mysen and Griffin 1973) is a typical feature of the retrograde metamorphism.

The eclogites in gneiss vary widely in chemical composition. In terms of basaltic norms, they range from quartz-tholeiitic to moderately alkaline (Fig. 2) and are generally tholeiitic rather than calc-alkaline. Much of the variation seen in Fig. 2 can be found within single bodies, expressed as mineralogical layering on scales of centimetres to metres. Alkali-olivine basalt compositions are largely confined to the Sunnfjord area, where many also have high contents of TiO_2 and P_2O_5 (Krogh 1980a).

Microstructural evidence and geochemical studies suggest that the protoliths of many of the eclogites are igneous in origin; high values of Ni and Cr (>100 ppm), and positive Eu anomalies in aluminous varieties (Griffin and Garmann in preparation; Brastad this volume; Brastad and Brunfelt in preparation) are especially diagnostic. Some of the chemical variation may therefore represent primary differences in magma composition, or the effects of cumulate processes (Mysen and Heier 1972).

However, detailed studies by Krogh (1980a) and Krogh and Brunfelt (1981) showed that eclogites in the Sunnfjord area were extensively contaminated by Si, Al, Na, K, Ba, and REE from the surrounding gneisses either before or during the eclogite metamorphism. Later amphibolitization has further disturbed the chemistry, mainly by introduction of H_2O, K, Sr, Rb, and Ba. The Sunnfjord eclogites have equilibrated under high f_{H_2O} (Krogh 1980a) and have rare-earth patterns enriched in LREE. Other eclogites in gneiss commonly show strong *depletion* of La, Ce, and Sm (Garmann *et al.* 1975). This may be a primary feature, but also can be modelled in terms of loss of H_2O at high P and T (Griffin and Garmann in preparation). Such dehydration–depletion may also explain the very low K and Rb of many eclogites. While such detailed data are only available for a few eclogites, it is now clear that the composition of a given eclogite sample, particularly one from a small body, may not be simply related to its original magmatic composition.

Although eclogites in gneiss generally have very low Rb/Sr ratios (<0.2), they typically have high $^{87}Sr/^{86}Sr$ ratios (>0.705); this is especially true of the orthopyroxene eclogites (up to 0.723; Brueckner 1977b; Griffin and Brueckner 1980, 1983). This implies either massive exchange of Sr with the (Sr-poor) gneisses, or a crustal prehistory with a higher Rb/Sr ratio. Similarly, Nd isotopic compositions suggest that at least some eclogites have had a long premetamorphic history with a Sm/Nd ratio unlike possible mantle ratios (Griffin and Brueckner 1980; 1983). Oxygen-isotope studies (Vogel and Garlick 1970; Griffin, Ahmad and Perry in preparation) show that while many eclogites have 'igneous' $\delta^{18}O$ values, others range from much lower to much higher. The high values in particular suggest isotopic exchange with the surrounding gneisses at high T, and imply the presence of a fluid phase during metamorphism.

Gabbros/Dolerites and associated eclogites

Dolerite dykes and bodies of fine/medium-grained gabbro occur throughout the WGR, but are especially abundant in a belt extending eastward from Ålesund. They commonly have a lensoidal form, concordant with the gneiss foliation, though some of the larger bodies can be followed for 2–3 km (Gjelsvik 1952). They are generally olivine-tholeiitic in composition (Fig. 2), with local variations from olivine-rich to feldspar-rich types. The primary magmatic minerals have reacted, in varying degrees, to assemblages of high-P granulite facies and eclogite facies (Gjelsvik 1952; Griffin and Råheim 1973; Griffin and Heier 1973). These corona-forming reactions have commonly led to the formation of eclogites in the fine-grained margins of the dolerites (Griffin and Råheim 1973). An excellent example has been described from the interior of the Flatraket 'mangerite' (Bryhni *et al.* 1977; Griffin and Mørk 1981). In Vindøldalen on the south side of the Surnadal syncline (loc. 50), a zone of coarse grained eclogite has developed on the rim of a gabbro that otherwise shows relatively little corona development (Tørudbakken 1981). On Flemsøy (loc. 35), north of Ålesund, coronitic gabbros have been isochemically converted to eclogites with obvious pseudomorphs after gabbroic textures, and with gradual transitions into irregular bodies of medium-grained eclogite (Mørk 1983). The large Naustdal eclogite in Sunnfjord (loc. 8) contains relict subophitic textures, in which magmatic pyroxene is replaced by omphacite + amphibole + garnet, and plagioclase by clinozoisite + paragonite + phengite + qtz. In the Gjølanger area of Sunnfjord (loc. 2), anorthositic gabbros are pseudomorphed by eclogite; magmatic

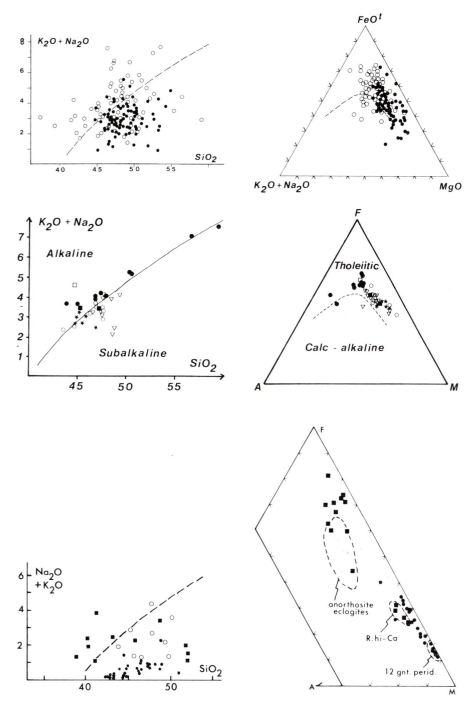

Fig. 2 Chemical data on Norwegian eclogites. Top: eclogites in gneiss. Rings, Sunnfjord area; dots, other areas. Centre: dolerites, variably metamorphosed to eclogite. Full circles, loc. 50; open circles, Møre region (Gjelsvik 1952); triangles, Kristiansund area; open squares, loc. 54; filled squares, loc. 13; stars, Nordøyene (loc. 34–36). Bottom: eclogitic rocks in ultramafic bodies. Squares. Raudkleivane; rings, Bjørkedalen; dots, Almklovdalen, Tajford, Otrøy, Sandvik. AFM diagram shows fields for garnet-peridotes, Raudkleivane high-Ca type, and eclogites in anorthosite

assemblages, coronites and eclogites are intimately intermixed on a scale of metres (Cuthbert 1981).

Griffin and Råheim (1973) ascribed the coronitic dolerites, and the apparent 'coexistence' of gabbro and eclogite on a regional scale, to convergent metamorphism during post-intrusion cooling of the dolerites toward the ambient geotherm, represented by the eclogites. However, recent radiometric evidence (see below) suggests that the gabbros are Precambrian, and the metamorphism Caledonian. This would require that the gabbroic rocks are metastable relics that survived the eclogite-facies metamorphism. Fig. 2 shows that all dolerite compositions are represented by equivalent eclogites. This fact, and the field and textural evidence for a close relation between the two rock types (Mørk 1983), suggest that many of the common country-rock eclogites were derived by metamorphism of pre-existing dolerites or gabbros.

Ultramafic rocks

Bodies of ultramafic rocks from a few metres to several kilometres across occur throughout the WGR. Many are intimately associated with anorthosite. The foliation of the gneisses typically bends smoothly around them; Brueckner (1977a), Medaris (1980) and Cordellier et al. (1980) have concluded that the peridotites have shared all the deformational phases recognized in the gneisses. The ultramafic rocks are mainly dunitic and lherzolitic, with lesser amounts of wehrlite and harzburgite.

The earliest ultramafic mineral assemblages recognized in the western part of the WGR are garnetiferous lherzolites, wehrlites and websterites, often complexly interlayered. The minerals are highly magnesian (olivine > Fo_{90}) and both clinopyroxene and garnet are Cr-rich; cpx is Na-poor (sodic augite) (Medaris 1980; Carswell 1968a,b, 1973b; O'Hara and Mercy 1963). The garnet lherzolite assemblages are overprinted locally by a symplectitic ol + cpx + spinel assemblage (O'Hara and Mercy 1963; Carswell 1968b). This is in turn replaced by the ubiquitous amphibolite-facies assemblage ol + opx + amph + chlor ± spinel, the 'chlorite harzburgites' of earlier writers. These chlorite peridotites are commonly strongly folded, and orthopyroxene megacrysts define an E–W lineation in some bodies (Medaris 1980). The chlorite–amphibole-bearing assemblages are variably serpentinized.

Only chlorite–peridotite and serpentinite assemblages have been recognized in the eastern part of the WGR and in the nappes (Moore and Qvale 1977). Some of the eastern bodies show large (2–20 cm) opx crystals in a remarkable radiate arrangement which Moore (1977) has interpreted as being derived from prograde metamorphism of serpentinites. Alteration of these rocks produces opx–talc–carbonate peridotites, similar to the sagvandites of the Tromsø area (Ohnmacht 1974; Schreyer et al. 1972). Fe-rich garnetiferous ultramafic rocks (oliv ~Fo_{80}) have been reported from several localities (Carswell and Gibb 1980b; Carswell and Harvey 1981). These strongly resemble ultramafic layers within the large body of hypersthene eclogite at Eiksundal on Hareidland (loc. 53) (Schmidt 1963).

The magnesian ultramafic rocks enclose two distinct types of 'internal' or 'ultramafic' eclogites. One,

the classic Rødhaugen type of Eskola (1921) is essentially bimineralic, with magnesian sodic augite + pyropic garnet ± orthopyroxene. This type is distinguished from similar rocks within the layered ultramafic sequences by lack of olivine, higher Fe/Mg in both pyroxene and garnet, lower Cr contents, and somewhat higher Na_2O contents (Carswell 1981; Fig. 2). These rocks typically occur as bands from centimetres to decimetres in thickness, and may often be traced for many metres along strike. Their chemistry is basaltic only in a general sense. Normatively, most of them contain very calcic plagioclase (An_{80-90}), <30 per cent olivine, and varying amounts of both cpx and opx.

The other type of 'internal' eclogite occurs as lenses and layers from 1 to 30 m across, apparently resulting from the disruption of tabular bodies. This 'Raudkleiva type' has been described from Almklovdalen (loc. 24) (Lappin 1973, 1974; Griffin and Qvale this volume) and from Bjørkedalen (loc. 21) (Brastad 1984). Lappin (1973, 1974) regarded these bodies as layers, cogenetic with the garnet peridotites and Rødhaugen-type eclogites in the same outcrop; Griffin and Qvale (this volume) interpret them as originally basaltic dykes in the ultramafic body. These eclogites contain jadeite-rich omphacites and almandine-rich garnets, as well as abundant rutile and apatite. Some occurrences have concentrations of low-Na clinopyroxenite that may have formed local central zones or inclusions prior to disruption of the dykes (Lappin 1974; Griffin and Qvale this volume). The eclogites are characterized by high Fe/Fe + Mg and by much higher Na_2O and Al_2O_3 than the Rødhaugen type (Fig. 2). They are comparable in chemistry to amphibolitized dykes in Caledonian 'alpino-type' serpentinites (Type I of Moore and Qvale 1977), to the ferrogabbros of the Mid-Atlantic ridge and to the superferrian Voltri Group eclogites of the Ligurian ophiolites (Griffin and Qvale this volume).

Anorthosites and associated eclogites

Anorthosites are abundant in the area north of Nordfjord (Bryhni 1966); they typically occur as concordant layers in gneiss, up to 300 m thick, and are continuous over distances up to 50 km. Some are intimately associated with ultramafic rocks; anorthosites form a nearly continuous collar around the large Bjørkedalen peridotite (loc. 21) (Brastad 1981, this volume). Two larger, more massive bodies at Fiskå and on Sandøy (Gjelsvik 1952; Griffin and Mørk 1981) preserve primary magmatic mineralogy and structures: coarse-grained euhedral to subhedral plagioclase (An_{45-60}) and megacrysts of Al-rich opx up to 30 cm across. The opx crystals contain abundant exsolution lamellae of garnet, and are rimmed by thin cpx + gnt coronas (Griffin and Mørk 1981; Carswell *et al.* this volume). These early assemblages are overprinted by amphibolite- and greenschist-facies mineralogy (plagioclase, hornblende, biotite, epidote, muscovite, margarite) that characterize the meta-anorthosites.

Eclogites occur as fine-grained, conformable layers and lenses 0.3–5 m thick within the meta-anorthosites (Brastad this volume); some of the eclogites show gradational contacts to the anorthosite. They consist mainly of omphacite + gnt ± zoisite, and typically contain abundant kyanite. Chemically they resemble the superferrian eclogites in the ultramafic rocks, but with higher $Na_2O + K_2O$ (Fig. 2). Their REE patterns show low ΣREE, enrichment in LREE, and large positive Eu anomalies. The Eu anomalies, together with their high Al content, clearly relate them genetically to the surrounding anorthosites; some may have formed by metamorphism of opx + ilm−mgt concentrations in the anorthosite (Brastad and Brunfelt in preparation).

Summary

The Western Gneiss Region includes, in addition to supracrustal rocks and orthogneisses, a suite of peridotites, anorthosites, gabbros, and coarse-grained intermediate–acid igneous rocks ('rapakivi granites', 'mangerites', etc.). All of these rock types enclose eclogites. Some gneisses and peridotites contain early mineral assemblages compatible with those in the eclogites. Igneous mineral assemblages in anorthosites, gabbros and mangerites are overprinted by eclogite mineral assemblages or equivalent high-P granulite assemblages. The age of these granulite-facies assemblages is not always clear. In the Bergen Arcs, Precambrian granulite-facies assemblages are either preserved, or overprinted by eclogite-facies assemblages (Austrheim and Griffin 1982). Within the WGR, some felsic high-P granulite assemblages (cpx + gnt + plag + qtz) can be shown to be contemporaneous with eclogite (cpx + gnt ± qtz) assemblages, either by interlayering of the two on small scales (Midøy, Fjørtoft, Kristiansund area) or by coexistence of omphacite with plagioclase (Midøy, Eiksundal, Ulsteinvok). Retrograde granulite-facies assemblages in some eclogites (described above) suggest that some of the granulite-facies assemblages in the gneisses may also have formed during uplift *following* the eclogite-facies metamorphism. We cannot, therefore, unambiguously assume that all outcrops of felsic high-P granulites in the WGR were formed contemporaneously with the eclogites. However, it seems probable, on available evidence, that all rock types physically associated with eclogites also have suffered metamorphism under the same conditions as the eclogites, and probably at the same time. In most rocks, including many eclogites, these early high-P mineral assemblages have been partly or completely obliterated by later metamorphism in amphibolite- and greenschist-facies.

Age of Metamorphism

On the basis of field evidence, the emplacement or metamorphism of the eclogites in the Gneiss Region has generally been related to the 'major metamorphism' of the gneisses, assuming that only one such event has occurred. This 'major metamorphism' was long regarded as Caledonian on structural grounds, an idea supported by early K–Ar dates on minerals. As Rb–Sr whole-rock isochrons of Precambrian age became available, they were widely interpreted as metamorphic ages, and the importance of the Caledonian orogeny was correspondingly reduced (e.g. Krogh 1977; Bryhni *et al.* 1977).

Rb–Sr whole rocks dates and U–Pb dates on zircons (upper discordia intercepts) from the WGR cluster in the period 1600–1800 Ma for various types of gneisses. Studies of adjacent palaeosome and neosome in migmatites suggest that this period includes both protolith formation and high-grade metamorphism (Råheim 1977). Similar data from rapakivi granites/augen gneisses give dates clustered in the 1400–1600 Ma range, consistent with field evidence for an intrusive origin (Lappin *et al.* 1979; Carswell and Harvey 1981; Harvey 1983). A gabbro in Vindøladalen (loc. 50) with eclogitic margins has yielded a Rb–Sr whole-rock date of 1517 ± 60 Ma (Tørudbakken 1981). Some Sveconorwegian dates (*ca.* 1000 Ma) have been reported from the Gneiss Region; many seem to be related to small granitic intrusions (Brueckner 1972, 1979). Sm–Nd whole-rock data suggest an age of *ca.* 550 Ma for the protolith of the Naustdal eclogite in Sunnfjord (loc. 8) (Griffin and Brueckner 1983).

Krill and Griffin (1981) have argued that whole-rock Rb–Sr isochrons cannot *a priori* be expected to give the age of a metamorphic event superimposed on older gneisses. Within the WGR essentially all *mineral* dates (Rb–Sr, K–Ar) from gneisses are Caledonian.

Sm–Nd dates on cpx–gnt pairs from five eclogites average 425 ± 12 Ma (Griffin and Brueckner 1980, 1983); one older date was interpreted as disequilibrium due to preservation of pre-eclogite garnet. Mearns and Lappin (1982) give similar Sm–Nd mineral ages for a country-rock eclogite and a garnet-bearing gneiss from Almklovdalen (408 ± 8 and

414 ± 31 Ma, respectively). In contrast, three cpx–gnt–whole-rock Sm–Nd ages from one small garnet peridotite outcrop at Lien, Almklovdalen, show a wide range (1477 ± 7 Ma, Jacobsen and Wasserburg 1980; 1316 ± 138 and 1029 ± 34 Ma, Mearns and Lappin 1982), suggesting considerable disequilibrium. Zircons from the large Ulsteinvik eclogite have yielded nearly concordant U–Pb dates of *ca*. 410 Ma (Krogh *et al*. 1974).

Similar zircon U–Pb dates, interpreted as the age of eclogitization, were obtained from five eclogites in the Selje area by Gebauer *et al*. (1982); they also obtained upper-intercept ages (protolith ages) of *ca*. 1500 Ma and *ca*. 1760 Ma for two of these bodies.

Rb–Sr dates on 'primary' micas from eclogites group around 395 Ma (Griffin and Brueckner 1983). Mineral isochrons from pegmatites that cut eclogites suggest that pegmatite intrusion continued well up into Devonian time (Griffin and Krill in preparation). Available data therefore suggest that most of the Precambrian rocks of the WGR were recrystallized during the Caledonian orogeny, and that at least local anatexis occurred during the late stages of this orogeny.

Reymer *et al*. (1980) obtained a Rb–Sr whole-rock date of *ca*. 1000 Ma for quartzitic gneisses surrounding eclogites in the Seve Nappe of the Västerbotten area, Sweden. This date is based on regional sampling, which suggests that it is a 'primary' age of some sort. Claesson (1981) has demonstrated that Seve rocks in the Åreskutan area of Sweden have suffered granulite-facies metamorphism and anatexis at *ca*. 415 Ma. Acid metavolcanics interlayered with the eclogite-bearing schists of the Tromsø Nappe Complex have given a Rb–Sr whole-rock date of 433 ± 12 Ma (A. Andresen personal communication). The eclogite-bearing mica schists of the Blåhø Nappe in the Oppdal area (east of loc. 51) yield no Rb–Sr isochron, but their maximum possible model age is *ca*. 700 Ma, suggesting that their metamorphism to amphibolite/eclogite facies is Caledonian. Intermediate gneisses interlayered with the micaschists give a Rb–Sr whole-rock date of 583 ± 69 (Krill and Röshoff in press).

Eclogites like those of the Caledonides are rare rocks on a worldwide scale, and their formation appears to require unusual tectonic conditions. No eclogites have been found on the Baltic Shield outside of the area affected by Caledonian orogenesis. The simplest, and therefore most likely, model would involve *one* eclogite-forming event in the Scandinavian Caledonides. The available mineral ages on eclogites, and the whole-rock Rb–Sr data from the Tromsø and Blåhø Nappes, require in that case that the metamorphism be Caledonian.

Eclogite Protoliths: Mantle or Crust?

The controversy surrounding the origin of the Norwegian eclogites boils down to one question: Are they fragments of the mantle, or are they crustal rocks subjected to unusually high P and T? The answer obviously has great implications for tectonic models of the Caledonides. A mantle origin was favoured by Eskola (1921), who first recognized the high-P nature of the eclogites. He was followed by O'Hara and his coworkers (1963, 1967, 1971) and by Lappin (1966, 1974), Lappin and Smith (1978), and Smith (1980). These later studies were strongly influenced by the assumption that *extremely* high pressures (30–40 Kb) were necessary. A crustal origin was first argued strongly by Gjelsvik (1952) and later by Bryhni (1966), Bryhni *et al*. (1969), Carswell (1973a), Griffin and Råheim (1973), and Krogh (1977). These workers have been impressed by several types of evidence for a crustal, prograde metamorphic origin. This evidence includes (1) partial or complete conversion of clearly low-P protoliths to eclogites, (2) amphibolite-facies relicts enclosed in eclogite garnets, and (3) prograde zoning in eclogite minerals.

1. The complete, often pseudomorphous, conversion of gabbros to eclogites, via a transitional coronite stage, is described above. These reactions are evidence of a P increase; the primary olivine + plagioclase assemblage is not stable above *ca*. 8 Kb (Herzberg 1978). Mysen and Heier (1972) showed on geochemical grounds that gnt–cpx layering in the large Ulsteinvik eclogite (loc. 31) could not have originated by segregation of garnet and clinopyroxene, but could be explained by sorting of plagioclase, clinopyroxene, and olivine. The intimate association of eclogites with anorthosites in the Bergen Arcs and in the Nordfjord area (Griffin 1972; Brastad this volume) is impossible to explain in terms of a mantle origin. The primary high-Al opx megacrysts seen in some anorthosites are in any case not stable above 10 kb at magmatic temperatures (Maquil 1978). The high positive Eu anomalies of the eclogites enclosed in anorthosites, and of several other kyanite eclogites (Garmann *et al*. 1975; Griffin and Brueckner 1983; Brastad and Brunfelt in preparation) clearly imply the involvment of plagioclase and thus a low-P origin for the protoliths. The occurrence of eclogites as thin layers in supracrustal sequences (Griffin and Råheim 1973; Hernes 1954; Carswell and Harvey 1981) also seems to be evidence that at least some eclogites had supracrustal protoliths.

2. Euhedral to subhedral inclusions of amphibolite-facies minerals are relatively common in the cores of large eclogite garnets, where they apparently have been shielded from later reaction. Among the phases recognized are various types of amphibole, plagioclase, diopside, microcline, biotite, epidote, paragonite, phengite, magnetite, ilmenite, pyrite, and quartz. Single garnet grains have been observed to contain up to six of these phases as discrete and polymineralic inclusions (Krogh 1982; Bryhni and Griffin 1971; Bryhni et al. 1977).

In many cases the inclusion phases are not found in the matrix, even where the eclogite is partially retrograded. For example, in the Verpeneset eclogite (loc. 14) (Bryhni and Griffin 1971; Krogh 1982) paragonite and hornblende occur only as inclusions in garnet, while the equivalent assemblage omphacite + kyanite + quartz occurs in the matrix; phengite is the only mica in the matrix. We conclude that these inclusions in garnet record either an early stage of the eclogite metamorphism, or a relict metamorphic assemblage from a previous orogenic period. In either case, they demonstrate that many eclogites evolved by prograde breakdown of a garnet–amphibolite assemblage.

3. Chemical zoning is commonly observed in the eclogite garnets, though less commonly in other phases. Two general types of garnet zoning have been recognized: (a) a decreasing Fe/Mg ratio from core to rim, usually accompanied by a decrease in Mn, (b) rims enriched in Fe/Mg, and sometimes in Mn, relative to the inner zone. The former type (prograde zoning) was related to growth with increasing T by Bryhni and Griffin (1971) and by Råheim and Green (1975) on the basis of experimental studies.

The most pronounced zoning observed in the eclogite garnets appears to coincide with the transition from garnet cores with amphibolite-facies inclusions, to rims with omphacite inclusions. This portion of the zoning patterns probably reflects the change in the garnet/matrix partitioning accompanying the breakdown of the low-grade assemblage. The similar, but less pronounced, zoning in the outer parts of such garnets is interpreted as due to growth of garnet in equilibrium with pyroxene (±amphibole) as T continued to increase. This interpretation has been confirmed by detailed studies of Fe/Mg partitioning between omphacite inclusions and the enclosing garnet, as a function of distance from the core of the garnet crystal (Bryhni and Griffin 1971; Bollingberg and Bryhni 1972; Krogh 1980a, 1982; Carswell et al. 1982). Where matrix

pyroxenes are zoned, their compositional variations mirror those of the garnets (e.g. Krogh 1980a; Carswell et al. 1982). Most commonly, however, matrix clinopyroxenes are essentially homogeneous.

The second type (retrograde zoning) occurs either as thin rims on otherwise prograde-zoned garnets, or as broad zones around homogeneous garnet cores. Omphacite inclusions in the latter type are commonly more Fe-rich and more Jd-rich than those in the matrix (Mysen 1972; Krogh 1982). In general, garnets containing amphibolite-facies inclusions are prograde-zoned, while retrograde-zoned garnets contain only eclogite phases as inclusions. This suggests that complete chemical equilibrium is only attained during complete recrystallization of the garnets. Griffin and Brueckner (1980, 1983) reached a similar conclusion from studies of Sm–Nd distribution in Norwegian eclogites. Calculated T values (see below) suggest that temperatures over 700°C are necessary for this recrystallization.

Metamorphic Conditions

T-gradient

Values of K_D = (Fe/Mg gnt)/(Fe/Mg cpx) change consistently across the WGR with low values near the coast and higher values inland; this suggests a corresponding westward increase in the maximum metamorphic temperature (Krogh 1977). We have expanded and revised Krogh's original data base (Table 1); all analyses are by electron microprobe. The use of K_D to determine temperatures is not straightforward. In all samples where zoning was found, we have taken the garnet composition with the *lowest* Fe/Mg, corresponding to the lowest K_D, as being in equilibrium with cpx. Wherever retrograde zoning is present, the resulting value for T_{MAX} is likely to be too low. Calculation of a meaningful K_D requires knowledge of the Fe^{3+}/Fe^{2+} ratio in the minerals. We have assumed that all Fe is divalent in garnets; even where stoichiometry requires small amounts of Fe^{3+}, the difference in K_D is trivial. Fe^{3+} in cpx has been calculated assuming charge balance (Neumann 1976); this procedure is very sensitive to errors in the analysis of SiO_2 (Mysen and Griffin 1973), and is a particular problem in pyroxenes with low FeOᵀ. For most samples (K_D = 4–8), we estimate the probable error in K_D to ca. ± 1 unit (±30°C), with a bias toward too *high* K_D values (low T), due to overestimation of Fe^{3+} in cpx.

K_D can vary considerably at constant T, as a function of the Ca content of garnet. We have therefore

Table 1   Condensed analytical data on Norwegian eclogites and granulites

Locality (Fig. 3)	K_D	X_{Ca}^{gnt}	X_{Jd}^{cpx}	X_{Jd+Ac}^{cpx}	T, 15 Kb (°C)	Comments
1. Sellevoll	7.2	0.19	0.47	0.47	700	
2. Tyssedalsvatn	9.4	0.19	0.45	0.50	640	
3. Flekke	13.5	0.24	0.38	0.44	605	$n = 3$ (585–620)
4. Breivatn	17.2	0.14	0.31	0.47	490	
5. Hestnes	12.6	0.21	0.42	0.51	590	
6. Sande	19.1	0.22	0.42	0.59	520	$n = 2$ (515–525)
7. Kvineset	12.9	0.15	0.38	0.54	550	$n = 9$ (520–590)
8. Naustdal	10.2	0.22	0.44	0.49	645	$n = 5$ (610–665)
9. Redal	10.8	0.17	0.43	0.53	595	
10. Gjelsvik	8.5	0.18	0.48	0.53	655	
11. Davik	6.9	0.22	0.35	0.37	735	
12. Oppedal	6.6	0.21	0.42	0.44	740	
13. Måløy	5.5	0.23	0.14	0.21	810	Ksp+cpx+gnt+qtz gneiss
14. Verpeneset	7.0	0.24	0.40	0.42	735	$n = 2$ (730–740)
15. Kvalneset*	7.2	0.15	0.10	0.10	640	opx eclogite
16. Bryggja	12.6	0.30	0.35	0.43	655	
17. Tasken	7.1	0.22	0.41	0.44	730	$n = 2$ (715–745)
18. Levdalseid	7.1	0.19	0.43	0.47	705	
19. Stryn	9.5	0.19	0.26	0.35	640	$n = 2$ (625–650)
20. Hornindal	7.1	0.15	0	0.03	675	pegm. opx eclogite
21. Bjørkedalen	8.0	0.21	0.34	0.39	705	$n = 4$ (680–730)
22. Høydalsneset	5.5	0.12	0.04	0.08	720	
23. Straumshavn	7.5	0.18	0.37	0.43	685	$n = 2$, incl in gnt
24a. Duen	7.2	0.23	0.28	0.31	735	
24b. Raudkleivane	8.7	0.25	0.28	0.39	700	
25. Grytingvåg*	4.8	0.11	0.08	0.12	765	$n = 5$ (685–800) pegm. opx ecl
26. Årviksnes*	6.1	0.15	0.22	0.31	715	
27. Hellesylt	7.0	0.11	0.36	0.41	645	
28. Flåskjær	6.8	0.11	0.19	0.25	650	
29. Gjøneset	7.9	0.24	0.25	0.31	710	$n = 2$ (700–720)
30. Nerlandsøy	6.5	0.10	0.21	0.32	655	
31. Ulsteinvik†	5.0	0.20	0.45	0.49	810	$n = 4$ (780–840) incl. in gnt
32. Hjørungavåg	4.6	0.10	0.08	0.18	750	opx eclogite
33. Stordal	9.8	0.15	0.01	0.11	600	oxydized: $T = T_{MIN}$
34. Vigra	3.6	0.10	0.26	0.27	825	
35. Flemsøy	4.2	0.10	0.30	0.39	775	$n = 3$ (750–825)
36. Fjørtoft	7.0	0.33	0.25	0.32	820	$n = 2$. 6 HP granul. aver. 805°
37. Uglevik*	4.5	0.10	0.13	0.17	760	$n = 2$ (740–785) gnt websterite
38. Skarshaugen*	3.9	0.11	0.08	0.18	750	opx eclogite
39. Solholmen*	5.3	0.15	0.08	0.08	760	opx eclogite
40. Reset	4.6	0.10	0.24	0.29	750	
41. Harøysund	8.5	0.17	0.31	0.42	645	Gnt zoned, inclusion-rich
42. Bud	5.2	0.20	0.22	0.29	800	
43. Kolmanskog	5.0	0.15	0	0.10	770	Fe-rich gnt period.
44. Bolia	8.8	0.16	0.41	0.48	630	
45. Visnes	6.0	0.24	0.33	0.37	790	layer in marble
46. Frie (K-11)‡	6.7	0.24	0.03	0.09	760	
47. Frei§	5.8	0.23	0.14	0.19	795	$n = 6$ (755–855)
48. Frei (K-6)‡	6.1	0.16	0.20	0.26	720	
49. Magnhildberget‡	7.4	0.29	0.05	0.15	780	$n = 2$ (770–790):HP granulites
50. Vindøladalen	8.7	0.18	0.27	0.47	645	
51. Grøvudalen	12.9	0.24	0.38	0.48	610	
52. Fylkesgrensa	11.9	0.23	0.25	0.39	620	
53. Eiksundal	8.2	0.10	0.24	0.37	605	eclogite
54. Romsdalshorn	7.9	0.15	0.16	0.28	635	
55. Frøyset	9.0	0.20	0.08	0.14	655	
56. Engesetvatn	7.1	0.28	0.04	0.12	780	High-P granulite
57. Vinjefjord	7.7	0.23	0.36	0.41	715	
58. Lesjaskog	13.8	0.24	0.25	0.40	595	
59. Skardsøy bru	4.7	0.17	0.25	0.33	810	

*Carswell et al. (1982); †Mysen and Heier (1972); ‡Krogh (1981); §Griffin and Råheim (1973, unpubl. data).

recalculated our K_D values to temperatures (Table 1, Fig. 3), normalized to the average P = 15 Kb, using the calibration of Ellis and Green (1979). This procedure probably raises the lowest T values by 10–15°C, and reduces the highest values by a similar amount; the variations are within the errors caused by the other factors discussed above.

The isotherms in Fig. 3 represent the most easterly point at which the metamorphic *peak* reached a

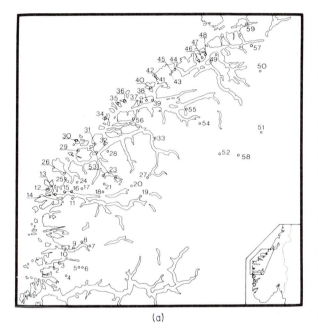

(a)

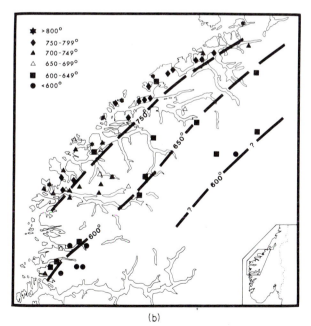

(b)

Fig. 3 Thermometry of eclogites in the Western Gneiss Region. (a) Locations of analysed samples; data are given in Table I. (b) Pattern of T values (normalized to P = 15 Kb) calculated from cpx + gnt pairs (Ellis and Green 1979)

given temperature. The calculated T values define a gradient of *ca.* 200°C across the WGR, even though large areas still lack suitable samples. This regional T variation is greater than the probable errors by a factor of 3–5x. Scattered lower-T values occur throughout the high-T zones; these low values may represent (a) disequilibrium at the peak of metamorphism, (b) samples in which retrograde zoning has 'wiped out' the zone of lowest Fe/Mg in the garnets, (c) overestimation of Fe^{3+} in cpx. In a few cases (c) can be eliminated since there is almost no Fe^{3+} present. Furthermore, several of the anomalously low-T eclogites show strong prograde zoning of the garnets. Such disequilibrium suggests that the 'compression' of the 750° and 800° isotherms along the coast is artificial, and that a true 750° isotherm should lie farther east than shown on Fig. 3.

The absolute T values calculated here are of course dependent on the chosen geothermometer, and may change drastically (±50–100°C) if other calculation methods are used (see discussions by Carswell and Griffin 1981 and Carswell *et al.* this volume). However, the existence of a *gradient* in T across the WGR persists, using any of the published gnt–cpx thermometers.

The Sunnfjord area (loc. 1–10, Fig. 3a) is separated from the rest of the region by major Devonian grabens and at least one major E–W fault. It shows both lower temperatures (in the eastern part) and a steeper westward T gradient (3–4°/km) than the region farther north. Glaucophane occurs in the low-T eclogites of the Naustdal area (Krogh 1980a). The lower T values in this area are from eclogites in the basal Jostedal complex and the higher ones from the overlying Fjordane complex (Table 1, Fig. 3), but the data are too few to evaluate the tectonic implications of this distribution.

The northern part of the WGR seems to show a steeper T gradient than the area farther south, but the data are too few to be sure. The best-documented areas are along Nordfjord (loc. 11–26, Fig. 3a), and from inner Romsdal to the Molde Peninsula and Nordøyene (loc. 35–42, 51, 52, 54, 55). Both of these traverses show average gradients of *ca.* 2°/km.

P estimates

Minimum values of metamorphic pressures were estimated for the bimineralic eclogites in Table 1, by assuming coexistence of cpx with albite and using the model of Holland (1980) for the activity of Jd in omphacite (Fig. 4). The common occurrence of omphacites with *ca.* 40 per cent Jd suggests that pressures were well above the $Di_{80}Jd_{20}$ curve (2) in Fig. 4, throughout the area in which eclogites occur.

Many of the high-T eclogites have Na-poor bulk compositions, and the P_{MIN} estimates are therefore anomalously low. In the Sunnfjord eclogites, which commonly contain paragonite, the P_{MIN} estimates may be close to the true P, and all of the samples fall within the stability field of Pg + Zo + Qz + $DiJd_{40}$ (Franz and Althaus 1977; Holland 1979b) in agreement with the observed assemblages. The Verpeneset eclogite, in which paragonite occurs only as inclusions in garnet, plots just outside this field at higher T and P (Table 1; Krogh 1982). Smith et al. (1980) have reported nearly pure jadeite from the Nybø eclogite; this sample gives the highest P_{MIN} (19 Kb) recorded for a bimineralic eclogite.

Opx-bearing eclogites can in principle be used to estimate both T and P. The P estimate is based on the Al content of opx, and several lines of evidence suggest that disequilibrium between opx and gnt is the rule rather than the exception. Carswell et al. (this volume) show that earlier estimates of very high pressures (30–45 Kb; Lappin 1974; Lappin and Smith 1978) result from use of unequilibrated mineral pairs; they give an average of selected P, T values for 13 opx eclogites from the WGR of 700–750°C, 17–19 Kb. This is within error of similar estimates for the 'internal' opx-bearing eclogites and for garnet lherzolite assemblages in the ultramafics (Carswell et al. this volume).

Krogh (1980a,b) has estimated P and T of metamorphism for the gneisses associated with eclogites at two localities, one in Sunnfjord and the other southeast of Kristiansund. In both cases, these estimates are similar to those for eclogites from the same area. Griffin and Carswell (this volume) calculated p = 19–21 Kb at 750°C for gnt + plag + ky + qtz assemblages from felsic veins in eclogites on Midøy.

An *upper* limit on the metamorphic pressures can only be a guess. No jadeite, nor textural evidence of its earlier presence, has yet been recognized from the felsic rocks of the WGR, and the most jadeite-rich clinopyroxene reported from the eclogites (Smith et al. 1980) does not coexist with quartz. The association of eclogite with anorthosites, as described above, also suggests that the Ab = Jd + Qtz equilibrium curve in Fig. 4 was not crossed during the metamorphism.

The data of Figs. 3 and 4 show that a regional gradient in P and T existed across the WGR during the eclogite-forming event. Maximum conditions of metamorphism increased from *ca.* 500°C, 12 Kb in the inner parts of Sunnfjord to *ca.* 800°C, 18–20 Kb along the coast of Vestlandet. The evidence from mineral inclusions and prograde mineral zoning shows that these conditions were reached by prograde metamorphism at each locality. However, the available data do not allow detailed determination of the P–T trajectory followed by any given rock on its way to the metamorphic peak.

The highest-T eclogites are characterized by retrograde zoning of garnets, and cpx inclusions in these garnets are commonly more Jd-rich than the matrix cpx (Mysen 1972; Krogh 1982). This suggests that the high-T eclogites in Fig. 4 have equilibrated to lower P, and somewhat lower T, than the peak metamorphic conditions. This is in accord with England and Richardson's (1977) models for the P–T evolution of overthrust terranes, and again suggests that P_{MAX} was close to 20Kb at the high-T end of the pieziothermic array.

Retrograde metamorphism

Many eclogites show retrograde reaction to granulite-facies assemblages, others only to amphibolite-facies assemblages. P and T can be calculated for the granulite-facies assemblages; they group around 700–750°C, 8–13 Kb (Krogh 1977, 1982; Carswell et al. 1982). These conditions can be interpreted as reflecting essentially isothermal, and therefore relatively rapid, uplift of the high-T parts of the terrane. Similar models have been suggested for eclogites from subduction zones (e.g. Ernst 1977). The amphibolite-facies assemblages are more difficult to constrain, but appear to have formed between 600–700°C, 6–10 Kb (*cf.* Lappin and Smith 1978). Uplift paths for various parts of the terrane, consistent with petrographic observations, are suggested in Fig. 4.

Fluid phase

The abundance of 'primary' micas and amphiboles in the Norwegian eclogites, the common occurrence of irregular quartz-rich segregations with coarse-grained eclogite phases, and the pegmatitic nature of some eclogites (Grytingvåg type of Eskola 1921) are evidence for a fluid phase being involved in the formation of many eclogites (*cf.* Green and Mysen 1972; Krogh 1980; Holland 1979a). The equilibration temperatures of the high-T eclogites lie well over the 'wet' eclogite solidus of Hill and Boettcher (1970). If the hbl + plag pegmatites described by Green and Mysen (1972) formed near the peak of metamorphism, they may be evidence for anatexis in the presence of a fluid phase. However, the timing of these pegmatites relative to retrograde metamorphism is not clear. Bryhni et al. (1970) and Green and Mysen (1972) have pointed out that anatexis in the surrounding gneisses may have provided a local 'sink' for H_2O, and thus helped to dehydrate the eclogites; this water would still be available for retrograde metamorphism during uplift and cooling.

The availability of a fluid phase may have been the major control on the kinetics of the eclogite-forming reactions (*cf.* Ahrens and Schubert 1975). Originally anhydrous rocks such as gabbros commonly have reacted only on their margins or along other zones where water could gain access. Breakdown of pre-existing amphibolite-facies assemblages (locally preserved as inclusions in garnets) would, on the other hand, free H_2O and speed the recrystallization of these rocks to eclogites. The survival of the high-P assemblages probably depends on whether a fluid phase was driven off at the peak of metamorphism. Garnet amphibolites are common in the WGR; many can be recognized as meta-eclogites by the preservation of omphacite inclusions in their garnets.

Discussion

In situ metamorphism

Evidence summarized here indicates that a variety of low-P protoliths, including both supracrustal and intrusive rocks, have been converted to eclogites over large areas of the WGR. The ubiquitous occurrence of eclogites in these areas suggests that the enclosing gneisses have gone through the same P, T conditions. It has been argued that the gneisses show only low-P amphibolite-facies metamorphism, and that the eclogites must therefore be exotic blocks (Lappin and Smith 1978, with references). Other observations, however, have shown widespread evidence for an early high-P metamorphic event in the gneisses as well. These high-P granulite assemblages in the felsic rocks may be the contemporaneous metamorphic equivalents of the eclogite assemblage in basic rocks (*cf.* experimental work by Green and Lambert 1965) but this can only be proven where the coexistence of omphacite + plagioclase + quartz in these rocks can be demonstrated.

Phengite-breakdown textures like those described by Heinrich (1982) from the Adula Nappe occur at many places in the WGR, and suggest that high-P metamorphism of the gneisses was much more widespread than previously recognized. It seems reasonable to conclude that the formation of most, if not all, eclogites occurred *in situ*. Contact relations often appear tectonic, because the large ductility contrast between eclogite and gneiss during later deformation has caused boudinage (Lappin 1966). Griffin and Carswell (this volume) have presented detailed evidence of *in situ* metamorphism at one locality.

Garnet-peridotite problem

The occurrence of the garnet-peridotites in the WGR has been a major argument for the tectonic introduction of mantle fragments into the gneiss terrane (O'Hara and Mercy 1963; Lappin 1974; Lappin and Smith 1978; O'Hara *et al.* 1971; Carswell 1973a). Like the eclogites, they occur in both large and small bodies. However, they are only found in the western parts of the WGR. The T, P values at the upper end of the metamorphic gradient across the WGR lie above the spinel lherzolite–garnet lherzolite transition of O'Hara *et al.* (1971). Jenkins and Newton (1979) have shown that this transition may lie at even lower P (15–13 Kb) in complex natural systems. The garnet-peridotite assemblages may have been present in these ultramafic rocks when they were tectonically emplaced from the upper mantle. This is consistent with the available Precambrian Sm–Nd gnt/cpx ages. However, it is not strictly necessary to invoke such an origin to explain the garnetiferous assemblages. The similar P, T conditions calculated for external eclogites and gneisses on the one hand, and internal opx eclogites and garnet lherzolites on the other, link the garnet-peridotites to the regional metamorphism. These P, T conditions may reflect metamorphic reequilibration of older garnet-peridotite assemblages.

Alternatively, the garnet-peridotites may have formed by prograde metamorphism of serpentinites, which are common in the eastern parts of the region, or of alpine-type spinel lherzolites similar to the Feragen complex (Moore 1977; Moore and Qvale 1977; Moore and Hultin 1980). This suggestion is supported by the similarity of the *superferrian* internal eclogites (Raudkleiva type) to basic dykes in Caledonian serpentinites and in alpine ophiolite complexes (Griffin and Qvale this volume). Furthermore, Carswell *et al.* (1982) have demonstrated the existence of relict spinel in some garnet-peridotites, and the existence of possible metarodingites, now garnet-pyroxenites, in others. Similar models, based on similar data, have been proposed for the formation of garnet-peridotites in the Alps (Evans and Trommsdorff 1978).

Bryhni (1966) has proposed that the eclogites, ultramafites, and anorthosites of the Nordfjord area present fragments of a layered complex similar to those of the Bergen Arcs and the Jotun Nappe. This is supported by the intimate associations of the three rock types in Bjørkedalen (Brastad this volume) and elsewhere; it also implies formation of the garnet-lherzolites by prograde metamorphism.

Tectonic Model

The high-P, high-T metamorphism of the WGR requires a rapid increase in crustal thickness over a belt >100 km across. The only obvious modern analogies are the Zagros and Himalaya belts of continent–continent collision. In the Zagros region, crustal thickening appears to occur in a wide belt of imbricate faulting; actual subduction of continental crust into the mantle is relatively limited (Bird *et al.* 1975). The P–T regime along the top of the overridden plate in the Zagros collision zone is shown in Fig. 4; it coincides with the P–T values derived for eclogite from the low-T part of the WGR, but lies at higher P in the high-T region. However, the models of England and Richardson (1977) imply that P_{MAX} in the high-T region lay *ca.* 3 Kb above our estimates. In that case, the P–T distribution across the WGR at the metamorphic peak may have been very close to that calculated for the Zagros collision zone.

We suggest, therefore, that the Western Gneiss Region was overridden by the American/Greenland plate during the Caledonian orogeny. Radiometric evidence suggests that this occurred at a rather late stage in the development of the orogen. The belt affected by this 'subduction' was at least 300 km long and 150 km wide (Figs. 1 and 3). Both P and T apparently increased toward the west and north, and this gradient was accompanied by deformation and recrystallization. The widespread metamorphic disequilibrium in the WGR suggests that the high-pressure episode was relatively short-lived (Ahrens and Schubert 1975), but available data can only limit this period to less than *ca.* 30 Ma.

The presence of similar eclogites in the nappes of northern Norway and Sweden, and the Hekla Hoek eclogites in Svalbard, implies that similar continental-subduction processes have occurred along the whole length of the Scandinavian Caledonides (van Roermund this volume). The orogenic belt involved was at least 2000 km long, quite comparable to the present Himalayan belt.

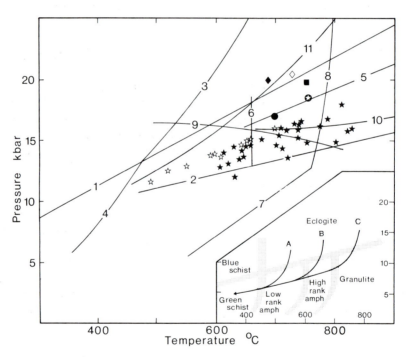

Fig. 4 Calculated P, T values for eclogites and related rocks from the WGR. Curves: 1 and 2: Lower stabilities of $Jd_{100} + Q$ and $Jd_{20} + Q$, respectively (Holland 1980). 3 and 4: $Lw + Jd + Zo + Pg + Q + Vp$ and $Lw + Ab + Zo + Pg + Q + Vp$, respectively (Holland 1979b). 5: Plag out in high-Al basalt/andesite, based on data for relevant compositions at 1100°C, extrapolated at 20 bars/°C (Green and Ringwood 1972). 6: $Pg + Zo + Q + L$ (Franz and Althaus 1977). 7 and 8: $Zo + Ky + Q + An + Vp$ and $Zo + Ky + Q + Vp + L$, respectively (Boettcher 1970). 9: $Pg + DiJd_{40} + Ky + Vp$ (Holland 1979b). 10: Spinel lherzolite/garnet lherzolite transition (O'Hara *et al.* 1971). 11: P, T path, Zagros, top of slab (Bird *et al.* 1975)

Symbols		Source
☆	Opx-free eclogites, Sunnfjord	Table 1, this volume
★	Opx-free eclogites, Nordfjord, Møre, Romsdal	Table 1, this volume
✪ & ●	External cpx-eclogites with Fe^{2+} in cpx calculated and analysed, respectively	Carswell *et al.* 1982
◆	Lien ultramafic body (rim compositions of gnt, cpx, and opx)	Medaris 1980
◇	Crustal garnet lherzolites	Carswell and Gibb 1980b
■	Felsic veins in eclogites, Midøy (gnt + plag + ky + qtz)	Griffin and Carswell 1982

Inset: Suggested retrograde P,T paths (deduced from petrographical evidence) for west Norwegian eclogites equilibrated at low (A), intermediate (B), and high (C) temperatures. Metamorphic facies division from Ernst (1976, 1977)

Uplift of the terrane

The calculated P, T conditions for the granulite-facies breakdown assemblages in eclogites from the WGR (Figs. 3 and 4) imply that uplift was rapid relative to cooling, following the peak of metamorphism. Two general mechanisms can be suggested: erosional removal of the overthrust plate, or 'bouyant return' by back-thrusting up the subduction zone (England and Holland 1979; Ernst 1977). A rough estimate for the uplift rate may be made by assuming closure temperatures of *ca.* 700°C for the Sm–Nd system in cpx and gnt, and of *ca.* 350°C for the Rb–Sr system in micas, and estimating pressures using thermal gradients of 10–15°/km. The average age difference for these two systems in the WGR is *ca.* 30 Ma. This translates into uplift rates of *ca.* 1 mm/year, well within the range expectable from erosion alone. Stripping of nappes from above the WGR would imply much faster uplift rates in the beginning of this period, and correspondingly slower uplift rates would be required later.

The 'bouyant return' mechanism has been used primarily to explain the occurrence of relatively small 'packets' of eclogite-bearing material, commonly with a mélange structure (Ernst 1977; *cf.* Smith 1980). The major arguments against this mechanism in the present case are the general continuity of the prograde metamorphism across the region, and the fact that both major units and single horizons (anorthosite, quartzite) may be followed for long distances (10–60 km) along strike, across the trend of the isotherms in Fig. 3b (Råheim 1972; Griffin and Mørk 1981). However, N–S to NE–SW faults have been observed within the WGR, and may have played a role in the uplift of the coastal areas relative to the inland areas.

The formation of large fault-bounded Devonian clastic basins, and Rb–Sr mineral dates on late granite pegmatites, indicate that the uplift continued into late Devonian time (Griffin and Krill in preparation). The differences in the *apparent* metamorphic gradients between Sunnfjord and the Nordfjord area (Fig. 3), suggest that uplift of these two blocks may have occurred independently, separated by the faults that controlled the sinking of the Devonian basins at Hornelen and Kvamshesten.

Tectonostratigraphic position

Eclogites are most abundant in the western part of the WGR, where the tectonostratigraphy is least understood. In the Oppdal area (east of loc. 51) an early nappe sequence is folded together with the basement (Holtedahl 1938; Krill 1980). Krill (this volume) has traced these rocks westward well into the WGR, where they are infolded with the basement rocks. The metamorphic grade of the Palaeozoic mica schists of the Blåhø Nappe increases westward, and amphibolites enclosed in the mica schists are metamorphosed to eclogite within 25 km of the eastern margin of the WGR. These observations suggest that the continental-collision stage of the orogeny followed the earliest nappe emplacement, in the southernmost part of the orogen.

On the other hand, the Seve Nappes of Sweden contain eclogites (van Roermund this volume) which may suggest that the continental–subduction stage began earlier in some parts of the orogen than in others, or that the Seve Nappes have gone through two stages of transport to reach their present positions. The eclogites of the Bergen Arcs nappes appear to have formed at moderate T (700°C) and high P (*ca.* 15 Kb) (Griffin 1972; Austrheim 1981; Austrheim and Griffin 1982). The eclogites of the Tromsø Nappe Complex are higher-T types, comparable to the middle of the WGR, and it is probably significant that they occupy the highest levels of the local tectonostratigraphy.

Conclusions

1. Most, if not all, of the eclogites in the Western Gneiss Region (WGR) of Norway have been derived from a variety of low-P protoliths, both supracrustal and intrusive, by prograde metamorphism to high P and T. By implication, the same is true of similar eclogites in the Caledonian nappes.

2. The eclogites define a regional metamorphic gradient across the WGR. Maximum metamorphic conditions varied from *ca.* 500°C, 10–12 Kb in Sunnfjord to *ca.* 800°C, 18–20 Kb along the coast of Møre og Romsdal.

3. High-pressure metamorphic mineral assemblages in the gneisses, peridotites, anorthosites, and 'rapakivi granites' are compatible with P–T conditions deduced from their enclosed eclogites. This suggests that all of these rocks were metamorphosed together at high P, T and during later retrograde events.

4. The garnet-peridotites may have inherited their mineral assemblages from the upper mantle. Alternatively, they may have formed by metamorphism of either serpentinites or spinel lherzolites, in the higher-grade parts of the regional metamorphic gradient.

5. The regional metamorphic gradient was established during subduction of a continental plate in a Himalaya-type continent–continent collision zone. Sm–Nd mineral dates on eclogites suggest that the metamorphic peak (T_{max}) may have been at *ca.* 425 Ma or earlier.

6. Eclogites in the Caledonian nappes give calculated P, T values similar to those from the low to middle parts of the gradient over the WGR. Some early nappes were apparently involved in the subduction *after* their emplacement, while later ones (Seve and higher) may represent pieces of the subducted plate, scaled off as part of the obduction process.

7. The uplift of the WGR may have been controlled largely by erosional removal of the overlying plate. Uplift, anatexis, and local deformation continued into the later Devonian, associated with relative subsidence of large clastic basins along E–W faults.

Acknowledgements

This work was supported by Norges Almenvitenskapelige Forskningsråd (D 48.22–008) and Nansenfondet. Many of our colleagues have helped us with critical discussions; we are particularly grateful for contributions from D. A. Carswell, S. Cuthbert, G. Medaris, A. Thompson, J. Hunziker, Igor Koons, B. Yardley, and other participants in both the Zürich/ Bern excursion (July 1981) and Excursion B1 (Sept. 1981).

References

Ahrens, T. J. and Schubert, G. 1975. Rapid formation of eclogite in a slightly wet mantle. *Earth Planet. Sci. Lett.,* **27**, 90–94.

Andréasson, P.-G., Gee, D. G. and Sukotjo, S. this volume. Seve eclogites in the Norrbotten Caledonides, Sweden.

Andresen A., Fareth, E., Bergh, S., Kristensen, S. A. and Krogh, E. J. 1981. Overview of the Caledonian lithotectonic units of western and central Troms, North Norway. Abstract. *Terra Cognita,,* **1**, 33.

Austrheim, H. 1981. Bergen Arcs. *In*: Griffin, W. L. and Mørk, M.B.E. (Eds), *Excursion B1, Excursions in the Scandinavian Caledonides.* Geologisk Museum, Oslo.

Austrheim, H. and Griffin, W. L. 1982. Shear deformation and eclogite formation within granulite-facies anorthosites of the Bergen Arcs, western Norway. Abstract. *Terra Cognita,* **2**, 315.

Binns, R. A. 1967. Barroisite-bearing-eclogite from Naustdal, Sogn og Fjordane. *J. Petrol.,* **8**, 349–371.

Bird, P., Toksöz, M. N. and Sleep, N. H. 1975. Thermal and mechanical models of continent–continent convergence zones. *J. Geophys. Res.,* **80**, 4405–4416.

Boettcher. A. L. 1970. The system $CaO–Al_2O_3–SiO_2–H_2O$ at high pressures and temperatures. *J. Petrol.,* **11**, 337–379.

Bollingberg, H. and Bryhni, I. 1972. Minor element zonation in an eclogite garnet. *Contr. Miner. Petrol.,* **36**, 113–122.

Brastad, K. 1981. Bjřkedalen. In Griffin, W. L. and Mørk, M.B.E. (Eds), *Excursion B1, Excursions in Scandinavian Caledonides.* Geologisk Museum, Oslo.

Brastad, K. this volume. Relations between anorthosites, eclogites, and ultramafics in Bjørkedalen, Western Norway.

Brueckner, H. K. 1972. Interpretation of Rb–Sr ages from the Precambrian and Paleozoic rocks of southern Norway. *Am. J. Sci.,* **272**, 334–358.

Brueckner, H. K. 1977a. A structural, stratigraphic and petrologic study of anorthosites, eclogites, and ultramafic rocks and their country rocks, Tafjord area, western south Norway. *Norges geol. Unders.,* **332**, 1–53.

Brueckner, H. K. 1977b. A crustal origin for eclogites and a mantle origin for garnet peridotites: Sr isotopic evidence from clinopyroxenes. *Contr. Miner. Petrol.,* **60**, 1–15.

Brueckner, H. K. 1979. Precambrian ages from the Geiranger–Tafjord area of the Basal Gneiss Region, west Norway. *Norsk geol. Tidsskr.,* **59**, 141–154.

Bryhni, I. 1966. Reconnaissance studies of gneisses, ultrabasites, eclogites and anorthosites in outer Nordfjord, western Norway. *Norges geol. Unders.,* **241**, 1–68.

Bryhni, I., Bollingberg, H. J. and Graff, P. R. 1969. Eclogites in quartzo-Feldspathic gneisses of Nordfjord, western Norway. *Norsk geol. Tidsskr.,* **49**, 193–225.

Bryhni, I., Fyfe, W. S., Green, D. H. and Heier, K. S. 1970. On the occurrence of eclogite in Western Norway. *Contr. Miner. Petrol.,* **26**, 12–19.

Bryhni, I. and Griffin, W. L. 1971. Zoning in eclogite garnets from Nordfjord, West Norway. *Contr. Miner. Petrol.,* **32**, 112–125.

Bryhni, I., Krogh, E. J. and Griffin, W. L. 1977. Crustal derivation of Norwegian eclogites: a review. *N. Jb. Miner. Abh.,* **130**, 49–68.

Carswell, D. A. 1968a. Picritic magma–residual dunite relationships in garnet peridotite at Kalskaret near Tafjord, southern Norway. *Contr. Miner. Petrol.,* **19**, 97–124.

Carswell, D. A. 1968b. Possible primary upper mantle peridotite in Norwegian basal gneiss. *Lithos,* **1**, 322–355.

Carswell, D. A. 1973a. The age and status of Basal Gneiss Complex of north-west southern Norway. *Norsk geol. Tidsskr.,* **53**, 65–78.

Carswell, D. A. 1973b. Garnet pyroxenite lens within Uglevik layered garnet peridote. *Earth Planet. Sci. Lett.,* **20**, 347–352.

Carswell, D. A. 1974. Comparative equilibration temperatures and pressures of garnet lherzolites in Norwegian gneisses and in kimberlite. *Lithos,* **7**, 113–121.

Carswell, D. A. 1981. Clarification of the petrology and occurrence of garnet lherzolites, garnet websterites and eclogite in the vicinity of Rødhaugen, Almklovdalen, W. Norway. *Norsk geol. Tidsskr.,* **61**, 249–260.

Carswell, D. A. and Gibb, F. G. F. 1980a, Geothermometry of garnet lherzolite nodules with special reference to those from the kimberlites of Northern Lesotho. *Contr. Miner. Petrol.,* **74**, 403–416.

Carswell, D. A. and Gibb, F. G. F. 1980b. The equilibrium conditions and petrogenesis of European crustal garnet lherzolites. *Lithos,* **13**, 19–29.

Carswell, D. A. and Griffin, W. L. 1981. Calculation of equilibrium conditions for garnet granulite and garnet websterite nodules in African kimberlite pipes. *Tscher. Min. Petr. Mitt.* **28**, 229–244.

Carswell, D. A. and Harvey, M. 1981. Molde–Otrøy–Midøy. In Griffin, W. L. and Mørk, M. B. E. (Eds), *Excursion B1. Excursions in Scandanavian Caledonides.*

Carswell. D. A. and Harvey, M. this volume. Intrusive history and tectonometamorphic evolution of the basal gneiss complex in the Moldefjord region, west Norway.

Carswell, D. A., Harvey, M. A. and Al-Samman, A. 1982. The petrogenesis of garnetiferous peridotites and related rocks in the high grade gneiss complex of western Norway. Abstract. *Terra Cognita, 2,* 327.

Carswell, D. A., Krogh, E. J. and Griffin, W. L. this volume. Norwegian orthopyroxene eclogites: calculated equilibration conditions and petrogenetic implications.

Claesson, S. 1981. Caledonian metamorphism of Proterozoic Seve rocks on Mt. Åreskutan, southern Swedish Caledonides. *Geol. Fören. Stockh. Förh., 103,* 291–304.

Cordellier, F., Boudier, F., and Boullier, A. M. 1981. Structural study of the Almklovdalen massif (Southern Norway). *Tectonophysics, 77,* 257–281.

Currie, K. L. and Curtis, L. W. 1976. An application of multicomponent solution theory to jadeitic pyroxenes. *J. Geol., 84,* 179–194.

Cuthbert, S. 1981. Gjørlangen–Sørdal area. *In:* Griffin, W. L. and Mørk, M. B. E. (Eds), *Excursion B1. Excursions in Scandinavian Caledonides.* Geolisk Museum, Oslo.

DeWaard, D. 1965. The occurrence of garnet in the granulite-facies terrane of the Adirondack highlands. *J. Petrol., 6,* 165–191.

Ellis, D. J. and Green, D. H. 1979. An experimental study of the effect of Ca upon garnet–clinopyroxene Fe–Mg exchange equilibria. *Contr. Miner. Petrol., 71,* 13–22.

England, P. C. and Holland, T. J. B. 1979. Archimedes and the Tauern eclogites: the role of buoyancy in the preservation of exotic eclogite blocks. *Earth Planet. Sci. Lett., 44,* 287–294.

England, P. C. and Richardson, S. W. 1977. The influence of erosion upon the mineral facies of rocks from different metamorphic environments. *J. geol. Soc. London, 134,* 201–219.

Ernst, W. G. 1973. Interpretive synthesis of metamorphism in the Alps. *Bull Geol. Soc. Amer., 84,* 2053–2078.

Ernst, W. G. 1976. Mineral chemistry of eclogites and related rocks from the Voltri Group, western Liguria, Italy. *Schweiz. Min. Ret. Mitt., 56,* 293–343.

Ernst, W. G. 1977. Tectonics and prograde versus retrograde P–T trajectories of high-pressure metamorphic belts. *Rend. Sco. Italiana Miner. Petrol., 33,* 191–220.

Eskola, P. 1921. On the eclogites of Norway. *Skr. Vidensk. Selsk. Christiania, Mat.–Nat. Kl. I, 8,* 1–118.

Evans, B. W. and Trommsdorff, V. 1978. Petrogenesis of garnet lherzolite, Cima di Gagnone, Lepontine Alps. *Earth Planet. Sci. Lett., 40,* 333–348.

Franz, G. and Althaus, E. 1977. The stability relations of the paragenesis paragonite–zoisite–quartz. *Neues Jahrb. Miner. Abh., 130,* 159–167.

Garmann, L. B., Brunfelt, A. O., Finstad, K. G. and Heier, K. S. 1975. Rare-earth element distribution in basic and ultrabasic rocks from West Norway. *Chemical Geol., 15,* 103–116.

Gebauer, D., Lappin, M., Gruenenfelder, M., Koestler, A. and Wyttenbach, A. 1982. Age and origin of some Norwegian eclogites: a U–Pb zircon and REE study. *Terra Cognita, 2,* 323.

Gee, D. G. 1966. A note on the occurrence of eclogites in Spitsbergen. *Norsk Polarinst. Årb., 1964,* 240–241.

Gjelsvik, T. 1952. Metamorphosed dolerites in the gneiss area of Sunnmøre on the West Coast of Southern Norway. *Norsk geol. Tidsskr., 30,* 33–134.

Green, D. A. and Lambert, I. B. 1965. Experimental crystallization of anhydrous granite at high pressures and temperatures. *J. Geophys. Res., 70,* 5259–5268.

Green, D. H. and Mysen, B. O. 1972. Genetic relationship between eclogite and hornblende + plagioclase pegmatite in western Norway. *Lithos, 5,* 147–161.

Green, D. H. and Ringwood, A. E. 1972. A comparison of recent experimental data on the gabbro–eclogite transition. *Jour. Geol., 80,* 277–288.

Griffin, W. L. 1971. Genesis of coronas in anorthosites of the Upper Jotun Nappe, Indre Sogn, Norway. *J. Petrol., 12,* 219–243.

Griffin, W. L. 1972. Formation of eclogites and the coronas in anorthosites, Bergen Arcs, Norway. *Geol. Soc. Am. Mem., 135,* 37–63.

Griffin, W. L. and Brueckner, H. K. 1980. Caledonian Sm–Nd ages and a crustal origin for Norwegian eclogites. *Nature, London, 285,* 319–321.

Griffin, W. L. and Brueckner, H. K. in press. REE, Rb–Sr and SM–Nd studies of Norwegian eclogites. *isotope Geoscience.*

Griffin, W. L. and Carswell, D. A. this volume. *In situ* metamorphism of Norwegian eclogites: an example.

Griffin, W. L. and Heier, K. S. 1973. Petrological implications of some corona structures. *Lithos, 6,* 315–335.

Griffin, W. L. and Mørk, M. B. E. 1981. Eclogites and basal gneisses in western Norway In Griffin, W. L. and Mørk, M. B. E. (Eds), *Excursion B1, Excursion in Scandinavian Caledonides.* Geologisk Museum, Oslo.

Griffin, W. L. and Qvale, H. this volume. Superferric eclogites and the crustal origin and garnet peridotites.

Griffin, W. L. and Råheim, A. 1973. Convergent metamorphism of eclogites and dolerites, Kristiansund area, Norway. *Lithos, 6,* 21–40.

Harris, L. D. and Bayer, K. C. 1979. Sequential development of the Appalachian orogen above a master decollement—a hypothesis. *Geology, 7,* 568–572.

Harvey, M. 1983. A geochemical and Rb–Sr study of the Proterozoic augen gneisses on the Molde Peninsula, west Norway. *Lithos, 16,* 325–338.

Heinrich, C. A. 1982. Kyanite–eclogite to amphibolite facies of mafic and pelitic rocks in the Adula nappe, central Alps. *Terra Cognita, 2,* 307.

Hernes, I. 1954. Eclogite–amphibolite on the Molde peninsula, Southern Norway. *Norsk geol. Tidsskr., 33,* 163–184.

Herzberg, C. I. 1978. Pyroxene geothermometry and geobarometry: experimental and thermodynamic evaluation of some subsolidus relations involving pyroxenes in the system $CaO–MgO–Al_2O_3–SiO_2$. *Geochim. Cosmochim. Acta, 42,* 945–957.

Hill, R. E. T. and Boettcher, A. L. 1970. Water in the earth's mantle: melting curves of basalt–water and basalt–water–carbon dioxide. *Science, 167,* 980–982.

Holland, T. J. B. 1979a. High water activities in the generation of high pressure kyanite eclogites of the Tauern-window, Austria. *J. Geol., 87,* 1–27.

Holland, T. J. B. 1979b. Experimental determination of the reaction paragonite = jadeite + kyanite + H_2O, and internally consistent thermodynamic data for part of the system $Na_2O–Al_2O_3–SiO_2–H_2O$, with applications to eclogites and blueschists. *Contr. Miner. Petrol., 68,* 293–301.

Holland, T. J. B. 1980. The reaction albite–jadeite + quartz determined experimentally in the range

600–1200°C. *Am. Min.,* **65**, 129–134.

Holtedahl, O. 1938. Geological observations in the Opdal–Sundal–Trollheimen district. *Norsk geol. Tidsskr.,* **18**.

Jacobsen, S. B. and Wasserburg, G. T. 1980. Nd and Sr isotopes of the Norwegian garnet peridotites and eclogites. *EOS,* **61**, 389.

Jenkins, D. M. and Newton, R. C. 1979. Experimental determination of the spinel peridotite to garnet peridotite inversion at 900°C and 1000°C in the system $CaO–MgO–Al_2O_3–SiO_2$, and at 900°C with natural olivine and garnet. *Contr. Miner. Petrol.,* **68**, 407–419.

Krill, A. G. 1980. Tectonics of the Oppdal area, Central Norway. *Geol. Fören. Stockh. Förh.,* **102**, 523–530.

Krill, A. G. this volume. Relationships between the Western Gneiss Region and the Trondheim Region: Stockwerttectonics reconsidered.

Krill, A. G. and Griffin, W. L. 1981. Interpretation of Rb–Sr dates from the Western Gneiss Region: a cautionary note. *Norsk geol. Tidsskr.,* **61**, 83–86.

Krill, A. G. and Röshoff, K. 1981. Basement–cover relationships in the central Scandinavian Caledonides, *Excursion in Scandinavian Caledonides.*

Krogh, E. J. 1977. Evidence for a Precambrian continent–continent collision in western Norway. *Nature, London,* **267**, 17–19.

Krogh, E. J. 1980a. Geochemistry and petrology of glaucophane-bearing eclogites and associated rocks from Sunnfjord, western Norway, *Lithos,* **13**, 355–380.

Krogh, E. J. 1980b. Compatible P–T conditions for eclogites and surrounding gneisses in the Kristiansund area, western Norway. *Contr. Miner. Petrol.,* **75**, 387–393.

Krogh, E. J. 1982. Metamorphic evolution deduced from mineral inclusions and compositional zoning in garnets from Norwegian country-rock eclogites. *Lithos,* **15**, 305–321.

Krogh, E. J. and Andresen, A. 1979. Geokjemi og petrologi av eklogitter og omliggende bergarter i øvre dekkeenhet i Troms. *Geolognytt,* **13**, 38.

Krogh, E. J. and Brunfelt, A. O. 1981. REE, Cs, Rb, Sr and Ba in glaucophane-bearing eclogites and associated rocks, Sunnfjord, western Norway. *Chem. Geol.,* **33**, 295–305.

Krogh, E. J. and Råheim, A. 1978. Temperature and pressure dependence of Fe–Mg partitioning between garnet and phengite, with particular reference to eclogites. *Contr. Miner. Petrol.,* **66**, 75–80.

Krogh, T. E., Mysen, B. O. and Davis, G. L. 1974. A paleozoic age for the primary minerals of a Norwegian eclogite. *Ann. Rep. Geophys. Lab. Carnegie Inst. Washington,* **73**, 575–576.

Kushiro, I. 1969. Clinopyroxene solid solution formed by reactions between diopside and plagioclase at high pressures. *Miner. Soc. Am. Spec. Pap.,* **2**, 179–191.

Landmark, K. 1973. Beskrivelse til de geologiske kart Tromsø og Målselv, del 2. *Tromsø Mus. Skr.,* **15**, 263 pp.

Lappin, M. A. 1966. The field relationships of basic and ultrabasic masses in the basal gneiss complex of Stadlandet and Almklovdalen, Nordfjord, southwestern Norway. *Norsk geol. Tidsskr.,* **46**, 439–495.

Lappin, M. A. 1973. An unusual clinopyroxene with complex lamellar intergrowths from an eclogite in the Sunndal–Grubse ultramafic mass, Almklovdalen, Nordfjord, Norway. *Miner. Mag.,* **39**, 313–320.

Lappin, M. A. 1974. Eclogites from the Sunndal-Grubse ultramafic mass, Almklovdalen, Norway and the T–P

history of the Almklovdalen mass. *J. Petrol.,* **15**, 567–601.

Lappin, M. A., Pidgeon, R. T. and Breemen, O. van 1979. Geochronology of basal gneisses and mangerite syenites of Stadlandet, west Norway. *Norsk geol. Tidsskr.,* **59**, 161–181.

Lappin, M. A. and Smith, D. C. 1978. Mantle-equilibrated orthopyroxene eclogite pods from the basal gneisses in the Selje district, western Norway. *J. Petrol.,* **19**, 530–584.

Maquil, R. 1978. Preliminary investigation on giant orthopyroxenes with plagioclase exsolution lamellae from the Egersund–Ogna anorthositic massif (S. Norway). In *Progress in Experimental Petrology,* N.E.R.C. Series D, **11**, 144–146.

Mearns, E. W. and Lappin, M. A. 1982. A Sm–Nd isotopic study of 'internal' and 'external' eclogites, garnet lherzolite and gray gneiss from Almklovdalen, western Norway. *Terra Cognita,* **2**, 324.

Medaris, Jr. L. G. 1980. Petrogenesis of the Lien peridote and associated eclogites, Almklovdalen, western Norway. *Lithos,* **13**, 339–353.

Moore, A. C. 1977. The petrography and possible regional significance of the Hjelmkona ultramafic body (sagvandite), Nordmøre, Norway. *Norsk geol. Tidsskr.,* **57**, 55–64.

Moore, A. C. and Hultin, I. 1980. Petrology, mineralogy and origin of the Feragen ultramafic body, Sør-Trøndelag, Norway. *Norsk geol. Tidsskr.,* **60**, 235–254.

Moore, A. C. and Qvale, H. 1977. Three varieties of alpine-type ultramafic rocks in the Norwegian Caledonides and Basal Gneiss Complex. *Lithos,* **10**, 149–161.

Mørk, M. B. E. 1983. A gabbro–eclogite transition on Flemsøy, Sunnmøre, Western Norway. *Terra Cognita,* **2**, 316.

Mykkeltveit, S., Husebye, E. S. and Oftedahl, C. 1980. Subduction of the Iapetus Ocean crust beneath the Møre Gneiss Region, southern Norway. *Nature, London,* **288**, 473–475.

Mysen, B. O. 1972. Five clinopyroxenes in the Hareidlandet eclogite, western Norway. *Cont. Miner. Petrol.,* **34**, 315–325.

Mysen, B. O. and Griffin, W. L. 1973. Pyroxene stoichiometry and the breakdown of omphacite. *Amer. Miner.,* **58**, 60–63.

Mysen, B. O. and Heier, K. S. 1972. Petrogenesis of eclogites in high grade metamorphic gneisses, exemplified by the Hareidlandet eclogite, western Norway. *Contr. Miner. Petrol.,* **36**, 73–94.

Neumann, E.-R. 1976. Two refinements for the calculations of structural formulae for pyroxenes and amphiboles. *Norsk geol. Tidsskr.,* **36**, 1–6.

O'Hara, M. J. 1967. Mineral paragenesis in ultrabasic rocks. In Wyllie, P. J. (Ed.), *Ultramafic and Related Rocks,* J. Wiley, New York.

O'Hara, M. J. and Mercy, E. L. P. 1963. Petrology and petrogenesis of some garnetiferous peridotites. *Trans. Roy. Soc. Edinburgh,* **65**, 251–314.

O'Hara, M. J., Richardson, S. W. and Wilson, G. 1971. Garnet peridotite stability and occurrence in crust and mantle. *Contr. Miner. Petrol.,* **32**, 48–68.

Ohnmacht, W. 1974. Petrogenesis of carbonate-orthopyroxenites (sagvandites) and related rocks from Troms, northern Norway. *J. Petr.,* **15**, 303–323.

Ohta, Y. 1979. Blueschists from Motalafjellet, Western

Spitsbergen. *Norsk Polarinst. Skr.,* **167**, 171–217.

Reymer, A. P. S., Boelrijk, N. A. I. M., Hebeda, E. H., Priem. H. N. A., Verdurmen, E. A. Th. and Verschure, R. H. 1980. A note on Rb–Sr whole-rock ages in the Seve nappe of the central Scandinavian Caledonides. *Norsk geol. Tidsskr.,* **60**, 139–148.

Roermund, H. L. M. van. this volume. Eclogites of the Seve Nappe, central Scandinavian Caledonides.

Rosenqvist, I. Th. 1956. Granulitt–mangeritt–ecklogitt. *Norsk geol. Tidsskr.,* **36**, 41–47.

Råheim, A. 1972. Petrology of high grade metamorphic rocks of the Kristiansund area. *Norges geol. Unders.,* **270**, 1–75.

Råheim, A. 1977. A Rb–Sr study of the rocks of the Surnadal Syncline. *Norsk geol. Tidsskr.,* **57**, 193–204.

Råheim, A. and Green, D. H. 1975. P, T paths of natural eclogites during metamorphism—a record of subduction. *Lithos,* **8**, 317–328.

Schmidt, H. H. 1963. *Petrology and Structure of the Eiksundal Eclogite Complex, Hareidlandet, Sunnmøre, Norway.* Unpubl Ph.D. Thesis, Harvard Univ. U.S.A., 323 pp.

Schreyer, W., Ohnmacht, W. and Mannchen, J. 1972. Carbonate-orthopyroxenites (sagvandites) from Troms, northern Norway. *Lithos*, **5**, 345–364.

Smith, D. C. 1980. A tectonic melange of foreign eclogites and ultramafics in West Norway. *Nature, London,* **287**, 366–367.

Smith, D. C. 1982. The essence of an eclogite: Al^{VI}, as exemplified by the crystal-chemistry and petrology of high-pressure minerals in Norwegian eclogites. *Terra Cognita,* **2**, 300.

Smith, D. C., Mottana, A. and Rossi, G. 1980. Crystal chemistry of a unique jadeite-rich acmite-poor omphacite from the Nybø eclogite pod. Sørpollen, Nordfjord, Norway. *Lithos,* **13**, 227–236.

Stern, C. R., Huang, W. L. and Wyllie, P. J. 1975. Basalt—andesite–rhyolite–H_2O: crystallization intervals with excess H_2O-undersaturated liquidus surfaces to 35 kilobars, with implications for magma genesis. *Earth Planet. Sci. Lett.,* **28**, 189–196.

Sørensen, H. 1955. A preliminary note on some peridotites from northern Norway. *Norsk geol. Tidsskr.,* **35**, 93–104.

Tørudbakken, B. 1981. Vindøladalen dolerite. In Griffin, W. L. and Mørk, M. B. E. (Eds), *Excursion B1. Excursions in Scandinavian Caledonides.* Geologisk Museum, Oslo.

Vogel, D. E. and Garlick, G. D. 1970. Oxygen-isotope ratios in metamorphic eclogites. *Contr. Miner. Petrol.,* **28**, 183–191.

The Caledonide Orogen—Scandinavia and Related Areas
Edited by D. G. Gee and B. A. Sturt

Superferrian eclogites and the crustal origin of garnet peridotites, Almklovdalen, Norway

W. L. Griffin and **H. Qvale**

Mineralogisk–Geologisk Museum, Sarsgate 1, Oslo 5, Norway

ABSTRACT

Basic rocks of superferrian composition (MgO < 8 per cent; FeO^T > 14 per cent) occur as dykes and lenses in ultramafic rocks at several localities in the Norwegian Caledonides. In serpentinites and amphibolite-facies chlorite + orthopyroxene (± carbonate) peridotites, the basic rocks are amphibolites.

At Raudkleivane, Almklovdalen, the basic bodies are eclogite, while relict garnet peridotite assemblages occur in the enclosing ultramafic rock. Garnets in the eclogites show strong prograde zoning of Fe, Mg, and Mn, and garnet cores contain abundant inclusions of a Cl-rich ferropargasite, as well as K-feldspar, apatite, ilmenite, and pyrite, none of which are now observed in the eclogite matrix. The similarity of the primary amphibole compositions to the whole rock compositions suggests that the rocks once consisted of amphibole + minor plagioclase + magnetite. The Cl content may link the amphibolitization to serpentinization of the dunite.

Our observations suggest that amphibolitization of the basic rocks preceded prograde metamorphism to eclogite facies. Estimated metamorphic conditions for the eclogites and the surrounding terrain are 700–750°C, 15–18 Kb, which is also sufficient to convert spinel lherzolite to garnet lherzolite. There is thus no need to regard the garnet lherzolite assemblages as inherited from the upper mantle.

Introduction

Garnet peridotites occur as relict assemblages in chlorite–amphibolite harzburgites and serpentinites at several localities in the Western Gneiss Region of Norway (Fig. 1). They have long been regarded as fragments of the upper mantle, tectonically introduced into the surrounding gneisses (O'Hara and Mercy 1963; Lappin 1974; Lappin and Smith 1978; O'Hara et al. 1971).

Recent studies of the gneisses and the eclogites enclosed in them (Bryhni et al. 1977; Krogh 1977, 1980a,b; Griffin et al. 1983) have demonstrated the existence of early high-pressure mineral assemblages in a wide variety of rock types in the Gneiss Region. These high-P assemblages define a metamorphic gradient increasing from ca. 500°C, 12 Kb in the south and east, to ca. 800°C, 18 Kb in the north and west. All of the known localities of garnetiferous ultramafic rocks in the Gneiss Region lie in areas that were subjected to T ≥ 650°C, P ≥ 13 Kb (Fig. 1).

Recent experimental studies (Jenkins and Newton 1979) show that these conditions lie above the spinel lherzolite–garnet lherzolite transition. This suggests that the garnetiferous ultramafic rocks could have formed by metamorphism of low-P protoliths, rather than being unmodified fragments of the upper mantle. Similar origins have been proposed for some garnet peridotites in the Alps (Evans and Trommsdorff 1978; Evans et al. 1981) and the Pamirs (Budanova and Budanov 1975). This possibility has led us to reexamine the available evidence on the garnet peridotites of western Norway.

This paper presents data on one type of eclogite within the (locally garnetiferous) ultramafic body in Almklovdalen, Møre and Romsdal county, Norway (Fig. 1). The chemistry of these bodies is compared

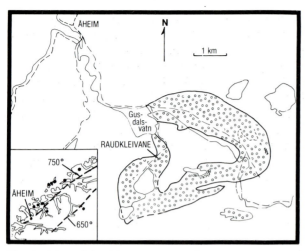

Fig. 1   Map of the Almklovdalen area with outline of the ultramafic mass. Inset: Location of known garnet peridotites, and isotherms derived from eclogite assemblages (Griffin *et al*. 1982)

with that of amphibolitized dykes in Caledonian serpentinites, and with eclogites within the ophiolites of the Ligurian Alps. Petrographic and chemical evidence suggests that the eclogites described here formed by prograde metamorphism of amphibolites.

Geological Setting

The early high-pressure mineral assemblages of Almklovdalen ultramafic body (Fig. 1) occur as relics within a strongly folded chlorite + amphibole + orthopyroxene + olivine peridotite, irregularly distributed within dunite (Medaris 1980). Two general categories of eclogite-facies rocks can be distinguished:

1. Garnet lherzolite, garnet websterite, and 'eclogite' (garnet clinopyroxenite) commonly interlayered and isoclinally interfolded on a scale of centimetres to decimetres (Lappin 1966; Medaris 1980). The eclogites ('Rødhaug type' of Eskola 1921) have Fe/Fe + Mg < 0.5; clinopyroxenes are Na-poor (diopside to sodic augite) and garnets are pyropes with variable, generally high, Cr contents. These rocks are found at Rødhaugen (Eskola 1921), Lien (Lappin 1966; Medaris 1980) and Raudkleivane ('Sunndal–Grubse' locality of Lappin 1973, 1974).

2. Fe-rich eclogite pods, apparently formed by boudinage of continuous layers. These consist of Na-rich omphacites + almandine–grossular–pyrope garnet + rutile + apatite. They occur in a small area at Raudkleivane, where Lappin

(1974) described them as 'layers in garnet peridotite'.

This locality is shown in more detail in Fig. 2; two trains of eclogite pods appear to be related to an isoclinal fold system in the chlorite peridotite. The form of the pods leaves little doubt that they are boudins: the entire array may have formed by boudinage of a single layer, or a series of closely-spaced, parallel layers. Individual boudins commonly show a cm-scale layering of clinopyroxene-rich *vs* garnet-rich types, especially toward their margins. Pods 5 and 6 consist partly of strongly lineated cpx + opx + amphibole rock with local clinopyroxene megacrysts. Layering in both rock types is conformable to the contacts on two sides of each pod, but often truncated by one side. All of the pods are amphibolitized, some heavily, along their contacts with the chlorite peridotite. Two of these pods, apparently our numbers 6 and 19, were described by Lappin (1974).

Garnetiferous ultramafic rocks, with thin layers of garnet clinopyroxenite, occur in several places at Raudkleivane (Fig. 2). One of the outcrops shows a broad open folding, apparently preceding the isoclinal folding of the surrounding chlorite peridotite. The garnetiferous rocks show all gradations into chlorite peridotite, but no boudinage. The two types of eclogite-facies rocks thus differ not only in composition, but in their rheological behaviour relative to the peridotites.

Petrography

The clinopyroxenite found in pods 5 and 6 (Fig. 2) has been well described by Lappin (1973, 1974). It consists mainly of clinopyroxene with minor

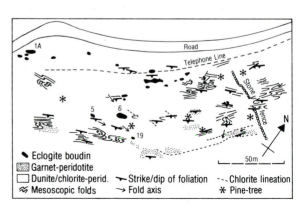

Fig. 2   Detailed map of the Raudkleivane locality

orthopyroxene and garnet, and variable amounts of secondary (?) amphibole, and commonly is well lineated parallel to the elongation of the boudins. Rare 1-cm megacrysts of clinopyroxene contain abundant blebs of orthopyroxene and garnet. Lappin (1973) interpreted these as due to exsolution, though he also demonstrated that the 'exsolution-planes' were non-rational. Rost and Brenneis (1978) and Obata and Morten (in preparation) have described very similar textures in clino- and ortho-pyroxenes from the Ultental spinel peridotites. These have clearly formed by exsolution from primary high-Al pyroxenes that originally coexisted with spinel + olivine. A similar origin, perhaps with more extensive recrystallization, may explain the complex pyroxene megacrysts in the clinopyroxenites at Raudkleivane.

The eclogitic rocks consist in general of omphacite + garnet + rutile + apatite in a medium-grained mosaic texture. Pyroxene-rich bands are usually somewhat coarser-grained and well lineated, with varying amounts of a pale, apparently primary amphibole. Most samples show extensive symplectitic breakdown of omphacite to augite + plagioclase (An_{20}); replacement of these symplectites and of garnet by dark green hornblende is also common.

In several of our samples the garnets are visibly zoned, with dark orange cores and pale red–pink rims. The cores of zoned garnets contain numerous inclusions. The most abundant inclusion phase is a pale blue–grey amphibole, clearly different from the secondary green hornblendes that replace both garnet and cpx. The inclusions are subhedral in form; many have negative-crystal shapes within the garnet. Other inclusion phases identified by microprobe are apatite, pyrite, ilmenite, and K-feldspar. Clinopyroxene and rutile occur as inclusions in the rims of zoned garnet grains, and in the cores of some unzoned garnets.

Sample 5 contains rounded areas consisting of a fine granular mosaic of a cloudy Na-zeolite, apparently analcime. These *areas* appear to be pseudomorphs after a phase of the same grain size as the garnet and omphacite. This phase might have been albite. The rock analysis is relatively low in Na, suggesting that this thin section is not representative of the analysed material.

Chemistry

The selection of samples for analysis is complicated by the existence of the small-scale banding. We have attempted to use homogeneous rocks. Samples were taken along traverses across several of the pods, to study effects of contamination and mixing. Analyses were done by wet-chemical methods for FeO, Na_2O, K_2O, P_2O_5, H_2O, and CO_2 and by XRF for all others, at Norges Geologiske Undersøkelse, Trondheim. Results are given in Table 1. Several of the analyses show low analytical totals, and in most samples there is poor agreement between loss on ignition (corrected for oxidation of Fe) and the sum of analysed ($H_2O + CO_2$). We suspect that the volatile analyses, especially H_2O^+, are *generally* too low, but this does not affect the conclusions drawn below.

FeO/FeO + MgO variation diagrams (Fig. 3) show that the total variation in the rocks (including analyses from Lappin (1974)), can be described in terms of limited mixing between two end members, the clinopyroxenites and the Fe-rich eclogites. The normative composition of the pyroxenites is, as expected, dominated by clinopyroxene, with minor orthopyroxene and a few per cent olivine. The low FeO/(FeO + MgO) and Na_2O of the pyroxenites is clear evidence that they have not originated simply by a large-scale segregation of eclogitic pyroxene and garnet.

The true eclogites have such high Fe contents that they fall into the *superferrian* group of Mottana (1977) and Mottana and Bocchio (1975). Normatively they are dominated by olivine, plagioclase, and clinopyroxene. However, several of them (1, 2, 10, 11) have extremely An-rich plagioclase together with an Fe-rich olivine. This suggests that the high Al in these samples may not be a primary feature (see below). Several of the samples show normative Ne, but this is largely the effect of a high Fe^{3+}/Fe^{2+}, which also may be secondary. All of the rocks have moderate Na contents and very little K (except no. 5—see below). Comparison of fresh and amphibolitized samples, and the mass balance between garnet and clinopyroxene, suggest that little, if any, alkalies have been added during the amphibolitization. Comparison of amphibolitized boudin margins *vs* fresher boudin cores suggests that the following elements *may* have been added during amphibolitization: Na, Sr, Ba, Ni. It seems obvious that the high Fe/Fe + Mg ratios, and the high Na contents that distinguish these superferrian eclogites from the garnet clinopyroxenites ('internal eclogites') in the ultramafic rocks, were established prior to eclogite-facies metamorphism.

Both Cr and Ni are high in the clinopyroxenites, low in the superferrian eclogites, and high again in the peridotite. This also suggests that the differentiation between pyroxenites and eclogites did not occur by segregation of garnet and clinopyroxene, since Cr would then be concentrated in the eclogites. The

Table 1 Rock analyses

	1	2	3	4	5	6	7	8	9	10	11	BP	BH	BR	BE	VE	RE
SiO$_2$	40.30	40.23	41.30	45.79	43.31	48.68	51.75	52.00	52.19	41.05	38.86	48.3	40.2	27.4	47.02	45.18	42.23
TiO$_2$	1.15	1.32	1.29	0.79	0.92	0.55	0.31	0.27	0.25	0.68	1.34	2.69	3.28	2.69	1.05	3.91	0.87
Al$_2$O$_3$	18.37	17.63	15.90	13.31	14.60	12.22	4.69	3.98	3.23	17.53	16.96	12.8	10.8	15.5	14.50	11.61	16.44
Fe$_2$O$_3$	1.85	2.71	5.04	1.99	2.18	1.94	2.25	3.29	2.08	2.77	2.95	3.1	9.0	13.9	3.18	4.50	2.46
FeO	17.72	17.37	12.94	14.03	15.70	8.79	5.45	3.82	4.34	18.17	17.89	11.4	8.9	13.0	8.67	13.90	16.59
MnO	0.21	0.25	0.22	0.25	0.29	0.11	0.12	0.09	0.09	0.32	0.31	0.28	0.25	0.2	0.17	0.23	0.26
MgO	5.76	6.41	6.71	8.37	6.36	9.69	16.45	17.97	18.23	7.46	7.07	5.6	9.6	25.2	9.16	5.86	6.93
CaO	10.39	9.36	9.35	12.00	10.86	12.40	15.73	16.60	17.71	9.91	9.80	10.32	13.28	1.3	11.80	8.66	10.27
Na$_2$O	1.87	2.28	3.65	2.19	1.92	3.32	1.90	1.39	1.04	1.04	1.21	3.69	2.28	—	2.47	3.59	2.10
K$_2$O	0.03	0.06	0.15	0.03	0.45	0.10	0.10	0.11	0.02	0.03	0.05	0.25	0.31	—	0.21	0.09	0.05
P$_2$O$_5$	0.12	0.10	0.07	0.04	0.05	0.04	0.02	0.01	0.03	0.04	0.10	0.28	0.23	0.6	0.21	0.14	0.06
H$_2$O$^+$	0.89	1.35	2.33	0.31	0.66	0.85	0.70	1.24	0.88	0.65	0.99	1.8	1.9	—	n.a.	1.84	0.81
Co$_2$	0.08	0.09	0.10	0.09	0.11	0.07	0.07	0.11	0.07	0.06	0.08	n.a.	n.a.	n.a.	n.a.	n.a.	0.0
	98.74	99.16	99.05	99.19	97.41	98.76	99.54	100.96	100.16	99.71	97.61	100.6	100.2	(100)	98.42	99.51	99.07
Zr	27	14	30	26	40	33	24	24	22	28	20	202	227			46	27
Y	21	24	19	20	26	18	11	9	9	35	36	73	89				25
Sr	67	46	209	102	77	125	269	331	200	127	420	171	295				120
Rb	<5	<5	<5	<5	<5	<5	<5	<5	<5	<5	<5	1	1				
Zn	93	89	85	74	88	79	45	38	36	71	70						77
Cu	21	11	32	102	217	100	116	46	38	18	155					37	67
Ni	79	101	154	309	318	377	537	476	776	100	224	48	90			55	277
V	154	185	241	171	195	290	200	186	173	179	172					610	199
Ba	18	45	227	60	81	37	62	43	47	145	848	145	80				83
Co	62	54	55	51	46	51	52	51	42	59	66					75	53
Cr	17	48	270	578	295	769	1300	1900	2000	137	80	70	153			110	274
Molecular norms																	
Or	0.15	0.29	0.71	0.14	2.24	0.48	0.45	0.46	0.08	0.14	0.24	1.23	1.9	0.3			
Ab	9.07	10.69	12.97	15.14	13.38	18.40	12.89	8.87	6.65	7.17	9.01	27.46	3.2	—			
An	34.19	30.43	21.15	21.52	25.24	14.67	3.00	3.05	2.91	33.10	33.75	14.66	19.3	2.8			
Ne	4.71	5.83	13.12	0.89	1.18	5.70	—	—	—	—	—	—	10.9	—			
Di	2.76	2.60	8.23	14.24	8.61	24.02	48.46	51.43	53.76	1.86	2.50	14.77	38.5	—			
Hd	4.32	3.47	6.70	12.10	10.78	10.48	7.08	3.51	5.41	2.33	3.11	11.64	—	—			
En	—	—	—	—	—	—	4.53	14.61	14.45	6.32	0.13	2.24	—	—			
Fs	—	—	—	—	—	—	0.66	1.00	1.45	7.90	0.17	1.76	—	—			
Fo	14.94	16.55	14.33	16.43	14.22	15.02	16.40	11.06	10.67	15.69	18.91	7.58	16.0	42.7			
Fa	23.38	22.05	11.66	13.96	17.80	6.55	2.39	0.76	1.07	19.63	23.52	5.79	—	—			
Ilm	3.29	3.71	3.58	2.24	2.70	1.55	0.82	0.67	0.62	1.82	3.87	7.79	4.8	3.8			
Mgt	2.65	3.81	6.99	2.83	3.21	2.73	2.96	4.07	2.58	3.71	4.26	4.49	5.3	14.4			
Ap	0.13	0.11	0.07	0.04	0.06	0.04	0.02	0.01	0.03	0.04	0.11	0.30	0.5	1.2			

Notes: 1–3: Lens 1A; 40, 20, and 0 cm from contact. 4–6: 5B; 70, 30, and 20 cm from contact. 7–8: lens 5A; 75, and 0 cm from contact. 9–11: Lens 6, 175, 50, and 0 cm from contact. Samples 7–9 are clinopyroxenites.
BP, Baldersheim 'prasinite' (aver. of 9); BH, Baldersheim hornblendfels (aver. of 4); Br, Baldersheim, chlorite schists (aver. (anhydrous) of 2); BE, Bjørkedalen eclogite (aver.); VE, Voltri eclogites (aver. of analyses from Bocchio and Mottana 1975; Cortesogno et al. 1977; Mottana and Bocchio 1975); RE, Raudkleivane eclogites, aver. incl. anal. by Lappin 1974.

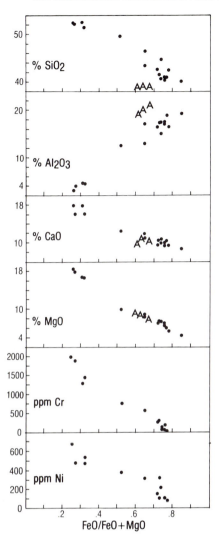

Fig. 3 MgO variation diagrams for Raudkleivane eclogites. Data from Table I and Lappin 1974. A, compositions of amphibole inclusions in garnet from Table II

duced a hornblende fels with higher Fe but lower Fe/Mg (Table 1), and the iron-rich compositions are regarded as primary. Enrichment of Al occurs in an outermost, chlorite + mgt-rich zone. Relicts of diopsidic clinopyroxene in this zone suggest that minor Ca-enrichment (rodingitization) may have occurred prior to development of the chlorite zone (*cf.* Bøe 1980). Amphibolites of superferrian composition also occur in serpentinites and chlorite–enstatite peridotites at Sjona in Nordland and Lesja near Dombås (Qvale unpublished data; Fig. 4).

Superferrian eclogites have been described from the serpentinites of the Ligurian ophiolites (Mottana and Bocchio 1975; Cortesogno *et al.* 1977; Morten *et al.* 1979; see Table 1). These authors have shown that the Ligurian eclogites are chemically similar to ferrogabbros dredged from the mid-Atlantic ridge. The superferrian eclogites and amphibolites known so far from Norwegian ultramafic rocks differ from these rocks mainly in lower average Ti and higher Al contents, but the two groups overlap even in these parameters, and are similar in, for example, AFM proportions (Fig. 4). FeTi basalts, and ferrogabbros ($FeO^T \sim 13$ per cent) with lower TiO_2 contents (1–2 per cent) are common at Pacific spreading centres (Clague and Bunch 1976), and they develop toward superferrian compositions by clinopyroxene fractionation, at shallow depths. The superferrian composition of the Raudkleivane eclogites may therefore be regarded as evidence of a low-P protolith.

high Ni contents suggest that olivine was originally an important phase in the cores, since distribution coefficients cpx/liq for Ni are generally small (Jensen 1973).

The large difference in Fe/Fe + Mg between the clinopyroxenites (clinopyroxene + olivine cumulates?) and the eclogites raises the question of whether the superferrian rocks represent magmatic compositions. Similar eclogites occur in a thick continuous layer, also with local Ca + Mg-enriched zones, in the ultramafic mass at Bjørkedalen, 25 km from Almklovdalen (Brastad 1977, 1983). Qvale (1978, in preparation) has described very similar Fe-rich compositions in the cores of amphibolitized basic dykes in Caledonian serpentinite at Baldersheim in Hordaland (Table 1). In this locality marginal alteration, related to serpentinization, has pro-

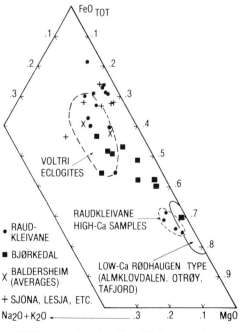

Fig. 4 AFM variation in Raudkleivane eclogites and comparable rocks from other bodies

Sr Isotope Data

The Sr isotope data are summarized in Table 2. The value for the NBS 987 standard for this laboratory is 0.71029 ± 4 (2SE), based on analyses done over the same time period.

The data show a good inverse correlation of $^{87}Sr/^{86}Sr$ with per cent CaO, corresponding to the mixing lines shown in Fig. 3. The spread of the data suggests again that the analysed rocks represent the mixing of the two end members: the clinopyroxenites with $^{87}Sr/^{86}Sr$ ratios of *ca.* 0.7036, and the eclogites with ratios of *ca.* 0.7065. These two groups of rocks are clearly not comagmatic; the data strengthen our interpretation that the pyroxenites are accidentally included in the eclogites. The eclogites are strikingly different in $^{87}Sr/^{86}Sr$ from the classic 'internal eclogites' of Almklovedalen; Brueckner (1977) has shown that these 'eclogites' and the garnet peridotites have $^{87}Sr/^{86}Sr$ values (at very low Rb/Sr ratios) of 0.702–0.703. On the other hand, the Sr isotope ratios of the Raudkleivane eclogites are similar to those of many 'external' eclogites (Griffin and Brueckner in preparation). The Raudkleivane pyroxenites may also be isotopically distinct from the classic 'internal eclogites', but more data on both groups are necessary to be certain.

Mineral Chemistry

Analyses of the phases in two clinopyroxenites are given by Lappin (1974). Analyses of minerals in the superferrian eclogites are given in Table 3 (*cf.* Lappin 1974).

The matrix clinopyroxenes are omphacites, with Jd + Acm = 38–40 per cent. Clinopyroxene inclusions in the garnet of sample 11 are lower in Jd + Acm, and higher in Ts than this. Inclusions of

clinopyroxenes in the rims of other garnets are indistinguishable from the matrix omphacite.

The zoning profile of a typical garnet from sample 5 (Fig. 5) shows a marked decrease in Fe/Mg from the core toward the rim, followed by a small increase at the outer margin. This trend is seen in garnets from many of the so-called 'external' or 'country-rock' eclogites, and is generally believed to reflect growth during prograde metamorphism, followed by retrograde equilibration during uplift (Bryhni and Griffin 1971; Råheim 1974; Krogh 1977, 1982). The garnets of samples 10 and 11 show the same zoning pattern, while sample 6 contains unzoned garnets with Fe/Mg roughly equivalent to the rims of the zoned garnets in the other samples. No zoning was observed in the omphacite.

The amphiboles trapped as inclusions in garnet are low-Ca, Cl-rich ferropargasites. They are clearly distinct in chemistry from secondary amphiboles replacing clinopyroxene, garnet, or even clinopyroxene inclusions within garnet (Table 3). The extremely low Si and high Al^{IV} in these amphiboles, and their common negative-crystal shapes, suggest that they have at least partially equilibrated with the surrounding garnet, both texturally and chemically. Amphiboles with these Si/Al ratios are rare, but might be expected to form in SiO_2-poor environments at high T (Ungaretti personal communication).

The K-feldspar inclusions in garnet are low in Na;

Table 2   Raudkleivane Rb/Sr data

Sample	ppm Rb	ppm Sr	$^{87}Rb/^{86}Sr$	$^{87}Sr/^{86}Sr$ ± 2SE
1	0.400	59.25	0.01953	0.70676 ± 12
3	2.842	211.3	0.03891	0.70709 ± 10
4	0.497	102.6	0.01401	0.70511 ± 12
7	0.981	209.5	0.01354	0.70498 ± 10
9	2.893	213.8	0.03913	0.70360 ± 10
11	1.271	460.7	0.00798	0.70613 ± 10

Concentration data by isotope dilution, precision better than ±0.5 per cent (2SE).

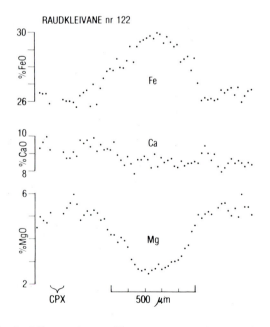

Fig. 5   Microprobe profile across a zoned garnet from a Raudkleivane eclogite. Note position of cpx inclusion in rim

Table 3   Microprobe analyses of minerals

Sample:	5 Gnt core (5)	5 Gnt rim (7)	5 Cpx inclus. (6)	5 Amph. inclus. (3)	5 Kspar. inclus. (2)	6 Gnt (5)	6 Cpx (5)	6 Amph. inclus. (2)	10 Gnt core (4)	10 Gnt rim (3)	10 Cpx (4)	10 Amph. inclus. (7)	10 Amph. second. (2)	11 Gnt core (3)	11 Gnt rim (2)	11 Cpx inclus. (2)	11 Amph. inclus. (4)	11 Amph. on cpx (2)	11 Amph. on gnt (2)
SiO_2	37.49	38.46	54.25	38.95	63.92	39.08	54.22	38.67	38.58	38.44	52.89	38.58	42.25	37.93	38.54	52.97	38.31	44.22	38.93
TiO_2	0.0	0.0	0.10	0.24	0.0	0.0	0.14	0.0	0.0	0.0	0.24	0.31	0.38	0.0	0.0	0.18	0.36	0.91	0.10
Al_2O_3	20.62	21.97	7.74	21.10	18.45	22.41	9.50	19.14	21.63	21.64	5.82	20.89	12.57	21.34	21.48	4.70	20.73	12.51	19.62
Fe_2O_3*	2.38	0.31	4.01	—	—	1.20	4.29	—	2.50	1.78	5.03	—	—	2.32	1.63	5.03	—	—	—
FeO	28.20	26.09	4.92	16.75	0.24	22.84	1.42	15.03	23.13	23.92	4.39	15.20	16.44	22.75	23.21	3.18	15.58	12.57	19.28
MnO	0.99	0.60	0.0	0.14	—	0.46	0.0	0.0	0.49	0.39	0.0	0.11	0.0	0.58	0.53	0.0	0.10	0.12	0.74
MgO	2.47	5.40	8.84	8.03	—	8.05	9.92	9.73	6.68	5.49	9.93	8.80	11.03	5.82	6.27	12.32	8.68	12.46	8.25
CaO	8.76	8.48	15.02	10.10	—	7.11	15.06	9.84	8.27	9.26	17.56	10.92	10.37	9.09	8.71	20.03	10.32	10.45	9.98
Na_2O	—	—	5.40	3.61	0.15	—	5.73	4.03	—	—	4.07	2.54	2.95	—	—	2.74	3.04	2.44	3.40
K_2O	—	—	—	0.0	16.55	—	—	0.0	—	—	—	0.0	0.11	—	—	—	0.0	0.0	0.14
Cl	—	—	—	0.65	—	—	—	1.24	—	—	—	0.82	0.21	—	—	—	0.98	0.0	0.0
	100.91	101.31	100.28	99.57	99.30	101.13	100.28	97.67	101.28	100.92	99.93	98.08	96.31	99.83	100.37	101.14	98.10	95.64	99.94
	8 cat.	8 cat.	4 cat.	23 0	8 0	8 cat.	4 cat.	23 0	8 cat.	8 cat.	4 cat.	23 0	23 0	8 cat.	8 cat.	4 cat.			
Si	2.985	2.969	1.967	5.743	2.980	2.972	1.937	5.825	2.970	2.978	1.943	5.732	6.413	2.970	2.987	1.922	5.727	6.576	5.762
Al^{IV}	0.015	0.031	0.033	2.257	0.020	0.028	0.063	2.175	0.030	0.022	0.057	2.268	1.587	0.030	0.013	0.078	2.273	1.424	2.238
Al^{VI}	1.920	1.958	0.298	1.410	0.994	1.981	0.337	1.222	1.933	1.953	0.195	1.398	0.662	1.939	1.949	0.123	1.379	0.768	1.184
Ti	—	—	0.003	0.027	—	—	0.004	—	—	—	0.007	0.035	0.043	—	—	0.005	0.040	0.102	0.011
Fe^{3+}	0.142	0.018	0.109	—	—	0.068	0.115	—	0.144	0.103	0.139	—	—	0.136	0.095	0.137	—	—	—
Fe^{2+}	1.867	1.661	0.149	2.066	0.008	1.449	0.042	1.893	1.481	1.543	0.135	1.893	2.087	1.481	1.499	0.096	1.948	1.563	2.386
Mn	0.067	0.039	—	0.017	—	0.030	—	—	0.032	0.026	—	0.014	—	0.038	0.035	—	0.013	0.015	0.030
Mg	0.293	0.621	0.478	1.765	—	0.913	0.528	2.184	0.767	0.634	0.544	1.953	2.496	0.679	0.724	0.666	1.934	2.762	1.820
Ca	0.747	0.702	0.583	1.596	—	0.579	0.576	1.588	0.682	0.769	0.691	1.742	1.687	0.763	0.723	0.779	1.653	1.665	1.583
Na	—	—	0.380	1.032	0.013	—	0.397	1.177	—	—	0.290	0.733	0.868	—	—	0.193	0.881	0.703	0.976
K	—	—	—	0.013	0.985	—	—	—	—	—	—	—	0.021	—	—	—	—	—	0.026

they apparently formed at low T, and have since been isolated within the garnet.

The prograde zoning of the garnets, their abundant inclusions of amphibole and other phases, and the low-Jd, high-Ts clinopyroxene inclusions in the garnet of sample 11, all suggest that the superferrian eclogites have formed by progressive metamorphism of amphibolite. The bulk composition of the amphiboles included in the garnets is similar to that of the bulk rocks (Fig. 3), only the addition of a few per cent of plagioclase and ilmenite–magnetite is needed to make the correspondence complete.

Pressure and Temperature Estimates

Temperatures of metamorphic equilibration can be calculated from the Fe/Mg distribution in gnt/cpx pairs (Ellis and Green 1979) provided that the Fe^{3+}/Fe^{2+} ratio in these phases is known. We have assumed that all $Fe = Fe^{2+}$ in garnets and amphiboles, and calculated Fe^{3+} by charge balance in the clinopyroxene. The Fe^{3+}/Fe^{2+} of pyroxenes 6094 (Lappin 1974), 5, 10 and 11 is reasonable (< 2), and comparable with that found in other omphacites of the Gneiss Region. Those from 6 and 6096 (Lappin 1974) have very high calculated Fe^{3+}/Fe^{2+}, perhaps as a result of small analytical errors (especially in SiO_2).

$K_{D\,gnt/cpx}^{Fe/mg}$ values have been calculated using the lowest observed values of Fe/Mg_{gnt}. This gives the maximum temperature, but if any retrograde zoning is present, as in samples 5 and 11, this will be lower than the metamorphic peak. The temperatures obtained, normalized to $P = 15$ Kb, are from 690–760°C. Calculation of Fe^{3+} in garnets raises these values by up to 30°C.

The clinopyroxene inclusion in sample 11, taken together with the adjacent garnet, yields $K_D = 16.6$, much higher than the matrix cpx + gnt rim pairs, even though the Fe^{3+}/Fe^{2+} of the clinopyroxene is reasonable. This K_D gives T = 570° at P = 15 Kb, and further suggests that the garnets grew under a regime of increasing T.

Pressures can be calculated from the assemblage opx + cpx + gnt in the clinopyroxenites, using the barometer of Wood (1974), based on Al_{opx}. Carswell et al. (this volume) used data from Lappin (1974) to calculate conditions of 730°, 22 Kb (using the two-pyroxene thermometer of Wood) or 650°, 17 Kb (using the thermometer of Ellis and Green). Similar calculations on country-rock eclogites from the surrounding area (Medaris 1980; Griffin et al. this volume) yield conditions of 700–750°C, 15–18 Kb. Calculated P–I values for the garnet lherzolite

assemblages, using the same methods, average 725°, 20 Kb (Carswell et al. this volume). These values suggest that the superferrian 'internal' eclogite and the enclosing peridotite were metamorphosed under tha same conditions as the eclogites of the surrounding gneiss terrain.

Discussion

The data presented here suggest that the Raudkleivane superferrian eclogites were originally basic bodies (dykes?) in a low-P ultramafic body, similar to those now found as serpentinites at high levels of the Caledonides and in ophiolite fragments of the Ligurian Alps. These basic rocks were apparently converted to amphibolites at some stage of their evolution. In the high-level Caledonian peridotites such metamorphism is related to serpentinization of the ultramafic rocks (Qvale 1978, in preparation).

Regardless of the origin of the basic bodies or their magmatic affinities, the petrographic data presented here suggest that they were converted to amphibolites before being metamorphosed to eclogite. Thus neither the dykes, nor the enclosing peridotite, can be unaltered fragments of the upper mantle, as has been so often claimed.

Bryhni et al. (1977) and Griffin et al. (this volume) have summarized the evidence for a high-pressure metamorphic event in the Western Gneiss Region. This event has produced eclogites from a variety of low-P protoliths, and apparently was shared by gneisses, anorthosites and late acid/intermediate intrusive rocks. The garnet peridotite localities all lie in the high-T part of this metamorphic gradient, as defined by the high-P assemblages. If they were present during this metamorphism, they would have been subjected to conditions of T = 650–800°C, P = 13–18 Kb.

Jenkins and Newton (1979) have shown that these conditions are sufficient to convert spinel lherzolite to garnet lherzolite. The evidence presented here suggests that in at least one locality, rocks enclosed by peridotite have been prograde metamorphosed to these conditions. If this is correct, there is no need to call on an upper mantle origin for the garnet peridotite assemblages, and there is direct evidence against such a model in the one case illustrated here.

Medaris (1980) in a detailed study of the garnet peridotites at Lien, within the same body studied here, found only minor retrograde zoning (rimward decrease in Mg/Fe) in the Mg-rich garnets. This accords with our experience elsewhere in the Gneiss Region, and suggests that Mg-rich garnets homogenize and react more easily than Fe-rich ones (Krogh

1982). Medaris also observed zoning of Al_2O_3 in orthopyroxene, with the highest values near garnet grains. He combines these two zoning trends to calculate a P–T evolution from *ca.* 28 Kb, 750°C to 17 Kb, 700°C.

Carswell *et al.* (this volume) have presented a model of opx/gnt equilibration that relates the Al zoning seen in the orthopyroxenes of garnet lherzolites and garnet websterites to disequilibrium during prograde metamorphism. This model implies that only the P, T values given by garnet and orthopyroxene in mutal contact can approach reality, and that the higher P values given by the low Al_2O_3 contents in orthopyroxene cores are spurious (*cf.* Carswell and Griffin 1981). This model thus removes the apparent contradiction between Medaris' 'retrograde' history and our 'prograde' history.

The presence of Cl-rich amphiboles is especially interesting with regard to the possibility of a low-P history for the ultramafic body. Rucklidge and Patterson (1977) and Miura *et al.* (1981) have drawn attention to the presence of Cl in ultramafic rocks undergoing serpentinization, and proposed that the conversion of olivine to serpentine involves an intermediary $Fe_2(OH)_3Cl$ phase. This phase continuously precipitates and redissolves; all Cl will eventually disappear when serpentinization is complete since there will be no stable phase that the Cl can enter. We suggest that under some conditions a basic (amphibolite) layer might serve as a sink for the Cl being transported in the fluid phase. For example, Honnorez and Kirst (1975) report Cl-bearing amphiboles from partially rodingitized olivine gabbros dredged from the Mid-Atlantic fracture zones. Thus the Cl content of the amphiboles included in the garnets described here may record the earlier serpentinization of the enclosing ultramafic rocks.

Lappin (1974) and Lappin and Smith (1978) have suggested that the eclogites described here originated at T > 1500°C, P ~ 25 Kb, and passed through *ca.* 1100°C, 20 Kb on their way to a crustal equilibration at *ca.* 700°C, 10 Kb. The early stages of this trajectory are based largely on the assumption that the complex high-Al pyroxene relics in the pyroxenites coexisted with orthopyroxene and garnet at some stage. There is no textural evidence to support this assumption. Rost and Brenneis (1978) have shown that such complex pyroxenes are associated with spinel peridotites, and thus are relatively low-P phases. Experimental studies (Green and Ringwood 1968; Thompson 1974) show that moderately Al-, Na-rich clinopyroxenes comparable to samples 7–10 (Table I) can precipitate as liquidus or subliquidus phases from tholeiitic basalts at moderate pressures (< 10 Kb). We suggest therefore, that pyroxenites seen at Raudkleivane may represent cumulates related to the differentiation of basaltic magma.

The pyroxenites (± spinel) may have been emplaced into the peridotite as inclusions in the superferrian basaltic magmas. Alternatively they may have formed much earlier as lenses or vein fillings, so that their relation to the superferrian magma is entirely accidental. Their present cpx + opx + gnt mineralogy might have formed by isobaric cooling at P ≳ 11 Kb (Irving 1974; Herzberg 1978); or, like the Ultenthal peridotites, the Almklovdalen bodies may have been tectonically emplaced with low-P, high-T mineralogy and then subjected to high-P metamorphism together with the surrounding terrain. The prograde metamorphic history suggested here would provide conditions suitable for the exsolution and recrystallization of complex pyroxene (± spinel) cumulates to eclogite-facies minerals.

In the model outlined here there is no need to postulate any high-P stage other than that reacted in the Caledonian prograde metamorphism. The model assumes tectonic emplacement of low-P spinel peridotite protoliths, as seen in many mountain belts. It thus removes the necessity to contrive a mechanism for the vertical transport of large bodies of garnet peridotite many kilometres through the mantle and their introduction into a much less dense continental crust.

Acknowledgements

Field and laboratory work were supported by Norges Almenvitenskapelige Forskningsråd (D 48.22–008). We are grateful to M. B. E. Mørk and A. Krill for their contribution to the mapping of the locality, and to D. A. Carswell, A. Mottana, and L. Morten for useful discussions.

References

Bocchio, R. and Mottana, A. 1975. Trace-element abundances of iron-rich eclogites, with implications on the geodynamical evolution of the Voltri Group (Pennidic Belt). *Chem. Geol.*, **15**, 273–283.

Brastad, K. 1977. *En Geologisk Undersøkelse av Anorthositter og Andre Bergarter på Grensen Mellom Nordfjord og Sunnmøre.* Unpubl. Cand. Real. Thesis, Univ. of Oslo.

Brastad, K. this volume. Relations between anorthosites, eclogites and ultramafics in Bjørkedalen, Western Norway.

Brueckner, H. K. 1977. A crustal origin for eclogites and a mantle origin for garnet peridotites: Sr isotopic evidence from clinopyroxenes *Centr. Miner. Petrol.*, **60**, 1–15.

Bryhni, I. and Griffin, W. L. 1971. Zoning in eclogite garnets from Nordfjord, West Norway *Contr. Mineral. Petrol.*, **32**, 112–125.

Bryhni, I., Krogh, E. J. and Griffin, W. L. 1977. Crustal derivation of Norwegian eclogites: a review. *N. Jb. Miner. Abh.*, **130**, 49–68.

Budanova, K. T. and Budanov, V. I. 1975. The southwest Pamirs, a new province of garnetiferous ultramafic rocks. *Doklady Akad. Nauk SSSR*, **222**, 190–193.

Bøe, R. 1980. Rodingitt fra Lindås, Hordaland (abstract). *Geolognytt*, **15**, 13.

Carswell, D. A. and Griffin, W. L. 1981. Calculation of equilibration conditions for garnet granulite and garnet websterite nodules in African kimberlite pipes. *Tscher. Min. Petr. Mitt.*, **28**, 229–244.

Carswell, D. A., Krogh, E. J. and Griffin, W. L. this volume. Norwegian orthopyroxene eclogites: calculated equilibration conditions and petrogenetic implications.

Clague, D. A. and Bunch, T. E. 1976. Formation of ferrobasalt at East Pacific midocean spreading centers. *J. Geophys. Res.*, **81**, 4247–4258.

Cortesogno, L., Ernst, W. G., Galli, M., Messiga, B., Pedemonte, G. M. and Piccardo, G. B. 1977. Chemical petrology of eclogite lenses in serpentinite, Gruppo di Voltri, Ligurian Alps. *J. Geol.*, **85**, 255–277.

Ellis, D. J. and Green, D. H. 1979. An experimental study of the effect of Ca upon garnet–clinopyroxene Fe–Mg exchange equilibria. *Contr. Miner. Petrol.*, **71**, 13–22.

Eskola, P. 1921. On the eclogites of Norway. *Skr. Vidensk. Selsk. Christiania, Mat. Naturv. Kl. I*, **8**, 1–118.

Evans, B. W. and Trommsdorff, V. 1978. Petrogenesis of garnet lherzolite, Cina di Gagnone, Lepontine Alps. *Earth Planet. Sci. Lett.*, **40**, 333–348.

Evans, B. W., Trommsdorff, V. and Goles, G. G. 1981. Geochemistry of high-grade eclogite and metarodingites from the Central Alps. *Contr. Miner. Petrol.*, **76**, 301–311.

Green, T. H. and Ringwood, A. E. 1968. Genesis of the calc-alkaline igneous rock suite. *Contr. Miner. Petrol.*, **18**, 105–162.

Griffin, W. L., Austrheim, H., Brastad, K., Bryhni, I., Krill, A., Mørk, M. B. E., Qvale, H. and Tørudbakklen, B. this volume. High-pressure metamorphism in the Scandinavian Caledonides.

Herzberg. C. R. 1978. The bearing of phase equilibria in simple and complex systems on the origin and evolution of some well-documented garnet–websterites. *Contr. Miner. Petrol.*, **66**, 375–382.

Honnorez, J. and Kirst, P. 1975. Petrology of rodingites from the equatorial mid-Atlantic fracture zones and their geotectonic sequence. *Contr. Miner. Petrol.*, **49**, 233–257.

Irving, A. J. 1974. Geochemical and high-pressure experimental studies of garnet pyroxenites and pyroxene granulite xenoliths from the Delegate basic pipes, Australia. *J. Petrol.*, **15**, 1–40.

Jenkins, D. M. and Newton, R. C. 1979. Experimental determination of the spinel peridotite to garnet peridotite inversion at 900°C and 1000°C in the system CaO–MgO–Al$_2$O$_3$–SiO$_2$, and at 900°C with natural olivine and garnet. *Contr. Miner. Petrol.*, **68**, 407–419.

Jensen, B. B. 1973. Patterns of trace element partitioning. *Geochim. Cosmochim. Acta*, **37**, 2227–2242.

Krogh, E. J. 1977. Evidence for a Precambrian continent–continent collision in western Norway. *Nature, London*, **267**, 17–19.

Krogh, E. J. 1980a. Geochemistry and petrology of glaucophane-bearing eclogites and associated rocks from Sunnfjord, western Norway. *Lithos*, **13**, 355–380.

Krogh, F. J. 1980b. Compatible P–T conditions for eclogites and surrounding gneisses in the Kristiansund area, western Norway. *Contr. Miner. Petrol.*, **75**, 387–393.

Krogh, E. J. 1982. Metamorphic evolution deduced from mineral inclusions and compositional zoning in garnets from Norwegian country-rock eclogites. *Lithos*, **15**, 305–321.

Lappin, M. A. 1966. The field relationships of basic and ultrabasic masses in the basal gneiss complex of Stadlandet and Almklovdalen, Nordfjord, southwestern Norway. *Norsk geol. Tidsskr.*, **46**, 439–495.

Lappin, M. A. 1973. An unusual clinopyroxene with complex lamellar intergrowths from an eclogite in the Sunndal–Grubse ultramafic mass, Almklovdalen, Nordfjord, Norway. *Min. Mag.*, **39**, 313–320.

Lappin, M. A. 1974. Eclogites from the Sunndal–Grubse ultramafic mass, Almklovdalen, Norway and the T–P history of the Almklovdalen mass, *J. Petrol.*, **15**, 567–601.

Lappin, M. A. and Smith, D. C. 1978. Mantle-equilibrated orthopyroxene eclogite pods from the basal gneisses in the Selje district, Western Norway. *J. Petrol.*, **19**, 530–584.

Medaris, L. G. Jr. 1980. Petrogenesis of the Lien peridote and associated eclogites, Almklovdalen, western Norway. *Lithos*, **13**, 339–353.

Miura, Y., Rucklidge, J. and Nord, G. L. 1981. The occurrence of chlorine in serpentine minerals. *Contr. Miner. Petrol.*, **76**, 17–23.

Morten, L., Brunfelt, A. O. and Mottana, A. 1979. Rare earth abundances in superferrian eclogites from the Voltri Group (Pennidic Belt, Italy). *Lithos*, **12**, 25–32.

Mottana, A. 1977. Petrochemical characteristics of the western Alps eclogites. *N. Jb. Miner. Abh.*, **130**, 78–88.

Mottana, A. and Bocchio, R. 1975. Superferric eclogites of the Voltri group (Pennidic belt, Apennines). *Contr. Miner. Petrol.*, **49**, 201–210.

O'Hara, M. J. and Mercy, E. L. P. 1963. Petrology and petrogenesis of some garnetiferous peridotites. *Trans. R. Soc. Edinburgh*, **65**, 251–314.

O'Hara, M. J., Richardson, S. W. and Wilson, G. 1971. Garnet peridotite stability and occurrence in crust and mantle. *Contr. Miner. Petrol.*, **32**, 48–68.

Qvale, H. 1978. *Geologisk Undersøkelse av et Kaledonsk Serpentinittfelt ved Baldersheim, Hordaland*. Unpubl. Cand. Real. Thesis, Univ. of Oslo, 252 pp.

Rost, F. and Brenneis, P. 1978. Die Ultramafitite in Bergzug südlich des Ultentales, Provinz Alto Adige (Oberitalien). *Tscherm. Min. Petr. Mitt.*, **25**, 257–286.

Rucklidge, J. C. and Patterson, G. C. 1977. The role of chlorine in serpentinization. *Contr. Miner. Petrol.*, **65**, 39–44.

Råheim, A. 1974. *Pressure, Temperature, and Time Relationships in Eclogite Metamorphic Terranes*. Unpubl. Ph.D. Thesis, Aust. Nat. Univ, Canberra.

Thompson, R. N. 1974. Some high-pressure pyroxenes. *Min. Mag.*, **39**, 768–787.

Wood, B. J. 1974. The solubility of alumina in orthopyroxene coexisting with garnet. *Contr. Miner. Petrol.*, **46**, 1–15.

The Caledonide Orogen—Scandinavia and Related Areas
Edited by D. G. Gee and B. A. Sturt
© 1985 John Wiley & Sons Ltd

In situ metamorphism of Norwegian eclogites: an example

W. L. Griffin* and **D. A. Carswell†**

*Mineralogisk–Geologisk Museum, Sarsgate 1, Oslo 5. Norway
†Dept. of Geology, Sheffield Univ., Sheffield S1 3JD, England

ABSTRACT

Field relations and petrographic observations on the west coast of Midøy, W. of Molde, show the following sequence of events: 1. Intrusion of the igneous protoliths of the augen orthogneiss complex; 2. Intrusion of dolerite dykes; melting of wall rocks, back-veining of dolerites by anatectic melt (intrusive relations of gneiss into eclogite are preserved locally); 3. Metamorphism to *ca.* T = 750°C, P = 17–21 Kb; dolerites boudinaged and converted to eclogites, felsic veins to Ca-rich garnet + kyanite + plagioclase + quartz ± K-feldspar ± omphacite; 4. Extreme flattening, ENE–WSW shearing and development of blastomylonite in gneisses; further boudinage of eclogites, formation of granitic pegmatites, amphibolite-facies mineral assemblages in gneiss and along eclogite margins.

This sequence is now recognized over much of the Western Gneiss Region. Radiometric data suggest that (1) occurred *ca.* 1500 Ma ago, while (3) and (4) are Caledonian (450–380 Ma). The high-pressure metamorphism of the eclogites must have occurred *in situ*; the surrounding segment of continental crust was depressed to 60–70 km depths. This 'subduction' was probably related to continent–continent collision during the Caledonian orogeny.

Introduction

The origin of the eclogites that occur as lenses in the Western Gneiss Region of Norway is still a matter of debate, sixty years after the classic descriptions of Eskola (1921). Many investigators see compelling evidence for *in situ* metamorphism of the eclogite lenses enclosed in gneiss ('country-rock' or 'external' eclogites; Bryhni *et al.* 1977; Krogh 1977; Griffin *et al.* 1984). Others, noting the presence of apparently mantle-derived garnet peridotite bodies in the gneisses, also interpret the 'country-rock' eclogites as mantle fragments (Lappin 1966; Lappin and Smith 1978; O'Hara *et al.* 1971). If eclogites can be shown to have formed by *in situ* metamorphism at high P and T, this would imply an overthickening of the continental crust, with important tectonic ramifications.

The main argument generally used against an *in situ* formation of the eclogites is that the surrounding gneisses are usually in amphibolite facies. The eclogites are thus 'out of equilibrium' with their country

rocks, as evidenced by the zone of marginal amphibolitization that is typically present. Models have been presented to show that this difference in facies could be due to local variations in f_{H_2O} (Bryhni *et al.* 1977; Green and Mysen 1972) and these models have experimental support (Fry and Fyfe 1969). However, our experience suggests that *in situ* metamorphism of the eclogites and their country rock has been followed by extensive deformation and retrograde metamorphism, involving relative movement (especially along eclogite/gneiss contacts) and introduction of water. The observed eclogites therefore, in this interpretation, represent relics of an earlier high-pressure regional metamorphism. Many eclogites have been completely retrograded to garnet amphibolites, and are recognizable only by the occasional relics of omphacite found as inclusions in garnet grains. Similar relict assemblages are also found in the gneisses. Omphacite relics occur in gneiss near the Ulsteinvik eclogite, and omphacite – pyroxene–garnet–K-feldspar–quartz-kyanite assemblages occur in mangeritic rocks at several localities

(Mysen 1972; Bryhni 1966; Griffin and Mørk 1981). Krogh (1980a, b) has demonstrated in two localities that the metamorphic mineral assemblages of the gneisses are compatible with those of the associated eclogites.

Field work on Otrøy and Midøy, southwest of Molde has recently revealed rare examples of preserved 'primary' eclogite/gneiss contacts, showing relations that not only imply the *in situ* metamorphism of some country-rock eclogites, but which give detailed evidence of the relative age relations in the gneiss complex there.

Field Relations

The islands of Midøy and Otrøy (Fig. 1) expose several zones of eclogite boudins within a massive intermediate gneiss, mapped by Carswell (see Carswell and Harvey 1984) as part of an 'augen orthogneiss complex'. This complex, which includes rocks with original rapakivi-type feldspar megacrysts and relict high-P granulite mineralogy, intrudes an older supracrustal sequence. Eclogites occur in elongate concentrations, resembling basic dyke swarms, within the complex (Carswell and Harvey 1984). We will discuss one locality in detail: Litledigerneset, map sheet 1220 III, series M711, coordinates 521/769. Relations similar to those described here have also been observed at several other localities on Otrøy, Midøy, and other islands to the west (Carswell unpublished; Mørk and Griffin unpublished).

The gneiss at Litledigerneset contains abundant feldspathic lenses 1–5 cm thick and 10–100 cm long,

which are isoclinally folded about a marked foliation (S_1). The origin of these lenses is ambiguous. They may be anatectic segregations, formed during high-grade metamorphism of the gneiss, or they may be of essentially tectonic origin. In other localities on Otrøy and Midøy, deformation of coarse-grained rapakivi-type structures (S_1) has flattened feldspar megacrysts into elongate ovoid lenses. More extreme deformation locally results in coalescence and segregation of this feldspathic material into the cores of tight small-scale folds. The less strongly deformed augen gneiss on Litledigerneset point, between two main eclogite zones, does not have this 'migmatitic' appearance.

The eclogites at this locality, as elsewhere on Otrøy and Midøy, occur in elongated zones of lensoid boudins, typically from 1–20 m long and $\frac{1}{2}$–10 m across, enclosed in strongly flattened gneiss. The boudins are often bordered by zones of intense amphibolitization 1–10 cm thick. Both the gneisses and the eclogites are cross-cut by coarse-grained granite pegmatites up to several decimetres thick. In the gneisses, these pegmatites are usually folded, stretched, and finally disrupted into trains of feldspar and quartz augen, easily distinguishable from the feldspathic lenses mentioned above. Where the pegmatites cut the eclogites, they are bordered by decimetre-wide amphibolite zones. Some of the pegmatites within the eclogites show very little deformation, but others have developed a strong foliation parallel to their length.

The Midøy locality is interesting because several of the eclogite boudins *do* preserve the original contact relations. In these cases the blastomylonitic foliation (S_2) swings around the eclogite bodies in such a

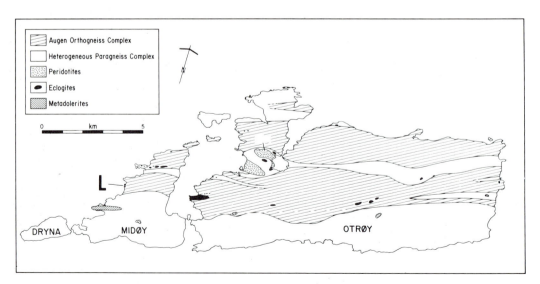

Fig. 1   Geological map of Otrøy and Midøy

way as to leave some of the surrounding gneiss relatively little affected by S_2, in a 'pressure shadow' area adjacent to the eclogite. These eclogites also have the form of irregular pods, but they are bordered by a 20–60 cm zone of massive, fine-grained quartz-dioritic gneiss lacking the feldspathic lenses common in the rest of the gneiss. This zone grades outward into the 'migmatitic' gneiss through the appearance of scattered coarser feldspathic lenses in the fine-grained matrix (Fig. 2).

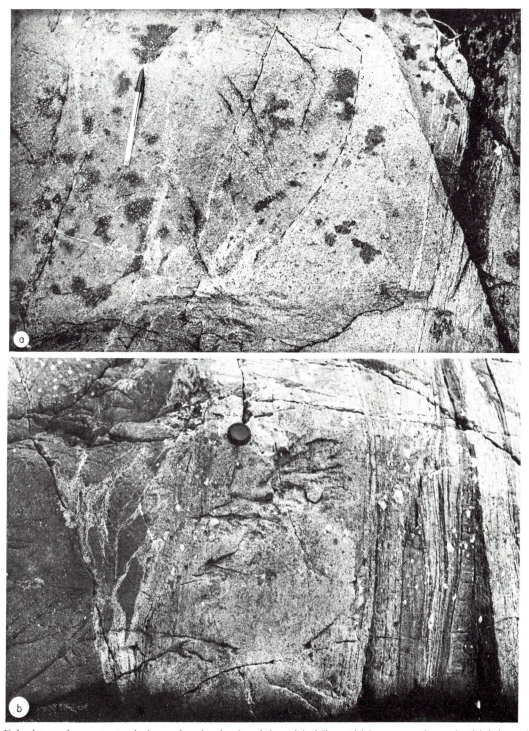

Fig. 2   Eclogite-gneiss contact relations, showing back-veining. (a) 'Migmatitic' augen orthogneiss (right) grades into fine-grained quartz dioritic gneiss toward contact with eclogite (left) and material from the fine-grained zone intrudes eclogite as irregular veins. Contact outlined for visibility. (b) Similar relations to (a), but showing blastomylonite development in migmatitic gneiss outside the fine-grained quartz-dioritic zone

Fig. 3   Irregular, zoned felsic veins in eclogite. Note absence of any amphibolitization along veins

The contact between the fine-grained massive zone and the eclogite is generally sharp, but in several cases the former appears to intrude the eclogite in irregular veins 1–50 cm thick (Fig. 2). These felsic veins, unlike the granite pegmatites, are *not* bordered by amphibolite (Fig. 3), but contain pyroxene, kyanite and garnet (see below). The fine-grained quartz-dioritic zone, the unretrograded eclogite, and the felsic veins in the eclogite are cut by a weak foliation. This foliation appears to be the same (S_1) as that which is parallel to the axial planes of isoclinal folds in the migmatitic gneiss.

The field relations strongly suggest that the fine-grained quartz-dioritic zone and the veins in the eclogites were formed by local anatexis of the gneiss adjacent to the eclogites. We interpret this as an example of the 'Sederholm effect'–melting of the country rock adjacent to the basic intrusion, and back-veining of the chilled basic magma by the anatectic melt. The relatively wide zone of melting suggests that the country rock was at rather high T at the time of intrusion. Comparable relations, in similar rocks, have been described from the Raftsund mangerite in Lofoten (Griffin *et al.* 1974).

Petrography and Mineral Chemistry

The eclogites at Litledigerneset, where least amphibolitized, show a retrograded-eclogite mineral assemblage, consisting of garnet and symplectitic masses of clinopyroxene + plagioclase (An_{23}), with minor secondary hornblende. Quartz and rutile are locally present. The pyroxene–plagioclase symplectites are common in Norwegian eclogites and are widely recognized as the result of exsolution of the jadeite (Jd) component from high-pressure omphacite pyroxene, probably at lower pressure. The chemistry of this breakdown has been described in detail elsewhere (Mysen and Griffin 1973; Mysen 1972). The larger grains in the pyroxene–plagioclase symplectites contain up to 20 per cent Jd (Table 1); the original pyroxene apparently contained *ca.* 50 per cent Jd. The eclogite garnets are zoned to higher Fe/Mg at the rims, suggesting partial reequilibration at lower T (Table 1). Adjacent to the felsic veins, the eclogite may be enriched in garnet and/or biotite; the latter appears to have coexisted with both clinopyroxene and garnet.

The thin (*ca.* 1.5 cm) anatectic veins, where not subjected to later amphibolitization, consist of quartz, K-feldspar ($Or_{85}Ab_{15}$), plagioclase (An_{20-25}), biotite, kyanite, garnet, and clinopyroxene. The clinopyroxene occurs in pyroxene + plagioclase symplectites texturally similar to those in the adjacent eclogites. This symplectite occurs in patches that appear to be pseudomorphs after omphacite grains (Fig. 4). Microprobe analyses show that the cpx is a sodic augite with *ca.* 10% Jd (Table 1). Point-counting of photomicrographs yielded Jd con-

Table 1 Mineral analyses: eclogite and veins

	Eclogite				Vein G1			Vein U502		Vein 521G		Vein 521C		Vein U518×6	
	cpx cores	cpx rims	gnt cores	gnt rims	cpx	gnt cores (4)	gnt rims (4)	biotite (3)	Kspar (3)	gnt cores (4)	gnt rims (4)	gnt cores (2)	gnt rims (4)	gnt cores (3)	gnt rims (3)
SiO_2	51.0	50.9	40.0	39.2	51.6	39.6	39.6	37.66	64.03	39.05	39.24	38.77	38.30	39.12	39.21
TiO_2	0.38	0.46	—	—	0.19	—	—	2.79	0.30	0.0	0.0	0.0	0.0	0.0	0.0
Al_2O_3	10.7	5.1	23.0	22.8	4.6	23.0	23.0	20.24	18.61	21.89	22.07	21.94	21.37	21.64	22.05
$Fe_2O_3^*$	1.9	1.3	—	—	0.0	—	—	—	—	—	—	—	—	—	—
FeO	5.9	7.4	19.6	21.2	9.3	19.2	18.8	8.65	0.0	19.53	21.90	19.13	23.80	19.64	21.37
MnO	0.10	0.15	0.30	0.30	0.25	0.30	0.35	0.0	0.0	0.10	0.21	0.10	0.24	0.20	0.20
MgO	9.3	11.0	7.6	6.6	11.8	7.7	7.6	16.74	0.10	4.98	6.97	5.14	5.94	5.30	6.46
CaO	17.3	20.1	10.9	10.7	20.3	11.4	11.9	0.0	0.15	14.82	10.16	14.64	9.75	14.30	11.15
Na_2O	3.6	1.8	—	—	1.2	—	—	0.36	1.52	—	—	—	—	—	—
K_2O	—	—	—	—	—	—	—	9.91	14.36	—	—	—	—	—	—
	100.18	98.21	101.4	100.8	99.24	101.2	101.25	96.35	98.97	100.37	100.55	99.72	99.40	100.20	100.44
Si	1.861	1.922	2.989	2.971	1.937	2.971	2.967	5.375	2.972	2.994	2.994	2.987	2.989	3.003	2.997
Al^{IV}	0.139	0.078	0.011	0.029	0.063	0.029	0.033	0.625	0.028	0.006	0.006	0.013	0.011	—	0.003
Al^{VI}	0.319	0.148	2.020	2.013	0.139	2.003	2.013	2.779	0.990	1.972	1.979	1.979	1.954	1.958	1.983
Ti	0.010	0.013	—	—	0.005	—	—	0.299	0.010	—	—	—	—	—	—
Fe_3	0.051	0.035	—	—	—	—	—	—	—	—	—	—	—	—	—
Fe_2	0.180	0.233	1.225	1.348	0.291	1.201	1.181	1.032	—	1.252	1.398	1.233	1.553	1.261	1.366
Mn	0.003	0.005	0.025	0.025	0.010	0.020	0.021	—	—	0.006	0.014	0.007	0.016	0.013	0.013
Mg	0.508	0.622	0.846	0.745	0.659	0.862	0.848	3.561	0.007	0.569	0.793	0.590	0.691	0.606	0.736
Ca	0.676	0.813	0.874	0.872	0.813	0.917	0.952	—	0.007	1.217	0.831	1.208	0.815	1.176	0.913
Na	0.252	0.131	—	—	0.081	—	—	0.100	0.137	—	—	—	—	—	—
K	—	—	—	—	—	—	—	1.804	0.850	—	—	—	—	—	—
Plag:	An_{23} (symplectite)	An_{25} (symplectite)			An_{25} (symplectite, grains)			An_{15}		An_{30}(28–35)		An_{35}(33–43)		An_{35}(33–40)	

Analyses by energy-dispersive microprobe techniques (LINK system, ZAF–4 reduction program, ARL–EMX probe).
*Fe_2O_3 calculated by charge balance for cpx (4 cations); all Fe as FeO in garnets (12 oxygen) and biotites (22 oxygen).

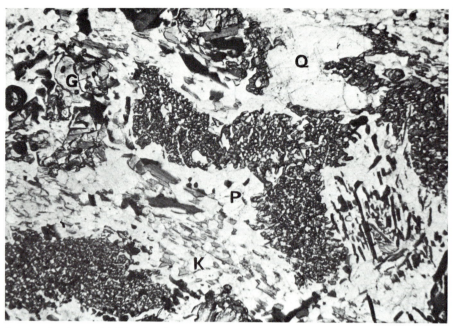

Fig. 4   Photomicrograph of sample G1, showing cpx + plag (P) symplectites, biotite + K-feldspar (K) + kyanite
intergrowths and minor garnet (G) and quartz (Q)

tents of 30–40 per cent for the original omphacite. The presence of large homogeneous plagioclase grains, not in contact with symplectite, suggests that this pyroxene may have coexisted with plagioclase, but this is difficult to establish with certainty.

Hornblende + plagioclase + quartz symplectites are common in samples that lack pyroxene, and appear to have formed by amphibolitization of pre-existing cpx + plag symplectites, or of primary omphacite. Biotite occurs mainly in skeletal or radial aggregates, intergrown with K-feldspar + quartz + kyanite. These are interpreted as pseudomorphs after phengite, according to the reaction:

$$6K_2Al_{4-x}(Mg,Fe)_xSi_{6+x}Al_{2-x}O_{20}(OH)$$
$$\text{Phengite}$$

$$\rightarrow XK_2(Mg,Fe)_6Si_6Al_2O_{10}(OH)_4$$
$$\text{Biotite}$$

$$+ 12-2x\ Al_2SiO_5 + 12-2x\ KAlSi_3O_8$$
$$\text{Kyanite} \qquad\qquad \text{K-feldspar}$$

$$+ (12x-12)SiO_2 + 12-2x\ H_2O$$
$$\text{Quartz}$$

Thicker veins commonly show a marked zoning, with dark outer zones and a felsic central zone. The adjacent eclogite is typically enriched in garnet and/or biotite. The dark outer zones are similar in mineralogy and texture to the thinner veins, but with higher biotite contents. Again, much of the biotite, and all of the K-feldspar appear to be breakdown products of phengite flakes. The felsic cores contain a coarse-grained (1–3 mm) equilibrium assemblage

of gnt + kyanite + plag + qtz. Kyanite blades are commonly, but not always, separated from quartz by a collar of plagioclase grains. Biotite + kspar and cpx/hbl + plag intergrowths occur in some samples. The garnets are rich in Ca; they are typically zoned from *ca*. $Gros_{40-45}$ in the cores to $Gros_{30-33}$ or less at the rims. Fe/Mg typically decreases rimward as well (Table 1). Plagioclase varies from *ca*. An_{28} to An_{40}, with more calcic varieties (aver. An_{35}) adjacent to the garnets.

The fine-grained quartz-dioritic zones consist dominantly of antiperthitic plagioclase (An_{18-25}) and minor quartz. The foliation is defined by dark green–brown hornblende and minor very dark brown biotite in ragged flakes. Small grains of garnet are common, and minor cpx was seen in one section. The fine-grained zone is thus depleted in K_2O compared to the surrounding gneisses (Table 2), which contain abundant microcline and biotite, but rarely garnet or hornblende. This depletion is ascribed to removal of the anatectic melt now represented by the veins in the eclogites.

Rock Chemistry

One retrograded eclogite from Litledigerneset (U517) has a Ne normative composition, whereas the others are tholeiitic (Table 2). Similar, better-preserved eclogites elsewhere on Midøy and Otrøy have the composition of olivine tholeiites (Carswell and Harvey 1984).

Analyses of two thin veins (one from Otrøy), and three samples taken across a thicker one, are presented in Table 3. Microprobe analyses have been made on glasses prepared by fusing rock powders on a Mo strip in an Ar atmosphere. Tests on standard rocks have shown that the method produces no loss of alkalies, and that the results are comparable in precision and accuracy with XRF data, for concentrations > 0.5 per cent.

One of the thin veins (U502), and the outer part of a thicker one, are granitic in composition. The other thin vein (G1) is more mafic; this vein has a diffuse margin, and appears to have been contaminated by basic material, now represented by the eclogite. The other analyses show that the central parts of the zoned veins are higher in Al and Ca, and much lower in K, than the marginal zones.

These relations suggest that originally granitic to quartz-monzonitic veins have interacted with the surrounding basic material. Mg has migrated into the veins, while K and Na have moved toward the vein margins. These variations are now seen as enrichment of clinopyroxene, phengite, and/or biotite in the vein margins, and of biotite and/or garnet in the

Table 3   Analyses of felsic veins in eclogite

	G1	U502	U518 ×4	U518 ×5	U518 ×6
SiO$_2$	53.95	70.69	78.94	78.52	79.12
TiO$_2$	0.91	0.39	0.22	0.37	0.27
Al$_2$O$_3$	18.50	15.67	10.82	12.40	12.30
Fe$_2$O$_3$	1.01	n.a.	n.a.	n.a.	n.a.
FeO	6.48	2.82	2.17	2.93	2.57
MnO	0.16	0.0	0.0	0.0	0.0
MgO	5.34	2.09	1.56	0.88	0.89
CaO	5.30	1.78	1.48	2.60	2.13
Na$_2$O	3.83	2.84	1.86	1.26	1.58
K$_2$O	3.30	3.98	3.09	0.17	0.52
P$_2$O$_5$	0.49	n.a.	n.a.	n.a.	n.a.
H$_2$O$^\pm$	0.88	n.a.	n.a.	n.a.	n.a.
	100.15	100.25	100.13	99.14	99.53

G1, wet-chemical analysis by B. Bruun. Other vein analyses by microprobe on fused samples.

G1: 2-cm wide irregular vein, abundant cpx + plag symplectite.

U502: 2.5-cm felsic vein, scattered large garnets, bio + kyan + ksp symplectite.

U518: 20-cm wide zoned felsic vein.

×4: Mica-rich outer zone with rare garnets, bio + kyan + ksp symplectite.

×5: Outer part of central zone, with large gnt and kyanite grains.

×6: Central zone with gnt + kyan + plag + qtz, and minor hbl + plag symplectite.

Table 2   Whole rock analyses

	Eclogite U514	Average oliv. thol. eclogite ($n = 4$)	Average augen orthogneiss ($n = 12$)
SiO$_2$	46.65	46.42	65.02
TiO$_2$	2.45	1.30	0.85
Al$_2$O$_3$	12.16	15.82	15.11
Fe$_2$O$_3$	0.45	2.22	1.48
FeO	9.85	11.14	4.08
MnO	0.18	0.20	0.11
MgO	9.81	10.56	1.80
CaO	12.98	9.10	3.62
Na$_2$O	2.65	2.39	3.14
K$_2$O	0.08	0.08	3.64
P$_2$O$_5$	0.80	0.18	0.32
S	0.44	0.03	0.05
H$_2$O$^+$	0.45	0.31	0.65
Total	99.03	100.08	99.91
ppm			
Ni	184	259	4
V	263	182	53
Cr	814	301	13
Zn	104	112	85
Cu	45	61	18
Rb	5	2	44
Sr	534	231	264
Y	40	23	42
Zr	346	89	345
Pb	9	40	16
Ba	254	69	1151

XRF analyses by D. A. Carswell.

adjacent eclogite. The removal of K and Na from the centres of larger veins resulted in an excess of Al and the presence of abundant kyanite. These contamination and differentiation processes may have occurred either during intrusion of the veins, or during later metamorphism. The relatively small size (20–50 cm) of the zoned veins suggests small-scale, subsolidus diffusion; we therefore prefer the latter interpretation.

Metamorphic Conditions

The original presence of the metamorphic omphacite + garnet + feldspar + quartz assemblage in the veins corroborates the field evidence that these veins were metamorphosed together with the surrounding eclogite. The vein assemblage may have included phengite at the metamorphic peak, or this phase have formed at an earlier stage of prograde metamorphism and broken down at high-T conditions. An accurate estimate of the metamorphic T is difficult, since no omphacite is preserved well enough to be considered in equilibrium with garnet. However, studies of eclogites on a regional basis have documented a metamorphic gradient across the Western Gneiss Region (Griffin *et al.* 1984), and many of the data points are from the area around Midøy. These data suggest that the temperature at the metamorphic peak here lay near 750°C.

The assemblage calcic garnet + plagioclase + kyanite + quartz is indicative of relatively high P, and several attempts have been made to calibrate the reaction:

$$An \leftrightharpoons Gros + kyan + qtz \qquad (2)$$

for P and T. We have used the calibration of Schmid and Wood (1976) and Wood (1977), with the value of ω_{Ca}^{gnt} suggested by Jenkins and Newton (1979) and $\gamma_{An}^{plag} = 1.28$ (Orville 1972). This method gives results that are identical (within the probable error) to those obtained by the method of Ghent (1976). We have assumed that the rims of the analysed garnet ($Gros_{30}$) equilibrated with the adjacent plagioclase (An_{35}). This leads to P = 19–21 Kb at 750°C for the four samples of this assemblage (Table 1). This P estimate lies near the upper end of the P–T array for the eclogites of the Western Gneiss Region (Griffin et al. this volume) and is similar to the average P estimate (17–18 Kb at 700–750°C) derived from orthopyroxene-bearing eclogites by Carswell et al. (this volume). An error of 50° in T leads to a positively correlated error in P of ca. 1.5 Kb. This estimate, therefore, represents an independent confirmation of the high pressures reached during the metamorphism of the Western Gneiss Region.

The zoning of the garnets in the veins suggests growth during increasing T (reflected by a rimward decrease in Fe/Mg). The rimward decline in CaO could be interpreted as reflecting a moderate drop in P as T increased; England and Richardson (1977) predict such a thermal evolution for subducted or overthrust terranes. However, we have no evidence on the composition of the plagioclase that may have coexisted with garnet cores. It is therefore possible that the rimward drop in CaO seen in the garnets mainly reflects depletion of Ca from the coexisting plagioclase.

Discussion

The field relations described here, and the evidence of high-P mineral assemblages in the anatectic back-veins, clearly show that the eclogites and the surrounding gneiss have been metamorphosed together at high P and T (F_1). Both gneiss and eclogite have suffered a later, lower-grade hydrous metamorphism, accompanied by extreme local flattening (F_2), boudinage, and local anatexis. The preservation of the eclogite and high-P granulite has depended heavily on the size of the eclogite bodies

and the large difference in competence between the basic bodies and the more ductile gneiss.

The Litledigerneset locality shows a rather detailed sequence of events.

1. Formation of the host rock, probably as an intermediate plutonic intrusive.
2. Intrusion of the basic magma, probably as dykes: marginal anatexis, back-veining of the dykes.
3. Metamorphism at high T, P: development of S_1 foliation, 'migmatization', either anatectic or tectonic, of orthogneiss, boudinage of basic dykes, formation of eclogite/high-P granulite mineral assemblages.
4. Strong ca. N–S compression (ENE–WSW shearing): blastomylonite formation, further boudinage of basic bodies, formation of coarse-grained granitic pegmatites around and locally cross-cutting the eclogite boudins.

The oldest foliation that can be recognized is that which forms the axial plane to isoclinal folds in feldspathic lenses in the gneiss (S_1). This foliation is also found in the eclogites and in the fine-grained quartz-dioritic zone around the eclogites.

If the feldspathic lenses in the gneiss represent deformed feldspar megacrysts, their absence from the quartz-dioritic marginal zones can be a result of the local anatexis associated with intrusion of the basic dykes. If these lenses were formed by anatexis during (F_1) regional metamorphism they would not have formed in the quartz-dioritic zones, which were already depleted in granitic components. Thus we cannot be sure that the formation of these lenses post-dates the intrusion of the dykes; the sequence given here is the simplest that fits the observed relations.

Several workers (Bryhni et al. 1969; Green and Mysen 1972) have suggested that the formation of pegmatites by anatexis was important in lowering f_{H_2O} locally and thus promoting the formation of eclogites. The field relations at Litledigerneset show that on the contrary the granitic pegmatites there clearly post-date the high-grade metamorphic event. They have been important in introducing water, leading to amphibolitization of the eclogites. This retrograde step occurred either prior to, or during the latest intense deformation phase.

Several of the steps in this sequence can be dated, at least tentatively, on the basis of available radiometric data. The formation of the augen orthogneiss complex has been dated as 1478 ± 42 Ma by Rb–Sr whole-rock studies, along the strike from the present locality (Carswell and Harvey 1984). This agrees with U–Pb and Rb–Sr data from similar rocks in the Nordfjord area (Lappin et al. 1979; Malinconico personal communication). These ages are significantly younger than those on the surrounding supracrustal rocks (1600–1800 Ma: Tørudbakken and Ilebekk in prepartion).

The intrusion of the basic magma has not been dated directly, but is believed to have occurred soon after the intrusion of the orthogneiss complex, as discussed above. Tørudbakken (1981) has obtained a Rb–Sr whole-rock age of 1517 ± 60 Ma on a partially eclogitized gabbro of similar composition, in the Trollheimen area.

The high-P metamorphism may be dated by Sm–Nd ages of cpx + gnt pairs, which cluster tightly around 425 Ma (Griffin and Brueckner 1980; in preparation). U–Pb data on zircons from several Norwegian eclogites also indicate metamorphism in this time range (Krogh *et al.* 1974; Gebauer personal communication). Rb–Sr dates on micas and other phases from eclogites cluster in the period 380–430 Ma (Griffin and Brueckner in preparation). It is not clear whether these ages are related only to cooling and uplift, or represent a real metamorphic episode—step (4) of the above sequence. Similar Rb–Sr ages are given by mica + whole-rock pairs from various types of gneiss (Brueckner 1972; Krill and Griffin in preparation).

Brueckner (1972) has given a biotite-rock Rb–Sr date of 396 Ma on a granite pegmatite cutting a retrograded (oliv + opx + chlor + amph) garnet peridotite on Gurskøy. Feldspar–mica pairs from similar dykes in eclogites yield ages as low as 355 Ma (Krill and Griffin in preparation), suggesting that intrusion and local deformation continued into Devonian time.

These radiometric data suggest that all of the *metamorphic* events recognized at the Litledigerneset locality occurred during the Caledonian orogeny, while the igneous protoliths date from the end of the Svecofennian period. This sequence of events is recognizable over larger areas of the Western Gneiss Region, wherever eclogites occur within massive or migmatitic orthogneisses. A metamorphic event older than step (1) may be recognized where metasupracrustal rocks occur; we interpret this as a Svecofennian amphibolite-facies metamorphism. The garnets of eclogites in metamorphosed supracrustal rocks commonly contain inclusions of amphibolite-facies minerals (Bryhni *et al.* 1977; Krogh 1982) that may derive from this metamorphic event.

The relations between gneiss and eclogite described here leave little doubt that the formation of the eclogites has occurred *in situ*. Other evidence, summarized by Bryhni *et al.* (1977) and Griffin *et al.* (this volume) suggests that this conclusion is valid for many, if not all, of the eclogites in the Western Gneiss Region. This requires in turn that this edge of the Baltic Shield was depressed to depths of *ca.* 60–70 km, probably in a continent–continent collision zone, during the Caledonian orogeny (Bryhni *et al.* 1977; Krogh 1977; Griffin *et al.* this volume; Cuthbert *et al.* 1981).

Acknowledgements

This work has been supported by NAVF (grant D 48–22.08) and NERC (grant GR3/3085). Discussions with M. B. E. Mørk have contributed much to our understanding of the field relations.

References

Brueckner, H. K. 1972. Interpretation of Rb–Sr ages from the Precambrian and Paleozoic rocks of southern Norway. *Am. Journal of Science*, **272**, 334–358.

Bryhni, I. 1966. Reconnaissance studies of gneisses, ultrabasites, eclogites and anorthosites in outer Nordfjord western Norway *Norges geol. Unders.*, **241**, 1–68.

Bryhni, I., Fyfe, W. S., Green, D. H. and Heier, K. S. 1970. On the occurrence of eclogite in Western Norway. *Contr. Mineral. Petrol.*, **26**, 12–19.

Bryhni, I., Krogh, E. J. and Griffin, W. L. 1977. Crustal derivation of Norwegian eclogites: a review. *N. Jb. Miner. Abh.*, **130**, 49–68.

Carswell, D. A. and Harvey, M. 1981. Molde–Otrøy–Midøy. In Griffin, W. L. and Mørk M.-B. E. (Eds), *Excursion B1. Excursions in the Scandinavian Caledonides.* Geologisk Museum, Oslo.

Carswell, D. A. and Harvey, M. this volume. The intrusive history and tectonometamorphic evolution of the basal gneiss complex in the Moldefjord region. W. Norway.

Carswell, D. A., Krogh, E. J. and Griffin, W. L. this volume. Norwegian orthopyroxene eclogites: calculated equilibration conditions and petrogenetic implications.

Cuthbert, S., Harvey, M. and Carswell, D. A. 1981. A plate tectonic model for the metamorphic evolution of the Basal Gneiss Complex in Western Norway. *Terra Cognita*, **1**, 41.

England, P. C. and Richardson, S. W. 1977. The influence of erosion upon the mineral facies of rocks from different metamorphic environments. *J. geol. Soc. London*, **134**, 201–219.

Eskola, P. 1921. On the eclogites of Norway. *Skr. Vidensk. Selsk. Christiania, Mat.–Natv. KII*, No 8, 1–118.

Fry, N. and Fyfe, W. S. 1969. Eclogites and water pressure. *Contr. Mineral. Petrol*, **24**, 1–6.

Ghent, E. D. 1976. Plagioclase–garnet–Al_2SiO_5–quartz: a potential geobarometer–geothermometer. *Am. Mineral*, **61**, 710–714.

Green, D. H. and Mysen, B. O. 1972. Genetic relationship between eclogite and hornblende + plagioclase pegmatite in western Norway. *Lithos*, **5**, 147–161.

Green, D. H. and Ringwood, A. E. 1967. An experimental investigation of the gabbro to eclogite transformation and its petrological implications. *Geochim. Cosmochim. Acta.*, **31**, 767–833.

Griffin, W. L., Austrheim, H., Brastad, K., Bryhni, I., Krill, A., Mørk, M.-B. E., Qvale, H. and Tørudbakken,

B.' this volume. High-pressure metamorphism in the Scandinavian Caledonides.

Griffin, W. L., Heier, K. S., Taylor, P. N. and Weigand, P. W. 1974. General geology, age and chemistry of the Raftsund mangerite intrusion, Lofoten–Vesterålen. *Norges geol. Unders.*, **312**, 1–30.

Griffin, W. L. and Brueckner, H. K. 1980. Caledonian Sm–Nd ages and a crustal origin for Norwegian eclogites. *Nature, London*, **285**, 319–321.

Griffin, W. L. and Mørk, M.-B. E. 1981. Eclogites and basal gneisses in western Norway. *Excursion B1. Excursions in the Scandinavian Caledonides.* Geolisk Museum, Oslo.

Jenkins, D. M. and Newton, R. C. 1979. Experimental determination of the spinel peridotite to garnet peridotite inversion at 900°C and 1000°C in the system CaO–MgO–Al$_2$O$_3$–SiO$_2$, and at 900°C with natural olivine and garnet. *Contr. Miner. Petrol.*, **68**, 407–419.

Krogh, E. J. 1977. Evidence for a Precambrian continent–continent collision in western Norway. *Nature, London*, **267**, 17–19.

Krogh, E. J. 1980a. Geochemistry and petrology of glaucophane-bearing eclogites and associated rocks from Sunnfjord, western Norway. *Lithos*, **13**, 355–380.

Krogh, E. J. 1980b. Compatible P–T conditions for eclogites and surrounding gneisses in the Kristiansund area, western Norway. *Contrib. Mineral. Petrol.*, **75**, 387–393.

Krogh, E. J. 1982. Metamorphic evolution deduced from mineral inclusions and compositional zoning in garnets from Norwegian country-rock eclogites. *Lithos*, **15**, 305–321.

Krogh, T. E., Mysen, B. O. and Davis, G. L. 1974. A Paleozoic age for the primary minerals of a Norwegian eclogite. *Ann. Rep. Geophys. Lab. Carnegie I not. Washington*, **73**, 575–576.

Lappin, M. A. 1966. The field relationships of basic and ultrabasic masses in the basal gneiss complex of Stadlandet and Almklovdalen, Nordfjord, northwestern Norway. *Norsk geol. Tidsskr.*, **46**, 439–495.

Lappin, M. A. and Smith, D. C. 1978. Mantle-equilibrated orthopyroxene eclogite pods from the basal gneisses in the Selje district, western Norway. *J. Petrol.*, **19**, 530–584.

Lappin, M. A., Pidgeon, R. T. and Breemen, O. Van. 1979. Geochronology of basal gneisses and mangerite syenite of Stadlandet, west Norway. *Norsk geol. Tidsskr.*, **59**, 161–181.

Mysen, B. O. 1972. Five clinopyroxenes in the Hareidlandet eclogite, western Norway. *Cont. Mineral. and Petrol.*, **34**, 315–325.

Mysen, B. O. and Griffin, W. L. 1973. Pyroxene stoichiometry and the breakdown of omphacite. *Am. Miner.*, **58**, 60–63.

O'Hara, M. J., Richardson, S. W. and Wilson, G. 1971. Garnet peridotite stability and occurrence in crust and mantle. *Contr. Mineral. Petrol.*, **32**, 48–68.

Orville, P. M. 1972. Plagioclase cation exchange equilibria with aqueous chloride solution: results at 700°C and 2000 bars in the presence of quartz. *Am. J. Sci.*, **272**, 234–272.

Schmid, R. and Wood, B. J. 1976. Phase relationships in granulitic metapelites from the Ivrea–Verbano Zone (Northern Italy). *Contr. Miner. Petrol.*, **54**, 255–279.

Tørudbakken, B. 1981. Vindøldalen dolerite. In Griffin, W. L. and Mørk, M.-B. E. (Eds), *Excursion B1. Excursions in the Scandinavian Caledonides.* Geologisk Museum, Oslo.

Wood, B. E. 1977. The activities of components in clinopyroxene and garnet solid solutions and their application to rocks. *Phil. Trans. R. Soc. London*, **A286**, 331–342.

Yoder, H. S. and Tilly, C. E. 1964. Origin of basalt magmas: an experimental study of natural and synthetic rock systems. *J. Petrol.*, **3**, 342–532.

The Caledonide Orogen—Scandinavia and Related Areas
Edited by D. G. Gee and B. A. Sturt

Norwegian orthopyroxene eclogites: calculated equilibration conditions and petrogenetic implications

D. A. Carswell[*] **E. J. Krogh**[†] and **W. L. Griffin**[‡]

[*]*Department of Geology, University of Sheffield, Mappin Street, Sheffield S1 3JD, England*
[†]*Institute of Biology & Geology, University of Tromsø, Tromsø, Norway*
[‡]*Mineralogisk–Geologisk Museum, Sarsgate 1, N–Oslo 5, Norway*

ABSTRACT

Within the Western Gneiss Region of Norway, orthopyroxene-bearing eclogites occur as part of garnetiferous ultramafic rocks ('internal' or 'A' type) and as layers and lenses enclosed in gneisses ('external' or 'B' type). Calculation of metamorphic pressure–temperature conditions is complicated by apparent disequilibrium among orthopyroxene, clinopyroxene, and garnet. Garnets commonly show minor retrograde Fe/Mg zoning, while in some external orthopyroxene eclogites, garnets are prograde zoned and preserve amphibolite facies mineral assemblages as inclusions. Orthopyroxene grains are commonly zoned from low Al in cores to maximum Al in rims in contact with garnet and clinopyroxene. Some orthopyroxene eclogites show phase intergrowths which may reflect exsolution from primary high Al pyroxenes, comparable to those found elsewhere in anorthosites or spinel peridotites. There is no evidence that these high-Al pyroxenes ever have coexisted with garnets; they are high temperature, but not necessarily high pressure, phases. The low Al contents of some orthopyroxene grain cores may reflect disequilibrium during prograde metamorphism. These low-Al compositions cannot be paired with particular garnet compositions as meaningful mineral equilibria. Hence the high pressure values (30–50 Kbars) calculated for many orthopyroxene–garnet composition pairs are probably spurious.

Calculated pressure–temperature conditions using maximum values of Al in orthopyroxenes, average: external orthopyroxene eclogites *ca*. 700–740°C, 17–18 Kbars; internal orthopyroxene eclogites *ca*. 710°C, 20 Kbars; garnet lherzolites, *ca*. 725°C, 20 Kbars. As similar pressure–temperature values have been calculated for associated granulite facies gneisses and orthopyroxene free external eclogites, it is considered that all of these rock types equilibrated together at high pressures. This may have occurred during a transient subduction event, prior to the extensive retrograde metamorphism which occurred during the ensuing uplift of these rocks towards the surface.

Introduction

The Gneiss Region of western Norway, extending from Bergen to Kristiansund, is a classic area for the study of eclogites. Within this area, two general types are commonly distinguished:

1. 'Internal' or Type A eclogites are layers or lenses enclosed within ultramafic rocks, sometimes garnet-bearing. The peridotites are dominated by the amphibolite facies assemblage olivine + orthopyroxene + amphibole + chlorite, which overprints the garnetiferous assemblages (Medaris 1980).

2. 'External' or Type B eclogites are layers, lenses, and pods from centimetres to kilometres in size, enclosed within gneisses of both sedimentary and igneous origin. The gneisses locally contain relict high-pressure granulite assemblages, but are dominated by the amphibolite facies mineral assemblages that also overprint the eclogites.

It appears obvious from field and petrographic data that the eclogite facies assemblages are relics of

Table 1 Summary of field relations and petrography of analysed samples

Sample no.	Locality	Field type	Primary mineral assemblage	Secondary minerals	Special features
KR1	Hornindal	B	Gnt+cpx+opx+amph+magnetite+spinel?	Sec. green amph around cpx	Uncertain if green spinel associated with magnetite is primary or secondary. Sub- to euhedral gnt inclusions in pyroxenes may reflect exsolution from original higher T. Al-rich pyroxenes.
KR6	Hellesylt	B	Gnt+cpx+opx+(colourless)amph+rutile	none	Rounded irregular gnt grains enclosed in larger opx, cpx and amph grains.
KR2A KR2B	Kvalneset	B	Gnt+cpx+opx+amph+rutile	none	Large opx grains contain sub- to euhedral grains of gnt. Primary amph is colourless in thin sections in contrast to inclusions of green amph in some gnt.
KR3	Hjørungvåg	B	Gnt+cpx+opx+amph+qtz+rutile	Secondary reactions as follows: (a) CpxI+gnt → CpxII+plag +opxII	Cpx grains contain thin spindles of quartz // to cleavage planes—perhaps exsolved during oxidation.
KR3	Hjørungvåg	B	Gnt+cpx+opx+amph+qtz+rutile	(b) CpxI+gnt+H_2O → cpxII+amphII+plag (c) OpxI+gnt+H_2O → amphII (d) Gnt+qtz → opxII+plag	
KR4	Kvalvåg	B	Gnt+cpx+opx+qtz+rutile	Considerable sec. cpx after both gnt and cpx. Sec. reaction (d) Gnt+qtz → OpxII+plag	Resembles KR3 but sec. reaction products not so abundant. Green amph inclusions in some gnt.
U206 U243	Solholm Otrøy	B	Gnt+cpx+opx+mica+rutile	Sec. amph after cpx and gnt especially in U206; also extensive alteration of opx	Red–brown mica in apparent textural equil. with primary phases and also as inclusions in gnt.
U19	Skarshaugen Otrøy	B	Gnt+cpx+opx+mica?+qtz+rutile	Minor sec. amph	Numerous gnt inclusions in pyroxenes. Pale brown mica is enigmatic—sometimes in text. equil. with primary phases elsewhere of apparent replacement origin.
KR5 A44A			Gnt+cpx+opx+rutile		Pegmatitic opx bearing variants.
A44B	Grytingvåg	B	Gnt+cpx+opx+rutile	Sec. amph abundant in broad zones towards margin of lens	Contains contact of coarse opx eclogite with finer grained, foliated, opx free eclogite.
A45			Gnt+cpx+mica+rutile		Interlayered opx free eclogite with pale brown mica. Many cpx+mica inclusions in gnt.
V80/19	Høydalsneset	B	Gnt+cpx+opx+amph+qtz+rutile	Some sec. amph	Gnt contain inclusions of pyroxenes and amph., foliation defined by pyroxenes and amph.
U95	Ugelvik (Harbour)	A	Gnt+opx+cpx	none	All cpx and much (all?) gnt exsolved from original high Al–opx. Details and data in Carswell (1973).
U47	Ugelvik (Raudhaugene)	A	Gnt+opx+cpx+rutile	Pyroxenes extensively hydrated	Sharply bounded layer in serp. gnt peridotite body. Porphyroclastic texture with clasts up to 6 mm in recryst. mosaic of 0.2–0.5 mm grains.

an early high-pressure metamorphic event, now largely obliterated by later amphibolite facies metamorphism. The nature, tectonic setting, and timing of the high-P metamorphism have been a matter of heated controversy for at least 50 years. One group (O'Hara and Mercy 1963; O'Hara 1967; O'Hara et al. 1971; Lappin 1966, 1974; Lappin and Smith 1978), regard the ultramafic rocks as tectonic slices of the subcontinental mantle, and have extended this interpretation to include the external eclogites. Others (Bryhni 1966; Bryhni et al. 1970, 1977; Carswell 1974; Griffin 1972; Griffin and Råheim 1973; Krogh 1977a,b; Mysen and Heier 1972; Råheim 1972) regard the external eclogites as an integral part of the gneiss terrane and implicitly or explicitly consider a dual origin: essentially in situ metamorphism of the external eclogites, and tectonic introduction of the ultramafic rocks and their associated internal eclogites. Recent geotectonic models relate both processes to limited transient subduction of the west Norwegian continental crust in a Himalaya-type collision belt, probably in an early stage of the Caledonian orogeny (Bryhni et al. 1977; Krogh, 1977a; Cuthbert et al. 1983; Griffin et al. this volume).

One approach to the resolution of this controversy is the determination of the P–T conditions involved in the metamorphism of both types. Garnet websterites are especially pertinent to this problem, since the assemblage orthopyroxene + clinopyroxene + garnet allows estimation of both P and T by comparison with experimental work, assuming that these phases have equilibrated under metamorphism. Garnet websterites occur in both 'internal' and 'external' situations in the Western Gneiss Region. Although these rocks contain clinopyroxene, they have traditionally been called 'orthopyroxene eclogites', and we will follow this practice. 'Internal' orthopyroxene eclogites may contain olivine and grade into garnet lherzolites. 'External' orthopyroxene eclogites may contain quartz and rutile, as well as hydrous phases.

Mineralogical data and P–T estimates for the internal orthopyroxene eclogites and garnet lherzolites are summarized by Carswell and Gibb (1980a), and Medaris (1980). Data on external orthopyroxene eclogites are less extensive. Lappin (1974) and Lappin and Smith (1978) presented data from several bodies in the Selje area; they concluded that these originated deep in the upper mantle, and have been tectonically emplaced into the crust ($\Delta P \sim 20$ Kbars). This paper presents data on a wider range of orthopyroxene eclogite occurrences, mostly of the external type. Special attention is paid to evidence of chemical disequilibrium and the difficulty of calculating meaningful P–T estimates for these rocks.

The locations of the analysed samples are given in Griffin et al. (this volume), and field relations, petrography, etc. are summarized in Table 1.

Mineral Chemistry

Analytical techniques

Mineral separates were analysed by D. A. Carswell at the University of Sheffield using a combination of atomic absorption, flame photometry, spectrophotometric, and titration techniques. All other mineral analyses were by electron microprobe using either the ARL–EMX microprobe at Oslo (analysts: A. J. Krogh and W. L. Griffin) or the Microscan IX microprobe at Sheffield (analyst: D. A. Carswell). The Oslo analyses include both WDS and some EDS (LINK system, ZAF–4 reduction program) analyses; Sheffield analyses are all WDS data. Analyses of low-Al orthopyroxene standards are shown in Table 2.

Microprobe point analyses of the primary mineral

Table 2 Microprobe analyses of 'standard' orthopyroxenes

| | R 2537 | | | R 1742 | | | Shallow water | |
	Oslo	nominal		Oslo	nominal		Oslo	nominal
SiO$_2$	51.91	52.2		49.92	50.2		59.13	59.98
Al$_2$O$_3$	1.78	1.8	(2.2)	1.40	1.1	(1.8)	0.11	0.13
TiO$_2$	0.11	0.1		0.12	0.1		—	—
MgO	22.45	22.8		13.25	13.8		40.04	39.96
FeO	21.91	21.3		32.97	33.3		—	—
MnO	0.61	0.6		0.78	0.7		—	—
CaO	0.54	0.6		0.69	0.6		0.09	0.10
	99.31			99.13			99.37	

Nominal values for R 2537, R 1742 are probe analyses by Howie and Smith (1966); wet chemical values in parentheses. Shallow water enstatite, probe analysis by K. Fredriksson (pers. comm. 1977). Oslo analyses done using ARL–EMX manual probe, mixed synthetic and natural mineral standards, Bence–Albee correction methods. Analyst: E. J. Krogh.

Table 3   Analyses of primary orthopyroxenes

	KR 1	KR 2A-1	KR 2B-1	KR 2B-2	KR 3-1	KR 3-2	KR 3-3	KR 4-1	KR 4-2	KR 4-3	KR 6-1	KR 6-2	U 206	U 243-1	U 243-2	U 19-1	U 19-2
	No zoning Cores	Rims ADJ gnt	Rims ADJ gnt	Cores	Min. Al	Max. Al	Aver. Al	Min. Al	Max. Al	Aver. Al	Rims ADJ gnt	Cores	No zoning	Large grains	Inc. in gnt	Min. Al	Max. Al
SiO_2	55.3	55.5	54.4	55.3	56.3	54.7	55.8	55.2	55.2	54.9	55.6	55.9	55.4	55.7	54.6	56.4	55.8
TiO_2	n.d.	n.d.	0.03	0.02	n.d.	n.d.	n.d.	n.d.	n.d.	n.d.	0.06	0.09	n.d.	n.d.	n.d.	0.02	0.02
Al_2O_3	0.88	0.57	0.78	0.33	0.77	2.58	1.16	0.90	2.41	1.53	0.69	0.35	0.86	0.73	1.11	1.24	1.99
Cr_2O_3	n.d.	n.d.	0.06	0.02	n.d.	n.d.	n.d.	n.d.	n.d.	n.d.	0.05	0.02	n.d.	n.d.	n.d.	0.04	0.05
FeO^*	14.4	15.8	15.4	15.0	13.5	13.3	13.6	11.4	10.7	11.8	12.1	12.2	13.0	12.7	13.2	8.40	8.04
MnO	0.21	n.d.	0.14	0.13	0.34	0.34	0.34	0.16	0.10	0.17	0.10	0.15	0.18	n.d.	n.d.	0.15	0.15
NiO	n.d.	n.d.	n.d.	n.d.	n.d.	n.d.	n.d.	n.d.	n.d.	n.d.	n.d.	n.d.	n.d.	n.d.	n.d.	n.d.	n.d.
MgO	29.6	27.7	28.2	28.8	29.1	28.5	28.9	32.0	30.8	30.8	30.8	31.1	30.5	31.1	30.4	34.0	34.1
CaO	0.14	0.26	0.25	0.10	0.23	0.14	0.21	0.23	0.17	0.37	0.16	0.15	0.29	0.28	0.27	0.21	0.13
Na_2O	n.d.	n.d.	n.d.	n.d.	n.d.	n.d.	n.d.	n.d.	n.d.	n.d.	n.d.	n.d.	0.02	n.d.	n.d.	0.01	0.02
Total	100.5	99.9	99.3	99.7	100.2	99.6	100.0	99.9	99.4	99.5	99.6	100.0	100.3	100.5	99.6	100.4	100.3
Structural formulae O = 6																	
Si	1.968	1.996	1.971	1.988	1.998	1.953	1.986	1.953	1.951	1.951	1.976	1.980	1.965	1.966	1.954	1.954	1.934
Al^{IV}	0.032	0.004	0.029	0.012	0.002	0.047	0.014	0.038	0.049	0.049	0.024	0.015	0.035	0.030	0.046	0.046	0.066
Al^{VI}	0.005	0.020	0.004	0.002	0.030	0.066	0.035	0.000	0.051	0.015	0.005	0.000	0.001	0.000	0.001	0.005	0.015
Ti	—	—	0.001	0.001	—	—	—	—	—	—	0.002	0.002	—	—	—	0.001	0.001
Cr	—	—	0.002	0.001	—	—	—	—	—	—	0.001	0.001	—	—	—	0.001	0.001
Fe	0.428	0.477	0.467	0.451	0.400	0.398	0.404	0.337	0.315	0.351	0.360	0.361	0.387	0.375	0.395	0.243	0.233
Mn	0.006	—	0.004	0.004	0.010	0.010	0.010	0.005	0.003	0.005	0.003	0.005	0.005	—	—	0.004	0.004
Ni	—	—	—	—	—	—	—	—	—	—	—	—	—	—	—	—	—
Mg	1.568	1.485	1.523	1.542	1.537	1.516	1.534	1.686	1.622	1.631	1.631	1.642	1.613	1.637	1.617	1.757	1.763
Ca	0.005	0.010	0.010	0.004	0.009	0.005	0.008	0.009	0.006	0.014	0.006	0.006	0.011	0.011	0.010	0.008	0.005
Na	—	—	—	—	—	—	—	—	—	—	—	—	0.001	—	—	0.001	0.001
Total	4.013	3.992	4.011	4.006	3.986	3.992	3.990	4.028	3.998	4.017	4.008	4.012	4.018	4.019	4.023	4.020	4.024

* All Fe as FeO; analytically determined FeO contents in mineral separates are U19: 8.61 wt%, U47: 4.41 wt%, A44: 8.42 wt%.

Table 3 (continued)

	U 19-3 Aver. Al	U 19-4 Min. Sep.	U 95-1 Min. Al	U 95-2 Max. Al	U 95-3 Aver. Al	U 47-1 Aver. clasts	U 47-2 Aver. recrysts.	U 47-3 Min. sep.	KR 5-1 Rims ADJ gnt	KR 5-2 Core	KR 5-3 Rims ADJ cpx	A 44-1 Aver. cores	A 44-2 Rims ADJ gnt	A 44-3 Rims ADJ cpx/amp	A 44-4 Min. sep.	V80/17 1 High-Al	V80/17 2 Min. Al
SiO_2	56.0	53.5	58.4	57.4	58.0	57.7	58.1	58.0	56.4	56.6	56.8	57.2	56.9	57.0	56.7	55.9	55.7
TiO_2	0.02	0.05	n.d.	n.d.	n.d.	n.d.	n.d.	0.08	n.d.	n.d.	n.d.	0.02	0.02	0.02	0.03	n.d.	n.d.
Al_2O_3	1.52	6.66	0.54	2.03	1.22	0.15	0.38	0.75	0.78	0.53	0.38	0.35	0.68	0.48	0.82	1.18	0.59
Cr_2O_3	0.05	0.13	0.15	0.28	0.24	n.d.	n.d.	0.04	0.03	0.03	0.03	0.03	0.02	0.02	0.02	n.d.	n.d.
FeO*	8.17	8.88	4.14	4.18	4.26	4.93	5.11	4.96	10.0	10.4	10.1	9.67	10.0	9.82	8.84	12.1	11.9
MnO	0.15	0.20	0.12	0.14	0.13	n.d.	n.d.	0.08	0.14	0.15	0.15	0.06	0.09	0.10	0.08	0.0	0.14
NiO	n.d.	0.07	n.d.	n.d.	n.d.	n.d.	n.d.	0.11	n.d.	n.d.	n.d.	0.21	0.17	0.19	0.21	n.d.	n.d.
MgO	34.0	27.2	36.0	35.6	35.4	37.3	37.3	35.4	31.7	30.8	30.6	32.3	32.0	32.0	32.8	31.0	31.0
CaO	0.18	2.15	0.20	0.20	0.20	0.17	0.15	0.39	0.17	0.14	0.18	0.13	0.13	0.14	0.63	0.15	0.25
Na_2O	0.02	0.39	n.d.	n.d.	n.d.	0.01	0.02	0.09	n.d.	n.d.	n.d.	n.d.	n.d.	n.d.	0.06	n.d.	n.d.
Total	100.1	99.2	99.6	99.8	99.5	100.3	101.0	99.9	99.2	98.7	98.2	100.0	100.1	99.8	100.1	100.3	99.6

Structural formulae O = 6

	U 19-3	U 19-4	U 95-1	U 95-2	U 95-3	U 47-1	U 47-2	U 47-3	KR 5-1	KR 5-2	KR 5-3	A 44-1	A 44-2	A 44-3	A 44-4	V80/17 1	V80/17 2
Si	1.946	1.890	1.999	1.961	1.987	1.972	1.971	1.989	1.982	2.009	2.020	1.999	1.990	1.997	1.976	1.979	1.978
Al^{IV}	0.054	0.110	0.001	0.039	0.013	0.006	0.015	0.011	0.011	0.000	0.000	0.001	0.010	0.003	0.024	0.021	0.022
Al^{VI}	0.008	0.167	0.021	0.043	0.036	0.000	0.000	0.019	0.021	0.022	0.016	0.013	0.018	0.017	0.010	0.003	0.002
Ti	0.001	0.001	—	—	—	—	—	0.002	—	—	—	0.001	0.001	0.001	0.001	—	—
Cr	0.001	0.004	0.004	0.008	0.006	—	—	0.001	0.001	0.001	0.001	0.001	0.001	0.001	0.001	—	—
Fe	0.238	0.262	0.118	0.120	0.122	0.141	0.145	0.142	0.296	0.309	0.302	0.283	0.294	0.288	0.258	0.353	0.353
Mn	0.004	0.006	0.004	0.004	0.004	—	—	0.002	0.004	0.005	0.005	0.002	0.003	0.003	0.002	0.004	0.004
Ni	—	0.002	—	—	—	—	—	0.003	—	—	—	0.006	0.005	0.005	0.006	—	—
Mg	1.763	1.432	1.834	1.813	1.810	1.899	1.886	1.807	1.666	1.628	1.622	1.683	1.670	1.673	1.703	1.639	1.641
Ca	0.007	0.081	0.007	0.007	0.007	0.006	0.005	0.014	0.006	0.005	0.007	0.005	0.005	0.005	0.024	0.010	0.010
Na	0.001	0.027	—	—	—	0.001	0.001	0.006	—	—	—	—	—	—	0.004	—	—
Total	4.023	3.982	3.988	3.994	3.985	4.025	4.023	3.997	3.995	3.980	3.972	3.993	3.995	3.992	4.008	4.009	4.010

*All Fe as FeO; analytically determined FeO contents in mineral separates are U19: 8.61 wt%, U47: 4.41 wt%, A44: 8.42 wt%.

Table 4   Analyses of

	KR 1 Not zoned	KR 2A-1 Cores	KR 2A-2 Rims	KR 2B-1 Cores	2B-2 Rims	KR 3-1 Cores	KR 3-2 Rims	KR 4-1 Cores	KR 4-2 Rims	KR 6 Not zoned	U 206 No zoning	U 243 No zoning	U 19-1 No zoning
SiO_2	54.9	55.3	55.1	53.9	54.0	54.7	53.7	54.3	54.2	54.1	54.4	53.8	54.5
TiO_2	n.d.	n.d.	n.d.	0.03	0.03	n.d.	n.d.	n.d.	n.d.	0.13	0.04	n.d.	0.08
Al_2O_3	1.11	1.91	2.25	1.48	1.49	3.96	4.18	5.41	5.82	2.45	1.83	1.78	2.99
Cr_2O_3	n.d.	n.d.	n.d.	0.15	n.d.	n.d.	0.24	n.d.	0.28	n.d.	n.d.	n.d.	n.d.
FeO^*	5.01	4.65	4.06	5.56	4.18	5.63	5.78	4.51	4.59	4.09	4.20	4.16	3.88
MnO	0.11	n.d.	n.d.	0.03	0.08	0.10	0.12	0.07	0.08	0.04	0.05	n.d.	0.08
MgO	16.3	14.9	15.3	14.3	15.4	13.7	13.9	12.6	12.6	15.1	16.01	15.9	15.2
CaO	22.8	22.1	22.3	21.8	22.7	18.6	19.1	19.2	19.1	21.6	23.1	22.6	21.1
Na_2O	1.10	1.54	1.55	1.78	1.30	2.86	2.37	3.37	3.15	1.88	1.10	1.15	2.20
Total	101.3	100.5	100.5	99.1	99.3	99.5	99.2	99.5	99.6	99.7	100.7	99.5	100.4

Structural formulae O = 6

Si	1.987	2.007	1.995	1.998	1.988	1.999	1.975	1.979	1.971	1.979	1.974	1.976	1.975
Al^{IV}	0.013	0.000	0.005	0.002	0.012	0.001	0.025	0.021	0.029	0.021	0.026	0.024	0.025
Al^{IV}	0.034	0.082	0.091	0.063	0.053	0.170	0.156	0.211	0.221	0.085	0.052	0.053	0.103
Ti	—	—	—	0.001	0.001	—	—	—	—	0.004	0.001	—	0.002
Cr	—	—	—	0.007	0.004	—	—	—	—	0.008	—	—	0.011
Fe	0.152	0.141	0.123	0.172	0.129	0.172	0.178	0.138	0.140	0.125	0.128	0.128	0.118
Mn	0.003	—	—	0.001	0.002	0.003	0.004	0.002	0.002	0.001	0.002	—	0.002
Mg	0.879	0.807	0.824	0.790	0.845	0.744	0.759	0.683	0.686	0.823	0.867	0.872	0.821
Ca	0.883	0.860	0.864	0.866	0.896	0.726	0.754	0.751	0.746	0.847	0.897	0.891	0.821
Na	0.077	0.108	0.109	0.128	0.093	0.203	0.169	0.238	0.222	0.133	0.077	0.082	0.155
Total	4.028	4.006	4.012	4.028	4.023	4.017	4.019	4.024	4.016	4.026	4.024	4.026	4.032

*All Fe as FeO; analytically determined FeO contents in mineral separates are U19: 2.71 wt%, U47: 1.49 wt%, A44: 2.10 wt%.

phases in many of the analysed samples have shown significant within-grain compositional variations. As the zoning is important to interpretations of the metamorphic evolution of these rocks, averages for both grain core and rim analyses are given in these instances (Tables 3–7).

Orthopyroxene compositional variability

Al_2O_3 contents of orthopyroxenes are crucial to the calculation of equilibration pressures for these rocks but often vary significantly within single grains. Our observations indicate that the Al_2O_3 content of the orthopyroxene is typically highest adjacent (20–50 μm distance) to garnet grains and appreciably lower in the cores of grains and adjacent to clinopyroxenes. Secondary amphibolitization in some of the orthopyroxene eclogites results in a notable decrease in the Al_2O_3 contents of both orthopyroxenes and clinopyroxenes. Inclusions of pyroxenes within garnets, which are shielded from the effects of secondary amphibolitization, typically have higher Al_2O_3 contents than matrix pyroxene grains. In Table 3 several different orthopyroxene analyses are given for most samples. In most instances we have elected to give analyses with both maximum and minimum Al_2O_3 contents in addition to an 'average' analysis, that is probably not very meaningful.

Lappin and Smith (1978) reported very low values for Al_2O_3 (0.24–0.65 wt. per cent) in orthopyroxenes from a number of orthopyroxene eclogite lenses directly enclosed in gneisses. However, our Al_2O_3 determinations have indicated a much wider range (0.28–2.58 wt. per cent) for the orthopyroxenes of such bodies. It is of particular interest to compare our various analyses for orthopyroxenes in samples from the Grytingvåg locality with the microprobe determination of 0.39 wt. per cent Al_2O_3 given by Lappin and Smith (1978). This latter value lies within our microprobe-determined ranges of 0.38–0.78 wt. per cent (sample KR5—analyst: Krogh, Oslo microprobe) and 0.35–0.68 wt. per cent (sample A44—analyst: Carswell, Sheffield microprobe). However, it is clearly more in line with the values that we have measured in orthopyroxene grain cores (or rims adjacent to clinopyroxene or amphibole) than with the higher values of 0.78 wt. per cent (Krogh) and 0.68 wt. per cent (Carswell) for orthopyroxene rims adjacent to garnets. Likewise in other orthopyroxene eclogites the values for Al_2O_3 in orthopyroxenes given by Lappin and Smith (1978) are much more comparable to our determinations for grain cores than for rims adjacent to garnets. Our microprobe studies thus suggest that equilibrium between orthopyroxene, garnet, and clinopyroxene has been attained over very short dis-

primary clinopyroxenes

	U 19–2 Min. sep.	U 95 No zoning	U 47–1 Aver. clasts	U 47–2 Aver. recryst.	U 47–3 Min. sep.	KR 5–1 Cores	KR 5–2 Rims ADJ gnt	A 44–1 Cores	A 44–2 Rims	A 44–3 Min. sep.	A 45–1 Cores	A 45–2 Rims	U80/19 Not zoned
SiO_2	54.2	55.1	54.9	54.8	54.7	55.1	55.1	55.3	55.4	54.7	55.2	55.1	54.0
TiO_2	0.20	n.d.	n.d.	n.d.	0.26	n.d.	n.d.	0.34	0.03	0.12	0.02	0.04	0.17
Al_2O_3	3.22	4.04	3.29	3.10	3.76	3.34	2.94	2.96	2.50	2.72	1.82	2.22	1.43
Cr_2O_3	0.33	1.67	n.d.	n.d.	0.37	0.19	0.20	0.26	0.18	0.18	0.11	0.10	n.d.
FeO^*	3.67	1.12	2.06	1.93	2.49	3.39	3.50	3.07	3.32	2.76	2.16	2.24	3.65
MnO	0.08	0.10	n.d.	n.d.	0.05	0.11	0.08	0.03	0.03	0.04	0.01	0.00	0.10
MgO	15.3	15.2	16.2	15.9	16.1	14.8	15.5	14.9	15.2	16.7	16.0	15.7	16.3
CaO	20.0	20.8	21.6	21.4	19.9	21.1	21.4	20.9	21.2	20.2	22.6	22.3	22.7
Na_2O	2.54	2.45	2.30	2.33	2.37	1.97	1.80	2.06	1.86	2.13	1.31	1.36	1.20
Total	99.7	100.5	100.3	99.5	100.0	100.0	100.5	99.8	99.7	99.6	99.4	99.1	99.6

Structural formulae O = 6

Si	1.976	1.973	1.975	1.986	1.970	1.993	1.987	2.000	2.008	1.981	2.008	2.006	1.979
Al^{IV}	0.024	0.027	0.025	0.014	0.030	0.007	0.013	0.000	0.000	0.019	0.000	0.000	0.021
Al^{VI}	0.114	0.143	0.115	0.118	0.130	0.136	0.112	0.126	0.107	0.097	0.078	0.095	0.041
Ti	0.005	—	—	—	0.007	—	—	0.009	0.001	0.003	0.001	0.001	0.005
Cr	0.010	0.047	—	—	0.010	0.005	0.006	0.007	0.005	0.005	0.003	0.003	—
Fe	0.112	0.034	0.062	0.058	0.075	0.103	0.106	0.093	0.101	0.084	0.066	0.068	0.112
Mn	0.002	0.003	—	—	0.002	0.003	0.002	0.001	0.001	0.001	0.000	0.000	0.003
Mg	0.830	0.810	0.868	0.857	0.865	0.797	0.834	0.803	0.821	0.904	0.866	0.851	0.890
Ca	0.780	0.796	0.831	0.832	0.766	0.819	0.825	0.810	0.823	0.783	0.882	0.872	0.891
Na	0.180	0.170	0.160	0.164	0.165	0.138	0.126	0.144	0.131	0.150	0.092	0.096	0.085
Total	4.034	4.003	4.035	4.030	4.020	4.002	4.011	3.992	3.998	4.026	3.997	3.992	4.028

*All Fe as FeO; analytically determined FeO contents in mineral separates are U19: 2.71 wt%, U47: 1.49 wt%, A44: 2.10 wt%.

tanccs at best. This being the case, it is clearly important to calculate P–T values, based on element distribution considerations, on analyses of adjacent grain rims rather than the grain cores.

Al_2O_3 contents determined in orthopyroxene bulk mineral separates are invariably higher (as are the CaO contents) than in microprobe determinations for the same sample. Values determined for example on an orthopyroxene mineral separate from the Grytingvåg eclogite are 0.82 wt. per cent Al_2O_3 and 0.63 wt. per cent CaO, which compare closely with the values of 0.90 wt. per cent Al_2O_3 and 0.55 wt. per cent CaO independently determined by Green (1969). These significantly higher values undoubtedly reflect minor contamination by both garnet and clinopyroxene. Certainly this is the case in the orthopyroxene of sample U19. The contents of 6.66 wt. per cent Al_2O_3 and 2.15 wt. per cent CaO in the orthopyroxene separate reflect the presence of abundant small inclusions of both garnet and clinopyroxene, which prohibit preparation of a pure orthopyroxene separate.

Sample U47 is the only one with a notably porphyroclastic texture. Al_2O_3 contents are low in both the large strained porphyroclasts (0.15 wt. per cent) and the much smaller recrystallized grains (0.38 wt. per cent), but none the less significantly higher in the latter.

Clinopyroxene compositions (Table 4)

Unlike the orthopyroxenes these show little in the way of significant within-grain compositional variation of Al_2O_3 content. There is no consistent variation in the Na_2O content between grain cores and rims, so that Jd contents are fairly uniform. Similarly there is a little significant difference between the analyses of the large clinopyroxene porphyroclasts and smaller recrystallized grains in sample U47.

Garnets and $K_{D\,Fe-Mg}^{Gnt-Cpx}$ values

The metamorphic evolution of eclogite facies assemblages is often conveniently monitored by changes in $Fe^{2+}-Mg^{2+}$ partitioning between the coexisting garnets and clinopyroxenes—usually expressed by $K_D = (Fe^{2+}/Mg^{2+}$ garnet$)/(Fe^{2+}/Mg^{2+}$ clinopyroxene$)$. K_D values are strongly temperature dependent (Banno 1970) but are also influenced by pressure and compositional factors, notably the Ca content of the garnet (Råheim and Green 1974; Ellis and Green 1979). The appreciable compositional zonation observed in eclogite minerals from the Nordfjord and Sunnfjord regions (Bryhni and Griffin 1971; Bryhni et al. 1977 Krogh 1977a, 1980a, 1982) provides a convincing record of the prograde metamorphic history suffered by these

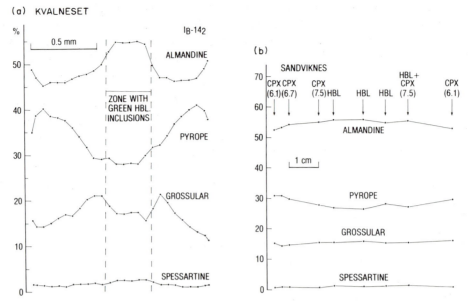

Fig. 1   Zoning profiles across garnet grains in two orthopyroxene bearing eclogite bodies. (a). Kvalneset, orthopyroxene eclogite, sample KR2A (IB–14$_2$). (b) Sandviknes, quartz eclogite, sample V80/33. Note radial variation in inclusions and K_D

rocks. This evolution is further documented by the existence of amphibolite facies inclusion suites in the cores of many eclogite garnets (Bryhni *et al*. 1977; Krogh 1982).

The compositional zonation of the garnets in the orthopyroxene eclogites is not usually so pronounced (Table 5). The most extreme prograde zonation that we have observed occurs in sample

KR2A from Kvalneset (Fig. 1) in which the Fe/Mg of the garnet (with all Fe taken as Fe^{2+}) decreases from 1.81 (cores) to 1.08 (rims). Comparison with the coexisting clinopyroxene suggests that the zoning represents a decrease in K_D from 10.3 to 7.3 as the garnets grew. A less extreme example, from the Sandviknes body described by Lappin and Smith (1978), is shown in Fig. 1B. More typically, any sig-

Table 5   Analyses

	KR 1 Not zoned	KR 2A–1 Cores	KR 2A–2 Rims	KR 2B–1 Cores	KR 2B–2 Rims	KR 3–1 Cores	KR 3–2 Rims	KR 4–1 Cores	KR 4–2 Rims	KR 6 Not zoned	U 206 No zoning	U 243 No zoning	U 19– No zoni
SiO$_2$	39.7	38.8	39.4	38.1	39.6	40.0	39.6	40.2	40.1	40.2	39.9	39.5	40.8
TiO$_2$	n.d.	n.d.	n.d.	0.17	0.09	n.d.	n.d.	n.d.	n.d.	0.08	0.02	n.d.	0.0
Al$_2$O$_3$	23.3	22.4	22.9	22.3	22.4	23.2	23.0	22.8	22.9	23.1	21.8	22.0	22.8
Cr$_2$O$_3$	n.d.	n.d.	n.d.	0.30	0.19	n.d.	n.d.	n.d.	n.d.	0.21	n.d.	n.d.	0.4
FeO*	20.5	25.6	22.0	25.5	21.8	18.3	19.9	17.4	19.0	19.0	18.5	18.1	13.3
MnO	0.58	n.d.	n.d.	1.17	0.75	1.39	1.32	0.65	0.84	0.68	0.82	n.d.	0.7
MgO	12.0	7.96	11.4	7.27	10.9	13.3	11.9	14.1	12.5	13.9	13.1	13.2	17.5
CaO	5.22	5.07	4.30	6.17	5.43	3.67	3.90	4.32	4.43	3.98	5.66	5.63	4.2
Total	101.2	99.8	100.0	100.9	101.2	99.8	99.6	99.5	99.8	101.2	99.9	98.5	99.9

Structural formulae O = 12

Si	2.947	2.987	2.966	2.930	2.968	2.975	2.981	2.984	2.993	2.959	2.988	2.982	2.97
Al^{IV}	0.053	0.013	0.034	0.070	0.032	0.025	0.019	0.016	0.007	0.04	0.012	0.018	0.03
Al^{VI}	1.983	2.019	2.002	1.952	1.947	2.013	2.016	1.985	2.011	1.962	1.915	1.946	1.92
Ti	—	—	—	0.010	0.005	—	—	—	—	0.004	0.001	—	0.0
Cr	—	—	—	0.018	0.011	—	—	—	—	0.012	—	—	0.0
Fe	1.273	1.648	1.384	1.640	1.366	1.139	1.252	1.081	1.185	1.169	1.160	1.146	0.81
Mn	0.036	—	—	0.076	0.048	0.088	0.084	0.041	0.053	0.042	0.052	—	0.0
Mg	1.325	0.913	1.283	0.833	1.218	1.474	1.335	1.564	1.396	1.52	1.465	1.486	1.89
Ca	0.416	0.418	0.347	0.510	0.436	0.293	0.314	0.344	0.354	0.314	0.454	0.456	0.3
Total	8.035	7.997	8.016	8.039	8.031	8.006	8.002	8.015	7.998	8.02	8.048	8.036	8.0

*All Fe as FeO; analytically determined FeO contents in mineral separates are: U19: 11.99 wt%, U47: 8.17 wt%, A44: 13.15 wt%.

nificant zoning is of a retrograde nature and is confined to narrow zones at the grain rims. Retrograde zoning also appears to be the norm in the orthopyroxene eclogites enclosed within the peridotite bodies (see Medaris 1980).

'Primary' amphiboles and micas

The primary or secondary status of the amphiboles and micas in these eclogites is often enigmatic from textural evidence alone. Some grains of these minerals are in apparent textural equilibrium with the primary anhydrous eclogite facies phases (garnets and pyroxenes) or occur as inclusions within these minerals. Other grains (even in the same rock) show obvious replacement relationships. While the latter are undoubtedly secondary, in the sense that they post-date and replace the primary eclogite facies mineral assemblages the former would appear to be primary. This is especially obvious in the case of hornblende inclusions in the cores of garnets, which apparently are relicts of pre-eclogite amphibolite assemblages (Krogh 1982). However, Lappin and Smith (1978) considered them to have developed later than the primary anhydrous eclogite facies assemblage and referred to such hydrous phases as 'early' rather than primary.

Like Lappin and Smith (1978) we have found it impossible to distinguish on chemical grounds between the 'primary' (or 'early') amphiboles in these rocks and the undoubted secondary replacement amphiboles (Table 6), as both show wide compositional ranges, especially in their Al_2O_3 and Na_2O contents. It seems reasonable to expect that different generations of amphibole may have similar compositions if they formed under similar P–T conditions on either the prograde or retrograde metamorphic paths (Krogh 1982). However, amphiboles replacing garnet clearly have higher Al_2O_3 contents than those replacing pyroxenes.

The pale brown micas in these rocks all appear to be Ti-rich phlogopites, regardless of their textural relationships. Primary phengitic micas were not found in these particular rocks and appear to be restricted to more aluminous eclogites (Krogh and Råheim 1978; Lappin and Smith 1978; Krogh 1980a).

Secondary pyroxenes (Table 7)

Breakdown of the primary garnet–clinopyroxene assemblage in the Hjørungavåg orthopyroxene eclogite, according to reactions (a) and (b) listed in Table I, results in the development of a secondary labradoritic plagioclase (An_{50-63}). The secondary orthopyroxene has a higher Fe/(Fe+Mg) ratio and an Al_2O_3 content similar to the minimum value recorded in the primary orthopyroxene grains. Breakdown of the garnet + quartz assemblage according to reaction (d) in both the Hjørungavåg

primary garnets

	U 19–2 Min. sep.	U 95 No zoning	U 47–1 No zoning	U 47–2 Min. sep.	KR 5–1 Cores	KR 5–2 Rims ADJ opx	A 44–1 Cores	A 44–2 Rims	A 44–3 Min. sep.	A 45–1 Cores	A 45–2 Rims	V80/19 1 Cores	V80/19 2 Rims
SiO_2	41.2	42.2	41.9	41.6	40.3	40.9	40.8	40.4	40.9	41.7	41.5	40.3	40.4
TiO_2	0.10	n.d.	n.d.	0.14	n.d.	n.d.	n.d.	0.02	0.16	0.02	0.02	n.d.	n.d.
Al_2O_3	23.3	22.7	23.3	23.2	22.5	22.5	22.5	22.5	22.4	23.1	23.1	22.9	22.7
Cr_2O_3	0.28	1.85	n.d.	0.37	0.25	0.27	0.23	0.38	0.24	0.09	0.12	n.d.	n.d.
FeO^*	12.0	7.73	9.75	10.0	18.3	16.5	16.9	16.3	15.1	13.9	14.3	17.8	18.0
MnO	0.50	0.35	n.d.	0.41	0.61	0.57	0.41	0.38	0.37	0.28	0.30	0.37	0.55
MgO	17.5	21.2	20.4	20.1	14.7	15.6	15.0	15.4	16.0	17.1	16.9	14.4	14.2
CaO	4.02	4.37	3.85	3.79	4.28	4.23	4.19	4.23	4.46	4.15	4.37	4.66	4.73
Total	100.1	100.4	99.5	99.5	100.9	100.5	100.1	99.7	99.6	100.4	100.6	100.4	100.6

Structural formulae O = 12

Si	2.985	2.983	2.989	2.978	2.969	2.993	3.002	2.982	3.001	3.012	3.000	2.972	2.980
Al^{IV}	0.015	0.017	0.011	0.022	0.031	0.007	0.000	0.018	0.000	0.000	0.000	0.028	0.020
Al^{VI}	1.972	1.876	1.951	1.935	1.922	1.936	1.955	1.940	1.941	1.964	1.964	1.962	1.953
Ti	0.005	—	—	0.008	—	—	—	0.001	0.009	0.001	0.00	—	—
Cr	0.016	0.103	—	0.021	0.014	0.016	0.013	0.022	0.014	0.005	0.007	—	—
Fe	0.791	0.457	0.582	0.601	1.127	1.010	1.042	1.008	0.925	0.840	0.862	1.098	1.110
Mn	0.031	0.021	—	0.024	0.038	0.035	0.026	0.024	0.023	0.017	0.018	0.023	0.034
Mg	1.882	2.232	2.202	2.146	1.609	1.699	1.646	1.698	1.748	1.842	1.823	1.583	1.561
Ca	0.312	0.331	0.294	0.291	0.338	0.332	0.330	0.334	0.351	0.231	0.338	0.368	0.374
Total	8.008	8.020	8.030	8.026	8.048	8.028	8.014	8.027	8.013	8.002	8.013	8.033	8.033

*All Fe as FeO; analytically determined FeO contents in mineral separates are: U19: 11.99 wt%, U47: 8.17 wt%, A44: 13.15 wt%.

Table 6　Partial analyses of amphiboles and micas

	Amphiboles																Micas		
	KR 1-1	KR 1-2	KR 2A	KR 2B-1	KR 2B-2	KR 3	KR 4	KR 6	U 206	U 243	U 19-1	U 19-2	U 95-1	U 95-2	A 44	V80/19	U 206	U 19	A 45
	Prim.	Sec.	Inc. in gnt	Inc. in gnt	Prim.	Prim.	Inc. in gnt	Prim.	Sec.	Inc. in gnt	Inc. in gnt	Sec.	Sec. ADJ pyrox.	Sec. ADJ gnt	Sec.	Prim.	Prim.?	Prim.	Prim.
SiO_2	48.5	53.7	40.8	49.8	56.5	53.4	51.2	55.5	48.2	47.6	54.0	52.5	47.6	43.4	55.0	51.8	38.4	40.2	41.4
TiO_2	n.d.	n.d.	n.d.	0.24	0.07	n.d.	n.d.	0.03	0.42	n.d.	0.09	0.11	n.d.	n.d.	0.09	0.14	1.24	1.42	1.11
Al_2O_3	9.70	6.41	18.2	8.12	1.51	4.43	8.79	2.97	10.2	10.1	3.53	5.96	12.8	18.6	3.58	5.67	17.1	14.0	14.6
Cr_2O_3	n.d.	n.d.	n.d.	n.d.	n.d.	n.d.	n.d.	0.08	n.d.	n.d.	0.17	0.22	2.02	1.52	0.08	n.d.	n.d.	0.30	0.08
FeO^*	8.58	11.4	12.8	7.89	5.70	7.86	5.26	4.88	6.14	5.46	4.11	4.49	1.73	2.07	3.96	5.42	8.16	4.69	3.84
MnO	0.09	0.10	n.d.	0.17	0.09	0.18	0.08	0.12	0.02	n.d.	0.10	0.13	0.09	0.12	0.02	0.05	n.d.	n.d.	0.02
NiO	n.d.	n.d.	n.d.	n.d.	n.d.	n.d.	n.d.	n.d.	n.d.	n.d.	n.d.	n.d.	n.d.	n.d.	0.23	n.d.	n.d.	n.d.	n.d.
MgO	17.7	13.2	12.1	18.0	21.6	19.9	19.5	21.6	18.1	18.2	22.7	21.4	18.6	17.2	21.14	20.7	19.5	24.0	22.4
CaO	11.9	12.2	10.5	11.8	11.4	11.7	9.72	11.4	11.7	11.5	11.0	11.1	11.6	10.8	10.92	11.7	0.03	0.04	n.d.
Na_2O	0.84	0.46	3.13	1.85	0.72	1.14	2.22	1.16	1.52	1.85	1.03	1.44	2.65	3.04	1.19	1.19	1.59	0.32	0.24
K_2O	n.d.	n.d.	n.d.	0.17	0.13	n.d.	n.d.	0.18	n.d.	n.d.	0.58	0.44	n.d.	n.d.	n.d.	0.12	9.28	9.32	9.43
Total	97.4	97.4	97.6	98.0	97.7	98.6	96.8	97.9	96.3	94.7	97.3	97.8	99.2	96.8	96.2	96.8	95.4	94.3	93.1

*All Fe as FeO.

Table 7   Analyses of phases in reaction coronas

Sample No. Phase	KR 3 cpx II	KR 3 opx II	KR 3 cpx II	KR 3 amph IIA	KR 3 amph IIB	KR 3 amph IIA	KR 3 amph IIB	KR 3 opx II	KR 4 opx II
Reaction type (see text)	(a)	(a)	(b)	(b)	(b)	(c)	(c)	(d)	(d)
SiO_2	55.1	54.9	52.7	51.0	44.7	53.2	45.5	55.3	55.6
TiO_2	n.d.	n.d.	n.d.	n.d.	n.d.	n.d.	n.d.	n.d.	n.d.
Al_2O_3	2.47	0.79	3.89	7.58	13.2	6.20	14.4	0.69	0.75
FeO	7.37	16.8	7.40	7.75	9.03	6.52	8.30	18.4	14.6
MnO	0.16	0.29	0.14	0.17	0.17	0.14	0.15	0.44	0.20
MgO	14.6	26.5	16.2	18.7	17.0	20.0	16.7	25.3	28.3
CaO	19.2	0.36	18.4	11.0	11.3	10.4	10.6	0.10	0.20
Na_2O	0.94	n.d.	0.85	1.34	2.25	1.27	2.52	n.d.	0.20
Total	99.8	99.7	99.6	97.5	97.6	97.7	98.2	100.2	99.6

and Kvalvåg samples gives secondary orthopyroxenes of similar composition coexisting with oligoclase (An_{23-29}).

Pressure–Temperature Estimates

A number of experimentally calibrated mineralogical geothermometers/barometers based on element partition relationships are now available; together they allow calculation of absolute P–T estimates for the equilibration of garnet + orthopyroxene + clinopyroxene assemblages. Extensive discussions of these methods exist in the literature (Carswell 1980; Carswell and Gibb 1980a, b; Carswell and Griffin 1981) mainly in connection with their application to garnet lherzolite assemblages. From these considerations it appears that either the Wells (1977) 'best fit' calibration of the two-pyroxene solvus geothermometer or the Ellis and Green (1979) calibration of the $K_{D Fe-Mg}^{Gnt-Cpx}$ geothermometer, used in conjunction with the garnet–orthopyroxene geobarometer of Wood (1974), can be expected to yield the most reliable P–T estimates for orthopyroxene eclogite assemblages. There are nevertheless problems with the application of both geothermometers to these particular rocks. The two-pyroxene solvus geothermometer is relatively insensitive below about 1000°C and use of the $K_{D Fe-Mg}^{Gnt-Cpx}$ geothermometer is hampered by lack of knowledge of the $Fe^{2+}/(Fe^{2+} + Fe^{3+})$ ratios in the minerals in the case of microprobe analyses. However, the two independent methods do at least provide a cross check on the temperature estimates.

Our calculated P–T estimates for the various orthopyroxene eclogite samples analysed by us or reported in the literature are given in Table 8. In the case of the Ellis and Green/Wood combination alternative P–T values are given, depending on whether (a) all Fe is taken as Fe^{2+} in both garnets and clinopyroxene, (b) Fe^{3+} calculated on a charge balance basis (Neumann 1976) is excluded from the K_D calculation, or (c) K_D calculation is based on Fe^{2+} values determined in mineral separates.

For the majority of samples there is reasonably good agreement between the P–T values given by the Wells/Wood and Ellis and Green/Wood combinations, especially when all Fe is taken as Fe^{2+} in the latter case (Table 8). However, in certain instances the Ellis and Green/Wood combination yields much higher P–T estimates. For these samples the correspondence with the Wells/Wood values is considerably improved if wet-chemical Fe^{2+} values are used in the K_D calculation. On the other hand the Ellis and Green/Wood values which exclude calculated Fe^{3+} contents from the K_D calculation are rather erratic. Calculation of the $Fe^{2+}/(Fe^{2+}+Fe^{3+})$ ratio in these magnesian garnets and clinopyroxenes is apparently too sensitive to analytical errors (notably in SiO_2) to yield consistently reliable results. The Wells/Wood values for mineral separates tend to be higher than those based on microprobe analyses of the same sample. Even very small amounts of contamination in pyroxene mineral separates (or the existence of exsolution lamellae) will result in erroneously high temperature estimates from the two-pyroxene solvus geothermometer.

The profound influence that intrasample variations in orthopyroxene composition have on the P–T estimates is also illustrated in Table 8. In the case of the Wells/Wood values only the P value is affected as this has been obtained after calculation of the T by the Wells (1977) geothermometer. However, with the Ellis and Green/Wood combination both the T and P values are affected, and decrease as the Al_2O_3 contents of the orthopyroxene increase, since these values were obtained by simultaneous solution of the two P–T dependent equations. In several of

Table 8 P–T estimates for Norwegian orthopyroxene eclogites

Locality	Sample no.	Features	Data ref.	T°C Wells	P Kbars Wood	all Fe as Fe^{2+} T°C Ellis & Green*	all Fe as Fe^{2+} P Kbars Wood	calc. Fe^{3+} excl. T°C Ellis & Green†	calc. Fe^{3+} excl. P Kbars Wood	analyz. Fe^{3+} excl. T°C Ellis & Green‡	analyz. Fe^{3+} excl. P Kbars Wood
					'External' opx eclogites						
Hornindal	KR 1	Unzoned minerals	1	770	21.0	740	19.1	545	8.0		
Hellesylt	KR 6	Max. Al opx against gnt	1	674	19.3	752	24.1	499	8.5		
		Min. Al opx in cores	1	674	26.1	781	33.6	515	15.2		
Kvalneset	KR 2A	Grain cores	1	739	31.5	600	21.8	567	19.3		
		Grain rims		726	23.8	653	19.2	570	14.2		
	KR 2B	Grain rims Max. Al opx	1	600	11.5	659	15.0	501	5.7		
		Grain cores Min. Al opx	1	529	13.7	649	21.8	511	12.4		
Hjørungavåg	KR 3	Min. Al opx Gnt & cpx cores	1	833	27.2	910	31.8	767	23.1		
		Max. Al opx Gnt & cpx rims	1	853	13.9	782	10.6	652	4.5		
		Aver. Al opx Gnt & cpx cores	1	834	22.4	888	25.3	750	17.6		
Kvalvåg	KR 4	Min. Al opx Gnt & cpx cores	1	566	10.2	910	30.2	657	15.8		
		Max. Al opx Gnt & cpx rims	1	711	8.4	786	11.9	652	5.6		
		Aver. Al opx Gnt & cpx cores	1	568	5.5	881	22.0	638	9.4		
Solholm	U 206	Unzoned minerals	1	713	17.7	759	20.3	564	9.1		
	U 243	Large opx Opx inc. in gnt	1	730	21.0	779	23.9	548	10.2		
				730	16.4	761	18.0	538	5.9		
Skarshaugen	U 19	Min. Al opx	1	694	15.2	955	29.6	954	29.3	755	18.5
		Max. Al opx		696	10.6	928	22.2	482	0.2	736	12.5
		Aver. Al opx		695	13.2	943	26.4	488	2.5	747	15.8
		Min. separates		814	3.9	847	5.1	—		766	2.0
Grytingvåg	KR 5	Opx rims adj. gnt Gnt & cpx rims	1	808	26.4	780	24.7	689	19.1	674	18.1
		Grain cores		776	28.6	743	26.5	777	28.4	688	22.5
	A 44	Grain cores	1	799	35.3	759	32.6	775	33.6	707	28.6
		Opx rims adj. gnt Gnt & cpx rims		803	27.9	777	26.3	810	28.2	684	20.3
		Min. separates		882	30.5	717	20.6	—		678	18.1
	2539	Min. separates	2	735	10.6	737	20.7	—		652	15.6
	8G	Min. sep. analy. except Ca & Al in opx	3	880	39.0	807	34.1	759	30.4	714	27.6
Liseter	D 171	Aver. probe Anal.	3	774	27.2	835	31.0	721	24.0		
Sørpollen	C 411	Aver. probe Anal.	3	722	33.2	820	40.3	713	32.5		
Sandviknes	66/87/ 11	Aver. probe Anal.	3	685	22.2	961	40.2	612	17.4		
Ganges–Kardneset	127 M	Aver. probe Anal.	3	660	24.4	684	26.1	360	3.9		
Høydalsneset	V80/19	High Al opx Gnt rims	1	690	14.5	710	15.5	423	−0.4		
		Low Al opx Gnt cores	1	689	21.5	743	24.9	743	24.9		
					'Internal' opx eclogites						
Ugelvik	U 95	Min. Al opx	1	757	28.8	783	30.5	707	25.4		
		Max. Al opx		761	14.6	729	13.0	661	9.5		
		Aver. Al opx		760	20.0	749	19.4	678	15.3		
		Min. separates	4	904	35.8	735	25.2	—		759	26.6
	U 47	Opx & cpx clasts		605	31.4	966	60.1	—		896	54.5
		Opx & cpx recryst.	1	533	17.8	890	42.6	—		856	40.1
		Min. separates		881	32.8	936	36.2	—		803	28.1

Table 8 (*continued*)

Locality	Sample no.	Features	Data ref.	TC Wells	P Kbars Wood	all Fe as Fe^{2+} T°C Ellis & Green*	all Fe as Fe^{2+} P Kbars Wood	calc. Fe^{3+} excl. T°C Ellis & Green†	calc. Fe^{3+} excl. K Bars Wood	analyz. Fe^{3+} excl. T°C Ellis & Green‡	analyz. Fe^{3+} excl. P Kbars Wood
Sunndal	6912 { Aver. probe anal.		5	<u>722</u>	<u>23.9</u>	<u>714</u>	<u>23.4</u>	—		<u>640</u>	<u>18.5</u>
Grubse	6095 { Aver. probe anal.			<u>752</u>	<u>21.1</u>	<u>731</u>	<u>19.8</u>	—		<u>657</u>	<u>15.2</u>
Lien	1 J	Grain cores	6	623	20.2	801	31.8	—		797	31.3
		Grain rims		<u>608</u>	<u>19.7</u>	<u>734</u>	<u>28.1</u>	—		<u>730</u>	<u>27.6</u>
	N71	Min. separates	7	686	16.0	675	15.4	—		647	14.7

* All Fe taken as Fe^{2+} in K_D calculation
† Calculated Fe^{3+} excluded from K_D calculation
‡ K_D calculation excludes Fe^{3+} indicated by analytical Fe^{2+} determination
Data references: 1—This paper
             2—Green (1969)
             3—Lappin and Smith (1978)
             4—Carswell (1973)
             5—Lappin (1974)
             6—Medaris 1980
             7—O'Hara and Mercy (1963)

our analysed samples orthopyroxenes have highest Al$_2$O$_3$ contents adjacent to garnets, and the lower P–T estimates yielded by such analyses (calculated in conjunction with garnet and clinopyroxene rim analyses) are probably the most appropriate to the garnet + orthopyroxene + clinopyroxene assemblages.

We have underlined in Table 8 those P–T values which we judge from the foregoing considerations to give the 'best' indication of the actual equilibration conditions for the primary eclogite facies assemblages in these rocks. For the orthopyroxene eclogites directly enclosed in the gneisses these values range from 674–853°C and 8.4–27.9 Kbars (mean values 743°C and 18.2 Kbars) with the Wells/Wood combination and from 653–786°C and 10.6–26.3 Kbars (mean values 737°C and 18.6 Kbars) with the Ellis and Green/Wood combination, assuming all Fe as Fe^{2+} in the K_D calculation.

Although the ranges for the different samples are considerable the mean values given by these different calculation procedures are remarkably close. This may be fortuitous, as taking all Fe as Fe^{2+} in K_D yields P–T estimates, using the Ellis and Green/Wood pair, which are effectively maxima for this particular method. We consider the somewhat lower values, ranging from 674–736°C and 12.5–20.3 Kbars (mean values 698°C and 17.0 Kbars) and which utilize K_D based on analytically determined Fe^{2+} values, to be more accurate. We suspect that the variation in P–T estimates between the different analysed orthopyroxene eclogite samples essentially reflects analytical errors, Fe^{2+}–Fe^{3+} uncertainties, or disequilibrium effects. These data cannot be used to demonstrate any regional P–T variation over the Sunnmøre–Romsdal–Nordmøre area even though

such a gradient is apparent in data from other types of eclogites (Krogh 1977a,b; Griffin *et al*. this volume).

Our 'best' estimates for the equilibration conditions of the 'external' orthopyroxene eclogites (Table 8) are substantially lower than the equivalent estimates of 700–850°C and 30–45 Kbars given by Lappin and Smith (1978) for their 'Regime B'. The differences arise partly from the fact that they used the obsolete Wood and Banno (1973) calibration of the garnet–orthopyroxene geobarometer instead of the improved Wood (1974) calibration, and partly from the lower analytical values for Al$_2$O$_3$ in the orthopyroxenes given by Lappin and Smith (values which in our experience do not represent equilibrium with garnet.).

We have also calculated T values for garnet–phlogopite in the three samples containing apparently primary mica, using the data of Ferry and Spear (1978). Assuming P = 17 Kbars, these temperatures are: U206, 855°C; U19, 789°C; A45, 690°C. The value for A45 is within the range of other estimates, while the other two are considerably higher. Errors may well result from failure to take account of the influence of the significant TiO$_2$ contents in the micas on K_{Fe-Mg}^{Gnt-Bi}.

Olivine-free 'internal' orthopyroxene eclogites, enclosed within the west Norwegian peridotite bodies, are rare and P–T estimates for these, based on currently available data (Table 8), present a somewhat more confused picture (especially as sample U47 yields consistently high values). We have again underlined those P–T values that we judge to be the 'best' indication of the eclogite facies equilibration conditions. These P–T estimates range from 608–761°C and 13.0–28.1 Kbars for the different methods, with overall mean 'best' values of 707°C and 20.4 Kbars.

When realistic allowances are made for errors in the various P–T estimates, we feel justified in concluding that the eclogite facies assemblages in both the 'internal' and 'external' orthopyroxene eclogites equilibrated in similar P–T regimes. Our preferred mean P–T estimates for the West Norwegian garnet lherzolite assemblages are likewise closely similar at 726°C and 20.3 Kbars, although again this disguises a substantial range of estimates (651–834°C and 11.3–31.8 Kbars) for individual samples. These estimates for the garnet lherzolite assemblages are slightly lower than those given recently by Carswell and Gibb (1980a). This is the result of incorporation of additional data, provided by the new experimental calibrations for $K_{DFe-Mg}^{Gnt-Cpx}$ and for K_{DFe-Mg}^{Gnt-Ol} (Ellis and Green 1979); O'Neill and Wood 1979, 1980), and the use of analytically determined $Fe^{2+}/(Fe^{2+}+Fe^{3+})$ ratios in the calculations.

Discussion

Prograde metamorphism of external eclogites

There is abundant evidence that many of the 'external' or Type B eclogites of the Western Gneiss Region have formed through prograde metamorphism of crustal rocks (see summaries by Bryhni *et al.* 1977; Griffin *et al.* this volume). The evidence includes (1) observation of partial to complete eclogitization in rocks of demonstrably low-P origin (sediments, anorthosites, gabbros, dolerites); (2) preservation of relict amphibolite facies mineral assemblages in the garnet grains of many eclogites; (3) marked prograde zoning in eclogite minerals; and (4) a regional gradient in the maximum T calculated from $K_{DFe-Mg}^{Gnt-Cpx}$, reflecting increasing grade of metamorphism toward the W and NW. Maximum P–T conditions are estimated at 750–800°C, 16–18 Kbars (Krogh 1977; Griffin *et al.* this volume; this paper).

In the high-grade part of this regional gradient, homogenization appears to have occurred in both garnets and clinopyroxenes. Prograde zoning is generally found only in coarse-grained rocks, while many samples show some retrograde zoning near grain contacts. Most of the external orthopyroxene eclogites in this study are from this high-grade area, and show only minor retrograde zoning. However, some evidence of prograde metamorphic evolution is found locally. As noted above (see Fig. 1), prograde zoning is obvious in the minerals of the Kvalneset orthopyroxene eclogite, suggesting an increase in T, and presumably of P, with time.

The 'Sandviksnes' orthopyroxene eclogite at Årheimneset on Stadlandet has been described by

Lappin and Smith (1978), who considered it to be a mantle derived fragment. A restudy of this strongly layered body shows layers of coarse-grained (1 cm) clinopyroxene + garnet + quartz eclogite, in which the garnets show a clear prograde zoning (Fig. 1B). The cores contain abundant euhedral hornblende inclusions; the rims contain only omphacite inclusions (aver. $Jd_{25}Acm_7Ts_1$). $K_{DFe-Mg}^{Gnt-Cpx}$ calculated for clinopyroxene inclusions and adjoining garnet declines from 7.5 to 6.1 across the outer rim. The lowest K_D implies a T of *ca.* 675° at an assumed P of 18 Kbar, and is thus compatible with the calculated T for the orthopyroxene eclogite in the same body (Table 7). The zoning, and the preservation of hornblende grains in the garnet cores (none is seen in the matrix) of this layer, strongly suggest that the orthopyroxene-bearing layers of this body also have had a crustal, prograde history.

Disequilibrium model

As noted above, most of the orthopyroxene eclogites, both internal and external, show a minor retrograde zoning (Fe/Mg of garnet increasing toward the rim, Fe/Mg of clinopyroxene decreasing), suggesting a partial adjustment to declining T. Many also show a marked zoning in Al_2O_3 content of the orthopyroxene, from low values in grain cores to high values near garnet. These two trends could be interpreted as reflecting *large* drops in P, together with modest drops in T (as, for example, in the 'internal' garnet pyroxenites studied by Medaris (1980)).

However, it is not necessary to postulate a variety of tectonic/metamorphic histories to reconcile these data with the (apparently conflicting) evidence for prograde metamorphism of other eclogites. There is abundant evidence that *disequilibrium* was common during the eclogite-forming event in western Norway; the existence of zoned minerals and relict inclusion assemblages is in itself such evidence. Partial reaction to eclogite (corona structures) is common in the basic rocks of the region (Griffin and Heier 1973). Several examples are known of gabbros in which apparently unmetamorphosed portions alternate with completely eclogitized zones, and in which the gabbro textures are pseudomorphed by garnet and omphacite (Mørk in prep.; Tørudbakken 1981; Cuthbert 1981).

We can suggest two ways in which disequilibrium during prograde metamorphism could produce the commonly observed zoning of orthopyroxene from low-Al cores to high-Al rims.

1. The orthopyroxene + clinopyroxene + garnet assemblage forms at relatively low T (and P),

where the orthopyroxene in equilibrium with garnet would have low Al. During a relatively rapid T increase, Al_{opx} equilibrates slowly compared to equilibration of Fe/Mg between clinopyroxene and garnet.

2. The orthopyroxene cores are relics from a lower-P, T assemblage, such as orthopyroxene + amphibole ± chlorite. Such rocks are seen in lower-grade parts of the area and as retrograde assemblages in the high-T part. With increasing metamorphic grade this assemblage would react to orthopyroxene + clinopyroxene + garnet, but the original low-Al orthopyroxenes might not equilibrate with the new assemblage, especially in coarse-grained rocks.

In either case, the use of the low-Al orthopyroxene core compositions to calculate P, after T has been calculated from the garnet–clinopyroxene pair, will lead to spurious P–T estimates (Fig. 2). In gen-

eral, the use of the core compositions from zoned grains to calculate P–T must be regarded with suspicion, unless other evidence can show which of these compositions originally were in mutual equilibrium. In our view, the use of such data has misled Lappin and Smith (1978) to propose (erroneously) a very high-P origin for the orthopyroxene eclogites.

Origin of orthopyroxene eclogites

Most orthopyroxene eclogites, unlike the Sandviknes and Kvalneset bodies discussed above, retain no structural or textural clues to the nature of the protolith. Many of the 'external' orthopyroxene eclogites may be high-pressure equivalents of the Mg(–Al–Si) rich rocks found in lower-P gneiss terranes as cordierite + anthophyllite assemblages. It is striking that *all* analysed external orthopyroxene eclogites, but only a few orthopyroxene-free ones, contain very high unsupported $^{87}Sr/^{86}Sr$ ratios (Brueckner 1977; Griffin and Brueckner 1980, in prep.), implying that they have at least undergone chemical exchange with the surrounding gneisses.

However, certain orthopyroxene eclogites contain pyroxenes apparently showing extensive exsolution, notably of garnet. Lappin (1973, 1974) and Lappin and Smith (1978) have 'reconstituted' some of these pyroxenes from modal and microprobe analyses. By assuming that these reconstituted phases coexisted with other pyroxenes, they have calculated very high temperatures (ca. 1000°C). By assuming coexistence of the reconstituted orthopyroxene with garnet at these temperatures, they have then calculated very high pressures (30–45 Kbars) which have been used as evidence for a mantle origin. Likewise Carswell (1973) argued for an upper mantle P–T regime for the crystallization of the primary aluminous orthopyroxene of the orthopyroxene eclogite (garnet pyroxenite) lens within the garnet peridotite body at Ugelvik, Otrøy. However, high primary crystallization pressures (35–40 Kbars) are only required on the assumption that some garnet coexisted with the original high temperature orthopyroxene, and the petrographic evidence for that is equivocal.

The exsolution textures described by Lappin and Smith (1978) in the Sandviknes orthopyroxene eclogite are difficult to reconcile with our described zoning and solid-inclusion evidence for a prograde metamorphism history in the associated layers of quartz eclogite. However, the petrographic features attributed by these authors to exsolution are somewhat enigmatic, and in the case of the amphibole lamellae may reflect parallel growth or replacement rather than exsolution. Evidence for garnet and pyroxene exsolution in the pyroxenites at Sunn-

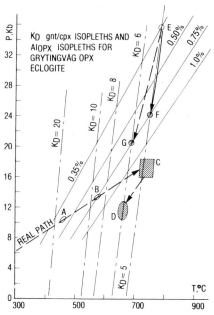

Fig. 2 K_D and Al_{opx} isopleths for Grytingvåg eclogite, calculated from relations of Ellis and Green (1979) and Wood (1974). Path A–B–C is a possible regional P–T development derived for the West Norwegian eclogites by Krogh (1977a). Assume that (1) low Al cores of orthopyroxenes formed in equilibrium with amphibole + chlorite ± clinopyroxene at T < 600°C, (2) this assemblage reacted to opx + cpx + gnt at higher T, and (3) equilibration of Al between cpx and gnt lagged behind equilibration of Fe/Mg between cpx and gnt. When rock is at C, opx has Al_2O_3 content appropriate to, for example, B. Uplift is accompanied by minor retrograde readjustment of Fe/Mg to D. Calculated P/T conditions will be: E for cores of opx + cpx + gnt; F for rim opx + core cpx + core gnt; G for rims of all phases. E–F and E–G illustrate the types of spurious P–T paths that might be deduced from particular combinations of mineral composition data

dal–Grubse (Lappin 1973, 1974) and Ugelvik (Carswell 1973) is more convincing.

Accepting that exsolution textures in certain orthopyroxene eclogites imply a high temperature origin for the protoliths of those particular rocks, the following considerations indicate that very high (mantle) pressures may not be necessary.

1. Many anorthosites contain high-Al orthopyroxenes as primary magmatic phases, even though these cannot have formed at mantle depths. Morse (1975) concluded that the high Al content is controlled by kinetic factors rather than being a pressure effect. In the layered anorthosites of the Bergen Arcs, the primary low-pressure mineral assemblages of olivine + plagioclase + clinopyroxene ± orthopyroxene have been metamorphosed to granulite facies, then to eclogite facies conditions (Griffin 1972; Austrheim 1981). Pods of coarsely granular 'primary' orthopyroxene and clinopyroxene show intragrain and interstitial exsolution of garnet. The least-exsolved (not 'reconstituted') orthopyroxene bears a strong compositional resemblance to Lappin and Smith's (1978) 'reconstituted' orthopyroxene from an external eclogite pod (Table 9).

The Fiskå anorthosite, north of Åheim in Sunnmøre, contains large (10–20 cm) primary orthopyroxene crystals bordered by clinopyroxene + garnet coronas and choked with garnet and clinopyroxene exsolution lamellae (Griffin and Mørk 1981). This pyroxene is again similar to those proposed by Lappin and Smith (1978) as protoliths for the orthopyroxene eclogites. Obviously extreme pressures are not necessary for the production of such pyroxenes, since the primary olivine + plagioclase assemblage of the Bergen anorthosites (Griffin 1972) limits crystallization of the anorthosites to P < 8–10 Kbars (Griffin and Heier 1973). Maquil (1978) has experimentally produced such pyroxenes in anorthosite compositions at 1000°C, 10 Kbars.

2. T. H. Green and Ringwood (1968) showed experimentally that subliquidus clinopyroxenes coexisting with garnet in tholeiitic melts contain up to 13.6 per cent Al_2O_3 at 18 Kbars, 1230–1300°C. Subsolidus clinopyroxenes coexisting with plagioclase, orthopyroxene, and amphibole at 920–1040°C and 9–10 Kbars contained 9.9–11.3 per cent Al_2O_3, while coexisting orthopyroxene contains 7.3–8.4 per cent Al_2O_3. Thompson (1974) has produced similar pyroxenes at liquidus to subliquidus T and intermediate P, from a range of basalts. Na and Al contents rise rapidly with falling T or rising P for each

Table 9   Exsolved pyroxenes in anorthosites—comparison with 'reconstituted' pyroxenes from opx eclogites

	Bergen–cpx POD 113 A				Bergen–opx POD 119 A			
	cpx	cpx by gnt	exs. gnt	gran. gnt	opx	exs. cpx	exs. gnt	gran. gnt
SiO_2	49.8	53.4	39.2	39.3	51.4	50.4	38.9	39.5
TiO_2	1.2	0.2	—	—	0.1	0.7	—	—
Al_2O_3	9.2	6.7	22.3	22.9	3.8	5.7	21.9	21.9
FeO^+	7.2	6.0	21.7	17.6	20.6	9.3	24.3	22.5
MnO	—	—	0.8	0.6	—	—	1.1	0.9
MgO	11.1	12.4	9.7	11.3	21.8	11.7	7.5	9.6
CaO	19.3	19.8	5.1	7.3	0.3	20.3	5.5	5.3
Na_2O	2.2	1.9	—	—	—	1.7	—	—
	100.0	100.4	98.8	99.0	98.0	99.8	99.2	99.7

	Fiskå–opx megacryst		'Reconstituted' eclogite pyroxenes		
	opx	exs. gnt	opx 68/88*	cpx 68/87*	6912†
SiO_2	53.1	39.2	51.83	55.20	52.3
TiO_2	0.16	0.0	0.02	0.07	0.0
Al_2O_3	3.00	22.5	3.58	4.50	4.2
FeO	16.3	21.0	17.45	10.17	7.80
MnO	0.18	0.97	0.17	0.06	0.1
MgO	27.0	10.7	25.38	14.93	16.3
CaO	0.18	5.86	0.72	12.04	17.0
Na_2O	0.0	0.0	0.07	3.44	1.18
	99.9	100.2	99.22	100.51	99.30

*Lappin and Smith (1978).
†Lappin (1974).

composition. Extrapolation of Thompson's data also suggests that at T = 1100–1200°C, P = 15–18 Kbars, many basalts will precipitate pyroxenes with high Na/Ca ratios and high Al contents.

3. Rost and Brenneis (1978) have described exsolution of garnet in the pyroxenes of spinel peridotites in the Ultenthal, and suggested that the associated garnet peridotites and garnet pyroxenites have evolved by crustal metamorphism of the Al-rich primary pyroxenes of the spinel lherzolites and pyroxenites.

We conclude that cumulates of various igneous pyroxenes, formed at moderate pressures, may well have been the protoliths for at least some orthopyroxene eclogites, both internal and external. These may stem from magmas intruded into the crust during the eclogite metamorphism. Alternatively, they may represent much older rocks, carried to depth and recrystallized during the eclogite forming event (*cf.* Griffin and Qvale this volume). Isotopic data on some external orthopyroxene eclogites (Griffin and Brueckner 1980) tend to support the latter interpretation. These older rocks may include both pyroxenites, related to the ultramafics, and the anorthosites, and metasupracrustal rocks; further analytical data are obviously necessary.

Internal vs. external eclogites

Most discussions of the garnetiferous ultramafic rocks and associated internal 'eclogites' (± orthopyroxene) in the Western Gneiss Region have concluded that these rocks represent tectonic slices of the subcontinental mantle. This conclusion has been based on (a) the assumption (based on early experimental work) that the garnet–peridotite mineral assemblage is only stable at very high P, (b) high calculated P–T values, (c) a decompression history, derived both from observation of pyroxene + spinel symplectites around garnet (O'Hara and Mercy 1963; Carswell 1968), and from compositional zoning in constituent minerals (Medaris 1980).

We have shown here that careful evaluation of both the analytical data and of published barometers and thermometers points to substantially lower pressures of formation for all the orthopyroxene eclogites than previously indicated. We have also shown how the zoning profiles can be reconciled with a prograde, rather than retrograde, metamorphic history (Fig. 2). Recent experimental evidence (Jenkins and Newton 1979) demonstrates that a garnet–peridotite-like assemblage is stable to pressures as low as 13 Kbars at 800°C. Our work shows that, when appropriate error brackets are attached to the various calculated values, there is no resolv-

able difference in the average P and T of equilibration among the garnet peridotites, internal orthopyroxene eclogites, external orthopyroxene eclogites, and external orthopyroxene-free eclogites in the coastal parts of the Western Gneiss Region. Finally, the widespread high-pressure granulite facies relics within the gneisses of the region (Mysen and Heier 1972; Krogh 1980a,b; Griffin and Carswell this volume; Griffin and Mørk 1981; Griffin *et al.* this volume) suggest that the gneisses have had a metamorphic history similar to that of the eclogites. Krogh (1980a,b) has demonstrated this point in detail for two localities at extremes of the regional P–T gradient.

All the available data are therefore consistent with a unified model for the genesis of all the observed eclogite facies assemblages during a major, regional scale, high pressure metamorphic event. This may have resulted from transient subduction (Bryhni *et al.* 1977; Krogh 1977a; Cuthbert *et al.* 1983) of a large continental lithospheric slab during a major continent–continental collision event. Recent geochronological data suggest that this event was part of the Caledonian orogeny (Griffin and Brueckner 1980, in prep.; Griffin and Carswell this volume).

There are two possible origins for the garnetiferous ultramafic rocks and 'internal' eclogites in this geotectonic scenario: (a) tectonic introduction of mantle slices into the descending slab; (b) progressive metamorphism of older 'alpine-type' peridotites or serpentinites already lodged in the crust during earlier orogenic episodes (Griffin and Qvale this volume). In either case the data and arguments presented here imply final 'equilibration' of all the eclogite facies types under similar conditions. Further evidence of the metamorphic and tectonic history of the ultramafic association will have to come from studies of their field associations and geochemistry, especially isotopic geochemistry.

Acknowledgements

This research has been supported by grants from N.E.R.C. (grant GR3/3085), N.A.V.F. (grant D48–22.08) and Nansenfondet. The manuscript was considerably improved after a perceptive review by Simon Harley.

References

Austrheim, H. 1981. Bergen Arc. In Griffen, W. L. and Mørk, M. B. E. (Eds), *Excursion B1, Excurs. in Scand. Caledonides*, 74–82.

Banno, S. 1970. Classification of eclogites in terms of physical conditions of their origin. *Phys. Earth Planet. Inter.*, **3**, 405–421.

Brueckner, H. K. 1977. A crustal origin for eclogites and a mantle origin for garnet peridotites: Strontium isotopic evidence from clinopyroxenes. *Contr. Miner. Petrol.*, **60**, 1–15.

Bryhni, I. 1966. Reconnaissance studies of gneisses, ultrabasites, eclogites and anorthosites in outer Nordfjord, western Norway. *Norges geol. Unders.*, **241**, 1–68.

Bryhni, I., Fyfe, W. S., Green, D. H. and Heier, K. S. 1970. On the occurrence of eclogite in western Norway. *Contr. Miner. Petrol.*, **26**, 12–19.

Bryhni, I. and Griffin, W. L. 1971. Zoning in eclogite garnets from Nordfjord, West Norway. *Contr. Miner. Petrol.*, **32**, 112–125.

Bryhni, I., Krogh, E. J. and Griffin, W. L. 1977. Crustal derivation of Norwegian eclogites: a review. *N. Jb. Miner. Abh.*, **130**, 46–68.

Carswell, D. A. 1968. Possible primary upper mantle peridotite in Norwegian basal gneiss. *Lithos*, **1**, 322–355.

Carswell, D. A. 1973. Garnet pyroxenite lens within Uglevik layered garnet peridotite. *Earth Planet. Sci. Lett.*, **20**, 347–352.

Carswell, D. A. 1974. Comparative equilibration temperatures and pressures of garnet lherzolites in Norwegian gneisses and in kimberlite. *Lithos*, **7**, 113–121.

Carswell, D. A. 1980. Mantle derived lherzolite nodules associated with kimberlite, carbonatite and basalt magmatism: a review. Lithos, **13**, 121–138.

Carswell, D. A. 1982. Clarification of the petrology and occurrence of garnet lherzolites, garnet websterites and eclogites in the vicinity of Rødhaugen, Almklovdalen, west Norway. *Norsk geol. Tidsskr.*, **62**, 249–260.

Carswell, D. A. and Gibb, F. G. F. 1980a. The equilibration conditions and petrogenesis of European crustal garnet lherzolites. *Lithos*, **13**, 19–29.

Carswell, D. A. and Gibb, F. G. F. 1980b. Geothermometry of garnet lherzolite nodules with special reference to those from the kimberlites of Northern Lesotho. *Contr. Miner. Petrol.*, **74**, 403–416.

Carswell, D. A. and Griffin, W. L. 1981. Calculation of equilibration conditions for garnet granulite and garnet websterite nodules in African kimberlite pipes. *Tscher. Min. Petr. Mitt.*, **28**, 229–244.

Cuthbert, S. J. 1981. Gørlangen–Sørdal area. In Griffin, W. L. and Mørk, M. B. E. (Eds), *Excursion B1, Excurs. in Scand. Caledonides*, 70–73.

Cuthbert, S. J., Harvey, M. A. and Carswell, D. A. 1983. A tectonic model for the metamorphic evolution of the Basal Gneiss Complex, Western South Norway—a working hypothesis. *J. Metamorphic Geol.*, **1**, 63–90.

Ellis, D. J. and Green, D. H. 1979. An experimental study of the effect of Ca upon garnet–clinopyroxene Fe–Mg exchange equilibria. *Contr. Miner. Petrol.*, **71**, 13–22.

Eskola, P. 1921. On the eclogites of Norway. *Skr. Vidensk. Selsk. Christiania, Mat.–Naturv. Kl.*, **I**, 8, 1–118.

Ferry, J. M. and Spear, I. S. 1978. Experimental calibration of the partitioning of Fe and Mg between biotite and garnet. *Contr. Miner. Petrol.*, **66**, 113–117.

Green, D. H. 1969. On the mineralogy of two Norwegian eclogites. *Contr. to Physico-chemical Petrology*, **1**, 37–44.

Green, T. H. and Ringwood, A. E. 1968. Genesis for the calc-alkaline igneous rock suite. *Contr. Miner. Petrol.*, **18**, 105–162.

Griffin, W. L. 1972. Formation of eclogites and the coronas in anorthosite, Bergen Arcs, Norway, *Mem. Geol. Soc. Am.*, **135**, 37–63.

Griffin, W. L., Austrheim, H., Brastad, K., Bryhni, I., Krill, A. G., Krogh, E. J. Mørk, M.|B. E., Qvale, H. and Tørudbakken, B. this volume. High-Pressure metamorphism in the Scandinavian Caledonides.

Griffin, W. L. and Brueckner, H. K. 1980. Caledonian Sm–Nd ages and a crustal origin for Norwegian eclogites. *Nature*, London, **285**, 319–321.

Griffin, W. L. and Carswell, D. A. this volume. *In situ* metamorphism of Norwegian eclogites: an example. this volume.

Griffin, W. L. and Heier, K. S. 1973. Petrological implications of some corona structures. *Lithos*, **6**, 315–335.

Griffin, W. L. and Mørk, M. B. E. 1981. Eclogites and basal gneisses in western Norway. *Excursion B1, Excurs. in Scand. Caledonides*.

Griffin, W. L. and Råheim, A. 1973. Convergent metamorphism of eclogites and dolerites, Kristiansund area, Norway. *Lithos*, **6**, 21–40.

Griffin, W. L. and Qvale, H. this volume. Superferric eclogites and the crustal origin of garnet peridotites.

Howie, R. A. and Smith, J. V. 1966. X-ray emission microanalysis of rock-forming minerals. V. Orthopyroxenes. *J. Geol.*, **74**, 443–462.

Jenkins, D. M. and Newton, R. C. 1979. Experimental determination of the spinel peridotite to garnet peridotite inversion at 900°C and 1000°C in the system $CaO–MgO–Al_2O_3–SiO_2$, and at 900°C with natural olivine and garnet. *Contr. Miner. Petrol.*, **68**, 407–419.

Krogh, E. J. 1977a. Evidence for a Precambrian continent–continent collision in western Norway. *Nature*, London, **267**, 17–19.

Krogh, E. J. 1977b. Crustal and *in situ* origin of Norwegian eclogites. Reply. *Nature*, London, **269**, 730.

Krogh, E. J. 1980a. Geochemistry and petrology of glaucophane-bearing eclogites and associated rocks from Sunnfjord, western Norway. *Lithos*, **13**, 355–380.

Krogh, E. J. 1980b. Compatible P–T conditions for eclogites and surrounding gneisses in the Kristiansund area, western Norway. *Contr. Miner. Petrol.*, **75**, 387–393.

Krogh, E. J. 1982. Metamorphic evolution of Norwegian country rock eclogites, as deduced from mineral inclusions and compositional zoning in garnets. *Lithos*, **15**, 305–321.

Krogh, E. J. and Råheim, A. 1978. Temperature and pressure dependence of Fe–Mg partitioning between garnet and phengite, with particular reference to eclogites. *Contr. Miner. Petrol.*, **66**, 75–80.

Lappin, M. A. 1966. The field relationships of basic and ultrabasic masses in the basal gneiss complex of Stadlandet and Almklovdalen, Nordfjord, southwestern Norway. *Norsk geol. Tidsskr.*, **46**, 439–495.

Lappin, M. A. 1973. An unusual clinopyroxene with complex lamellar intergrowths from an eclogite in the Sunndal–Grubse ultramafic mass, Almklovdalen, Nordfjord, Norway. *Miner. Mag.*, **39**, 313–320.

Lappin, M. A. 1974. Eclogites from the Sunndal–Grubse ultramafic mass, Almklovdalen, Norway, and the T–P history of the Almklovdalen mass. *J. Petrol.*, **15**, 567–601.

Lappin, M. A. and Smith, D. C. 1978. Mantle-equilibrated orthopyroxene eclogite pods from the basal gneisses in the Selje district, western Norway. *J. Petrol.*, **19**, 530–584.

Maquil, R. 1978. Preliminary investigation on giant

orthopyroxenes with plagioclase exsolution lamellae from the Egersund–Ogna anorthositic massif (S. Norway). In *Progress in Experimental Petrology. N.E.R.C. Series D, no. 11*, 144–146.

Medaris, L. G. Jr. 1980. Petrogenesis of the Lien peridotite and associated eclogites, Almklovdalen, western Norway. *Lithos*, **13**, 339–353.

Morse, S. A. 1975. Plagioclase lamellae in hypersthene, Tikkoatokhakh Bay, Labrador. *Earth Planet. Sci. Lett.*, **26**.

Mysen, B. O. and Heier, K. S. 1972. Petrogenesis of eclogites in high grade metamorphic gneisses, exemplified by the Hareidlandet eclogite, western Norway. *Contr. Miner. Petrol.*, **36**, 73–94.

Mørk, M. B. E. 1982. A gabbro–eclogite transition on Flemsøy, Sunnmøre, Western Norway. *Proc. 1st Int. Eclogite Conf.* (Abst.).

Neumann, E.-R. 1976. Two refinements for the calculations of structural formulae for pyroxenes and amphiboles. *Norsk geol. Tidsskr.*, **56**, 1–6.

O'Hara, M. J. 1967. Garnetiferous ultrabasic rocks of orogenic zones. In *Ultramafic and Related Rocks* (Ed. P. J. Wyllie), Wiley, New York, 167–172.

O'Hara, M. J. and Mercy, E. L. P. 1963. Petrology and petrogenesis of some garnetiferous peridotites. *Trans. Roy. Soc. Edinburgh*, **65**, 251–314.

O'Hara, M. J., Richardson, S. W. and Wilson, G. 1971. Garnet peridotite stability and occurrence in crust and mantle. *Contr. Miner. Petrol.*, **32**, 48–68.

O'Neill, H. St. C. and Wood, B. J. 1979. An experimental study of Fe–Mg partitioning between garnet and olivine and its calibration as a geothermometer. *Contr. Miner. Petrol.*, **70**, 59–70.

O'Neill, H. St. C. and Wood, B. J. 1980. Corrections. *Contr. Miner. Petrol.*, **72**, 337.

Råheim, A. 1972. Petrology of high grade metamorphic rocks of the Kristiansund area. *Norges geol. Unders.*, **270**, 1–75.

Råheim, A. and Green, D. H. 1974. Experimental determination of the temperature and pressure dependence of the Fe–Mg partition coefficient for coexisting garnet and clinopyroxene. *Contr. Miner. Petrol.*, **48**, 179–203.

Rost, F. and Brenneis, P. 1978. Die Ultramafite in Bergzug südlich des Ultenthales, Provinz Alto Adige (Oberitalien). *Tscherm. Min. Petr. Mitt.*, **25**, 257–286.

Thompson, R. N. 1974. Some high-pressure pyroxenes. *Mineral Mag.*, **39**, 768–787.

Tørudbakken, B. 1981. Vindøladalen dolerite. In *Excursion B1. Excurs. in Scand. Caledonides*.

Wells, P. R. A. 1977. Pyroxene thermometry in simple and complex systems. *Contr. Miner. Petrol.*, **62**, 129–139.

Wood, B. J. 1974. The solubility of alumina in orthopyroxene coexisting with garnet. *Contr. Miner. Petrol.*, **46**, 1–15.

Wood, B. J. and Banno, S. 1973. Garnet–orthopyroxene and orthopyroxene–clinopyroxene relationships in simple and complex systems. *Contr. Miner. Petrol.*, **42**, 109–124.

The Caledonide Orogen—Scandinavia and Related Areas
Edited by D. G. Gee and B. A. Sturt
© 1985 John Wiley & Sons Ltd

The intrusive history and tectonometamorphic evolution of the Basal Gneiss Complex in the Moldefjord area, west Norway

D. A. Carswell and **M. A. Harvey**

Department of Geology, University of Sheffield, Mappin Street, Sheffield S1 3JD, U.K.

ABSTRACT

Field, petrographic and geochemical data are presented which indicate that the voluminous quartz monzonitic augen gneisses in this area, together with subordinate granitic gneisses, are orthogneisses originally emplaced in the mid-Proterozoic (Rb–Sr isochron age of 1506 ± 22 Ma) but extensively deformed and recrystallized during the Caledonian orogeny. Least deformed samples retain porphyritic igneous textures with coronitic development of garnet and scarce relict clinopyroxene, mineralogical features considered to reflect the influence of an early Caledonian high pressure metamorphism more obviously evidenced by the associated metabasic eclogite lenses. These lenses are interpreted as tectonically disrupted intrusive sheets, mostly with olivine tholeiite basaltic chemistry. Original intrusive back-veining relationships have occasionally been preserved between these metabasic intrusives and the enclosing gneisses. A few of the larger metabasic bodies also retain igneous textures with impressive coronitic development of the eclogite facies mineral phases. However, both the eclogites and the augen orthogneisses show extensive deformation which induced retrogression to late Caledonian amphibolite facies mineral assemblages.

A more variable sequence of quartzo-feldspathic gneisses, pelites, calc silicate gneisses, marbles and metabasites (the Paragneiss Complex) has not been dated isotopically but is considered most likely to represent an older Svecofennian rock sequence intruded by the mid-Proterozoic Orthogneiss Complex. Metabasic rocks in the Paragneiss Complex only rarely retain eclogite facies mineralogies and are mostly garnet granulites (as on Tverrfjella) or amphibolites (as on Bolsøy and other islands in Moldefjord). The lower metamorphic grade of the rocks along Molde-fjord is associated with the imposition of a near pervasive protomylonitic fabric and is considered to reflect the more intensive late Caledonian deformation and recrys-tallization of these rocks. However, the occasional preservation of kyanite in pelites and of eclogite facies mineralogies in metabasic rocks indicate that the rocks of the Paragneiss Complex have also witnessed the earlier high pressure metamorphism.

Introduction and Previous Studies

This paper summarizes the results of a study of the rocks in the Moldefjord area of west Norway (62°52′N, 7°W) in the so-called Basal Gneiss Complex (Holtedahl 1944; Carswell 1973).

The rocks of this Complex (Fig. 1) have been the subject of several conflicting interpretations. The earliest workers (e.g. Reusch 1881) regarded the gneisses as Precambrian basement to the overlying Palaeozoic sediments. Later it was considered that part (e.g. Holtedahl 1944; Gjelsvik 1951; Kolderup 1960) or all (e.g. Hernes 1967) of the gneisses were derived by feldspathization of 'Caledonian' or late Precambrian/Cambro-Silurian sediments, respectively. The situation was further complicated by the suggestion that perhaps the basement was composed of two distinct elements. These were first defined by Bryhni (1966) as the homogeneous Jostedal Complex, of reworked Precambrian gneisses and the overlying heterogeneous Fjordane Complex. Strand (1969), Brueckner (1977a), and Skjerlie (1969) discerned similar units at Grotli, Tafjord, and Sunnfjord, respectively.

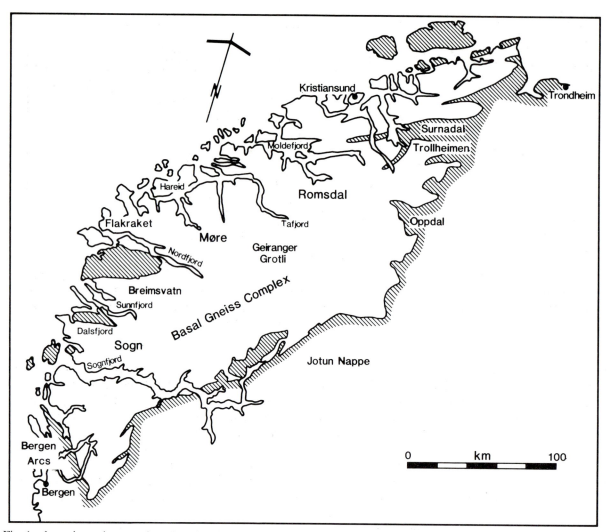

Fig. 1   Location of area of study within the Basal Gneiss Complex, of western Norway. Basal Gneiss Complex unshaded, higher lithological units shaded

In the northern part of the Complex the gneisses are directly overlain by a sequence of heterogeneous rocks of late Precambrian to Ordovician age, with intercalated portions of the basement, shown to be a series of allochthonous nappes with a supposed origin at least 500 km to the west (Gee 1978, 1980). These are separated from the basement Complex by a thin veneer of sediment considered to have acted as a décollement surface for the movement (Gee 1980). The nature of the nappe/basement contact here is crucial to the interpretation of the Moldefjord rocks.

The southern part of the Complex is bordered by further allochthonous material in the Jotun nappe and the Bergen arcs. The origin of the former, consisting of anorthosites and granulites, is equivocal; often thought to be a far-travelled thrust nappe (e.g. Hossack *et al.* 1981). Gravity studies have suggested a deep dense root, and a local derivation from the lower crust as an 'Ivrea-type flake' (Smithson *et al.*

1974; Banham *et al.* 1979). The Bergen Arcs, consisting of a series of arcuate belts of lower Palaeozoic metasediments and metavolcanics, and gneisses with migmatites and anorthosites, have been interpreted as a series of nappes thrust onto a Precambrian gneiss complex of probable autochthonous nature (Sturt *et al.* 1975; Sturt and Thon 1976). The anorthosite portion contains eclogites of postulated Caledonian or possibly Sveconorwegian age and may correlate with the Jotun Nappe (Austrheim and Råheim 1981).

Field studies at Moldefjord have so far been concentrated (Fig. 2) between Moldefjord and Langvatnet–Nåsvatnet, just north of Tverrfjella on the Molde peninsula and on the nearby islands of Otrøy, Midøy, and Dryna to the west and Bolsøy to the south.

Bugge (1934) described the discovery of greenschists, micaschists, quartz schists, and marble on Bolsøy, Saeterøy, and Hjertøy in Moldefjord. Later

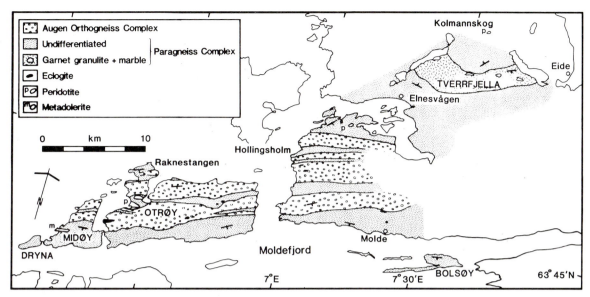

Fig. 2   Generalized geological map of the northern Moldefjord region, only the largest observed lenses of eclogite, peridotite, and metadolerite are indicated

Hernes (1954a, 1956) mapped a narrow synclinal outcrop of similar rocks along the south side of Fannefjord to the east of Molde. These rocks were thought to be contained in a discontinuous westward extension, along the regional strike of the Surnadal syncline which contains comparable rocks (Strand 1952); they were shown as such on the map of Hernes (1955). Furthermore, as Strand considered that the Surnadal rocks could be traced eastwards directly into the Cambro-Silurian rocks of the Trondheim region (as shown on the geological map of Norway—Holtedahl and Dons 1960) it was presumed that the greenschists and associated rocks in the Moldefjord area were of similar age. Gjelsvik (1953) has described rocks near Brattvåg to the southwest of Moldefjord; here micaschists associated with greenschists contain porphyroblasts of feldspar which increase in number over a distance of 50 m across the strike so that the schists pass into an augen gneiss and then into uniform grey gneisses. Thus Gjelsvik, like Hernes (1954a, 1956) apparently considered the contact between the 'basal gneisses' and the supracrustal metasedimentary and metavolcanic rocks to represent a Caledonian migmatite front—the basal gneisses therefore being interpreted as extensively migmatized Eocambrian–Lower Palaeozoic rocks.

Hernes (1955) published a map of the geology between Moldefjord and Kristiansund, including Surnadalen, and constructed a tectonostratigraphy based on a eugeosynclinal sequence of assumed late Precambrian age (Hernes 1967). From his studies near Kristiansund Råheim (1972) expanded upon Hernes's sequence but questioned the Caledonian

age for the formation of the gneisses. Subsequently, Pidgeon and Råheim (1972) demonstrated that at least some of the rocks (Tingvoll Group) had a Svecofennian age (1708 ± 60 Ma). The apparent lack of any structural or metamorphic break between these rocks and those of the Surnadal synform suggested that the latter (Røros Group) were of a similar age which Råheim (1977) showed to be so, although the isochron was not well constrained. In addition he demonstrated the existence of a metamorphic and structural break between these Svecofennian-aged rocks in the synform and those higher in the sequence, which he equated with the Støren nappe of Gale and Roberts (1974) and Råheim (1979).

This suggested that the Surnadal synform was not an integral part of the Palaeozoic rocks of the Trondheim Basin. However, Krill (1981) regarded this tectonic break as a zone separating an infrastructure of high-grade metamorphism from a superstructure of low-grade metamorphism as found at Oppdal (Krill 1980). He also correlated an allochthonous Caledonian-aged unit at Oppdal (Blåhø unit) with part of the Surnadal sequence, thereby contradicting the Svecofennian age given for them by Råheim (1977). Furthermore Gee (1980) has correlated the Blåhø unit with the Seve nappe thereby implying that the cover rocks of the Surnadal syncline were allochthonous and Caledonian in age. These contrasting interpretations of the rocks at Surnadal have cast serious doubt over the age and nature of the supracrustal rocks in Moldefjord.

According to the previous mapping and interpretation of Hernes (1956) and Råheim (1972) in the area between Molde and Kristiansund, the rocks of the actual Basal Gneiss Complex on the northern side of Moldefjord should correspond to Hernes' Frei Group. This lithologically heterogeneous group of rocks was reported to include migmatites, augen gneisses, amphibole and diopside bearing schists and gneisses, marbles and calc-silicate gneisses, garnet amphibolites, eclogites, metaperidotites, metapyroxenites, and metadolerites. On his map, Hernes (1955) indicated that a lithological boundary exists within the Frei Group rocks about 3 km north of Molde which can be followed for tens of kilometres northeast along strike. Our mapping has shown (Fig. 2) that this boundary can also be followed southwestwards across the islands of Otrøy and Midøy and separates contrasting lithological units which we refer to as the Heterogeneous Paragneiss Complex and the more homogeneous Augen Orthogneiss Complex. However, the rocks of the Paragneiss Complex are not restricted to the southern side of this boundary but also occur further north, notably at Raknestangen (the most northerly point at Otrøy) in the area north of Hollingsholm, and on Tverrfjella about 18 km north of Molde (see Fig. 2).

The Augen Orthogneiss Complex

The dominant lithology in this unit is an impressive coarse augen-textured gneiss. The conspicuous pink feldspar augen are usually somewhat flattened and are typically about 1 cm across, although they may be up to 5 cm long. However, dependent upon the degree of deformation, the size, shape, and apparent proportion of feldspar augen may be quite variable, as therefore is the overall appearance of the rock. Nevertheless, where least deformed, the rock has the appearance of a consistently porphyritic igneous rock. An igneous origin is also indicated by the relative uniformity in chemical composition of the augen gneiss (Table 1, column E). On a normative feldspar

Table 1   Averaged whole rock analyses

	A $\bar{x}$	A σ_n	B $\bar{x}$	B σ_n	C $\bar{x}$	C σ_n	D $\bar{x}$	D σ_n	E $\bar{x}$	E σ_n	F $\bar{x}$	F σ_n	G $\bar{x}$
SiO_2	46.11	1.98	50.06	1.33	48.59	1.31	46.77	1.20	66.39	2.93	77.54	0.78	66.03
TiO_2	1.76	1.23	0.29	0.04	1.23	0.27	1.78	0.59	0.76	0.23	0.09	0.07	0.71
Al_2O_3	14.87	1.54	8.50	0.78	14.72	0.63	16.43	0.79	14.84	0.74	12.10	0.27	15.88
Fe_2O_3	2.35	0.82	1.75	0.26	3.53	0.76	1.47	0.51	1.75	0.74	0.39	0.32	1.87
FeO	12.23	2.78	7.62	1.14	7.38	1.21	11.41	1.15	3.33	1.55	0.53	0.39	3.24
MnO	0.20	0.05	0.19	0.01	0.17	0.04	0.32	0.33	0.09	0.03	0.02	0.02	0.08
MgO	9.72	2.19	18.52	1.02	8.60	1.98	8.47	2.21	1.55	0.48	0.30	0.13	1.70
CaO	10.11	1.75	10.12	0.18	11.09	1.09	8.81	0.94	3.23	0.72	0.99	0.27	3.00
Na_2O	2.15	0.56	0.76	0.11	3.11	0.84	2.70	0.41	3.13	0.43	2.88	0.54	2.80
K_2O	0.06	0.06	0.42	0.25	0.37	0.28	0.81	0.33	4.12	0.75	4.85	0.90	4.21
P_2O_5	0.18	0.12	0.34	0.37	0.10	0.04	0.33	0.16	0.27	0.10	0.02	0.02	0.15
S	0.02	0.03	0.02	0.01	0.01	0.02	0.03	0.05	0.02	0.04	0.01	0.01	—
H_2O^+	0.35	0.12	0.89	0.60	0.74	0.20	0.65	0.47	0.58	0.19	0.35	0.23	0.90
TOTAL	100.11		99.48		99.64		99.98		100.06		100.07		100.57
100Na/(Na + Ca)	27.8		11.9		33.7		35.7		63.7		84.0		62.8
100Mg/(Mg + Fe)	57.4		78.3		64.8		56.1		42.1		45.6		44.7
ppm													
Ni	217	132	498	41	163	95	155	93	3	2	0	0	
V	341	315	139	3	303	44	220	48	45	17	5	5	
Cr	279	199	1960	55	412	178	101	37	12	3	11	4	
Zn	109	28	86	13	76	16	103	22	85	20	18	10	
Cu	67	39	29	11	67	52	70	14	18	10	7	3	
Rb	2	2	23	4	6	4	21	6	80	20	139	60	
Sr	206	123	102	38	105	40	368	49	231	96	82	74	
Y	23	8	8	6	29	6	30	10	38	12	21	17	
Zr	90	38	30	8	70	21	132	43	339	85	65	44	
Ba	77	27	149	83	84	30	403	185	1123	403	253	106	

A—Average of 11 essentially unmetasomatized olivine normative tholeiitic eclogites
B—Average of 3 somewhat metasomatized 'eucritic' orthopyroxene eclogites
C—Average of 11 garnet granulites from Tverrfjella
D—Average of 5 metadolerites
E—Average of 19 augen orthogneisses
F—Average of 9 granites in orthogneiss complex
G—Average mangerite syenite, Flatraket (Lappin et al. 1979)

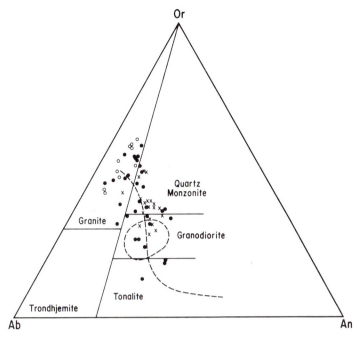

Fig. 3   Normative feldspar plot of whole rock compositions for coarse augen orthogneisses (crosses) and associated granitic sheets (open circles) and miscellaneous augen free gneisses (closed circles). The labelled classification fields outlined by solid lines are those of O'Conner (1965), the field encircled by a broken line the granodiorite field of Streckeisen (1976), and the curved dashed line corresponds to the chemical trend recognized by de Waard (1969) for anorthosite–norite–jotunite–mangerite–farsundite–opdalite–charnockite suite rocks

classification basis (Fig. 3) the augen gneiss compositions straddle the quartz monzonite and granodiorite fields of O'Conner (1965) and lie partly within the granodiorite field of Streckeisen (1976).

Where relatively undeformed it is apparent that the feldspar augen are of rapakivi type with cores of finely perthitic microcline (Or_{90-94} Ab_{5-10}) mantled by oligoclase (An_{20-30}). This plagioclase is commonly antiperthitic enclosing coarse patches of microcline and is often myrmekitic adjacent to the potash feldspar core. A further conspicuous feature of the augen gneiss is the presence of appreciable fine-grained garnet developed in a crude corona fashion between the feldspar augen and the clots of mafic minerals which include rare clinopyroxene relicts, more common dark green hornblende, and significant amounts of both magnetite and sphene. The fabric of these predominantly quartz monzonitic augen gneisses is invariably to some extent porphyroclastic with 'ribbons' of finely granulated quartz. With extreme deformation the feldspar augen clasts are drawn out into long 'pencils' and the content of garnet clearly diminishes whilst that of biotite increases as the rocks are transformed into finely banded, grey, hornblende–biotite gneisses.

Locally the augen-gneisses are sheeted by more nearly equigranular pink granites (Table 1, column F). Despite their low mafic content these granites sometimes contain scarce pyralspite garnet.

Isotopic studies have shown that the augen gneisses and associated granitic rocks give a Rb–Sr isochron age of 1506 ± 22 Ma (Fig. 4). In view of the low initial $^{87}Sr/^{86}Sr$ ratio of 0.7035 and the lack of disturbance of the LIL elements in general (including Rb) we consider this to represent essentially the magmatic crystallization age of these rocks rather than a metamorphic recrystallization age (cf. Krill and Griffin 1981). It is of interest that this age is essentially identical to the U–Pb magmatic crystallization age obtained by Lappin et al (1979) for the mangerite syenite intrusion within the Basal Gneiss Complex at Flakraket, Nordfjord and to the age for other rapakivi-like orthogneisses in Scandinavia (Point 1975; Röshoff 1978; Krill 1980). The fairly high M.S.W.D. of 14.78 is considered to be the result of subsequent high grade metamorphism during the Caledonian (see below).

We also note the close similarity in chemical composition between the Flakraket mangerite syenite and the Moldefjord quartz monzonitic augen gneisses (compare columns E and G, Table 1), and the fact that together with the associated granitic rocks and miscellaneous augen free gneisses the augen-textured orthogneisses define a chemical trend (Fig. 3) which closely corresponds to that recognized by de Waard (1969) for the anorthosite–norite– jotunite– mangerite– farsundite– opdalite– charnockite suite of rocks. Although anorthosites appear to be

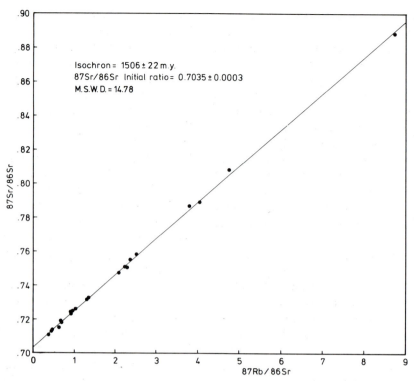

Fig. 4   Rb–Sr isochron diagram for augen textured gneisses and associated granitic rocks from the Augen Orthogneiss Complex on the north side of Moldefjord

absent in the Moldefjord area, they are not uncommon elsewhere in the Basal Gneiss Complex (Bryhni 1966; Brueckner 1977a). We suggest that the anorthosite–augen orthogneiss (mangerite)–granite suite of plutonic intrusives within the Basal Gneiss Complex may be related to the voluminous emplacement of such anorogenic plutonic igneous rocks which appears to have occurred between 1450–1700 Ma, in a broad belt across Scandinavia and North America (Emslie 1978, Bridgewater and Windley 1973).

Lenses of essentially bimineralic eclogite frequently occur throughout the Augen Orthogneiss Complex in this area. Only the outcrops of the largest lenses are shown on Fig. 2. When unaffected by later metasomatism associated with their transformation into biotite amphibolites, the majority of these lenses have broadly similar basaltic rock chemistries (Table 1, columns A and B). In normative terms the dominant eclogites are olivine tholeiites, whilst the scarcer orthopyroxene bearing eclogites (garnet pyroxenites) (Carswell *et al.* this volume) appear to have had a eucritic (i.e. hypersthene–olivine gabbro) parentage. There seems little doubt that these eclogite types represent metamorphosed basic igneous rocks which originally sheeted the intermediate–acid plutonic complex. Typically the eclogite facies metabasites occur as tectonically bounded 'strings' of lenses indicating extensive

boudinage of the original intrusive sheets or dykes. This tectonic disruption clearly post-dated the development of the eclogite facies assemblages as witnessed by disorientated internal layering of such assemblages in certain lenses. Particularly large eclogite lenses at Kolmannskog and Raknestangen illustrate internal layering from bimineralic eclogite through garnet websterite to iron-rich garnet–spinel lherzolite, similar to the layering in the Eiksunddal eclogite complex further south on Hareidlandet (Schmitt 1963). Some degree of marginal amphibolitization of all eclogite lenses against the surrounding gneisses is normally apparent.

In rare instances where, as at Litledigerneset on the west coast of Midøy, the contacts between the eclogite bodies and the gneisses have escaped tectonism, original intrusive backveining relationships have been preserved (Griffin and Carswell this volume). Field evidence demonstrates that the higher temperature basic sheets have caused partial remelting of the igneous precursors of the augen gneisses into which they were intruded, leading to the injection of acid backveins into the basic rocks. Quartz and alkali feldspars are the dominant mineral phases in these veins but the preservation of small amounts of symplectitic clinopyroxene, garnet, and kyanite strongly suggests that the emplacement of these veins predated the high pressure eclogite facies metamorphism. These early quartzo-feldspathic veins contrast with the much more conspicuous post-eclogite facies granitic pegmatites which lack high pressure mineral phases and produce amphibolite haloes where they cross-cut eclogite lenses.

The presence of almandine garnets with significant grossular contents (Table 3, columns H and J) in the intermediate–acid plutonic rocks of the Augen Orthogneiss Complex, is taken to indicate pressures of at least 9 Kbars at the likely 700–800°C metamorphic recrystallization temperatures (Green and Mysen 1972). Likewise the occasional preservation of symplectitic clinopyroxenes in the augen gneisses and the general observation that eclogites have developed in basaltic sheets intruded into the igneous precursors of the augen gneisses, requires that these intermediate–acid plutonic rocks were subjected to the high pressure metamorphic conditions responsible for the development of the eclogite facies assemblages in the metabasic rocks.

Utilizing the garnet–symplectitic clinopyroxene–plagioclase–quartz assemblage equilibria (Newton and Perkins 1981) and the Fe^{2+}/Mg^{2+} partitioning between garnet and symplectitic clinopyroxene (Ellis and Green 1979) pressure and temperature conditions for this assemblage both in augen gneisses and in retrograded eclogites (garnet granulites) were calculated as ~800°C and 9–10 Kbars. The original omphacitic clinopyroxene in these rocks must therefore have been formed at P > 9 Kbars.

The Heterogeneous Paragneiss Complex

This rock complex includes migmatitic gneisses (sometimes garnetiferous but invariably with hornblende and/or biotite), diopside ± scapolite-bearing calc-silicate gneisses, marbles, garnet–mica pelites (sometimes with conspicuous kyanite), and garnet and/or epidote bearing amphibolites. The undoubted metasedimentary parentage of many of these rocks and their dominant almandine–amphibolite facies mineralogy led one of us (Carswell 1973) to suggest previously that the rocks of this Complex, which outcrop on the southern sides of Otrøy and Midøy and on the smaller islands nearer Molde, represent a sequence of Caledonian supracrustal rocks deposited on an older Precambrian gneiss basement within which the high grade eclogite facies assemblages were thought to be restricted. However, this interpretation is now recognized to be at least partly incorrect. Firstly, recent Sm–Nd dating of garnet–clinopyroxene mineral pairs from several eclogite lenses within the Basal Gneiss Complex (including samples from the Moldefjord area) has indicated a Caledonian age of around 425 Ma for the eclogite facies metamorphism contrary to the earlier interpretations by

Pidgeon and Råheim (1972) and Krogh (1977) of a Svecofennian age (1700–1800 Ma) for this event. Secondly, although the majority of metabasic rocks within the Paragneiss Complex in the near vicinity of Moldefjord are amphibolites, it is now appreciated that eclogite assemblages are in fact occasionally preserved within such amphibolite lenses. Moreover, there are large numbers of eclogite lenses enclosed within migmatitic metasedimentary gneisses around Raknestangen on the extreme north of Otrøy and abundant partly retrograded eclogitic rocks associated with marbles on Tverrfjella (Hernes 1954b). In addition we now recognize that the Paragneiss Complex as mapped (Fig. 2) in places includes conspicuous volumes of possible orthogneisses which may be related to comparable rocks in the more homogeneous Augen Orthogneiss Complex.

So far the Paragneiss Complex around Molde has failed to give a suitably constrained whole-rock Rb–Sr isochron. However, the rocks dated by Pidgeon and Råheim (1972) as 1708 ± 60 Ma near Kristiansund were from the Frei Group, whilst Brueckner (1979) dated eclogite-bearing supracrustal rocks at Tafjord as 1775 ± 57 Ma. These dates for possibly equivalent rocks to those of the Paragneiss Complex suggest that they might be of similar age.

As previously mentioned the Paragneiss Complex on Bolsøy has been related to rocks in the Surnadal syncline (Strand 1952; Hernes 1955; Råheim 1972) but no agreement exists over the interpretation of the latter as either Svecofennian basement (Råheim 1979) or allochthonous Caledonian rocks (Krill 1981; Gee 1980).

However, we question the validity of Råheim's (1977) dating of the Røros Group at Surnadal as Svecofennian on three points. Firstly, no actual isochron was produced for this age. Secondly, the samples described were not of the same lithology and therefore were unlikely to possess the same $^{87}Sr/^{86}Sr$ initial ratio. Thirdly, the 1700 Ma age was suggested because some of the Røros Group rocks fell on the Tingvoll Group isochron, but again the lithologies were different and this superimposition may be purely fortuitous. Furthermore, the concordance of the Røros and underlying Tingvoll groups, taken to indicate a single lithological sequence (Råheim 1977) may well be an artifact of the intense flattening suffered by these rocks during the Caledonian orogeny (cf. Carswell 1973). On the other hand the interpretation of Krill and Gee for the Surnadal rocks seems better founded and is preferred on present evidence.

If the lithostratigraphic correlation of the paragneisses on Bolsøy and Tverrfjella eastwards with the rocks at Surnadal is accepted, this would suggest that the former may also be allochthonous Caledonian rocks. However, as such an interpretation is at odds with the Svecofennian whole rock Rb–Sr isochron age for the Frei Group rocks to the south of Kristiansund (Pidgeon and Råheim 1972), it appears that the key question of the age of the Paragneiss Complex rocks in the Moldefjord area cannot be satisfactorily resolved at present.

Due to this uncertainty over their age, two interpretations are possible for their relationship to the Orthogneiss Complex. The simplest explanation is that the igneous precursors to the augen gneisses intruded and partially migmatized a sequence of pre-existing sediments. Such a model explains the appearance of apparent orthogneisses within the Paragneiss Complex and vice versa, although these relationships have undoubtedly been complicated by later (Caledonian) crustal imbrication. The more complex alternative model requires one, or possibly both, of the complexes to be allochthonous. The augen gneisses could correlate with the Tännäs Augen Gneiss nappe in Sweden which contains Proterozoic rapakivi-like augen orthogneisses with anorthosites and anorthositic gabbros in the Oppdal area (Krill 1980), although the last two rock types are absent at Moldefjord, whilst the Paragneisses could possibly correlate with allochthonous Caledonian rocks of the overlying nappes (e.g. the Seve nappe). To achieve the close intercalation of Orthogneisses and Paragneiss Complexes large scale folding and/or faulting movements would clearly have been required.

It is feasible that the Paragneiss Complex in the Moldefjord area (Fig. 2) may in fact contain both Svecofennian (basement) and Caledonian (cover) lithostratigraphic units. However, the currently available data do not enable us to discriminate between two such rock units, if indeed both do exist in this area.

Metabasic eclogite lenses within the Paragneiss Complex on northern Otrøy have undergone deformation-induced retrogression which has converted the rocks into plagioclase-bearing garnet-granulites. The sporadic preservation of inclusions of omphacitic clinopyroxene within garnets indicates the original presence of eclogite facies assemblages in these granulites. Jadeite-poor clinopyroxenes now occur in equilibrium with oligoclase in the same rocks (Table 2, columns B–D). The conspicuous metabasaltic rocks associated with the marbles on Tverrfjella, described by Hernes (1954b) as eclogite–amphibolites, contain clinopyroxenes with low jadeite contents, intergrown with oligoclase as a symplectite (Table 2, column E) and thus are also garnet-granulites. Further retrogression of such garnet-granulites, ultimately to garnet-free epidote

amphibolites would have depended upon the accessibility of water in a declining P/T environment. Metabasic rocks on Bolsøy and adjacent islands in Moldefjord, previously described as greenschists (Bugge 1934; Gjelsvik 1953) are better termed epidote amphibolites. Our recent discovery of relict kyanite in garnetiferous pelites on Bolsøy suggests that these paragneisses may have suffered the same high metamorphic pressures as experienced further north. It would appear that the appearance of fine-grained, well-foliated epidote amphibolites on Bolsøy and adjacent islands and the paucity of eclogite facies assemblages in the Paragneiss Complex along Moldefjord are related to the imposition of a late Caledonian protomylonitic fabric on the rocks in that area.

One of us (MAH) has recently been particularly concerned with the interpretation of the relationships between garnet-granulites (eclogite-amphibolites), marbles, and associated schists and gneisses in the Tverrfjella area. From his earlier study, Hernes (1954b) was uncertain as to whether the conspicuous mafic rocks there (his eclogite-amphibolites) represented metabasaltic rocks or some form of reaction skarn between the marbles and the surrounding gneisses. Since the marbles occur as megaboudins and can be traced along presumed original sedimentary horizons, a further possibility to be considered is that they are of metasedimentary origin and have been derived from impure carbonate rocks.

From geochemical studies the Tverrfjella garnet-granulites display a basaltic chemistry (compare columns A and C in Table 1). These are plotted on a Pearce and Cann (1973) diagram (Fig. 5a), together with various eclogites, amphibolites, and other garnet-granulites from elsewhere in the region. The Tverrfjella garnet-granulites are fairly well separated from the eclogites, the majority of which form a compact group although there are a few with anomalously high or low titanium contents. The majority of the eclogites lie in or near the field defined as ocean island or continental basalts by Pearce and Cann (1973) which is consistent with our hypothesis of them originally being basic dykes intruded into continental crust rocks. By contrast the Tverrfjella garnet-granulite samples lie predominantly in the field of ocean floor basalts as do the majority of the amphibolite and garnet-granulite samples from elsewhere, suggesting that all of these rocks are of an igneous origin. However, another scattered group of amphibolites and garnet-granulites appear towards the Zr apex and are interpreted as having a different, possibly sedimentary, parentage with the high Zr resulting from concentrations of detrital zircon.

Table 2   Representative electron microprobe clinopyroxene analyses

Analysis Sample No.	A U484	B U160 Inc.	C U160 Symp. Max. Al	D U160 Symp. Min. Al	E 1410 Symp.	F U263 Prim.	G U263 Corona	H U262 Inner corona	J U262 Outer corona	K U332
SiO_2	55.90	55.39	49.64	50.87	49.86	52.09	52.64	53.16	53.28	50.71
TiO_2	0.12	0.31	0.32	0.51	0.42	0.25	0.17	0.07	0.08	0.12
Al_2O_3	8.24	10.08	5.71	1.78	5.07	2.81	3.22	2.15	5.59	2.21
FeO^T	3.97	8.11	10.35	9.38	11.42	8.41	8.78	5.78	5.67	15.08
MnO	0.02	0.09	0.24	0.22	0.11	0.17	0.14	0.14	0.13	0.30
MgO	11.15	8.32	10.79	13.02	10.74	13.52	13.08	14.19	13.51	9.45
CaO	16.26	14.34	20.43	22.00	20.52	21.94	20.35	22.75	18.78	20.89
Na_2O	4.86	6.05	1.68	1.09	1.22	0.99	1.86	1.43	2.63	0.63
TOTAL	100.53	99.69	99.15	99.32	99.36	100.18	100.24	99.67	99.67	99.39
Si	1.989	1.960	1.956	1.931	1.892	1.937	1.952	1.968	1.947	1.950
Ti	0.003	0.008	0.010	0.015	0.012	0.007	0.005	0.002	0.002	0.004
Al	0.346	0.421	0.107	0.080	0.227	0.123	0.141	0.094	0.241	0.101
Fe	0.118	0.240	0.341	0.312	0.362	0.262	0.272	0.179	0.173	0.490
Mn	0.001	0.003	0.008	0.007	0.004	0.005	0.005	0.004	0.004	0.010
Mg	0.591	0.439	0.633	0.736	0.608	0.749	0.723	0.783	0.736	0.547
Ca	0.620	0.544	0.862	0.895	0.834	0.874	0.809	0.902	0.735	0.870
Na	0.335	0.415	0.128	0.080	0.090	0.072	0.134	0.103	0.186	0.048

A   —Unaltered grains in large eclogite body at Torveneset, W. Otrøy.
B   —Omphacite inclusions in garnets from partly retrograded eclogite lens in roadcut between Rakvåg and Ugelvik, Otrøy.
C, D —Symplectitic grains intergrown with plagioclase An$_{16}$. Same sample as Analysis B but showing the range of alumina contents observed in different spot analyses.
E   —Symplectitic grains in garnet–granulite from Tverrfjella.
F   —Primary 'igneous' grains in metadolerite from Drynasund Lighthouse, W. Midøy.
G   —Corona grains around clusters of orthopyroxene grains in same sample as F.
H   —Inner part of corona adjacent to corona orthopyroxene around primary olivines (Fo$_{67}$) in metadolerite from Drynasund Lighthouse, W. Midøy.
J   —Outer part of corona adjacent to garnet in same sample as H.
K   —Relict grains intergrown with plagioclase in augen gneiss from Sundsbø roadcut, N.E. Otrøy.
FeO^T—Total Fe as FeO.

Table 3   Representative electron microprobe garnet analyses

Analysis Sample No.	A U484	B U160	C 1410	D U263 Inner corona	E U263 Outer corona	F U262 Inner corona	G U262 Outer corona	H U332 High Fe + Mn	J U332 Low Fe + Mn
SiO_2	40.89	38.65	38.03	38.68	38.64	39.04	38.66	37.32	37.67
TiO_2	0.02	0.05	0.02	0.05	0.05	0.07	0.06	0.02	0.04
Al_2O_3	23.10	21.31	20.92	21.12	21.18	22.37	22.68	20.58	20.63
FeO^T	18.49	23.88	27.59	25.34	26.04	20.95	14.82	29.64	27.81
MnO	0.29	0.24	0.80	1.14	1.12	0.81	0.70	2.33	0.83
MgO	13.94	7.88	4.57	7.36	6.90	10.32	4.70	2.14	3.26
CaO	4.01	7.34	8.25	6.27	5.88	6.33	17.56	7.95	9.44
Na_2O	0.02	0.02	n.d.	0.02	0.01	0.03	0.02	n.d.	n.d.
TOTAL	100.76	99.87	100.18	99.98	99.82	99.94	99.40	99.98	99.68
Si	6.003	5.973	5.877	5.977	6.015	5.923	5.933	5.989	5.993
Ti	0.004	0.008	0.004	0.008	0.008	0.012	0.010	0.004	0.007
Al	3.998	3.882	3.811	3.848	3.888	4.000	4.104	3.894	3.870
Fe	2.270	3.087	3.566	3.275	3.390	2.658	1.902	3.978	3.701
Mn	0.001	0.097	0.105	0.150	0.148	0.104	0.091	0.316	0.112
Mg	3.050	1.814	1.052	1.695	1.600	2.334	1.075	0.512	0.773
Ca	0.631	1.216	1.366	1.038	0.981	1.029	2.887	1.367	1.609
Na	0.005	0.006	—	0.006	0.004	0.009	0.006	—	—

Sample locations as in Table 2.
FeO^T—total Fe as FeO.

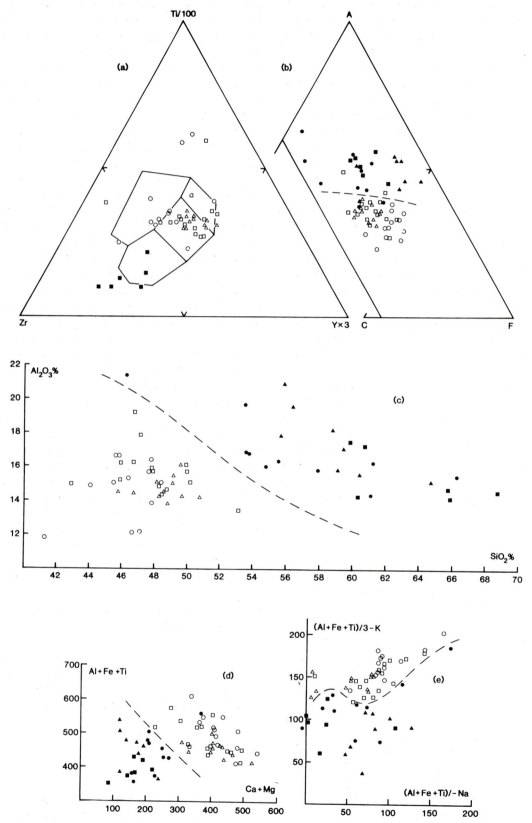

Fig. 5   (a) to (e): various discrimination diagrams showing plots of whole rock analyses of samples from the Moldefjord region as follows: open circles—eclogites; open triangles—garnet granulites from Tverrfjella; open squares—ortho-amphibolites and granulites; solid circles—various calc-silicate bearing and semipelitic gneisses; solid triangles—pelites; solid squares—para-amphibolites and granulites

In an attempt to clarify further the origin of the various garnet-granulites and amphibolites, several discrimination plots were constructed including analyses of pelites and calc-silicate bearing rocks from the Tverrfjella and Molde sequences.

In all these figures (5b–e) the meta-igneous basaltic rocks lie in a cluster (i.e. eclogites, garnet-granulites, and amphibolites) which does not include the apparent metasedimentary amphibolites. Rather, these occur with the undoubted metasedimentary rocks (pelites and calc-silicates). The para-amphibolites are rather more alkali-rich (Fig. 5b) and siliceous, often extremely so (Fig. 5c). The diagrams of Moine and of de la Roche (1968) (Figs. 5d and 5e) were devised specifically to discriminate between meta-igneous and metasedimentary rocks. Although the data do not fall exactly within the fields defined by de la Roche we feel that there is sufficient consistancy of pattern to define two contrasting groups of amphibolites.

From the relative compositional unconformity of the Tverrfjella garnet-granulite and the evidence of the discrimination plots we conclude that this rock may represent submarine basalt flows interbedded with sedimentary material now represented by the associated marbles; calc-silicate gneisses (composed of garnet, diopside, scapolite, feldspar, and sphene) appearing as bands within the marble; amphibolites (often feldspathic and rarely garnetiferous), kyanite-bearing garnet–biotite pelites and quartzo-feldspathic schists. These rocks appear as a definite sequence passing from quartzofeldspathic schists, through the garnet-granulite with marbles, amphibolites, to pelites at the top of the Tverrfjella sequence.

Petrogenetic Significance of Metaperidotites and Metadolerites

Variably sized bodies of both of these rocks types are conspicuous within the area which we have studied. Particularly noteworthy are the garnetiferous peridotites outcropping between Ugelvik and Raknes on the island of Otrøy and the large metadolerite body at Drynasund on western Midøy.

The metadolerites are characterized by the fact that for the most part they retain an igneous fabric but with obvious development of coronas of fine-grained garnet and secondary pyroxenes around primary orthopyroxenes and, if present, olivines. There are considerable variations in composition within these corona phases (Tables 2 and 3) but it is clear that the overall metamorphic trend is towards the development of high pressure eclogite facies assemblages with pyropic garnets and omphacitic clinopyroxenes (cf. Griffin and Råheim 1973). Although these dolerites have a basaltic rock chemistry (Table 1, column D) similar to the common bimineralic eclogite lenses, there are certain mincr chemical differences, notably in K, Rb, Ba, Zr, P, and Cr contents. Two splits of hand-picked clinopyroxenes from a coarse metadolerite lens exposed at Bolgavatnet within the Frei Group (Råheim 1972), some 45 km northeast of Molde and just south of Kristiansund, have yielded ^{39}Ar–^{40}Ar weighted mean plateau ages on stepwise degassing, of 428 ± 23 Ma and 421 ± 25 Ma (Lynch 1976) with little evidence from the Ar release pattern that these rocks have had an older origin. Hence, we consider that it is more likely that these metadolerites represent a suite of later Caledonian intrusives (probably emplaced close in time to the attainment of the metamorphic 'high') rather than them being older (~ 1500 Ma) rocks of the extensive early basaltic dyke suite (now represented by the thoroughly recrystallized coarse grained eclogites).

For kinetic reasons these metadolerites have partly retained their parent igneous mineralogy through having remained as essentially passive bodies during the subsequent tectonic events. What is certain is that their intrusion and high pressure corona development predates the emplacement of the common Caledonian granite pegmatite suite observed throughout both the Augen Orthogneiss and the Paragneiss Complexes. As in eclogite lenses, amphibolite haloes are developed around such pegmatites where they cut the metadolerite bodies. In addition the metadolerites are converted into foliated amphibole schists where intersected by ENE-striking late Caledonian shear zones.

Geochemical studies of the garnet peridotite bodies in the Basal Gneiss Complex suggest an ultimate mantle origin for these rocks (Carswell 1968; Brueckner 1977b). However, it is now considered (Carswell and Gibb 1980; Medaris 1980) that these peridotite bodies were tectonically intercalated into the gneiss complex prior to the equilibration of the garnet peridotite assemblages. Indeed these eclogite facies assemblages are now thought to have developed contemporaneously with the high pressure assemblages sporadically preserved in metabasic, pelitic, and gneissic rocks of the surrounding Basal Gneiss Complex.

Tectonometamorphic Implications of Deduced P/T Equilibrium Conditions for the High Pressure Metamorphic Assemblages

Judicious interpretation of P/T values obtained from various experimentally calibrated mineral thermometers and barometers applicable to garnet + clinopyroxene ± orthopyroxene ± olivine assemblages indicates values of around $740 \pm 50°C$ and 22 ± 5 Kbars for the P/T equilibration conditions responsible for the high grade metabasic and metaperidotitic assemblages outcropping in this area (Krogh 1977; Carswell and Gibb 1980; Carswell *et al* this volume). In view of the evidence for early high pressure assemblages in a wide range of rock types (including the extensive augen orthogneisses) we envisage that these conditions were operative over a wide area in this part of Møre and Romsdal region. However, observations of regional variations in $Fe^{2+}–Mg^{2+}$ partition coefficients for garnet–clinopyroxene pairs in eclogites indicate that the P/T conditions responsible for the metamorphic culmination in the Basal Gneiss Complex were at a maximum around Moldefjord and adjacent areas along the Norwegian coast and decreased significantly to the southeast (Krogh 1977). Taking into consideration both this and recent isotopic age evidence (Griffin and Brueckner 1980) of a Caledonian age for the metamorphic culmination responsible for the generation of the eclogite facies assemblages, we propose a revised plate tectonic model for the evolution of the Basal Gneiss Complex as outlined more fully in Cuthbert *et al.* (1981, 1983).

We envisage that, following collision of the Greenland and Baltic plates during the Caledonian orogeny, the Baltic plate was underthrust beneath the Greenland plate down a northwesterly dipping Benioff Zone (relative to present geography). Imbrication of mantle peridotite into crustal gneisses situated close to the nose of the downgoing plate is thought to have preceded imposition of the P/T conditions for the metamorphic 'high' which may be expected to have occurred shortly after cessation of subduction and attainment of maximum crustal thickness. Subsequent rapid recovery of these rocks to near surface by mid-Devonian times is considered to have led to varying degrees of retrogression of the high grade assemblages depending upon the extent of strain induced recrystallization.

It is apparent throughout the Moldefjord area that rocks of both the Augen Orthogneiss and Paragneiss Complexes show widespread development of a protomylonitic fabric associated with which the early high grade mineral assemblages are extensively retrograded to lower almandine amphibolite facies assemblages. This deformation attenuates and progressively eliminates early fabrics and structures and is responsible for the dominant steeply dipping ENE striking foliation and prominent rodding lineation in many of the rocks. We consider that this deformation occurred late in the Caledonian orogenic cycle, during the final stages of the recovery of these rocks

to the surface. It is our thesis that the virtual elimination of early high grade metamorphic assemblages within the Paragneiss Complex rocks in the southern part of the area can be attributed to the fact that the intensity of this late deformation was at a maximum roughly along the present line of Moldefjord itself.

Summary of the Geological Evolution of the Basal Gneiss Complex

Due to the uncertainty over the age and nature of the Paragneiss Complex two alternatives are given in the following sequence of events for its role in the geological evolution of the Basal Gneiss Complex in the Moldefjord area.

1. Possible deposition and subsequent almandine amphibolite facies metamorphism and migmatization of the sedimentary and volcanic rocks of the Paragneiss Complex at around 1700–1800 Ma.

2. Intrusion into the Paragneiss Complex at around 1500 Ma of the rapakivi-textured quartz monzonites/granodiorites with subsidiary granitic and doleritic sheets now seen as the Orthogneiss Complex. This plutonic igneous activity may have been related to the extensive suite of anorthosite–mangerite–rapakivi granite intrusions emplaced between 1450–1700 Ma in a broad belt across Scandinavia and North America connected with the incipient break-up of the old North American–European continent prior to the Grenville (Sveconorwegian) orogeny.

3. Although we have no definite evidence it seems unlikely that the rocks in this area entirely escaped the tectonometamorphic effects of the Grenville (Sveconorwegian) orogeny at around 900–1200 Ma. Certainly they lie within the Grenville front in Scandinavia as depicted by Shackleton (1979).

4. Our model for the closure and collision of the Baltic and Greenland plates, resulting in the Caledonian orogeny, is presented in Cuthbert *et al.* (1983); only a synopsis relevant to the rocks in Moldefjord is given here but is tenable for both possible suggested relationships between the Paragneiss and Orthogneiss Complexes. During collision the leading edge of the Baltic plate underthrust the Greenland plate and suffered intense telescoping and imbrication of both basement and cover giving a crustal thickness of around 80 km. The possible juxtapositioning of the two Complexes by thrusting could have occurred at that time, with the Orthogneisses thrust over basement or perhaps a parautochthonous unit, and the Paragneisses in turn thrust

over the Orthogneisses. The alternative intrusive relationship of the two Complexes would require that the two behaved as a single unit of crust which escaped much of this imbrication. The metaperidotites are considered to have been incorporated during this imbrication perhaps due to thrusts transecting the subcontinental upper mantle. There also appears to have been an episode of emplacement of mantle derived basaltic magma, resulting in the metadolerite bodies, at around the time of maximum crustal thickness, approximately 450 Ma.

5. Imposition of the P/T conditions ($740 \pm 50°C$, 22 ± 5 Kbars) for the Caledonian metamorphic culmination in the Basal Gneiss Complex occurred shortly afterwards, following thermal 'recovery' in the imbricated pile (Richardson and England 1979). During this high pressure metamorphic event eclogite facies mineral assemblages were developed in 'dry' metabasic rocks and in the metaperidotite bodies and garnet (and rare jadeitic clinopyroxene) formed in the intermediate–acid orthogneisses. By contrast, almandine amphibolite assemblages may have remained stable (Bryhni et al. 1977) in certain 'wetter' metasedimentary/metavolcanic rocks of the Paragneiss Complex. Partial melting of the latter rocks under the high P_{H_2O} conditions may have been responsible for the generation of the extensive suite of Caledonian granite pegmatites.

6. During the ensuing rapid, isostatically induced, recovery of these rocks towards the surface, the early high pressure metamorphic assemblages suffered varying degrees of retrogression to low almandine amphibolite and ultimately greenschist facies assemblages, depending upon the extent to which the rocks were affected by associated deformation. Where eclogite facies assemblages remained essentially 'dry', but deformation prompted recrystallization, retrogression proceeded only as far as the development of granulite facies assemblages. The pervasive imposition of a low grade protomylonitic fabric in the paragneisses in the vicinity of Moldefjord suggests that the final transfer of these rocks to close to the surface by mid-Devonian times (~ 385 Ma) involved extensive shearing, thrusting, and isoclinal folding presumably associated with continued compression in the orogen.

Acknowledgements

We wish to thank R. Kanaris-Sotiriou and F. G. F. Gibb for assistance and advice with X-ray fluorescence and electron microprobe analyses performed at the University of Sheffield; W. L. Griffin for his cooperation and encouragement with Rb–Sr isotopic analyses conducted at the Mineralogisk–Geologisk Museum in Oslo; and P. Mellor for cheerfully typing several versions of the manuscript of this paper. In addition, D. A. Carswell acknowledges financial support for field studies in Norway from the University of Sheffield and the Natural Environment Research Council, and M. A. Harvey the receipt of a research studentship from the latter body.

References

Austrheim, H. and Råheim, A. 1981. Age relationships within the high grade metamorphic rocks of the Bergen Arcs, western Norway. Abstr. Uppsala Caledonide Symp. Terra Cognita, 1, 33.

Banham, P. H., Gibbs, A. D. and Hopper, F. W. M. 1979. Geological evidence in favour of Jotunheimen Caledonian suture. Nature, London, 277, 289–291.

Bridgewater, D. and Windley, B. F. 1973. Anorthosites, post-orogenic granites, acid volcanic rocks, and crustal development in the North Atlantic Shield during the mid-Proterozoic. In Lister, L. A. (Ed.), Symposium on Granites, Gneisses and Related Rocks, Geol. Soc. S. Africa Sp. Publ., 3, 307–318.

Brueckner, H. K. 1977a. A structural, stratigraphic and petrologic study of anorthosites, eclogites and ultramafic rocks and their country rocks, Tafjord area, western south Norway. Norges geol. Unders., 332, 1–53.

Brueckner, H. K. 1977b. A crustal origin for eclogites and a mantle origin for garnet peridotites: strontium isotopic evidence from clinopyroxenes. Contrib. Mineral. and Petrol., 60, 1–15.

Brueckner, H. K. 1979. Precambrian ages from the Geiranger–Tafjord–Grotli area of the Basal Gneiss region, west Norway. Norsk geol. Tidsskr., 59, 141–153.

Bryhni, I. 1966. Reconnaissance studies of gneisses, ultrabasites, eclogites and anorthosites in outer Nordfjord, western Norway. Norges geol. Unders., 241, 1–60.

Bryhni, I., Krogh, E. and Griffin, W. L. 1977. Crustal derivation of Norwegian eclogites: a review. N. Jb. Miner. Abh., 130, 49–68.

Bugge, C. 1934. Grønne trondhjemsskifre på øyene ved Molde. Norges geol. Unders., 43, 167–175.

Carswell, D. A. 1968. Possible primary upper mantle peridotite in Norwegian basal gneiss. Lithos, 1, 322–355.

Carswell, D. A. 1973. The age and status of the basal gneiss complex of north-west southern Norway. Norsk geol. Tidsskr., 53, 65–78.

Carswell, D. A. and Gibb, F. G. F. 1980. The equilibrium conditions and petrogenesis of European crustal garnet lherzolites. Lithos, 13, 19–29.

Carswell, D. A., Krogh, E. J. and Griffin, W. L. this volume. Norwegian orthopyroxene eclogites: calculated equilibration conditions and petrogenetic implications.

Claesson, S. 1981. Caledonian metamorphism of Proterozoic Seve rocks in Mt. Åreskutan, southern Swedish Caledonides. Geol. För. Stockholm Förh., 103, 291–304.

Cuthbert, S. J., Harvey, M. A. and Carswell, D. A. 1981. A plate tectonic model for the Basal Gneiss Complex of

Western South Norway—a working hypothesis. *Abstr. Uppsala Caledonide Symp. Terra Cognita*, **1**, 41.

Cuthbert, S. J., Harvey, M. A. and Carswell, D. A. 1983. A tectonic model for the metamorphic evolution of the Basal Gneiss Complex, Western South Norway. *J. Metam. geol.*, **1**, 63–90.

de Waard, D. 1969. The anorthosite problem: the problem of the anorthosite–charnockite suite of rocks. In Isachesen, Y. W. (Ed.), *Origin of Anorthosite and Related Rocks*, pp. 466.

Ellis, D. J. and Green, D. H. 1979. An experimental study of the effect of Ca upon garnet–clinopyroxene Fe–Mg exchange equilibria. *Contrib. Mineral. Petrol*, **71**, 13–22.

Emslie, R. F. 1978 Anorthosite massifs, rapakivi granites and late Proterozoic rifting of North America. *Precambrian Res.*, **7**, 61–98.

Gale, H. G. and Roberts, D. 1974. Trace-element geochemistry of Norwegian lower Palaeozoic basic volcanics and its tectonic implications. *Earth Planet. Sci. Lett.*, **22**, 380–390.

Gee, D. G. 1978. Nappe displacement in the Scandinavian Caledonides. *Tectonophysics*, **47**, 393–419.

Gee, D. G. 1980. Basement–cover relationships in the central Scandinavian Caledonides. *Geol. För. Stockholm Förh.*, **102**, 455–474.

Gjelsvik, T. 1951. Oversikt over bergartene i Sunnmøre og tilgrensende deler av Nordfjord. *Norges geol. Unders.*, **179**, 1–45.

Gjelsvik T. 1953. Det nordvestlige gneis-område i det sydlige Norge, aldersforhold og tektonisk–stratigrafisk stilling. *Norges geol. Unders.*, **184**, 71–94.

Green, D. H. and Mysen, B. O. 1972. Genetic relationship between eclogite and hornblende plagioclase pegmatite in western Norway. *Lithos*, **5**, 147–161.

Griffin, W. L. and Råheim, A. 1973. Convergent metamorphism of eclogites and dolerites, Kristiansund area, Norway. *Lithos*, **6**, 21–40.

Griffin, W. L. and Brueckner, H. K. 1980. Caledonian Sm–Nd ages and crustal origin for Norwegian eclogites. *Nature*, London, **285**, 319–321.

Griffin, W. L. and Carswell, D. A. this volume. In-situ metamorphism of Norwegian eclogites: an example.

Hernes, I. 1954a. Trondhjemsskifrene ved Molde. *Norsk geol. Tidsskr.*, **34**, 123–137.

Hernes, I. 1954b. Eclogite amphibolite on the Molde peninsula, southern Norway. *Norsk geol Tidsskr.*, **33**, 163–184.

Hernes, I. 1955. Geologisk oversikt over Molde–Kristiansundsområdet. *Det. Kongl. Norske Vidensk. Selsk. Skrifter.*, **5**, 16.

Hernes, I. 1956. Surnadalssynklinalen. The Surnadal syncline, central Norway. *Norsk geol. Tidsskr.*, **36**, 35–39.

Hernes, I. 1967. The late Pre-Cambrian stratigraphic sequence in the Scandinavian mountain chain. *Geol. Mag.*, **104**, 557–563.

Holtedahl, O. 1944. On the Caledonides of Norway with some scattered local observations. *Skr. Norske Vidensk-Akad. i Oslo, Mat.-Naturv. Kl.*, **4**, 31.

Holtedahl, O. and Dons, J. A. 1960. Geologiske kart over Norge, Bergrunnskart 1: 1,000,000. *Norges geol. Unders.*, **208**.

Hossack, J. R., Nickelsen, R. P. and Garton, M. 1981. The geological section from the foreland up to the Jotun Sheet in the Valdres area, South Norway. *Abstr. Uppsala Caled. Symp. Terra Cognita*, **1**, 52.

Kolderup, N.-H. 1960. Origin of Norwegian eclogites in gneisses. *Norsk geol. Tidsskr.*, **40**, 73–76.

Krill, A. G. 1980. Tectonics of the Oppdal area, central Norway. *Geol. För. Stockholm Förh.*, **102**, 523–530.

Krill, A. G. 1981. 'Stockwerk' tectonic relationships between the Trondheim synclinorium and the Western Gneiss Region of Norway. *Abstr. Uppsala Caledonides Symp. Terra Cognita*, **1**, 56.

Krill, A. G. and Griffin, W. L. 1981. Interpretation of Rb–Sr dates from the western gneiss region: a cautionary note. *Norsk geol. Tidsskr.*, **61**, 83–86.

Krogh, E. J. 1977. Evidence of Precambrian continent–continent collision in western Norway. *Nature*, London, **267**, 17–19.

Lappin, M. A., Pidgeon, R. T. and van Breemen, O. 1979. Geochronology of basal gneisses and mangerite syenites of Stadlandet, west Norway. *Norsk geol. Tidsskr.*, **59**, 161–181.

Lynch, M. C. F. 1976. *Application of the ^{39}Ar–^{40}Ar technique to terrestrial rocks and minerals with special reference to the Lewisian of N.W. Scotland*. Unpublished Ph.D Thesis, University of Sheffield, U.K.

Medaris, L. G. 1980. Petrogenesis of the Lien peridotite and associated eclogites, Almklovdalen, Western Norway. *Lithos*, **13**, 339–353.

Moine, B. and de la Roche, H. 1968. Nouvelle approche de probleme de l'origine des amphibolites a partir de leur composition chimique. *Comptes Rendues Acad. Sci. Paris, seire D*, **267**, 2084–2087.

Newton, R. C. and Perkins, D. III 1981. Thermodynamic calibration of geobarometers based on the assemblages garnet–plagioclase–orthopyroxene (clinopyroxene)–quartz. *Am. Mineral.*, **67**, 203–222.

O'Conner, J. T. 1965. A classification for quartz-rich igneous rocks based on feldspar ratios. *U.S. geol. Surv. Prof. Paper*, **525–B**, 79–84.

Pearce, J. A. and Cann, J. 1973. Tectonic setting of basic volcanic rocks using trace element analysis. *Earth Planet. Sci. Lett.*, **19**, 290–300.

Pidgeon, R. T. and Råheim, A. 1972. Geochronological investigation of the gneisses and minor intrusive rocks from Kristiansund, west Norway. *Norsk geol. Tidsskr.*, **52**, 241–256.

Point, R. 1975. Mylonites et orogenèse tangentielle nature, geochemie, origine et âge des gneiss oeillés dans les nappes calédoniennes externes. *Bulletin de la Societe Geologique de France*, **7**, 17, 664–679.

Råheim, A. 1972. Petrology of high grade metamorphic rocks of the Kristiansund area. *Norges geol. Unders.*, **279**, 1–75.

Råheim, A. 1977. A Rb, Sr study of the rocks of the Surnadal syncline. *Norsk. geol. Tidsskr.*, **57**, 193–204.

Råheim, A. 1979. Structural and metamorphic break between the Trondheim basin and the Surnadal synform. *Norsk geol. Tidsskr.*, **59**, 195–198.

Reusch, H. H. 1881. Konglomerat–Sandstenfelterne i Nordfjord, Søndfjord og Sogn. *Nytt. Mag. Naturvitensk*, **26**, 93–170.

Richardson, S. W. and England, P. C. 1979. Metamorphic consequences of crustal eclogite production in overthrust orogenic zones. *Earth Planet. Sci. Lett.*, **42**, 183–190.

Roberts, D., Thon, A., Gee, D. G. and Stephens, M. B. 1981. Scandinavian Caledonides-tectonostratigraphy map, scale 1:1,000,000. *Uppsala Caledonide Symp.*

Röshoff, K. 1978. Structure of the Tännäs Augen Gneiss Nappe and its relation to under- and overlying units in the central Scandinavian Caledonides. *Sver. geol. Unders.*, **C739**, 1–35.

Schmitt, H. H. 1963. *Petrology and Structures of the Eiksundal Eclogite Complex, Hareidland, Sunnmøre, Norway.* Unpublished|Ph.D.|thesis,|Harvard University.

Shackleton, R. M. 1979. The British Caledonides: comment and summary. In Harris, A. L. Holland, C. H. and Leake, B. E. (Eds), *The Caledonides of the British Isles, Reviewed*, pp. 768.

Skjerlie, F. J. 1969. The Pre-Devonian rocks in the Askvoll–Gaular area and adjacent districts, Western Norway. *Norges geol. Unders.*, **28**, 325–359.

Smithson, S. B., Ramberg, I. B. and Grønlie, G. 1974. Gravity interpretation of the Jotun Nappe of the Norwegian Caledonides. *Tectonophysics.*, **22**, 205–222.

Strand, T. 1952. The relation between the basal gneiss and the overlying meta-sediments in the Surnadal district (Caledonides of Southern Norway). *Norges geol. Unders.*, **184**, 100–123.

Strand, T. 1969. Geology of the Grotli area. *Norsk geol. Tidsskr.*, **49**, 341–360.

Streckeisen, A. 1976. Classification of the common igneous rocks by means of their chemical composition. A provisional attempt. *N. Jb. Miner. Mh.*, **1**, 1–15.

Sturt, B. A. and Thon, A. 1976. The age of orogenic deformation in the Swedish Caledonides. *Am. J. Sci.*, **276**, 385–390.

Sturt, B. A., Skarpenes, O., Ohanian, A. T. and Pringle, I. R. 1975. Reconnaissance Rb/Sr isochron study in the Bergen Arc system and regional implications. *Nature*, London, **253**, 595–599.

The Caledonide Orogen—Scandinavia and Related Areas
Edited by D. G. Gee and B. A. Sturt
© 1985 John Wiley & Sons Ltd

Relationships between peridotites, anorthosites and eclogites in Bjørkedalen, western Norway

K. Brastad

Mineralogisk–Geologisk Museum, Sarsgt. 1, Oslo 5

ABSTRACT

Anorthosites, peridotites, and eclogites show close spatial associations in Bjørkedalen. The country rocks are micaceous paragneisses with quartzitic layers or tonalitic to granitic orthogneisses. Eclogites are found both in ortho- and paragneisses, being far more abundant in the latter. Eclogites are also enclosed in anorthosites and in the largest peridotite; these eclogites show compositional layering while the eclogites in gneiss do not. The eclogites in anorthosite are quartz-noritic in composition; the others are olivine-gabbroic. On the bases of REE-data and comparison with equivalent rocks elsewhere in the Western Gneiss Region, both types are considered to have low-P protoliths. Maximum equilibration temperatures for the eclogites are compatible with those of other eclogites from this district, except for the eclogites in peridotite, which show clearly higher maximum equilibration conditions (assuming all Fe = FeO). Paragneisses show mineral assemblages which are equivalent to the eclogite assemblages in the mafic rocks thus showing that they were subjected to the eclogite metamorphism together with their enclosed eclogites. It is proposed that the whole association of rocks in Bjørkedalen was brought together before the Caledonian eclogite-forming metamorphism, and that the Caledonian orogeny only imposed local tectonic displacements along the boundaries of the rock units.

Introduction

The Bjørkedalen valley is situated in the Precambrian Western Gneiss Region (WGR) of Norway (Fig. 1) which consists of partially Caledonized Precambrian ortho- and paragneisses, anorthosites, peridotites, and a variety of basic rocks. Bjørkedalen was chosen as a suitable area for studying relationships between the various rock types because most elements of the Western Gneiss Region are present there. This study mainly deals with field relationships, possible origin and genetic relations of the anorthosites, peridotites, amphibolites, and eclogites, and their relations to the eclogite-forming event. The Bjørkedalen rocks are strongly tectonized and have suffered at least three major metamorphic events. The original mineralogy is almost totally obliterated and cannot be used to draw conclusions as to the original compositions. A final discussion of the origin of these rocks must be based on a broad knowledge of the chemical composition of both rocks and minerals. This study may, however, throw light on similarities with equivalent rocks elsewhere and thereby give an indication of the possible origin of the anorthosite–peridotite–eclogite–gneiss association.

This knowledge will have important consequences for the interpretation of the eclogite-forming event which in turn has consequences for interpretation of the Caledonian tectonic picture in this area.

The mode of formation of the eclogites in the WGR is controversial. One group of geologists postulates an *in situ* high-P metamorphism of low-P protoliths (summarized by Griffin *et al.* this volume); another group argues for a foreign origin of the eclogites and postulates that they were subjected to a high-P metamorphism elsewhere and then brought into this part of the crust by a tectonic mixing process (Lappin 1966; Lappin and Smith 1977; Smith 1981). The relations in Bjørkedalen indicate that high-P metamorphism has been superimposed on an anorthosite–peridotite–mafic rock-gneiss associa-

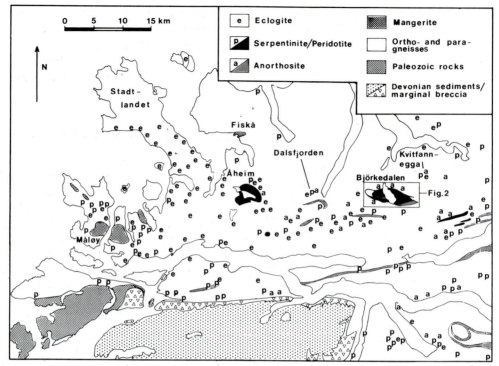

Fig. 1   Geological and spatial setting of the Bjørkedalen area. Main serpentinite/peridotite, anorthosite, and eclogite occurrences within the Nordfjord/southern Sunnmøre parts of the Western Gneiss Region are shown

tion with low-P protoliths, thus supporting the *in situ* model.

Field Relations and Mineral Compositions of the Bjørkedalen Rocks

The structural pattern in Bjørkedalen is complex and is dominated by the major peridotite (Fig. 2) with the neighbouring gneisses folded around it. The peridotite has the form of a trough with E–W elongation, and probably has the same general structural evolution as the Almklovdalen peridotite (Cordellier *et al.* 1981). The rocks on the northern side of the peridotite are migmatitic and are probably orthogneisses, while the rocks on the southern side are dominated by paragneisses with layers of quartzites, amphibolites, anorthosites, eclogites, peridotites, and subordinate orthogneisses. The paragneiss association is exposed as an E–W striking

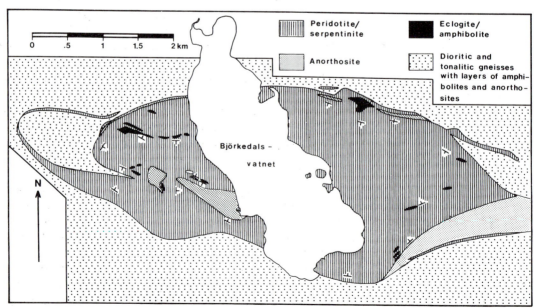

Fig. 2   Simplified geological map over the central peridotite and its surrounding rocks in Bjørkedalen

complex with steep inclination and with very variable thicknesses of the different units. The structural history of this area has not been worked out in detail but it is easily seen that the rocks have been repeatedly isoclinally folded.

Description of the Rocks

Gneisses

The orthogneisses are mainly tonalitic to granitic in composition with an amphibolite facies mineralogy: quartz + plagioclase + K-feldspar + hornblende + biotite + muscovite ± epidote ± sphene. The paragneisses range from micaceous to very quartz-rich rocks also including layers of quartzites. Locally, in the vicinity of some eclogites an older assemblage with quartz + phengite + kyanite + garnet ± plagioclase ± rutile can be recognized. This is overprinted by the younger amphibolite facies assemblage with quartz + biotite + muscovite + plagioclase + K-feldspar ± hornblende + ilmenite. The most important reactions here are:

A. Phengite → biotite + muscovite + plagioclase.

B. Garnet + H_2O → biotite + plagioclase.

According to Heinrich (1982) the first assemblage is equivalent to the eclogite assemblages in mafic rocks and the retrograde reactions are characteristic for the transition from eclogite facies to amphibolite facies conditions.

Ortho- and paragneisses are found as alternating zones, striking east–west. They are intimately folded together and it is often impossible to decide whether the rocks belong to one or the other category.

Peridotites

The central parts of Bjørkedalen valley are dominated by a major peridotite body and a number of smaller ones (Fig. 2). The latter are mostly dunitic, but strongly serpentinized along fractures and contacts with other rocks. At least 90% of the central peridotite consists of dunite/lherzolite and serpentinite. The rest are composed of more or less retrograded eclogitic rocks with minor amounts of garnet orthopyroxenite and garnet clinopyroxenite which grade into eclogites (Tables 1 and 3). Garnet peridotites have not been found, but garnet + serpentine + clinopyroxene and garnet + tremolite + serpentine assemblages suggest that such rocks have been present prior to retrograde metamorphism.

The least retrograded peridotites are pure dunites with minor amounts of orthopyroxene, chromite, and chlorite and usually without any distinct layering. Layering is best seen where chromite is abundant. Olivine and orthopyroxene are Mg-rich with 90–91% of the forsterite and the enstatite component respectively (Table 2). This is in harmony with most of the other peridotites in this district (Gjelsvik 1951; Mercy and O'Hara 1965; Carswell 1968; Medaris 1980).

Anorthosites and related amphibolites

Anorthositic rocks envelop the major peridotite, and are also found folded into it and as thin layers within it. The anorthositic rocks are mainly in contact with orthogneisses and amphibolites. Anorthositic layers or lenses are also found elsewhere in Bjørkedalen within the paragneiss areas. These are commonly associated with lenses of orthogneisses and minor peridotites. The anorthositic rocks include anorthosite *sensu stricto (s.s.)*, amphibolites with varying femic/mafic ratio, and eclogites. Amphibolites and eclogites are enclosed within the anorthosite bodies as layers or lenses or are found along their contacts and their extension along strike.

All these rocks have been extensively metamorphosed and the only relics of the original igneous parageneses are dark-coloured very coarse-grained plagioclase crystals (10–20 cm) which are found in two exposures near the major peridotite within anorthosite (*s.s.*). The metamorphosed anorthosites (*s.s.*) (metaanorthosites) exhibit an amphibolite-facies mineralogy with plagioclase + hornblende + cpidote/clinozoisite ± biotite ± muscovite ± quartz ± rutile. This assemblage is in turn overprinted by a very low grade assemblage with local epidote, chlorite, and zeolite growth. Light amphibolites have more or less the same mineral assemblages as the metaanorthosites, but with less plagioclase and epidote/clinozoisite in the amphibolite facies paragenesis. Dark amphibolites usually have metaeclogite textures and are clearly retrograded eclogites (see eclogites below). Plagioclase from metaanorthosites and light amphibolites lie most commonly in the An_{40} to An_{60} range, while the relic igneous plagioclase gives An_{55}–An_{65} (Table 2). Thin ilmenite/magnetite-rich layers within the largest anorthosite lens have recrystallized plagioclase with An_{80-93} (Brastad 1980). These data are also in accordance with feldspar analyses and normative plagioclase composition for anorthositic rocks from the same and from adjoining areas (Brastad 1980; Griffin and Mørk 1981; Bryhni personal communication). The general decrease of the An content in the plagioclase during recrystallization is due to extensive epidote and hornblende crystallization.

The original igneous assemblage of these rocks is difficult to establish in Bjørkedalen as primary plagioclase only is preserved as relics. Anorthosite

Table 1 Whole rock analyses and norms of peridotite (274, 774, 2174), anorthosite (E73, 2873, 2974), and eclogites; in peridotite (7174, 7963, 7964), in anorthosite (E73, 2873, 2974), and eclogites; in peridotite (7174, 7963, 7964), in anorthosite (3573, 8033, 8062), in orthogneiss (8037, 8065), and in paragneiss (4374, 8674, 9074)

	274	774	2174	E73	2873	2974	7174	7963	7964	3573	8033	8062	8037	8065	4374	8674	9074
SiO_2	42.69	43.69	51.03	54.49	53.52	54.52	46.55	45.65	45.22	52.59	52.96	52.02	51.53	46.08	48.58	45.55	47.13
TiO_2	0.05	0.03	0.05	0.13	0.12	0.11	0.89	1.21	1.25	1.13	0.28	0.43	0.37	1.00	0.55	0.86	0.55
Al_2O_3	0.80	1.93	3.44	27.65	27.94	28.32	7.32	13.08	18.54	16.44	21.06	18.04	10.48	16.50	15.43	14.93	18.30
Fe_2O_3	0.96	0.07	1.02	0.58	0.35	0.24	3.52	2.36	3.80	1.72	7.30	9.18	0.49	1.29	1.96	2.83	1.13
FeO	6.92	8.48	4.87	0.36	1.15	0.57	9.55	10.60	8.33	8.05	4.67	5.57	8.07	14.08	4.53	10.89	6.52
MnO	0.11	0.13	0.12	0.01	0.20	0.01	0.23	0.18	0.14	0.18	0.11	0.17	0.15	0.23	0.12	0.20	0.11
MgO	47.45	43.49	33.70	0.12	0.74	0.10	18.55	12.34	6.34	6.03	7.81	8.42	16.35	8.68	10.12	14.80	11.07
CaO	0.54	1.49	1.92	11.25	10.56	11.09	11.37	12.02	11.09	8.19	6.56	9.32	9.88	7.94	16.18	7.08	12.02
Na_2O	0.18	0.20	0.14	4.49	4.52	4.86	1.06	1.28	2.53	2.79	2.74	2.90	1.17	3.19	1.98	1.33	1.76
K_2O	0.01	0.01	—	0.57	0.47	0.17	0.11	0.08	0.31	0.68	0.07	0.07	1.07	0.20	—	0.11	0.17
P_2O_5	0.01	0.01	0.03	0.14	0.13	0.13	0.07	0.06	0.28	0.26	0.04	0.06	0.22	0.17	0.03	0.16	0.17
Cr_2O_3	0.28	0.28	0.35	n.d.	n.d.	n.d.	n.d.	n.d.	n.d.	n.d.	n.d.	n.d.	n.d.	n.d.	n.d.	n.d.	n.d.
H_2O	n.d.	n.d.	2.73	n.d.	n.d.	n.d.	n.d.	n.d.	n.d.	1.43	n.d.	n.d.	n.d.	n.d.	n.d.	1.72	1.27
Total	100.00	99.81	99.40	99.79	99.70	100.12	99.22	98.96	97.83	99.49	98.93	100.61	99.78	99.36	99.48	100.46	100.20
CIPW—norms																	
Q	—	—	—	2.43	1.05	1.10	—	—	—	4.36	6.30	4.17	—	—	—	—	—
C	—	—	—	—	—	—	—	—	—	—	5.09	—	—	—	—	—	—
Or	0.05	0.05	—	3.32	2.73	0.98	0.23	0.48	1.89	4.10	0.40	0.40	5.95	1.19	—	0.66	1.00
Ab	1.43	1.61	1.20	39.35	39.90	43.39	10.76	11.63	23.40	25.70	24.65	25.80	9.91	28.76	14.48	12.06	15.70
An	1.19	3.90	8.40	52.98	50.68	52.89	8.23	30.07	39.47	31.13	32.40	35.70	19.00	30.20	32.96	34.41	41.27
Di	0.83	1.94	0.41	0.55	—	—	41.44	17.17	8.18	7.04	—	7.76	20.70	6.76	31.48	—	10.51
Hd	0.06	0.21	0.03	0.33	—	—	8.14	6.74	4.08	—	—	—	—	—	5.72	—	—
En	7.59	10.98	63.62	0.05	2.01	0.27	9.66	7.25	2.21	14.68	21.62	19.19	23.36	—	—	21.10	3.00
Fs	0.59	1.21	4.75	0.03	1.66	0.56	1.90	2.85	1.10	8.84	8.14	0.19	7.10	—	—	7.60	1.79
Ol	87.32	77.97	20.44	—	—	—	15.47	19.47	13.19	—	—	—	12.62	29.96	10.62	20.56	18.15
Il	0.01	0.04	0.07	0.18	0.16	0.15	1.24	1.71	1.79	1.62	0.40	0.60	0.48	1.39	0.76	1.21	0.76
Mt	0.89	0.07	1.02	0.29	0.36	0.24	2.82	2.50	4.09	1.85	0.83	5.93	0.53	1.49	2.03	1.93	1.17
Hm	—	—	—	0.20	—	—	—	—	—	—	—	—	—	—	—	—	—
Ap	0.02	0.02	0.06	0.29	0.27	0.27	0.12	0.13	0.60	0.56	0.08	0.13	0.39	0.25	0.06	0.34	0.35

n.d. = not detected.

Table 2   Analyses of olivine and orthopyroxene from peridotite, and plagioclase from anorthosite

	374 Ol	774 Ol	Opx	E73 Plag	2873 Plag
SiO_2	40.76	40.73	57.29	52.96	55.73
Al_2O_3	—	0.04	1.18	30.14	28.14
TiO_2	0.04	—	0.08	—	—
FeO	9.27	9.59	6.11	—	—
MnO	0.04	0.16	0.21	—	—
MgO	49.33	49.00	34.70	—	—
CaO	—	—	0.18	12.04	10.12
Na_2O	—	—	—	4.60	5.66
K_2O	—	—	—	0.18	0.21
Cr_2O_3	0.02	0.09	0.29	—	—
NiO	0.42	0.38	0.14	—	—
Total	99.92	99.99	100.18	99.92	99.84
Cation proportions					
Si	0.997	0.997	1.968	2.398	2.510
Al	—	0.001	0.032	0.602	0.490
Al	—	—	0.016	1.006	1.002
Ti	0.001	—	0.002	—	—
Fe	0.190	0.196	0.176	—	—
Mn	0.001	0.003	0.006	—	—
Mg	1.800	1.788	1.777	—	—
Ca	—	—	0.007	0.584	0.488
Na	—	—	—	0.404	0.494
K	—	—	—	0.010	0.012
Cr	—	0.002	0.008	—	—
Ni	0.008	0.007	0.004	—	—

bodies with preserved igneous assemblages are very rare in the Western Gneiss Region. The best example is that at Fiskå (Griffin and Mørk 1981). This body has igneous textures with megacrysts (5–25 cm) of andesine and hypersthene, and with local ilmenite concentrations. Another body in the Oppdal area consists also of plagioclase + orthopyroxene (Krill, personal communication 1981). The norms for these rocks from Bjørkedalen are plagioclase + orthopyroxene-dominated (Table 1); the dominant mafic igneous mineral probably was orthopyroxene, and the original rocks were anorthosite, leuconorite, and norite depending on the femic/mafic ratio. A crude estimate for the mean composition of the anorthositic rocks is leuconorite.

Eclogites

Eclogites are found within ortho- and paragneisses, peridotites and anorthosites. They are scarce in the orthogneisses, peridotites, and anorthosites, and very common in the paragneisses, and are even found within quartzites. The eclogites form pods or lenses ranging from a few decimetres to some hundreds of metres across.

The eclogites in orthogneiss are medium-grained rocks without any distinct layering. Usually they have a pronounced foliation defined by omphacite and phlogopite orientation. The eclogite paragenesis is usually omphacite + garnet + phlogopite + quartz ± rutile. More massive varieties lack phlogopite and have well-defined atoll garnets + omphacite + quartz ± orthopyroxene ± rutile.

Eclogites in the paragneisses are medium to coarse-grained and normally without any clear layering or foliation. The largest eclogites in this group seem, however, to have a large-scale layering only visible from a distance in a cliff-wall. This layering is probably defined by variations in modal garnet and omphacite content. This eclogite paragenesis includes garnet + omphacite ± orthopyroxene ± amphibole + quartz + kyanite ± zoisite ± phengite ± rutile.

Eclogites within and associated with anorthosites are medium- to fine-grained rocks and have a poorly defined compositional layering with alternating 0.3–1.5 cm thick garnet-, omphacite-, and kyanite-rich layers. Usually they show a pronounced foliation defined by the orientation of omphacite. The mineral assemblage of these eclogites is: Omphacite + garnet + kyanite + quartz ± rutile ± zoisite.

Eclogites within peridotite are medium-grained and show a pronounced layering with alternating 1–10 cm thick clinopyroxene- and garnet-rich layers, but without any distinct foliation. Eclogite parageneses for these rocks are:

1. Diopsidic clinopyroxene + garnet + orthopyroxene ± rutile.

2. Omphacite + garnet ± orthopyroxene ± quartz ± rutile.

3. Orthopyroxene + garnet.

All eclogites are more or less affected by retrograde metamorphism but those in peridotites are more strongly influenced than the others. Their central parts usually show well preserved eclogite mineralogy while margins and small bodies are strongly retrograded. This retrogradation is best described by a series of assemblages: eclogite → clinopyroxene + plagioclase symplectite + relic garnet → amphibolite (hornblende + plagioclase) ± relic garnet → amphibolite with hornblende + plagioclase, and without any textural evidence for the earlier existence of omphacite and garnet. Whether all the amphibolites are retrograded eclogites or have survived the eclogite metamorphism as amphibolites is not clear. There are clear mineralogical differences between the different groups of eclogites which probably reflect different modes of origin. Eclogites from paragneisses and anorthosites locally contain abundant kyanite, zoisite, and phengite. This sug-

Table 3A Garnet analyses in eclogites from anorthosite (3474, 8062, 8063), paragneiss (4374, 8674, 9074), and orthogneiss (8037, 8065). Fe^{3+} values are calculated

	3474	8062		8063		4374		8674	9074		8037		8065
	core	core	rim	core	rim	core	core	core	core	rim	core	rim	core
SiO_2	39.40	39.64	39.78	39.71	39.02	40.66	40.35	40.86	39.93	41.24	40.44	40.83	38.88
Al_2O_3	22.17	22.38	22.64	22.45	21.98	22.75	23.25	23.48	22.90	23.30	22.96	23.18	22.33
TiO_2	0.03	0.10	—	0.01	—	0.07	0.12	0.05	0.14	—	0.02	0.03	0.05
Fe_2O_3	1.47	0.73	0.86	1.20	2.41	1.15	1.64	—	1.35	0.72	1.88	1.69	0.70
FeO	21.96	19.70	19.15	20.97	21.22	15.51	15.22	19.62	19.38	14.07	16.85	17.64	23.58
MnO	0.45	0.63	0.51	0.57	0.79	0.44	0.09	0.19	0.42	0.13	0.47	0.34	0.49
MgO	9.25	10.14	9.88	9.48	9.61	12.16	11.71	12.21	9.70	14.57	14.74	14.83	8.80
CaO	6.43	7.09	8.03	7.06	5.86	8.63	9.50	5.49	8.41	7.14	3.73	3.46	5.29
Cr_2O_3	—	—	—	—	—	—	—	—	—	—	0.16	0.08	0.07
Total	101.16	100.40	100.86	101.45	100.89	101.36	101.88	101.90	102.24	101.16	101.25	102.08	100.19
Cation proportions													
Si	5.943	5.963	5.954	5.949	5.903	5.963	5.895	5.986	5.911	5.972	5.908	5.920	5.939
Al	0.057	0.037	0.046	0.051	0.097	0.037	0.105	0.014	0.089	0.028	0.092	0.080	0.061
Al	3.884	3.932	3.948	3.914	3.822	3.895	3.898	4.040	3.907	3.949	3.862	3.880	3.960
Ti	0.003	0.011	—	0.001	—	0.008	0.013	0.006	0.016	—	0.002	0.003	0.006
Fe^{3+}	0.167	0.082	0.097	0.135	0.274	0.127	0.180	—	0.151	0.078	0.207	0.184	0.081
Fe^{2+}	2.770	2.478	2.398	2.628	2.685	1.902	1.860	2.404	2.400	1.704	2.059	2.139	3.012
Mn	0.057	0.080	0.065	0.072	0.101	0.055	0.011	0.024	0.053	0.016	0.058	0.042	0.063
Mg	2.080	2.274	2.204	2.117	2.167	2.658	2.550	2.666	2.140	3.145	3.210	3.205	2.004
Ca	1.039	1.143	1.288	1.133	0.950	1.356	1.487	0.862	1.334	1.108	0.584	0.537	0.866
Cr	—	—	—	—	—	—	—	—	—	—	0.018	0.009	0.008

Table 3B Cpx analyses in eclogites. $Fe^{2+} = Fe^{tot}$

	3474 inc.	8062 core	rim	8063 core	rim	4374 core	inc.	8674 core	9074 core	rim	8037 core	rim	8065 core
SiO_2	56.50	55.88	55.50	55.99	55.64	55.04	54.34	56.65	55.60	55.64	55.06	55.10	54.90
Al_2O_3	11.91	10.53	10.62	10.03	11.06	7.27	7.29	8.63	8.01	8.05	4.25	4.12	10.37
TiO_2	0.13	0.06	—	0.02	0.09	0.01	0.11	0.11	0.08	0.05	0.09	—	0.10
FeO	3.56	2.95	2.83	3.10	3.66	2.42	2.57	3.66	2.22	2.38	3.63	3.67	5.23
MnO	0.01	0.10	0.01	0.07	0.04	0.02	0.01	—	0.12	0.16	0.04	0.15	—
MgO	8.84	9.59	9.34	9.44	8.88	12.48	12.78	11.34	12.42	12.11	14.00	13.69	8.63
CaO	13.00	14.49	14.39	14.69	14.09	19.00	18.92	15.66	18.08	18.18	19.93	19.78	13.04
Na_2O	7.13	6.50	6.33	6.29	6.74	4.06	4.00	5.53	4.30	4.29	2.96	3.03	6.93
Cr_2O_3	—	—	—	—	—	—	—	—	—	—	0.17	0.25	0.12
Total	101.08	100.10	99.02	99.63	100.20	100.30	100.02	101.58	100.83	100.86	100.13	99.79	99.02
Cation proportions													
Si	1.980	1.982	1.987	1.996	1.976	1.967	1.951	1.990	1.968	1.971	1.989	1.997	1.985
Al	0.020	0.018	0.013	0.004	0.024	0.033	0.049	0.010	0.032	0.029	0.011	0.003	0.015
Al	0.471	0.423	0.435	0.417	0.439	0.273	0.260	0.347	0.302	0.307	0.169	0.173	0.427
Ti	0.003	0.002	—	0.001	0.002	—	0.003	0.003	0.002	0.001	0.002	—	0.003
Fe^{2-}	0.104	0.088	0.085	0.092	0.109	0.072	0.077	0.108	0.066	0.070	0.110	0.111	0.158
Mn	—	0.003	—	0.002	0.001	0.001	—	—	0.004	0.005	0.001	0.005	—
Mg	0.462	0.507	0.498	0.502	0.470	0.665	0.684	0.594	0.655	0.639	0.754	0.740	0.449
Ca	0.488	0.551	0.552	0.561	0.536	0.728	0.728	0.589	0.686	0.690	0.771	0.768	0.505
Na	0.484	0.447	0.439	0.435	0.464	0.281	0.278	0.377	0.295	0.295	0.207	0.213	0.486
Cr	—	—	—	—	—	—	—	—	—	—	0.005	0.007	0.003
% Jad	45.5	40.4	41.9	41.3	41.6	23.9	21.4	34.0	27.4	27.9	16.1	16.9	41.2

gests that they had Al-rich protoliths such as plagioclase-rich rocks. On the other hand, eclogitic rocks from peridotite, with diopsidic clinopyroxene and orthopyroxene + garnet assemblages and without quartz, kyanite, etc., are probably formed from plagioclase-poor or plagioclase-free protoliths. No primary igneous minerals have been recognized in any of the Bjørkedalen eclogites.

Peridotite/Anorthosite/Eclogite Relations at other Localities in the Sunnmøre District

The peridotite/anorthosite/eclogite association is not restricted to Bjørkedalen but is regionally distributed. In order to show this, and to demonstrate the intimate peridotite/anorthosite association in this district, descriptions of three other localities are given below.

Kvitfannegga

In the mountain Kvitfannegga (Fig. 1) anorthosites and serpentinized peridotite are folded together in a tight antiform. On the eastern and western side of the mountain the following sections are found, outwards from the core of the fold.

Rock type	Eastern side Thickness	Western side Thickness
Peridotite	1–2 m	>20 m
Anorthosite	1 m	3–5 m
Peridotite	2 m	10 m
Anorthosite	1–2 m	2–3 m
Peridotite	2–3 m	5 m
Anorthosite	100–200 m	100–200 m
Peridotite	0–5 m	0–5 m
Amphibolite/ Eclogite	0–ca 200 m	0–ca 200 m
Orthogneiss		

The distance between these localities is about 1.5 km along the strike which gives the minimum continuous length of these layers. The lack of mafic rocks intermediate between the peridotites and the anorthosites in both sections is a striking feature though there are amphibolites and eclogites in the outer zone. With the exception of the eclogites, the metamorphic grade is amphibolite facies.

Valldal

This locality is situated outside the area of Fig. 1, about 80 km east–northeast of Bjørkedalen. This is one of the largest anorthositic bodies of the district and a schematic section through it is given in Fig. 3.

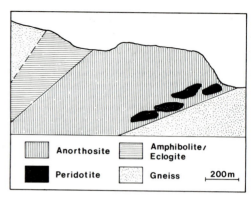

Fig. 3 A simplified section across the anorthosite–peridotite–amphibolite/eclogite series at the mountain Sandfjellet at Valldal, Sunnmøre

The peridotites occur there as lens-shaped bodies near the structural base of the anorthosite and appear to be a boudinaged layer. Again there is a complete lack of intermediate mafic rocks between the olivine and plagioclase rocks. Metamorphic grade is the same as for the Kvitfannegga locality.

Dalsfjorden

Dalsfjorden (Fig. 1) was mentioned by Eskola in his work on Norwegian eclogites (Eskola 1921). His section through the rock units at the bottom of the cliff at Åmelfot is given below, together with a cross-section recorded by the author at the top of the cliff.

Bottom (Eskola 1921)	*Top* (this work)
Gneiss	Gneiss
Peridotite	Eclogite
Eclogite	Peridotite
Anorthosite	Amphibolite
	Anorthosite
	Eclogite

From a distance it is easy to see why the two sections are so different (Fig. 4). The peridotite cuts through a series of anorthosites and partially

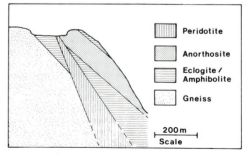

Fig. 4 A schematic drawing of the SE-side of the mountain Sølvsbergknausen at Dalsfjord, Sunnmøre. The section which is shown is simplified

amphibolitized eclogites which appear to have mutually gradational contacts. Metamorphic grade is the same as at the two other localities.

Contact Relationships

Contacts between the rock types described here are in most cases so disturbed by late retrograde reactions and shearing that it has not been possible to demonstrate original relations. A summary of the best preserved contacts, however, gives some indications of the relations between these rocks before hydration reactions started.

Anorthosite–peridotite

The contacts between anorthositic and peridotitic rocks are more or less tectonized. Those that are strongly tectonized have a well-developed late foliation, defined by chlorite and epidote, and probably belong to the very last low-grade metamorphic episode. The least tectonized ones are found at Kvitfannegga (Fig. 1) where small serpentinites are strongly folded together with anorthosites. Here the following sequence of minerals and mineral assemblages is found outwards from the serpentinite: serpentine → actinolite → hornblende + plagioclase intergrowth → actinolitic hornblende → clinozoisite + actinolite → clinozoisite/epidote + actinolite + plagioclase. The total thickness of the sequence is 5 cm.

This section is probably comparable with those found in the Dalsfjord area in Sunnfjord at anorthosite/peridotite contacts (Cuthberth personal communication 1981). The differences are that in Sunnfjord, red garnets appear instead of the hornblende + plagioclase intergrowths and the epidote mineral is zoisite instead of clinozoisite/epidote. Hornblende + plagioclase intergrowths are common as garnet breakdown textures elsewhere in Bjørkedalen and there can be little doubt that the intergrowths observed in the reaction zone at Kvitfannegga are derived from pre-existing garnets. These zones have two possible origins: high pressure reaction between peridotite and anorthosite or rodingitization of an anorthosite in contact with a serpentinite.

High pressure reaction between olivine and plagioclase commonly produces the following sequence (Griffin 1971): olivine–orthopyroxene–clinopyroxene–garnet–plagioclase. Rodingitization is essentially a Ca-metasomatism which produces a variety of Ca-minerals such as hydrogrossular, grossular/andradite, diopside, prehnite, zoisite, vesuvianite etc. (Anhaeusser 1979). This contact rock is not a true rodingite but shows some characteristics of rodingites, especially the zoisite/clinozoisite bearing zones with a Ca-content of 17–25% CaO—far higher than normal basic rocks. It is not yet clear whether the Ca-enrichment occurred prior to the eclogite metamorphism, or has overprinted a corona assemblage formed by the olivine/plagioclase reactions described by Griffin.

Peridotite–eclogite

The eclogites within the peridotites are usually strongly hydrated and tectonized along their margins with the formation of hornblende/actinolite + chlorite zones. This might be due to the fact that the internal eclogites seem to lie near the strongly tectonized margins of the main peridotite. However, the least retrograded eclogites show no extra disturbance of the contacts towards the peridotites.

Anorthosite–eclogite

The contacts between the anorthosites and their enclosed eclogites are mostly gradational. Eclogites in anorthosite may also have sharp contacts with the anorthosites and are then usually amphibolitized along the margins. Gradational contacts (1–8 m thick) consist of amphibolites with leuconoritic and noritic normative composition. In these contact zones no remnants of eclogite metamorphism are seen. Anorthosite–eclogite contacts are less sheared than eclogite/gneiss and anorthosite/peridotite contacts.

Chemistry

Bulk chemical composition

Bulk chemical analyses and normative calculations for peridotitic, anorthositic, and eclogitic rocks are presented in Table 1. The analyses were performed by XRF on glass pellets (sample/$Na_2B_4O_7$ = 1/9), except for Na and low Mg-analyses which were done by atomic adsorption, and Fe^{2+} which was determined by titration. Peridotite analyses correspond to dunite/lherzolite except for sample 2174 which was taken from an orthopyroxene-rich part near the border of the peridotite. The anorthosite analyses show quartz + orthopyroxene-normative compositions and rather constant Ca/Na-ratios. These analyses are representative for the most common type of anorthosite in Bjørkedalen (Brastad 1980).

Eclogite analyses show basaltic compositions and large variations in Mg/Fe-ratios within each group. The eclogites within anorthosite are lower in MgO and higher in Al_2O_3 than the others, and have quartz

+ plagioclase + orthopyroxene-dominated norms, as do the enclosing anorthosites. The three other groups of eclogites have plagioclase ± clinopyroxene ± orthopyroxene + olivine in the norms. Eclogites within peridotite show the largest variations in MgO and Al_2O_3-content. These variations are probably best explained by accumulation of plagioclase + clinopyroxene-dominated mixtures with different proportions of these two minerals (Brastad 1982).

Eclogites within orthogneiss and paragneiss show no clear compositional differences which would separate them into two groups. The most striking feature—which might be important—is the difference in normative plagioclase compositions which is An_{50-55} for the first category and An_{58-74} for the second.

REE-patterns for eclogites from peridotite, anorthosite, and paragneiss show clear differences (Fig. 5). Eclogites from peridotite are enriched in light or intermediate REE, and with small positive Eu anomalies for the first group. Eclogites within anorthosite are usually enriched in light REE and show large positive Eu anomalies. Eclogites within paragneiss have flat REE patterns with consistent small but distinct positive Eu anomalies. These patterns suggest low pressure (plagioclase-bearing) protoliths for all the eclogites of these groups.

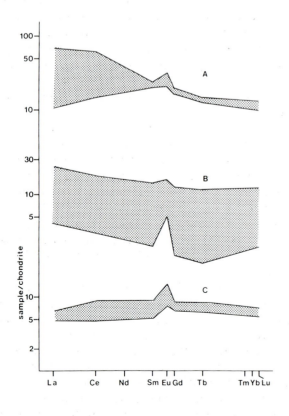

Fig. 5   Generalized REE-patterns for eclogites in: (a) peridotite, (b) anorthosite, (c) paragneiss. Data from Brastad and Brunfelt (in prep.)

Mineral chemistry and equilibration conditions

Minerals from peridotite, anorthosite, and eclogites were analysed using an ARL–EMX microprobe analyser with LINK energy dispersive equipment (Table 2 and 3). Omphacite and garnet pairs from eclogites in anorthosite, orthogneiss, and paragneiss (Table 3) have been used for calculation of equilibration temperatures, following the method of Ellis and Green (1979) with the revisions of Krogh (1982). The calculated temperatures (Table 4) are maximum temperatures ($Fe^{2+}_{cpx} = Fe^{tot}$ and $Fe^{2+}_{gnt} = Fe^{tot} - Fe^{3+calc}$). Omphacite analyses are low in Fe and calculation of Fe^{3+} will therefore introduce unreasonably large variations in Fe^{2+}/Fe^{3+}-ratios which in turn give large variations in calculated temperatures. Equilibration temperatures (Table 4) show good agreement for these three groups of eclogites with averages of 711°C and 722°C for core/core and rim/rim combinations respectively, assuming 15 kb for the equilibration pressure. These temperatures are in accord with the general prograde trend of the eclogites in WGR, but slightly higher than temperatures from neighbouring eclogites (Griffin et al. this volume). As the temperatures given here are maximum values, these differences are not significant. Equilibration pressures are not yet calculated for these eclogites.

Equilibration conditions for the eclogites found within the main peridotite are clearly different from those given above. Average maximum temperature for $core_{cpx}/core_{gnt}$ is 860°C at an average maximum

Table 4   Equilibrium temperatures for eclogites in anorthosite (3474, 8062, 8063), paragneiss (4374, 8674, 9074), and orthogneiss (8037, 8065). The methods used are described by Ellis and Green (1979) and Harvey and Green (1982). For calculations: $Fe^{2+}_{cpx} = Fe^{tot}$, $Fe^{3+}_{gnt} = $ calc.

Sample	K_D	T°C(15 kb)	T°C(25 kb)
3474 Core	5.92	681	714
8062 Core	6.20	687	719
Rim	6.37	707	739
8063 Core	6.77	662	694
Rim	4.34	771	807
4374 Core	6.61	705	737
Core	6.48	729	762
8674 Core	4.96	706	741
9074 Core	4.95	756	791
8037 Core	4.40	704	740
Rim	4.45	689	726
8065 Core	4.27	766	803
Average Core	5.13	711	745
Rim	5.05	722	756

Equilibration conditions for eclogite in peridotite:
Average Core	860°C, 27.6 kb	(K_D = 3.95)
Rim	700°C, 13.0 kb	(K_D = 5.36)

pressure of 27.6 kb while rim_{cpx}/rim_{gnt} yields 700°C, 13 kb–still maximum values (Brastad 1983). The first P, T conditions are clearly too high, and have to be lowered when the real Fe^{3+}/Fe^{2+} ratios in clinopyroxene, which control the temperatures, can be established. An explanation for the differences in equilibration temperatures shown in Table 4 might be that omphacites from eclogites in peridotite have higher Fe^{3+}/Fe^{2+} ratios than omphacites from the other eclogites. Whole rock Fe^{3+}/Fe^{2+} ratios (Table 1 and unpublished analyses) indicate that this is true, and with few exceptions.

Pressure estimates are based on the solubility of Al in orthopyroxenes in equilibrium with garnets (Harley and Green 1982). The orthopyroxenes used for calculation are clearly zoned with respect to Al (Brastad 1983), and the interpretation of this zoning has strong influence on the pressure estimates which are ca. 30 kb and 17–18 kb respectively for core and rim.

The contrast in P, T conditions between eclogitic rocks from peridotites in the WGR and from outside have been discussed by a number of authors. Brueckner (1977), Carswell (1968), Medaris (1980, 1982), and Mercy and O'Hara (1965) concluded that the eclogites within peridotite were originally equilibrated under higher P, T conditions than those outside. Carswell (1981) and Griffin and Qvale (this volume) have demonstrated that the eclogitic rocks from within peridotite have experienced metamorphic conditions compatible with those from outside. Most of these authors do, however, agree that the existing data on P, T conditions on eclogites demonstrate that the same high-P metamorphism was imposed on all the eclogites.

Origin of Peridotites, Anorthosites and Eclogites in Bjørkedalen

Peridotite and eclogites within peridotite

The generally accepted hypothesis for the origin of the Western Gneiss Region peridotites is that they are fragments of depleted upper mantle (Carswell 1968; Brueckner 1977; Medaris 1980). This is based on the consistently high forsterite content of the olivines (Fo ≈ 90%), of the dunitic composition, and of the assumed very high equilibration pressures of the eclogitic assemblages. All these characteristics are present in Bjørkedalen, and the peridotites there will therefore fit easily into this model. Whether these arguments are valid or not is however still an open question. In a recent study, Carswell (1983) has shown that a number of peridotites with lower forsterite content (Fo > 67) exist in the gneiss region, and that these are interlayered with mafic rocks.

The peridotite in Bjørkedalen is mostly dunitic, but has clearly a more mafic character near the structural top, with interlaying of eclogites, amphibolites, and even meta-anorthosite. Whether this interlaying also is accompanied by a decrease in modal or normative olivine content and a decrease in Fo content of the olivines has not yet been studied.

This interlayering might have been formed by intrusion of mafic rocks or tectonic introduction of neighbouring rocks, but might also have been formed by accumulation of alternating olivine-rich and olivine-poor mixtures. Repeated accumulation of peridotites, olivine-bearing (normative) mafic rocks, and quartz-bearing (normative) anorthosites seem to be highly unlikely and most probably indicates that the anorthosites are tectonically emplaced into a series consisting of the first two types of rocks. The eclogites and amphibolites of this series show large variations in $Mg/Mg + Fe^{2+}$ ratios (0.35–0.67). Variations of this range are common in large layered mafic complexes such as the Muskox intrusion (Irvine 1970; Irvine and Baragar 1972), but not in 10–40 m thick intrusions like these. The observed $Mg/Mg + Fe^{2+}$ variations are comparable with those in an olivine-gabbro layer in the upper part of the Muskox intrusion. The contrast in $Mg/Mg + Fe^{2+}$ ratios between the enclosing peridotite and the gabbroic rock is, however, somewhat larger in Bjørkedalen than in the Muskox intrusion because of the higher Fo content of the olivine here. Whether the measured Fo content in Bjørkedalen is consistent throughout the peridotite body or is confined to the analysed massive dunite is still an open question and cannot be used as a conclusive argument.

A second process which might yield chemical variations like those observed in Bjørkedalen is intrusion of a ferrobasaltic magma into an existing system of peridotites with a resulting mixing and differentiation. Processes like this are common at ocean floor spreading centres (Clague and Bunch 1976) and have been proposed as explanation for the origin of some superferrian eclogites in the neighbouring peridotite at Åheim described by Griffin and Qvale (this volume).

The first model of formation, 'the layered complex process' will relate the internal eclogites and amphibolites directly to the peridotite by the same differentiation and accumulation process. The second model relates them partially to the peridotite as being formed in the same environments, but by different processes. With the existing knowledge of these rocks, it seems impossible to favour either of these models. An important conclusion is, however, that both theories require low pressure protoliths for the eclogites as well as for their enclosing peridotites.

Anorthosites and their enclosed eclogites

A short consideration of other anorthosite occurrences may help in understanding the relations between the rock types dealt with here. There are three major groups of anorthosites.

1. Layers in basic plutons of Stillwater–Bushveld type (Hess 1960; Visser and von Gruenewaldt 1970). These have mainly a bytownitic plagioclase, and are associated with large amounts of intermediate mafic rocks together with relatively small amounts of peridotites with intermediate to magnesian (Fo 50–90)-olivines. The mean compositions of the intrusions are gabbroic.
2. Intermediate anorthosites of Michikaman type (Olmsted 1969). Here the plagioclase is labradorite with mean composition ca. An 60. These anorthosites are clearly differentiated rocks with small amounts of relatively iron-enriched peridotites near the base, associated with olivine gabbros, and with anorthosites and gabbroic anorthosites as the principal rocks. The estimated mean composition is that of an anorthositic gabbro.
3. Massive anorthosites of Adirondack–Egersund type (Buddington 1969; Michot and Michot 1969). The main constituent of these is andesine (An 38–45%). Field and chemical relations have been interpreted in terms of (a) genetic relations to relatively small amounts of mafic rocks (Simmons and Hanson 1978; Wiebe 1980; Morse 1982), or (b) genetic relations to small amounts of mafic rocks and large amounts of intermediate to acid rocks (Michot and Michot 1969; de Waard and Romey 1969). The mean compositions are considered to be gabbro–anorthositic and intermediate respectively.

In Bjørkedalen the close field and chemical relationships between anorthosites and their associated amphibolites and eclogites makes a genetic relation between them through magmatic processes probable. Bearing this in mind and considering the mean composition (leuconoritic), the bulk feldspar composition (An 50–65) and the relative amounts of the different rock types within this series of Bjørkedalen rocks, the greatest similarities are to the intermediate anorthosites of Michikaman type.

Eclogites in gneiss

The country-rock eclogites of the WGR are suggested by Bryhni et al. (1970) and Griffin et al. (this volume) to have originated by metamorphism of a variety of basic rocks. The question of the origin of the eclogites is intensely debated (see summary by Griffin et al. this volume). It is not the aim of this study to repeat that discussion, but to indicate the most probable origin of the Bjørkedalen eclogites. The latter can be subdivided into eclogites in ortho- and paragneisses, the first being scarce and the second very common. The two associations of eclogites might have had the same type of protoliths, but it is more likely that they have different origins, corresponding to their different modes of occurrence. The bulk compositions are olivine-gabbroic for both associations (Table 1). Bulk REE-analyses on eclogites from paragneiss (Fig. 5) give horizontal patterns comparable with tholeiitic lavas. Eclogites from orthogneiss are not yet analysed for REE, but analyses from elsewhere in the WGR on eclogitized olivine-gabbro from orthogneiss give LREE-enriched patterns (Krogh and Brunfelt 1981; Mørk personal communication.) Olivine-gabbros (dolerites) are common in the WGR, but all the known localities are found within rocks which are supposed to be orthogneisses.

The eclogite parageneses are comparable for both associations except that eclogites in orthogneiss may have abundant phlogopite while those in paragneiss commonly have amphibole. Metamorphosed olivine-gabbros from elsewhere in the WGR commonly have abundant phlogopite as a part of the eclogite assemblages (Mørk 1983; Brastad in preparation), but may also have amphibole.

From these arguments it is inferred that the eclogites within the orthogneisses at Bjørkedalen most probably originated as olivine-gabbros like other eclogites within orthogneisses in the WGR. The eclogites within paragneisses are clearly different from those in orthogneiss, suggesting different protoliths such as tholeiitic lavas.

Discussion

The intimate association of peridotite and anorthosite seen in Bjørkedalen and in neighbouring localities is common in the Western Gneiss Region (Bryhni 1966; Brueckner 1977). The question of a genetic relationship between these two groups of rocks is relevant in the light of their field association and has been proposed by a number of authors, e.g. Eskola (1921), Bryhni (1966). The interlayering of magnesian peridotites and intermediate An anorthosites is unusual with regard to the three main classes of anorthosites mentioned above. Moreover, this interlayering most commonly includes peridotites s.s. and anorthosites s.s. as seen in a number of occurrences in Bjørkedalen, in the Kvitfannegga mountain, in Valldal and in other localities in the WGR. Basic rocks—amphibolites with normative quartz + opx and with layers of metaanorthosites—are in fact present next to the main peridotite in Bjørkedalen, but these show field and whole rock chemical relations to the anorthosites and not to the peridotites. The basic rocks within the peridotites, which might be related to them, occur in very small volumes. The peridotites at Bjørkedalen have more magnesian olivines than peridotites associated with any of the three main classes of anorthosites. A metamorphic event (serpentinization followed by metamorphism) might produce very magnesian secondary olivines (Qvale 1978) but it seems difficult to generate large masses of dunitic peridotites, like the main peridotite in Bjørkedalen, with a supposed constant $Fo \approx 90\%$ by this mechanism. With the present data there is no obvious genetic relationship between the anorthositic and peridotitic rocks of Bjørkedalen.

Although genetic relationships between the Bjørkedalen rocks cannot be demonstrated, they may still have a long common history. This is of particular interest with respect to the question of whether they have been subjected to a Caledonian eclogite metamorphism (Griffin and Brueckner 1980). All the eclogitic rocks have probably been equilibrated under the same high-P metamorphism as the other eclogites in the WGR (Griffin *et al.* this volume). The paragneisses show mineral assemblages which are equivalent to the eclogite assemblages in the basic rocks. Equivalent assemblages have not yet been found in the orthogneisses, but have most probably been obliterated by the retrograde metamorphism. Anorthosites and their enclosed eclogites show the same chemical characteristics and their protoliths have most probably been formed by the same magmatic process. In Bjørkedalen anorthositic and peridotitic rocks have been folded together at least twice, and in Tafjord and in Almklovdalen these rocks share all the fold episodes with the eclogites and gneisses (Brueckner 1977; Medaris 1980; Cordellier *et al.* 1981).

These relationships strongly suggest that the association of rocks described here was brought together before the Caledonian eclogite-forming event. Whether this happened prior to or at an early stage of the Caledonian orogeny has not yet been determined. This implies also that the late crushing and shearing at the contacts between rock units are of only local importance and not related to large scale tectonic movements.

Acknowledgements

Financial support for this study were given by the Norwegian Council for Science and the Humanities (NAVF) and by the Nansen foundation. Thanks are given to Drs. William L. Griffin and Inge Bryhni for revising the manuscript.

References

Anhaeusser, C. R. 1979. Rodingite occurrences in some Archaean ultramafic complexes in the Barberton Mountain Land. *Precambrian Res.*, **8**, 49–76.

Brastad, K. 1977. *Anorthositter og andre bergarter på grensen mellom Nordfjord og Sunnmøre.* Unpubl. Cand. Real. Thesis. Univ. of Oslo, 198 pp.

Brastad, K. 1980. Beskrivelse av anorthosittforekomstene i Bjørkedalen, Ljosurda, Bauvann: Eid, Sogn og Fjordane, Volda, Møre og Romsdal. *Norges geol. Unders.*, unpublished report nr. 1560/19, 11 pp.

Brastad, K. 1982. Eclogites within the Bjørkedalen Peridotite, western Norway. *Terra Cognita*, **2**, 327.

Brastad, K. 1983. Petrology of eclogitic rocks within the Bjørkedalen peridotite, W. Norway. *Bull. Minéralogie*, (in press).

Brueckner, H. K. 1977. A structural, stratigraphic and petrologic study of anorthosites, eclogites, and ultramafic rocks and their country rocks, Tafjord area, western south Norway. *Norges geol. Under.*, **332**, 1–53.

Bryhni, I. 1966. Reconnaissance studies of gneisses, ultrabasites, eclogites and anorthosites in outer Nordfjord, western Norway. *Norges geol. Under.*, **241**, 1–68.

Bryhni, I., Fyfe, W. S., Green, D. H. and Heier, K. S. 1970. On the occurrence of eclogite in Western Norway, *Contr. Miner. Petrol.*, **26**, 12–19.

Buddington, A. F. 1969. Adirondack anorthositic series, *N.Y. State Mus. and Sci. Serv. Mem.*, **18**, 215–232.

Carswell, D. A. 1968. Possible primary upper mantle peridotite in Norwegian basal gneiss. *Lithos*, **1**, 322–355.

Carswell, A. D. 1981. Clarification of the petrology and occurrence of garnet lherzolites, garnet websterites and eclogite in the vicinity of Rødhaugen, Almklovdalen, W. Norway. *Norsk geol. Tidsskr*, **61**, 249–260.

Carswell, D. A. 1983. The petrogenesis of contrasting Fe–Ti and Mg–Cr garnet peridotites in the high grade gnesis complex of western Norway. *Bull. Minéralogie*, (in press).

Clague, D. A. and Bunch, T. E. 1976. Formation of ferrobasalt at East Pacific midocean spreading centers, *J. Geophys. Res.*, **81**, 4247–4258.

Cordellier, F., Boudier, F. and Boullier, A. M. 1981. Structural study of the Almklovdalen massif (Southern Norway). *Tectonophysics*, **77**, 257–281.

Ellis, D. J. and Green, D. H. 1979. An experimental study of the effect of Ca upon garnet–clinopyroxene Fe–Mg exchange equilibria. *Contr. Miner. Petrol.*, **71**, 13–22.

Eskola, P. 1921. On the eclogites of Norway. *Skr. Vidensk. Selsk. Christiania, Mat.–Nat. kl. 1*, **8**, 1–118.

Gjelsvik, T. 1951. Oversikt over bergartene i Sunnmøre og tilgransende deler av Nordfjord. *Norges geol. Unders.*, **179**, 1–45.

Griffin W. L. 1971. Genesis of coronas in anorthosites of the Upper Jotun Nappe, Indre Sogn, Norway. *J. Petrol.*, **12**, 219–243.

Griffin, W. L. and Brueckner, H. K. 1980. Caledonian Sm–Nd ages and a crustal origin for Norwegian eclogites. *Nature*, London, **285**, 319–321.

Griffin, W. L. and Mørk, M. B. E. 1981. Eclogites and

basal gneisses in western Norway. *Excursion B1, Excurs. in Scand. Caled.*

Griffin, W. L., Austrheim, H., Brastad, K., Bryhni, I., Krill, A., Mørk, M. B. E., Qvale, H. and Tørudbakken. B. this volume. High pressure metamorphism in the Scandinavian Caledonides.

Griffin, W. L. and Qvale, H. this volume. Superferric eclogites and the crustal origin of garnet peridoties.

Harley, S. L. and Green, D. H. 1982. Garnet–orthopyroxene barometry for granulites and peridoties. *Nature, 300,* 697–701.

Heinrich, C. A. 1982. Kyanite-eclogite to amphibolite facies evolution of hydrous mafic and pelitic rocks, Adula Nappe, Central Alps. *Contr. Miner. Petrol., 81,* 30–38.

Hess, H. H. 1980. Stillwater Igneous Complex, Montana. *Geol Soc. Am. mem., 80,* 230 pp.

Irvine, T. N. 1970. Crystallization sequences in the Muscox intrusion and other layered intrusions. I. Olivine–pyroxene–plagioclase relations. *Spec. Publs. Geol. Soc. S. Africa, 1,* 441–476.

Irvine, T.N. and Baragar, W. R. A. 1972. The Muscox intrusion and Coppermine River lavas, Northwest Territories, Canada. *Int. Geol. Congr., 24th, Montreal, Field Excursion A29, Guidebook, 70.*

Krogh, E. J. 1982. Granat-klinopyroksen geotermometri—en retolkning av eksisterende eksperimentelle data. *Geolognytt, 17,* 33.

Krogh, E. J. and Brunfelt, A. O. 1981. REE, Cs, Rb, Sr and Ba in glaucophane-bearing eclogites and associated rocks, Sunnfjord, western Norway. *Chem. Geol., 33,* 295–305.

Lappin, M. A. 1966. The field relationships of basic and ultrabasic masses in the basal gneiss complex of Stadlandet and Almklovdalen, Nordfjord, southwestern Norway. *Norsk geol. Tidsskr., 46,* 439–495.

Lappin, M. A. and Smith, D. C. 1978. Mantle-equilibrated orthopyroxene eclogite pods from the basal gneisses in the Selje district, western Norway. *J. Petrol., 19,* 530–584.

Medaris, L. G. Jr, 1980. Petrogenesis of the Lien peridote and associated eclogites, Almklovdalen, western Norway. *Lithos, 13,* 339–353.

Medaris, L. G. Jr, 1982. A review of garnet peridotites within gneiss in western Norway. *Terra Cognita, 2,* 303.

Mercy. E. L. P. and O'Hara, M. J. 1965. Chemistry of some garnet-bearing rocks from south Norwegian peridotites. *Norsk geol. Tidsskr., 45,* 323–332.

Michot, J. and Michot, P. 1969. The problem of anorthosites: the South-Rogaland igneous complex, southwestern Norway. *N.Y. State Mus. & Sci. Serv. Mem., 18,* 339–410.

Morse, S. A. 1982. A partisan review of Protoerozoic anorthosites. *Am. Mineral., 67,* 1087–1100.

Mørk, M. B. E. 1983. A gabbro–eclogite transition on Flemsøy, Sunnmøre, western Norway. *Chem. Geol* (in press).

Olmsted, J. F. 1969. Petrology of the Mineral Lake Intrusion, northwestern Wisconsin. *N.Y. State Mus. and Sci. Serv. Mem., 18,* 149–161.

Qvale, H. 1978. *Geologisk undersøkelse av et kaledonsk serpentinittfelt ved Baddersheim, Hordaland.* Unpubl. Cand Real. Thesis, Univ. of Oslo, 252 pp.

Simmons, E. C. and Hanson, G. N. 1978. Geochemistry and origin of massif-type anorthosites. *Contr. Miner. Petrol., 66,* 119–135.

Smith, D. C. 1981. A reappraisal of factual and mythical evidence concerning the metamorphic and tectonic evolution of eclogite-bearing terrain in the Caledonides. *Terra Cognita, 1,* 73–74.

Visser, D. R. L. and von Gruenewaldt, (Eds) 1970. Symposium on the Bushveld Igneous Complex and other layered intrusions. *Geol. Soc. S.Afr. Spec. Publ., 1,* 763 pp.

de Waard, D. and Romey, W. D. 1969. Petrogenetic relationships in the anorthosite–charnockite series of Snowy Mountain Dome, south–central Adirondacks. *N.Y. State Mus. and Sci. Serv. Mem., 18,* 307–315.

Wiebe, R. A. 1980. Anorthositic magmas and the origin of proterozoic anorthosite massifs. *Nature, 286,* 564–567.

The Caledonide Orogen—Scandinavia and Related Areas
Edited by D. G. Gee and B. A. Sturt
© 1985 John Wiley & Sons Ltd

Eclogites of the Seve Nappe, central Scandinavian Caledonides

H. van Roermund

Mineralogisk–Geologisk Museum, Sarsgt. 1, N-Oslo 5, Norway

ABSTRACT

Eclogites occur at two tectonic levels in the Seve Nappe, Jämtland, Central Scandinavian Caledonides, i.e., the Central and the Eastern Belts. Both eclogite types have a comparable tholeiitic bulk composition but differ in their mineral chemistry. Seve eclogites from the Central Belt contain garnets with 22–35% pyrope and omphacites with 28–41% jadeite. Eastern Belt Seve eclogites contain garnets with 8–22% pyrope and omphacites with ~48% jadeite. The eclogites also contain quartz, phengite, rutile, apatite, and zircon, while zoisite is confined to Central Belt eclogites. The physical conditions of eclogite formation have been estimated to be 18.0 ± 1.0 Kb and $780 \pm 50°$C for the Central and 14.0 ± 1.5 Kb and $550 \pm 70°$C for the Eastern Belt eclogites. These data, with a reconstruction of the prethrusting position of the nappes, support a northeasterly continuation of the linear NE-trending regional pattern of minimum distribution coefficients of Mg and Fe between garnet and clinopyroxene in eclogites from the Basal Gneiss Complex of southwest Norway.

Introduction

Since Eskola's classical work *'On the Eclogites of Norway'* the Norwegian eclogites have become one of the best known eclogite occurrences in the world (Eskola 1921). Most of the Norwegian eclogites occur in a wide belt between Bergen and Trondheim but further north in Norway scattered occurrences of eclogitic rocks appear again along the trend of the Caledonides (Bryhni *et al.* 1977). Less well known are the eclogites of the Seve Nappe, Jämtland (Zwart 1974) and Nasafjäll, Central Scandinavian Caledonides (Andreasson *et al.* 1981). The aim of the present paper is to describe the field appearance, the whole rock chemistry, and the primary mineralogy of the Jämtland Seve eclogites. The origin of the rocks will be discussed and a tentative comparison with the eclogite occurrences in the Western Norwegian Gneiss Belt of Central Norway will be made.

The Seve Nappe, which is the structurally lowermost part of the Seve–Köli Nappe Complex, forms one of the major nappe units in the Central Scandinavian Caledonides. It rests upon lower grade nappes (Särv, Offerdal, Olden, and Blaik Nappes)

and it is overlain by the lower grade Köli Nappe. The Seve Nappe is commonly subdivided from top to bottom into a Western, a Central, and an Eastern Belt (Trouw 1973; Zwart 1974). The Central Belt consists of upper amphibolite to granulite facies rocks which have subsequently been reworked under amphibolite and greenschist facies conditions (Zwart 1974; Williams and Zwart 1977). From the Central Belt upwards and downwards the metamorphic grade decreases (lower to middle amphibolite facies) but the plurifacial and/or polymetamorphic evolution of the Eastern and Western Belts is still rather poorly understood. The Seve eclogites described here are found in the Central and in the Eastern Belt of the Seve Nappe (Fig. 1) and are called Central Belt or type C and Eastern Belt or type E eclogites. Type localities of the Jämtland Seve eclogites are Sipmikken Creek for the Central Belt and the Tjeliken Mountain for the Eastern Belt eclogites. (Sverige Fältkarta 22E Frostviken $14°24'$–$26'/71°81'$–$83'$ and $14°48'/71°62'$ respectively.) Recent age determinations on eclogite-bearing country rocks from both units have yielded Precambrian ages of 1000–1100 Ma (Reymer *et al.* 1980).

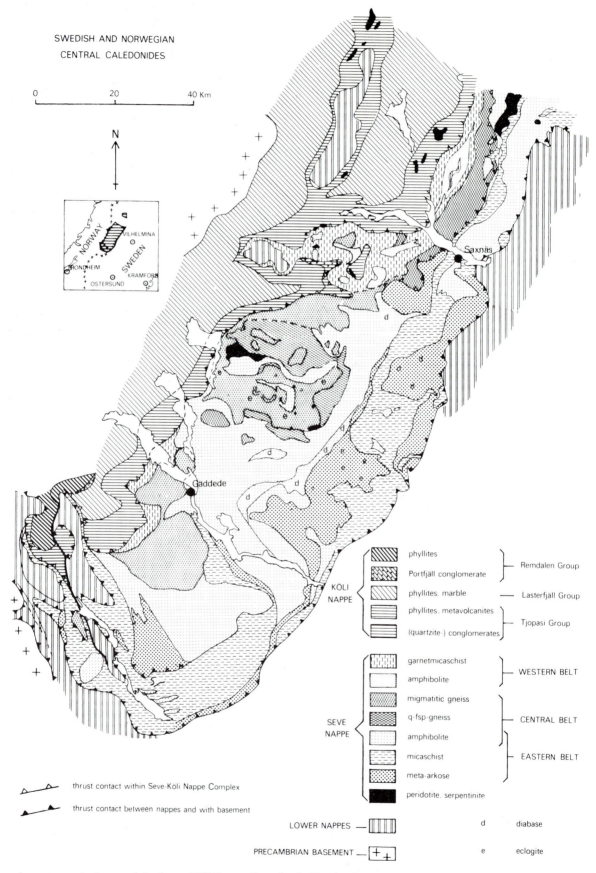

Fig. 1   Geological map of the Seve–Köli Nappe Complex in Jämtland and Västerbotten (Sweden) and Nord Trøndelag (Norway). After Williams and Zwart 1977

Analytical techniques

Wet chemical analyses of major elements were performed at the petrochemical laboratory of the Institute of Geology and Mineralogy in the State University of Leiden. The single solution method of Shapiro (1967) was used. Out of 25 samples, 18 were made in duplicate. Mean values of duplicates are used in the tables. Minor elements have been analysed by neutron activation analysis at the Interuniversity Reactor Institute (IRI) at Delft Technical University. Electron microprobe analyses (EMP analyses) on primary rock-forming minerals were carried out in the electron microprobe laboratory at the Institute of Earth Sciences, Free University of Amsterdam with a Cambridge Geoscan and/or Microsan 9. Operating conditions: 20 Kv, sample current 25 nA, ZAF correction program: microscan 9. In each sample spot analyses have been made of 3 to 7 grains. Homogeneity was checked in at least two grains by spot analyses in core and rim or occasionally by continuous line scans. Mean values are listed in the tables.

Petrography

The dominant rock type of the Central Seve Belt is a migmatitic kyanite/sillimanite potash-feldspar gneiss. Minor intercalations of quartzites, amphibolites, garnet–mica schists and marbles occur. The dominant eclogite-bearing country rock of the Eastern Belt is an unmigmatized quartzofeldspathic gneiss with minor intercalations of amphibolites, garnet–mica schists and marbles. The eclogite bodies occur as layers or lenses within the country rocks. Some occurrences strongly suggest disruption of originally continuous layers. The maximum diameter of the eclogitic inclusions varies from a few decimetres to several tens of metres.

The eclogite facies mineral assemblage of Eastern Seve Belt eclogites is omphacite, garnet, quartz, ±phengite, rutile, zircon and apatite. Central Belt Seve eclogites contain zoisite in addition.

The eclogites are generally strongly affected by post-eclogite facies transformation products, i.e. red garnets lie in a foliated matrix which is dominated by light green plagioclase–pyroxene or plagioclase–amphibole symplectites. Garnets commonly have rims (up to 0.5 cm) of black amphibole ± plagioclase ± oxide (hercynite or ilmenite). Secondary amphiboles and epidote group minerals generally constitute 25–50% of the matrix mineralogy. However, the degree of post-eclogite facies alteration is more pronounced in the Central than in the Eastern Seve Belt eclogites. At Tjeliken

Mountain omphacite still forms one of the major matrix phases which is more the exception than the rule at Sipmikken Creek. The eclogites may show a compositional layering due to both mineralogical and grain-size variations. The compositional layering is parallel to a foliation defined by the shape preferred orientation of omphacite or zoisite. This primary foliation becomes mimetically replaced by secondary minerals, such as amphiboles and epidotes. The contact between (retro) eclogite and country rock may be gradational or sharp. The rocks at the contact vary from coarse grained with only traces of secondary transformation products to fine or medium grained and fully amphibolitized and/or mylonitized.

Whole Rock Chemistry

Averages of the analyses of 25 samples from different outcrops (18 type C and 7 type E) are presented in Table 1 and Fig. 2. From these data it is concluded that the bulk chemistry of the two eclogite types is similar. A magmatic origin for the metabasic rocks is inferred from the techniques outlined by Leake (1964) and Moine and La Roche (1968) or alternatively from the positive correlation of MnO, P_2O_5, and TiO_2 versus FeO^*/MgO, while MgO and Al_2O_3 are negatively correlated (Fig. 2), which is suggestive of a magmatic trend. On the normative Ne-Ol-Di-Hy-Qtz diagram (Fig. 3) the chemical composition of the eclogites varies from slightly normative Ne-bearing (<5.0%) to slightly quartz normative

Table 1 Chemical composition of Seve eclogites. Major element oxides in wt%, minor elements in ppm. $FeO = FeO + Fe_2O_3$

Component	Central belt Mean	Sigma	Eastern belt Mean	Sigma
SiO_2	49.50	1.80	49.51	1.22
TiO_2	1.27	0.33	1.44	0.46
Al_2O_3	15.08	0.64	14.90	0.72
Fe_2O_3	2.39	1.16	1.31	0.73
FeO	8.62	1.23	10.12	1.40
MnO	0.22	0.05	0.22	0.03
MgO	7.78	0.78	6.89	0.81
CaO	11.22	1.46	11.28	0.98
Na_2O	2.43	0.55	2.90	0.38
K_2O	0.42	0.26	0.37	0.21
P_2O_5	0.13	0.04	0.18	0.09
H_2O	1.00	0.38	0.74	0.20
CO_2	<0.1		<0.1	
Total	100.16		99.96	
FeO^*	10.76	1.65	11.31	1.18
V	320	45	291	68
Cr	278	80	202	58

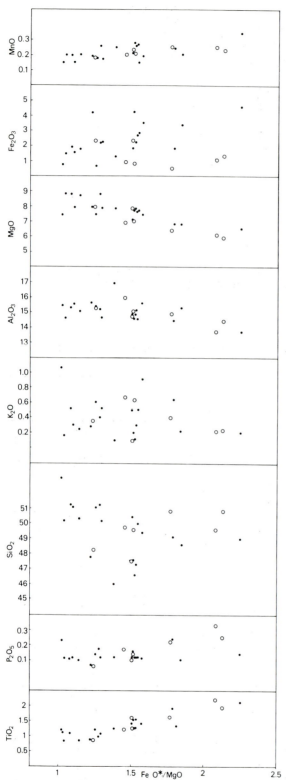

Fig. 2 Diagram showing the variation of each element against FeO*/MgO. Sample numbers are the same as used for the compilation of Table I. Symbols: dots—type C eclogites, open circles—type E eclogites

(<13%). This spread in the Ne-Ol-Di-Hy-Qtz diagram is at least partly due to secondary alteration processes as is seen from the arrows in Fig. 3 which corresponds to the variation in normative composition between eclogite cores and amphibolitized rims. Other geochemical criteria such as the AFM and Miyashiro's FeO/MgO versus SiO_2 diagrams reveal a tholeiitic affinity for both eclogite types (Miyashiro 1975). A tholeiitic composition is also inferred from plots of FeO* (=total iron as FeO), V, Ti, and Cr versus FeO*/MgO. With advancing fractional crystallization, i.e. with increasing FeO*/MgO ratio, FeO*, TiO_2 and V show a clear positive-Cr a negative trend (Miyashiro and Shido 1975). In terms of geotectonic setting the majority of eclogites plot in the field of ocean floor basalts (OFB's) according to the discriminant F_1F_2 function field and the Ti–Cr diagram of Pearce (1975, 1976). This is in agreement with Miyashiro and Shido's (1975) abyssal tholeiitic field using their FeO*, V, Cr versus FeO*/MgO classification diagrams and with the oceanic field in the TiO_2–K_2O–P_2O_5 diagram (Pearce 1975). Field evidence, however, indicates that the quantities of metabasites are volumetrically insignificant relative to the quartzo-feldspathic country rock gneisses. The eclogites, in terms of tectonic setting, can therefore best be considered to represent dykes or sills intruded during the early stages of a continental break-up.

Mineralogy

Clinopyroxenes

Optical observations

Most omphacites from anhedral, elongated grains, 1 to 7 mm long. The shape preferred orientation of omphacites defines a foliation. A crystallographic fabric, c//l and b⊥s, as defined by Binns (1967), is generally present. Most of the original grain and interphase boundaries are obliterated because of incomplete, post-eclogite facies alteration. The intracrystalline defect structure in the omphacite consists of undulatory extinction, infrequent twinning and subgrain boundaries (van Roermund and Boland 1981).

Chemistry

Analysed clinopyroxenes appear to be chemically (and optically) homogeneous. Representative EMP analyses and structural formulae are given in Table 2. The sum of cations, based on six oxygens, is generally higher than 4. This indicates the presence of trivalent iron (Fe^{3+} calculation method according to

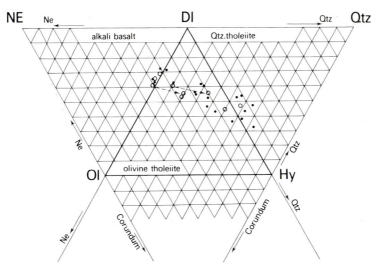

Fig. 3 C.I.P.W. norms of Seve eclogites plotted in the Ne–Ol–Di–Hy–Qtz diagram. Symbols as in Fig. 2. Arrows show direction of chemical variation between eclogitic cores and amphibolitized rims

Table 2 EMP analyses and calculated structural formulae, based on six oxygens, assuming all Fe to be ferrous, of clinopyroxenes from Seve eclogites

Sample	Eastern Belt					Central Belt				
	85	77	78	84	84*	1	8	38	38*	49
SiO_2	55.63	55.40	55.80	55.66	55.70	54.71	54.00	54.68	54.65	53.85
TiO_2	0.11	0.13	0.14	0.11	0.10	0.15	0.15	0.20	0.20	0.25
Al_2O_3	10.98	11.60	11.48	11.45	11.35	10.23	9.30	10.20	10.40	9.40
V_2O_5	0.03	0.10	0.08	0.10	0.10	—	—	—	—	—
FeO	5.27	4.55	5.23	4.78	4.90	5.20	4.45	4.78	4.90	4.40
MnO	0.03	—	0.05	—	—	0.10	—	0.05	0.05	0.05
MgO	7.63	7.63	7.52	7.50	7.70	8.85	10.40	9.27	9.20	10.20
CaO	12.87	12.60	12.54	12.79	12.85	14.86	16.85	16.06	15.95	17.40
Na_2O	7.17	7.58	7.23	7.24	7.10	5.87	4.65	5.30	5.35	4.50
Cr_2O_3	0.03	—	—	—	—	0.03	—	—	—	—
Total	99.75	99.59	100.07	99.63	99.80	99.95	99.80	100.54	100.70	100.05
Si	1.99	1.98	1.99	1.99	1.99	1.97	1.95	1.95	1.95	1.94
Al^{4+}	0.01	0.02	0.01	0.01	0.01	0.03	0.05	0.05	0.05	0.06
T	2.00	2.00	2.00	2.00	2.00	2.00	2.00	2.00	2.00	2.00
Al^{6+}	0.46	0.47	0.47	0.47	0.47	0.40	0.32	0.38	0.39	0.34
Ti	0.003	0.003	0.004	0.003	0.003	0.004	0.005	0.005	0.005	0.01
V	—	0.003	0.002	0.003	0.003	—	—	—	—	—
Fe	0.16	0.14	0.16	0.14	0.15	0.16	0.13	0.14	0.15	0.13
Mn	0.001	—	0.002	—	—	0.001	—	0.002	0.002	0.002
Mg	0.41	0.41	0.40	0.40	0.41	0.47	0.56	0.49	0.49	0.55
Cr	0.005	—	—	—	—	0.001	—	—	—	—
M_1	1.035	1.026	1.038	1.016	1.036	1.036	1.015	1.017	1.037	1.022
Ca	0.49	0.48	0.48	0.49	0.49	0.57	0.65	0.62	0.61	0.67
Na	0.50	0.52	0.50	0.50	0.49	0.41	0.33	0.37	0.37	0.31
M_2	0.99	1.00	0.98	0.99	0.98	0.98	0.98	0.98	0.98	0.99
Total	4.025	4.04	4.018	4.006	4.016	4.016	3.995	3.997	4.017	4.012
†Fe^{3+}	0.06	0.09	0.04	0.04	0.04	0.05	0.04	0.03	0.03	0.04
Fe^{2+}	0.10	0.04	0.11	0.10	0.11	0.10	0.09	0.12	0.11	0.10

* Clinopyroxene inclusion in garnet (rim).
†Ferric iron estimated following Neumann (1976)

Neumann 1976). As usual for pyroxenes formed under high pressure the amount of tetrahedrally coordinated Al is low (type E < 0.02 atoms per formula unit, type C < 0.06 a.f.u.). Consequently in the pyroxene end member diagram of White (1964) all pyroxenes plot in the eclogite field. Clinopyroxene compositions expressed in terms of the end member molecules jadeite, acmite, and augite are given in Fig. 4. In the calculation the method of Cawthorn and Collerson (1974) has been used since it is least affected by the uncertainties in trivalent iron content of the clinopyroxenes, but En, Fs, Wo, and tschermakitic molecules (Ca–Fe^{3+}, Ca–Ti, and Ca–Tschermaks) were taken together. Type E clinopyroxenes reveal a constant composition and plot in the field of omphacites. Due to Al < Na type E pyroxenes contain a small amount of Ca–Ti Tschermak's molecule component (<0.4%). Type E clinopyroxenes have a molecular composition around $Jd_{48}Ac_3Aug_{49}$. In contrast type C clinopyroxenes reveal a characteristic lower jadeite content (range 41.7–23%). In the triangular end member diagram of Fig. 4 type C clinopyroxenes, except for 8(2), plot also in the field of omphacite; 8(2) has a sodic augite composition. Type C clinopyroxenes have a molecular composition intermediate between $Jd_{41.7}Ts_{1.2}Aug_{57.1}$ and $Jd_{23.9}Ts_{7.2}Aug_{68.9}$. Omphacites incorporated in garnets have the same chemistry as those in the matrix. They are also chemically unzoned.

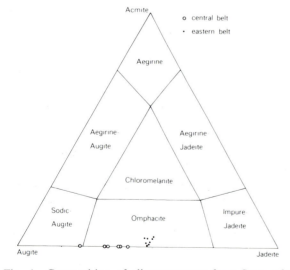

Fig. 4 Composition of clinopyroxenes from Seve eclogites in terms of the end member molecules jadeite (Jd), acmite (ac), diopside + hedenbergite + tschermakite (Aug). Clinopyroxenes from type E eclogites are represented by dots, clinopyroxenes from type C eclogites by open circles. Pyroxene composition field after Essene and Fyfe (1967)

Garnets

Optical observations

Most garnets form euhedral to subhedral grains or porphyroblasts 5 to 150 mm in diameter for the Central Belt and 1 to 10 mm for the Eastern Belt. Although inclusion-free garnets do occur, especially in the Eastern Belt, most garnets contain inclusions of other minerals which are generally concentrated in the garnet cores. The dusty inclusion-rich cores are surrounded by clear, inclusion-poor garnet rims. In the Central Belt garnet aggregates have been found; i.e. several dusty, inclusion-rich cores have been tied together by clear outer rims. In the Eastern Belt locally atoll and barrier reef garnet microstructures are most abundant. Euhedral or subhedral garnet cores are surrounded by euhedral to subhedral garnet rims with quartz between the two (atoll garnet). Often in the atoll garnet microstructures the cores and rims are connected by bridges of garnet. If in the atoll garnet microstructures the central euhedral garnet cores are lacking the microstructure is called 'barrier reef'. Within the atoll and barrier reef garnet microstructures phengitic mica or less frequently brown–green amphibole may occur locally instead of quartz.

Quartz and rutile (needles and grains) are the most abundant inclusions in garnet. Locally the quartz can be surrounded by a thin plagioclase film. Other inclusions are amphiboles, epidote group minerals, calcite, phengite, biotite, zircon, apatite, and omphacite. If primary amphiboles and omphacites are incorporated in a single garnet grain then omphacite is found in the rim and amphiboles in the cores, while a zone rich in quartz and/or rutile may mark the boundary between the two. In the Central Belt elongated inclusions in garnet are locally aligned and define an internal foliation (S_i). S_i has been found parallel as well as oblique to the external foliation in the eclogite matrix (S_e). S_i consists of amphiboles, epidote group minerals, quartz and calcite. S_i is restricted to the garnet cores and is bounded by an almost continuous circular inclusion-rich zone. These garnet cores, due to a high Ca content, may have a pronounced yellowish colour in transmitted light.

Chemistry

EMP analyses and structural formulae, based on 12 oxygens, are given in Table 3. Core and rim analyses are given since some of the analysed garnets are chemically zoned. The sum of cations of analysed Seve garnets is generally higher than 8. This is accompanied by Al values lower than 2 (>1.95 a.f.u) which indicates the presence of trivalent iron.

Table 3 EMP analyses and calculated structural formulae, based on 12 oxygens, of garnets from Seve eclogites

	Central belt							
	1		8		38		49	
	Core	Rim	Core	Rim	Core	Rim	Core	Rim
SiO_2	38.80	39.25	39.70	39.95	39.20	39.20	40.00	39.80
Al_2O_3	21.30	22.00	22.10	22.00	21.80	22.00	22.00	22.25
TiO_2	—	—	0.10	0.05	—	—	0.05	0.05
FeO	23.00	21.95	19.65	18.95	22.05	21.30	18.95	18.75
MnO	0.90	0.60	0.55	0.40	1.75	0.40	0.45	0.45
MgO	5.75	7.30	9.00	10.10	6.40	8.05	9.80	9.95
CaO	9.65	8.45	8.95	8.35	9.75	8.90	9.00	8.80
Total	99.40	99.55	100.05	99.80	100.95	99.85	100.25	100.05
Si	3.02	3.02	3.01	3.01	3.00	3.00	3.01	3.00
Al^{4+}	—	—	—	—	—	—	—	—
Al^{6+}	1.96	2.00	1.97	1.96	1.97	1.98	1.96	1.98
Ti	—	—	0.01	0.005	—	—	0.01	0.005
Fe	1.50	1.41	1.25	1.20	1.41	1.36	1.19	1.18
Mn	0.06	0.04	0.04	0.03	0.12	0.03	0.03	0.03
Mg	0.67	0.84	1.02	1.14	0.73	0.92	1.10	1.12
Ca	0.81	0.70	0.72	0.68	0.80	0.73	0.73	0.71
Total	8.02	8.01	8.02	8.025	8.03	8.02	8.03	8.025

	Eastern belt							
	77		78		84		102	
	Core	Rim	Core	Rim	Core	Rim	Core	Rim
SiO_2	39.25	38.20	38.45	38.65	38.40	38.35	38.20	38.55
Al_2O_3	21.55	21.95	21.80	21.80	21.70	21.65	21.55	21.55
TiO_2	0.10	—	0.05	0.10	—	—	—	—
FeO	26.70	25.95	26.15	26.35	25.80	25.95	27.55	26.00
MnO	0.45	0.45	0.65	0.70	0.60	0.55	1.65	0.40
MgO	5.80	5.80	4.95	5.35	4.85	5.50	2.05	5.05
CaO	7.90	7.70	7.60	7.25	8.45	7.35	9.10	8.00
Total	101.75	100.05	99.65	100.20	99.80	99.35	100.10	99.55
Si	3.01	2.97	3.01	3.01	3.00	3.01	3.03	3.02
Al^{4+}	—	0.03	—	—	—	—	—	—
Al^{6+}	1.95	1.98	2.01	2.00	2.00	2.00	1.99	1.99
Ti	0.01	—	0.01	0.01	—	—	—	—
Fe	1.72	1.69	1.71	1.72	1.69	1.70	1.83	1.70
Mn	0.03	0.03	0.05	0.05	0.04	0.04	0.11	0.03
Mg	0.67	0.67	0.58	0.62	0.57	0.64	0.24	0.59
Ca	0.65	0.64	0.64	0.60	0.71	0.62	0.78	0.67
Total	8.04	8.01	8.01	8.01	8.01	8.05	7.98	8.00

Table 3 clearly demonstrates that for both eclogite occurrences two distinct garnet types can be recognized:

1. Chemically-homogeneous garnets.
2. Chemically-zoned garnets.

Examples of microprobe stepscan profiles across chemically-zoned garnets from both eclogite types are given in Fig. 5. In cases of obvious chemical zoning Ca decreases and Mg increases from core to rim, Mn remains constant or decreases. Fe, depending on the behaviour of the other elements, behaves more capriciously, but commonly Fe also decreases from core to rim. As is illustrated in Fig. 5 the shape of the chemical profiles in the garnets can vary from gently bell shaped to more abrupt. In the triangular pyrope–almandine + spessartine–grossular + andradite end member diagram this chemical zonation is also quite evident and a marked increase in the pyrope content from core to rim can be recognized (Fig. 6). Furthermore Fig. 6 illustrates a clear separation of garnets from the two eclogite occurrences. Eastern Belt garnets belong to the group C eclogites of Coleman et al. (1965), i.e. eclogites associated with glaucophane and lower amphibolite facies rocks. Central Belt garnets tend to plot in the group B eclogite field; i.e. eclogites associated with migmatites, granulites, and amphibolite facies rocks.

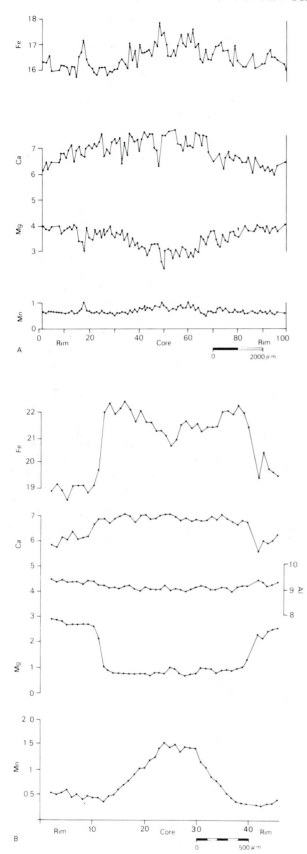

Fig. 5　Microprobe stepscan profiles across garnets of Seve eclogites. A: Central Belt, Sipmikken Creek, step distance 100 μm. B: Eastern Belt, Tjeliken Mountain, step distance 50 μm

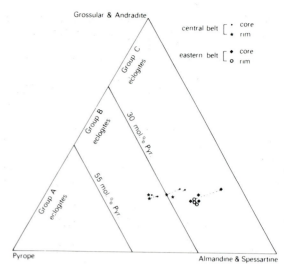

Fig. 6　Triangular garnet end member diagram showing the compositional variation of garnets from Seve eclogites. Group A, B, and C eclogites fields after Coleman *et al.* (1965)

Amphiboles

Optical observations

The amphiboles in both eclogite occurrences can be classified into four types:

Type 1 is found as fine-grained anhedral to sub-hedral, green or greyish green pleochroic grains incorporated in garnet (cores or less frequently rims) omphacite and quartz.

Type 2 is a matrix component. Its post-eclogite facies nature is demonstrated by the absence of intracrystalline defect structures such as subgrain boundaries and undulating extinction at places where these are also commonly observed features in the omphacites. Type 2 amphiboles contain inclusions of omphacite, plagioclase–pyroxene symplectites and of a concentric pattern in which the inclusions from core to rim consist of quartz, clinopyroxene and plagioclase.

Type 3 is the blue–green amphibole commonly occurring in kelyphitic rims around garnet or as pseudomorphs after omphacite or plagioclase–pyroxene symplectites. Because it forms rims around type 2 amphiboles type 3 is considered to be younger.

Type 4 is a pale bluish-green actinolitic amphibole, homoaxially intergrown with the bluish-green (type 3) or brownish-green (type 2) ones.

Only types 1 and 2 will be discussed here.

Chemistry

EMP amphibole analyses and structural formulae, based on 23 oxygens, are given in Table 4. The total

Table 4 EMP analyses and calculated structural formulae, based on 23 oxygens, of amphiboles from Seve eclogites

	Amphibole			
	Central Belt		Eastern Belt	
Sample	1	1	84	84
Type	I	II	I	II
SiO₂	38.45	40.85	43.05	45.15
Al₂O₃	19.40	17.10	17.00	14.05
TiO₂	—	0.35	0.70	0.65
FeO	14.20	12.95	13.40	11.00
MnO	0.10	0.20	—	—
MgO	9.00	10.70	9.75	12.85
CaO	10.00	9.35	8.25	7.80
Na₂O	4.20	4.25	4.95	5.10
K₂O	—	—	0.50	0.60
Total	95.35	95.75	97.50	97.25
Si	5.85	6.12	6.31	6.57
Al	3.48	3.02	2.94	2.41
Ti	—	0.04	0.08	0.07
Fe²⁺	1.81	1.63	1.64	1.34
Mn	0.01	0.03	—	—
Mg	2.04	2.39	2.13	2.78
Ca	1.63	1.50	1.30	1.22
Na	1.23	1.24	1.41	1.44
K	—	—	0.09	0.11
Total	16.05	15.97	15.90	15.93
O	23.00	23.00	23.00	23.00

cations, excluding Ca, Na, and K, always exceed 13, which indicates the presence of trivalent iron. In the calculation of the standard amphibole formulae, expressed as $A_{0-1}B_2C_5^{6+}T_8^{4+}O_{22}(OH, F, Cl)_2$, the procedure proposed by Leake (1978) has been followed. According to this nomenclature all analysed amphiboles of the Central Seve Belt can be grouped as calcic amphiboles of class B (Fig. 7C), amphiboles of Eastern Belt eclogites belong either to the calcic–((Class B) Fig. 7B), or to the sodic–calcic amphibole group with $(Na)_A > 0.50$ (Fig. 7A). Summarizing one can say that the sodium content of types 1 and 2 amphiboles from the Eastern Belt eclogites exceeds in general that of amphiboles from eclogites from the Central Belt. This is clearly demonstrated in Fig. 8.

Phengite

Phengites are found as infrequent matrix components or as inclusions in the cores of garnets. Due to post-eclogite facies alteration processes phengites are commonly replaced by fine-grained biotite–feldspar intergrowths. In the Central Belt this process has gone to completion and therefore only phengites from Eastern Belt eclogites have been analysed (Table 5). Type E phengites are faintly zoned. From

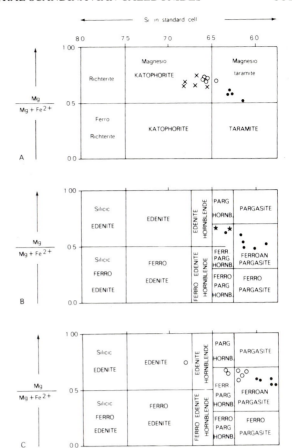

Fig. 7 Classification of amphiboles according to Leake (1978). 7a: Sodic–calcic amphiboles from Eastern Seve Belt eclogites in which $(Na + K)_A > 0.50$; 7b and c: Calcic amphiboles (class B) from Eastern (7b) and Central (7c) Seve Belt eclogites. Symbols: dots inclusions in garnets, open circles matrix amphiboles, crosses inclusions in quartz, stars inclusions in omphacite

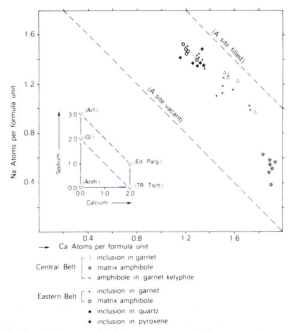

Fig. 8 Sodium and calcium atoms per 23 oxygens in analysed amphiboles from Seve eclogites. Symbols as in Fig. 7

Table 5 EMP analyses and calculated structural formulae of phengite and zoisite

Phengite Eastern Belt			Zoisite Central Belt	
78 Core	78 Rim		38 Core	38 Rim
49.95	48.45	SiO_2	39.25	39.35
28.10	30.45	Al_2O_3	30.65	31.40
0.90	0.90	TiO_2		
1.75	1.20	Fe_2O_3	6.22	4.56
—	—	MnO		
2.95	2.10	MgO	0.20	0.05
—	—	CaO	23.90	24.25
0.80	0.45	Na_2O		
10.50	10.80	K_2O		
94.25	94.35	Total	100.20	99.61
6.69	6.52	Si	5.92	5.95
4.44	4.83	Al	5.46	5.59
0.10	0.10	Ti	—	—
0.20	0.14	Fe^{3+}	0.71	0.53
—	—	Mn	—	—
0.59	0.43	Mg	0.05	0.001
—	—	Ca	3.86	3.92
0.21	0.12	Na	—	—
1.80	1.86	K	—	—
14.03	14.00	Total	16.00	15.98
22.00	22.00	O	25.00	25.00

core to rim there is a decrease in the Si^{4+} content (6.70 → 6.50) and in the FeO_{tot}/MgO ratio. The substitution of Na for K in the phengites is up to 15% but generally does not exceed 10% (of total). The chemical composition of phengites included in garnet corresponds to the core-compositions of matrix phengites.

Zoisite

Zoisite is found in the matrix of Central Seve Belt eclogites as isolated crystals or as elongated aggregates. In both cases the shape preferred orientation of zoisite defines a foliation. Some zoisites are chemically zoned, with c. 6.2 wt.% Fe_2O_3 in the cores and 4.6 wt.% in the rims (Table 5).

Metamorphic Conditions of Eclogite Formation

The following methods have been used to obtain information about the metamorphic conditions of eclogite formation:

1. The thermal and compositional dependence of the Fe^{2+}–Mg distribution coefficient between coexisting garnet and clinopyroxene, hereafter called K_D. Method: Ellis and Green 1979.

2. The P–T dependence of the jadeite content of the omphacites. Method: Gasparik and Lindsley 1980.

Results

In the Central Belt, K_D values range between 5.8 and 7.6 and the maximum jadeite percentage of clinopyroxenes is 41.7%. In the Eastern Belt L_D values range between 9.5 and 25.8 and the maximum jadeite percentage of clinopyroxenes is 49. Intersection of these curves define temperatures and minimum pressures of eclogite formation (Fig. 9), which corresponds for the Eastern Belt eclogites to 14.0 ± 1.5 Kb and 550 ± 70°C and for the Central Belt eclogites to 18.0 ± 1.0 Kb and 780 ± 50°C The inferred temperature difference between both eclogite occurrences is consistent with the migmatitic character of the country rocks in the Central Belt and the lack of migmatization in the Eastern Belt.

Discussion

Microprobe stepscan profiles across garnets from metabasic rocks have been discussed extensively in the literature, among others Råheim (1975), Miller (1977), and several mechanisms have been proposed to explain chemically zoned garnets, e.g. Rayleigh fractionation (Hollister 1966), segregation model (Atherton 1965), zoning due to diffusion (Anderson and Buckley 1973) but the most widely accepted

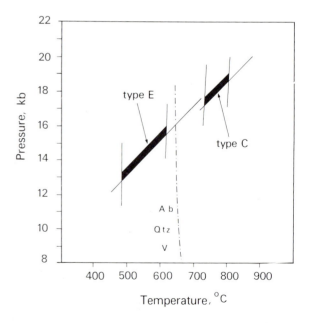

Fig. 9 Estimated PT conditions of eclogite formation. Data after Gasparik and Lindsley (1980) and Ellis and Green (1979). Symbols: E Eastern Belt eclogites, C Central Belt eclogites

interpretation is that zoning is due to changing P,T conditions (among others Råheim and Green 1975). Based on the assumption that the observed chemical zoning of garnet (and/or phengite) reflects the failure of cores of crystals to reequilibrate to the changed P,T conditions, the Seve eclogites can be interpreted as being the result of the prograde metamorphism of lower grade rocks (amphibolites) of tholeiitic composition. Such an interpretation is consistent with the observed zonal distribution of inclusions in garnet, i.e. amphiboles, epidote group minerals in the cores, and clinopyroxenes in the rims. This prograde metamorphism requires in this case geothermal gradients of about 12°C/km, which is characteristic for subduction zone metamorphism. Descent along a subduction zone could account for the observed progressive metamorphism of the eclogites.

Recently eclogite occurrences in the Norwegian Scandinavian Caledonides have been reviewed by Bryhni et al. (1977) and a slightly modified version of their map of the regional distribution of eclogitic rocks in the Scandinavian Caledonide Belt is given in Fig. 10. Since recent age determinations of the eclogite bearing country rocks from the Seve Nappe, Jämtland, Sweden, reveal a Sveconorwegian or Grenvillian age (Reymer et al. 1980), it becomes interesting to compare the Seve eclogites with the classical and well studied eclogite occurrences from the west coast of Central Norway (Eskola 1921), which occur in supracrustal rocks of comparable age (1000 Ma, Brueckner (1972) and 1700 Ma, Mysen and Heier (1972). It is claimed that the Norwegian country rock eclogites are metamorphosed in situ (Råheim 1972, 1975; Mysen and Heier 1972; Bryhni et al. 1977; Griffin and Råheim 1973; Råheim and Green 1975) rather than being (solid) introductions from the mantle (Lappin 1974). Inferred metamorphic conditions of eclogite formation vary regionally in a systematic way (500–800°C and 8–22 Kbar (Bryhni et al. 1977) or 600–800°C at 15 Kbar (Griffin et al. this volume)). This is illustrated in the regional $K_D(Fe^{2+}/Mg;gt–cpx)$ distribution pattern of Fig. 10, which indicates an increasing P–T trend from the Norwegian inland towards the Atlantic coastline. Glaucophane-bearing eclogites occur in the Sunnfjord area (Krogh 1980) where the $K_D(Fe^{2+}/Mg;gt–cpx)$ value exceeds 9.3. Towards the Atlantic coast glaucophane incorporated in garnets is absent, P–T increases and the early post-eclogite-facies cooling history is marked by the transformation of garnet + quartz into orthopyroxene + plagioclase ± spinel (Bryhni et al. 1977). Good examples of the latter are represented by the Hareidland (Mysen and Heier 1972), Kristiansund (Råheim 1972; Griffin and Råheim 1973), and Molde peninsula (Hernes 1954) eclogite occurr-

ences. Inferred prograde P–T paths for the development of eclogites in the Basal Gneiss region of Norway correspond to the proposed geothermal gradients of Seve eclogites.

Recently several authors claim a northward extension of the Sveconorwegian province, Fig. 10, which is defined here as an area which has suffered a tectonometamorphic event characterized by folding, high grade metamorphism and granitic and anorthositic plutonism between 1200 Ma and 850 Ma ago (Kratz et al. 1968; Reymer et al. 1980; Zwart and Dornsiepen 1978). In doing so the eclogite-bearing basement gneisses of the Central Norwegian west coast can be traced further northwards along the Atlantic coast immediately west of Norway towards the Lofoten where basement gneisses of comparable ages are exposed (1000 Ma, Heier and Compston 1969; Griffin et al. 1978). This northward extension of the Sveconorwegian province is important for the present study since nappe displacement studies in the Scandinavian Caledonides around the Grong culmination indicate that the Seve rocks have their roots immediately west of the present Norwegian coast (Gee 1978). Results obtained by Gee (1978) are indicated in Fig. 10. Also in this figure the presently studied Seve rocks are translated backwards towards their inferred positions before the onset of the Caledonian thrusting. This implies that the Central and structurally highest Seve unit comes from a position further west of the present Norwegian west coast than the Eastern or lower Seve unit. From this reconstruction it can be seen that the regional K_D distribution pattern obtained from eclogite occurrences from the west coast of Central Norway can be traced further northwards and Eastern Belt eclogites (K_D = 9.5–25.8) can be correlated with the Sunnfjord (or Eastern Gneiss Region) occurrences (Krogh 1980). Central Belt eclogites (K_D = 5.8–7.6) can be correlated with the higher P–T eclogite occurrences closer to the Atlantic coastline (Hareidland, Molde peninsula, Kristiansund).

From the present observations it might be concluded that, despite their present geological setting, the Seve eclogites differ only from their Norwegian country rock equivalents in their (late) post-eclogite-facies geological history, which is obviously related to Caledonian thrusting. A thorough comparative study of the post-eclogite facies microstructures developed in the eclogites of Norway and Sweden seems therefore to be promising.

Conclusions

1. The Central and Eastern Seve Belt eclogites have a comparable tholeiitic bulk chemistry.

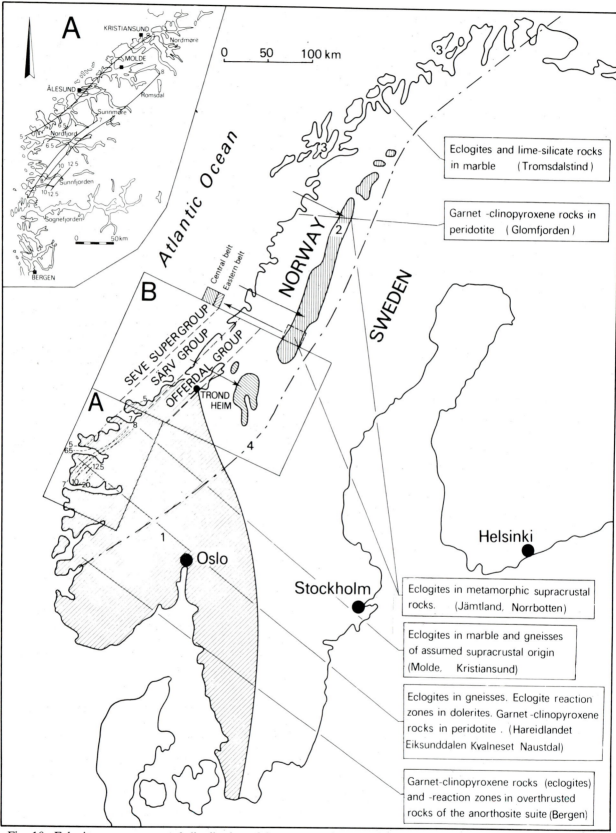

Fig. 10   Eclogite occurrences and distribution of Sveconorwegian ages in the Scandinavian Caledonides. Data after Bryhni *et al*. (1977) and Reymer *et al*. (1980). Legend: 1. Sveconorwegian province of SW Scandinavia. 2. Seve Nappe, area under consideration is indicated with broken lines. 3. Sveconorwegian rocks of the Lofoten and Seiland province. 4. Caledonian front Square A and inset A illustrate the regional pattern of minimum distribution coefficients of Mg and Fe between coexisting garnet and clinopyroxene from eclogites of the Basal Gneiss Region (after Bryhni *et al*. 1977). Square B illustrates the inferred root zones of the Seve and lower nappes (after Gee 1978)

2. Rimwards changing garnet chemistry and zonal distribution of inclusions in garnets indicate that some eclogites started to form as foliated garnet-amphibolites and subsequently became transformed into garnet-omphacite assemblages as a result of increasing temperature and pressure.

3. Following the trivalent iron calculation method of Neumann (1976) minimum pressure and temperature of eclogite formation for Central Belt eclogites are 18.0 ± 1.0 Kb and $780 \pm 50°C$, for Eastern Belt eclogites 14.0 ± 1.5 Kb and $550 \pm 70°C$.

4. Paleogeographic reconstructions of the Seve Nappe eclogites indicate the northeasterly extension of the regional pattern of minimum apparent distribution coefficients of Mg and Fe between coexisting garnet and clinopyroxene as has been mapped along the Basal Gneiss Region in the west coast of Norway.

Acknowledgements

I am indebted to Profs. H. J. Zwart and W. L. Griffin, and Drs. J. N. Boland, M. Otten, P. Sauter, and P. Verhoef for critical reading and discussion of some of the manuscript. Electron microprobe analyses reported in this paper were carried out by Dr. C. Kieft and W. J. Lustenhouwer at the Free University of Amsterdam. Whole rock chemical analyses were performed by V. B. Kwee and F. Sanchez at the State University of Leiden. Both institutions have financial and personnel support by Z.W.O.–W.A.C.O.M. (Research Group for analytical Chemistry of Minerals and Rocks subsidized by the Netherlands Organisation of Pure Research). Also grants from the Molengraaf- and Schurmannfonds are gratefully acknowledged.

References

Anderson, D. E. and Bickley, G. R. 1973. Zoning in garnets—diffusion models. *Contrib. Mineral. Petrol.*, **40**, 87–104.

Atherton, M. P. 1968. The composition of garnet in regionally metamorphosed rocks. In Pitcher, W. S. and Flinn, G. W. (Eds), *Controls of Metamorphism*, 281–291.

Andreasson, P. G., Gee, D. G. and Sukotjo, S. 1981. Lithological association and structural setting of amphibolite eclogites in the Seve Nappe of Norbotten, Sweden. *Terra cognita*, **1**, 32.

Binns, R. A. 1967. Barroisite-bearing eclogite from Naustdal, Sogn og Fjordane, Norway. *J. Petrol.*, **8**, 349–371.

Brueckner, H. K. 1972. Interpretation of Rb–Sr ages from the Precambrian and Paleozoic rocks of southern Norway. *Am. J. Sci.*, **272**, 334–358.

Bryhni, I., Krogh, E. and Griffin, W. L. 1977. Crustal

derivation of Norwegian eclogites: a review. *N. Jb. miner. Abh.*, **130**, 49–68.

Cawthorn, R. G. and Collerson, K. D. 1974. The recalculation of pyroxene end-member parameters and the estimation of ferrous and ferric iron content from electron microprobe analyses. *Am. Mineral.*, **59**, 1203–1208.

Coleman, R. G., Lee, D. E., Beatty, L. B. and Brannock, W. W. 1965. Eclogites and eclogites: their differences and similarities. *Bull. Geol. Soc. Am.*, **76**, 483–508.

Ellis, D. J. and Green, D. H. 1979. An experimental study of the effect of Ca upon garnet–Cpx. Fe–Mg. exchange equilibria. *Contrib. mineral. Petrol.*, **71**, 13–22.

Eskola, P. 1921. On the eclogites of Norway. *Skr. Norske Vidensk. Akad. Oslo, Kl,1*, **8**, 118 pp.

Essene, E. J. and Fyfe, W. S. 1967. Omphacite in Californian metamorphic rocks. *Contrib. mineral. Petrol.*, **15**, 1–23.

Gasparik, T. and Lindsley, D. H. 1980. Phase equilibria at high pressure of pyroxenes containing monovalent and trivalent ions. In Prewitt, C. T. (Ed.). *Pyroxenes: Review in Mineralogy. Min. Soc. Am. Washington*, **7**, 309–339.

Gee, D. G. 1978. Nappe displacement in the Scandinavian Caledonides. *Tectonophysics*, **47**, 393–419.

Griffin, W. L., Austraheim, A., Brastad, K., Bryhni, I., Krill, A., Krogh, E., Mørk, M. B. E., Qvale, H., and Tørudbakken, B. this volume. High-pressure metamorphism in the Scandinavian Caledonides.

Griffin, W. L. and Råheim, A. 1973. Convergent metamorphism of eclogites and dolerites, Kristiansund area, Norway. *Lithos*, **6**, 1973, 21–40.

Griffin, W. H., Taylor, P. N., Hakkininen, J. W., Heier, K. S., Iden, I. K., Krogh, E. J., Malm, O., Olsen, K. L., Ormaasen, P. R. and Tveten, E. 1978. Archaean and proterozoic crustal evolution in Lofoten Vesterlen, N. Norway. *J. Geol. Soc. London*, **135**, 629–647.

Heier, K. S. and Compston, W. 1969. Interpretation of Rb–Sr age patterns in high-grade metamorphic rocks, Norway. *Norsk geol. Tidsskr.*, **57**, 257–283.

Hernes, I. 1954. Eclogite–amphibolite on the Molde Peninsular Southern Norway. *Norsk geol. Tidsskr.*, **33**, 163–182.

Hollister, L. S. 1966. Garnet zoning: an interpretation based on the Rayleigh fractionation model. *Science*, **154**, 1647–1680.

Kratz, K. O., Gerling, E. K. and Lobach-Zhuchenko, S. B. 1968. The isotope geology of the Precambrium of the Baltic Shield. *Can. J. Earth Sci.*, **5**, 657–660.

Krogh, E. J. 1980. Compatible P, T conditions for eclogites and surrounding gneisses in the Kristiansund area, Western Norway. *Contrib. mineral. Petrol.*, **75**, 387–393.

Lappin, M. A. 1974. Eclogites from the Sunndall–Grubse ultramafic mass, Almklovdalen, Norway and the T–P history of the Almklovdalen Masses. *J. Petrol.*, **15**, 567–601.

Leake, B. E. 1964. The chemical distinction between ortho- and para-amphibolites *J. Petrol.*, **5**, 238–254.

Leake, B. E. 1978. Nomenclature of amphibolites. *Canadian Mineral*, **16**, 501–520.

Miller, C. 1977. Mineral parageneses recording the P.T. history of Alpine eclogites in the Tauern Window, Austria. *N. Jb. miner. Abh.*, **130**, 69–77.

Miyashiro, A. 1975. Volcanic rock series and tectonic setting. *Ann. Rev. Earth Plan. Sciences*, **3**, 251–269.

Miyashiro, A. and Shido, F. 1975. Tholeiitic and calc-alkalic series in relation to the behaviours of Ti, V, Cr and Ni. *Am. J. Sci.*, **175**, 265–277.

Moine, B. and La Roche, H. 1968. Nouvelle approche du probleme de l'origine des amphiboles a partir de leur composition chimique. *Compt. Rend.*, **267** (Acad. Sci., Paris) 2084–2087.

Mysen, B. O. and Griffin, W. L. 1973. Pyroxene stoichiometry and the breakdown of omphacite. *Am. Mineral.*, **58**, 60–63.

Mysen, B. O. and Heier, K. S. 1972. Petrogenesis of eclogites—high grade metamorphic gneisses, exemplified by the Hareidland eclogite, Western Norway. *Contrib. mineral. Petrol.*, **36**, 73–94.

Neumann, E. R. 1976. Two refinements for the calculation of structural formulae for pyroxenes and amphiboles. *Norsk geol. Tidsskr.*, **56**, 1–6.

Pearce, J. A. 1975. Basalt geochemistry used to investigate past tectonic environments on Cyprus. *Tectonophysics*, **25**, 41–67.

Pearce, J. A. 1976. Statistical analysis of major elements patterns in basalts. *J. Petrol.*, **17**, (1), 15–43.

Priem, H. N. A., Boelrijk, N. A. I. M., Hebeda, E. H., Verdurmen, E. A. Th. and Verschure, R. H. 1973. Rb–Sr investigations on Precambrian granites, granitic gneisses and acidic meta-volcanics in Central Telemark: metamorphic resetting of Rb–Sr whole-rock systems. *Norges geol. Unders.*, **289**, 37–53.

Råheim, A. 1972. Petrology of high grade metamorphic rocks of the Kristiansund area. *Norges geol. Unders.*, **279**, 1–75.

Råheim, A. 1975. Mineral zoning as a record of P, T history of Precambrian eclogites and schists in Western Tasmania. *Lithos*, **8**, 221–236.

Råheim, A. and Green, D. H. 1975. P, T paths of natural eclogites during metamorphism—a record of subduction. *Lithos*, **8**, 317–328.

Reymer, A. P. S., Boelrijk, N. A. I. M., Hebeda, E. H., Priem, H. N. A., Verdurmen, E. A. Th. and Verschure, R. H. 1980. A note on Rb–Sr whole-rock ages in the Seve Nappe of the Central Scandinavian Caledonides. *Norsk geol. Tidsskr.*, **60**, 139–147.

Roermund, H. L. M. van and Boland, J. N. 1981. The dislocation substructures of naturally deformed omphacites. *Tectonophysics*, **78**, 403–418.

Shapiro, L. 1967. Rapid analysis of rocks and minerals by a single solution method. *U.S. geol. Surv. Prof. Pap.*, B187–B191.

Trouw, R. 1973. Structural geology of the Marsfjällen area, Caledonides of Västerbotten, Sweden. *Sver. geol. Unders.*, *Serie C.*, **689**. 155 pp.

White, A. J. R. 1964. Clinopyroxenes from eclogites and basic granulites. *Am. Mineral.*, **49**, 883–888.

Williams, P. F. and Zwart, H. J. 1977. A model for the development of the Seve–Köli Caledonian Nappe Complex. In Saxena, S. K. and Bhattacharji, S. (Eds), *Energetics of Geological Processes*, Springer-Verlag, New York, 170–187.

Zwart, H. J. 1974. Structure and metamorphism in the Seve–Köli Nappe Complex and its implications concerning the formation of metamorphic nappes. *Cent. Soc. Geol. Belg.*, 129–144.

Zwart, H. J. and Dornsiepen, U. F. 1978. The tectonic framework of Central and Western Europe. *Geologie en Mijnbouw*, **57**, 627–654.

The Caledonide Orogen—Scandinavia and Related Areas
Edited by D. G. Gee and B. A. Sturt
© 1985 John Wiley & Sons Ltd

Seve eclogites in the Norrbotten Caledonides, Sweden

P.-G. Andréasson*, D. G. Gee† and S. Sukotji†

*Department of Mineralogy and Petrology, University of Lund, Sölvegatan 13, S–223 62 Lund, Sweden
†Geological Survey of Sweden, Box 670, S–751 28 Uppsala, Sweden

ABSTRACT

The well-known Scandinavian eclogites occur in the lowest tectonic units in the
Caledonides, being confined to the basal gneisses of southwestern Norway. Recent
isotopic age-determinations suggest that the high pressure metamorphism required
for their origin occurred during Silurian orogenesis. This study describes the
lithological association and structural setting of amphibolitic eclogites occurring
within the upper allochthonous units (Seve Nappes) in northern Sweden. They
occur as trails of small boudins and attenuated bands in psammitic and pelitic
schists. In close association with the eclogites occur thicker psammitic units with
coherent concordant metabasic bands preserving an ophitic texture and chilled
contacts, but showing mineralogical and microchemical adjustment to high pressure
conditions.

The metasediments that host the metabasic rocks show primary depositional
features in some rare localities. The metadolerite-bearing and the eclogite-bearing
associations suffered the same deformational history which included early recumbent isoclinal folding on NNW–SSE axes and subsequent stretching in an ESE
direction, parallel to the main direction of nappe displacement. This stretching
resulted in abundant non-cylindrical folds with all transitions to tubular folds.

In the underlying tectonic units, included by previous authors in the Seve Nappes, dolerite-intruded sandstones occur, metamorphosed in greenschist facies.
Lithological correlation with the overlying eclogite-bearing associations favours an
origin of the latter by high pressure Caledonian metamorphism related to subduction of the Baltoscandian margin, involving build up of a tectonic overburden,
depression of the miogeocline and underlying Precambrian crystalline basement,
and subsequent thrusting to higher tectonic levels on the Baltoscandian Platform.

Introduction

Scandinavian eclogites have been a focus of research
since the classical studies of Eskola (1921). Most of
them are located within the Precambrian crystalline
basement beneath the Caledonian allochthon in
southwestern Norway. As in other parts of the fragmented Caledonide orogen around the North Atlantic, they only occur in the interior of the orogen. They
are not present in Precambrian terrains marginal to the orogen or even in the autochthonous–parautochthonous basement beneath the front
range thrusts. In Scandinavia, they were thought
until recently to be confined to the deepest levels of
the orogen.

The mineral chemistry of at least some of the
eclogites requires exceptionally high pressures of
crystallization and this has led some authors
(O'Hara *et al.* 1971; Lappin and Smith 1978) to
favour derivation from the mantle and tectonic
intercalation of the eclogites into their present
gneissic environment. Most authors, however, prefer
an alternative hypothesis of *in situ* metamorphism
(Gjelsvik 1953; Bryhni *et al.* 1970; Bryhni *et al.*
1977) and accept very deep depression of the crust
during eclogite crystallization. Whereas the spatial
relationships of the eclogites within the Caledonides
led supporters of the former hypothesis to favour
their Caledonian emplacement as some kind of tectonic mélange (Smith 1980), until very recently

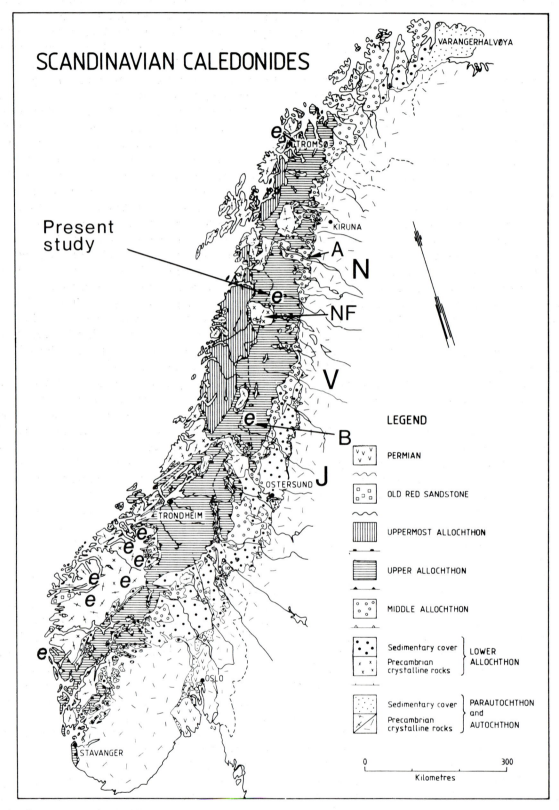

Fig. 1   Occurrences of eclogitic rocks in the Scandinavian Caledonides. Tectonostratigraphy from the IGCP–CO Tectonostratigraphic Map, this volume. Eclogite occurrences in Norway from Bryhni *et al.* 1977; in Jämtland from Williams
and Zwart 1977, and Sjöstrand 1978. N = Norrbotten province. V = Västerbotten province. J = Jämtland province. A =
Akkajaure Complex. NF = Nasafjäll Window. B = Blåsjön area. e = eclogite occurrence

(Griffin and Brueckner 1980; Griffin *et al.* 1981 and this volume) those supporting the alternative hypothesis regarded the high grade metamorphism as a Precambrian phenomenon (*cf.* however, Krogh *et al.* 1974).

The presence of eclogites at high tectonostratigraphic levels within the orogen has been reported during the last decade (Landmark 1973; Zwart 1974; Gee and Zachrisson 1979); these occurrences are less well known. This paper refers briefly to their distribution in the Caledonian allochthon in Sweden and describes one area of occurrence where investigations are in progress.

Within the successions of nappes in Scandinavia (Fig. 1), metamorphic rocks of amphibolite facies and higher grade occur in the Upper Allochthon (*cf.* the IGCP–CO Tectonostratigraphic Map, this volume). The high grade of metamorphism is unambiguously of Precambrian origin in some areas (e.g. Jotun Nappe); in others, the evidence favours high temperatures and pressures during Caledonian deformation of lithologies that may well be in part of Precambrian origin (Claesson 1981). In the Swedish Caledonides, the basal tectonic units in the Upper Allochthon are referred to as the Seve Nappes; these generally amphibolite facies and higher grade rocks compose the lower part of the Seve–Köli Nappe Complex, a unit that dominates the Upper Allochthon throughout the Swedish part of the Caledonian mountains. The overlying Köli Nappes are generally of lower metamorphic grade and made up of volcano-sedimentary sequences including fossiliferous Ordovician and Silurian strata. Metamorphic grade increases downwards from the Köli Nappes into the Seve Nappes; in some areas this occurs transitionally, in others abruptly, with or without the record of tectonic discontinuity.

The general character of the Seve units in Sweden has been summarized by Zachrisson (1973). They have been described, along with other units, in regional descriptions of the various provinces of the mountain belt by Kulling (1955, 1964, 1982) and have been subject to several local studies by Helfrich (1967), Zachrisson (1969), Trouw (1973), Zwart (1974), Williams and Zwart (1977), Sjöstrand (1978), Arnbom (1980), Sjöström (in press), and a variety of other unpublished theses. These descriptions show the Seve to be a complex of nappes. Lithologies vary and calcareous schists are also characteristic. Solitary ultramafic rocks are frequent in some areas. Some of the basic rocks are of intrusive origin but most are probably extrusive. Various local studies of the Seve have shown that the tectonic contacts between the various units are characterized by amphibolite facies ductile shear zones.

Early investigations of the Seve rocks (Svenonius

1881; Backlund 1935, 1936) mention the occurrence of eclogites and 'eclogite amphibolites'. Backlund referred to the extensive development of garnet amphibolites, compared them with similar lithologies occurring in East Greenland (Liverpool Land), where omphacite–garnet assemblages occur, and considered the Seve amphibolites to be of similar affinities. However, he did not identify pyroxene in the Swedish rocks and provided no supporting evidence for the hypothesis that the garnet–hornblende assemblages might be related to eclogite. In 1968, A. Wikström (personal communication) reported eclogites in the Norrbotten Caledonides. Subsequently, in 1971, T. Sjöstrand found eclogites in the Blåsjön area of Västerbotten (Sjöstrand 1978, p. 13). High grade gneisses of the central Seve unit occur there in tectonic contact with overlying Köli schists and phyllites. Some of the gneisses are migmatitic and contain a K-feldspar + Kyanite + sillimanite-bearing mineralogy (Lillfjället Gneiss). Pods of amphibolite occur frequently; ultramafic lenses are also present. The (amphibolitized) eclogite (Table 1) occurs as lenses in a unit referred to as the Avardo Formation (Sjöstrand 1978).

Subsequently, pyroxene–garnet assemblages from a variety of localities in the Seve have been analysed. Only two areas have so far been recognized, where the metabasic rocks carry omphacitic pyroxene. The southern of these areas (Jämtland–southern Västerbotten) is treated by van Roermund (1981, this volume); the other (southern Norrbotten) is the subject of this paper. In the former area, eclogites occur both in the Avardo gneisses and associated with lower-grade psammites and amphibolites in the lower part of the Seve in association similar to that described here.

This paper concentrates on the lithological and structural setting of the Norrbotten eclogites. Studies of their metamorphic history and age are in progress (M. B. E. Mørk, W. L. Griffin, Mineralogisk–Geologisk Museum, Oslo). Our study has involved mapping of numerous small eclogite lenses. This was greatly facilitated by a detailed photogeological interpretation, using infrared colour aerial photographs (scale 1:60 000). The interpretation and projection of geological information on to a basemap (1:20 000) has been conducted with a Wild Aviograph B8S in which an undistorted model is obtained, fixed in scale and orientated in space. Good exposure and inhomogeneity of lithologies facilitated mapping of most elements of the late structural geometry of the rocks. The colour and competency differences between the metabasites and the metasediments allowed the identification of most eclogite and metadolerite lenses greater than 2 m in diameter.

Regional Geological Setting

The Seve-Köli Nappe Complex can be followed northwards from Blåsjön in northern Jämtland through Västerbotten County (Zachrisson 1969; Zwart 1974) to Norrbotten (see Stephens *et al.* this volume). In southern Norrbotten (Kulling in Strand and Kulling 1972 and Kulling 1982), greenschist facies, fossiliferous successions correlatable with the stratigraphy in the Köli Nappes farther south have been described. These successions are underlain by a higher grade complex of amphibolites, schists, and gneisses correlated with the Seve. Kulling (1982) did not attempt to distinguish Seve from Köli Nappes; although the latter overlies the former, the 'transition-zone' is highly variable, apparently due to tectonic intercalation of the two major units (Stephens personal communication 1981). The Seve Nappes wedge out westwards, as is apparent in the Akkajaure profile (Fig. 1) and along the eastern margin of the Nasafjäll Window. This geometry is part of the very large-scale pinch-and-swell tectonics that characterize the Scandinavian Caledonides. The units may well reappear further west in Norway at the same tectonostratigraphic level.

The geology of the area around Jäckvik is shown on Fig. 2. The map is based on Kulling (1982) and amended in accord with our recent reconnaissance. This paper is not concerned with the lower tectonic units (Autochthon, Parautochthon, Lower Allochthon) and these are therefore not separated in Fig. 2. Within the overlying Middle Allochthon have been included the arkosic sandstones and Precambrian crystalline rocks of Kulling's 'Middle Thrust Rocks'. Also included in this complex are some of the lower units in Kulling's 'Upper Thrust Rocks' where these have been interpreted by us to be similar to the underlying units.

South of Jäckvik, Kulling (1982) reported the presence of dolerites in Seve schists. Closer examination has shown that these dykes cut and contact metamorphose cross-bedded feldspathic sandstones. Composition of both the dykes (Kulling 1982; Table 1) and the sediments favours correlation with the lithologies of the Särv Nappes of Jämtland and southern Norrbotten which occur at the same tectonostratigraphical level.

North of Jäckvik, apparently in the same tectonic unit, dolerites occur extensively in the schistose psammites. Higher in the structural succession, the

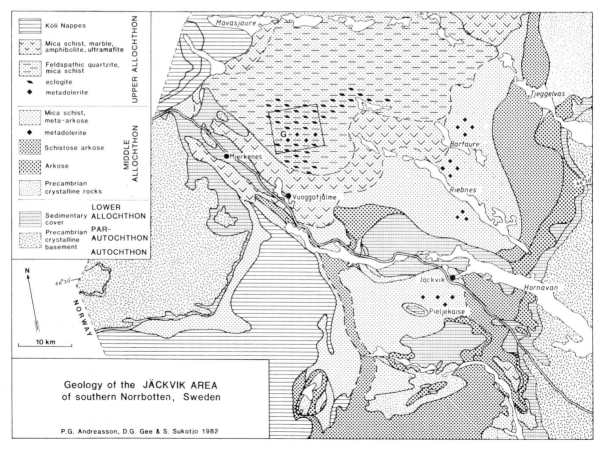

Fig. 2   Simplified geological map of the area around Jäckvik, southern Norrbotten, based on Kulling 1982; Thelander *et al.* 1980, and our own reconnaissance. Outlined field is map area of Fig. 3. G = Mt. Grapesvare

Table 1 Chemical composition of garnet and clinopyroxene from the Avardo eclogite. Fe^{3+} contents calculated as differences between ideal and calculated charges. Analyst: C. Ålinder, Laboratories of the Geological Survey of Sweden, Uppsala. Instrument: ARL–SEMQ operated at 15 kV. Specimens provided by T. Sjöstrand, Luleå

	Garnet T72:53C	Clinopyroxene T71:85		T72:53C
SiO_2	39.40	38.8	SiO_2	52.91
Al_2O_3	20.60	22.0	Al_2O_3	8.17
TiO_2	0.06	—	TiO_2	0.23
MgO	9.98	10.5	MgO	10.02
FeO	19.77	21.4	FeO	4.59
MnO	0.43	0.5	CaO	18.90
CaO	9.61	6.3	Na_2O	4.19
Si	5.96	5.90	Si	1.92
Al^{IV}	0.04	0.10	Al^{IV}	0.08
Al^{VI}	3.64	3.85	Al^{VI}	0.27
Fe^{3+}	0.35	0.15	Ti	0.01
Ti	0.01	—	Fe^{3+}	0.09
Mg	2.25	2.38	Fe^{2+}	0.05
Fe	2.15	2.57	Mg	0.58
Mn	0.06	0.07	Mg	0.01
Ca	1.56	1.03	Ca	0.74
			Na	0.30
Almandine	36	43		
Pyrope	37	39	Acmite	9
Spessartine	1	1	Jadeite	21
Grossular	17	13	Augite	70
Andradite	9	4		

metamorphic grade increases into the Seve Nappes; psammitic lithologies give way into mica schists with marbles, amphibolites and occasional ultramafites. Within the regional W-dipping foliation these are overlain by schistose psammites and semipelites containing both dolerites and eclogites. This unit was referred to by Kulling (1982) as the Juron Quartzites. It is in turn overlain by more mica schists, amphibolites, and marbles similar to the rock associations in the underlying unit. Köli phyllites and volcanites cap the succession.

Although the Juron Quartzites do not occur at the base of the Upper Allochthon, Kulling (1982) regarded them as the lowermost unit, appearing in the core of a major recumbent antiform, overlain and underlain by the schists, marbles, and amphibolites. Lithological similarities between the Juron quartzite–dolerite–eclogite association and Jäckvik sandstones and dolerites suggest that the differences between these tectonic units are essentially related to style of deformation and grade of metamorphism. Part of the area in which the eclogites are preserved is treated in more detail below.

The Grapesvare Area

The eclogites occur in the mountains north of Vuoggatjålme. They have been studied by us on Grapesvare (Figs. 3 and 4) where they occur in psammitic and pelitic schists. South of the map area, they are also present in marbles.

Host rocks

Psammitic, semipelitic, and pelitic compositions dominate the metasediments that host the metabasic rocks. Pure quartzite is subordinate and although feldspar contents may be low, the rocks are generally *feldspathic quartzites*. They contain dark bands of varying thickness and spacing. In the field, some of these bands can be observed to originate from extreme thinning of metabasites. Calcareous psammites occur interbanded with the quartzites and a regular compositional repetition of quartzites, calcareous psammites, and pelites (Fig. 5) apparently reflects a cyclic depositional environment. Regular gradational changes in mineralogy (Fig. 5) are interpreted as graded bedding; they are particularly well preserved in fold hinges. Calcareous bands are thin (usually <10 cm) and contain subordinate calcite, poikiloblastic garnet (grossular <30 molar %), plagioclase (An <45 molar %) and rarely, actinolite. In the northern part of the mapped area (Fig. 3), a belt of intensely folded, banded quartzite occurs. With increasing mica content, feldspar quartzites grade into flagstones, which as a rule occur in the border zone between psammitic and pelitic units (Fig. 3). The compositional banding is eventually transposed into a closely spaced, flaggy foliation (Fig. 5).

Pelites and semipelites vary from very fine-grained to medium-grained *mica schists*; the former are rich in quartz and feldspar and the latter contain conspicuous quartz–feldspar segregations. Phengite is often the dominating mica of the schist. It is rimmed by greenish brown biotite. It is overgrown by subidioblastic garnet and locally accompanied by kyanite. Quartz and quartz–feldspar veins are closely spaced and run strictly parallel to the pervasive schistosity. Veins are frequent where deformation was intense, as in the vicinity of metabasite boudins. Here, mica schists may also become phyllonitic.

Typical host rock parageneses (Table 2) indicate lower amphibolite facies metamorphism (Bryhni and Andréasson this volume). Despite favourable bulk compositions, minerals diagnostic of higher grade (e.g. sillimanite) were not found in any of the over one hundred thin sections examined. The number of phases is low and a bulk compositional control of

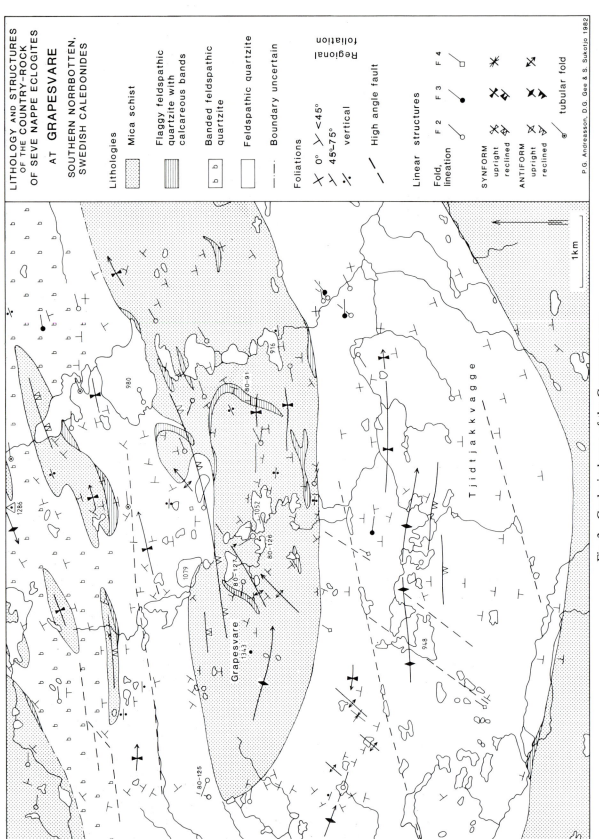

Fig. 3   Geological map of the Grapesvare area

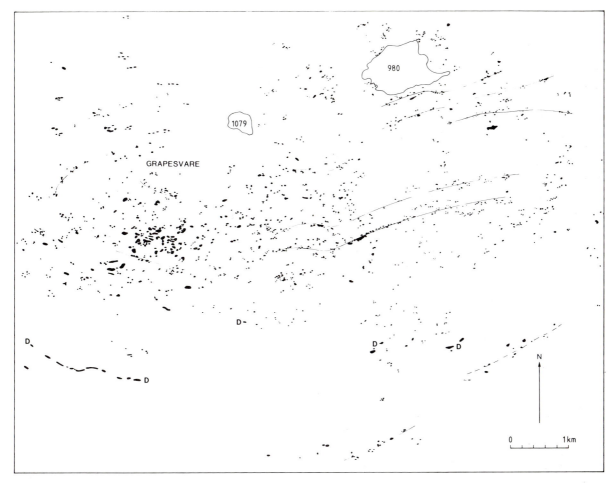

Fig. 4   Distribution of metabasite bands and boudins in the Grapesvare area, based on photogeological interpretation and ground survey

mineral composition is therefore probable and is, in fact, suggested by the strong variation of Mg in biotites of samples taken from one and the same outcrop but from different bulk compositions (Table 3). The often greenish colour of biotite and the coexistence of albite and oligoclase are other criteria suggesting low temperatures of recrystallization.

Analysed white micas are phengites or phengitic muscovites with Si^{4+} contents between 6.40–6.70 and (Fe + Mg) contents between 0.55–0.75 (Table 4). Garnets are poor in Mg and rich in Ca (Table 5). In general, Mg-contents increase towards the rims of crystals, though often with a drop in the outermost rim. Plagioclase crystals display zoning from An_{20} (core) to An_5 (rim) in the analysed samples.

High contents of Ca, Mn, and Ti, textural disequilibrium, and other unfavourable textural relationships prevented the use of garnet–biotite geothermometers employing Fe–Mg distribution. A few analysed garnet–phengite pairs fulfilled the requirements of the garnet–phengite geothermometer of Green and Hellman (1982) for low Ca-systems

with mg (of phengites) close to 0.67 and 0.30 respectively. The following temperatures were obtained by this method for P = 15 kb: 520, 530, 545, 640, 680°C. Since the Fe^{3+} contents of phengites have not been taken into account, these values are maximum temperatures.

The mineralogical composition and morphology of quartz–feldspar veins indicate formation by metamorphic segregation (White 1966; Vidale 1974; Yardley 1978). The minerals of the veins occur also in the host rock and there appears to be no difference between vein and host rock as regards plagioclase composition and zoning. Veins are coarse grained in comparison with the host rock, planar and strictly parallel to the pervasive foliation. There is an obvious concentration of veins to zones of high and heterogeneous strain.

The metamorphic parageneses of host rocks indicate a single major episode of prograde recrystallization, followed by minor retrogression. Both schistose and non-schistose units have closely similar parageneses. Phengite, biotite, and epidote define the axial plane foliation of inferred early folds. Garnets

Fig. 5   Host rock psammites of the Grapesvare area. A: Compositional banding in calcareous feldspathic quartzite. The banding is seen close above the snow. Upwards, it is transposed into the tight foliation which contains the eclogite boudin (to the right of the man). B: The outlined area of Fig. 5A, showing the primary compositional banding folded by F1 folds. C: Transposition of compositional banding into the foliation of the flagstone. D: Flagstone at the exposure shown in Fig. 5C. Locs. 80–126–127 (A–B) and 80–91 (C–D)

are, with few exceptions, syn to post-kinematic with idioblastic rims.

Metadolerite

In the Grapesvare area, metadolerites are restricted to the Tjidtjakkvagge valley, where they are hosted by feldspathic quartzites. On the northern slope of the valley, the metadolerites gradually pass into eclogites, with disappearance of plagioclase phenocrysts and growth of garnet. In addition, one and the same metadolerite band may exhibit an ophitic texture which passes laterally into garnetiferous metadolerite. Metadolerites occur as large boudins or semiconnected bands. Chilled contacts and ophitic textures are frequent; discordant intrusive relationships have not been observed. The primary

mineralogy is characterized by plagioclase, augite, and olivine in descending order of modal abundance. Completely fresh igneous assemblages have not been found. Under the microscope, garnet coronas accentuate the ophitic texture of the best preserved metadolerites. Microprobe analyses of the primary clinopyroxenes have demonstrated a gradual increase in jadeite content from core to rim (up to 10 molar %). There is also a gradual increase in exsolution of rutile needles towards the rim. The plagioclase is heavily saussuritized. In some cases, a narrow rim of green amphibole separates primary pyroxene from plagioclase.

Eclogite

Eclogitic mineral assemblages are best preserved in the cores of larger boudins, where omphacite and

Table 2 Parageneses of host rock metasediments in the Grapesvare area. Selection based on examination of c. 100 specimens. Brackets indicate that a crystal is wrapped in the regional (pervasive) foliation; italics that a crystal post-dates this foliation. Specimen numbers refer to the Swedish IGCP–CO Collection File A

Feldspathic quartzite

78–167A   quartz + microcline + plagioclase + albite + greenish biotite + white mica ± zoisite ± epidote ± *garnet* ± sphene

78–170C   quartz + oligoclase + microcline + phengitic white mica + greenish biotite + *almandine* ± epidote ± sphene

Calcareous band in feldspathic quartzite

78–165B   quartz + plagioclase + biotite + white mica + garnet (Gros 25–30) + epidote-zoisite + hematite ± sphene ± (garnet)

79–80A   quartz + zoisite-clinozoisite + plagioclase + (garnet) + actinolite ± sphene

80–127   carbonate + quartz + epidote + sphene + chlorite + (garnet)

Flaggy feldspathic quartzite

78–165A   quartz + plagioclase (An 5–15) + phengitic white mica + greenish biotite + *almandine* ± sphene

79–213   quartz + biotite + plagioclase + albite + *garnet* + sphene ± carbonate

Banded quartzite with quartz–feldspar segregations

80–119   *host rock*: quartz + microcline ≫ white mica

             segregation: quartz + microcline + biotite + white mica

Fine-grained mica schist

78–163B   quartz + biotite + phengite + albite + plagioclase + almandine + hematite

Coarse-grained mica schist

81–132   quartz + phengite + feldspar + *garnet* + *kyanite*

Mica schist with quartz–feldspar segregations

78–165C   *host rock:* quartz + plagioclase (An 5–20) + biotite + phengite + *almandine*

             segregation: quartz + plagioclase (An 5–20) ≫ biotite + white mica

78–166C   host rock: quartz + plagioclase (An 10–20) + biotite + phengitic white mica + *almandine* ± zoisite

79–67, −97   segregation: quartz + plagioclase + biotite + white mica + *garnet* ± zoisite

Shield of mica schist around eclogite boudin

78–171A   phengitic white mica + quartz + greenish biotite + Fe–Ti–oxides + *almandine (Gros 12–16)*

Table 3 Chemical composition of biotite from the Grapesvare metasediments. Analyst and instrument: see under Table 1

	78–163B	78–165A	78–165B	78–165C	78–166C	78–170C	78–171A
SiO_2	35.61	35.66	37.70	35.53	35.46	34.94	36.23
Al_2O_3	16.87	16.19	14.97	16.70	16.69	16.07	15.43
TiO_2	2.78	3.26	2.05	2.63	2.29	3.63	2.72
FeO	24.25	23.10	17.98	21.28	20.74	25.62	26.23
MnO	0.03	0.08	0.18	0.01	0.01	0.20	0.11
MgO	7.48	7.61	11.24	9.36	9.97	5.73	4.92
K_2O	9.21	9.16	9.15	9.28	9.29	9.01	8.97
Si	5.49	5.54	5.81	5.49	5.49	5.50	5.73
Al^{IV}	2.51	2.46	2.19	2.51	2.51	2.50	2.27
Al^{VI}	0.56	0.51	0.52	0.53	0.54	0.49	0.61
Ti	0.32	0.38	0.24	0.31	0.27	0.43	0.32
Fe	3.13	3.01	2.32	2.75	2.68	3.37	3.47
Mn	—	0.01	0.03	—	—	0.03	0.01
Mg	1.72	1.76	2.59	2.16	2.30	1.34	1.16
K	1.82	1.82	1.80	1.83	1.84	1.81	1.81
$X_{Mg/Mg+Fe}$	0.35	0.37	0.53	0.44	0.46	0.28	0.25

garnet are totally dominant. Fresh rutile occurs as aggregates as do biotite and quartz. The composition of garnet and clinopyroxene in the Grapesvare eclogite is presented in Table 6. Similar compositions have recently been obtained by other analysts (M. B. E. Mørk, Oslo, personal communication 1982 and T. Widmark, Lund, personal communication 1982).

The clinopyroxene is omphacitic (Essene and Fyfe 1967) with 25 per cent jadeite. Garnets are Mg-poor but Ca-rich and correspond to 'garnets from eclogites within glaucophane schists' (Coleman *et al*. 1965). While the garnets of some boudins indicate retrogression (i.e. the Mg/Fe ratio falls from core towards rim of the crystal: Table 6 and M. B. E. Mørk, per-

Table 4 Chemical composition of phengite and phengitic muscovite from the Grapesvare metasediments. Analyst and instrument: see under Table 1

	78–163B	78–165A	78–165C	78–166C	78–170C	78–171A
SiO_2	50.50	48.55	50.22	48.07	48.45	47.95
TiO_2	0.90	0.89	1.80	1.14	1.22	0.83
Al_2O_3	29.06	31.83	28.66	30.45	31.79	31.93
FeO	3.18	2.74	2.75	2.49	2.81	3.18
MgO	1.63	1.56	2.35	2.07	1.26	1.17
CaO	0.05	—	—	0.03	0.02	—
Na_2O	0.30	0.38	0.36	0.51	0.38	0.32
K_2O	10.06	10.75	10.49	10.72	10.24	10.81
Si	6.71	6.41	6.63	6.46	6.42	6.39
Al^{IV}	1.29	1.59	1.37	1.54	1.58	1.61
Al^{VI}	3.26	3.37	3.10	3.27	3.40	3.40
Ti	0.09	0.09	0.18	0.11	0.12	0.08
Fe	0.35	0.30	0.30	0.28	0.31	0.35
Mg	0.32	0.31	0.46	0.41	0.25	0.23
Ca	0.01	—	—	—	—	—
Na	0.08	0.10	0.09	0.13	0.10	0.08
K	1.70	1.82	1.77	1.83	1.73	1.84

sonal communication 1982), other boudins carry garnets with unambiguous evidence of prograde growth (T. Widmark, personal communication 1982). The cores of the latter garnets contain inclusions of randomly oriented, prismatic epidote, and of quartz and sphene. The rims carry inclusions of omphacite and rutile. The Mg/Fe ratio of these garnets increases from core to rim. Calculations of eclogitization temperatures employing the garnet–clinopyroxene geothermometer of Ellis and Green (1979) have yielded values between 550°C and 750°C, for the Grapesvare eclogites.

Retro-eclogites

Retrogression of the eclogites was coeval with the formation of the regional foliation developed in the host rocks. The first evidence of amphibolization is associated with development of a NW–SE trending stretching ('transverse') lineation (Fig. 6A). A common variety of retrograde eclogite consists of subidioblastic garnets embedded in a very fine-grained plagioclase–hornblende symplectite. Where penetrative deformation reached into the core of a boudin, a spaced cleavage defined by amphibole and biotite porphyroblasts, quartz streaks and trails of magnetite formed. With the onset of amphibolitization, garnets lost their sharp outlines and recrystallized to a granular intergrowth of plagioclase and hornblende. Sphene grew on rutile and fresh, single grains of albite–oligoclase formed.

In better preserved eclogites, retrogression is characterized by the breakdown of omphacite into plagioclase–clinopyroxene symplectite, passing into coarser plagioclase–amphibole symplectite; the amphibole sometimes carrying a small Na_B component.

Amphibolite

The retrograde eclogites described above pass into amphibolites with granoblastic textures and with epidote, biotite, and sphene as important constituents. Such amphibolites characterize smaller boudins, bands connecting boudins, and margins to eclogites. An investigated 30 m long and less than 30 cm thick band consists essentially of a hornblende–epidote–biotite–sphene paragenesis, but a few samples contain relics of coarse plagioclase–hornblende symplectite and garnet. This band was traced continuously laterally into a centimetre-thick mica schist composed of phengite, epidote, and sphene.

Structures

The eclogite-bearing rocks of the Grapesvare area occur within the central part of the wedge of Seve rocks located northeast of Nasafjäll (Fig. 2). The internal geometry of this Seve wedge is under study and the description of structures that follows here is limited to those occurring in the immediate association of the metabasic rocks. As mentioned earlier, Kulling (1982) suggested that the eclogite-bearing Juron quartzites occur in the core of a major recumbent antiform.

The metabasite bands and boudins lie within the gently dipping regional foliation (S2) of the Seve wedge. On the western slope of Grapesvare (Fig. 3),

Table 5 Chemical composition of garnets from the Grapesvare metasediments. Fe^{3+} contents calculated as differences between ideal and calculated charges. Analyst and instrument: see under Table 1. Specimen numbers refer to the Swedish IGCP-CO Collection File A

| | 78-163B | | 78-165A | | 78-165B | | 78-165C | | 78-166C | | 78-170C | | 78-171A | | | |
| | | | | | | | | | | | | | core | | rim | |
	core	rim	core	rim	core	rim	core	rim	core	rim	core	rim	I	II	I	II
SiO_2	36.76	36.99	36.97	36.36	38.42	36.60	36.49	35.16	37.77	37.55	37.45	37.33	37.44	37.32	37.47	36.69
Al_2O_3	20.94	21.03	21.54	21.17	21.40	21.09	21.61	21.60	21.66	21.82	21.13	21.37	20.61	20.77	20.60	21.23
TiO_2	0.02	0.04	0.14	0.13	0.05	—	0.11	0.16	0.16	0.21	0.05	0.08	0.17	0.07	0.14	0.01
MgO	2.92	2.35	1.61	1.77	3.10	2.04	0.92	1.02	1.18	1.33	1.01	1.83	1.24	1.27	1.25	1.29
FeO	33.42	33.33	32.95	33.71	24.43	25.15	23.69	25.25	27.94	28.50	25.69	29.96	31.41	33.65	33.46	34.24
MnO	5.01	3.88	5.59	4.97	3.84	3.82	7.00	5.21	1.88	1.85	5.90	1.39	3.63	2.97	2.74	2.78
CaO	1.03	2.67	2.47	2.40	9.82	10.47	10.94	10.92	10.55	10.05	8.83	8.41	5.66	4.48	4.81	4.80
Si	5.95	5.97	5.93	5.90	6.00	5.89	5.83	5.72	5.96	5.94	6.00	5.97	6.04	6.02	6.04	5.94
Al^{IV}	0.05	0.03	0.07	0.10	—	0.11	0.17	0.28	0.04	0.06	—	0.03	—	—	—	0.06
Al^{VI}	3.94	3.97	4.00	3.95	3.94	3.89	3.90	3.86	3.99	4.01	3.99	4.00	3.92	3.95	3.91	3.99
Fe^{3+}	0.06	0.03	—	0.03	0.05	0.11	0.09	0.12	—	—	—	—	0.03	0.03	—	0.01
Ti	—	—	0.02	0.02	0.01	—	0.01	0.02	0.02	0.03	0.01	0.01	0.02	0.01	0.02	—
Mg	0.71	0.56	0.38	0.43	0.72	0.49	0.22	0.25	0.28	0.31	0.24	0.43	0.30	0.31	0.30	0.31
Fe	4.46	4.47	4.41	4.54	3.14	3.28	3.07	3.32	3.69	3.77	3.44	3.98	4.20	4.51	4.51	4.63
Mn	0.69	0.53	0.76	0.68	0.51	0.52	0.95	0.72	0.25	0.18	0.80	0.19	0.50	0.41	0.37	0.38
Ca	0.18	0.46	0.43	0.42	1.64	1.81	1.87	1.90	1.78	1.70	1.52	1.43	0.98	0.77	0.83	0.70
Almandine	73	74	74	74	52	53	50	53	61	61	57	66	70	75	75	77
Pyrope	12	9	6	7	12	8	4	4	5	5	4	7	5	5	5	5
Spessartine	11	9	13	11	8	8	15	11	4	3	13	3	8	7	6	6
Grossular	3	8	7	7	27	29	30	30	30	31	26	24	16	13	14	12
Andradite	1	—	—	1	1	2	1	2	—	—	—	—	1	—	—	—

Table 6 Chemical composition of garnet and clinopyroxene from the Grapesvare eclogite. Fe^{3+} contents calculated as the difference between ideal and calculated charges. Analyst and instrument: see under Table 1. Specimen provided by A. Wikström, Uppsala

	Garnet W68–76		Clinopyroxene W68–76	
	core	rim		
SiO_2	38.68	38.20	SiO_2	52.75
Al_2O_3	21.06	21.00	Al_2O_3	8.65
TiO_2	0.27	0.29	TiO_2	0.25
MgO	3.79	3.31	MgO	9.09
FeO	24.35	27.13	FeO	7.26
MnO	4.08	0.86	CaO	16.43
CaO	9.22	10.51	Na_2O	4.62
Si	6.01	5.96	Si	1.95
Al^{IV}	—	0.04	Al^{IV}	0.05
Al^{VI}	3.86	3.82	Al^{VI}	0.30
Fe^{3+}	0.02	0.07	Ti	0.01
Ti	0.03	0.03	Fe^{3+}	0.05
Mg	0.88	0.77	Fe^{2+}	0.18
Fe	3.13	3.47	Mg	0.48
Mn	0.54	0.11	Mg	0.03
Ca	1.54	1.76	Ca	0.66
			Na	0.31
Almandine	51	56		
Pyrope	14.5	12.5		
Spessartine	9	2	Acmite	5
Grossular	25	28.5	Jadeite	26
Andradite	—	1	Augite	69

the outcrop pattern of mica schists and eclogite boudins (Fig. 4) is controlled by an open F3 synform with a general southerly dip of the regional schistosity in the northern part of the map area and a general dip towards the north in the Tjidtjakkvagge valley. F3 folds indicated on the map (Fig. 3) are parasitic to this major F3 synform. West of the map area, F3 fold axes swing northwest to align with the Vuoggatjålme–Mierkenes system of NW–SE trending synforms and antiforms, the latter exposing lower allochthonous cover and basement in their cores (Fig. 2).

Systematic mapping of minor folds suggests that the major F3 synform refolded a large-scale recumbent, southwards closing (F2) fold. The central mica schist belt (Fig. 3) with abundant boudins occurs in the upper limb, and the southernmost mica schist belt in the lower limb of this fold. The Tjidtjakkvagge quartzite belt with well preserved metabasite bands is situated in the hinge zone of the recumbent fold.

Non-cylindrical folding of early foliations (at least locally related to F1) characterizes the Grapesvare area. Fig. 6A illustrates how a metabasite band was isoclinally folded on a NW–SE trending axis and later refolded by two nearly coaxial folds. The presence of sheath and tubular folds (Fig. 6B,C) bear witness to locally extreme deformation during F2. Fig. 6B shows how a large recumbent fold with average NW–SE trend was stretched in an easterly direction. Tubular folds occur, in which the 'tube' has constant dimensions (i.e. parallel limbs) for several tens of metres. Reflecting ductility contrasts between host rocks, the metabasites of the pelite belts are invariably strongly boudined and disrupted, while those of the psammite belt are often only weakly necked. Boudins in the psammite belts are, moreover, usually larger than those of the pelite belts.

The earliest mesofolds recognized are recumbent and isoclinal with approximate NW–SE axial trends (F1; Fig. 6). They were apparently the first structures to fold the primary stratification preserved in the calcareous feldspathic quartzites (Fig. 5).

Metasediments, metadolerites, and eclogites share all recognized phases of deformation. Dolerite bands and eclogite trails never transect each other (Fig. 4). While the cores of some eclogite boudins and most of the metadolerite bands lack deformational fabrics, the foliation of amphibolitized margins is concordant with the pervasive S2 foliation of the country rock.

Origin of the Eclogites

The eclogites occurring in the Scandinavian Caledonides show considerable variation in mineralogy, reflecting varying regional metamorphic environments. They occur in both low tectonostratigraphical levels towards the interior of the orogen (southwest Norway) and high levels in the long-transported nappes.

Three lines of evidence favour their Caledonian origin:

1. Their occurrence exclusively within the orogen.
2. Their development in both basement and Caledonian supracrustal successions.
3. Sm/Nd age-determination evidence of c. 425 Ma crystallization (Griffin and Brueckner 1980).

An eclogite association very similar to that described here occurs in the lower part of the Seve Nappes in Västerbotten (van Roermund this volume), where temperatures of c. 600°C (at c. 15 kb) were obtained for the eclogite crystallization. Low-temperature eclogites have also been reported from the basal gneisses and 'parautochthonous' cover of the Sunnfjord area (Krogh 1980) of southwestern Norway.

In Norrbotten (Grapesvare area), the occurrence of the eclogites within the Seve metasediments indi-

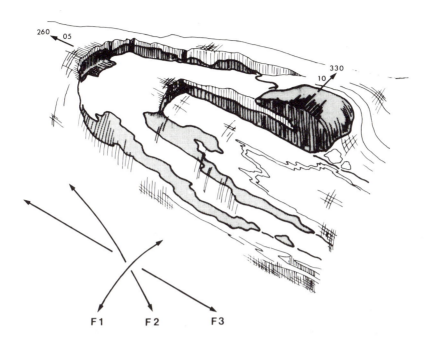

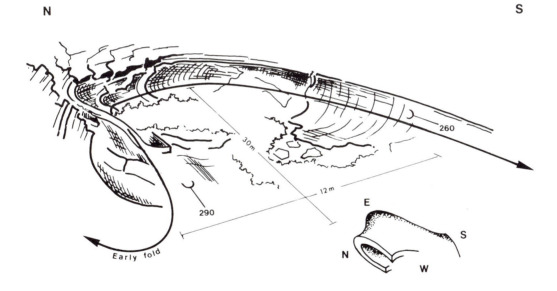

Fig. 6   Structures in the Grapesvare area. Drawings from photographs and field sketches. A: A metabasite band was folded by an early, isoclinal and recumbent fold with approximate NNW–SSE axial trend (inferred F1) and later refolded by two nearly coaxial folds (F2 with stretching and boudinage and F3 with open, concentric folding). A lineation defined by hornblende and plagioclase (330°) may be related to the first phase of folding. Loc. 79–112. B: An early, recumbent and isoclinal fold (F1?) with average N–S axial trend was stretched in easterly direction (F2). The size of the structure is indicated by figures on the floor of the cave. Loc. 80–125. C: Tubular fold. The length of the broken 'tube' is 1.5 m Loc. 79–139

cates their origin during regional metamorphism. Both corona dolerites and eclogites have been influenced by the same sequence of deformation. The metasedimentary association of the Seve eclogites is reminiscent of that of the underlying Särv Nappes, where dolerites intrude feldspathic sandstones. Whole rock chemical analysis for major and immobile elements has failed to detect any differences between eclogites, metadolerites, and amphibolites of the Grapesvare area (Solyom personal communication 1982). These basic rocks are tholeiites with a chemistry identical to that of the dolerites in the Särv Nappes. The latter dolerites were probably related to the opening of the Iapetus Ocean (Gee 1975; Solyom *et al.* 1979; Kumpulainen 1980). The eclogite-bearing Seve units contain substantial pelitic and semi-pelitic units and subordinate marble together with feldspathic quartzites. Stratigraphy within the Seve Nappes has not been established. However, the frequent development of thick psammite sequences in the lower tectonic units of the Seve throughout the Swedish Caledonides suggests that they originated in a depositional basin closely related to that of the underlying sandstone successions (Kumpulainen 1980) of the Särv Nappes. The hypothesis favoured here is that the Särv and Seve Nappes differ in terms of metamorphic grade and deformational history but underwent a related pretectonic development (Gee 1975). Superposition of the Seve on the Särv requires their original location to the west of the Särv, forming part of the depositional basin that developed along the Baltoscandian margin immediately prior to and during rifting of the Laurasian continent.

The structural and metamorphic history recorded in the Seve Nappes suggests that the late Precambrian miogeoclinal prism was subject to deep depression during the Caledonian cycle prior to thrust emplacement onto the Baltoscandian Platform. It is of interest in this context to note that the eclogites occurring higher in the Seve tectonostratigraphy in the central gneiss complex (Avardo Formation) in Jämtland are indicative of higher temperatures and pressures than those in the underlying units in that area (van Roermund this volume) and in Norrbotten.

We have no age-determination evidence from the Seve in Norrbotten to indicate when in the Caledonian cycle the high pressure metamorphism occurred. Scandian ages (mid-Silurian to early Devonian) have been obtained by the U–Pb method on zircons and monazite and by the Rb–Sr method on migmatite neosomes in the granulite facies Seve units of central Jämtland (Claesson 1981). On the basis of this evidence and that obtained by the dating of the eclogites of the parautochthonous basement in

southwest Norway (Griffin and Brueckner 1980), we prefer the hypothesis that the Grapesvare eclogites formed during Scandian collisional orogeny (Gee 1975). The eclogites of southwest Norway crystallized in response to the deep depression achieved during build up of the nappe pile in that area. The Grapesvare eclogites differ from them in that they were thrust at least 200 km eastwards after regional metamorphism.

These two lines of evidence suggest subduction of the Baltoscandian margin, involving build up of a tectonic overburden, depression of the miogeocline and underlying Precambrian crystalline basement, eclogite crystallization, and subsequent thrusting to higher tectonic levels on the Baltoscandian Platform. The eclogites provide an estimate of the depth of crustal depression during the collisional process and, in the case of southwestern Norway, also may allow an estimate of the amount of tectonic thinning (vertical shortening) and lateral spreading that accompanied the final uplift of the orogen.

Acknowledgements

This study was supported by grants from the Swedish Natural Sciences Research Council to the IGCP project Caledonide Orogen, and was in part sponsored by Swedish funds to Project Nordkalotten. Drs. A. Wikström and T. Sjöstrand kindly contributed unpublished material. Drs. C. Ålinder, T. Widmark, and Z. Solyom carried out the microprobe work. We acknowledge constructive criticism from anonymous referees.

References

Arnbon, J.-O. 1980. Metamorphism of the Seve Nappes at Åreskutan Swedish Caledonides. *Geol. Fören. Stockh. Förh.*, **102**, 359–371.

Backlund, H. G. 1935. Zur Tektonik des nordschwedischen Hochgebirges. *Geol. Rundschau*, **26**.

Backlund, H. G. 1936. Zur genetischen Deutung der Eklogite. *Geol. Rundschau*, **27**, 47–61.

Bryhni, I. and Andréasson, P. G. this volume. Metamorphism in the Scandinavian Caledonides.

Bryhni, I., Fyfe, W. S., Green, D. H. and Heier, K. S, 1970. On the occurrence of eclogite in western Norway. *Contr. Mineral. Petrol.*, **26**, 71–94.

Bryhni, I., Krogh, E. J. and Griffin, W. L. 1977. Crustal derivation of Norwegian eclogites: a review. *N. Jb. Miner. Abh.*, **130**, 49–68.

Claesson, S. 1981. Caledonian metamorphism of Proterozoic Seve rocks on Mt. Åreskutan, southern Swedish Caledonides. *Geol. Fören. Stockh. Förh.*, **103**, 1–14.

Coleman, R. G., Lee, D. E., Beatty, L. B. and Brannoch, W. W. 1965. Eclogites and eclogites. Their differences and similarities. *Geol. Soc. Am. Bull.*, **76**, 483–508.

Ellis, D. J. and Green, D. H. 1979. An experimental study of the effect of Ca upon garnet–clinopyroxene Fe–Mg exchange equilibria. *Contr. Mineral. Petrol.*, **71**, 13–22.

Eskola, P. 1921. On the eclogites of Norway. *Skr. Vidensk. Selsk., Christiania, Mat.–Nat. Kl.I*, **8**, 1–118.

Essene, E. J. and Fyfe, W. S. 1967. Omphacite in Californian metamorphic rocks. *Contr. Mineral. Petrol.*, **15**, 1–23.

Gee, D. G. 1975. A tectonic model for the central part of the Scandinavian Caledonides. *Am. J. Sci.*, **275A**, 468–515.

Gee, D. G. and Zachrisson, E. 1979. The Caledonides in Sweden. *Sver. geol. Unders.*, C **796**, 1–48.

Gjelsvik, T. 1953. Det nordvestlige gneisområde i det sydlige Norge, aldersforhold og tektonisk-stratigrafisk stilling. *Norges geol. Unders.*, **184**, 71–94.

Green, T. H. and Hellman, P. L. 1982. Fe–Mg partitioning between coexisting garnet and phengite at high pressure, and comments on a garnet–phengite geothermometer. *Lithos*, **15**, 253–266.

Griffin, W. L. and Brueckner, H. K. 1980. Caledonian Sm–Nd ages and a crustal origin for the Norwegian eclogites. *Nature, London*, **285**, 319–321.

Griffin, W. L., Austrheim, H., Brastad, K., Bryhni, I., Krill, A., Mørk, M.-B. E. Qvale, H. and Tørudbakken, B. 1981. High-pressure metamorphism in the Scandinavian Caledonides. *Terra Cognita*, **1**, 48 (Abstract).

Helfrich, H. K., 1967. Ein Beitrag zur Geologie des Åregebietes aus dem centralen Teil der Schwedischen Kaledoniden. *Sver. geol. Unders.*, C **612**, 35 pp.

Krogh, E. J. 1980. Geochemistry and petrology of glaucophane-bearing eclogites and associated rocks from Sunnfjord, western Norway. *Lithos*, **13**, 355–380.

Krogh, T. E., Mysen, B. O. and Davies, G. L. 1974. A Palaeozoic age for the primary minerals of a Norwegian eclogite. *Ann. Rep. Geophys. Lab. Carnegie Inst, Washington*, **73**, 575–576.

Kulling, O. 1955. Den kaledoniska fjällkedjans berggrund inom Västerbottens län. Beskrivning till berggrundskarta över Västerbottens län 2. *Sver. geol. Unders.*, **Ca 37**, 101–296.

Kulling, O. 1964. Översikt över norra Norrbottensfjällens **Kaledonberggrund**. *Sver. geol. Unders.*, **Ba 19**, 166 pp.

Kulling, O. 1982. Översikt över södra Norrbottensfjällens Kaledonberggrund. *Sver. geol. Unders.*, **Ba 26**, 295 pp.

Kumpulainen, R. 1980. Upper Proterozoic stratigraphy and depositional environments of the Tossåsfjället Group, Särv Nappe, southern Swedish Caledonides. *Geol. Fören. Stockh. Förh.*, **102**, 531–550.

Landmark, K. 1973. Beskrivelse til det geologiske kart 'Tromsø' og 'Måselv'. Et snitt gjennom Fjellkjeden i midt-Troms. Del II. *Tromsø Mus. Skr.*, **15**, 263 pp.

Lappin, M. A. and Smith, D. C. 1978. Mantle-equilibrated orthopyroxene eclogite pods from the basal gneisses in the Selje district, western Norway. *J. Petrol.*, **19**, 330–384.

O'Hara, M. J., Richardson, S. W. and Wilson, G. 1971.

Garnet peridotite stability and occurrence in crust and mantle. *Contr. Mineral. Petrol.*, **32**, 48–68.

Roermund, H. L. M. van. this volume. Eclogites of the Seve Nappe, central Scandinavian Caledonides.

Sjöstrand, T. 1978. Caledonian geology of the Kvarnbergsvattnet area, northern Jämtland, central Sweden. *Sver. geol. Unders.*, C **735**, 107 pp.

Smith, D. C. 1980. A tectonic mélange of foreign eclogites and ultramafites in West Norway. *Nature, London*, **287**, 366–368.

Solyom, Z., Gorbatschev, R. and Johansson, I. 1979. The Ottfjäll Dolerites. Geochemistry of the dyke swarm in relation to the geodynamics of the Caledonide Orogen in central Scandinavia. *Sver. geol. Unders.*, C **756**, 38 pp.

Stephens, M. B., Gustavson, M., Ramberg, I. B. and Zachrisson, E. this volume. The Caledonides of central–north Scandinavia—a tectonostratigraphic overview.

Strand, T. and Kulling, O. 1972. Scandinavian Caledonides, Wiley, London, 302 pp.

Svenonius, F. 1881. Om den s.k. Sevegruppen i nordligaste Jämtland och Ångermanland samt dess förhållande till fossilförande lager. *Geol. Fören. Stockh. Förh.*, **5**, 484–497.

Thelander, T., Bakker, E. and Nicholson, R. 1980. Basement cover relationships in the Nasafjället Window, Central Swedish Caledonides. *Geol. Fören. Stockh. Förh.*, **102**, 569–580.

Trouw, R. 1973. Structural geology of the Marsfjällen area, Caledonides of Västerbotten, Sweden. *Sver. geol. Unders.*, C **689**, 155 pp.

Vidale, R. J. 1974. Vein assemblages and metamorphism in Dutchess County, New York. *Geol. Soc. Am. Mem.*, **74**, 153 pp.

White, A. J. R. 1966. Genesis of migmatites from the Palmer region of South Australia. *Chem. Geology*, **1**, 165–200.

Williams, P. F. and Zwart, H. J. 1977. A model for the development of the Seve–Köli Caledonian Nappe Complex. In Saxena, S. K. and Bhattacharji, S. (Eds), *Energetics of Geological Processes*, Springer–Verlag, New York, pp. 170–187.

Yardley, B. W. D. 1978. Genesis of the Skagit Gneiss migmatites, Washington, and the distinction between possible mechanisms of migmatization. *Geol. Soc. Am. Bull.*, **89**, 941–951.

Zachrisson, E. 1969. Caledonian geology of northern Jämtland—southern Västerbotten. *Sver. geol. Unders.*, C **644**, 33 pp.

Zachrisson, E. 1973. The westerly extension of Seve rocks within the Seve–Köli Nappe Complex in the Scandinavian Caledonides. *Geol. Fören. Stockh. Förh.*, **95**, 243–251.

Zwart, H. J. 1974. Structure and metamorphism in the Seve Köli Nappe Complex and its implication concerning the formation of metamorphic nappes. *Cent. Soc. geol. Belg. Liège*, **2**, 129–144.

The Caledonide Orogen—Scandinavia and Related Areas
Edited by D. G. Gee and B. A. Sturt
© 1985 John Wiley & Sons Ltd

Geology and metamorphism of the Krutfjellet mega-lens, Nordland, Norway

M. B. E. Mørk

Mineralogisk–Geologisk Museum, Sarsgt. 1, Oslo 5, Norway

ABSTRACT

The Krutfjellet area forms a mega-lens (a separate nappe) of medium-grade polymetamorphic rocks situated within the low-grade Köli Nappe in Hattfjelldal near the Swedish border. The Krutfjellet Group consists of metasediments and metavolcanics intruded by the Krutfjellet gabbro–metagabbro complex. The gabbro was intruded syntectonically, before the nappe emplacement into the low-grade Köli rocks. The gabbro intrusion is associated with contact metamorphism and anatexis of abundant xenoliths and of the country rocks in a narrow contact zone.

Textural evidence of multistage mineral reactions is preserved both in the Krutfjellet Group and in the gabbro complex. Contact metamorphism, and thus time of intrusion, closely follows the main medium-grade regional metamorphic event. A later retrograde metamorphism is characterized by extensive garnet growth both in the contact zone and outside it. This metamorphism is associated with regional gar + biot equilibration at $450° \pm 50°C$, and was continued even after the thrusting with emplacement into the pile of low-grade Köli rocks.

Introduction

The nappes of eastern central Nordland can be divided into two main complexes: the 'highest Nordland nappe units' (Rødingsfjell and Helgeland nappes: Ramberg and Riis 1979), and the underlying Köli nappe complex. Whilst the former consists of rock sequences containing basement/cover relationships typically metamorphosed at high grade, the underlying Köli rocks are of lower Palaeozoic age and are usually of much lower metamorphic grade. However, interest has recently been focused on the occurrence of gabbros and associated higher-grade metamorphic rocks occurring as tectonic mega-lens structures *within* the Köli succession on both sides of the Norwegian–Swedish border (Mørk 1979; Senior 1978; Stephens 1979; Häggbom 1980). The gabbro-intruded mega-lenses have been correlated and assigned to a separate nappe (Häggbom 1980) forming the lower part of the Upper Köli Nappes (Stephens 1980).

The Krutfjellet mega-lens is situated on the eastern side of lake Rösvatnet, close to the imbricated thrust zones of the overlying Rödingsjället and

Helgeland Nappe Complexes (figs. 6 and 21 in Ramberg and Stephens 1981). The mega-lens appearance is related to extreme pinch-and-swell deformation after the emplacement of the nappes (Ramberg 1981).

The central and northern part of the Krutfjellet mega-lens is occupied by the Krutfjellet gabbro complex (40 km^2). This gabbro was intruded syntectonically before the nappe emplacement. Due to the characteristic association of gabbro intrusion and high metamorphism, the effect of contact metamorphism is of special interest. In order to study metamorphic gradients towards the gabbro intrusion both the prograde metamorphic mineral assemblages and the retrograde mineral equilibrations have been considered. Replacement textures, mineral reactions, and element exchange data are discussed with emphasis on the pelitic assemblages.

The diagnostic occurrence of syntectonic gabbros and granites and high metamorphism are not found in the Köli nappes below (Ramberg and Stephens 1981), but can be compared with the highest nappe complexes. The metamorphic conditions calculated for the gabbro emplacement are also very similar to those described for the Sulitjelma gabbro (Mason 1980).

General Geology

The high-grade part of the Krutfjellet area consists of a series of polymetamorphic metasedimentary and metavolcanic rocks constituting the Krutfjellet Group (Mørk 1979). The Krutfjellet Group is separated from the surrounding Köli rocks both by a break in metamorphism and by thrust zones appearing as broad zones of strong deformation and retrogression to green-schist facies. The contact is folded. In the north the high grade rocks are overlain by north-dipping calcareous phyllites of the upper Köli nappes, while at the southern and eastern contacts a sequence of phyllites, trondhjemites, and greenstones intruded by serpentinites dip NW under the Krutfjellet Group (Fig. 1).

The Krutfjellet Group consists of layered biotite–muscovite mica schists and gneisses, with alternating pelitic, calc-pelitic, and psammitic layers and several conglomerate horizons. These rocks are interlayered with amphibolites and hornblende-rich gneisses which dominate in the western part of the area.

The hornblende-rich gneisses occur both in layers isoclinally folded with the pelitic layers, and as inclusions in them. While the amphibole-bearing rocks represent basic volcanics, primary sedimentary structures are locally preserved in the quartz–mica-rich layers. In the northern part of the area quartz–epidote-rich gneisses and calcareous conglomerate are intimately associated and are intruded by both gabbro and granites (Fig. 1).

The Krutfjellet intrusive complex consists of gabbro, metagabbro, diorite, quartz diorite, and granite with clear intrusive contacts to the Krutfjellet Group. The intrusives contain numerous xenoliths of various Krutfjellet Group rock types, affected by contact metamorphism, anatexis, and formation of magmatic breccia structures. Mixtures of partly melted xenoliths and contaminated gabbro are common. The dioritic and quartz-dioritic rocks in the complex are always related to xenolith occurrences and are interpreted as hybrid rocks (Mørk 1979).

The gabbro ranges from augite-norite to olivine gabbro. Olivine and plagioclase are cumulus phases followed by orthopyroxene and/or augite, producing ophitic to subophitic textures. Hornblende and ilmenite are intersitial in some samples. The gabbro is subalkaline according to chemical classification schemes, and shows a tholeiitic affinity by the occurrence of orthopyroxene and by the clinopyroxene chemistry (low Ti-augite).

The granites are chemically very heterogeneous. They are of limited occurrence, closely associated with the gabbro intrusion in space and time, and show no continuity from the gabbro–diorite in two-element variation diagrams. The granites are therefore interpreted as the products of local anatexis.

Deformation

The rocks are usually strongly deformed with two phases of tight to isoclinal folding in the Krutfjellet Group, followed by more open and crenulated folds which also affected the gabbro body. In the F_2 phase the folding style varies and folds are often asymmetrical especially in the southeastern and northeastern parts of the Krutfjell massif. The planar structures S_0, S_1, S_2 together define a good schistosity, and generally are parallel due to the pattern of tight folding. S_1 represents fine-scale metamorphic layering such as rhythmic alternation of hornblende-rich, mica-rich, or quartz-rich layers or 'quartz trains', parallel to the primary layering occurring on a scale of 0.1–10 m.

The S_2-foliation represents reorientation of biotite and muscovite crystals parallel to axial planes of the *tight* F_2 folds. The later crenulation-folding is developed as sinus-folded or more asymmetrically folded S_2-mica foliation and has not developed good cleavage, but is associated with reorientation of retrogressed mineral assemblages.

The main foliation and fold axis are usually parallel to (bending around) the gabbro contact, and the gabbro complex thus forms the core of a regional dome structure. The main schistosity has, however, a persistent N–S direction in the western part of the mega-lens.

Relation between gabbro intrusion and deformation

The gabbro body is folded into open folds, but foliation is only locally developed, usually near the contact and near xenoliths. The part of the intrusion close to the northern thrust contact is the most deformed and retrogressed, and here the WNW–ESE elongation of the body follows the axes of late mesoscopic folds in the NE-contact area, and defines an elongated dome structure. On the western side, however, the gabbro contact is roughly parallel to steep F_2 axial planes, and its emplacement may have been facilitated by such planes of weakness.

The occurrence of foliated xenoliths, and especially of the 8 km long conglomerate horizon with strongly elongated boulders, shows that the intrusion

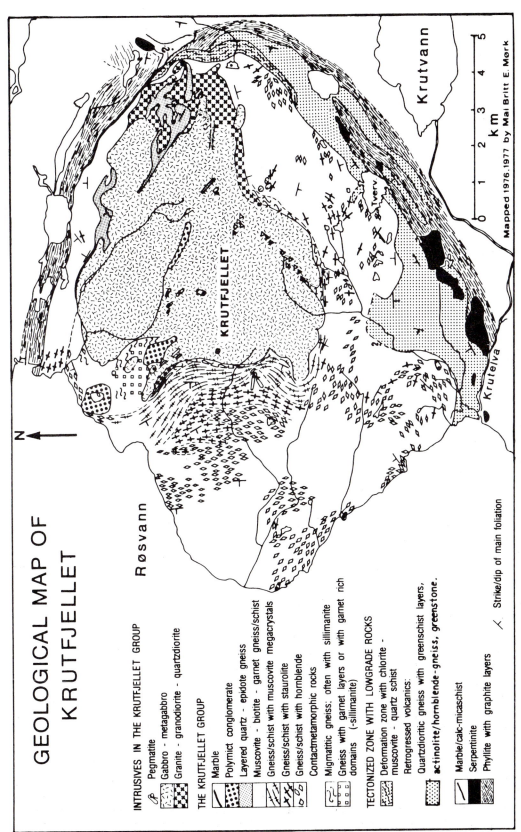

Fig. 1   Geological map of Krutfjellet. Distribution of the index-minerals staurolite, hornblende, and of contact metamorphic rocks in the Krutfjellet Group

clearly post-dates F_1. The late F_3 deformation certainly affected the gabbro; and is strongest near the thrust zone (N). The F_2 deformation phase probably is closely connected to the gabbro emplacement, and the high strain features (extensive boudinage) recorded in the magmatic breccias probably reflect the syntectonic character of the intrusion which is therefore interpreted as being emplaced during late F_2.

The granites (N and NE) are more strongly deformed than the gabbro with mesoscopic double folds and blastomylonitic fabric in distinct zones. The high strain in the granites is probably due to their proximity to the northern thrust zone and to the limited size of the granite bodies.

Metamorphism

In order to study the extent of contact metamorphism, a large number of samples were selected to cover all the rock types at different distances from the gabbro contact. No change in mineral paragenesis recorded with distance from the gabbro contact outside the few hundred metres wide contact zone shown on the map (Fig. 1). The diagnostic parageneses staurolite + biotite + garnet occurs in pelitic layers all over the area, within 4 km of the gabbro. The examination of metamorphic gradients, however, is complicated by polymetamorphism. Staurolite is frequently replaced by retrograde assemblages. Even when the contact areas mentioned above are excluded, polymetamorphism is evidenced by (a) a larger number of mineral phases than predicted by the phase rule, (b) inclusion and emplacement relations, and (c) different relations to foliation for different mineral generations. Determination of successive mineral parageneses has therefore been based on careful microfabric studies using the criteria of contemporaneousness proposed by Zwart (1962) and more general equilibrium criteria summarized in Spry (1969) and Vernon (1976).

Regional metamorphism: multistage mineral growth

Mineral growth stages in relation to the development of the foliation of the pelitic schists are summarized in Fig. 2a–ii; three stages M1–M3 of metamorphism are recognized.

M1 is defined by the assemblage biotite–muscovite–staurolite–garnet, marking the peak of regional metamorphism. Staurolite usually forms porphyroblasts with inclusions of quartz and less abundant inclusions of graphite and mica. The inclusions either cluster in the core-domains or less commonly show strong preference to crystallographic faces in the staurolite 'host' reflecting expulsion from the growing crystal according to Hollister and Bence (1967). In the most common case however inclusion-free domains are recorded with a tabular shape after mica ('negative pseudomorphs') indicating crystal growth by chemical replacement. The outer rims are usually inclusion-free. Only in a few cases is an older foliation defined by graphite or elongated quartz grains preserved as inclusions in staurolite. The main foliation (S_2) of biotite and muscovite is always seen to bend around the staurolite porphyroblasts with recrystallized quartz and micas in pressure shadows. Using the criteria for growth by chemical replacement (Zwart 1962) the peak of staurolite growth had been reached before the F_2-flattening process.

The garnet porphyroblasts clearly demonstrate crystal growth by chemical replacement, resulting in a garnet fabric that reflects both the minerals replaced during garnet growth (local fabric) and time of growth relative to deformation. Multiple growth stages of garnet are preserved in the mica schists with the most extensive garnet growth represented by helicitic post-S_2 garnets. These garnets often define a second growth stage rimming inclusion-free, inclusion-discordant, or rotated garnet cores. Mutual contact between staurolites and garnet is not common but textures (Fig. 2a–iii) indicate that the two minerals coexisted during M1.

During the retrograde metamorphism M2, staurolite breaks down into fine-grained muscovite aggregates and simultaneously chlorite nucleates in the surrounding biotite foliation. The following breakdown reaction is recorded all over the area:
M2: staur + biot + H_2O → musc + chlor.
The staurolite shape is often excellently preserved, as is the mica foliation which bends around the staurolite pseudomorphs. Since the pseudomorphic muscovite aggregates are recrystallized and oriented only in the most crenulated rock specimens, the textures indicate that the widespread staurolite breakdown occurred post-F_2 and pre-F_3.

A third metamorphic event (M3) is defined by the widespread growth of garnet, hornblende and less abundant epidote in the pelitic layers. Both euhedral garnet and hornblende have nucleated across or within the fine-grained muscovite aggregate in the pseudomorphs after staurolite. The calcic minerals (hornblende, epidote) are not recorded in contact with fresh staurolite, demonstrating significant chemical migration with Ca-transport into the pseudomorphs. From Figs. 2a iii–iv, the euhedral garnets are correlated with the helicitic, post-S_2 outer garnet rims, and from Fig. 3 it can be seen that

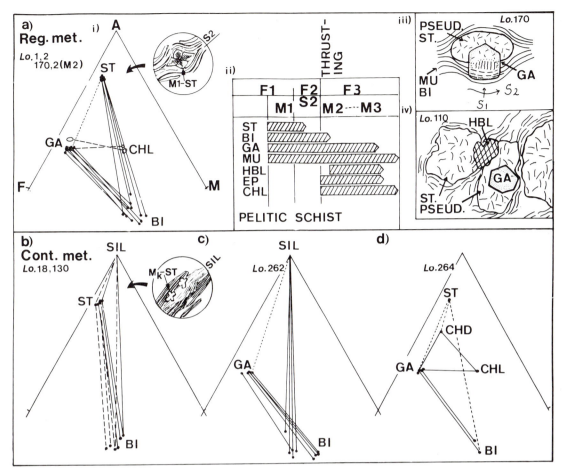

Fig. 2   AFM projections for pelitic rocks from Krutfjellet, with tie lines (═══) connecting coexisting mineral pairs. Earlier parageneses are indicated by dotted lines. (a) Regional metamorphism. (ii)–(iv) Evidence for multistage mineral growth described in the text. (b)–(d) Contact metamorphism

this latest stage of garnet growth is associated with an increase in Ca.

For the pelitic schists the following parageneses occur.

M1 (pre-F_2):
   P_{1a}: staur + gar + biot + musc + qtz ± plag
   P_{1b}: gar + biot + musc + qtz ± plag

M2 (post-F_2, pre-syn-F_3):
   P_{2a}: gar + biot + chlor + musc + qtz ± plag

M3 (post-M2?):
   P_3:   gar + hbl + musc + chlor + epid
       + qtz ± plag.

M2 and M3 may be closely related in time, but there is textural evidence in places that the growth of garnet and hornblende occurred after the general staurolite breakdown by reaction (1). In one sample without the M2 replacement of staurolite only, garnet was seen to replace staurolite directly. Since Ca-phases are very rare in pelitic M1-paragenesis, the M3-event must involve local mobilization of Ca

into the pelitic layers. The adjacent alternating more calcareous layers (described below) are likely to be the source, since there are gradations of hornblende content between these layers.

Hornblende mica-schists and hornblende–plagioclase gneisses show a simpler metamorphic evolution, with foliation parallel to the compositional banding. The common paragenesis is:

   P_{1c}: hbl + gar + biot + musc + plag(An_{45})
       + qtz ± epid with gradual transitions
       between the layers:
   P_{1d}: hbl + gar + musc + plag + qtz ± epid
   P_{1e}: biot + gar + musc + plag + qtz ± epid

The most mafic layers show extreme hornblende enrichment but the parageneses found are strongly retrogressed during M2–M3:

   P_{2b}: hbl + chlor ± musc ± qtz ± calcite
   P_{2c}: hbl + musc + qtz ± chlor

Generally epidote and chlorite are most abundant in

samples with strong crenulation folding (F_3), suggesting that they are retrograde in the mafic rocks also.

Pelitic schist and gneiss from the contact zone

The rocks in the contact zone are distinguished from the others by a larger grain size (locally), appearance of 5–20 mm long aluminosilicates, muscovite megacrystals, or feldspar porphyroblasts. Schistosity is often less planar and muscovite is less dominant as an S_2-forming mineral (in contrast to biotite) but occurs more frequently in poikiloblasts and porphyroblasts growing across the main foliation. Coarse-grained muscovite commonly replaces and overgrows aluminosilicate porphyroblasts. At the southwestern contact tabular muscovite replaces sillimanite and is oriented at a steep angle to the main foliation (biotite, quartz foliation), thus defining a S_3-direction, probably related to the later F_3 folding. This particular muscovite growth clearly appeared after the main deformation (F_2) but still reflects a contact metamorphic temperature gradient, as shown by the large grain size and the generally high degree of recrystallization of muscovite in aluminosilicate pseudomorphs close to the gabbro.

The sillimanite–biotite pair defines the peak of contact metamorphism at the southern and western gabbro contact, while the assemblage sillimanite–garnet–K–feldspar–biotite was only recorded in the northwestern contact zone. The contact metamorphic rocks preserve retrograde replacement textures indicating higher metamorphic conditions than those described for the regional metamorphism.

At the *southern and western gabbro contact* sillimanite occurs in idiomorphic porphyroblasts or as lenticular fibrolite aggregates, but is most frequently retrogressed to pseudomorphs of muscovite. The porphyroblasts show no preferred orientation, while fibrolite often subparallels the pre-existing biotite foliation.

In strongly deformed rocks near the *western* gabbro contact fibrolite relics are elongated parallel to the main biotite foliation, but since sillimanite preferentially *nucleates* in biotite (Chinner 1961) the growth of fibrolite might have occurred after the main deformation.

The sillimanite is associated with a type of staurolite which differs from the regional metamorphic M1-staurolite both in size, chemistry, and relation to the main foliation. These contact metamorphic staurolites were formed very locally as minute grains overgrowing the main foliation. They are orientated across the boundary between fibrolite and biotite

grains, and preserve the inclusion pattern after fibrolite. Thus at least this late staurolite growth occurs after the main deformation and S2-foliation.

The common contact metamorphic paragenesis is

$$P_{k1} \text{ sill} + \text{biot} + \text{musc} + \text{plag} + \text{qtz}$$

Staurolite is formed in P_{kl} at the expense of sillimanite and biotite. Together with the extensive muscovite formation and recrystallization in sillimanite pseudomorphs, this relation demonstrates the following retrograde reaction:

$$M_{k1a} \text{ sill} + \text{biot} \rightarrow \text{staur} + \text{musc}$$

Similarly, appearance of idiomorphic garnets in recrystallized muscovite megacrysts with inclusion trails after fibrolite indicates the reaction:

$$M_{k1b} \text{ sill} + \text{biot} \rightarrow \text{gar} + \text{musc}$$

In the western and southern contact zone late formed garnet and staurolite do not coexist. In other less recrystallized pseudomorphs from the contact zone, chlorite appears in the fine-grained muscovite aggregates.

At a location at the western gabbro contact (loc 86) the porphyroblast habits indicate that sillimanite replaced earlier-formed andalusite. During retrograde metamorphism sillimanite broke down to fine-grained muscovite at the margins and in cracks where staurolite and chloritoid formed locally (fig. 30 a, b in Mørk 1979).

The *northwestern contact zone* (Fig. 1) consists of a very heterogeneous complex of gneisses, veined and intruded by granitic to dioritic rocks and metagabbro. High-T mineralogy reveals evidence of multistage metamorphic breakdown reactions in a special rock that was only recorded in this area. The rock is characterized by garnet-enrichment in layers and spots (loc 262, 264). Sillimanite occurs in the paragenesis:

$$P_{k2}: \text{sill} + \text{biot} + \text{gar} + \text{plag} + \text{K-fsp} + \text{qtz}$$

Sillimanite also occurs as relics in muscovite-rich shear zones where biotite is almost totally broken down.

At loc 262b a microlayering is defined by sillimanite–biotite and garnet–quartz enrichments. The same type of garnet–quartz layers occurs in an adjacent rock (264) but here they alternate with layers enriched in euhedral chloritoid crystals in a fine-grain matrix of muscovite and chlorite, indicating much lower metamorphic grade. However, a part of the metamorphic evolution of this rock can be read from inclusion relationships. Whilst chloritoid contains inclusions of relic staurolite and even shows an example of pseudomorphism with the preservation of a typical staurolite cross, biotite is preserved only as minute inclusions in garnet crystals. Presuming that relic staurolite and biotite coexisted before the retrograde reactions, the following parageneses may represent the two metamorphic stages:

$$P_{k3a}: \text{gar + biot + staur + musc + qtz}$$
$$\pm \text{(ep?)} \pm \text{(opaque)}$$
$$P_{k3b}: \text{chld + gar + musc + chlor + qtz + epid}$$
$$\pm \text{(opaque)}$$

The lack of strong preferential mica orientation in the retrograde assemblages indicates that the retrogression occurred after the main deformation.

Cordierite–orthopyroxene xenolith

This xenolith, a few metres across, is found in the gabbro in the central southern part of Krutfjellet. Most of the cordierite is broken down to pseudomorphic aggregates of kyanite and chlorite. Orthopyroxene is locally retrogressed to biotite and serpentine. The garnets (poikiloblasts, *ca.* 1 cm diameter) are often rimmed by quartz but also share contacts with the other minerals. Microstructures indicate a step-wise transition from the paragenesis:

1. cord + gar + opx + plag + qtz ± bio ± ilm to
2. cord + gar + biot + kya + plag + qtz ± ilm
   down to the local associations of
3. ky + chlor + biot in the pseudomorphs after cordierite.

Most of the xenoliths preserved in the northern part of the intrusion are calcareous rocks. These are fine-grained and 'hornfelsic' in appearance when mica poor, and are dominated by epidote, actinolite, or diopside. Marble xenoliths are associated with grossular skarn formation, and other types of skarn have formed in mushes of partly melted country rock in magmatic breccias (Mørk 1981), demonstrating the high temperature conditions of country rock during the gabbro intrusion.

The contact metamorphic gneisses contain many garnet bearing anatectic veins, very often quartz dioritic. The garnets may achieve a diame r of up to 1 cm.

Metamorphism of the gabbro

The Krutfjellet gabbro has a very heterogeneous appearance due to the abundant xenoliths associated with contamination and retrogression of the gabbro, and to abundant alternations between gabbro and a range of metagabbro variants. Retrogression of the gabbro started at high-T conditions with corona-forming reactions similar to those described from the Sulitjelma gabbro (Mason 1968) producing orthopyroxene, pale green (Fe-pargasitic) hornblende + spinel coronas between primary olivine and plagioclase. Other high-T reactions are the replacement of magmatic clinopyroxene by brown, Ti-rich hornblende. These high-T reactions are very incomplete and are interpreted as an early retrogression during cooling of the gabbro.

Various stages of retrogression have been recorded down to low-grade assemblages that formed in the more Mg-rich samples in the northern part of the complex. The following mineral replacements are seen:

$$\text{cpx + ilm} \rightarrow \text{hbl (brown)}$$
$$\text{ol + plag} \rightarrow \text{hbl (pale green), opx, spin, plag}$$
$$\text{hbl (brown)} \rightarrow \text{hbl (green) + rutile + sphene}$$
$$\text{hbl (brown)} \rightarrow \text{biot}$$
$$\text{plag}(An_{69-48}) \rightarrow \text{plag}(An_{40-18}) + \text{epid + musc}$$
$$\text{hbl (brown, green)} \rightarrow \text{actinolite}$$
$$\text{cpx, opx, ol} \rightarrow \text{actinolite}$$

Usually magmatic (cumulate) textures are preserved, and the highest degree of metamorphic recrystallization is reached in fine-grained samples with extensive development of blue–green hornblende.

The lack of metamorphic equilibrium is evident from the mineral fabric, strong colour-zoning of amphiboles, and relict of multistage mineral replacements. Mixtures of magmatic, high, medium, and low-grade mineralogy occur in the same thin section.

A tempting division of the metagabbro types would be to distinguish the medium-grade samples dominated by brown and green hornblende from samples with the low-grade actinolite–chlorite parageneses. But there is no relation between amphibole type, plagioclase breakdown, and chlorite occurrence. The fine-grained partly recrystallized metagabbro with green hornblende would classify as andesine amphibolite, representing medium grade metamorphism (Winkler 1976). Texturally however, the two phases seem to be in equilibrium with chlorite, and chlorite is diagnostic for lower grade. Thus the assemblage is transitional between medium and low grade.

In the most 'retrograded' samples, the actinolite–chlorite–epidote–muscovite assemblages are indicative of low-grade metamorphism, though even the most saussuritic plagioclase commonly has a composition of *ca.* An$_{60}$, implying that plagioclase has not re-equilibrated to the same extent as the mafic phases.

Mineral Chemistry and Equilibrium in the Pelites

Garnet zoning

Garnet compositions range between Alm$_{80-62}$ Pyr$_{30-8}$Spes$_{16-1,5}$Gros$_{20-2}$ with the most Pyr-rich types in the contact metamorphic rocks. Zoning is pronounced, with greatest variation in zoning patterns for the elements Mg, Ca, Mn (Fig. 3). Generally the greatest zoning is recorded in the outer rim.

Outside the contact zone, garnet zoning was examined in samples with M1-staurolite preserved and in rocks with pseudomorphs after staurolite. These garnets can be divided into two groups:

1. Composite matrix garnets with helicitic post-S$_2$ rims, and
2. Garnet idioblasts overgrowing/replacing pseudomorphic mica aggregates after staurolite.

The general zoning pattern for the matrix garnets is a rimwards decrease in Mg and Mn, and an increase in Ca (Fig. 3). The small irregularities in the curves either reflect chemical equilibration towards

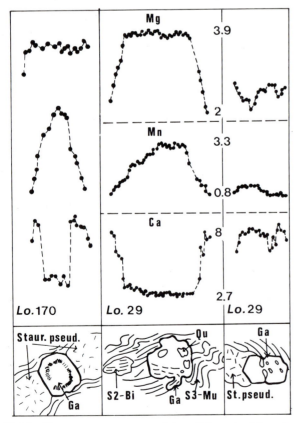

Fig. 3   Two stage garnet growth in pelitic rocks. Note the Ca-enrichment in post-S$_2$ rims and in M3-garnets

inclusions in the garnet, or alternatively, garnet growth starting from different nuclei. In the latter case the garnet growth may be described by the Hollister (1966) fractionation model with restricted diffusion in garnet.

The texturally deduced multistage garnet growth is shown by abrupt changes of the slopes of zoning curves in the matrix garnets. In contrast the type (2) garnets, which crystallized entirely after the staurolite breakdown, have smooth zoning profiles with bell shaped Mn-curves, reflecting one growth stage with regular Mn-fractionation (Hollister 1966).

Texturally the type (2) garnets were correlated with the helicitic rims of the type (1) garnets, and this is supported also by the zoning patterns, as indicated below.

Zoning directions are compared for type (1) and (2) garnets from two locations (170 and 29; 0, 3 and 4 km from the gabbro contact). In both places the type (1) garnets show a marked Ca-increase in the rim zones; in loc 29 this is associated with a marked Mg-decrease. Type (2) garnets from the same thin sections show less zoning but in the opposite sense, and are generally more Ca-rich and Mg-poor. However, the Ca and Mg (and also Mn) contents in the helicitic rims of type (1) garnets approximate the main levels of type (2). To summarize, the matrix garnets (type 1) reflect two growth stages. The second occurred post-S$_2$ under P and T approaching the retrograde conditions reflected by the type (2) garnets.

Garnets from the contact metamorphic rocks are either almost unzoned or show strong retrograde zoning in the outer margin, with a decrease in Mg/Fe and an increase in Ca. The greatest rimwards Mg/Fe decrease is recorded in garnets from the anatectic quartz dioritic veins inside or close to the gabbro complex. Here the garnets have reached the largest size and are characterized by concentric fracturing and somewhat rounded shapes. The Mg decrease and Ca increase is most pronounced.

However, in contrast to the typical bell-shaped Mn-curves recorded in the other rocks, Mn is almost unzoned except for a marked Mn-increase in the outer rims.

Interpretation of the garnet zoning

Garnet zoning reflects variations in element preferences with changing metamorphic conditions (Miyashiro 1953; Lambert 1959; Sturt 1962) or with rock chemistry (Lyons and Morse 1970) but may equally well be caused by kinetic factors and various fractionation and diffusion models have been proposed (Hollister 1966; Atherton 1968; Anderson and Buckley 1973).

In the present case the retrograde (multistage) evolution is evident from mineral reactions, and the retrograde character of the garnet zoning is well established. However, the reactions observed in the pelites do not explain the general Ca-increase observed. It was earlier argued that Ca was mobilized and transported into the pelitic layers during the retrograde metamorphism, and this general Ca-increase would also explain the Ca-patterns in the garnets.

The Mn-zoning is most consistent with the bell-shaped fractionation pattern of Hollister (1966), and occurs in spite of retrograde mineral growth. Mn-garnet is stablized under decreasing metamorphic conditions (Miyashiro 1953), but only in the migmatite veins is an Mn-increase observed in the outer rims. Thus the Mn-patterns are more likely to reflect the availability of Mn. According to Grant and Weiblen (1971) an Mn-increase may be caused by a drastic increase in the Mn content of the reservoir, due to resorption of a Mn-rich phase. Two alternative Mn-sources exist for the anatectic veins:
1. garnet resorption, and
2. Mn-release due to the breakdown of ilmenite to rutile in the gabbro during retrogression.

In either case the high-T, Mg-rich, Mn-poor garnets became unstable during retrogression.

Zoning of other minerals

The regional metamorphic M1-staurolite only shows weak zoning with increasing Mg/Fe rimwards, while chloritoid shows weak decrease in Mg/Fe. Generally the mafic phases vary in composition depending on which mineral they share contact with, and this is illustrated by Thompson's (1957) AFM-projection (Fig. 2b). The AFM-diagrams are also used to compare mineral chemistry for the different rocks, and to evaluate equilibrium and metamorphic reactions in the pelites.

Comparison of regional metamorphic (M1) and contact metamorphic parageneses

The M1-staurolite–biotite pairs have higher Mg/Fe ratios than the later-formed contact metamorphic staurolite–biotite pairs, but the staur/biot Mg/Fe distribution coefficient is higher in the latter. The tie line patterns for staurolite–biotite reflect mosaic equilibrium in the contact metamorphic rocks (loc 18, 130; Fig. 2b).

Garnet–biotite pairs are, however, most magnesian in the contact metamorphic rocks which also have the highest garnet/biotite Mg/Fe distribution coefficient (0.17–0.20 in xenoliths, 0.13–0.18 in contact zone, 0.10–0.12 outside contact zone). From textural evidence and garnet zoning patterns the garnet–biotite equilibration outside the contact zone is retrograde and these phases are not in equilibrium with staurolite.

Retrograde equilibration in the contact zone

The lack of general equilibrium is apparent in the garnet-banded contact metamorphic gneiss (Fig. 2c) where biotite is strongly zoned with increase in F and A values towards adjacent sillimanite grains. Biotite in the sillimanite-rich layers has lower Mg/Fe-contents than in the garnet layers with garnet-biotite two-phase mosaic equilibrium. From the AFM configurations the garnet–biotite pairs are not in equilibrium with the sillimanite–biotite pairs. The least magnesian garnet–biotite pair shown represents a different rock (loc 78) with a retrograde contact metamorphic assemblage; the garnets have strong retrograde zoning and overgrow pseudomorphic fibrolite. The K_D of this retrograde garnet rim and adjacent biotite is similar to that in the above mentioned garnet–biotite layers. By comparison, the biotite–garnet equilibrium in the latter is also interpreted as retrograde.

Similar garnet–biotite tie lines are recorded for the chloritoid–garnet gneiss; but in this rock retrogression went further. The replacement of the garnet–biotite–staurolite assemblage by chloritoid–chlorite–garnet is illustrated in Fig. 2d.

Zoning of high grade mineralogy: cordierite–opx xenolith

Determination of coexisting minerals is difficult for the gar–cord–opx xenolith due to the multistage retrogression (described earlier). Opx is best preserved in domains with plagioclase and cordierite (usually cord pseudomorphs) shielded from garnets. Cordierite pseudomorphs are, however, common also in garnet-rich domains, and judging from the microfabric the minerals probably coexisted before the cordierite breakdown. The garnets show no significant zoning, but have a total Mg/(Mg + Fe) spread 0.19–0.32 for different grains. The few cordierites preserved are more constant in composition with a Mg/(Mg + Fe) range 0.73–0.76. Orthopyroxene is often strongly zoned for Al with a maximum rimward Al-decrease from 7.2–4.6% Al_2O_3 while Mg/(Mg + Fe) is quite constant. The opx-zoning may either reflect crystal growth during cooling or a change in the availability of Al due to garnet crystallization.

Mineral Reactions and Metamorphic Conditions

Staurolite is a common index-mineral for medium grade metamorphism of pelitic rocks (Winkler 1976), and the P–T dependence of various staurolite-forming reactions has been studied experimentally by Hoschek (1967, 1969) and Ganguly (1968) (Fig. 4). The lower stability limit of staurolite is defined by the reactions

1. chlor + musc = staur + biot + qtz + fluid
2. chld + Al₂SiO₅ (ky, and) = staur + qtz + fluid
3. chld + qtz = staur + gar + fluid
4. chld + chlor + musc = staur + biot + qtz + fluid

while the upper stability limit is defined by

5. sill + biot = staur + qtz + musc

In the Krutfjell area the reactions suggested from the textural relations are all *retrograde*, but still the prograde reactions (1)–(5) set a lower limit for the peak of the regional metamorphism and the contact metamorphism.

Staurolite breakdown by reaction (1) is most common, and defines the lower limit of the regional metamorphism M1. The *prograde* reaction (1) sug-

gests temperatures of 565 and 540°C (± 15°C) at P_{H_2O} = 7 and 4 Kb, respectively.

Chloritoid was formed by retrograde metamorphism in the contact zone where the microfabric suggests that chloritoid formed by reaction (4) and possibly by reaction (3). Equilibrium conditions for these reactions are close to the P–T line of reaction (1). Lack of biotite in the chloritoid–garnet gneiss (loc 264) except as shielded inclusions in garnet, suggests that reaction continued beyond the biotite-chloritoid stability limit, defined by

6. chlor + musc + qtz = chld + biot

Reaction (6) has equilibrium conditions at slightly lower T than (1) and lies within the upper greenschist facies.

The upper stability of staurolite is defined by (5) which has been calibrated to 675 ± 15°C at 5.5 Kb. (Hoschek 1969). Again, the appearance of the retrograde reaction in the contact zone demonstrates that this is a lower limit for equilibrium conditions for the contact metamorphic formation of sillimanite.

The sillimanite–K-feldspar assemblage is only recorded in one location (loc 262), and suggests still higher temperatures according to the equilibrium conditions for the reaction (7).

7. musc + qtz = K-fsp + sill + H₂O

The xenolith assemblage cordierite–almandine–hypersthene represents high-grade metamorphism and is related to the alm–biot–hyp assemblage by the discontinuous reaction:

8. alm (gar) + biot → cord + hyp (Reinhardt 1968).

In the gneiss xenolith 6B, garnet–opx equilibrium could not be established. But garnet may have coexisted with *some* opx composition at an earlier stage and then metamorphic conditions may be inferred from reaction (9) which was experimentally calibrated by Henson and Green (1972, 1973):

9. cord + hyp = ga + qtz

The stability curve for this reaction is strongly dependent on the Mg/(Mg + Fe) ratios of garnet and cordierite.

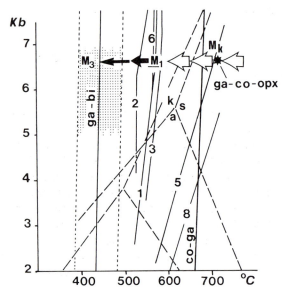

Fig. 4 P_{H_2O}–T diagram for the pelites with stability curves for staurolite according to reactions 1, 2, 3, 6 and 5, and Al₂SiO₅-phase boundaries (Richardson *et al*. 1969 and Holdaway 1971). Retrograde gar–biot equilibration (Ferry and Spear 1978) outside the contact zone (filled arrows). P–T fixed from ga–cord–opx (asterisk). Retrograde change of contact metamorphism from sillimanite into kyanite and staurolite stability fields (open arrows). Gar–cord coincides with gar–biot for the contact metamorphic xenolith loc 6B (calibration of Thompson 1976)

Pressure estimate, garnet–cordierite–opx

Application of the P–T-composition (garnet, cordierite) diagram of Hensen and Green (1973, fig. 3) fixes both P and T for given garnet and cordierite compositions according to reaction (9). The cordierite in the relevant xenolith has fairly constant composition so that P–T estimate depends most on the garnet composition used. This will strongly influence the calculated temperatures, while the effect on the pressure calculation is insignificant. The mean gar–cord composition values from the garnet-rich domains of the xenolith plot outside the P–T diagram (less than 650°C). The more Mg-rich garnets give 750°C, 6.5 Kb. The total pressure estimates for crystallization of the gabbro (cpx-composition and crystallization order in relation to magma chemistry) restrict the pressure to the interval 5–7 Kb (Mørk 1979). Other evidence for the relatively high pressure is retrograde replacement of cordierite by kyanite in the contact zone.

Garnet–biotite equilibration temperatures

The extensive garnet formation occurred late in the metamorphic history, and the garnet–biotite pairs studied involve S_2-biotite and post-S_2 garnets. Thus, except for a few biotite inclusions in garnet cores, the garnet–biotite equilibria are retrograde relative to the peaks of both regional and contact metamorphism. Even in the garnet–sillimanite gneiss (loc 262) the garnet–biotite pairs were not in equilibrium with the adjacent sillimanite–biotite pairs (Fig. 2c). Garnet–biotite temperatures were calculated by use of four different geothermometers (Perchuk 1969; Thompson 1976; Holdaway and Lee 1977; Ferry and Spear 1978) which are in close agreement in the relevant temperature interval. Garnet–biotite pairs from the pelites outside the contact zone cluster at about 450°C (±50°C) with a total range between 420–500°C. Within the contact zone garnet–biotite pairs from garnet–sillimanite gneiss (loc 262) and garnet–chloritoid gneiss (loc 264, biotite as inclusions in garnet only) average 550 ± 50°C with total range 480–560°C. Similar temperatures are derived for the rim compositions of the strongly zoned garnets in gneiss and anatectic veins in the contact zone, but pairs of garnet plus biotite inclusions give higher values (585–624°C, loc 118). The highest garnet–biotite equilibrium temperatures are derived in the garnet–opx–cordierite xenolith (loc 6B, 660–680°C) within the gabbro complex. But even here mineralogy, opx-zoning, and garnet–cordierite temperatures demonstrate that the garnet–biotite equilibration is retrograde.

In all samples, calculation using garnet *cores* would raise the temperatures, and the garnet zoning shows that these higher temperatures are minimum temperatures if garnet and biotite *equilibrated* towards the retrograde temperatures. Outside the contact zone this temperature difference is generally slight. The highest values (~550°C) are derived by using the inclusion-free garnet cores together with matrix biotite in two samples with staurolite

pseudomorphs. The more strongly zoned garnets in the anatectic veins would give 660–750°C by application of garnet-core compositions.

Discussion of the T-estimates

The metamorphic peaks (M1, Mk) are inferred from the mineralogy while the highest temperature values are derived from the equilibrium curves of the reactions (1)–(9). The experimental data are based on the condition $P_{H_2O} = P_{total}$. However the presence of graphite in some of the samples suggests that P_{H_2O} was less than P_{total}, and the stability temperatures for the staurolite forming reactions have been lower than those of the experiments (Hoschek 1969). Since the mole fraction of H_2O is usually $0.7 \leq X–H_2O < 1$ for graphite-bearing pelites under most metamorphic conditions, the T-decrease for reaction (1) will probably be less than 50°C (by interpolation of fig. 9 in Hoschek 1969). From their study of X_{H_2O}, P and T relations for lower sillimanite-zone graphitic pelites ($X_{H_2O} = 0.75 \pm 10$) Novak and Holdaway (1981) conclude that the available experimental results for staurolite may be 25–50 °C too high.

When the effect of $P_{H_2O} < P_{total}$ is considered, there is still a temperature gap between the metamorphic peaks (M1, Mk) and the T calculated from late garnet–biotite equilibria.

Application of temperatures in the polymetamorphic evolution

1. The M1-parageneses have minimum temperatures of about 500–550 °C (dependent on local X_{H_2O}). The main garnet-growth was after the M1 metamorphic peak, and after the main deformation (F1, F2). These garnets equilibrated with S_2-matrix biotite at 450 ± 50°C. Some garnet cores may be relics of the M1-parageneses, both from the textural, mineral zoning, and temperature data discussions. The temperature data therefore agree with the petrographic interpretations that helicitic garnets were not in equilibrium with M1-staurolite and that garnet grew in several stages.

2. Within the contact zone, garnet–biotite equilibration is also retrograde, but about 100°C higher temperatures are preserved here than outside the contact. The highest garnet–biotite temperatures preserved are in shielded areas (loc 6B) where the gabbro is little deformed. But still higher temperatures are indicated by the composition of relict garnet cores in anatectic veins (750°C) and by opx-zoning (loc 6B).

3. Neither mineral paragenesis nor garnet core nor garnet rim/biotite compositions in the Krutfjellet Group show gradients toward the gabbro outside the inner high-T contact aureole. Close to the gabbro and within xenoliths both high-T mineralogy and anatexis are due to relatively high-P (5–7 Kb) contact metamorphism. The intrusion post-dated the M1-regional metamorphism. It was followed by retrograde metamorphism under low grade conditions with extensive garnet growth occurring after the main deformation and thrusting.

Widespread recrystallization of muscovite pseudomorphing staurolite and sillimanite in the contact zone, and the high-T retrograde reactions there, suggest that the cooling gabbro body shielded this zone from deformation and maintained a higher temperature during retrograde metamorphism and deformation.

Conclusion

The Krutfjellet mega-lens is interpreted as having been tectonically emplaced into the pile of low-grade Köli rocks. The medium-grade metamorphism is interpreted as a regional metamorphism and that the temperatures were locally raised even higher due to gabbro intrusion. The emplacement of the Krutfjellet Complex into the Köli rocks occurred after the metamorphic peak, after the gabbro intrusion and after the S_2-flattening process. This emplacement or thrusting was followed by low-grade metamorphism M2, M3, and the thrust contact was folded (F_3). Thus the growth of M3 garnet and hornblende continued at low-grade conditions after the thrusting.

Acknowledgements

This paper is based on parts of my cand. real. thesis at the University of Oslo 1979. Profs. W. L. Griffin and I. B. Ramberg are acknowledged for advice during the thesis work. I would also like to thank the staff at the Geological Museum, Oslo, for valuable help and discussions.

References

Anderson, D. E. and Buckley, G. R. 1973. Zoning in garnets—diffusion models. *Contrib. mineral. Petrol.*, **40**, 87–104.

Atherton, M. P. 1968. The variation in garnet, biotite and chlorite composition in medium grade pelitic rocks from the Dalradian, Scotland, with particular reference to the zonation in garnet. *Contrib. mineral. Petrol.*, **18**, 347–371.

Chinner, G. A. 1961. The origin of sillimanite in Glen Clova, Angus. *J. Petrol*, **2**, 312–323.

Ferry, J. M. and Spear, F. S. 1978. Experimental calibration of the partitioning of Fe and Mg between biotite and garnet. *Contrib. mineral. Petrol.*, **66**, 113–117.

Ganguly, J. 1968. Analysis of the stabilities of chloritoid and staurolite and some equilibria in the system $FeO-Al_2O_3-SiO_2-H_2O-O_2$. *Am. J. Sci.*, **266**, 277–299.

Grant, J. A. and Weiblen, P. W. 1971. Retrograde zoning in garnet near the second sillimanite isograd. *Am. J. Sci.*, **270**, 281–296.

Hensen, B. J. and Green, D. H. 1972. Experimental study of cordierite and garnet in pelitic compositions at high pressures and temperatures. II Composition without excess alumino-silicate. *Contrib. mineral. Petrol.*, **35**, 331–354.

Hensen, B. J. and Green, D. H. 1973. Experimental study of the stability of cordierite and garnet in pelitic compositions at high pressures and temperatures. III Synthesis of experimental data and geological applications. *Contrib. mineral. Petrol.*, **38**, 151–166.

Holdaway, M. J. 1971. Stability of andalusite and the aluminium silicate phase diagram. *Am. J. Sci.*, **271**, 97–131.

Holdaway, M. J. and Lee, S. M. 1977. Fe–Mg cordierite stability in high-grade pelitic rocks based on experimental, theoretical, and natural observations. *Contrib. Mineral. Petrol.*, **63**, 175–198.

Hollister, L. S. 1966. Garnet zoning: an interpretation based on the Rayleigh fractionation model. *Science*, **154**, 1647–1650.

Hollister, L. S. and Bence, A. E. 1967. Staurolite: sectoral compositional variations. *Science*, **158**, 1053–1056.

Hoschek, G. 1967. Untersuchungen zum Stabilitätsbereich von Chloritoid und Staurolith. *Contrib. mineral. Petrol.*, **14**, 123–162.

Hoschek, G. 1969. The stability of staurolite and chloritoid and their significance in metamorphism of pelitic rocks. *Contrib. mineral. Petrol.*, **22**, 208–232.

Häggbom, O. 1980. Polyphase deformation of a discontinuous nappe in the central Scandinavian Caledonides. *Geol. Fören. Stockholm Förh.*, **100**, 249–354.

Lambert, R., St. J. 1959. The mineralogy and metamorphism of the Moine schist of the Morar and Knoydant districts of Inverness-shire. *Roy. Soc. Edinburgh*, **63**, 553–588.

Lyons, J. B. and Morse, S. A. 1970. Mg/Fe partitioning in garnet and biotite from some granitic, pelitic and calcic rocks. *Am. Mineral.*, **55**, 231–246.

Mason, R. 1968. Electron-probe microanalysis of coronas in a troctolite from Sulitjelma, Norway. *Mineral. Mag.*, **36**, 504–514.

Mason, R. 1980. Temperature and pressure estimates in the contact aureole of the Sulitjelma gabbro, Norway. Implications for an ophiolite origin. In Panayiotou, A. (ed.), *Ophiolites. Proceedings International Ophiolite Symposium, Cyprus 1979.*

Miyashiro, A. 1953. Calcium-poor garnet in relation to metamorphism. *Geochim. Cosmoch. Acta.*, **4**, 179–208.

Mørk, M. B. E. 1979. Metamorf utvikling og gabbrointrusjon på Krutfjellet, Nordland. Abstract. *Geolognytt*, **12**, p. 16.

Mørk, M. B. E. 1979. *Metamorf utvikling og gabbrointrusjon på Krutfjell, Nordland. En petrografisk–petrologisk undersøkelse*, Cand. real. thesis (unpublished), University of Oslo.

Mørk, M. B. E. 1981. The Krutfjellet traverse. In

Ramberg, I. and Stephens, M. B. (Eds), *The Central Scandinavian Caledonides—Storuman to Mo i Rana*, 1981, UCS Excursion guide.

Novak, J. M. and Holdaway, M. J. 1981. Metamorphic petrology, mineral equilibria, and polymetamorphism in the Augusta quadrangle, south–central Maine. *Am. Mineral.*, **66**, 51–69.

Perchuk, L. L. 1969. The effect of temperature and pressure on the equilibrium of natural iron–magnesium minerals. *Int. Geol. Rev.*, **11**, 8, 875–901.

Ramberg, I. B. 1967. Kongsfjell-områdets geologi, en petrografisk og strukturell undersøkelse i Helgeland, Nord-Norge. *Norges geol. Unders.*, **240**, 1–152.

Ramberg, I. B. 1981. The Brakfjellet tectonic lens: evidence of pinch-and-swell in the Caledonides of Nordland, north central Norway. *Norsk geol. Tidsskr.*, **61**, 87–92.

Ramberg, I. B. and Riis, F. 1979. Et snitt gjennom Nordlands øvre dekke-kompleks sør for Ranafjorden. Abstract. *Geolognytt*, **12**, p. 17.

Ramberg, I. B. and Stephens, M. B. 1981. The Central Scandinavian Caledonides–Storuman to Mo i Rana. *Uppsala Caledonide Symposium. Excursion A3*. Unpublished excursion guide.

Reinhardt, E. W. 1968. Phase relations in cordierite-bearing gneisses from the Gananoque area, Antario. *Can. J. Earth Sci.*, **5**, 455–482.

Richardson, S. W., Gilbert, M. E. and Bell, P. M. 1969. Experimental determination of kyanite–andalusite and andalusite–sillimanite equilibria, the aluminum silicate triple point. *Am. J. Sci.*, **267**, 259–272.

Senior, A. 1978. The Artfjäll gabbro, Västerbotten län. Abstract. *XIII Nordiske geologiske Vintermøde*, København, p. 64.

Spry, A. 1969. Metamorphic Textures. Pergamon Press, Oxford, 350 pp. 350 p.

Stephens, M. B. 1979. Stratigraphy and relationship between folding, metamorphism and thrusting in the Tärna–Björkvattnet area, Northern Swedish Caledonides, *Sver. geol. Unders., Ser. C.*, **726**, 146 pp.

Stephens, M. B. 1980. Occurrence, nature and tectonic significance of volcanic and high-level intrusive rocks within the Swedish Caledonides. In Wones, D. R. (Ed.) *The Caledonides of the USA*, Virginia Polytechnic Institute and State University, Department of Geological Science, Mem. 2, 289–298.

Sturt, B. A. 1962. The composition of garnets from pelitic schists in relation to the grade of regional metamorphism. *J. Petrol.*, **3**, 181–191.

Thompson, A. B. 1976. Mineral reactions in pelitic rocks: II Calculations of some $P-T-X_{Fe,Mg}$ phase relations. *Am. J. Sci.*, **276**, 425–454.

Thompson, J. B. 1957. The graphical analyses of mineral assemblages in pelitic schists. *Am. Mineral.*, **42**, 842–852.

Vernon, R. H. 1976. Metamorphic Processes, George Allen & Unwin Ltd, London, 247 pp.

Winkler, H. G. F. 1976. Petrogenesis of Metamorphic Rocks (Four ed.), Springer-Verlag, New York, 244 pp.

Zwart, H. J. 1962. On the determination of polymetamorphic mineral associations, and its application to the Bosost Area (Central Pyrenees). *Geol. Rundsch.*, **52**, 38–65.

Tectonic Evolution

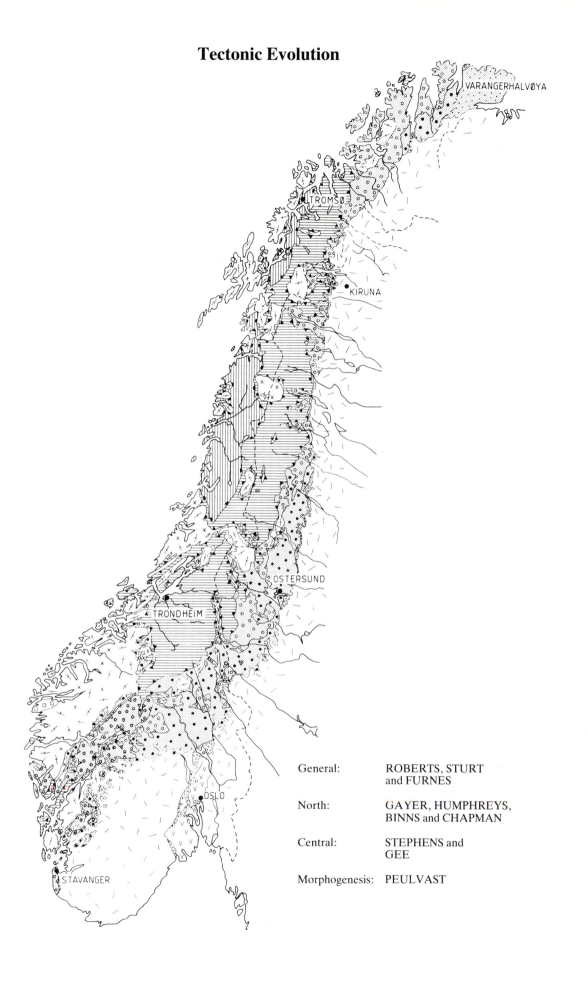

General: ROBERTS, STURT and FURNES

North: GAYER, HUMPHREYS, BINNS and CHAPMAN

Central: STEPHENS and GEE

Morphogenesis: PEULVAST

The Caledonide Orogen—Scandinavia and Related Areas
Edited by D. G. Gee and B. A. Sturt
© 1985 John Wiley & Sons Ltd

Volcanite assemblages and environments in the Scandinavian Caledonides and the sequential development history of the mountain belt

D. Roberts[*] **B. A. Sturt**[†] and **H. Furnes**[†]

[*]*Norges geologiske Undersøkelse, Postboks 3006, 7001 Trondheim, Norway*
[†]*Geologisk Institutt Avd. A, Universitetet i Bergen, Allégt. 41, 5014 Bergen, Norway*

ABSTRACT

Geochemical signatures of many Scandinavian Caledonide, Late Precambrian–Early Palaeozoic, volcanite and associated plutonic assemblages, when viewed in relation to local internal stratigraphy and regional tectonostratigraphy, have provided important indications of palaeotectonic setting. Magmatic associations recognized include those generated as oceanic crust along major spreading ridges or in marginal basin situations, as well as some deriving from volcanic arc constructions and others denoting oceanic or continental within-plate affiliation. Fragmented and dismembered ophiolites occurring over a 1500 km strike length of the mountain belt in many instances provide evidence signifying that initial tectonic transport, or obduction, was completed by approximately Caradoc time; in one case, faunal evidence favours a pre-Middle Arenig obduction, while radiometric data from another area denote pre-Mid Ordovician thrusting. This points to a temporal correspondence between ophiolite dissection and thrust emplacement, and the Finnmarkian and Grampian orogenic events of northern Norway and the British/Irish orthotectonic belt, respectively. Although direct control on obduction direction is lacking, fold facing and other criteria denote an east to southeastward vergence and translation during the latest Cambrian to early Ordovician Finnmarkian deformation.

Clastic sequences of shallow-marine to locally continental facies deriving from the obducted ophiolite and early arc slices and their substrates initiated an Ordovician–Silurian volcanosedimentary cycle which in central districts of Norway and Sweden includes marginal basin and mature magmatic arc environments related to eastward subduction of Iapetus crust. The terminal plate-collisional process, which led to the thermodynamic transformation and thrusting (in some cases by several hundred kilometres) of these heterogeneous sequences, began in Wenlock or even Late Llandovery time in internal zones of the orogen, and transmitted its energies diachronously southeastward through to Early Devonian time in the frontal detachment zone of southern Norway and Sweden. At this time the highest thrust sheets in the nappe stack further west were shedding thick molase sediments into fault-controlled Old Red Sandstone basins. These intermontane basinal sequences were themselves subject to Late Devonian folding and thrust deformation.

Introduction

The emergence of volcanite petrochemistry as an invaluable implement in our assessment of the early stages of evolution of the Scandinavian Caledonide orogen stems from the initial investigations in central and western Norway, with indications of the presence of ocean floor (Gale and Roberts 1974; Prest-vik 1974) and island arc environments (Gale and Roberts 1972; Furnes and Færseth 1975). More recent studies have led to the discovery of almost complete though fragmented ophiolite assemblages in some areas, considered to represent vestiges of the Lower Palaeozoic Iapetus oceanic crust (Sturt and Thon 1978; Sturt et al. 1979; Furnes et al. 1980; Furnes et al. this volume; Grenne et al. 1980; Prestvik 1980), while similar geochemical work in the Swedish part of the metamorphic allochthon has outlined the occurrence of a Late Proterozoic continental rift sequence (Solyom *et al.* 1979a, b) as well as Early Palaeozoic rifted arc complexes (Stephens 1977, 1980, 1982).

Important though this work has been, parallel research throughout the orogen has increasingly been pinpointing the fundamental contributions of basic stratigraphy and tectonostratigraphy, major unconformities and tectonometamorphic breaks, and regional patterns of basement/cover relationships in chronicling a history of Caledonide evolution (Gee 1975; Roberts and Sturt 1980). That the clast petrography of certain polymict conglomerates, for example, is telling us an important story (Sturt and Thon 1976; Minsaas and Sturt this volume; Sturt *et al.* this volume) has not really been emphasized strongly enough up to the present time; yet this relates to the recognition on structural and metamorphic grounds in many cases, of distinctive orogenic events which are essentially reflecting changes in Lower Palaeozoic palaeogeography.

In any attempt at piecing together the threads of Scandinavian Caledonide evolution, one must therefore deliberate over a series of features which are now part and parcel of a multicomponent nappe pile, and which have in some cases suffered two principal, and several minor, tectonic deformations. In plate tectonic modelling, palinspastic reconstructions are thus, at best, difficult, and more especially for the older volcanostratigraphic sequences.

Taking the above criteria into consideration, the aims of this paper are to present a brief review of Caledonide volcanite assemblages and their likely palaeotectonic settings, with some emphasis on the significance of local stratigraphy. These important facets of the geological history will then be viewed in the context of the evolving orogen, from its rift beginnings in Late Precambrian time to its ultimate dissection in the Siluro-Devonian period. Full details of regional stratigraphies, tectonostratigraphy, magmatic associations and geochemistry are contained in other articles in this volume.

Principal Volcanite Associations

In terms of palaeogeographic setting it is convenient to divide the volcanite associations into specific groups distinguished on the bases of geochemical signature and geological characteristics, including the presence of consanguineous plutonic rocks. The principal associations are those of ophiolites, island arcs and within-plate volcanites, with a minor component of early, rift-related magmatic activity.

Ophiolite assemblages

A fairly comprehensive survey of the fragmented ophiolite assemblages present in the metamorphic allochthon is that of Furnes *et al.* (1980). A near-complete ophiolite pseudostratigraphy in the sense of the Penrose Conference (Anon. 1972) or Coleman's (1977) definition is seen only on Karmøy (Sturt and Thon 1978). Elsewhere in Scandinavia, ophiolitic associations are dismembered and considerably fragmented, in some cases a result of two major time-separated thrust-deformations. Nevertheless, in several instances there are recognizable relics of the tripartite 'stratigraphy' of gabbro–sheeted dyke complex–pillow lavas, with the lavas showing clear ocean floor basalt trace-element chemistries.

A classification of many of the presently known ophiolitic assemblages into two major groups is given elsewhere in this volume (Furnes *et al.*) together with a map showing their locations. In brief, group I is considered to be representative of either a major ocean or a mature, arc-remote, marginal basin setting, whereas group II sequences relate to younger, small, back-arc marginal basins. Other ophiolite slices are difficult to classify, in most cases because of either severe fragmentation, later structural disruption, or a lack of definitive features characteristic of the two main types. Group I complexes include those of Karmøy, Lykling, Gullfjell, Leka, and Lyngen (see Furnes *et al.* this volume, Fig. 1). Of these, Karmøy presents the most comprehensive pseudostratigraphy from a thin development of deformed serpentinized peridotites through cumulate and static gabbros, a 2 km-thick sheeted dyke complex, pillow lavas, and a cap of hemipelagic sediments and volcaniclastics which contains intercalated units of alkaline, within-plate pillow lavas (Sturt *et al.* 1980a). The Karmøy ophiolite appears to have developed partly in a transform fault/oceanic fracture zone (Solli, 1981; Furnes and Sturt work in progress). The Lykling ophiolite complex is unconformably overlain by an ensimatic island arc assemblage (Amaliksen 1983) which has been Rb–Sr-dated to 535 ± 46 Ma (Furnes *et al.* 1983).

The Leka occurrence exposes the best developed tectonized ultramafic complex yet recorded in the Scandinavian Caledonides, but other members of the standard pseudostratigraphy are also represented, though in part strongly deformed. Associated with the Leka ophiolite are hawaiitic ocean-island type lavas together with oceanic sediments (Prestvik 1974, 1980). The Støren slice is composed largely of a thick, OFB pillow lava sequence with rare basic dykes (Gale and Roberts 1974). Juxtaposed against its tectonized base is a reported mélange unit (Horne 1979; Roberts 1980a). In northern Norway the Lyngen mafic complex comprises tectonized layered gabbro, greenstones, basic dykes, and lensoid ultramafite bodies.

Significant in the context of palaeogeographic modelling are the strong indications that the group I ophiolites are older than those of group II, and suffered their initial obduction in earliest Ordovician time. The Støren, for example, was obducted, internally deformed, uplifted and eroded prior to deposition of shallow-water sediments carrying a Middle Arenig fauna (Furnes et al. 1980, 1982a; Hardenby et al. 1981. On Karmøy, constraints on obduction timing are provided by a Rb–Sr isochron age of 450 Ma on late acidic dykes in a major igneous complex, which cuts the ophiolite and carries xenoliths of both the ophiolite and the immediately underlying gneissic basement on to which obduction had taken place. At Lykling on Bømlo, volcanics of the post-ophiolite Siggjo Complex have yielded Rb–Sr whole-rock ages of 468 ± 23 Ma and 464 ± 16 Ma (Furnes et al. 1983). The cap sequence to the Karmøy ophiolite was deformed and metamorphosed in upper greenschist facies prior to the deposition of the Skudenes Group (Upper Ordovician) and emplacement of the West Karmøy 'granite' suite (Solli 1981), a deformation which is presumed to broadly coincide with Finnmarkian orogenesis (Ryan and Sturt 1981). In the north, the Lyngen complex was thrust and metamorphosed prior to deposition of Middle to Upper Ordovician continental sediments (Minsaas 1981; Minsaas and Sturt this volume). This evidence, of obduction and deformation age, from Støren, Karmøy, Lykling, and Lyngen fits quite neatly into the time frame of the major Finnmarkian orogenic phase first established in northern Norway (Sturt et al. 1978) and now recognized increasingly, and fairly widely, in central and southern parts of the Scandinavian Caledonides (Guezou 1978; Roberts 1978; Claesson 1980).

Group II ophiolitic complexes, of marginal ocean basin character, are closely associated with arc-derived volcaniclastics and also siliclastic sediments. They are post-Finnmarkian and mostly of Middle Ordovician age, and occur solely in the Upper Allochthon of the Trondheim region; at Vassfjell, Forbordfjell, Jonsvatn, Løkken, Grefstadfjell, and Snåsavatn. Massive and pillowed basaltic greenstones at Forbordfjell and Jonsvatn show LREE-depleted MORB characteristics in their thicker central portions, and transitional to LREE-enriched within-plate chemistry in their basal and upper members (Grenne and Roberts 1980). The Grefstadfjell occurrence has faunal control both below

and above (Middle Arenig to Early Llanvirn) and the volcanites pass laterally into Late Arenig volcanoclastic sediments (Ryan et al. 1980). Gabbros and a sheeted dyke complex have also been recorded here. The Snåsavatn basaltic greenstones are largely of OFB geochemical character, although the succession contains a calc-alkaline basal portion.

Fauna in limestones adjacent to the Snåsa occurrence denote a Middle Ordovician age. Another fragmented ophiolite unit which most probably belongs to the group II back-arc basin ophiolites is that of Sulitjelma (Boyle 1980). In contrast to the group I ophiolites, those in the group II category were initially deformed and thrust during the principal Scandinavian, or Scandian, orogenic phase in Middle to Late Silurian time. Ophiolite fragments which have been less easy to categorize occur in the following places; Stord, Solund, Stavfjord, Skålvær, Velfjord, and Terråk. Details are given in Furnes et al. (1980 and this volume). Local deformation patterns and stratigraphies provide clues which are relevant to the topic of palaeotectonic modelling, and the evidence accrued to date suggests that all the above-mentioned fragments belong to the group I category.

Island arc associations

Volcanite sequences of magmatic arc affinity have been recorded from throughout the mountain belt, though principally from central and southern districts (Gale and Roberts 1974; Halls et al. 1977; Stephens 1977, 1980, 1982; Furnes et al. 1980; Roberts 1980b). Arc constructions are of varied plate tectonic settings—ensimatic, transitional (oceanic-to-continental) and Andean types have been reported—which to some extent relate to their actual time of development.

Initial closure of the Iapetus Ocean commenced in Late, or even Middle, Cambrian time (Gale and Roberts 1974), and resulted in the construction of an ensimatic arc, with bimodal volcanite character, above what has been assumed to have been an eastward-directed subduction zone. Such immature arc associations are represented in the Gjersvik–Skorovatn and Fundsjø–Røros districts (Halls et al. 1977; Grenne and Lagerblad this volume; also Furnes et al. 1980, Fig. 1), and one has recently been reported from Lykling (Amaliksen 1983). At Norra Storfjället, in Sweden, the volcanites and associated sediments were probably accumulated on the trench side of the arc (Stephens and Senior 1981).

Arc associations developed during the post-Finnmarkian Ordovician period are on the whole more evolved, although that at Stekenjokk which involved arc-splitting is reportedly ensimatic

(Stephens 1977, 1982). The island of Smøla in west–central Norway provides perhaps the best example of a mature volcanic arc, with products from high-alumina basalt through andesite to rhyolite defining a typical calc-alkaline trend (Roberts 1980b). Intercalated limestones carry a fauna of Late Arenig to Llanvirn age (Bruton and Bockelie 1979). Open folding and faulting deformed the sequence, which was then intruded by Late Ordovician quartz diorite and granodiorite plutons. Some 200 km northeastwards along strike, calc-alkaline volcanites are present structurally beneath the marginal basin type Snåsavatn metabasalts, again in association with Middle Ordovician limestones. Limestones with a very similar fauna occur 100 km east-southeast of Smøla in the Hølonda–Løkken district, together with local andesites (Bruton and Bookelie 1980), here in a back-arc basinal environment (Roberts et al. in press). This is generally considered to be in the same nappe (the Støren Nappe, part of the Upper Allochthon) as the succession at Snåsa.

In west Norway, the supposed Mid to Late Ordovician Siggjo volcanic unit on Bømlo, which lies unconformably upon the deformed ophiolite (Lykling), comprises mainly subaerial lavas, ash-falls, and flows of basaltic of rhyolite composition. Geochemical data denote a comparison with Andean margin type magmas (Nordås et al., this volume). The diorite/granite suite of the West Karmøy Igneous Complex (Ledru 1980) probably represents the deeply eroded part of a magmatic arc (Ryan and Sturt this volume). In northern Norway, our knowledge of the thin volcanite sequences is more scanty. Ordovician conglomerates above the Lyngen ophiolite fragment contain clasts of undeformed arc-type plutonics and volcanites (Minsaas 1981), of uncertain provenance. Gayer and Humphries (this volume) have data which they interpret as suggesting that both island arc volcanites and more easterly marginal basin ocean floor tholeiites are present within the Ordovician rocks of the Upper Allochthon in Troms; these, they suggest, indicate eruption in an Andean-type continental margin situation.

Within-plate associations

Within-plate ocean island volcanites occur in association with the cap-rock sequences to the Karmøy and Leka ophiolites. Elsewhere in the Caledonide allochthon intraplate volcanic units are ensialic, generated in basins developed on thick continental crust (Furnes et al. 1980), and occur in the Bømlo–Stord, Joma and Sagelvvatn districts. On the islands of Bømlo and Stord in southwest Norway, subaerial basalts and andesites give way upwards to rhyodacitic lavas and ignimbrites with intercalated fluviatile sediments, and Ashgill limestones higher up (Lippard 1976; Nordås et al., this volume). A Rb–Sr isochron on Stord rhyolites has yielded a date of 455 Ma (Priem and Torske 1973). In the Joma area in central Norway, metabasalts and tuffs are associatd with shallow-marine sediments of uncertain, though probably Late Ordovician age (Gale and Roberts 1974; Olsen 1980).

The Sagelvvatn greenstones of Troms in northern Norway interdigitate with a carbonate evaporite tidal sequence containing corals and other fossils of Late Llandovery age (Bjørlykke and Olaussen 1981). The environment of volcanism was that of a supratidal carbonate platform succeeding fluvial arenites which developed upon continental crust of the Finnmarkian orogenic zone; just to the north, these sediments pass into those lying unconformably upon the Lyngen ophiolite (Minsaas and Sturt this volume).

Early rift-related magmatism

Important in any synopsis of magmatic environments and essential to the evolutionary plate model are the basaltic amphibolites and doleritic dykes which occur in the Middle and Upper Allochthons, mainly in the Sårv and Seve Nappes of central Sweden and their Norwegian equivalents. The dykes are most common in the sandstone-dominated Särv Nappe (which also includes tillites), and were initially Rb–Sr-dated to 720 ± 260 Ma (Claesson 1976). A more recent $^{40}Ar–^{39}Ar$ and K–Ar study has yielded an age of 665 ± 10 Ma (Claesson and Roddick 1983), which is probably closer to reality in view of the fact that the dykes transect tillites. Geochemically, the dykes are of transitional ocean floor to within-plate continental character (Andréasson et al. 1979; Solyom et al. 1979a), and may be consanguineous with tholeiitic amphibolites of comparable chemistry occurring in the Seve Nappe (Solyom et al. 1979b). Similar dykes and with similar, transitional, OFB–WPB chemistries occur in the Finnmarkian-deformed Middle Allochthon in northernmost Norway (D. Roberts, unpublished data). In the sparagmite basin area of the Lower Allochthon in southern Norway, evidence of basaltic volcanism has been reported from below the tillite (Bjørlykke 1978; Nystuen 1981).

In a plate tectonic context, these magmatic products are generally considered to relate to a continental break-up or rifting setting, prior to or at an early stage in the Caledonian orogenic cycle (Gale and Roberts 1974; Gee 1975; Bjørlykke et al. 1976).

Tectonostratigraphic Position of the Volcanite Assemblages

The tectonostratigraphic location of the various magmatic associations described above and their pre-orogenic restoration are major points to be considered in any attempt at palinspastic reconstruction. Division of the Caledonian metamorphic allochthon into four main complexes, the Lower, Middle, Upper, and Uppermost Allochthons, has been outlined elsewhere (Roberts and Gee, this volume). Within this tectonic framework, the rift-related dolerite dyke swarms are restricted to the Middle Allochthon. A few such dykes do occur in the Lower Allochthon in northern areas, but in the frontal Dividal Group autochthon they are absent.

Earlier in this account it was noted that, where faunal control is available, it can be shown that the ophiolite fragments occur at two principal stratigraphic levels. The older group I ophiolites and the early insular arc rocks were involved in the latest Cambrian to earliest Ordovician Finnmarkian orogenic phase, which included their initial thrusting or obduction. In northern Norway this deformation phase encompasses rocks from the autochthon up to the Middle Allochthon. The only possible remnant of Vendian–Cambrian oceanic crust yet postulated there, however, is that of the Lyngen complex which now occurs as local 'basement' within a higher Silurian nappe of the Upper Allochthon. This situation, of an originally lower-level ophiolitic (or early arc) allochthon now rethrust as 'basement' within higher-level Scandian (or Scandinavian) phase nappes, is also in evidence in central and western Norway. In Sweden, recognition of the Finnmarkian event in the Middle Allochthon is only slowly emerging, although there are now indications of this in Rb–Sr ages on mylonites (Claesson 1980). Initial detachment of some of the lowermost nappes of the Upper Allochthon may also carry a Finnmarkian, or at least Early Ordovician, signature. This appears to have been the case for the Dalsfjord Nappe (Skjerlie 1969 of west Norway) and possibly also for the Jotun (Hossack *et al*. this volume); and if the amphibolite sheets and dykes in the Särv and Seve Nappes are cogenetic (Solyom *et al*. 1979b) then it is not inconceivable that the latter may have begun its movement in the Ordovician, as with the Särv (Claesson 1980).

In contrast to this pattern of events, the younger group II ophiolites, most of the mature magmatic arc sequences, and the ensialic within-plate volcanites all have adequate biostratigraphic control signifying that their tectonic translation can be ascribed fairly safely, in most cases, to the Mid to Late Silurian, Scandian event. These volcanosedimentary associations, which are commonly floored by major unconformities, all occur within the Upper Allochthon. The Uppermost Allochthon, positioned above the Ordovician–Silurian sequences and containing large areas of granitoid plutons, must clearly have been translated eastwards into that situation in Late Silurian time (or later?), yet there are indications in these highest nappes of components of Finnmarkian deformation (e.g. Graversen *et al*. 1981; Riis and Ramberg 1981). Again, this is a feature of some significance in palaeotectonic modelling.

Aspects of Evolutionary History of the Orogen

Implicit in the above descriptions of the Caledonide volcanoplutonic associations is the view that these different rock complexes and their environments of formation can readily be compared with magmatic products and regimes known to occur in present-day oceans and smaller oceanic basins. This mobilistic concept, and in particular the cycle of ocean opening and destruction has been propounded by Wilson (1966), and the cumulative evidence favouring these ideas for the Caledonian–Appalachian orogen documented by Roberts and Gale (1978). Here, we will simply outline a few of the constraints on modelling and briefly attempt to integrate the data into the framework of the evolving orogen.

Tectonic fragmentation of the stratigraphic record in the Scandinavian segment of the orogen during Late Silurian time appears to have been more advanced than elsewhere in the mountain chain. Coupled with the evidence of earlier, Finnmarkian dissection, there are fundamental problems for reconstruction of palaeoenvironments. Basic restorations, i.e. of nappe units, are now possible, but strain data are minimal and balanced sections are few. Because of these constraining factors it is essential to keep in mind the basic features, the stratigraphies and tectonometamorphic histories, of neighbouring segments of the Caledonides, in particular East Greenland and the British Isles. Without a consideration of these areas, the value of any attempt at plate tectonic reconstruction for Scandinavia would be reduced.

Another critical and in some ways restricting factor is that of faunal provinciality, which by and large has demanded that, in Cambrian to Early Ordovician time, faunas of e.g. North American affinity were restricted to the environs and marginal oceanic waters of the North American–Laurentian plate, whereas European or Baltic faunas flourished close to the Baltoscandian plate along the eastern side of a wide oceanic (Iapetus) barrier (Cocks and Fortey 1982). The immediate conclusion from this, in many cases, and a factor influencing the drafting of several pre-drift maps and models, has been that sediments containing, for example, North American faunas must therefore have accumulated along the shelf or rise marginal to the North American plate. We now know that this may not necessarily have been the case; faunal distributions have more plausible explanations (Skevington 1974; Bruton and Bockelie 1980), and there is the attractive hypothesis of effective and ready migration of faunas to and from volcanic islands far removed from the continental margins (Neuman 1972; Bruton and Harper 1982). Whatever the primary reasons for provincialism, its significance gradually waned from Middle Ordovician time onwards. This is generally taken as indicating a progressive closing of the Iapetus Ocean.

Late Precambrian

The stages of Iapetus oceanic development, from initial crustal distension and rifting through to oceanic maturity and progressive contraction and extinction, are all considered to be represented in the Scandinavian Caledonides (Gale and Roberts 1974; Gee 1975; Roberts and Gale 1978; Furnes et al. 1980). In Late Riphean to Vendian time, fault-controlled sedimentation charaterized the early rifting phase with graben-type tectonics in central and southern regions (Bjørlykke et al. 1976) and thick continental margin clastic prisms further north (Figs. 1 and 2). Dolerite dykes of swarm proportions occur only in the higher parts of the Middle Allochthon (Strömberg 1969; Gee 1975; Andreasson et al. 1979). Moving down the tectonostratigraphy, dykes become scarce and are absent in the autochthon, which thickens appreciably in northeasternmost Norway. This denotes a fundamental stability for the Baltoscandian platform during this early period. At a slightly later stage, in the time range 590–565 Ma, an important carbonatite–alkaline intrusive event referable to an extensive rift/aulacogen system has been traced from Laurentia to northwestern Europe (Doig 1970), including the Fen area of southern Norway. It was during this latest Vendian to basal Cambrian period that the Iapetus Ocean truly developed and expanded. Evidence for this also comes from the extensive marine transgression at the Vendian–Cambrian boundary, recorded widely within East Greenland and Scotland as well as in Scandinavia; and palaeomagnetic data from several of these areas are unanimous in showing that apparent polar-wander paths diverge markedly at the base of the Cambrian, signifying the break-up and drifting apart of major shield areas (Briden et al. 1973; Piper 1981).

Cambrian to earliest Ordovician

While the Early and Middle Cambrian period in the orogenic interior was thus clearly one of oceanic expansion, there are very few clues as to the precise palaeogeography within the oceanic tract and its marginal continental rise prisms. Data from the Karmøy and Lykling ophiolites signify close proximity to an oceanic fracture zone or ridge–ridge transform system (Amaliksen 1979; Solli 1981). Variolitic lavas and the association of pelagic chert–lutite sequences denote comparatively deep oceanic conditions, and alkalic volcanics in the Karmøy cap-rock unit point to ocean island effusions or fracture zone magmatism analogous to those in the present-day Pacific and Atlantic. In the far north the evidence is more scanty, but palaeotransforms active during this period have been inferred (Kjøde et al. 1978). The thick Vendian to Early Cambrian arenites of the Finnmarkian Kalak Nappe Complex (Middle Allochthon) lie unconformably upon Proterozoic gneisses of the Baltic Shield (Ramsay and Sturt 1977; Ramsay et al. 1979), a cratonic basement which must have extended westwards for approximately 400 km (Ramsay et al. this volume). In the Middle Cambrian, pelite and limestone deposition gave way to deeper marine turbidites. Ophiolite is not present in this actual nappe complex, but comes in as Finnmarkian basement in higher nappes (Lyngen); the Cambrian oceanic crust must therefore have existed even farther to the west.

The initial closure of Iapetus, in latest Middle to the first part of Late Cambrian time is indicated by the immature, ensimatic volcanic arc construction and associated trench–forearc mélange above a subduction zone of uncertain polarity; in Figs. 1c and 2c, drawn in 1981, an easterly dipping Benioff zone is indicated, but argument favouring westward-dipping subduction has been presented in a more recent view article (Sturt et al. 1984). This development continued at least up into the Tremadocian in the Upper Allochthon of central Norway, as shown by the presence of a graptolite fauna in pelites in one area (Gale and Roberts 1974; Gee 1981); but this situation was soon destroyed by a major tectonism broadly equivalent to the later stages of the Finnmarkian, and involving ophiolite and ensimatic arc obduction, and nappe emplcement, arising from arc/Baltoscandian margin collision (Fig. 2D). Further south, detachment and southeastward transport of the Dalsfjord and Jotun Nappes and other thrust-sheets may have begun at this time (Skjerlie 1969; Hossack et al., this volume).

In the north, abundant radiometric dating has shown that evorogenic deformation with eastward-directed folds and thrusts began in Middle to early

NORTH NORWEGIAN CALEDONIDES

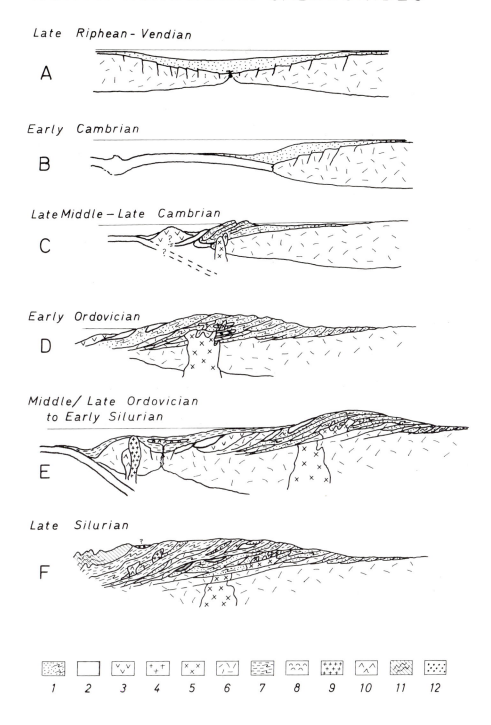

Fig. 1   Schematic c. E–W sections depicting stages in evolution of the Caledonian orogen in northern Norway from Late Riphean to latest Silurian time. For description of the geological evidence, see text. Parts of sections C–E are modified from Sturt *et al*. (1978). Legend: 1. Late Riphean to Cambrian (Tremadocian in east) sediments, subsequently deformed in Finnmarkian orogenic event. 2. Ocean floor magmatic rocks. 3. Possible volcanic arc (not exposed today). 4. Early synorogenic (Finnmarkian) plutons. 5. Later synorogenic (Finnmarkian) plutons. 6. Pre-cambrian basement (Karelian and pre-Karelian). 7. Orodovian to Lower Silurian sediments, subsequently deformed in the Scandian/Scandinavian event. 8. Within-plate tholeiitic volcanites. 9. Granitoid plutons. 10. Gabbroic plutons. 11. Higher nappes (rocks deformed in Sveconorwegian or earlier, Finnmarkian and Scandian orogenic events). 12. Possible Old Red Sandstone basinal deposits (not present today)

CENTRAL AND SOUTHERN SCANDINAVIAN CALEDONIDES

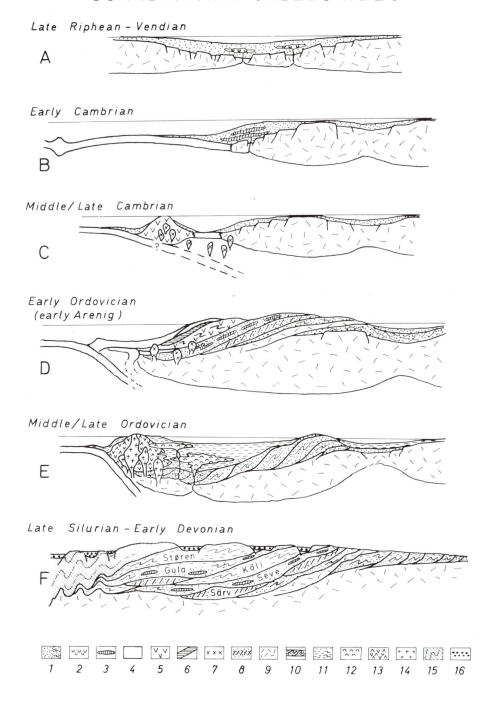

Fig. 2 Schematic c. E–W sections depicting stages in development of the Caledonian orogen of central and south–central Scandinavia from latest Riphean to Early Devonian time. Description of the geology is given in text. Note that in E, for simplification, the nappe units shown in D are combined in a composite, undifferentiated allochthon beneath the unconformably overlying Middle/Upper Ordovician volcanosedimentary pile. In F, the principal nappe units/complexes of central Scandinavia are indicated. Legend: 1. Late Riphean to latest Cambrian/Tremadocian sediments deformed, in most areas, in approx. Early Arenig time. 2. Alkali basalts of early rifting stage. 3. Tholeiitic basalts. 4. Ocean floor magmatic rocks (type I ophiolite). 5. Cambro-Tremadocian ensimatic arc volcanites. 6. Mélange unit. 7. Trondhjemitic to dioritic plutons. 8. Dyke-intruded sediments of the Särv Nappe (Sweden) and equivalents. 9. Precambrian basement (deformed in Svecofennian and Sveconorwegian events). 10. Undifferentiated Early Ordovician-deformed nappes, and sediments in east. 11. Middle Arenig and younger sediments and volcanites. 12. Marginal basin ophiolites (type II). 13. Ordovician arc volcanites. 14. Diorite, granodiorite and granite plutons. 15. Higher nappes (e.g. Helgeland). 16. Old Red Sandstone intermontane basinal sediments

Late Cambrian time in the west (in the highest Finnmarkian nappes), broadly coeval with initial magmatism, of mantle plume character (Sturt *et al.* 1980b), in what is now the Seiland igneous province. This Finnmarkian deformation was diachronous, from west to east, affecting the foreland autochthon in Early Ordovician (post-Tremadoc) time (Sturt *et al.* 1978) (Fig. 1C–D). It is uncertain if a volcanic arc was ever present in Cambrian time in these northern regions of Iapetus, and the precise cause of this tectonometamorphic event is therefore conjectural. Nevertheless, it seems clear that initial oceanic closure and deformation occurred earlier in Finnmark than in southern Scandinavia, possibly in part related to oblique collision and the presence of a westward-bulging cratonic promontory along the Baltoscandian margin. In east and northeast Greenland it is significant that a major non-sequence interrupts the stratigraphic record of the autochthon and lower nappes in Middle to Upper Cambrian time (Haller 1971); distant reflections of Finnmarkian/Grampian orogenesis (Roberts and Gale 1978). In southwest Norway, pre-Middle Ordovician SE-facing folds are thought to correlate in time with southeastward obduction of the fragmenting ophiolites upon a sialic Baltoscandian basement (Andresen and Færseth 1981; Færseth 1982). In Scotland, on the northwestern margin of Iapetus, northwestward obduction of the Ballantrae ophiolitic complex is considered to date to Middle Arenig time (Bluck *et al.* 1980), correlating with the Grampian orogenesis which affected the Dalradian ensialic basinal sediments and volcanics further to the northwest.

Ordovician–Silurian

The stratigraphic record in the metamorphic allochthon shows that the Ordovician period from Middle to Late Arenig time onwards, i.e. after initial ophiolite detachment, deformation and unroofing, was characterized by accumulation of sediments and volcanic products from a variety of environments. The volcanites, as described earlier, relate to mature magmatic arcs, back-arc marginal basins, local intra-arc basins and ensialic basinal regimes. These associations are all located within the Upper Allochthon.

In northern Norway, Gayer and Humphries (this volume) and Minsaas and Sturt (this volume) argue for the probable construction of a volcanic arc adjacent to an ensialic marginal basin (Fig. 1E), in the manner of the present-day Andean situation. The tholeiitic greenstones in the ensialic basin are of Llandovery age. Moving south, the Köli Nappes of central Sweden provide evidence of Ordovician arc development with subsequent arc splitting and younger small oceanic basin formation (Stephens

1980, 1982). The problem of tectonic fragmentation makes it difficult here to decide upon subduction direction. In Lower to Middle Silurian time, greywacke sequences in the Köli Nappes denote closure of the ocean basins and probable initial erosion of the uplifting and advancing nappes. Within Norway, the marginal basin ophiolite fragment of Sulitjelma is too isolated to allow establishment of a coherent palaeogeography.

Further south, the Trøndelag district of central Norway, and the Støren Nappe in particular, exposes perhaps the best example of an Ordovician mature arc/marginal basin couplet in Scandinavia. Here, there is the question of the provenance of the Støren Nappe sediments, relating to their content of fossils which are largely, though not exclusively, of North American affinity. The group II ophiolite assemblages are well developed here, in the time-range Late Arenig to approximately Llandeilo (Fig. 2e). In the Snåsavatn area, geochemical data show that calc-alkaline metabasalts pass up into OFB tholeiites. These calc-alkaline rocks, and adjacent limestones, can be traced southwestwards along strike towards the mature magmatic arc of Smøla. The change of magma type at Snåsa may signify an oceanward migration of the main arc with time, OFB volcanism taking over up-column; subduction is considered to have been eastward, with the trench west of the arc. Evidence for the polarity, and differences between the geology seen on Smøla and at Hølonda, have been outlined by Roberts (1980b). The continental margin directly east of the extensive marginal basin may be part of an elongate microcontinent with an additional small oceanic basin between this and the main Baltoscandian margin. Summing up, it is doubtful if the Ordovician rocks of the Støren Nappe were accumulated close to the North American continental margin, as the fauna might lead one to suggest. Alternative solutions to the problem have been presented by Gee (1975), Roberts (1980b), and Bruton and Harper (this volume).

In southern Norway, the Siggjo volcanites on Bømlo and the West Karmøy Igneous Complex provide further examples of the Andean margin situation. In many ways this is similar to the postulated Ordovician–Silurian setting for north Norway and, interestingly, also for the Welsh area in Britain. It would thus seem as if the marginal basin extensional regime, of Japan Sea type, with group II ophiolite assemblages to the fore, was selectively developed in the central portion of the Scandinavian orogenic belt (Roberts and Gale 1978). It is possble that this was related to an original sinuosity of the continental margin immediately following the Finnmarkian orogenesis. Alternatively, the main Middle Ordovician arc, and subduction zone, may have bulged oceanward in this central segment of the belt.

Terminal events

The terminal plate-collisional process, which led to the thermodynamic transformation, extreme dissection and thrusting of these varied Late Precambrian to Silurian volcanosedimentary sequences, began in Wenlock or possibly late Llandovery time in internal zones of the orogen and transmitted its energies diachronously southeastward through to Early Devonian time in the frontal décollement zone (Figs. 1 and 2). By latest Silurian time, the highest thrust sheets in the nappe pile were being eroded and shedding thick molasse sediments into fault-controlled Old Red Sandstone basins (Fig. 2F).

Cover shortening, as expressed by estimated displacement of individual nappes, was considerable in southern and central areas, in the order of several hundred kilometres (Gale and Roberts 1974; Gee 1975); even the Lower Allochthon in the south has been transported by over 150 km (Nystuen 1981). While details of the features of this Scandian orogenic phase and the possible mechanisms of deformation have been given in many articles in recent years, and will therefore not be repeated here, there are one or two points worthy of special mention in concluding this account. The first is that the continent–continent collision between Baltica and Laurentia was clearly a violent one with extreme telescoping of the thick continental rise and shelf successions, the intra-oceanic arc and back-arc associations, and the microcontinental highs which may have disturbed this oceanic regime; and the SE–E nappe translations of Scandinavia are mirrored by westward to northwestward thrusting in East Greenland, Scotland, and the Appalachians. Results from work on eclogites in west Norway have shown that these rocks crystallized c. 425 Ma ago and were subjected to temperatures of 700–800°C and pressures of 14–18 kbar, considered by Griffin and Brueckner (1980) to reflect an underthrusting of the Baltoscandian margin beneath the Laurentian following the initial continent–continent impact (see also Gee 1975). The effect of this on cover and oceanic sequences must inevitably have been catastrophic. In addition, there is the probability that major, orogen-parallel, strike-slip dislocations may have disturbed the precollisional plate distributions, and thus juxtaposed terrains of diverse palaeogeographic origin as in the Appalachians (e.g. Keppie, this volume). Future palaeomagnetic studies will be helpful in this respect.

It is also of interest to note that the Uppermost Allochthon in Norway, derived from further 'west' than the other nappes, consists of Silurian-emplaced thrust sheets containing Precambrian crystallines bearing evidence of Grenvillian deformation. With the Finnmarkian event also probably represented in these rocks, and the fact that these uppermost nappes contain much granitoid plutonic material of probable Late Silurian age (Priem *et al*. 1975), it is tempting to suggest that the rocks of these highest thrust sheets may derive, in part, from a northeasterly extension of the Grampian-deformed and granite-intruded, continental rise to ensialic basinal environment in which the Moine and Dalradian assemblages of Scotland are believed to have accumulated.

Acknowledgements

David Bruton, George Gale and Tor Grenne are thanked for their critical reading of the original 1981 manuscript. Many other colleagues in Scandinavia, Britain and Ireland have read preprints of the manuscript over the last two years and we thank them all for access to, and discussions of their latest data, which have led to small refinements in our general thesis.

References

Amaliksen, K. G. 1979. Lykling-ophiolittens breksjer—en indikasjon på en fossil fracture zone? Abstract, 14. Nordiske geol. v. møte. *Geolognytt*, **13**, 4.

Amaliksen, K. G. 1983. *The Geology of the Lykling Ophiolitic Complex, Bømlo, SW Norway*. Unpubl. Cand. Real. thesis, Univ. of Bergen.

Andréasson, P. G., Solyom, Z. and Roberts, D. 1979. Petrochemistry and tectonic significance of basic and alkaline–ultrabasic dykes in the Leksdal Nappe, northern Trondheim region, Norway. *Norges geol. Unders.*, **348**, 47–71.

Andresen, A. and Færseth, R. B. 1981. An evolutionary model for the southwest Norwegian Caledonides. *Amer. J. Sci.* (in press).

Anonymous 1972. Penrose field conference on ophiolites. *Geotimes*, **17**, 24–25.

Bjørklykke, A. and Olaussen, S. 1981. Silurian sediments, volcanics and mineral deposits in the Sagelvvatn area, Troms, North Norway. *Norges geol. Unders.*, **365**, 1–38.

Bjørlykke, K. 1978. The eastern marginal zone of the Caledonide orogen in Norway. *Geol. Surv. Canada, Paper*, **78–13**, 49–55.

Bjørlykke, K., Elvsborg, A. and Hoy, T. 1976. Late Precambrian sedimentation in the central sparagmite basin of South Norway. *Norges geol. Tidsskr.*, **56**, 233–290.

Bluck, B. J., Halliday, A. N., Aftalion, M. and Macintyre, R. M. 1980. Age and origin of Ballantrae ophiolite and its significance to the Caledonian orogeny and Ordovician time scale. *Geology*, **8**, 492–495.

Boyle, A. P. 1980. The Sulitjelma amphibolites, Norway: part of a Lower Palaeozoic ophiolite complex? *Proc. Int. Ophiolite Symp. Cyprus, 1979*, 567–575.

Briden, J. C., Morris, W. A. and Piper, J. D. A. 1973. Palaeomagnetic studies in the British Caledonides—VI Regional and global implications. *Geophys. J. R. astr. Soc.*, **34**, 107–134.

Bruton, D. L. and Bockelie, J. F. 1979. The Ordovician

sedimentary sequence on Smøla, west Central Norway. *Norges geol. Unders.*, **348**, 21–31.

Bruton, D. L. and Bockelie, J. F. 1980. Geology and paleontology of the Hølonda area, western Norway—a fragment of North America? *Proc. IGCP Symp. Blacksburg, Virginia*, 41–47.

Bruton, D. L. and Harper, D. A. T. this volume. Early Ordovician (Arenic–Llanvirn) faunas from oceanic islands in the Appalachian–Caledonide orogen.

Claesson, S. 1976. The age of the Ottfjället dolerites of Sårv Nappe, Swedish Caledonides. *Geol. Fören. Stockh. Förh*, **98**, 370–374.

Claesson, S. 1980. A Rb–Sr isotope study of granitoids and related mylonites in the Tännäs augen gneiss nappe, southern Swedish Caledonides. *Geol. Fören. Stockh. Förh*, **102**, 403–420.

Claesson, S. and Roddick, J. C. 1983. $^{40}Ar/^{39}Ar$ data on the age of metamorphism of the Ottfjället dolerites, Särv Nappe, Swedish Caledonides. *Lithos*, **16**, 61–73.

Cocks, L. R. M. and Fortey, R. A. 1082. Faunal evidence for oceanic separations in the Palaeozoic of Britain. *J. geol. Soc. London*, **139**, 465–478.

Coleman, R. G. 1977. Ophiolites. Springer-Verlag, Berlin, 229 pp.

Doig, R. 1970. An alkaline rock province linking Europe and North America. *Can. J. Earth Sci.*, **7**, 22–28.

Furnes, H. and Færseth R. B. 1975. Interpretation of preliminary trace element data from the Lower Palaeozoic greenstone sequences on Stord, West Norway. *Norsk geol. Tidsskr.*, **55**, 157–169.

Furnes, H., Roberts, D., Sturt, B. A., Thon, A. and Gale, G. H. 1980. Ophiolite fragments in the Scandinavian Caledonides. *Proc. Int. Ophiolite Symp. Cyprus, 1979*, 582–600.

Furnes, H., Austrheim, H., Amaliksen, K. G. and Nordås, J. 1983. Evidence for an incipient early caledonian orogenic phase in SW Norway. *Geol. Mag.* **120**, 607–612.

Furnes, H., Ryan, P. D., Grenne, T., Roberts, D., Sturt, B. A. and Prestvik, T. this volume. Geological and geochemical classification of the ophiolitic fragments in the Scandinavian Caledonides.

Færseth, R. B. 1982. Geology of southern Stord and adjacent islands, southwest Norwegian Caledonides. *Norges geol. Unders.*, **371**, 57–112.

Gale, G. H. and Roberts, D. 1972. Palaeogeographical implications of greenstone petrochemistry in the southern Norwegian Caledonides. *Nature Phys. Sci.*, **238**, 50–61.

Gale, G. H. and Roberts, D. 1974. Trace element geochemistry of Norwegian Lower Palaeozoic basic volcanics and its tectonic implications. *Earth Planet. Sci. Lett.*, **22**, 380–390.

Gayer, R. A. and Humphreys, R. J. Binns, R. E. and Chapman, T. J. this volume. Tectonic modelling of the Finnmark and Troms Caledonides based on high-level igneous rock geochemistry.

Gee, D. G. 1975. A tectonic model for the central part of the Scandinavaian Caledonides. *Amer. J. Sci.*, **275-A**, 468–515.

Gee, D. G. 1981. The Dictyonema-bearing phyllites at Nordaunevoll, eastern Trøndelag, Norway. *Norsk geol. Tidsskr.*, **61**, 93–95.

Graversen, O., Marker, M. and Søvegjarto, U. 1981. Precambrian and Caledonian nappe tectonics in the Central Scandinavian Caledonides, Nordland, Norway. *Terra Cognita*, **1**, 47.

Grenne, T. and Roberts, D. 1980. Geochemistry and vol-

canic setting of the Ordovician Forbordfjell and Jonsvatn greenstones, Trondheim region, central Norwegian Caledonides. *Contrib. Mineral. Petrol.*, **74**, 374–386.

Grenne, T. and Roberts, D. 1981. Fragmented ophiolite sequences in Trøndelag, Central Norway. *Excursion guide B12, Uppsala Caledonide Symposium, 1981*, 40 pp.

Grenne, T. and Lagerblad, B. this volume. The Fundsjø Group, central Norway—a Lower Palaeozoic island arc sequence: geochemistry and regional implications.

Grenne, T., Grammeltvedt, G. and Vokes, F. M. 1980. Cyprus-type sulphide deposits in the western Trondheim district, central Norwegian Caledonides. *Proc. Int. Ophiolite Symp. Cyprus, 1979*, 727–743.

Griffin, W. L. and Brueckner, H. K. 1980. Caledonian Sm–Nd ages and a crustal origin of Norwegian eclogites. *Nature*, London, **285**, 419–321.

Guezou, J. C. 1978. Geology and structure of the Dombås–Lesja area, southern Trondheim region, South–central Norway. *Norges geol. Unders.*, **340**, 1–34.

Haller, J. 1971. Geology of the East Greenland Caledonides. Wiley–Interscience, London, 413 pp.

Halls, C., Reinsbakken, A., Ferriday, I., Haugen, A. and Rankin, A. 1977. Geological setting of the Skorovass orebody within the allochthonous volcanic stratigraphy of the Gjersvik Nappe. *Geol. Soc. London Spec. Publ.*, **7**, 128–151.

Hardenby, C., Lagerblad, B. and Andréasson, P. G. 1981. Structural development of the northern part of the Trondheim Nappe Complex central Scandinavian Caledonides. *Terra Cognita*, **1**, 50.

Horne, G. S. 1979. Mélange in the Trondheim Nappe suggests a new tectonic model for the central Norwegian Caledonides. *Nature*, London, **281**, 267–270.

Hossack, I. R., Garton, M. R. and Nickelsen, R. P. this volume. The geological section from the foreland up to the Jotun sheet in the Valdres area, South Norway.

Keppie, J. D. this volume. The Appalachian collage.

Kjøde, I., Storetvedt, K. M., Roberts, D. and Gidskehaug, A. 1978. Palaeomagnetic evidence for large-scale dextral movement along the Trollfjord–Komagelv Fault, Finnmark, North Norway. *Phys. Earth Planet. Interiors*, **16**, 132–144.

Ledru, P. 1980. Evolution structurale et magmatique du complexe plutonique de Karmøy (sud-ouest de Calédonides norvegiennes). *Bull. Soc. geol. mineral. Bretagne*, **12**, 1–106.

Lippard, S. J. 1976. Preliminary investigations of some Ordovician volcanics from Stord, western Norway. *Norges geol. Unders.*, **237**, 41–66.

Minsaas, O. 1981. *Lyngenhalvøyas geologi, med spesiell vekt på den sedimentologiske utvikling av de Ordovisisk–Siluriske klastiske sekvenser som overligger Lyngen gabbrokompleks.* Unpubl. Cand. Real. thesis, Univ. of Bergen, 294 pp.

Minsaas, O. and Sturt, B. A. this volume. The Ordovician–Silurian clastic sequence overlying the Lyngen Gabbro Complex and its environmental significance.

Neuman, R. B. 1972. Brachiopods of early Ordovician volcanic islands. *24th. Int. Geol. Congr. Montreal. Sec.*, **7**, 297–302.

Nordås, I., Amaliksen, K. G., Brekke, H., Suthern, R., Furnes, H., Sturt, B. A. and Robins, B. 1981. Lithostratigraphy and petrochemistry of Caledonian rocks on Bømlo, SW Norway. *Terra Cognita*, **1**, 61–62.

Olsen, J. 1980. Genesis of the Joma stratiform sulfide deposit, central Norwegian Caledonides. *Proc. 5th. IAGOD Symp.*, 745–757.

Nystuen, J. P. 1981. The Late Precambrian 'Sparagmites' of southern Norway: a major Caledonian allochthon—the Osen–Røa Nappe Complex. *Am. J. Sci.*, **281**, 69–94.

Piper, J. D. A. 1981. Late Precambrian–Cambrian palaeomagnetism and initial stages in Caledonian orogenesis. *Abstract, Eur. Geophys. Soc. meeting, Uppsala*, 1981.

Prestvik, T. 1974. Supracrustal rocks of Leka, Nord-Trøndelag. *Norges geol. Unders.*, **311**, 65–87.

Prestvik, T. 1980. The Caledonian ophiolite complex of Leka, north central Norway. *Proc. Int. Ophiolite Symp. Cyprus, 1979*, 555–566.

Priem, H. N. A. and Torske, T. 1973. Rb–Sr isochron age of Caledonian acid volcanics from Stord, western Norway. *Norges geol. Unders.*, **300**, 83–85.

Priem, H. N. A., Boelrijk, N. A. I M., Hebeda, E. H., Verdurman, E. A. T. and Verschure, R. H. 1975. Isotopic dating of the Caledonian Bindal and Svenningdal granitic massif, central Norway. *Norges geol. Unders.*, **319**, 29–36.

Ramsay, D. M. and Sturt, B. A. 1977. A sub-Caledonian uncoformity within the Finnmarkian nappe sequence and its regional significance. *Norges geol. Unders.*, **334**, 107–116.

Ramsay, D. M., Sturt, B. A. and Andersen, T. B. 1979. The sub-Caledonian unconformity on Hjelmsøy—new evidence of primary basement/cover relations in the Finnmarkian nappe sequence. *Norges geol. Unders.*, **351**, 1–12.

Ramsay, D. M., Sturt, B. A., Zwaan, K. B. and Roberts, D. this volume. Caledonides of northern Norway.

Riis, F. and Ramberg, I. B. 1981. The Uppermost Allochthon—the Rödingsfjället and the Helgeland Nappe Complexes in a segment south of Ranafjorden, Norway. *Terra Cognita*, **1**, 69.

Roberts, D. 1978. Caledonides of South central Norway. *Geol. Surv. Can. Paper*, **78–13**, 31–37.

Roberts, D. 1980a. Mélange in the Trondheim Nappe, central Norwegian Caledonides. *Nature*, London, **285**, 593.

Roberts, D. 1980b. Petrochemistry and palaeogeographic setting of the Ordovician volcanic rocks of Smøla, Central Norway. *Norges geol. Unders.*, **359**, 43–60.

Roberts, D. and Gale, G. H. 1978. The Caledonian–Appalachian Iapetus Ocean. In Tarling, D. H. (Ed.), *Evolution of the Earth's Crust*, Academic Press, London, 255–342.

Roberts, D. and Gee, D. G. this volume. An introduction to the structure of the Scandinavian Caledonides.

Roberts, D. and Sturt, B. A. 1980. Caledonian deformation in Norway. *J. geol. Soc. London*, **137**, 241–250.

Roberts, D., Grenne, T. and Ryan, P. D. in press. Ordovician marginal basin development in the central Norwegian Caledonides. *J. geol. Soc. London*.

Ryan, P. D. and Sturt, B. A. 1981. Early Caledonian orogenesis in northwestern Europe. *Terra Cognita*, **1**, 71.

Ryan, P. D., Williams, D. M. and Skevington, D. 1980. A revised interpretation of the Ordovician stratigraphy of Sør-Trøndelag. *Proc. IGCP Symp. Blacksburg, Virginia*, 99–103.

Skevington, D. 1974. Controls influencing the composition and distribution of Ordovician graptolite faunal provinces. *Spec. Paper Palaeont.*, **13**, 59–73.

Skjerlie, F. J. 1969. The pre-Devonian rocks in the Ask-voll–Gaular area and adjacent districts, western Norway. *Norges geol. Unders.*, **258**, 325–359.

Solli, T. 1981. *The Geology of the Torvastad Group, the Cap Rocks to the Karmøy Ophiolite*. Unpubl. Cand. Real. thesis, Univ. of Bergen.

Solyom, Z., Gorbatschev, R. and Johansson, I. 1979a. The Ottafjäll dolerites: geochemistry of the dyke swarm in relation to the geodynamics of the Caledonide orogen of central Scandinavia. *Sver. geol. Unders.*, **C756**, 1–38.

Solyom, Z, Andréasson, P. G. and Johansson, I. 1979b. Geochemistry of amphibolites from Mt. Sylarna, central Scandinavian Caledonides. *Geol. Fören. Stock. Förhl.*, **101**, 17–25.

Stephens, M. B. 1977. The Stekenjokk volcanites—segment of a Lower Palaeozoic island arc complex. In Bjørlykke, A., Lindahl, I. and Vokes, F. M., (Eds), *Kaledonske malmforekornster:* Trondheim, Bergverkenes Landssammenslutnings Industrigruppe, 24–36.

Stephens, M. B. 1980. Occurrence, nature and tectonic significance of volcanic and high-level intrusive rocks within the Swedish Caledonides. *Proc. IGCP Symp. Blacksburg, Virginia*, 289–298.

Stephens, M. B. 1982. Field relationships, petrochemistry and petrogenesis of the Stekenjokk volcanites, Central Swedish Caledonides. *Sver. geol. Unders.*, **786C**, 1–111.

Stephens, M. B. and Senior, A. 1981. The Norra Storfjället lens—an example of forearc basin sedimentation and volcanism in the Scandinavian Caledonides. *Terra Cognita*, **1**, 76–77.

Stephens, M. B., Furnes, H., Robins, B. and Sturt, B. A. this volume. Igneous activity within the Scandinavian Caledonides.

Strömberg, A. G. B. 1969. Initial Caledonian magmatism in the Jämtland area, Sweden. *Am. Assoc. Petrol. Geol. Mem*, **12**, 275–387.

Sturt, B. A. and Thon, A. 1976. The age of orogenic deformation in the Swedish Caledonides. *Amer. J. Sci.*, **276**, 385–390.

Sturt, B. A. and Thon, A. 1978. An ophiolite complex of probable early Caledonian age discovered on Karmøy. *Nature*, London, **275**, 538–539.

Sturt, B. A., Pringle, I. R. and Ramsay, D. M. 1978. The Finnmarkian phase of the Caledonian orogeny. *J. geol. Soc. London*, **135**, 597–610.

Sturt, B. A., Thon, A. and Furnes, H. 1979. The Karmøy ophiolite, southwest Norway. *Geology*, **7**, 316–320.

Sturt, B. A., Thon, A. and Furnes, H. 1980a. The geology and preliminary geochemistry of the Karmøy ophiolite, S. W. Norway. *Proc. Int. Ophiolite Syhmp. Cyprus, 1979*, 538–554.

Sturt, B. A., Speedyman, D. L. and Griffin, W. L. 1980b. The Nordre Bumandsfjord ultramafic pluton, Seiland, North Norway. Part I: field relations. *Norges geol. Unders.*, **358**, 1–30.

Sturt, B. A., Andersen, T. B. and Furnes, H. this volume. The Skei Group, Leka: an unconformable clastic sequence overlying the Leka Ophiolite.

Sturt, B. A., Roberts, D. and Furnes, H. 1984. A conspectus of Scandinavian Caledonian ophiolites. *J. geol. Soc. London, Spec. Publ.*, **13**, 381–391.

Wilson, J. T. 1966. Did the Atlantic close and then re-open? *Nature*, London, **271**, 676–681.

The Caledonide Orogen—Scandinavia and Related Areas
Edited by D. G. Gee and B. A. Sturt
© 1985 John Wiley & Sons Ltd

Tectonic modelling of the Finnmark and Troms Caledonides based on high level igneous rock geochemistry

R. A. Gayer[*], **R. J. Humphreys**[*], **R. E. Binns**[†] and **T. J. Chapman**[‡]

[*]*Dept. of Geology, P.O. Box 78, University College, Cardiff, CF1 1XL*
[†]*Strindveien 64, 7000 Trondheim, Norway*
[‡]*Geology Division, Department of Environmental Sciences, Plymouth Polytechnic, Drake Circus, Devon, PL4 8AA*

ABSTRACT

Within the Finnmark Caledonides of northernmost Norway pre- (or early syn-) tectonic dykes intrude both the metasediments of the Laksefjord and Kalak Nappe Complexes and also basement slices within the Laksefjord N.C. The dyke geochemistry, using immobile major and trace elements, suggests that they are subalkaline tholeiitic basalts. In general they show affinities to MORB but with a tendency towards transitional within plate continental basalts. They may have developed during main Iapetus rifting or in a marginal basin related to subduction along its southeastern margin. Evidence for subduction generated magma may be present in the West Finnmark Seiland Igneous Province, radiometric dating of which indicates a 540–490 Ma age range.

Farther south, metaigneous rocks occur within the Skibotn, Ullsfjord, and Tromsø Nappe Complexes of the Troms Caledonides, in part as high level intrusive basic sheets and in part as pillow basalts and more acidic volcanic rocks. Associated sediments within the Skibotn and Ullsfjord N.C. contain late Ordovician/early Silurian faunas; thus this igneous activity post-dates that of the Finnmark area but may be the equivalent of volcanicity in the Køli and higher nappes farther south. Major and trace element geochemistry indicates an ocean floor tholeiitic affinity for the pillowed metabasites within the Skibotn N.C. Concordant amphibolitic sheets within a thrust isolated unit in the Skibotn N.C. have a more transitional to continental geochemistry and probably belong to an earlier, possibly Finnmarkian, sequence. Metaigneous rocks in the overlying Ullsfjord N.C. show somewhat ambiguous geochemistry but generally indicate transitional features between continental and MORB or island arc tholeiites. Continued syn-orogenic magma intrusion involved later, subduction-related calc-alkaline and granodioritic magmas. A marginal basin adjacent to and east of an island arc environment is suggested for the whole of this sequence, probably developed just within the earlier Finnmarkian continental margin comparable with present day Andean tectonic regimes.

Introduction

The Caledonian nappes of northern Norway have been thrust southeastwards across the Precambrian Baltic Shield, and are considered to result from gravitational spreading (Chapman *et al.* this volume). The nappes are thought to have developed in two distinct orogenic episodes: an earlier Finnmarkian phase (Sturt *et al.* 1978), dated at 540–490 Ma; and a younger Scandian phase (Gee 1975a), dated at *ca.* 420 ma.

The nappes formed during the Finnmarkian phase are best developed in northernmost Norway, in Finnmark, and in N. Troms (Fig. 1). The earlier farthest travelled, higher nappes (Kalak N.C. and Laksefjord N.C.), exhibit polyphase deformation and metamorphism. They have been thrust over lower nappes (Gaissa Nappe) and parautochthonous cover rocks of the foreland in which Late Riphean, Vendian, and Cambrian sequences are preserved (e.g. Gayer and Roberts 1973; Sturt and Roberts 1978). These nappes virtually lack volcanic rocks. Syn-orogenic plutonic igneous rocks, however, are present in the upper nappes of the Kalak N.C. (Seiland petrogenic province, e.g. Robins and Gardner 1975) and pre- to early syn-orogenic high-level

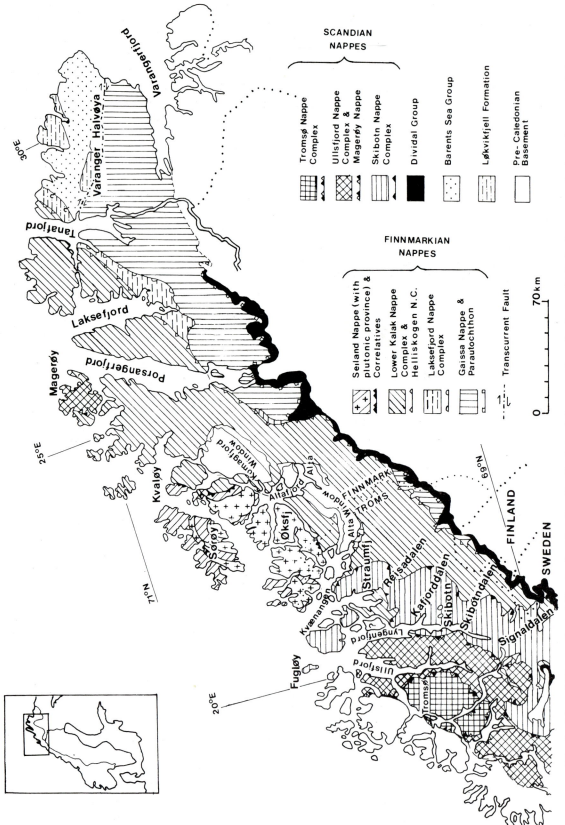

Fig. 1 Map of Finnmark and Troms showing the distribution of the tectonic units from which the igneous rocks have been sampled

intrusive dykes and sheets are present throughout the nappe pile (e.g. Føyn 1960; Gayer and Roberts 1971).

The nappes formed during the Scandian phase have their maximum development farther south in Scandinavia (e.g. Gee 1975a,b) where the evidence for earlier Caledonian events has been largely obscured by Scandian phase reworking. The Scandian nappes extend northwards into Troms where three major nappe units have been recognized (Tromsø N.C., Ullsfjord N.C., and Skibotn N.C.) (Binns 1978), the latter two containing polyphasally deformed and metamorphosed Ordovician/Silurian shelf facies sequences. The lowest of these nappes, the Skibotn N.C. is thrust over the Helligskogen N.C., a correlative of the Finnmarkian Kalak N.C. Binns (1978) envisaged a complex reworking of the Finnmarkian units during later Scandian nappe emplacement. Recently Ramsay *et al.* (this volume) have described an apparent unconformity between the Ordovician/Silurian sequence and an underlying previously deformed Finnmarkian sequence. Farther north, in Finnmark, the only representative of the younger nappes is the Magerøy Nappe (Ramsay and Sturt 1976) which outcrops as a klippe above a passive Kalak N.C. basement on Magerøy (Ramsay and Sturt 1976; Sturt and Roberts 1978). The Scandian nappe units contain major plutonic intrusive complexes (e.g. the tectonically-emplaced Lyngen Gabbro in the Ullsfjord N.C. and the syn-orogenic intrusive Honningsvag Gabbro in the Magerøy Nappe). However, volcanic rocks only occur in the Troms region.

Various plate tectonic models have been suggested to account for the two phase Caledonian development in northern Scandinavia. Harland and Gayer (1972) proposed southeastward subduction of Iapetus oceanic crust beneath an irregular Baltic Shield continental margin to produce late Cambrian orogenesis in the north and island arc/marginal basin development in the south. Contraction of the Iapetus Ocean by continued subduction caused a gradual migration of the zone of orogenesis southwards. This model failed to account for the Scandian Magerøy Nappe in Finnmark and the presence of two distinct orogenic phases in Troms. Ramsay (1973) argued that the Finnmarkian phase could be the northward equivalent of the marginal ocean basins developed farther south but without the formation of a marginal basin. The deep-seated expression of this 'still-born' marginal basin could be the massive intrusion of the syn-orogenic Seiland plutonic province which Robins and Gardner (1975) argued was derived from a steeply dipping subduction zone. Evidence for such a model should be obtainable from the geochemistry of igneous rocks,

particularly volcanic rocks and high-level intrusions, within the various nappe units. The presence of igneous rocks of broadly similar age within nappe units thrust from widely separated localities across the belt should allow palaeogeographic and tectonic reconstructions to be made.

In this paper we present major and trace element analyses for 181 samples of lavas and high-level intrusions selected from most of the nappe units. These represent the first published geochemical analyses for such rocks in all but the Ullsfjord N.C. for which one previous set of analyses is available (Bjørlykke and Olaussen 1981). The results are plotted on discriminant diagrams, using elements thought to have been largely immobile during subsequent metamorphism, to show both magma type and tectonic environment of emplacement.

Geological Setting

Finnmarkian nappes

Intrusive igneous rocks occur in all the main nappe complexes developed during the Finnmarkian phase. Volcanic rocks are, however, absent with the possible exception of hornblende–garnet garbenschiefers in the Hellefjord Group of the highest exposed nappe of the Kalak N.C. (Roberts 1968). The Seiland igneous province represents a major synorogenic intrusive phase within the Kalak N.C. of west Finnmark. Extreme fractionation has taken place in this province (Robins and Gardner 1975) and although it is likely that early phases of intrusion were tholeiitic, it is not possible to use the geochemical analyses to discriminate the tectonic environment of these intrusions. Robins and Gardner (1975) argued that the magmatism was generated above a steeply SE-dipping subduction zone although Sturt (personal communication 1981) suggested a deep mantle source, possibly related to plume activity.

In this investigation sampling was restricted to the high-level intrusive sheets within: (1) the Kalak N.C., (2) the Laksefjord N.C., and (3) the Gaissa Nappe; the field relations in each of these tectonic units are shown diagrammatically in Fig. 2.

Kalak Nappe Complex

All the intrusions investigated occur on either side of Porsangerfjord (Fig. 1) where three distinct nappes can be recognized. The majority of the intrusions occur in a pronounced swarm within the middle Kalak Nappe. The sheets are all metabasites and vary in field relationships from clearly discordant dykes intruded into massive competent units to almost concordant 'sill-like' sheets frequently

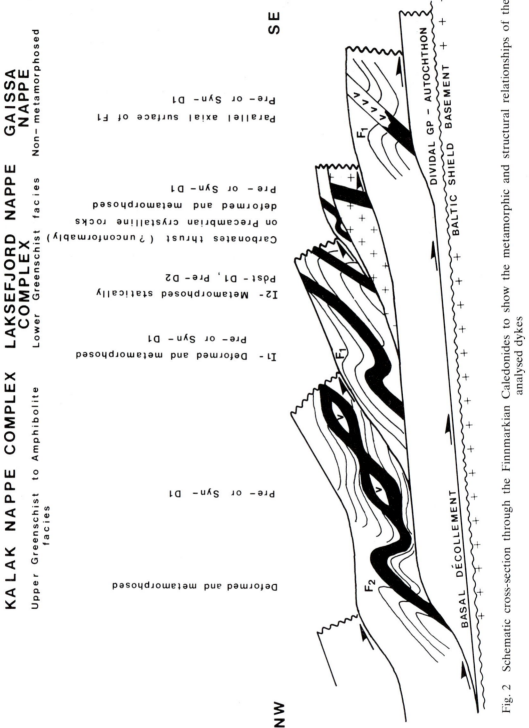

Fig. 2 Schematic cross-section through the Finnmarkian Caledonides to show the metamorphic and structural relationships of the analysed dykes

deformed into discrete boudins occurring in foliated interbanded psammite/semipelite and pelites. In some dykes multiple intrusion has occurred. Studies of the metamorphism and deformation of these metabasites (Gayer *et al.* 1978) showed that porphyroblastic phases, mainly amphibole and garnet, preserved helicitic S1 fabrics, indicating that dyke intrusion took place either before the onset of deformation or very early during the first deformation phase. Since metamorphism reached upper greenschist to lower amphibolite facies within the Kalak N.C. after the close of the first deformation phase, metamorphism will have caused element mobility, particularly within the sheared margins of the metabasites. Thus the least affected cores have been used in the discussion of original igneous geochemistry below.

Laksefjord Nappe Complex

Metabasites occur in four distinct tectonic situations within the nappe complex (Fig. 2). Two of these intrude the Laksefjord Group sediments of the Middle Laksefjord Nappe (Chapman 1980). One set (L_1) is affected by the earliest deformation and metamorphism recognized in the surrounding sediments and are interpreted, as with the similarly situated Kalak dykes, as pre- to early syn-tectonic intrusions. They are metamorphosed to lower greenschist facies and contain the S_1 foliation at least in their margins, but commonly throughout. Subophitic textures are only rarely preserved in their cores. The other set (L_2) are unfoliated, preserving subophitic textures, although they are metamorphosed to lower greenschist facies. It is possible that the two sets represent the same intrusive suite affected by varying levels of subsequent strain. The outcrops of the two sets are mutually exclusive with L_2 dykes intruding massive quartzites that have suffered less strain than the more pelitic lithologies into which the L_1 dykes are intruded. On the other hand one L_2 dyke has been mapped intruding across a major F_1 fold axial plane, suggesting that at least this L_2 dyke post-dates the F_1 folding and L_1 set.

The other two groups of metabasites occur in the Lower Laksefjord Nappe which comprises a slice of crystalline Precambrian basement overthrust by highly-deformed carbonate metasediments. Intrusions in the carbonate metasediments (L_3) are affected by the earliest deformation and metamorphism recognized in the sediments and are interpreted as pre- or early syn-tectonic. The thick dykes in the crystalline basement (L_4) post-date most of the Precambrian events seen in the country rocks but are metamorphosed to lower greenschist facies in common with all the rocks of the Laksefjord N.C.

Gaissa Nappe

Rare basic dykes occur within the Gaissa Nappe south of Porsangerfjord and south of Laksefjord. The dykes parallel F_1 fold axial surfaces but are affected by shearing related to the fold development. The primary ophitic texture has been largely obliterated by epidote and chlorite growth, which may have resulted from D_1 shearing. Thus, as with the dykes in the higher Finnmarkian nappes, those intruding the Gaissa Nappe are either pre- or early syn-D_1.

Scandian nappes

Igneous rocks occur in a number of tectonic situations in both the Ullsfjord and Skibotn N.C. (Fig. 3) as described below. Geochemical data for the Lyngen gabbro, a major tectonically-emplaced igneous body within the lower part of the Ullsfjord N.C., have been given by Munday (1970) and the mean of his results has been plotted in Fig. 7 for comparison.

Ullsfjord Nappe Complex

This unit chiefly consists of the Balsfjord Supergroup (Binns and Matthews 1981), a sequence of metamorphosed pelites, carbonates, psammites, psephites, and some volcanic and volcano-sedimentary rocks formed in ?middle Ordovician to ?early Silurian time (Olaussen 1977; Binns and Matthews 1981; Bjørlykke and Olaussen 1981; Humphreys 1981). Ultramafic, gabbroic, and granodioritic–trondhjemitic intrusions of ?pre- syn- orogenic age occur. Local and more regional breaks in sedimentation probably reflect the changing configuration of a marginal basin. In broad terms, sedimentation started with the shallow marine Jøvik Formation, continued with the chiefly shallow marine, periodically/locally intertidal to subtidal Sandøyra and Lakselvdal Formations, and concluded with the largely continental, fluviatile (locally marine) Vardtind Formation (Binns and Matthews 1981; Minsaas and Sturt this volume). All the lithologies were polyphasally deformed and regionally metamorphosed in low greenschist to upper amphibolite facies during the Scandian event. This deformation also involved movement along several thrust zones within as well as at the base and top of the nappe complex, and the previously folded, layered Lyngen gabbro was tectonically emplaced (e.g. Munday 1970; Randall 1971; Binns 1978) but *cf.* Minsaas and Sturt (this volume).

Concordant metabasite sheets are particularly associated with specific horizons in the Jøvik and Sandøyra Formations. Although generally thin

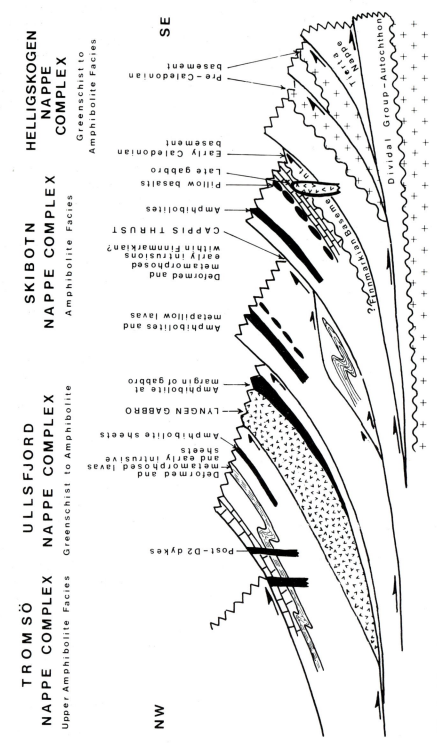

Fig. 3 Schematic cross-section through the Troms Caledonides showing the tectonic situation of the basic igneous rocks sampled. The main unconformity beneath the late Ordovician/Silurian sequence in the lower Skibotn N.C. is after Ramsay et al. (this volume)

(10 mm–3 m), many sheets show a considerable lateral continuity, 2 km or more not being atypical. Precise extrapolation is often hindered by multiple fold repetitions.

Metabasites are typically deep green to black, usually aphanitic, with an amphibolite-dominated composition. Mineralogies vary from greenschist to amphibolite facies assemblages with typical associations being amphibole–albite–epidote and chlorite–albite–carbonate. Garnet is present at higher grades. Saussuritization almost universally affects plagioclase (An_{5-10}).

Weakly schistose fabrics are usually pervasive through sheets except in the thickest bodies (>10 m). This fabric is strongest at sheet margins which are sometimes characterized by greater modal chlorite or biotite, and is coplanar with the main regional, low-angle foliation. Both upper and lower contacts of sheets show sharp boundaries although much of the adjacent semipelite is also believed to have an associated volcanogene origin (Humphreys 1981). Pillow structure, jasper, and agglomerate are found in E. Ullsfjord and southwest of Balsfjord (Binns and Matthews 1981). At least some sheets are therefore interpreted to represent lava flows or ash bands. In contrast, a shallow intrusive origin cannot be dismissed for certain more massive-textured bodies in upper parts of the complex.

Similar amphibolitic sheets were described by Munday (1970) in metasediments (≡ Jøvik Formation) structurally above the Lyngen Gabbro. He considered a thick, more massive body (his Kjosen Formation) immediately below to be distinct however and likely to have a tectonic derivation from the Lyngen Gabbro itself (a branded hypersthene norite). Binns (1978) suggested a correlation between this and the Vardtind Formation; both are probably predominantly sedimentary and volcano-sedimentary in origin and both contain imbricate slices derived from the Lyngen Gabbro.

Occasional thin, stratiform layers of a garbenschiefer-like lithology also occur in both formations. Their more leucocratic mineralogy usually with a grossular-rich garnet even at the lowest grades may have formed from a mixed ash/tuff, sediment precursor. Analyses of these rocks have been included in this paper to allow better control of chemical variation to be expected from sediment/volcanic admixtures. Rare, late, truly discordant dykes with a fine-grained aphanitic but greenschist mineralogy intrude the middle and upper levels of the nappe complex. A few outcrops of rhyolite and andesite porphyry ?flows are found in the lower and upper parts of the volcano-sedimentary sequence; these have not been analysed.

Skibotn Nappe Complex

Lithostratigraphic descriptions of this amphibolite facies lower nappe complex have been given by Padget (1955), with revisions by Binns (1978), Quenardel (1978), and Zwaan and Roberts (1978). Large scale folding and a tectonic break, the Cappis Thrust, have been shown to duplicate the primarily shallow marine Ordovician–Silurian sequence (Binns and Gayer 1980) and a major unconformity separating this sequence from an underlying Finnmarkian complex has been described (Ramsay et al. this volume). Between Skibotndalen, Signaldalen, and the head of Lyngenfjord, the Cappis Thrust bifurcates apparently forming a major tectonic lens, partly containing possible Finnmarkian complex material (Binns 1978, and in prep.).

Up to at least 300 m of concordant, schistose amphibolite occur at several horizons in the nappe (the 'Green Beds' of Padget (1955)), locally showing traces of pillow structure, volcanic breccia and agglomerate. A hornblende, plagioclase (oligoclase–andesine), epidote, and garnet mineralogy is typical with accessory ilmenite. Subordinate, more acid lava derivatives also occur (Binns and Gayer 1980).

Thinner, boudined, concordant amphibolite sheets are locally common, especially in those portions of the nappe complex which radiometric dating (Quenardel et al. 1981; Binns and Taylor in preparation) suggests may have been deformed and metamorphosed in an early, pre-Silurian (?Finnmarkian) orogenic phase. Although less demonstrably extrusive, deformation usually being more intense in these horizons, their modal mineralogy is similar to the other amphibolites.

Several small, late tectonically-emplaced gabbros as well as basic dykes and sheets occur at different levels in the nappe complex. These cut all the main Caledonian structures of the complex but parallel the dominant joint trend. In addition a troctolitic complex with relatively undeformed doleritic intrusions occurs at the top of the underlying Helligskogen N.C. Hausen (1942) and Bøe (1976) described slightly alkaline normative mineralogies for these sill-like bodies with modal plagioclase, clinopyroxene, hornblende, and biotite.

Sampling and Analytical Techniques

Sampling has been concentrated in well exposed areas with good representation of igneous rocks and where the regional geology was sufficiently understood. In general the rocks are deformed and affected by regional metamorphism (see below).

Samples were chosen to represent the range of petrographic variation present in the rocks and care was taken to collect unweathered and unsheared samples that contained no secondary veining.

The samples were analysed for 10 major and up to 12 trace elements using X-ray fluorescence spectrometers at the Department of Geology, Bristol University, the Department of Geology, Durham University, and the Department of Mineral Exploitation, U.C. Cardiff. Tests were carried out for inter-laboratory variation. Major elements were analysed on glass beads prepared by fusion with a borax flux. Rock powders were ignited prior to fusion to remove volatiles so that major element concentrations are presented on a volatile free, anhydrous basis. This follows the suggestion by Irvine and Baragar (1971) and Thirlwall (1981) who argued that the volatile content in metamorphic rocks has little to do with the original igneous geochemistry and is largely introduced during metamorphic alteration. Trace elements were analysed on pressed powder pellets, with matrix corrections calculated from sample major element composition. Both international and synthetic standards were used to calibrate each element.

The results of the chemical analyses are given in Tables 1 and 2.

Metamorphism apart from affecting volatile content is likely to have mobilized other elements. In an attempt to assess the extent of such mobility samples were collected at 1m intervals across two dykes from the amphibolite facies Kalak N.C. and the analyses plotted in Fig. 4. The first is a metadolerite with amphibolitized margins, although the central portions still partially retain an igneous mineralogy. The margins preserve a porphyritic plagioclase (An_{35}) fabric with laths up to 3 mm long in a fine-grained granulose mosaic of blue-green hornblende with subordinate quartz. The central portion is a partially amphibolitized metadolerite in which original subophitic texture is preserved with two pyroxenes marginally altered to blue–green hornblende. The plagioclase grains preserve complex twinning in their cores but are altered from An_{50} to An_{35} around the edges. The second is a marginally biotitized amphibolite dyke in which no trace of the original igneous mineralogy or texture has been preserved. The margins have been recrystallised to a biotite schist with a foliation parallel to the margins, whilst the centre is a coarse-grained schistose quartz-amphibolite with subordinate biotite. Untwinned, recrystallized plagioclase (An_{20}) is present throughout the dyke.

Fig. 4a shows considerable variation in the alkalis K_2O and Rb with marginal enrichment. Very slight variation occurs in Na_2O and MgO with a tendency for marginal depletion. The other major and most of the minor elements show little variation across the dyke suggesting relative immobility during deformation and metamorphism. Much greater variation occurs in the biotitized dyke (Fig. 4b) with mobility of MgO, CaO, Na_2O, K_2O, and most traces. The petrographic study suggests that the central portions of the dykes are more likely to preserve original igneous chemistry. Thus analyses from the dyke cores have been used in the following discussion and, in general, discriminant diagrams have been chosen to avoid the use of mobile elements.

Whole-Rock Chemistry

Finnmarkian nappes

Magma type

The mobility of the alkaline elements precludes the use of the standard alkali v. silica plot to determine the alkalinity of these metabasic rocks. However, the use of an A–F–M diagram to show Fe and Mg variations can be justified, particularly as alkali mobility is at a minimum in the least altered dyke centre samples used in this analysis. The Middle Kalak dykes (Fig. 5a) show a fairly well-defined pattern of Fe enrichment, i.e. a tholeiitic trend. The Lower Kalak dykes, however, show a cluster generally within the field of the tholeiitic fractionation trend; the lack of definition probably results from alkali mobility. The tholeiitic trend is also seen using either Zr (Fig. 5g) or $FeO_{(TOT)}/MgO$ (Fig. 5d and e) as indices of fractionation (e.g. Allegre *et al.* 1977; Miyashiro and Shido 1975). Y/Nb (Fig. 5b) also indicates a tholeiitic magma, but the Zr/P_2O_5 ratios (Fig. 5h) show transitional tholeiitic/alkaline characteristics.

In the case of the Laksefjord N.C., the L_1 and L_2 dykes of the Middle Laksefjord Nappe together with the L_3 intrusions into the carbonates of the Lower Laksefjord Nappe show similar well defined patterns of Fe enrichment on the A–F–M diagram (Fig. 6a), with the exception of four intrusions that fall within the calc-alkaline field. These intrusions are acidic granophyres. The L_4 intrusions in the basement slice of the Lower Laksefjord Nappe show a broad scatter on Fig. 6a, transitional between tholeiitic and calc-alkaline trends. This scatter may have resulted from an original transitional magma chemistry or from strong alkaline element mobility. The L_1, L_2, and L_3 intrusions also show very similar and strong tholeiitic trends using both $FeO_{(TOT)}/MgO$ and Zr as indices of fractionation (Fig. 6c and e) and also using Y/Nb ratios (Fig. 6b). The L_4 dykes, intruding the Precambrian basement, show less distinct but definite tholeiitic tendencies, particularly using Y/Nb as a discriminant (Fig. 6b). The low Zr/P_2O_5 ratio of

Table 1 Mean compositions ($\bar{x}$) and standard deviations (s) for samples of basic dykes from the Finnmark Caledonides. Column 1: Upper Kalak N.C. dykes, 2: Middle Kalak N.C. dykes, 3: Lower Kalak N.C. dykes, 4: Middle Laksefjord N.C. dykes, 5: Lower Laksefjord N.C. dykes intruding carbonates, 6: Lower Laksefjord N.C. dykes intruding granitoid basement, 7: Gaissa Nappe dykes

	1	2		3		4		5		6		7	
	$\bar{x}$	$\bar{x}$	s	$\bar{x}$	s	$\bar{x}$	s	$\bar{x}$	s	$\bar{x}$	s	$\bar{x}$	s
SiO_2	48.05%	47.66%	1.75	50.23%	2.19	48.56%	2.35	50.38%	2.32	48.45%	0.99	54.29%	2.72
Al_2O_3	12.63	13.06	0.78	13.05	1.40	14.24	1.51	14.78	1.02	15.66	0.77	14.00	1.04
TiO_2	1.49	2.30	0.42	2.54	1.17	2.62	0.82	1.72	0.43	1.89	0.12	1.60	0.39
FeO_T	12.54	14.62	1.03	13.09	2.69	15.29	2.59	12.70	1.93	13.47	0.37	13.52	1.95
MgO	11.74	8.12	0.72	7.62	1.99	6.90	0.61	8.38	1.20	7.90	0.29	5.61	2.19
CaO	11.23	10.68	0.49	8.74	2.54	9.93	1.00	8.64	2.23	9.48	1.08	6.99	2.54
Na_2O	1.80	2.00	0.68	1.39	1.17	1.69	0.36	1.94	0.56	2.15	0.47	2.62	0.96
K_2O	0.16	1.11	1.57	2.84	1.80	0.22	0.27	1.04	0.72	0.67	0.33	0.91	0.72
MnO	0.22	0.24	0.01	0.24	0.11	0.26	0.03	0.24	0.03	0.20	0.04	0.16	0.03
P_2O_5	0.14	0.21	0.06	0.28	0.12	0.29	0.20	0.18	0.07	0.13	0.02	0.30	0.14
Total	100.00	100.00		100.02		100.00		100.00		100.00		100.00	
Ba	28 ppm	102 ppm	65	627 ppm	583	n.d.		n.d.		n.d.		n.d.	
Rb	4	34	56	96	69	13 ppm	7	56 ppm	39	10 ppm	8	29 ppm	12
Sr	115	126	88	190	139	744	353	691	466	145	91	290	203
Zr	61	135	35	185	104	171	74	143	91	98	29	172	13
Y	58	48	11	51	26	41	10	25	3	29	11	24	4
Nb	3	4	1	9	7	6	3	7	6	4	2	11	3
La	7	6	2	9	5	n.d.		n.d.		n.d.		n.d.	
Ce	2	12	7	26	19	n.d.		n.d.		n.d.		n.d.	
Cu	2	50	32	107	283	n.d.		n.d.		n.d.		n.d.	
Ni	74	73	14	70	48	n.d.		n.d.		n.d.		n.d.	
Cr	537	202	37	175	96	n.d.		n.d.		n.d.		n.d.	
Zn	98	121	18	122	32	n.d.		n.d.		n.d.		n.d.	
Pb	n.d.	n.d.		n.d.		16	8	13	7	33	51	22	27
No. of Samples	1	26		22		11		18		11		12	

Table 2   Mean composition ($\bar{x}$) and standard deviation (s) for samples of basic igneous rocks from the Troms Caledonides. Column 8: Ullsfjord N.C. metabasites from the Jøvik and Sandøyra Formation, 9: Ullsfjord N.C. late orogenic dyke, 10: Ullsfjord N.C. amphibolites immediately east of and structurally below the Lyngen Gabbro, 11: Skibotn N.C. amphibolite sheets (?Finnmarkian), 12: Skibotn N.C. pillow lavas and amphibolites, 13: Skibotn N.C. late orogenic intrusive

	8		9		10		11		12		13	
	$\bar{x}$	s	$\bar{x}$	s	$\bar{x}$	s	$\bar{x}$	s	$\bar{x}$	s	$\bar{x}$	s
SiO_2	50.80%	3.25	54.90%		48.84%	2.95	52.15%	3.29	50.26%	2.71	52.62%	1.68
Al_2O_3	13.67	1.79	13.94		13.18	2.16	14.37	1.70	13.74	1.90	19.81	3.78
TiO_2	1.58	0.67	0.74		0.38	0.12	2.08	0.52	1.53	0.45	0.98	0.61
FeO_T	13.99	2.11	10.16		8.44	2.34	12.96	2.02	10.02	1.61	6.28	1.05
MgO	6.97	1.95	7.00		11.36	2.47	5.29	1.55	8.33	1.72	5.45	1.10
CaO	9.66	3.90	9.01		12.90	3.01	9.73	1.67	12.52	2.92	11.09	2.22
Na_2O	2.34	0.93	2.55		1.78	0.59	2.63	0.47	2.88	0.56	2.94	0.84
K_2O	0.67	0.65	1.53		0.34	0.48	0.54	0.24	0.41	0.34	0.69	0.73
MnO	0.24	0.07	0.19		0.16	0.05	0.25	0.03	0.17	0.04	0.14	0.01
$P_2O_5{}^*$	0.21	0.13							0.18	0.05		
Total	100.13	—	100.02		100.00	—	100.00		100.04		100.00	—
Ba	175 ppm	332	710 ppm		32 ppm	33	125 ppm	63	78 ppm	47	145 ppm	35
Rb	28	27	92		11	15	20	20	12	12	29	31
Sr	287	271	278		242	66	261	58	264	131	330	40
Zr	135	72	145		38	22	176	60	133	44	127	42
Y	28	8	21		28	28	29	6	31	7	17	1
Nb	19	10	16		15	15	21	7	9	5	11	2
La	23	13	51		7	7	28	13	12	11	51	42
Ce	35	14	57		7	7	44	14	33	11	54	18
Cu	36	32	31		68	35	37	15	35	30	24	16
Ni	53	39	53		151	98	37	15	59	22	47	32
Cr	176	111	251		325	211	177	53	241	67	158	17
Zn	155	68	84		103	103	151	56	68	23	64	3
No. of Samples	29(*9)		1		5		11		31		2	

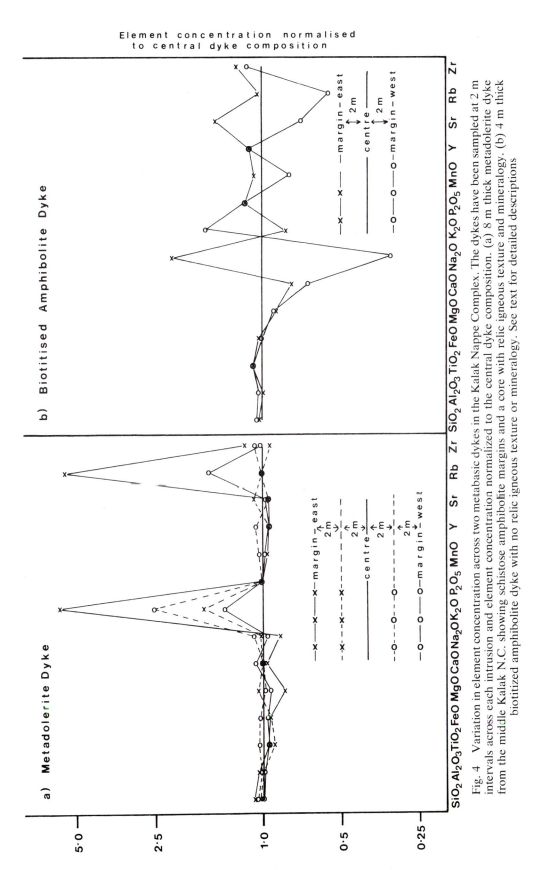

Fig. 4 Variation in element concentration across two metabasic dykes in the Kalak Nappe Complex. The dykes have been sampled at 2 m intervals across each intrusion and element concentration normalized to the central dyke composition. (a) 8 m thick metadolerite dyke from the middle Kalak N.C. showing schistose amphibolite margins and a core with relic igneous texture and mineralogy. (b) 4 m thick biotitized amphibolite dyke with no relic igneous texture or mineralogy. See text for detailed descriptions

KALAK NAPPE COMPLEX

- ■ Upper Kalak Dykes
- ● Mid Kalak Dykes
- ▲ Lower Kalak Dykes

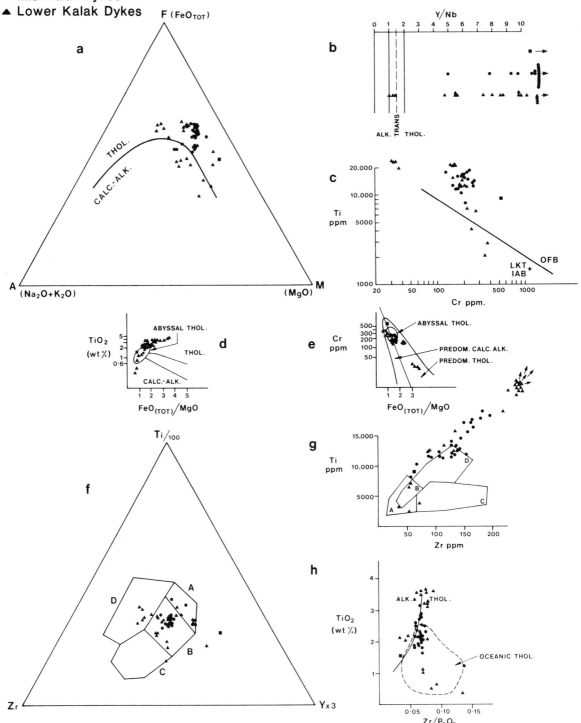

Fig. 5   Kalak Nappe Complex dykes plotted on: (a) AFM plot; the boundary between tholeiitic and calc-alkaline trends is after Irvine and Baragar (1971); (b) Y/Nb discrimination between tholeiitic and alkaline magmas after Pearce and Cann (1973); (c) Ti–Cr plot of Pearce (1975) distinguishing ocean-floor basalts (MORB) and subduction generated magmas (LKT + IAB); (d) and (e) Miyashiro and Shido (1975) plots of TiO_2 (d) and Cr (e) against $Fe_{(TOT)}/MgO$ separating calc-alkaline from tholeiitic trends with the field of abyssal tholeiites (MORB) indicated; (f) Ti–Zr–Y discrimination diagram of Pearce and Cann (1973), field A—low potassium tholeiite, field B—ocean-floor basalts (MORB) and island arc basalts; field C—calc-alkaline basalts, field D-within-plate basalts; (g) Ti–Zr variation plot after Pearce and Cann (1973), field A + B—low potassium tholeiites, field C + B—calc-alkaline basalts, field D + B—ocean-floor basalts (MORB); (h) TiO_2–Zr/P_2O_5 variation plot of Floyd and Winchester (1975) separating alkaline basalt from tholeiitic trends with the field of oceanic tholeiites (MORB) indicated

LAKSEFJORD & GAISSA NAPPE

- ● MID LAKSEFJORD Dykes
- ▲ LR LAKSEFJORD dykes in carbonates
- ♦ LR LAKSEFJORD dykes in basement
- ■ GAISSA Dykes

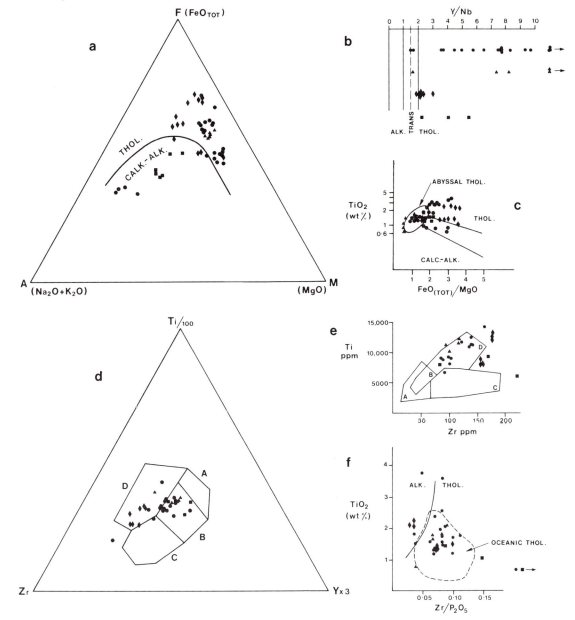

Fig. 6   Laksefjord and Gaissa Nappe dykes plotted on variation diagrams. See caption to Fig. 5 for explanation of fields and sources of the diagrams

some of these L_4 intrusions (Fig. 6f) suggest transitional chemistry between tholeiitic and alkaline basalt.

The dykes from the Gaissa Nappe show no tendency towards Fe enrichment on the A–F–M diagram (Fig. 6a) but rather fall on a calc-alkaline trend. This, however, could be a product of alkali mobility. No clear fractionation trend is seen using $FeO_{(TOT)}/MgO$ (Fig. 6c) or Zr (Fig. 6e) as indices but the Y/Nb and Zr/P_2O_5 ratios (Fig. 6b and f) suggest a tholeiitic magma.

Tectonic affinity

The Ti–Y–Zr diagram of Pearce and Cann (1973) (Fig. 5f) shows a tight grouping in field B for the majority of the Middle Kalak dykes but a more scattered grouping in and around field B for the upper and Lower Kalak intrusions, extending into the within-plate (field D) and low K tholeiite (field A). When plotted on the Ti–Zr diagram (Fig. 5g) these tholeiites fall within the MORB field, extending beyond this field to show the strongly fractionated characteristics of high Ti and Zr of within plate basalts. The fractionation diagrams using $FeO_{(TOT)}/MgO$ (Fig. 5d and e) and Zr/P_2O_5 (Fig. 5h) and the Ti–Cr plot of Pearce (1975) (Fig. 5c) also show a tendency for more highly fractionated magmas than is typical for MORB. All the discriminant diagrams show very similar relationships for the three sets of intrusions into the Laksefjord N.C. cover sediments (L_1, L_2, and L_3). These relationships are clearly shown by the Ti–Y–Zr diagram (Fig. 6d) which together with the Ti–Zr plot (Fig. 6e) show strong MORB tendencies with a slight scatter into the within plate (field D) in Fig. 6d. These intrusions show a similar chemistry to the Kalak dykes but without the pronounced enrichment of Zr and Ti shown by some of the Lower Kalak dykes (Fig. 5g).

The less strongly tholeiitic intrusions into the crystalline basement (L_4) show different chemical relationships. They group in the within plate (field D) on the Ti–Y–Zr plot (Fig. 6f) and show greater enrichment of $FeO_{(TOT)}$ and Zr (Fig. 6c and g) than the other Laksefjord intrusions suggesting a more strongly fractionated tholeiite from a within plate setting. It seems unlikely that these intrusions can represent feeders to the intrusions into the overlying cover since they would appear to have been derived from a different magma source, thus supporting the field evidence for a thrust contact between the basement and cover. It is suggested that the L_4 intrusions represent an entirely pre-Caledonian, possibly early to middle Proterozoic phase of igneous activity.

Unfortunately trace element analyses are available for only a few of the Gaissa Nappe dykes and these do not give a very definite picture of tectonic setting. The Ti–Zr–Y plot (Fig. 6d) indicates a MORB chemistry whilst the Ti–Zr (Fig. 6e) and TiO_2–Zr/P_2O_5 (Fig. 6f) plots are indeterminate. Thus, tentatively, the Gaissa dykes are correlated with the main Laksefjord dykes and the Kalak dykes.

Scandian nappes

Magma type

Moderately tholeiitic A–F–M (Fig. 7a) affinities appear from the main group of metabasite sheets of the Ullsfjord N.C. The two analyses located apart from the main data well into the calc-alkaline field represent a garbenschiefer body and metavolcanic agglomerate, and indicate the effects of sedimentary contamination. Confirmation of related, igneous trends and this tholeiitic chemistry can also be seen using either Zr or $FeO_{(TOT)}/MgO$ as indices of fractionation. Trends for the main group of concordant metabasic sheets in the Jøvik and Sandøyra Formations involving progressive enrichment of Ti and depletion of Cr are notable (Fig. 7c, d, e, and g). Additionally, where available P_2O_5 data (Fig. 7h) also support such a character. The considerably weaker tholeiitic or more transitional character indicated by Y/Nb ratios (Fig. 7b) is related to the notably wide scatter of data shown on the Zr–Y–Ti plot (Fig. 7f).

The Lyngen Gabbro (mean) and the majority of the Kjosen Formation amphibolites (data from Munday 1970) appear to represent very poorly fractionated, primitive, magma. They show very little iron enrichment on the A–F–M diagram (Fig. 7a); titanium and zirconium levels are low (Fig. 7c, d, e, and g); whereas concentrations of the compatible elements chromium and nickel are high (Fig. 7c and d and Table II). One sample of the Kjosen amphibolites appears to show definite tholeiitic characteristics on most diagrams (Fig. 7a, d, and e) but it is difficult to assess the significance of this solitary analysis. The possible distinction of a true calc-alkaline character for a late orogenic dyke sample from the Ullsfjord N.C. is upheld by most of the commonly employed discrimination systems (Figs. 7a, d, f, and h), despite the uncertainty from only one analysis.

Fig. 8a and b show a clear difference in basic chemistry between the groups of metabasite sheets of the Skibotn N.C. The hornblende schists/pillow lavas are strongly tholeiitic by Y/Nb ratios (Fig. 8b) but less clearly so on the A–F–M diagram (Fig. 8a), perhaps as a result of high grade metamorphic alkali enrichment. Higher Ti and lower Cr contents of the amphibolite sheets (Fig. 8c) are in agreement with a greater degree of fractionation implied by Fig. 8a although this tholeiitic character is less clearly shown by Y/Nb ratios (Fig. 8b).

Data, although more limited, for the late basic intrusions support a calc-alkaline affinity even in the major element statistical discriminant plots of Pearce (1976).

Tectonic affinity

The main group of metabasites of the Ullsfjord N.C. are rather difficult to interpret since they give a wide spread of data on most of the currently used discriminant diagrams (e.g. Fig. 7f). If the Ullsfjord data

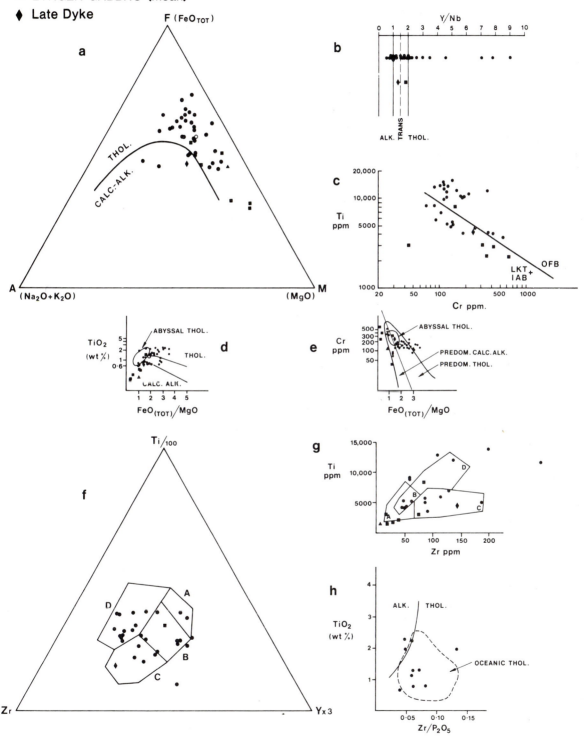

Fig. 7   Variation diagrams of igneous rocks from the Ullsfjord Nappe Complex. See caption to Fig. 5 for explanation of fields and sources of diagrams

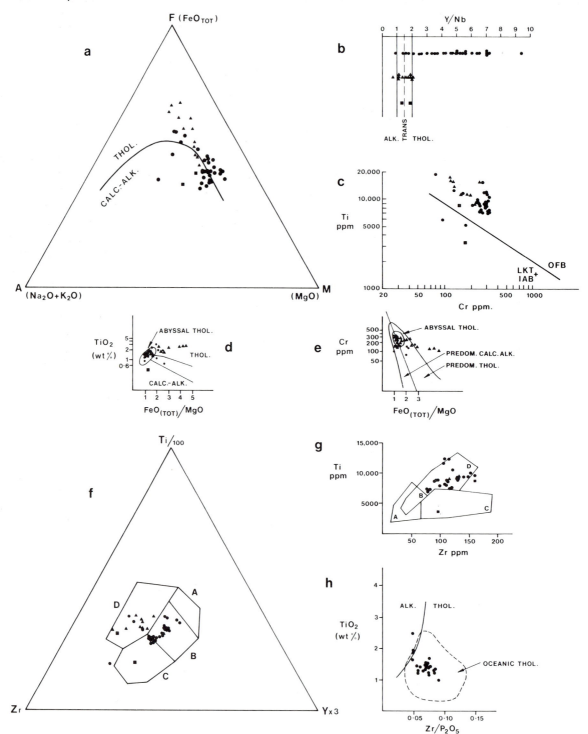

Fig. 8   Variation diagrams of basic igneous rocks from the Skibotn Nappe Complex. See caption to Fig. 5 for explanation of fields and sources of the diagrams

are compared with modern basalts at 'typical fractionation degrees' (data from Pearce and Cann 1973 and Pearce 1973), the mean of the main group metabasites (Table III, col. 8) resembles neither MORB nor IAT. The spread of data into the within plate basalt field is therefore likely to imply a magma source transitional between WBP and either IAT or MORB characteristics. The fractionation range of Cr and Zr related to Ti contents (Fig. 7c, g, and h) suggests the presence of minor true calc-alkaline lavas. The data would thus appear to resemble most closely a transitional magma between IAT and continental tholeiite, perhaps generated above a subduction zone in areas where continental back-arc rifting was also taking place (see e.g. Pearce 1980). This environment is also indicated for the Jøvik (Svensby) amphibolite sheets, analysed by Munday (1970) who used the Karoo dolerite for comparison.

The Lyngen Gabbro and Kjosen amphibolites represent poorly fractionated magma. It is thus not possible to use the commonly employed discriminant diagrams to determine tectonic setting for these rocks since such diagrams assume moderately fractionated basaltic liquid. The only analysis of a well fractionated Kjosen amphibolite appears to indicate a MORB setting (Fig. 7f, g, and d) but the reliability of one analysis must be suspect. However, the primitive nature of the gabbros is entirely consistent with an ophiolite interpretation for the Lyngen Gabbro (Minsaas and Sturt this volume) whereby primary mantle melts could rise rapidly to high levels in the lithosphere.

The late syn-tectonic dyke shows subduction related island arc basalt or continental margin affinities in many of the discriminant plots (e.g. Fig. 7d, f, and g), in agreement with their calc-alkaline composition.

Fractionation trends (Fig. 8d, e, g, and h), Cr–Ti values (Fig. 8c), the Zr–Y–Ti ratios (Fig. 8f) all suggest a MORB character for analyses of the hornblende schist/pillow samples. The tight clustering of data within the MORB field of the latter two plots suggests major element mobility has not been significant in these rocks.

Although distinctly tholeiitic, the tectonic environment of the probably older amphibolite sheets is clearly different, with a within-plate setting indicated by Ti–Y–Zr (Fig. 8f) perhaps transitional to MORB as indicated in the Cr–Ti plot (Fig. 8c). The higher levels of incompatible elements in these transitional basalts suggest a less depleted underlying mantle (Jamieson and Clarke 1970), possibly resulting from undepleted subcontinental mantle during initial stages of back-arc rifting (e.g. Saunders et al. 1979). Compared with the possibly similarly aged Kalak and Laksefjord N.C. dykes, the closest resemblance

is with the L_4 dykes intruding the basement slice, although there is also a similarity with the Lower Kalak dykes in the high fractionation trends seen in the TiO_2–$FeO_{(TOT)}$/MgO (cf. Figs. 4d and 8d).

The late emplaced gabbros and dolerites have many features suggesting island arc related calc-alkaline basalts (Fig. 8c, f, and g) and they may well be comparable with the late syn-tectonic subduction generation intrusions in the Ullsfjord N.C.

Tectonic models

The tholeiitic dykes intruding the Finnmarkian Kalak, Laksefjord, and Gaissa Nappes and the older amphibolite sheets of the Skibotn Nappe, showing transitional MORB/within plate characteristics, are interpreted as mantle derived magma intruded through thin continental crust. The dykes have a very similar composition to the Ottfjäll dolerites of the Central Scandinavian Caledonides which have been related to Iapetus opening (Solyom et al. 1979) and it is possible that the Finnmark dykes represent a northward continuation of the Iapetus rift zone (Fig. 9a). It is not possible to determine whether such a rift zone was isochronous since the age of the Finnmark intrusions is uncertain; no isotopic data are available and the field relationships can only suggest age limits between early Vendian and late Mid Cambrian. The dykes are very similar geochemically to the Kongsfjord dyke swarm (Roberts 1975) which have been dated at 640 ± 19 Ma (Beckinsale et al. 1976), comparable with that of the Ottfjäll dolerites (Claesson and Roddick this volume) although the isotopic chemistry of the latter is complex. Roberts (1972) however has shown that the Kongsfjord dykes were intruded during the first phase of deformation in the country rocks which makes an extensional rift zone origin for these dykes less likely. The field relationships of the analysed Finnmark dykes cannot preclude an early D_1 intrusion age (see Fig. 3). However, it is difficult to reconcile their transitional MORB characteristics with intrusion through the thickened continental crust required for the gravitational spreading model suggested for the Finnmarkian deformation (Chapman et al. this volume).

The Finnmarkian deformation is broadly coeval with the major plutonism of the Seiland Province, the individual components of which intrude the highest Kalak nappes of west Finnmark syntectonically. Robins and Gardner (1975) documented the magmatic evolution of the complex, starting with the development of low K tholeiitic basalts during D_1, passing through high K calc-alkaline magmas into alkaline olivine basalts and highly differentiated alkaline magmas and carbonatites in D_2.

They argued that this sequence of magmas originated from successively deeper levels in the mantle and suggested magma generation above a subducted slab of oceanic lithosphere, penetrating progessively deeper into the mantle. They required a steepening subduction zone with time to explain the constant position of successive intrusions into the overlying Finnmarkian nappes. An alternative model to explain both the deep mantle source and the constant location of the Seiland Province has been invoked by Sturt (personal communication 1981) who suggested a deep mantle plume beneath the continental margin. Both models satisfactorily explain the Finnmarkian deformation; the rise of magma increasing the geothermal gradient, causing thickening of the overlying continental crust and thus initiating nappe translation by a gravity spreading or fluid welt process (Fig. 9b).

Following the Finnmarkian deformation, the continental margin appears to have become stabilized with considerable uplift and erosion of the Finnmarkian nappes so that the younger Ordovician/Silurian shallow marine clastic and carbonate sequences were deposited unconformably on both Finnmarkian complex basement in ensialic basins—the Skibotn N.C., and, at least partially on oceanic crust, the Ullsfjord N.C. (Minsaas and Sturt this volume). This study suggests that the Lyngen Gabbro and associated Kjosen amphibolites represent primitive mantle melts consistent with an interpretation by Minsaas and Sturt (this volume) that they are a fragment of oceanic crust underlying the Ullsfjord sediments.

Although the Silurian succession of the Magerøy Nappe of north Finnmark contains no volcanic rocks, the Ordovician–Silurian sequences of Troms are interbedded with lavas and pyroclastic sediments showing a variety of chemical affinities. Those in the ensialic Skibotn sequence show close similarities to MORB, whereas those in the partially ensimatic Ullsfjord sequences show only weakly tholeiitic chemistry with transitional island arc/within-plate basalt characteristics. A palaeogeographic model to explain these relationships can be attempted by restoring the nappes to their relative pre-Scandian deformation positions. The Ullsfjord succession will lie to the west of the Skibotn succession separated by the oceanic/continental margin. The model (Fig. 9c) invokes a destructive plate margin to the west with both lava sequences generated above a subduction zone in areas of back-arc rifting (cf. Pearce 1980). In the case of the Ullsfjord lavas, the varied chemistry of individual sheets suggests more than one magma source, possibly with some material derived from early phases of subduction generated magma whilst other lavas were related to magmas formed in a back-arc rifting process. Back-arc rifting within the continental margin would be expected to produce rapid facies changes and thickness variations. P. R. Johnson (personal communication 1978) has interpreted the Silurian Sagelvvatn marble as a dolomite carbonate sabkha similar to present day Red Sea depositional systems. The presence of local sulphide mineralization may suggest hydrothermal brine systems (Emery et al. 1969), leaching from new oceanic crustal basalt or evaporite associated stratiform deposits (Renfro 1974).

The island arc complex appears to be poorly preserved but evidence of its existence comes from rare exposures of metarhyolite and partially recrystallized andesite porphyry along eastern Ullsfjord (Binns in preparation), as well as from clasts of typical island arc igneous material in the conglomerates said to overly the Lyngen Gabbro (Minsaas and Sturt this volume).

The calc-alkaline nature of late syn-tectonic dykes in both the Ullsfjord and Skibotn N.C. suggests that whilst early igneous activity was transitional in character, later activity developed a definite calc-alkaline island arc/continental margin affinity perhaps indicating the evolution of a mature subduction zone in association with early stages of Scandian deformation. Destruction of ocean crust along this subduction zone would have closed Iapetus and brought about the continent/continent collision that generated the main Scandian deformation.

The significance of the ?early Caledonian Tromsø N.C. is uncertain. It may represent a microcontinent separated from the Baltic Shield by marginal ocean spreading following the Finnmarkian orogenic phase, or it may represent part of the Laurentian margin emplaced after Iapetus closure during the main Scandian event. The latter interpretation is implicit in Fig. 9c since Iapetus and the subduction zone along which it was destroyed would have lain to the west of the Tromsø N.C. microcontinent in the former. However, we have very little evidence to favour one solution over the other.

The Ordovician/Silurian successions of the Troms nappes can be followed southwards around the Rombak window into Nordland. Here, in the Helgeland, N.C. Gustavson (1978) has described a number of greenstone units showing chemical characteristics of island arc and marginal basin development. The underlying Køli N.C. also displays a wide range of volcanic units that have been related to island arc, marginal basin, and arc splitting environments (e.g. Stephens this volume). At present it is not possible to correlate the individual tectonic units of Nordland, Norbotten, and Västerbotten with those of Troms at more than the most general level. However, it is clear that the complex development of late Ordovician/Silurian island arcs and related mar-

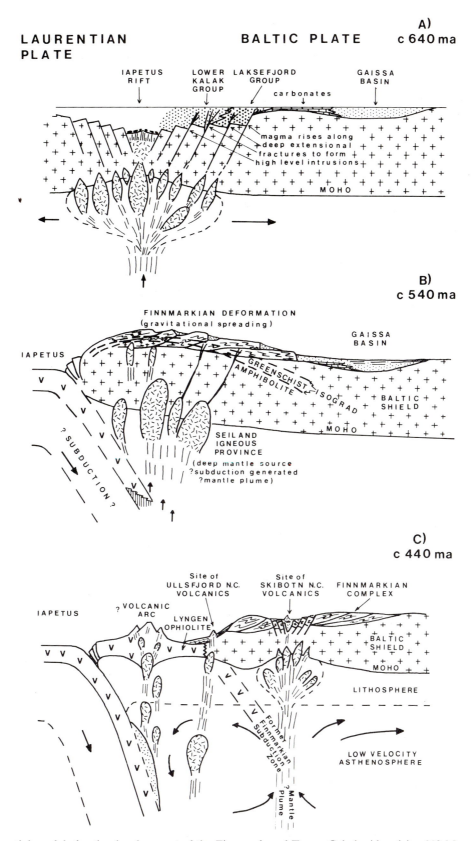

Fig. 9   Sequential models for the development of the Finnmark and Troms Caledonides. (a) *c*.640 Ma—rifting of the Laurentian/Baltic continental shield with dyke injection of MORB/WPB into distended continental crust and into the overlying Kalak and Laksefjord sediments. (b) *c*. 540 Ma—plutonic emplacement of the Seiland Igneous Province during the continuing Finnmarkian Phase deformation. The presence of the subduction zone to the west is speculative after Robins and Gardner (1975), the Seiland Province magmas may have been generated above a deep mantle plume. (c) *c*. 440 Ma—the development of a second subduction zone west of the continent (Baltic Shield)/ocean (Iapetus) passive margin. The Finnmarkian Complex has been incorporated within the Baltic Shield and the associated Finnmarkian S.Z. is defunct. See text for discussion of the sites of the Skibotn and Ullsfjord N.C. volcanites and of the Lyngen ophiolite and volcanic arc

ginal basins recognized in the Central Scandinavian Caledonides continued northwards to at least the Troms area along the Iapetus margin of the Balto-scandian Plate.

Acknowledgements

The work in Finnmark and N. Troms was funded by research grants to RAG from NERC and the Royal Society. T. J. C. received an NERC research studentship and R. J. H. wishes to acknowledge financial assistance from U.C. Cardiff. We are grateful to Drs. R. E. Bevins and S. Scott for helpful discussion arising from an early draft. We wish to thank Drs. M. Bennett, J. G. Holland, and D. Robinson for assistance with XRF analyses and Margaret Millen for drafting the geochemical diagrams.

References

Allegre, C. J., Treuil, M., Minster, J. F., Minster, J. B. and Albarede, F. 1977. Systematic use of trace elements in igneous processes. I, Fractional crystallization in volcanic suites. *Contrib. Mineral. Petrol.*, **60**, 57–75.

Beckinsale, R. D., Reading, H. G. and Rex, D. C. 1976. Potassium–argon ages for basic dykes from East Finnmark: stratigraphical and structural implications. *Scott. J. Geol.*, **12**, 51–65.

Binns, R. E. 1978. Caledonian nappe correlation and orogenic history in Scandinavia north of lat. 67°N. *Bull. geol. Soc. Am.*, **89**, 1475–1490.

Binns, R. E. and Gayer, R. A. 1980. Silurian or Upper Ordovician fossils at Guolasjav'ri Troms, Norway. *Nature*, London, **284**, 53–55.

Binns, R. E. and Matthews, D. W. 1981 Stratigraphy and structure of the Ordovician–Silurian Balsfjord Supergroup, Troms, North Norway. *Norges geol. Unders.*, **365**, 39–54.

Binns, R. E. and Taylor, P. N. In preparation. Caledonian orogenic events in Troms, Northern Norway.

Bjørlykke, A. and Olaussen, S. 1981. Silurian sediments and volcanites in the Sagalvvatn area, Balsfjord, Northern Norway. *Norges geol. Unders.*, **365**,

Bøe, P. 1976. Geology of the troctolite complex of Raisduodar–Hal'di, Troms, northern Norway. *Norges geol. Unders.*, **324**, 29–46.

Chapman, T. J. 1980. *The Geological Evolution of the Laksefjord Nappe Complex, Finnmark, N. Norway.* Unpub. thesis, Univ. of Wales.

Chapman, T. J., Gayer, R. A. and Williams, G. D. this volume. Structural cross-sections through the Central Finnmark Caledonides and timing of the Finnmarkian event..

Claessòn, S. and Roddick, J. C. 1983. ^{40}Ar–^{39}Ar data on the age and metamorphism of the Ottfjället dolentes, Särv Nappe, Swedish Caledonides. *Lithos*, **16**, 61–73.

Emery, K. O., Hunt, J. M. and Hayes, E. E. 1969. Summary of hot brines and heavy metal deposits in the Red Sea. In Degeus, E. T. and Ross, D. A. (Eds), *Hot Brines and Recent Heavy Metal Deposits in the Red Sea:*

a Geochemical and Geophysical Account, Springer-Verlag, New York, 557–571.

Floyd, P. A. and Winchester, J. A. 1975. Magma type and tectonic setting discrimination using immobile elements. *Earth planet. Sci. Lett.*, **27**, 211–218.

Føyn, S. 1960. Tanafjord to Laksefjord. In Dons, J. A. (Ed.), *Aspects of the Geology of Northern Norway, Guide to Excursion A3, Int. geol. Cong. XXI*, pp. 45–55.

Gayer, R. A., Powell, D. B. and Rhodes, S. 1978. Deformation against metadolerite dykes in the Caledonides of Finnmark, Norway. *Tectonophysics*, **46**, 99–115.

Gayer, R. A. and Roberts, J. D. 1971. The structural relationships of the Caledonian nappes of Porsangerfjord, W. Finnmark, N. Norway. *Norges geol. Unders.*, **269**, 21–67.

Gayer, R. A. and Roberts, J. D. 1973. Stratigraphic correlation of the Finnmark Caledonides and possible tectonic implications. *Proc. geol. Assoc.*, **84**, 405–428.

Gee, D. G. 1975a. A tectonic model for the central part of the Scandinavian Caledonides. *Am. J. Sci.*, **274A**, 468–515.

Gee, D. G. 1975b. A geotraverse through the Scandinavian Caledonides—Östersund to Trondheim. *Sver. geol. Unders.*, **C717**, 1–66.

Gustavson, M. 1978. Geochemistry of the Skalvaer greenstone, and a geotectonic model for the Caledonides of Helgeland, north Norway. *Norsk geol. Tidsskr.*, **58**, 161–174.

Hausen, H. 1942. Das Halditjokkomassiv. *Acta Acadeimiae Aboensis, Math et Physica*, **13** (14).

Harland, W. B. and Gayer, R. A. 1972. The Arctic Caledonides and earlier oceans. *Geol. Mag.*, **109**, 289–314.

Humphreys, R. J. 1981. *The Geology of Northwest Ullsfjord, Troms, Norway and its Regional Caledonian Implications.* Unpub. thesis, Univ. of Wales.

Irvine, T. N. and Baragar, W. R. A. 1971. A guide to the chemical classification of the common volcanic rocks. *Can. J. Earth Sci.*, **8**, 523–548.

Jamieson, B. J. and Clarke, D. B. 1970. Potassium and associated elements in tholeiitic basalts. *J. Petrol.*, **11**, 183–204.

Minsaas, O. and Sturt, B. A. this volume. The Ordovician–Silurian clastic sequence overlying the Lyngen Gabbro Complex and its environmental significance.

Munday, R. J. C. 1970. *The Geology of the Northern Part of the Lyngen Peninsula, Troms, Norway.* Unpubl. thesis Univ. Newcastle.

Miyashiro, A. and Shido, F. 1975. Tholeiitic and calc-alkaline series in relation to the behaviour of titanium vanadium, chromium and nickel. *Am. J. Sci.*, **275**, 265–277.

Olaussen, S. 1977. Palaeozoic fossils from Troms, Norway. *Norsk geol. Tidsskr.*, **56**, 457–459.

Padget, P. 1955. The geology of the Caledonides of the Birtavarre region, Troms, northern Norway. *Norges geol. Unders.*, **192**, 1–107.

Pearce, J. A. 1973. *Some Relationships Between the Geochemistry and Tectonic Setting of Basic Volcanic Rocks.* Unpubl. thesis Univ. East Anglia.

Pearce, J. A. 1975. Basalt geochemistry used to investigate past environment on Cyprus. *Tectonophysics*, **25**, 41–67.

Pearce, J. A. 1976. Statistical analysis of major element patterns in basalts. *J. Petrol.*, **17**, 15–43.

Pearce, J. A. 1980. Geochemical evidence for the genesis

and eruptive setting of lavas from Tethyan ophiolites. *Proc. Int. Ophiolite Symp. Cyprus 1979.*

Pearce, J. A. and Cann, J. R. 1973. Tectonic setting of basic volcanic rocks determined using trace elements analysis. *Earth planet. Sci. Lett.*, **19**, 290–300.

Pearce, J. A. and Gale, G. H. 1977. Identification of ore deposition environments from trace element geochemistry of associated host rocks. In *Volcanic Processes in Ore Genesis, Inst. Mining and Metallurgy and Geol. Soc. London*, 14–24.

Pearce, J. A. and Norry, M. J. 1979. Petrogenetic implications of Ti, Zr, Y, and Nb variations in volcanic rocks. *Contrib. Mineral. Petrol.*, **69**, 33–47.

Quenardel, J-M. 1978. Géologie de la rive orientale du Lyngenfjord (Caledonides de Norvège du Nord). *103ᵉ Congres national des societes savantes, Nancy, Sciences*, **4**, 55–65.

Quenardel, J.-M, Ploquin, A., Dangla, P. and Sonet, J. 1981. Geochronological evidence for Precambrian and Silurian elements in the northern Scandinavian Caledonides. *Terra Cognita*, **1**, 66.

Ramsay, D. M. 1973. Possible existence of a stillborn marginal ocean in the Caledonian orogenic belt of North-West Norway. *Nature Phys. Sci.*, **245**, 107–109.

Ramsay, D. M. and Sturt, B. A. 1976. The synmetamorphic emplacement of the Magerøy Nappe. *Norsk geol. Tidsskr.*, **56**, 291–307.

Ramsay, D. M., Sturt, B. A., Zwaan, K. B. and Roberts, D. this volume. Caledonides of northernmost Norway.

Randall, B. A. O. 1971. An outline of the geology of the Lyngen peninsula, Troms, Norway. *Norges geol. Unders.*, **269**, 68–71.

Renfro, A. R. 1974. Genesis of evaporite-associated stratiform metalliferous deposits—a sabkha process. *Econ. Geol.*, **69**, 33–45.

Roberts, D. 1968. The structural and metamorphic history of the Langstrand Finfjord area, Sørøy, Northern Norway. *Norges geol. Unders.*, **253**, 1–160.

Roberts, D. 1975. Geochemistry of dolerite and metadolerite dykes from Varanger Peninsula, Finnmark, northern Norway. *Norges geol. Unders.*, **322**, 55–72.

Robins, B. and Gardner, P. M. 1975. The magmatic evolution of the Seiland Province, and Caledonian plate boundaries in northern Norway. *Earth planet. Sci. Lett.*, **26**, 167–178.

Saunders, A. D., Tarney, J. Stern, C. R. and Dalziel, I. W. D. 1979. Geochemistry of Mesozoic marginal basin floor igneous rocks from southern Chile. *Bull. geol. Soc. Am.*, **90**, 237–258.

Solyom, Z., Gorbatschev, R. and Johansson, I. 1979. The Ottfjäll Dolerites: geochemistry of the dyke swarm in relation to the geodynamics of the Caledonide orogen of Central Scandinavia. *Sver. geol. Unders.*, **C756**, 1–38.

Stephens, M. B. 1982. Field relationships, petrochemistry and petrogenesis of the Stekenjokk volcanites, central Swedish Caledonides. *Sver. geol. Unders.*, **C786**, 111 pp.

Stephens, M. B. Furnes, H., Robins, B. and Sturt, B. A. this volume. Igneous activity within the Scandinavian Caledonides.

Sturt, B. A., Pringle, I. R. and Ramsay, D. M. 1978. The Finnmarkian phase of the Caledonian orogeny. *J. geol. Soc. London*, **135**, 597–610.

Sturt, B. A. and Roberts, D. 1978. Caledonides of Northernmost Norway (Finnmark). In Tozer, E. T. and Schenck, P. E. (Eds), *Caledonian—Appalachian Orogen of the North Atlantic Region. Geol. Surv. Can.*, **78–13**, 17–23.

Thirlwall, M. F. 1981. Implications for Caledonian plate tectonic models of chemical data from volcanic rocks of the British Old Red Sandstone. *J. geol. Soc. Lond.*, **138**, 123–138.

Zwaan, K. B. and Roberts, D. 1978. Tectonostratigraphic succession of the Finnmarkian nappe sequence, North Norway. *Norges geol. Unders.*, **343**, 53–71.

The Caledonide Orogen—Scandinavia and Related Areas
Edited by D. G. Gee and B. A. Sturt

A tectonic model for the evolution of the eugeoclinal terranes in the central Scandinavian Caledonides

M. B. Stephens and **D. G. Gee**

Sveriges geologiska undersökning, Box 670, S–75128 Uppsala, Sweden

ABSTRACT

The structure of the Caledonides in Scandinavia is dominated by thrust nappes emplaced from west to east onto the Baltoscandian platform during Silurian–Devonian (Scandian) orogeny. In the central part of the Scandes, biostratigraphic control of the successions in the nappes is more complete than elsewhere in the mountain belt, providing particularly favourable conditions for interpreting the tectonic evolution. The nappes can be readily divided into two groups. Lower units (including the Lower and Middle Allochthons and probably the Seve Nappes of the Upper Allochthon) contain successions derived from the Baltoscandian platform and miogeocline and provide evidence for the late Proterozoic and Cambrian development of a passive continental margin, followed by Ordovician and Silurian instability. The passive margin during the Cambrian extended at least 400 km west of the present thrust front. The higher allochthonous units were derived from terranes that lay outboard of the Baltoscandian miogeocline; their relationships to Baltica are uncertain. These suspect terranes represent both eugeoclinal (Köli Nappes of the Upper Allochthon) and miogeoclinal (Uppermost Allochthon) environments, the former including (from east to west) the Virisen, Gjersvik, and Hølonda terranes and the latter being inferred to have been a part of Laurentia.

The opening of the Iapetus Ocean during the Cambrian and probably also the Vendian was followed by closure in the Ordovician; the eugeoclinal successions provide evidence for development of both oceanic crust and subsequent volcanic arcs and related basins. The Hølonda terrane is composite, being composed essentially of an ophiolitic basement that was obducted to high structural levels close to or on a continental margin in the earliest Ordovician and thereafter overlain by a calc-alkaline arc complex and back-arc basin succession. The Arenig–Llanvirn faunas in this terrane suggest that this margin belonged to Laurentia. The Gjersvik terrane is dominated by a tholeiitic arc complex built on an ensimatic basement and with both fore- and back-arc facies represented. This arc is inferred to have developed as a result of easterly-directed subduction probably related to the obduction of the ophiolites in the Hølonda terrane along the western margin of the Iapetus Ocean. The Gjersvik terrane is thought to have been accreted to the Laurentian margin prior to the intrusion of mid-late Ordovician granitoids. The deformation and metamorphism that preceded this intrusion is thought to be related to the accretion of the Gjersvik and Hølonda terranes to Laurentia. The Virisen terrane of arc volcanites and craton-derived clastic sediments lay outboard of the Baltoscandian miogeocline on the eastern side of the Iapetus Ocean and may have accreted to Baltica during early and mid Ordovician time. Thus, closure of the Iapetus Ocean occurred by B-subduction outboard of both margins followed by A-subduction involving continent-arc collision along the Laurentian and possibly also the Baltoscandian margins.

By the beginning of the Silurian, successions along the Baltoscandian margin and in the Virisen and Gjersvik terranes, are so similar that proximity is inferred. This sharing of a depositional basin suggests the elimination of the Iapetus Ocean, prior to collision of Baltica and Laurentia and the onset of Scandian deformation. During the mid Silurian to Devonian, the accreted Laurentian and western eugeoclinal terranes with their early Palaeozoic intrusive complexes were thrust onto the Virisen terrane and Baltoscandian miogeocline and platform. High-pressure metamorphism ensued in the western part of the Baltoscandian platform. These relationships suggest A-subduction of Baltica westwards beneath an active Laurentian margin. During the final stages of this emplacement, an Old Red Sandstone facies was deposited in both extra- and intramontane basins.

Introduction

The Scandinavian Caledonides flank the western margin of the Fennoscandian Shield over a distance of c. 1800 km. Within the orogen, as it is now exposed in the Scandes, a cycle of tectonic evolution is recorded that started in the late Proterozoic, culminated in the Ordovician and Silurian and ended in the Devonian. The structure of the orogen is dominated by a succession of thrust nappes, transported eastwards onto the Baltoscandian platform in the late Silurian and early Devonian (Scandian deformation). Older units occur in the orogen, both as crystalline basement beneath the younger successions and as allochthonous sheets incorporated in the nappes.

The last decade has witnessed a rapid advance in the general understanding of the orogen, an advance as fundamental and revolutionary as the introduction of the nappe hypothesis nearly one hundred years ago. Törnebohm (1888, 1896) provided evidence that the higher units now occurring in the Scandian front in Sweden were allochthonous; he inferred that they were derived by thrusting from areas c. 100 km further west in Norway. This hypothesis met with considerable opposition, but found support in subsequent mapping and correlation. It provided the basis for a model that regarded the Norwegian west coast gneisses as the core of the orogen, a model that persisted into the 1960's (Ramberg 1966) with the concept of a buoyant mobile core rising and spreading laterally to provide, by extension, the long-transported nappes in the mountain front.

A more radical hypothesis, first entertained by Högbom (1910), suggests that the higher Scandian Nappes are derived from unknown distances west of the Norwegian coast. This hypothesis was supported by Holtedahl (1936) for the displacement of the Jotun Nappe and by Asklund (1938) in the general tectonic context of the central Scandinavian Caledonides. Further north, in the Lofoten–Akkajaure profile, Kautsky (1946, 1953) likewise provided evidence favouring several hundreds of kilometres of transport for the higher tectonic units. Nevertheless, it was not until the 1970's that compelling evidence favouring Högbom's hypothesis was assessed and ventilated in sufficient detail to become generally accepted. With its acceptance came the realization that the rock units in the Scandes can be separated into two categories, those that are directly relatable to the development of the Baltoscandian continental margin and those (occurring in the higher nappes) that were derived from environments west of this margin.

Early attempts to interpret the late Proterozoic and early Palaeozoic evolution in Scandinavia in terms of ocean development and subsequent closure (Wilson 1966; Dewey 1969) provided an invaluable stimulus for work in the 1970's. These models raised fundamental questions concerning regional structure, igneous rock affinities, and faunal provincialism, subjects that were extensively debated during the following decade. Initial attempts to outline evolutionary models of this kind are inevitably open to criticism for oversimplification. Provided they are anchored in the rocks, they nevertheless provoke rethinking of old problems and the search for relevant new data.

Significant advances during the 1970's occurred in five main fields:

1. Mapping and correlation in various parts of the mountain belt established Högbom's (1910) hypothesis that the higher nappes were derived from west of the Norwegian coast. This led to the production of the new tectonic map of the Scandes which accompanies this volume. The tectonostratigraphy (Roberts and Gee, this volume), includes four major nappe complexes, the Lower, Middle, Upper, and Uppermost Allochthons, each representing related environments that are distinguishable in terms of history of sedimentation, deformation, metamorphism, and, in some cases, igneous activity.

   Evidence is presented here indicating that the Baltoscandian continental margin is represented by the upper Proterozoic and lower Palaeozoic rocks occurring in the Autochthon, Parautochthon, Lower and Middle Allochthons, and probably also the lower part (Seve Nappes) of the Upper Allochthon. Terranes developed outboard of this platform and miogeocline occur in the Uppermost and upper part (Köli Nappes) of the Upper Allochthon and are of highly 'suspect' (Coney et al. 1980; Williams and Hatcher 1982; Keppie, this volume) character. The rocks in the Uppermost Allochthon represent a continental margin related to a western continent (Laurentia?) while the rocks in the Köli Nappes of the Upper Allochthon comprise the eugeocline.

2. Intensive field analysis of the volcanosedimentary complexes of the Upper Allochthon, coupled with geochemical investigations of basic igneous rocks, particularly their trace and REE chemistry, resulted in the identification of several ophiolite complexes and a variety of volcanic island-arc associations.

3. Major unconformities were identified within the Upper Allochthon, with deposition of upper Ordovician successions on deformed and metamorphosed volcanosedimentary complexes (including ophiolites). The age of this deformation remains to be established but, in several cases, it probably occurred in the early Ordovician.

4. Precambrian crystalline sheets were identified within the amphibolite facies and higher grade units in the Upper and Uppermost Allochthons. Radiometric age-determination work has provided much of the evidence.

5. Palaeontological investigations, particularly of the faunas in the Upper Allochthon, have resulted in the discovery of new fossils and allowed a reassessment of the evidence for faunal provincialism. The North American affinities of the faunas in the lower and middle Ordovician successions deposited on the ophiolites of the Støren Nappe at the top of the Upper Allochthon have been reaffirmed. The fauna in the Otta detrital serpentinites, lying in a lower tectonic unit in the middle of the Upper Allochthon, has been shown to be of mixed North American and Baltoscandian affinities.

These various lines of evidence have persuaded us to reject the hypothesis (Gale and Roberts 1974; Furnes et al. 1976; Gustavson 1978a; Roberts 1980) that assumes Baltoscandian affinities for the entire nappe pile. The model presented here expands on that presented by one of us (Gee 1975a, 1978) by summarizing briefly the evidence for the Baltoscandian platform and miogeocline and analysing in considerably more detail the evidence from the terranes inferred to have been located outboard of this miogeocline prior to thrust emplacement on to the platform. Fundamental to our model is the convic-

tion that no assessment of the pre-Silurian development of the orogen can usefully be attempted without first restoring the various allochthonous units to their positions prior to Scandian deformation.

We restrict our evidence to the central part of the Scandes, presenting data from north and south of the Grong–Olden Culmination (Fig. 1). General descriptions of these areas are to be found in Gee et al. and Stephens et al. (a) elsewhere in this volume. In this paper, we often have found it convenient to refer to the location of terranes in terms of their present geographical orientation (e.g. the eugeocline lay to the west of the Baltoscandian platform). This practice should not influence interpretation of the early Palaeozoic relationships. All Rb–Sr whole-rock isochron ages quoted here have been recalculated to $\lambda^{87}Rb = 1.42 \times 10^{-11} a^{-1}$. Discussions of the age-determinations in relation to the early Palaeozoic timescale are based on McKerrow et al. (1980).

The Baltoscandian Platform and Miogeocline

Thin Vendian and/or lower Cambrian quartzites and middle and upper Cambrian uraniferous black shales occur in the autochthonous sedimentary veneer overlying the Precambrian crystalline basement throughout most of the Caledonian thrust front in the Scandes. Locally, the black shales are overlain by Ordovician limestones. This succession (Thorslund 1960; Bergström 1980) passively underlies the sole thrust of the Lower Allochthon; it has been traced locally by drilling, at least 30 km west of the thrust front (Gee et al. 1978 and unpublished SGU data). Further west, in the windows, similar successions have been identified (Gee 1974, 1980; Tirén in Gee and Kumpulainen, 1980; Sjöström in Gee and Kumpulainen 1980) and it is apparent that, during the Vendian and Cambrian, the Baltoscandian platform extended westwards at least to coastal Norway.

Successions in the Lower Allochthon, included within the Jämtland Supergroup (Gee 1975b), likewise contain Cambrian uraniferous black shales underlain by quartzites (Fig. 2). The latter are thicker than in the underlying tectonic units and, in some areas, are themselves underlain by tillites and continental and shallow marine sandstones (Kumpulainen 1982) of pre-Vendian, probably late Proterozoic age. The Cambrian shales give way upwards via limestones to westerly-derived greywackes of mid Ordovician age. The instability reflected in the deposition of the latter apparently ceased in the late

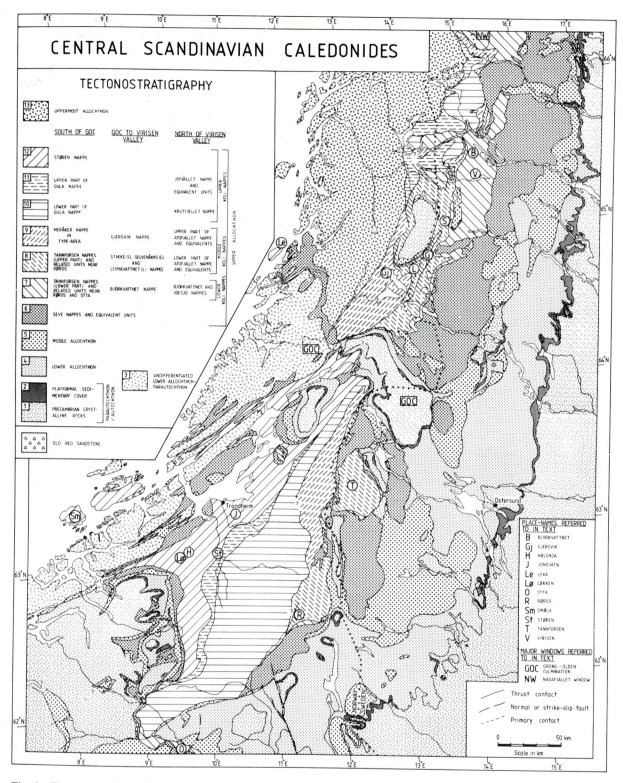

Fig. 1 Tectonostratigraphic map of the central Scandinavian Caledonides (based on Gee *et al.*, this volume and Stephens *et al.*, this volume, a)

Ordovician with general regression and shallow marine deposition, prior to renewed transgression in the Llandovery. Deepening of the basin in the late Llandovery with deposition of black shales and then greywackes apparently reflected renewed instability

further west, in this case with the advance of the Scandian Nappes.

Comparison of the successions in the Lower Allochthon with those deposited on the parautochthonous basement in the deepest tectonic levels of the windows, suggests that the allochthonous sequences were deposited in westernmost Jämtland and Trøndelag. However, exposure of the Jämtland Supergroup is generally poor and has not allowed assessment of stratal shortening based on the construction of balanced sections (cf. Hossack *et al.*, this volume). The extent to which the parautochthonous basement may have been thrust eastwards also remains uncertain. Hence, there is considerable uncertainty in the palinspastic reconstructions. However, it is important to note that in the Bottenviken Nappe, referred by Tirén (in Gee and Kumpulainen 1980) to the upper part of the Lower Allochthon, uraniferous black shales occur underlain by quartzites. These shales, of probable Cambrian age, have been thrust from westernmost Trøndelag or further west and testify to the existence, during the Cambrian, of a passive continental margin extending to west of the present Norwegian coast.

In the overlying units of the Middle Allochthon, the successions (Tossåsfjället Group, Kumpulainen 1980) include tillites of probable Vendian age but do not extend upwards into a black shale facies and are therefore thought to be probably of pre-mid Cambrian age. Elsewhere in the orogen, radioactive black shales have been reported (Bergström and Gee this volume) in this tectonic unit. The Tossåsfjället Group successions testify to subsidence and rifting of the Baltoscandian margin during the late Proterozoic. The highest unit in the Middle Allochthon (the Sarv Nappes, Strömberg 1961, 1969) contains an extensive dolerite dyke swarm of largely tholeiitic mid-ocean ridge basalt (MORB) but partly alkaline affinities (Solyom *et al.* 1979a) and probable Vendian age (Claesson 1976; Claesson and Roddick 1983). This basic intrusive activity is thought to be related to incipient development of the Iapetus Ocean (Gee 1975a; Solyom *et al.* 1979a). Palinspastic restoration of the Lower and Middle Allochthons suggests (Gee 1978) that the Baltoscandian miogeocline extended off-shore the present coast of Norway. Outer units were emplaced distances of at least 300 km onto the platform.

Within the central Scandes, there is a prominent tectonic break between the Särv Nappes of the Middle Allochthon and the overlying Seve Nappes (Zachrisson 1973) of the Upper Allochthon. The metamorphic grade increases upwards into the Seve Nappes to amphibolite and even granulite facies (Arnbom 1980). Claesson's age-determination work (1982) on high-grade rocks within the middle part of the Seve Nappes suggests that the metamorphism is essentially of Caledonian (Scandian) age (c. 415 Ma), but that Proterozoic material is present in the nappes. Both the lower and upper parts of the Seve

Nappes are composed of schistose feldspathic psammites and amphibolites, the latter being similar in composition (Solyom *et al.* 1979b) to the dolerites in the underlying Särv Nappes. It has been suggested (Gee 1975a) that the outer margin of the Baltoscandian miogeocline may be represented in the Seve units. This hypothesis finds support in evidence from northern Jämtland (van Roermund this volume) and further north (Norrbotten) in the orogen (Andréasson *et al.* this volume), where there appears to be a transition upwards from the Särv into the Seve Nappes. In Norrbotten, dolerite-intruded sandstones pass upwards via schistose psammites and metadolerites into Seve Nappe psammites, corona dolerites, and amphibolites, and finally into psammites, corona dolerites, and eclogites. These relationships have been interpreted (Andréasson *et al.*, this volume) to be the result of depression of the miogeocline during orogenesis, high pressure metamorphism and subsequent thrust emplacement onto the Baltoscandian platform.

The Seve Nappes locally contain ultramafites, usually in the upper parts close beneath the Köli Nappes, but also occasionally in the lower units. They generally occur as solitary peridotites or dunites (Stigh 1979) marginally serpentinized, and apparently occurring randomly in the surrounding metasediments. They were emplaced prior to the high-grade metamorphism of the Seve Nappes and thrusting onto the Baltoscandian platform. Their mode of emplacement is treated further below.

The Eugeoclinal Development of the Köli Nappes

General

The Seve Nappes are overlain by greenschist facies volcanosedimentary units of Silurian, Ordovician, and perhaps Cambrian age, occurring in the Köli Nappes. Törnebohm (1896) referred to the Köli all the units above the Seve that are now included in the Upper Allochthon and this practice is followed here. The Köli Nappes are extensively developed north and south of the Grong–Olden Culmination. Correlation of individual tectonic units over this basement high remains a subject of some controversy but there is general agreement that the lower nappes are best developed north of the culmination in northern Jämtland and Västerbotten and the upper units most fully represented in the south in Sør Trøndelag.

South of the Grong–Olden Culmination, the Köli Nappes occur in the Tännforsen area (Beckholmen 1978, 1982) and in Trøndelag in the Trondheim Nappe Complex (Wolff and Roberts 1980). The latter has been subdivided into the Meråker, Gula, and

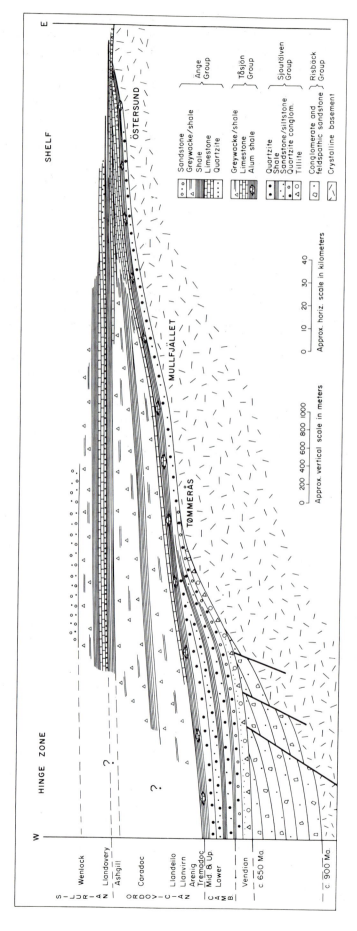

Fig. 2 Pre-tectonic reconstruction of the Jämtland Supergroup stratigraphy

Støren Nappes. The internal structure of the Trondheim Nappe Complex is much disputed (Gee 1975b; Gee et al., this volume); whereas some authors have interpreted the major internal structure in the complex to be antiformal and have correlated the volcanosedimentary successions of the Meråker and Støren Nappes (Roberts and Wolff 1980), others regard these three major units as independent allochthons (Gee and Zachrisson 1974; Dyrelius et al. 1980), the island-arc complex of the Meråker Nappe (Grenne and Lagerblad, this volume) at the base being separated from the ophiolite-bearing Støren Nappe (Grenne et al. 1980) by the largely metasedimentary Gula complex. This interpretation has been supported by experience north of the Grong–Olden Culmination where much of our recent work on the Köli Nappes has been concentrated.

In the area straddling the border between Norway and Sweden, from the Grong–Olden Culmination to the Nasafjället Window, a variety of nappes containing metamorphosed igneous and sedimentary associations have been recognized (Foslie and Strand 1956; Zachrisson 1969; Halls et al. 1977; Stephens 1977a; Häggbom 1978; Ramberg 1981) that fall readily into three groups, referred to as the Lower, Middle, and Upper Köli Nappes (Stephens 1980). The general geology has been described elsewhere in this volume (Stephens et al., this volume, a) as has the geochemistry of the igneous rocks (Stephens et al., this volume, b) and this contribution treats briefly the evidence from the different tectonic units that is particularly relevant to the interpretation of the tectonic evolution.

Diagnostic fossils are seldom preserved in the successions of the Köli Nappes although crinoids are not infrequent in limestones. Only in the lowest (Björkvattnet Nappe and probable equivalents at Otta) and highest (Støren Nappe) units are the faunas well enough preserved to control the stratigraphic interpretation decisively. Deformation and metamorphism are polyphase. Lithostratigraphic successions have been established, the interpretation of the way up of which usually depending on local sedimentological criteria, the preservation of an occasional fossil, isolated isotopioc age-determinations, studies of metal zonation in massive sulphide deposits, etc. Stratal thicknesses vary greatly, very often the result of tectonic repetition or thinning. Nevertheless, it is possible to interpret and correlate the various stratigraphies in the nappes and on this basis to build up alternative models for the evolution of the eugeocline.

The Köli Nappes contain a complex diversity of island-arc, back-arc, fore-arc, and oceanic environments. In the Middle and Upper Köli Nappes, various segments across a major island-arc complex have been recognized. Without the evidence of faunal affinities, the rocks in these higher units could represent environments from any part of the Iapetus Ocean and its marginal islands and basins. Our interpretation is guided first and foremost by the evidence that obduction of the ophiolites in the highest Köli unit, the Støren Nappe, occurred to a location that was exclusively inhabited by North American faunas and, therefore, inferred to be closely related to the Laurentian margin. Subsequent build-up of a volcanic arc environment on that margin occurred with an oceanic segment still providing considerable separation from the Baltoscandian margin (Bruton and Harper, this volume). The mixed faunas occurring in the isolated environments of the serpentinite islands (Otta conglomerates, probably in the Lower Köli Nappes) also witness to the existence of this ocean, separating the rocks within the Støren Nappe from the Baltoscandian miogeocline.

In the following treatment of the Köli Nappes, the tectonic units are described (Figs. 3 and 4) from the base upwards, as they occur in the area north of the Grong–Olden Culmination. Their correlation southwards (Fig. 1) is briefly treated thereafter in as far as it influences interpretation of the age of the successions and the tectonic model.

Lower Köli Nappes

The contact between the Seve and Köli Nappes is a thrust in Jämtland and southern Västerbotten (Zachrisson 1969; Trouw 1973; Sjöstrand 1978). Further north in central Västerbotten, Stephens (1977a) described a prograde metamorphic sequence from phyllites and metavolcanites typical of the Köli Nappes downwards into psammitic schists and amphibolites typical of the upper Seve and no significant tectonic break was recognized in this area. Thus, the importance of the tectonic contact between the Seve and the Köli Nappes remains uncertain; it may exist only locally or it may be of fundamental regional importance and obscured in some areas by subsequent metamorphism. Another possibility is that it is located in certain areas such that lithologies typical of the Seve Nappes occur within the overlying Köli Nappes; such a solution also limits the regional significance of this tectonic contact.

The stratigraphy (Fig. 3) of the lowest tectonic units in the Köli (Kulling 1933; Zachrisson 1969; Stephens 1977a; Sjöstrand 1978; and Kollung 1979) is most completely preserved in Kulling's classical Björkvattnet–Virisen area where the fossiliferous successions are located. The biostratigraphical control occurs in the upper part of the sequence in which

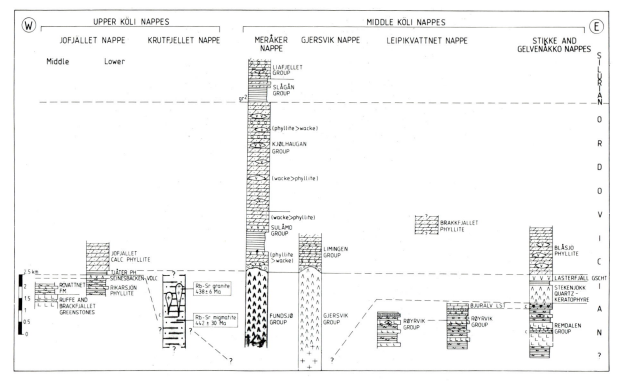

Fig. 3   Summary of stratigraphy and fossil localities in the Köli Nappes between the Grong–Olden Culmination and the Nasafjället Window, and the Meråker Nappe in its type-area south of the Grong–Olden Culmination. The sequences are shown in their inferred order of deposition and relative west–east position (distance between sequences unspecified) prior to Scandian displacement. Data sources for each nappe are based on the authors' own observations and published work referred to in the respective section in the text. Thicknesses indicated are only approximate

shallow-marine Ashgill limestones (Slätdal Formation) pass up into Llandovery black phyllites (Broken Formation) overlain by turbidites and conglomerates (Lövfjäll, Vesken, and Viris Formations). The only evidence of volcanic activity in this younger part of the succession is the presence of greenstones (locally pillow lavas) with predominantly MORB affinities (Stephens *et al.*, this volume, b) in the Llandovery black phyllites.

The Ashgill limestones usually pass down into calcareous sandstones, quartzites, and quartzite conglomerates (Vojtja Formation) which rest on a varied succession of turbidites, conglomerates, and volcanites as well as solitary and detrital serpentinites (Tjopasi Group in the west, Seima and Gilliks Formations in the east). In these pre-Ashgill successions, age control is limited to the occurrence of poorly preserved gastropods (Holmqvist 1980) in thin limestones within the detrital serpentinites and a U/Pb (zircon) age-determination (488±5 Ma) of an albitè trondhjemite associated with predominantly acid volcanites (Claesson *et al.* 1983). Further north at this tectonic level in Norrbotten, similar successions below Ashgill limestones contain tabulate corals (Kulling in Strand and Kulling 1972) suggesting a mid Ordovician age.

The detritus in the mid Ordovician turbidites and

conglomerates of the Gilliks Formation indicates erosion of a continental margin with local influx of material from a basic volcanic–ultramafic source. Wacke-phyllite ratios and wacke bed thickness in the pre-Ashgill succession attain maximum development in eastern areas towards the top of the sequence; various phyllites, locally with polymict conglomerates, dominate westwards. The presence of both acid and basic tuffs in the Gilliks Formation and the distribution of serpentinite and associated serpentinite conglomerate suggest that the subduction-related (low-K tholeiitic, LKT) igneous activity in the Tjopasi Group to the west is younger than the transitional within-plate basalts (WPB) of the Seima Formation to the east (Stephens *et al.*, this volume, b and Fig. 3).

South of the Grong–Olden Culmination (Fig. 1), equivalents to the Lower Köli Nappes occur in the western and southern parts of the Tännforsen area (Beckholmen 1978, 1982). Further south, they apparently occur in southeastern Trøndelag and in the Otta area (see discussion in Holmqvist 1980, 1982, and Bruton and Harper 1982). Bruton and Harper (1981) have redescribed a well-preserved shelly fauna (Hedström 1930) of Llanvirn age from detrital serpentinites at Otta. They have shown that this fauna exhibits mixed North American and Bal-

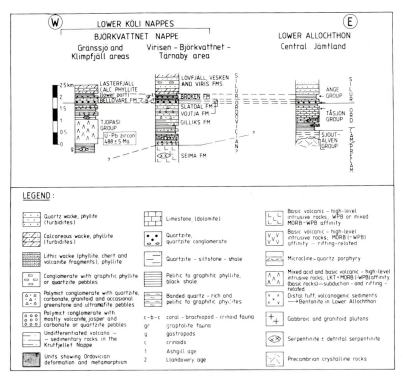

Fig. 3 *continued*

toscandian affinities. An isolated depositional environment is indicated both by the fauna and the monomict character of the detritus.

Middle Köli Nappes

A variety of nappes, at least four in number in Jämtland and Nord Trøndelag north of the Grong–Olden Culmination, are thrust over the Lower Köli units (Stephens 1982). Descriptions of the stratigraphy (Fig. 3) in the Stikke, Gelvenåkko and Leipikvattnet Nappes are provided by Zachrisson (1964, 1969), Nilsson (1964), Sjöstrand (1978) and Kollung (1979) and, in the Gjersvik Nappe, by Oftedahl (1956), Halls *et al.* (1977), Kollung (1979), and Lutro (1979).

Petrochemical data on volcanic and high-level intrusive rocks have been presented by Stephens (1977b, 1980, 1982) and Stephens *et al.* (this volume, b) for units in the Stikke and Gelvenåkko Nappes, by Gale and Roberts (1974), Olsen (1980) and Stephens *et al.* (this volume, b) for the Leipikvattnet Nappe and by Gale and Roberts (1974), Halls *et al.* (1977), Lutro (1979) and Grenne and Reinsbakken (1981) for the Gjersvik Nappe. Further north, in Västerbotten, the stratigraphy in the tectonostratigraphically equivalent and probably composite Atofjället Nappe (Fig. 1) has been described by Beskow (1929), Kieft (1952), Stephens (1977a), and Häggbom (1978). These five nappes are grouped together as the Middle Köli. They are separated from underlying and overlying units by containing similar volcanite or mixed volcanite–phyllite sequences associated with calcareous turbidites, the latter in most cases containing basic extrusions or pretectonic high-level intrusions. In Västerbotten, they are overlain, along a late- or post-metamorphic thrust, by a variety of schists, subordinate metavolcanites and syntectonic intrusions showing a distinctive history of deformation, high-grade metamorphism and intrusion which are included in the Upper Köli Nappes (see following section). Way-up criteria in the Stikke (Juve 1977; Sundblad 1980; Zachrisson 1982) and Gjersvik (Oftedahl 1956; Halls *et al.* 1977) Nappes suggest that the calcareous turbidites are younger than the volcanite and mixed volcanite–phyllite complexes and that the sequences in the individual slices of the Middle Köli are mostly inverted. Apart from the rare occurrence of crinoids in the Stikke Nappe, no fossils have been found in this group of nappes.

In eastern sequences (Stikke and Gelvenåkko Nappes), predominantly fine-grained phyllites associated with transitional basalts showing mixed

MORB–WPB affinities (Remdalen Group) pass upwards via subduction- and later rifting-related volcanic and subvolcanic rocks (Stekenjokk Quartz-Keratophyre and Lasterfjäll Greenschist) into calcareous turbidites (Blåsjö Phyllite) intruded by pretectonic, rifting-related basic rocks. Pyroclastic rocks and subvolcanic intrusions dominate the rifted low-K tholeiitic arc association of the Stekenjokk Quartz-Keratophyre. Westwards, in the overlying Leipikvattnet Nappe, there occurs a basic volcanite–phyllite complex (Røyrvik Group and Bjurälv Limestone) and calcareous turbidites (Brakkfjället Phyllite) which may be correlated with the Remdalen Group and the Blåsjö Phyllite, respectively. Lithic wacke and polymict conglomerate, the latter containing volcanite and jasper pebbles (Nilsson 1964), occur in the inferred, upper part of the Røyrvik Group (Fig. 3). Further westwards, in the overlying Gjersvik Nappe, polymict conglomerates, calcareous turbidites, basic lavas and, at least locally, alluvial sediments (Lutro 1979) stratigraphically overlie a complex dominated by basic lavas whose geochemistry suggests a rifted low-K tholeiitic arc affinity (Limingen and Gjersvik Groups). The volcanites of the Gjersvik Group have a plutonic infrastructure consisting of gabbros, diorites, trondhjemites, and granodiorites. The debris within the conglomerates of the Limingen Group can be matched directly with the igneous rocks in the Gjersvik Group; erosion of a carbonate sequence, possibly deposited around the emerging volcanic complex, is also suggested from the conglomerate detritus. The strikingly similar volcanosedimentary histories recorded in the Gjersvik Nappe and the post-Remdalen Group rocks in the Stikke and Gelvenåkko Nappes suggest the correlations shown in Fig. 3. The facies differences between equivalent units in these two sequences are thought to be controlled by a closer proximity to volcanic centres and sediment source in the western succession.

South of the Grong–Olden Culmination, Beckholmen (1982) has correlated the calcareous phyllites and garbenschiefer in the higher thrust slices of the Tännforsen area with the Middle Köli (Fig. 1). Further south towards Røros (Fig. 1), similar lithologies beneath the Tydal Thrust (Rui and Bakke 1975) belonging to the Røros and Røsjø Formations (Rui 1972; Rui and Bakke 1975) probably also lie at this tectonostratigraphic level. The overlying volcanosedimentary sequence of the Meråker Nappe (Wolff 1967; Wolff and Roberts 1980), which has been correlated with similar lithologies in the Gjersvik Nappe (Grenne and Reinsbakken 1981; Gee et al., this volume), provides a better control on the age of the volcanites and calcareous turbidites in the Middle Köli (see Fig. 3). In the Meråker Nappe, volcanic and subvolcanic rocks (Fundsjø Group) showing a rifted low-K tholeiitic arc affinity (Grenne and Lagerblad, this volume) stratigraphically underlie calcareous turbidites and conglomerates containing basic extrusions and high-level intrusions (Sulåmo and Kjølhaugan Groups, see Chaloupsky and Fediuk 1967; Siedlecka 1967; Hardenby 1980). According to Grenne and Lagerblad (this volume), the Fundsjø Group was deformed and metamorphosed prior to uplift and erosion and subsequent deposition of the Sulåmo Group. By contrast to the Gjersvik Nappe, the stratigraphy of the Meråker Nappe continues upwards (Fig. 3), the Kjølhaugan Group being overlain by the Llandovery (Getz 1890) pelitic and graphitic phyllites of the Slågån Group and turbidites and conglomerates of the Liafjellet group (Hardenby 1980). The diagnostic fossil occurrence in the Slågån Group combined with the correlations shown in Fig. 3 suggest that calcareous turbidite sedimentation in the Middle Köli Nappes was of Ordovician age. This sedimentation differs markedly from the pre-Silurian clastic lithofacies development (Gilliks Formation, Tjopasi Group) in the Lower Köli Nappes. However, the Silurian rocks in the Meråker Nappe are reminiscent of the Silurian sequences in the Lower Köli Nappes and the Lower Allochthon (see Fig. 3).

Upper Köli Nappes

Krutfjellet and Jofjället Nappes and their equivalents south of the Grong–Olden Culmination

North of the Grong–Olden Culmination, the Upper Köli Nappes have been divided (Häggbom 1978) into a lower complex (Krutfjellet Nappe), metamorphosed in amphibolite facies and preserving evidence for Ordovician deformation and metamorphism prior to Scandian nappe emplacement, and an upper complex (Jofjället Nappe) metamorphosed in greenschist facies. The stratigraphy in the Jofjället Nappe resembles that in the Stikke Nappe of the Middle Köli.

Although various petrological studies have been carried out on the amphibolite facies rocks of the Krutfjellet Nappe (Beskow 1929; Kulling 1938; Mørk 1979; Senior and Otten, this volume; various University of Amsterdam unpublished theses from the 1950's), the internal structural and stratigraphical relationships remain uncertain. The Krutfjellet Nappe is dominated by graphitic, quartz-rich and pelitic schists, locally displaying sedimentary structures indicative of turbidite deposition. Conglomerates, containing pebbles of quartzite, limestone (locally with crinoids, Gee and Zachrisson 1979), and abundant volcaniclastic debris occur at various structural levels together with marble (locally with crinoids, Kulling 1938) and predominantly basic volcanites. The latter, comprising both pyroclastic rocks and massive or pillowed lavas, occur together with high-level intrusions and show affinity to both subduction-related magmas of low-K tholeiitic arc type (including high-Mg basaltic andesite) and alkaline basalts of WPB type (Stephens and Senior 1981). On the basis of the significant volcaniclastic component in the sediments and the nature of the subduction-related volcanites, Stephens and Senior (1981) suggested that deposition had occurred in a fore-arc basin. It is possible that this basin was fed by both a distal terrestrial source as well as the arc itself and includes both pre-arc and syn-arc sequences. This may explain the association of quartz-rich and volcaniclastic sediments and the contrasting geochemical characteristics of the basic rocks. This volcanosedimentary succession, partly of post-Cambrian age, suffered penetrative deformation prior to migmatitization, intrusion of layered gabbro, diorite, trondhjemite, and granite, and later Scandian thrust emplacement onto underlying units. Rb–Sr whole-rock radiometric dating of the Vilasund Granite (Gee and Wilson 1974) and the migmatites (Reymer 1979) suggest late Ordovician ages. When combined with the scanty fossil evidence, an early and/or mid Ordovician age is indicated for the earlier deformation phase in this nappe.

Based on lithological association, tectonic position, and the relative timing of intrusion and deformation phases, the Krutfjellet Nappe shows significant similarities to the higher-grade part of the Gula Nappe (Wolff and Roberts 1980), south of the Grong–Olden Culmination (Fig. 1). At the base of this complex, radioactive (Gee 1981) graphitic schist of Tremadoc age (Vogt 1889; Størmer 1941) is intimately associated with pillow lava and is intruded by dolerites (Nilsen 1971); contemporaneity of volcanism and black shale deposition has been inferred. The basic volcanic rocks were included in the Hersjø Formation and correlated with the Fundsjø Group in the Meråker Nappe (Rui 1972); their relationship to the latter remains unclear. Such a correlation suggests a link between the stratigraphies in the Middle and Upper Köli Nappes and an early Ordovician (Tremadoc) age for the arc volcanism in these units. Rb–Sr and U–Pb (zircon) whole-rock radiometric studies on syntectonic trondhjemites in the Gula Nappe (Klingspor and Gee 1981) indicate intrusion during the early and possibly mid Ordovician. Since the fossiliferous Tremadoc rocks have probably been influenced by the deformation which commenced

prior to the intrusion of the trondhjemites, an early and possibly also mid Ordovician age is inferred for this deformation, supporting correlation with the Krutfjellet Nappe.

The low-grade rocks in the overlying Jofjället Nappe can be divided into three sub-units. The internal structure and stratigraphy of the lower and middle parts (Fig. 3) have been described in the type-area in Västerbotten by Sandwall (1981a). In this part of the nappe, fine-grained calcareous turbidites (Jofjället Calcareous Phyllite) pass upwards into various pelites (Tjåter Phyllite), distal tuffs, volcanogenic sediments, and conglomeratic dolo-ankeritestone (Seinesbäcken Volcanite) and subsequently quartz-rich and pelitic phyllites with minor basic volcanite intercalations (Rikarsjön Phyllite). This sequence is comparable to the inferred inverted succession in the Stikke Nappe (Middle Köli Nappes) and provides further support for a link between the stratigraphies in the Middle and Upper Köli Nappes. The sequence described above is repeated at higher structural levels in a series of minor thrust sheets. The middle part of the nappe is dominated by basaltic greenstones (Ruffe and Brakkfjället Greenstones) showing mixed MORB–WPB and WPB affinities (Sandwall 1981b), and fine-grained phyllites (Rövattnet Formation) similar to the Rikarsjön Phyllite. The upper unit (Akfjell Nappe) contains carbonate rocks, phyllites, and volcanites, reminiscent of the lower part of the Jofjället complex, as well as serpentinites. It cuts discordantly across underlying units and possesses a mega lens (Brakfjället Lens) of higher-grade mica gneiss, quartzite, and cross-cutting basic dykes at its base reminiscent of the overlying rocks in the Uppermost Allochthon (Ramberg 1981).

South of the Grong–Olden Culmination, the sequence in the lower part of the Jofjället Nappe has been compared to the greenschist facies rocks in the western, upper part of the Gula Nappe (Sandwall 1981a and Fig. 1). Horne (1979) identified a clast-bearing facies in these low-grade Gula rocks and interpreted it as mélange. Early deformation structures in these phyllites are cut by later trondhjemite intrusions in a similar manner to those in the underlying higher-grade Gula rocks. The Ruffe and Brackfjället Greenstones have been correlated (Sandwall 1981b) with the basalts of the Støren Group in the immediately overlying Støren Nappe (Gale and Roberts 1974). Since other characteristic units in the Støren Nappe appear to be lacking in the Jofjället area, the Støren Nappe is treated here as a separate tectonic complex lying above the Jofjället Nappe and restricted to areas south of the Grong–Olden Culmination (Fig. 1).

Støren Nappe

The Støren Nappe overlies the Gula Nappe throughout much of western Trøndelag. Early work (Vogt 1945; Carstens 1960) established a stratigraphy in the Støren–Horg–Hølonda–Løkken area composed of thick pillow basalts (Støren Group), overlain by an Ordovician volcanosedimentary sequence containing andesites and rhyolites (Lower and Upper Hovin Groups). An overlying unfossiliferous sedimentary unit (Horg Group) is thought to cap the succession. Recently, Oftedahl (in Wolff *et al.* 1980) and Bruton and Bockelie (1980) have suggested that this succession occurs in at least three separate thrust sheets.

Interpretation of the internal structure and stratigraphy of the Støren Nappe remains controversial. Some of the basalts earlier correlated with the Støren Group have, in recent years, been interpreted (Roberts 1975; Wolff 1976; Grenne and Roberts 1980; Ryan *et al.* 1980) to lie at higher stratigraphic levels within the early–mid Ordovician succession.

As discussed below, the biostratigraphic and structural controls are insufficient to constrain the alternative interpretations. The basalts all show predominantly MORB petrochemical affinities (Gale and Roberts 1974; Grenne *et al.* 1980; Furnes *et al.* 1980), some contain layered gabbros, sheeted-dyke complexes, and a chert–argillite cap-rock and all are regarded as fragments of ophiolites (Furnes *et al.*, this volume). If all these basic complexes (the Bymark, Løkken, Støren, Jonsvatn, and Forbordfjell Greenstones) are fragmented ophiolites, then we think it likely that their occurrence at different tectonostratigraphic levels within the Støren Nappe is the result of Scandian tectonic imbrication and the traditional lithostratigraphic interpretation remains viable (Fig. 4). However, some units, such as Jonsvatn and Forbordfjell, are composed only of basalts and may indeed by younger extrusions (see discussion below).

The basic igneous complexes and associated cherts and argillites were emplaced to high structural levels and deeply eroded prior to deposition of the overly-

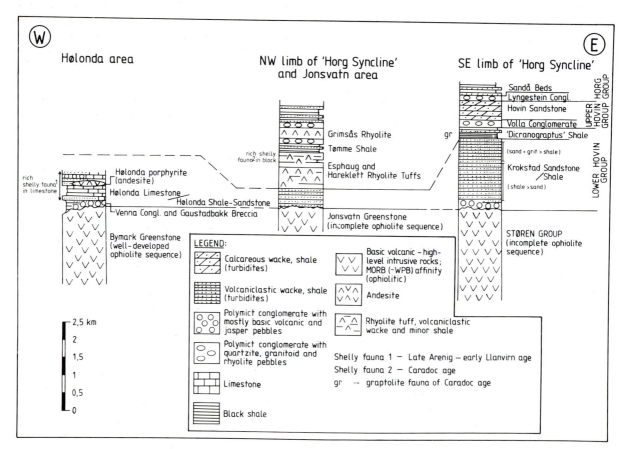

Fig. 4   Summary of stratigraphy and fossil localities in the Støren–Horg–Hølonda part of the Støren Nappe. The stratigraphy in each of the three sub-nappes (Oftedahl in Wolff *et al.* 1980) is based on Vogt (1945); the facies variations in the Hølonda area are taken from Bruton and Bockelie (1980). The Jonsvatn Greenstone is portrayed here as part of an ophiolite fragment (Furnes *et al.*, this volume) and correlated with the Støren Group and Bymark Greenstone (Carstens 1960). An alternative interpretation of the stratigraphy in the Lower Hovin (upper part), Upper Hovin and Horg Groups has been put forward by Chaloupsky (1970). Thicknesses indicated are only approximate

ing successions. Basal units of the Lower Hovin Group are generally conglomeratic and dominated by greenstone and chert clasts. The overlying successions in the various thrust sheets can be treated in two general associations, a western and eastern, the former probably being thrust from a region west of the latter.

The eastern successions are illustrated in Fig. 4 from Vogt's (1945) 'Horg Syncline'. The greenstone-boulder conglomerates, overlying the Støren Group pillow lavas, pass up via volcaniclastic turbidites into black shales and rhyolites and then into turbiditic sandstones and polymict conglomerates. The only diagnostic fossils collected in these sequences are of mid Ordovician (Caradoc) age.

The western succession has been interpreted in recent years by Bruton and Bockelie (1980) in the Hølonda type area (Fig. 4) where shales and limestones occur together with a suite of penecontemporaneous extrusions and intrusions of porphyritic andesite. In the Hølonda area and further to the west towards Løkken and Grefstadfjell, the shales and limestones have yielded a variety of fossils of Arenig and Llanvirn age, which show distinct affinities with North American platform faunas, (Neuman and Bruton 1974; Bergström 1979; Bruton and Bockelie 1980; Bruton and Harper, this volume; Spjeldnaes, this volume). The oldest faunas, occurring in black shales intercalated with basic volcanites, are reported to be of mid-late Arenig age (Ryan *et al.* 1980); the youngest are Caradoc (Kiaer 1932) or possibly Ashgill (Bruton and Bockelie 1980), in a succession of shales, sandstones, rhyolitic tuffs, and limestones. Silurian fossils have not been found in the Støren Nappe.

About 100 km west of the Hølonda area on the island of Smøla, fossiliferous limestones occur, of similar age and faunal affinity to those at Hølonda (Holtedahl 1915; Strand 1932; Bruton and Bockelie 1979). According to Roberts (1980), these limestones interdigitate with and are overlain by subduction-related volcanites ranging from basalts and andesites to dacites and rhyolites of medium-K calc-alkaline character. These volcanites are intruded by a suite of gabbros, quartz diorites, granodiorites, and granites. Rb–Sr whole-rock dating of a quartz diorite yielded an age of 436±7 Ma (Sundvoll and Roberts 1977) suggesting that intrusion occurred in the late Ordovician or early Silurian. The Smøla igneous and sedimentary rocks have been assumed (Roberts 1980) to compose an essential westerly component of the Støren Nappe. Further north along the Trøndelag coast, on the island of Leka, Prestvik (1972, 1974, 1980) has described ophiolites, which may also be contained in the Støren Nappe. In the case of both these isolated island associations, their level within the tectonostratigraphy remains uncertain.

Discussion

Disrupted arc complex in the Middle and Upper Köli Nappes

Beneath the ophiolite-bearing igneous and sedimentary associations of the Støren Nappe, the Middle and Upper Köli Nappes are dominated by subduction-related volcanic and subvolcanic rocks of low-K, mildly tholeiitic character. The lithofacies within the various nappes include a more distal succession largely composed of pyroclastic rocks, subvolcanic intrusions and sedimentary rocks in the east (Stikke and Gelvenåkko Nappes) and a main magmatic arc complex including plutonic infrastructure further west (Gjersvik Nappe). Volcanites and volcaniclastic sediments of possible fore-arc basin association occur in the highest units (Krutfjellet Nappe). The coherence of this facies variation from east to west and the related stratigraphies in the Middle and Upper Köli Nappes suggest that these rocks represent a single disrupted island arc complex of early Ordovician (Tremadoc) age (note correlation lines in Fig. 3). Interpreted in this manner, the direction of subduction appears to have been to the east.

The basement to this arc is thought to be preserved in the back-arc (Stikke, Gelvenåkko and Leipikvattnet Nappes) and extreme fore-arc (Jofjället Nappe) situations (see Fig. 3). The character of the pre-arc rocks in these nappes suggests build-up of the arc upon ensimatic crust modified by the input of a distal terrestrial source. The occurrence of coarse volcaniclastic successions in the upper (?) part of the Røyrvik Group in the Leipikvattnet Nappe suggests that part of this unit, like at least some of the rocks in the Krutfjellet Nappe, may be a time equivalent to the main magmatic arc. The post-arc history is dominated by incipient rifting of the arc and the deposition of calcareous turbidites in all tectonic units except the Krutfjellet Nappe (Fig. 3). The basaltic extrusions and high-level intrusions providing the evidence for rifting occur in the main arc and back-arc situations, in the Middle Köli Nappes. The Ordovician deformation and high-grade metamorphism, combined with an absence of calcareous turbidites in the Krutfjellet Nappe, are suggested to be related to the same tectonic event involving: (1) depression of the Krutfjellet rocks beneath the main magmatic arc; (2) progressive deformation, metamorphism and intrusive activity at depth during the Ordovician; (3) uplift and contemporaneous erosion of the arc rocks and associated carbonates at higher levels providing igneous and calcareous detritus to flanking sedimentary basins. The evidence from south of the Grong–Olden Culmination suggests that this early deformation also affected at least part of the main arc (Meråker Nappe) and the extreme fore-arc facies (upper part of the Gula Nappe).

Palaeotectonic significance of faunas in the Støren Nappe

Within the eugeoclinal successions in the Scandinavian Caledonides, early–mid Ordovician faunas have been found in only two tectonic units, the Støren Nappe and at Otta, probably in the Lower Köli Nappes. As mentioned earlier, the faunas in the Støren Nappe contain many genera of North American affinities. This incompatability with faunas in the lower nappes and autochthon is generally interpreted (see Bruton and Harper, this volume; Spjeldnaes, this volume) to be related to substantial separation at the time of deposition (McKerrow and Cocks 1976). In view of the unconstrained 'suspect' character of the volcanosedimentary successions in the Støren Nappe, we accept this interpretation.

Gale and Roberts (1974), Roberts and Gale (1978), and Roberts (1980) have interpreted the Arenig–Llanvirn and younger Ordovician development in the Støren Nappe in terms of a back-arc basin situated east of a main magmatic arc (Smøla). In their tectonic models, these authors have consistently placed this arc/back-arc basin on the southeast side of the Iapetus Ocean adjacent to the Baltoscandian margin, thus regarding the faunal provincialism as related to factors other than physical separation from this margin (see discussion in Roberts 1980). On the assumption that the early–mid Ordovician, subduction-related igneous activity on Smøla lay west of and was related to the contemporaneous volcanicity in the Hølonda area, Roberts (1980) inferred an eastwards-directed subduction system. Clearly, a variety of alternative interpretations are possible but this is the simplest on present data and it is adopted here with the proviso that the arc/back-arc system developed nearer the Laurentian than the Baltoscandian margin.

Correlation of ophiolites

Within the Støren Nappe, only the ophiolites in the Hølonda–Løkken–Grefstadfjell area are closely related to the Arenig–Llanvirn faunas. In the Hølonda area (Bruton and Bockelie 1980), the fossils provide unambiguous control of the mid Arenig or earlier age of the underlying ophiolite complex (Bymark Greenstone). However, in the Løkken area, Ryan *et al.* (1980) described graptolitic shales of mid–late Arenig age to both underlie and overlie pillow basalts and claimed the entire igneous complex to be of late Arenig age. The structural interpretation upon which this conclusion was based requires remarkable primary variations in thickness of the ophiolite. Until details of both the faunal assemblages and the structural relationships are presented, we follow Vogt (1945) and Carstens (1960) in correlating the Løkken ophiolite with the other major ophiolite fragments in the area.

As far as the remaining basic complexes in the Støren Nappe are concerned, biostratigraphic control is lacking. The basalts of the Støren Group are of pre-Caradoc age while the interpretation of the Jonsvatn and Forbordfjell Greenstones as early–mid Ordovician (Roberts 1975; Wolff 1976) is based solely on their occurrence within lithologies of the Lower Hovin Group. Grenne and Roberts (1980) presented geochemical data for the Jonsvatn and Forbordfjell Greenstones, demonstrating their close similarity with the basalts of the Støren Group (Gale and Roberts 1974). The chemical criteria for inferring the oceanic affinity of the Støren basalts were used by Gale and Roberts (1974) to claim vast (200–250 km) displacement of these rocks onto the Baltoscandian margin during Silurian orogeny. However, in 1980, Grenne and Roberts did not entertain the possibility that the basal contacts of the Jonsvatn and Forbordfjell Greenstones may be thrusts, thus allowing correlation of these greenstones with the Støren Group as proposed by earlier workers. Such an interpretation is favoured here (Fig. 4) if, indeed, the Jonsvatn and Forbordfjell Greenstones are parts of ophiolite complexes.

In view of the discussion above, the ophiolite fragments and basalt complexes interpreted as parts of ophiolites are regarded here as the disrupted members of a single thrust sheet of oceanic crust of pre-late Arenig age. This oceanic crust was obducted to high structural levels adjacent to Laurentia in the early Ordovician and subsequently formed the basement to the early–mid Ordovician arc/back-arc complex. This interpretation of the stratigraphy of the Støren Nappe suggests that the post-ophiolite arc was stationary (Dickinson 1975) with respect to its back-arc region. Alternatively, if the Jonsvatn and Forbordfjell Greenstones are younger extrusions, then back-arc spreading and a migratory (Dickinson 1975) arc situation is envisaged (Grenne and Roberts 1980).

Miogeoclinal Rocks Related to a Western Continent (Laurentia?)

The Rödingsfjället and Helgeland Nappe Complexes in the Uppermost Allochthon (Stephens *et al.*, this volume, a) are dominated by gneisses, in part migmatitic, and a variety of psammitic, pelitic, and calcareous schists with associated subordinate amphibolites. Thick dolomite and calcite marbles are also characteristic; they are in part conglomeratic and contain sedimentary iron ore deposits. These medium- and high-grade rocks were penetratively deformed and metamorphosed prior to intrusion of gabbros and abundant granitoids, and later (Scandian) thrust emplacement onto the underlying low-grade Köli Nappes. Several age-determination studies of the granitoids (Priem *et al.* 1975; Claesson 1979; Ramberg *et al.* in press) have provided Rb–Sr whole-rock isochrons suggesting mid Ordovician to Silurian ages. Thus, both the rocks and much of the deformation and metamorphism in the Uppermost Allochthon appear to be pre-mid Ordovician in age. Riis and Ramberg (1979) have suggested that, locally, the Uppermost Allochthon can be divided into 'basement' and 'cover' units, the former consisting of various gneisses and the latter composed of schist–marble complexes with associated iron ores. The 'basement' gneisses appear to have suffered a more complex history of deformation and metamorphism than the 'cover' sequences. Rb–Sr whole-rock dating of an augen gneiss near the base of the Helgeland Nappe Complex yielded a mid Proterozoic age (Ramberg *et al.* in press).

Along the Norwegian west coast, isolated remnants of low-grade rocks in the western part of the Helgeland Nappe Complex (Gustavson 1975, 1978b) include basaltic lavas on Skålvaer showing MORB affinities (Gustavson 1978a; Furnes *et al.*, this volume) and interpreted as part of an incomplete ophiolite sequence (Furnes *et al.* 1980). It is possible that the ophiolitic rocks and overlying volcanosedimentary sequence on Leka described by Prestvik (1974, 1980) also belong to the Helgeland Nappe Complex.

Stephens *et al.* (this volume, a) have interpreted the higher-grade rocks in the Uppermost Allochthon as a continental margin sequence deposited in an ensialic environment. The tectonostratigraphic setting of these rocks indicates that this miogeocline lay west of the Köli eugeocline. In the Helgeland Nappe Complex, there is an apparently incompatible association of these continental margin rocks and the low-grade ophiolite-bearing slices. Stephens *et al.* (this volume, a) have suggested that this association may be due to a Scandian imbrication of a more continuous sheet of obducted ophiolite and overlying post-ophiolite volcanosedimentary sequences together with the continental margin rocks.

Suspect Terranes

The development of the Baltoscandian platform and miogeocline during the late Proterozoic and early Palaeozoic can be readily traced in the successions of the Autochthon and Lower and Middle Allochthons (Fig. 5). The upper units in the latter have been transported (Gee 1978) from areas lying at least 300 km west of the Caledonian thrust front.

The overlying Seve Nappes contain sedimentary and basic igneous associations which are directly comparable with those in the underlying Särv Nappes of the Middle Allochthon. However, these Seve units are of higher metamorphic grade and contain solitary ultramafites. Interpretation of the terranes from which the Seve Nappes have been derived is therefore more problematical. Tectonic intercalation of the ultramafic rocks implies that the Seve Nappes may be of composite terrane character. Nevertheless, we prefer the hypothesis that the Seve units composed an essentially outboard component of the Baltoscandian miogeocline, comprising the transition zone from a passive continental margin to oceanic environments during the early development of the Iapetus Ocean. On this interpretation, the Baltoscandian miogeocline extended to at least 400 km west of the present Caledonian thrust front and the higher tectonic units were derived from suspect terranes that were located outboard of this miogeocline.

Miogeocline sequences related to a western continent (Laurentia?) are inferred for the successions in the Uppermost Allochthon. This terrane (Fig. 5) is unrelated to the Baltoscandian miogeocline from which it is separated by the eugeoclinal terranes represented in the Köli Nappes. It differs from the Baltoscandian miogeocline in three respects:

1. The sedimentary successions are more variable than in the Seve and Särv Nappes; carbonates are particularly conspicuous and the associated iron ores are not known in other tectonic units.
2. The successions in this terrane were intruded by syntectonic gabbro and granitoid plutons, the latter in some cases of batholithic dimensions; ages of intrusion range at least from mid Ordovician to Silurian.
3. Penetrative deformation and amphibolite facies metamorphism of the rocks in this terrane occurred prior to the intrusions referred to above.

It should be noted that the subordinate, low-grade ophiolite-bearing slices in the Uppermost Allochthon have not been distinguished in Fig. 5; this terrane is clearly of composite character. Its position in the nappe pile above the Støren Nappe favours the interpretation that the rocks of miogeoclinal charac-

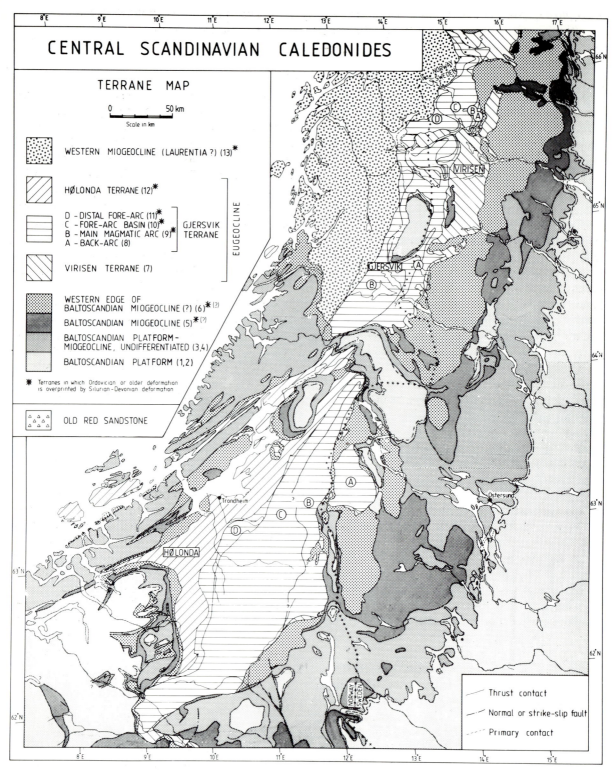

Fig. 5   Suspect terrane map of the central Scandinavian Caledonides. Numbers in brackets (1–13) refer to tectono-stratigraphic subdivisions in Fig. 1

ter in this tectonic unit originated along the Laurentian margin.

The eugeoclinal successions preserved in the Köli Nappes are sandwiched between the two miogeoclinal complexes referred to above. They were derived from terranes which are suspect both with respect to each other and the adjacent miogeoclinal rocks. Three terranes are recognized within the eugeoclinal belt (Fig. 5), being distinguished on the basis of differences in stratigraphy, history of deformation, metamorphism, and intrusion, and evidence of faunal affinities. They are referred to as the Virisen, Gjersvik, and Hølonda terranes.

Virisen terrane

The volcanosedimentary complex of the Virisen terrane occurs in the Lower Köli Nappes, the rocks ranging in age from the early Ordovician (perhaps Cambrian) into the early Silurian. The pre-Ashgill successions contain arc-related (low-K tholeiitc) volcanites of early to mid Ordovician age as well as mid Ordovician clastic rocks; rift-related basalts occur in the Llandovery part of the sequence following deposition of Ashgill shallow-water sediments. The direction of pre-Ashgill subduction is not known. The Lower Köli terrane was located outboard of the Baltoscandian miogeocline, proximity to which is suggested by:

1. The apparent lack of a significant tectonic break between the Köli Nappes and the underlying Seve Nappes in some areas.
2. The nature of the detritus in the mid Ordovician clastic rocks, suggesting derivation from an eastern platformal/cratonic source.
3. The presence of a concentration of solitary ultramafites in the Seve and Lower Köli Nappes, showing similar primitive petrochemical characteristics (Stigh 1979).

However, this evidence for proximity is difficult to reconcile with the character of the detrital serpentinites which occur frequently in the Lower Köli Nappes mantling solitary serpentinites. The monomict nature of the sediments and the mixed faunal affinities of the fossils at Otta suggest an isolated depositional environment separated from the Baltoscandian margin. If the ultramafic rocks in the Seve and Lower Köli Nappes are related and emplaced by a protrusive mechanism (Stigh 1979), then the sedimentary and faunal evidence suggest that the Virisen terrane was not located immediately adjacent to the miogeocline. An alternative hypothesis accepts that the detrital serpentinites existed in the early Ordovician in truly ensimatic environments remote from the Baltoscandian margin (Bruton and Harper, this volume). It requires that they were transported towards this margin during closure of Iapetus and in the mid Ordovician were emplaced tectonically as olistoliths into the depositional basin of the Virisen terrane, immediately outboard of the Baltoscandian miogeocline. Such an explanation involves separate mechanisms of emplacement of the ultramafites in the Lower Köli and subjacent Seve Nappes, those in the latter being perhaps

related to protrusion in a continent–ocean transition zone (cf Boillot et al. 1980).

Gjersvik terrane

The volcanosedimentary complex of the Gjersvik terrane occurs in the Middle and Upper Köli Nappes. It is inferred to be of Ordovician to early Silurian age and is characterized by an early Ordovician (Tremadoc), low-K tholeiitic arc complex, related to easterly-directed subduction, overlain by volcaniclastics as well as calcareous turbidites and locally, alluvial sediments. The eastern back-arc facies and the main magmatic arc show evidence of post-arc rifting, while the western fore-arc facies and part of the main magmatic arc underwent deformation and metamorphism during the early and, possibly, mid Ordovician.

The Gjersvik terrane is distinguished from the Virisen terrane by the character of the mid Ordovician clastic sequence and the evidence of post-arc rifting. The presence of Llandovery black shales passing up into a coarsening-upwards greywacke succession is, however, similar to the early Silurian development in the Virisen terrane and the Baltoscandian miogeocline.

Hølonda terrane

The Hølonda terrane is identified in the Støren Nappe. It is composed of pre-late Arenig ophiolites, overlain by an early–mid Ordovician volcaniclastic succession and medium-K calc-alkaline arc complex thought to be related to easterly-directed subduction. The post-ophiolite succession ranges into the Caradoc and perhaps Ashgill. Mid Ordovician calcareous turbidites are comparable with sediments probably of similar age in the Gjersvik terrane. However, in the Hølonda terrane, these sediments lack the ubiquitous basic intrusions of the eastern part of the Gjersvik terrane. Faunas in the Hølonda terrane suggest Laurentian affinities.

Tectonic Model

The evidence presented above allows restoration of the various allochthonous units in the central part of the Scandinavian Caledonides to their relative positions prior to Scandian deformation. Baltoscandian platform and miogeoclinal successions have been recognized. The higher nappes were derived from areas that lay outboard of this Baltoscandian miogeocline and an attempt has been made to reconstruct the palaeotectonic setting of these suspect terranes. A Laurentian margin has been tentatively identified and evidence has been presented for interpreting the accretionary history of the intervening eugeoclinal terranes. The latter is outlined below as a basis for constructing the tectonic model.

Accretionary history

Within the eugeoclinal successions, ophiolites have only been recognized in the Hølonda terrane; they are interpreted to have been derived from an early Caledonian ocean (Iapetus). Whereas the various other ophiolites (Furnes et al. 1980) in the Scandinavian Caledonides may have been derived from any part of this ocean or its marginal basins, the ophiolites in the Hølonda terrane are thought to have been located near the Laurentian margin in the latest Cambrian and to have been emplaced to high structural levels on or adjacent to that margin in the earliest Ordovician. Proximity to the Laurentian margin is inferred from the faunal assemblages in the overlying Arenig–Llanvirn strata (evidence which is lacking in the case of the other Scandinavian ophiolites). The timing of ophiolite obduction is constrained only by the age of the cap-rock successions; it may have occurred prior to the Arenig.

The Gjersvik terrane was located between the Hølonda terrane and the Baltoscandian margin in the early Ordovician. Proximity of the Gjersvik and Hølonda terranes in the mid Ordovician is inferred from the development in both areas of similar calcareous turbidites. It is therefore suggested that the build-up of the arc complex within this terrane was related to the obduction of the ophiolites in the Hølonda terrane and that both these terranes were accreted to the Laurentian margin at least by the mid Ordovician. Support for this interpretation is found in the Uppermost Allochthon and Upper Köli Nappes in areas north of the Grong–Olden Culmination where the strata in both these complexes were intruded by granitoid plutons and dykes of mid–late Ordovician age. The deformation and metamorphism that preceded this intrusion is inferred to be related to accretion of the Hølonda and Gjersvik terranes to Laurentia.

The faunal evidence is interpreted to imply that a large expanse of early to mid Ordovician ocean (McKerrow and Cocks 1976) separated the accretion of the Hølonda and Gjersvik terranes along the Laurentian margin from the development of the Virisen terrane and the Baltoscandian margin. Evidence for basin sharing between the Virisen and Gjersvik terranes only becomes apparent in the early Silurian. Furthermore, the similarity in the latest Ordovician and early Silurian successions of the Virisen terrane and the Baltoscandian miogeocline suggest proximity of the latter to the Laurentian margin by the end

of the Ordovician. This interpretation finds support in the breakdown of faunal provincialism in most fossil groups by the early Silurian (Cocks and Fortey 1982).

Late Proterozoic and Cambrian

Interpretation of the pre-Ordovician history of the suspect terranes is inhibited by the lack of faunal evidence and isotope age-determination data. The pre-late Arenig ophiolites of the Hølonda terrane are assumed to be of Cambrian and possibly Vendian age on the basis of wider considerations of the early history of the Iapetus Ocean (Harland and Gayer 1972; Roberts and Gale 1978). The late Proterozoic and Cambrian history of the Baltoscandian miogeocline, with the development of a passive continental margin to Baltica and the extensive intrusion of tholeiitic magma into the outer parts of the miogeocline during the Vendian, supports the inferred age of the ophiolites. Likewise, it suggests that development of a volcanic arc outboard of the Baltoscandian miogeocline did not start until after the Cambrian. The pre-Ordovician history has been treated previously (Gee 1975a, Fig. 8) and is not expanded on here. Closure of the Iapetus Ocean is thought to have started in the Tremadoc and the discussion that follows concentrates on the inferred convergence of Laurentia with Baltica during the Ordovician and Silurian. The tectonic evolution of the central Scandinavian Caledonides during the Ordovician and early Silurian prior to Scandian orogenesis is illustrated in a series of west–east schematic sections (Fig. 6).

Early Ordovician

Easterly-directed subduction leading to the build-up of the low-K, tholeiitic arc in the Gjersvik terrane (Fig. 6a) was established in an ensimatic setting during the early Ordovician (Tremadoc). As argued earlier, the different segments of this arc are preserved in the Middle and Upper Köli Nappes. The ophiolites in the Støren Nappe are thought to represent the oceanic basement in the western, trench-side part of this arc, i.e. to have been a part of the overriding plate. Thus, the terranes represented in the Støren Nappe and in the underlying units of the Upper and Middle Köli Nappes are thought to have been in juxtaposition prior to the mid Ordovician. Progressive elimination of the oceanic segment between the Laurentian continental margin and the arc complex in the Gjersvik terrane led subsequently to continent–arc collision. This compressional phase is

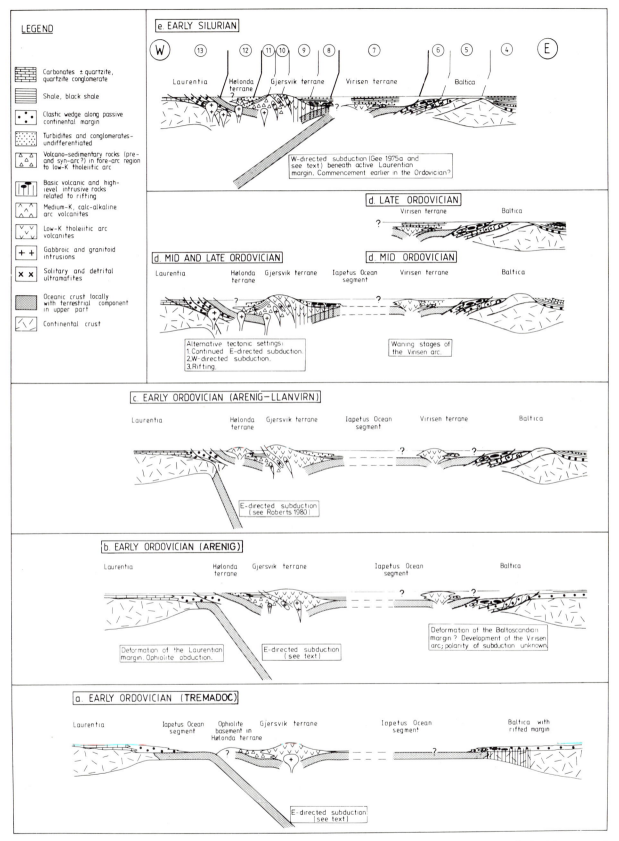

Fig. 6   Schematic sections illustrating progressive elimination of the Iapetus Ocean during early Ordovician to early Silurian time in the central part of the Scandinavian Caledonides. Numbers in circles (4–13) in the upper figure refer to the tectonostratigraphic subdivisions in Fig. 1. They are shown here above the respective source areas for these nappe complexes

thought to have resulted in:

1. Obduction of the oceanic basement in the outer part of the arc *westwards* onto the *Laurentian margin*, and
2. Deformation of the western continental margin, the rocks of the fore-arc and part of the main magmatic arc to the east.

By late or possibly even mid Arenig time, the ophiolites were in a sufficiently high position to be eroded and to supply debris to the basin in which the younger strata in the Hølonda terrane were deposited (Fig. 6b). This model for the earliest Ordovician contrasts with previous interpretations (Gale and Roberts 1974; Roberts 1980) which assume derivation of the ophiolites from the downgoing plate and their emplacement eastwards onto the Baltoscandian margin in the Silurian (Gale and Roberts 1974) or early Ordovician (Furnes *et al.* 1980).

During Arenig–Llanvirn time, the locus of arc volcanism shifted westwards and resulted in the build-up of the medium-K, calc-alkaline arc in the Hølonda terrane (Fig. 6c). As argued earlier, this arc is thought to have been constructed along the Laurentian margin on a basement composed either of tectonically thickened oceanic crust or, as Roberts (1980) has suggested, crust with both oceanic and continental components. In either case, the nature of the arc volcanism indicates sub-arc crustal thicknesses in excess of c. 25 km (Miyashiro 1974). Roberts (1980) has suggested easterly-directed subduction for this arc, based, in part, on higher contents of K_2O in contemporaneous andesites at Hølonda in the east relative to Smøla in the west. Although reservations were expressed above concerning the spatial relationship between the Smøla and Hølonda volcanites prior to Scandian deformation, this subduction polarity has been tentatively adopted in Fig. 6c. Such a model envisages deposition of the Hølonda sequence in an inter-arc basin between the active arc to the west and the remnant arc complex in the Gjersvik terrane to the east (Fig. 6c). As argued earlier, this basin may have been stationary with respect to the active arc. Such an interpretation does not favour the westwards shift of the arc volcanism as being due to actual migration of the subduction zone. Instead, a model involving steepening of the subduction zone related to tectonic activity and crustal thickening further east is proposed.

During the early Ordovician, an oceanic segment is thought to have separated the active Laurentian margin from the Baltoscandian margin. The tectonic evolution outboard of and along the Baltoscandian margin can be gleaned from the Lower Köli Nappes and the successions in the Lower Allochthon. Outboard of the miogeocline, there is evidence for car-bonate deposition together with detrital serpentinites around serpentinite plugs in Llanvirn time and the eruption of low-K tholeiitic, arc-related volcanites, the latter ranging in age from early to mid Ordovician. Alternative interpretations of the mode of emplacement of the ultramafites have been discussed above; they allow radically different models for the early development of the Virisen terrane and its relationship to the Baltoscandian margin. In addition, constraints on the polarity of subduction are lacking.

A basement high situated along the Baltoscandian margin (see Fig. 6c–d) has been postulated by several authors (Asklund 1960; Gee 1975a) to have developed in the early–mid Ordovician, providing a source for the detritus in the greywackes of the Jämtland Supergroup (Lower Allochthon) and the Virisen terrane (Gee 1975a). The instability reflected in the deposition of these sediments started during the Llanvirn (Karis in Gee and Kumpulainen 1980) in the Jämtland basin and continued throughout the mid Ordovician. This overlaps in time with the subduction-related igneous activity in the Virisen terrane. Thus, one possibility is that the basement high was a passive feature which rose in response to increased heat-flow and instability in the back-arc region (cf. Watanabe *et al.* 1977) above an easterly-directed subduction zone. A second possibility, prompted by the evidence in northernmost Scandinavia for late Cambrian to early Ordovician deformation (Finnmarkian phase of Sturt *et al.* 1978), envisages a more active basement high, established in the early Ordovician in response to thrusting of the outer margin of the miogeocline and crustal thickening (Fig. 6b–d). The latter model is favoured by Hossack *et al.* (this volume), who relate the early movement of the Jotun Nappe in southern Norway to the development of a foredeep and deposition of a mid–late Ordovician clastic wedge in front of the advancing thrust sheets. The dating of a mylonite (c. 485 Ma) at the base of the Särv Nappes in central Jämtland (Claesson 1980) provides supporting evidence for this hypothesis. This alternative allows the possibility that the Virisen volcanism was related to westerly-directed subduction and that early deformation of the outer edge of the Baltoscandian miogeocline resulted from collision of the Virisen arc with Baltica. Fig. 6 (b–d) portrays early to mid Ordovician deformation of the Baltoscandian margin but leaves the question of subduction polarity open.

Mid and late Ordovician

The mid and late Ordovician development of the western terranes is difficult to reconstruct due to the scanty age-control. Detritus in the post-arc sedimentary cover suggests that a landmass was established during this interval towards the fore-arc region of the remnant arc in the Gjersvik terrane and was flanked by sedimentary basins (Fig. 6d). The calcareous turbidites and conglomerates of the Middle Köli Nappes were deposited in a basin to the east of this high while, to the west, the calcareous turbidites of the Jofjället Nappe were laid down in a separate marginal basin. Paucity of age data in the eastern basin allows this deposition to have started in the early Ordovician, a possibility indicated in Fig. 6c.

To the east of the high, there is evidence for a change-over to a distensional tectonic régime with the injection and extrusion of rifting-related basaltic magmas in what were the main-arc and back-arc regions of the Gjersvik arc (Fig. 6d). Once again, the poor age control does not preclude inception of this rifting event in the early Ordovician. Further west, layered gabbros and granitoids, the latter ranging in age from early Ordovician to Silurian, intrude already deformed rocks in both the relic fore-arc region of the Gjersvik arc and the western miogeocline itself (Fig. 6b–e); furthermore, rhyolites of mid and possibly late Ordovician age erupted in the western basin. The tectonic significance of this predominantly intrusive igneous activity along and to the west of the high is not yet understood. One possibility is that it represents the waning stages of subduction on the Gjersvik–Smøla arc system. An alternative possibility is that after accretion of the Gjersvik and Hølonda terranes to the Laurentian margin, easterly-directed subduction gave way to shallow westerly-dipping subduction. A further possibility is that the plutonism was unrelated to subduction, being perhaps triggered by incipient rifting in a tectonically thickened crust and being related to the rifting further east. A combination of these possibilities is also viable.

Along the Baltoscandian margin, the instability which commenced in the early Ordovician continued through the mid Ordovician. The basement high separating the Jämtland basin from the Virisen terrane supplied detritus both to east and west. Bentonites in the Caradoc shales and greywackes of the Jämtland successions witness to volcanic activity further to the west. By the mid Ordovician, the Virisen terrane lay directly outboard of the Baltoscandian miogeocline, and was receiving craton-derived conglomerates and greywackes. The subordinate occurrence of ultramafic clasts in these conglomerates indicate that ultramafites were in a position for erosion close to the Baltoscandian margin at least by the mid Ordovician. Low-K tholeiitic, subduction-related volcanicity continued.

The late Ordovician successions of both the Virisen terrane and the Baltoscandian miogeocline indicate a return to more stable conditions. In the former, marked regression was followed by deposition of a transgressive sequence of quartz sandstones and quartzite-bearing conglomerates passing upwards into limestones (Fig. 6d). Further east, the Caradoc greywackes pass up into Ashgill shales. During deposition of the latter, evidence diminishes for the existence of a basement high separating the Virisen terrane from areas of deposition further to the east.

Early Silurian

Llandovery black shales, coarsening upwards into greywackes and conglomerates occur in the Gjersvik and Virisen terranes and along the Baltoscandian miogeocline (Fig. 6e). Minor, distension-related basalts occur in the Lower Köli black shales and bentonites are present further east. The Llandovery limestones and quartzites underlying the black shale facies of the Baltoscandian miogeocline are reminiscent of the Ashgill strata of the Virisen terrane further west. These relationships suggest proximity of the Gjersvik and Virisen terranes to the Baltoscandian margin by the early Silurian, implying closure of the Iapetus Ocean. The marine transgression that progressed from west to east across these terranes, starting in the latest Ordovician and continuing through the Llandovery, testifies to a deepening basin and increasing instability further west—the onset of Scandian deformation.

Mid Silurian to Devonian

The subsequent history of Scandian collisional orogeny involving emplacement of the accreted Laurentian and western eugeoclinal terranes onto the Virisen terrane and Baltoscandian miogeocline, as well as thrusting of the entire nappe pile onto the platform, has been treated earlier (Gee 1975a, 1978). The Scandian evolution has been inferred to be related to westerly-dipping, A-subduction. Underthrusting of Laurentia by Baltica led to high-pressure metamorphism with crystallization of eclogites in the Precambrian gabbros and dolerites of the western part of the Baltoscandian platform (Griffin and Brueckner 1980). Gabbros and granitoids of Silurian age occur in the terranes of the overthrust plate. Emplacement of the nappe pile was accompanied by rapid uplift with the deposition of an Old Red Sandstone facies in intramontane and extramontane basins (Steel *et al.*, this volume).

Concluding Remarks

The model presented here is an attempt to synthesize a broad spectrum of Caledonian data into a coherent evolutionary history. We are persuaded that plate movements similar to those prevailing today occurred in the latest Proterozoic and early Phanerozoic and that processes of plate accretion and destruction were active then as now. The Caledonian history involved development of an ocean (Iapetus) between the continents of Laurentia and Baltica and subsequent closure of that ocean. Elements from both sides of Iapetus are thought to be represented in the Scandes. The ocean opened in the Vendian and Cambrian, reached its greatest width in the late Cambrian and closed during the Ordovician. This closure involved B-subduction with development of island-arc systems outboard of both continental margins leading to A-subduction, involving thrusting of the arcs onto the continental margins. The latter initiated during the early Ordovician at the western margin and may have occurred during early and mid Ordovician time at the Baltoscandian margin. Subsequent sedimentation was dominated by deposition of turbidites and conglomerates. Final elimination of the ocean was accompanied by Scandian continental collision (A-type subduction) and thrusting of the Laurentian and eugeoclinal elements onto the Baltoscandian miogeocline and platform. In this paper, we have tried to spell out the basis for our interpretations in some detail, in the hope that this will promote closer analysis and provoke concrete alternative solutions.

Correlations of events in time and space is fundamental to all models of orogenic evolution. Unfortunately, early Phanerozoic histories are plagued by the paucity of the time constraints and this is particularly the case in the Scandinavian Caledonides. The model presented here and the alternatives that have been discussed require a more rigorous time-frame; priority needs to be given to palaeontological and isotopic age-determination studies. The latter, by the application of different methods, should seek more specifically than before to distinguish the age of rock units from the time of their metamorphism and subsequent uplift.

On-going regional mapping of the mountain belt will revise aspects of the structural geometry and influence correlation of the nappe units both within and between the major complexes. This survey mapping needs to be supplemented by detailed studies of local critical tectonostratigraphical relationships, these providing the skeleton on which the model is built.

During the last decade, basic igneous suites have been the focus of interest in the Scandes, particularly basalts and related high-level intrusions. Trace element and REE studies have supplemented the field data to provide new insight into the evolution of the eugeocline. These aspects have been highlighted in this model and will no doubt continue to contribute to reinterpretation of the Caledondian evolution. By contrast, there is a notable paucity of geochemical data on the granitic and related intrusions in the higher nappe units.

In this deeply eroded mountain belt only fragments of the tectonostratigraphy are preserved. Judging by the high grade of Caledonian metamorphism, vast thicknesses of volcanosedimentary successions have been removed by erosion and little of the record preserved in successor basins. The fragmentary character of the tectonostratigraphy provides an incomplete record which inevitably is reflected in the oversimplifications of the model. Nevertheless, we trust that this will not inhibit continuous reassessment of the details of mountain geology and their relevance for models of the tectonic evolution.

Acknowledgements

We thank R. Mason for comments on an early draft of this manuscript. MBS wishes to acknowledge special funding by the Geological Survey of Sweden under the auspices of the IGCP Project No. 60, *Correlation of Caledonian Stratabound Sulphides*. DGG acknowledges support via the IGCP Project No. 27, *Caledonide Orogen* by the Swedish Natural Science Research Council.

References

Andréasson, P. -G., Gee, D. G. and Sukotjo, S. this volume. Seve eclogites in the Norrbotten Caledonides, Sweden.

Arnbom, J.-O. 1980. Metamorphism of the Seve Nappes at Åreskutan, Swedish Caledonides. *Geol. Fören. Stockh. Förh.*, **102**, 359–371.

Asklund, B. 1938. Hauptzüge de Tektonik und Stratigraphie der mittleren Kaledoniden in Schweden. *Sver. geol. Unders.*, **C 417**, 99 pp.

Asklund, B. 1960. The geology of the Caledonian mountain chain and of adjacent areas in Sweden. In *Description to Accompany the Map of the Pre-Quarternary Rocks of Sweden*. *Sver. geol. Unders.*, **Ba 16**, 126–149.

Beckholmen, M. 1978. Geology of the Nordhallen–Duved–Greningen area in Jämtland, central Swedish Caledonides. *Geol. Fören. Stockh. Förh.*, **100**, 335–347.

Beckholmen, M. 1982. Mylonites and pseudotachylites associated with thrusting of the Köli Nappes, Tännforsfältet, central Swedish Caledonides. *Geol. Fören. Stockh. Förh.*, **104**, 23–32.

Bergström, J. 1980. The Caledonian Margin of the Fennoscandian Shield during the Cambrian. In Wones, D. R. (Ed.), *The Caledonides in the USA. Virginia Polytechnic Inst. and State Univ., Dept. Geol. Sci., Mem.*, **2**, 149–156.

Bergström, J. and Gee, D. G. this volume. The Cambrian in Scandinavia.

Bergström, S. M. 1979. Whiterockian (Ordovician) conodonts from the Hølonda Lime stone of the Trondheim region, Norwegian Caledonides. *Norsk geol. Tidsskr.*, **59**, 295–307.

Beskow, G. 1929. Södra Storfjället im südlichen Lappland. *Sver. geol. Unders.*, **C 350**, 335 pp.

Boillot, G., Grimaud, S., Mauffret, A., Mougenot, D., Kornprobst, J., Mergoil-Daniel, J. and Torrent, G. 1980. Ocean–continent boundary off the Iberian margin: a serpentinite diapir west of the Galicia Bank. *Earth Planet. Sci. Lett.*, **48**, 23–34.

Bruton, D. L. and Bockelie, J. F. 1979. The Ordovician sedimentary sequence on Smøla, west Central Norway. *Norges geol. Unders.*, **348**, 21–31.

Bruton, D. L. and Bockelie, J. F. 1980. Geology and palaeontology of the Hølonda area, western Norway—a fragment of North America? In Wones, D. R. (Ed.), *The Caledonides in the USA*. Virginia Polytechnic Inst. and State Univ., Dept. Geol. Sci., Mem., **2**, 41–47.

Bruton, D. L. and Harper, D. A. T. 1981. Brachiopods and trilobites of the early Ordovician serpentine Otta Conglomerate, south central Norway. *Norsk geol. Tidsskr.*, **61**, 153–181.

Bruton, D. L. and Harper, D. A. T. 1982. Ordovician gastropods from Vardofjället, Swedish Lapland, and the dating of Caledonian serpentinite conglomerates: a discussion. *Geol. Fören. Stockh. Förh.*, **104**, 189–190.

Carstens, H. 1960. Stratigraphy and volcanism of the Trondheimsfjord area, Norway. In *Guide to Excursions No. A4 and C1*, Internat. Geol. Cong., 21st, Norden 1960. *Norges geol. Unders.*, **212b**, 23 pp.

Chaloupsky, J. and Fediuk, F. 1967. Geology of the western and north-eastern part of the Meråker area. *Norges geol. Unders.*, **245**, 9–21.

Claesson, S. 1976. The age of the Ottfjället dolerites of the Särv Nappe, Swedish Caledonides. *Geol. Fören. Stockh. Förh.*, **98**, 370–374.

Claesson, S. 1979. Pre-Silurian orogenic deformation in the north–central Scandinavian Caledonides. *Geol. Fören. Stockh. Förh.*, **101**, 353–356.

Claesson, S. 1980. A Rb–Sr isotope study of granitoids and related mylonites in the Tännäs Augen Gneiss Nappe, southern Swedish Caledonides. *Geol. Fören. Stockh. Förh.*, **102**, 403–420.

Claesson, S. 1982. Caledonian metamorphism of Proterozoic Seve rocks on Mt. Åreskutan, southern Swedish Caledonides. *Geol. Fören. Stockh. Förh.*, **103**, 291–304.

Claesson, S. and Roddick, J. C. 1983. $^{40}Ar/^{39}Ar$ data on the age and metamorphism of the Ottfjället dolerites, Särv Nappe, Swedish Caledonides. *Lithos*, **16**, 61–73.

Claesson, S., Klingspor, I. and Stephens, M. B. 1983. U–Pb and Rb–Sr isotopic data on an Ordovician volcanic/subvolcanic complex from the Tjopasi Group, Köli Nappes, Swedish Caledonides. *Geol. Fören. Stockh. Förh.* **105**, 9–15.

Cocks, L. R. M. and Fortey, R. A. 1982. Faunal evidence for oceanic separations in the Palaeozoic of Britain. *J. geol. Soc. London*, **139**, 465–478.

Coney, P. J., Jones, D. L. and Monger, J. W. H. 1980. Cordilleran suspect terranes. *Nature*, London, **288**, 329–333.

Dewey, J. F. 1969. Evolution of the Appalachian/Caledonian Orogen. *Nature, London*, **222**, 124–129.

Dickinson, W. R. 1975. Potash–depth (K–h) relations in continental margin and intra-oceanic magmatic arcs. *Geology*, **3**, 53–56.

Dyrelius, D., Gee, D. G., Gorbatschev, R., Ramberg, H. and Zachrisson, E. 1980. A profile through the central Scandinavian Caledonides. *Tectonophysics*, **69**, 247–284.

Foslie, S. and Strand, T. 1956. Namsvatnet med en del av Frøyningsfjell. *Norges geol. Unders.*, **196**, 82 pp.

Furnes, H., Skjerlie, F. J. and Tysseland, M. 1976. Plate tectonic model based on greenstone geochemistry in the Late Precambrian–Lower Palaeozoic sequence in the Solund–Stavfjorden areas, west Norway. *Norsk geol. Tidsskr.*, **56**, 161–186.

Furnes, H., Roberts, D., Sturt, B. A., Thon, A. and Gale, G. H. 1980. Ophiolite fragments in the Scandinavian Caledonides. In Panayiotou, A. (Ed), *Ophiolites. Proc. Int. Ophiolite Symp. Cyprus 1979*, 582–600.

Furnes, H., Ryan, P. D., Grenne, T., Roberts, D., Sturt, B. A. and Prestvik, T. this volume. Geological and geochemical classification of the ophiolite fragments in the Scandinavian Caledonides.

Gales, G. H. and Roberts, D. 1974. Trace element geochemistry of Norwegian Lower Palaeozoic basic volcanics and its tectonic implications. *Earth Planet. Sci. Lett.*, **22**, 380–390.

Gee, D. G. 1974. Comments on the metamorphic allochthon in northern Trøndelag, central Scandinavian Caledonides. *Norsk geol. Tidsskr.*, **74**, 435–440.

Gee, D. G. 1975a. A tectonic model for the central part of the Scandinavian Caledonides. *Am. J. Sci.*, **275–A**, 468–515.

Gee, D. G. 1975b. A geotraverse through the Scandinavian Caledonides—Östersund to Trondheim. *Sver. geol. Unders.*, **C 717**, 66 pp.

Gee, D. G. 1978. Nappe displacement in the Scandinavian Caledonides. *Tectonophysics*, **47**, 393–419.

Gee, D. G. 1980. Basement–cover relationships in the central Scandinavian Caledonides. *Geol. Fören. Stockh. Förh.*, **102**, 455–474.

Gee, D. G. 1981. The Dictyonema-bearing phyllites at Nordaunevoll, eastern Trøndelag. *Norsk geol. Tidsskr.*, **61**, 93–95.

Gee, D. G. and Kumpulainen, R. 1980. An excursion through the Caledonidan mountain chain in central Sweden from Östersund to Storlien. *Sver. geol. Unders.*, **C 774**, 66 pp.

Gee, D. G. and Wilson, M. R. 1974. The age of orogenic deformation in the Swedish Caledonides. *Am. J. Sci.*, **274**, 1–9.

Gee, D. G. and Zachrisson, E. 1974. Comments on stratigraphy, faunal provinces and structure of the metamorphic allochthon, central Scandinavian Caledonides. *Geol. Fören. Stockh. Förh.*, **96**, 61–66.

Gee, D. G. and Zachrisson, E. 1979. The Caledonides in Sweden. *Sver. geol. Unders.*, **C 769**, 48 pp.

Gee, D. G., Kumpulainen, R. and Thelander, T. 1978. The Tåsjön Décollement, central Swedish Caledonides. *Sver. geol. Unders.*, **C 742**, 35 pp.

Gee, D. G., Guezou, J. C., Roberts, D., and Wolff, F. C. this volume. The central–southern part of the Scandinavian Caledonides.

Getz, A. 1890. Graptolitførende skiferzoner i det trondhjemske. *Nyt. Mag. Naturv.*, **3**, 31–42.

Grenne, T. and Lagerblad, B. this volume. The Fundsjø Group, central Norway—a Lower Palaeozoic island arc sequence: geochemistry and regional implications.

Grenne, T. and Reinsbakken, A. 1981. Possible correlations of island arc greenstone belts and related sulphide deposits from the Grong and eastern Trondheim districts of the central Norwegian Caledonides. *Trans. Inst. Min. Metall. (Sect. B: Applied earth sci.)*, **90**, B59 (Abstr.).

Grenne, T. and Roberts, D. 1980. Geochemistry and volcanic setting of the Ordovician Forbordfjell and Jonsvatn Greenstones, Trondheim Region, Central Norwegian Caledonides. *Contr. Mineral. Petrol.*, **74**, 375–386.

Grenne, T., Grammeltvedt, G. and Vokes, F. M. 1980 Cyprus-type sulphide deposits in the western Trondheim district, central Norwegian Caledonides. In Panayiotou, A. (Ed.), *Ophiolites. Proc. Int. Ophiolite Symp. Cyprus 1979*, 727–743.

Griffin, W. L. and Brueckner, H. K. 1980. Caledonian Sm–Nd ages and a crustal origin for Norwegian eclogites. *Nature, London*, **285**, 319–321.

Gustavson, M. 1975. The low-grade rocks of the Skålvaer area, S. Helgeland and their relationship to high-grade rocks of the Helgeland Nappe Complex. *Norges geol. Unders.*, **322**, 13–33.

Gustavson, M. 1978a. Geochemistry of the Skålvaer greenstone, and a geotectonic model for the Caledonides of Helgeland, north Norway. *Norsk geol. Tidsskr.*, **58**, 161–174.

Gustavson, M. 1978b. Caledonides of north–central Norway. In Tozer, E. T. and Schenk, P. E. (Eds), *IGCP Project 27, Caledonian—Appalachian Orogen of the North Atlantic Region. Geol. Surv. Canada, Paper*, **78–13**, 25–30.

Halls, C., Reinsbakken, A., Ferriday, I., Haugen, A. and Rankin, A. 1977. Geological setting of the Skorovas orebody within the allochthonous volcanic stratigraphy of the Gjersvik Nappe, central Norway. In *Volcanic Processes in Ore Genesis. Spec. Pap. No. 7, Inst. Min. Metall.—Geol. Soc. London*, 128–151.

Hardenby, C. 1980. Geology of the Kjølhaugen area, eastern Trøndelag, central Scandinavian Caledonides. *Geol. Fören. Stockh. Förh.*, **102**, 475–492.

Harland, W. B. and Gayer, R. A. 1972. The Arctic Caledonides and earlier oceans. *Geol. Mag.*, **109**, 289–314.

Hedström H. 1930. Om Ordoviciska fossil från Ottadalen i det centrala Norge. *Avhandl. Nor. Vidensk. Akad. i Oslo Mat.–naturv. klasse*, **1930–10**, 10 pp.

Holmqvist, A. 1980. Orodovician gastropods from Vardofjället, Swedish Lapland, and the dating of Caledonian serpentinite conglomerates. *Geol. Fören. Stockh. Förh.*, **102**, 493–497.

Holmqvist, A. 1982. Ordovician gastropods from Vardofjället, Swedish Lapland, and the dating of Caledonian serpentinite conglomerates: a reply. *Geol. Fören. Stockh. Förh.*, **104**, 191–192.

Holtedahl, O. 1915. Fossiler fra Smølen. *Norges geol. Unders.*, **69**, 1–14.

Holtedahl, O. 1936. Trekk av det Scandinaviske fjelkjedestrøks historie. *Nordiska (19. Skand.) Naturforskaremötet i Helsingfors 1936*, 129–145.

Horne, G. S. 1979. Mélange in the Trondheim Nappe suggests a new tectonic model for the central Norwegian Caledonides. *Nature, London*, **281**, 267–270.

Hossack, J. R., Garton, M. R. and Nickelsen, R. P. this volume. The geological section from the foreland up to the Jotun thrust sheet in the Valdres area, South Norway.

Häggbom, O. 1978. Polyphase deformation of a discontinuous nappe in the central Scandinavian Caledonides. *Geol. Fören. Stockh. Förh.*, **100**, 349–354.

Högbom, A. G. 1910. Studies in the post-Silurian thrust region of Jämtland. *Geol. Fören. Stockh. Förh.*, **31**, 289–346.

Juve, G. 1977. Metal distribution at Stekenjokk: primary and metamorphic patterns. *Geol. Fören. Stockh. Förh.*, **99**, 149–158.

Kautsky, G. 1946. Neue Gesichtspunkte zu einigen nordskandinavischen Gebirgsproblemen. *Geol. Fören. Stockh. Förh.*, **68**, 589–602.

Kautsky, G. 1953. Der geologische Bau des Sulitelma–Salojauregebietes in den nordskandinavischen Kaledoniden. *Sver. geol. Unders.*, **C 528**, 232 pp.

Keppie, J. D. this volume. The Appalachian collage.

Kiær, J. 1932. The Hovin Group in the Trondheim area. *Skr. Nor. Vidensk. Akad. i Oslo Mat.–naturv. klasse*, 1932–4, 175 pp.

Kieft, C. 1952. *Geology and Petrology of the Tärna Region, Southern Swedish Lapland*, Diss., Univ. of Amsterdam, 98 pp.

Klingspor, I. and Gee, D. G. 1981. Isotopic age-determination studies of the Trøndelag trondhjemites. *Terra Cognita*, **1**, 55 (Abstr.).

Kollung, S. 1979. Stratigraphy and major structures of the Grong District, Nord–Trøndelag. *Norges geol. Unders.*, **354**, 1–51.

Kulling, O. 1933. Bergbyggnaden inom Björkvattnet–Virisen–området i Västerbottensfjällens centrala del. *Geol. Fören. Stockh. Förh.*, **55**, 167–422.

Kulling, O. 1938. Grönstenarnas placering inom Västerbottensfjällens kambrosilurstratigrafi. *Geol. Fören. Stockh. Förh.*, **60**, 153–176.

Kumpulainen, R. 1980. Upper Proterozoic stratigraphy and depositional environments of the Tossåsfjället Group, Särv Nappe, southern Swedish Caledonides. *Geol. Fören. Stockh. Förh.*, **102**, 531–550.

Kumpulainen, R. 1982. The upper Proterozoic Risbäck Group, northern Jämtland and southwestern Västerbotten, central Swedish Caledonides. *Uppsala Univ. Dept. Min. Pet., res. rep.* **28**, 60 pp.

Lutro, O. 1979. The geology of the Gjersvik area, Nord–Trøndelag. *Norges geol. Unders.*, **354**, 53–100.

McKerrow, W. S., Lambert, R. St. J. and Chamberlain, V. faunal migration across the Iapetus Ocean. *Nature, London*, **263**, 305–306.

McKerrow, W. S., Lambert, R. St. J. and Chamberlain, V. E. 1980. The Ordovician, Silurian and Devonian time scales. *Earth Planet. Sci. Lett.*, **51**, 1–8.

Miyashiro, A. 1974. Volcanic rock series in island arcs and active continental margins. *Am. J. Sci.*, **274**, 321–355.

Mørk, M. B. E. 1979. *Metamorf Utvikling og Gabbrointrusjon på Krutfjell*. Thesis, Univ. of Oslo, 307 pp.

Neuman, R. B. and Bruton, D. L. 1974. Early Middle Ordovician fossils from the Hølonda area, Trondheim Region, Norway. *Norsk geol. Tidsskr.*, **54**, 69–115.

Nilsen, O. 1971. Sulphide mineralization and wall rock alteration at Rødhammaren Mine, Sør–Trøndelag, Norway. *Norsk geol. Tidsskr.*, **51**, 329–354.

Nilsson, G. 1964. Berggrunden inom Blåsjöområdet i nordvästra Jämtlandsfjällen. *Sver. geol. Unders.*, **C 595**, 70 pp.

Oftedahl, C. 1956. Om Grongkulminasjonen og Grongfeltets skyvedekker. *Norges geol. Unders.*, **195**, 57–64.

Olsen, J. 1980. Genesis of the Joma stratiform sulphide deposit, central Norwegian Caledonides. *Proc. 5th IAGOD symposium, Alta, Utah 1978*, **1**, 745–757.

Prestvik, T. 1972. Alpine-type mafic and ultramafic rocks of Leka, Nord–Trøndelag. *Norges geol. Unders.*, **273**, 23–34.

Prestvik, T. 1974. Supracrustal rocks of Leka, Nord-Trøndelag. *Norges geol. Unders.*, **311**, 65–87.

Prestvik, T. 1980. The Caledonian ophiolite complex of Leka, north central Norway. In Panayiotou, A. (Ed.), *Ophiolites. Proc. Int. Ophiolite Symp. Cyprus 1979*, 555–566.

Priem, H. N. A., Boelrijk, N. A. I. M., Hebeda, E. H., Verdurmen, E. A. T. and Verschure, R. H. 1975. Isotopic dating of the Caledonian Bindal and Svenningdal granitic massifs, central Norway. *Norges geol. Unders.*, **319**, 29–36.

Ramberg, H. 1966. The Scandinavian Caledonides as studied by centrifuged dynamic models. *Uppsala Univ. Geol. Inst. Bull.*, **43**, 72 pp.

Ramberg, I. B. 1981. The Brakfjellet tectonic lens: evidence of pinch-and-swell in the Caledonides of Nordland, north central Norway. *Norsk geol. Tidsskr.*, **61**, 87–91.

Ramberg, I. B., Tørudbakken, B. O. and Råheim, A. in press. Rb–Sr geochronological evidence for the involvement of Precambrian rocks within the Uppermost Allochthon, Tustervatn area, N. Norway. *Norges geol. Unders.*

Reymer, A. P. S. 1979. *Investigations into the Metamorphic Nappes of the Central Scandinavian Caledonides on the Basis of Rb–Sr and K–Ar Age Determinations*. Thesis, Univ. of Leiden, 123 pp.

Riis, I. and Ramberg, I. B. 1979. Nyere data fra Nordlands høymetamorfe dekkekompleks. *Geolognytt*, **13**, 60–61 (Abstr.).

Roberts, D. 1975. The Stokkvola conglomerate—a revised stratigraphical position. *Norsk geol. Tidsskr.*, **55**, 361–371.

Roberts, D. 1980. Petrochemistry and palaeogeographic setting of Ordovician volcanic rocks of Smøla, Central Norway. *Norges geol. Unders.*, **359**, 43–60.

Roberts, D. and Gale, G. H. 1978. The Caledonian–Appalachian Iapetus Ocean. In Tarling, D. H. (Ed.), *Evolution of the Earth's Crust*, Academic Press, London, New York, 255–342.

Roberts, D. and Gee, D. G. this volume. An introduction to the structure of the Scandinavian Caledonides.

Roermund, H. van. this volume. Eclogites of the Seve Nappe, central Scandinavian Caledonides.

Rui, I. J. 1972. Geology of the Røros district, southeastern Trondheim region with a special study of the Kjøliskarvene–Holtsjøen area. *Norsk geol. Tidsskr.*, **52**, 1–21.

Rui, I. J. and Bakke, I. 1975. Stratabound sulphide mineralization in the Kjøli area, Røros district. Norwegian Caledonides. *Norsk geol. Tidsskr.*, **55**, 51–75.

Ryan, P. D., Williams, D. M. and Skevington, D. 1980. A revised interpretation of the Ordovician stratigraphy of Sør Trøndelag, and its implications for the evolution of the Scandinavian Caledonides. In Wones, D. R. (Ed.), *The Caledonides in the USA. Virginia Polytechnic Inst. and State Univ., Dept. Geol. Sci., Mem.*, **2**, 99–105.

Sandwall, J. 1981a. Caledonian geology of the Jofjället area, Västerbotten county, Sweden. *Sver. geol. Unders.*, **C 778**, 105 pp.

Sandwall, J. 1981b. Greenstones related to rifting and ocean basin opening in the Jofjället area, central Swedish Caledonides. *Geol. Fören. Stockh. Förh.*, **103**, 421–428.

Senior, A. and Otten, M. T. this volume. The Artfjället gabbro and its bearing on the evolution of the Storfjället Nappe, central Swedish Caledonides.

Siedlecka, A. 1967. Geology of the eastern part of the Meråker area. *Norges geol. Unders.*, **245**, 22–58.

Sjöstrant, T. 1978. Caledonian geology of the Kvarnbergsvattnet area, northern Jämtland, central Saeden. *Sver. geol. Unders.*, **C 735**, 107 pp.

Solyom, Z. Gorbatschev, R. and Johansson, I. 1979a. The Ottfjället Dolerites. Geochemistry of the dyke swarm in relation to the geodynamics of the Caledonide orogen in central Scandinavia. *Sver. geol. Unders.*, **C 756**, 38 pp.

Solyom, Z., Andréasson, P. G. and Johansson, I. 1979b. Geochemistry of amphibolites from Mt. Sylarna, Central Scandinavian Caledonides. *Geol. Fören. Stockh. Förh.*, **101**, 17–27.

Spjeldnaes, N. this volume. Biostratigraphy of the Scandinavian Caledonides.

Steel, R., Siedlecka, A. and Roberts, D. this volume. The Old Red Sandstone basins of Norway and their deformation: a review.

Stephens, M. B. 1977a. Stratigraphy and relationship between folding, metamorphism and thrusting in the Tärna–Björkvattnet area, northern Swedish Caledonides. *Sver. geol. Unders.*, **C 726**, 146 pp.

Stephens, M. B. 1977b. The Stekenjokk volcanites—segment of a Lower Palaeozoic island arc complex. In Bjørlykke, A., Lindahl, I. and Vokes, F. M. (Eds.), *Kaledonske Malmforekomster. BVLI's Tekniske Virksomhet, Trondheim*, 24–36.

Stephens, M. B. 1980. Occurrence, nature and tectonic significance of volcanic and high-level intrusive rocks within the Swedish Caledonides. In Wones, D. R. (Ed.), *The Caledonides in the USA. Virginia Polytechnic Inst. and State Univ., Dept. Geol. Sci., Mem.*, **2**, 289–298.

Stephens, M. B. 1982. Field relationships, petrochemistry and petrogenesis of the Stekenjokk volcanites, central Swedish Caledonides. *Sver. geol. Unders.*, **C 786**, 111 pp.

Stephens, M. B. and Senior, A. 1981. The Norra Storfjället lens—an example of fore-arc basin sedimentation and volcanism in the Scandinavian Caledonides. *Terra Cognita*, **1**, 76–77 (Abstr.).

Stephens, M. B., Gustavson, M., Ramberg, I. B. and Zachrisson, E. this volume. (a). The Caledonides of central–north Scandinavia—a tectonostratigraphic overview.

Stephens, M. B., Furnes, H., Robins, B. and Sturt, B. A. this volume (b). Igneous activity within the Scandinavian Caledonides.

Stigh, J. 1979. Ultramafites and detrital serpentinites in the central and southern parts of the Caledonian Allochthon in Scandinavia. *Geol. Inst., Chalmers Tekniska Högskola och Göteborgs Univ., Publ.* **A 27**, 222 pp.

Strand, T. 1932. A Lower Ordovician fauna from the Smøla Island, Norway. *Norsk geol. Tidsskr.*, **11**, 356–366.

Strand, T. and Kulling, O. 1972. Scandinavian Caledonides. John Wiley and Sons Ltd., London, 302 pp.

Strömberg, A. 1961. On the tectonics of the Caledonides in the south-western part of the county of Jämtland, Sweden. *Uppsala Univ. Geol. Inst. Bull.*, **39**, 92 pp.

Strömberg, A. 1969. Initial Caledonian magmatism in Jämtland area, Sweden. In *North Atlantic Geology and Continental Drift. Am Assoc. Petroleum Geologists, Mem.*, **12**, 375–387.

Sturt, B. A., Pringle, I. R. and Ramsay, D. M. 1978. The Finnmarkian phase of the Caledonian Orogeny. *J. geol. Soc. London.*, **135**, 547–610.

Størmer, L. 1941. Dictyonema shales outside the Oslo Region. *Norsk geol. Tidsskr.*, **20**, 161–170.

Sundblad, K. 1980. A tentative 'volcanogenic' formation model for the sediment-hosted Ankarvattnet Zn–Cu–Pb massive sulphide deposit, central Swedish Caledonides. *Norges geol. Unders.*, **360**, 211–227.

Sundvoll, B. and Roberts, D. 1977. Framgangsrapport på datering og geokjemi av eruptivbergarter på Smøla og Hitra. *Norges geol. Unders.*, internal report (unpubl.).

Thorslund, P. 1960. The Cambro–Silurian. In *Description to Accompany the Map of the Pre-Quaternary Rocks of Sweden. Sver. geol. Unders.*, **Ba 16**, 69–110.

Trouw, R. A. J. 1973. Structural geology of the Marsfjällen area, Caledonides of Västerbotten, Sweden. *Sver. geol. Unders.*, **C 689**, 115 pp.

Törnebohm, A. E. 1888. Om fjällproblemet. *Geol. Fören. Stockh. Förh.*, **10**, 328–336.

Törnebohm, A. E. 1896. Grunddragen af det centrala Skandinaviens bergbyggnad. *Kgl. svenska vetensk, akad. Handl.*, **28 (5)**, 212 pp.

Vogt, J. H. L. 1889. Funn av Dictyonema ved Holtsjøen, Holtålen. Foredrag 16 nov. 1888. *Förh. Vitenskapsselsk. Christiania för 1888*. Oversikt over møder, 12 pp.

Vogt, T. 1945. The geology of part of the Hølonda–Horg district, a type area in the Trondheim region. *Norsk geol. Tidsskr.*, **25**, 449–527.

Watanabe, T., Langseth, M. G. and Anderson, R. N. 1977. Heat flow in back-arc basins of the Western Pacific. In Talwani, M. and Pitman, W. C. III. (Eds), *Island Arcs, Deep Sea Trenches, and Back-Arc Basins*. AGU, 137–162.

Williams, H. and Hatcher, R. D. 1982. Suspect terranes and accretionary history of the Appalachian orogen. *Geology*, **10**, 530–536.

Wilson, J. T. 1966. Did the Atlantic close and then re-open? *Nature*, London, **211**, 676–681.

Wolff, F. C. 1967. Geology of the Meråker area as a key to the eastern part of the Trondheim region. *Norges geol. Unders.*, **245**, 123–142.

Wolff, F. C. 1976. Geologisk kart over Norge, berggrunnskart Trondheim 1:250 000. *Norges geol. Unders.*

Wolff, F. C. and Roberts, D. 1980. Geology of the Trondheim region. *Norges geol. Unders.*, **356**, 117–128.

Wolff, F. C., Roberts, D., Siedlecka, A., Oftedahl, C. and Grenne, T. 1980. Guide to excursions across part of the Trondheim Region, Central Norwegian Caledonides. *Norges geol. Unders.*, **356**, 129–167.

Zachrisson, E. 1964. The Remdalen Syncline. *Sver. geol. Unders.*, **C 596**, 53 pp.

Zachrisson, E. 1969. Caledonian geology of Northern Jämtland–Southern Västerbotten. *Sver. geol. Unders.*, **C. 644**, 33 pp.

Zachrisson, E. 1973. The westerly extension of Seve rocks within the Seve–Köli Nappe Complex in the Scandinavian Caledonides. *Geol. Fören. Stockh. Förh.*, **95**, 243–251.

Zachrisson, E. 1982. Spilitization, mineralization and vertical metal zonation at the Stekenjokk strata-bound sulphide deposit, central Scandinavian Caledonides. *Trans. Inst. Min. Metall. (Sect. B: Applied earth sci.)*, **91**, B192–B199.

Post-orogenic morphotectonic evolution of the Scandinavian Caledonides during the Mesozoic and Cenozoic

J.-P. Peulvast

Laboratoire de Géographie Physique (L.A. 141), et Université de Paris–Sorbonne, 191 rue Saint-Jacques, 75005 Paris, France

ABSTRACT

The marginal bulge of the Baltic Shield includes high Caledonian structural reliefs and large basement areas. This bulge incorporates an old mountain chain, eroded over a long period; it is the result of uplift and renewed differential erosion along an anteclise approximately coinciding with the axis of the Caledonides. Morphological and sedimentological evidence show that the uplift represents the late strengthening of an old positive trend and that it follows a now well-known period of taphrogenic activity, with vertical movements restricted to the sides of the Oslo rift (Permo-Trias) and of the North Atlantic North Sea rift system (Trias–Jurassic). The epeirogenic phase is associated with subsidence of the neighbouring areas of the continental margin and of the North Sea; it is thought to be of essentially Neogene age, after a Palaeogene initiation. The anteclise is supposed to be an indirect result of the opening of the Norwegian Sea, according to the most recent geophysical theories. The influence of the Caledonian structures on the style of uplift and on the morphological development is analysed through identification of the successive morphogenic sequences.

Introduction

The Caledonian orogeny ended during the Middle Devonian. In spite of this antiquity and long erosion, the Scandinavian Caledonides constitute the main part of the Scandes range, and its highest mountains. Classically considered as the result of Tertiary 'oblique uplift' (O. Holtedahl 1953, 1960), this mountain range forms the western marginal 'bulge' of the Baltic Shield east of the Norwegian Sea. An analysis of the main morphostructural features, and of the uplift style and history, allows a discussion of the role of deep and shallow Caledonian structures in the tectonic and morphological evolution of the present mountain range.

The Caledonides and the Marginal Bulge

Morphostructural features

The Scandian mountain range is 2000 km long and 200–300 km wide. Its higher peaks range from 1000 to 2470 m. A smoothing of its topographic features and drawing of a subenvelope map (Gjessing 1967, Fig. 1 and Fig. 2) give the picture of a huge asymmetric bulge, with a long and gentle eastern slope and a short and steep western side.

Geographically, the Scandes do not coincide exactly with the Caledonides, though their directions are roughly parallel (Fig. 1). In southern Norway, they incorporate a large outcrop of the Baltic Shield (Hardangervidda), whereas the Caledonian cover is restricted to the oblique belt of the Hardanger–Ryfylke klippes. Further north, there is a good fit between the Scandes summital zone and the Caledonian nappes, except in western Norway where it includes large areas of basal gneisses. In eastern Finnmark, the elevations decrease slowly towards the east, obliquely within the Gaissa nappe and the Kalak nappe complex.

The western and northwestern slope of the Scandes decreases in width towards the north, with increasing asymmetry of the bulge. It is composed of plateaus (around the fjords of Vestlandet from the Vest Agder to the outer Møre, Trøndelag, Finnmark) that are cut by a high and steep cliff above the strandflat

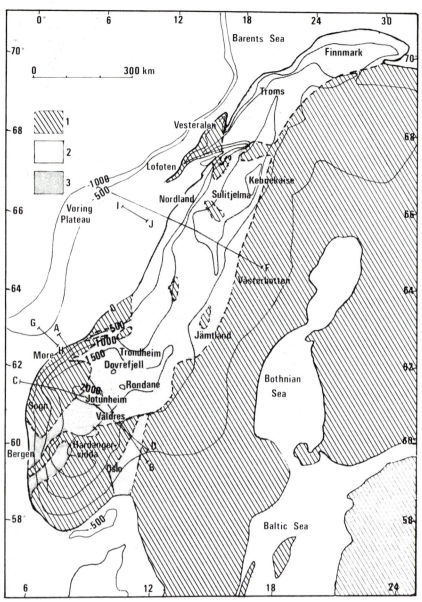

Fig. 1   The enveloping surface of the Scandian Bulge and the Caledonides. In part from Gjessing (1967), with location
of Fig. 2, 4, 6, 8, 10. 1: Baltic Shield; 2: Caledonides; 3: Oslo Graben

(from Stavanger to Tromsø) or above the continental shelf (Finnmark). Its general direction, turning from north in Vestlandet to northeast and east towards northern Scandinavia is often transverse or oblique to the Caledonian structures and the western basement areas. But it is dissected by fjords and basins related to the strandflat or overdeepened below sea-level and which often follow Caledonian or Precambrian weakness zones (Hardangerfjord along phyllites: H. Holtedahl 1975; fjords between Sogn and Nordfjord along Middle Devonian E–W structures; Trondheimsfjord; Lofoten–Vesterålen: Peulvast 1977; Hamarøy–Sagfjord: Fig. 3; etc.).

The Scandian range itself includes two long dome-like areas of high mountains and plateaus in southern and northern Norway (Jostedalsbre–Jotunheim–Møre, Nordland–Troms), and a lower central zone in Trøndelag. In spite of the general asymmetry, the summital zone is often widened towards and above the eastern gentle slope (Jotunheim–Rondane; Sarek–Kebnekaise), owing to the presence of major structural reliefs (Rudberg 1962; Peulvast 1978). In the southern culmination, mountainous reliefs rise above a high complex peneplain gently sloping towards the periphery, and which truncates basement rocks of both igneous and supracrustal origin, except some quartzites (Gaustatoppen, in Telemark), as well as parts of the external Caledonian structures (Valdres, South Trøndelag). Some of these mountains bear remnants

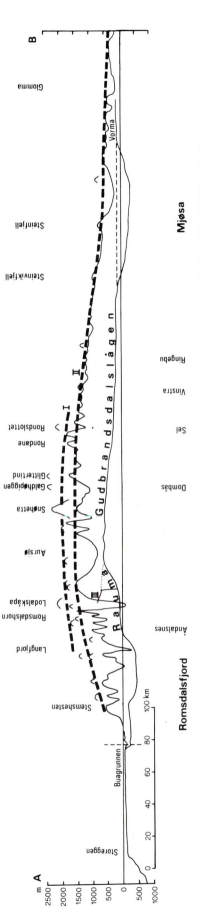

Fig. 2 Schematic profile of the Scandian Bulge, between Romsdalsfjord and Mjøsa. Redrawn from Holtedahl (1960) and Gjessing (1967), modified. I: distinct remnants of the pre-Palaeocene surface. II: post-Palaeocene peneplain. III: pre-glacial valleys

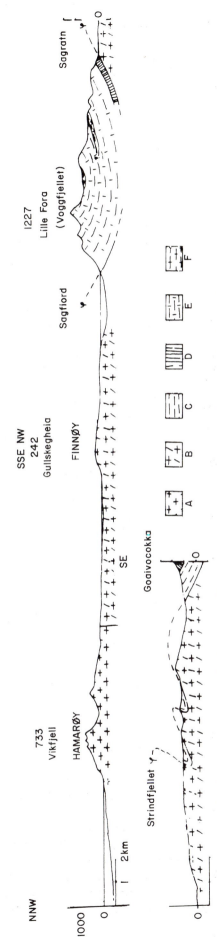

Fig. 3 Profile within Hamarøy–Sagfjord area. The western edge of the Scandes is the result of differential erosion within a broad flexure zone. Basement: A: mangerites; B: monzonitic orthogneisses. Caledonian nappes: C: micaschists (Seve–Köli nappes); D: marbles; E: micaceous gneiss; F: granitoid rocks

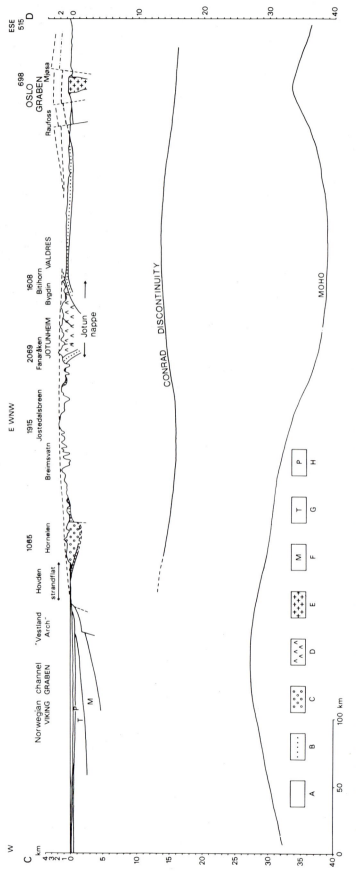

Fig. 4   Profile across the Scandian Bulge from the North Sea to the Oslo Graben. Partly compiled from Sellevoll (1973), Seguin (1971), Ziegler (1978). A: basement; B: metasediments of the Caledonian nappes; C: Jotun rocks; D: Devonian sediments; E: Permian intrusive rocks; F: Mesozoic sediments; G: Tertiary sediments; H: Pleistocene sediments

of higher rolling peneplains. All of them coincide with the most resistant rocks of the Caledonidian units intensely folded together with the basement (Fig. 4): the high orthogneissic plateau of Jostedalsbre (1800 to 2083 m), the jagged peaks and high cupolas of Jotunheimen (1800 to 2470 m) carved into pyroxene gneisses, the huge isolated massifs of Dovrefjell and Rondane (2000 to 2286 m). They are separated by broad hanging depressions or 'palaeic valleys' extending the eastern and southern plateau across the mountains, along upper Gudbrandsdal and Valdres; those lower areas are hollowed into various metasediments (phyllites, Valdres sparagmites, etc.) and mylonitic rocks surrounding the Jotun nappe. The southwestern anorthositic part of this nappe is itself truncated together with gneisses south of the Jostedalsbre by the same peneplain (Main Sogn surfce: Peulvast, in preparation) which slopes gently towards the western edge of the highlands. These morphological units and the valleys or fjords that are deepened below are often surrounded by monoclinal escarpments which coincide with the present erosional Caledonian nappe fronts or with minor folded or faulted structures, e.g. around the Jotun nappe (Fig. 5), giving to the reliefs of the main topographical levels an Appalachian-like style.

The same complex Appalachian-like morphology exists in the northern highlands, below at least two main superposed peneplains. Mountains rise above the western coast, from Helgeland to Troms and West Finnmark; the highest of them coincide with parts of the Caledonian upper nappes and with some very resistant rocks of the basement (e.g. some parts

of the mangerites and migmatites of the Lofoten–Versterålen that have not undergone the Grenvillian retrograde recrystallization: Griffin *et al.* 1978; Peulvast 1977). But the crest of the Scandes is located further east; some of the highest summits rise above the Swedish side. The reason is that the location of all these mountains is controlled by the distribution of the most resistant rocks and by the Caledonian structures, e.g. the basic and ultramafic intrusive rocks incorporated in the upper and middle nappes (Lyngen, Sulitjelma, Seiland Province), the migmatites and certain gneisses of these nappes (Beiarn, Tromsø), the amphibolites of the Seve nappes (Kebnekaise, Sarek, Abisko mountains: Rudberg 1954, 1962), the basement in several windows (Store Børgefjell, Rombak). Metasedimentary units are truncated by lower plateaus (Finnmark, Eastern Troms) or furrowed by longitudinal depressions and valleys (e.g. Namdal) between narrow structural ridges of folded schists, quartzites, or limestones (e.g. Håfjell, Ofoten).

Parts of the *eastern and southern boundary of the Scandes* coincide with the erosional Caledonian thrust front above the basement and a fringe of autochthonous soft sediments (e.g. Hallingskarvet–Hallingdal, parts of Sweden: Fig. 6; Finnmark). But some parts of this 'glint line' (Rudberg 1962) are located at the erosional thrust front of higher nappes (e.g. the eastern edge of Jotunheimen above Valdres: Fig. 7), whereas other parts of the mountain edge are located within the autochthonous basement (Västerbotten) above low peneplains gently sloping towards the Bothnian Sea (Rudberg 1954).

Differential erosion accompanying the develop-

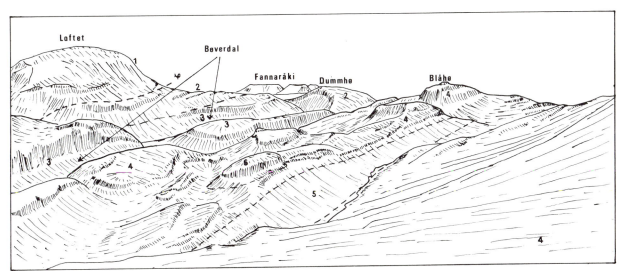

Fig. 5 Bøverdal, as seen from Nettoseterfjell. Appalachian relief carved on the NW edge of Johunheimen below the Palaeogene (?) uncomplete peneplain and a lower (Neogene?) basin in the NW edge of the Faltungsgraben. Jotunheimen (Loftet, Fannaråki) bears remnants of a higher rolling peneplain. 1: Jotun pyroxene gneisses; 2: Valdres sparagmite; 3: limestones; 4: phyllites and quartzites; 5: greenschists; 6: greenstones

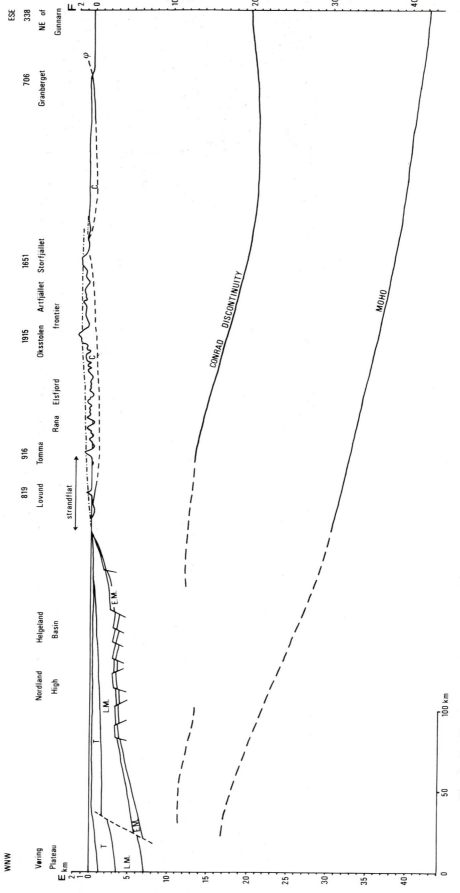

Fig. 6   Profile across the Scandinavian Bulge, from Vøring Plateau to Sweden (NE of Gunnarn). Partly compiled from Sellevoll (1973) and Jørgensen with Navrestad (1981). C: Caledonian units; M: Mesozoic sediments; T: Tertiary sediments

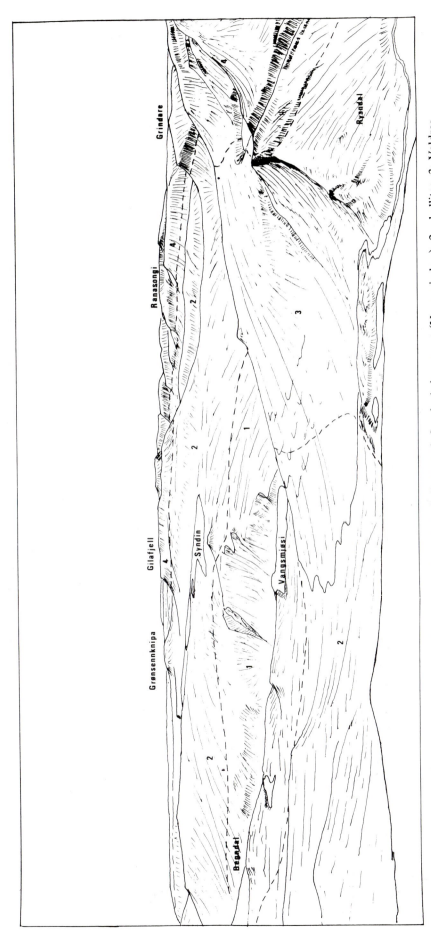

Fig. 7   Southeastern edge of Jotunheimen as seen from Slettefjell, towards South. 1: basement (Vang window); 2: phyllites; 3: Valdres sparagmite; 4: Jotun rocks

ment of successively lower peneplains and the excavation of valleys explains the details of the configuration of the ridge and its edges. Obviously, such a pattern cannot be explained only by glacial erosion and Pleistocene evolution (Peulvast 1978). But the high altitudes of plateaux and mountains, the general asymmetry and the approximate coincidence with the various rocks and folded structures of an old chain preserved above the basement in spite of a greater intensity of erosion in this high zone must be explained by a relatively recent uplift. Such an epeirogenic movement must be distinguished from the post-glacial isostatic rebound, which reaches its maximum intensity in the more eastern area of the Bothnian Sea.

The asymmetric uplift

Tectonic style

In spite of the rugged morphology, the enveloping surface and the main plateaus are disposed within a broad (300 km) asymmetric but simple bulge, antiform, or anteclise (Mattauer 1973). Between the southern and northern higher zones or domes (Torske 1975), the lower altitudes of the Trondheim region may be explained by erosion of predominantly pelitic rocks between more resistant units, but also by a less important uplift, as shown by the preservation of Devonian deposits on relatively low plateaus near Røros and of downfaulted Jurassic deposits near the Trondheimsfjord (Vigran 1970).

The western steeper slope of this bulge is mainly flexured, as indicated by Gjessing (1977), but the present topography, with the strandflat and a high cliff carved into the western plateaus or mountains is obviously the result of erosion east of the hinge line. Only a few normal faults around Vestfjord (Dekko 1975) and Andfjord (Dalland 1975), between the coast and the Lofoten horst, directly control parts of this western edge. Elsewhere, the coastal areas show a differentiated and irregular relief carved into a flexural zone along old weakness lines, without cliffs or escarpments which could be demonstrated as directly created by faulting (Peulvast 1978). In southwestern Norway, N–S or NNW–SSE faults, ascribed to Mesozoic extension around the North Sea (Nilson 1973; Indrevaer and Steel 1975) are truncated by the strandflat. But the west coast of Norway is a seismic zone (Husebye *et al.* 1978); even if the post-glacial isostatic rebound is mainly responsible for this seismicity, this fact can also reveal recent faulting along old fractures of the coastal area (Meløy: Gabrielsen and Ramberg 1979). However, the longitudinal trenches of the shelf, which had been considered as excavated by glaciers along fault lines associated with the recent

oblique uplift (O. Holtedahl 1953, 1960; H. Holtedahl 1955) are now interpreted as furrows excavated by erosion along the contact between the basement and Cenozoic sediments (Sellevoll and Sundvor 1974).

This flexured pattern therefore looks like the one of Eastern Greenland, a symmetrical area where Eocene basalts are flexured in the same way (Torske 1972; Surlyk 1977).

The flexure hinge line is thought to be located off the outer edge of the strandflat, where the basement disappears under sediments. Offshore, the pre-Pleistocene sediments show dips of 2–9°, with maxima west of the two culminations and lower angles offshore Trøndelag (Sellevoll 1975). The later sedimentation seems to have been restricted to the western part of the shelf (Fig. 8), showing the influence of eustatic events (low Mid-Chattian sea-level and a late Oligocene to Miocene sea-level rise which did not reach the pre-Chattian level: Jørgensen and Navrestad 1981) and perhaps a shift of the hinge line towards the west during Miocene and later. Faults may be more important along the Barents Sea, where the disposition of rivers passing from internal low areas (e.g. Finnmarkvidda, Inari) through higher coastal areas could be explained by antecedence to a recent uplift (Gjessing 1977).

The eastern side is locally bounded by distal longitudinal faults which do not seem to be related to the uplift, along Skagerrak and Oslo Graben. But information about recent faulting remains rather scarce (Ramberg and Smithson 1975) and the morphology shows only furrows and escarpments carved by differential erosion along these weakness zones, within a gently sloping coastal area (Barth 1939). In spite of the locally steep eastern edge of the Caledonian mountains, the general appearance is that of a broad flexured zone bounding the wide and shallowly inclined eastern plateaus. Its hinge line is rather indefinite, and is probably located east of the mountain edge above the depressed areas of the Skagerrak, central Sweden and the Baltic–Bothnian basins. The preservation of Cambro-Silurian sediments (South Sweden, Baltic Sea) and of older weathered deposits (Talvitie 1981) shows for these low areas a probably epeirogenic origin (stability or old and slight tendency to subsidence) in spite of a present tendency to uplift (Mörner 1981). Recent faulting in northern Sweden (the Pärve fault: Lundqvist and Lagerbäck, 1976), adjacent to the flexured area, is related to this uplift and to the post-glacial isostatic rebound rather than to the bulge-synclise disposition of the Scandes and the Baltic area. But some abnormal rivers in northern Sweden may be antecedent to faulting and block tilting in the basement (Rudberg 1976).

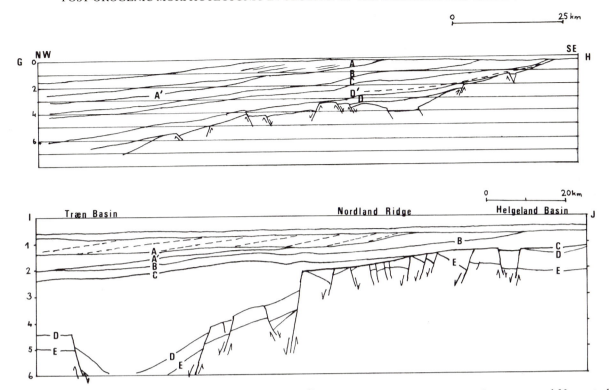

Fig. 8 Profiles across the Norwegian continental shelf. Reproduced with permission from Jørgensen and Navrestad (1981; lines B–2/75 and B–6/78). Age of the horizons: A: Lower Miocene; A': Mid-Chattian (Mid-Oligocene); B: top Lower Eocene; C: base Tertiary; D': Mid Cretaceous (lowermost Cenomanian); D: Late Cimmerian unconformity (top Jurassic); E: Middle Jurassic (top Bathonian)

Within the bulge, recent faulting along old lineaments cannot be proved, even around the relatively depressed Trondheim area, between the lineaments of Ringebu–Vågå–Nordfjord in the south (Prost 1975) and of Jämtland in the north (Strömberg 1976); high gneissic and quartzitic reliefs extend the higher adjoining mountains across these lineaments (Rondane, Dovrefjell). Elsewhere, the old faults control mainly transverse valleys rather than major escarpments of recent tectonic origin. Except in some distal areas the long wavelength fold formed without significant faulting.

Structural environment

The Scandian bulge is partially surrounded by the old rifted structures of Permian to late Cimmerian age which between 62 and 70°N are separated from their Greenlandian part by the line of oceanic opening of the Norwegian Sea (Fig. 9).

South of Stadlandet (62°N), the bulge is *entirely intracratonic*. The distal part of the eastern side incorporates the peneplanated rift structures of the Oslo–Skagerrrak area, which were formed during the Permian and lower Mesozoic and perhaps later (Ramberg and Spjeldnaes 1978). Offshore the western edge, the Viking Graben and the associated

structures (e.g. the Vestland Arch), formed from the Trias to Late Cretaceous times (Whiteman *et al.* 1975; Ziegler 1978) have subsided; they are covered by Tertiary sediments which are disposed in a broad synclinal basin between the Scotland–Shetland platform and Norway, and show no faulting, except during Palaeocene.

Between Stadlandet and Tromsø, the western side of the bulge is formed by a flexure which arches the basement and the sediments of the continental shelf; this part borders an Atlantic-type continental margin (Heezen 1974). Below the Tertiary sediments of this long basin lie the northern extensions of the North Sea rift structures (Rønnevik and Navrestad 1977; Jørgensen and Navrestad 1981). These rift structures were formed during the Late Carboniferous or Permian to Late Jurassic (Early and Late Cimmerian) times, but distinct longitudinal basins continued to subside during the Cretaceous, before the oceanic opening. Except in the Lofoten–Vesterålen area (Vestfjord, Andfjord), they have been incorporated in the broad subsiding shelf along the Scandes, and the line of oceanic opening is roughly parallel to them (Eldholm and Thiede 1980).

The northeastern part, like the southern one, is entirely intracratonic. It is bordered to the north by the shallow Barents platform, where old age grabens

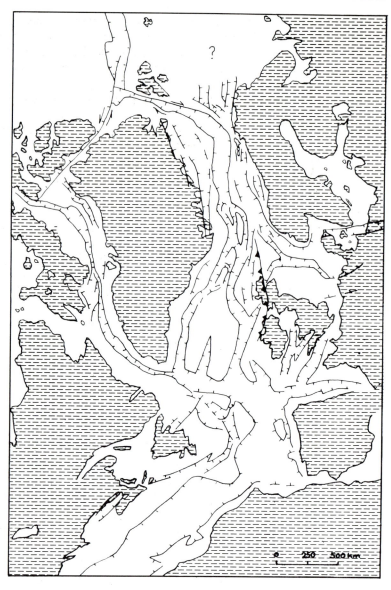

Fig. 9 Structural environment of NW Europe at the beginning of Cretaceous. Reproduced with permission from Ziegler (1978)

are fossilized below sediments further from the coast (Hammerfest Basin, with the Troms–Finnmark fault: Rønnevik 1981).

Therefore, the Scandinavian bulge appears as a new structure parallel to late Palaeozoic and Mesozoic rift structures, but related to large flexural movements which incorporated these structures on the outer edges. It appears to be geometrically related, for its major part, to an Atlantic-type continental margin.

Epeirogeny and its timing

As the Scandes is un uplifted area, practically devoid of post-Caledonian sediments, interpretation of the vertical movements must be based upon study of the neighbouring areas, their sedimentation history, and deformation.

Though parts of the bulge are zones of old positive trends (South Norway, probably since Grenvillian: Prost 1977), it incorporates Caledonian and older structures in a broad antiform without any documented reactivation of these structures (e.g. the large late-Caledonian synforms of the Trondheim and Bergen areas and the antiformal windows); thus the Caledonian orogeny was followed by an increased crustal rigidity. Subsidence and associated crustal thinning occurred during Devonian or Carboniferous times (Sturt *et al.* 1979; Jørgensen and Navrestad 1981), probably around a large zone of wrench faults of late-Caledonian origin between Greenland and Fennoscandia, but the location and sedimentology of the Devonian basins (Kvale 1975; Steel 1978) show that the pattern of uplifted and depressed areas is different from the present ones.

Torske (1975) has suggested a Mesozoic dome-like uplift of the Scandian bulge in Jotun–Møre and Lofoten–Troms. Indeed, late Palaeozoic and Mesozoic vertical movements were important along the northwest European rift structures; they were accompanied in southwestern Norway by intrusion of basic dykes (Faerseth *et al.* 1976). But the documented movements involved tilted blocks along tensional grabens (Ziegler 1978; Jørgensen and Navrestad 1981) rather than domes or anteclises. According to Bott (1976), the width of uplifted rift borders cannot exceed 50 or 79 km if the graben is 30 to 60 km wide, with a 10 to 20 km thick rigid upper crust (this was probably the case with the Viking graben and with the 'Arctic–North Atlantic rift'). As those rifts are now more than 50 to 100 km from the edge of the bulge (except in the Lofoten–Vesterålen area, where they are incorporated in it), the main part of the present uplifted area may be thought to have remained stable or only affected by a slow positive trend together with the Baltic Shield (Ziegler 1978; Meissner 1979). Its westernmost part probably subsided together with the rift zone (except the Lofoten–Vesterålen area: Eldholm *et al.* 1977), but the onlapping of Cretaceous sediments on the basement (Ziegler personal communication 1981) may be more likely explained as the result of an important eustatic sea level rise (Vail *et al.* 1977). In the Trondheim area, Mesozoic grabens and related vertical movements may also have occurred (Vigran 1970; Oftedahl 1980). The reliefs associated with most of the Cimmerian rifts provided considerable quantities of clastic rocks (sandstones, shales: Jurassic; Sturt *et al.* 1979; Deegan and Scull 1977) and all of them were worn down by erosion and peneplained (Oslo region; Nordland ridge as early as Cretaceous: Jørgensen and Navrestad 1981).

Morphological arguments are not sufficient for demonstrating a Mesozoic uplift of large parts of the Caledonides. Such an early uplift would probably have been followed by erosion and would not explain the preservation of the Caledonian structures in the uplifted areas, next to the depressed eastern area where the basement is laid bare (southeast Norway, Sweden, Finnmark). Moreover, it seems unlikely that elements of a Mesozoic radial hydrographic system are preserved on the supposed domes as suggested by Torske (1975). These valleys include subsequent segments (fault line valleys of southwestern Norway; Sogn–Jotun, Nordland), and owing to the development of peneplains until the Tertiary, their disposition and patterns are preferably explained by more recent differential erosion

under wet temperature or cold conditions (Peulvast 1978 and in preparation). Some of the Mesozoic climates, as shown by Vigran (1970) were warm and dry, and probably rather favourable to peneplanation.

The limits between the now-uplifted zone and the subsident areas of the Norwegian Shelf, North Sea, and Danish Basin (Surlyk 1980) had been established by the beginning of the Mesozoic period next to the present coast line, from North Norway to Skagerrak. But the morphological effects of this Mesozoic taphrogenic phase in the present landscapes are probably as weak as in East Greenland, where similar structures incorporated in the Atlantic slope of the bulge are completely truncated (Surlyk 1977). No signs of strong erosion are shown by the sediments of the North Sea during the Late Cretaceous (chalk; Ziegler 1978). A relative tectonic stability may have allowed this peneplanation.

The opening of the Norwegian Sea during the Palaeocene (Eldholm and Thiede 1980) was accompanied by vertical movements near the initial fracture zone. Subsidence and basaltic eruptions occurred on the outer edge of the shelf (Vøring plateau), though some parts were subject to uplift and subaerial erosion (e.g. the North Sea: Burke 1976; the continental margin in the Lofoten–Vesterålen area, where a post-uplift downwarping could occur along the present limit between the land and the inner shelf area: Eldholm *et al.* 1977; Rokoengen *et al.* 1977). But a clastic terrigenous sedimentation on the shelf edge and a broad marginal flexure of the continent east of the northern North Sea (Eldholm and Thiede 1980) show that uplift began to affect at least a part of the western margin of Fennoscandia. *The first important vertical movements in the Scandes probably began during Palaeocene*, along a flexure zone close to the present coast line.

During the Eocene and Oligocene, the continental shelf and Denmark received fine grained, argillaceous sediments (Holtedahl *et al.* 1974; Spjeldnaes 1975): no strong erosion seems to have occurrred on the continent, which has certainly undergone a slow uplift, while the shelf was gently subsiding. But extension, faulting, and even volcanism still occurred along the Skagerrak coast, on the southeastern side of the bulge (Äm 1973). Northwards, the position of the eastern hinge line is not known.

An important change is thought to have occurred during the Late Oligocene, which is also a period of sea level fall (Mid-Chattian). Clastic sediments, with fresh micas, appear in Denmark and become more abundant during the Miocene and Pliocene; they come from the north, therefore from the southern part of the Scandinavian Peninsula and from the South Baltic area (Spjeldnæs 1975). On the continental margin of Norway, terrigenous sedimentation and subsidence increased abruptly during the Miocene or Late Miocene (Kossovskaya *et al*. 1978), and progradation structures can be seen on the outer part of the shelf, on profiles published by Jørgensen and Navrestad (1981; Fig. 8); but the Lofoten area appeared as a stable or positive area incorporated in the western side of the bulge. *A Neogene phase of reinforced uplift along the Scandes anteclise* may be one of the causes of this phenomenon, together with more aggressive climatic conditions and erosion. The unconformity between the undeformed Pleistocene sediments and the inclined and eroded Tertiary sediments may be explained by a Pre-Quaternary interruption of the flexural movement along the western side of the bulge, followed by erosion (Talwani and Eldholm 1977), though in some areas faulting seems to have occurred up to the present time (e.g. Vestfjord, Dekko 1975; Meløy area: Gabrielsen and Ramberg 1979).

Discussion: The Caledonides and the Formation of the Scandes

Caledonides and uplift

According to Theilen and Meissner (1979) and Meissner (1979), there is no distinct root under the Scandinavian Caledonides. But maps of the Moho (Sellevoll 1973) show that the continental crust is slightly thickened along the eastern edge of the bulge (45 km under Jämtland) and that there is a crustal thinning around the southern tip of the bulge (under the Oslo region, the Skagerrak, and the Norwegian Trench (32–36 km and 28 km) and along the continental shelf (28 to 26 km, and 15 km under the Vøring Plateau); the maximum uplift zone is located between these respectively thickened and thinned zones of the crust (Figs. 4 and 6). Therefore, it is difficult to explain the uplift by the influence of a light root, though recent geophysical evidence suggests that such an influence cannot be completely excluded (Zadelhoff *et al*. 1981). Moreover, indications of a more or less complete peneplanation of the Caledonian structures as early as in Carboniferous times (e.g. Andøy and neighbouring areas of Northern Norway: Sturt *et al*. 1979; Jørgensen and Navrestad 1981; Oslo area: Oftedahl 1960) may be interpreted as the result of a relative tectonic stability after Devonian vertical movements. The Caledonian roots were perhaps already reduced in the Late Palaeozoic, after isostatic compensation and erosion of the Caledonian superstructures, and were just able to create a slight positive trend during the Mesozoic, especially in the eastern part of the range,

far from the zone of rifting. This old positive tendency might explain the exhumation of the basement in front of the Caledonides. With this hypothesis, it would not be necessary to suppose a Tertiary 'scraping off' of the Caledonide roots (Theilen and Meissner 1979); crustal thinning may have occurred earlier in the western areas nearer to the Mesozoic rift structures.

Gravimetric maps show a zone of negative Bouguer anomalies which coincides with the bulge and the Caledonides. As they are stronger under zones of maximum thickening (eastern border of the Caledonides), a crustal origin cannot be excluded. Negative anomalies are less distinct under zones of weaker uplift (e.g. Trøndelag and Finnmark). Lower zones outside the bulge and especially areas of thinned crust associated with old rifts (e.g. the Oslo region) and with zones of subsidence in the North Sea and the continental margin show distinct positive anomalies.

According to Theilen and Meissner (1979), negative anomalies within the bulge must be explained by the presence of an underlying low density asthenosphere. This hypothesis would explain the large wave length of the uplift. Unfortunately, it does not fit with the steep gravimetric gradients on both sides of the bulge; the question remains unresolved and some of the implications of this theory must be considered.

As the marginal bulge appeared *at the time of oceanic opening*, a correlation between the two events may be suggested, though parts of the anteclise are intracratonic and may be thought to be more recently uplifted (Finnmark: Gjessing 1977; Peulvast 1978). Compressive stresses within the Caledonides and the Baltic Shield (Hast 1969) are interpreted as being caused by oceanic expansion at lithospheric and asthenospheric levels (Theilen and Meissner 1979). Creep of a hot and low viscosity asthenosphere, or a kind of non-seismic subduction, especially at the time of opening, is then supposed by the authors to be the main cause of uplift. This advection and the density heterogeneities in the upper mantle cause the destabilization of the thick old continental lithosphere deprived of its cold root. Preliminary models (Fleitout and Peulvast 1982; Fleitout *et al*. 1983) show that a thermal perturbation of 600°C at a depth of 100 km would cause an uplift of 1170 m; the geothermal flow after 50 Ma would be hardly higher in the uplifted zone than in the adjoining areas, as it is now in the Caledonides (Grønlie *et al*. 1977). Moreover, it may explain the delay between the oceanic opening and the period of maximum uplift, and the asymmetry of the bulge, the phenomenon being more important near the oceanic area. The asthenospheric creep may be temporarily strengthened by the flow associated with the post-glacial isostatic rebound (Mörner 1981). The Palaeocene vertical movements, and faulting in the Lofoten–Versterålen area until the Palaeocene (Sturt *et al.* 1979) might be explained as the result of the thermal event directly related with the oceanic opening. Later, overloading of the continental margin and of the adjoining areas (North Sea) by sediments, and the mechanism of flexural loading (Watts and Ryan 1976), probably strengthened subsidence induced by crustal thinning and thermal contraction. The associated continental uplift may have been strengthened by erosive unloading: increased erosion during the Neogene can therefore be the result and one of the causes of a reactivation of the vertical movements. A later and weaker subsidence of the Barents Sea is thought to be correlated with later and weaker uplift along the Finnmark coast.

Oceanic opening along the Caledonian lineaments is probably the cause of the parallelism and coincidence between the bulge and the Caledonides. There is only an indirect correlation between the Caledonian structures and the uplift caused by oceanic opening. This would explain why not only the Caledonides but also some large basement areas are involved in the uplift.

Morphostructural expression of the Caledonides

The influence of the Caledonian structures on the morphology of the present mountain range has been demonstrated. It can be explained by differential erosion, itself induced by the uplift. I have previously mentioned a probable peneplanation of most of the Caledonian range as early as in Carboniferous or Permian times. But it is not known whether some very resistant units (e.g. quartzites or Jotun pyroxene gneisses) were completely truncated, as no peneplain is preserved on most of the highest summits (e.g. Dovrefjell, Rondane, western Jotunheimen), except on the Lyngen massif (O. Holtedahl 1960), on the Jotesdalsbre, and around summits which look like huge residual hills (massifs of northern Jotunheim). The peneplanation of the Oslo rift and of the Arctic–North Atlantic rift structures during the Mesozoic suggests that the Late Paleozoic peneplanation was continued during post-Palaeozoic times, owing to slow vertical movements in the present uplifted area. Erosion probably reached the basement in the slow uplift zone east of the Caledonides, whereas on the margin of the western rifting zone, after Late Caledonian deep erosion and peneplanation, the positive movements became restricted to narrow and tilted blocks faulted within a henceforth entirely subsiding area. Between these two zones, the Caledonian nappes were more or less truncated, but never completely eroded.

The timing of morphogeny within the Caledonides remains somewhat speculative, owing to the lack of post-Palaeozoic sediments (Fig. 10). However, the Palaeocene uplift probably initiated the development of a lower generation of peneplains, inside and on both sides of the Caledonides, especially within soft Caledonian metasediments (e.g. in Valdres and Trøndelag) and other higher cleaved and jointed rocks of pre-Caledonian origin within several nappes (e.g. the psammites and gneisses of West Finnmark). These peneplains extend also within the basement, in southeastern Norway and in Sweden (Västerbotten: Rudberg 1954) where erosion had only to wear down the autochthonous cover or small thicknesses of basement rocks below the sub-Cambrian peneplain, tilted by the Caledonian orogeny. Their width is less important in the basement areas of the western side of the bulge, especially in the areas of maximum coastal uplift where they can be difficult to correlate with the other peneplains (e.g. the narrow peneplain later raised to 300–400 m a.s.l. in the Lofoten region, which truncates Svecofennian paragneisses, retrograded mangeritic rocks, and basic rocks: Peulvast 1977). But correlations between the peneplains of the western and eastern slopes are possible in some of these areas (Sogn–Jotun: Peulvast in preparation). In Hardangervidda, as in parts of Sweden and in Finland, these peneplains are nearly parallel to the exhumed and weakly deformed sub-Cambrian peneplain (Licstøl 1960; Rudberg 1970; Simonen 1980). This peneplanation probably took place during most of the Palaeogene. It implies an epeirogenic movement which involved the Caledonides together with their foreland within a broad bulge between the west coast and the more stable Baltic–Bothnian area. It was allowed by slowness of the vertical movements after the initial uplift and by relatively warm climates which had induced deep weathering and might still allow pediplanation, except in the most resistant rocks (poorly porous rocks, some quartzites: Peulvast 1977, 1978 and in preparation) which constitute the present mountains. The present limits of most of the main reliefs were certainly already established above the Palaeogene peneplains. These surfaces, namely a more or less narrow fringe on the western side, immense plains on the Baltic Shield and broad basins within the Caledonides (e.g. the main Sogn surface) are some of the main elements of the 'Palaeic surface' of Ahlmann (1919) and Gjessing (1967).

Cooler climates, as shown by the sediments in Denmark (Spjeldnæs 1975) probably explain further deepening of consequent and subsequent narrower valleys or basins below these forms, along particu-

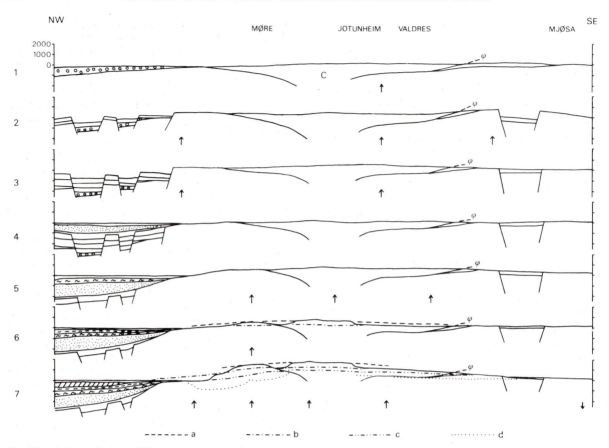

Fig. 10   Schematical profiles across Norway from Møre to Mjøsa, showing a speculative reconstruction of the morphotectonic evolution of the southern part of the Scandinavian Bulge. 1: Carboniferous; 2: Permian to Trias; 3: Late Cimmerian; 4: Cretaceous; 5: Palaeocene; 6: Eocene–Oligocene; Neogene. Topographic surfaces: a: pre-Palaeocene peneplain; b: Palaeogene peneplain; c: Neogene valleys and basins; d: glacial valleys and fjords

larly weak zones such as soft metasediments or highly tectonized and jointed rocks (the mylonites of the Jotun nappe for instance). But this fact is also the result of the assumed Late-Oligocene–Neogene uplift. Broad basins were carved 300 to 500 m under the Palaeogene surface, in the central zone of the bulge (e.g. in Sogn: low plateaus around Kanpanger) and also on the distal parts of its western and eastern sides (pre-strandflat erosion surface and basins of the Lofoten–Versterålen: Peulvast 1978; Västerbotten near the coast of the Bothnian Sea: Rudberg 1954). On the eastern side, this disposition indicates that the positive movement, though maximum in the Caledonide area, involved also parts of the Baltic area. The Baltic and Bothnian basins, as in the case of the Barents Sea, north of the Scandes, were probably lowered later, owing to slight differential subsidence along flexures or even fault lines (Winterhalter 1972). The development and subsequent tilting of the lowest 'cycles' of eastern Västerbotten (Rudberg 1954) could be explained in this way. Uncertainties about the initial longitudinal profiles of old drainage systems do not allow a good estimate of the amplitude

of this phase of uplift, which seems to have ended before the Pleistocene.

The excavation of broad basins together with relatively deep weathering was quickly followed by deeper and deeper dissection which achieved the morphological differentiation within the Caledonides. The organization of hydrographic systems on both sides of the bulge and the carving of low forms and valley sides with their own structural forms were controlled by the mechanical properties of the rocks (Peulvast 1978). The glaciers, which probably appeared at the end of the Pliocene (Ruddiman 1977) have emphasized or modified this dissection and the longitudinal profiles overdeepened the fjords on the western side and perhaps helped antecendence processes (Finnmark: Gjessing 1977). They altered a water divide the western initial position of which coincided with the axis of the bulge within the two culmination zones (Rudberg 1976; Peulvast 1978 and in preparation). Glacial influence on the morphology of plateaux and lower areas was probably less important, as the mean glacial erosion of Scandinavia is calculated to be only 16 m (Ruddiman 1977).

Conclusion

The post-Caledonian evolution of western Fenno-scandia was essentially controlled by rifting around the Scandinavian Caledonides from the Oslo region to Western Norway and the Barents Sea and then by opening of the Norwegian Sea. The influence of deep Caledonian structures on the vertical movements has been somewhat indirect, in spite of the reactivation of some lineaments. But the Caledonian superstructure controls the most striking features of the relief, to such an extent that it resuscitates in the present landscapes parts of this old mountain range.

Acknowledgements

This paper is based upon geomorphological studies carried out by the author in Norway (Sogn–Jotun and Lofoten–Vesterålen) for a 'Thèse de Doctorat d'Etat' during the years 1971 to 1979, under the direction of Professor P. Birot. This research has been supported by the 'Laboratoire de Géographie Physique' (L.A 141, C.N.R.S., Paris). I wish to thank Prof. P. Birot, Prof. A. Godard, Dr A. E. Prost (B.R.G.M.) and Dr P. A. Ziegler (S.I.P.M.) for critical reading of previous versions of the manuscript.

References

Ahlmann, H. W. Jnr. 1919. Geomorphological studies in Norway. *Geografiska Annaler.*, **1–2**, 3–205 and 193–252.

Åm, K. 1973, Geophysical indications of Permian and Tertiary igneous activity in the Skagerrak. *Norges geol. Unders.*, **287**, 1–25.

Barth, T. F. W. 1939. Geomorpholgy of Vest-Agder fjord-land. *Norsk geogr. Tidsskr.*, **7**, 290–305.

Bott, M. H. P. 1971, Evolution of young continental margins and formation of shelf basins. *Tectonophysics*, **11**, 319–327.

Bott, M. H. P. 1976. Sedimentary basins of continental margins and cratons. *Development in Geotectonics*, **12**, Elsevier publ. Co., Amsterdam. 314 pp.

Burke, K. 1976. Development of graben associated with the initial ruptures of the Atlantic Ocean. In Bott, M. H. P. (Ed.), *Sedimentary Basins of Continental Margins and Cratons. Development in Geotectonics*, **12**, Elsevier, Amsterdam, 93–112.

Dalland, A. 1975. The Mesozoic rocks of Andøy, northern Norway. *Norges geol Unders.*, **316**, 271–287.

Deegan, C. E. and Scull, B. J. 1977. A proposed standard lithostratigraphic nomenclature for the Mesozoic of the Central and Northern North Sea. In N. P. F. *Mesozoic Northern North Sea Symposium.* Oslo, 3, 24 pp.

Dekko, T. 1975. Refleksjonsseismiske undersøkelser i Vestfjorden 1972. *NTNF's Kontinentalsokkelkontor*, 77.

Eldholm, I., Sundvor, E. and Myhre, A. 1977. Continen-tal margin off Lofoten–Vesterålen, Northern Norway. *Inst. for Geol. Univ. i Oslo, Intern skriftserie*, **13**.

Eldholm, O. and Thiede, J. 1980. Cenozoic continental separation between Europe and Greenland. *Palaeogeogr., Palaeoclim., Palaeoecol.*, **30**, 243–259.

Faerseth, R. B., McIntyre, R. M. and Naterstad, J. 1976. Mesozoic alkaline dykes in the Sunnhordaland region, western Norway: ages, geochemistry and regional significance. *Lithos*, **9**, 331–345.

Fleitout, L., Froidevaux, C., Peulvast, J. P. and Yuen, D. A. 1983. Surface tectonics, lithospheric instabilities deep density anomalies. *Europ. Union Geosc. Meeting*, Strasbourg, abstracts.

Fleitout, L. and Peulvast, J. P. 1982. Les bourrelets marginaux des hautes latitudes. L'exemple du bourrelet scandinave. *Bull. Assoc. Géogr. Fr.*, **489–490**, 245–253.

Gabrielsen, R. H. and Ramberg, I. B. 1979. Tectonic analysis of the Meløy earthquake area based on Landsat lineament mapping. *Norsk geol. Tidsskr.*, **59**, 2, 183–187.

Gjessing, J. 1967. Norway's paleic surface. *Norsk geogr. Tidsskr.*, **20**, 8, 69–133.

Gessing, J. 1977. Landformene. In *Norges Geografi*, Universitetsforlaget, Oslo, Bergen, Tromsø, 15–42.

Griffin, W. L., Taylor, P. N., Hakkinen, J. W., Heier, K. S., Iden, I. K., Krogh, E. J., Malm, O. A., Olsen, K. I., Ormaasen, D. E. and Tveten, E. 1978. Crustal evolution in Lofoten, Norway: 3500–400 MY BP. *J. Geol. Soc.*, **135**, 6, 629–647.

Grønlie, G., Heier, K. S. and Swanberg, C. A. 1977. Terrestrial heat-flow determinations from Norway. *Norsk geol. Tidsskr.*, **57**, 153–162.

Hast, N. 1969. The state of stress in the upper part of the earth's crust. *Tectonophysics*, **8**, 169–211.

Heezen, B. C. 1974. Atlantic-type continental margins. In *The Geology of Continental Margins*, Springer-Verlag, 13–24.

Holtedahl, H. 1955. On the Norwegian continental terrace, primarily outside Møre–Romsdal: its morphology and sediments. *Årb. Univ. Bergen, Naturvid. rk.* **14**, 209 pp.

Holtedahl, H. 1975. The geology of the Hardangerfjord, West Norway. *Norges geol. Unders.*, **323**, 1–84.

Holtedahl, H., Haldorsen, S. and Vigran, J. O. 1974. A study of two sediment cores from the Norwegian continental shelf between Haltenbanken and Frøyabanken (64°06'N, 7°38'E). *Norges geol. Unders.*, **304**, 1–20.

Holtedahl, O. 1953. On the oblique uplift of some northern lands. *Norsk geogr. Tidsskr.*, **14**, 1–4, 132–139.

Holtedahl, O. 1960. Features of the geomorphology. In Holtedahl, O. (Ed.), *Geology of Norway. Norges geol. Unders.*, **208**, 507–531.

Husebye E. S., Bungum, H., Fyen, J. and Gjøystdal, H. 1978. Earthquake activity in Fennoscandia between 1497 and 1975 and intraplate tectonics. *Norsk geol. Tidsskr.*, **58**, 68.

Indrevaer, G. and Steel, R. J. 1975. Some aspects of the sedimentary and structural history of the Ordovician and Devonian rocks of the Westernmost Solund Islands, West Norway. *Norges geol. Unders.*, **317**, 23–32.

Jørgensen, F. and Navrestad, T. 1981. The geology of the Norwegian shelf between 62°N and the Lofoten Islands. In *Petroleum Geology of the Continental Shelf of North west Europe*, Inst. of Petroleum, London, 407–413.

Kossovskaya, A. G., Timofeev, P. P. and Shutov, V. D. 1978. The lithology and genesis of the sedimentary deposits in the Norwegian Basin and western part of the

Lofoten Basin. *Initial Rep. of the DSDP*, suppl. to vol. XXXVIII, XXXIX, XL and XLI. National Science Foundation, Washington, 67–72.

Kvale, A. 1975. Caledonides in Scandinavia compared with East Greenland. *Bull. geol. Soc. Denmark*, **24**, 129–160.

Lagerbäck, R. 1978. Neotectonic structures in northern Sweden, *Geol. Fören. Stockh. Förland.*, **100**, 3, 263–269.

Liestøl, O. 1960. Det subkambriske peneplan i omradet Haukelifjell Suldalsheine. *Norsk geol. Tidsskr.*, **40**, 69–72.

Lundqvist, J. and Lagerbäck, R. 1976. The Pärve fault: a late-glacial fault in the Precambrian of Swedish Lapland, *Geol. Fören. Stockh. Förland.*, **98**, 45–51.

Matttauer, M. 1973. Les Déformations des Matériaux de l'écorce terrestre, Hermann, Paris, 493 pp.

Meissner, R. 1979. Fennoscandia. A short outline of its geodynamic development. *Geol. J.*, **3**, 3, 227–233.

Mörner, N. A. 1981. Crustal movements and geodynamics in Fennoscandia. *Tectonophysics*, **71**, 1–4. Special issue: Recent crustal movements, 1979, 241–252.

Nilsen, T. H., 1973. The relation of joint patterns to the formation of fjords in Western Norway. *Norsk geol. Tidsskr.*, **53**, 2, 183–194.

Oftedahl, C. 1960. Permian rocks and structures of the Oslo region. In Holtedahl, O. (Ed.), *Geology of Norway. Norges geol. Unders.*, **208**, 298–343.

Oftedahl, C. 1980. Norway. In *Geology of the European Countries. Denmark, Finland, Iceland, Norway, Sweden.* C.N.F.G., Dunod, Paris 346–456.

Peulvast, J. P. 1977. L'érosion différentielle et ses implication dans les roches cristallines: exemple de la Norvège du Nord (Flakstadøy, îles Lofoten). *Rev. de Géo. Phys. Géol. syn.*, **19**, 2, 149–163.

Peulvast, J. P. 1978. Le bourrelet scandinave et les Calédonides: un essai de reconstitution des modalités de la morphogénèse en Norvège. *Géogr. Phys. Quart.*, **32**, 4, 295–320.

Peulvast, J. P. in prep. *Recherches Géomorphologiques en Norvège Septentrionale et Centre-sud*, Thèse.

Prost, A, E. 1975. Un accident rhegmatique en Scandinavie: la discontinuité de Ringebu-Vågå–Bremanger. *Rev. Géogr. Phys. Géol. Dyn.*, **17**, 4, 361–374.

Prost, A. E. 1977. Répartition et évolution géodynamique des Éxternides calédoniennes scandinaves. *Rev. de Géo. Phys. Géol. syn.*, **19**, 5, 421–432.

Ramberg, I. B. and Smithson, S. B. 1975. Geophysical interpretation of crustal structure along the southeastern coast of Norway and the Skagerrak. *Bull. Geol. Soc. Am.*, **86**, 769–775.

Rambert, I. B. and Spjeldnaes, N. 1978. The tectonic history of the Oslo region. In Ramberg, I. B. and Neumann, E. R. (Eds), *Tectonics and Geophysics of Continental Rifts. Nato ASI Series*, Ser. C, D. Reidel Publ. Co., Dordrecht, Boston, London, 167–194.

Rokoengen, K., Bell, G., Bugge, T., Dekko, T., Gunleiksrud, T., Lien, R. L., Løfaldi, M. and Os Vigran, J. 1977. Prøvetaking av fjellgrunn og løsmasser utenfor deler av Nord–Norge i 1976. *Inst. Kont. Unders.*, **91**.

Rudberg, S. 1954. Västerbottens berggrundsmorfologi, ett försök till rekonstruktion av preglaciala erosionsgenerationer i Sverige. *Geographica*, **25**, 457p.

Rudberg, S. 1962. Geology and morphology of the 'fjells'. *Biul. Peryglac.*, **11**, 173–186.

Rudberg, S. 1970. The sub-Cambrian peneplain in Sweden and its slope gradient. *Z. für Geomorph. N.F. Suppl. Bd.*, **9**, 157–167.

Rudberg, S. 1976. River valleys anomalies. One approach to the study of Fennoscandian bedrock relief. *Norsk geogr. Tidsskr.*, **30**, 3, 83–92.

Ruddiman, W. F. 1977. Late Quarternary deposition of ice-rafted sand in the subpolar North Atlantic (lat. 40° to 65°N). *Bull. Geol. Soc. Am.*, **88**, 12, 1813–1827.

Rønnevik, H. C. 1981. Geology of the Barents Sea. In *Petroleum Geology of the Continental Shelf of North west Europe*, Inst. of Petroleum, London, 395–406.

Rønnevik, H. and Navrestad, T. 1977. Geology of the Norwegian Shelf between 62°N and 69°N. *Geol. J.*, **1**, 33–46.

Sellevoll, M. A. 1973. Mohorovicic discontinuity beneath Fennoscandia and adjacent parts of the Norwegian Sea and the North Sea. In Mueller, S. (Ed.), *The Structure of the Earth's Crust, Based on Recent Data. Tectonophysics*, **20**, 1–4, 359–366.

Sellevoll, M. A. 1975. Seismic refraction measurements and continuous seismic profiling on the continental margin off Norway between 60°N and 69°N. *Norges geol. Unders.*, **316**, 219–235.

Sellevoll, M. A. and Sundvor, E. 1974. The origin of the Norwegian Channel. A discussion based on seismic measurements. *Can. J. Earth Sci.*, **11**, 2, 224–231.

Simonen, A. 1980. The Precambrian in Finland. In *Geology of the European Countries. Denmark, Finland, Iceland, Norway, Sweden.* C.N.F.G. Dunod, Paris, 55–108.

Spjeldnaes, N. 1975. Paleogeography and facies distribution in the Tertiary of Denmark and surrounding areas. *Norges geol. Unders.*, **316**, 289–311.

Steel, R. J. 1978. Late-orogenic Devonian basin formation in the western Norwegian Caledonides. In Tozer, E. T. and Schenck, P. E. (Eds), *Caledonian–Appalachian Orogen of the North Atlantic Region. Geol. Survey of Canada Pap¯.*, **78–13**, Ottawa, 57–62.

Strömberg, A. G. B. 1976. A pattern of tectonic zones in the western part of the East European platform. *Geol. Fören. Stockh. Förland.*, **98**, 227–243.

Sturt, B. A., Dalland, A. and Mitchell, J. L. 1979. The age of the sub-Mid-Jurassic tropical weathering profile of Andøy, Northern Norway, and the implications for the Late Palaeozoic palaeogeography in the North Atlantic Region. *Geol. Rdsch.*, **68**, 2, 523–542.

Surlyk, F. 1977. Mesozoic faulting in East Greenland. In Frost, R. T. C. and Diickers, A. J. (Eds), *Fault Tectonics in NW Europe. Geol. Mijnbouw*, **56**, 311–327.

Surlyk, F. 1980. Denmark. In *Geology of the European Countries. Denmark, Finland, Iceland, Norway, Sweden.* C.N.F.G., Dunod, Paris, 1–54.

Talvitie, J. 1981. Deformation phases in the Blue Road Geotraverse region of Finland. *Earth Evol. Sci.*, **1**, 58–60.

Talwani, H. and Eldholm, O. 1977. Evolution of the Norwegian–Greenland Sea. *Bull. geol. Soc. Amer.*, 969–999.

Theilen, F. and Meissner, R. 1979. A comparison of crustal and upper mantle features in Fennoscandia and the Renisch Shield, two areas of recent uplift. *Tectonophysics*, **61**, 1–3. Special issue, Plateau uplift: mode and mechanism, 227–242.

Torske, T. 1972. Tertiary oblique uplift of Western Fennoscandia; crustal warping in connection with rifting and break-up of the Laurasian continent. *Norges geol. Unders.*, **273**, 43–48.

Torske, T. 1975. Possible Mesozoic mantle plume activity beneath the continental margin of Norway. *Norges geol. Unders.*, **322**, 73–90.

Vail, P. R., Mitchum, R. M. and Todd, R. G. 1977. Eustatic model for the North Sea during the Mesozoic. In *N.P.F. Mesozoic Northern North Sea Symposium*, Oslo, 12, 35 p.

Vigran, J. O. 1970. Fragments of a Middle Jurassic flora from northern Trøndelag, Norway. *Norsk geol. Tidsskr.*, 50, 193–214.

Watts, A. B. and Ryan, W. B. F. 1976. Flexure of the lithosphere and continental margin basins. In Bott, M. H. P. (Ed.), *Sedimentary Basins of Continental Margins and Cratons. Development in Geotectonics*, 12, Elsevier, Amsterdam, 25–44.

Whiteman, A., Rees, C., Naylor, D. and Pegrum, R. M.

1975. North Sea troughs and plate tectonics. *Norges geol. Unders.*, 316, 137–161.

Winterhalter, B. 1972. On the geology of the Bothnian Sea, an epeiric sea that has undergone Pleistocene glaciation. *Bull. Geol. Surv. Finland*, 258, 1–66.

Zadelhoff, K. V., Gorling, L. and Vogel, A. 1981. Optimization of crustal density models along the Blue Road Geotraverse by use of gravity data and additional constraints. *Earth Evol. Sci.*, 1, 34–37.

Ziegler, P. A. 1978. North-Western Europe: Tectonics and basin development. In van Loon, A. J. (Ed.), *Keynotes of the MEGS-II* (Amsterdam, 1978). *Geol. Mijnbouw*, 57, 589–626.

Related Caledonian Areas

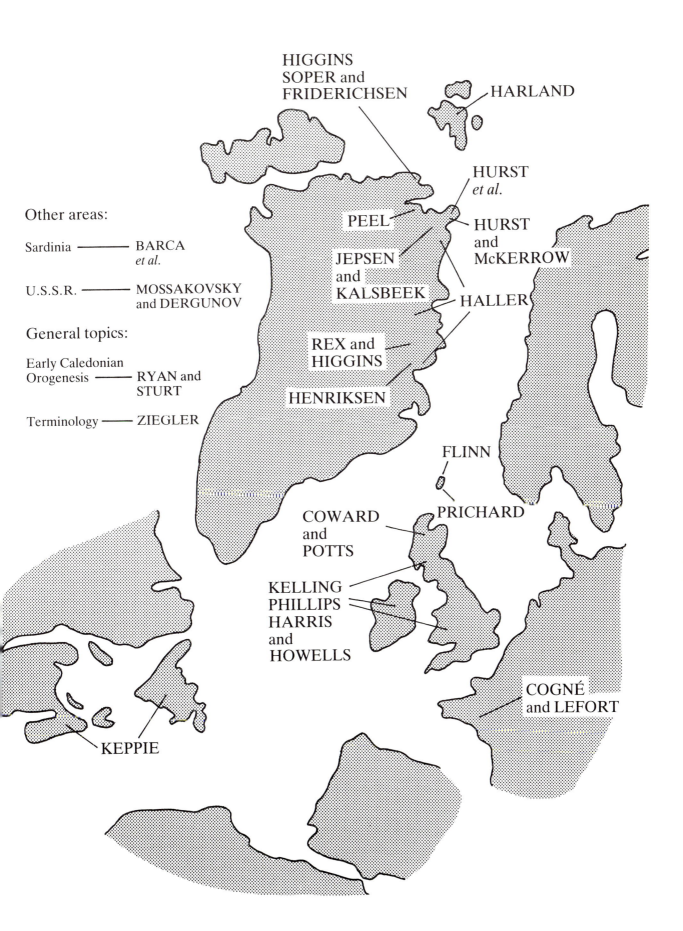

Other areas:

Sardinia ———— BARCA *et al.*

U.S.S.R. ———— MOSSAKOVSKY and DERGUNOV

General topics:

Early Caledonian Orogenesis ———— RYAN and STURT

Terminology ——— ZIEGLER

HIGGINS SOPER and FRIDERICHSEN

HARLAND

HURST *et al.*

PEEL

HURST and McKERROW

JEPSEN and KALSBEEK

HALLER

REX and HIGGINS

HENRIKSEN

FLINN

PRICHARD

COWARD and POTTS

KELLING PHILLIPS HARRIS and HOWELLS

COGNÉ and LEFORT

KEPPIE

The Caledonide Orogen—Scandinavia and Related Areas
Edited by D. G. Gee and B. A. Sturt
© 1985 John Wiley & Sons Ltd

Caledonide Svalbard

W. B. Harland

Department of Earth Sciences, University of Cambridge, Cambridge, CB2 3EQ England

ABSTRACT

The Svalbard archipelago was affected by more than one mid-Palaeozoic orogenic phase, and it is generally considered that the whole area is Caledonide. However, westernmost Svalbard may have formed from a distinct Holtedahl Geosyncline situated off North Greenland, outside the Caledonian Orogen. It exhibits thick Vendian through Silurian strata of mobile facies rich in tillites and volcanics. Its deformation could be more closely related to the Ellesmerian Orogeny.

The classic Caledonian terrane (Eastern Province, i.e. Ny Friesland and Nordaustlandet) began with the Hecla Hoek Geosyncline forming over a very long time span, deformed after Llanvirn time and reaching a climax in the Ny Friesland Orogenic phase.

An intermediate (Central) province formed from the Hornsundian Geosyncline and extended into northwest Svalbard. The mid-Cenozoic West Spitsbergen Orogeny affected the Palaeozoic structures which are difficult to interpret. The structure of the Eastern Province (not so affected) throws light on transpression as a major orogenic process. These three provinces were juxtaposed by Svalbardian (Late Devonian) sinistral strike-slip whose magnitude is debated but should allow the Hecla Hoek Geosyncline to have formed near central East Greenland.

Introduction

The Younger rocks

There is a clear distinction between the post-Devonian and the pre-Carboniferous rocks of Svalbard. The post-Devonian strata mostly belong to what has been referred to as a platform or epicontinental sequence from Tournaisian through Albian, and Palaeogene. Strata up to 8 km thick define the central basin of Spitsbergen.

The contrast between the platform sequence and the underlying rocks is accentuated by the lack of recognizable Late Palaeozoic (or Hercynian) diastrophism. Indeed tectonic instability decreased in intensity from late Devonian through (Early Carboniferous) Dinantian time so that by Permian time Svalbard in common with much of the western Arctic (N. Greenland and the Sverdrup Basin) was exceedingly stable and in effect remained so till Cretaceous time, when igneous activity and gentle warp-ing resulted in a widespread hiatus between Albian and Palaeogene strata.

Post-Caledonian tectonism (with dextral transpression) is most evident in the West Spitsbergen Orogen of mid-Eocene age. It caused overfolding and overthrusting to the east of rocks along the western coastal belt to the south of Kongsfjorden, so obscuring much of the earlier, albeit more intense mid-Palaeozoic deformation.

The Older rocks

Pre-Carboniferous rocks may be divided into Old Red Sandstone (roughly Early and Middle Devonian and possibly latest Silurian) and pre-Devonian rocks (see Figure 1).

The Old Red Sandstone has suffered *Haakonian* diastrophism (latest Silurian or earliest Devonian) and Late Devonian *Svalbardian* folding with extensive faulting. These have been regarded as late Caledonian tectonic phases.

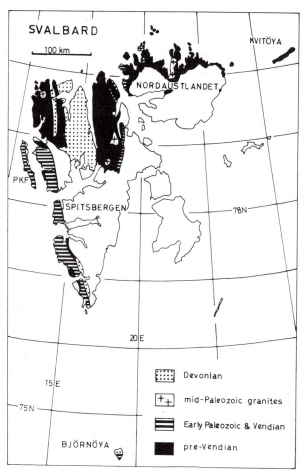

Fig. 1   Map of Svalbard with all the major islands of the archipelago shown and those with pre-Carboniferous rocks named. PKF = Prins Karls Forland. The classification into four rock outcrop ornaments is simplified and the outcrops are schematized

The pre-Devonian rocks have generally been more intensely tectonized as in a major orogeny that has usually been regarded as the main Caledonian phase. However, perhaps only eastern Svalbard records the typical main Caledonian phase (*Ny Friesland* Orogeny) that deformed the Hecla Hoek geosyncline.

In the extreme northwestern, northeastern and southern corners of Svalbard and in the older rocks throughout Svalbard, metamorphosed and/or tectonized facies are the rule. Older maps show that the whole of Svalbard lay within the Caledonian belt (e.g. Holtedahl 1920; Hurley 1968). Radiometric determinations throughout Svalbard have yielded apparent ages ranging from 450 Ma to 350 Ma and so confirm that mid-Palaeozoic metamorphism was general.

However, the recognition of a pre-Caledonian, i.e. Precambrian, metamorphosed basement remains a major question. Basement rocks, if they exist, have almost certainly been remobilized in mid-Palaeozoic

time and so in the better known areas are difficult to recognize. In the less accessible areas of eastern Nordaustlandet, however, it has been claimed that there is an Archean basement. An unconformity has yet to be demonstrated and so an alternative possibility is that rocks representing another zone of intense Caledonian tectogenesis are exposed to view. An intermediate explanation is that the terrane consists largely of Caledonian remobilized Precambrian basement. Similar, though weaker arguments could be made for remobilized basement in all the areas of highest metamorphic facies.

The Caledonides

In a broad sense the mid-Palaeozoic tectogenesis affected the whole width of Svalbard and the resulting orogenic structure was planed down and covered by post-Devonian rocks. Uplift to the north and west has again exposed these older rocks beneath the cover.

The traditional view is that the whole pre-Carboniferous structure of Svalbard should be regarded as 'Caledonian' and an integral part of the great Caledonides extending throughout the Appalachians, British Isles, Scandinavia, and Greenland to Svalbard (e.g. Holtedahl 1920, 1925; Bailey and Holtedahl 1938). On this unified view the Caledonian Orogeny deformed the Hecla Hoek Geosyncline and in the later stages of the orogeny Old Red Sandstone was generated elsewhere in the Caledonian Belt.

This view will be examined. An alternative model is considered in which Svalbard consists of three disparate tectonic provinces (exotic or allochthonous terranes analogous to recent Californian models) of which one is typically Caledonian and one at least formed outside the Caledonian realm.

Faults

The geologic map of Svalbard has been variously drawn with faults dominating the structure. Early notions that the fjords were delineated by faults bounding their coastlines (e.g. de Geer 1909) are not supported by detailed studies; but certainly most fjords are occupied by faults that have to some extent determined their locus. Another older view that faults separate more and less metamorphosed strata, as Archaean faulted against Caledonian terranes (e.g. within Ny Friesland and Nordaustlandet) has been retained by Soviet authors (e.g. Sokolov *et al.* 1968). But these boundaries have been shown to result largely from erosion of dipping strata.

The view that Svalbard as a whole was bounded by a Late Devonian major sinistral transcurrent fault system was advanced by the author (Harland 1965). The model of orogenic transpression leading to longitudinal transcurrence in an orogen was based on Svalbard studies (Harland 1971). In 1972 I suggested that some of the major faults within Svalbard could represent Early Palaeozoic plate boundaries and this concept was more fully documented (Harland *et al.* 1974) for the Billefjorden Fault Zone (BFZ). Evidence within Svalbard suggested a minimal sinistral transcurrence of 200 km. This had the effect of forming the present Old Red Sandstone Graben so that it was bounded in the west by fault systems of possibly minor strike-slip component, and in the east by the BFZ that post dated the 'graben' sediments. External stratigraphic evidence suggested larger values of the order of 1000 km or more. BFZ then seemed to divide Svalbard into Eastern and Western tectonic provinces. Another major sinistral feature—the Central Western Fault Zone (CWFZ), also acting as a plate boundary—was postulated by Harland and Wright (1979). It could not be easily observed because its supposed line was obscured by the Palaeogene West Spitsbergen Orogenic Front which it controlled. The BFZ and CWFZ would then divide Svalbard up into three distinct provinces or terranes (East, Central, and West) that were formed in widely separated regions and juxtaposed in late Devonian time. Major faults of all ages in Svalbard were reviewed by the author (Harland 1979) and Fig. 2 and Table 1 are abstracted from that paper.

Some possible vindication of these extreme views applying to the whole length of the Caledonian–Appalachian orogens appeared to come from palaeomagnetic work in the Appalachian system (Van der Voo 1980; Harland 1980) where suggestions of up to 2000 km sinistral transcurrence were indicated. These matters continue to be investigated.

Most minor faults appear to be of Palaeogene age but a number of oblique faults with sinistral motion that cut the older rocks are consequences of late Caledonian compression and transpression.

Tectonic provinces

According to the AGI *Glossary of Geology* (1980) a province is 'any large area or region as a whole, all parts of which are characterized by similar features or by a history differing significantly from that of adjacent areas'. Terrane is often used in this sense but is more valuable as an exploratory term.

Until about 1972 it was customary to consider the Caledonian geology of Svalbard as a whole, i.e. as a single tectonic province. Accordingly attempts were made at correlation of the older rocks throughout Svalbard and some of these correlations were accepted mainly because, until lately, few geologists were familiar with all the rock groups. Difficulty in elucidating the stratigraphy of western Svalbard according to the well displayed sequence of North Eastern Svalbard may stem from a false assumption that both belonged to the same province. An example of such an assumption with a unifying scheme was illustrated by Birkenmajer in 1975 and again in 1981. Part of his scheme was questioned on internal stratigraphic evidence (Harland 1979).

The hypothesis of substantial strike-slip, i.e. of ancient plate boundaries within Spitsbergen (analogous to those in late Phanerozoic California) led to the postulation of at least three tectonic provinces or terranes: Eastern, Central, and Western. However, this is controversial and is not assumed in the following description—which nevertheless adopts the same areas for descriptive convenience.

The evidence of sinistral transpression, and then

Table 1   Major faults of Svalbard (abbreviations refer to Fig. 2)

Name of fault	(Abbreviation as on Fig. 1)	Comments
Lady Franklinfjorden Fault	(LFF)	Age and significance unknown
Hinlopenstredet Fault Zone	(HFZ)	Age and significance unknown
Lomfjorden Fault Zone	(LFZ)	Post-Permian fault probably Palaeogene
Billefjorden Fault Zone	(BFZ)	Late Devonian transcurrence
Bockfjorden Fault Zone	(BKFZ)	
Breibogen Fault	(BBF)	
Raudfjorden Fault Zone	(RFZ)	
West Spitsbergen Orogenic Front·and	(WSOF)	Palaeogene thrust
Central Western Fault Zone	(CWFZ)	Front postulated late
including Kongsfjorden Fault	(KF)	Devonian transcurrence
Forlandsundet Graben	(FG)	Palaeogene structure could conceal ancient fault zone
Sutorfjella Fault	(SF)	Postulated from ? Silurian conglomerates

transcurrence, between and within these provinces is based on structural evidence to be outlined in the following accounts of the region. The argument for substantial displacement (i.e. of hundreds of kilometres on each fault) derives from stratigraphic comparisons between Svalbard rocks and (most notably) those of Greenland (e.g. Harland 1965, 1969, 1979, and especially by Harland and Wright 1979). In brief the eastern Svalbard sequence is very similar to that of East Greenland (as confirmed independently by Swett 1981), it contrasts with the central and western sequences in a number of respects, the latter having more in common with Northeast and North Greenland. This will be taken further in the last section of this paper.

Ages

Radiometric ages have been adjusted to the 1976 convention on decay constants and the time scale adjusted here is essentially that of Harland *et al.* 1982.

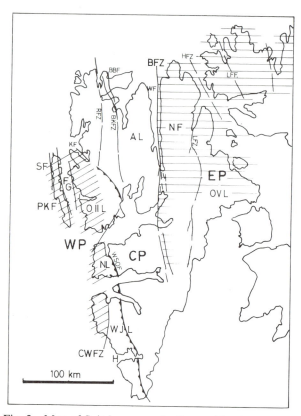

Fig. 2 Map of Spitsbergen and adjacent islands to show: (1) the three tectonic provinces, EP (shaded horizontally) = Eastern, CP = Central, and WP (shaded obliquely) = Western; (2) the principal fault lineaments, the key to these symbols is in Table I; (3) some place names mentioned in the text namely: Al = Andrée Land; H = Hornsund; J = St. Jonsfjorden; NL = Nordenskiöld Land; OIIL = Oscar II Land; OVL = Olav V. Land; PKF = Prins Karls Forland; WF = Wijdefjorden; WJL = Wedel Jarlsberg Land; NF = Ny Friesland

Northeastern Svalbard

The Northeastern Province comprises western Nordaustlandet and northeastern Spitsbergen (Ny Friesland and parts of Olav V Land), indicated on Fig. 1, Fig. 2 (by EP), and Fig. 3. Too little is known to separate or relate eastern Nordaustlandet. It is the classic Hecla Hoek area of the initial descriptions of the last century (e.g. Nordenskiöld 1863) and of this century (e.g. Kulling 1934) whose stratigraphic sequence will first be outlined—then the tectonic sequence of its deformation.

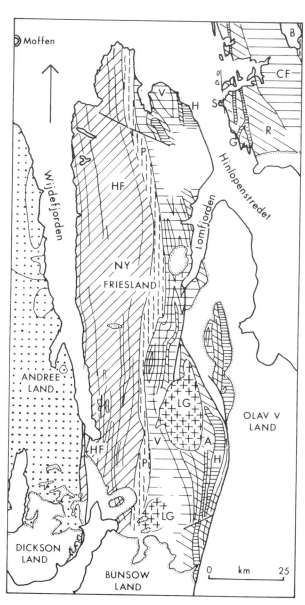

Fig. 3 Schematic geological map of Ny Friesland and surrounding lands based mainly on Harland 1959. The key to the ornament is via the letter symbols to Table 2. LG = Late granite

The Hecla Hoek Geosyncline

As a result of a series of investigations over many years a geosynclinal sequence has been established in Nordaustlandet (e.g. Kulling 1934; Flood *et al*. 1969) and in Ny Friesland (e.g. Harland and Wilson 1956; Harland, Wallis, and Gayer 1966). These sequences are proving to have considerable sedimentological and biostratigraphical interest (e.g. Swett 1981; Knoll 1982a, 1982b). A Varangian climatic sequence is now well established by distinctive early and late unequivocal tillites as in Norway and East Greenland (Hambrey 1982, 1983). The geosynclinal sequences are shown in outline on Table 2.

Structure

The structure of *western Nordaustlandet* was outlined by Kulling (1934) and by Flood *et al*. (1969).

Essentially the pattern is of upright folds, open to tight, with N–S axes plunging gently. The overall dip is westerly so that rocks are successively exposed to the east.

The structure of *Ny Friesland* (e.g. outlined by Harland 1959) is, in the above respect, opposite to that of Nordaustlandet. The youngest strata (Llanvirn) are exposed on the west side of the strait nearest to Nordaustlandet and the oldest rocks occur in a zone running N–S along the west of Ny Friesland.

In eastern Ny Friesland upright folds trend N–S—open or tight according to competence. The incompetent Polarisbreen Group commonly occurs in tight synclines in both eastern Ny Friesland and western Nordaustlandet.

In central Ny Friesland folds are tighter and wrap around the two or three large unfoliated granite (porphyritic adamellite) batholiths that in part trun-

Table 2   The Hecla Hoek sequences of northeastern Svalbard (abbreviations refer to Fig. 3)

Northeast Spitsbergen Ny Friesland		West Nordaustlandet Gustav V Land	
Hinlopenstretet Supergroup	H	Kap Sparre Fm (limestones)	
Oslobreen Group (1.2 km)		*Gotia Group* (0.6 km)	
Valhallfonna Fm (limestone) Arenig-Llanvirn		Klackberget Fm (shales with sandstones and marls)	
Kirtonryggen Fm (limestone and dolomites) Arenig		Sveanor Fm (tillites)	
Tokommane Fm (dolomites and sandstones) Caerfai		Backaberget Fm (shales with sandstones and marls)	
Polarisbreen Group (0.8 km)		*Murchisonfjorden Supergroup*	
Dracoisen Fm (shales)		*Roaldtoppen Group*)1.4 km)	R
Wilsonbreen Fm (tillites) Late Varangian		Ryssö Fm (dolomites with limestones, cherts and shales)	
Elbobreen Fm (shales, tillites) Early Varangian		Hunnberg Fm (limestones)	
		Celsius Group (2.1 km)	CF
Lomfjorden Supergroup		Raudstup-Sälodd Fm (dolomites and sandstones)	
Akademikerbreen Group (2 km)	A	Norvik Fm (sandstones with shales)	
Backlundtoppen Fm (dolomites and shales)		Flora Fm (sandstones)	
Draken Conglomerate Fm Early Sturtian		*Franklinsundet Group* (1.8 km)	CF
Svanbergfjellet Fm (limestones and dolomites)		Kapp Lord Fm (mudstones with sandstones and limestones)	
Grusdievbreen Fm (limestones)		Westmanbukta Fm (mudstones with siltstones)	
Veteranen Group (3.8 km)	V	Persberget Fm (quartzites with shales)	
Oxfordbreen Fm (shales)		*Botniahalvøya Group* (>6 km)	B
Glasgowbreen Fm (greywackes and quartzites)		Kapp Platen Fm (quartzites and shales)	
Kingbreen Fm (quartzites and shales with greywackes and carbonates)		Austfonna Fm (quartzites and limestones)	
Kortbreen Fm (quartzites and limestones)		Brennevinsfjorden Fm (quartzites, siltstones and shales)	
Stubendorffbreen Supergroup		Kap Kansteen Fm (acid volcanics)	
Planetfjella Group (4.8 km)	P		
Vildalen Fm (semipelites, psammites and quartzites)			
Flåen Fm (semipelites, psammites and quartzites with acid pyroclastics)			
Harkerbreen Group (4.1 km)	HF		
Sørbreen Fm (quartzites and psammites)			
Vassfaret Fm (semipelites, psammites and amphibolites)			
Bangenhuk Fm (feldspathites, psammites and amphibolites)			
Rittervatnet Fm (psammites, semipelites and amphibolites)			
Polhem Fm (quartzites and amphibolites)			
Finlandveggen Group (2.7 km)	HF		
Smutsbreen Fm (semipelites and marbles)			
Eskolabreen Fm (feldspathites, semipelites and amphibolites)			

cate and in part shoulder aside the strata with noticeable thinning adjacent to the intrusions.

In western Ny Friesland the rocks are all highly tectonized, and are typically of amphibolite facies up to sillimanite grade. They show three or more phases of homoaxial folding, generally with bedding schistosity and large recumbent folds apparently overfolded to the west and themselves antiformal in the south. Throughout there is remarkably constant N–S strike, and boudinage is ubiquitous with N–S extension and E–W compression. The only stones observed occur in a tilloid formation (Rittervatnet) and are strongly elongated. Throughout this western zone rock fabrics are conspicuously lineated, whether amphibolites or augengneisses. The boudinage and other structures often show sinistral displacement as seen from above and from the south. Tectonization is so extreme that in the deepest rocks it would be difficult to see if original struc-

tures were discordant or not. West of Ny Friesland the few outcrops available reveal a zone, possibly a few kilometres wide, of chloritic rocks originally with lineated amphibolite textures. Discordant strikes and late brittle fractures are also more evident in this zone.

Tectonic sequence

Within the Hecla Hoek sequence no stratigraphic break is evident. Archaean basement could be infolded with the cover rocks in the lowest formations but the tectonism to make them parallel would then post-date the cover rocks. On superficial examination many works have assumed that there must be an unconformable relationship between the unmetamorphosed and fossiliferous Ordovician strata and the extremely tectonized rocks 18 or more stratigraphic kilometres lower down. But no evi-

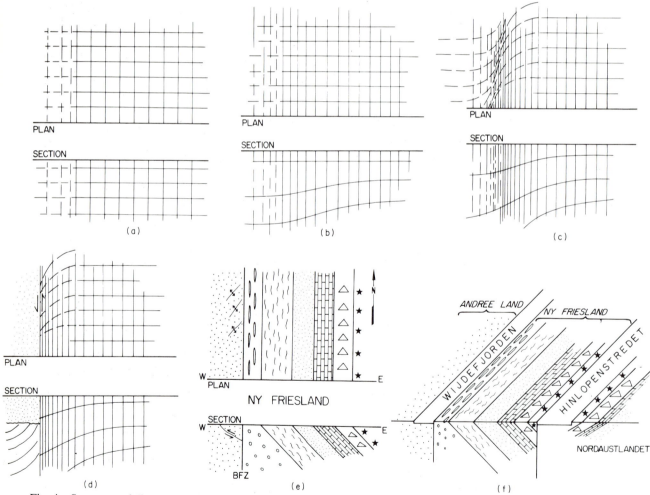

Fig. 4   Sequence of diagrams to show possible evolution of Caledonian structure of Ny Friesland (a) undistorted grid in plan and section (b) grid showing the effects of compression with horizontal shortening and vertical thickening (c) zone of sinistral transpression accentuates compression (d) truncation by sinistral transcurrence (e) similar sections with rock groups showing extreme elongation in west of Ny Friesland and relatively undistorted stones and fossils in the east (f) the same seen as block diagram. (Figure drawn by K. E. Fancett)

dence for such a break has yet been identified and we must take seriously the view that the prime tectogenesis is post-Llanvirnian. This was named the Ny Friesland orogenic phase (Harland 1959). It was far advanced but not finally completed with the emplacement of the granites dated at 413–392 Ma and the metamorphic ages were 444–372 Ma both recalculated K/Ar. The oldest overlying rocks are Tournaisian.

A tectonic sequence had first been argued as continuous compression with a deep consequent axial elongation (e.g. Harland 1959; Harland and Bayly 1958). But this failed to account for the lack of internal deformation throughout eastern Ny Friesland that is continuous and transitional with the western rocks that show such marked elongation. So the concept of transpression (Harland 1971) was introduced and appears still to be adequate (see Fig. 4).

According to this hypothesis a geosyncline, gripped between converging plates, was first squeezed intensely and normally. If this took place along a subduction zone, it has not been identified and need not have coincided with the present BFZ. Normal compressions increasingly gave place to sinistrally oblique compression or transpression which would cause the elongation along the western tectonized zone but need not tectonize the areas away from it. If there was a proto-Iapetus Ocean in this area it should be possible to identify its relationship to the Hecla Hoek and Greenlandian geosynclinal sedimentation and this is the subject of a current investigation.

In due course with a great sialic welt formed by compression and transpression, and virtual completion of the orogen, the strike-slip component would take over completely and give a transcurrent regime. This would have taken place with continuing uplift and erosion. So that the disturbance would have provided an environment for retrogressive metamorphism. Strike-slip motion along the Billefjorden Fault Zone with local secondary transpression produced structures is also evident in the Old Red Sandstones of Andrée Land to the west (see Northwest Svalbard).

External evidence suggests that between 1000 and 2000 km sinistral displacement occurred between Laurentia and Baltica (e.g. Van der Voo 1980). How much of this was affected by early transpressive subduction in an ocean basin or by collisional transpression followed by intraorogenic transcurrence is not clear, nor on which mobile zones such a sequence operated and in what degree. Indeed the degrees of freedom allow a distribution of the required motion in any or all of three or four fault zones and in each case active at different times over a span of at least 100 Ma.

Northwest(ern) Svalbard (North Central and Northwest Spitsbergen)

This province is shown on Figs. 1, 2, and 5. In previous reviews (e.g. Harland 1972, 1978(a), 1979) I kept open the possibility that the Raudfjorden fault, while demonstrably of post-Gedinnian age, could also have occupied the line of an earlier major fault and if this had been transcurrent then the two areas east and west of it might have initially belonged to

Fig. 5. Schematic geological map of northwest Spitsbergen based mainly on Gee and Hjelle 1966. The key to the ornament is via the letter symbols to Table 3

different provinces. After careful rereading of the few descriptions of this area and especially of Gee and Hjelle (1966) I accept their map (p. 33) which indicates no continuity of the fault to the south even though a nearly straight line could be drawn through glaciers. The whole area will thus be treated together and according to the thesis of this paper might well form the northern part of the same Central Province, the southern part of which (South Central Svalbard) will be treated next.

The province is separated to the east by Wijde-fjorden and the Billefjorden Fault Zone, to the south-east by overlying Carboniferous rocks. The other boundaries are coastal—to the southwest in Kongs-fjorden where difficulty in correlating the rocks on either side has led to the postulation of a major fault; there is no way of knowing how far the terrane may have extended westwards, and to the north there is the Yermak plateau, Moffen, and (separated by the Arctic Ocean Basin) the Lomonosov Ridge.

The rock sequence

The area exposes rocks of three main kinds informally referred to here as supergroups: Old Red Sandstone, Late Tectonic Granites, and metamorphic basement. These are most conveniently summarized in Table 3 which is based for Old Red Sandstone on Friend 1961; Friend and Moody-Stuart 1972; Gee and Moody-Stuart 1966; and Gee 1972; for granites and for Krossfjorden and Svitjodbreen Groups on Gee and Hjelle 1966; and for the Biskayerhuken Group on Gee 1972.

Structure

The Krossfjorden Group occupies the southwest of the province where low grade metamorphism (biotite) obtains. The Liefdefjorden, Biskayerfonna, and Richarddalen Groups occupy the east and northeast of the province and are of higher grade (garnet is general and feldspathic gneisses develop locally).

Table 3   The pre-Carboniferous rocks of northwestern Svalbard (abbreviations refer to Fig. 5)

Old Red Sandstone (*Supergroup*)		
Andrée Land Group (5 km)	AL	
Mimer Valley Fm in S and Wijde Bay Fm		Givetian–Eifelian
(0.5 km) in N		
Grey Hoek Fm (1 km)		Late Emsian grey siltstones
Wood Bay Fm (2.9 km)		Early Emsian–Siegenian
Red Bay Group (0.2 km)	RB	
Ben Nevis Fm		Gedinnian
Fraenkelryggen Fm		
Andréebreen Sst Fm		
Raudfjorden Conglomerate Fm		Unconformity
Siktefjellet Group (1.6 km)	SG	
Siktefjellet Sst Fm		
Lilljeborgfjellet Cgl Fm		Unconformity
Granite (*Supergroup*)		
Hornemantoppen Batholith	HB	Previously referred to as Smeerenburgbreen Batholith
a quartz monzonite		Fm, the name of the glacier that extends throughout
		its outcrop (Harland 1961)
Syntectonic granite group	g	Granite granodiorites, aplites and pegmatites occur
		intermittently from N to S of area
Northwest Metasediment Complex		
Krossfjorden Group (in SW)		
Generalfjella Fm (2 km)	GF	Marbles and interbanded marbles, pelites and
		quartzites
Signehamna Fm (2–2.5 km)	SE	Pelites with psammites and subordinate quartzites
Nissenfjella Fm (3 km)	NF	Pelites with subordinate amphibolites and psammites
		passing down into migmatites
Svitjodbreen migmatite group (in centre)		
Eastern sequence from D. G. Gee		(Personal communication. Informal nomenclature
Liefdefjorden Group 1.2–1.5 km (in E)	L	pending fuller description)
Pteraspistoppen marbles		
Erikbreen pelites		
Wulffberget marbles		
Biskayerfonna Group 4.5–5.5 km (in NE)	B	
Biskayerhuken Fm		Pelites, psammites and quartzites
Montblanc Fm		Feldspathic with amphibolites
Richarddalen Group (in NE)	R	
amphibolites, feldspathites, eclogites and marbles		

The centre of the province is mapped as migmatites and forms a narrow zone about 3 km wide in the south to a broad coastal exposure 30 to 40 km across in the north (Gee and Hjelle 1966). They are referred to here as the Svitjodbreen migmatites, and would appear to lie at a deeper level than the Krossfjorden Group because the fold structures plunge southwards in the south and the migmatization is transitional in the lower Nissenfjella Formation. The migmatite zone is characterized by local grey syntectonic granites (referred to by Gee and Hjelle as post-tectonic) which occur in belts and patches from north to south. Also there is a single large, late tectonic, intrusion of distinctive pink quartz-monzonite—the Horneman Batholith, that occurs in the middle part of this axial migmatite zone.

The successions on either side of the Svitjodbreen migmatite zone (Krossfjorden Group to the southwest and the Liefdefjorden, Biskayerfonna, and Richarddalen Groups to the northeast) are not similar. The fact that Gee and Hjelle (1966) knowing both successions did not attempt to correlate them lent support to my earlier view that the Raudfjorden Fault might be a major transcurrent displacement. But in such a large and complex terrane it is likely that more than the 7 km of recorded strata have been tectonized and that the strata varied somewhat, therefore the failure to correlate isolated successions on either side of a broad migmatite zone need not imply that either is exotic.

Folds and faults have a generally N–S or NNW–SSE trend—they are open to steep, often overturned and even recumbent. Overfolding and overthrusting towards the west is the rule.

Tectonic sequence

Within such a large complex it would be surprising if some Precambrian basement were not present though it could be difficult to recognize. A few apparent Precambrian radiometric ages were obtained but these are too uncertain except to speculate that the rocks of the Richarddalen Group would best qualify as Precambrian basement.

Most metamorphic rocks cooled at 450–460 to 387 Ma implying a Caledonian age for the main tectogenesis. The Horneman Batholith gave 347 to 324 Ma values and, if true, these ages would imply that uplift and granite emplacement was proceeding, in the northwest of the province, concurrently with Old Red Sandstone deposition to the east of it.

However before 400 Ma and probably in later Silurian time the Siktefjellet Group was deposited on the deeply eroded high grade rocks and was folded and thrust in the Haakonian diastrophic phase (Gee 1972).

There was further erosion and uplift with sediments pouring down from the west to form the remarkable Red Bay conglomerates and sandstones (Friend and Moody-Stuart 1972). This was followed further east, at least, by the quieter fluviatile Old Red Sandstone facies of the Andrée Land Group of Early and Middle Devonian ages. They were then folded and faulted in the well known Svalbardian diastrophic phase which was completed before Tournaisian time.

It is worth recalling that on his own evidence Gee (1972) argued that the Haakonian and Svalbardian structures required NW–SE compression across N–S lineaments (i.e. sinistral transpression) and he gave evidence that the three faults (Raudfjorden, Hannabreen–Rabotdalen, and Breibogen faults) all show characteristics of sinistral strike-slip in the early stages but later dip-slip motion in both Devonian and post-Carboniferous time. Gee concluded that it is 'probable that they represent major crustal fractures'. This evidence, taken in conjunction with the joint drag phenomenon, suggests that these fundamental faults originated as transcurrent phenomena prior to Siktefjellet Group deposition and that by the time these post-orogenic sediments had been deposited strike-slip movements were ceasing and giving way to dip-slip movement'.

He also related the strike-slip motion to the conspicuous oblique chevron folding in the Biskayerhuken pelites and to retrogressive metamorphism throughout this belt.

This accords well with my hypotheses, based elsewhere, of transpressive through to transcurrent regimes and substantial strike-slip (e.g. Harland 1965, 1971). This is further evidence that transpression, which concentrated in fault zones was general throughout Svalbard in 'Caledonian times'. But for the reason stated above it is not proposed that these particular faults were transcurrent fractures with more than a few kilometres of displacement (in contrast to the BFZ and CWFZ).

South Central Svalbard

Without attempting to give precise boundaries to this province we may include the areas to the north and south of Hornsund (Southern Wedel Jarlsberg Land and Sørkapp Land respectively) where both Norwegian and Polish studies have predominated e.g. Major and Winsnes (1955), Birkenmajer from 1958 onwards (e.g. 1975).

It is not easy to correlate north and south of Hornsund so the possibility of some faulting in the fjord cannot be ruled out but there is sufficient similarity for both sides to be included in this account.

There is a significant geosyncline and the name *Hornsundian* referring to the pre-Caledonian basin development of Birkenmajer 1975 has been selected here as a name for the geosyncline in this region.

It might also be added that the whole of the pre-Carboniferous outcrop in the province has been enveloped by Cenozoic tectonism of the West Spitsbergen Orogen and so the original dispositions even after mid-Palaeozoic deformation have been disrupted and so complicates interpretation.

Hornsundian Geosyncline

The first modern investigation of the area south of Hornsund was by Major and Winsnes (1955). Then a series of Polish expeditions not only studied the areas around their base building north of Hornsund, but also integrated Norwegian and Polish results into a unified system (Birkenmajer 1958, 1959, 1960, 1972, 1975). Table 4 shows the succession as synthesized by Birkenmajer. A tentative alternative based on Harland 1979 is also added because recent observations have led me to question parts of this sequence. For example the evidence for the unconformities above and below the Deilegga rocks are not easy to observe. To some extent Birkenmajer's gaps in the sequence follow from the correlation made between Ny Friesland and Hornsund which I

was initially in agreement with. However having seen the rocks in all the areas, and especially in the Western Province, I suggest that the greater similarity is not with Ny Friesland but with the much younger (mainly Vendian) eugeosynclinal sequence of the west coast. We (Birkenmajer and I) correlated the Rittervatnet tilloid of Ny Friesland with the Vimsodden tillite north of Hornsund. Later observations led Winsnes and I to view the Vimsodden tillite as part of the Vendian sequence.

The tectonic sequence

Birkenmajer's investigations were synthesized in a general tectonic paper (1975) in which he set out a sequence of phases of crustal deformation that separated his lithostratigraphical groups and related them to phases of basin formation. Table 4 is abstracted from this for Hornsund. The Jarlsbergian, Torellian, and Werenskiöldian tectonic phases are inferred from the supposed gaps in the stratal sequence and the recurrence of conglomerates. If my suspicions are correct the 'Slyngfjellet conglomerate' is not evidence of tectonism but of glaciation and the evidence for the stratigraphic gaps is difficult to see. In 1981 Birkenmajer correlated his Torellian phase with Haller's Carolinidian phase in Greenland. Good evidence for either is still awaited.

Table 4 Carboniferous and earlier sequence of Hornsund area in south Spitsbergen

Unit (thickness km)	Age	Facies	Tectonic Phase	Alternative possibilities
	Late & Middle Carbonif.	'molasse'		
			— Adriabukta	
	Early Carboniferous	'molasse'		
			— Svalbardian	
	Devonian intramontaine	'molasse'		
			— Ny Friesland	
Sørkapp Land Gp (1.7)	Early Ordovician	carbonates		
			— Hornsundian	
Sofiekammen Gp (0.8)	Early Cambrian	carbonates etc.		
			— Jarlsbergian	
Sofiebogen Gp (2.5)	Vendian Riphean	Gåshamna phyllite Fm .... Höferpynten dolomite Fm Slyngfjellet conglomerate Fm		Late Vendian Varangian tilloids
			— Torellian	
Deilegga Gp (3.5)	Upper Fm Middle Fm Lower Fm	slates, phyllites & quartzites slates, phyllites & dolomites dolomites and conglomerates		Höferpynten Gp (Vendian–Late Riphean)
			— Werenskiöldian	
Eimfjellet Gp (1.5)	Vimsodden Fm Skålfjellet Fm Gulliksenfjellet Fm	variable facies .... feldspathic rocks etc. .... quartzites		Varangian tilloid metavolcanics
Isbjørnhamna Gp (1.5)	Revdalen Fm Ariekammen Fm Skoddefjellet Fm	garnetiferous pelites garnetiferous pelites and marbles garnetiferous pelites		

Southern Svalbard—Bjørnøya (Bear Island)

Bjørnøya is a small island of which only a small part exposes older rocks—insufficient for significant conclusions to be drawn. While it can be asserted that it does not match closely any of the above successions, at a distance of about 250 km from the nearest it would not necessarily be expected to do so. However, the facies and thicknesses have more affinity with those of South Central Svalbard than of Western Svalbard.

Table 5 Pre-Carboniferous sequence of Bjørnøya. Succession after Krasil'shchikov and Livshits 1974, with older nomenclature after Horn and Orvin 1928)

Roedvika Fm }	(Ursa Sandstone)		Famennian
	————unconformity————		
Ymerdalen Fm Upper Member	(*Tetradium* Limestone) 240 m	(Chazy)	Llandeilo
Middle Member Lower Member Sørhamna Fm	(Younger Dolomite) 400 m (Slate–Quartzite) 175 m		
	————unconformity————		
Russehamna Fm	(Older Dolomite)	Late	Riphean
	400 m		

The overlying *Roedvika Formation* is a continental sandstone formation with plants of late Famennian to Tournaisian age and is part of the 'Ursa Sandstone'. There is no major unconformity in the island from Famennian to Triassic time so the Svalbardian movements, which in Spitsbergen are demonstrably post-Givetian and pre-part-early Tournaisian, may well have preceded the Roedvika Formation and therefore be constrained in age as pre-part-Famennian. There is a related question whether the youngest underlying *Ymerdalen Formation* of which the upper part ('Tetradium Limestone') is of Chazy facies and age (Ordovician) was deformed at the same time as the Svalbardian, the Haakonian, or the Ny Friesland diastrophism in North Spitsbergen.

The *Ymerdalen Formation* rests unconformably on the *Sørhamna Formation* (175 m of pelites and quartzites) which have quite speculatively been considered as Vendian in age (e.g. Harland and Wilson 1956 *et seq.*) and the *Russhamna Formation* (400 m of dolostone) which may be Vendian or Late Riphean in age. Correlation of these strata with particular Spitsbergen units is not as inevitable as it once seemed because of the widely divergent successions now known in Spitsbergen.

Inspection of Table 5 suggests an alternative correlation. The Middle and Lower Ymerdalen Members could well be Canadian and the *Sørhamna Formation* may be Cambrian. On this basis there is some similarity with the Ny Friesland sequence and the unconformity would correspond to Vendian strata generally including tillites. It could then possibly correspond with Birkenmajer's Jarlsbergian diastrophic phase.

Western Svalbard, Central West and Southwest Svalbard

In common with the South Central Province (Hornsund area) the whole of the Western Province has been subject both to mid-Palaeozoic deformation (Caledonian or Ellesmerian orogenies) and to mid-Cenozoic deformation of the West Spitsbergen Orogeny. In short the earlier tectogenesis led to low and medium grade metamorphism with recumbent structures overfolded from east to west while the later orogeny on a smaller and more superficial scale caused folding and thrusting towards the east. The resulting complexity of structures is one reason why the older strata of this province have taken so long to elucidate. Another reason is that the sedimentary facies of the belt is highly mobile with evidence of rapid deposition. For some years attempts were made to interpret the western succession in terms of the better known Ny Friesland sequence and this tended to force correlations that have since proved to be untenable.

The rocks clearly belong to a geosyncline that has little in common with the Central and Eastern Provinces and a separate name was given to it in honour of the scientist who first made substantial progress in the investigation of the rocks in their regional setting—Olav Holtedahl.

The Holtedahl Geosyncline

This was defined by Harland *et al.* for the 1975 Symposium (1979), and based on the succession as interpreted then in Oscar II Land and Prins Karls Forland. It is possible to relate these two successions but they could have originally been far distant and separated by a fault along the present Forlandsundet Graben.

Since 1975 further work has led to significant modifications e.g. Harland 1978(b), Waddams (in press), and Winsnes and Harland (in press). Because this new information is not yet published and there is not space here to give it in definitive detail the 1975 sequence is tabulated with some modifications. Tables 6 and 7 are based on sequences towards the northern and southern end of the west coast outcrops of the postulated Western Province. There are still

Table 6　Pre-Carboniferous sequences of central western Spitsbergen

Prins Karls Forland	*Oscar II Land*
Harland, W. B., Horsfield, W. T., Manby, G. M., and　Morris, A. P. 1979	Harland *et al*. 1979
	Modified from Waddams (1983)
Grampian Group (3.6 km)	*Bullbreen Group*
Geddesflya Fm (turbidites)	Holmesletfjella Fm
	(Bulltinden Mbr) + mid Silurian
Fugelhuk Fm (quartzite, shales)	Motalafjella Fm
Barents Fm (turbidites)	
Conqueror Fm (quartzites)	
Utnes Fm (slates, quartzites)	*Comfortlessbreen Group*
Scotia Group (1.0 km)	
	Aavatsmarkbreen Fm
	(part of Sarsøyra of Harland *et al*.)
Roysha Fm (siltstones, slates)	
Kaggan Fm (slates)	
Baklia Fm (siltstones, slates, limestones)	mainly dark phyllite with subsidiary marble, psammite,　conglomerate and volcanics
Peachflya Group (1.3 km)	
Knivodden Fm (phyllites)	
Hornnes Fm (phyllites, sandstones)	
Alasdairhornet Fm (volcanics)	
Fisherlaguna Fm (phyllites)	Annabreen Fm
	2 km (quartzites, schists and phyllites)
Geikie Group (0.77 km)	
Rossbukta Fm (sandstones)	
Gordon Fm (limestones)	
Ferrier Group (0.75 km)	
Neupiggen Fm (tillites)	Haakon Fm (tillites) + 2–3 km
Peterbukta Fm (greywackes)	
Hardiefjellet Fm (tillites)	
Isachsen Fm (tillites, schists, volcanics)	
	St. Jonsfjorden Group
	Alkhorn Fm (dk laminated limestone) 1 km
	Lovliebreen Fn (Qi, phyllites and volcanics) 1 km
	Moefjellet Fm (dolostone) 0.8 km
	Trondheimfjella Fm (tillite) +
	Kongsvegen Group
	⎧ Nielsenfjellet Fm
	N ⎨ Steenfjellet Fm
	⎩ Bogegg (dolostone and semipelites)
Pinkie Fm (metavolcanics)	
	s ⎰ Müllerneset (amphibolite)
	⎱ Vestgotabreen (glaucophane schist)

outstanding problems but the following generalizations can be made:

1. Early Palaeozoic strata are thick and of clastic facies that are not easy to date except for Silurian (? Wenlock) rocks in the Bullbreen Group. The youngest strata in the Hornsundian and Hecla Hoek Geosynclines are early mid-Ordovician in age. So there is a contrast in age and facies.

2. Recent work, notably by P. Waddams, has shown that an earlier tillite, almost certainly equivalent to early Varangian (Smalfjord tillite formation), occurs in the sequence at a lower level than was previously thought. Thus the whole of the Comfortlessbreen, St. Jonsfjorden, Kapp Lyell Groups are certainly Varangian; while the Dunderbukta, Recherchefjorden, and Konglomerat-

fjellet groups are also probably Varangian. Varangian strata in the Eastern Province are approximately 1 km in thickness, with no recorded volcanic component. In the Central Province about 1½ km of Vendian strata include green and purple phyllites. In the west these Varangian strata vary in thickness from 6 to 10 km of highly mobile and variable facies including extrusive volcanic facies. The rocks outcropping in areas between those shown in Tables 6 and 7 (i.e. Southern Oscar II Land and Western Nordenskiöldland) are not described here—they add little to the story being almost entirely Vendian.

3. A pre-Varangian conformable sequence may be seen in the Nordbukta Group.

4. Scattered outcrops of more metamorphosed

Table 7   Late Precambrian sequences south of Bellsund in south Spitsbergen

Western After Harland 1978 Winsnes and Harland in press Waddams 1983	Eastern After Harland 1978	Southern After Hjelle 1969
	Kapp Lyell Group Lyellstranda Fm	
		Bellsund–Dunderdalen (tillite)
Kolvebekken Fm (0.7 km)		
	Renardbreen (1–2 km)	
Logna Fm (0.16 km)		*Konglomeratfjellet Gp*
Dundrabeisen Fm (3 km) (tillites)	Vestervagen (0.34 km)	shale and quartzite fm
Dundabukta Group	*Recherchefjorden Gp*	
Dunderdalen Fm	Chamberlindalen Fm	volcanic fm
Floykalven Fm (tillite)	Solhögda Fm	calcareous fm
	Gaimardtoppen Fm	
	Foldnutane Fm (conglomerate)	conglomerate fm
Nordbukta Group	*Magnethogda Gp*	
	= Revdalen	
	Hornsund area belongs to South	
	Central Province	

strata are difficult to relate at this stage. They are presumed to be older and are tectonically separated from the rocks described. These include:

(a) The Kongsvegen Group at the northern extremity of the Western Province (in Brøggerhalvøya).
(b) Pinkie Formation of Prins Karls Forland
(c) Müllerneset beds just south of St. Jonsfjorden
(d) Vestgotabreen blue schists still further south in Oscar II Land (e.g. Ohta 1979).

Structure

The generally N–S trending structure of this coastal area, combining as it does the effects of mid-Palaeozoic and mid-Cenozoic diastrophism, is complex. The West Spitsbergen Orogen (Harland and Horsfield 1974) exposes easterly directed overfolds and thrusts in a small scale dextral transpressive structure (Lowell 1972). Its basement reveals earlier and altogether larger features that appear as disconnected and somewhat reorientated fragments of the major orogenic zone of the Western Province. The fragments are separated by oblique faults, some being inferred within glaciers and fjords. In the north there is evidence of a substantial westerly directed nappe zone with roots near eastern St. Jonsfjorden, and allochthonous apex in southern Prins Karls Forland. In northern Wedel Jarlsberg Land is the enormous fold structure whose axis is the Kapp Lyell northerly plunging syncline. These and other structures are manifest. Two cryptic features are worth noting: (1) the Forlandsundet Graben (a Ter-

tiary structure between Oscar II Land and Prins Karls Forland) may well occupy a mid-Palaeozoic sinistral transpression and transcurrent zone. How much strike-slip has separated the eastern and western successions in Table 6 is uncertain. (2) To the west of Prins Karls Forland at Sutorfjella is a remarkable synsedimentary breccia with blocks many metres across suggestive of contemporaneous faulting in Early Palaeozoic times. This could have been another (splay of an) ancient sinistrally active fault system.

Tectonic sequence

The oldest rocks exposed may well not be so old as in the Central and Eastern Provinces. The record becomes readable for Vendian time with thick mobile sedimentation, volcanism, and glaciation which continued, apparently without interruption into early Palaeozoic time when facies did not favour a fossil record.

Then there is stratigraphic evidence for at least two Palaeozoic tectonic episodes (in addition to the superimposed Palaeogene diastrophism). Highly mobile environments yielded molasse-like facies, including conglomerates, themselves part of the pre-Carboniferous deformed succession. They contain slate and other weakly metamorphosed fragments of Vendian rock types. In St. Jonsfjorden such is the Bullbreen Group with its conglomerate which is probably Silurian being demonstrably Wenlock or pre-Wenlock. Unfortunately there is a tectonic hiatus between those and probable Vendian strata. In Prins Karls Forland the Sutorfjella Conglomerate of the Barents Formation (Grampian Group) of similar facies could be of similar age but there is no direct evidence. In Prins Karls Forland, however, a more complete succession is preserved so that about 5 km of variable strata come between the conglomerates and the latest Vendian tillite. It is arguable from comparisons elsewhere that the first tectonic episode was late Ordovician or earliest Silurian and the second was post-part Wenlock probably Pridoli or earliest Devonian. No Devonian strata are known from the western province. Without elaborating further, there is marked contrast with, say, the Eastern and Central Provinces and the timing of diastrophism would better fit that of the Ellesmerian (and presumably) the North Greenland Fold Belt, than, say, that of Ny Friesland or Central East Greenland Caledonides.

As to the resulting structure we observe a similar effect to that seen elsewhere in Svalbard, with at least localized zones showing extreme N–S elongation of tillite stones, and also oblique folding of earlier structures. This is consistent with a sequence of compression, followed by transpression, and probably then by transcurrence, all completed between about Wenlock and Tournaisian time.

Conclusions

Sidestepping the problems which abound, here are some conclusions that I believe can be established. However, because there is no room for critical evidence or discussion they may well be regarded as questions.

The composite nature of Svalbard

There may be more than three terranes but the Eastern, Central, and Western Provinces outlined above constitute a minimum number that formed at far greater distances than now found and were juxtaposed by mid-Palaeozoic time (Figs. 6 and 7). A minimum of 200 km sinistral strike-slip was argued for the Billefjorden Fault Zone (Harland *et al.* 1974) but a greater displacement was subsequently suggested (Harland and Wright 1979). Birkenmajer (1975, 1981), however, accepted the 200 km as a maximum not a minimum value. The effect of this is to place the Hornsund sequence 200 km nearer to the Ny Friesland sequence i.e. almost adjacent. The contrasts between the two sequences become more difficult to explain if they were adjacent. However, the evidence for the hundreds of kilometres of strike-slip claimed is based mainly on stratigraphic similarity of the Hecla Hoek sequence of the Eastern Province with the Eleonore Bay and associated

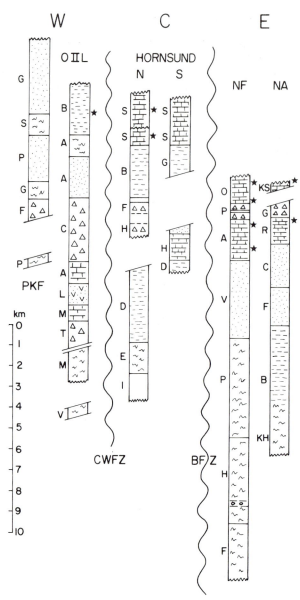

Fig. 6 Diagrammatic representation of the Eastern (E), Central (C), and Western Provinces (W) with two representative stratigraphic sections from each. The major rock units are indicated by the initial letter of the name which will be found in Tables 2, 3 and 6 respectively. PKF = Prins Karls Forland; OIIL = Oscar II Land; NF = Ny Friesland; NA = Nordaustlandet. (Figure drawn by K. E. Fancett)

rocks of central East Greenland and the difference of the intervening rocks of the Central and Western Province. Arguments developed (e.g. Harland and Wright 1979) might support 1000 km strike-slip distributed variously in different fault zones (Fig. 7), and support comes also from palaeomagnetic studies (e.g. Van der Voo 1980; Van der Voo and Scotese 1981). Thus the hypothesis, ignored in 1975 when I first proposed it, is now widely accepted (e.g. Ziegler 1982).

Significance of transpression

It is generally more rewarding to interpret mesoscopic and microscopic structures from regional and
global tectonics than to attempt the reverse investigation. This applies to the phenomena of transpression. Transpression is an inevitable consequence of
plate tectonics—it being impossible to manipulate
plates without a significant proportion of mobile
zones at any one time obliquely converging and
diverging. The effects of such oblique compression
or transpression have commonly escaped notice
partly because of the tendency to depict and con-
ceive orogenic belts in terms of cross-sections only,
and partly because significant transpression pro-

duces parallel from discordant structures as well as
elongations that are not conspicuously dextral or
sinistral. There is much to learn here but the conclusion is inescapable, in Svalbard, that zones of
marked axial direction and elongation (with
homoaxial folding) are the product of a phase of
transpression transitional between compression and
transcurrence—in this case sinistral (see Fig. 4). It is
believed that although Svalbard clearly exhibits the
phenomena, the process must commonly apply in
orogenic belts anywhere.

Caledonian and Ellesmerian Provinces

The structure of eastern North Greenland shows a
meeting of the Caledonian Fold Belt passing out to
sea with NE trend and the North Greenland Fold
Belt converging with it from an E–W trend (Fig. 7).
Permian reconstructions typically show Spitsbergen
with its dominant mid-Palaeozoic strike truncating
the Greenland trend projected out beyond the Wandel Sea. That configuration would make difficult the
interpretation of the main Caledonian kinematics. It
got there later, in Late Devonian time. Earlier the
Western Province of Spitsbergen was probably not
far from its Permian position and belongs to the
North Greenland Fold Belt or the extreme north of
the Caledonides. The Eastern Province cannot have
been far from central East Greenland. So we can
only be sure that the Eastern and possibly Central
Provinces of Svalbard are Caledonian (*ss*). The
Western Province in timing, facies, and location has
more affinity with the Ellesmerian Orogen.

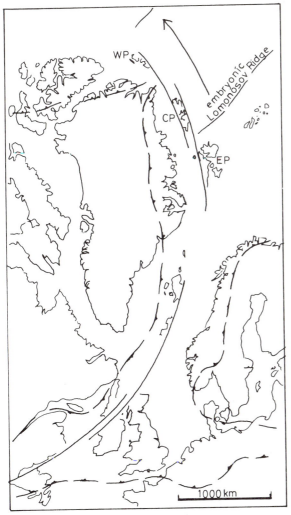

Fig. 7 Schematic mid-Caledonian map to show relative
positions of Svalbard's tectonic provinces in relation to
Greenland and Europe adapted from Harland and Gayer
1972, Harland and Wright 1979, and Ziegler (1982).
The map is much simplified to illustrate only one aspect. It
depicts a situation with Iapetus nearly closed and about to
intensify the late Caledonian (up to Svalbardian) sinistral
strike-slip motion that juxtaposed the parts of Svalbard,
British Isles, and Newfoundland

Hecla Hoek, Hornsundian and Holtedahl Geosynclines

In consequence the three geosynclines corresponding to the Eastern, Central, and Western Provinces
should be clearly distinguished (by name) until
shown to be continuous. The habit of referring to
any older rocks in Svalbard as Hecla Hoek (which
first referred to the succession in the east) obfuscates
an essential truth and method of getting at it.

Iapetus and Proto-Iapetus

Iapetus was originally drawn so as to separate
Greenland with Spitsbergen and much of the Barents Shelf from Scandinavia (Harland and Gayer
1972). On this basis the three geosynclines mentioned above would not have been separated by
ocean floor from Greenland. It is possible that at an
earlier stage there was a proto-Iapetus, possibly an
aulacogen to separate (slightly) these areas and
allow subsidence and sedimentation. The matter is
currently under investigation by means of the infer-

red source and direction of flow of Varangian continental ice sheets.

Lomonosov and Arctic tectonics

The consequences of the significant (i.e. 100 km or more) mid-Palaeozoic sinistral strike-slip motion has repercussions for distant areas now in the Arctic. It was proposed in 1965 that the Lomonosov Ridge was a product of Svalbardian and possibly earlier motion producing a compression (? subduction) zone along the leading edge of the Barents Plate—and now separated by the spreading of the Gakkel Ridge. In that sense the Lomonosov Ridge would be the converging end of a transform. This could be tested by estimating the amount of sinistral movement relative to Greenland on either side of the Lomonosov Ridge allowing that it might itself have been displaced since formation.

Acknowledgement

Such a general paper is indebted to countless strands that have contributed to my experience. Nevertheless I must acknowledge substantial support for renewed field work in Ny Friesland in 1981 (and 1982) provided through the Natural Environment Research Council (GR/4342).

Postscript

It was remarked by a referee that describing the older rocks of Svalbard separately according to area emphasized the differences and possibly obscured the similarities between them, so giving spurious support to my hypothesis of tectonic juxtaposition of originally widely separate successions. In response to this I would offer two considerations.

1. It is philosophically sounder to describe different objects independently so leaving it to consider how similar or how different they may seem to be. The more popular alternative of lumping objects into the same class prejudges the issue at the outset. Subsequently it is easier to discuss whether A is or is not equivalent to B than whether A (with or without some qualification) is or is not equivalent to A (with or without another qualification).
2. The only sure Precambrian correlation between the older rocks in the different terranes is by the ubiquitous tilloids which in *some* cases have been shown unequivocally to be tillites (Hambrey 1983) and are *generally* taken here to be tillites,

so affording a long distance correlation. It is now confirmed that in the North Atlantic Arctic region two main groups of Varangian (Vendian) tillites can be correlated (Hambrey 1983). The earlier one (equivalent to the Smalfjord tillite of North Norway and the lower tillite of East Greenland) is generally devoid of feldspathic stones which are almost entirely limestones, dolostones, sandstones, and quartzites matching immediately underlying strata. The upper tillite horizon (equilvalent to the Mortensnes tillite of North Norway and the upper tillite of East Greenland) is characterized by a much greater variety of stones including many pink granitoid rocks of unknown provenenace.

These two distinctive facies can be traced not only in the three main provinces of Svalbard but extending thousands of kilometres beyond. This tells us that we have a widespread constant climatic facies component superimposed on the variable local facies component, and so these distinctive yet widespread facies cannot be used to argue original proximity of deposition. There is little else that is similar between the provinces in Svalbard.

A more detailed paper (Harland, Hambrey, and Waddams) is in preparation setting out the correlation of Vendian strata throughout Svalbard based not only on tillites but on biostratigraphy and other characters. In this respect the tables in this paper were simply abstracted in 1981 from published work and do not attempt the more precise correlation possible in 1983 when going to press.

References

American Geological Institute. 1980. Bates, R. L. and Jackson, J. A. (Eds), Glossary of Geology, ACI, Fall Church, Virginia, USA, 749 pp.

Bailey, E. B. and Holtedahl, O. 1938. Northwestern Europe, Caledonides. In *Regionale Geologie der Erde*, 2, *Palaeozoische Tafeln und Gebirge*, Absch. II. Leipzig.

Birkenmajer, K. 1958. Preliminary report on the stratigraphy of the Hecla Hoek formation in Wedel–Jarlsberg Land, Vestspitsbergen. *Bull. Acad. Polonaise Sci. Ser. Chim. Geol. Geog.*, 6 (2), 143–150.

Birkenmajer, K. 1959. Report on the geological investigations of the Hornsund Area, Vestspitsbergen in 1958, Part I—The Hecla Hoek Formation, Part II—The Post-Caledonian Succession, Part III—The Quaternary Geology. *Bull. Acad. Polonaise Sci. Ser. Chim. Geol. et Geog.* 7 (2), 129–136, (3) 191–196 and 197–202.

Birkenmajer, K. 1960. Geological sketch of the Hornsund area. *I.G.C. 21st Session*, Norden 1960. Supplement to the guide for excursion A16. (*Aspects of the Geology of Svalbard*) 12 pp.

Birkenmajer, K. 1972. Cross bedding and stromatolites in the Precambrian Høferpynten Dolomite Formation of Sørkapp land, Spitsbergen. *Norsk Polarinst. Årbok 1970*, 128–145.

Birkenmajer, K. 1975. Caledonides of Svalbard and plate tectonics. *Bull. Geol. Soc., Denmark*, **24**, (1–2) 1–19.

Birkenmajer, K. 1981. The geology of Svalbard, the western part of the Barents sea, and the continental margin of Scandinavia. In Nairn, A. G. M., Churki, M., Jr., and Stehli, F. (Eds), *The Ocean Basins and Margins, Vol. 5 The Arctic Ocean*, 265–329.

De Geer, G. 1909. Some leading lines of dislocation in Spitsbergen. *Geol. Fören. Stockh. Förh.*, **31**, H.4, 199–208.

Flood, B., Gee, D. G., Hjelle, A., Siggerud, T. and Winsnes, T. 1969. The geology of Nordaustlandet, northern and central parts. *Norsk Polarinst. Skr.*, **146**, 139 pp + geol. map 1:250,000.

Friend, P. F. 1961. The Devonian stratigraphy of north and central Vestspitsbergen. *Proc. Yorks. Geol. Soc.*, **33**, Part I, No. 5, 77–118.

Friend, P. F. and Moody-Stuart, M. 1972. Sedimentation of the Wood Bay Formation (Devonian) of Spitsbergen: regional analysis of a late orogenic basin. *Norsk Polarinst. Skr.*, **157**, 77 pp.

Gee, D. G. 1972. Late Caledonian (Haakonian) movements in northern Spitsbergen. *Norsk Polarinst. Årbok 1970*, 92–101.

Gee, D, G. and Hjelle, A. 1966. on the crystalline rocks of north-west Spitsbergen. *Norsk Polarinst. Årbok 1964* 31–45.

Gee, D. G. and Moody-Stuart, M. 1966. The base of the Old Red Sandstone in central north Haakon VII Land, Spitsbergen. *Norsk Polarinst. Årbok 1964*, 57–68.

Hambrey, M. H. 1983. Correlation of Late Proterozoic tillites in the North Atlantic region and Europe. *Geol. Mag.*, **120**(3), 209–320.

Harland, W. B. 1959. The Caledonian sequence in Ny Friesland, Spitsbergen. *Q. J. geol. Soc. London*, **114**, 307–342.

Harland, W. B. 1965. Critical evidence for a great Infra-Cambrian glaciation. *Geol. Rdsch.*, **54**(1), 45–61.

Harland, W. B. 1969. Contribution of Spitsbergen to understanding of tectonic evolution of North Atlantic region. in *North Atlantic—Geology and Continental Drift, A.A.P.G. Mem.*, **12**, 817–851.

Harland, W. B. 1971. Tectonic transpression in Caledonian Spitsbergen. *Geol. Mag.*, **108**(1), 27–42.

Harland, W. B. 1972. Early Palaeozoic faults as margins of Arctic plates in Svalbard. In *24th Internat. Geol. Congress, (Montreal 1972)*, Section 3, 230–237.

Harland, W. B. 1978a. The Caledonides of Svalbard. In *Caledonian—Appalachian Orogen of the North Atlantic Region, I.G.C.P. Project 27*, Geol. Surv. Can. Paper, **78–13**, 3–11.

Harland, W. B. 1978b. A reconsideration of Late Precambrian stratigraphy of South Spitsbergen. *Polarforschung*, **48**(1932) 61.

Harland, W. B. 1979. A review of major faults in Svalbard. *U.S.G.S. Open File Report 79–1239*, 157–80 (Proc. of Conference 8. Analysis of actual fault in bedrock. Convened under the auspices of the National Earthquake Hazards Reduction Program, 1–5 April, 1979).

Harland, W. B. 1980. Comment on 'A paleomagnetic pole position from the folded upper Devonian Catskill red beds, and its tectonic implications'. *Geology*, **8**, (6) 258.

Harland, W. B. and Bayly, M. B. 1958. Tectonic regimes. *Geol. Mag.*, **95**, 89–104.

Harland, W. B., Cox, A., Llewellyn, P. G., Pickton, C. A.

G. and Walters, R. 1982. A Geologic Time Scale, Cambridge, xii + 132 pp.

Harland, W. B. and Gayer, R. A. 1972. The Arctic Caledonides and earlier oceans. *Geol. Mag.*, **109**, 289–384.

Harland, D, W. B. and Horsfield, W. T. 1974. West Spitsbergen Orogen. In Spencer, A. M. (Ed.), *Mesozoic–Cenozoic Orogenic Belts. Data for Orogenic Studies Geol. Soc. London Special Publ.* **4**, 747–755.

Harland W. B. and Wilson, C. B. 1956. The Heckla Hoek Succession in Ny Friesland, Spitsbergen. *Geol. Mag.*, **93**(4), 265–286.

Harland, W. B. and Wright, N. J. R. 1979. Alternative hypothesis for the pre-Carboniferous evolution of Svalbard. *Norsk Pet. Skr.*, **167**, 89–117.

Harland, W. B., Wallis, R. H. and Gayer, R. A. 1966. A revision of the Lower Heckla Hoek Succession in Central North Spitsbergen and correlation elsewhere. *Geol. Mag.*, **103**(1), 70–97.

Harland, W. B., Cutbill, J. L., Friend, P. F., Cobbett, D. J., Holliday, D. W., Maton, P. I., Parker, J. R. and Wallis R. H. 1974. The Billefjorden Fault Zone, Spitsbergen. The long history of a major tectonic lineament. *Norsk Polarinst. Skr.*, **161**.

Harland, W. B., Hoesfield, W. T., Manby, G. M. and Morris, A. P. 1979. An outline pre-Carboniferous stratigraphy of central western Spitsbergen. *Norsk Polarinst. Skr.*, **167**, 89–117.

Holtedahl, O. 1920. Paleogeography and diastrophism in the Atlantic–Arctic region during Paleozoic time. *Am. J. Sci.*, **49**(289), 25 pp.

Holtedahl, O. 1925. Some points of structural resemblance between Spitsbergen and Great Britain and between Europe and North America. *Avh. Norske Vidensk. Akad.*, **4**, 1–20.

Horn, G. and Orvin, A. K. 1928. Geology of Bear Island. *Skr. Svalbard og Ishavet*, **15**, 152 pp

Hurley, P. M. 1968. The confirmation of continental drift. *Scientific American*, **218**, No. 4, 53–64.

Knoll, A. H. 1982a. Microfossils from the Late Precambrian Draken Conglomerate, Ny Friesland, Svalbard. *J. Palaeontol.*, **56**(3), 755–790.

Knoll, A. H. 1982b. Microfossil-based biostratigraphy of the Precambrian Hecla Hoek Sequence, Nordaustlandet, Svalbard. *Geol. Mag.*, **119**, 269–279.

Krasil'shchilov, A. A. and Livshits, Yu. Ya. 1974. Testf Bjørnøya. *Geotectonics 4*, Academy of Sciences USSR, Moscow, pp. 39–51 [in Russian].

Kulling, O. 1934. Scientific results of the Swedish–Norwegian Arctic Expedition in the summer of 1931 (vol. II) Part XI: 'The Hecla Hoek Formation' round Hilopenstredet. *Geogr. Annaler Stock.*, **4**, 161–254.

Nordenskiöld, A. E. 1863. Geografisk och geognostisk beskrifning över de nordöstra delarna of Spetsbergen och Hinlopen Strait *K. Svenska Vetensk. Akad.*, **4**(7), 1–25.

Ohta, Y. 1979. Blue schists from Motalafjella, Western Spitsbergen. *Norsk Polarinst. Skr.*, **167**, 171–217.

Sokolov, V. N., Mrasilschikov, A. A. and Livshits, Yu. Ya. 1968. The main features of the tectonic structure of Spitsbergen. *Geol. Mag.*, **105**(2), 95–115. (2), 95–115.

Swett, K. 1981. Cambro-Ordovician strata in Ny Friesland, Spitsbergen and their palaeotectonic significance. *Geol. Mag.*, **118**(3), 225–336.

Van der Voo, R. 1980. Reply on 'A paleomagnetic pole

position from the folded Upper Devonian Catskill red beds, and its tectonic implications'. *Geology,* **8**, 259.

Van der Voo, R. and Scotese, C. 1981. Paleomagnetic evidence for a large (200 km) sinistral offset along the Great Glen Fault during Carboniferous time. *Geology,* **9**, 583–589.

Waddams, P. 1983. The late Precambrian succession in north-west Oscar II Land, Spitsbergen. *Geol. Mag.,* **120**, 232–252.

Winsnes, T. S., and Harland, W. B. in press. Further observations on Precambrian stratigraphy of south Spitsbergen. *Geol. Mag.*

Ziegler, P. A. 1982. Faulting and graben formation in western and central Europe. *Geol. Rdsch.,* **71**, 747–772.

The Caledonide Orogen—Scandinavia and Related Areas
Edited by D. G. Gee and B. A. Sturt
© 1985 John Wiley & Sons Ltd

North Greenland fold belt in eastern North Greenland

A. K. Higgins*, **N. J. Soper**† and **J. D. Friderichsen***

**Grønlands Geologiske Undersøgelse, Øster Voldgade 10, DK–1350 Copenhagen K, Denmark*
†*Department of Geology, The University of Sheffield, Mappin Street, Sheffield S1 3JD, UK*

ABSTRACT

The North Greenland fold belt is younger than the East Greenland Caledonian fold belt. The North Greenland fold belt developed on the site of an Upper Proterozoic–Lower Palaeozoic trough, an eastward extension of the Hazen trough of northern Ellesmere Island. A simplified comprehensive stratigraphy for the trough sequence is briefly described. The main deformation took place in Devonian–Carboniferous time and corresponds to the Ellesmerian orogenic deformation of the Innutian tectonic province in Arctic Canada. In North Greenland the Ellesmerian deformation and metamorphic grade increase in intensity northwards, and six tectonic–metamorphic zones can be distinguished. Three phases of coaxial folding are developed in northern areas, where metamorphism locally reaches amphibolite facies grade. Post-Ellesmerian events are briefly summarized.

Introduction

The North Greenland fold belt is part of the Innutian tectonic province of Arctic Canada, which in its eastern part extends through northern Ellesmere Island and North Greenland (Fig. 1). The North Greenland fold belt is 600 km long from west to east and up to 100 km wide. The age of deformation is younger than in the East Greenland Caledonian fold belt; the two fold belts have trends that if continued offshore would intersect in the Wandel Sea. Apart from differences in trend, the two fold belts contrast in that the East Greenland fold belt is thrust westwards over the Lower Palaeozoic rocks of the foreland (Hurst *et al.* this volume), whereas the southern margin of the North Greenland fold belt is transitional.

The North Greenland fold belt developed on the site of a Lower Palaeozoic flysch trough, though deposition probably began in the Upper Proterozoic. The flysch trough is part of the Franklinian Geosyncline of Arctic Canada and Greenland (Schuchert 1923). Throughout Cambro-Ordovician time there was in the area of present North Greenland a fairly well-defined boundary between shelf carbonate deposition of the platform to the south, and flysch deposition in the trough. The deformation associated with the North Greenland fold belt is almost confined to the flysch trough sediments, but in some areas extends sufficiently far south as to affect the northernmost platform carbonates. In Silurian time flysch sedimentation was more widespread, and extended southwards over former platform areas. A similar Lower Palaeozoic depositional history is known in northern Ellesmere Island, Arctic Canada (Trettin and Balkwill 1979).

The main deformation of the North Greenland fold belt took place in Devonian–Carboniferous time (Ellesmerian orogeny), and is responsible for the strong E–W alignment of structures. The intensity of metamorphism and deformation increases northwards, and in parts of extreme north Peary Land (Johannes V. Jensen Land) amphibolite facies is reached in highly deformed rocks. Three main phases of coaxial folding are distinguished and folds are generally overturned northwards, in contrast to the situation in Ellesmere Island where the general sense of overturning is southwards. There are no Ordovician volcanic rocks, Ordovician granites or *in situ* Proterozoic crystalline rocks in the North Green-

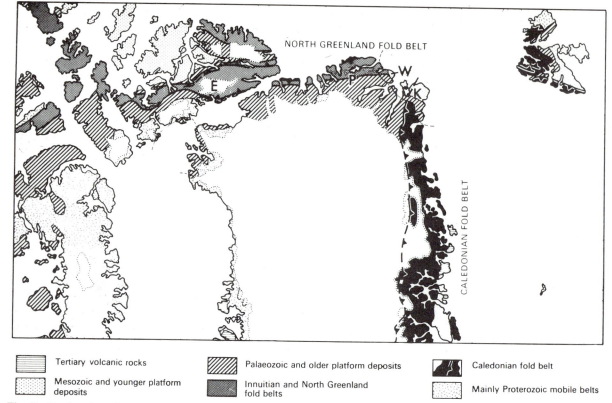

| | Tertiary volcanic rocks | | Palaeozoic and older platform deposits | | Caledonian fold belt |
| | Mesozoic and younger platform deposits | | Innuitian and North Greenland fold belts | | Mainly Proterozoic mobile belts |

Fig. 1   Regional setting of the North Greenland fold belt and East Greenland Caledonian fold belt. E: Ellesmere Island, P: Peary Land W: Wandel Sea, K: Kronprins Christian Land

land fold belt, whereas all these elements occur in the folded strata of the extreme north of Ellesmere Island.

Carboniferous–Palaeogene sediments (Wandel Sea Basin) overlying the folded rocks were deformed during the Tertiary Eurekan orogeny, and Eurekan orogenic events locally overprint the older structures. Folds, faults, thrusts, and local retrogressive metamorphism are recognized in North Greenland, and presumably related magmatic events include Cretaceous dyke swarms and late Cretaceous peralkaline volcanism.

This account is mainly concerned with the eastern half of the North Greenland fold belt, north Peary Land (Johannes V. Jensen Land) and adjacent islands, where new systematic geological investigations were carried out by the Geological Survey of Greenland (GGU) between 1978 and 1980. It is based to some extent on earlier accounts of the North Greenland fold belt (Dawes 1971, 1973, 1976; Dawes and Soper 1973, 1979). The most recent reviews of the Innuitian tectonic province in Canada are those of Trettin and Balkwill (1979) and Christie (1979).

?Upper Proterozoic–Lower Palaeozoic

The ?Upper Proterozoic and Lower Palaeozoic sediments of North Greenland were deposited in an eastern extension of the Franklinian basin of Arctic Canada. Evolution of the platform-trough system was probably initiated in the late Precambrian with development of an E–W trending trough crossing what is now North Greenland; this was the eastern extension of the Hazen trough of Ellesmere Island. The Lower Palaeozoic rocks of the platform are dominantly carbonates and reach a cumulative thickness of several kilometres. The equivalent deep-water trough facies is dominantly turbiditic and considerably thicker. Surlyk et al. (1980) envisage the trough to have formed by rifting with the southern margin (and hypothetical northern margin) controlled by major faults. The trough is considered to have expanded in several episodes, with collapse of the platform margin in the Silurian (latest Llandovery) such that considerably larger areas became part of the trough (Hurst and Surlyk 1980; Hurst et al. 1983).

Various stratigraphic terms have been used in the

past for the trough sequence of different parts of the fold belt (Dawes and Soper, 1973, 1979), but recent systematic mapping in Peary Land (Soper *et al.* 1980; Hurst and Surlyk 1980; Pedersen 1979, 1980; Higgins *et al* 1981), and especially the discovery of diagnostic fossils in the central part of the fold belt in north Peary Land (Surlyk *et al.* 1980), have made possible a simplified, comprehensive stratigraphic scheme for the entire trough sequence (Friderichsen *et al.* 1982). The six lithostratigraphic groups are described below, and their distribution in eastern North Greenland is shown on Fig. 2.

The lowest division, the *Skagen Group*, consists of structureless quartzitic sandstones and phyllitic mudstones of late(?) Proterozoic or early Cambrian age. It is at least 500 m thick, but has a very limited occurrence, mainly in the extreme east of the fold belt; its base is not exposed. The Skagen Group is probably younger than different developments of quartzites with dolerite intrusions in fault contact south of the main outcrop area, which are correlated with the Independence Fjord Group (Collinson 1980). The Independence Fjord Group has yielded isotopic ages of c. 1350 Ma on clays (Larsen and Graff-Petersen 1980) and 1230 Ma on an intrusive granophyre sheet (Jepsen and Kalsbeek 1979).

The *Paradisfjeld Group* is dominated by dark grey, impure carbonates (lime mudstones), calcareous phyllites and phyllites, with light-coloured limestone conglomerates characterizing the upper levels. It is at least 1000 m thick. Fragmentary fossils from near the top of the group indicate a post-Precambrian age (Peel and Higgins 1980), but on regional grounds it cannot be younger than Cambrian. The carbonates of the Paradisfjeld Group are widely distributed in northern parts of the fold belt, and even in complexly folded and metamorphosed areas form distinctive mappable units (Fig. 2).

The *Polkorridoren Group* consists of thick turbiditic sandstone and mudstone units with a total thickness of 2–3 km. The group can be described as 'classical flysch', and reflects deeper water sedimentation with a greater differentiation of shelf and trough. Sedimentary transport was mainly from the south and the east. The uppermost unit of the Polkorridoren Group (as redefined by Friderichsen *et al.* 1982) is the *Frigg Fjord mudstone*, a distinctive purple and green coloured sequence of mudstones from c. 400 to 1000 m thick, whose incompetent nature has played a significant role in the subsequent deformation. The Polkorridoren Group is of Cambrian age.

The *Vølvedal Group* is of Cambrian to earliest Ordovician age, and comprises 600–700 m of chert, grey mudstones, turbiditic sandstones (often silicified), and carbonate and chert conglomerates. The succeeding *Amundsen Land Group* comprises similar rocks, 350–500 m thick, and is of early Ordovician to early Silurian age. These deposits are considered to reflect a general decrease in the supply of clastics in the late Cambrian, with relatively slow deposition during the Ordovician, punctuated by debris flows derived from the platform margin to the south which gave rise to the carbonate and chert conglomerate units (Surlyk *et al.* 1980; Hurst and Surlyk 1980).

The uppermost division of the trough sequence, the *Peary Land Group*, is dominated by turbiditic sandstones and mudstones. It marks a new phase of turbidite sedimentation at the end of Ordovician and beginning of Silurian time, and is widely distributed throughout the fold belt. With the foundering of the platform margin in the Silurian (latest Llandovery), the upper turbidites of the Peary Land Group spread for several hundred kilometres to the south, overlapping the carbonates of the former platform (Peel 1980; Hurst and McKerrow 1981; Hurst *et al.* 1983). Paleocurrent directions in the Peary Land Group in Peary Land indicate derivation from the east along the axis of the trough, and the turbidites are interpreted as the products of erosion of the rising Caledonide mountains to the east of North Greenland (Hurst *et al.* 1983). The youngest parts of the turbidite sequence are latest Silurian (possibly lowest Devonian) strata found in Hall Land in the western part of the fold belt (Bendix-Almgreen and Peel 1972; Berry *et al.* 1974).

The Lower Palaeozoic successions of North Greenland provide no direct evidence of the existence of a northern source area comparable to the 'Pearya geanticline' of northern Ellesmere Island (Trettin 1971). However, the numerous indications of transport from east to west along the assumed axis of the trough indirectly suggest the existence of a northern barrier (Hurst and Surlyk 1980).

No *in situ* basement rocks occur in the North Greenland fold belt. However, numerous xenoliths of a variety of gneisses and granitic rocks occur in volcanic centres in the southern part of the fold belt in Johannes V. Jensen Land, east of Frigg Fjord (Soper *et al.* 1980; Parsons 1981). Numerous granitic gneiss blocks also occur in a Cretaceous dyke in the extreme north of the fold belt south of Benedict

---

Fig. 2   Geological map of north Peary Land (Johannes V. Jensen Land)—the eastern half of the North Greenland fold belt. See Fig. 5 for cross-sections. Modified after Higgins *et al.* (1981)

Fig. 3   Structural map of the eastern half of the North Greenland fold belt

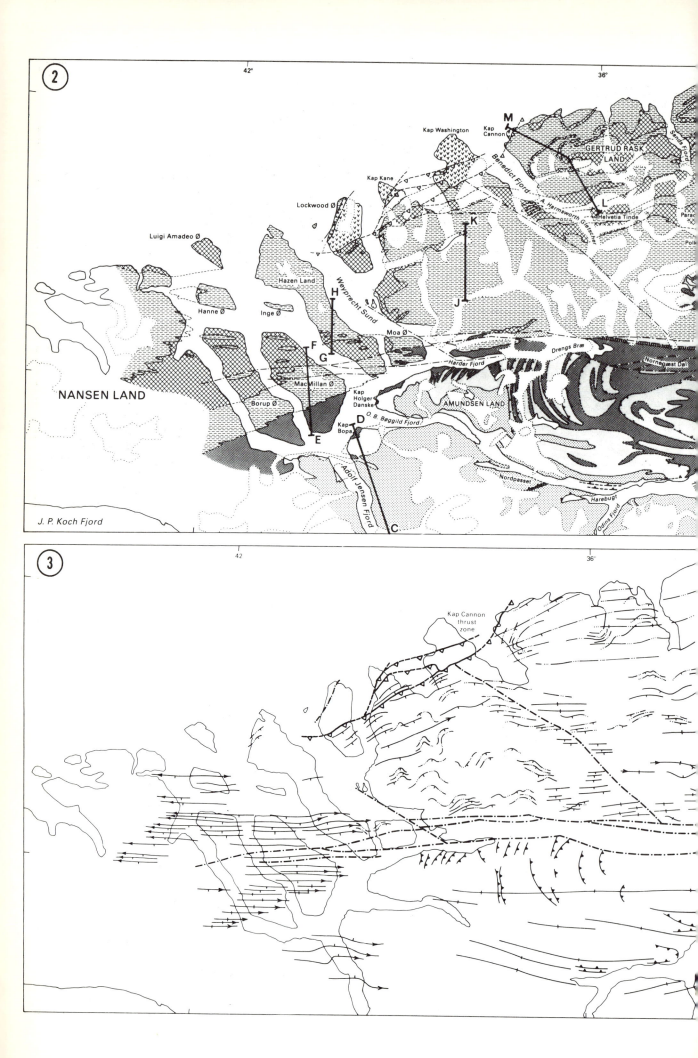

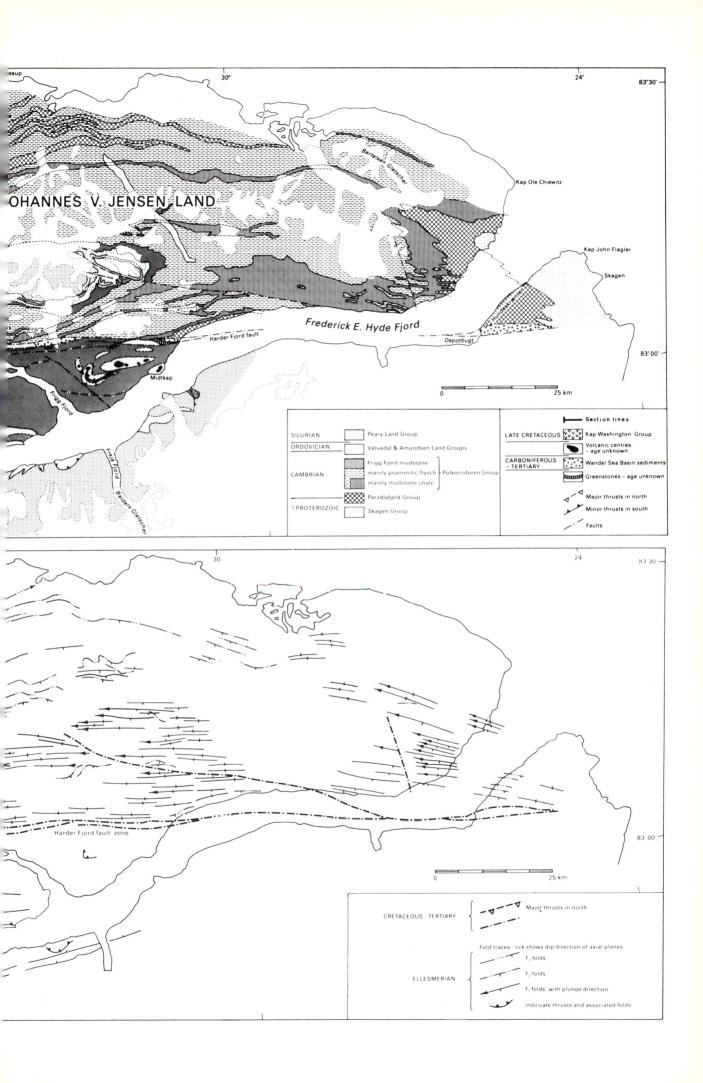

SILURIAN Peary Land Group

ORDOVICIAN Vølvedal & Amundsen Land Groups

CAMBRIAN
Frigg Fjord mudstone
mainly psammitic flysch } Polkorridoren Group
mainly mudstone shale

Paradisfjeld Group

? PROTEROZOIC Skagen Group

LATE CRETACEOUS Kap Washington Group

Volcanic centres - age unknown

CARBONIFEROUS – TERTIARY Wandel Sea Basin sediments

Greenstones - age unknown

Section lines

Major thrusts in north

Minor thrusts in south

Faults

OHANNES V. JENSEN LAND

Bertelsen Gletscher

Kap Ole Chiewitz

Kap John Flagler

Skagen

Frederick E. Hyde Fjord

Hundesk

Depotbugt

Harder Fjord fault

Midtkap

Frigg Fjord

Freja Fjord

Balders Gletscher

30° 24° 83°30'

83°00'

0 25 km

CRETACEOUS – TERTIARY Major thrusts in north

Fold traces - tick shows dip direction of axial planes

ELLESMERIAN
F₁ folds
F₂ folds
F₃ folds, with plunge direction
Imbricate thrusts and associated folds

Harder Fjord fault zone

30 24 83 30

83 00

0 25 km

Fjord. These are clear indications that the Franklinian basin in North Greenland is underlain by continental crust. These inferences are in agreement with those deduced from the more completely exposed section in northern Ellesmere Island where it is also considered that the Franklinian basin developed on continental crust (Trettin 1971; Trettin and Balkwill 1979).

Ellesmerian Orogeny in North Greenland

The North Greenland fold belt is dominated by rocks of the ?Upper Proterozoic–Lower Palaeozoic trough sequence, which were deformed and metamorphosed during the Ellesmerian orogeny. The fold belt is characterized by a strong east–west trending grain, roughly parallel to the north coast of Greenland and the continental margin (Figs 3 and 4). The fold belt is at least 100 km wide in Johannes V. Jensen Land, where it is most widely exposed and where the northern margin lies an unknown distance offshore.

The main sense of tectonic transport in the fold belt is northwards towards the Arctic Ocean. Northerly overturned fold structures are particularly evident in northern parts of Johannes V. Jensen Land, where all planar structural elements have a southerly dip (Fig. 5). Deformation increases in intensity northwards, as does the metamorphic grade; the lat-

ter reaches amphibolite facies on the north coast of Johannes V. Jensen Land. The Tertiary Kap Cannon thrust cuts obliquely across the metamorphic zones (Fig. 4), and brings metasedimentary rocks of the fold belt over non-metamorphosed Carboniferous–Permian sediments and late Cretaceous volcanic rocks of the Kap Washington Group.

Structures attributable to three major episodes of deformation were recognized by Dawes and Soper (1973, 1979). Recent work in north Peary Land has confirmed the interpretation (Soper et al. 1980; Higgins et al. 1981), and has defined more closely certain areas of anomalous structures (Pedersen 1980). The major folds of the three deformation episodes are effectively coaxial with gentle plunges and trend E–W over large areas (Fig. 3). The southernmost folds affect the northern edge of the carbonate platform (Fig. 4); farther south in the platform the strata are gently inclined, with a regional northward dip so that the carbonates disappear northwards under the Silurian clastic cover. The first folds appear initially as open structures which intensify northwards and then become refolded by second folds. Farther north, with increasing metamorphic grade, these develop an axial planar fabric which in the very north of Johannes V. Jensen Land is deformed by third structures (Fig. 5). These systematic changes have enabled six tectonic–metamorphic zones to be defined (Dawes and Soper, 1973, 1979; Dawes 1976) which, with some modifi-

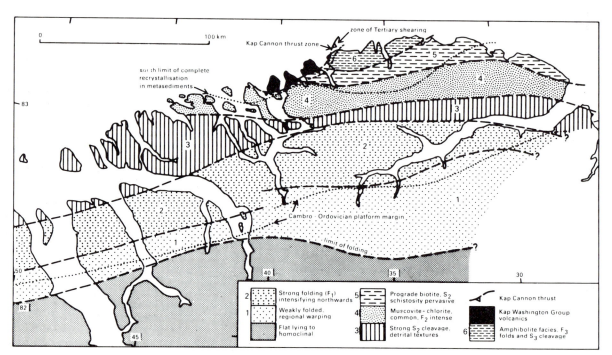

Fig. 4   Tectonic–metamorphic zones in the eastern half of the North Greenland fold belt. Modified after Dawes (1976). Numbers 1 to 6 correspond to zones 1 to 6 described in the text

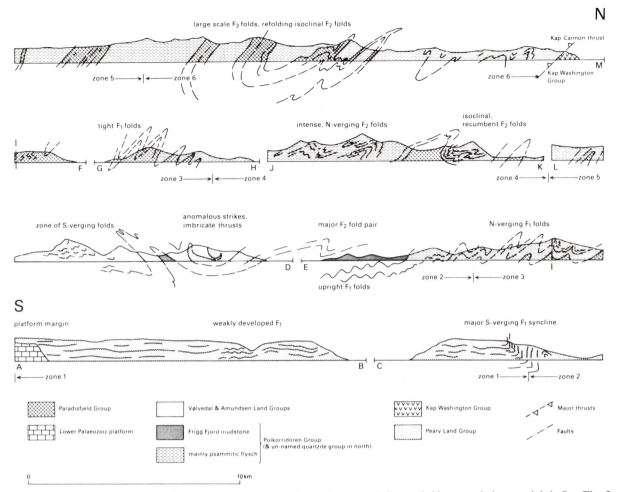

Fig. 5 Cross-sections through the North Greenland fold belt from south (lower left) to north (upper right). See Fig. 2 for section lines. Section A–B is south of the map limit of Fig 2, in continuation of section C–D

cations arising from recent work, are still a convenient means of description (Fig. 4). In the following account comments on metamorphic grade are based on mineral assemblages in pelitic and semipelitic units of the Polkorridoren Group. The three main fold phases are referred to as F1 to F3, and the associated planar fabrics as S1 to S3.

Zone 1

This zone (Fig. 4) is a transitional one of gentle folding between the unfolded carbonate platform to the south and the more severe deformation of zone 2. It is characterized by large and medium scale, open, small amplitude folds. These are referred, in the absence of evidence to the contrary, to the F1 phase. Fold axes generally trend ENE to E, but south of Frederick E. Hyde Fjord structural trends swing towards the southeast (Pedersen 1979).

Zone 2

The southern boundary of zone 2 is here taken as the

margin of the North Greenland fold belt. The boundary is abrupt, but involves no thrusting. In western North Greenland it is marked by the appearance of major E–W trending symmetrical folds and monoclines, which in the north became overturned towards the north. Box-folds are evident in some areas (Fig. 6). In eastern North Greenland the abrupt nature of the fold belt margin is well seen in several N–S trending fjord sections between J. P. Koch Fjord and Freja Fjord. There, over an E–W distance of 100 km, several small undulations of box-fold style are followed abruptly by a large south-verging syncline with a vertical to inverted north limb (Fig. 5) (Dawes and Soper 1979, fig. 10; Pedersen 1979, fig. 18).

In northern parts of zone 2 in Peary Land, both north- and south-facing major folds occur. The south-facing structures were thought to be early (F1) structures by Dawes and Soper (1979), whereas north-facing folds were assigned to the second phase (F2).

An extensive area between O. B. Bøggild Fjord

Fig. 6   Box folds in Silurian turbidites on the east side of Castle Ø, in the western part of the North Greenland fold belt. Cliff is 350 m high. Castle Ø is located 40 km west of the map limit of Fig. 4

and Frigg Fjord is characterized by spectacular imbricate thrusting and an anomalous, arcuate strike pattern ranging from E–W through NW–SE to N–S trends (Fig. 3). The thrusts continuously repeat the boundary between the Frigg Fjord mudstone and the overlying turbidites of the Vølvedal Group. Displacement is reported by Pedersen (1980) to be consistently westwards, and he regards the thrusts and their associated folds to be entirely earlier than the F1–F3 structural sequence of the remainder of the fold belt. The anomalous area is everywhere underlain by the Frigg Fjord mudstone, and it is evident that the anomalous structural pattern is associated with the presence of this thick incompetent unit, and specifically that the imbricate thrusting is due to contact strain effects at its upper boundary.

Zone 3

The southern edge of zone 3 is defined by the appearance of cleavage. In pelitic rocks the cleavage is often penetrative, forming good slates. Coarser lithologies occasionally show weaker, spaced cleavages. The cleavage is axial planar to northerly overturned F2 folds which occur on minor to major

scales. The F1 structures also begin to show northward overturning (Fig. 7), presumably due to the superimposition of F2 strain on originally upright structures.

In thin section the pelitic slates are seen to contain prograde sericitic material aligned parallel to the cleavage, and in the coarser clastic rocks some recrystallization of the matrix is generally apparent. Otherwise the detrital textures are little disturbed.

Zone 4

This zone is defined where the argillaceous matrix of silicoclastic rocks becomes completely recrystallized, and where prograde chlorite coexists with muscovite in many lithologies. In coarse psammitic rocks detrital textures are still discernible, though strongly modified. A chlorite isograd evidently lies close to the southern boundary of zone 4 (Fig. 4), but the data are as yet insufficient to map the isograd. The northern boundary of zone 4 is taken at the first appearance of biotite in appropriate lithologies. The degree of recrystallization in the metasediments increases northwards, but the line representing the development of completely recrystallized fabrics

Fig. 7   Northerly overturned F1 folds on the west side of Hazen Land. The light bands repeated by folding are the upper units of the Paradisfjeld Group. The ridge is about 800 m high

shown on Fig. 4 does not quite coincide with the first appearance of biotite. Thus along the northeast coast of Johannes V. Jensen Land prograde biotite occurs in the matrix of incompletely recrystallized metasediments, whereas farther west, to the south of Sands Fjord, totally recrystallized rocks with muscovite and chlorite (but no biotite) are present.

F2 folds (Fig. 8) are widely developed in zone 4, and the S2 fabric affects many lithologies. F1–F2 interference is well displayed and is approximately coaxial. It results in a generally northward vergence of both fold sets. In the northern part of this zone both fold sets become long-limbed, almost isoclinal, and their hinges are difficult to discern. A third set of medium to small scale folds makes its appearance in the north; these folds are accompanied by a crenulation fabric.

Zone 5

The S2 fabric of zone 5 is represented by a thoroughly penetrative schistosity in pelitic and semipelitic metasediments, and is normally the dominant fabric in psammitic rocks. In zone 5 'way-up' criteria (cross-bedding, grading, etc.) are usually obliterated by deformation and recrystallization, and the stratigraphic sequence becomes uncertain. However, distinctive lithologies provide good mapping units, and

some of these, such as the marbles of the Paradisfjeld Group, can be traced continuously southwards into less deformed terrain.

F1 and F2 folds are effectively isoclinal, recumbent, and northward verging, and it is frequently impossible to distinguish between the two sets. The F3 structures become increasingly important, with the S3 schistosity inclined to the south more steeply than the S2 schistosity.

Zone 6

A relatively small area adjacent to the north coast of Johannes V. Jensen Land is characterized by amphibolites facies mineral assemblages. The southern boundary of this area in Fig. 4 is placed where garnet appears in appropriate lithologies. Staurolite, cordierite, and rare andalusite are also present in some parts of zone 6. These assemblages indicate a cover of 10–13 km at the peak of metamorphism in this, the deepest part of the fold belt at present exposed. Mineral textures in thin sections suggest the thermal maximum accompanied and outlasted the F2 deformation.

Major F3 folds are present in this region (Fig. 5). They refold the two earlier fold sets and the composite S2 fabric, and they control the gross outcrop pattern (Figs. 2 and 3). Insufficient data are available

Fig. 8   Major, recumbent F2 folds in Polkorridoren Group turbidites in western Johannes V. Jensen Land. Section is oblique to the fold axes and exaggerates the limb lengths. Cliff is about 1000 m high

to determine whether they also fold the isograd boundaries. The composite S2 fabric is heterogeneously developed; the rocks in certain zones have a coarse, almost gneissose character, while elsewhere strain is sufficiently low for sedimentary structures to be preserved in turbidites. The gneissose zones are thought to be associated with high F2 shear strains.

Zones 5 and 6 are truncated by the Tertiary Kap Cannon thrust zone. Associated with this structure is a belt a few kilometres wide in which shearing and retrogressive metamorphism are superimposed on the Palaeozoic mineral fabrics. Close to the thrust the rocks become very platy and develop blastomylonitic fabrics (Dawes and Soper 1973), and Cretaceous basic dykes are altered to greenstones (Higgins *et al.* 1981). This zone narrows westwards as it passes into calcareous rocks of the Paradisfjeld Group.

Retrogressive mineral assemblages are commonly developed throughout zone 6, outside the zone of Tertiary shearing. The development of chlorite and muscovite in association with S3 shows that some of these retrogressive effects are of late Ellesmerian age, but there is also K–Ar isotopic evidence for a period of reheating in Tertiary time (Dawes and Soper 1971).

Discussion

The most characteristic feature of the fold belt in north Peary Land (Johannes V. Jensen Land) is the northerly vergence and northerly intensification of the folds—in a direction away from the stable platform to the south. Dawes (1973) interpreted this fold pattern as a consequence of southerly underthrusting of the fold belt by crustal material. Pronounced northerly vergence is characteristic of only the easternmost part of the fold belt; in Nansen Land, where mapping is not yet complete, folds tend to be more upright, while farther west in northern Ellesmere Island southerly vergence predominates (Trettin 1971). It is evident that the northerly vergence in Johannes V. Jensen Land is due essentially to the superimposition of F2 deformation on more upright F1 structures. Important F2 folding and cleavage development are confined to Johannes V. Jensen Land north of the boundary between zones 2 and 3 (Fig. 4), and this coincides with the region of northerly vergence. The original attitude of F1 in the area of overprinting is unknown, but upright and even a few south-verging first folds are preserved locally.

Metasediments correlated with the Polkorridoren

Group are exposed in several areas to the north of the Kap Cannon thrust zone, where they are overlain unconformably by marine Carboniferous–Permian sediments and the late Cretaceous Kap Washington Group volcanics. They are of a lower tectonic–metamorphic grade than nearby stratigraphic equivalents that now lie above the thrust, the latter presumably having lain at a deeper tectonic level prior to the Tertiary thrusting.

Older Palaeozoic intrusions

A few deformed mafic layers in the metamorphosed terrain of the North Greenland fold belt represent minor intrusions into the trough sequence (Dawes 1976). Some are apparently concordant sills, while others have an irregular form. The rock types vary from foliated metadolerite and metagabbro to amphibolite and crystalline schist, and have evidently suffered the same general degree of deformation and metamorphism as their host rocks. In the southern part of the fold belt Pedersen (1980) distinguishes an older generation of basic dykes emplaced after a first and before a second phase of (Palaeozoic) deformation.

Age of orogenesis

The youngest sediments in the North Greenland sector of the Franklinian basin are latest Silurian, possibly earliest Devonian (Berry et al. 1974; Bendix-Almgreen and Peel 1972; Surlyk et al. 1980). The oldest sediments of the Wandel Sea Basin sequence unconformably overlying deformed rocks of the fold belt are Upper Carboniferous clastics, limestones, and cherts. Thus the Ellesmerian diastrophism can only be loosely dated as Devonian to Middle Carboniferous (early Pennsylvanian).

Rb–Sr isotopic studies of whole rock and clay fractions (Springer 1981) suggest a tectonic–metamorphic event in Upper Devonian to Lower Carboniferous (Lower Mississippian) time, but these results are regarded as provisional. K–Ar ages (Dawes and Soper 1971) testify to a later Cretaceous–Tertiary thermal event.

In Ellesmere Island the most extensive and intensive deformation of the Ellesmerian orogen took place before the Middle Devonian in northern areas and was completed before the late Early Pennsylvanian (Trettin and Balkwill 1979). However, here there is a distinct indication that deformation progressed from north to south, and in the south deformation probably began in latest Devonian or early Carboniferous time.

Post-Ellesmerian Events

Sedimentary rocks ranging in age from early Carboniferous to Palaeogene occur as outliers in eastern North Greenland where they rest unconformably on the Palaeozoic platform and on the deformed rocks of the East and North Greenland fold belts. These cover sequences are referred to the Wandel Sea Basin, of which the most recent descriptions are those of Håkansson (1979) and Håkansson et al. (1981). In the North Greenland fold belt occurrences are limited to thrust slices in the Tertiary Kap Cannon thrust zone and downfaulted blocks within the Harder Fjord fault zone (Croxton et al. 1980; Soper et al. 1980; Higgins et al. 1981; Wagner et al. 1982).

Magmatic events recorded in the North Greenland fold belt include the emplacement of alkalic basic dyke swarms in the Cretaceous, and eruption of peralkaline volcanic rocks (Kap Washington Group) in the late Cretaceous (Dawes and Soper 1973; Larsen et al. 1978; Brown and Parsons 1981; Batten et al. 1981). The dykes occur in several trends, the most intense swarm being aligned roughly N–S, normal to the present north coast (Fig. 9). Both the dykes and extrusive rocks are regarded as products of continental rift magmatism associated with an extensional tectonic regime, perhaps a precursor to the opening of the Eurasian Basin. The main dyke swarm aligns with the initial position of the Nansen spreading axis (Soper et al. 1982).

Metasediments of the fold belt were displaced northwards over the Kap Washington Group volcanics and late Palaeozoic Wandel Sea Basin sediments along the Tertiary Kap Cannon thrust zone. In the vicinity of the thrust the Cretaceous dykes are deformed and altered to greenstones, while the metasediments are locally mylonitized. Widespread retrogression of the higher grade Palaeozoic mineral assemblages in the northern part of the fold belt may be in part of Tertiary age; K–Ar mineral ages indicate a period of reheating in the Tertiary (Dawes and Soper 1971). The northward thrusting is considered by Soper et al. (1982) to be associated with anticlockwise rotation of Greenland due to the opening of the Labrador Sea, though delayed possibly until Eocene time when the Wandel Sea and Barents Sea continental margins became 'unpinned' across the Spitsbergen fracture zone.

The E–W trending Harder Fjord fault zone can be traced for over 200 km within the eastern half of the fold belt (Fig. 9). Suggestions that it may have been one of the faults controlling the southern margin of the Lower Palaeozoic flysch trough (Surlyk et al. 1980; Hurst and Surlyk 1980) are not yet substantiated (Higgins et al. 1981). Substantial vertical uplift

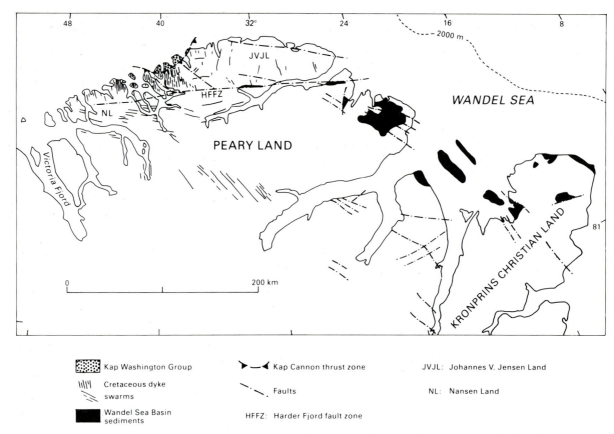

Kap Washington Group

Cretaceous dyke
swarms

Wandel Sea Basin
sediments

Kap Cannon thrust zone

Faults

HFFZ: Harder Fjord fault zone

JVJL: Johannes V. Jensen Land

NL: Nansen Land

Fig. 9   Pattern of Cretaceous–Tertiary magmatic and tectonic events in eastern North Greenland

of the northern side of the fault took place in Tertiary time. There is no evidence of major wrench displacement.

Folds and faults that affect the Wandel Sea Basin sediments and trend parallel to the present continental margin have been described by Håkansson (1979) and Håkansson *et al.* (1981). Soper *et al.* (1982) suggest that this Tertiary deformation resulted from oblique compression followed by weak extension due to interaction of the Wandel Sea and Barents Sea continental margins along the Spitsbergen fracture zone during the opening of the North Atlantic Ocean and Eurasian Basin. The structures are thus seen as weaker counterparts of the West Spitsbergen orogen and Forlandsundet graben (Harland and Horsfield 1974).

Acknowledgements

This paper is published with the permission of the Director of the Geological Survey of Greenland. P. R. Dawes and J. M. Hurst kindly made suggestions for improvements of the manuscript, and we also acknowledge the helpful comments of two anonymous referees.

References

Batten, D. J., Brown, P. E., Dawes, P. R., Higgins, A. K., Eske, Koch, B., Parson, I. and Soper, N. J. 1981. Peralkaline volcanicity on the Eurasia Basin margin. *Nature London*, **294**, 150–152.

Bendix-Almgreen, S. E. and Peel, J. S. 1972. Early Devonian vertebrates from Hall Land, North Greenland. *Rapp. Grønlands geol. Unders.*, **65**, 13–16.

Berry, W. B. N., Boucot, A. J., Dawes, P. R. and Peel, J. S. 1974. Late Silurian and early Devonian graptolites from North Greenland. *Rapp. Grønlands geol. Unders.*, **65**, 11–13.

Brown, P. E. and Parsons, I. 1981. The Kap Washington Group volcanics. *Rapp. Grønlands geol. Unders.*, **106**, 65–68.

Christie, R. L. 1979. The Franklinian geosyncline in the Canadian Arctic and its relationship to Svalbard. *Norsk Polarinst. Skr.*, **167**, 263–314.

Collinson, J. D. 1980. Stratigraphy of the Independence Fjord Group (Proterozoic) of eastern North Greenland. *Rapp. Grønlands geol. Unders.*, **99**, 7–24.

Croxton, C. A., Dawes, P. R., Soper, N. J. and Thomsen, E. 1980. An occurrence of Tertiary shales from the Harder Fjord Fault, North Greenland fold belt. *Rapp. Grønlands geol. Unders.*, **101**, 61–64.

Dawes, P. R. 1971. The North Greenland fold belt and environs. *Bull. geol. Soc. Denmark*, **20**, 197–239.

Dawes, P. R. 1973. The North Greenland fold belt: a clue to the history of the Arctic Ocean Basin and the Nares Strait Lineament. In Tarling, D. H. and Runcorn, S. K.

(Eds), *Implications of Continental Drift to the Earth Sciences*, **2**, Academic Press, London and New York, 925–948.

Dawes, P. R. 1976. Precambrian to Tertiary of northern Greenland. In Escher, A. and Watt, W. S. (Eds), Geology of Greenland, Geological Survey of Greenland, Copenhagen, 248–303.

Dawes, P. R. and Soper, N. J. 1971. Significance of K/Ar age determinations from northern Peary Land. *Rapp. Grønlands geol. Unders.*, **35**, 60–62.

Dawes, P. R. and Soper, N. J. 1973. Pre-Quaternary history of North Greenland. In Pitcher, M. G. (Ed.), *Arctic Geology. Mem. Am. Assoc. Petrol. Geol.*, **19**, 117–134.

Dawes, P. R. and Soper, N. J. 1979. Structural and stratigraphic framework of the North Greenland fold belt in Johannes V. Jensen Land, Peary Land. *Rapp. Grønlands geol. Unders.*, **93**, 40 pp.

Friderichsen, J. D., Higgins, A. K., Hurst, J. M., Pedersen, S. A. S. Soper, N. J. and Surlyk, F. 1982. Lithostratigraphic framework of the Upper Protrozoic and Lower Palaeozoic deep water clastic deposits of North Greenland. *Rapp. Grønlands geol. Unders.*, **107**, 19 pp.

Håkansson, E. 1979. Carboniferous to Tertiary development of the Wandel Sea Basin, eastern North Greenland. *Rapp. Grønlands geol. Unders.*, **88**, 73–83.

Håkansson, E., Heinberg, C. and Stemmerik, L. 1981. The Wandel Sea Basin from Holm Land to Lockwood Ø, eastern North Greenland. *Rapp. Grønlands geol. Unders.*, **106**, 47–63.

Harland, W. B. and Horsfield, W. T. 1974. West Spitzbergen orogen. In Spencer, A. M. (Ed.), *Mesozoic–Cenozoic Belts—Data For Orogenic Studies. Special Publ. geol. Soc. Lond.*, **4**, 747–755.

Higgins, A. K., Friderichsen, J. D. and Soper, N. J. 1981. The North Greenland fold belt between central Johannes V. Jensen Land and eastern Nansen Land. *Rapp. Grønlands geol. Unders.*, **106**, 35–45.

Hurst, J. M. and McKerrow, W. S. 1981. The Caledonian nappes of Kronprins Christian Land, eastern North Greenland. *Rapp. Grønlands geol. Unders.*, **106**, 15–19.

Hurst, J. M. and Surlyk, F. 1980. Notes on the Lower Palaeozoic clastic sediments of Peary Land, North Greenland. *Rapp. Grønlands geol. Unders.*, **99**, 73–78.

Hurst, J. M., McKerrow, W. S., Soper, N. J. and Surlyk, F. 1983. The relationship between Caledonian nappe tectonics and Silurian turbidite deposition in North Greenland. *J. geol. Soc. London.*

Jepsen, H. F. and Kalsbeek, F. 1979. Igneous rocks in the Proterozoic platform of eastern North Greenland. *Rapp. Grønlands geol. Unders.*, **88**, 11–14.

Larsen, O. and Graff-Petersen, P. 1980. Sr-isotopic studies and mineral composition of the Hagen Bræ Member in the Proterozoic clastic sediments at Hagen Bræ, eastern North Greenland. *Rapp. Grønlands geol. Unders.*, **99**, 111–118.

Larsen, O., Dawes, P. R. and Soper, N. J. 1978. Rb/Sr age of the Kap Washington Group, Peary Land, North Greenland, and its geotectonic implication. *Rapp. Grønlands geol. Unders.*, **90**, 115–119.

Parsons, I. 1981. Volcanic centres between Frigg Fjord and Midtkap, eastern North Greenland. *Rapp. Grønlands geol. Unders.*, **106**, 69–75.

Pedersen, S. A. S. 1979. Structural geology of central Peary Land, North Greenland. *Rapp. Grønlands geol. Unders.*, **88**, 55–62.

Pedersen, S. A. S. 1980. Regional geology and thrust fault tectonics in the southern part of the North Greenland fold belt, North Peary Land. *Rapp. Grønlands geol. Unders.*, **99**, 79–87.

Peel, J. S. 1980. Geological reconnaissance in the Caledonian foreland of eastern North Greenland with comments on the Centrum Limestone. *Rapp. Grønlands geol. Unders.*, **99**, 61–72.

Peel, J. S. and Higgins, A. K. 1980. Fossils from the Paradisfjeld Group, North Greenland fold belt. *Rapp. Grønlands geol. Unders.*, **101**, 28.

Schuchert, C. 1923. Sites and nature of the North American geosynclines. *Bull. geol. Soc. Amer.*, **34**, 151–229.

Soper, N. J., Dawes, P. R. and Higgins, A. K. 1982. Cretaceous–Tertiary magmatic and tectonic events in North Greenland and the history of adjacent ocean basins. In Dawes, P. R. and Kerr, J. W. (Eds), *Nares Strait and the Drift of Greenland: A Conflict in Plate Tectonics. Meddr Grønland, Geosci.*, **8**, 205–220.

Soper, N. J., Higgins, A. K. and Friderichsen, J. D. 1980. The North Greenland fold belt in eastern Johannes V. Jensen Land. *Rapp. Grønlands geol. Unders.*, **99**, 89–98.

Springer, N. 1981. Preliminary Rb–Sr age determinations from the North Greenland fold belt, with comments on the metamorphic grade. *Rapp. Grønlands geol. Unders.*, **106**, 77–84.

Surlyk, F., Hurst, J. M. and Bjerreskov, M. 1980. First age-diagnostic fossils from the central part of the North Greenland fold belt. *Nature, London*, **286**, 800–803.

Trettin, H. P. 1971. Geology of lower Palaeozoic formations, Hazen Plateau and southern Grant Land Mountains, Ellesmere Island, Arctic Archipelago. *Geol. Surv. Can. Bull.*, **203**, 134 pp.

Trettin, H. P. and Balkwill, H. R. 1979. Contributions to the tectonic history of the Innuitian Province, Arctic Canada. *Can. J. Earth Sci.*, **16**, 748–769.

Wagner, R. H., Soper, N. J. and Higgins, A. K. 1982. A late Permian flora of Pechora affinity in North Greenland. *Rapp. Grønlands geol. Unders.*, **108**, 5–13.

The Caledonide Orogen—Scandinavia and Related Areas
Edited by D. G. Gee and B. A. Sturt
© 1985 John Wiley & Sons Ltd

The East Greenland Caledonides—reviewed

J. Haller[*]

Department of Geological Sciences, Harvard University, Cambridge, Mass. 02138 USA

ABSTRACT

Common 'pre-drift' reconstructions of the North Atlantic Caledonides are not in accord with pre-Permian palaeogeography, especially not with the relationship between the British–Scandinavian and the East Greenland Caledonides. The palinspastic position of the stratotectonic elements, combined with evidence of longitudinal strike-slip movement within the Caledonian system, make the East Greenland fold belt a markedly separate branch of the Caledonides, more peripheral to, and exclusive of, the simatic vestiges of the late Proterozoic-early Palaeozoic Iapetus Ocean. Moreover, due to their ensialic setting, the East Greenland Caledonides have preserved a good record of successive Proterozoic regimes. This record includes the geodynamic process that led to the novel trend of Iapetus by sharply truncating the tectonic fabric of its bounding craton.

The closing and suturing of Iapetus brought about the linear Caledonian orogenic trend, marked by substantial Silurian tectonic translations, westward in Greenland and eastward in Scandinavia. In Devonian time, diversification of structural provinces and general uplift of the folded ranges resulted in emergence of the 'Old Red Continent', which included Britain and Scandinavia, but not Greenland. The emergent East Greenland fold belt (together with the ranges of Svalbard?) presumably was separated from the 'Old Red Continent' by marine successor basins which had evolved west of the Iapetus suture.

By late Devonian, but mainly in Carboniferous time, rifting accompanied by supposedly significant left-lateral transcurrent faulting heralds an incipient opening of the Scandic Sea, which thus would have developed from the sites of episutural Caledonian basins. Structural evolution of the modern Atlantic seaboard of East Greenland commenced in Jurassic time. The step-faulted Mesozoic shelf sequences were exhumed by crustal arching that preceded the eruption of early Tertiary plateau basalts, and by subsequent uplift in the Neogene.

Large-scale three-dimensional exposures of Caledonian core complexes in Central East Greenland gave rise to classic concepts in petrogenesis; and these can now be refined in the light of modern isotope studies.

Introduction

That the conveners of the Uppsala Caledonide Symposium have invited me and my colleagues from the Geological Survey of Greenland to talk about East Greenland is geologically and historically befitting. Were it not for Swedish ingenuity, and a Swedish tradition to venture into the unknown, perhaps even today we should be spared the trouble of deciphering how East Greenland relates to what the Romans called *Sevo mons*, and what we consider to be the 'backbone' of the Scandinavian Highlands. Was it not A. E. Törnebohm who made himself unpopular by suggesting that these pristine mountains were merely imports from the Far West (see emblem of this symposium)? Törnebohm spoke of tectonic allochthony, of windows and of thrust sheets, at a time when even in the Alps the existence of *nappes* was far from having been settled.

Törnebohm's bold departure from the fixistic perception of his contemporaries about the structure of the Earth was matched, in a different way, by the

* Deceased.

Swedish engineer S. A. Andrée, who broke with traditional means of reaching the North Pole. He built a balloon that was to carry him and two companions across the ice of the Arctic Ocean. He and his companions perished north of Spitsbergen, and remains of the expedition were found much later (Sundman 1967). But it was the search for Andrée—at least officially—that caused his countryman A. G. Nathorst to mount the first geological expedition to East Greenland in 1899. Among other results, Nathorst (1901) was able to demonstrate that in the region of Kong Oscar Fjord (72°–73°N) folded Lower Palaeozoic rocks are overlain unconformably by Old Red deposits. This was like the fold belts of Britain and Scandinavia, which Eduard Suess (1888) had called *Caledonian* mountain chain, and which constitute his *Old Red Continent*. And so, after Nathorst's discovery Greenland was naturally annexed to the Old Red Continent. Without hesitation, the East Greenland fold belt was tied into the same chain of mountains that once stretched from Britain to Scandinavia. This became part of the authenticity of the modern paradigm.

I shall take issue with this and other vexed questions, although almost a quarter of a century has passed since last I visited Greenland. I had then spent a winter and nine summers of field work between 70° and 81°N with the Danish expeditions conducted by Lauge Koch. It was during these post-war years that the 'heroic age' of polar exploration came to an end, which meant that blank spots disappeared from topographic maps as did rigged steamships, dog-team travel, and strenuous backpacking. Modern methods of air transport began to open the Arctic wilderness and to allow the geologist easy access to hitherto unknown terrane.

Although foreign expeditions made some important contributions to the knowledge of the geology of East Greenland, the surveys mounted by the Danish Government were paramount. Their activity falls into three distinct periods: (1) Lauge Koch's expeditions 1926 to 1938 carried out topographic mapping and geologic reconnaissance between 70° and 78°N; (2) Geologic mapping of the region 72° to 76°N and reconnaissance to 81°N by Koch's expeditions 1947 to 1958; (3) Geologic mapping of the region 70° to 72° by the Geological Survey of Greenland from 1968 to 1972. The results of the first two periods were summarized by Haller (1970, 1971) and by Koch and Haller (1971); those of the last period by Henriksen and Higgins (1976) and Higgins and Phillips (1979).

When I concluded the writing of my synthesis of the East Greenland Caledonides in 1968, the geological sciences had just reached the threshold of a new era. In the wake of geophysical exploration of oceanic ridges and basins, the English-speaking world in particular all at once tried to catch up with the mobilistic view that Alpine geologists had propounded for decades. The emerging theory of *plate tectonics* scored its banner year and, consequently, geotectonic interpretations took frantic turns; too quickly changing to be included in my synthesis. I had trouble enough with what were already entrenched indoctrinations, such as the Stillean phases of orogeny and Kay's geosynclines. Moreover, the structure of the Caledonides did not resemble that of the Alps, which unsettles the mind of any native Swiss. Now, a decade later, the initial commotion in geotectonic interpretations has given place gradually to more critical assessments of geophysical and structural data. A return to objectivity, combined with worldwide studies in regional tectonics, has been leading us to a better perception of *palinspastic problems*, including the questions raised by three generations of students of the North Atlantic Caledonides. With regard to East Greenland, moreover, geochronological sampling carried out by the Geological Survey of Greenland has resulted in a much clearer picture of the pre-Caledonian history of the region. In view of these more recent developments, and of my own studies in Britain, Scandinavia, the Appalachians, Cordilleran, and Alpine systems, I am reassessing in this study the structural anatomy and geodynamic setting of the East Greenland Caledonides.

Synopsis of Structural Evolution

The East Greenland Caledonides extend over 1,400 km, from 70° to 82°N, with a dominance of metamorphic rocks of Precambrian and Palaeozoic age over most of this length (Fig. 1). The main orogenic movements and metamorphic events appear to be of Silurian age. It was then that westward translations of thrust sheets and folding of meridional trend established the major tectonic grain that is roughly parallel to the course of the present coast. The western edge of the fold belt is marked by overthrusts and by foreland folds, except for the region between 72° and 76°N, where the Caledonian front is obscured by inland ice. Moreover, between 74° and 76°N the fold belt is dominated over its entire width by transverse tectonic translations of Devonian age. In the coastal region the superimposed structures trend northwesterly, while along the inland ice a separate system trends northeasterly. Within the zones of Devonian movements, syn- and post-kinematic intrusions of granite are widespread.

North of latitude 76°, the Caledonian allochthon

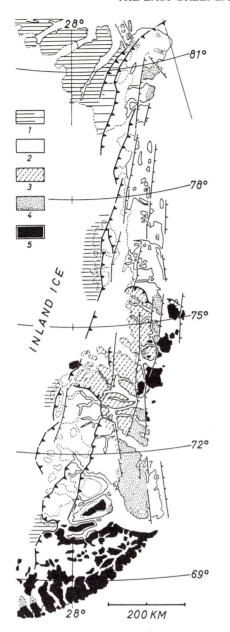

Fig. 1 Sketch map of the East Greenland Calcdonides, with areas of (1) Foreland, partly folded, (2) Silurian main movements, (3) Devonian movements, (4) Carboniferous through Tertiary strata, (5) Palaeogene plateau basalt (From Haller, 1983; reproduced by permission of the Commission for Scientific Research in Greenland)

consists of Precambrian sediments and partially Caledonized basement complexes. South of latitude 76°, the more thoroughly Caledonized basement is overlain by Proterozoic and Lower Palaeozoic strata. In the region of Kong Oscar Fjord (72°N) and Kejser Franz Joseph Fjord (73°N) shallow-water shelf deposits of late Proterozoic age exceed 10,000 m in thickness and are followed by some

3,000 m of Cambro-Ordovician platform deposits. Except for possible occurrences in the allochthon of Kronprins Christian Land (80°N), no Silurian strata were found within the entire East Greenland fold belt. Their absence is part of a hiatus that extends from the Upper Ordovician to the Lower Devonian.

South of latitude 75°, continental Devonian (Old Red molasse) accumulated in intermontane basins, which were subjected intermittently to compressional and extensional deformation of varying trends. The latest phase of compression occurred in the Early Carboniferous. Silicic and mafic volcanism affected various parts of the intermontane basin region.

In Upper Carboniferous time (Moscovian), the sea transgressed across the northernmost part of the Caledonian belt, while its central and southern portions still formed a belt of highlands that stretched along an extensive, north-northeasterly trending fracture pattern. South of latitude 79°, continental Upper Carboniferous and Lower Permian sediments are found in fault-bounded, linear troughs. The formation of these epi-orogenic grabens initiated extensional fault-block tectonics and subsidence that turned the eastern parts of the presently exposed Caledonian terrane into an unstable shelf from the Late Permian (Zechstein Sea) through all of Mesozoic time.

At the end of the Cretaceous, a wide region south of latitude 76° rose above sea-level, with the Mesozoic shelf being denuded by subaerial erosion so that its block-faulted Caledonian basement became partly exhumed. A thick pile of Palaeocene–Eocene plateau basalts (Brito-Arctic province) then buried the region, which was affected by renewed pulses of uplift in the Neogene. The latter led to the present elevations of 2,000 to 3,000 m in Central East Greenland and of 1,000 to 1,500 m in Northeast Greenland.

In comparison to the Scandinavian Caledonides it is quite remarkable that the East Greenland fold belt and its foreland contain a depositional record of practically the entire Phanerozoic, with the distinct exception of Lower Devonian strata. But the price for this completeness has been a severe fragmentation of the Caledonian structural pattern by late Palaeozoic and Mesozoic block-faulting, as is obvious from the generalized structure sections shown in Haller (1971, p. 334). Unlike Scandinavia, where the Caledonian structure usually can be shown in continuous sections, post-Caledonian faulting and differential vertical movements in Greenland have virtually mutilated the orogen. That is to say, today we find side by side segments that represent quite different structural levels of the orogen. Ordinary structure sections from Greenland, therefore, cannot convey a coherent picture of that region's Caledonian tectonics.

The North Atlantic Caledonides and the Evolution of the Atlantic Ocean

In 1965, E. Bullard and coworkers excited the profession by their 'fit of the continents', which matched the 500-fathom contours on both sides of the Atlantic Ocean over a length of 12,000 km (Bullard *et al.* 1965, Fig. 8). For the first time, a palinspastic reconstruction of global dimension was computed properly as relative rotation on the Earth's spherical surface, thus recognizing the 200-year-old theorem of Euler. In the eyes of many, Bullard's professionally executed 'fit' was more convincing than earlier attempts (Wegener 1915; Choubert 1935; Carey 1958). At the same time the process of sea floor spreading (Hess 1962) had become a credible mechanism for continental dispersal, and so, after half a century of ridicule, credence was lent also to Wegener's Permian supercontinent, *Pangaea*. On the basis of the new data from the sea floor, it was assumed the Mid-Atlantic Ridge would mark the line of breakup from which the Americas moved relatively westward, Africa and Eurasia relatively eastward. To Bullard (1975, p. 21) it occurred only later that his reconstruction should have included also the crest of the Mid-Atlantic Ridge, the outline of which deviates considerably from the line of the 'fit' as was pointed out by Dillon (1974, p. 175). But in spite of this and other flaws, pre-drift reconstructions of the Bullard-type have become common, and it is the tacit acceptance of such 'fits' that must be blamed for some prevailing misconceptions about the palaeogeography of the North Atlantic region in Palaeozoic time.

The Atlantic Ocean of today appears to be a simple longitudinal megastructure, symmetrically spaced about the axis of the Mid-Atlantic Ridge. Accepting marine magnetic anomalies as virtual increments of sea floor growth, and the pivots to Bullard's 'fit' as poles of finite rotation, it is then tempting to extrapolate the opening of the Atlantic from the present setting backward to the time of continental breakup (Pitman and Talwani 1972). This simple and widely held view, however, fails to recognize the distinctly different evolution of the major segments of the Atlantic Ocean (Fig. 2). In particular, the history of the Equatorial and South Atlantic differs strikingly from that of the North Atlantic. The southern segments appear to have formed only in Late Mesozoic through Cenozoic time by way of separation of South America from Africa. The North Atlantic, on the other hand, has a

Fig. 2 Global view of North Atlantic Ocean and adjacent continents, including Palaeozoic (thin lines) and Mesozoic–Cenozoic orogenic belts (heavy lines). In the Atlantic, the crest of the Mid-Atlantic Ridge and transverse fractures are shown. Major physiographic-tectonic segments are: (A) Equatorial Atlantic, (B) 'Tethyan' or Central North Atlantic, (C) Northern North Atlantic, including Greenland and Norwegian Seas

long and complex history of recurrent opening and closing. Moreover, the segment between latitudes 10° and 50°N is thought to have opened along the sutures of Late Palaeozoic fold belts as part of the Tethys. It therefore appears that this segment, referred to as the Central North Atlantic (Lancelot 1980), only became an independent 'Atlantic' basin in late Cretaceous–early Tertiary time when the Equatorial Atlantic gradually widened and the Alpine orogen, in turn, underwent its major phase of foreshortening.

North of latitude 50°, within the realm of the North Atlantic Caledonides, the history of the Atlantic Ocean, including the Greenland and Norwegian Seas, is intimately related to the evolution of the Caledonides and their post-orogenic fragmentation (Ziegler 1978). Moreover, the eruption of huge quantities of basaltic magma in the early Tertiary led to the unusual buildup of extensive plateau basalts between latitudes 55° and 75°N as part of the *Brito-Arctic province*. Thick piles of subaerial flood basalts are exposed in Greenland, Iceland, the Faeroes, Scotland, and Ireland. Other wide areas, still covered by plateau basalt, subsided in mid-Tertiary time and lie now beneath the sea at depths of up to several kilometres (Deep Sea Drilling Project: Leg 12, sites 116, 117; Leg 38, sites 336, 337, 338, 343, 350).

The vast extent and tonnage of these eruptions,

together with the vertical movements that accompanied and followed them, is likely to have resulted in the discomposure of pre-Tertiary crust. One should expect therefore that over this region part of the Palaeozoic and Mesozoic record has been concealed and virtually destroyed in the course of this early Tertiary igneous activity. From this point of view, it is conceivable that some of the oceanic-type crust within the Norwegian Sea did form by *in situ* transformation rather than by accretionary spreading, which is commonly assumed.

The Caledonian Trend: A case of Tectonic Heredity

That controlling structural patterns may persist through geological time was long considered a highlight in Suess's perception of global tectonics. Nevertheless, when Argand (1924) pondered the question of randomness versus order in the development of orogenic belts he came to abandon Suess's axiom with respect to its connotation that hereditary trends would ensue only from the structure of continents, that is from sialic 'Rahmen'. Having traced the major structural elements of the Western Alps to developing stages of the Tethyan realm in a paper of 1916, Argand now suggested that orogenic hereditary trends evolve rather from simatic domains. He felt that this assumption might even overcome apparent conflicts between his mobilistic interpretation of global tectonics and the fixists' principle of permanence because he recognized crustal dispositions which seemingly had persisted through time. For example, by calling the Caledonian 'geosyncline' an earlier design (*vieulle ébauche*) of the present-day Atlantic Ocean, Argand considered Caledonian folding and thrusting to be merely the inverse of continental dispersal. To him, the Caledonian orogeny was but the suturing of an older simatic scar that was to sunder again.

Argand's notion about the opening and closing of simatic domains was pursued by Staub (1928) who elaborated on a geodynamic model for the regeneration of simatic sea floor as a result of the reversal in the direction of orogenic translations. Van der Gracht (1928) once more noticed that the Caledonian trend had predetermined the disposition of the present North Atlantic. But otherwise the topic was largely neglected until Wilson (1966) revived the subject because of the striking alignment of the Appalachians and the Caledonides on Bullard's 'fit'. Ironically, the geophysicist had now reminded the palaeontologists that in the Cambro-Ordovician strata of both Appalachian and Caledonian orogenic systems a dichotomy of faunal correlation had been noticed for quite some time. Specifically, Lower

Fig. 3   J. Tuzo Wilson's perception of an early Palaeozoic Atlantic Ocean that would have formed a complete barrier between two faunal provinces, the American-Pacific (horizontal ruling) and the Acado-Baltic (vertical ruling). As was to be expected, this simple sketch of 1966 provoked excitement and criticism, which both helped to modify and improve the notion of a Caledonian–Appalachian ocean, now referred to as Iapetus. (Reprinted by permission from *Nature*, Vol. 211, No 5050, p. 678, 1966 and J. Wilson. Copyright © 1966 Macmillan Journals Limited)

Cambrian trilobites of the Olenellid realm appear to outline a persistent pair of contrasting faunal provinces along the shores of a Lower Palaeozoic Atlantic (Fig. 3). Because this seaway preceded the Atlantic Ocean by one generation the name *Iapetus*, in Greek mythology the father of Atlas, was then introduced by Harland and Gayer (1972).

It may be fortuitous that the provinciality of Lower Cambrian shelf faunas should have exaggerated so greatly the perception of the Iapetus Ocean. During the preceding latest Precambrian (Vendian) widespread glaciation presumably extended from high to low palaeolatitudes (Sokolov 1973). Related geographic and drastic climatic shifts are thought to have acted upon marine life at a time of its fundamental physiological modification. Glacioeustatic lowering of the sea-level during glaciation, and worldwide transgressions at its close, would be expected consequences. Thus, the flooding of littoral sites by the Cambrian sea may not seem overly significant. But when looking at the evolving palaeogeographic pattern, a surprisingly uniform meridional trend of the newly forming coastlines (palaeoslopes) is discernible along both sides of Iapetus. Prior to the Cambrian transgression, the prevalent structural grain of the terranes involved was markedly different; it ran roughly transverse to what was to become the Caledonian trend. It is only in the Appalachian region, that a similar crosscutting relationship had been established already in the Proterozoic by the pervasive Grenvillian trend (Fig. 4).

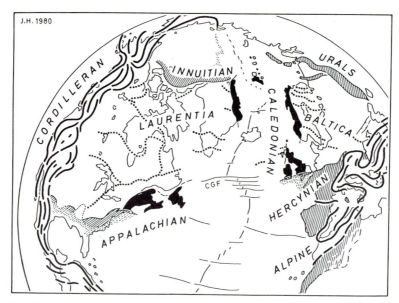

Fig. 4   Map showing the global tectonic setting of the North Atlantic Caledonides. Areas affected by Ordovician and/or Siluro-Devonian orogenic movements are shown in black. Hatchure indicates belts of Upper Palaeozoic orogenies, except for the Southern Appalachians where Ordovician and Late Palaeozoic movements are discernible, and the Innuitian system where Palaeozoic and Cenozoic movements occurred. Heavy lines mark the trends of the Mesozoic–Cenozoic belts. On the Precambrian cratons Laurentia and Baltica, the major structural trends are indicated by dotted lines. Note that in Geenland, Scotland, and Scandinavia these trends are transverse to the Caledonian trend, while the Appalachians parallel the Proterozoic Grenvillian trend. In the Atlantic Ocean, the crest of the Mid-Atlantic Ridge and major transverse fractures are shown, namely the Charlie-Gibbs Fracture Zone (CGF), which appears to relate to both the Appalachian grain (Haworth 1977) and the Hercynian front (Cherkis *et al.* 1973)

In Greenland and Scandinavia, however, the linear Caledonian trend is a *novum*. It ensued from the foreshortening and closing of Iapetus by 'delamination' and splintering of the crust. These effected large-scale translations, eastward in Scandinavia, westward in Greenland. The linearity of the Caledonian trend seemingly resulted from a deep-seated anisotropy of the lithosphere that guided the dynamics of the North Atlantic Caledonides (Wegmann 1959; Haller 1980).

Suggestions from Studies in Comparative Tectonics

Subsequent to my field work in Greenland I visited many parts of the British and Scandinavian Caledonides, and of the Appalachians, in order to get a better understanding of the structural evolution of the entire system. But having spent much time in this deeply truncated orogen I also felt the need to learn more about the anatomy of recent orogenic belts, namely the North American Cordillera and the Alpine–Himalayan system (Haller 1979). It is from these comparative studies that I have derived new thoughts about some of the tectonic principles involved in the structure of the North Atlantic Caledonides.

In comparison to Mesozoic–Cenozoic mountain ranges, or for that matter in comparison to the

Variscides of Northwestern Europe, the Urals, or the Southern Appalachians, the North Atlantic Caledonides are devoid of significant foredeep structures, except for southwest Britain. It is true that the sparse outcrops of Devonian in the Oslo Fjord region may be the remainder of a modest foredeep fill in Scandinavia and, based on erratics of questionable age, one might speculate about a rudimentary foredeep structure in East Greenland, beneath the inland ice between latitudes 72° and 73°N. But otherwise tangible evidence for such structures is missing. Of course, foredeep basins which themselves were not involved in thrust tectonics may rebound later so that their fill will be removed by erosion, and the record of subsidence so destroyed. A case in point would be the Colorado Plateau which was buried beneath the Cordilleran foredeep in Cretaceous time but then was uplifted and stripped of most of that overburden during the Cenozoic. Nevertheless, in the case of the Caledonides, both in Scandinavia and Greenland, I am inclined to believe that no significant foredeep structures ever existed. The reason for this belief lies in the principle of the succession of thrust emplacement, derived from comparative studies, and illustrated in Fig. 5 by two contrasting patterns.

The first pattern (type A) shows a *progression* of active thrust movements from the mobile belt toward the cratonic foreland. The outward migration

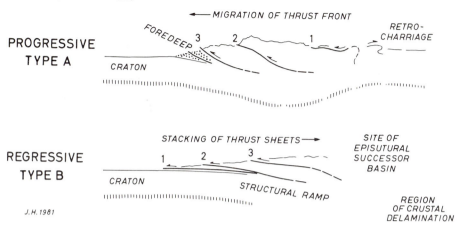

SUCCESSION OF THRUSTING

(NOT TO SCALE)

Fig. 5   Diagrammatic presentation of successions of thrust emplacement in orogenic belts: (A) The *progressive type* usually involves thick thrust units, with the activity migrating from 1 (oldest) to 3 (youngest). When thrust zone 2 was in motion the adjacent craton became isostastically depressed and detraction products accumulated in the ensuing fore-deep. The youngest thrust zone 3 eventually involved the foredeep itself. At the same time, retrocharriage (back-folding) developed inside the thrust belt. (B) The *regressive type* consists of a stack of relatively thin thrust sheets. The oldest unit (1) forms the front of thrust belt; the youngest unit (3) tops the accretionary wedge which amassed over a structural ramp. There is no foredeep in this case. But in the region from where the thrust units originated, episutural successor basins may form. (Distances and crustal thickness are not to scale)

of the thrust front appears to give rise to a foredeep, which itself eventually may become affected by fore-shortening. Progressive thrust patterns of this kind are observed, for example, in the Himalayas, the Canadian Rocky Mountains, and the Southern Appalachians.

The second pattern (type B) shows a *regression* of active thrust zones from the cratonic foreland into the mobile belt. The result is a stacking of successively younger thrust sheets over a structural ramp (the position of which presumably is controlled by the edge of the adjacent craton). In this case no foredeep basin will form, but in the region from which the translated crustal flakes 'delaminated', where 'decoupling' and 'suturing' took place, *successor basins* are likely to ensue.

There is another distinctive characteristic in which the two patterns differ: the regressive type of thrust belt (B) tends to follow Argand's axiom according to which major thrust units do reflect sedimentary facies belts and hence palaeogeography, whereas in the case of the progressive type (A) the thrusts usually cut across facies belts and palaeogeographic elements are therefore not conformable to structural trends.

The regressive pattern does not preclude a later progression of thrust belts. In fact, progressive patterns often are the sequel to regression, such as in the Central Alps. I suspect also that in the case of Central East Greenland there is a rudimentary regime of progressive thrusts (type A) of different

trend superimposed over an older regime of regressive thrusts (type B). This would explain some of the drastic changes in tectonic style that make the Devonian NW–SE structures so distinct from the older N–S pattern. Nevertheless, it is my perception that the principles of regressive thrust patterns (type B) played an important role in both East Greenland and Scandinavia. I must admit though that from neither belt is there much published data in support of my view.

With respect to Scandinavia my view, in fact, appears to run counter to the widely recognized pattern of thrust faults younging towards the cratonic foreland (Gee 1975). But this pattern seemingly relates to the gravitational collapse (Ramberg 1966, 1981), stretching and eastward spreading (Gee 1978) of an already existing stack of nappes. It is important, therefore, to distinguish between such late movements and the major translations that had caused the build up of the nappe pile in the first place (Nicholson 1974; Roberts 1978). With regard to those major translations, however, an indication of the regressive principle of stacking is seen in (1) the geometrical superposition of thrust sheets (Roeder 1973; Gee 1977; Sturt and Roberts 1978), and (2) the close correlation between thrust units and palaeogeographic entities (Oftedahl 1966; Størmer 1967; Prost 1977; Gee 1978) and/or pre-existing belts of regional metamorphism (Kulling 1972; Prost 1972; Bryhni and Brastad 1980).

In East Greenland, large-scale thrust systems were

postulated early by Parkinson and Whittard (1931). But it was not until the 1950s that the Caledonian allochthon was documented along the western edge of the orogen, namely in Kronprins Christian Land (80° to 81°N), Dronning Louise Land (76° to 77°N), and Paul Stern Land (70°N). Although inland ice obscures the edge of the orogen between 72° and 76°N there is evidence that thrust sheets also mark this segment (see Fig. 1). The sedimentary nappes of Kronpins Christian Land (Fränkl 1955; Hurst and McKerrow 1981) owe their emplacement mainly to gravitational sliding as is indicated by an inversion of stratigraphic sequences (*diverticulation*). South of 79°N, however, the thrust front is marked by actively shoved thrust plates, which consist mainly of partially Caledonized basement rocks. For the region around 70°N, Higgins and Phillips (1979) assumed a tectonic transport of some 130 km. No attempt has been made to analyse the kinematics of thrust emplacement except for the area around 71°N, from where Homewood (1973) described a distinctly regressive pattern. In this southern region, the stack of thrust sheets subsequently was affected by folding and arching. Moreover, Caledonian metamorphism appears to have affected the thrust sheets prior to translation (Homewood 1973), in places during the major movement (Steck 1971), or afterwards. In the latter case Chadwick (1975) reported progressive overprinting from the interior of the orogen, while Phillips *et al.* (1973) reported retrogressive overprinting from the edge.

In the region of the inner fjords between 71° and 73°N, a belt of mylonite parallels the trend of the marginal thrusts (see Fig. 1). Although this linear zone was the site of normal faulting in late and post-Caledonian time (Haller 1971), the formation of the mylonite probably is due to Caledonian thrust movements. In fact, it was the occurrence of such mylonite at Junctiondal, Kejser Franz Joseph Fjord (73°N) that led Parkinson and Whittard to suggest their thrust-belt model of the orogen. At Junctiondal there is indeed a significant thrust. But this structure is part of the wide belt of transverse tectonic translations of Devonian age that overprinted the Caledonian main structures (Haller 1970). Aside from these Devonian transverse thrusts and the well developed major thrusts along the edge of the orogen, the discernment of additional allochthonous entities within the orogen becomes naturally speculative. In contrast to Scandinavia, the cover rocks preserved within the East Greenland Caledonides cannot be attributed to different facies belts. Therefore, the lithologic distinction between possibly separate tectonic entities has to be based merely on the plutonometamorphic characteristics

of former basement terranes. But with regard to these aspects and the few observations on thrust kinematics referred to above, I assume that the major Caledonian translations in East Greenland followed a broad pattern of regressive thrusting. Except for the superposed Devonian structures, I have not noticed any indication of a progression of thrusting.

It is my postulate of regressive thrust tectonics in both Scandinavia and East Greenland that makes me wonder also whether in the late stages of the Caledonian orogeny the palaeogeography had not been quite different from that which is commonly assumed on the basis of Bullard-type reconstructions. First of all, there is simply no evidence for the widely held assumption that Britain–Scandinavia and Greenland were consolidated into one coherent landmass by Old Red time. Secondly, the East Greenland fold belt was probably much farther north at that time and therefore did not 'rub its back' with Scandinavia (Bailey and Holtedahl 1938; Harland 1966; Irving 1977). For the intervening regions I envisage marine rather than terrestrial conditions, most likely an environment of intraorogenic successor basins, which would have evolved west of the Iapetus suture (Fig. 6). The implied separate position of East Greenland concurs with its stratotectonic record that relates the Late Precambrian evolution to Svalbard, and the Cambro-Ordovician to the Hebridean platform, that is the cratonic forelands of the British Caledonides.

The Devonian was a time of great structural change in all parts of the North Atlantic Caledonides (Friend 1981). There is a remarkable diversification of structural provinces that reflects a major reorientation of regional stresses (regroupment of crustal dynamics). Extensional tectonics with intermittent phases of compression prevailed through Middle and Late Devonian time in East Greenland, where the last compressive movements occurred in the earliest part of the Carboniferous (Haller 1971). The Silurian main movements were followed by general uplift and erosional unroofing of the folded ranges. In the course of this process huge quantities of rock material were removed. At places in Central East Greenland, for example, over 8,000 m of strata already had been eroded before the earliest intramontane molasse accumulated in Middle Devonian time. Similarly, huge rock quantities 'tracelessly' disappeared from Scandinavia, which then emerged together with Britain to form the classic 'Old Red Continent'. But the Greenland ranges, together (?) with those of Svalbard, seemingly were separated by episutural, Western Mediterranean-type depressions, which thus would have served as depocentres for that initial Devonian

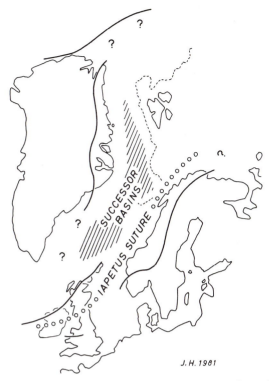

J.H. 1981

Fig. 6 Palinspastic map showing possible configuration of the North Atlantic Caledonides in Devonian time, including the postulate of marine successor basins west of the Iapetus suture. This reconstruction also implies that the Caledonian front of Britain (Flinn 1979) does not relate to the edge of the East Greenland fold belt (Haller 1969; Brooks *et al.* 1976)

debris that so far has not been explained. The separate position of the Greenland ranges also would explain why *Ichthyostegalia*, the celebrated earliest genus of the Tetrapoda, remained as endemic as it did in the Upper Devonian of East Greenland.

By Late Devonian, but mainly in Carboniferous time, rifting (Haller 1971) and significant left-lateral strike-slip movement paralleling the Caledonian trend (Harland 1969, 1978), appear to have initiated the opening of the northern North Atlantic (Ziegler 1978), which thus would have evolved directly from episutural Caledonian basins. In support of my conclusion about Greenland's palinspastic setting within the North Atlantic Caledonides, there also are recent palaeomagnetic data (Van der Voo and Scotese 1981). But as a sceptic in these matters I will not take advantage of the situation. Rather, I prefer to admit how, after having drawn my reconstruction (Fig. 6), I noticed that, stimulated by Wegener's ideas, O. Holtedahl already had come up with approximately the same configuration in 1936. Moreover, the existence of Late Palaeozoic successor basins in the boreal region has been postulated by Christie (1979), although his palinspastic map showed North Greenland facing Svalbard.

Finally, to consider the Lomonosov Ridge as part of the Caledonian scenario (Trettin 1969) is currently favoured by many, and I am afraid that such consensus soon may make a fact out of what should remian conjecture.

In the context of profound structural changes during the late stages of Caledonian orogeny, it is important to point out that both in Greenland and in Scandinavia, the Devonian structures tend to mimic the Precambrian trends of the Caledonian forelands (*cf.* Fig. 4). Similar trend lines, transverse to the dominant grain of the Caledonides, apparently still control major fractures in the Norwegian–Greenland Seas, the Jan Mayen and Greenland Fractures in particular.

Proterozic 'Blueprints' in the East Greenland Caledonides

Transverse trends and pre-Caledonian stratigraphy

The linear trend of the North Atlantic Caledonides may distract our attention from the fact that the internal structure of these belts is complex. In Scandinavia, it was mainly Wegmann (1959) who pointed out that the pervasive trend of Caledonian tectonic translations has obscured effectively many important changes that occur along the strike. Similar changes are known from East Greenland, and for that matter also from Britain. Each of these belts includes Proterozoic depositional basins of greatly varying thickness and, more significant, basin axes that are oblique or normal to the later Caledonian trend. A particular case in point is Central East Greenland, where the late Proterozoic Eleonore Bay Group makes up the bulk of pre-Caledonian stratigraphy (Fig. 7).

The division referred to as 'Lower' Eleonore Bay Group is of variable thickness but otherwise rather monotonous. It consists mostly of quartzitic schists that represent pelitic and psammitic material, which in part had been cannibalized from a lithologically similar but older substratum (Krummedal Series). A notable exception, however, is seen in the Alpefjord region (72°N) where a stack of distinctly deltaic beds reaches a thickness of over 9,000 m. According to Caby (1975) eastward current directions dominate the sequence. The volume and layout of these shallow-water deposits resemble the infill of the Mississippi Embayment, whose distal clastic wedge was formed in the early Tertiary when the Mississippi and its tributaries transported an enormous volume of rock-waste from the rising Rocky Mountains. By analogy, I suspect that the Alpefjord region represents a Proterozoic embayment into which detritus of westerly provenance was deposited by what one

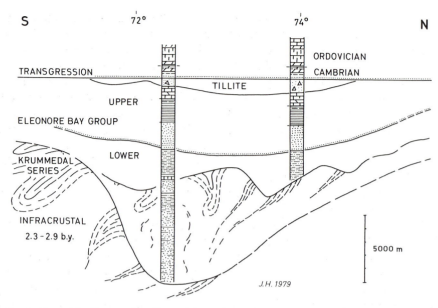

Fig. 7   Palinspastic stratigraphic diagram for Central East Greenland, with the Cambrian transgression as datum. The Lower Eleonore Bay Group varies in thickness, from a few thousand meters in the inner fjord and nunatakker region (backward section) to over 9,000 m in the Alpefjord area (front section). The older basement includes infracrustal rocks (blank) as well as supracrustal rocks (dashed lines). The horizontal scale is indicated by the geographic latitudes

might call *secundum naturam*, the 'Alpeppippi'. Supposing that palaeodrainage pattern comparable to the modern Mississippi, the divide would have been about halfway to British Columbia, where in the Purcell and Windermere sequences similar accumulations were supplied from equally unknown, but eastern sources.

As is the case in the modern analogue, the elongation of the subsiding embayment apparently was controlled by the pre-existing tectonic fabric, which was E–W to NW–SE as is evident from both earlier supracrustal belts and infracrustal basement complexes (Haller 1970).

The next cycle of sedimentation, the 'Upper' Eleonore Bay Group, begins with a disconformity of regional extent that heralds an orderly, uniform shelf accumulation that extended N–S for at least 500 km. In the Late Precambrian of East Greenland, this is the first accumulation to be so extremely regular and to change with time from clastic to carbonate platform facies. But with the onset of the Vendian tillites the pattern reverts once more to the former transverse trend, which apparently directed the axis of the 'Tillite Basin' between latitudes 72° and 74°N. The transverse trend of the basin is indicated by the trend of sedimentary facies boundaries (Fränkl 1953). From the phenoclasts included in these ground moraine and glaciomarine strata as well as from the sporadic occurrence of tillites in the Caledonian foreland, it is concluded that the region west of the basin had been affected by widespread block-faulting and igneous activity. In both East and North

Greenland, the magnitude of this Vendian tectonic event appears to be much larger than hitherto recognized. Although Stille (1958) had stressed the transmuting aspect (*Umbruch*) of such tectonic events in the latest Precambrian, their significance in terms of continental fragmentation (Doig 1970) and the initial expansion of Iapetus (Roberts and Gale 1978) has been acknowledged only recently.

With the commencement of the last cycle of sedimentation, the transgressing Cambrian sea established from Central to Northeast Greenland the remarkable linear trend that was first indicated by the disposition of the Upper Eleonore Bay Group. Moreover, the lateral uniformity of the subsequently developing Cambro-Ordovician carbonate platform is truly amazing as there are no transitions from littoral to basin facies. But over the entire region there is a remarkable hiatus in Upper Cambrian deposition, and a similar break is known from Svalbard. Roberts and Gale (1978) consider this oscillation of the carbonate platform a reaction to the early orogenic movements in Finnmark.

The stratigraphic and metamorphic dates from East Greenland refer the major tectonic and metamorphic events to Silurian time, although in the southernmost region some early plutons appear to be Ordovician in age. But the ensialic setting of this orogenic belt has preserved also the imprints of Proterozoic structural and metamorphic events. At first it was merely on the basis of petrogenetic characteristics that Precambrian infracrustal complexes were identified within the Caledonian orogen (Mittelholzer 1941; Haller and Kulp 1962; Koch and Haller 1971). Although in the northern part of the orogen metasedimentary sequences were attributed to Proterozoic metamorphism, it was not realized at that time that pre-Caledonian cycles of deposition and orogeny had also affected some supracrustal sequences in the southern part. The significance of such Proterozoic events came to light mainly through subsequent mapping and geochronological studies (Higgins 1976; Henriksen and Higgins 1976; Higgins *et al.* 1981). The new data naturally gave rise to a reassessment of the nature of Caledonian tectonism and metamorphism in East Greenland, which had been the subject of a heated controversy in the 1930s (Haller 1971). I do not criticize the renewed debate because it sheds new light on questions that I have been pondering ever since I completed my work in Greenland. Nevertheless, at this point I should like to recapitulate some of the classic principles of petrogenesis in the following section and then also offer my current opinion on the subject.

Caledonian metamorphism and basement reactivation

In the late 1940s when I pursued the study of petrology and began my work in Greenland, I was caught in the middle of the 'granite controversy'. There was the *magma* (about which Paul Niggli lectured in Zurich) and there was the *migma* (a term introduced by my teacher Max Reinhard in Basel). The only thing I understood at that time was that 'magmatists' and 'transformists' were not on speaking terms. And it was especially the term *front* that in the late 1930s had become one of those 'explosive terms that rouse dark passions in the most staid of geologists', as Read (1948) noticed.

The concept of 'front' has its roots in the French school (Termier 1912). It ran counter to the then fashionable doctrine of *Tiefenstufe* (depth zone), which related regional metamorphism to load and to passive geosynclinal burial (Becke 1913; Grubenmann and Niggli 1924). Instead, the 'front' concept postulates the active upward motion of metamorphic processes as well as their selective placement within the orogen. The principles involved were best formulated by E. Wegmann, who had learned tectonics from E. Argand, and his petrology from J. J. Sederholm, and who had worked with H. G. Backlund at the deciphering of the 'Central Metamorphic Complex' in East Greenland. And so Wegmann, in his celebrated paper of 1935, applied the 'front' concept to the formation of migmatites. These are rocks composed of pre-existing material, the *palaeosome*, and a quartzo-feldspathic mobile phase, the *neo-*

some. These two parts are intimately mingled so that a new kind of rock is formed (Sederholm 1907). Thus, by definition, much of a migmatitic rock consists of old material, nor need the mobile phase come from distant sources. This kind of rock was not once a melt or crystal mush; it is believed to be the product of metasomatic changes in a macroplastic environment. Wegmann then emphasized that the ascendant front of metasomatism would have a geometry distinctly different from that of the subsequently receding heat dome (Fig. 8). Moreover, the degree of retrogressive equilibration would be dependent largely on the rate of cooling.

The 'front' concept appertains to dynamic processes within the orogenically warmed sialic crust. It relates metasomatism to the interaction of movement and metamorphism. While isochemical changes are conveniently presented by isogradic surfaces, that is, by the first occurrence of index minerals, allochemical changes encompass whole rock complexes because the effect of metasomatism is volumetric and therefore best ascribed to a 'front'. Migmatization hence takes place behind the 'migmatite front'. The position of that front is defined by mineral paragenesis; its geological age depends on the time of equilibration of the mineral phases. The ascendant front reflects a new set of physicochemical conditions; it transforms pre-existing rock material by rising fluids (Backlund 1936) and by thermal convective effects (Talbot 1979). This is quite different from a plutonic intrusion, where molten rock material moves into place by replacing or displacing the host rock.

During the upward and lateral progression of metamorphism, both pressure and temperature will change and so does the *rheological* behaviour of the affected rock material, that is, the manner in which rocks respond mechanically to deformation. It is the rheological behaviour that determines the tectonic style, which principally arises from contrasts in duc-

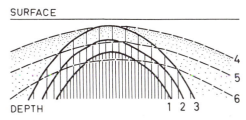

Fig. 8 Diagrammatic vertical section showing E. Wegmann's perception of an ascendant front and the subsequent stages of cooling. Solid lines (1–3) indicate the prograde vertical and lateral spreading of the metamorphic transformation, which may lead to migmatization in the centre of the copula. The dashed lines, in turn, reflect retrogressive stages (4–6) of cooling, the rate of which determines the degree of retrograde equilibration. (Reproduced by permission of Geologische Rundschau)

tility. Albert Heim (1878) was the first to recognize that in a mountain range the style of rock deformation proceeds from brittle (fracturing) to ductile (folding) with increasing structural depth. Within the aureole of regional metamorphism, particularly in the realm of migmatization, the ductility of rock units changes drastically, so that a mobile 'infrastructure' with flow folding will develop beneath a zone of less ductile fold style, the 'superstructure'. This contrast in rheologic behaviour is mainly vertical but also lateral, and it has been called 'stockwerk tectonics' by Wegmann.

Furthermore, the rock complexes that make up the anatomy of a mountain range have to be examined as to their stratotectonic setting, that is, as to the pre-orogenic lithological characteristics of basement complexes versus cover strata.

The three geological principles of grouping rock material of the core complexes of orogenic belts are compared in Fig. 9, which illustrates how each of these principles is based on entirely different observational data, namely lithology, petrochemistry, and rheology. In an orogenically heated and moving crust the three principles interact and interfere in the most complex way. Wegmann had assumed that the 'migmatite front' would define also the rheological boundary condition that makes the infrastructure so distinctly different from the superstructure. Originally I followed him in his conclusion, but now I think that this was an oversimplification. Because what is readily recognizable in the field is only the style of deformation, that is the rheological behaviour of the rock material. This was, of course the main criterion I had been using for the mapping of infrastructural elements (Haller 1955, 1970).

Vertical ductility contrasts—*stockwerks*—may develop at different levels of the stratotectonic scheme. Thus, with regard to the subdivision into basement and cover, there are three possible posi-

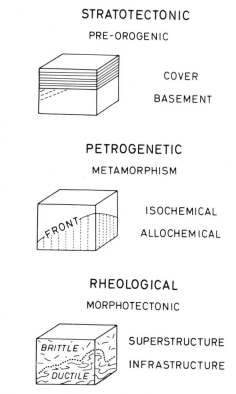

Fig. 9 Diagrammatic illustration of interferential principles in Caledonian petrogenesis in East Greenland. The three pairs of contrasting phenomena are based on entirely different kinds of field observation, namely lithology (stratotectonic subdivision), petrochemistry (metamorphism, migmatization), and rheology (ductility contrasts)

tions of the boundary of the mobile infrastructure: (a) inside the basement complex, (b) at the basal unconformity, and (c) in the cover strata (Fig. 10).

In the first case, which comes close to the situation in Northeast Greenland, pervasive shearing of regional trend correlates with the westward translation of thrust sheets of basement rocks. The zones of

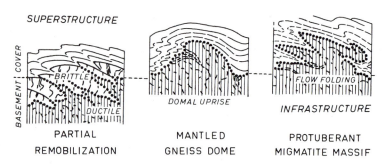

Fig. 10 Diagrammatic presentation of synorogenic ductility contrast, showing boundary of mobile infrastructure within basement complex (left), at basal unconformity (centre), and protuberant into cover (right)

shearing thereby guided the advance of metasomatic and thermal conductive processes. As a result, strips of intact old basement are now found adjacent to belts of entirely new mould. In the course of this process the new infrastructure had destroyed the old grain of the basement.

The second and third cases relate to Central East Greenland, where Caledonian regional metamorphism reached high up into the late Proterozoic cover (around 73°N close to the base of the Vendian tillites). In these cases the effects upon the old basement can be envisaged as a physicochemical reactivation of selected rock groups or structures. And so, old basement structures received new life. This aspect, of course, poses the interesting question of how much of these structures is old and how much is new? The question is particularly tricky where the infrastructure forms ductile protuberant massifs, the dynamics of which are most difficult to interpret in any way. As an illustration, I shall refer to two cases of infrastructural protuberance, where in either instance the final emplacement of the migmatitic rock body was outlasted by growth of spectacular garnet porphyroblasts close to the contorted migmatite sheets. The first example, the giant ptygmatic injection at Isfjord (73°24′N) obviously represents a hypermobile migmatite that emplaced by way of plasticoviscous flow (see Haller 1971, phot. 50). In the second example, exposed at the head of Kejser Franz Joseph Fjord (73°08′N), the protuberance is featured by highly contorted lamellae of pre-Caledonian gneiss. A detailed study of these vertical exposures (Haller 1970, pl. 42, 1971, phot. 46) suggests that part of the 'mise en place' of these migmatite sheets occurred in the form of solid thrust wedges. This mechanical emplacement—protrusion—clearly predates the ductile contortion, which is Caledonian. The protrusion, however, is seemingly of Proterozoic vintage. In fact, the structural design looks like the 'blueprint' of a former thrust belt but now reshuffled, plastically folded, and associated with cover strata that reached staurolite–kyanite grade in Caledonian times.

In East Greenland, as elsewhere, there has been and there will be much debate about the 'mise en place' of *in situ* granites and associated migmatites, because the field geologist now has available new tools with which to look at old problems. The new analytical techniques of *isotope chemistry* in particular help us to unravel the history and derivation of igneous and metamorphic complexes (Rex and Gledhill 1981).) To know about the subtle ratios obtained from mineral or bulk rock samples can be of great help. But it is an 'invisible' property, which by its plain number may not tell us unambiguously how it relates to the complex history of the rock itself, much less of the whole orogen (Steiger *et al.* 1979). For example, Rb–Sr isochrons are likely to reflect the stratotectonic age of the rock while mineral ages refer to crystallization and cooling, that is to morphotectonic events. The first may give information about the origin of the rock material while the second more likely reflects the anatomy and the evolution of the orogenic belt. In either case, it is now our challenge to relate the 'invisible' data to the visible history of the entire rock complex.

Conclusion

In this study I have reviewed a variety of topics. To recapitulate, I shall conclude as follows:

1. Palinspastic reconstructions make the East Greenland Caledonides a markedly separate branch of the North Atlantic Caledonides, more peripheral to and exclusive of the simatic vestiges of the Iapetus Ocean.
2. Due to their ensialic setting, the East Greenland Caledonides have preserved a good record of successive Proterozoic regimes. This record includes the onset of the novel trend of Iapetus that sharply truncated the Proterozoic tectonic fabric.
3. The closure of Iapetus brought about the linear Caledonian trend, which is outlined by a stack of thrust sheets that were shoved westward in Silurian time.
4. In Northeast Greenland, Caledonian core complexes developed a new tectonic fabric, guided by and conformable to the Silurian thrust pattern. In the core complexes of Central East Greenland, the 'blueprints' of Proterozoic structures are discernible throughout the Caledonized basement.
5. In Devonian time, Central East Greenland was affected by folding and thrusting of transverse trend that roughly mimics the Proterozoic fabric.
6. Uplift of the folded ranges resulted in erosional unroofing and the removal of vast quantities of Lower Devonian detritus which is not accounted for. Because the emergent East Greenland fold belt, together with the ranges of Svalbard, were distinctly separate from those of Britain and Scandinavia (Suess's 'Old Red Continent') the existence of intraorogenic, marine successor basins is postulated. These basins developed west of the Iapetus suture and became disrupted by dextral strike-slip movements in Devono-Carboniferous time.
7. The opening of the northern North Atlantic began in the late Palaeozoic and developed from the sites of these episutural Caledonian basins.

References

Argand, E. 1916. Sur l'arc des Alpes Occidentales. *Eclog. geol. Helv.*, **14**, (1), 143–191.

Argand, E. 1924. La tectonique de l'Asie. *Int. geol. Congr., 13th Belgique, 1922, Rep 1*, 171–332.

Backlund, H. G. 1936. Der 'Magmaaufstieg' in Faltengebirgen. *C. R. Soc. geol. Finl.*, **9**, 293–347.

Bailey, E. B. and Holtedahl, O. 1938. Northwestern Europe. Caledonides. In Andrée, K. *et al.* (Eds), *Regionale Geologie der Erde, vol. 2, Palaeozoische Tafel und Gebirge, pt. 2*, Akademische Verlagsgesellschaft, Leipzig, 84 pp.

Becke, F. 1913. Über Mineralbestand und Struktur der kristallinischen Schiefer. *Denkschr. k. Akad. Wiss. Wien, Mat.–Nat. Kl.*, **75**, I, 1–53.

Brooks, C. K., Fawcett, J. J. and Gittins, J. 1976. Caledonian magmatic activity in southeastern Greenland. *Nature, London*, **260**, 694–696.

Bryhni, I. and Brastad, K. 1980. Caledonian regional metamorphism in Norway. *J. geol. Soc. London*, **137**, 251–259.

Bullard, E. 1975. The emergence of plate tectonics: a personal view. *Ann. Rev. Earth Planet. Sci.*, **3**, 1–30.

Bullard, E. C., Everett, J. E. and Smith, A. G. 1965. The fit of the continents around the Atlantic. *Phil. Trans. R. Soc. (A)*, **258**, 41–51.

Caby, R. 1976. Investigations on the Lower Eleonore Bay Group in the Alpefjord region, central East Greenland. *Rapp. Grønlands geol. Unders.*, **80**, 102–106.

Carey, S. W. 1958. The tectonic approach to continental drift. In *Continental Drift—A Symposium*, Dept. Geol. Univ. Tasmania, Hobart, 177–355.

Chadwick, B. 1975. The structure of South Renland, Scoresby Sund, with special reference to the tectonometamorphic evolution of a southern internal part of the Caledonides of East Greenland. *Medd. Grønland*, **201** (2), 1–67.

Cherkis, N. Z., Fleming, H. S. and Massingill, J. V. 1973. Is the Gibbs Fracture Zone a westward projection of the Hercynian Front into North America? *Nature Phys. Sci., Lond.*, **245**, 113–115.

Choubert, B. 1935. Récherches sur la genèse des chaînes paléozoïque et anté-cambriennes. *Revue Géogr. phys. géol. dyn.*, **8** (1), 5–50.

Christie, R. L. 1979. The Franklinian Geosyncline in the Canadian Arctic and its relationship to Svalbard. *Norsk Polarinst. Skr.*, **167**, 263–314.

Dillon, L. S. 1974. Neovolcanism: a replacement for the concepts of plate tectonics and continental drift. *Am. Ass. Petrol. geol. Mem.*, **23**, 167–239.

Doig, R. 1970. An alkaline rock province linking Europe and North America. *Can J. Earth Sci.*, **7**, 22–28.

Flinn, D., Frank, P. L., Brook, M. and Pringle, I. R. 1979. Basement–cover relations in Shetland. In Harris, A. L., Holland, C. H. and Leake, B. E. Eds), *The Caledonides of the British Isles—reviewed*, Scottish Academic Press, Edinburgh, 109–115.

Fränkl, E. 1953. Geologische Untersuchungen in Ost-Andrées Land (Nordostgrönland). *Meddr. Grønland*, **113** (4), 1–160.

Fränkl, E. 1955. Weitere Beiträge zur Geologie von Kronprins Christians Land (Nordostgrönland, zwischen 80° und 80°30′N). *Meddr. Grønland*, **103** (7), 1–35.

Friend, P. F. 1981. Devonian sedimentary basins and deep faults of the northernmost Atlantic borderlands. In Kerr, J. W. and Fergusson, A. J. (Eds), *Geology of the North Atlantic Borderlands. Can. Soc. Petrol. geol. Mem.*, **7**, 149–165.

Gee, D. G. 1975. A tectonic model for the central part of the Scandinavian Caledonides. *Am. J. Sci.*, **275–A**, 468–515.

Gee, D. G. 1977. Reply [to Discussion: a tectonic model for the central part of the Scandinavian Caledonides]. *Am. J. Sci.*, **277**, 657–665.

Gee, D. G. 1978. Nappe displacement in the Scandinavian Caledonides. *Tectonophysics*, **47**, 393–419.

Grubenmann, U. und Niggli, P. 1924. *Die Gesteinsmetamorphose, I. Allgemeiner Teil*, (3rd edition), Borntraeger, Berlin, 539 pp.

Haller, J. 1955. Der 'Zentrale Metamorphe Komplex' von Nordostgrönland, Teil I. Die geologische Karte von Suess Land, Gletscherland und Goodenoughs Land. *Meddr. Grønland*, **73**, I (3), 1–174.

Haller, J. 1969. Tectonics and neotectonics in East Greenland—review bearing on the drift concept. In Kay, M. (Ed), *North Atlantic—Geology and Continental Drift, A Symposium, Am. Ass Petrol. geol. Mem.*, **12**, 852–858.

Haller, J. 1970. Tectonic map of East Greenland (1:500, 000). An account of tectonism, plutonism, and volcanism in East Greenland. *Meddr. Grønland*, **171** (5), 1–286.

Haller, J. 1971. Geology of the east Greenland Caledonides, Wiley–Interscience, London, xxiii + 413 pp.

Haller, J. 1979. Himalayan orogenesis in perspective (Preface), pp. i–xxxiv. In Verma, P. K. (Ed.), *Metamorphic Rock Sequences of the Eastern Himalaya*, K. P. Bagchi Co., Calcutta, 166 pp.

Haller, J. 1980. Geodynamic significance of hereditary trends in orogenesis. *Int. geol. Congr., 26th Paris, 1980. Abstracts I*, **347**.

Haller, J. 1983. Geological map of Northeast Greenland 75°–82°N. Lat. (1:1,000,000). *Meddr. Grönland*, **200** (5), 1–22.

Haller, J. and Kulp, J. L. 1962. Absolute age determinations in East Greenland. *Meddr. Grønland*, **171** (1), 1–77.

Harland, W. B. 1966. A hypothesis of continental drift tested against the history of Greenland and Spitsbergen. *Cambridge Res.*, **2**, 18–22.

Harland, W. B. 1969. Contribution of Spitsbergen to understanding of tectonic evolution of North Atlantic Region. In Kay, M. (Ed.), *North Atlantic—Geology and Continental Drift, A Symposium, Am. Ass. Petrol geol. Mem.*, **12**, 817–851.

Harland, W. B. 1978. The Caledonides of Svalbard. In Tozer, E. T. and Schenk, P. E. (Eds), *Caledonian–Appalachain Orogen of the North Atlantic Region, Geol. Surv. Canada Paper*, **78–13**, 3–11.

Harland, W. B. and Gayer, R. A. 1972 The Arctic Caledonides and earlier oceans. *Geol. Mag.*, **109** (4), 289–314.

Haworth, R. T. 1977 The continental crust northeast of Newfoundland and its ancestral relationship to the Charlie Fracture Zone. *Nature*, London, **266**, 246–249.

Heim, A. 1878. *Untersuchungen über den Mechanismus der Gebirgsbildung* (im Anschluss an die geologische Monographie der Tödi–Windgällen-Gruppe), 3 vols. (I, xiv + 346 pp.; II, 246 pp., Atlas), Benno Schwabe, Basel.

Henriksen, N. and Higgins, A. K. 1976. East Greenland Caledonian fold belt. In Escher, A. and Watt, W. S.

(Eds), *Geology of Greenland*, Geol. Surv. Greenland Copenhagen, 182–246.

Hess, H. H. 1962. History of ocean basins. In Engel, A. E. J., James, H. L. and Leonard, B. F. (Eds), *Petrologic Studies: a volume in honor of A. F. Buddington*, Geol. Soc. Am., New York, 599–620.

Higgins, A. K. 1976. Pre-Caledonian metamorphic complexes within the southern part of the East Greenland Caledonides. *J. geol. Soc. London*, **132**, 289–305.

Higgins, A. K., Fiderichsen, J. D. and Thyrsted, T. 1981. Precambrian metamorphic complexes in the East Greenland Caledonides (72°–74°N)—their relationships to the Eleonore Bay Group, and Caledonian orogenesis. *Rapp. Grönlands geol. Unders.*, **104**, 5–46.

Higgins, A. K. and Phillips, W. E. A. 1979. East Greenland Caledonides—an extension of the British Caledonides. In Harris, A. L., Holland, C. H. and Leake, B. E. (Eds), *The Caledonides of the British Isles—reviewed*, Scottish Academic Press, Edinburgh, 19–32.

Holtedahl, O. 1936. Trekk av det skandinaviske fjellkjede-strøks historie. *Nordiska (19 skandinaviska) naturfors-karmötet i Helsingfors Eripainos*, Helsinki, 129–145.

Homewood, P. 1973. Structural and lithological divisions of the western border of the East Greenland Caledonides in the Scoresby Sund region between 71°00′ and 71°22′N. *Rapp. Grønlands geol. Unders.*, **57**, 1–27.

Hurst, J. M. and McKerrow, W. S. 1981. The Caledonian nappes of eastern North Greenland. *Nature, London*, **290**, 772–774.

Irving, E. 1977. Drift of the major continental blocks since the Devonian. *Nature, London*, **270**, 304–309.

Koch, L. and Haller, J. 1971. Geological map of East Greenland 72°–76°N. Lat (1:250, 000). *Meddr. Grønland*, **183**, 1–26. (13 map sheets in portfolio).

Kulling, O. 1972. The Swedish Caledonides. In Strand, T. and Kulling, O., *Scandinavian Caledonides*, Wiley–Interscience, London, 147–285.

Lancelot, Y. 1980. Birth and evolution of the 'Atlantic Tethys' (Central North Atlantic). Colloque C 5, Géologie des chaînes alpines issues de la Téthyl. Int geol. Congr. 26th Paris, 1980, *Bureau Récherches géol. Min. Mem.*, **115**, 215–223.

Mittelholzer, A. E. 1941. Die Kristallingebiete von Clavering Ø and Payer Land (Ostgrönland), vorläufiger Bericht über die Untersuchungen im Jahre 1938/39. *Meddr. Grønland*, **114** (8), 1–42.

Nathorst, A. G. 1901. Bidrag till nordöstra Grönlands geologi. *Geol. För. Stockh. Förh.*, **23** (207), 275–306.

Nicholson, R. 1974. The Scandinavian Caledonides. In Nairn, A. E. M. and Stehli, F. G. (Eds), *The Ocean Basins and Margins, vol. 2, The North Atlantic*, Plenum Press, New York, 161–203.

Oftedahl, C. 1966. Note on the main Caledonian thrusting in northern Scandinavia. *Norsk geol. Tiddskr.*, **46** (2), 237–244.

Parkinson, M. M. L. and Whittard, W. F. 1931. The geological work of the Cambridge Expedition to East Greenland in 1929. *Q. J. geol. Soc. London*, **87**, 650–674.

Phillips, W. E. A., Stillman, C. J., Friderichsen, J. D. and Jemelin, L. 1973. Preliminary results of mapping in the western gneiss and schist zone around Vestfjord and inner Gåsefjord, southwest Scoresby Sund. *Rapp. Grønlands geol. Unders.*, **58**, 17–32.

Pitman, W. C. and Talwani, M. 1972. Sea floor spreading in the North Atlantic. *Bull. geol. Soc. Am.*, **83** (3), 619–646.

Prost, A. E. 1972. Aperçu synthétique sur les Calédonides scandinaves. *Sci. Terre*, **17** (3), 217–233.

Prost, A. E. 1977. Répartition et évolution géodynamique des externides calédoniennes scandinaves. *Rev. Géogr. phys. géol. dyn.*, **19** (5), 421–432.

Ramberg, H. 1966. The Scandinavian Caledonides as studied by centrifuged dynamic models. *Uppsala Univ. geol. Inst. Bull.*, **43**, 1–72.

Ramberg, H. 1981. The role of gravity in orogenic belts. In McClay, K. R. and Price, N. J. (Eds), *Thrust and Nappe Tectonics*, Geol. Soc. London; Blackwell Sci. Publ., Oxford, 125–140.

Read, H. H. 1948. A commentary on place in plutonism. *Q. J. geol. Soc. London*, **104**, 155–206.

Read, H. H. 1975. The Granite Controversy, Murby, London, xix + 430 pp.

Rex, D. C. and Gledhill, A. R. 1981. Isotopic studies in the East Greenland Caledonides (72°–74°N)—Precambrian and Caledonian ages. *Rapp. Grønlands geol. Unders.*, **104**, 47–72.

Roberts, D. 1978. Caledonides of South Central Norway. In *Caledonian—Appalachian Orogen of the North Atlantic Region. Geol. Surv. Canada, Paper*, **78–13**, 31–37.

Roberts, D. and Gale, G. H. 1978. The Caledonian–Appalachian Iapetus Ocean. In Tarling, D. H. (Ed.), *Evolution of the Earth's Crust*, Academic Press, London, 255–342.

Roeder, D. H. 1973. Subduction and orogeny. *J. Geophys. Res.*, **78** (23), 5005–5024.

Sederholm, J. J. 1907. Om granit och gneis (deras uppkomst, uppträdande och utbredning inom urberget i Fennoskandia). *Bull. Comm. géol. Finl.*, **23**, 1–110.

Sokolov, B. S. 1973. Vendian of Northern Eurasia. In Pitcher, M. G. (Ed.), *Arctic Geology, Am. Ass. Petrol. geol. Mem.*, **19**, 204–218.

Staub, R. 1928. *Der Bewegungsmechanimus der Erde*, Gebr. Borntraeger, Berlin, 270 pp.

Steck, A. 1971 Kaledonische Metamorphose der Praekambrischen Charcot Land Serie, Scoresby Sund, Ost-Grönland. *Meddr. Grønland*, **192** (3), 1–69.

Steiger, R. H., Hansen, B. T., Schuler, Ch., Bär, M.T. and Henriksen, N. 1979. Polyorogenic nature of the southern Caledonian fold belt in East Greenland: an isotopic age study. *J. geol.*, **87** (5), 475–495.

Stille, H. 1958. Die assyntische Tektonik im geologischen Erdbild. *Beih. geol. Jb.*, **22**, 1–255. Hannover.

Sturt, B. A. and Roberts, D. 1978. Caledonides of Northernmost Norway (Finnmark). In *Caledonian–Appalachian Orogen of the North Atlantic Region. Geol. Surv. Canada Paper*, **78–13**, 17–24.

Størmer, L. 1967. Some aspects of the Caledonian geosyncline and foreland west of the Baltic Shield. *Q. J. geol. Soc. London*, **123**, 183–214.

Suess, E. 1888. *Das Antlitz der Erde*, vol. 2, Tempsky/Freytag Publ., Vienna/Leipzig, 704 pp.

Sundman, P. O. 1967. Ingenjör Andrées Luftfärd, Norstedt & Söners Förlag, Stockholm, 343 pp.

Talbot, C. J. 1979. Infrastructural migmatitic upwelling in East Greenland interpreted as thermal convective structures. *Precambrian Rs.*, **8**, 77–93.

Termier, P. 1912. Sur la genèse des terrains cristallophylliens. *Int. geol. Congr., 11th Stockholm, 1910*, **Rep. 1**, 587–595.

Trettin, H. P. 1969 A Paleozoic–Tertiary fold belt in

northernmost Ellesmere Island aligned with the Lomonosov Ridge. *Bull. geol. Soc. Am.*, **80** (1), 143–148.

Van Der Gracht, W. A. and van Waterschoot, J. M. 1928. Introduction. In *Theory of Continental Drift*, Am. Ass. Petrol. Geol., Tulsa, Okla., 1–75.

Van Der Voo, R. and Scotese, C. 1981. Paleomagnetic evidence for a large (2,000 km) sinistral offset along the Great Glen fault during Carboniferous time. *Geology*, **9** (12), 583–589.

Wegener, A. 1915. Die Enstehung der Kontinente und Ozeane. *Sammlung Vieweg. Tagesfragen aus den Gebieten der Naturwissenschaften und det Techiik. Vieweg, Braunschweig*, **23**, 94 pp.

Wegmann, C. E. 1935. Zur Deutung der Migmatite. *Geol. Rdsch.*, **26**, 305–350.

Wegmann, C. E. 1959. La Flexure axiale de la Driva et quelques problèmes structuraux des Caledonides Scandinaves. *Norsk geol. Tidsskr.*, **39** (1), 25–74.

Wilson, J. T. 1966. Did the Atlantic close and re-open? *Nature*, London, **211**, 676–681.

Ziegler, P. A. 1978. North-western Europe: Tectonics and basin development. In Van Loon, A. J. (Ed.), *Keynotes of the MEGS–II (Amsterdam, 1978)*. *Geol. Mijnb.*, **57** (4), 589–626.

The Caledonide Orogen—Scandinavia and Related Areas
Edited by D. G. Gee and B. A. Sturt

The geology of the northern extremity of the East Greenland Caledonides

J. M. Hurst[*][1], **H. F. Jepsen**[*], **F. Kalsbeek**[*], **W. S. McKerrow**[†] and **J. S. Peel**[*]

Grønlands Geologiske Undersøgelse, Øster Voldgade 10, DK-1350 Copenhagen K, Denmark
†*Department of Geology and Mineralogy, Parks Road, Oxford OX1 3PR, England*

ABSTRACT

The Caledonian fold belt of Kronprins Christians Land, eastern North Greenland, has four main elements. In the eastern coastal regions, crystalline basement, consisting of gneisses, migmatized mica schists and amphibolites, is of possible early Proterozoic age. To the west of this, a thick sequence of late Proterozoic sandstones with basic intrusions comprising thick sills, discordant sheets, and subvertical dykes (Independence Fjord Group) is present. The base of the Independence Fjord Group is interfolded with the crystalline basement. Still further to the west, a 20 to 30 km wide belt of allochthonous Proterozoic and Lower Palaeozoic clastics and carbonates occurs in the Vandredalen, Sæfaxi Elv, and Finderup Land nappes. In the extreme westernmost part of the area, late Proterozoic to Silurian platform carbonates and clastics form the foreland to the fold belt. Throughout all zones the sequence of structural events and the style of deformation appears the same. The nappes are displaced westwards. Folds are consistently overturned to the west or northwest and have gentle to moderately dipping eastern limbs. The nappes were emplaced prior to the deformation producing the major westerly-directed folds. Thrusting in the Independence Fjord Group accompanied folding. Thrusting and faulting in the foreland could be related to either nappe emplacement, the later folding or a combination of both. A series of northwest to southeast normal faults cuts folds and thrust planes. These could be Caledonian or younger in age. A precise minimum age for diastrophism cannot be established, but mid Wenlock sediments, affected by the thrusts, provide a maximum age for the nappe emplacement in Kronprins Christian Land. There is no evidence of the late Precambrian Carolinidian Orogeny.

Introduction

The Caledonian fold belt of Kronprins Christian Land, eastern North Greenland, is, at the present day, the most northerly expression of both the whole Caledonide orogen and also its East Greenland segment (Fig. 1). As with the remainder of East Greenland, the western margin of the orogenic belt is preserved (*cf.* Henriksen 1978; Higgins and Phillips 1979); nappes of Proterozoic and Lower Palaeozoic clastic and carbonate rocks are displaced westward over Proterozoic and Lower Palaeozoic platform sediments (Fig. 2). The fold belt is cored by crystalline basement of possible early Proterozoic age but, unlike the rest of the East Greenland Caledonides, sediments are not strongly metamorphosed and there is a total lack of plutonic rocks. In most Caledonian areas there is controversy concerning the extent of Caledonian orogenic events, and Kronprins Christian Land is no exception. Haller (1961, 1970, 1971) claimed that a Proterozoic orogenic event, the Carolinidian Orogeny, predated the Caledonian orogeny. Part of the evidence on which this orogeny was based, was derived from the Prinsesse Caroline-Mathilde and Prinsesse Elisabeth Alper (Fig. 2).

The Caledonides of Kronprins Christian Land extend 80 km by approximately 200 km from latitudes 79° to 82°N, and occupy almost the whole area between the eastern coast and Danmark Fjord

---

[1]Present address: B. P., Petroleum Development Ltd, Britannic House, Moor Lane, London EC2Y 9BU, England.

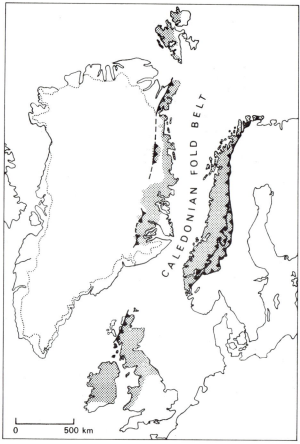

Fig. 1   Relationship of the Kronprins Christian Land
Caledonides (northernmost stippled area in Greenland) to
East Greenland, Scandinavia, and Britain in a pre-drift
configuration

(Fig. 2). Carboniferous to Tertiary marine and continental rocks overlie the parautochthonous basement in the eastern coastal regions (Håkansson *et al*. 1981). Four main tectonostratigraphic elements can be differentiated from east to west in the Caledonian folded units of Kronprins Christian Land: zone 1, parautochthonous early Proterozoic crystalline basement in the eastern coastal regions; zone 2, parautochthonous Proterozoic sandstones; zone 3, allochthonous Proterozoic and Lower Palaeozoic clastics and carbonates; and zone 4, parautochthonous Lower Palaeozoic platform carbonates (Fig. 2).

The stratigraphy and structure of these four elements of the Caledonian fold belt are described below. Finally, they are compared briefly to the rest of the East Greenland and North Atlantic Caledonides.

Zone 1: Crystalline Basement

The crystalline basement occupies most of the eastern part of Lambert Land, Hovgaard Ø, and Holm Land as well as the northeastern corner of Amdrup Land. The zone is up to 50 km wide (Fig. 2). The basement consists predominantly of variable gneiss types, migmatized mica schists, quartzites, and amphibolites. Discordant basic dykes, now amphibolites, are common at several localities.

U–Pb isotopic dating on zircons from two gneiss samples (see Jepsen and Kalsbeek this volume) indicates that the basement was formed at least 2000 Ma ago and later reactivated during the Caledonian orogeny at c. 400 Ma.

On the south coast of Hovgaard Ø and on two islands near the front of the glacier in Nioghalvfjerds fjorden (Fig. 2), augen gneisses, presumably belonging to the basement, occur interfolded with the overlying Independence Fjord Group sandstones.

Zone 2: Sandstones of the Independence Fjord Group

Proterozoic sandstones with numerous dolerite intrusions occur in a 20 to 50 km wide zone that runs

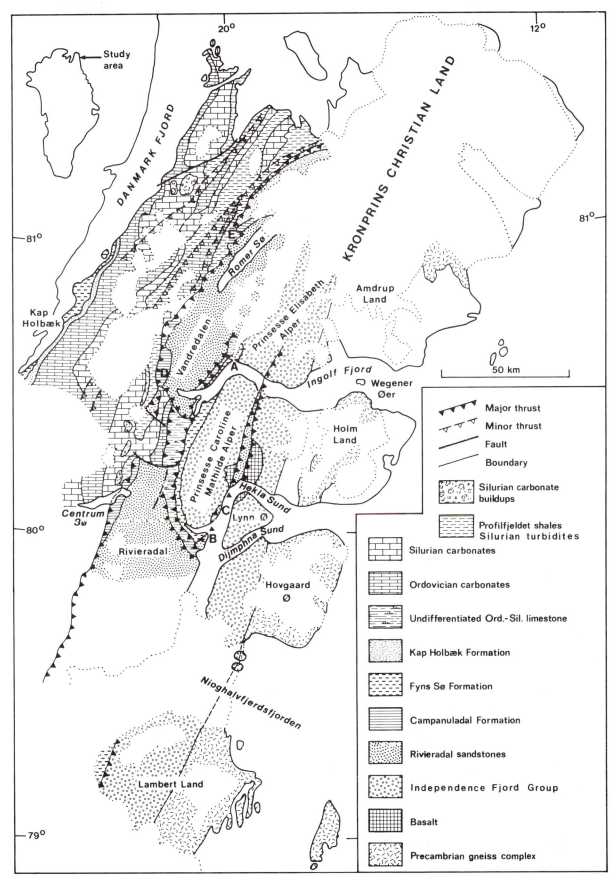

Fig. 2   Simplified geological map of the sediments affected by Caledonian folding in Kronprins Christian Land. A: Hjørnegletscher; B: Marmorvigen; C: west Hekla Sund; D: Nøglefjeldet; E: west of Romer Sø

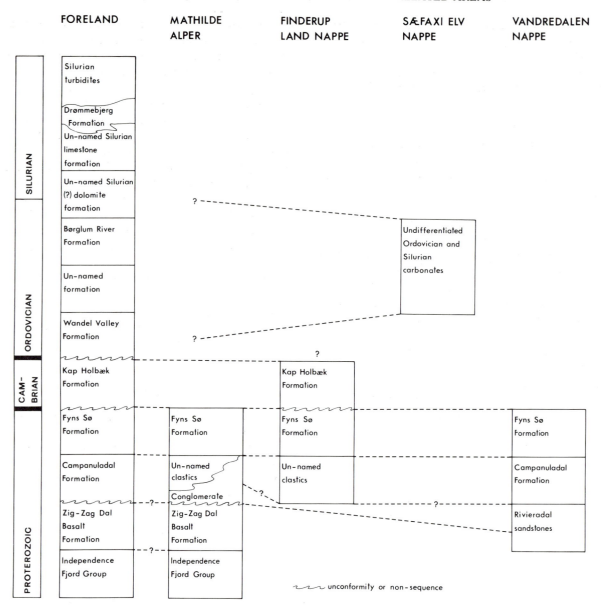

Fig. 3   Simplified schematic stratigraphies and correlation of both the parautochthonous sediments, platform carbon-ates of the foreland and the Proterozoic sandstone swell of the Mathilde Alper, and allochthonous sediments (Van-dredalen, Sæfaxi Elv and Finderup Land Nappes) incorporated in the Caledonian earth movements in Kronprins Christian Land

from central Kronprins Christian Land to Lambert Land in the south (Fig. 2). These rocks are now referred to the Independence Fjord Group (Collison 1979, 1980) of the platform area to the northwest, where the sandstones have a thickness of more than 1750 m. The sandstones are capped by isolated out-crops of basalt, and are also overlain by the Rivieradal sandstone and other late Precambrian sediments (see later section on the Caledonian nappes).

Stratigraphy (Fig. 3)

The sandstones

Haller (1970) referred to the sandstones in this zone as the 'Thule Group'.

According to Haller (1970, p. 52) this group could be split into two:

1. 'Lower Thule Group' consisting of a basal 200 m thick quartzite and limestone unit, overlain by

some 3000 m of semipelitic rocks which accumulated in a geosynclinal trough. Haller (1971, p. 56) suggested that parts of this succession were composed of basic extrusive rock.

2. 'Upper Thule Group' consisting of 3000 m of psammitic strata, partially in red bed facies.

The 'Lower Thule Group' was purported to be widespread in Kronprins Christian Land (Haller 1971, Figs 18 and 24), but no evidence of the presence of such sediments has been found (Jepsen and Kalsbeek 1981).

The 'Upper Thule Group' is now referred to the Independence Fjord Group of the platform area. The sequence is over 1000 m thick and consists of medium to coarse-grained sandstones of variable composition from feldspathic to quartzitic, and pink, yellow, or white in colour. The sandstones are intruded by dolerites (now metadolerites) in the form of thick sills to discordant sheets and subvertical dykes (Fig. 4).

The age of the sandstones is not known, but is probably approximately the same as that of similar deposits to the west, where Larsen and Graff-Petersen (1980) have obtained a Rb–Sr clay-mineral age of 1380 Ma for a siltstone horizon within the sandstones which they regard as a minimum age for the deposition of the rocks.

Basalts

To the west of Danmark Fjord (Fig. 2) the sandstones of the Independence Fjord Group are overlain conformably by the Zig-Zag Dal Basalt Formation (Jepsen, Kalsbeek and Suthren 1980; Fig. 3), which in all probability is contemporaneous with the instrusions in the underlying sandstones.

A sequence of basaltic rocks also occurs in Kronprins Christian Land where it attains a minimum thickness of about 500 m. At one locality the basalts are overlain by sandstone intruded by dolerites. It is not clear whether this overlying sandstone is brought into its present position by thrusting or whether the stratigraphy in Kronprins Christian Land differs from the stratigraphy in the platform area west of Danmark Fjord. The suggestion of Haller (1971, p. 56) that basic extrusive rocks probably occur in Kronprins Christian Land may stem from his flight over this local succession.

Intrusions in the Independence Fjord Formation from the platform area have given Rb–Sr whole-rock isochron ages of c. 1250 Ma (Jepsen and Kalsbeek 1979), and it is probable that the basalts are of the same general age.

Late Proterozoic sediments

The sandstones of the Independence Fjord Group are unconformably overlain by a sequence of shallow marine late Proterozoic clastics and carbonates (Fränkl 1954, 1956; Hurst and McKerrow 1981a,b; Jepsen and Kalsbeek 1981). The contact with the underlying Independence Fjord Group is a profound erosional unconformity of regional extent, which has been recognized over large parts of the North

Fig. 4   Dolerite (now metadolerite) intrusions in the Proterozoic sandstones of eastern Ingolf Fjord (Fig. 2). Cliff height approximately 1000 m

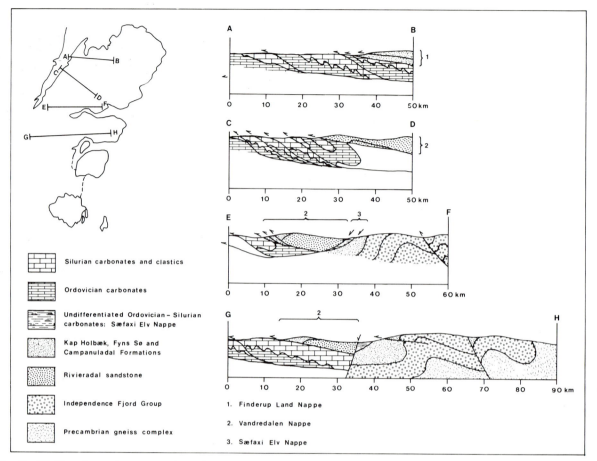

Fig. 5 Highly schematic cross-sections through Kronprins Christian Land. Vertical scale is distorted and only relative scale is implied

Greenland platform. Local beds of conglomerate occur at the contact, sometimes filling in an irregular erosional topography in the underlying sandstones and intrusions.

This late Proterozoic sequence in Kronprins Christian Land is comparable to the Campanuladal Formation west of Danmark Fjord (Fig. 3; Clemmensen 1979) in that it is overlain by limestone (strongly sheared in Kronprins Christian Land) that grades upward into stromatolitic dolomites of the Fyns Sø Formation (Fig. 3).

The precise age of these sediments is not known, but they predate the Lower Cambrian Portfjeld Formation (O'Connor 1979; Peel 1980a) which unconformably overlies the Fyns Sø Formation. The conglomerates may be equivalent in age to the upper parts of the Rivieradal sandstones (see section on the Caledonian nappes) in which acritarchs suggest a Late Riphean age (approximately 800 Ma).

Rivieradal sandstones

(Fränkl 1954, 1955, 1956; Hurst and McKerrow 1981a). This late Proterozoic turbidite sequence is entirely allochthonous. In zone 2 the unit has only been found in a tightly compressed syncline that separates the sandstones of the Independence Fjord Group into two elongate outcrop areas. The Rivieradal sandstones dominate the area of Caledonian nappes (zone 3) and a full description of them is presented below.

Structure (Fig. 5)

The quartzites and sandstones of the Independence Fjord Group and the associated dolerite intrusions are strongly folded, on scales varying between tens of metres and a few kilometres, with north to northeast trending, often subhorizontal, axes (Fig. 6). East to southeast dipping axial planes are most common; the folds have gentle to moderately dipping eastern limbs and steep to overturned western limbs (Fig. 5). This is the same style of folding as that found in the younger, late Proterozoic to Palaeozoic sediments lying above the quartzites.

Lack of good marker horizons, poorly developed bedding as well as the presence of commonly irregular dolerite intrusions make the study of the struc-

Fig. 6   In the foreground large-scale folds in the Proterozoic sandstones (P). F, assorted clastics and Fyns Sø Formation (Fig. 3). Note in the background fold in Fyns Sø Formation (F), with overlying 'Centrum Limestone' (C) on the Sæfaxi Elv Nappe and finally overlying Rivieradal sandstones (R) on the Vandredalen Nappe. Foreground locality, Hjørnegletscher (A, on Fig. 2) t–t, thrust planes, Note that the thrust plane between Fyns Sø Formation and 'Centrum Limestone' is folded. Cliff height approximately 1000 m

tures difficult. The presence of strongly sheared rocks in places suggests that the folding was accompanied by westward thrusting of the rocks, but the amount of displacement along individual shear zones could not be determined. In the Ingolf Fjord region (Fig. 2) and south of Ingolf Fjord, the rocks of the Independence Fjord Group have locally overthrust the Fyns Sø Formation. Apparently the overthrusted rocks did not move far, as the thrust planes pass latterly into folded sedimentary contacts.

Shearing in the sandstones increases in intensity towards the east. Here, steep belts of strongly sheared sandstone, in which the dolerites are commonly heavily deformed and boudinaged, alternate with zones of less deformed sandstone. Such shear belts become less common towards the west where large-scale folds predominate.

Zone 2 provides one of the few keys to a chronology of structural events in the Kronprins Christian Land Caledonides. The Rivieradal sandstones west of zone 2 lie on a large-scale westerly thrust sheet, and the Rivieradal sandstones found within the central syncline in the Prinsesse Caroline-Mathilde Alper appear to represent the trailing edge of this nappe. Subsequent to the thrusting, the sediments were folded into their synclinal form. Thus, the major episode of westerly directed nappe emplacement would appear to predate the folding and thrusting in the sandstones of the Independence Fjord Group. As a result the thrust planes are often steeply inclined and in the Marmorvigen region dip up to 70° to the east (B, Fig. 2).

Carolinidian orogeny

Details concerning this supposed orogenic event can

be found in Jepsen and Kalsbeek (this volume). In short, Haller (1961, 1970, 1971) proposed a late Proterozoic orogeny, post-dating the deposition of the Independence Fjord Group and prior to the Caledonian, which he called the Carolinidian orogeny after the Prinsesse Caroline-Mathilde Alper in Kronprins Christian Land. Recent work by Jepsen and Kalsbeek (1981) has shown that there is no evidence of a pre-Caledonian orogeny in Kronprins Christian Land. For example, it was found that there is no angular discordance between the Independence Fjord Group and the overlying late Proterozoic sediments as thought by Haller (1961, 1970, 1971).

It should, however, be pointed out that there is probably a considerable time span between deposition of the Independence Fjord Group and the overlying younger Proterozoic sediments equivalent to the Campanuladal and Fyns Sø Formations. Regional considerations indicate the gap to be more than 500 Ma. The lack of the widespread basalt unit in this area, equivalent to the Zig-Zag Dal Basalt Formation further west (Jepsen *et al*. 1980) is probably due to erosion in this time period.

Zone 3: Caledonian Nappes

To the west of the Proterozoic sandstones of zone 2 is a belt of allochthonous non-metamorphic carbonate and terrigenous clastic sediments of Proterozoic and Lower Palaeozoic age. This zone varies in width from 30 to 50 km and extends for the length of Kronprins Christian Land (Fig. 2).

Fränkl (1954, 1955, 1956) considered that all the sediments occurring west of the Proterozoic sandstones and east of the Lower Palaeozoic platform

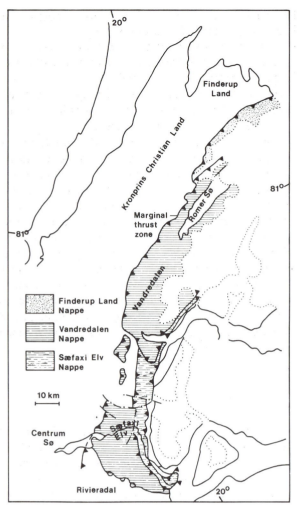

Fig. 7 Simplified map showing the distribution of sediments belonging to the three main Caledonian nappes

thin laminated quartz sand units, up to 100 m thick. These are overlain by approximately 300 m of red and yellow laminated and stromatolitic limestones (Fig. 3). Both units are comparable to the Campanuladal Formation west of Danmark Fjord. Approximately 100 m of tectonized stromatolitic dolomite, the Fyns Sø Formation, follow conformably. The top stratum of this nappe consists of green bioturbated siltstones and white quartzites with *Skolithos*, assigned to the Kap Holbæk Formation. This is probably of late Proterozoic age, as noted by Peel (this volume), and it rests without any angular discordance on the Fyns Sø Formation of similar age.

Structure

The sole of the nappe is flat or inclined at an angle of 10° to the southeast. In places along the southwestern margin, a pronounced imbricate zone is developed and the succession, Fyns Sø and Kap Holbæk Formation, is repeated at least three times (Hurst and McKerrow 1981b).

The sediments are strongly folded, wavelengths ranging from a few metres to several hundreds of metres. Fold axes which trend between northeast and east-northeast are often subhorizontal but plunges up to 45° to the northeast are recorded. Some large folds, up to 200 m amplitude, affect the Kap Holbæk Formation. All the folds have gentle to moderately dipping eastern limbs and steep to vertical (but not overturned) western limbs consistent with a westward displacement direction.

Penetrative deformation

Deformation is frequently penetrative but varies considerably in intensity. Sediments of the Campanuladal and Fyns Sø Formations, which are always at the base of the thrust nappe, are strongly sheared. In some areas low domal stromatolites typical of the Campanuladal Formation are compressed by a factor of 10:1 in width and almost obliterated by vertical axial plane cleavage. Stromatolites in the Fyns Sø Formation are completely sheared and have recrystallized into either a homogeneous limestone or a laminated friable talc dominated unit. Sediments of the Kap Holbæk Formation are less deformed and in many places, where only sandstones outcrop, a cleavage is difficult to discern.

carbonate rocks belonged essentially to one huge nappe, which he termed the Main Nappe. Recently, Hurst and McKerrow (1981a,b) have recognized three large nappes in this allochthonous zone (Fig. 7). Two, the Vandredalen and Sæfaxi Elv Nappes, are present south of Romer Sø and the other, the Finderup Land Nappe, is confined to Finderup Land in the north of the area (Fig. 7).

Finderup Land Nappe (Fig. 7)

This nappe is the most areally restricted, occurring only in northern Kronprins Christian Land where it rims the western edge of the ice sheet. As a consequence it was not recognized by Fränkl (1954) who in fact probably only saw its southern margin and thus referred it to his Main Nappe.

Stratigraphy (Fig. 3)

The lowest sediments are laminated mudstones with

Vandredalen Nappe (Fig. 7)

This is the largest and most widespread nappe in Kronprins Christian Land (Fig. 7).

Stratigraphy (Fig. 3)

Within this nappe Fränkl (1954, 1955) erected a stratigraphy which from below comprised: Stenørkenen phyllites (+1000 m); Sydvejdal marbles with chloritic shales 100–400 m); Taagefjeldene greywackes (+700 m) with a layer of alum shales at the base (approximately 150 m); Rivieradal sandstones (1000 to 200 m) consisting of light coloured sandstones alternating with dark shales; Ulvebjerg sandstones and tillites (20–35 m); red Campanuladal limestones (150 m) and Fyns Sø dolomite (250 m).

All sediments prior to the Campanuladal limestones were thought to be of 'Greenlandian' age, that is, late Precambrian. The younger sediments were assumed to be either late Precambrian or Cambrian and until comparatively recently (Poulsen 1978) it was still suggested that the Campanuladal and Fyns Sø Formations were of early Cambrian age.

Based on this work of Fränkl, Haller (1970, 1971) assigned all the post-Proterozoic sandstones (Independence Fjord Group) and pre-early Cambrian sediments to the Hagen Fjord Group. It is quite clear that the purpose of erecting the Hagen Fjord Group was to assemble all these post-Carolinidian sediments into one lithostratigraphic unit, and this was based mainly on the sediments occurring in the Vandredalen Nappe. Unfortunately, it is now apparent that the Hagen Fjord Group is not well founded, since it not only contains sediments which are supposedly post-Carolinidian in age, but also basalts of the Zig-Zag Dal Basalt Formation (Jepsen and Kalsbeek 1979; Jepsen, Kalsbeek, and Suthren 1980). The latter are demonstrably related to the intrusions in the Proterozoic sandstones of the Independence Fjord Group and are therefore supposedly pre-Carolinidian in age.

Recently, Hurst and McKerrow (1981a,b) demonstrated that the Stenørken phyllites, Sydvejdal marbles, Taagefjeldene greywackes, Rivieradal sandstones, and Ulveberg sandstones and tillites are part of a single unit of deep water turbidites (partly calcareous), mud and resedimented conglomerates, here collectively referred to as Rivieradal sandstones. The resedimented conglomerates were thought by Fränkl (1954, 1955) to be tillites. The Rivieradal sandstones are about 2.5 km thick, with the lower part dominated by mudstones (now shales and phyllites), passing up into interbedded sandstones and mudstones, massive thick-bedded sandstones and resedimented conglomerates in two distinct horizons (Fig. 3). The conglomerates are particularly interesting in that they consist of 95–98 per cent rounded and spherical quartzite clasts and 2–5 per cent rounded and spherical dolerite clasts, indicating derivation from the Proterozoic sandstones of the Independence Fjord Group (Fig. 3).

Conformably overlying the deep water succession are approximately 600 m of quartz sands, silts, variegated shales, silts, and red limestones (in part Campanuladal Formation) and, finally, stromatolitic dolomites (Fyns Sø Formation), indicating a transition into shallow shelf environments.

Dating this sequence is difficult. Acritarchs are commonly thermally altered, but several comparatively well preserved specimens from the upper part of the Rivieradal sandstones can be identified with dominantly upper Proterozoic taxa. A single specimen of *Chuaria circularis* is in agreement with an upper Riphean age (G. Vidal, written communication 1980). The absolute age of upper Riphean acritarch assemblages in Scandinavia and North America which contain *C. circularis* is approximately 800 Ma (Vidal 1979; G. Vidal, written communication 1980). Thus, the overlying Campanuladal and Fyns Sø Formations are of late Riphean or Vendian age.

Structure (Fig. 5)

In northern Kronprins Christian Land, to the west of Romer Sø, the sole thrust of the nappe dips at about 20° to the east. Along the west side of Vandredalen the sole appears to be flat (Fig. 8), but around Centrum Sø it again dips gently to the east. Where the Vandredalen Nappe rests on the Sæfaxi Elv Nappe the sole is flat. At Marmorvigen (B on Fig. 1) the basal thrust is steeply dipping to subvertical. This high inclination may be related to post thrust folding as apparently was the case with the wedge of Rivieradal sandstones caught within the Independence Fjord Group (Fig. 2).

Complex imbricate zones occur at the base of the Vandredalen Nappe, at Marmorvigen (B on Fig. 1) and Nøglefjeldet (D on Fig. 1). At Nøglefjeldet a sequence of at least 10 imbricated slices has been produced by décollement along the variegated shales of the Campanuladal Formation (Fig. 9). Other thrusts and imbricate zones are present in the thick muddy turbidites sequence on the lower part of the Vandredalen Nappe. However, due to a lack of internal marker beds it has not been possible to trace these.

Sediments of the Vandredalen Nappe are strongly folded. Along the western side of Vandredalen folds, up to 500 m in amplitude, have gently to moderately dipping eastern limbs and steep to overturned western limbs. Fold axes are horizontal or plunge gently to the northeast. Areas of intense isoclinal folding, with folds up to 300 m in amplitude and covering

s   n

Fig. 8   Flat to gently lying Rivieradal sandstone (R) on the Vandredalen Nappe, lying on a thrust slice of 'Unnamed Silurian (?) dolomite formation' (Sd), which is thrust on top of the stratigraphic succession 'Unnamed Silurian (?) dolomite formation' (Sd) and 'Unnamed Silurian Limestone formation' (S1). t–t thrust planes. Location the west side of Vandredalen (Fig. 2). Cliff height approximately 800 m

many square kilometres are present in Rivieradal itself (Fig. 2).

All sediments are affected by penetrative deformation. Shales almost invariably show an easterly dipping fracture cleavage and in the folds, when bedding can be seen, it is evident that it is axial plane cleavage. Bedding is often lost in the carbonate sediments, but ubiquitous east dipping cleavage or jointing is considered to be of tectonic origin. The axial planar cleavage is steep to subvertical in the eastern part of the outcrop area of the Independence Fjord Group, becoming gradually flatter towards the west. This trend is similar to that in the Independence Fjord Group.

Sæfaxi Elv Nappe (Fig. 7)

This nappe is restricted to the area south of Ingolf Fjord. Fränkl (1954, 1955) recognized the unit but considered it to be partly in fault contact and partly thrust against younger and older units.

Stratigraphy (Fig. 3)

The lowest sediments on the Sæfaxi Elv Nappe are pale grey dolomites of uncertain age with vertical sandstone fissures between 50 and 100 cm thick. The sandstone fissures led Fränkl (1954, 1955) to infer the presence of the Kap Holbæk Formation unconformably overlain by dolomite. The dolomites were assigned to the 'Danmarks Fjord dolomite' of western Kronprins Christian Land which was considered to be of Cambrian–Ordovician age (Adams and Cowie 1953; Cowie 1961). This unit in western Kronprins Christian Land is now considered to be of Early Ordovician age and is not longer correlated with outcrops in the Sæfaxi Elv Nappe (Peel *et al.* 1981). The dolomites at Sæfaxi Elv are conformably overlain by Ordovician to ?Silurian thin bedded black limestone and shales which form a deeper marine equivalent to the western platform carbonates (Peel *et al.* 1981).

Structure (Fig. 5)

The sole thrust is poorly exposed but appears flat or

w   e

Fig. 9   Sediments on the Vandredalen Nappe, to the east of the basal thrust (t–t) overlying vertical to overturned Silurian carbonates and turbidites to the west. Note folding and imbricate thrusts on the nappe. Location Nøglefjedet (Fig. 1)

dipping gently eastwards, south of Ingolf Fjord. West of Hjørnegletscher (A on Fig. 1) the sole dips 0 to 50° to the west. Apparently, the thrust has been folded and the degree of inclination of the thrust plane can be directly related to the proximity of large westward directed folds in the Proterozoic sandstones. No imbricate zones have been recorded. Small-scale folding, on the scale of tens of metres, is present. Folds again have shallow dipping eastern limbs and steep western limbs, but no overturning was seen. Fold axes trend northeast.

The carbonates are penetratively deformed so that original sedimentary structures and bedding are often difficult to identify. Cleavage is widespread and apparently of an axial plane nature.

Relationship between the nappes

In southern Kronprins Christian Land the relationship between the nappes is clearly exposed. Near Marmorvigen (B on Fig. 1) and along the southwestern side of Sæfaxi Elv (Fig. 7) the Vandredalen Nappe rests on the Sæfaxi Elv Nappe. Both nappes are thrust over a Proterozoic parautochthonous sequence of Independence Fjord Group clastics and conglomerates and finally Fyns Sø Formation. The relationships between the nappes are less clear in the north. Both the Vandredalen and Finderup Land Nappes are seen to lie above parautochthonous Silurian platform carbonates and turbidites. The outcrop distribution suggests that the Vandredalen Nappe overlies the Finderup Land Nappe, but the region of overlap is largely obscured by ice.

Derivation of the nappes

Fränkl (1954, 1955) derived his Main Nappe, i.e. Vandredalen Nappe, from east of Prinsesse Caroline-Mathilde Alper (Fig. 2) from a 'Wurzelzone' (root zone) which he termed the Hekla Sund Basin. It is now clear that this 'Wurzelzone' corresponds precisely to the thrusted tight syncline of Rivieradal sandstones sandwiched between sandstones of the Independence Fjord Group which is exposed between Hekla Sund and Ingolf Fjord (Fig. 2). Later Haller (1970) considered that the sediments in the 'Wurzelzone' belonged to the 'Lower Thule Group' and were not involved in the thrusting. Haller (1970, p. 64) thus concluded that the sedimentary pile contained in the Vandredalen Nappe was presumably originally deposited east of the present-day coastline of Kronprins Christian Land.

Hurst and McKerrow (1981a,b) consider that all three nappes were derived from east of the Prinsesse Caroline-Mathilde Alper and probably east of the present coastline. Evidence for this can be summarized as follows:

1. In the Prinsesse Caroline-Mathilde Alper the Fyns Sø Formation lies with normal stratigraphic contact upon the Independence Fjord Group sandstones. On the Vandredalen Nappe the Fyns Sø Formation lies with normal stratigraphic contact upon the Campanuladal Formation and Rivieradal sandstones. Likewise on the Finderup Land Nappe the Fyns Sø Formation lies with normal stratigraphic contact on the Campanuladal Formation. Thus, it is evident that the nappes could not have been derived from the 'Wurzelzone' of Fränkl (1954, 1955) i.e. the Prinsesse Caroline-Mathilde Alper. In consequence they must have been derived from at least east of this area.

2. The sediments of the Vandredalen Nappe represent a shallowing sequence from deep basin turbidites and resedimented conglomerates of the Rivieradal sandstones, up into shelf clastics (Campanuladal Formation) and finally shallow marine carbonates of the Fyns Sø Formation. It is quite clear that the Independence Fjord Group sandstones in the alps (Fig. 2) acted as a swell region at least until the deposition of the Fyns Sø Formation. Therefore, deposition of the sediment pile of the Vandredalen Nappe must have taken place in a deep trough to the east of the present outcrop of the Independence Fjord Group.

These points are compelling evidence for derivation of the Finderup Land Nappe and Vandredalen Nappe from the east. Evidence for derivation of the Sæfaxi Elv Nappe is not as clear-cut. The Ordovician–Silurian carbonate unit in this nappe was deposited in relatively deep water, probably slope environments, but is equivalent to the shallow water platform carbonate deposits 5 km to the west (Peel 1980b; Peel et al. 1981). It is unlikely that the change from platform to slope could have occurred in such a short distance without a distinctive sequence of platform edge facies (cf. Hurst 1980a,b; Surlyk, Hurst, and Bjerreskov 1980), of which there is no evidence. The Sæfaxi Elv Nappe must therefore have been derived from the east, but this need not necessarily have involved as great a distance of transport as the other two nappes.

Zone 4: Platform Carbonate Foreland

This zone covers the largest area of Kronprins Christian Land. It varies from between 50 and 100 km in width, and accounts for the whole western part of the land region (Fig. 2).

Stratigraphy (Fig. 3)

Lower Palaeozoic platform stratigraphy in the foreland region is discussed more fully by Peel (this volume).

Adams and Cowie (1953) were the first to erect a stratigraphy in the foreland sequence around the head of Danmark Fjord. They recognized a basal Norsemandal sandstone of at least 300 m thickness, which has subsequently been redefined as the Norsemandal Formation of the Independence Fjord Group (Collinson 1980). Overlying this they recognized the Campanuladal sandstones and limestones (250 m), Fyns Sø dolomite (324 m), and Kap Holbæk sandstone (135 m), which are now known as the Campanuladal Formation, Fyns Sø Formation, and Kap Holbæk Formation and which also occur on the nappes to the east (Fig. 3).

The uppermost unit recognized by Adams and Cowie (1953) was an approximately 2500 m thick sequence of platform carbonates to which they gave the name 'Centrum Limestone'. Later workers (Fränkl 1955; Cowie 1971; Scrutton 1975) have subdivided the 'Centrum Limestone' into several formations, namely, from base to top, Danmark Fjord Formation, Amdrup Formation, *Opikina* Limestone, Centrum Formation, and Drømmebjerg Formation. Clastics overlying the carbonates have been termed Profilfjeldet shales or, later, the Profilfjeldet Formation.

Ambiguities surrounding this stratigraphy were noted by Peel (1980) who recognized that the carbonate stratigraphy of Kronprins Christian Land could be described in terms of the better known Peary Land succession (Troelsen 1949; Christie and Peel 1977). The 'Centrum Limestone' based stratigraphy was not only ill defined, but proved to be severely complicated due to repetition by thrusting not recognized by Adams and Cowie (1953). As a consequence, use of the terms 'Centrum Limestone' and 'Centrum Formation' was abandoned by Peel *et al*. (1981).

The Wandel Valley Formation of early to middle Ordovician age directly overlies the Kap Holbæk Formation (Peel 1980b; Peel *et al*. 1981). The size and extent of this unconformity was first recognized in Peary Land (Peel 1979). Peel *et al*. (1981) have reduced the Danmarks Fjord Dolomite of Fränkl (1955) and the Amdrup Formation of Cowie (1971) to members of the Wandel Valley Formation (Fig. 3). The formation consists of up to 400 m of monotonous stromatolitic dolomites and lime mudstones.

A new, as yet unnamed, dolomite formation (c.

100 m) has been recognized between the Wandel Valley Formation and the overlying middle to upper Ordovician Børglum River Formation (Fig. 3). The latter formation is at least 400 m thick and consists essentially of subtidal lime mudstones (Peel *et al*. 1981).

The unnamed (?)Silurian dolomite formation and unnamed Silurian limestone formation of Christie and Peel (1977) are both present in Kronprins Christian Land (Peel 1980b; Peel *et al*. 1981). The former is a shallow to marginal marine stromatolitic unit in Peary Land (Fürsich and Hurst 1980) but, in Kronprins Christian Land, it is more marine and attains a thickness in excess of 200 m. The latter is a shallow subtidal lime mudstone unit, some 400 m thick. Carbonate buildups, the Drømmebjerg Formation, occur within the Silurian limestone and are typical of Silurian buildups across North Greenland (Hurst 1980a).

The uppermost Lower Palaeozoic strata are the turbidites of the Profilfjeldet shales (Llandovery to middle Wenlock age). They are identical to the unnamed Silurian turbidites of Peary Land (Christie and Peel 1977) and represent a deep water trough facies overlying the shallow platform carbonates (Surlyk, Hurst, and Bjerreskov 1980).

Structure (Fig. 6)

Haller (1971, fig. 88) recognized four principal structural zones in Kronprins Christian Land of which the platform carbonates and clastics and the overlying turbidites were incorporated in a western foreland zone. The eastern boundary was drawn at a prominent northeast to southwest Quaternary filled valley, but in fact it occurs some 5 km to the east (Peel 1980b; Hurst and McKerrow 1981a,b; Fig. 10).

At the present-day nappe front the Lower Palaeozoic succession is folded into a tight overturned syncline with a north to northeast axis (Figs. 5 and 9) north of Romer Sø (E on Fig. 2) and at Nøglefjeldet (D on Fig. 2). This syncline may be a regional feature between Nøglefjeldet and Romer Sø.

Immediately west of the syncline the foreland deposits are disrupted by a complex sequence of high angle reverse faults and low angle thrusts. The transport distances of the blocks involved are not great and all sediments are essentially autochthonous (Fig. 10). The thrusts trend north to northeast, but up in the northeastern part of Kronprins Christian Land they turn to almost due east. Most thrusts and reverse faults are of limited extent, up to 10 km long, suddenly appearing and disappearing in the homogenous carbonates.

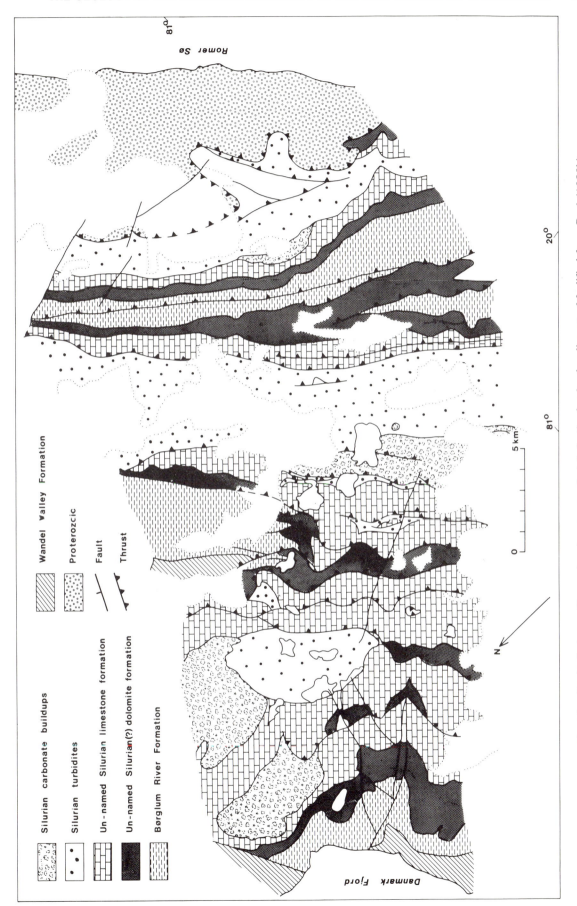

Fig. 10   Geological map illustrating the structural style in the foreland sediments. Modified from Peel (1980b)

At least three thrusts are of regional extent (Fig. 2). The most easterly one adjacent to the nappe front has the largest displacement, bringing up the Ordovician Wandel Valley Formation, and occupies the large valley, mentioned above, in which Haller (1971) placed his nappe front. Ordovician carbonates are also thrust over the Silurian along the westernmost of these three thrusts (Fig. 11). In this case, overthrusting is most substantial in the northeast, but is reduced to some few tens of metres to the southwest. Displacements on the thrusts and reverse faults generally decrease westward. Reverse faults extend to Kap Holbæk on the western bank of Danmark Fjord (Fig. 2), which can be taken as the western limit of deformation attributable to Caledonian movements.

Both the carbonates and clastics are folded, especially in the northeast (Fig. 12), but folds with an amplitude of more than 100 or 200 m are rare. Fold styles conform with other areas. In general, fold axes are subhorizontal or plunge up to 45° to the north or northeast. Folds have gentle to moderately dipping eastern limbs and steep or overturned western limbs (Fig. 12). In the carbonates dislocation commonly occurs along the overturned limb of tight folds. This is not necessarily the case in the clastics.

In northeastern Kronprins Christian Land the foreland sediments immediately in front of the Finderup Land Nappe are buckled into a series of tight folds. The carbonates form a series of 'whalebacks' pushed through the turbidite cover of the Profilfjeldet shales. In Finderup Land the fold axes in the foreland are orientated northeast to east.

A series of small normal faults trends approximately perpendicular to strike. Many of these cut thrust planes and high angle reverse faults as well as the thrusts of the main nappes to the east. Clearly, the NW–SE normal faults post-date the main thrusting and deformation, but it is not yet known how young they are. Of further note is a SW–NE orientated wrench fault cutting the northwestern corner of Kronprins Christian Land (Fig. 2). Its trend is slightly oblique to that of the thrusts, but its relationship to the structural chronology of the area is unknown.

The carbonates are jointed but no cleavage has been seen. In contrast, the turbidites of the Profilfjeldet shales are jointed and cleaved, the cleavage being axial planar when seen with folds.

Comparison with East Greenland

The basement to the Caledonides of Kronprins Christian Land is composed of a variable suite of gneisses and schists of possible early Proterozoic age. These are similar to the Archaean basement complexes of East Greenland (Henriksen and Higgins 1969, 1976). There is a thick Proterozoic marine clastic sequence in both areas, although in East Greenland the Eleonore Bay Group is apparently a fluvial to shallow shelf sequence (Caby 1976), whilst the Rivieradal sandstones of Kronprins Christian Land are of deep water origin. Also, the Eleonore Bay Group is autochthonous.

The foreland sequence of Kronprins Christian Land is far more extensively developed than in central East Greenland. Furthermore, a complete Lower Silurian sequence is represented in the north, whilst in East Greenland no sediments involved in Caledonian folding are younger than middle Ordovician (cf. Frykman 1979; Smith 1982).

The Caledonides of Kronprins Christian Land record a full transition across the western margin of the orogen. In contrast, in East Greenland central parts of the fold belt are better represented. Together the two areas provide a fairly complete traverse of the orogenic zone.

The oldest post-Caledonian sediments in Kronprins Christian Land are Lower Carboniferous (Håkansson et al. 1981), whilst in East Greenland thick Devonian molasse occurs. No plutons are present in Kronprins Christian Land, unlike East Greenland.

A distinct marginal thrust system is present in both areas. Indeed, Higgins and Phillips (1979) infer that the marginal thrust zone in Kronprins Christian Land is the same as that in East Greenland. They also suggest that the East Greenland marginal thrust system is the logical continuation of the Moine thrust belt of Scotland which implies that the Kronprins Christian Land marginal thrust system is also a correlative of the Moine thrust. This is a tempting correlation, but the amount of water and ice between these two extremes calls for caution. In East Greenland speculation as to the original position of the rock units within the thrust sheets has led to suggestions of 100 km displacements (Homewood 1973; Higgins 1974). These figures are in accord with displacements envisaged in Kronprins Christian Land (Hurst and McKerrow 1981a,b).

Conclusions

It is possible to recognize four tectoniostratigraphic zones in the Caledonides of Kronprins Christian Land, western North Greenland. These are, from southeast to northwest: (1) a crystalline basement complex of possible early Proterozoic age in the eastern coastal regions; (2) a belt of Proterozoic sandstones of the Independence Fjord Group; (3) a zone of nappes; and (4) platform carbonates and clastics of the imbricate zone and foreland.

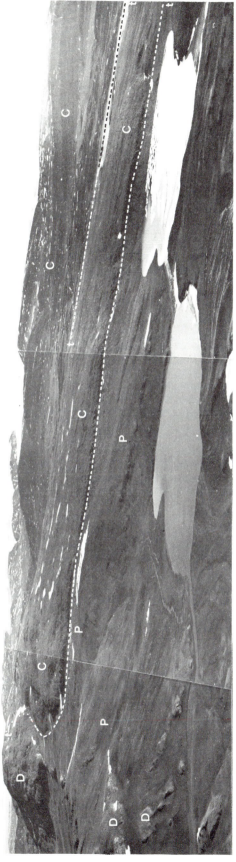

Fig. 11 Ordovician and Silurian carbonates (C) thrust (t–t) over Profilfjeldet shales (P) and carbonate mounds of the Drømmebjerg Formation (D). Foreland zone, looking northeast

Fig. 12   Rivieradal sandstones (R) on the Vandredalen nappe overlying the 'Unnamed Silurian (?) dolomite formation' (Sd) of the platform carbonate foreland sequence. Note the steep or overturned western limbs of the westerly directed folds, which in some cases are disrupted. The carbonates are in a tight overturned syncline immediately below the nappe. Cliff height approximately 800 m. t–t, thrust plane

Throughout all four zones the sequence of structural events and the style of deformation appears the same. Thrusting has displaced units westwards and northwestwards. Folds are all overturned to the west or northwest and have gently to moderately dipping eastern limbs. The intensity of folding, thrusting and penetrative deformation progressively decreases to the west. Essentially, areas to the west of Danmark Fjord have not been affected by the folding and thrusting. The lack of any discordant earlier structures in zones 1 and 2 is evidence that an earlier Proterozoic Carolinidian orogeny did not exist.

It seems probable that the Vandredalen Nappe, Finderup Land Nappe, and Sæfaxi Elv Nappe were derived from the western margins of the Iapetus Ocean (Harland and Gayer 1972). If so, they would have originated east of the gneiss outcrops in eastern Kronprins Christian Land, requiring a total displacement distance of at least 100 to 150 km.

A basic three-fold structural chronology is recognizable. Emplacement of the Vandredalen Nappe (and probably the other two) preceded the main folding. This is indicated by the folded syncline of Rivieradal sandstone within the fold Independence Fjord Group in zone 2. Thrusting also occurred in association with the folding in the Independence Fjord Group. Therefore, there are at least two periods of thrusting, one long distance and the other shorter. Whether the thrusting and reverse faulting in the foreland are contemporaneous with one or both of these events or indeed represent another event is not known. It is possible that the deformation in the foreland was related to nappe emplacement (Hurst and McKerrow 1981a). Finally, a series of NW–SE faults cuts both the folds and thrust planes. The age of these is unknown so they may not be related to the Caledonian events.

A precise minimum age of the thrusting in Kronprins Christian Land cannot be established as no late Silurian or Devonian rocks are known. The profilfjeldet shale extends into the Wenlock and it is possible that diastrophism started in late Wenlock or Ludlow time, if not earlier to the east.

Acknowledgements

J. R. Inesone, P. D. Lane and H. A. Armstrong participated in field work during 1980. A. K. Higgins provided useful comments. Published with the permission of the Director or the Geological Survey of Greenland.

References

Adams, P. J. and Cowie, J. W. 1953. A geological reconnaissance of the region around the inner part of Danmarks Fjord, Northeast Greenland. *Meddr Grønland*, **11** (7), 1–24.

Caby, R. 1976. Reconnaissance on the Lower Eleonore Bay Group in the Alpefjord region, central East Greenland. *Rapp. Grønlands geol. Unders.*, **80**, 102–106.

Christie, R. L. and Peel, J. S. 1977. Cambrian–Silurian stratigraphy of Børglum Elv, Peary Land, eastern North Greenland. *Rapp. Grønlands geol. Unders.*, **82**, 1–48.

Clemmensen, L. B. 1979. Notes on the palaeogeographical setting of the Eocambrian tillite-bearing sequence of southern Peary Land, North Greenland, *Rapp. Grønlands geol. Unders.*, **88**, 15–22.

Collison, J. D. 1979. The Proterozoic sandstones between Heilprin Land and Mylius-Erichsen Land, eastern North Greenland. *Rapp. Grønlands geol. Unders.*, **88**, 5–10.

Collison, J. D. 1980. Stratigraphy of the Independence Fjord Group (Proterozoic) of eastern North Greenland. *Rapp. Grønlands geol. Unders.*, **99**, 7–23.

Cowie, J. W. 1961. The Lower Palaeozoic geology of Greenland. In Raasch, G. O. Ed), *Geology of the Arctic*, Toronto U.P., pp. 160–169.

Cowie, J. W. 1971. The Cambrian of the North American Arctic regions. In Holland, C. H. (Ed.), *Cambrian of the New World*, Interscience, London, pp. 325–383.

Frykman, P. 1979. Cambro-Ordovician rocks of C. H. Ostenfeld Nunatak, northern East Greenland. *Rapp. Grønlands geol. Unders.*, **91**, 125–132.

Fränkl, E. 1954. Vorläufige Mitteilung über die Geologie von Kronprins Christian Land (NE-Grönland). *Meddr Grønland*, **116** (2), 1–85.

Fränkl, E. 1955. Weitere Beiträge zur Geologie von Kronprins Christian Land (NE-Grönland). *Meddr Grønland*, **103** (7), 1–35.

Fränkl, E. 1956. Some general remarks on the Caledonian chain of East Greenland. *Meddr Grønland*, **103** (11), 1–43.

Fürsich, F. P. and Hurst, J. M. 1980. Euryhalinity of Palaeozoic articulate brachiopods. *Lethaia*, **13**, 303–312.

Haller, J. 1961. The Carolinides: an orogenic belt of late Precambrian age in Northeast Greenland. In Raasch, G. O. (Ed.), *Geology of the Arctic*, Toronto U.P., pp. 155–159.

Haller, J. 1970. Tectonic map of East Greenland (1:500,000). An account of tectonism, plutonism, and volcanism in East Greenland. *Meddr Grønland*, **171** (5), 1–286.

Haller, J. 1971. Geology of the East Greenland Caledonides, Interscience. New York.

Harland, W. B. and Gayer, R. A. 1972. The Arctic Caledonides and earlier oceans. *Geol. Mag.*, **109**, 289–314.

Henriksen, N. 1978. East Greenland Caledonian fold belt. *Pap. Geol. Surv. Can.*, **78–13**, 105–109.

Henriksen, N. and Higgins, A. K. 1969. Preliminary results of mapping in the crystalline complex around Nordvestfjord, Scoresby Sund, East Greenland. *Rapp. Grønlands geol. Unders.*, **21**, 5–20.

Henriksen, N. and Higgins, A. K. 1976. East Greenland Caledonian fold belt. In Escher, A. and Warr, W. S. (Eds), *Geology of Greenland*, Geo. Surv. Greenland, Copenhagen, pp. 182–246.

Higgins, A. K. 1974. The Krummedal supracrustal sequence around inner Nordvestfjord, Scoresby Sund, East Greenland. *Rapp. Grønlands geol. Unders.*, **67**, 1–34.

Higgins, A. K. and Phillips, W. E. A. 1979. East Greenland Caledonides—an extension of the British Caledonides. In Harris, S. L., Holland C. H. and Leake, B. E. (Eds), *The Caledonides of the British Isles—reviewed*, Scottish Academic Press, Edinburgh, pp. 19–32.

Homewood, P. 1973. Structural and lithological divisions of the western border of the East Greenland Caledonides in the Scoresby Sund region between 71°00′ and 71°22′N. *Rapp. Grønlands geol. Unders.*, **57**, 1–27.

Hurst, J. M. 1980a. The palaeogeographic and stratigraphic differentiation of the Silurian carbonate buildups and biostromes of North Greenland. *Bull. Am. Ass. Petrol. Geol.*, **64**, 527–548.

Hurst, J. M. 1980b. Silurian stratigraphy and facies distribution in Washington Land and western Hall Land, North Greenland. *Bull. Grønlands geol. Unders.*, **138**, 1–95.

Hurat, J. M. and McKerrow, W. S. 1981a. The Caledonian nappes of eastern North Greenland. *Nature*, London, **290**, 772–774.

Hurst, J. M. and McKerrow, W. S. 1981b. The Caledonian nappes of Kronprins Christian Land, eastern North Greenland. *Rapp. Grønlands geol. Unders.*, **106**, 15–19.

Håkansson, E., Heinberg, C. and Stemmerik, L. 1981. The Wandel Sea Basin from Holm Land to Lockwood Ø, eastern North Greenland. *Rapp. Grønlands geol. Unders.*, **106**, 47–63.

Jepsen, H. F. 1971. The Precambrian, Eocambrian and early Palaeozoic stratigraphy of the Jørgen Brønlund Fjord area, Peary Land, North Greenland. *Meddr Grønland*, **192** (2), 1–42.

Jepsen, H. F. and Kalsbeek, F. 1979. Igneous rocks in the Proterozoic platform of eastern North Greenland. *Rapp. Grønlands geol. Unders.*, **88**, 11–14.

Jepsen, H. F. and Kalsbeek, F. 1981. Non-existence of the Carolinidian orogeny in the Prinsesse Caroline Mathilde Alper of Kronprins Christian Land, eastern North Greenland. *Rapp. Grønlands geol. Unders.*, **106**, 7–14.

Jepsen, H. F. and Kalsbeck, F. this volume. Evidence for non-existence of a Carolinidian fold belt in eastern North Greenland.

Jepsen, H. F., Kalsbeek, F., and Suthren, R. J. 1980. The Zig-Zag Dal Basalt Formation, North Greenland. *Rapp. Grønlands geol. Unders.*, **99**, 25–32.

Larsen, O. and Graff-Petersen, P. 1980. Sr-isotopic studies and mineral composition of the Hagen Bræ Member in the Proterozoic clastic sediments at Hagen Bræ, eastern North Greenland. *Rapp. Grønlands geol. Unders.*, **99**, 111–118.

O'Connor, B. 1979. The Portfjeld Formation (?early Cambrian) of eastern North Greenland. *Rapp. Grønlands geol. Unders.*, **88**, 23–28.

Peel, J. S. 1979. Cambrian–Middle Ordovician stratigraphy of the Adams Gletscher region, south-west Peary Land, North Greenland. *Rapp. Grønlands geol. Unders.*, **88**, 29–39.

Peel, J. S. 1980a. Early Cambrian microfossils from the Portfjeld Formation, Peary Land, eastern North Greenland. *Rapp. Grønlands geol. Unders.*, **100**, 15–17.

Peel, J. S. 1980b. Geological reconnaissance in the Caledonian foreland of eastern North Greenland with comments on the Centrum Limestone. *Rapp. Grønlands geol. Unders.*, **99**, 61–62.

Peel, J. S. this volume. Cambrian–Silurian platform stratigraphy of Northeastern Greenland.

Peel, J. S., Ineson, J. R., Lane, P. D. and Armstrong, H. A. 1981. Lower Palaeozoic stratigraphy around Danmark Fjord, eastern North Greenland. *Rapp. Grønlands geol. Unders.*, **106**, 21–27.

Poulsen, V. 1978. The Precambrian–Cambrian boundary in parts of Scandinavia and Greenland. *Geol. Mag.*, **115**, 131–136.

Scrutton, C. T. 1975. Corals and stromatoporoids from the Ordovician and Silurian of Kronprins Christian Land, Northeast Greenland. *Meddr Grønland*, **171** (4), 1–43.

Smith, P. 1982. Ordovician conodont biostratigraphy of central East Greenland. *Paleont. Contr. Univ. Oslo*, **280**, 48.

Surlyk, F., Hurst, J. M. and Bjerreskov, M. 1980. First age-diagnostic fossils from the central part of the North Greenland foldbelt', *Nature, London*, **286**, 800–803.

Troelsen, J. C. 1949. Contributions to the geology of the area round Jørgen Brønlunds Fjord, Peary Land, North Greenland. *Meddr Grønland*, **149** (2), 1–29.

Vidal, G. 1979. Acritarchs from the Upper Proterozoic and Lower Cambrian of East Greenland. *Bull. Grønlands geol. Unders.*, **134**, 1–40.

The Caledonide Orogen—Scandinavia and Related Areas
Edited by D. G. Gee and B. A. Sturt
© 1985 John Wiley & Sons Ltd

Origin of the Caledonian nappes of eastern North Greenland

J. M. Hurst* and **W. S. McKerrow**†

**Grønlands Geologiske Undersøgelse, Øster Voldgade 10, DK–1350 Copenhagen K, Denmark*
†*Department of Geology and Mineralogy, Parks Road, Oxford OX1 3PR, UK*

ABSTRACT

The coastal area of Kronprins Christian Land, eastern North Greenland, is composed of gneisses, sandstones, and quartzites that are at least 1380 Ma old. These are overlain unconformably by a thin sequence of late Proterozoic sediments, including a discontinuous conglomerate with quartzite clasts below carbonates of the late Proterozoic Fyns Sø Formation. To the west of these parautochthonous Precambrian rocks, in a belt 20 and 30 km wide, we recognize three large nappes made up of latest Proterozoic and Lower Palaeozoic shelf, slope, and basin sediments. The youngest allochthonous sediments are present on the lowest nappe, and the oldest sediments are on the highest nappe. To the west of the nappes, shelf rocks include the Fyns Sø Formation and older Precambrian sediments and a Lower Palaeozoic sequence ranging up to Middle Silurian (Wenlock) age. Because the Fyns Sø Formation is present on the shelf to the west, on two of the three large nappes, and on the parautochthon to the east, the nappes are interpreted as having travelled from east of the present outcrops of the Fyns Sø Formation; this would involve transport over distances of more than 100 km. Perhaps the source area of the nappes was in the wide continental shelf still preserved to the east of Kronprins Christian Land. It cannot be proved whether the nappes resulted from gravity sliding or gravity spreading.

Introduction

In contrast to Norway, East Greenland appears to have been a passive margin throughout the Lower Palaeozoic. There are no structures related to any previous subduction history (Henriksen and Higgins 1976). The Caledonian Orogeny in this area is thus confined to events associated with continental collision during the closure of the Iapetus Ocean.

The geology of the northern extremity of the East Greenland Caledonian fold belt is described by Hurst *et al.* (this volume). In this contribution we compare the stratigraphy of each of the three large nappes in Kronprins Christian Land (Hurst and McKerrow 1981) with the parautochthonous area of the Prinsesse Caroline-Mathilde Alper to the east of the thrust belt and with the North Greenland autochthonous Palaeozoic and Proterozoic shelf sediments to the west (Fig. 1). We also briefly discuss the possible origin of the nappes.

Stratigraphy

Autochthonous shelf sediments to the west of the thrust belt (Fig. 1) include latest Proterozoic carbonates and clastics of the Campanuladal and Fyns Sø Formations (Clemmensen 1979) which rest disconformably on earlier Proterozoic basalts which overlie sandstones and quartzites of the Independence Fjord Group (Fig. 2); the latter are possibly around 1380 Ma old (Larsen and Graff-Petersen 1980). The distinctive yellow stromatolitic dolomites of the Fyns Sø Formation are followed by Late Proterozoic *Skolithos* sandstones of the Kap Holbæk Formation, which in turn are followed by Ordovician and Silurian shallow marine shelf carbonates. In latest Llandovery times, a sudden deepening occurred (Surlyk, Hurst, and Bjerreskov 1980; Hurst and Surlyk 1982; Surlyk and Hurst 1983) prior to deposition of turbidites and shales which range up into the Wenlock and include the youngest beds

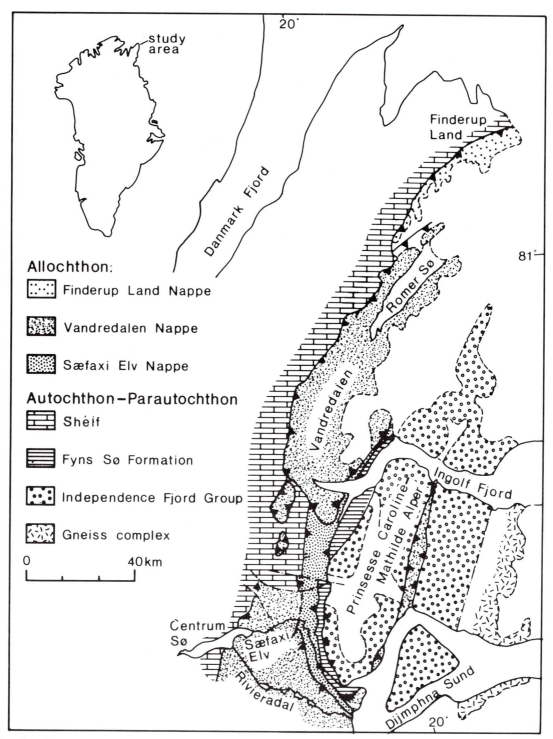

Fig. 1   Simplified geological map of the Caledonian nappes and sediments affected by their emplacement. The shelf is a convenient term covering Lower Palaeozoic carbonates, and the Independence Fjord Group corresponds to the area of the 'Proterozoic Swell' of Hurst and McKerrow (1981)

affected by Caledonian thrusting in eastern North Greenland.

In the thrust belt, the most widespread of the three large nappes is the highest one: the Vandredalen Nappe (Fig. 1). The Finderup Land Nappe is only present in the north of Kronprins Christian Land, while the Sæfaxi Elv Nappe is only exposed in the south. The key to the stratigraphy and structure of the region (Fig. 2) is the presence of the yellow stromatolitic carbonates of the Fyns Sø Formation on the autochthonous shelf to the west in the Vandredalen and Finderup Land Nappes, and on the

Shelf	Vandredalen Nappe	Finderup Land Nappe	Sæfaxi Elv Nappe	Parautochthon to east
Silurian turbidites 400m				
Silurian carbonates 300m			Silurian & Ordovician carbonates 550m	
Ordovician carbonates 600m				
Late Proterozoic clastics 200m		Late Proterozoic clastics 200m		
Fyns Sø Formation 300m	Fyns Sø Formation 300m	Fyns Sø Formation 300m		Fyns Sø Formation 400m
Campanuladal Formation 850m	Campanuladal Formation 600m	Campanuladal Formation 300m		Local clastics with conglomerate 200m
Basalts 1300m	Rivieradal sandstones 2500m			Local basalts
Independence Fjord Group 1800m				Independence Fjord Group
				Gneisses

Fig. 2   Simplified stratigraphies of the shelf, nappes and eastern parautochthon. The shelf is also autochthonous. Note the presence of the distinctive Fyns Sø Formation which is a key for correlation. Thicknesses for the nappe sediments are estimates on the maximum side

parautochthon to the east. The widespread occurrence of the Fyns Sø Formation is not merely useful in correlating the allochthonous and autochthonous successions, but it indicates that the nappes must have originated outside the region where the Fyns Sø Formation is now seen.

The Ordovician and Silurian dark limestones and shales of the Sæfaxi Elv Nappe are of deeper water facies than the equivalent beds of the shelf; similarly the Fyns Sø and Kap Holbaek Formations of the Finderup Land Nappe contain thicker shale units, which suggest intermittent deeper water facies than the stromatolitic dolomites and *Skolithos* dominated facies of the western shelf. However, the greatest differences with the shelf are seen in the large Vandredalen Nappe, where the Fyns Sø and Campanuladal Formations are underlain by the thick (2 to 3 km) turbidite sequence, the Rivieradal sandstones, which have no known equivalents on the shelf to the west (Peel 1980). The Rivieradal sandstones contain a late Riphean acritarch assemblage (Vidal personal communication 1981), and their stratigraphic position coincides with the long period of time represented by the stratigraphic break below the Campanuladal Formation on the platform (Fig. 2).

Several thick conglomerate beds are present within the upper parts of the Rivieradal sandstones. These are resedimented quartzite cobble to boulder conglomerates and may represent the reworking of shallow water deposits, as the clasts (98 per cent quartzite, 2 per cent dolerite) are well rounded. One of these conglomerates was previously thought to be a tillite (Fränkl 1954, 1955), but this possibility is now discounted.

In the parautochthonous region to the east of the thrust belt, the Proterozoic sandstones and quartzites of the Independence Fjord Group are intruded by 1,250 Ma dolerite dykes (Jepsen and Kalsbeek 1979) and are overlain in places by related basalts (Jepsen and Kalsbeek 1981). In the Prinsesse Caroline-Mathilde Alper, immediately east of the thrust belt, the Independence Fjord Group sandstones and quartzites are locally overlain by a discontinuous conglomerate which contains clasts of the underlying quartzites and dolerites. These conglomerates occur in depressions (valley-fill deposits) up to 50 m thick; the rounding of the clasts is very variable: some depressions are filled with large (1 m) angular blocks of quartzite, while others contain smaller, rounded clasts.

At Ingolf Fjord (Fig. 1), the beds above these

conglomerates are shallow marine sandstones and shales, followed by the stromatolitic Fyns Sø Formation. Further south, on the west coast of Dijmphna Sund, limestones of the Fyns Sø Formation rest either on the conglomerate or directly on the sandstones and quartzites of the Independence Fjord Group. The similarity in stratigraphic position (Fig. 2) of the deep water conglomerates in the upper parts of the Rivieradal sandstones and the shallow water conglomerates on the parautochthon to the east (both occur below the Fyns Sø Formation), and the similarity of the clasts (dominantly quartzite with subsidiary dolerite) suggest that the resedimented conglomerates of the Rivieradal sandstones may have been derived from the shallow water environments present on the swell formed by the Independence Fjord Group of the Prinsesse Caroline-Mathilde Alper. This swell probably lay to the west of the region where the Rivieradal sandstones were originally deposited, though it now lies east of the present position of the Vandredalen Nappe where they now lie.

The Origin of the Nappes

The nappes in Kronprins Christian Land were first described by Fränkl (1954, 1955), who considered that his 'Main Nappe' (equivalent in part to our Vandredalen Nappe) was derived from a 'Wurzelzone' immediately east of the Prinsesse Caroline-Mathilde Alper (Fig. 1). Haller (1970, 1971) thought that the sediments in this 'Wurzelzone' were not involved in the thrusting, so he (Haller 1970, p. 64) concluded that the allochthonous sediments were originally deposited east of the modern coastline of Kronprins Christian Land. We now recognize (Hurst et al. this volume) that Fränkl's 'Wurzelzone' is a klippe of Rivieradal sandstones downfaulted into the quartzites of the Independence Fjord Group, but (though for different reasons) we agree with Haller that the original site of deposition of the allochthonous sediments was to the east.

The presence of the Fyns Sø Formation in the Vandredalen and Finderup Land Nappes, on the shelf to the west and on the parautochthon to the east of the thrust belt, shows that the nappes must have been derived from outside the region. As the shelf extends for many hundreds of kilometres westwards, we conclude that the nappes came from the east of the Prinsesse Caroline-Mathilde Alper. This would entail displacements of over 100 km. An easterly derivation is also supported by the orientation of the folds, minor thrusts, and imbricate structures near the soles of the large nappes. In particular, the Wenlock turbidites, which underlie the

thrusts on the platform to the west, are repeatedly folded into recumbent synclines which close to the east and have the upper limbs inverted below the basal thrust.

In places, the Wenlock turbidites contain conglomerates with rounded quartzite clasts, very similar in their petrology to the conglomerates in the upper Rivieradal sandstones and which are likewise also derived from the Independence Fjord Group sandstones and quartzites. These conglomerates and the sandstone turbidite sequence suggest uplift of an area to the east of the Caledonian thrust belt during the Wenlock. This uplift was very probably a direct precursor to the emplacement of the nappes, as the Wenlock beds are the youngest to be affected by the Caledonian movements in eastern North Greenland.

Conclusions

Many of the allochthonous sequences contain deep shelf sediments which were probably deposited originally on the site of the present-day continental shelf to the east of Kronprins Christian Land. The Rivieradal sandstones and the resedimented conglomerates in their upper part were probably deposited on a continental slope and deep water basin on the margin of the Iapetus Ocean. If this conclusion is correct, then the clasts in these sediments were transported eastwards from a swell on the Prinsesse Caroline-Mathilde Alper in the late Riphean long before the westward movement (in the late Silurian) of the nappes to their present position.

The Sæfaxi Elv Nappe has the youngest sediments (Fig. 2), and yet it is the lowest nappe, while the highest (Vandredalen) nappe contains the oldest allochthonous sediments: the late Riphean Rivieradal sandstones. The nappes thus appear to be in reverse stratigraphic sequence.

As the nappes include incompetent sediments and are also very extensive it is possible that they were formed by gravity gliding. The only indications of compression in the thrust belt are folds, striking north or northeast, which affect both the nappes and the parautochthonous region to the east and are thus clearly later than the napper emplacement.

Bally (1981) has suggested that gravitational gliding can only occur by sliding off an uplifted basin and notes that no such tectonically denuded region is known to him, but such an uplifted region would surely continue to suffer erosion after the nappes had slipped off. The continental shelf east of Kronprins Christian Land is up to 250 km wide. If it were underlain by an eastward continuation of the Precambrian rocks exposed along the coast, it could have provided an ample source for the three large nappes.

Alternatively, some of the nappes may result from gravity spreading. For instance, the stratigraphic differences between the Finderup Land and Sæfaxi Elv Nappes may indicate that the thrusts cut up section along the strike. If this interpretation is correct these two nappes do not have to have moved that far from the east.

At present it is not proven whether the nappes are the result of gravity sliding or spreading. The inherent differences between the three nappes could suggest both processes were operative.

Acknowledgements

We thank D. G. Gee, A. K. Higgins, and an anonymous referee for critically reading the manuscript and the Director of the Geological Survey of Greenland for permission to publish.

References

Bally, A. W. 1981. Thoughts on the tectonics of folded belts. In Price, N. J. and McClay, K. R. (Eds), *Thrust and Nappe Tectonics, Geol. Soc. Lond. Special Pub.*, 9, 13–32.

Clemmensen, L. B. 1979. Notes on the palaeogeographical setting of the Eocambrian tillite-bearing sequence of southern Peary Land, North Greenland. *Rapp. Grønlands geol. Unders.*, 88, 15–22.

Fränkl, E. 1954. Vorläufige Mitteilung über die Geologie von Kronprins Christian Land (NE-Grönland). *Meddr Grønland*, 116(2), 1–85.

Fränkl, E. 1955. Weitere Beiträge zur Geologie von Kronprins Christian Land (NE-Grönland). *Meddr Grønland*, 103(7), 1–35.

Haller, J. 1970. Tectonic map of East Greenland (1:500,000). An account of tectonism, plutonism, and volcanism in East Greenland. *Meddr Grønland*, 171(5), 1–286.

Haller, J. 1971. Geology of the East Greenland Caledonides. Interscience, New York.

Henriksen, N. and Higgins, A. K. 1976. East Greenland Caledonian fold belt. In Escher, A. and Watt, W. S. (Eds), *Geology of Greenland*, Geol. Surv. Greenland, Copenhagen, pp. 182–246.

Hurst, J. M., Jepsen, H. F., Kalsbeek, F., McKerrow, W. S. and Peel, J. S. this volume. Geology of the northern extremity of the East Greenland Caledonides.

Hurst, J. M. and McKerrow, W. S. 1981. The Caledonian nappes of eastern North Greenland. *Nature*, London, 290, 772–774.

Hurst, J. M. and Surlyk, F. 1982. Stratigraphy of the Silurian turbidite sequence of North Greenland. *Bull. Grønlands geol. Unders.*, 145, 121 pp.

Jepsen, H. F. and Kalsbeek, F. 1979. Igneous rocks in the Proterozoic platform of eastern North Greenland. *Rapp. Grønlands geol. Unders.*, 88, 11–14.

Jepsen, H. F. and Kalsbeek, F. 1981. Non-existence of the Carolinidian orogeny in the Prinsesse Caroline-Mathilde Alper of Kronprins Christian Land, eastern North Greenland. *Rapp. Grønlands geol. Unders.*, 106, 7–14.

Larsen, O. and Graff-Petersen, P. 1980. Sr-isotopic studies and mineral composition of the Hagen Bræ Member in the Proterozoic clastic sediments at Hagen Bræ, eastern North Greenland. *Rapp. Grønlands geol. Unders.*, 99, 111–118.

Peel, J. S. 1980. Geological reconnaissance in the Caledonian foreland of eastern North Greenland with comments on the Centrum Limestone. *Rapp. Grønlands geol. Unders.*, 99, 61–72.

Surlyk, F., Hurst, J. M. and Bjerreskov, M. 1980. First age-diagnostic fossils from the North Greenland fold belt. *Nature*, London, 286, 800–803.

Surlyk, F. and Hurst, J. M. 1983. Evolution of the early Paleoic deep-water basin of North Greenland—Aulacogen or narrow ocean? *Geology*, 11, 77–81.

The Caledonide Orogen—Scandinavia and Related Areas
Edited by D. G. Gee and B. A. Sturt
© 1985 John Wiley & Sons Ltd

Evidence for non-existence of a Carolinidian fold belt in eastern North Greenland

H. F. Jepsen and **F. Kalsbeek**

The Geological Survey of Greenland, Østervoldgade 10, DK–1350 København K

ABSTRACT

Descriptions of the Caledonian fold belt of eastern North Greenland have suggested the existence of late Proterozoic (Carolinidian) deformation predating the Caledonian orogeny. The existence of a Carolinidian fold belt was inferred mainly on the basis of a postulated angular unconformity within the Proterozoic sedimentary sequence and from the suggested presence of two generations of Proterozoic basic intrusions, the one apparently deformed during both the Carolinidian and Caledonian orogenies and the other only affected by the latter. During our field work in the Kronprins Christian Land area, the type area of the Carolinidian orogeny, we have been unable to find evidence either for the angular unconformity or for the pre-Caledonian deformation of the intrusions.

Introduction

Eastern North Greenland (Fig. 1) is divided into two structural units: (1) the undeformed Proterozoic and Lower Palaeozoic platform between Independence Fjord and Danmark Fjord, and (2) the NNE–SSW trending Caledonian fold belt located between Danmark Fjord and the east coast (Hurst *et al*. this volume).

The oldest rocks in the area are composed of early Proterozoic crystalline basement exposed only along the east coast. The crystalline rocks are unconformably overlain by the middle Proterozoic Independence Fjord Group, composed of quartzitic and feldspathic sandstones that are intruded by numerous basic sills and dykes, and this is overlain in the platform region by the Zig-Zag Dal Basalt Formation. The sandstones with associated basic rocks are equivalent to Haller's 'Thule Group' (Haller and Kulp 1962), which he divided into a lower mainly semipelitic group and an upper psammitic group. Only the psammitic group is present in eastern North Greenland. Areas which according to Haller (1971, Fig. 18) are underlain by the semipelitic group were found to belong either to the crystalline basement, the psammitic group, the basaltic rocks, or the younger Hagen Fjord Group (Jepsen and Kalsbeek 1981). The name 'Thule Group' is now restricted to the Thule basin, northwestern Greenland, and is no longer in use in eastern North Greenland (Dawes 1976).

The Independence Fjord Group or the Zig-Zag Dal Basalt Formation (in the platform) are overlain by the late Proterozoic Hagen Fjord Group and Lower Palaeozoic sediments.

The Carolinidian disturbance of late Proterozoic age was defined by Haller (1961) and later (Haller and Kulp 1962; Haller 1970) renamed as the Carolinidian orogeny. It was supposed to have affected the crystalline basement and the middle Proterozoic Independence Fjord Group in eastern North Greenland and possibly northern Ellesmere Island. The Hagen Fjord Group was considered to have not been influenced by the Carolinidian orogeny.

The concept of a late Proterozoic Carolinidian orogeny was based on observations made by Haller and co-workers during field work and aerial reconnaissance mapping carried out under extremely difficult conditions in North and Northeast Greenland in the period 1952–1958 (Fränkl 1954, 1955, 1956; Peacock 1956, 1958; Haller 1956). These observations convinced Haller (1961) that there was a major unconformity between the Independence Fjord Group and the Hagen Fjord Group, requiring an

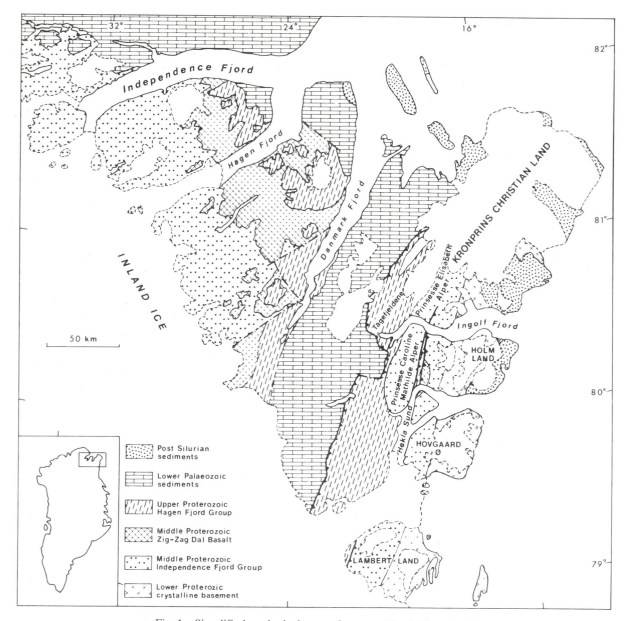

Fig. 1   Simplified geological map of eastern North Greenland

orogenic event to have taken place before the deposition of the late Proterozoic Hagen Fjord Group.

Field work carried out by the authors in 1980 in the Kronprins Christian Land area of eastern North Greenland between latitudes 79°–81°N clearly demonstrates that, although the contact between the Independence Fjord Group and the Hagen Fjord Group is an important erosional unconformity, there is complete structural continuity across that boundary (Jepsen and Kalsbeek 1981).

The authors do not claim to have solved all problems involving the Caledonian deformational and metamorphic history, but with respect to the Carolinidian orogeny as defined by Haller the field relationships seem conclusive.

A brief resumé is given below of the stratigraphy and structure of the Proterozoic rocks of the Kronprins Christian Land area.

Stratigraphy

Crystalline basement

Crystalline rocks are exposed in a 30 km wide zone along the eastern coast of Kronprins Christian Land, Holm Land, Hovgaard Ø, and Lambert Land (Fig. 1). They are composed of a variety of gneisses, migmatized mica schists, and amphibolitic rocks. The gneisses and migmatized mica schists are cut by discordant basic dykes which themselves are folded but not migmatized.

U–Pb isotopic analyses on zircon fractions from two samples of the gneisses have yielded discordant U–Pb isotopic trends, which indicate formation or recrystallization of this mineral during the generation of the gneisses c. 2000 Ma ago, followed by a major isotopic disturbance during the Caledonian orogeny c. 400 Ma ago (R. T. Pidgeon, personal communication 1981).

Independence Fjord Group

The crystalline basement is unconformably overlain by a sequence of middle Proterozoic sandstones with associated basic intrusions and effusive rocks, which make up the elevated mountain chain stretching from the Prinsesse Elizabeth Alper in the north, through the Prinsesse Caroline-Mathilde Alper, to western Lambert Land in the south. This zone, trending NNE–SSW, is from 20 to 40 km wide.

Intense Caledonian deformation of the sandstones prevents precise measurement of their total thickness, but it probably exceeds 1000 m. Similarity of lithology and stratigraphic context strongly suggest a correlation with the Independence Fjord Group of the platform region west of Danmark Fjord (Collinson 1980). There, Rb–Sr isotopic ages on clay minerals from the Hagen Bræ Member (Larsen and Graff-Petersen 1980) give an approximate age of 1380 Ma. Rb–Sr whole rock isochrons on the intrusive rocks in the Independence Fjord Group of the platform (Jepsen and Kalsbeek 1979) give ages of about 1250 Ma. The Zig-Zag Dal Basalt Formation (Jepsen et al. 1980), also on the platform, is probably contemporaneous with the intrusive rocks.

The top of the Independence Fjord Group with its intrusive and effusive rocks is an erosional surface, which in Kronprins Christian Land marks a time gap in the stratigraphical succession of more than 450 Ma. This unconformity is known not only from the Kronprins Christian Land area but can be followed from Dronning Louise Land (Peacock 1956) c. 200 km south of Lambert Land to the platform region west of Danmark Fjord (Jepsen 1971), and further southwest to Inglefield Land in Northwest Greenland (Yochelson and Peel 1980, Fig. 1).

Hagen Fjord Group

A sequence of late Proterozoic clastic rocks and carbonates (Hurst and McKerrow 1981) unconformably overlies the Independence Fjord Group and was referred to by Haller (Haller and Kulp 1962; Haller 1970) as the Hagen Fjord Group. The oldest

strata, the Rivieradal Sandstone (Fränkl 1954), are only known in an allochthonous position in nappe structures displaced westwards over the Proterozoic sandstones and younger Proterozoic and Palaeozoic strata. The Rivieradal Sandstone is a sequence of turbidites and some conglomerates and has a thickness of at least 1000 m (Hurst and McKerrow 1981). An acritarch assemblage from the upper half of the formation suggests a late Riphean age (c. 800 Ma) (G. Vidal, personal communication 1981). The Rivieradal Sandstone, which is overlain by the Campanuladal Formation, is thought to have been deposited in a marine, deep-water basin located east of the Prinsesse Caroline-Mathilde Alper (Hurst and McKerrow 1981).

In the central parautochthonous region around Ingolf Fjord and Hekla Sund, the Independence Fjord Group is overlain by a few metres of basal conglomerate, which is in turn overlain by the Campanuladal Formation, composed of shallow water mudstone with some quartzitic sandstones. The Campanuladal Formation varies in thickness from 0–80 m and rests on the Independence Fjord Group and overlying conglomerate without angular discordance.

Concordantly overlying the Campanuladal Formation follows a characteristic stromatolitic dolomite named the Fyns Sø Dolomite (Adams and Cowie 1953). At Hekla Sund the dolomite rests directly on the basal conglomerate overlying the Independence Fjord Group. The thickness of the Fyns Sø Formation at the type locality is 320 m, but due to tectonic thinning in the Kronprins Christian Land area the thickness may be reduced to a few metres. The ages of the Campanuladal and Fyns Sø Formations are not known precisely but in the platform area around Hagen Fjord the Fyns Sø Formation is overlain by the Portfjeld Formation of probably earliest early Cambrian age (Peel 1980), so it is likely that both formations are of very late Proterozoic age.

The Hagen Fjord Group in the Kronprins Christian Land area is overlain by Lower Palaeozoic carbonates and clastic rocks, including Kap Holbæk Sandstone, Danmark Fjord Formation, and Centrum Sø Limestone (Hurst and McKerrow 1981).

Structure

The structures developed during the Caledonian deformations in the Kronprins Christian Land area are well shown in sections across the Prinsesse Caroline-Mathilde Alper along Ingolf Fjord and Hekla Sund. This area was the 'type area' of the

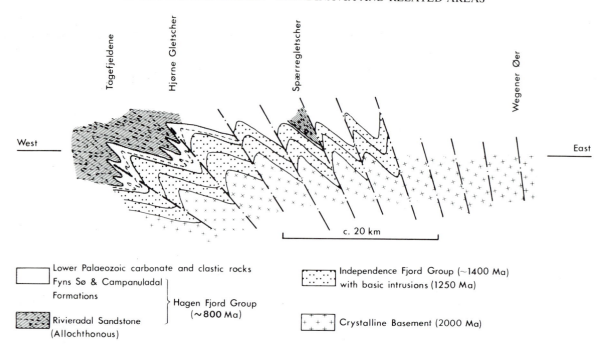

Fig. 2   Schematic cross-section through Prinsesse Caroline-Mathilde Alper along the northern shore of Ingolf Fjord

Carolinidian orogeny (Haller and Kulp 1962; Haller 1970).

Figure 2 illustrates how we interpret the structures in the Ingolf Fjord region. The deformation is characterized by E–W compression resulting in folding and thrusting with displacement towards the west. The trends of fold axes and strike of axial planes vary between north and northeast. Because of differences in competency between the massive sandstones of the Independence Fjord Group and the sediments of the Hagen Fjord Group the structures vary between the two groups, but the basic style of deformation is the same.

Towards the east, near the contact with the basement, the sandstones with intrusions are folded into near isoclinal folds with steeply dipping axial planes. Towards the west, the sandstones and the overlying beds of the Hagen Fjord Group are more openly folded into asymmetric folds with axial planes dipping horizontally to 50 degrees eastwards. Amplitudes of folds ranging up to several hundred metres are observed in many places. Shear zones, subparallel to the axial planes, traverse the section at several localities; they are several hundred metres wide in the western and central parts, and may be up to several kilometres wide in the eastern part of the Prinsesse Caroline-Mathilde Alper. The shear zones mark locations of thrusts or zones of extreme flattening in the eastern part of the section where the lack of good marker horizons in the Independence Fjord Group prohibits demonstration of displacement across the zones. The rocks in the shear zones have been strongly mylonitized and, towards the east,

they are transformed into fine-grained quartz-phyllites with abundant development of sericite. Intrusive bodies in the shear zones are often boudin-aged and metamorphosed into greenschists.

The sandstones located between the shear zones are not much disturbed by the deformation and primary structures such as cross-bedding and ripple marks are well preserved. In contrast to the sandstones the less competent rocks of the Hagen Fjord Group often show well developed axial plane cleavage often completely obliterating primary sedimentary structures.

The allochthonous Rivieradal Sandstone, which is thrust over the younger beds of the Hagen Fjord Group, occurs in nappes displaced westwards during an early stage of the Caledonian deformation. The Rivieradal Sandstone was later involved in the main phase of deformation as described above (Fig. 2). It is now exposed mainly west of the Prinsesse Caroline-Mathilde Alper (Hurst and McKerrow 1981) but is also found in a narrow syncline east of that mountain range in the topographic depression formed by Spærregletscher and Hekla Sund (Figs. 1 and 2).

No evidence of a late Proterozoic orogeny

In his definition of the Carolinidian orogeny, Haller (1961, 1970; Haller and Kulp, 1962) postulated that the sandstones of the Independence Fjord Group were folded and metamorphosed prior to the deposition of the Hagen Fjord Group, and that there was a distinct angular unconformity between the two groups. He also considered that the Independence Fjord Group was intruded by two generations of basic rocks. The first generation, according to Haller, was folded during the Carolinidian deformation, and the second, post-Carolinidian generation was not deformed but intruded before the deposition of the Hagen Fjord Group.

As already mentioned above, the boundary between the Independence Fjord Group and the Hagen Fjord Group is a regional erosional unconformity; however, at the three localities in the Kronprins Christian Land area where it has been examined, the bedding in the two groups is parallel or subparallel and there is complete structural continuity across the boundary.

The sandstones of the Independence Fjord Group are intruded by numerous basic rocks forming dykes, sheets, sills, and other more irregular discordant bodies. One type often passes laterally or vertically into another type. We have examined the basic rocks at many localities, including those referred to by Haller (1970, Figs. 17, 19 and plate 12) but we have not been able to distinguish two generations of intrusions and we are convinced that they have all been deformed by only one phase of deformation.

In the platform region, around the head of Independence Fjord, c. 200 km west of the Kronprins Christian Land area, Haller (1961) described very open folds with wavelengths of up to 25 km. These folds, which were thought to have affected only strata of the Independence Fjord Group, were ascribed to the Carolinidian orogeny. Field work carried out by the authors in 1978 in the Independence Fjord area has not confirmed their existence.

Conclusions

In the Kronprins Christian Land area of eastern North Greenland the oldest exposed rocks comprise a crystalline basement, probably of early Proterozoic age. This basement is overlain by the mid-Proterozoic Independence Fjord Group, a succession of sandstones intruded by basic rocks. After a period of erosion and non-deposition lasting c. 450 Ma, the late Proterozoic Hagen Fjord Group and Lower Palaeozoic rocks were deposited.

The area was deformed during the Caledonian orogeny in mid-Silurian time. Large nappes were thrust westwards, followed by folding and reverse faulting with continued displacement towards the west.

The Carolinidian orogeny, as defined by Haller (1961) has not affected the rocks of eastern North Greenland; in an area stretching over a distance of 400 km from Lambert Land to Independence Fjord,

no evidence has been found which supports the concept of this late Proterozoic orogenic episode.

Acknowledgements

We thank D. G. Gee and A. K. Higgins for critically reading the manuscript and the Director of the Geological Survey of Greenland for permission to publish.

References

Adams, P. J. and Cowie, J. W. 1953. A geological reconnaissance of the region around the inner part of Danmarks Fjord, Northeast Greenland. *Meddr Grønland*, **111** (7), 24 pp.

Collinson, J. D. 1980. Stratigraphy of the Independence Fjord Group (Proterozoic) of eastern North Greenland. *Rapp. Grønlands geol. Unders.*, **99**, 7–23.

Dawes, P. R. 1976. Precambrian to Tertiary of northern Greenland. In Watt, W. S. and Escher, A. (Ed.), *Geology of Greenland*, Copenhagen, Geol. Surv. Greenland, 249–303.

Fränkl, E. 1954. Vorlaufige Mitteilung über die Geologie von Kronprins Christians Land. *Meddr Grönland*, **116** (2), 86 pp.

Fränkl, E. 1955. Weitere Beiträge zur Geologie von Kronprins Christians Land. *Meddr Grønland*, **103** (7), 36 pp.

Fränkl, E. 1956. Some general remarks on the Caledonian mountain chain of East Greenland. *Meddr Grønland*, **103** (11), 43 pp.

Haller, J. 1956. Die Strukturelemente Östgrönlands zwischen 74° und 78°N. *Meddr Grønland*, **154** (2), 27 pp.

Haller, J. 1961. The Carolinides: an orogenic belt of late Precambrian age in Northeast Greenland. In Raasch, G. O. (Ed), *Geology of the Arctic 1*, Toronto U.P., 155–159.

Haller, J. 1970. Tectonic map of East Greenland (1:500,000). *Meddr Grønland*, **171** (5), 286 pp.

Haller, J. 1971. *Geology of the East Greenland Caledonides*, New York, Interscience Publishers, 423 pp.

Haller, J. and Kulp, J. L. 1962. Absolute age determinations in East Greenland. *Meddr Grönland*, **171** (1), 77 pp.

Hurst, J. M. and McKerrow, W. S. 1981. The Caledonian nappes of Kronprins Christian Land, eastern North Greenland. *Rapp. Grønlands geol. Unders.*, **106**, 15–19.

Jepsen, H. F. 1971. The Precambrian, Eocambrian and early Palaeozoic stratigraphy of the Jørgen Brønlund Fjord area, Peary Land, North Greenland. *Meddr Grønland*, **192** (2), 42 pp.

Jepsen, H. F. and Kalsbeek, F. 1979. Igneous rocks in the Proterozoic platform of eastern North Greenland. *Rapp. Grønlands geol. Unders.*, **88**, 37–56.

Jepsen, H. F. and Kalsbeek, F. 1981. Non-existence of the Carolinidian orogeny in the Prinsesse Caroline-Mathilde Alper of Kronprins Christian Land, eastern North Greenland. *Rapp. Grønlands geol. Unders.*, **106**, 7–14.

Jepsen, H. F., Kalsbeek, F. and Suthren, R. J. 1980. The

Zig-Zag Dal Basalt Formation, North Greenland. *Rapp. Grønlands geol. Unders*., **99**, 25–32.

Larsen, O. and Graff-Petersen, P. 1980. Sr-isotopic studies and mineral composition of the Hagen Bræ Member in the Proterozoic clastic sediments at Hagen Bræ, eastern North Greenland. *Rapp. Grønlands geol. Unders*., **99**, 111–118.

Peacock, J. D. 1956. The geology of Dronning Louise Land, N.E. Greenland. *Meddr Grønland*, **137** (7), 38 pp.

Peacock, J. D. 1958. Some investigations into the geology and petrography of Dronning Louise Land, N.E. Greenland. *Meddr Grønland*, **157** (4), 139 pp.

Peel, J. S. 1980. Early Cambrian microfossils from the Portfjeld Formation, Peary Land, eastern North Greenland. *Rapp. Grønlands geol. Unders*., **100**, 15–18.

Yochelson, E. L. and Peel, J. S, 1980. Early Cambrian Salterella from North-west Greenland. *Rapp. Grønlands geol. Unders*., **101**, 29–36.

The Caledonide Orogen—Scandinavia and Related Areas
Edited by D. G. Gee and B. A. Sturt
© 1985 John Wiley & Sons Ltd

Cambrian–Silurian platform stratigraphy of eastern North Greenland

J. S. Peel

Grønlands Geologiske Undersøgelse, Øster Voldgade 10, DK–1350 Copenhagen K, Denmark

ABSTRACT

Cambrian–Silurian platform sediments in eastern North Greenland form part of a Precambrian–Silurian sequence occurring in the angle between the Innuitian foldbelt of North Greenland and the Caledonian East Greenland foldbelt.The Cambrian–Silurian sequence is relatively uniform and it is now possible to integrate older local stratigraphic schemes into a regionally applicable nomenclature. About 1.5 km of Lower, Middle, and Upper Cambrian carbonates, shales, and sandstones are subdivided into 16 formations in the northwestern part of the region. This sequence is progressively overstepped by carbonates of late Early Ordovician (late Canadian) age, such that Cambrian strata are probably absent in the southeast. The Ordovician consists principally of a lower formation of pale, recessive dolomites (Wandel Valley Formation; 320 m) and an upper, massive-weathering limestone unit (Børglum River Formation; 430 m) recognized throughout the area. An intervening unnamed formation (100 m) of intermediate lithology occurs to the southeast. The Ordovician–Silurian boundary occurs within an unnamed dolomitic unit (95–200 m) which is itself overlain by an unnamed lower Silurian limestone formation (325 m). Carbonate mounds (Drømmebjerg Formation) with a relief of over one hundred metres and a diameter of several kilometres are rooted within this limestone formation. Silurian carbonate deposition ceased at about the end of the Llandovery with the accumulation of up to 100 m of black shales and siltstones (Wulff Land Formation). Calcareous turbidite deposition (Lauge Koch Land Formation) subsequently overwhelmed the entire platform area at about the Lower Silurian–Upper Silurian boundary, reflecting the collapse of the carbonate platform and the rising Caledonides to the east and southeast.

Introduction

Eastern North Greenland contains the most northerly preserved part of the Caledonian mountain chain in Greenland (Fig. 1; see Hurst *et al.* this volume). It is also here that the N–S trending Caledonian foldbelt converges upon the W–E extension across North Greenland of the Innuitian foldbelt. In Greenland, the two foldbelts are referred to as the East Greenland and North Greenland foldbelts, respectively (Fig. 1), and are fully described elsewhere in this volume (Henriksen, this volume; Higgins *et al.*, this volume; Hurst *et al.*, this volume). The relationship between the two foldbelts is not precisely known, since changes in the Caledonian trend in northern Kronprins Christian Land (Fig. 1)

indicate that the junction, if present, occurs offshore; land exposures of the relevant Lower Palaeozoic strata are also obscured by Upper Palaeozoic–Tertiary deposits of the Wandel Sea Basin (Dawes and Soper 1973; Håkansson *et al.* 1981; Dawes and Peel 1981) and by significant ice cover.

The onshore area of eastern North Greenland in the angle between the North Greenland foldbelt and the East Greenland foldbelt is occupied by a sequence of Precambrian–Silurian sediments which were deposited on a broad platform lying to the south of the North Greenland trough (Dawes 1976; Dawes and Peel 1981; Surlyk *et al.* 1980).

The platform area between the two foldbelts has a long history, extending from the Precambrian

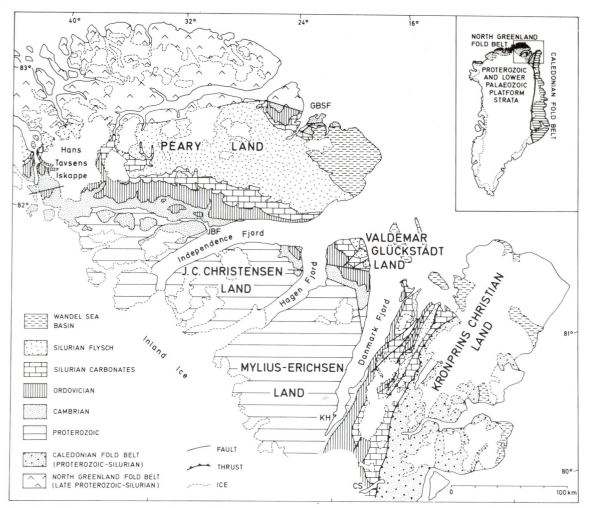

Fig. 1   Simplified geological map of eastern North Greenland. GBSF: G.B. Schley Fjord; JBF: Jørgen Brønlund Fjord; KH: Kap Holbæk; JPKF: J.P. Koch Fjord (compiled with permission from maps accompanying *Rapp. Grönlands geol. Unders.*, **88**, (1979) and **106** (1981)). Geological Survey of Greenland standard practice refers areas to the northwest of Hagen Fjord in Fig. 1 to 'central North Greenland', and areas to the southeast to 'eastern North Greenland'. The combination under the latter, in the present context, is purely a matter of convenience

(Riphean: late Proterozoic) until the end of the Early Silurian (Llandovery). Early geological work was scattered; hence there is a duplication of stratigraphic nomenclature which has obscured the fact that the platform is essentially a single structural and sedimentological entity. There is inevitably variation in deposition and geological history across the platform, as might be expected in an area of this size (500 km × 500 km).

Notwithstanding the coincidence in terms of timing of the main orogenesis (latest Silurian–Devonian), there is little geological similarity between the North Greenland foldbelt and the East Greenland foldbelt in eastern North Greenland. The former contains a well developed continuous trough sequence of Precambrian to Silurian age (summarized by Higgins *et al.*, this volume), while trough deposition in the East Greenland foldbelt is only represented by

the Precambrian Rivieradal sandstones (Hurst *et al.*, this volume) and a Silurian flysch sequence. The Rivieradal sandstones are without direct equivalents in the North Greenland foldbelt. The Silurian flysch was derived from the rising Caledonides; it is found everywhere because the platform collapsed at the end of the Early Silurian and trough deposition then spread throughout the region (Surlyk 1982). Hurst and McKerrow (this volume) propose derivation of the main Caledonian nappes, with the Rivieradal sandstones, from an area located some 150 km to the east of their present location. Lower Palaeozoic trough sediments were presumably also deposited in this source area but are not preserved in the allochthonous Caledonian sequences anywhere in Greenland.

Thus, in the context of the North Greenland foldbelt, Lower Palaeozoic strata form a deep water trough sequence, which was the locus of orogenic activity (Higgins *et al.*, this volume), and part of an essentially undeformed, carbonate-dominated platform sequence. Variably deformed strata of this same platform sequence comprise the Lower Palaeozoic record in the East Greenland foldbelt and are overridden by the Caledonian nappes.

Eastern North Greenland is accessible only during a short summer season. Geologists have visited the area only sporadically during the last 65 years and often under physically demanding conditions associated with primary geographical exploration. Most of this earlier geological work was concentrated in widely separated localities because of logistic difficulties. Summaries are given by Dawes (1971, 1976), Haller (1971), Christie and Peel (1977), Dawes and Peel (1981), and Peel (1982b). The scattered nature of earlier field work inevitably resulted in the establishment of separate stratigraphic schemes. Thus, in terms of the platform areas, a Peary Land school (Troelsen 1949; Jepsen 1971; Christie and Peel 1977) and a Danmark Fjord school (Adams and Cowie 1953; Fränkl 1955; Cowie 1971; Haller 1971; Scrutton 1975) have developed.

Recent expeditions by the Geological Survey of Greenland have resulted in geological investigation throughout the region (preliminary results are given in *Rapport Grønlands Geologiske Undersøgelse* numbers 88, 99, and 106, published in 1979, 1980, and 1981 respectively). Field work during 1978–80 demonstrated the essential continuity of the Precambrian to Silurian platform sequence throughout the region and established an integrated stratigraphic scheme. The Precambrian history of the platform has been described by Clemmensen (1979), Collinson 1980), and Jepsen *et al.* (1980). The present paper summarizes Lower Palaeozoic platform stratigraphy in eastern North Greenland, although formal description of many of the units has not yet been accomplished.

Cambrian

The Cambrian in eastern North Greenland is most fully represented in southern Peary Land (Figs. 1, 2), where a sequence nearly 1.5 km in thickness has yielded faunas of Early, Middle, and Late Cambrian age. The sequence has been described by Troelsen (1949), Jepsen (1971), Dawes (1976), Christie and Peel (1977), Peel (1979), and Ineson (1980), while recent summaries are given by Ineson and Peel (1980) and Peel (1982b). Fifteen principal formations are recognized (Fig. 3), although most of these have not yet been formally named. An additional

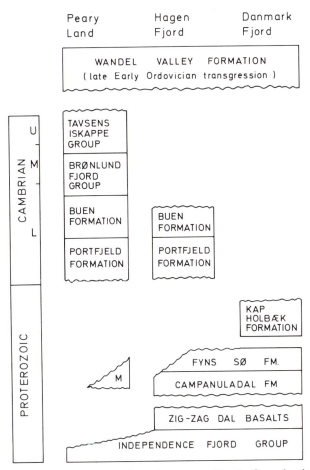

Fig. 2 Cambrian sections in eastern North Greenland showing the relationship with underlying Proterozoic and overlying Early Ordovician strata

formation, the Schley Fjord shale, is equivalent in part to the sandstone and shale dominated Buen Formation of southern Peary Land. The lowest formation, the predominantly dolomite Portfjeld Formation, unconformably overlies the supposed tillite-bearing Morænesø Formation in Peary Land (Jepsen 1971; Clemmensen 1979); the latter fills hollows in the irregular, eroded, upper surface of the Inuiteq Sø Formation of the Independence Fjord Group (Collinson 1980). To the southeast, around Hagen Fjord, the Portfjeld Formation unconformably overlies the faulted and eroded Campanuladal and Fyns Sø Formations (Fig. 2).

At Kap Holbæk and in western Kronprins Christian Land (Fig. 1), siltstones and sandstones of the Kap Holbæk Formation (maximum thickness 150 m) occur between the Proterozoic Fyns Sø Formation and the Ordovician Wandel Valley Formation (Fig. 2). The formation also occurs in the Caledonian nappes of eastern Kronprins Christian Land described elsewhere in this volume by Hurst and McKerrow and Hurst *et al.* Peel *et al.* (1981) and Peel (1982b) correlated the Kap Holbæk Formation

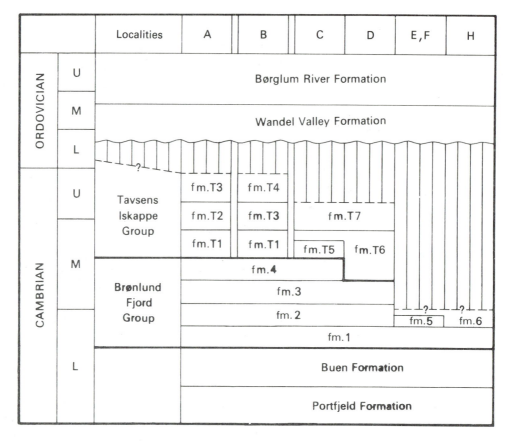

Fig. 3   Cambrian and Ordovician stratigraphy in southern Peary Land, eastern North Greenland. Localities are indicated in Fig. 4 (Reproduced with permission from Ineson and Peel 1980)

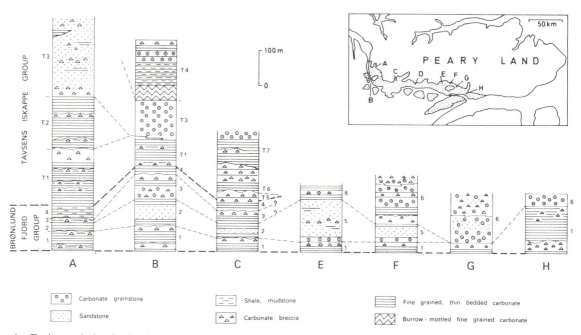

Fig. 4   Facies variation in the Cambrian Brønlund Fjord and Tavsens Iskappe Groups of southern Peary Land, eastern North Greenland. Numerals refer to formations identified in Fig. 3 (Reproduced with permission from Ineson and Peel 1980)

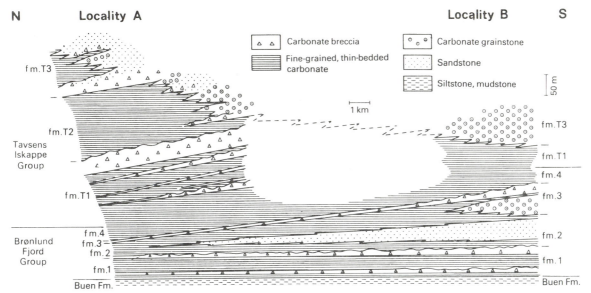

Fig. 5   Generalized facies relationships of south–north prograding sediments of the Cambrian Brønlund Fjord and Tavsens Iskappe Groups. West of Hans Tavsens Iskappe, southern Peary Land, eastern North Greenland. Localities A and B are identified in Fig. 4 (after Ineson 1980), with permission)

with the Lower Cambrian Buen Formation of Peary Land on account of their considerable lithological similarity and stratigraphic position beneath the Wandel Valley Formation. However, preliminary examination of acritarchs from near the base of the formation at Kap Holbæk suggests a late Proterozoic age (G. Vidal written communication 1982). Strata in Valdemar Glückstadt Land which Peel (1980b) referred to the Kap Holbæk Formation are re-assigned to the Buen Formation.

Lower to upper Cambrian strata assigned to the Brønlund Fjord and Tavsens Iskappe Groups form a complex of limestones, dolomites, sandstones, and shales showing substantial facies variation and dia-chronism (Figs. 4, 5). The sequence is characterized by carbonate and some sandstone breccia sheets, carbonate grainstones, and sandstones which pro-grade south to north toward the North Greenland trough, and are interbedded with dark, thin-bedded dolomites, limestones and shales (Ineson 1980).

Portfjeld Formation

This formation varies in thickness from 200 m to about 280 m in southern Peary Land (O'Connor 1979), but Christie and Ineson (1979) record more than twice this thickness at G. B. Schley Fjord (Fig. 1). The formation is greatly reduced in thickness to the southeast and absent in Kronprins Christian Land (Fig. 2). The Portfjeld Formation is principally composed of grey or buff coloured, cliff-forming dolomites which are generally well bedded, often cross-bedded, and may contain beds of oolite,

intraformational conglomerate, and stromatolites. A chert bed (10–15 m) composed of dark grey bituminous dolomite with lenses and beds of black chert overlying pale dolomites forms a conspicuous mapping horizon throughout the region (Fig. 6). A large scale cross-bedded quartz sandstone (10–30 m) is prominent in the upper part of the formation in southern Peary Land.

Pedersen (1976) recorded filamentous microfos-sils from the Portfjeld Formation in southern Peary Land which he suggested were of youngest Precam-brian age. Peel (1980a) reported irregularly coiled phosphatic tubes, considered indicative of an Early Cambrian age.

Buen Formation

The Buen Formation (Jepsen 1971) consists of about 425 m of sandstones, siltstones, and shales at its type locality near Jørgen Brønlund Fjord (Fig. 1). Jepsen noted that two sedimentary cycles made up the sequence, each beginning with quartzitic sand-stones and ending with interbedded shales, silt-stones, and impure sandstones. In more westerly areas of Peary Land, however, the formation can be subdivided into a lower sandstone unit and an upper, recessive, shale, siltstone, and thin sandstone alter-nation. The sandstones may be cross-bedded, ripple marked, and sometimes heavily bioturbated.

At G. B. Schley Fjord (Fig. 1), equivalent strata have been given the name Schley Fjord shale (Troelsen 1956; Poulsen 1974; Christie and Ineson 1979).

Fig. 6   Lower Cambrian stratigraphy, southern Peary Land. View eastward along Øvre Midsommersø showing carbonates of the Brønlund Fjord Group (BFG) overlying recessive shales and prominent sandstones of the Buen Formation (BUEN) and carbonates of the Portfjeld Formation (PF). A prominent dark chert bed occurs in the Portfjeld Formation just above its unconformable junction with the Precambrian Inuiteq Sø Formation (ISF) which contains dark weathering dolerite intrusions. The cliff is about 700 m high. Reproduced by permission of the Geological Survey of Greenland

A nevadiid, occurring together with *Olenellus hyperboreus* (Poulsen 1974), hyolithids, *Pelagiella*, and archaeocope ostracods, near the middle of the formation, suggests that the Buen Formation has a medial and late Early Cambrian age (Palmer and Peel 1979). *Olenellus* cf. *O. svalbardensis* Kielan, 1964, occurs at G.B. Schley Fjord in the uppermost beds of the formation (Poulsen 1974; Christie and Ineson 1979).

Brønlund Fjord Group

Cambrian carbonates overlying the Buen Formation in the Jørgen Brønlund Fjord area of southern Peary Land (Fig. 1) were named Brønlund Fjord dolomite by Troelsen (1949) and have subsequently been recognized as a formation (Cowie 1971; Christie and

Peel 1976). Peel (1979) gave the formation group status after field work west of Hans Tavsens Iskappe (Figs. 1, 7). Ineson and Peel (1980) discussed the stratigraphy of the Brønlund Fjord Group throughout southern Peary Land and established the subdivision into six formations employed here (Figs. 3, 4, 5).

The Brønlund Fjord Group ranges in age from late Early Cambrian to late Middle Cambrian (Palmer and Peel 1979). As a result of regional overstep by the Wandel Valley Formation, the group is not present to the southeast of Independence Fjord (Figs. 1, 2).

A typical section through formations 1–4 of the Brønlund Fjord Group in the Hans Tavsens Iskappe area is given in Fig. 8. Formations 5 and 6 are only recognized in more easterly outcrops in southern

Fig. 7   The Brønlund Fjord and Tavsens Iskappe Groups (Lower Cambrian–earliest Ordovician?) of southern Peary Land. View of east side of J.P. Koch Fjord, immediately west of Hans Tavsens Iskappe (Fig. 1). The cliffs (height 1 km) are cut by basic dykes and capped by Ordovician carbonates of the Wandel Valley Formation (WV). BFG: Brønlund Fjord Group; TIG: Tavsens Iskappe Group; BF: Buen Formation. Reproduced by permission of the Geological Survey of Greenland

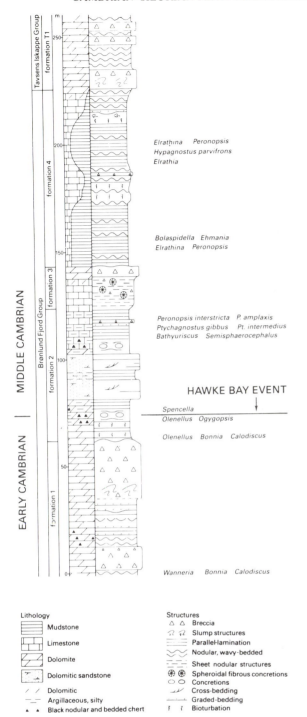

Fig. 8 Stratigraphy and palaeontology of the Brønlund Fjord Group. Hans Tavsens Iskappe area, southern Peary Land, eastern North Greenland. The Early–Middle Cambrian boundary is marked by a hiatus, the Hawke Bay Event of Palmer and James (1980). Trilobite identifications principally by R. A. Robison and A. R. Palmer. Stratigraphy from Ineson (1980)

breccia interpreted as debris flows (Ineson 1980). Well-sorted structureless or locally cross-bedded sandstones are present in formation 2, although more characteristic of formation 5 in more easterly outcrops.

Formation 1 forms conspicuous cliffs of pale weathering dolomite above the dark, recessive Buen Formation, throughout southern Peary Land (Fig. 7). Fossils are confined to the lower few metres, a glauconite and phosphorite rich member (Frykman 1980), where *Bonnia, Calodiscus*, fragments of *Wanneria, Chancelloria*, molluscs, and problematic fossils indicate a late Early Cambrian age (Palmer and Peel 1979).

Formation 2 is a dark, recessive unit of dolomites, shales, and limestones, although pale dolomitic sandstones become prominent to the southeast as the formation interdigitates with formation 5 (Figs. 3, 4, 7). Formation 2 is richly fossiliferous, yielding a wide variety of trilobites, molluscs, brachiopods, and problematic fossils of Early and Middle Cambrian ages. *Bonnia, Olenellus, Ogygopsis*, and *Kootenia* in the lower part of the formation indicate a late Early Cambrian age (Palmer and Peel 1979). Immediately overlying finely laminated dolomites, indistinguishable from those below which yield Early Cambrian faunas, yield medial Middle Cambrian trilobites. The inferred unconformity, removing early Middle Cambrian strata, can be correlated with the Hawke Bay Event of Palmer and James (1980) documented along the eastern seaboard of North America. In western North Greenland, some 500 km to the west of Peary Land, at least part of this missing interval is present (Poulsen 1964). Higher beds of formation 2 yield trilobites indicative of the *Ptychagnostus gibbus* Zone.

Formation 3 is mainly composed of cliff-forming nodular limestones and cross-bedded grainstones, although a dolomite breccia sheet often forms a conspicuous top to the sequence. The formation has produced no age diagnostic fossils.

Formation 4 consists of grey shales and lenticular limestones in northwestern exposures (Figs. 3, 4 localities A, B), but the shale content decreases toward the south and east. Lower beds contain *Bolaspidella, Elrathina*, and *Peronopsis*, while *Elrathina, Elrathia, Hypagnostus*, and *Peronopsis* occur at higher levels, indicating a general correlation with the *Ptychagnostus gibbus* and *P. atavus* Zones of the Middle Cambrian (R. A. Robison written communication 1981).

Formation 5 is dominated by pale cream, fine-grained, dolomitic sandstones, interbedded with dark, bioturbated siltstones. The formation interdigitates with formation 2, to the west, and thins out to the southeast (Figs. 3, 4). Palmer and Peel (1979) noted an abundant fauna of late Early Cambrian age.

Peary Land (Figs. 3, 4). At locality B (Figs. 3, 4), the group is dominated by dark, thin or medium-bedded dolomites and micritic limestones which are often argillaceous, bituminous, or cherty. About 25 per cent of the sequence is formed by beds of carbonate

Formation 6 is a cliff-forming dolomite unit which directly overlies formation 1 in eastern areas of the Cambrian outcrop of southern Peary Land (Figs. 3, 4). A variety of bioclastic and oolitic grainstones are present, often associated with slumped horizons. Dark, thin bedded, graded and laminated dolomites become more common in northern outcrops and yield a rich fauna of archaeocyathids, trilobites, brachiopods, molluscs, and *Chancelloria* indicative of a late Early Cambrian age.

Tavsens Iskappe Group

The Tavsens Iskappe Group (Peel 1979) is a sequence of carbonates and sandstones occurring between the Brønlund Fjord Group and the sub Wandel Valley Formation unconformity in the area immediately west of Hans Tavsens Iskappe (Fig. 1). The group is restricted to western parts of the Cambrian outcrop in Peary Land (Fig. 1) and was redefined by Ineson and Peel (1980). Four formations (T1–T4) are recognized to the west of Hans Tavsens Iskappe, but these outcrops cannot be precisely correlated with outcrops on the eastern side, where three additional formations are recognized (Figs. 3, 4).

The group attains a thickness of about 700 m and is of Middle and Late Cambrian (earliest Ordovician?) age (Peel 1982a). Lower formations are mainly dolomites and limestones similar to those of the underlying Brønlund Fjord Group, upon which the Tavsens Iskappe Group rests with apparent conformity. However, the uppermost formations (T3, T7) are prograding units of carbonate grainstones, cross-bedded sandstones, and breccias. The pronounced diachronism of most formations (Fig. 4, compare localities A, B, C) reflects the strong south–north prograding character of the sequence (Fig. 5).

Formation T1 consists of cliff-forming limestones and dolomites overlying the recessive formation 4 of the underlying Brønlund Fjord Group. West of Hans Tavsens Iskappe, at locality A (Figs. 4, 7), the formation is composed of thick dolomite breccia sheets interbedded with darker, thin-bedded dolomites and argillaceous limestones. A thickness of 175 m is attained, but less than half this amount is present at locality B, some 20 km to the south (Fig. 4). Abundant agnostids are all indicative of the *Ptychagnostus punctuosus* Zone of the Middle Cambrian (R. A. Robinson written communication 1981).

Formation T2 is composed of about 150 m of dark, recessive, thinly-bedded limestones, dolomites, and shales at locality A (Figs. 4, 7) but thins rapidly to the south and is absent at locality B. The formation is richly fossiliferous, yielding agnostids

indicative of the latest Middle Cambrian *Lejopyge laevigata* Zone including a possible ancestor to *Agnostus pisiformis*. Of particular note is the occurrence of *Cedaria* together with *L. laevigata*. *Cedaria* is the zonal index of the lowest Dresbachian stage, the base of which has been employed as the base of the Late Cambrian in North America. It is evident that the base of the Dresbachian falls within the Middle Cambrian, in European terms.

Formation T3 is a cliff-forming unit of thick-bedded dolomitic sandstones, carbonate grainstones, and quartzites, with abundant quartzitic and dolomitic breccia sheets. Individual beds wedge out northwards into darker carbonates, with depositional dips of up to 20 degrees (Fig. 9). At least 400 m of strata are present at locality A (Fig. 4), but the formation is progressively overstepped by the unconformably overlying Wandel Valley Formation to the south. At locality B, the formation consists of about 120 m of oolitic dolomite (Fig. 4). Late Cambrian molluscs (*Proplina*) occur rarely, but conodonts of the *Cordylodus proavus* Zone, which narrowly spans the Cambrian–Ordovician boundary, occur in the upper beds (Peel 1982a).

Formation T4 is restricted to the area around locality B (Fig. 4) and consists of 170–200 m of bioturbated dolomites and greenish siltstones. Thin algal laminates occur near the top of the formation which is capped by a prominent quartzitic breccia sheet.

Formations T5–T7 occur to the east of Hans Tavsens Iskappe (Figs. 3, 4, localities C, D) and can be correlated only tentatively with sequences to the west. Formation T5 is a prominent pale dolomite breccia overlying the darker, recessive formation 4 of the Brønlund Fjord Group. The overlying formation T6 (100 m) consists of dark, recessive thin-bedded dolomites and limestones yielding rich trilobite faunas of Middle Cambrian age, and is thus

Fig. 9 Prograding Upper Cambrian carbonates and sandstones of the Tavsens Iskappe Group, east side of J. P. Koch Fjord, immediately west of Hans Tavsens Iskappe. Section height 1 km; TIG; Tavsens Iskappe Group; WV: Wandel Valley Formation (Ordovician). Reproduced by permission of the geological Survey of Greenland

older than the lithologically similar formation T2, to the west of the Iskappe. Formation T7 is at least 200 m in thickness and contains a variety of cliff-forming, pale-weathering dolomites which interdigitate to the north with wedges of darker, bioturbated dolomite. Thick breccia sheets in the lower part of the formation contain huge blocks of dolomitic grainstones, tens of metres in diameter.

Ordovician

The Ordovician platform sequence in eastern North Greenland is composed of dolomites and limestones, principally referred to the Wandel Valley Formation and the overlying Børglum River Formation (Figs. 1, 10); these were originally described from the Jørgen Brønlund Fjord area of southern Peary Land (Troelsen 1949; Christie and Peel 1977). However, an as yet unnamed Ordovician formation is recognized in the Danmark Fjord region, between the Wandel Valley and Børglum River Formations (Peel *et al*. 1981). An overlying sequence of dolomites and dolomitic limestones, the unnamed Silurian (?) dolomite formation of Christie and Peel (1977), is now known to span the Ordovician–Silurian boundary (Peel 1980b; Armstrong and Lane 1981).

An alternative nomenclature has been used in Kronprins Christian Land where Adams and Cowie (1953) referred the entire Lower Palaeozoic carbonate sequence to the Centrum Limestone. The terminology was modified by Fränkl (1955), Cowie (1971), and Scrutton (1975). Recent field work (Peel 1980b; Peel *et al*. 1981) demonstrated the essential identity of the Peary Land and Kronprins Christian Land sequences and a regionally more uniform stratigraphy is now employed (Fig. 10). The Danmarks Fjord Dolomite, variously considered to be a formation of Cambrian or Ordovician age (see discussion in Peel *et al*. 1981), is now known to contain late Canadian (late Early Ordovician) conodonts (M. P. Smith written communication 1981). It is reduced in status to a member within the Wandel Valley Formation, as is the immediately overlying Amdrup Formation of Cowie (1971). The term *Opikina* Limestone of Scrutton (1975) is abandoned; it is apparently equivalent to part of the Wandel Valley Formation and the overlying unnamed Ordovician Formation.

The term Centrum Formation is the much modified remnant of the Centrum Limestone of Adams and Cowie (1953). In view of the lack of precision surrounding its definition and content, and the failure of Adams and Cowie to recognize the widespread effects of repetition by thrusting upon the stratigraphic sequence in Kronprins Christian Land, Peel *et al*. (1981: 21) recommended that this name also be abandoned. The concept of the Centrum Formation used by Scrutton (1975) corresponds to the Børglum River Formation.

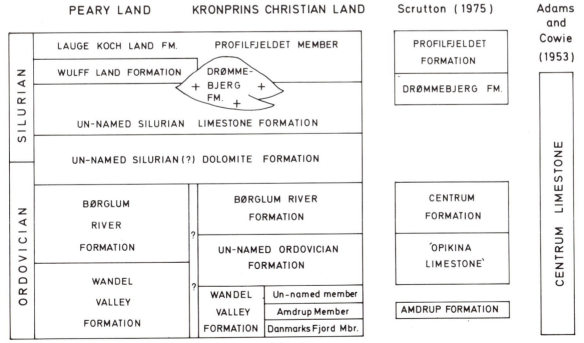

Fig. 10 Ordovician and Silurian stratigraphy in the platform sequences of eastern North Greenland. Early stratigraphic schemes employed by Scrutton (1975) and Adams and Cowie (1953) in Kronprins Christian Land are shown for comparison

Wandel Valley Formation

This formation was named by Troelsen (1949) in southern Peary Land (Fig. 1) and redefined by Christie and Peel (1977). A profound hiatus separates the Wandel Valley Formation from underlying strata throughout the region (Fig. 2), although angular discordance or other evidence of unconformity is rarely discernible at individual outcrops. At Hans Tavsens Iskappe (Figs. 1, 3, 9), the formation overlies Upper Cambrian clastics and carbonates of the Tavsens Iskappe Group, but this is gradually overstepped to the south and southeast so that the Wandel Valley Formation overlies the Lower Cambrian Brønlund Fjord Group at Jørgen Brønlund Fjord (Figs. 1, 3). South of Independence Fjord, the formation lies directly on clastics of the Buen or Kap Holbæk Formations (Figs. 2, 11). The Wandel Valley Formation has a gradational contact with the overlying Børglum River Formation, in Peary Land, and with the unnamed Ordovician formation in the Danmark Fjord region (Fig. 12).

In southern Peary Land, Christie and Peel (1977) recorded a thickness of 320 m distributed between three members. A lower member (45 m) of thin bedded, pale grey, cherty, laminated dolomite, with thin beds and pockets of intraformational conglomerate, contains darker beds of mottled dolomite with silicified fossils. A middle member (75 m) of dark grey, fossiliferous, sugary dolomite is in marked contrast to underlying and overlying pale weathering members. The upper member (200 m) is similar in character to the lower member, but generally unfossiliferous.

Three members are also recognized in the Danmark Fjord region, but no precise correlation is attempted between these and the Peary Land mem-

Fig. 11 The sub-Ordovician unconformity in the Danmark Fjord area. Ordovician carbonates of the Danmarks Fjord and Amdrup Members of the Wandel Valley Formation (WV) overlying sandstones and siltstones of the Kap Holbæk Formation (KH). About 50 m of Wandel Valley Formation are exposed. View of west side of the headland at Kap Holbæk (Fig. 1), looking south. Reproduced by permission of the Geological Survey of Greenland

bers. The Danmarks Fjord Member (10–12 m) is readily recognized at Kap Holbæk (Fig. 1), the type locality, but in Kronprins Christian Land it grades into the overlying Amdrup Member (200 m). The former consists of brown weathering cherty dolomite; the latter is mainly dark, medium-bedded limestones and limey dolomites. The upper member (unnamed) of the Wandel Valley Formation is of similar lithology and thickness to the upper member in Peary Land.

Abundant *Ceratopea* in the lower beds of the Wandel Valley Formation indicate a late Canadian (late Early Ordovician) age. Yochelson and Peel (1975) identified *C. ankylosa* and *C. unguis* from the lower and middle members in Peary Land. In Kron-

Fig. 12 Ordovician carbonates in Kronprins Christian Land. Pale and dark weathering dolomites of the Wandel Valley Formation (WV) are overlain by the 100 m thick unnamed Ordovician Formation (O), the lower member of which is a cliff-forming dark dolomite, and dark cliff-forming limestones of the Børglum River Formation (BR). East side of Danmark Fjord, looking eastward. Reproduced by permission of the Geological Survey of Greenland

prins Christian Land, *C. billingsi* is locally abundant in the lower beds of the Amdrup Member (Peel 1980b); this member has also yielded a number of trilobites and may be more marine in character than its counterpart in Peary Land. Macrofossils are rare in the upper part of the Wandel Valley Formation, but conodonts indicate that the highest beds are of Chazy (Middle Ordovician) age (M. P. Smith and R. J. Aldridge written communication 1981).

Unnamed Ordovician formation

Peel *et al*. (1981) recognized a new, as yet unnamed Ordovician formation between the Børglum River Formation and the Wandel Valley in the Danmark Fjord region (Figs. 10, 12). The formation is about 100 m thick and consists of a lower, cliff-forming, generally dark-weathering dolomite member, reminiscent of the Børglum River Formation, and an upper member of pale grey weathering, recessive dolomite similar in character to the upper member of the Wandel Valley Formation. Brachiopods (*Oepikina*) and trilobites occur in thin beds of intraformational conglomerate at the base of the formation and it is probably to part of this formation, together with the underlying upper member of the Wandel Valley Formation, that the ill-defined term *Opikina* Limestone has been applied (Scrutton 1975). The unnamed Ordovician formation is tentatively correlated with upper beds of the Wandel Valley Formation in Peary Land but may also be equivalent to the lowest part of the Børglum River Formation (Fig. 10).

Børglum River Formation

The Børglum River Limestone was first described in southern Peary Land by Troelsen (1949) and subsequently extended by Troelsen (in Fränkl 1956:19) and Mayr (1976) to include overlying Silurian carbonates. However, Christie and Peel (1977) restricted the name to the lowest of three formations which they recognized in this stratigraphic interval: Børglum River Formation, unnamed Silurian (?) dolomite formation, unnamed Silurian limestone formation.

The Børglum River Formation is a thick, competent unit of brownish-grey weathering limestones which attains a thickness of about 430 m in southern Peary Land. This thickness may be substantially increased in southern Kronprins Christian Land but the effects of repetition by thrusting in the stratigraphically monotonous unit cannot be ascertained. Christie and Peel (1977) recognized three members in southern Peary Land but the regional applicability of this subdivision is not known.

A lower limestone member (260 m) weathers thick-bedded but often shows thin-bedding or even lamination on closer inspection. Bioturbation produces a pronounced mottling on bedding surfaces with chert nodules often lying within burrows. Scoriaceous weathering is characteristic, with poorly silicified corals (*Catenipora*), gastropods (*Maclurites*), and orthocone cephalopods conspicuous. A darker weathering limestone and dolomite member (40 m) is mottled and apparently poorly fossiliferous. An upper limestone member (130 m) resembles the lower member, but tends to form rubbly outcrops on account of a silty lamination.

The Børglum River Formation is locally richly fossiliferous, particularly in the upper part where large molluscs and corals are conspicuous. *Gonioceras*, *Bathyurus*, and *Oepikina* near the base of the formation suggest a medial Middle Ordovician age. Upper beds yield *Kochoceras*, *Paleofavosites*, *Catenipora*, *Calapoecia*, *Lobocorallium*, gastropods, trilobites, and brachiopods indicating a Late Ordovician age. Condont studies in progress suggest that the base of the Børglum River Formation in southern Peary Land is probably of latest Chazy (medial Middle Ordovician) age, while uppermost beds are of late Richmondian (Late Ordovician) age (M. P. Smith and R. J. Aldridge written communication 1981).

Unnamed Silurian (?) dolomite formation

This formation was first recognized in southern Peary Land (Peel and Christie 1975; Christie and Peel 1977) where a lower member of pale weathering dolomite was seen to be overlain by darker dolomites with thin intraformational conglomerates. Armstrong and Lane (1981) measured a thickness of about 95 m, rather less than the original estimates of Christie and Peel (1977). The formation is about 200 m thick in Kronprins Christian Land and forms a conspicuous mapping horizon on account of its alternating beds of very pale and dark dolomite.

The Ordovician–Silurian boundary in eastern North Greenland lies within the unnamed Silurian (?) dolomite formation. Ordovician corals and molluscs occur at the base of the formation in Kronprins Christian Land while thin-shelled pentamerid brachiopods high in the sequence are indicative of an earliest Silurian age. Preliminary conodont examination suggests that about 40 m of the formation in southern Peary Land are of Late Ordovician age (Armstrong and Lane 1981).

Silurian

Carbonate deposition persisted in the platform areas of eastern North Greenland until the end of the Early Silurian, when turbidite deposition overwhelmed the platform and terminated a period of almost continuous carbonate accumulation extending back to the Early Cambrian.

As with the Ordovician, described in the previous section, two schemes of nomenclature have been built up in southern Peary Land and Kronprins Christian Land. Stratigraphic relationships are best seen in Peary Land which is little affected by tectonism and forms the model for the present description. Much of the Silurian in Kronprins Christian Land was not recognized by Adams and Cowie (1953) and is consequently absent from earlier stratigraphic schemes (Fig. 10). However, names given in Kronprins Christian Land are maintained.

A different stratigraphy has been recognized by Hurst (1979) in the area west of Hans Tavsens Iskappe (Fig. 1). The sequence shows affinities to western North Greenland (Hurst and Peel 1979; Hurst 1980a,b) and is not further discussed here.

Unnamed Silurian (?) dolomite formation

This formation, discussed above, spans the Ordovician–Silurian boundary in eastern North Greenland (Christie and Peel 1977; Peel 1980b; Armstrong and Lane 1981; Peel et al. 1981).

Unnamed Silurian limestone formation

The unnamed Silurian limestone formation, as described by Christie and Peel (1977), includes strata referred by Mayr (1976) to the Børglum River Formation, although this name is now restricted to purely Ordovician carbonates occurring lower in the sequence. A thickness of about 325 m in southern Peary Land is subdivided into six members, many of which appear to be recognizable throughout eastern North Greenland. The limestones are generally pale grey or yellowish grey weathering and appear thick bedded and prominent when compared with the underlying dolomites and overlying clastics (Fig. 13).

Members A–D (175 m) consist of a variety of pale, dark, mottled, and slightly silicified limestones. Member E (100 m) is a thick-bedded, cherty, nodular or mottled biostromal limestone with characteristic yellowish-grey scoriaceous weathering; a distinctive poorly silicified fauna includes large subspherical stromatoporoids and tabulate corals. Member F (50 m) is a recessive unit of dark rubbly limestones overlain abruptly by black shales in southern Peary Land.

The unnamed Silurian (?) dolomite formation and

the unnamed Silurian limestone formation are not recognizable in stratigraphic descriptions built on the work of Adams and Cowie (1953) around Danmark Fjord. Scrutton (1975) noted that many supposed Silurian corals from the Centrum Limestone were in fact Ordovician species, while undoubted Silurian species were restricted to the Drømmebjerg Formation, see below. Both the unnamed Silurian formations are well developed in Kronprins Christian Land (Peel 1980b; Peel et al. 1981), but no attempt has been made to recognize the subdivision into members proposed in southern Peary Land. The distinctive stromatoporoidal limestones of member E of the limestone formation are, however, well represented.

Fig. 13   Silurian sequence in southern Peary Land. Lower Silurian (Llandovery) limestones (SL) are overlain by dark weathering shales (100 m) of the Wulff Land Formation (WL) yielding Llandovery–Wenlock graptolites, and turbidites of the Lauge Koch Land Formation (LKL). View southeast, downstream along Børglum Elv, north of Jørgen Brønlund Fjord (Fig. 1). Reproduced by permission of the Geological Survey of Greenland

Preliminary examination of pentamerid brachiopods (Boucot in Christie and Peel 1977; J. M. Hurst personal communication 1981) and conodonts (Aldridge 1979) suggests an Early–Late Llandovery (Early Silurian) age for the unnamed Silurian limestone.

Drømmebjerg Formation

Silurian carbonate mounds (Fig. 14) rooted in the unnamed Silurian limestone formation were reported from southern Peary Land by Peel and Christie (1975) but are now known to occur throughout eastern North Greenland (Mayr 1976; Christie and Peel 1977; Lane and Thomas 1979; Mabillard 1980; Peel 1980b; Peel *et al.* 1981). The mounds form the eastern continuation of a complex of Silurian carbonate mounds stretching across North Greenland in a variety of stratigraphic and environmental settings (Dawes 1976; Hurst 1980a, 1980b; Dawes and Peel 1981).

Limestones now known to form one such mound were given the name Drømmebjerg Limestone (Fränkl 1955) and this name is maintained as a formation to include the mounds of Peary Land, Valdemar Glückstadt Land, and Kronprins Christian Land (Fig. 1). Individual mounds range in size from a few tens of metres to several kilometres in diameter; they may be more than a hundred metres thick and on account of their original relief penetrate into the overlying turbidite sequence. Pale grey crinoid and stromatoporoid bioclastic limestones dominate. A rich fauna is locally abundant (e.g. Lane 1972; Lane and Peel 1980; Christie and Peel 1977) and this, together with preliminary conodont examination (H. A. Armstrong personal communication 1981), suggests an age around the Llandovery–Wenlock (Early–Late Silurian) boundary.

Wulff Land Formation

A distinctive unit of black shales and grey siltstones of turbiditic origin with a maximum thickness of about 100 m overlies the Silurian carbonates (Fig. 13) in southern Peary Land (Christie and Peel 1977; Hurst and Surlyk 1980, 1982). The shales are referred to the Wulff Land Formation and together with the overlying Lauge Koch Land Formation form part of the Peary Land Group of Hurst (1980a) and Hurst and Surlyk (1982). In the Danmark Fjord region, the formation is much reduced in thickness and appears to be absent in the eastern part of Kronprins Christian Land (Peel 1980b; Hurst and McKerrow 1981), such that the recognition of a separate unit is not practicable. In this area, the shale and the overlying turbidite sequence are grouped into the Profilfjeldet Member of the Lauge Koch Land Formation. In Fig. 1 all three units are referred to as Silurian flysch.

Graptolites from the Wulff Land Formation indicate a latest Llandovery–Wenlock age (J. M. Hurst and M. Bjerreskov personal communication 1981).

Lauge Koch Land Formation

The shales of the Wulff Land Formation grade sharply up into a sequence of turbidites, the Lauge Koch Land Formation of Hurst and Surlyk (1982), which outcrop over large areas of central Peary Land, forming distinctive terraced hills (Fig. 14).

Fig. 14   Silurian carbonate mound of the Drømmebjerg Formation, Samuelsen Høj, north of Jørgen Brønlund Fjord (Fig. 1), southern Peary Land. Terraced hills in the distance produced by turbiditic sandstone beds within the Lauge Koch Land Formation. (After Lane and Thomas 1979). Reproduced by permission of the Geological Survey of Greenland

The turbidites are generally rather fine grained and slightly calcareous, but chert pebble conglomerates occur higher in the sequence. Hurst and Surlyk (1982) reported at least 1,500 m of strata.

The turbidites lie within a basin extending some 800 km across North Greenland and into Ellesmere Island (Dawes 1976; Dawes and Peel 1981; Surlyk 1982). Similar deposits were deposited in the North Greenland trough throughout the Early Silurian (Hurst and Surlyk 1980, 1982; Surlyk *et al*. 1980; Soper *et al*. 1980), but in latest Llandovery to Wenlock times turbidite deposition flooded the areas of earlier carbonate sedimentation to the south. Palaeocurrent directions indicate that much of the Silurian turbidite sequence was derived from the east, probably from the rising Caledonides of East Greenland (Hurst and Surlyk 1980, 1982; Surlyk 1982).

Profilfjeldet Member

Caledonian tectonism in Kronprins Christian Land prevents a precise correlation of Silurian flysch sequences in this area with those of Peary Land, although there is no doubt about the continuity of the deposits. In consequence, it is convenient to maintain Nielsen's (1941) name Profilfjeldet Shales for the combined black shale unit and overlying turbidites in the Danmark Fjord region. Hurst and Surlyk (1982) gave the shales member status within the Lauge Koch Land Formation of the Peary Land Group (Fig. 10). Black shales may be present at the base of the sequence in this area, but it is not practicable to separate the few metres of sediments in question from the overlying turbidites. Turbidites of the Profilfjeldet Member are indistinguishable from their counterparts in Peary Land, although graptolites at the base of the clastic sequence indicate a slightly older, Late Llandovery age (J. M. Hurst and M. Bjerreskov personal communication 1981). Peel (1980b) noted the presence of Ordovician and Silurian carbonates, sandstones, metamorphic, and crystalline rock types as clasts within thin beds of conglomerate near the base of the formation, reflecting erosion of the rising Caledonides to the east (Hurst *et al*. this volume).

Correlation with East Greenland Caledonides

A number of small outcrops of Cambrian and Ordovician platform sediments occur within the Caledonian fold belt of northern East Greenland (see Henriksen this volume), but Silurian strata are not present (Fig. 15). The sequence is best known through the work of Cowie and Adams (1957), summarized by Henriksen and Higgins (1976), with more recent developments noted by Peel and Cowie (1979), Frykman (1979), and Peel (1982b). The close relationship between the Cambrian and Ordovician sequences in East Greenland, Svalbard, Scotland, and Newfoundland is discussed by Swett and Smit (1972a,b), and Swett (1981).

The East Greenland Cambrian begins with clastics of the Kløftelv and Bastion Formations (c. 210 m) overlain by late Early Cambrian carbonates of the Ella Island (= Ella Ø) and Hyolithus Creek Formation (c. 300 m). The upper part of the clastic sequence and these carbonates may be correlated in terms of age with the Buen Formation and the lower part of the Brønlund Fjord Group of Peary Land, but the lower clastics are of older Cambrian age (Vidal 1979); carbonates equivalent to the Portfjeld Formation are seemingly absent.

Dolomites of the overlying Dolomite Point Formation (c. 400 m) are not readily correlated with the northeastern Greenland sequence. A pre-Early Ordovician age has previously been assumed with Cowie's (1971) tentative suggestion of Middle Cambrian requiring a profound unconformity below the Early Ordovician Antiklinalbugt Formation. Early Ordovician conodonts are now known from the uppermost beds of the formation (Frykman 1979; V. E. Kurtz and J. Miller written communication 1980), but the Middle and Late Cambrian are not proven. Swett (1981: 242) suggested that the absence of fossil evidence for the Middle and Late Cambrian probably reflected slow sedimentation in a schizohaline environment. On the basis of general age, the Dolomite Point Formation is equivalent to all or part of the upper Brønlund Fjord Group and Tavsens Iskappe Group of Peary Land together with earliest Ordovician strata not preserved in Peary Land.

The Ordovician sequence in northern East Greenland attains a thickness in excess of 3 km, more than three times the thickness preserved in Peary Land. The Antiklinalbugt Formation (the renamed Cass Fjord Formation of East Greenland, see Peel and Cowie 1979) is of Early–Middle Canadian (early to middle Early Ordovician) age (Cowie and Adams 1957) and represents a time interval not preserved below the sub-Wandel Valley Formation unconformity in the platform regions of eastern North Greenland. However, graptolites of early Arenig (middle Early Ordovician) age are reported from the clastic sequences in the North Greenland foldbelt (Bjerreskov and Poulsen 1973; Dawes and Soper 1979).

It is uncertain how much of the overlying 1 km thick Cape Weber (= Kap Weber) Formation of East Greenland has a time equivalent unit in eastern

North Greenland. Cowie and Adams (1957) suggested a middle Canadian (middle Early Ordovician) to Chazy (early Middle Ordovician) age, and in many ways the formation is reminiscent of the Amdrup Member of the Wandel Valley Formation in Kronprins Christian Land. However, the Amdrup Member has yielded *Ceratopea billingsi* (Peel 1980c) described by Yochelson (1964) from the Narwhale Sound Formation (= Narhval Sund Formation) of East Greenland. Thus, while the Cape Weber Formation probably ranges up into the late Canadian (Smith 1982), where it may be equivalent to the lowest beds of the Wandel Valley Formation, it is probable that lower beds of the formation have no equivalent in the platform sequence of eastern North Greenland.

The Narwhale Sound Formation (460 m) is reminiscent of, and apparently equivalent to, much of the Wandel Valley Formation. A late Canadian (late Early Ordovician) to late Whiterock (early Middle Ordovician) age is suggested (Smith 1982). The Heim Bjerge Formation consists of thick bedded limestones, lithologically equivalent to the Børglum River Formation of northeastern Greenland. Cowie and Adams (1957) gave a minimum thickness of 320 m, but this has been increased to at least 1200 m (Frykman 1979). Smith (1982) noted that conodonts indicate that the Heim Bjerge Formation is entirely of Middle Ordovician age (late Whiterock–Chazy), and thus older than the Børglum River Formation.

Correlation with Western North Greenland and Ellesmere Island

Peel and Christie (1982) and Peel (1982b) compared Cambrian and Ordovician platform sequences in eastern Ellesmere Island, western North Greenland, and eastern North Greenland, pointing out the close similarity between the two former areas (Fig. 15). Correlation eastward across North Greenland to Peary Land is less precise, but a gradual transition is evident from intermediate sections (Hurst and Peel 1979; Peel 1980d).

The Portfjeld Formation is overstepped by late Lower Cambrian clastics equivalent to the Buen Formation in the inner platform regions of western North Greenland, such that the Humboldt Formation of Washington Land (or the equivalent Dallas Bugt Formation of Inglefield Land and Bache Peninsula, see Peel *et al.* 1982) rests unconformably upon

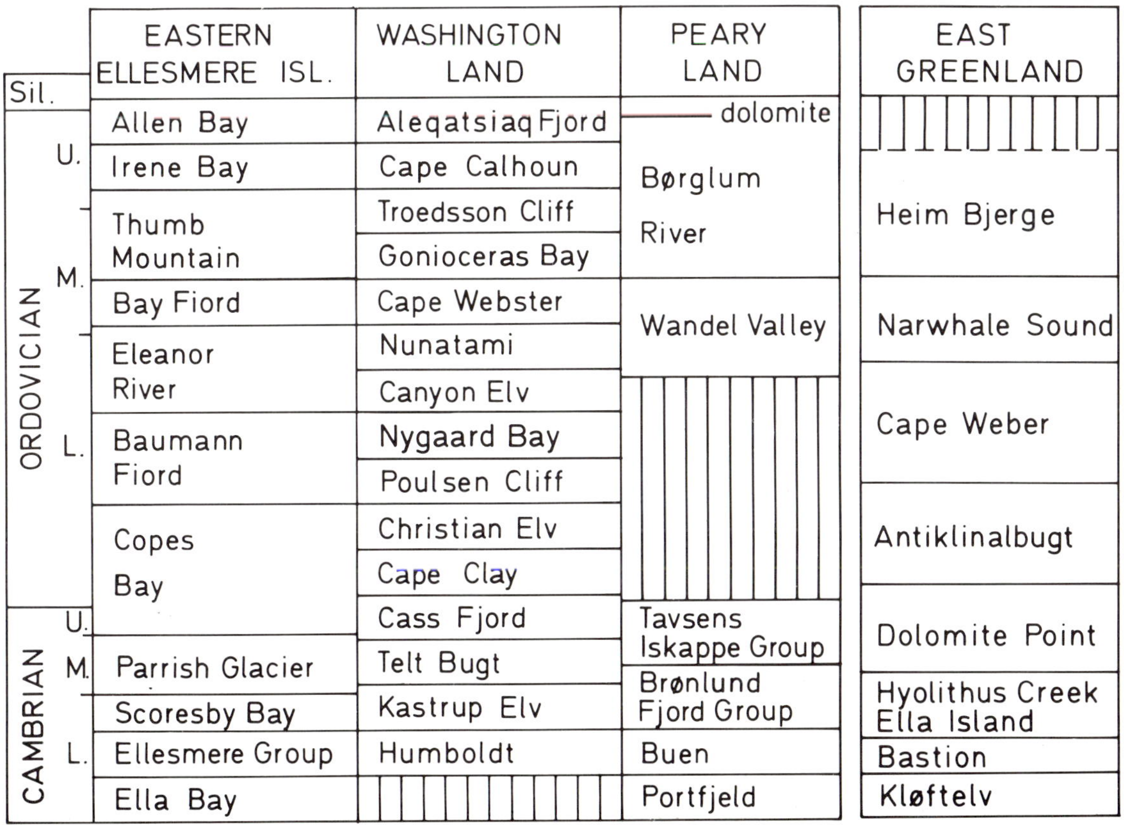

		EASTERN ELLESMERE ISL.	WASHINGTON LAND	PEARY LAND	EAST GREENLAND
Sil.				— dolomite	
ORDOVICIAN	U.	Allen Bay	Aleqatsiaq Fjord	Børglum River	
		Irene Bay	Cape Calhoun		
	M.	Thumb Mountain	Troedsson Cliff		Heim Bjerge
			Gonioceras Bay		
		Bay Fiord	Cape Webster	Wandel Valley	Narwhale Sound
		Eleanor River	Nunatami		
	L.		Canyon Elv		Cape Weber
		Baumann Fiord	Nygaard Bay		
			Poulsen Cliff		
		Copes Bay	Christian Elv		Antiklinalbugt
			Cape Clay		
CAMBRIAN	U.		Cass Fjord	Tavsens Iskappe Group	Dolomite Point
	M.	Parrish Glacier	Telt Bugt	Brønlund Fjord Group	Hyolithus Creek Ella Island
		Scoresby Bay	Kastrup Elv		
	L.	Ellesmere Group	Humboldt	Buen	Bastion
		Ella Bay		Portfjeld	Kløftelv

Fig. 15   Correlation of Cambrian and Ordovician sequences in eastern Ellesmere Island, western North Greenland (Washington Land), eastern North Greenland (Peary Land) and East Greenland

Precambrian strata. Carbonates of the Ella Bay Formation occurring below the clastic Ellesmere Group in the outer platform of Ellesmere Island could correlate with the Portfjeld Formation. The Kastrup Elv Formation of Washington Land is a dolomite unit equivalent to Lower and Middle Cambrian fossiliferous limestones in Inglefield Land to the southwest (C. Poulsen 1927, 1958; V. Poulsen 1964; Peel and Christie 1982). This, and the succeeding shallow water limestone sequences of the Telt Bugt and Cass Fjord Formations show little lithological similarity with strata of similar age in Peary Land. Of particular note, however, is the general absence of early Middle Cambrian strata in Peary Land, while at least some strata of this age are present in western North Greenland. The Dresbachian, Franconian, and Trempealeauan Stages of the Late Cambrian are identified within the Cass Fjord Formation in western North Greenland, although faunal evidence for the Franconian is not yet available in eastern North Greenland (Palmer and Peel 1981).

Strata of early Early Ordovician age are absent in eastern North Greenland, although some 400 m of carbonates, with evaporites, shales and sandstones are represented in the west by the latest Cass Fjord to Nyegaard Bugt Formations (Fig. 15). Limestones of the Canyon Elv and Nunatami Formations and the evaporite-bearing Cape Webster Formation show a gradual west to east transition into the Wandel Valley Formation. The widespread development of evaporites in Ellesmere Island (Baumann Fjord and Bay Fjord Formations) and western North Greenland (Poulsen Cliff and Nyegaard Bay Formations, Cape Webster Formation) is not seen in eastern North Greenland.

The incoming of cliff-forming, thick-bedded limestones above recessive dolomites or evaporites is a conspicuous event in the Middle Ordovician of Greenland and adjacent Ellesmere Island. The facies change is diachronous, coinciding with the base of the Heim Bjerge Formation in East Greenland and the base of the Børglum River Formation in eastern North Greenland. To the west, the limestones have been given the name Morris Bugt Group (Gonioceras Bay–Aleqatsiaq Fjord Formations in Fig. 15) by Peel and Hurst (1980), equivalent to the Thumb Mountain, Irene Bay, and Allen Bay Formations of Ellesmere Island. In eastern North Greenland, the Ordovician–Silurian boundary lies within an unnamed dolomite formation; in western North Greenland, the boundary lies in the upper part of the Aleqatsiaq Fjord Formation (Hurst 1980a).

Silurian flysch occurs across all North Greenland and into Ellesmere Island (Dawes 1976; Dawes and Peel 1981; Hurst and Surlyk 1980; Surlyk 1982). However, Silurian carbonates in western North Greenland show a much greater differentiation in terms of lithology and environment of deposition than is seen in the eastern area (Hurst and Peel 1979; Hurst 1980a,b). Carbonate deposition persisted into the Upper Silurian in western North Greenland, suggesting that the carbonate platform was overwhelmed at a later date in this area than in eastern North Greenland, where carbonate deposition effectively ceased at the end of the Lower Silurian.

Acknowledgements

P. R. Dawes, A. K. Higgins, and J. M. Hurst reviewed the manuscript, which is published with the permission of the Director of Grønlands Geologiske Undersøgelse.

References

Adams, P. J. and Cowie, J. W. 1953. A geological reconnaissance of the region round the inner part of Danmarks Fjord, Northeast Greenland. *Meddr Grønland*, **111**(7), 1–24.

Aldridge, R. J. 1979. An upper Llandovery conodont fauna from Peary Land, eastern North Greenland. *Rapp. Grønlands geol. Unders*, **91**, 7–23.

Armstrong, H. A. and Lane, P. D. 1981. The un-named Silurian(?) dolomite formation, Børglum Elv, central Peary Land. *Rapp. Grønlands geol. Unders*, **106**, 29–34.

Bjerreskov, M. and Poulsen, V. 1973. Ordovician and Silurian faunas from northern Peary Land, North Greenland. *Rapp. Grønlands geol. Unders*, **55**, 10–14.

Christie, R. L. and Ineson, J. R. 1979. Precambrian–Silurian geology of the G.B. Schley Fjord region, eastern Peary Land, North Greenland. *Rapp. Grønlands geol. Unders*, **88**, 63–71.

Christie, R. L. and Peel, J. S. 1977. Cambrian–Silurian stratigraphy of Børglum Elv, eastern North Greenland. *Rapp. Grønlands geol. Unders*, **82**, 1–48.

Clemmensen, L. B. 1979. Notes on the palaeogeographical setting of the Eocambrian tillite-bearing sequence of southern Peary Land, North Greenland. *Rapp. Grønlands geol. Unders*, **88**, 15–22.

Collinson, J. D. 1980. Stratigraphy of the Independence Fjord Group (Proterozoic) of eastern North Greenland. *Rapp. Grønlands geol. Unders*, **99**, 7–23.

Cowie, J. W. 1971. The Cambrian of the North American Arctic regions. In Holland, C. H. (Ed.), *Cambrian of the New World*, Interscience, London, pp. 325–383.

Cowie, J. W. and Adams, P. J. 1957. The geology of the Cambro-Ordovician rocks of central East Greenland. 1. *Meddr Grønland*, **153**,1, 1–193.

Dawes, P. R. 1971. The North Greenland fold belt and environs. *Bull. geol. Soc. Denmark*, **20**, 197–239.

Dawes, P. R. 1976. Precambrian to Tertiary of northern Greenland. In Escher, A. and Watt, W. S. (Eds), *Geology of Greenland*. Geol. Survey Greenland, Copenhagen, pp. 248–303.

Dawes, P. R. and Peel, J. S. 1981. The northern margin of

Greenland from Baffin Bay to the Greenland Sea. In Nairn, A. G. M., Churkin, M., Jr., and Stehli, F. G. (Eds), *The Ocean Basins and Margins*, vol. 5, Plenum, New York, pp. 201–264.

Dawes, P. R. and Soper, N. J. 1973. Pre-Quaternary history of North Greenland. In Pitcher, M. G. (Ed.), *Arctic Geology*. *Mem. Am. Assoc. Petrol. Geol.*, **19**, 117–134.

Dawes, P. R. and Soper, N. J. 1979. Structural and stratigraphic framework of the North Greenland fold belt in Johannes V. Jensen Land, Peary Land, *Rapp. Grønlands geol. Unders*, **93**, 1–40.

Fränkl, E. 1955. Weitere Beiträge zur Geologie von Kronprins Christian Land (NE-Grönland). *Meddr Grønland*, **103**(7), 1–35.

Fränkl, E. 1956. Some general remarks on the Caledonian Mountain chain of East Greenland. *Meddr Grønland*, **103**(11), 1–43.

Frykman, P. 1979. Cambro-Ordovician rocks of C.H. Ostenfeld Nunatak, northern East Greenland. *Rapp. Grønlands geol. Unders*, **91**, 125–132.

Frykman, P. 1980. A sedimentological investigation of the carbonates at the base of the Brønlund Fjord Group (Early–Middle Cambrian), Peary Land, eastern North Greenland. *Rapp. Grønlands geol. Unders*, **99**, 51–55.

Håkansson, E., Heinberg, C. and Stemmerik, L. 1981. The Wandel Sea Basin from Holm Land to Lockwood Ø, eastern North Greenland. *Rapp. Grønlands geol. Unders*, **106**, 47–63.

Haller, J. 1971. Geology of the East Greenland Caledonides. Interscience, New York, 411 pp.

Henriksen, N. this volume. The Caledonides of central East Greenland 70–76°N.

Henriksen, N. and Higgins, A. K. 1976. East Greenland Caledonian fold belt. In Escher, A. and Watt, W. S. (Eds), *Geology of Greenland*, Geol. Survey Greenland, Copenhagen, pp. 183–246.

Higgins, A. K., Soper, N. J. and Friderichsen, J. D. this volume. North Greenland fold belt in eastern North Greenland.

Hurst, J. M. 1979. Uppermost Ordovician and Silurian geology of north-west Peary Land, North Greenland, *Rapp. Grønlands geol. Unders*, **88**, 41–49.

Hurst, J. M. 1980a. Silurian stratigraphy and facies distribution in Washington Land and western Hall Land, North Greenland. *Bull. Grønlands geol. Unders*, **138**, 1–95.

Hurst, J. M. 1980b. The palaeogeographic and stratigraphic differentiation of the Silurian carbonate buildups and biostromes of North Greenland. *Bull. Am. Ass. Petrol. Geol.*, **64**, 527–548.

Hurst, J. M., Jepsen, H. F., Kalsbeek, F., McKerrow, W. S. and Peel, J. S. this volume. The geology of the northern extremity of the East Greenland Caledonides.

Hurst, J. M. and McKerrow, W. S. 1981. The Caledonian nappes of eastern North Greenland. *Nature*, London, **290**, 772–774.

Hurst, J. M. and McKerrow, W. S. this volume. Origin of the Caledonian nappes of eastern North Greenland.

Hurst, J. M. and Peel, J. S. 1979. Late Proterozoic(?) to Silurian stratigraphy of southern Wulff Land, North Greenland. *Rapp. Grønlands geol. Unders*, **91**, 37–56.

Hurst, J. M. and Surlyk, F. S. 1980. Notes on the Lower Palaeozoic clastic sediments of North Greenland. *Rapp. Grønlands geol. Unders*, **99**, 73–78.

Hurst, J. M. and Surlyk, F. S. 1982. Stratigraphy of the Silurian turbidite sequence of North Greenland. *Bull. Grønlands geol. Unders*, **145**, 1–125.

Ineson, J. R. 1980. Carbonate debris flows in the Cambrian of south-west Peary Land, eastern North Greenland. *Rapp. Grønlands geol. Unders*, **99**, 43–49.

Ineson, J. R. and Peel, J. S. 1980. Cambrian stratigraphy in Peary Land, eastern North Greenland. *Rapp. Grønlands geol. Unders*, **99**, 33–42.

Jepsen, H. F. 1971. The Precambrian, Eocambrian and early Palaeozoic stratigraphy of the Jørgen Brønlund Fjord area, Peary Land, North Greenland. *Bull. Grønlands geol. Unders*, **96**, 1–42.

Jepsen, H. F., Kalsbeek, F. and Suthren, R. J. 1980. The Zig-Zag Dal Basalt Formation, North Greenland. *Rapp. Grønlands geol. Unders*, **99**, 25–32.

Lane, P. D. 1972. New trilobites from the Silurian of north-east Greenland. *Palaeontology*, **15**, 336–364.

Lane, P. D. and Peel, J. S. 1980. Trilobites and gastropods from Silurian carbonate mounds in Valdemar Glückstadt Land, eastern North Greenland. *Rapp. Grönlands geol. Unders*, **101**, 54.

Lane, P. D. and Thomas, A. T. 1979. Silurian carbonate mounds in Peary Land, North Greenland. *Rapp. Grønlands geol. Unders*, **88**, 51–54.

Mabillard, J. E. 1980. Silurian carbonate mounds of south-east Peary Land, eastern North Greenland. *Rapp. Grønlands geol. Unders*, **99**, 57–60.

Mayr, U. 1976. Middle Silurian reefs in southern Peary Land, North Greenland. *Bull. Can. Petrol. Geol.*, **24**, 440–449.

Nielsen, E. 1941. Remarks on the map and the geology of Kronprins Christian Land. *Meddr Grønland*, **126**,2, 1–34.

O'Connor, B. 1979. The Portfjeld Formation (?Early Cambrian) of eastern North Greenland. *Rapp. Grønlands geol. Unders*, **88**, 23–28.

Palmer, A. R. and James, N. P. 1980. The Hawke Bay Event: a circum-Iapetus regression near the lower Middle Cambria boundary. In Wones, D. R. (Ed.), *The Caledonides in the U.S.A.*, Mem. Virginia Polytechnic Instit. & State Univ., Blacksburg, pp. 15–18.

Palmer, A. R. and Peel, J. S. 1979. New Cambrian faunas from Peary Land, eastern North Greenland. *Rapp. Grønlands geol. Unders*, **91**, 29–36.

Palmer, A. R. and Peel, J. S. 1981. Dresbachian trilobites and stratigraphy of the Cass Fjord Formation, western North Greenland. *Bull. Grønlands geol. Unders*, **141**, 1–46.

Pedersen, K. R. 1976. Fossil floras of Greenland. In Escher, A. and Watt, W. S. (Eds), *Geology of Greenland*, Geol. Survey Greenland, Copenhagen, pp. 519–535.

Peel, J. S. 1979. Cambrian–Middle Ordovician stratigraphy of the Adams Gletscher region, south-west Peary Land, North Greenland. *Rapp. Grønlands geol. Unders*, **88**, 29–39.

Peel, J. S. 1980a. Early Cambrian microfossils from the Portfjeld Formation, Peary Land, eastern North Greenland. *Rapp. Grønlands geol. Unders*, **100**, 15–17.

Peel, J. S. 1980b. Geological reconnaissance in the Caledonian foreland of eastern North Greenland, with comments on the Centrum Limestone. *Rapp. Grønlands geol. Unders*, **99**, 61–72.

Peel, J. S. 1980c. *Ceratopea billingsi* (Gastropoda) from the Early Ordovician of Kronprins Christian Land, eastern North Greenland. *Rapp. Grønlands geol. Unders*, **101**, 68.

Peel, J. S. 1980d. Cambrian and Ordovician geology of Warming Land and southern Wulff Land, central North Greenland. *Rapp. Grønlands geol. Unders*, **101**, 55–60.

Peel, J. S. 1982a. The age of the Tavsens Iskappe Group, central North Greenland. *Rapp. Grønlands geol. Unders*, **108**, 30.

Peel, J. S. 1982b. The Lower Palaeozoic of Greenland. In Embry, A. F. and Balkwill, H. R. (Eds), *Proc. Third Arctic Geology Symp. Can. Soc. Petrol. Geol. Mem.* **8**, 309–330.

Peel, J. S. and Christie, R. L. 1975. Lower Palaeozoic stratigraphy of southern Peary Land, eastern North Greenland. *Rapp. Grønlands geol. Unders*, **75**, 21–25.

Peel, J. S. and Christie, R. L. 1982. Cambrian–Ordovician stratigraphy, correlations around Kane Basin. In Dawes, P. R. and Kerr, J. W. (Eds), *Nares Strait and the Drift of Greenland: a Conflict in Plate Tectonics. Meddr Grønland, Geosci.*, **8**, 117–135.

Peel, J. S. and Cowie, J. W. 1979. New names for Ordovician formations in Greenland. *Rapp. Grønlands geol. Unders*, **91**, 117–124.

Peel, J. S., Dawes, P. R., Collinson, J. D. and Christie, R. L. 1982. Proterozoic–basal Cambrian stratigraphy across Nares Strait: correlation between Inglefield Land and Bache Peninsula. In Dawes, P. R. and Kerr, J. W. (Eds), *Nares Strait and the Drift of Greenland: a Conflict in Plate Tectonics. Meddr Grønland, Geosci.*, **8**, 105–115.

Peel, J. S. and Hurst, J. M. 1980. Late Ordovician and early Silurian stratigraphy of Washington Land, western North Greenland. *Rapp. Grønlands geol. Unders*, **100**, 18–24.

Peel, J. S., Ineson, J. R., Lane, P. D. and Armstrong, H. A. 1981. Lower Palaeozoic stratigraphy around Danmark Fjord, eastern North Greenland. *Rapp. Grønlands geol. Unders*, **106**, 21–27.

Poulsen, C. 1927. The Cambrian, Ozarkian and Canadian faunas of Northwest Greenland. *Meddr Grønland*, **70**,1,2, 233–343.

Poulsen, C. 1958. Contribution to the Palaeontology of the Lower Cambrian Wulff River Formation. *Meddr Grønland*, **162**,2, 1–25.

Poulsen, V. 1964. Contribution to the Lower and Middle Cambrian paleontology and stratigraphy of Northwest Greenland. *Meddr Grønland*, **164**,6, 1–105.

Poulsen, V. 1974. Olenellacean trilobites from eastern North Greenland. *Bull. geol. Soc. Denmark*, **23**, 79–101.

Scrutton, C. T. 1975. Corals and stromatoporoids from the Ordovician and Silurian of Kronprins Christian Land, Northeast Greenland. *Meddr Grønland*, **171**,4, 1–43.

Smith, M. P. 1982. Conodonts from the Ordovician of East Greenland. *Rapp. Grønlands geol. Unders*, **108**, 14.

Soper, N. J., Higgins, A. K. and Friderichsen, J. D. 1980. The North Greenland fold belt in eastern Johannes V. Jensen Land. *Rapp. Grønlands geol. Unders*, **99**, 89–98.

Surlyk, F. 1982. Nares Strait and the down-current termination of the Silurian turbidite basin of North Greenland. In Dawes, P. R. and Kerr, J. (Eds), *Nares Straits and the Drift of Greenland: a Conflict in Plate Tectonics. Meddr Grønland, Geosci.*, **8**, 147–150.

Surlyk, F., Hurst, J. M. and Bjerreskov, M. 1980. First age-diagnostic fossils from the central part of the North Greenland fold belt. *Nature*, London, **286**, 800–803.

Swett, K. 1981. Cambro-Ordovician strata in Ny Friesland, Spitsbergen and their paleotectonic significance. *Geol. Mag.*, **118**, 225–250.

Swett, K. and Smit, D. E. 1972a. Palaeogeography and depositional environments of the Cambro-Ordovician shallow marine facies of the North Atlantic. *Bull. Geol. Soc. Am.*, **83**, 3223–3248.

Swett, K. and Smit, D. E. 1972b. Cambro-Ordovician shelf sedimentation of western Newfoundland, northwest Scotland, and central East Greenland. *Internatl. geol. Congr. 24, Montreal*, **6**, 33–41.

Troelsen, J. C. 1949. Contributions to the geology of the area round Jørgen Brønlund Fjord, Peary Land, North Greenland. *Meddr Grønland*, **149**,2, 1–29.

Troelsen, J. C. 1956. The Cambrian of North Greenland and Ellesemere Island. *Internatl. geol. Congr. 20, Mexico*, **3**,1, 71–90.

Vidal, G. 1979. Acritarchs from the Upper Proterozoic and Lower Cambrian of East Greenland. *Bull. Grønlands geol. Unders*, **134**, 1–40.

Yochelson, E. L. 1964. The Early Ordovician gastropod *Ceratopea* from East Greenland. *Meddr Grønland*, **164**,7, 1–12.

Yochelson, E. L. and Peel, J. S. 1975. *Ceratopea* and the correlation of the Wandel Valley Formation, eastern North Greenland. *Rapp. Grønlands geol. Unders*, **75**, 28–31.

Note Added in Proof:

Hurst (in press) has proposed a new lithostratigraphic nomenclature for the uppermost Ordovician–Lower Silurian of eastern North Greenland. The unnamed Silurian(?) dolomite formation (Fig. 10) is named Turesø Formation. The unnamed Silurian limestone formation (Figs 10, 13) is referred to the Odins Fjord Formation, with the exception of the former member A of Christie and Peel (1977) which becomes the Ymers Gletscher Formation. The latter formation is probably not present in Kronprins Christian Land. The member subdivision employed by Christie and Peel (1977) is abandoned. Two new members, Bure Iskappe and Melville Land, introduced by Hurst do not reflect the earlier member subdivision. The name Drømmebjerg Formation is replaced by Samuelsen Høj Formation after the type locality of the same name in Peary Land (Fig. 14). Hurst *et al.* (1983), Hurst and Surlyk (1983; 1984) and Surlyk and Hurst (1984) discuss aspects of the evolution of eastern North Greenland.

Hurst, J. M. in press. Upper Ordovician (Cincinnatian) and Silurian (Llandoverian) carbonate shelf stratigraphy, facies and evolution, eastern North Greenland. *Bull. Grønlands geol. Unders.*, **148**,

Hurst, J. M., McKerrow, W. S., Soper, N. J. and Surlyk, F. 1983. The relationship between Caledonian nappe tectonics and Silurian turbidite deposition in North Greenland. *J. geol. Soc. London*, **140**, 123–132.

Hurst, J. M. and Surlyk, F. 1983. Initiation, evolution, and destruction of an Early Paleozoic carbonate shelf, eastern North Greenland. *J. Geol.*, **91**, 671–691.

Hurst, J. M. and Surlyk, F. 1984. Tectonic control of Silurian carbonate-shelf margin morphology and facies, North Greenland. *Bull. Am. Assoc. Petrol. Geol.*, **68**, 1–17.

Surlyk, F. and Hurst, J. M. 1984. The evolution of the early Paleozoic deep-water basin of North Greenland. *Bull. geol. Soc. Am.*, **95**, 131–154.

The Caledonide Orogen—Scandinavia and Related Areas
Edited by D. G. Gee and B. A. Sturt
© 1985 John Wiley & Sons Ltd

The Caledonides of central East Greenland 70°–76°N

N. Henriksen

The Geological Survey of Greenland, ØsterVoldgade 10, DK-1350 Copenhagen K., Denmark

ABSTRACT

In central East Greenland, an Upper Riphean to Middle Ordovician sedimentary sequence up to 17 000 m in composite thickness has been affected by the Caledonian orogeny. These sediments, which are for the most part only moderately deformed and metamorphosed, rest on crystalline complexes which are believed to represent a pre-Caledonian basement. The crystalline complexes are polyorogenic and isotopic evidence suggests development during three orogenic phases 3000–2300 Ma, 2000–1800 Ma and 1200–1000 Ma ago.

Caledonian deformation of this sedimentary sequence is generally moderate, with N–S trending open folds as the dominant structural element. These structures indicate that only limited Caledonian crustal shortening took place in the area between 72°–76°N.

Westwards-directed thrusts of probable Caledonian origin form the marginal zone of the Caledonian fold belt in the west, south of 72°N. Thrust displacements are known to exceed 40 km, but may be much larger.

The Caledonian influence on the pre-Caledonian basement complex is very difficult to assess. Interpretations range from significant reactivation of the basement and formation of infracrustal units by transformation of the Caledonian sedimentary sequence, to only moderate reactivation of pre-existing basement complexes. Rb–Sr whole rock isotopic studies of the infracrustal complexes have yielded pre-Caledonian ages, whereas mineral age determinations on the same rocks are generally Caledonian.

A variety of late orogenic plutonic rocks were emplaced throughout the region. Some occur in the pre-Caledonian basement complexes, but many of them are confined to the border zone between the infracrustal complexes and the Caledonian supracrustals. Age determinations on the plutonic rocks show a range from c. 560 Ma to c. 380 Ma.

A sequence of more than 7000 m of Middle and Upper Devonian coarse clastic sediments and 4000–5000 m of Upper Carboniferous and Lower Permian continental deposits represents a late orogenic molasse deposition. This succession has been subjected to phases of superficial folding, marking the terminating stages of the Caledonian orogeny in central East Greenland.

Introduction

The Caledonian fold belt in East Greenland can be traced as a continuous coast parallel belt for more than 1200 kilometres from c. 70° to c. 82°N (Fig. 1). The southern continuation is obscured by Tertiary basalts, except for a possible small area around 67°30′N, where an intrusive complex giving a Caledonian age has been found (Brooks *et al*. 1976).

The exposed onshore part of the fold belt is more than 300 km wide, but it is assumed to have extended further eastwards representing a substratum to the younger sediments on the 100–300 km wide shelf region.

Systematic studies on the Caledonian geology of East Greenland were carried out during 'The Danish Expeditions to East Greenland 1926–58' led by Lauge Koch. Since 1968 the Geological Survey of

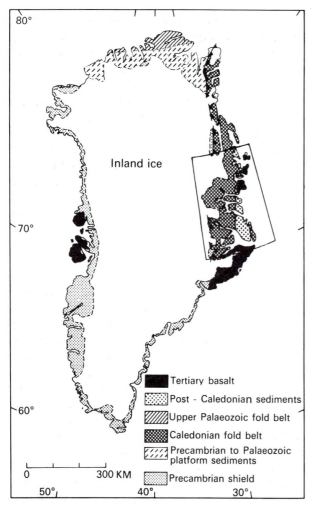

Fig. 1 Main geological divisions in Greenland with extent of Caledonian fold belt. The frame outlines the region between 70°–76°N

Map legend:
- Tertiary basalt
- Post - Caledonian sediments
- Upper Palaeozoic fold belt
- Caledonian fold belt
- Precambrian to Palaeozoic platform sediments
- Precambrian shield

Greenland has carried out regional studies both in East Greenland and in eastern North Greenland.

The region between 70°–76°N in East Greenland is almost completely covered by geological maps at scales of 1:250 000 (72°–76°N) and 1:100 000 (70°–72°N). From the same region an extensive geological literature exists, but in general terms the geological information level is substantially less detailed than that known from the North American and European parts of the orogen. A major regional synthesis was compiled by John Haller (1971), and more recently a series of review papers have been presented by Henriksen and Higgins (1976); Higgins (1976); and Higgins and Phillips (1979). An almost complete bibliography can be found in the above mentioned papers, some of which also contain a historical account of the development of geological investigations. Due to the existence of a number of recent review papers, the present contribution is brief and confined to a presentation of

the regional Caledonian geology in central East Greenland.

One of the main geological problems in the fold belt in East Greenland is to evaluate the character, intensity, and the extent of the Caledonian orogenesis. Two basically different viewpoints have been presented: one considers most of the fold belt with all its varying infracrustal and supracrustal rock units and structures to have essentially formed during the Caledonian orogeny (Haller 1971); the second view interprets the region affected by the Caledonian fold belt in East Greenland as of polyorogenic nature, including large areas of pre-Caledonian basement rocks which were only partly reworked during the Caledonian orogeny (Henriksen and Higgins 1976; Higgins et al. 1981).

Geological Main Units

Four main lithostructural units can be distinguished:

1. Riphean–Ordovician sedimentary sequence. This is an Upper Riphean to Middle Ordovician sequence of sediments with a cumulative thickness of more than 17 000 m. It is generally considered as one coherent sedimentary sequence of miogeosynclinal character deposited in a shallow water regime during a period of several hundred million years. The age of the lowest levels has not been determined, but may be only slightly older than the overlying, acritarch-bearing, Upper Riphean strata (Vidal 1981).

2. Caledonian and pre-Caledonian metamorphic crystalline complexes. These are the dominant exposed rock units and comprise both infracrustal rocks (gneisses, migmatites, granites) and supracrustal sequences (schists, quartzites, marbles). The infracrustal and supracrustal units are frequently deformed together in intricate fold patterns. At least three different age groups of supracrustal sequences can be distinguished. The youngest are the metamorphic representatives of the Upper Riphean sedimentary sequence. Older sequences include middle Proterozoic schists, a sequence of early Proterozoic metavolcanics and metasediments, and possibly still older minor supracrustal bands within Archaean gneiss units. The crystalline complexes preserve isotopic evidence of at least three pre-Caledonian orogenic episodes, as well as a varying degree of Caledonian reactivation with formation of new metamorphic rock units.

3. Caledonian intrusives. Late to post-kinematic acid to intermediate plutonic intrusions are widespread in central East Greenland. They are concentrated particularly along the boundary zone

between the Riphean–Ordovician sedimentary sequences and the adjacent crystalline complexes, but have a more widespread occurrence in the southern part of the fold belt. A number of intrusions can be dated by their cross-cutting relationships to sediments of known stratigraphic age, but others have only been identified as Caledonian using isotopic age determinations. These indicate that Caledonian intrusions were formed over a span of time from 560 to 380 Ma ago.

4. Late–post-Caledonian deposits. After the main phase of the Caledonian orogeny, probably in the early Silurian, uplift and denudation began and resulted in deposition of extensive clastic sedimentary sequences of molasse type. The Middle Devonian to Lower Permian sandstones and conglomerates have a composite thickness of more than 11 000 m, and accumulated in intramontane basins; the older units were subjected to superficial late orogenic folding and local thrusting.

Riphean–Ordovician Sedimentary Sequence

This Upper Riphean to Middle Ordovician sedimentary sequence (Fig. 2) outcrops mainly in a c. 250 km long N–S trending belt in the central fjord zone, while parts of the succession are preserved in other areas (Fig. 3). The sediments are folded mainly in large open folds, but otherwise are generally only slightly to moderately affected by Caledonian deformation and metamorphism. Only in the lowest part of the sequence, in the boundary zone with the metamorphic complexes, have the sediments been significantly metamorphosed and a schistosity developed.

Eleonore Bay Group

This comprises a lower unit up to 9000 m thick of dominantly psammitic and semipelitic sediments, and an upper unit up to 4000 m thick of well banded quartzite, mudstone, and carbonates. The sequence shows a grossly similar lithological development over a region 450 km from north to south and 200 km from east to west, and reflects an extensive, stable, depositional environment.

The subdivisions and main lithology are summarized in Fig. 2, which refers to the main outcrop area (Fränkl 1951, 1953a, b; Haller 1971). In other areas a different local nomenclature may be used, but this is not referred to in this review paper.

The age of the Eleonore Bay Group can only be inferred from studies of acritarchs and stromatolites.

Age-diagnostic acritarchs recovered from the Upper Eleonore Bay Group suggest a Late Riphean age (Vidal 1976), but stromatolites from some of the same units have been correlated with Vendian forms (Bertrand-Sarfati and Caby 1976).

The Lower Eleonore Bay Group reaches a maximum thickness of c. 9000 m in the southern part of the central fjord zone, but in other areas only 2000–4000 m of the sequence has been recorded. Sedimentological investigations in the southern part of the area (Caby 1976b) indicate that the sediments are very shallow water deposits, laid down on a subsiding and fluctuating, wide deltaic zone. The sequence is thus not a geosynclinal sequence of interbanded quartzites and greywackes as previously described (Haller 1971).

The Upper Eleonore Bay Group comprises three main divisions (Fig. 2) and many subdivisions (Eha 1953; Fränkl 1953a; Katz 1952). The sequence shows an exceptionally constant lithological development over the whole outcrop area (Fig. 3). The Quartzite 'series' comprises c. 2000 m of well sorted sandstones, quartzites, siltstone, and shales with sedimentary structures indicating shallow water deposition. The Multicoloured 'series' is a c. 1000 m thick succession of strongly coloured banded shales, mudstones, and carbonates probably deposited on a stable marine shelf. Stromatolites are abundant in certain levels throughout the whole outcrop area. The Limestone–Dolomite 'series' is up to 1000 m thick and exhibits sedimentary features suggesting shelf-type marine deposition. Stromatolitic biostromes imply that large areas were formed in a subtidal environment (Caby 1972).

An undisturbed lower contact of the Eleonore Bay Group is not exposed. Everywhere the lower levels are within the realm of Caledonian metamorphism and where the base is seen, it is marked by a movement or a décollement zone over older crystalline rocks. Various marble–greenschist associations found in the metamorphic complexes along the Inland Ice have been interpreted as a 'Basal Series' of the Eleonore Bay Group (Haller 1971), but at least some of these can be demonstrated as belonging to some of the older Proterozoic supracrustal sequences (Henriksen and Higgins 1976).

Tillite Group

This is a c. 500–800 m thick sequence of sediments with two distinct tillite levels of regional extent (Fig. 3). The lowest level rests with a slight angular unconformity on Eleonore Bay Group sediments. Five formations can be distinguished comprising the two tillite formations, inter-tillite sandstone, and two

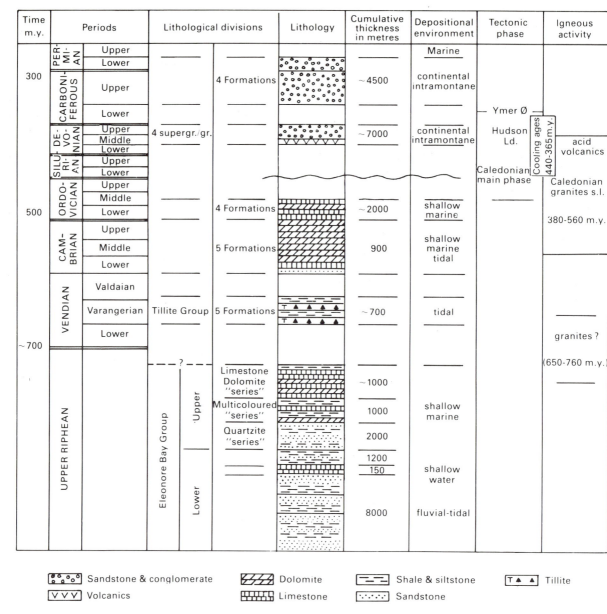

Fig. 2   Age and lithology of Upper Riphean to Permian stratigraphical units

overlying carbonate and mudstone formations. A review of the Tillite Group has recently been presented by Higgins (1981).

Both of the tillite formations may comprise several tillite horizons. The boulder assemblage in the lower tillite formation comprises mainly rocks from the underlying Upper Eleonore Bay Group whereas the upper tillite formation contains many crystalline blocks, probably derived from the metamorphic complexes. A Rb–Sr whole rock isochron on an assemblage of red granitic boulders, by P. J. Leggo (in Spencer 1978), indicates an age of 1830 Ma.

In tectonic windows in the inner Scoresby Sund region two tillite occurrences have been found (Fig.

3), resting on a crystalline basement (Phillips and Friderichsen 1981; Henriksen 1981). Although there are no firm stratigraphic indications it is most likely that these isolated tillite occurrences are of the same age as those in the central fjord zone.

Acritarchs recovered from the Tillite Group suggest a Varangerian age (Fig. 2), which implies the existence of a considerable hiatus between the Upper Riphean Eleonore Bay Group and the Lower Tillites (Vidal 1976, 1981). Comparison with other middle Vendian Tillites from the North Atlantic region suggests that the East Greenland tillite deposits were also formed by melting of a grounded ice sheet during the Varangerian Ice Age (Spencer 1978).

Cambro-Ordovician

The up to 3000 m thick Cambro-Ordovician sequence of dominantly limestones and dolomites was deposited on a shallow marine shelf. The sequence is divided into five Cambrian formations and four Ordovician formations (Cowie and Adams 1957; Cowie 1961; Peel and Cowie 1979; Frykman 1979). Previously it was supposed that a major hiatus spanned the period from Middle Cambrian throughout Upper Cambrian to Early Ordovician, but recently samples from a level previously thought to be Middle Cambrian have yielded Early Ordovician conodonts; there is no longer any evidence for a major hiatus (V.E. Kurtz, 1978 in Frykman, 1979). The uppermost part of the Lower Palaeozoic sequence, previously assumed to reach into the lowermost Cincinnatian, has recently yielded a conodont fauna indicating an upper Chazyan age (Smith 1982). There is thus no evidence of post-Middle Ordovician sedimentation in central East Greenland prior to Caledonian deformation.

The lowest Cambrian formation is a c. 125 m thick sequence of dominantly sandstones with trace fossils, interpreted as a tidal to shallow marine succession. The first body fossils occur in limestones just above the sandstone formation and include olenellids, brachiopods, and hyolithids. The succeeding sequence of c. 2800 m of alternating limestones and dolomites contain a diversified shelf type fauna referable to the Pacific province. The sedimentary structures indicate generally very shallow water deposition on a relatively stable shelf. There are no angular unconformities in the succession and this precludes any pre-Upper Ordovician folding in the Palaeozoic.

The Cambro-Ordovician strata are only exposed in the central fjord zone and in Canning Land (Fig. 3), but an originally much wider extent of Lower Cambrian strata in East Greenland is suggested by the occurrence of numerous erratic blocks of Skolithos quartzite, transported eastwards from occurrences hidden under the Inland Ice (Haller 1971, Fig. 48). Occasional erratic blocks of Ordovician limestones found near the Inland Ice also indicate an originally wider extent for these deposits (J. S. Peel, personal communication).

Caledonian and Pre-Caledonian Metamorphic Complexes

The Caledonian fold belt in central East Greenland is dominated by gneisses, migmatites, schists, and granites, which on a lithostructural basis can be grouped into a number of metamorphic complexes.

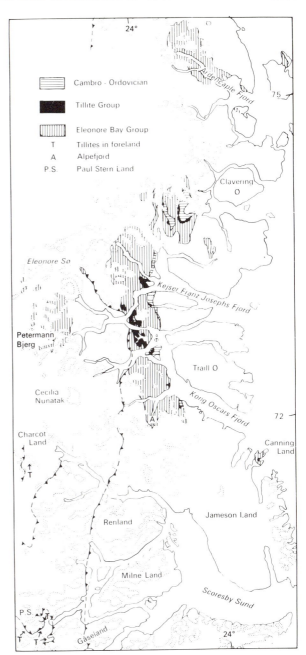

Fig. 3 Outcrops of Upper Riphean–Ordovician sedimentary sequence. After Higgins and Phillips (1979)

The oldest complexes are of Archaean origin, but were later influenced by one or more Proterozoic orogenic events and finally partially reactivated during the Caledonian orogeny. Younger metamorphic complexes were formed both during the early Proterozoic and the middle Proterozoic, and these were also partly reactivated by Caledonian events. Caledonian metamorphic complexes were formed partly by deformation and metamorphism of the lowest part of the Riphean–Ordovician sedimentary sequence, and also by strong reactivation, metamorphism, and migmatization of parts of the

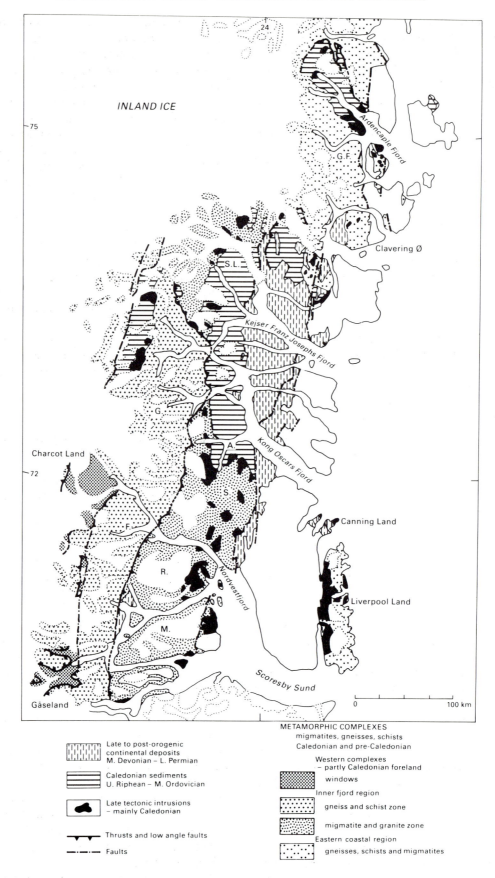

Fig. 4 Outline map of the Caledonian fold belt between 70°–76°N. Place names: M: Milne Land; R: Renland; F: Flyverfjord, S: Stauning Alper; A: Alpefjord; G: Gletscherland; S.L.: Strindberg Land and G.F.: Grandjean Fjord

pre-Caledonian crystalline complexes. This intense Caledonian reactivation is mainly concentrated in the eastern part of the inner fjord zone, where the Caledonian plutonic activity is also conspicuous (Fig. 4). All the different metamorphic complexes are welded together and at the present level of exposure form an extensive region of crystalline rocks characterized by a complex deformational pattern, which contrasts strongly with the simple deformation style in the adjacent belts of Riphean–Ordovician sedimentary rocks.

In the southwest, crystalline rocks occur in two tectonic windows exposing parts of the Precambrian Greenland shield beneath major thrust units derived from more easterly parts of the fold belt (Fig. 4). In the inner fjord zone a broad region of crystalline rocks, divisible into two distinct zones (Fig. 4), can be traced from c. 70° to c. 74°30′ north. This region has usually been referred to as the 'central metamorphic complex' (Haller 1955, 1971; Henriksen and Higgins 1976), because it is situated centrally between two belts of little or non-metamorphic Riphean–Ordovician sediments. In the eastern coastal regions further outcrops of crystalline complexes occur. These are geographically separated from the complexes found in the inner fjord zone, but their general lithological and structural character suggest that they may have a very similar geological history.

Tectonic windows

The Gåseland window in the southwest (Fig. 4) exposes gneissic basement rocks with abundant amphibolite bands and occasional augen granites. These reflect several phases of deep-seated folding, migmatization, and metamorphism, and are considered to be part of an Archaean–early Proterozoic complex. Above them occurs an autochthonous to parautochthonous sequence of marbles and chlorite schist with local tillites at the base (Phillips et al. 1973). The tillites have been correlated with the Varangerian tillites in central East Greenland, and the autochthonous to parautochthonous sediments therefore might form part of the Riphean–Ordovician sedimentary sequence. Major thrusts above this weakly metamorphosed unit outline the window and are overlain by major westwards directed thrust sheets dominated by high-grade middle Proterozoic schists (Krummedal supracrustal series).

The Charcot Land window (Fig. 4) comprises an older gneiss and granite complex of probable Archaean to early Proterozoic age, overlain by the Charcot Land supracrustal sequence of c. 2000 m of basic metavolcanics, marbles, semipelitic and quartz-itic metasediments. Both the older gneisses and granites and the Charcot Land supracrustals are cut by two post-kinematic granodioritic and granitic intrusions, which were emplaced c. 1840 Ma ago (Hansen et al. 1981). The supracrustal sequence and its basement were affected by prograde regional metamorphism, which attained low greenschist facies to amphibolite facies prior to the emplacement of the granites (Steck 1971). Isotopic age determinations suggest this metamorphic event took place c. 1900–1850 Ma ago. A local tillite occurrence, resting unconformably on the early Proterozoic granodiorite, is probably correlatable with the Varangerian tillites of central East Greenland (Henriksen 1981). Major thrusts arch over the above mentioned units, outlining the window, and are overlain by major thrust units of high grade schists (Krummedal supracrustal sequence) displaced westwards over the foreland window for at least 40 km.

If the correlation of the tillites in both windows with the Varangerian tillites is correct then the major westwards-directed thrusts must be of Caledonian age.

Metamorphic complexes of the inner fjord zone

These form the broad core region of the exposed part of the fold belt in central East Greenland (Fig. 4). This region of mainly high grade gneisses, migmatites and schists, can be traced more than 400 km from north to south between 70°–74°30′N, and it is up to 200 km wide.

The early investigators considered this region as mainly comprising pre-Caledonian basement rocks (Wordie 1927; Odell 1944), whereas others considered it to consist of Caledonian complexes formed as a result of deep-seated orogenic activity affecting the rock deposited in the Caledonian geosyncline (Wegmann 1935; Haller 1971). Recent mapping in the southern part of the fold belt (Henriksen and Higgins 1976), together with a large number of isotopic age determinations (Steiger et al. 1979; Hansen et al. 1978; Rex and Gledhill 1981), supplemented by various detailed investigations in key areas (Higgins et al. 1981), have led to the conclusion that this region of metamorphic complexes attained its present character during at least three pre-Caledonian orogenic episodes as well as during Caledonian reactivation. In the Caledonian orogeny the whole region was subjected to regional metamorphism and parts of it were mobilized and variously deformed.

The metamorphic complexes of the inner fjord zone can be divided into two zones separated by a major thrust or low angle fault zone, traceable from south to north for almost 400 km. To the west the metamorphic complexes can be characterized as a 'gneiss and schist zone', comprising an Archaean–early Proterozoic gneissic basement overlain by a middle Proterozoic cover sequence of quartzites and schists. To the east of the tectonic lineament occurs a 'migmatite and granite zone' dominated by migmatites and granites in southern areas but with more extensive occurrences of the middle Proterozoic supracrustal rocks in the north. This eastern migmatite and granite zone appears mainly to have formed as a result of middle Proterozoic orogenesis, but the zone was also the site of a substantial Caledonian reactivation.

Gneiss and schist zone

The oldest recognized unit in the gneiss and schist zone is the Flyverfjord infracrustal complex with outcrops between latitudes 70°–72°N. It forms a gneissic basement to the overlying younger cover series of mainly metasedimentary schists: the Krummedal supracrustal sequence. Both units are generally structurally conformable along their mutual boundaries and they are folded together with a varying degree of intensity.

The Flyverfjord infracrustal complex is a characteristic, amphibolite banded, gneiss with abundant ultrabasic lenses (Henriksen et al. 1980). It is intensely folded in a structural pattern reflecting several superimposed medium and large scale folds. A conspicuous feature is the widespread presence of swarms of folded discordant amphibolitic dykes, which post-date at least two phases of deformation, but which were themselves affected by a third phase of deformation. A Rb–Sr whole rock age of 2935 Ma has been obtained from the gneisses in Flyverfjord (Rex and Gledhill 1974; all Rb–Sr whole rock ages cited in the present paper are calculated, or recalculated, with a decay constant of λ_{87Rb} = 1.42×10^{-11} y; Rex and Gledhill 1982). Zircon ages on various foliated granites from the same region gave 2300–2630 Ma (Steiger et al. 1979), and also reflect an early Proterozoic to Archaean age for this complex.

The Flyverfjord infracrustal complex south of 72°N can be traced northwards into the so-called Gletscherland migmatite complex (Haller 1955). The latter has yielded a Rb–Sr whole rock age of c. 2450 Ma (Rex and Gledhill 1981). The Gletscherland migmatite complex comprises gneiss and migmatite units folded together with supracrustal rocks: marble–amphibolite associations and rusty red weathering metasediments. Gneisses and

dioritic rocks in the eastern part of the complex have given Rb–Sr whole rock isochrons of 1830 and 1705 Ma respectively (Rex and Gledhill 1981), which indicate that parts of the complex also evolved during an early Proterozoic orogenic episode, whose significance cannot at present be outlined precisely. The various supracrustal elements have not been dated. The structural pattern in the Gletscherland migmatite complex is dominated by major E–W trending tight and recumbent folds, which face both northwards and southwards. The transverse character of these folds (Fig. 5) has previously been interpreted as inherited from a pre-Caledonian basement (Wegmann 1935).

Between latitudes 73°–74°N there occur several areas of infracrustal rocks overlain and largely enveloped by later metasedimentary rocks. The former include the so-called Niggli Spids dome and Hagar sheet of Haller (1955) as well as other less well defined areas (Higgins et al. 1981). The Hagar sheet has yielded five Rb–Sr whole rock isochron ages of 1575–1725 Ma on strongly deformed granites and gneisses (Rex and Gledhill 1981). No isotopic data is available for the other infracrustal units, although a comparable age of development is suspected.

The Krummedal supracrustal sequence in its type area between 70°–72°N forms a cover on the Flyverfjord infracrustal complex. It is more than 2500 m thick and is dominated by rusty red weathering, well-banded, psammitic and pelitic metasediments (Higgins 1974). Locally the metasediments have preserved their primary features, but they are generally strongly deformed and schistose and have been recrystallized under amphibolite facies conditions.

Rusty red weathering metasediments similar to the Krummedal supracrustal sequence form a characteristic unit between 72°–74°N, where they are intensely folded together with the infracrustal gneissic rocks. In some regions the metasediments form envelopes around gneissic cores in major recumbent folds or nappes. Most of the characteristic rusty red metasediments throughout the region are probably equivalent in age to the Krummedal supracrustal sequence. Rb–Sr whole rock age studies on metasediments throughout the inner fjord zone (70°–74°N) suggest metamorphic ages between 1000–1250 Ma (Hansen et al. 1981; Rex and Gledhill 1981). Although the supracrustal rocks are generally conformable to the underlying infracrustal units, discordant relationships are locally preserved and interpreted as basement–cover contacts (Higgins et al. 1981).

The interfolding of the Archaean–early Proterozoic infracrustal units and the middle Proterozoic cover was most probably a middle Proterozoic orogenic event; it cannot be older, but a younger age is possible.

Migmatite and granite zone

The eastern complexes in the inner fjord zone (Fig. 4) are comprised largely of migmatitic metasedimentary rocks and granites described as the Gåseland–Stauning Alper migmatite and granite zone (Henriksen and Higgins 1976). The western boundary of this zone is a prominent N–S trending fault-thrust lineament. The migmatite and granite zone is dominated by gneissic migmatites and various synkinematic to post-kinematic plutonic rocks. Representatives of the Krummedal supracrustal sequence form the main part of the palaeosome bands, lenses, and schlieren in the migmatites, and occasionally occur as relatively little migmatized intact sequences. The migmatite and granite zone has a complex genesis, and is the net result of two major orogenic events both involving migmatization, deformation, and plutonism related to both. During the first of these orogenic events major sheets of augen granites were intruded into the middle Proterozoic supracrustal rocks, and both units folded together into major recumbent folds of nappe dimensions (Chadwick 1975). A major phase of migmatization, granite formation, and regional high grade metamorphism was related to the same orogenic event; Rb–Sr whole rock isochron and zircon determinations on the augen granites place this event at 987 Ma and 1053 Ma ago respectively (Steiger et al. 1979).

Late to post-kinematic plutonic intrusions are widespread mainly in the eastern part of the migmatite and granite zone. Age determinations indicate that many of these intrusions are Caledonian. A prominent hypersthene monzonite sheet outcropping in eastern Renland (Chadwick 1975) forms a major stratigraphical marker unit. The sheet is somewhat warped but generally not folded. It cuts through the strongly folded, highly migmatized, middle Proterozoic gneisses with augen granite sheets, but is itself affected by a later phase of migmatization and granite injection. The monzonite sheet was injected c. 475 Ma ago (Steiger et al. 1979) and it can be used to separate a middle Proterozoic first phase of migmatization and deformation from a Caledonian second phase of migmatization and granite formation. It is thus possible to recognize the occurrence of Caledonian migmatization and metamorphism, superimposed on the middle Proterozoic migmatites, though its extent can only be estimated indirectly. It seems to have been mainly concentrated in the eastern part of the migmatite and granite zone, where Caledonian plutonic intrusions are also concentrated (see below).

Metamorphosed Eleonore Bay Group sediments

Broad belts of low metamorphic to non-metamorphic Upper Riphean–Ordovician sedimentary rocks are found to the east and west of the metamorphic complexes at the inner fjord zone (Fig. 4). When traced towards the metamorphic complexes, and stratigraphically downwards, the sediments become progressively more metamorphosed and grade generally through a chlorite, a biotite, and a garnet-bearing zone. Adjacent to this follows, normally conformably, a zone dominated by rusty red kyanite-bearing schists and quartzites, which form the bulk of the supracrustal sequences (Krummedal supracrustal sequence) recognized in the metamorphic complexes. The apparent gradual transition from unmetamorphosed Eleonore Bay Group sediments into high grade schists infolded with infracrustal rocks in the metamorphic complexes, was the main evidence for the assumption that the metamorphic complexes were of Caledonian formation (Haller 1971). Recent reinvestigations of some of the key transition zone, however, have shown that the transition is not completely gradational (Higgins et al. 1981). Between the garnet grade schist and the underlying slightly migmatized kyanite schists, a tectonic décollement plane can usually be distinguished separating the overlying metamorphic Riphean sediments from the older metamorphic complexes below. The gradational transition is thus only apparent, and it is not possible to correlate stratigraphically the moderately deformed and metamorphosed Eleonore Bay Group metasediments with the bulk of the strongly deformed and metamorphosed schist sequences in the metamorphic complexes.

Metamorphic complexes in the eastern coastal regions

In the eastern regions of the fold belt two large areas of crystalline rocks occur, which are physically separated from the corresponding areas in the inner fjord zone (Fig. 4). In the southeast, Liverpool Land forms an isolated horst area bordered to the east by the sea and to the west by Mesozoic sediments. In the northeast, north of Clavering Ø (c. 74°N), crystalline rocks cover a large region between the sea and the Inland Ice.

Liverpool Land is a strongly deformed crystalline complex of various units of gneisses, migmatites, and supracrustal rocks. An extensive suite of plutonic intrusions occurs in the same region, and can be divided into an early synkinematic group of basic and granodioritic types and a later group of post-kinematic granitic types. The last group is probably mainly of Caledonian origin. However, so far only a few mineral age dates have been carried out in the Liverpool Land area (Hansen and Steiger 1971), and the intrusive age of the plutonic rocks is uncertain.

In Liverpool Land three metamorphic complexes can be distinguished (Coe and Cheeney 1972). The structurally lowest and probably oldest complex occurs in the central part of the area and comprises an at least 2500 m thick sequence of well banded flat-lying granitic gneisses with conformable marble and amphibolite bands. This complex is separated from the adjacent complexes by thrust and shear zones. A second metamorphic complex can be distinguished in northern areas, and this comprises a several thousand metres thick sequence of metasediments and their migmatized equivalents, the last forming the main part of the area. The metasediments are lithologically comparable with parts of the middle Proterozoic sequences known from the inner fjord zone. The third complex, outcropping in the south, comprises a structurally lower part of migmatites and metasediments and a higher part of gneissic granodiorites which eastwards grade into their more homogenous plutonic equivalents. Two mineral age dates from this third complex giving 1182 and 1125 Ma (Hansen et al. 1973) suggesting that it may be of Precambrian age. Of special interest in this complex is the occurrence of eclogitic rocks in isolated lenses (Smith and Cheeney 1981).

The general lithological similarities between some of the metamorphic units in Liverpool Land and those in the inner fjord zone suggest that the metamorphic complexes in Liverpool Land are also probably of Precambrian age. The few mineral ages from Liverpool Land are in agreement with this interpretation. However, it is probable that the Liverpool Land area constitutes a composite complex with elements from several orogenic events.

The crystalline region north of Clavering Ø is characterized by interfolded infracrustal and supracrustal rocks, which occur on both sides of a graben structure containing Eleonore Bay Group sediments at Ardencaple Fjord. The infracrustal rocks include migmatitic gneisses and foliated granites with various schist bands. The metamorphosed supracrustals are generally psammitic and pelitic schists, which have been interpreted as equivalents of the Eleonore Bay Group (Haller 1970, 1971). A large proportion of these metasediments are, however, lithologically very similar to the middle Proterozoic sequences known from the inner fjord zone, and it is therefore quite likely that those north of 74°N are of pre-Caledonian origin.

Two sets of interfering folds are recognized in the metamorphic complexes north of Clavering Ø: an older set of NNE-trending folds and a younger NNW-trending set. These are both, according to Haller (1970, 1971), Caledonian structures and both are associated with intense migmatization. The interpretation of the structures as Caledonian is clearly associated with Haller's assumption that the metamorphic supracrustal rocks are Eleonore Bay Group equivalents. If they are older the structures in the metamorphic complex might well be of pre-Caledonian origin, and the whole complex then essentially pre-Caledonian. At present there are no radiometric age determinations to support either view.

Caledonian Plutonic Rocks

A large group of mainly late- to post-kinematic plutonic rocks characterizes certain zones in the fold belt (Fig. 4). In central and northern areas the intrusions were mainly emplaced in the boundary zone between the Upper Riphean sediments and the adjacent metamorphic complexes, whereas in southern areas plutonic bodies are widespread within the crystalline complexes. The Caledonian age of most of these intrusions can be demonstrated by their discordant relationships to the Eleonore Bay Group sediments, and/or by Rb–Sr whole rock isochron age determinations, which indicate that the intrusions were emplaced between 560–380 Ma (Fig. 6). Most of the intrusions are large stocks or sheet-formed bodies of homogeneous to porphyritic biotite muscovite granite, with in some cases associated aplitic and pegmatitic veins. The plutonic activity has been most intense in the southern part of the area, where the intrusions reflect a more deep-seated level, and where some areas are characterized by associations of intrusions of various rock types. For example, in east Milne Land the oldest Caledonian intrusion is a sheet-like granodiorite giving a Rb–Sr whole rock isochron age of 453 Ma (Hansen and Tembusch 1979). This is cut by a coarse-grained mafic, quartz monzonite intrusion giving a zircon age of 425 Ma (B. T. Hansen, personal communication) and two different types of leucocratic biotite granites. The youngest granite has given a Rb–Sr whole rock isochron of 373 Ma (Hansen and Tembusch 1979).

Nearby in eastern Renland, a sheet of hypersthene monzonite up to 500 m thick forms a lopolith-like structure (Chadwick 1975), and has given a Rb–Sr whole rock isochron age of 475 Ma. It was followed by a late phase of granite formation with an associated phase of migmatization (see above).

The intrusions along the boundary between the Upper Riphean sediments and adjacent crystalline rocks include some which were involved in folding but most were emplaced after Caledonian deformation. They are mainly simple two-mica granites which form sheet-like bodies or more frequently major plutons. Rb–Sr whole rock determinations mainly give ages between 480–430 Ma, but examples of older (560–550 Ma) and younger intrusions (380 Ma) have been found (Rex and Gledhill 1981). A small granitic stock occurring in the Middle Devonian molasse sequence should also be referred to the Caledonian plutonic phase. It is associated with acid volcanics which occur locally in the eastern part of the Middle Devonian sequence.

Structures

Two distinct structural regimes characterize the Caledonian fold belt in central East Greenland. In the metamorphic complexes the fold patterns are composite, often with several superimposed phases of folding reflecting their polyorogenic development. Structures occur on all scales from immense disharmonic folds of nappe character, through mesoscopic folds to minor and intrafolial folds. The supracrustal mantle of Upper Riphean–Ordovician sediments is characterized by moderate deformation, and shows regular developments of N–S and NW–SE-trending open folds mainly formed during a single phase of deformation (Fig. 5). These folds are younger than the Middle Orovician sediments at the top of the sedimentary sequence and clearly of Caledonian age. A comprehensive account of the structural geology accompanied by a tectonic map at a scale of 1:500 000 has been presented by Haller (1970).

The age of the structures in the infracrustal complexes is a matter of debate. One view is that the structural development is essentially of Caledonian origin and was formed during formation of large migmatitic upwellings of infracrustal material which encroached upon and were intensely folded together with the overlying metasedimentary sequences; the latter were interpreted as transformed equivalents of the sediments deposited in a Caledonian geosyncline (Haller 1971). Another view is that the metamorphic complexes were to a large extent formed by pre-Caledonian rocks and that the structures reflect a pattern of superimposed sets of folds, which include Archaean, early and middle Proterozoic, and Caledonian elements (Henriksen and Higgins 1976; Higgins et al. 1981). This second viewpoint is largely adopted here. A later Caledonian deformation can be distinguished in some areas, but its regional extent and intensity is difficult to assess. In general it is now believed that the Caledonian deformation is of limited significance in the infrastructural levels.

The boundary zone between the infrastructure and the suprastructure is characterized by a shear zone or décollement zone. In both the above mentioned interpretations, it is considered as a zone of intense deformation, which separates two structural regimes and accommodates the differences in deformational style. In the field it is not everywhere possible to define precisely the décollement zone, but nevertheless it exists and is a zone of great structural importance.

The lateral shortening observed in the Riphean–Ordovician sediments is very limited. In the central part of the fjord zone figures of c. 5 per cent and c. 15 per cent have been calculated (Eha 1953; Higgins et al. 1981). In the southeastern part of the region the Caledonian structures have been considered as tensional features related to gliding tectonics (Caby 1976a). These observations indicate that crustal shortening in this part of the fold belt was restricted.

Caledonian crustal shortening is, however, considered to have taken place along the west margin of the fold belt. This is most marked in the southwestern border zone, where major westwards directed thrust units have been observed in a zone along the Inland Ice. The thrusts bring units of schists and gneisses from the interior parts of the inner fjord zone over the Precambrian foreland area, with outcrops in tectonic windows. The minimum displacement is 40 km, but thrust units may have been displaced for much larger distances (Homewood 1973; Phillips et al. 1973). The age of thrusting is generally considered from geological arguments to be Caledonian, though direct evidence is only seen in eastern North Greenland (Hurst and McKerrow 1981).

A series of N–S trending open folds and minor thrusts affect the Middle and Upper Devonian Old Red Sandstone sequence (Fig. 5). These structures are generally considered as the latest Caledonian phases of deformation, and locally also affect pre-Devonian rocks. Deformation took place intermittently through the Middle and Late Devonian, finally dying away in the earliest Carboniferous.

A major fault or steep thrust zone of supposed Caledonian age can be traced c. 400 km N–S through the metamorphic complexes in the inner fjord zone. The lineament is a major structural feature which separates some of the major lithostructural units. Its displacement is unknown but could be substantial. The northern end of the thrust zone cuts a Caledonian granite dated at 560 Ma, which gives a maximum age for the latest movements on the thrust (Rex and Gledhill 1981). Fault displacements along the same line in the Carboniferous influenced accumulation of the Røde Ø conglomerate in the

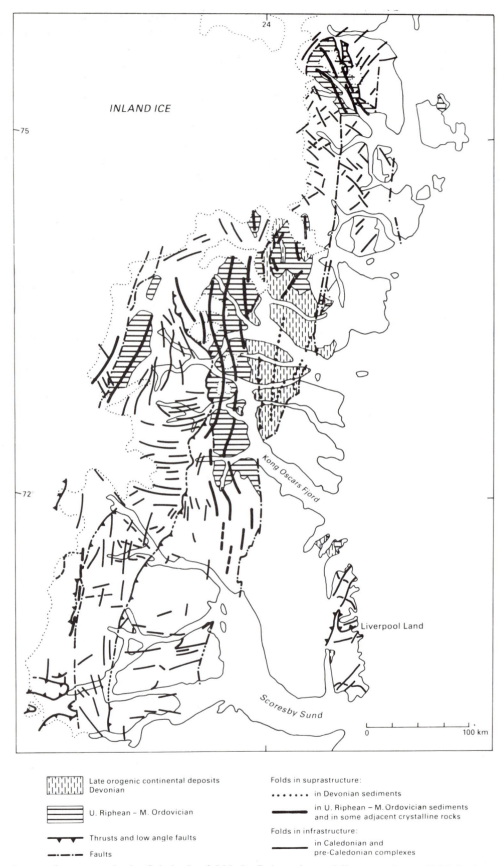

Fig. 5   Sketch map of structures in the Caledonian fold belt. Only major anticlines are shown. Folds in the suprastructure are Caledonian, but in the infrastructure the main pattern is probably of pre-Caledonian origin. Interpretation is based on maps by Haller (1970, 1971) and Henriksen and Higgins (1976)

Scoresby Sund region (Henriksen and Higgins 1976).

Faulting is widespread in both structural regimes and especially conspicuous in the Riphean–Ordovician and Devonian sediments. The fault systems are mainly of late Caledonian age, but with decreasing activity towards the upper part of the Palaeozoic. Most faults trend between NW and NE and are related to tensional movements in a general E–W direction i.e. perpendicular to the main trend of the fold belt.

Metamorphism

A regional study of the metamorphism has been carried out by T. Thyrsted (1979) whose work forms the basis for the following summary.

The metamorphic crystalline complexes are generally characterized by high-grade mineral parageneses, whereas the adjacent belts of Upper Riphean–Ordovician sedimentary sequences are characterized by a downwards increasing metamorphic grade varying from non-metamorphic to lowest amphibolite facies. The metamorphic patterns in the crystalline complexes are of a composite nature reflecting several pre-Caledonian and Caledonian events, which are difficult to distinguish individually due to lack of clear stratigraphical control. It is, however, assumed that the regional Caledonian metamorphism comprised both an intermediate and a low pressure facies series. The former reached into lower amphibolite facies, whereas the latter locally attained granulite facies conditions. Generally, however, the Caledonian metamorphism in the central and western part of the region only reached greenschist facies, and is seen most commonly as a retrogression of the earlier metamorphic assemblages. The pre-Caledonian metamorphic events include: (1) A middle Proterozoic (?) greenschist–upper amphibolite facies event of mainly intermediate pressure type; (2) An Archaean–early Proterozoic amphibolite facies. A prograde greenschist–amphibolite facies transition, probably related to an early Proterozoic event has been described from the Charcot Land window in the southwest of the region (Steck 1971; Hansen et al. 1981).

Late to Post-Caledonian Deposits

After the Caledonian orogenic main phase a thick sequence of continental clastic sediments was deposited in intermontane basins in central East Greenland (Fig. 4). The deposits comprise Middle to Upper Devonian non-marine conglomerates, sandstones, and arkoses with a cumulative thickness of c. 7000 m, and continental Upper Carboniferous to Lower Permian clastic deposits with an aggregate thickness of c. 4000–5000 m. A new lithostratigraphical subdivision of the Devonian strata supplemented by detailed sedimentological investigations has recently been presented by Friend et al. (1976), based on previous work by Bütler (1959, 1961a).

The Devonian sediments were deposited in a number of local intermontane basins to which the sediments were transported from north, south, east, and west at various times (Alexander-Marrack and Friend 1976). From the sedimentation patterns it is concluded that a substantial amount of material was derived from an eastern source reflecting the existence of an extensive uplifted landmass east of the present coastline of East Greenland. At the west margin of the Devonian outcrop area mainly sandstones were deposited by generally eastwards flowing river-fan systems, and in the centre of the area the deposits were laid down from mainly southwards flowing river systems (Yeats and Friend 1978).

Devonian volcanism of both acid and basic character took place intermittently with the deposition of the clastic sediments. Basic dykes and sills are widespread and can be related to four intrusive events. Acid volcanics with flows and pyroclastics were formed locally and reflect at least two extrusive events. Two Devonian granites, forming a small stock and a laccolith-shaped body, are probably related to the acid volcanic activity.

The Devonian deposits were affected by a series of late Caledonian disturbances (see above) forming sets of open folds and with minor thrusting and faulting (Haller 1971). The character of this structural pattern reflects intermittent compressional and tensional events.

Upper Carboniferous–Lower Permian sediments rest with an angular unconformity on the Upper Devonian indicating a hiatus and a phase of deformation between the two sequences. The continental Carboniferous–Lower Permian succession comprises clastic sediments with conglomerates, arkoses, and sandstone. There are strong lateral facies variations indicating numerous centers of accumulation. The stratigraphical division into at least four formations is based on works by Bütler (1961b) and Perch-Nielsen et al. (1972).

The Lower Permian continental strata are considered to be the latest developments in the late to post-Caledonian molasse sequences (Haller 1971). A regional marine transgression in Upper Permian time marked the beginning of the following cratogenic condition which characterizes Mesozoic and Tertiary development.

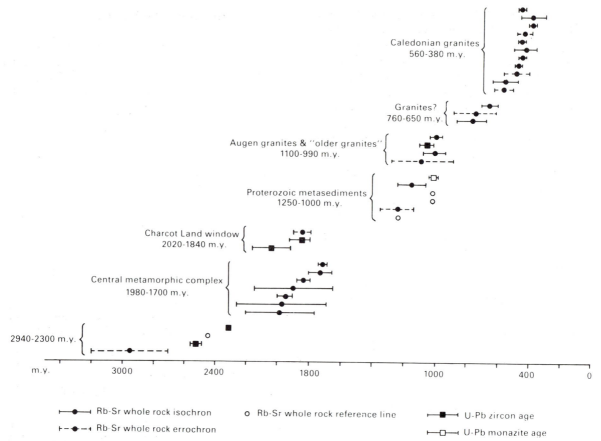

Fig. 6   Geochronological divisions in central East Greenland. Based on Rb–Sr whole rock, zircon, and monazite age determinations. Compiled after: Hansen *et al*. (1976, 1978, 1979, 1981), Rex and Gledhill (1974, 1981) and Steiger *et al*. (1979)

Age of the Caledonian Orogeny in Central East Greenland

On stratigraphical evidence the main phase of the Caledonian orogeny in central East Greenland can be bracketed between the Middle Ordovician and the Middle Devonian. The youngest Ordovician sediments are referred to the middle Middle Ordovician (Chazyan stage). The oldest deposits in the molasse sequence, following the orogenic main phase, are referred to the lower part of the Givetian (or Couvinian?).

After the main orogenic phase there followed a series of late Caledonian gentle fold phases with local thrusting, which affect the molasse sequence and adjacent older units. On a stratigraphical basis four minor deformation phases of mainly Late Devonian age can be distinguished. The latest deformations are referred to the Early Carboniferous (Haller 1971).

In eastern North Greenland the age of the main thrusting is somewhat younger than the main deformation phase in central East Greenland. There the large westwards directed thrusts rest on Upper Silurian shales on the foreland (Hurst and McKerrow 1981).

The substantial number of radiometric age determinations from central East Greenland clearly indicate that Caledonian plutonism, and presumably also metamorphism, took place over a long period. Caledonian granites were emplaced from c. 560–380 Ma with a main grouping between 480 and 430 Ma (see Fig. 6). The most precise evaluation at present of the age of the main deformation in the Caledonian suprastructure is indicated by the time span between deposition of the youngest sediments, which are Chazyan (corresponding to an age of c. 477 ± 11 Ma on a North American time scale; Barnes, Norford, and Skevington 1981) and the intrusive age of 445 ± 5 Ma for a post-tectonic granite, which clearly cuts the folded Eleonore Bay Group sediments (Rex and Gledhill 1981). Due to the lack of other stratigraphic evidence the age of the Caledonian main orogenesis in central East Greenland can only be evaluated from the existing age determinations. While most of the late to post-tectonic intrusions have been dated in the interval 480–430 Ma, many of the determinations have large uncertainties. They cannot, therefore, be considered to reflect precise intrusion ages and the results should be interpreted with prudence. As the intrusions are late to post-tectonic their ages should indicate minimum ages for the main phase of Caledonian diastrophism. It must therefore be assumed that the most likely time for the main orogeny is, very approximately, Late Ordovician. In this connection it is interesting to note that while Caledonian sedimentation lasted into the Middle Ordovician, no sign of synsedimentary tectonic disturbances have been recorded.

Rb–Sr and K–Ar mineral age dates from the region 70°–74°N fall in the range 440–365 Ma (Hansen et al. 1971, 1972, 1973; Rex and Higgins 1981). They are interpreted as cooling or uplift ages. Some other mineral ages are interpreted as an excess argon pattern. The age group between 440–365 Ma corresponds to the interval between earliest Silurian and earliest Carboniferous, and corresponds well with the beginning of uplift which generated the source area for the Middle Devonian to Lower Permian molasse deposits.

Summary and Conclusions

In central East Greenland the Caledonian fold belt is developed along the edge of the Precambrian Greenland shield, which also forms the basement of the fold belt. Upper Riphean to Middle Ordovician shallow water sediments form a sequence of regional extent several thousand metres thick overlying the crystalline basement, and both units were involved in the main phase of Caledonian orogeny, probably in the Late Ordovician. In the lower levels the basement crystalline rocks were partly reactivated during the Caledonian orogeny and plutonic activity led to formation of granites and in some areas migmatization. In the higher structural levels the Upper Riphean–Middle Ordovician sediments were very moderately folded and reflect only limited crustal shortening. In the western border zone large westwards directed thrusts of probably Caledonian age have a minimum displacement of 40 km.

The partly reactivated basement and the overlying Riphean–Ordovician exhibit two distinct structural regimes, separated by a décollement zone. The lower regime is of a composite polyorogenic nature. It is made up of elements from Archaean, early Proterozoic, middle Proterozoic and Caledonian deformation cycles (Fig. 7). The distinction between the pre-Caledonian and the Caledonian structural elements in the infrastructural levels is not clear and various interpretations have been suggested. Some authors regard most of the rocks and structures

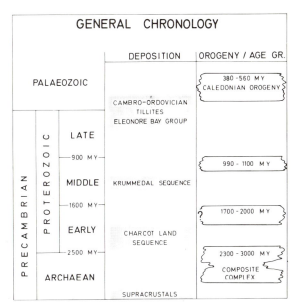

Fig. 7 General chronological scheme showing pre-Caledonian and Caledonian elements occurring in the Caledonian fold belt in central East Greenland

as mainly formed during the Caledonian orogeny (Haller 1971), whereas others consider the infrastructural regimes as dominantly of a pre-Caledonian origin, with rocks and structures primarily reflecting various Precambrian orogenic events (Henriksen and Higgins 1976). A substantial number of radiometric age determinations (Fig. 6) from rocks in the infracrustal complexes show that these rocks are of a pre-Caledonian origin, but the degree of Caledonian reworking is open to discussion. The age of formation of the several superimposed sets of major and minor folds in the infrastructural regime is difficult to determine. Some major nappe-like folds involve both rocks of middle Proterozoic and older ages, and the maximum age of these structures is therefore middle Proterozoic (c. 1000 Ma). They could, however, also be interpreted as Caledonian, although this is considered less likely.

The upper structural regime is clearly of Caledonian origin, as it affects mainly the sequence of Upper Riphean–Middle Ordovician sediments. In some areas it can also be traced into the adjacent metamorphic complexes, where it is superimposed on earlier structures. A pronounced characteristic is the limited lateral shortening of 5–15 per cent, estimated from the generally N–S trending gentle folding of the sediments. In one area the Caledonian deformation has even been explained as tensional and related to gliding tectonics.

A possible model for the two structural levels is that the suprastructure formed a crustal epidermis which was displaced eastwards and westwards off an uplifting crystalline infrastructure mainly composed of partly reactivated pre-

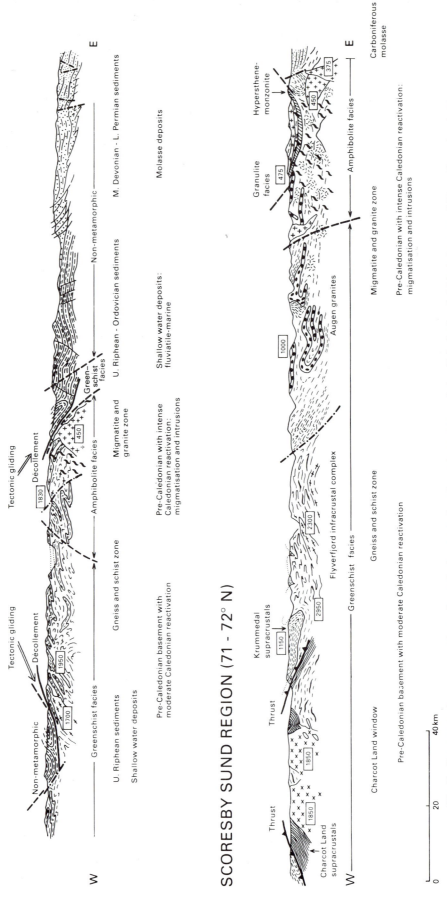

Fig. 8 Interpretative cross-sections through the central and southern parts of the Caledonian fold belt in central East Greenland. The intensity of Caledonian metamorphism and migmatization is indicated in the text below the profiles, where the corresponding pre-Caledonian features are not specified. Representative age determinations in Ma for major age groups are indicated in frames

Caledonian basement and Caledonian plutonic rocks (Fig. 8). Discovery of major recumbent folds on the eastern (Caby 1976b) and western (Higgins *et al.* 1981) side of the uplifted infrastructural region, in both cases just above the décollement zone, lend some support to this hypothesis. These two recumbent folds each face away from the central infrastructural region and imply tectonic transport in opposite directions. The model is thus in agreement with a limited crustal shortening, and the structures could even be tensional and related to gliding tectonics.

The Caledonian sediments were only slightly to moderately affected in their lower parts by Caledonian metamorphism, whereas in the underlying metamorphic complexes Caledonian metamorphism reached amphibolite facies. Caledonian plutonic activity was most intense in parts of the migmatite and granite zone (Fig. 4) and here metamorphic effects were also strongest and associated with extensive Caledonian migmatization. This migmatization in some areas was superimposed on pre-Caledonian migmatitic developments, and it is therefore very difficult to estimate more precisely the extent of the Caledonian migmatization. Generally it is only found below the décollement zone and only very locally have the lowermost parts of the Upper Riphean sediments been invaded by neosome material from the underlying migmatites. The highest levels of the Caledonian granite intrusions are sometimes surrounded by a rim of pegmatites and granite veins, which dissect the adjacent rocks including the Riphean sediments. This veining may grade into migmatitic neosome, and it is assumed that in general the Caledonian migmatization and granite emplacement are related.

The sedimentary sequence deposited prior to the Caledonian orogeny includes in its lowest part an Upper Riphean series of dominantly fluviatile deposits locally up to 9000 m thick. These are overlain by Upper Riphean mainly marine sediments, quartzites, shales, mudstones, and carbonates, deposited on a shallow water stable shelf covering an extensive region. After a pronounced hiatus a tillite-bearing succession of Varangerian age was deposited, followed by a new hiatus. The Lower Palaeozoic sequence begins with a thin basal sandstone of Early Cambrian age. The remaining 3000 m thick series of Lower Cambrian to Middle Ordovician sediments are all carbonates, deposited on a stable shelf and containing a shelly fauna related to the Pacific faunal province.

The general character of the Riphean–Ordovician sedimentary sequence shows no evidence of deep water deposits or oceanic crust material, and it must be concluded that all the exposed parts of the fold belt in central East Greenland were inside the Greenland shelf region of the Iapetus Ocean. The boundary between oceanic and continental crust during Iapetus Ocean time must have been situated east of the present coastline, and any Caledonian subduction zone must also have been situated in a more easterly position than the present outcrops.

The whole character of the Caledonian fold belt in central East Greenland therefore corresponds to a sector of a Cordilleran-type mountain belt, where only the zone underlain by continental crust is exposed.

Acknowledgements

Published with the permission of the Director of the Geological Survey of Greenland. Cooperation, advice, and criticism in connection with the preparation of this paper from my colleagues J. D. Friderichsen and A. K. Higgins are gratefully acknowledged. Special thanks are due to A. K. Higgins for improvements of both contents and language of the manuscript.

References

Alexander-Marrack, P. D. and Friend, P. F. 1976. Devonian sediments of East Greenland III. The eastern sequence, Vilddal Supergroup and part of the Kap Kolthoff Supergroup. *Meddr Grønland*, **206**(3), 122 pp.

Barnes, C. R., Norford, B. S. and Skevington, D. 1981. The Ordovician system in Canada. *Int. Union. Geol. Sci. Publ.* no. 8.

Bertrand-Sarfati, J. and Caby, R. 1976. Carbonates et stromatolites du sommet du Groupe d'Eleonore Bay (Précambrien terminal) au Canning Land (Groenland oriental). *Bull. Grønlands geol. Unders.*, **119**, 51 pp.

Brooks, C. K., Fawcett, J. J. and Gittins, J. 1976. Caledonian magmatic activity in south-eastern Greenland. *Nature*, London, **260**, 694–696.

Bütler, H. 1959. Das Old Red-Gebiet am Moskusoksefjord. Attempt at a correlation of the series of various Devonian areas in central East Greenland. *Meddr Grønland*, **160**(5), 188 pp.

Bütler, H. 1961a. Devonian deposits of central East Greenland. In Raasch, G. O. (Ed.), *Geology of the Arctic*, **1**, Toronto U.P, 188–196.

Bütler, H. 1961b. Continental Carboniferous and Lower Permian in central East Greenland. In Raasch, G. O. (Ed.), *Geology of the Arctic*, **1**, Toronto U.P, 205–213.

Caby, R. 1972. Preliminary results of mapping in the Caledonian rocks of Canning Land and Wegener Halvø, East Greenland. *Rapp. Grønlands geol. Unders*, **48**, 21–38.

Caby, R. 1976a. Tension structures related to gliding tectonics in the Caledonian superstructure of Canning Land and Wegener Halvø, central East Greenland. *Rapp. Grønlands geol. Under.*, **72**, 24 pp.

Caby, R. 1976b. Investigations on the Lower Eleonore Bay Group in the Alpefjord region, central East Greenland. *Rapp. Grønlands geol. Unders.*, **80**, 102–106.

Chadwick, B. 1975. The structure of south Renland, Scoresby Sund, with special reference to the tectonometamorphic evolution of a southern internal part of the Caledonides of East Greenland. *Bull. Grønlands geol. Unders.*, **112**, (also *Meddr Grönbland*, **201** (2), 67 pp.

Coe, K. and Cheeney, R. F. 1972. Preliminary results of mapping in Liverpool Land, East Greenland. *Rapp. Grønlands geol. Unders.*, **48**, 7–20.

Cowie, J. W. 1961. The Lower Palaeozoic of Greenland. In Raasch, G. O. (Ed.), *Geology of the Arctic*, Toronto U.P, 160–169.

Cowie, J. W. and Adams, P. J. 1957. The geology of the Cambro-Ordovician rocks of central East Greenland, 1. *Meddr Grønland*, **153**(1), 193 pp.

Eha, S. 1953. The Pre-Devonian sediments on Ymers Ø, Suess Land, and Ella Ø (East Greenland) and their tectonics. *Meddr Grønland*, **111**(2), 105 pp.

Fränkl, E. 1951. Die untere Eleonore Bay Formation im Alpefjord. *Meddr Grønland*, **151**(6), 15 pp.

Fränkl, E. 1953a. Geologische Untersuchungen in Ost-Andrées Land (NE-Grønland). *Meddr Grønland*, **113**(4), 160 pp.

Fränkl, E. 1953b. Die geologische Karte von Nord-Scoresby Land (NE-Grønland). *Meddr Grønland*, **113**(6), 56 pp.

Friend, P. F., Alexander-Marrack, P. D., Nicholson, J. and Yeats, A. K. 1976. Devonian sediments of East Greenland II. Sedimentary structures and fossils. *Meddr Grønland*, **206**(2), 91 pp.

Frykman, P. 1979. Cambro-Ordovician rocks of C.H. Ostenfeld Nunatak, northern East Greenland. *Rapp. Grønlands geol. Unders.*, **91**, 125–132.

Haller, J. 1955. Der 'Zentrale Metamorphe Komplex' von NE-Grönland. Teil I. Die geologische Karte von Suess Land, Gletscherland und Goodenoughs Land. *Meddr Grønland*, **73**(1) 174 pp.

Haller, J. 1970. Tectonic map of East Greenland (1:500 000). An account of tectonism, plutonism, and volcanism in East Greenland. *Meddr Grønland*, **171**(5), 286 pp.

Haller, J. 1971. Geology of the East Greenland Caledonides. Interscience Publishers, New York, 413 pp.

Hansen, B. T. and Steiger, R. H. 1971. The geochronology of the Scoresby Sund area, I: Rb/Sr mineral ages. *Rapp. Grønlands geol. Unders.*, **37**, 55–57.

Hansen, B. T., Steiger, R. H. and Henriksen, N. 1972. The geochronology of the Scoresby Sund area, 2: Rb/Sr mineral ages. *Rapp. Grønlands geol. Unders.*, **48**, 105–107.

Hansen, B. T., Frick, U. and Steiger, R. H. 1973. The geochronology of the Scoresby Sund area, 5: K/Ar mineral ages. *Rapp. Grønlands geol. Unders.*, **58**, 59–61.

Hansen, B. T. and Steiger, R. H. 1976. The geochronology of the Scoresby Sund region, central East Greenland. Progress report 7: Rb–Sr whole rock and U–Pb zircon ages. *Rapp. Grønlands geol. Unders.*, **80**, 133–136.

Hansen, B. T., Higgins, A. K. and Bär, M. T. 1978. Rb–Sr and U–Pb age patterns in polymetamorphic sediments from the southern part of the East Greenland Caledonides. *Bull. geol. Soc. Denmark*, **27**, part 1–2, 55–62.

Hansen, B. T. and Tembusch, H. 1979. Rb–Sr isochron

ages from east Milne Land, Scoresby Sund, East Greenland. *Rapp. Grønlands geol. Unders.*, **95**, 96–101.

Hansen, B. T., Steiger, R. H. and Higgins, A. K. 1981. Isotopic evidence for a Precambrian metamorphic event within the Charcot Land window, East Greenland Caledonian fold belt. *Bull. geol. Soc. Denmark*, **29**, 151–160.

Henriksen, N. 1981. The Charcot Land tillite, Scoresby Sund, East Greenland. In *Earth's Pre-Pleistocene Glacial Record*, Cambridge Earth Science series, 776–777.

Henriksen, N. and Higgins, A. K. 1976. East Greenland Caledonian fold belt. In Escher, A. and Watt, W. S. (Eds), *Geology of Greenland*, Copenhagen; The Geological Survey of Greenland, 182–246.

Henriksen, N., Perch-Nielsen, K. and Andersen, C. 1980. Descriptive text to 1:100 000 sheets Sydlige Stauning Alper 71 Ø.2 N and Frederiksdal 71 Ø.3 N, 46.pp. *Grønlands geol. Unders.*, Copenhagen.

Higgins, A. K. 1974. The Krummedal supracrustal sequence around inner Nordvestfjord. Scoresby Sund, East Greenland. *Rapp. Grønlands geol. Unders.*, **3**, 34 pp.

Higgins, A. K. 1976. Pre-Caledonian metamorphic complexes within the southern part of the East Greenland Caledonides. *J. geol. Soc. London*, **132**, 289–305.

Higgins, A. K. 1981. The Late Precambrian Tillite Group of the Kong Oscars Fjord and Kejser Franz Josefs Fjord region of East Greenland. In *Earth's Pre-Pleistocene Glacial Record*, Cambridge Earth Science series, 778–781.

Higgins, A. K. and Phillips, W. E. A. 1979. East Greenland Caledonides—an extension of the British Caledonides. In Harris, A. L., Holland, C. H. and Leake, B. E. (Eds), *The Caledonides of the British Isles—reviewed*, Edinburgh (Geol. Soc. London), 19–31.

Higgins, A. K., Friderichsen, J. D. and Thyrsted, T. 1981. Precambrian metamorphic complexes in the East Greenland Caledonides (72°–74°N)—their relationships to the Eleonore Bay Group, and Caledonian orogenesis. *Rapp. Grønlands geol. Unders.*, **104**.

Homewood, P. 1973. Structural and lithological divisions of the western border of the East Greenland Caledonides in the Scoresby Sund region. *Rapp. Grønlands geol. Unders.*, **57**, 27 pp.

Hurst, J. M. and McKerrow, W. S. 1981. The Caledonian nappes of Kronprins Christian Land, eastern North Greenland. *Rapp. Grønlands geol. Unders.*, **106**.

Katz, H. R. 1952. Zur Geologie von Strindbergs Land (NE-Grønland). *Meddr Grønland*, **111**(1), 150 pp.

Koch, L. and Haller, J. 1971. Geological map of East Greenland 72°–76°N.Lat. (1:250.000). *Meddr Grønland*, **183**, 26 pp.

Odell, N. E. 1944. The petrography of the Franz Josef Fjord region, North-East Greenland, in relation to its structures. *Trans. Roy. Soc. Edinburgh*, **61**(1), 221–246.

Peel, J. S. and Cowie, J. W. 1979. New names for Ordovician formations in Greenland. *Rapp. Grønlands geol. Unders.*, **91**, 117–124.

Perch-Nielsen, K., Bromley, R. G., Birkenmajer, K. and Aellen, M. 1972. Field observations in Palaeozoic and Mesozoic sediments of Scoresby Land and northern Jameson Land. *Rapp. Grønlands geol. Unders.*, **48**, 39–59.

Phillips, W. E. A., Stillman, C. J., Friderichsen, J. D. and Jemelin, L. 1973. Preliminary results of mapping in the western gneiss and schist zone around Vestfjord and inner Gåsefjord, south-west Scoresby Sund. *Rapp. Grønlands geol. Unders.*, **58**, 17–32.

Phillips, W. E. A. and Friderichsen, J. D. 1981. The Late

Precambrian Gåseland tillite. Scoresby Sund, East Greenland. In *Earth's Pre-Pleistocene Glacial Record*, Cambridge Earth Science series, 773–775.

Rex, D. C. and Gledhill, A. 1974. Reconnaissance geochronology of the infracrustal rocks of Flyverfjord, Scoresby Sund, East Greenland. *Bull. geol. Soc. Denmark*, 49–54.

Rex, D. C. and Gledhill, A. R. 1981. Isotopic studies in the East Greenland Caledonides (72°–74°N)–Precambrian and Caledonian ages. *Rapp. Grønlands geol. Unders.*, **104**.

Rex, D. C. and Higgins, A. K. 1981. Potassium–Argon mineral ages from East Greenland Caledonides between 72°N and 74°N. *Terra cognita*. **1**.

Smith, D. C. and Cheeney, R. F. 1981. A new occurrence of garnet–ultrabasite in the Caledonides: a Cr-rich chromite–garnet–lherzolite from Tværdalen, Liverpool Land, East Greenland. *Terra cognita*, **1**.

Smith, P. 1982. Ordovician conodont biostratigraphy of central East Greenland. *Paleont. Contr. Univ. Oslo*, **280**, 48.

Spencer, A. M. 1978. Late Precambrian glaciation in the North Atlantic region. In Wright, A. E. and Moseley, F. (Eds), *Ice Ages, Ancient and Modern. Geol. J. Special Issue*, **6**, 217–239.

Steck, A. 1971. Kaledonische metamorphose der praekambrischen Charcot Land Serie, Scoresby Sund, Ost-Grönland. *Bull. Grønlands geol. Unders.*, **97** (also *Meddr Grönland*, **192**(3)), 69 pp.

Steiger, R. H. and Jäger, E. 1977. Subcommission on geochronology: convention on the use of decay constants in geo- and cosmoschronology. *Earth. Planet. Sci. Lett.*, **36**, 359–362.

Steiger, R. H., Hansens, B. T., Schuler, C. H., Bär, M. T. and Henriksen, N. 1979. Polyorogenic nature of the southern Caledonian fold belt in East Greenland: an isotopic age study. *J. Geol.*, **87**, 475–495.

Thyrsted, T. 1979. The metamorphic zonation and the relationship between metamorphism and deformational events in the East Greenland Caledonides between 72°–74°N. *Grønlands geol. Unders. internal report*, 24 pp.

Vidal, G. 1976. Late Precambrian acritarchs from the Eleonore Bay Group and Tillite Group in East Greenland. *Rapp. Grønlands geol. Unders.*, **78**, 19 pp.

Vidal, G. 1981. Micropalaeontology and biostratigraphy of the Upper Proterozoic and Lower Cambrian Sequence in East Finnmark, Northern Norway. *Norges geol. Unders.*, **362**, 53 pp.

Wegmann, C. E. 1935. Preliminary report on the Caledonian orogeny in Christian X's Land (North-East Greenland). *Meddr Grønland*, **103**(3), 59 pp.

Wordie, J. M. 1927. The Cambridge expedition to East Greenland in 1926. V. Geology. *Geogr. J.*, **70**, 252–253.

Yeats, A. K. and Friend, P. F. 1978. Devonian sediments of East Greenland IV. The western sequence, Kap Kolthoff Supergroup of the western areas. *Meddr Grønland*, **206**(4), 112 pp.

The Caledonide Orogen—Scandinavia and Related Areas
Edited by D. G. Gee and B. A. Sturt
© 1985 John Wiley & Sons Ltd

Potassium–argon mineral ages from the East Greenland Caledonides between 72° and 74°N

D. C. Rex[*] and **A. K. Higgins**[†]

[*]*Department of Earth Sciences, The University, Leeds LS2 9JT, U.K.*
[†]*Grønlands Geologiske Undersøgelse, Øster Voldgade 10, DK–1350 Copenhagen K, Denmark*

ABSTRACT

207 K–Ar mineral ages from a variety of rock units within the East Greenland
Caledonides are presented. Most are biotite ages, which can be divided into three
age groups. The youngest ages (365–405 Ma) come from deep-seated infracrustal
units, presumed to be the last to cool through their blocking temperatures following
regional Caledonian metamorphism. The second group (405–440 Ma) come from
surrounding regions presumed to have cooled earlier. The oldest group of ages
(440–1171 Ma) come mainly from metasedimentary rocks, and are interpreted as
'excess argon' ages, since coexisting muscovites give a normal Caledonian range of
ages. The areal distribution and relationship of the age groups are discussed with
reference to cross-sections of thc foldbelt.

Introduction

The East Greenland Caledonian foldbelt extends
from latitude 72°–82°N, occupying most of the
50–250 km wide strip of land between the coast
and the Inland Ice (see Henriksen this volume,
Hurst *et al*. this volume). This study concerns the
region 72°–74°N in the southern part of the foldbelt
(Fig. 1), in which distinction is made between: (1)
the three infracrustal units described by Haller
(1971)—Gletscherland complex, Hagar sheet, Niggli Spids dome; (2) metasedimentary sequences
which envelope and overlie the infracrustal complexes; and (3) late Proterozoic sediments (Eleonore
Bay Group, Petermann Series) and Lower Palaeozoic sediments which flank the crystalline region to
the east and west. For reviews of the regional
geology, see Henriksen and Higgins (1976), Higgins
and Phillips (1979), and Higgins *et al*. (1981).

Geochronologic work on the Rb–Sr whole rock
and zircon isotopic systems in the infracrustal units
has given Archaean or early Proterozoic ages (Rex
and Gledhill 1974, 1981; Rex *et al*. 1976, 1977;
Steiger *et al*. 1979; Hansen *et al*. 1973; Higgins *et al*.
1978). Rb–Sr whole rock isotopic studies on the
enveloping metasedimentary rocks suggest they suffered a metamorphic event about 1000–1500 Ma

ago (Hansen *et al*. 1978; Rex and Gledhill 1981).
Late to post-tectonic granites which cut the late Proterozoic sediments have given Caledonian Rb–Sr
whole rock ages, mainly in the range of 480–430 Ma
(Rex and Gledhill 1981). Scattered K–Ar and
Rb–Sr mineral ages, obtained usually as a byproduct
of the isotopic studies quoted above, have generally
given Caledonian ages irrespective of the rock type
concerned.

The age of the main phase of the Caledonian
orogeny in this part of East Greenland can be
defined only loosely on stratigraphic grounds as
between Middle Ordovician and Devonian time.
The isotopic ages on the Caledonian granites suggest
the main deformation and metamorphism took place
in the early part of this timespan, most probably in
the Upper Ordovician (see discussion in Henriksen
this volume). The extent of Caledonian deformation
and metamorphism is easily traceable in the late Proterozoic and Lower Palaeozoic sediments, but very
difficult to define in the polyorogenic crystalline
basement rocks which underlie them. In an attempt
to define more closely the uplift and cooling patterns
related to Caledonian metamorphism and reactivation, especially in the older basement complexes,
K–Ar mineral isotopic studies were carried out on
extensive sample collections made during Geological

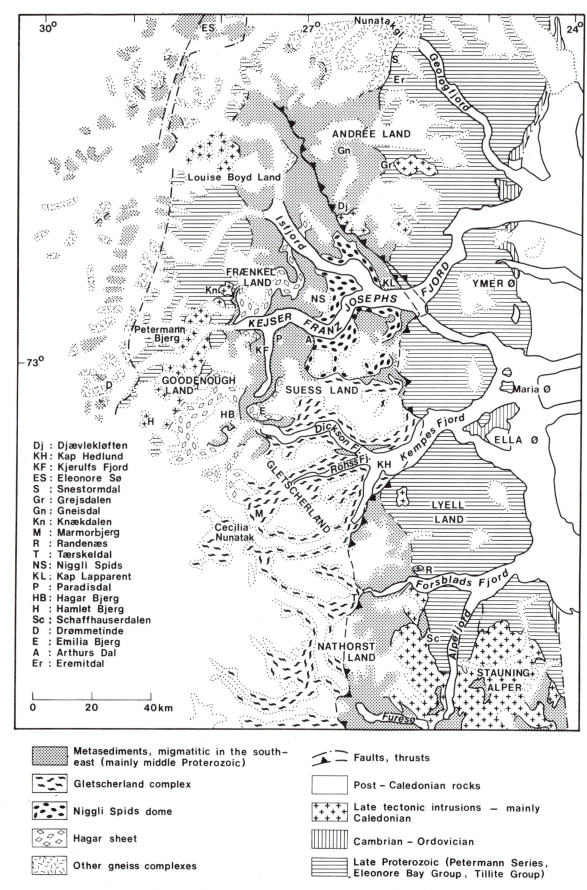

Dj : Djævlekløften
KH : Kap Hedlund
KF : Kjerulfs Fjord
ES : Eleonore Sø
S : Snestormdal
Gr : Grejsdalen
Gn : Gneisdal
Kn : Knækdalen
M : Marmorbjerg
R : Randenæs
T : Tærskeldal
NS : Niggli Spids
KL : Kap Lapparent
P : Paradisdal
HB : Hagar Bjerg
H : Hamlet Bjerg
Sc : Schaffhauserdalen
D : Drømmetinde
E : Emilia Bjerg
A : Arthurs Dal
Er : Eremitdal

0   20   40km

Metasediments, migmatitic in the south-east (mainly middle Proterozoic)

Gletscherland complex

Niggli Spids dome

Hagar sheet

Other gneiss complexes

Faults, thrusts

Post – Caledonian rocks

Late tectonic intrusions — mainly Caledonian

Cambrian – Ordovician

Late Proterozoic (Petermann Series, Eleonore Bay Group, Tillite Group)

Fig. 1.   Geological map of East Greenland (72°–74°N)

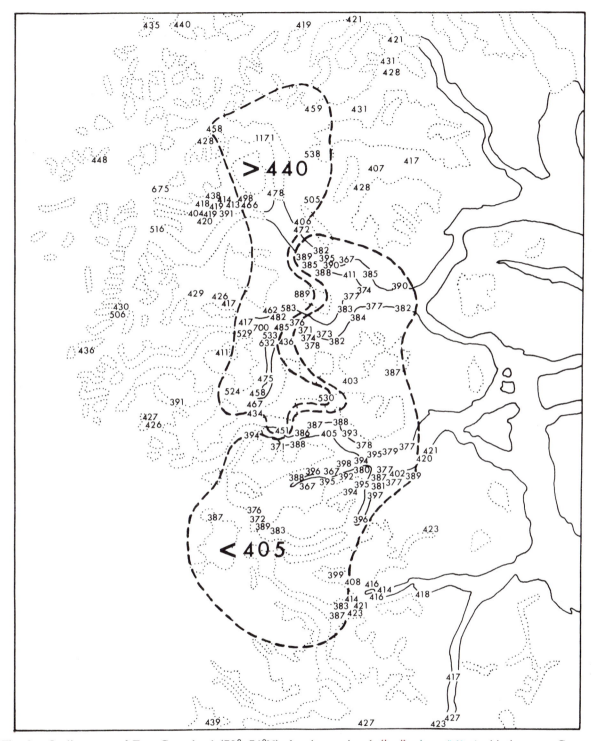

Fig. 2. Outline map of East Greenland (72°–74°N) showing regional distribution of K–Ar biotite ages. Contours outline areas giving ages less than 405 Ma, and greater than 440 Ma. Compare with geological map of Fig. 1

Survey of Greenland (GGU) expeditions in the period 1974–78. We report here 207 K–Ar mineral ages on biotite, muscovite, and hornblende.

It is widely accepted that radiometric ages determined on minerals from metamorphic belts reflect their cooling history rather than their primary crystallization. Armstrong (1966) first developed this concept in detail and Harper (1967) pioneered its application to the British Caledonides. Its general acceptance, however, stems largely from the work of E. Jäger and her colleagues in the Central Alps, which has been reviewed in Jäger and Hunziker (1979). Cooling history and mineral blocking temperatures have been further discussed by Harrison and McDougal (1980) from their detailed work on the Separatean Point batholith in New Zealand.

Results

The analytical data for the 207 mineral ages are listed in the appendix. The ages are mainly on biotite and muscovite, with a few hornblende ages, and come from a wide variety of rock types. The results are portrayed as histograms in Figs. 3 and 4. Biotite ages have concentrations at 390 and 425 Ma (Fig. 3), with a scatter of ages rising to 1171 Ma. The less numerous muscovite ages cluster around 415 Ma (Fig. 4). The biotite ages are plotted on the map of Fig. 2, and can be divided regionally into three age groups: 365–405 Ma, 405–440 Ma, and >440 Ma.

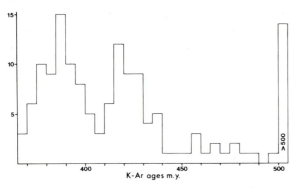

Fig. 3.   Histogram of K–Ar biotite ages

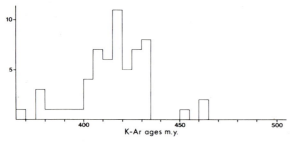

Fig. 4.   Histogram of K–Ar muscovite ages

365–405 Ma biotite ages (group I)

The youngest group of K–Ar biotite ages is represented by 69 analyses of samples collected mainly within two of the infracrustal units (Niggli Spids dome and Gletscherland complex), and indicate that these were the last to cool from the Caledonian orogeny. Ages for coexisting muscovite are available for nine samples of this group, and in general, the muscovites give older ages. This suggests that the rocks cooled slowly through their relative closure temperatures following Caledonian metamorphism (Purdy and Jäger 1976).

405–440 Ma biotite ages (group II)

Forty biotite ages fall in this range. They have a diffuse geographic distribution compared to the 365–405 Ma group, and have been obtained on samples from the infracrustal Hager sheet, metasedimentary rocks in Andrée Land and Lyell Land, and late Proterozoic rocks of the Eleonore Bay Group and Petermann Series. They do not form a simple zonal pattern with respect to the distribution of the group I ages. Ages for coexisting muscovite and biotite are available for 22 of the samples in this group. Approximately half record equivalent ages, which is indicative of rapid cooling (Purdy and Jäger 1976), while in most of the other sample biotite ages are older than the muscovites (possibly reflecting the presence of 'excess argon', see also below).

>440 Ma biotite ages (group III)

These ages come mainly from metasedimentary rocks believed to record the effects of both middle Proterozoic and Caledonian regional metamorphism (Higgins et al. 1981; Rex and Gledhill 1981), and from some rocks of the Hagar sheet. Areally these >440 Ma ages occur mainly in a N–S trending zone northwest of the broad area of the 365–405 Ma ages (Fig. 2). A few ages >440 Ma occur farther west in the nunatak region. The 29 biotite ages in this group range from 440–1171 Ma. Coexisting muscovite ages are available for 14 samples, and in all cases the ages are considerably younger (391–462 Ma), nearly all within the normal Caledonian range. This indicates that the biotites contain excess argon and are not reflecting a real event. The oldest biotite age (1171 Ma) is similar to that obtained by Rb–Sr whole rock studies (Rex and Gledhill 1981). However, this is likely to be a coincidence; the coexisting muscovite records one of the oldest muscovite ages obtained in this study (462 Ma), and may also reflect the presence of excess argon.

Previous work

Earlier K–Ar ages obtained from the area of this study (Haller and Kulp 1962) fall in with our age groupings (with one exception) when recalculated using the decay constants recommended by Steiger and Jäger (1977). K–Ar mica ages from the Scoresby Sund region (70°–72°N) fall mainly in the range 410–420 Ma (Higgins and Phillips 1979), but are too few and scattered to allow a systematic pattern to be discerned and they cannot be directly compared with the groupings of this study. Farther north at Danmarkshavn (76°N), K–Ar ages 355,

341, and 330 Ma have been reported by Steiger *et al.* (1976).

Interpretation

Infracrustal units

The three infracrustal units (Gletscherland complex, Hagar sheet, Niggli Spids dome) are all considered to be Archaean or early Proterozoic basement complexes partially reactivated in one or more subsequent orogenies, although Rb–Sr whole rock ages supporting this interpretation are so far known only for the Gletscherland complex and Hagar sheet (Rex and Gledhill 1981). The young group I ages centred on the Niggli Spids dome and Gletscherland complex are considered to record the time of cooling from regional Caledonian metamorphism. The distribution of these ages is consistent with late stage doming of the infracrustal units, as they represent the deepest orogenic levels and apparently were the last to cool. The older group II ages on samples from the Hagar sheet seem to indicate that it cooled significantly earlier than the Niggli Spids dome and Gletscherland complex.

Excess argon

The muscovite K–Ar ages show a well-defined concentration between 400–435 Ma, nearly corresponding with the group II biotite dates. All but three of the muscovite ages whose coexisting biotite ages are in group III also lie in this range. Thus the time-span 400–435 Ma likely defines the late Caledonian cooling age for all the rocks giving group II or group III biotite ages. This would suggest that all biotite ages greater than 435 Ma contain some component of excess argon.

Comparable differential accumulation of excess argon in biotite *vs.* muscovite has been described by several authors. In a study of migmatization at the Grenville front in Canada, Wanless *et al.* (1970) showed that the excess argon from the immediate environment was incorporated in the biotite, but not in the muscovite. In part of the eastern Austrian Alps, Brewer (1969) showed that biotites which contained excess argon coexisted with muscovites that either had less or none at all. He suggested that this was a result of build up in argon pressure from the outgassing of underlying old basement rocks, and that the biotites were subjected to a longer diffusion period because of their lower blocking temperature as cooling proceeded in this high partial pressure of argon. Roddick *et al.* (1980) further examined these and other biotites containing excess argon, and concluded that a greater solubility of argon in biotite compared with muscovite is the determining factor.

The mechanism by which the excess argon developed mainly in the metasedimentary rocks of the East Greenland Caledonides is uncertain. These metasedimentary rocks mainly occur structurally beneath the Hagar sheet (Fig. 5) but are also locally present above it. Resetting of argon systems in the early and middle Proterozoic rocks of the Hagar sheet and metasedimentary units during the Caledonian thermal event could have yielded a high partial pressure of argon in the environment. Although this could explain the excess argon in the neighbourhood of the Hagar sheet, there are also excess argon biotite ages from Andrée Land and parts of the nunatak zone (Fig. 2) which are spatially removed from the older terrains. This suggests that the source of excess argon must be more general. It may have evolved from older basement rocks suggested to occur in these areas. Higgins *et al.* (1981) suggest that some areas of gneissic rocks in the nunatak zone may be basement, while in Andrée Land Rex and Gledhill (1981) have obtained two *c.* 1000 Ma ages from granite bodies indicating that basement rocks are also present here.

The late Proterozoic Petermann Series and Eleonore Bay Group flanking the metamorphic complexes were locally metamorphosed in Caledonian time (see also below). K–Ar ages of biotite from these rocks are in group II and do not show any evidence of excess argon. The thickness of these late Proterozoic sequences, and the Lower Palaeozoic succession overlying them, varies from 7 km to about 17 km (Haller 1971; Henriksen and Higgins 1976). It is uncertain how thick and complete the late Proterozoic–Lower Palaeozoic cover was in the region of the present metamorphic complexes, but it could have some bearing on their cooling pattern.

Cross-section of the fold belt in Fraenkels Land and southern Andrée Land

The relationship between biotite age groups and the various lithostructural rock units of East Greenland is illustrated in Fig. 5. In the west, the metamorphic grade in the Petermann Series increases eastwards towards the metamorphic complexes. Biotite ages are all in group II. The contact between the Petermann Series and older metasediments is interpreted as a décollement surface (Higgins *et al.* 1981), but there is no significant change in the biotite ages across the contact suggesting that this structure formed prior to late Caledonian cooling below biotite blocking temperatures.

The infracrustal Hagar sheet has yielded early Proterozoic Rb–Sr whole rock isochron ages: 1725 Ma from a granitic gneiss at the west margin of the sheet; and 1950 Ma from a foliated granite at the

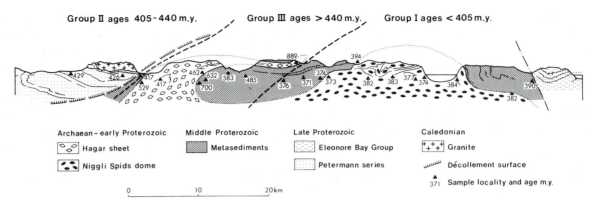

Fig. 5.   Cross-section of the fold belt in Frænkels Land and southern Andrée Land showing relationships of the biotite age groups to the lithostructural rock units

east margin of the sheet (Higgins *et al.* 1978). K–Ar biotite ages from the same units are 438 and 462 Ma, respectively. The latter apparently contains a component of excess argon as the corresponding muscovite age is 417 Ma. Metasedimentary rocks beneath the reverse limb of the Hagar sheet record group III biotite ages, thought to reflect an overpressure of argon.

All the K–Ar biotite ages from the Niggli Spids dome are in group I (ranging from 371–394 Ma). Muscovite and biotite from one sample both record a 383 Ma date. Young group I biotite ages have also been obtained from metasedimentary rocks from both the east and west sides of the Niggli Spids dome.

Southern Louise Boyd Land

Detailed sampling was carried out in an E–W valley in southern Louise Boyd Land which exposes a complete section through the Petermann Series and underlying metamorphic rocks (Fig. 6). This structural setting is comparable to that in Knækdalen at the west end of the cross-section of Fig. 5, with

metamorphic grade increasing systematically eastwards. A major recumbent Caledonian fold above the décollement surface is a notable feature (Fig. 6). The K–Ar biotite ages are typical group II ages, with—apart from one sample—muscovite ages agreeing closely with biotite dates. A mica schist below the décollement surface at the base of the Petermann Series also records group II ages for both micas. However, two augen gneiss samples from an infracrustal unit show excess argon, in one case for both muscovite and biotite. The generally systematic age pattern indicates that formation of the Caledonian folds and décollement structure was essentially completed prior to cooling of the micas below their blocking temperatures.

Discussion and Conclusions

The K–Ar biotite and muscovite ages presented here show that the entire region from which the samples were collected (72°–74°N) records the effects of Caledonian metamorphism. Ten hornblende ages from the area also yield Caledonian

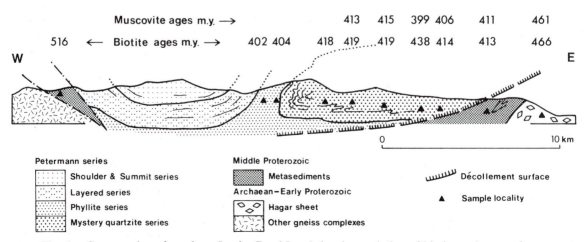

Fig. 6.   Cross-section of southern Louise Boyd Land showing variation of biotite and muscovite ages

dates. With biotite–hornblende pairs it is normally expected that the hornblende age would be equal to or older than the corresponding biotite age, which is the case with six out of the seven pairs analysed in this study. The hornblende from one metasedimentary rock with a group III biotite age records a younger date. However, the mineral age pattern on all rock units sampled is Caledonian, irrespective of the original age of the units as determined by other types of isotopic studies.

The group III biotite ages which scatter up to 1171 Ma are interpreted to reflect the presence of excess argon and not partially outgassed Proterozoic minerals. Muscovite ages of the same samples are in the range of 400–462 Ma, and are considered to be near the cooling age of these rocks. The single age peak on the muscovite histogram at 400–435 Ma could reflect sampling bias because muscovite-bearing rocks were not common in our collection from the Niggli Spids dome and related rocks. The few samples obtained gave muscovite ages very similar or slightly older than their corresponding biotite age.

The youngest ages are centred on the Niggli Spids dome and Gletscherland complex, with older ages in the surrounding region. This may reflect a simple concentric cooling pattern which was complicated by development of a broad zone of excess argon ages.

There is no obvious break in the age pattern across the supposed décollement surface between the Petermann Series and the assumed middle Proterozoic sediments. This is taken to indicate that the fold structures and movements on the décollement surface predate the cooling through the closure temperatures of the minerals.

The mineral ages from the Hagar sheet, where not affected by excess argon, are significantly older than those from the Niggli Spids dome and Gletscherland complex, indicating that it cooled earlier. Haller (1971) deduced from geological evidence that the Hagar sheet was the youngest of his three migmatite complexes. Although the early Proterozoic Rb–Sr whole rock isochron ages from the Hagar sheet and Gletscherland complex (Rex *et al.* 1976; Higgins *et al.* 1978; Rex and Gledhill 1981) demonstrate that the infracrustal units cannot simply be the product of deep-seated Caledonian mobilization, which was for many years considered to be the case (see e.g. Haller 1971), there is uncertainty as to the early Palaeozoic tectonic state within these metamorphic complexes. For example, Talbot (1979) reasoned that irrespective of the original age of the rock units involved, major structures of the infracrustal units could have formed during the Caledonian orogeny by processes of thermal convection. The mineral age patterns contribute only slightly to this discussion, in that they do support the idea of a late Caledonian dome-like uplift of the metamorphic complexes centred on the Niggli Spids dome and Gletscherland complex.

On the basis of the variability in ages across our area of study, it seems possible that the Danmarkshavn region at 76°N (Steiger *et al.* 1976) may be composed of deep-seated rocks uplifted by a late-stage event, comparable to the young ages of the Niggli Spids dome; the young Danmarkshavn K–Ar ages of 330–355 Ma need not necessarily mean that the Caledonian events lasted longer in northern areas.

Analytical Methods

Decay constants used are as recommended by Steiger and Jäger (1977). Potassium values quoted are the mean of three dissolutions. Concentrations were measured by flame photometry using a Corning Eel 450 Li internal standard flame photometer. Argon was extracted in a glass vacuum system using a pure 38-argon spike (approx. size 1×10^{-5} s.c.c.). Argon isotopes were measured on a modified A.E.I. MS10 fitted with peak switching and digital output (Rex and Dodson 1970).

Errors on the ages are ±3 per cent at the 2 sigma level, measured by replicate analyses. International standards were run at intervals throughout the period of the analytical work.

Acknowledgements

This paper is published with the permission of the Director of the Geological Survey of Greenland. Drs. M. Dodson and C. Roddick kindly made useful comments on a draft of this manuscript, and we also acknowledge the helpful suggestions of the two referees.

The K–Ar analyses were carried out in the isotope laboratories at Leeds University with support from GGU and NERC. The majority of the potassium analyses were carried out by A. R. Gledhill with assistance from Miss Kate Johnson. Mineral separations were performed by Walter Wilkinson and Tom Oddy.

References

Armstrong, R. L. 1966. K–Ar dating of plutonic and volcanic rocks in orogenic belts. In Schaeffer, O. A. and Zahrenger, J. (Eds), *Potassium Argon Dating*, Springer-Verlag, Berlin, 117–131.

Brewer, M. S. 1969. Excess radiogenic argon in metamorphic micas from the Eastern Alps, Austria. *Earth Planet. Sci. Lett.*, **6**, 321–331.

Haller, J. 1971. Geology of the East Greenland Caledonides, New York, Interscience Publishers, 413 pp.

Haller, J. and Kulp, J. L. 1962. Absolute age determinations in East Greenland. *Meddr Grønland*, **171**(1), 77 pp.

Hansen, B. T., Oberli, F. and Steiger, R. H. 1973. The geochronology of the Scoresby Sund area. 4: Rb/Sr whole rock and mineral ages. *Rapp. Grønlands geol. Unders.*, **58**, 55–58.

Hansen, B. T., Higgins, A. K. and Bär, M. T. 1978. Rb–Sr and U–Pb age pattern in polymethamorphic sediments from the southern part of the East Greenland Caledonides. *Bull. geol. Soc. Denmark*, **27**, 55–62.

Harper, C. T. 1967. The geological interpretation of potassium argon ages of metamorphic rocks from the Scottish Caledonides. *Scott. J. geol.*, **3**, 46–66.

Harrison, T. M. and McDougal, I. 1980. Investigations of an intrusive contact, northwest Nelson, New Zealand—I. Thermal, chronological and isotopic constraints. *Geochim. Cosmochim. Acta*, **44**, 1985–2003.

Henriksen, N. and Higgins, A. K. 1976. East Greenland Caledonian fold belt. In Escher, A. and Watt, W. S. (eds), *Geology of Greenland*, Copenhagen, Geol. Surv. Greenland, 182–246.

Higgins, A. K. and Phillips, W. E. A. 1979. East Greenland Caledonides—a continuation of the British Caledonides. In Harris, A. L., Holland, C. H. and Leake, B. E. (Eds), *The Caledonides of the British Isles—reviewed*, J. Geol. Soc. London, Spec. Publ., **8**, 19–32.

Higgins, A. K., Friderichsen, J. D., Rex, D. C. and Gledhill, A. R. 1978. Early Proterozoic isotopic ages in the East Greenland Caledonian fold belt. *Contrib. mineral. Petrol.*, **67**, 87–94.

Higgins, A. K., Friderichsen, J. D. and Thyrsted, T. 1981. Precambrian metamorphic complexes in the East Greenland Caledonides (72°–74°N)—their relationships to the Eleonore Bay Group, and Caledonian orogenesis. *Bull. Grønlands geol. Unders.*, **104**, 5–46.

Jäger, E. and Hunziker, J. C. 1979. Lectures in Isotope Geology, Springer-Verlag, Berlin, 329 pp.

Purdy, J. W. and Jäger, E. 1976. K–Ar ages on rock forming minerals from the Central Alps. *Memorie de Padova*, **30**.

Rex, D. C. and Dodson, M. H. 1970. Improved resolution and precision of argon analysis using an MS10 mass spectrometer. *Eclog. Geol. Helv.*, **63**, 275–280.

Rex, D. C. and Gledhill, A. R. 1974. Reconnaissance geochronology of the infracrustal rocks of Flyvefjord, Scoresby Sund, East Greenland. *Bull. geol. Soc. Denmark*, **23**, 49–54.

Rex, D. C. and Gledhill, A. R. 1981. Isotopic studies in the East Greenland Caledonides (72°–74°N)—Precambrian and Caledonian ages. *Bull. Grønlands geol. Unders.*, **104**, 47–72.

Rex, D. C., Gledhill, A. R. and Higgins, A. K. 1976. Progress report on geochronological investigations in the crystalline complexes of the East Greenland Caledonian fold belt between 72° and 74°N. *Bull. Grønlands geol. Unders.*, **80**, 127–133.

Rex, D. C., Gledhill, A. R. and Higgins, A. K. 1977. Precambrian Rb–Sr isochron ages from the crystalline complexes of inner Forsblads Fjord, East Greenland fold belt. *Bull. Grønlands geol. Unders.*, **85**, 122–126.

Roddick, J. C., Cliff, R. A. and Rex, D. C. 1980. The evolution of excess argon in Alpine biotites—Ar^{40} Ar^{39} analysis. *Earth Planet. Sci. Lett.*, **48**, 185–208.

Steiger, R. H. and Jäger, E. 1977. Subcommission on geochronology: convention on the use of decay constants in geo- and cosmochronology. *Earth Planet. Sci. Lett.*, **36**, 359–362.

Steiger, R. H., Harnik-Soptrajanova, G., Zimmermann, E. and Henriksen, N. 1976. Isotopic age and metamorphic history of the banded gneiss at Danmarkshavn, East Greenland. *Contrib. mineral. Petrol.*, **57**, 1–24.

Steiger, R. H., Hansen, B. T., Schuler, C., Bär, M. T. and Henriksen, N. 1979. Polyorogenic nature of the southern Caledonian fold belt in East Greenland. *J. Geol.*, **87**, 475–495.

Talbot, C. J. 1979. Infrastructural migmatitic upwelling in East Greenland interpreted as thermal convective structures. *Precambrian Res.*, **8**, 77–93.

Wanless, R. K., Stevens, R. D. and Loveridge, W. D. 1970. Anomalous isotopic relationships in rocks adjacent to the Grenville Front near Chibougamau, Quebec. *Eclog. Geol. Helv.*, **63**, 345–364.

Appendix: K–Ar analytical data

GGU no.	Latitude	Longitude	Rock type	Mineral	% K	Rad. 40Ar cm³/g⁻¹ × 10⁻⁴	% 40Ar rad.	Age m.y.
103422	72°00.9'N	28°02'W	gneiss	bio.	7.38	1.4273	97.3	439
108534	72°00.5'N	26°22'W	mica sch.	bio.	7.37	1.3784	96.0	427
133104	72°22.9'N	26°21'W	amphib.	hnbl.	0.70	0.1764	93.0	554
133112	72°22.9'N	26°21'W	gneiss	bio.	7.34	1.3512	96.0	421
133131	72°22.5'N	26°24'W	gneiss	bio.	7.30	1.3207	94.8	414
133135	72°24.0'N	26°25'W	amphib.	bio.	7.27	1.2950	97.8	408
				hnbl.	1.17	0.2930	96.3	416
133140	72°24.8'N	26°33'W	pegmatite	bio.	7.63	1.3228	97.0	399
133145	72°20.4'N	26°34'W	gneiss	bio.	7.14	1.2066	96.4	387
133152	72°20.4'N	26°34'W	amphib.	bio.	7.25	1.2275	97.1	423
133156	72°20.4'N	26°34'W	amphib.	hnbl.	0.70	0.2065	97.3	689
133160	72°20.3'N	26°35'W	amphib.	hnbl.	1.07	0.2446	95.3	440
133164	72°20.3'N	26°35'W	gneiss	bio.	7.07	1.1656	88.9	383
133169	72°24.2'N	26°16'W	gneiss	bio.	7.42	1.3023	94.4	414
133171	72°23.2'N	26°18'W	gneiss	bio.	7.20	1.3059	97.0	416
133175	72°23.3'N	26°19'W	mica sch.	bio.	6.92	1.2599	97.1	416
133184	72°42.8'N	26°05'W	hnbl. gn.	bio.	7.78	1.2684	96.7	377
133194	72°42.8'N	26°08'W	hnbl. gn.	bio.	7.64	1.3363	91.7	402
133202	72°43.4'N	26°11'W	hnbl. gn.	bio.	7.05	1.1531	85.7	377
133206	72°43.2'N	26°12'W	gneiss	bio.	7.54	1.2657	90.0	387
	72°40.6'N	26°15'W	hnbl. gn.	hnbl.	7.10	1.1722	93.5	381
133211	72°40.2'N	26°16'W	gneiss	hnbl.	1.15	0.2052	96.6	409
				bio.	5.76	0.9937	96.9	397
133214	72°34.8'N	26°25'W	gneiss	hnbl.	1.32	0.2263	87.0	395
				bio.	0.99	0.1967	93.0	450
133217	72°40.8'N	26°22'W	gneiss	bio.	7.09	1.3062	95.2	395
133220	72°40.8'N	26°26'W	gneiss	bio.	7.61	1.2187	82.5	394
133225	72°42.4'N	26°48'W	gneiss	bio.	7.12	1.3099	84.9	395
				hnbl.	1.46	0.2816	76.3	438
133228	72°42.9'N	26°01'W	granite	bio.	0.79	0.1331	96.4	389
				musc.	8.95	1.6186	89.1	414
133235	73°05.4'N	27°12'W	mica sch.	bio.	6.86	1.8239	94.8	436
133242	73°06.4'N	27°25'W	quartzite	bio.	7.57	2.1616	79.6	533
133244	73°06.4'N	27°25'W	quartzite	bio.	7.37	2.1732	98.1	632
133248	73°07.3'N	27°17'W	hnb. sch.	bio.	7.25	1.5648	98.1	485
133251	73°10.0'N	27°27'W	granite	hnbl.	0.79	0.1495	89.6	433
				bio.	6.06	1.2539	98.3	462
				musc.	8.62	1.2284	97.6	416
133257	73°10.7'N	27°18'W	quartzite	bio.	7.12	1.5504	96.4	583
133263	73°05.8'N	26°42'W	gneiss	bio.	7.51	1.5748	95.9	382
133267	73°12.3'N	26°37'W	gneiss	bio.	7.89	1.9042	94.6	383
133272	73°24.7'N	27°05'W	gneiss	musc.	8.88	1.2412	97.8	472
133275	73°21.3'N	26°54'W	granite	bio.	7.30	1.3084	94.1	382
133276	73°17.1'N	26°18'W	gneiss	bio.	7.84	1.4721	95.1	385
133289	73°23.2'N	25°48'W	granite	bio.	7.40	1.5300	97.9	418
				musc.	8.70	1.6065	92.2	422
133291	72°01.0'N	25°34'W	gneiss	bio.	7.41	1.3753	97.3	423
133294	72°01.2'N	25°32'W	gneiss	bio.	6.31	1.1798	94.0	427
133296	72°08.8'N	25°29'W	hnbl. gn.	bio.	7.44	1.3587	88.5	417
188964	73°14.6'N	25°52'W	mica sch.	bio.	7.57	1.2824	91.5	390
189117	72°50.8'N	26°37'W	gneiss	bio.	7.63	1.2997	95.4	393
189119	72°51.9'N	26°39'W	mica sch.	bio.	7.65	1.2882	94.9	388
189123	72°50.6'N	26°58'W	gneiss	bio.	7.66	1.2798	95.9	386
189130	72°49.3'N	26°42'W	mica sch.	bio.	7.08	1.2418	95.2	403
189154	72°44.7'N	26°38'W	gneiss	bio.	7.18	1.2443	96.0	398
189166	73°28.5'N	27°42'W	mica sch.	bio.	7.34	1.5178	95.7	466
				musc.	7.64	1.5581	92.4	461
189170	73°31.0'N	28°33'W	quartzite	bio.	7.15	2.2764	97.0	675
189171	73°40.2'N	28°03'W	granite	musc.	8.68	1.6281	92.1	428
189172	73°40.2'N	28°03'W	mica sch.	bio.	7.11	1.4402	97.2	458
189175	73°39.9'N	27°31'W	mica sch.	musc.	7.54	1.4211	95.6	430
	73°30.3'N	27°02'W		bio.	6.33	4.0598	98.3	1171
189233	73°09.1'N	26°46'W	mica sch.	musc.	7.59	1.5539	74.4	462
189260	73°20.7'N	26°53'W	granodio.	bio.	7.23	1.6228	95.6	505
189273	73°30.0'N	27°21'W	mica sch.	bio.	7.42	1.4748	93.8	450
189286	73°19.2'N	26°57'W	mica sch.	musc.	7.43	1.2727	94.0	394
189287	73°19.2'N	26°57'W	gneiss	bio.	7.63	1.3828	92.4	415
189289	73°17.6'N	26°51'W	mica sch.	bio.	7.06	1.1943	96.1	390
215015	72°49.8'N	27°31'W	quartzite	bio.	7.18	1.5278	95.6	478
215018	72°54.0'N	27°34'W	granite	bio.	8.27	1.4556	88.5	404
215022	73°12.2'N	27°50'W	mica sch.	bio.	7.26	1.2233	95.3	389
215030	73°12.4'N	27°43'W	mica sch.	musc.	7.54	1.2585	96.5	385
215034	73°13.6'N	28°05'W	phyllite	bio.	8.22	1.3826	96.2	388
215040	73°12.5'N	27°48'W	mica sch.	musc.	7.11	1.3870	95.6	394
215050	73°20.1'N	29°16'W	quartzite	bio.	9.02	1.2182	92.1	379
				musc.	7.06	1.4791	96.7	434
215095	73°31.9'N	26°27'W	mica sch.	bio.	9.02	1.3467	96.7	405
				musc.	7.57	1.5942	94.8	426
215096	73°36.2'N	25°58'W	granite	bio.	8.79	1.4136	95.2	408
				musc.	7.61	1.5636	96.0	429
215108	73°56.0'N	26°01'W	mica sch.	bio.	6.06	1.4314	93.0	421
				musc.	8.82	1.3560	96.1	436
215112	73°50.0'N	26°05'W	gneiss	bio.	7.43	1.6281	96.7	409
				musc.	7.60	1.4581	96.7	428
215119	73°45.0'N	26°32'W	granite	bio.	8.85	1.5796	96.5	415
				musc.	7.02	1.3191	98.1	417
215121	73°43.8'N	26°58'W	mica sch.	bio.	8.84	1.6037	95.9	421
				musc.	7.58	1.3830	96.1	424
215124	73°41.3'N	26°57'W	phyllite	musc.	8.63	1.6174	94.6	428
215126	73°37.2'N	26°58'W	quartzite	bio.	8.44	1.4031	96.8	419
215134	73°59.6'N	26°34'W	gneiss	bio.	7.22	1.5680	96.1	431
215136	73°59.3'N	27°05'W	quartzite	bio.	8.87	1.4321	96.5	415
215143	73°58.7'N	27°42'W	gneiss	musc.	7.58	1.6101	91.6	459
				bio.	8.88	1.4509	97.2	432
				musc.	7.14	1.5135	94.4	416
				musc.	7.98	1.5177	89.8	538
				bio.	7.38	1.3437	96.4	440
				bio.	7.47	1.8194	98.3	421
				musc.	7.83	1.5824	95.4	419
				musc.	8.59	1.3839	84.8	414

GGU no.	Latitude	Longitude	Rock type	Mineral	% K	Rad. 40Ar cm³ g⁻¹ ×10⁻⁴	% 40Ar rad.	Age m.y.
215161	73°59.6'N	28°18'W	gneiss	bio.	7.69	1.4896	86.4	440
215170	73°59.3'N	28°43'W	quartzite	bio.	7.17	1.3723	95.8	435
216522	72°55.3'N	26°50'W	gneiss	musc.	8.32	1.5062	94.4	414
216525	72°58.3'N	26°30'W	granite	bio.	8.12	1.9257	97.4	530
216529	72°58.8'N	25°55'W	gneiss	bio.	7.03	1.2334	95.5	403
216537	73°14.2'N	27°05'W	gneiss	bio.	6.14	1.0191	95.6	387
216537			gneiss	bio.	6.98	3.1173	97.8	889
216549	72°59.3'N	27°25'W	gneiss	musc.	7.73	1.4749	89.5	434
216549			gneiss	bio.	7.31	1.5450	96.9	475
228301	72°35.6'N	27°29'W	gneiss	musc.	8.45	1.4888	92.2	404
228302	72°34.4'N	27°29'W	gneiss	bio.	7.62	1.2245	95.8	372
228316	72°36.0'N	27°30'W	gneiss	bio.	8.18	1.3820	90.9	389
228322	72°34.2'N	27°24'W	gneiss	bio.	6.77	1.0011	96.1	376
228323	72°35.8'N	27°21'W	gneiss	bio.	6.40	1.0612	93.4	383
228325	72°50.8'N	28°39'W	granite	bio.	7.92	1.3305	95.7	387
228325			granite	bio.	7.57	1.4186	96.6	427
228333	72°50.8'N	28°39'W	granite	musc.	8.55	1.6053	95.5	431
228333			granite	bio.	7.50	1.3927	96.2	426
228334	72°54.6'N	28°07'W	gneiss	bio.	8.63	1.6389	98.7	433
228334			gneiss	bio.	7.81	1.3269	93.9	391
228337	72°56.2'N	27°43'W	gneiss	bio.	6.63	1.5664	94.7	524
228339	72°54.5'N	27°34'W	granite	musc.	7.36	1.2808	90.0	400
228339			granite	bio.	7.57	1.5676	97.2	467
228354	72°49.8'N	27°13'W	gneiss	musc.	8.83	1.6095	96.8	418
228355	72°48.4'N	27°16'W	gneiss	bio.	7.53	1.6173	95.7	451
228356	72°51.0'N	26°54'W	gneiss	bio.	8.44	1.4992	95.2	371
228357	72°48.3'N	27°07'W	gneiss	musc.	9.02	1.3511	94.5	375
228367	72°50.2'N	26°45'W	gneiss	bio.	7.68	1.4626	91.6	387
228372	72°46.5'N	26°25'W	mica sch.	bio.	7.49	1.2974	92.8	388
228373	72°43.4'N	26°45'W	diorite	bio.	6.30	1.2596	87.8	405
228381	72°43.3'N	26°38'W	diorite	bio.	7.85	1.1141	95.0	394
228386	72°47.0'N	26°09'W	gneiss	bio.	7.18	1.3447	92.5	367
228387	72°47.8'N	26°01'W	gneiss	bio.	7.19	1.1369	94.7	392
228388	72°47.7'N	26°21'W	gneiss	bio.	7.77	1.2234	91.6	379
228389	72°46.4'N	26°14'W	gneiss	bio.	7.81	1.2721	95.1	377
228390	72°44.3'N	26°24'W	gneiss	bio.	7.93	1.2959	93.6	378
228392	72°41.2'N	27°05'W	mica sch.	bio.	7.68	1.3190	90.1	395
228393	72°41.4'N	27°07'W	gneiss	bio.	6.70	1.1011	93.4	380
228394	72°48.1'N	26°55'W	mica sch.	bio.	7.06	1.1161	91.7	367
228397	73°11.2'N	25°58'W	gneiss	bio.	7.40	1.1171	91.7	388
228398	73°10.6'N	26°29'W	gneiss	bio.	7.79	1.2442	85.8	396
228398			gneiss	bio.	7.35	1.3405	93.0	382
228419	73°09.7'N	27°24'W	quartzite	musc.	9.05	1.4737	94.6	377
228419			quartzite	bio.	7.89	1.6960	95.4	482
228420	72°56.0'N	27°32'W	quartzite	musc.	9.04	1.6037	94.8	407
228420			quartzite	bio.	7.86	1.5977	95.8	458
228422	73°07.6'N	27°32'W	mica sch.	musc.	8.61	1.4620	79.2	391
228423	73°07.0'N	27°41'W	gneiss	bio.	7.54	2.5047	95.9	700
228424	73°08.6'N	27°40'W	granite	bio.	7.82	1.8696	95.0	529
228437	73°08.8'N	27°08'W	quartzite	bio.	6.74	1.2285	95.3	417
228437			quartzite	musc.	8.00	1.3000	96.6	376
228438	73°08.0'N	27°04'W	quartzite	bio.	8.98	1.5107	85.7	388
228439	73°06.9'N	26°57'W	quartzite	bio.	8.00	1.2806	93.5	371
228440	73°06.5'N	26°56'W	quartzite	bio.	7.92	1.2793	95.3	374
228440			quartzite	bio.	8.01	1.3042	94.3	378
228441	73°13.3'N	26°54'W	gneiss	bio.	7.76	1.3085	95.0	373
228442	73°13.3'N	26°32'W	gneiss	bio.	7.76	1.2497	94.7	377
228443	73°14.2'N	26°24'W	gneiss	bio.	7.94	1.2644	91.4	374
228445	73°17.0'N	26°30'W	gneiss	bio.	7.93	1.2829	93.1	411
228447	73°19.3'N	26°45'W	mica sch.	bio.	7.78	1.1243	92.7	367
228447			mica sch.	musc.	8.71	1.2298	95.0	368
228454	73°25.5'N	27°05'W	granite	bio.	7.51	1.3877	82.8	406
228454			granite	musc.	7.65	1.3303	90.7	411
228460	73°01.2'N	28°01'W	gneiss	bio.	8.83	1.3727	95.0	427
228461	73°10.0'N	28°53'W	gneiss	musc.	7.93	1.6546	95.5	430
228462	73°10.0'N	28°53'W	gneiss	bio.	9.08	1.4976	93.5	425
228464	73°24.5'N	28°34'W	mica sch.	musc.	7.78	1.6905	97.8	506
228464			mica sch.	bio.	8.94	1.7639	96.6	431
228465	73°27.3'N	27°52'W	granite	musc.	7.59	1.6921	98.1	516
228465			granite	bio.	8.37	1.7614	94.4	434
228476	73°36.5'N	26°03'W	granite	musc.	5.84	1.5969	90.5	391
228476			granite	bio.	8.98	0.9852	93.8	408
228480	73°34.9'N	26°20'W	granite	bio.	9.20	1.5975	93.3	404
228490	73°52.2'N	26°08'W	gneiss	bio.	6.65	1.6168	93.6	407
228491	73°35.6'N	29°15'W	mica sch.	musc.	8.98	1.6133	90.9	415
228491			mica sch.	bio.	7.66	1.4518	95.9	413
228502	73°39.9'N	28°03'W	granite	musc.	8.99	1.4272	97.2	448
228502			granite	bio.	7.21	1.5751	94.6	425
234401	73°27.9'N	27°58'W	mica sch.	musc.	8.46	1.4320	93.9	428
234401			mica sch.	bio.	7.63	1.6219	94.6	417
234402	73°28.0'N	27°57'W	quartzite	bio.	8.90	1.6410	93.4	419
234402			quartzite	musc.	9.09	1.4369	96.1	419
234403	73°28.1'N	27°55'W	quartzite	bio.	7.83	1.6290	95.6	415
234403			quartzite	musc.	8.99	1.4094	96.4	420
234404	73°28.3'N	27°54'W	mica sch.	musc.	7.67	1.5899	95.7	405
234404			mica sch.	bio.	9.00	1.3068	93.9	438
234405	73°28.5'N	27°53'W	quartzite	musc.	6.78	1.2513	92.7	399
234405			quartzite	bio.	7.20	1.4104	89.7	414
234406	73°27.5'N	28°00'W	quartzite	musc.	8.00	1.5917	95.5	406
234410	73°26.8'N	28°07'W	phyllite	bio.	7.77	1.4202	96.1	418
234411	73°27.0'N	28°08'W	quartzite	bio.	6.80	1.1976	92.0	404
234415	73°28.7'N	27°51'W	quartzite	bio.	6.60	1.1973	90.7	402
234416	73°28.8'N	27°49'W	gneiss	bio.	7.85	1.1560	75.0	413
234416			gneiss	musc.	8.97	1.4167	85.9	411
234465	72°33.5'N	25°42'W	quartzite	musc.	7.87	1.6073	92.0	498
234465			quartzite	bio.	9.02	1.6717	93.5	423
234484	72°46.0'N	25°46'W	quartzite	musc.	6.83	1.2677	96.1	423
234485	72°45.0'N	25°49'W	phyllite	bio.	7.31	1.3453	98.2	420
				bio.	5.05	0.9285	94.5	

The Caledonide Orogen—Scandinavia and Related Areas
Edited by D. G. Gee and B. A. Sturt
© 1985 John Wiley & Sons Ltd

The Caledonides of the British Isles: a review and appraisal

G. Kelling*, W. E. A. Phillips†, A. L. Harris‡ and M. F. Howells§

*University of Keele, Staffordshire, England
†Trinity College, Dublin, Ireland
‡University of Liverpool, England
§Institute of Geological Sciences, Aberstwyth, Wales

ABSTRACT

This paper summarizes the current state of knowledge of the sequence, structure, and evolution of the Late Precambrian and Early Palaeozoic rocks within the British portion of the Caledonide orogen, emphasizing recent results and interpretations. It is now evident not only that the northwestern and southeastern foreland terrains display significantly different histories but also that Precambrian rocks, such as the Moines, involved in the orthotectonic sector of the Caledonian foldbelt include assemblages belonging to a Grenville cycle. These 'older' Moines are succeeded by further Precambrian sequences which appear to have undergone at least local thermotectonic deformation and metamorphism prior to commencement of the major Caledonian orogenic cycle, represented by the Dalradian and Lower Palaeozoic sequences.

The Dalradian Supergroup is attributed to an early (Vendian) phase of rifting and ensialic sedimentation, followed in the Cambrian by enhanced subsidence within a northeast-widening trough which locally may have been floored by oceanic crust, probably marginal to the main Iapetus Ocean. Sedimentation in this northern Dalradian zone was brought to an end by the Grampian orogenic event (Cambro-Ordovician), the fundamental cause of which remains uncertain. However, evidence from northwest Ireland indicates possible collision of the volcanic arc flanking the northwestern margin of Iapetus with the North American plate. It is now apparent that during the late Cambrian and early Ordovician an array of small marginal and back-arc basins was interposed between the rising Dalradian cordillera and the active northwestern margin of Iapetus.

Development of this northern trench-arc complex and the localized growth of the accretionary prism was most marked in the late Ordovician and early Silurian, prior to final oceanic closure and collisional deformation in the late Silurian–Devonian (Caledonian event, *sensu stricto*).

The southeastern margin of the British Caledonides is characterized by an Avalon-type history, involving extensive late Precambrian bimodal volcanicity and continental or marine sedimentation in a series of calc-alkaline volcanic chains and possible rift-basins that were eliminated in a Cadomian event involving folding and high P/low T metamorphism. Whilst subsequent Cambrian sedimentation in the Welsh Basin appears to be of foundered passive margin character, evidence of unstable quasi-oceanic conditions, probably associated with the inception of south-easterly subduction and trench formation, is present in the late Cambrian and early Ordovician of the Lake District–Leinster zone. The effects of continued subduction in this zone are manifest in the development of a southeastern volcanic arc with maximum activity during the mid- to late Ordovician, and a subsequent diachronous cessation of volcanicity towards the southwest. In the Ordovician–Silurian back-arc basin of Wales, recent work suggests that the late Silurian collisional deformation resulted in limited southeasterly translation of slices bounded by listric thrust faults.

Although the geochronological framework is still somewhat tenuous, there is increasing evidence for the diachroneity of tectonometamorphic effects associated with the Caledonian, Grampian, and perhaps even the Grenvillian orogenic events. Furthermore, large-scale strike-slip displacements (many of which are poorly quantified, as yet) appear to have been important not only in the final phases of closure and collision but also during the stages of active sedimentation and volcanism along the mobile margins of Iapetus. Such features may be ascribed to oblique closure of this early Palaeozoic ocean, and this may also account for the differing behaviour and structural polarity of the Scandinavian Caledonides, corresponding to the northern zones of the British Isles. However, the Caledonian history of the southern zones is more readily equated with the evolution of the North German–Polish Caledonide belt.

Introduction

The following account is intended to provide a summary of the main aspects of the Caledonide geology and evolution of the British Isles, to enable fuller and more informed correlation with Scandinavia and other sectors of the orogen. The events described here span the interval of geological time only from the early Riphean to the early Devonian but several reviews of the Devonian post-orogenic history of various parts of the British Isles are now available, notably the accounts by Bluck (1980), Mykura (1983), Dineley (in press).

Moreover, the reader is referred to recent accounts of British Caledonide magmatism by Stillman and Francis (1979), Stillman (1980), Fitton *et al*. (1981), and Brown *et al*. (1980), for fuller treatment of that aspect. A number of general reviews of the British Caledonides and syntheses have been published in the past few years (e.g. Watson 1978; Harris *et al*. 1978; Kelling 1978; Phillips 1978; and Harris *et al*. 1979) and the following account emphasizes new data and interpretations and assumes a general acquaintance with these earlier reviews.

We have adhered to the two-fold division of the orogen into an orthotectonic northwestern and a paratectonic southeastern zone (Dewey 1974), while recognizing the gross simplification implicit in such a scheme.

The Basement

The basement on which the British Caledonian rocks were formed shows marked contrasts on either side of the Iapetus suture (Solway–Navan–Silvermines line).

To the northwest, the basement consists largely of late Archaean ortho- and paragneisses (Lewisian), that extend northwards from Donegal in Ireland (Bowes 1978) to form the Hebridean craton—the Caledonian foreland (Fig. 1). Similar rocks occur within the orogen in the extreme west of Mainland Shetland (Flinn *et al*. 1979), in the fold-cores of Morar and Glenelg, and emplaced along ductile thrusts within the Moine rocks of the Caledonian orogen in the Scottish Highlands (e.g. Tanner *et al*. 1970; Rathbone and Harris 1979, Rathbone *et al*. 1983). The oldest Lewisian granulites on the craton are transected by the Scourie basic dykes (c. 2200 Ma) and where subsequent Laxfordian orogenesis (c. 1800 Ma) has been severe this has brought country-rock gneisses and Scourie dykes into approximate parallelism. The presence of inherited Laxfordian and younger zircons in many of the Caledonian granites of the orogenic zone (Pidgeon and Aftalion 1978) suggests extensive Laxfordian basement below the orthotectonic Caledonides. Although Lewisian rocks south of the Great Glen Fault in Scotland are confined to Islay and Colonsay, the L.I.S.P.B. geophysical profile indicates that a layer of rocks with a seismic velocity (>6.4 km sec) comparable to that of the Lewisian (but see Shackleton 1979; Soper and Barber 1982) persists at depths varying from 6 to 14 km as far as the Southern Uplands Fault. The top of this 'layer' appears to be downthrown to the southeast at the Great Glen Fault but upthrown to the southeast at the Highland Boundary Fault, conferring a horst-type basement structure on the post-Caledonian Scottish Midland Valley graben and probably accounting for the granulitic xenoliths recovered from Carboniferous diatremes in this region (Graham and Upton 1978). It also may be the source of the coarse detritus of the Upper Dalradian metasediments near the Highland Border. In north Mayo in Ireland granites and paragneisses have undergone Grenville metamorphism and deformation (van Breeman *et al*. 1978; Long and Yardley 1979).

Southeast of the Iapetus suture (placed between the Southern Uplands and the Lake District on the basis of faunal and facies contrasts within the Lower Palaeozoic sequences), unequivocal pre-Caledonian basement is present in the Rosslare–Anglesey complexes (see below) and in the bordering platforms of south Wales and the Midlands, where the basement rocks are considerably younger than the Lewisian complex. These consist mainly of late Precambrian calc-alkaline volcanics (the Uriconian and Pebidian) and low-grade metasediments (e.g. Longmyndian). These formed part of a vigorous volcanic environment that generated new crust under much of this

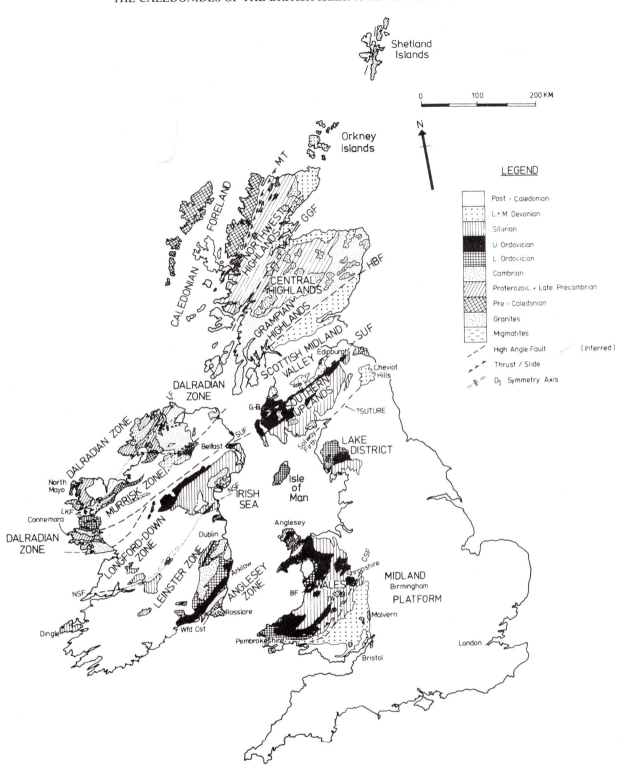

Fig. 1 Location map for Caledonides of the British Isles. The *orthotectonic* region lies between the Moine Thrust (MT) in the northwest and the Highland Boundary Fault (HBF) to the southeast; the Lower Palaeozoic rocks southeast of the Highland Boundary Fault constitute the *Paratectonic* region of the orogen. BF = Bala Fault; CSF = Church Stretton Fault; G–B = Girvan–Ballantrae complex; GGF = Great Glen Fault; LF = Leannan Fault; LKF = Leck Fault; NSF = Navan–Silvermines–Shannon Fault; SUF = Southern Uplands Fault; TA = Towy Axis; Wfd Cst = Waterford Coast

southern (? Avalonian) borderland (Thorpe 1979) in a c. 600 Ma thermotectonic event which conferred a broadly NW–SE tectonic grain on the Midlands basement, prior to unconformable deposition of the basal Cambrian.

The Orthotectonic Caledonides

The Northwestern Foreland

The Hebridean craton gneisses are overlain by virtually unmetamorphosed arkoses (Toridonian) and orthoquartzites and carbonates (Cambro-Ordovician).

The Torridonian Stoer Group rocks were deposited within rifts on an eroded and now partly exhumed terrain of Lewisian gneisses. The fault-determined margins of the rifts, according to Stewart (1982) not only controlled the sedimentary facies of the Stoer Group rocks but were later rejuvenated to influence sedimentation in the younger Torridonian Group, which is separated from the Stoer Group by an unconformity. Formerly believed to be 990 and 810 Ma respectively, the two Torridonian groups are now believed on palaeomagnetic grounds to be c. 1100 Ma and c. 1040 Ma (J. D. A. Piper, personal communication). The lower Cambrian Durness succession also rests unconformably on the Torridonian, overstepping westwards to lie on Lewisian. A further important unconformity has been reported by Palmer *et al.* (1980) and separates lower Cambrian carbonates of the Durness succession from Ordovician carbonates. Thus most of the time span of 1100 Ma to c. 500 Ma is not recorded by the foreland deposits and no precise stratigraphic correlation across the Moine Thrust into the metamorphic rocks of the orogenic zone can be convincingly sustained.

Moine

The main divisions of the Moine rocks are set out in Table 1 and their distribution is shown on Fig. 2. Whereas the general stratigraphic age of the Dalradian is known, and the environment of deposition of these rocks is increasingly understood, the status of the Moine rocks, and particularly those of the northern Scottish Highlands, is still a matter for debate. Many elements of the Moine were formed during the early Caledonian orogeny (Lambert and McKerrow 1976). These include; widespread deformation post-dating the 550 Ma Carn Chuinneag granite (Wilson and Shepherd 1979); biotite and muscovite cooling ages ranging from about 510 Ma to about 400 Ma (Dewey and Pankhurst 1970);

Table 1   The Moine Succession of the Scottish Highlands

Loch Eil Psammites	Loch Eil Division
—————— *Loch Quoich Line* ——————	
Glenfinnan Striped Schists	Glenfinnan
Lochailort Pelites	Division
————— *Sgurr Beag Slide* —————	
Upper Morar Psammites	
Lower Morar Psammites	Morar
Basal Pelites	Division
L E W I S I A N	

extensive pegmatite (c. 450 Ma) and granitic vein complexes (Fettes and McDonald 1978). Furthermore, the highest dislocation in the Moine Thrust Zone, the Moine Thrust itself, carries the Moine rocks about 77 km onto the foreland (Elliot and Johnson 1980) and must be a Caledonian structure, because it affects Cambro-Ordovician rocks (as well as Torridonian, Moine, and Lewisian) and displaces the c. 430 Ma Borolan intrusion. Fabrics in the Moine Thrust mylonites, characterized by a strong ESE-plunging stretching lineation are traceable not only into the Cambro-Ordovician but into the psammitic schists of the Moine Nappe above the thrust.

However, a growing body of evidence suggests that most, if not all, of the Moine rocks had a prolonged Precambrian history of polyphase deformation and high grade metamorphism before undergoing Caledonian reworking, reheating, and injection. Isotopic ages have been obtained for peak metamorphism in schists (Brook *et al.* 1977) (1024 ± 96 Ma) and the emplacement of granite gneiss (Brook *et al.* 1976) (1050 ± 46 Ma), while a foliated pegmatite in the Morar Division post-dates two fold episodes and yields a date of about 760 Ma (D. Powell, personal communication). In northwest Ireland the paragneisses of the Annagh Gneiss Complex of County Mayo (van Breeman *et al.* 1978) have undergone a comparable Grenville thermotectonic cycle. These rocks and the amphibolite–granulite facies metasediments of the Ox Mountains inlier may well be continuations of the 'old' Moine into Ireland (Fig. 1).

The divisions of the Moine rocks (Table 1) are structural/stratigraphic divisions and should not be confused with lithostratigraphic groups, in that they may well have greater structural than stratigraphic validity. Within the Morar Division several formations have been distinguished on the basis of a generalized psammitic or pelitic lithology. Abundant crossbedding in psammitic layers demonstrates the original way-up of this sequence, which youngs away

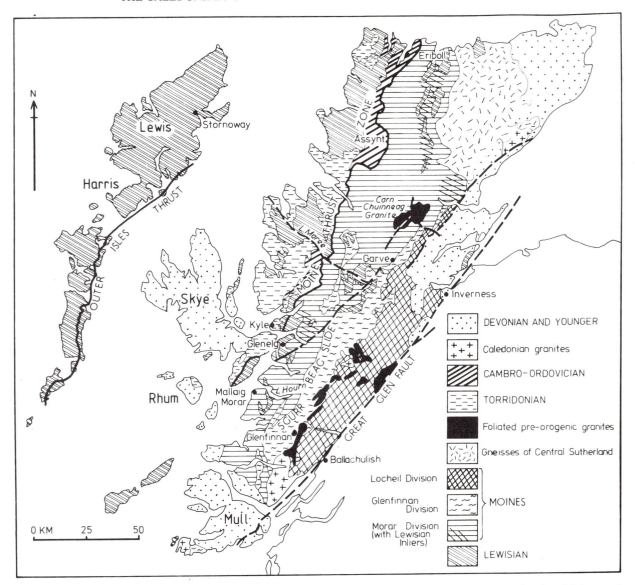

Fig. 2   Outline geology of the northwest Scottish Highlands, showing distribution of Lewisian, Torridonian, Moine, and associated rocks (after various authors)

from Lewisian slices forming the cores of isoclinal anticlines in the Morar and Glenelg areas.

The Glenfinnan Division lies regionally above and to the east of the Morar Division. Displacement across the Sgurr Beag slide, a ductile thrust which separates the two divisions, is probably many tens of kilometres. Although the rocks of both the Morar and Glenfinnan Divisions have had a prolonged history of folding and metamorphism it is likely that the Glenfinnan Division acquired many of its highly deformed and coarsely crystalline characteristics far to the east of, and at a lower crustal level than, the Morar Division. The relative age of the rocks in the two divisions remains unknown but they may have been deposited in contemporary though distant basins (Tanner *et al.* 1970). The Glenfinnan Division rocks are everywhere in the middle-to-upper

amphibolite facies of metamorphism and pelitic rocks are gneissose *lit-par-lit* migmatites and are kyanite- and fibrolite-bearing where composition permits.

The monotonously psammitic Locheil Division lies to the east of the Glenfinnan, from which it is separated by a zone of enigmatic status known as the Loch Quoich Line (Clifford 1957, p. 72). Psammites belonging to the Locheil Division occur to the west of the Loch Quoich Line at outliers within the Glenfinnan Division (Dalziel 1966, p. 132; A. M. Roberts personal communication). The Loch Quoich Line has been variously construed as a zone of high tectonic strain, a tectonically modified unconformity, or the steep limb of a late-stage asymmetric synform of regional extent trending approximately NNE. Over part of its course it is followed by a gneissic granite, the Ardgour gneiss, which has yielded a 1050 ± 46 Ma Rb:Sr date to Brook *et al.* (1976).

Southeast of the Great Glen Fault, in the Central Highlands of Scotland, rocks closely resembling the Moine of the northern Highlands have been described by Piasecki and van Breemen (1979) and included with their Central Highland Division. In the area to the northeast of the Foyers granite A. J. Highton and A. L. H. have examined many square kilometres of rock which have distinct affinities in lithology and metamorphic grade with the Glenfinnan Division of the type locality in the northern Highlands (see also Piasecki and van Breemen 1979, p. 140). Other Moine-like rocks which appear to pass up into the Dalradian are of widespread occurrence in the central Highlands. The relationship of these to the possibly very much older Moine rocks of northern Highland affinities is considered later.

Dalradian

Unlike the Moine rocks, the Dalradian Supergroup has suffered only Caledonian (Grampian) orogenesis). The Dalradian rocks (Table 2) range in age from Riphean to at least middle Cambrian. They have been divided into three groups by Harris and Pitcher (1975) (Table 2), the distribution of which is shown on Fig. 3. The base of the Dalradian is of unknown age but has a transition with Moine-like rocks. The Argyll/Appin boundary is marked by the Portaskaig Tillite (Spencer 1971). This is interpreted as being about 670 Ma old, but only by analogy with the age of sediments associated with the Varanger Tillite (Pringle 1972). The Argyll/Southern Highland boundary is marked by widespread basic volcanics and by a persistent metalimestone (the Tayvallich–Loch Tay Limestone) associated with black phyllites which have yielded an acritarch fauna of very early Cambrian age (Downie *et al.* 1971).

The Dalradian Supergroup basin of deposition

was probably ensialic, and witnessed increasing instability with time (Harris and Pitcher 1975; Harris *et al.* 1978; Anderton 1979). Laterally persistent shallow marine and tidal deposits—quartzites, carbonates, and black pelites—characteristic of the Appin Group gave way upwards after the 670 Ma glaciation to coarse turbidite sequences and carbonaceous pelites. In southeast Donegal there is a change from graphitic pelites with some coarser turbidites into coarse proximal turbidites as the Easdale Subgroup is traced southwards. In the same area, the overlying Crinan Subgroup shows a lateral transition from proximal turbidites to calcareous and quartzitic 'shelf' facies towards the south. This is one of the earliest indications of the emergence of a positive area to the south of the Dalradian. It is believed that block-faulting of the basement may have controlled the bathymetry of the Dalradian basin. Increased instability probably accompanied the widespread though sporadic basic volcanicity (Graham and Bradbury 1981) which is a feature of the Southern Highland Group.

The youngest 'Dalradian' rocks reported by Downie *et al.* (1971) are the Macduff Slates of northeast Scotland which have yielded possible lower Ordovician acritarchs. These rocks are possibly allochthonous (Ramsay and Sturt 1979) and hence their original stratigraphic relationships to the Dalradian elsewhere are now uncertain. Apart from the Macduff Slates the youngest fossils found are the *Pygetia* of high lower Cambrian age from the Leny Limestone (Stubblefield 1956) and the middle Cambrian *Protospongia hicksi* recorded by Rushton and Phillips (1973) from the Dalradian of Clare Island in western Ireland. It is possible, therefore, that 'traditional' Dalradian deposition did not extend much beyond the middle Cambrian.

The main outcrop of Dalradian in Ireland, excluding Connemara, lies to the northwest of a series of inliers of Tremadoc–Caradoc volcanic rocks and associated sediments which has been described as the Northwestern Volcanic Arc (Phillips *et al.* 1976). An important remaining question is to what extent these rocks have been involved in the Grampian orogeny; do they represent a conformable southern facies of the Dalradian or are they an unconformable cover? In Tyrone, the Tyrone Igneous Group (which has yielded a lower Caradoc graptolite; Hartley 1936) occurs on either flank of a central inlier of psammitic metasediments which resemble the '? Old Moine' of the Ox Mountains inlier. These metasediments of the central inlier appear to have been thrust northwards after their polyphase deformation and amphibolite facies metamorphism, and to have been retrogressed during the first deformation in the underlying Tyrone Igneous Group. These igneous

Table 2   The Dalradian Supergroup succession

Southern Highland Group (Upper Dalradian)

Argyll Group (Middle Dalradian)
  Tayvallich Subgroup
  Crinan Subgroup
  Easdale Subgroup
  Islay Subgroup

Appin Group (Lower Dalradian)
  Blair Atholl Subgroup
  Ballachulish Subgroup
  Lochaber Subgroup
  (Transition)

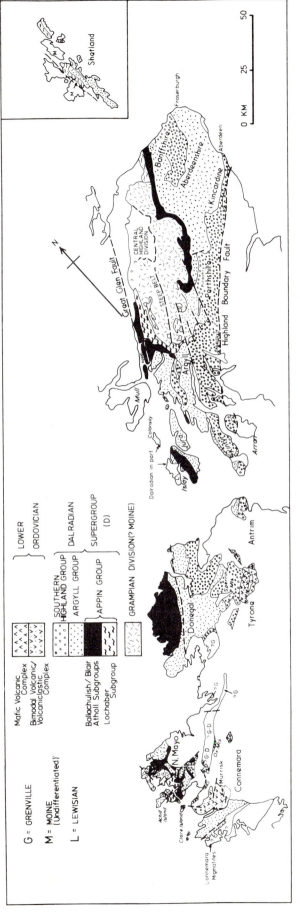

Fig. 3   Distribution of Dalradian and associated rocks in Scotland and Ireland. (After Harris and Pitcher 1975)

rocks display a structural and metamorphic history comparable to that seen in the tectonically overlying upper Dalradian to the north. To the south of the central inlier a further tract of basic lavas, metagabbros, and minor serpentinite resembles the Ballantrae Complex in the southwest Midland Valley of Scotland.

Phillips (1981) has suggested that these Tyrone rocks represent a lower Ordovician volcanic arc overriden to the north by Southern Highland Dalradian inverted on the lower limb of the Tay nappe, and overridden from the south by a slice of pre-Caledonian basement which in turn has been overthrust from the south by an ophiolite complex of Ballantrae type. The volcanic rocks appear to have shared at least much of the Grampian events of the Dalradian. The Tyrone Igneous Group is unconformably overlain by upper Caradoc sediments and hence its deformation occurred during the Caradoc.

Subsurface continuity between the volcanic rocks of Tyrone and the Charlestown inlier to the west (Fig. 3) is suggested by a zone of positive magnetic anomalies. Arenig graptolites of the Pacific faunal province have been obtained from argillites in a metabasalt and volcanic sequence at Charlestown.

In south Mayo and northwest Galway, Tremadoc–Llanvirn rocks are overlain with angular discordance by upper Llandovery sediments. The Ordovician rocks include tholeiitic basalts and andesite, overlain by calc-alkaline basalts of upper Tremadoc age (Ryan et al. 1980) and Llanvirn rhyolites associated with argillite and limestones containing graptolites of the Pacific faunal province and a rich assemblage of brachiopods and trilobites of American affinities (Williams 1969). Ryan et al. (1980) consider this volcanic arc to have developed during the Grampian orogeny as represented in the Dalradian of Connemara to the south. The main deformation (D$_2$) in Connemara is dated by U–Pb ages of 510 ± 10 Ma from zircons in syntectonic gabbros in the migmatite zone of south Connemara (Pidgeon 1969). In the northern limb of the Ordovician syncline in south Mayo at least 3.7 km of undated turbidites and pyroturbidites lie below the oldest Arenig graptolite fauna (D. extensus zone). Above this horizon, southerly and easterly derived turbidites and acid tuffs are followed by a transition through shallow marine Llanvirn sediments into fluviatile sediments which reach a thickness of at least 4 km and probably extend up into the Llandeilo.

In north Mayo, Kennedy (1980) has recognized a serpentinite bearing mélange in a high-level southerly derived tectonic slice of Southern Highland Group turbidites. This provides evidence for the emplacement of some 'oceanic' crust here during Cambrian sedimentation. The presence of an early

blueschist metamorphism in this tectonic unit (Gray and Yardley 1979) is also of interest as it may well be a relic of a more widespread blueschist metamorphism in the Dalradian, perhaps initiated during early D$_1$ thrusting.

It appears at present then that these Tremadoc–Caradoc volcanic and sedimentary sequences may be viewed as high level equivalents of the Connemara migmatite zone and its basic intrusions. They probably represent a magmatic arc along the southern side of the Dalradian basin, which shared at least part of the Grampian orogeny and which was associated locally with a shelf facies within the North American plate (Fig. 4F).

Some brief mention is required of the Highland Border Complex (H.B.C), a tectonically-thinned group of serpentinites, gabbros, spilites, amphibolites, cherts, clastics, and occasional limestone that recurs intermittently along the Highland Boundary Fault-zone across the breadth of Scotland. Despite a number of recent studies, the tectonic and palaeogeographic status of these rocks remains enigmatic. According to some workers the H.B.C. shared the same tectonic history as the adjacent Dalradian Supergroup (Johnson and Harris 1967), whereas fossil evidence now suggests an Arenig age for at least part of the Complex (Curry et al. 1982), which thus post-dates earlier deformation phases regionally affecting the Dalradian. Emplacement of the H.B.C. has also been variously ascribed to northwards subduction (van Breeman and Bluck 1981), or reverse faulting (Curry et al. 1982) of a marginal basin or back-arc sequence formed within the present Midland Valley. Alternatively, development of a small ocean basin within the Dalradian terrain has been suggested, with late Cambrian obduction towards the SE (Henderson and Robertson 1982). The role of major strike-slip faulting within the Midland Valley and its borders during the late Cambrian remains poorly quantified. However the tectonic removal by this means of a substantial southern microcontinent and the consequent juxtaposition of originally discrete terrains (Dalradian, H.B.C., Midland Valley) are possibilities that cannot be ignored (cf. Yardley et al. 1982).

Moine/Dalradian structure

Much of the Dalradian sequence of Scotland and NW Ireland is disposed in a major gravity nappe—the Tay nappe—that faces SE/SSE and was transported in that direction. This can be traced from the east coast of Scotland to the Sperrin Mountains of Tyrone in Ireland. A window into a lower nappe—the Tarfside nappe—has been described by Harte (1979) in the southeast Highlands. In the northeastern Highlands the Banff nappe, originally conceived by Read (1955), has recently been interpreted by Ramsay and Sturt (1979) as a high-level structure with allochthonous low-to-moderate grade metasediments resting on a discontinuous slice of pre-Caledonian gneissic basement and lying above the Tay nappe.

A steep '?root zone' of the Tay nappe, trending NE/SW, is found in the southwest Highlands, where it separates NE-facing primary folds overturned towards the northwest from the SE-facing Tay nappe.

In northern Ireland, such a root zone appears to extend westwards as far as the Leannan Fault (Fig. 1) where it is offset by at least 40 km (Pitcher et al. 1964), and possibly by as much as 160 km. In Donegal and Tyrone, the D_{1-2} deformation shows a dominance of thrusting.

However, the D_2 strain in the north Mayo Dalradian is dominated by a component of E–W dextral simple shear (Sanderson et al. 1980) which may explain the presence of the Tremadoc–Llanvirn arc sequence of south Mayo (and even the Connemara inlier) well south of the main Dalradian outcrop. The reduced component of thrusting could have preserved these units from being overridden by the main Dalradian zone to the extent which is evident in Tyrone (Fig. 3).

This hypothesis also could provide an explanation for the high temperature/low pressure Buchan metamorphism of the Dalradian of northeast Scotland, which could be the result of a still larger component of overthrusting to the northeast, emplacing the Banffshire Dalradian over the volcanic arc (Fig. 4G). This is a simpler explanation for the Buchan metamorphism than the hypothesis of ridge subduction suggested by Lambert and McKerrow (1976).

The position of the Dalradian inlier of Connemara is anomalous, lying well to the south of the main Dalradian outcrop of the British Isles. The stratigraphic succession in Connemara is typical of central parts of the main Dalradian outcrop, such as Perthshire in Scotland. However, the inlier lies to the south of a zone extending from Tyrone to south Mayo which shows evidence of being a southern margin of the Dalradian basin during the Cambrian and Lower Ordovician. This anomaly could be explained in several ways. Firstly, the distinctive Connemara succession could have been deposited before uplift of an internal horst within the Dalradian basin, represented by the Ox Moutains–Tyrone region. Secondly, the latter region may represent merely a basement slice thrust up within a wider Dalradian region, and the southern horst in late Dalradian times lay south of Connemara. Thirdly, the Connemara inlier may have been translated south of

the main outcrop belt by a component of sinistral strike-slip on the pre-Silurian faults which bound the inlier to the north, separating it from the Ordovician rocks. If the Tremadoc–Llanvirn volcanic rocks of south Mayo are correlated with the syntectonic plutonism of the migmatite zone in south Connemara, then the duplication of magmatic arcs might be best explained by this third alternative.

The age of the peak of metamorphism and that of the D_2–D_3 interval in one part of the Tay nappe is approximately indicated by the $514 \pm ^6_7$ Ma, post-D_2/pre-D_3 Ben Vuirich granite of the Pitlochry district of Perthshire (Bradbury et al. 1976). A similar age of 510 ± 10 Ma for D_2 in Connemara is given by Pidgeon (1969). However, if it is accepted that the rocks of the 'Northwestern Volcanic arc' were involved in some of the Grampian orogeny, then it follows that this event may well have been diachronous. Low grade metamorphism, mainly Barrovian in type, marks the peripheral parts of the Dalradian outcrop and is also a feature of structurally high levels (Fettes 1979). Grade locally rose to that of kyanite in the central parts of the belt. Buchan metamorphism is the feature of the northeastern Highlands—its type locality—where its thermal source was probably the c. 501 Ma gabbro sheets of Aberdeen and Banffshire.

Whereas the structure, metamorphism, and stratigraphy of the Dalradian form a fairly simple unified model, this is not the case for the Moine. Major structures of regional extent, such as the Sgurr Beag Slide, are traceable discontinuously from the south coast of Mull to Ross-shire and probably continue, or find their analogues, on the north coast of Scotland. Similar structures occur in the Moine-like rocks in Shetland where Lewisian-like slices mark the course of ductile thrusts in the western part of Mainland, to the east of the gneisses mentioned earlier. However, these persistent structures, of which the Loch Quoich line is another, are late in the orogenic history of the Moines. They were imposed on rocks already crystalline and complexly deformed and, in the case of the Sgurr Beag Slide in particular, have probably resulted in considerable foreshortening of the original Moine tract. A unifying model incorporating the stratigraphic, structural and thermal early history of the entire Moine as yet eludes the Highland geologists but the Moine may well come to be thought of as a fragment of a much larger tract of pre-Caledonian basement, heterogeneously reworked in the orogeny.

One aspect of this speculation is that an unconformity must occur between the Caledonian Dalradian (including the 'Young' Moine that passes upwards into the Dalradian) and the 'Old' Moine, whose orogenic history commenced pre-1000 Ma. If the 'Morarian' (780–720 Ma) dates obtained from syn-D$_2$ pegmatites in the Grampian Slide and associated slides are geologically meaningful (Piasecki *et al.* 1981) then a second unconformity may exist within the 'Young' Moine–Dalradian succession of the central Highlands. These unconformities remain elusive and the hypothesis which draws a sharp distinction between 'Old' and 'Young' Moine therefore relies heavily on radiometric and circumstantial geological evidence. It is possible that orogenesis commenced in Moine rocks at the root of a steadily accumulating sedimentary pile long before even the oldest Dalradian rocks had been deposited and that deformation, metamorphism, and igneous activity advanced upwards and outwards through the pile, imposing polyphase patterns at each successive level. While such a model of expanding diachronous orogenesis may be acceptable over even a few tens of millions of years, it is the apparent timespan of approximately 500 Ma in the case of the metamorphic Caledonides that makes it unlikely to apply here.

The Paratectonic Caledonides

Introduction

South of the Highland Boundary Fault of Scotland, the Caledonide sequences are of very low metamorphic grade and have undergone less intense deformation than in areas to the north. Although this pattern continues broadly across Ireland, it is complicated by the Dalradian inlier of Connemara lying within the paratectonic Caledonides. In the past there have been many fruitless attempts to trace across Ireland such major faults of the British Caledonides as the Great Glen Fault, Highland Boundary Fault, and Southern Uplands Fault, with scant regard for detailed evidence. A more useful approach to correlation has arisen from the recognition of common geotectonic zones, displaying distinctive stratigraphic sequences and structural/magmatic styles. Though the boundaries of these zones change character along strike, the zones themselves provide a natural framework for correlation. These zones of the paratectonic Caledonides are: (a) the Scottish Midland Valley–Murrisk zone; (b) the Southern Uplands–Longford–Down zone; (c) the Lake District–Leinster zone; (d) the Anglesey–Rosslare zone; (e) the Welsh Basin zone; (f) the Borderland–Midland Platform zone (Fig. 1).

Scottish Midland Valley–Murrisk zone

This northernmost zone is best defined in Scotland,

forming the region contained between the Highland Boundary Fault and the Southern Uplands Fault (Fig. 1). In this region the Caledonide units emerge through the prevalent Upper Palaeozoic cover as inliers. These are disposed along the northern border (the Highland Border Complex), and adjacent to the southern boundary fault (the early Ordovician Ballantrae ophiolite and succeeding Ordovician and Silurian sediments of the Girvan area). In several small inliers in the southern part of the Midland Valley, Silurian successions exhibit a concordant passage upwards from Llandovery marine turbidites to Wenlock and Ludlow brackish-water and terrestrial deposits. Most of these sediments were derived from a southerly source ('Cockburnland') that appears to have been an uplifted trench-slope break (Leggett *et al.* 1979).

The Girvan–Ballantrae Complex (Fig. 1) provides a tantalizing glimpse of an early Ordovician magmatic association that is more extensively preserved in Ireland. The Ballantrae rocks comprise an ophiolitic assemblage, including gabbros and trondhjemites dated at 484 ± 4 Ma (U–Pb age), and a thick pile of spilitic lavas and volcaniclastics, together with amphibolites and schists forming a thin aureole probably generated during obduction of hot ocean crust (*cf.* Spray and Williams 1980) on to a northern landmass, The age of the amphibolite (478 ± 4 Ma, by K/Ar method) is so close to that of the trondhjemite as to pose great difficulties if the ophiolite was generated at the median ridge of a large ocean (Lambert and McKerrow 1976). Bluck *et al.* (1980) have concluded that the ophiolite was formed in a small marginal basin flanked to the south by an early Arenig volcanic island arc which was subsequently obducted on to the Midland Valley continental margin. Shortly afterwards the region became a fore-arc basin in which olistostromes, coarse clastics, cherts, and black shales (containing Middle Arenig graptolites) accumulated.

The occurrence of abundant andesitic and rhyolitic clasts (accompanied by cobbles and pebbles of high-level granites yielding ages of 560–472 Ma) in the mid- to late Ordovician clastic sequences of southern Scotland and northern Ireland indicates the continued activity of this volcanic arc on the northwestern margin of the Iapetus Ocean (Longman *et al.* 1979; van Breemen and Bluck 1981).

In Ireland the equivalent Murrisk zone contains the Tremadoc–Lower Caradoc volcanic sequence forming a 'Northwestern Volcanic Arc' which has already been considered as a part of the Grampian orogen. An unconformable cover, as old as upper Caradoc in Tyrone but starting in the Upper Llandovery in Mayo rests upon this volcanic sequence and its ?pre-Caledonian basement. The cover shows a southward transition from shallow marine to deeper shelf and slope clastic sedimentation, with a turbidite fan developed during the peak of transgression in south Mayo (Piper 1972). The succeeding regressive sequence includes shallow marine and fluviatile red beds (probably upper Wenlock–Ludlow), comparable to the facies seen in the southern part of the Midland Valley of Scotland. However, in contrast to the Midland Valley there is no evidence for a southern 'Cockburnland' providing a source of clastic sediment in Ireland. This may result from the greater component of tectonic rotation found in the Longford–Down zone, as discussed below. Late Silurian deformation in south Mayo produced simple upright buckle folds and considerable faulting. The anomaly of more complex and intense polyphase deformation and metamorphism (up to garnet grade in pelites) intensifying northwards towards the basin margin in central Murrisk (Dewey and McManus 1964) is probably related to sinistral shear strains along the Leck–Leannan Fault which forms the northern boundary to the zone in Mayo and is a branch of the Great Glen Fault system (Fig. 1). Further deformation of Lower and Middle Devonian intermontane clastic sediments close to this fault in Mayo probably reflects continued sinistral shear in middle–upper Devonian times.

Southern Uplands–Longford–Down zone

This zone is bounded to the south by the postulated Iapetus suture, the Navan–Silvermines fault line in Ireland, and to the north by the Southern Uplands Fault. In the Scottish part of this zone the Caledonides are represented by Ordovician and Silurian rocks that attain a maximum apparent thickness of around 9 km (Walton 1965). This succession commences with Arenig–Llandeilo spilitic lavas and tuffs, black shales and cherts, passing up into thick piles of greywacke-turbidites (Caradoc–Wenlock). Sequences of these lithologies, usually occurring in the upward order given above, are disposed in perhaps a dozen broad, fault-defined slices trending parallel to the regional NE–SW strike.

Three main stratigraphic belts can be recognized in the Southern Uplands: the Northern Belt, entirely composed of Ordovician sequences, and the Central and Southern Belts which consist of Llandovery and Wenlock rocks respectively. Lambert et al. (1981) have shown that, traced southwards through the Northern Belt, the basal pillow lavas vary in composition from alkali basalts through ocean-floor tholeiites to primitive basalts. Detailed analysis also indicates that the turbidite facies appeared first in the north (early Caradoc) and spread progressively southwards with time.

The same general stratigraphic/structural pattern can be recognized throughout the Longford–Down sector of this zone. Kilometre-scale slumps and massive debris flows appear to be more abundant than in Scotland. A further new element, seen in the Irish Northern Belt, is that bedding is often flat-lying and inverted, facing down to the southeast on a steeper cleavage. This could be explained by longer duration of subduction at a faster rate towards the west, as suggested by Phillips et al. (1976), but limited rotation alone cannot explain why the Longford–Down sector overlies the granulite facies basement block in Ireland (Strogen 1974), a basement which seems to be south of the equivalent zone in Scotland. Work in progress in Longford–Down is expected to clarify the contribution of strike slip movements along this margin of Iapetus in conjunction with subduction-generated accretion.

The distinctive structural style of the whole zone involves a series of high-angle reverse faults with southwards translation, replacing the southern limbs of anticlines. This pattern, in conjunction with the general decrease in age towards the south of the change from pelagic to turbidite sedimentation, has led to the accretionary prism model being applied to the zone (Leggett et al. 1979) and a recent detailed structural study of the Silurian Hawick Rocks of southern Scotland indicates that the folding-style is consistent with the sequential accretion of sediments at a subduction zone (Stringer and Treagus 1981). A further test of the validity of this interpretation is to determine the times of deformation in individual fault-slices, for it is implicit in the model that deformation should become younger towards the ocean (SE in this zone). It is also interesting to note in Ireland that polyphase deformation appears to become more intense towards the southeast (Phillips et al. 1979; Anderson and Cameron 1979), and may result from collision and obduction strains being superimposed.

Lake District–Leinster zone

The Lake District (separated from the Southern Uplands by later Palaeozoic rocks believed to conceal the Iapetus suture-zone) displays sediments and volcanics of early Ordovician to late Silurian age. The oldest of these form the Skiddaw Group, a thick sequence of mud-turbidites of Arenig–Llanvirn age that is succeeded, with possible discordance, by nearly 5 km of Llandeilo–early Caradoc andesitic volcanics (the Eycott and Borrowdale Groups). An early Caradoc tectonic event generated broad, E–W trending folds (Simpson 1968; Soper and Moseley 1978), and the unconformably succeeding mid

Caradoc–Llandovery sequence is a relatively thin, southwards-transgressive group of mudstones with subordinate limestones and volcanics. The later Silurian is represented by thin graptolitic mudstones (Wenlock), succeeded by thick Ludlow turbidites that pass up into shallow marine shelly clastics with some red beds. The stratigraphic distribution of Lake District sequences crudely parallels that in the Southern Uplands, successively younger rocks being exposed to the southeast (apart from the *extensus* slates and volcanics of Black Combe, in the south-west Lake District), but the structure is dominated by broad complex folds of ENE–WSW trend, formed at the end of the Silurian.

Further to the southwest the Manx Group of the Isle of Man and the Ribband Group of the Leinster basin (Southeast Ireland) are lithologically comparable with the Skiddaw Group. Acritarch evidence (Molyneux 1979) indicates that most of the Manx Group (previously considered to be mainly Cambrian in age) is attributable to the Arenig, only the lowest part extending down into the uppermost Cambrian (Tremadoc). However, the base of the Ribband Group is probably Middle Cambrian (Smith 1977) and is conformably underlain by the early Cambrian Bray Group greywackes and quartzites that are in controversial (but probably unconformable) contact with the pre-Caledonian Rosslare gneisses of the Irish Sea Horst zone. In the northwest part of the Leinster Basin the Ribband Group is succeeded by the thick Kilcullen Group of turbidites (lower Ordovician–Wenlock; Bruck *et al.* 1979) but to the southeast, in Country Wexford, Mid-Ordovician faulting, tilting, and erosion led to deposition of shallow marine to subaerial Llandeilo–Ashgill sediments and acid volcanics unconformably upon the Ribband Group. The calc-alkaline basaltic, andesite, and rhyolitic volcanics form part of an arc extending westwards to the Dingle Peninsula (Fig. 1), where Wenlock volcanics pass up into marine shales, followed by late Ludlow to early Devonian fluviatile sediments.

In tracing this southeastern volcanic arc from the Lake District through southern Ireland, it diverges westwards from the suture zone, and a new geotectonic element appears to the northwest of the arc in central Ireland. This consists of a succession of Tremadoc–upper Wenlock sediments which include a series of isolated basaltic volcanic centres of Llanvirn–Caradoc age. Stillman (1980) interpreted these as a northern (oceanward) margin of the volcanic arc. After the Caradoc, sedimentation records the northwards migration of turbidites derived from the uplifted arc to the southeast (Bruck 1972). This geotectonic element is comparable to the forearc basins developed in arc–trench gaps in many of the

Mesozoic–Tertiary subduction zones of the Pacific (Seely *et al.* 1974).

The similar nature of the volcanism in the Lake District–Leinster–Dingle sequences, with their marked transition from tholeiitic to calc-alkaline character, is consistent with development of an island arc above a southeasterly dipping subduction zone. However, there is a notable absence of an accretionary prism on this southeastern margin of Iapetus. It is not clear whether this has been over-riden during collision or removed, either during subduction or by later transcurrent faulting.

Deformation throughout this zone involves notable mid-Ordovician strains in the north of England (Soper and Moseley 1978) which appear to have diminished southwestwards into Ireland. The main Caledonian strain was during the late Silurian in northern England and southeast Ireland, but further west in the Dingle peninsula it was post-Emsian (van der Zwan 1980), indicating diachronism towards the west. The deformation produced upright folds which form a complex arcuate pattern.

As in much of the Southern Uplands, cleavage commonly strikes clockwise of the axial planes of folds. A variety of strain models has been proposed to account for the development of this non-axial plane cleavage and for the alternations from down-dip to along-strike stretching (Sanderson *et al.* 1980; Treagus 1973). These involve dextral or sinistral transcurrent shear along strike during folding (Sanderson *et al.* 1980), or superposition of strain upon already tilted (non-coaxial) beds (Treagus 1973). The widespread distribution of this strain pattern on either side of the suture zone in Britain, Ireland, and Newfoundland makes it more probable that simple shear associated with collision was the controlling factor.

The Irish Sea Horst zone (Anglesey–Rosslare)

This geotectonic element is identified both from outcrops of pre-Caledonian crystalline basement in Anglesey (north Wales) and southeast Ireland (Rosslare), and also from the occurrence of Cambrian and Ordovician sequences on both flanks of the horst which yield evidence of local derivation from such crystalline sources, as in the region of the Ingleton area of the southern Lake District (Onions *et al.* 1973).

In Anglesey the generally accepted view involves a gneissic complex overlain by metasediments and metavolcanics (including an ophiolite body) of the late Precambrian Mona Complex, all conformably succeeded by Arenig to Caradoc shallow marine sediments (Shackleton 1975). However, there is now evidence (Barber and Max 1979) that the Bedded

Series of the Mona Complex is of Cambrian age and that most of the deformation and metamorphism of these rocks is Caledonian, implying a diminished role for the Irish Sea Horst in the Cambrian and an enhancement of the later tectonic activity in this region.

Recently the conspicuous differences in the Lower Palaeozoic histories of Anglesey and the adjacent North Wales mainland (including marked faunal contrasts in the Ordovician; Neuman and Bates 1978) have been attributed to post-Silurian dextral transcurrent faulting, of northeasterly trend, associated with closure of the Iapetus Ocean (Nutt and Smith 1981; Tegerdine et al. 1981; but see also Stewart 1982).

The Welsh Basin zone

This region is delimited to the northwest by the Irish Sea Horst and to the south and southeast by the shelf sequences of the Welsh Borderland and the cratonic block of central England and south Wales (now largely concealed below an Upper Palaeozoic and Mesozoic cover). Here the Lower Palaeozoic sediments and volcanics attain a maximum aggregate thickness of perhaps 10 km and are disposed in a series of broad, relatively open folds of general NE–SW trend, accompanied by major fractures that are best developed in the marginal tracts. These include major thrusts involving significant post-Silurian displacements along the boundary between the Central Wales basin and the peripheral platform to the east and south (Davies 1980; Cope 1979).

Apart from localized folding associated with block-faulting and warping in the early and middle Ordovician, the main deformation and low-grade metamorphism (up to prehnite–pumpellyite facies) is of post-Silurian date throughout the Welsh Basin. The phase of culminating deformation in the Welsh Basin may have been diachronous, varying from late Ludlow to mid-Devonian (McKerrow 1962). The structural relationships of the bordering crystalline rocks, together with limited geophysical evidence, strongly suggest that the major part of the Welsh Basin is floored by quasi-continental crust, and includes crystalline rocks emplaced in the interval 700–540 Ma (Dunning and Max 1975; Watson and Dunning 1979).

Two main areas can be distinguished within the Welsh Basin (Fig. 1), broadly separated by the Bala Fault: *The North Wales area* (Snowdonia–Harlech Dome–Denbighshire) is dominated by Cambrian and Ordovician clastic successions that display rapid lateral changes in thickness and lithology. To the northwest the basal Cambrian clastics are in near-conformable relationship with the Arvonian volcanics and pass up into thick sandy turbidites that

occupy the core of the Harlech Dome. Later Cambrian deposits are generally of shallow marine aspect and are overlain by Arenig sandstones which overstep northwestwards with increasing discordance across the Cambrian and older rocks. Near the Bala Fault the thick Arenig–Llandeilo sequence is dominated by basaltic–rhyolitic volcanics, related to contaminated mantle-derived magma, comparable with modern calc-alkaline island arcs (Kokelaar 1979). The later Ordovician history of North Wales is marked by post-Llanvirn uplift and extensive subaerial to shallow-water volcanicity, best developed in the thick Caradoc sequence of Snowdonia, and followed by widespread deposition of dark mudstones in the Ashgill and early Llandovery. During the Silurian an east-trending turbidite trough developed in northeast Wales (Cummins 1957, 1959).

The Central Wales trough is characterized by Ordovician and Silurian clastic marine sequences. The relatively thin Ordovician successions are dominated by dark, graptolitic mudstones, with localized thick submarine volcanics, especially common in the Llanvirn of southwest Wales (Dyfed) and the Caradoc of the Berwyns (northeast Wales). Thick bodies of sandy and pebbly turbidites were supplied from the south and southeast in the Silurian. The locus of maximum deep-marine deposition appears to have migrated eastwards with time (Ziegler 1970) as the basin became more shallow, culminating in the appearance of terrestrial facies in the late Ludlow of southeast Wales.

The quasi-continental basement character, and the nature and distribution of both the sedimentary and volcanic sequences, all indicate that the Lower Paleozoic Welsh Basin evolved as an ensialic marginal trough on the southeastern margin of the Iapetus Ocean. Most authors have interpreted the Welsh Region during the Ordovician as a back-arc extensional basin (e.g. Phillips et al. 1976). However, it has been suggested recently that the Silurian Welsh Basin may have been a fore-arc trough, created by renewed subduction following landward (southeastwards) migration of the Benioff zone (Okada and Smith 1980). The principal difficulties with this interpretation are the lack of evidence for the accretionary prism or mélange/olistostrome zone that might be expected on the oceanic margin of such a trough and the paucity of the volcanic effusions from the putative arc of the Pembrokeshire–Bristol region (Fig. 1).

The Borderland–Midland Platform zone

This region comprises the Shropshire–Pembrokeshire platform (bounded to the west by a zone of persisting tectonic activity, now represented by the

Towy axis or lineament, marking the shelf-edge through much of the Early Palaeozoic) and the more stable cratonic block of the English Midlands. This zone displays attenuated sequences of richly fossiliferous, shallow marine sediments ranging in age from early Cambrian to late Silurian and showing numerous hiatuses and internal disconformities, the most widespread of which occurs at the base of the Upper Llandovery. Localized volcanics, of basaltic to rhyolitic composition, are important in the Llanvirn and Caradoc of Shropshire and the Builth area of Powys (Wales), while sparse basic volcanics are known from the Llandovery of southwest Wales and the Bristol area.

Exposures of Cambrian quartzites and shales in the Coventry and Birmingham districts (Fig. 1) indicate an eastwards extension of the Shropshire–Pembrokeshire platform while the Tremadoc shales encountered in numerous boreholes in central England are part of an epicontinental mud blanket that covered the southeastern craton both in Britain and Scandinavia. Throughout this zone Caledonian deformation is restricted to localized folding associated with block-faulting that was most widespread in the early Silurian and early Devonian. However, boreholes in eastern and southeastern England have recovered steeply-dipping fossiliferous Ordovician and Silurian sediments (mainly mudstones) considered to form part of a Lower Palaeozoic fold-belt that may be an extension of the Brabant massif (Dunning and Watson 1977).

Summary and Speculations

Rocks assigned to the Caledonide orogen in the British Isles encompass a timespan of more than 300 million years and reveal a complex history of depositional, deformational, magmatic, and metamorphic events as outlined above. Nevertheless, the presently available evidence confirms that the Caledonide evolution of this region can be attributed to a single prolonged episode of ocean opening and closure, preceded by an extended period of intraplate rifting and deformation.

Fig. 4 is intended to convey schematically the salient features of British Caledonide evolution in plate tectonic terms, and emphasizes this two-stage approach. Thus Fig. 4A–C record late Proterozoic phases of ensialic crustal reworking, including a Grenville thermotectonic episode that affected the Lewisian and 'older' Moines. This was followed by a protracted phase of extension, accompanied by block-faulting and progressive disruption and subsidence of a shallow-marine shelf, and is represented by the lower and middle Dalradian successions (late

Riphean and Vendian). Towards the end of this period the southern flank of this rifted zone (the Baltic plate of Fig. 4) was marked by extensive calc-alkaline arc-type volcanicity which generated new crust beneath southeast Ireland, Wales, and central England, marking the inception of an Avalon-type terrane that may have become progressively separated from the Baltic area during the latest Precambrian and Cambrian by 'Tornquist's Sea' (Cocks and Fortey 1982). If the conventional interpretation of Anglesey stratigraphy is sustained, the Mona Complex ophiolites may be remnants of a small ocean basin of late Precambrian age, although an ensialic origin is not precluded for these rocks (see Maltman 1975). Much of the deformation, magmatism and metamorphism may then result from closure and compression of this small ocean basin as an early phase of the more widespread Cadomian or Avalonian orogeny (cf. Rast 1980; Rast and Skehan 1981).

Faunal provinciality and palaeomagnetic data are consistent with the concept of a relatively wide Iapetus Ocean in the British Isles by the early to middle Cambrian (McKerrow and Cocks 1976; Ziegler et al. 1977; Scotese et al. 1979; Smith et al. 1981). This ocean separated a northwestern region (Scotland, northern and central Ireland) from a southeastern region (England, Wales, and southeast Ireland). The date at which ocean spreading commenced in the intervening region remains uncertain, but it may well have been in the early Cambrian. The extensive mafic volcanicity (Tayvallichs etc.) within the Upper Dalradian of the northwestern may represent an abortive attempt at opening, coincident with the successful inception of Iapetus to the south (Fig. 4D; see also Anderton 1982). The presence of a substantial ophiolite complex in Shetland (Flinn this volume) suggests that oceanic crust developed locally here. This is consistent with the view that the original Moine–Dalradian basin may have widened to the northeast, as suggested by the deviation in the trends of the Moine Thrust belt and the Midland Valley–Murrisk zone (Fig. 1), even although some of this deviation can be attributed to later strike-slip faulting (especially along the Great Glen system). Further support for this idea derives from the evidence that the activity of the northwestern volcanic arc persisted longer towards the northeast, arc activity ceasing in the Llanvirn in Ireland and in the Caradoc in Tyrone, but continuing probably in the Llandovery in southern Scotland.

Evidence cited earlier suggests that the northwestern arc came into being during the middle to late Cambrian and was at least partly involved in the early Caledonian phase of orogenesis (late Cambrian to mid-Ordovician) that was responsible for the deformation, metamorphism, and 'early' plutonism of the Dalradian. Growth of the arc is plausibly attributed to the commencement of subduction and crustal consumption beneath the northwestern margin of Iapetus (Fig. 4E). However, the subsequent Grampian events (including initial outwards translation on the Outer Isles Thrust and Moine Thrust and formation of the gravity nappe complex of the Grampian Highlands) appear to require collision to the northwest and active participation by the North American Plate (Fig. 4F, G).

The southeastern margin of Iapetus remained essentially passive throughout the Cambrian and was marked by block-faulting (leading to uplift of the Irish Sea Horst and the North Leinster massif) and concurrent shallow to moderately deep marine sedimentation. Growth of the Southeastern Volcanic Arc is discernible in Wales from late Tremadoc times onwards and is inferred to mark the onset of southeasterly subduction of Iapetus crust (Dewey 1969). The general southeastwards change from tholeiitic to calc-alkaline character of these early to mid-Ordovician eruptives is consistent with such an interpretation (Fitton and Hughes 1970; Stillman 1980; Fitton et al. 1981), although the relative contributions of upper mantle melting, crustal fusion, and anatexis remain controversial (Stillman and Francis 1979; Hughes 1977).

Throughout most of the British Caledonides, the middle Ordovician (Llanvirn–Llandeilo) is characterized by a comparative lull in volcanic activity and widespread deposition of muds, prior to localized Llandeilo/early Caradoc deformation, especially marked in the Leinster–Lake District zone of the southeastern plate. In contrast, the late Ordovician saw a renewal of coarse clastic input on both margins of Iapetus, accompanied by intense volcanism, most conspicuous in the southeastern arc. The enhancement of coarse sediment supply on the northern margin is attributed to uplift of the Dalradian tectogene, with only a secondary contribution from the frontal arc-complex. The relative rates of sediment supply and plate convergence on this margin led to generation of a fore-arc accretionary prism within the Southern Uplands–Longford–Down zone. This prism prograded oceanwards throughout the Silurian and continued underthrusting and uplift of accreted slices ultimately gave rise to a well-developed fore-arc basin, partially preserved in the Scottish Midland Valley. Conversely, much of the coarse sediment influx on the southern margin of Iapetus was derived from the increased activity of the southeastern volcanic arc during the Caradoc. From the early Ashgill onwards there was an abrupt decline in volcanicity throughout this arc, with a generally diachronous cessation of activity towards

the southwest in the Lake District–Leinster zone (Phillips 1981).

The plate tectonic explanation for these mid- to late Ordovician events presents some problems, not the least of which is the width of Iapetus at this time, variously estimated at 2000–3000 km (McKerrow and Cocks 1976), 3300 km (Deutsch 1980), or 1300 km (Piper 1978). The changing behaviour of the southeastern margin may reflect rapid steepening of the southeast-dipping Benioff zone, and the subsequent development of a thermal dome near the leading edge of this Andean-type plate (cf. Stillman 1980) gave rise to the major Welsh and Leinster turbidite basins in the Silurian.

On the northwestern border of Iapetus the time of arrival of coarse detritus in the trench was probably determined by final structural elimination of the small marginal and back-arc basins which were previously interposed between the uplifted Dalradian terrain and the main trench. Major strike-slip faults appear to have been important in the destruction and displacement of these marginal basins. The contemporaneous activity of these features is difficult to prove, although some important faults of this type in northwest Ireland appear to have been initiated prior to the Silurian. Moreover the rotational effects seen in the late Ordovician and Silurian of the Longford–Down area, with accompanying mass-displacements, require significant strike-slip adjustments within the Southern Uplands accretionary wedge. Features of this type are consistent with the pattern of oblique closure of Iapetus inferred from structural and magmatic diachronism along the Caledonian orogen (Phillips et al. 1976).

Effective elimination of the Iapetus Ocean is indicated by the appearance of north-derived turbidites in the late Wenlock of the Lake District. However, continental collision and full suturing in this region did not occur until the early Devonian (Fig. 4H). In the vicinity of the suture this oblique collision involved tightening and some disruption of pre-existing structures, together with extensive dextral strike-slip displacements, succeeded by uplift, formation of molasse-basins, widespread calc-alkaline volcanicity, and injection of granite plutons into rocks which had undergone only burial metamorphism to prehnite–pumpellyite or local greenschist facies. The pattern of late Silurian–Devonian volcanicity in Scotland suggests that northwesterly subduction may have continued here until well into the Devonian (Lower Old Red Sandstone: Phillips et al. 1976; Thirlwall 1981). However, the occurrence of the Lower Devonian andesitic volcanics in both the fore-arc basin (Scottish Midland Valley) and the Southern Uplands subduction complex, and the abundance of post-collisional 'granitic' plutons in the

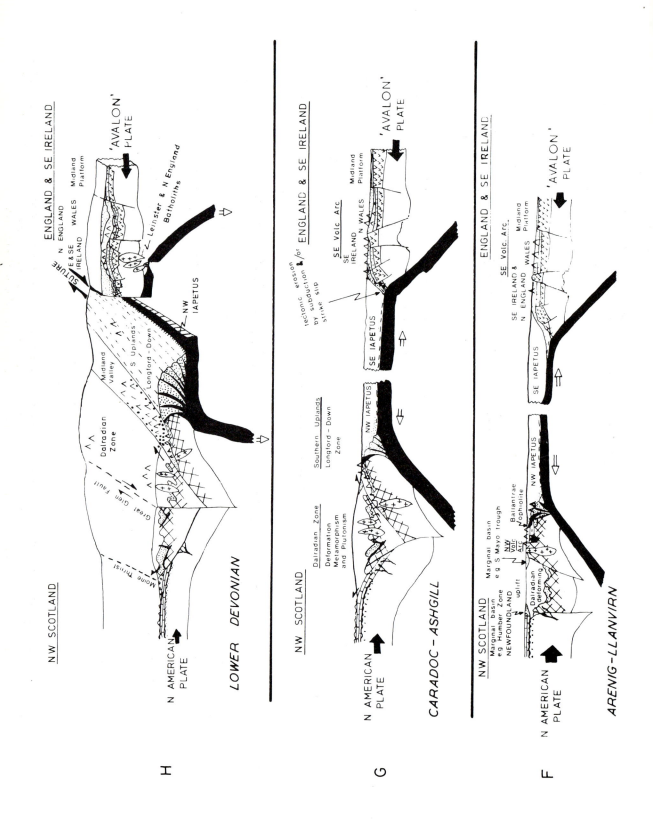

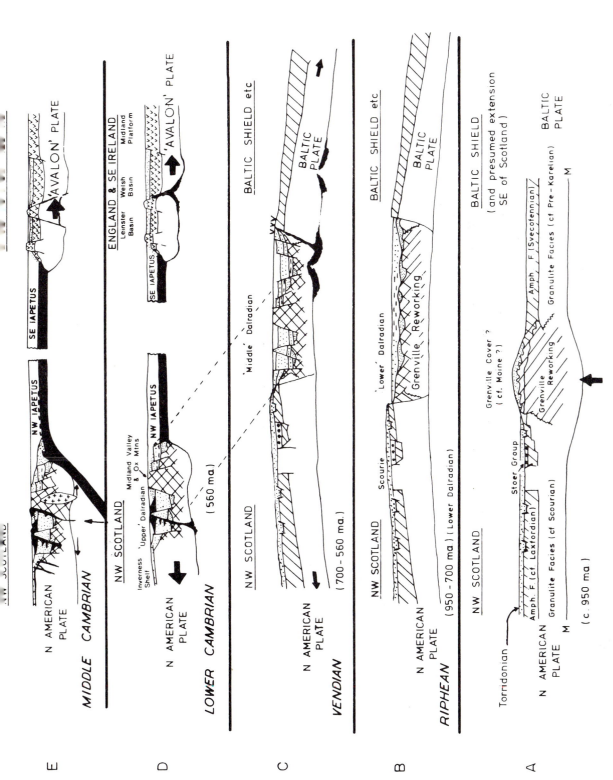

Fig. 4 Highly schematized profiles illustrating the postulated plate tectonic evolution of the British Caledonides. (See text for explanation.) With regard to profiles A and H, note that the boundaries between pre- and post-Grenville Moines are probably tectonic in the region northwest of the Great Glen Fault but may be unconformable in the Dalradian zone

Southern Uplands and the Lake District (south of the Solway suture) are difficult to explain simply in terms of continued subduction. A possible explanation, advocated by Leeder (1982), involves adoption of the model of Crook (1980) which postulates large-scale partial melting or 'ultrametamorphism' and mobilization of stationary oceanic lithosphere trapped beneath the accretionary prism in the immediate post-collisional phase.

Finally, what of the principal topic of this volume—the relationships with the Scandindavian Caledonides? There is little doubt that the orthotectonic Caledonides extend from Scotland beneath the North Sea to Norway. Continuity of the northwest foreland between the Lewisian of Scotland and the Lofoten Islands of western Norway, has been suggested by Talwani and Eldholm (1972). The correlation is based upon a nearly continuous belt of gravity and magnetic highs extending between these areas. Ziegler (1981, and this volume) has shown that isotopic age determinations on metamorphic and intrusive rocks from wells drilled in the North Sea indicate continuity between the Caledonides of Scotland and Norway. The question of making more detailed geological correlations has recently been reviewed by Nicholson (1979). A particularly interesting problem arises from the apparent migration of the Taconic–Grampian orogenic zone (late Cambrian–Ordovician orogenesis) from being entirely on the American margin of the Iapetus Ocean in the Appalachians and the British Isles to resting on the Baltic plate (Finnmarkian orogen, Roberts and Sturt 1980) in Scandinavia. A possible explanation for this is that the large-scale eastward thrusting in the Scandinavian Caledonides during the late Silurian and early Devonian may have translated part of the northwestern margin of Iapetus, with its Grampian orogen, on to the Baltic plate. This would require that an Iapetus suture zone lies below Finnmarkian elements in the Scandinavian Caledonides.

Thirlwall (1981) has shown that the chemistry of Old Red Sandstone volcanic rocks north of the Southern Uplands Fault is consistent with the existence of a subduction zone that changes strike from ENE in Ireland and southern Scotland to northwards in the North Sea. This provides further evidence that the southern edge of the North American plate curved northwards towards Norway. The conclusion also adds support to the hypothesis that collision between curved plate margins during the late Silurian–early Devonian closure of Iapetus produced a dominance of thrusting in Scandinavia and a major component of dextral strike-slip in the British Isles (Phillips 1981).

It is difficult to justify any direct correlations between the southeastern paratectonic Caledonides of the British Isles and the Scandinavian Caledonides and it seems more likely that continuations of Caledonian geology south of the Solway 'suture zone' in England are to be sought in the North German–Polish Caledonides (Brochwicz–Lewinski et al. 1981).

References

Anderson, T. B. and Cameron, T. D. J. 1979. A structural profile of Caledonian deformation in Down. In Harris, A. L., Holland, C. H. and Leake, B. E. (Eds), *The Caledonides of the British Isles–reviewed. Spec. Pub. geol. Soc. London No. 8*, 263–268.

Anderton, R. 1979. Slopes, submarine fans and syn-depositional faults: sedimentology of parts of the Middle and Upper Dalradian in the SW Highlands of Scotland. In Harris, A. L., Holland, C. H. and Leake, B. E. (Eds), *The Caledonides of the British Isles–reviewed. Spec. Pub. geol. Soc. London No. 8*, 483–488.

Anderton, R. 1982. Dalradian deposition and the late Precambrian–Cambrian history of the N. Atlantic region: a review of the early evolution of the Iapetus Ocean. *J. geol. Soc. London*, **139**, 423–434.

Bamford, D. 1979. Seismic constraints on the deep geology of the Caledonides of northern Britain. In Harris, A. L. Holland, C. H. and Leake, B. E. (Eds), *The Caledonides of the British Isles—reviewed. Spec. Pub. geol. Soc. London No. 8*, 93–96.

Barber, A. J. and Max, M. D. 1979. A new look at the Mona Complex (Anglesey, North Wales). *J. geol. Soc. London*, **136**, 407–432.

Bluck, B. J. 1980. Evolution of a strike-slip fault-controlled basin, Upper Old Red Sandstone, Scotland. In Ballance, P. F. and Reading, H. G. (Eds), *Sedimentation in Oblique-slip Mobile Zones. Spec. Pub. int. Ass. Sediment. No. 4*, 63–78.

Bluck, B. J., Halliday, A. N., Aftalion, M. and MacIntyre, R. M. 1980. Age and origin of Ballantrae ophiolite and its significance to the Caledonian orogeny and Ordovician time scale. *Geology*, **8**, 492–495.

Bowes, D. R. 1978. The absolute time scale and the subdivision of Precambrian rocks in north western Scotland, *Geol. Fören. Stöckholm Forhandl.*, **90**, 175–188.

Bradbury, H. J., Smith, R. A. and Harris, A. L. 1976. Older granites as time-markers in Dalradian evolution. *J. geol. Soc. London*, **132**, 677–684.

Brochwicz-Lewinski, W., Pozaryski, W. and Tomczyk, H. 1981. Strike-Slip fault movements along the SW edge of the Eastern European platform during the early Palaeozoic. *Przeglad Geol.*, **8**, 385–396. (In Polish with English Summary).

Brook, M., Powell, D. and Brewer, M. S. 1976. Grenville age for rocks in the Moine of northwestern Scotland. *Nature*, London, **260**, 515–517.

Brook, M., Powell, D. and Brewer, M. S. 1977. Grenville events in the Moine rocks of the Northern Highlands, Scotland. *J. geol. Soc. London*, **133**, 489–496.

Brown, G. C., Plant, J. A. and Thorpe, R. S. 1980. Plutonism in the British Caledonides: space, time and geochemistry. In Wones, D. R. (Ed.), *The Caledonides in the U.S.A.*, Virginia Polytechnic Institute, 157–166.

Bruck, P. M. 1972. Stratigraphy and sedimentology of the

Lower Palaeozoic greywacke formations west of the Leinster granite in counties Kildare and west Wicklow. *Proc. Roy. Irish Acad.*, **72B**, 25–53.

Bruck, P. M., Colthurst, J. R. J., Feely, M., Gardiner, P. R. R., Penney, S. R., Reeves, T. J., Shannon, P. M., Smith, D. G. and Vangeustaine, M. 1979. South-east Ireland: Lower Palaeozoic stratigraphy and depositional history. In Harris, A. L. *et al.* (Eds), *The Caledonides of the British Isles—reviewed. Spec. Pub. geol. Soc. London No. 8*, 533–544.

Clifford, T. N. 1957. The stratigraphy and structure of part of the Kintail district of northern Ross-shire: its relation to the northern Highlands. *Quart. J. geol. Soc. London*, **113**, 57–92.

Cocks, L. R. M. and Fortey, R. A. 1982. Faunal evidence for oceanic separations in the Palaeozoic of Britain. *J. geol. Soc. London*, **139**, 467–480.

Cope, J. C. W. 1979. Early history of the southern margin of the Tywi anticline in the Carmarthen area, South Wales. In Harris, A. L. *et al.* (Eds), *The Caledonides of the British Isles—reviewed. Spec. Pub. geol. London, No. 8*, 527–532.

Coward, M. and Siddons, A. W. B. 1979. The tectonic evolution of the Welsh Caledonides. In *The Caledonides of the British Isles—reviewed. Spec. Pub. geol. Soc. London, No. 8*, 187–198.

Crook, K. A. W. 1980. Fore-arc evolution and continental growth: a general model. *J. Struc. Geol.*, **2**, 289–303.

Cummins, W. A. 1957. The Denbigh Grits; Wenlock greywackes in Wales. *Geol. Mag.*, **94**, 433–451.

Cummins, W. A. 1959. The Lower Ludlow Grits in Wales. *L'pool. Manchr. geol. J.*, **2**, 168–179.

Curry, G. B., Ingham, J. K., Bluck, B. J. and Williams A. 1982. The significance of a reliable Ordovician age for some Highland Border rocks in Central Scotland. *J. geol. Soc. London*, **139**. 453–456.

Dalziel, I. W. D. 1966. A structural study of the granitic gneiss of Western Ardgour, Argyll and Inverness-shire, *Scott. J. geol.*, **2**, 125–152.

Davies, J. H. 1980. A suggested re-interpretation of the Lower Palaeozoic stratigraphy around Llanafan Fawr, central Wales. *Geol. J.*, **15**, 131–134.

Dewey, J. F. 1969. Evolution of the Appalachian/Caledonian orogen. *Nature*, London, **222**, 124–129.

Dewey, J. F. 1974. The geology of the southern termination of the Caledonides. In Nairn, A. E. M. and Stehli, F. (Eds), *The Ocean Basins and Margins, 2, The North Atlantic*, Plenum Press, New York, pp. 205–231.

Dewey, J. F. and McManus, J. 1964. Superposed folding in the Silurian rocks of Co. Mayo, Eire. *L'pool. Manchr. geol. J.*, **4**, 61–76.

Dewey, J. F. and Pankhurst, R. J. 1970. The evolution of the Scottish Caledonides in relation to their isotopic-age pattern. *Trans. R. Soc. Edinb.*, **68**, 361–389.

Deutsch, E. R. 1980. Magnetism of the Mid-Ordovician Tramore Volcanics, SE Ireland, and the question of a wide Proto-Atlantic Ocean. *J. Geomag. Geoelectr.*, **32**, Suppl. III., SIII 77–SIII 98.

Dineley, D. L. in press. The Devonian Period. In Duff, P. McL. D. and Smith, A. J. (Eds), *The Geology of England and Wales*, Scottish Academic Press.

Downie, C., Lister, T. R., Harris, A. L. and Fettes, D. J. 1971. A palynological investigation of the Dalradian rocks of Scotland. *Inst. geol. Sci. Rep. No. 71/9*, 29 pp.

Dunning, F. and Max, M. D. 1975. Explanatory notes to the geological map (Fig. 4) of the exposed and concealed Precambrian basement of the British Isles. In

Harris A. L. *et al* (Eds), *A Correlation of Precambrian Rocks in the British Isles. Spec. Rep. geol. Soc. London. No. 6*, 11–14.

Dunning, F. and Watson, J. V. 1977. Uber die mögliche Erstreckungder osteuropäischen Tafel bis England und Wales. *Zeit angewandte geol.*, **23(a)**, 465–470.

Elliot, D. and Johnson, M. R. W. 1980. Structural evolution in the northern part of the Moine thrust belt, NW Scotland. *Trans. R. Soc. Edinb. Earth Sci.*, **71**, 69–96.

Fettes, D. J. 1979. A metamorphic map of the British and Irish Caledonides. In Harris, A. L. *et al.* (Eds), *The Caledonides of the British Isles—reviewed. Spec. Pub. geol. London, No. 8*, 307–321.

Fettes, D. J. and McDonald, R. 1978. GlenGarry vein complex. *Scott. J. geol.*, **14**, 335–358.

Fitton, J. G. and Hughes, D. J. 1970. Volcanism and plate tectonics in the British Ordovician. *Earth planet Sci. Lett.*, **8**, 223–228.

Fitton, J. G., Thirwall, M. F. and Hughes, D. J. 1981. Volcanism in the Caledonian orogenic belt in Britain. In Thorpe, R. S. (Ed.), *Orogenic Andesites and Related Rocks*.

Flinn, D. this volume. The Caledonides of Shetland.

Flinn, D., Frank, P. L., Brook, M. and Pringle, I. R. 1979. Basement–cover relations in Shetland. In Harris, A. L. *et al.* (Eds), *The Caledonides of the British Isles—reviewed. Spec. Pub. geol. London, No. 8*, 109–115.

Graham, C. M. and Bradbury, H. J. 1981. Cambrian and late Precambrian basaltic igneous activity in the Scottish Dalradian: a review. *Geol. Mag.*, **118**, 27–37.

Graham, C. M. and Upton, B. G. J. 1978. Gneisses in diatremes, Scottish Midland Valley: petrology and tectonic implications. *J. geol. Soc. London*, **135**, 219–228.

Gray, J. R. and Yardley, B. W. D. 1979. A Caledonian blueschist from the Irish Dalradian. *Nature, London*, **278**, 736–737.

Harris, A. L. and Pitcher, W. S. 1975. The Dalradian Supergroup. In Harris, A. L. *et al.* (Eds), *A Correlation of Precambrian Rocks in the British Isles. Spec. Rep. geol. Soc. Lond. No. 6*, 52–75.

Harris, A. L., Baldwin, C. T., Bradbury, H. D., Johnson, H. D. and Smith, R. A. 1978. Ensialic basin sedimentation in the Dalradian Supergroup. In Bowes, D. R. and Leake, B. E. (Eds), *Crustal Evolution in Northwestern Britain and Adjacent Regions. Geol. J. Spec. Issue No. 10*, 115–138.

Harris, A. L., Holland, C. H. and Leake, B. E. (Eds), 1979. *The Caledonides of the British Isles—reviewed. Spec. pub. geol. Soc. London, No. 8*, 768 pp.

Harte, B. 1979. The Tarfside succession and the structure and stratigraphy of the eastern Scottish Dalradian rocks. In Harris, A. L. *et al.* (Eds), *The Caledonides of the British Isles—reviewed. Spec. Pub. geol. Soc. London, No. 8*, 221–228.

Hartley, J. J. 1936. The age of the igneous series of Slieve Gallion, Northern Ireland. *Geol. Mag.*, **73**, 226–228.

Henderson, W. G. and Robertson, A. H. F. 1982. The Highland Border rocks and their relation to marginal basin development in the Scottish Caledonides. *J. geol. Soc. London*, **139**, 435–452.

Hughes, D. J. 1977. *The Petrochemistry of the Ordovician igneous rocks of the Welsh Basin*. Ph.D. Thesis (Univ. of Manchester), unpublished.

Johnson, M. R. W. and Harris, A. L. 1967. Dalradian–?Arenig relations in parts of the Highland Border, Scotland and their significance in the chronology of the Caledonian orogeny. *Scott. J. Geol.*, **3**, 1–16.

Kelling, G. 1978. The paratectonic Caledonides of mainland Britain In *Caledonian–Appalachian Orogen of the North Atlantic Region, Geol. Surv. Can. Pap.*, **78–13**, 89–95.

Kennedy, B. 1980. Serpentine-bearing melange in the Dalradian of County Mayo and its significance in the development of the Dalradian basin. *J. Earth Sci. R. Dublin Soc.*, **3**, 117–126.

Kokelaar, B. P. 1979. Tremadoc to Llanvirn volcanism on the southeast side of the Harlech Dome (Rhobell Fawr), N. Wales. In Harris, A. L. *et al.* (Eds), *The Caledonides of the British Isles—reviewed. Spec. Pub. geol. Soc. London, No. 8*, 591.

Lambert, R. St. J. and McKerrow, W. S. 1976. The Grampian Orogeny. *Scott. J. Geol.*, **12**, 271–292.

Lambert, R. St. J., Holland, J. G. and Leggett, J. K. 1981. Petrology and tectonic setting of some Ordovician volcanic rocks from the Southern Uplands of Scotland. *J. geol. Soc. London*, **138**, 421–436.

Leeder, M. R. 1982. Upper Palaeozoic basins of the British Isles—Caledonide inheritance versus Hercynian plate margin processes *J. geol. Soc. London*, **139**, 481–494.

Leggett, J. K., McKerrow, W. S., Morris, J. H., Oliver G. J. H. and Phillips, W. E. A. 1979. The north-western margin of the Iapetus Ocean. In Harris, A. L. *et al.* (Eds), *The Caledonides of the British Isles—reviewed. Spec. Pub. geol.. Soc. London, No. 8*, 499–512.

Leggett, J. K., McKerrow, W. S. and Eales, M. H. 1979. The Southern Uplands of Scotland: A Lower Palaeozoic accretionary prism. *Jour. geol. Soc. London*, **136**, 755–770.

Long, C. B. and Yardley, B. W. D. 1979. The restricted distribution of pre-Caledonian basement in the Ox Mountains inlier, Ireland. In Harris, A. L. *et al.* (Eds), *The Caledonides of the British Isles—reviewed. Spec. Pub. geol. Soc. London, No. 8*, 153–156.

Longman, C. D., Bluck, B. J. and Van Breemen, O. 1979. Ordovician conglomerates and the evolution of the Midland Valley, *Nature, London*, **280**, 578–581.

McKerrow, W. S. 1962. The chronology of Caledonian folding in the British Isles. *Proc. nat. Acad. Sci. U.S.A.*, **68**, 1905–1913.

McKerrow, W. S. and Cocks, L. R. M. 1976. Progressive faunal migration across the Iapetus Ocean. *Nature, London*, **263**, 304–306.

Maltman, A. J. 1975. Ultramafic rocks in Anglesey: their non-tectonic emplacement. *J. geol. Soc. London*, **131**, 593–605.

Molyneux, S. G. 1979. New evidence for the age of the Manx Group, Isle of Man. In Harris, A. L. *et al.* (Eds), *The Caledonides of the British Isles—reviewed. Spec. Pub. geol. Soc. London, No. 8*, 415–422.

Mykura, W. 1983. The Old Red Sandstone. In Craig, G. Y. (Ed.), *The Geology of Scotland*, 2nd Ed., Scottish Academic Press.

Neuman, R. B. and Bates, D. E. B. 1978. Reassessment of Arenig and Llanvirn (early Ordovician) brachiopods from Anglesey, north-west Wales. *Palaeontology*, **21**, 571–613.

Nicholson, R. 1979. Caledonian correlations: Britain and Scandinavia. In Harris, A. L. *et al.* (Eds), *The Caledonides of the British Isles—reviewed. Spec. Pub. geol. Soc. London, No. 8*, 3–18.

Nutt, M. J. C. and Smith, E. G. 1981. Transcurrent faulting and the anomalous position of pre-Carboniferous Anglesey. *Nature, London*, **290**, 492–494.

Okada, H. and Smith, A. J. 1980. The Welsh 'geosyncline'

of the Silurian was a fore-arc basin. *Nature, London*, **288**, 352–354.

Onions, R. K., Oxburgh, R. K., Hawkesworth, C. J. and MacIntyre, R. M. 1973. New isotopic and stratigraphical evidence on the age of the Ingletonian: probable Cambrian of northern England. *J. geol. Soc. London*, **129**, 445–452.

Palmer, T. J., McKerrow, W. S. and Cowie, J. W. 1980. Sedimentological evidence for a stratigraphical break in the Durness Group. *Nature, London*, **287**, 720–722.

Phillips, W. E. A. 1978. The Caledonide Orogen in Ireland. In *Caledonian–Appalachian Orogen of the North Atlantic Region. Geol. Surv. Can. Pap.*, **78–13**, 97–103.

Phillips, W. E. A. 1981. Estimation of the rate and amount of absolute lateral shortening in an orogen using diachronism and strike-slipped segments. In McClay, K. R. and Price, N. J. (Eds), *Thrust and Nappe Tectonics. Spec. Pub. geol. Soc. London, No. 9*, 267–274.

Phillips, W. E. A., Stillman, C. J. and Murphy, T. 1976. A Caledonian plate tectonic model. *J. geol. Soc. London*, **132**, 579–609.

Phillips, W. E. A., Flegg, A. M. and Anderson, T. B. 1979. Strain adjacent to the Iapetus suture in Ireland. In Harris, A. L. *et al.* (Eds), *The Caledonides of the British Isles—reviewed. Spec. Pub. geol. Soc. London, No. 8*, 257–262.

Piasecki, M. A. J. and Van Breeman, O. 1979. The 'Central Highland Granulites': cover–basement tectonics in The Moinces. In Harris, A. L. *et al.* (Eds), *Caledonides of the British Isles—reviewed. Spec. Pub. geol. Soc. London, No. 8*, 139–144.

Piasecki, M. A. J., Van Breeman, O. and Wright, A. E. 1981. Late Precambrian geology of Scotland, England and Wales. *Can. Soc. Petrol. Geol. Mem.*, **7**, 57–94.

Pidgeon, R. T. 1969. Zircon U–Pb ages from the Galway granite and the Dalradian, Connemara, Ireland. *Scott. J. Geol.*, **5**, 375–392.

Pidgeon, R. T. and Aftalion, M. 1978. Cogenetic and inherited zircon U/Pb systems in granites: Palaeozoic granites of Scotland and England. In Bowes, D. R. and Leake, B. E. (Eds), *Crustal Evolution in Northwestern Britain and Adjacent Regions. Geol. J. Spec. Issue No. 10*, 183–220.

Piper, D. J. W. 1972. Sedimentary environments and palaeogeography of the late Llandovery and early Wenlock of North Connemara Ireland. *Jour. geol. Soc. London*, **128**, 33–51.

Piper, J. D. A. 1978. Palaeomagnetism and palaeogeography of the Southern Uplands block in Ordovician times. *Scott. J. Geol.*, **14**, 93–107.

Pitcher, W. S., Elwell, R. W. D., Tozer, C. F. and Cambray, F. W. 1964. The Leannan Fault. *Q. J. geol. Soc. London*, **120**, 241–274.

Pringle, I. R. 1972. Rb–Sr determinations on shales associated with the Varanger Ice Age. *Geol. Mag.*, **109**, 465–472.

Ramsay, D. M. and Sturt, B. 1979. The status of the Banff Nappe. In Harris, A. L. *et al.* (Eds), *The Caledonides of the British Isles—reviewed. Spec. Pub. geol. Soc. London, No. 8*, 145–152.

Rast, N. 1980. The Avalonian Plate in the Northern Appalachians and the Caledonides. In Wones, D. R. (Ed.), *The Caledonides in the U.S.A., Virginia Poly. Inst. Dept. Geol. Mem.*, **2**, 63–66.

Rast, N. and Skehan, J. W. S. J. 1981. Possible correlation of the Pre-Cambrian rocks of Newport, Rhode Island, with those of Anglesey Wales. *Geology*, **9**, 596–601.

Rathbone, P. A. and Harris, A. L. 1979. Basement–cover relationships at Lewisian inliers in the Moine rocks. In Harris, A. L. *et al.* (Eds), *The Caledonides of the British Isles—reviewed. Spec. Pub. geol. Soc. London, No. 8,* 101–107.

Rathbone, P. A., Coward, M. P., and Harris, A. L. 1983. Cover and basement: a contrast in style and fabrics. *Geol. Soc. Amer. Mem.* **158**, 213–223.

Read, H. H. 1955. The Banff Nappe: an interpretation of the structure of the Dalradian rocks of northeast Scotland. *Proc. geol. Assoc. London*, **66**, 1–29.

Roberts, D. and Sturt, B. A. 1980. Caledonian deformation in Norway. *J. geol. Soc. London*, **137**, 241–250.

Rushton, A. and Phillips, W. E. A. 1973. A Protospongia from the Dalradian of Clare Island, Co. Mayo, Ireland. *Palaeontology*, **16**, 231–237.

Ryan, P. D., Floyd, P. A. and Archer, J. B. 1980. The stratigraphy and petrochemistry of the Lough Na Fooey Group (Tremadocian), western Ireland. *J. geol. Soc. London*, **137**, 443–458.

Sanderson, D. J., Andrews, J. R., Phillips, W. E. A. and Hutton, D. H. W. 1980. Deformation studies in the Irish Caledonides. *J. geol. Soc. London*, **137**, 289–302.

Scotese, C., Bambach, R. K., Barton, C., Van der Voo, R., and Ziegler, A. 1979. Palaeozoic base maps. *J. Geol.*, **87**, 217–278.

Seeley, D. R., Vail, P. R. and Walton, G. G. 1974. Trench slope model. In Burk, C. A. and Drake, C. L. (Eds), *The Geology of Continental Margins*, Springer–Verlag, Berlin, pp 249–260.

Shackleton, R. M. 1975. Pre-cambrian rocks of Wales. In Harris, A. L. *et al.* (Eds), *A correlation of Precambrian Rocks in the British Isles. Spec. Rep. geol. Soc. London, No. 6,* 76–82.

Shackleton, R. M. 1979. The British Caledonides: Comments and Summary. In Harris, A. L. *et al.* (Eds), *The Caledonides of the British Isles—reviewed. Spec. Pub. geol. Soc. London, No. 8,* 229–304.

Simpson, A. 1968. The Caledonian history of the northeastern Irish Sea and its relation to surrounding areas. *Scott. J. geol.*, **4**, 135–163.

Smith, D. G. 1977. Lower Cambrian palynomorphs from Howth, County Dublin. *Geol. J.*, **12**, 159–168.

Smith, A. G., Hurley, A. M. and Briden, J. C. 1981. Phanerozoic palaeocontinental World Maps, Cambridge Univ. Press, 102 pp.

Soper, N. J. and Barber, A. H. 1982. A model for the deep structure of the Moine Thrust zone. *J. geol. Soc. London*, **139**, 127–138.

Soper, N. J. and Moseley, F. 1978. Structure. In Moseley, F. (Ed.), *Geology of the Lake District, Yorks. geol. Soc.*, 45–67.

Spencer, A. M. 1971. Late Precambrian glaciation in Scotland. *Mem. geol. Soc. London, No. 6.*

Spray, J. G. and Williams, G. D. 1980. The sub-ophiolite metamorphic rocks of the Ballantrae Igneous Complex, SW Scotland. *J. geol. Soc. London*, **137**, 359–368.

Stewart, A. D. 1982. Late Proterozoic rifting in northwest Scotland: the genesis of the 'Torridonian'. *J. geol. Soc. London*, **139**, 415–422.

Stillman, C. J. 1980. Caledonide volcanism in Ireland. In Wones, D. R. (Ed.), *The Caledonides in the U.S.A., Virginia Poly. Inst. Geol. Dept. Mem. 2,* 279–284.

Stillman, C. J. and Francis, E. H. 1979. Caledonide volcanism in Britain and Ireland. In Harris, A. L. *et al.* (Eds), *The Caledonides of the British Isles—reviewed. Spec. Pub. geol. Soc. London, No. 8,* 557–577.

Stringer, P. and Treagus, J. E. 1981. Asymmetrical folding

in the Hawick Rocks of the Galloway area, Southern Uplands. *Scott. J. Geol.*, **17**, 129–148.

Strogen, P. 1974. The sub-Palaeozoic basement in central Ireland. *Nature*, London, **250**, 562–563.

Stubblefield, C. H. 1956. Cambrian palaeogeography in Britain. In *The Cambrian System, its palaeogeography and the problem of its Base. CR. XX Int. Geol. Cong.*, Part I, pp. 1–43.

Talwani, M. and Eldholm, O. 1972. Continental margin off Norway: a geophysical study. *Bull. geol. Soc. Amer.*, **83**, 3575–3606.

Tanner, P. W. G., Johnstone, G. S., Smith, D. I. and Harris, A. L. 1970. Moine stratigraphy and the problem of the Central Ross-shire inliers. *Bull. geol. Soc. Amer.*, **81**, 299–306.

Tegerdine, G. D., Campbell, S. D. G. and Woodcock, N. H. 1981. Comments on Transcurrent faulting and pre-Carboniferous Anglesey. *Nature, London*, **293**, 760–761.

Thirwall, M. F. 1981. Implications for Caledonian plate tectonic models of chemical data from volcanic rocks of the British Old Red Sandstone. *J. geol. Soc. London*, **138**, 123–138.

Thorpe, R. S. 1979. Late Precambrian igneous activity in southern Britain. In Harris, A. L. *et al.* (Eds), *The Caledonides of the British Isles—reviewed. Spec. Pub. geol. Soc. London, No. 8,* 579–589.

Treagus, J. E. 1973. Buckling stability of a viscous single-layer system, oblique to the principal compression. *Tectonophysics*, **19**, 271–279.

Van Breemen, O. and Bluck, B. J. 1981. Episodic granite plutonism in the Scottish Caledonides. *Nature*, London, **291**, 113–117.

Van Breemen, O., Pidgeon, R. T. and Johnson, M. R. W. 1974. Precambrian and Palaeozoic pegmatites in the northern Moines of Scotland. *J. geol. Soc. London*, **130**, 493–508.

Van Breemen, O., Halliday, A. N., Johnson, M. R. W. and Bowes, D. R. 1978. Crustal additions in late Precambrian times. In Bowes, D. R. and Leak, B. E. (Eds), *Crustal evolution in northwestern Britain and adjacent regions. Geol. J. Spec. Issue 10,* 81–106.

Van der Zwan, C. J. 1980. Palynological evidence concerning the Devonian age of the Dingle Group, southwest Ireland. *Rev. Palaeobot. Palynol.*, **29**, 271–284.

Walton, E. K. 1965. Lower Palaeozoic rocks–palaeogeography and structure. In Craig, G. Y. (Ed.), *The Geology of Scotland*, Edinburgh, 201–227.

Watson, J. 1978. The basement of the Caledonide Orogen in Britain. In *The Caledonian–Appalachian Orogen of the North Atlantic Region. Geol. Surv. Can. Pap.*, **78–13**, 75–77.

Watson, J. and Dunning, F. W. 1979. Basement–cover relations in the British Caledonides. In Harris, A. L. *et al.* (Eds), *The Caledonides of the British Isles—reviewed. Spec. Pub. geol. Soc. London, No. 8,* 67–91.

Williams, A. 1969. Orodovician of British Isles. In Kay, M. (Ed.), *North Atlantic: Geology and Continental Drift. Am. Assoc. Petrol. Geol. Mem.*, **12**, 236–264.

Wilson, D. and Shepherd, J. 1979. The Carn Chuinneag granite and its aureole. In Harris, A. L. *et al.* (Eds), *The Caledonides of the British Isles—reviewed. Spec. Pub. geol. Soc. London, No. 8,* 669–675.

Yardley, B. W. D., Vine, F. J. and Baldwin, C. T. 1982. The plate tectonic setting of NW Britain and Ireland in late Cambrian and early Ordovician times. *J. geol. Soc. London*, **139**, 457–466.

Ziegler, A. M. 1970. Geosynclinal development of the British Isles during the Silurian period. *J. Geol.*, **78**, 445–479.

Ziegler, A. M., Scotese, C. R., McKerrow, W. S., Johnson, M. E. and Bambach, R. K. 1977. Palaeozoic biogeography of continents bordering the Iapetus (pre-Caledonian) and Rheic (pre-Hercynian) Oceans. In *Palaeontology and Plate Tectonics, Spec. Publ. Biol. Geol., Milwaukee Public Museum*, **2**, pp. 1–22.

Ziegler, P. A. 1981. Evolution of sedimentary basins in north-west Europe. In Illing, L. V. and Hobson, G. D. (Eds), *Petroleum Geology of the Continental Shelf of North-West Europe*, Institute of Petroleum, London, pp. 3–39.

Ziegler, P. A. this volume. Late Caledonian framework of western and central Europe.

The Caledonide Orogen—Scandinavia and Related Areas
Edited by D. G. Gee and B. A. Sturt
© 1985 John Wiley & Sons Ltd

Fold nappes: examples from the Moine Thrust zone

M. P. Coward and **G. J. Potts**

Department of Earth Sciences, Leeds University, Leeds LS2 9JT, England

ABSTRACT

Examples of folds within thrust sheets are taken from the Moine Thrust zone of northwest Scotland, in particular the Eriboll, Assynt, and Skye districts. Apart from the structurally necessary folds formed above or ahead of ramps in the thrust plane, most folds are formed by differential movement of the thrust slab. The folds may form in zones of compressional flow or zones of layer-normal differential shear, the first producing fold axes at high angles to the transport direction, the second producing oblique fold axes, often at a small angle to the transport direction. With progressive deformation these folds become asymmetrical and inclined to recumbent, and have intensely strained to sheared out, overturned limbs. Within a thrust stack there may be several phases of fold development and each successively underlying thrust may produce a fold which affects the overlying folds and thrusts. Such folds would not be regional but local. At Eriboll in the northern part of the Moine Thrust belt, more than five phases of folding have been produced locally by this mechanism. Hence there is no simple correlation of fold structures along or across the thrust zone as a whole.

Both basement and sedimentary cover rocks are involved in the folding. The fold size depends on the amount of thrust displacement and also the rheology and thickness of sediments involved. Thus, small fold nappes characterize the Cambrian sequence of the northern part of the thrust zone, while large recumbent nappes characterize the thick Torridonian sandstone sequence in the south.

Introduction

Rich (1934) recognized the association of rootless anticlines in the hanging wall of thrusts with the climb of these thrusts up ramps (Fig. 1). These folds may be explained without recourse to any shortening of the layers. Rich's (1934) concept originated from a study of cross-sections through the Pine Mountain–Cumberland Thrust block in Virginia, Kentucky, and Tennessee. The folds in the blocks are characterized by a large flat-topped anticline, the Powell Valley Anticline, separated from the hinge of the Middlesbrough Syncline by a short, steeply dipping limb (Fig. 1). Rich (1934) considered this fold to be a natural consequence of the transport of a thrust block over a ramp. Subsequent papers in Appalachian geology have verified the essential features of the Rich model (Harris 1970; Harris and Milici 1977) and comparable fold mechanisms have been applied to explain structures in the Rocky Mountains (Douglas 1950; Dahlstrom 1970). in the Pyrenees (Seguret 1970) and in the Moine Thrust

zone of Scotland (Elliot and Johnson 1980; Butler 1982).

The model of Rich (1934) has been modified by more recent workers (Berger and Johnson 1980) who noted that the northwest limb of the Powell Valley anticline dips steeply (around 70°) to the northwest, while the dip of the ramp is much more gentle (around 20°). If there is no deformation, other than that due to ramp climb, then the dip of the steep fold limb should be the same as that of the ramp. Berger and Johnson (1980) explain this steep dip, anomalous in the Rich model, by a component of drag on the ramp surface. This would produce an anticlinal fold in the hanging wall which may be highly asymmetric (Fig. 2).

In the Moine Thrust zone of Scotland, folds are associated with many of the thrusts. The zone involves the thrusting to the west–northwest of Upper Proterozoic Moine Schists over Lower Proterozoic Lewisian Gneisses with their cover of Upper Proterozoic Torridonian and Cambro–Ordovician sediments (Peach *et al.* 1907). The

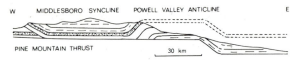

Fig. 1 Cross-section through the Middlesboro syncline–Powell Valley anticline, after Rich (1934) showing the form of fold development in the hanging wall of a thrust above and to the front of a ramp

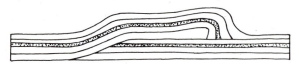

Fig. 2 Form of a fold in the hanging wall of a thrust subjected to high drag on the ramp surface (after Berger and Johnson 1980)

Torridonian is a several kilometres thick sequence of monotonous grits, sandstones, and shales, divided into a series of groups, many of which show only regional development (Stewart 1966). The Cambrian consists of a basal quartz sandstone (the Basal Quartzite), the upper part of which (the Pipe Rock) is bioturbated by worm tubes or pipes. This is overlain by a mixed series of dolomitic shales (the Fucoid Beds) and quartz-rich sandstones (the Serpulite Grit) and then by a thick sequence of limestones (the Durness Limestone). The bioturbation tubes of the Pipe Rock were originally normal to bedding and approximately circular in cross-section and make excellent strain markers as there is little to no ductility contrast between pipe and the host rock (Coward and Kim 1981).

The thrust sheets vary in thickness from several hundred metres, where they involve Lewisian Gneisses and their cover of Proterozoic and Phanerozoic rocks, to several metres or tens of metres where they involve thin zones of the Phanerozoic. Most thrust sheets contain folds which are similar in shape to the Pine Mountain structures described by Rich (1934). The hinges generally trend at 030°, normal to the thrust transport direction.

The aim of this paper is to discuss the shape and origin of these folds in the light of the models of Rich (1934) and Berger and Johnson (1980) using three particular areas in the Moine Thrust zone (Fig. 3A): (1) Eriboll on the north coast of Scotland, (2) Skye and the adjacent mainland at the southern end of the exposed thrust zone, and (3) the Assynt District.

The Eriboll District

The Eriboll District of the Moine Thrust zone (Fig. 3B) consists of four thrust sheets or nappes: a lower (Heilam) nappe containing finely imbricated Phanerozoic rocks, a middle (Arnaboll) nappe of Lewisian and Phanerozoic cover, a higher (Upper Arnaboll) nappe of mylonitized Lewisian and Phanerozoic rock, and an upper Moine Thrust Nappe. The lowest nappe is underlain by the sole or floor thrust which gradually climbs from the basement of the Lewisian Gneisses in the east to the Fucoid Beds and Durness Limestone in the west. Cross-sections through the folds associated with the thrusts are

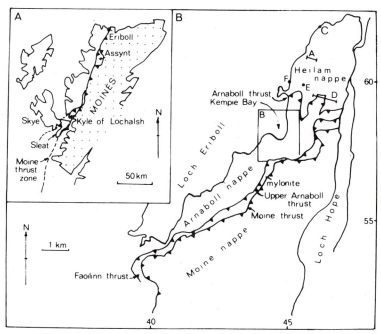

Fig. 3 (A) Sketch map of N.W. Scotland showing locations mentioned in the text. (B) Simplified location map of the east side of Loch Eriboll showing the main thrusts and nappes. A. location of section shown in Fig. 4. B. location map shown in Fig. 8. C. location of Fig. 5a. D. section line shown in Fig. 10. E. location of Fig. 11

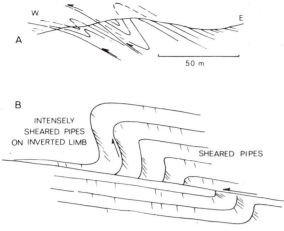

Fig. 4 (A) Cross-section through folded and thrust Pipe Rock from the Heilam nappe (for location see Fig. 3). (B) Sketch showing the form of the folds and the sheared nature of the pipes

shown in Fig. 4. These folds, which trend at about N30°, have the following characteristics.

1. They are characterized by long flat limbs, parallel to the thrust surface, separated by a steep, often inverted limb. The fold interlimb angle varies from 60° to about 120°. They are much tighter than may be expected from an unmodified fold formed solely by ramp climb.

2. The pipes show high shear strains on the inverted limbs and these strains are generally those expected for a flexural flow fold (Ramsay 1967, p. 392–395), though there are variations from this, as described by Fischer and Coward (1982). On the gentle fold limbs the pipes may show some shear strain but often they are normal to the bedding.

3. The pipes are not circular in cross-section on the bedding surface, but elliptical with long axes approximately parallel to the fold axes. Ellipse axial ratios range up to 1.5:1 even where the pipes are normal to bedding, suggesting a layer parallel shortening strain of up to 30–35 per cent, assuming no extension parallel to the fold axes, i.e. normal to the transport direction. This shortening varies; the greatest amount of layer parallel shortening is often seen on the flat limb, just to the east of the axial trace (Fischer and Coward 1982) but there are many areas of the Heilam Nappe, away from areas of folding, where layer parallel shortening is minimal.

4. Folds occur not only in the hanging wall of thrusts but also in the footwall. Sometimes there are several folds producing a fold train as shown in Fig. 5A.

Thus, though the shape of the folds in the hanging wall accords with the models of Berger and Johnson (1980), the variable amounts of layer parallel shortening and the presence of fold trains in the footwall suggest that shortening is an important mechanism in the production of folds. This suggests, therefore, a modification of the Berger and Johnson (1980) model, in that the layer-parallel shortening and folds were produced by stick not on the ramp but on the flat (Fig. 6). Indeed, the production of the thickened layers and the fold would produce the stress riser necessary to initiate the development of a thrust ramp (Wiltschko 1979a,b). This thrust ramp would then cut through the fold to carry the anticline–syncline pair of the hanging wall of the ramp some distance ahead of the associated structures in the footwall of the ramp.

Thrusts may also die out along strike (cf. Elliott 1976a) so that a thrust may pass laterally into a series of folds as shown in Fig. 7. As the thrust would propagate rapidly normal to the transport direction (cf. Elliott 1976a) then those folds produced at the lateral tips to the thrusts, would often be cut through by the faults as shown in sections B and C of Fig. 7.

A development of these types of fold structures is shown in the Arnaboll nappe of Kempie Bay and Arnaboll Hill. Here basement gneiss as well as Cambrian quartzites and the Pipe Rock are involved in the folds. At Kempie Bay (Fig. 8) the folds are asymmetric, with interlimb angles of approximately 70–80°. On the gentle limb the pipes show evidence of layer parallel shortening and there are localized zones of intense shearing in the Pipe Rock and Fucoid Beds. However, the Lewisian Gneisses preserve their amphibolite facies mineralogy and Lewisian structural trends. On the steep limb, the Pipe Rock and quartzites are variably sheared, presumably associated with flexural slip and flow folding. Angular shear strains of ψ greater than 85° are recorded from some well bedded parts of the Pipe Rock, while in the more massive quartzites ψ is low, often less than 5°. In the Lewisian Gneisses there is a new fabric of Caledonian age, formed by the reorientation of earlier Lewisian structures plus the new growth of greenschist facies minerals, in particular chlorite and epidote. Sometimes these greenschist facies minerals form a new closely-spaced cleavage but elsewhere they form on closely-spaced but irregular fracture planes in the gneiss. This Caledonian fabric in the gneiss generally dips steeply to the east–southeast and mineral lineations and slickensides plunge steeply down the dip of this cleavage.

A cross-section through the structures is shown in Fig. 9. It can be seen that there are several generations of thrusts, each producing its own fold pair. The high level thrusts are folded by the lower level structures confirming the argument for a general

Fig. 5   (A) Thrust in imbricate zone, western part of Heilam nappe showing folds in the footwall (for location see Fig. 3). The section shown by the photograph is some 10 m across. (B) Anticline–syncline pair above thrust plane at Na Tuadham, Assynt (for location see Fig. 14)

westward propagation of these structures in the thrust zone (*cf.* Elliott and Johnson 1980). Thus the Kempie structure (Fig. 9) is made up of not one simple syncline but a folded stack of thrusts, each with its own anticline–syncline pair.

A similar arrangement of folded thrusts and refolded folds occurs on the north side of Arnaboll Hill (Fig. 10). Here the Arnaboll Thrust has been folded round a large anticline by structures produced on subsequently developed lower thrusts. In fact, some five phases of thrusting and associated folding affect the rocks of Arnaboll as at least five of the

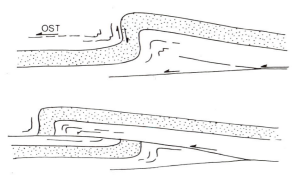

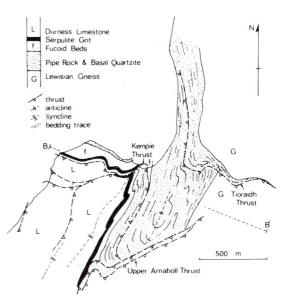

Fig. 6 Model to show the development of a fold by stick on a flat. Folding would be aided by flexural slip or shear on the steeply dipping limbs and by the generation of crumple zones and faults in the fold cores. Two such faults are shown: the thrust in the anticlinal arc and that in the syncline, the 'Out of Syncline-Thrust' (OST) of Dahlstrom (1970). This form of deformation was first described by Willis (1893) who suggested that the bending weakened the steep limb of the fold enabling it to be sheared through

Fig. 8 Map of Kempie Bay, Loch Eriboll (location B, Fig. 3) showing the main folds and thrusts. The Kempie thrust is the earliest, folded above the Tioradh thrust which is itself folded above the Arnaboll thrust. The Upper Arnaboll thrust is a later structure which slices through all earlier thrusts and folds

thrusts produce associated folds (Fig. 10). The Arnaboll folds are all upward facing, though Fig. 11 shows a small scale example of a downward facing fold, a structure which has been reorientated by structures developed on an underlying thrust. Figs. 10 and 11 illustrate the danger of fold correlation in areas of thrust tectonics. The rocks of Arnaboll are mainly Cambrian quartzites and so do not easily show cleavages. However, if the rocks had been more pelitic, several phases of folding and crenulation cleavage would be expected to have formed. These Arnaboll folds cannot be simply correlated with the structures at Kempie, though they were all formed by the same regional thrusting regime.

The Skye District

The Skye District, at the southern end of the Moine Thrust zone (Fig. 12) consists essentially of two major nappes (Bailey 1955). The lowermost nappe, the Kishorn nappe, contains very large recumbent

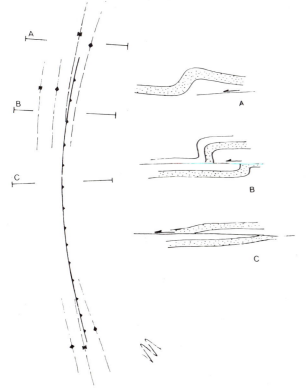

Fig. 7 Model to show the form of folds at the lateral tips to thrusts (cf. Elliott 1976a)

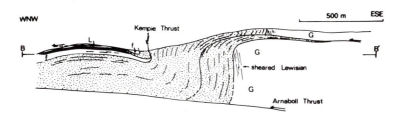

Fig. 9 Cross-section (BB' of Fig. 8) through the folds and thrusts of Kempie Bay

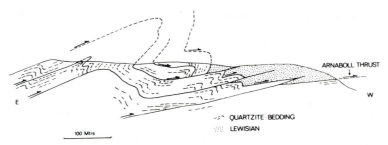

Fig. 10 Cross-section through the north side of Arnaboll Hill showing the form of the folds and thrusts in the Quartzite–Pipe Rock beneath the Arnaboll thrust

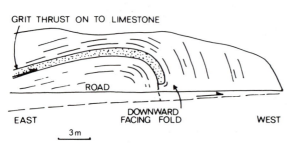

Fig. 11 Sketch of roadside outcrop (for location see Fig. 3) showing the structure of a downward facing fold in Durness Limestone

folds in Torridonian and Cambro-Ordovician rocks of low metamorphic grade. There is a small amount of Lewisian basement involved in the fold north of Lochalsh (Fig. 13A). The recumbent folds have limb lengths in the order of 10 kms, with interlimb angles which range from 90° to zero. The upper nappe, the Moine Nappe, consists of Lewisian Gneisses and the cover of Moine Schists of high metamorphic grade. Beneath the Moine Thrust are several nappes or groups of nappes which are of local extent. These include the Tarskavaig and Balmacara Nappes which consist of mylonitized Lewisian and Torridonian rocks. Within the Kishorn Nappe, the Lower Torridonian Sleat Group has its greatest development consisting of a thick sequence of grits, sandstone, and shales. Locally the recumbent folds of the Kishorn Nappe have been reorientated by later thrusting and folding but the main structure, the Lochalsh Syncline (Fig. 13A) trends at approximately 020° and is not truly normal to the transport direction. This syncline is a large recumbent fold which affects both the Sleat Group and the overlying Applecross Formation (Torridon Group). The normal limb of the fold is seen to the west of Kyle village and dips at about 20° to the northwest. The dip of the inverted limb gradually decreases from 70° to 20° but has a consistent east–southeasterly dip direction.

The hinge of the fold lies beneath the Moine Thrust and it is suggested that the fold formed as a footwall syncline to the Moine Thrust itself, following stick on the Moine Thrust. The interbedded shales and sandstones of the Sleat Group represent a highly anisotropic footwall to the thrust and folding was preferred to imbrication as seen further north in the Moine Thrust zone. As a result several large angular folds occur on the normal limb of the Lochalsh Syncline. On Sleat the main fold zone transfers to the more westerly and lower Eishort Anticline and Ord Syncline (Potts in press). It is not known whether these lower recumbent folds formed as part of a footwall fold train, or were produced as part of an anticline–syncline pair on the hanging wall of a lower thrust.

A strong mylonitic grain fabric is restricted to the overturned limb of the Lochalsh Syncline. The fabric dips to the east–southeast with a strong down-dip mineral and grain shape lineation. Within the Torridonian sandstones it is parallel to the bedding and in the overlying Lewisian Gneiss it replaces earlier structures and fabrics. The origin of this fabric is at present unknown. Throughout the Lochalsh Syncline, bedding-parallel slickensides are common and it is suggested that flexural slip, with movement taken up on bedding surfaces, is an important mechanism in the production of the fold.

Strain studies on the Pipe Rock of the Ord Syncline (Fig. 13B), the structurally lowermost recumbent fold, indicate no deformation of the bedding surface; the pipe sections are slightly elliptical on the bedding but with a random orientation. The pipes are normal to bedding and record no bedding-parallel shear strains (Potts 1982).

A palaeomagnetic direction for which a Torridonian age may be inferred (Potts unpublished data) is preserved in the Applecross Formation of the Eishort Anticline (Figs. 12 and 13B). This direction may be returned to a Torridonian position by simple rotation of the bedding about the strike. No component of strain is required to restore this direction to a Torridonian position (Potts 1982).

The absence of strain around the hinges of the Eishort Anticline and Ord Syncline suggests that flexural slip involving no internal deformation was the dominant folding mechanism. No evidence of layer-parallel shortening is seen in the folds of the Kishorn Nappe and it is thought that the highly

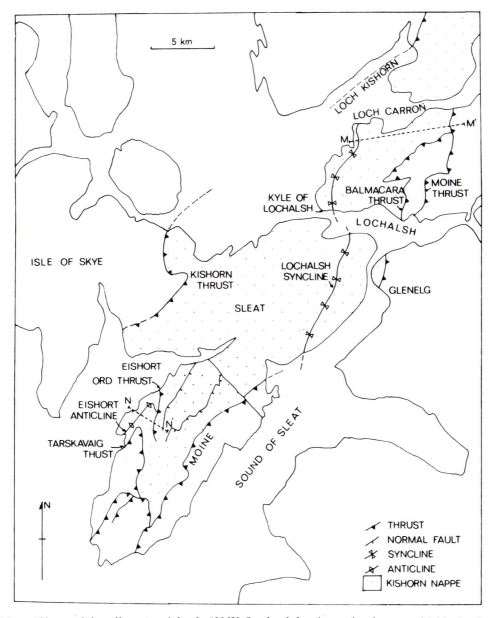

Fig. 12   Map of Skye and the adjacent mainland of N.W. Scotland showing major thrusts and folds. Section lines M–M′ and N–N′ are shown in Fig. 13

anisotropic Sleat Group preferred to fold rather than thicken homogeneously by layer-parallel shortening.

To the west of Ord village, the Eishort Anticline and overlying Tarskavaig Nappe are folded by the Tarskavaig synform (Fig. 13B). This is a relatively open structure which plunges gently to the southwest and clearly represents a footwall syncline to the Ord Thrust (Fig. 13B); it formed during the progressive development of successively lower thrusts, with associated folding as described in the Eriboll region. Several generations of folds and cleavage are seen within the Tarskavaig Nappes (Cheeney and Matthews 1965) and these may be related to successive folding and thrusting events.

The Assynt District

The north Assynt District consists of three major thrust nappes: a lower Nappe (un-named) made up of imbricated Cambrian rocks, a very thick Glencoul Nappe of imbricated Lewisian basement with its Cambrian cover, and the uppermost Moine Nappe. The map (Fig. 14) shows the major folds in north Assynt and their relationship to the thrusts; Fig. 15 shows a more detailed map of some of the folds at Cnoc an Droighinn. Fig. 5B shows a large oblique antiform in Na Tuadham. These folds have the following characteristics:

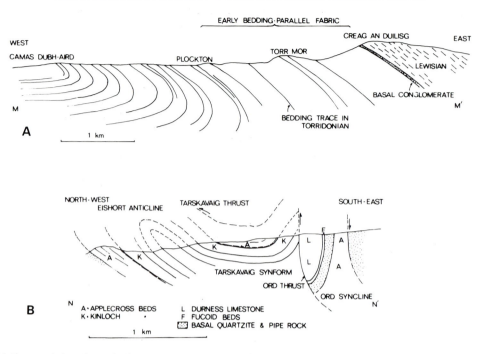

Fig. 13 (A) Cross-section through the Lochalsh syncline. The recumbent syncline lies beneath the Moine thrust which truncates the folded bedding. The Moine thrust outcrops approximately 2 kms to the east of the section. (B) Cross-section across part of Sleat, through a series of folds and thrusts. The early Eishort anticline and Tarskavaig thrust plane are both folded by the Tarskavaig synform which formed in the footwall to the Ord thrust

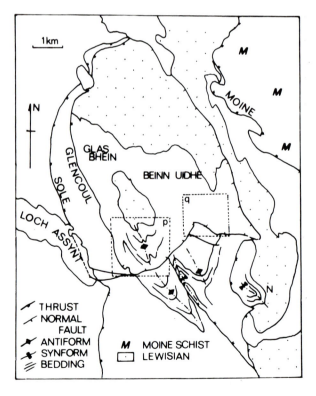

Fig. 14 Map of Assynt showing the position of the major faults and oblique folds. The transport direction is to N290°. The structure is complicated by the normal faults which cut down through and displace the earlier folds and thrusts. N = Na Tuadham

1. The anticlines are asymmetric with gently-dipping northeast limbs and steeply-dipping often overturned southwest limbs. The interlimb angles are between 60° and 100°. The fold hinges change their trend from N–S to NW–SE and from oblique to subparallel, with the thrust transport direction as determined from the sheared pipes and the orientation of Caledonian strike-slip faults.

2. The pipes are sheared by flexural flow processes on the steep limbs of the folds. On the bedding surfaces, the pipes are elliptical with long axes trending N–S to NW–SE (see Coward and Kim 1981). The ellipse long axes are not always parallel to the fold axes but often trend more N–S and thus change orientation around the folds.

3. As at Eriboll, folds occur both in the hanging wall and in the footwall to the thrusts.

4. The folds at Assynt deform not only the bedding but higher level thrusts, so that there are many generations of oblique trending asymmetric folds. This is shown by the folds near Loch nan Cuaran (Fig. 16).

The major difference between these folds and those described from Eriboll and Syke is their orientation; at Assynt the folds are oblique to lateral, footwall and hanging wall ramps. Coward and Kim (1981) have described how the finite strains at

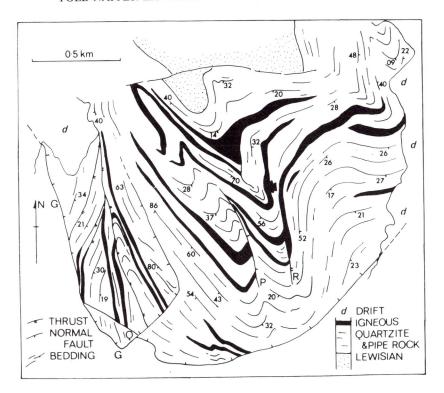

Fig. 15 Detailed map of the Cnoc an Droighinn (box p in Fig. 14) area showing the oblique folds. Faults P and R are back limb structures presumably due to accommodation problems in the folding. No structures are shown south of the Glencoul fault (G)

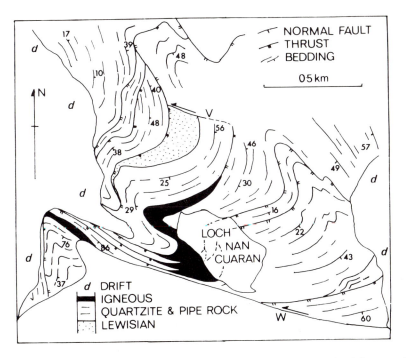

Fig. 16 Map of the area near Loch nan Cuaran (box q in Fig. 14) showing the stack of thrusts with oblique folds. The movement direction is given by the tear fault (w) south of which no structure is shown

Assynt, recorded by the pipes on the bedding surface, may be separated into two components: (a) layer-parallel shortening as described for the Eriboll area, and (b) shear strains on a plane normal to the main thrust plane but with the same transport direction. The two components result in ellipses which have long axes oblique to the general transport direction. Fischer and Coward (1982) have shown how the finite strains on the bedding surface in the Eriboll area may be factorized into the same components, but at Eriboll, the component of simple shear strain is much lower than at Assynt. It is suggested here that at Assynt the differential thrust movement that gave rise to the sinistral shear strains shown by the pipes was also responsible for the production of the oblique folds. Some component of stick on the thrusts in the Assynt area, allowing the Glencoul Nappe in the north to move further than in the south, caused the oblique folds to form. Further movement on the thrusts caused these folds to be decapitated so that the folds in the hanging wall were carried some distance from the structures in the footwall.

The majority of thrusts at Assynt carry oblique folds and these appear to have developed sequentially, as folds produced by the lower level thrusts deform the higher level folds and thrusts. Thus it is unlikely that the oblique folds formed by simple sticking on a thrust or the dying out of displacement as described by Elliot (1976a) from the Rocky Mountain examples. In such cases both sinistral and dextral oblique folds would be expected. However, at north Assynt almost all the folds show a sinistral component of shear strain. In southern Assynt the fold axes curve to trend northeast parallel to the regional trend.

Discussion

In many previous papers it has been assumed that thrust faults followed easy slip horizons and for some reason climbed through less easy slip layers to follow a higher level easy slip horizon. The fold formed above the ramp could be modified by resistance to movement on the thrust ramp. The reason for the fault climb has been attributed to the presence of some form of stress raiser, such as an earlier tectonic or sedimentary structure (Wiltschko and Eastman 1981). In the model proposed here, it is suggested that the stress raiser is produced by differential movement of the thrust on the flat, causing thickening and folding of the layers. For folding to take place above a decoupling horizon, there must have been detachment of the beds from the decoupling horizon to produce a more crumpled zone or, more likely, a secondary thrust, dipping at a low angle to

the decoupling horizon (Dahlstrom 1969, 1970). Without these structures, the fold section would not be balanced. If further thrust movement occurred, this detached zone, especially the secondary fault zone, would be an ideal site for ramp development and propagation. In some instances the folds produced above the decoupling horizon may be box shaped with axial surfaces and, possibly, secondary faults dipping in opposite directions. Further deformation of these structures may generate backward propagating thrusts as well as forward moving thrusts, to develop the structures known as 'pop ups' and 'triangle zones' (Elliot 1981).

Differential displacement on the thrusts, leading to the formation of folds may be due to:

1. Differential drag of thrust, possibly due to changes in the rock textures or structures. Major sedimentary facies changes as in the Lower Torridonian of Skye could produce differential resistance to fault movement which may have produced the major fold structures on Sleat Alternatively, variations in fluid pressure along a fault will cause variation in resistance to movement (Hubbert and Rubey 1959; Hsu 1969).
2. Variation in the rate of fault propagation. If the growth of the flat is prevented through variations in rock type or fluid pressure then thickening and folding of the detached slab is possible. The folds would be produced at thrust tips as shown in Fig. 17. In a similar manner to that described in section (1) above, changes in sedimentary facies could prevent further propagation of the thrust plane.
3. Variations in the driving forces behind the thrusts. This may be especially important in the production of the oblique and lateral folds seen at Assynt, where it appears that the Glencoul nappe north of Assynt consistently tried to move further to the northwest than did the thrust sheet to the south. Some differential push may be responsible for this but the most likely explanation involves gravitational forces. Elliott (1976b) has shown the importance of surface slope on the shear stress at the base of a thrust sheet. Though it is impossible to prove now, due to extensive post-Caledonian erosion, changes in erosion rate along the Moine Thrust, with the accompanying changes in surface topography, could lead to the production of differential shear strains and hence oblique and lateral folds and ramps.

Most of the folds produced are asymmetric, with the steep limbs more intensely deformed by flexural shear. The fold interlimb angles in deformed Basal Quartzite and Pipe Rock are generally fairly constant at $80° \pm 20°$ suggesting that the folding may

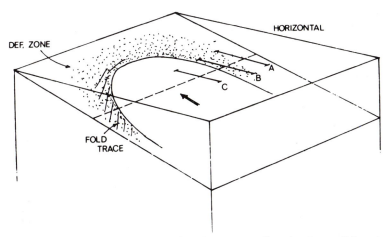

Fig. 17   Block diagrams showing the deformed region surrounding the thrust dislocation plane

have continued until the structure locked up (*cf.* Ramsay's 1974 discussion of chevron folds). In contrast to the Eriboll region where flexural flow processes produced large finite strains, the folds of Skye show no internal deformation and must be a consequence of flexural slip processes, producing the bedding parallel slickensides. Unless the Ord structures began as a large upright fold which later became inclined, the beds must have rolled round the hinges as the overturned limb grew. It is remarkable that no deformation microstructures are preserved. It is clear that whether the folding mechanism is one of flexural flow or flexural slip, the anticline–syncline pair decapitated by the associated thrust is the common feature. It is suggested that the contrasting fold mechanisms seen in the Moine thrust zone result from variations in the rock type in response to a common process, that of stick on flat. The thin Cambrian sequence of the northern part of the Moine Thrust zone produces small fold nappes as seen on Ben Arnaboll in contrast to the thick Torridonian sequence on Lochalsh and Skye which produces the large recumbent folds.

Finally, it is obvious that in zones of major shearing or thrusting, that is within most orogenic belts, care must be taken when attempting fold correlation across large areas. This work has shown that folds of the Moine Thrust zone develop sequentially, locally producing several fold phases in some areas during one part of the total thrust movement. If some form of differential movement was responsible for producing the folds, as at Assynt, then fold orientations cannot be used to distinguish deformation events. Fig. 18 shows how confusion and miscorrelation of folds may arise if two zones of differential movement should overlap, as may be possible in large orogenic belts. Fold phase correlation should therefore be confined to small areas only, or based on other criteria such as metamorphic or intrusion phases.

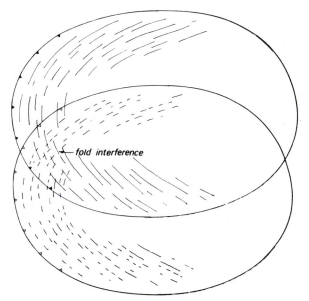

Fig. 18   Sketch showing how fold interference may be produced by two phases of differential movement within an orogenic belt

References

Bailey, E. B. 1955. Moine tectonics and metamorphism in Skye. *Trans. Edinb. geol. Soc.*, **16**, 93–166.

Berger, P. and Johnson, A. M. 1980. First order analysis of deformation of a thrust sheet moving over a ramp. *Tectonophysics, 70*, T9–T24.

Butler, R. W. H. 1982. A structural analysis of the Moine Thrust zone between Loch Eriboll and Fionaven, N.W. Scotland. *J. Struct. Geol.*, **4**, 19–29.

Cheeney, R. F. and Matthews, D. W. 1965. The structural evolution of the Tarskavaig and Moine nappes in Skye. *Scott. J. of Geol.*, **1**, 256–281.

Coward, M. P. and Kim, J. H. 1981. Strain within thrust sheets. In McClay, K. R. and Price, N. J. (Eds), *Thrust and Nappe Tectonics. Spec. Publ. geol. Soc. Lond.*, **9**, 275–292.

Dahlstrom, C. D. A. 1969. Balanced cross sections. *Can. J. Earth Sci.*, **6**, 743–757.

Dahlstrom, C. D. A. 1970. Structural geology of the eastern margin of the Canadian Rocky Mountains. *Bull. Can. Pet. Geol.,* **18**, 332–406.

Douglas, R. J. W. 1950. Callam Creek, Langford Creek, and Gap Map areas, Alberta. *Geol. Surv. Can. Mem.,* **225**, 124pp.

Elliott, D. 1976a. The energy balance and deformation mechanisms of thrust sheets. *Philos. Trans. R. Soc. Lond.,* Ser A, 283.

Elliott, D. 1976b. The motion of thrust sheets. *J. Geophys. Res.,* **81**, 949–963.

Elliott, D. 1981. The strength of rock in thrust sheets. *E.O.S.,* **62**, 397.

Elliott, D. and Johnson, M. R. W. 1980. Structural evolution in the northern part of the Moine thrust zone. *Trans. R. Soc. Edinburgh,* **71**, 69–96.

Fischer, M. W. and Coward, M. P. 1982. Strains and folds within thrust sheets: an analysis of the Heilam sheet, N.W. Scotland. *Tectonophysics,* **88**, 291–312.

Harris, L. D. 1970. Details of thin skinned tectonics in part of Valley and Ridge and Cumberland Plateau provinces of the southern Appalachians. In: Fischer, G. W. Pettijohn, F. J., Reed, J. C. Jr. and Weaver, K. N. (Eds), *Studies of Appalachian Geology: Central and Southern,* Wiley Interscience, N.Y. 161–173.

Harris, L. D. und Milici, R. C. 1977. Characteristics of thin skinned style of deformation in the southern Appalachians, and potential hydrocarbon traps. *U.S. Geol. Surv. Profess. Pap.,* **1018**, 40pp.

Hsu, K. J. 1969. Role of cohesive strength in the mechanics of overthrust faulting and landsliding. *Bull. geol. Soc. Am.,* **80**, 927–952. ·

Hubbert, M. K. and Rubey, W. W. 1959. Role of fluid pressure in mechanics of overthrust faults. I, Mechanics of fluid–filled porous solids and its application to overthrust faulting. *Bull. geol. Soc. Am.,* **70**, 115–206.

Peach, B. N., Horne, J., Gunn, W., Clough, C. T. and Hinxman, L. W. 1907. The geological structure of the northwest Highlands of Scotland. *Mem. geol. Surv. G.B.*

Potts, G. J. 1982. Strains within recumbent fold nappes. *Tectonophysics,* **88**, 313–319.

Potts, G. J. in press. Revision of the Ord Window, Sleat, Isle of Skye. *J. geol. Soc. London.*

Ramsay, J. G. 1967. Folding and Fracturing of Rocks. McGraw-Hill, N.Y., 568pp.

Ramsay, J. G. 1974. Development of chevron folds. *Bull Geol. Soc. Am.,* **85**, 1741–1754.

Rich, J. L. 1934. Mechanics of low angle overthrust faulting illustrated by Cumberland Thrust Block, Virginia, Kentucky and Tennessee. *Bull. Am. Assoc. Pet. Geol.,* **18**, 1584–1596.

Stewart, A. D. 1966. Torridonian rocks of Scotland reviewed. *Bull. Petrol, Geol.,* **18**, 332–402.

Seguret, M. 1970. Etude tectonique des nappes et série décollées de la partie centrale du versant sud des Pyrenées (caractères synsedimentaires, rôle de la compression et de la gravité). Thèse Sc. Montpellier, 162pp.

Willis B. 1893. Mechanics of Appalachian structure. *U.S. Geol. Surv. 13th Ann. Rept. Pt. 2.*

Wiltschko, D. V. 1979a. A mechanical model for thrust sheet deformation at a ramp. *J. Geophys. Res.,* **84**, 1091–1104.

Wiltschko, D. V. 1979b. Partitioning of energy in a thrust sheet and its implications concerning driving forces. *J. Geophys. Res.,* **84**, 6050–6058.

Wiltschko, D. V. and Eastman, D. B. 1981. The role of pre-existing basement faults and warps in localising thrust-sheet ramps. *E.O.S.,* **62**, 397.

The Caledonide Orogen—Scandinavia and Related Areas
Edited by D. G. Gee and B. A. Sturt

The Caledonides of Shetland

D. Flinn

Jane Herdman Laboratories of Geology, University of Liverpool, Liverpool L69 3BX

ABSTRACT

Shetland is part of an inlier of crystalline Caledonian and pre-Caledonian rocks surrounded by Devonian and younger sediments on the continental shelf north–northeast of Scotland. The islands lie equidistant from Scotland, Norway, and Greenland (in its pre-drift position) and close to any natural northeastward extrapolation of the Great Glen fault of Scotland. Since the islands are transected by a major transcurrent fault, which brings a segment of the Caledonian thrust front on the west side of the fault against Moine–Dalradian-type rocks on the east side of the fault, it is proposed that the fault in Shetland is a continuation of the Great Glen fault and that it shows a major sinistral displacement. The basement to the Caledonian front to the west of the fault is formed by a mass of orthogneiss giving a minimum age of 2900 Ma. This is overlain by steeply eastward-dipping alternating slices of schists and of blastomylonitized hornblende-banded gneisses interpreted as alternating slices of Caledonian and basement rocks. To the east of the fault, a succession of metamorphic rocks about 20 km thick dips steeply and strikes north–northeast. The western part is made of garnet-rich psammites and gneisses, while the eastern part is a more varied succession of schists, psammites, quartzites, limestones, metavolcanics, metaturbidites, and gneisses. These rocks have been correlated with the Moine and Dalradian of Scotland. On both sides of the fault, the rocks were deeply eroded before being covered by thick deposits of Middle Devonian sandstones which to the west of the fault include lavas, agglomerates, and ignimbrites. The the east of the fault, the Devonian sandstones post-dated the emplacement of large early Devonian granite complexes, while to the west, late Devonian granites and granophyres cut both the metamorphic rocks and the Devonian sandstones. In northeast Shetland, the metamorphic rocks are in thrust contact with an ophiolite complex in which nappes of banded ultrabasic rocks and gabbros are underlain by *mélanges* composed of metasediments containing ophiolite debris, and tectonic slices of ophiolite, or hornblende schist, and of the metamorphic rocks adjacent to the complex.

The Tectonic Setting of Shetland

Shetland lies half way between Scotland and Norway and would be the same distance from Greenland if Greenland were restored to its pre-drift position. It forms a large exposure of Caledonian rocks lying between the three parts of the Caledonian orogenic belt which are separated by post-Caledonian sediments and the sea (see Ziegler this volume, Fig. 1).

Shetland is an erosional monadnock-like remnant of Caledonian rocks rising above the continental shelf to the north of Scotland and is a part of the stable Shetland platform, the northward submarine continuation of Scotland. The platform is limited to the east by the north–south trending Viking graben and to the west by the north–northeast trending West Shetland basin, a half graben, fault-bounded to the east (Ziegler 1981). Both these structures and the platform itself came into being in Triassic times when a propagating rift started to form but failed to develop further. However, Hay (1978) proposes a different history of development. The grabens continued to develop and to fill with sediments until the end of the Cretaceous period when movement on the associated faults ceased, although subsidence and sedimentation centred on the two grabens continued. At about the end of the Mesozoic era the crust split to the west of the West Shetland basin and a propagating rift then formed, leading eventually to the creation of the present ocean between Scotland and Shetland in the east, and Greenland to the west.

Sedimentation continued throughout the Tertiary era in the East Shetland basin overlying the Viking graben, and less continuously and copiously above the West Shetland basin. During the Palaeocene period the Shetland Platform provided much sediment to each basin and to the Moray Firth area to the south (Ziegler 1981, p. 30), and must at that time have suffered uplift and erosion. Otherwise the platform seems to have remained stable for most of its existence despite its position a mere 60 km from the continental edge.

On the platform to the northeast of Unst lies the Unst basin, a half graben, fault-bounded on the east side and filled with Lower Cretaceous sediments (Day *et al.* 1982; Johnson and Dingwall 1981). The Fitful basin, another similar half-graben, fault-bounded to the east and filled with Permo-Triassic sediments, lies to the south–southwest of Shetland (Flinn 1969; Bott and Browitt 1975). The submarine platform and Shetland are cut by a system of north-trending transcurrent faults giving rise to post-Caledonian offsets (Flinn 1977). These post-Caledonian sediments and structures and the seas surrounding Shetland obscure and complicate the Caledonian geology of the area.

In the Viking graben, to the east of the platform, ten wells have reached crystalline rocks at a depth of several kilometres from which K–Ar ages of 430, 442, and 350 Ma have been obtained. These rocks vary from gneisses to low grade schists and include a sheared gabbro (Frost and Fitch 1981). Other wells have bottomed in Old Red sandstones. On the west side of the West Shetland basin twelve wells drilled into the Rona Ridge have revealed the presence of gneisses overlain by Old Red sandstones and Carboniferous sandstones. The gneisses contain hypersthene and have given K–Ar ages varying from 970 Ma (Ridd 1981, p. 420) to 2700 Ma (Ziegler this volume).

Within the Shetland Caledonian inlier the continuation of the Caledonian front from northwest Scotland can be seen in the northwest tip of Shetland as alternating layers of basement and cover-type rocks, overlain by Devonian sandstones and volcanics and cut by late Devonian granites. The east side of Shetland is underlain by a thick succession of schists and gneisses created or metamorphosed during the Caledonian orogeny, overthrust by an ophiolite complex, intruded by early Devonian granites,

and unconformably overlain by Middle Devonian sandstones. The two provinces of west and east Shetland are separated by the north-trending major transcurrent faults mentioned above, the Walls Boundary fault system (Flinn 1977).

The Caledonian Front

The metamorphic rocks to the west of the Walls Boundary fault in Shetland are an alternating series of slices of basement and cover rocks of different types (Flinn *et al.* 1979; Pringle 1970; Robinson 1983). There are two types of basement gneiss, the western and the eastern, and three types of cover rock, the Sand Voe, the Hillswick, and the Queyfirth (Fig. 1). The slices dip east at about 60% and there is no sign of thrusting at most of the contacts.

The *western gneisses* are acid, banded orthogneisses with accessory hornblende and some interbanded hornblendites. They are cut by foliated pegmatites. Two groups have been recognized. In the west (the Uyea Group) the hornblendites contain augite and the gneisses contain K-feldspar. The Uyea Group is separated from the other group lying to the east

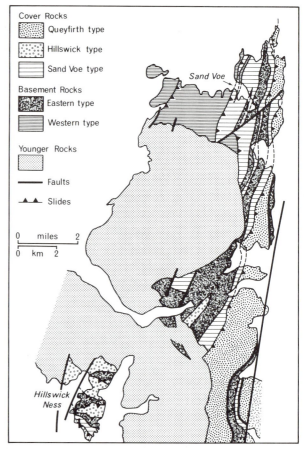

Fig. 1 Basement–cover relations in northwest Shetland (after Robinson 1983)

(the Wilgi Geo Group) by a major zone of shearing and mylonitization, and there is widespread shearing and even mylonitization in both groups of gneisses. The mylonitization within the Uyea Group appears to be older than that bounding the group. Hornblendes taken from a basic intrusion in the Uyea Group have given K–Ar ages of 2873 to 2661 Ma (Flinn *et al*. 1979).

The *eastern gneisses* are Lewisian inlier-like banded hornblendic gneisses. They are schistose, very feldspathic, psammitic looking rocks with fine stripes and bands of hornblendite and hornblende-rich rocks. It is apparent that they are blasto-mylonitized hornblende-banded orthogneisses, the chief minerals being plagioclase, green hornblende, garnet, and epidote. Brown hornblende with pyroxene occurs in places in these rocks and the garnets occasionally contain inclusions of pyroxene. Lenses of serpentinite ranging in size from a few centimetres to many tens of metres across occur frequently, as well as lenses of coarse hornblendite and steatite derived from them. One large mass of un-serpentinized ultrabasic rock gave K–Ar ages ranging from 2313 to 1043 Ma (Flinn *et al*. 1979).

The *Sand Voe cover rocks* are dominantly siliceous psammites with some garnet–mica psammites and semipelites with accessory hornblende. These rocks are generally highly schistose, especially near their contacts with the gneisses where they become platy, but what appear to be quartz and feldspar clastic grit grains occur in some thin sections and in a very few places current-bedding and conglomerate lenses have been found. Bands of pelitic garnet–mica schist, often extremely persistent, also occur especially adjacent to the slices of eastern gneiss.

The *Hillswick cover rocks* are largely siliceous psammites and semipelites, very similar to the Sand Voe cover rocks and like them containing biotite, garnet, and hornblende, as well as clastic grit grains. Rare conglomerate horizons and current-bedding have been recognized. Bands of pelitic garnet–mica schist also occur. These rocks differ from the Sand Voe rocks in the frequent presence of bands from 10 to 100 m in thickness which are rich in graphite and pyrite.

The *Queyfirth cover rocks* are composed of inter-banded quartzite beds, garnetiferous semipelitic schists, and green schists associated with siliceous limestones. The green schists contain chiefly green hornblende, biotite, chlorite, epidote, and plagioclase. The Queyfirth rocks are much cut up by faults and folded so that the true succession has not been established. They strike southwards into a poorly exposed area where occur graphitic phyllites, schistose conglomerates, laminated quartzite and

calcareous rocks with skarn–magnetite deposits that may belong to the Queyfirth cover rocks.

The eastern gneisses occur as a series of slices alternating with Sand Voe rocks to form the Sand Voe Unit, best, seen north of Sand Voe (Fig. 1). The unit also contains near its base a slice of strongly schistose western gneiss which in places has become a hornblende garbenscheifer. The Sand Voe Unit lies to the east of the western gneisses and in contact with them.

The eastern gneisses also occur as a series of slices alternating with Hillswick rocks to form the Hills-wick Unit, best seen at Hillswick Ness (Fig. 1). The junction between the Hillswick Unit and the other rocks is hidden by a later intrusion, but the Hillswick Unit appears to lie along the strike of the Sand Voe Unit. Both the Sand Voe and Hillswick Units are overlain to the east and south by Queyfirth rocks.

All the contacts between the gneisses and the cover rocks appear to be conformable but can be detected by the development of a platy schistosity or an enhancement of the local schistosity in the neighbourhood of the junction. A very sharp, approximately conformable junction can usually be detected by careful inspection. The junctions are all typical tectonic slides dipping between 40° and 60° to the east or south, parallel to a strong schistosity. A large-scale isoclinal folding with an axial plane parallel to this surface can be found in the cover rocks in the general areas of Sand Voe.

The contact between the Queyfirth rocks and the other rocks to the west is called the Virdibreck shear. In some places it is a slide but in others it has more the appearance of a late thrust. However, much of it lies in the shatter zone of the Walls Boundary fault and late movement may have taken place on it.

A further group of rocks including gneisses lies to the south of these rocks along the south shore of St. Magnus Bay (Fig. 2). It has no exposed contact with the rocks described above. These rocks, the Walls Metamorphic Series (Mykura and Phemister 1976; Frank 1977) are a series of nearly east–west striking quartzo-feldspathic gneisses, hornblendites, limestones, calc-silicate rocks, and semipelites, unlike any of the rocks to the north. A series of 24 K–Ar ages of hornblendes from the hornblendites gave ages ranging from 863 to 366 Ma associated with a change of hornblende colour from brown to green (Flinn *et al*. 1979). The brown hornblendes occur inside masses of hornblendite and the green hornblendes on the outsides. The rocks are interpreted as having experienced a high grade metamorphism more than 860 Ma ago and a later Caledonian overprinting. Micas and green hornblendes gave five K–Ar ages ranging from 400 to 435 Ma (Mykura and Phemister 1976).

This alternating succession of basement and cover type rocks, with basement gneisses in the west and Caledonoid schists alternating with slices of gneiss in the east has been interpreted as a segment of the Caledonian front (Miller and Flinn 1966, p. 98; Pringle 1970). Zones of late schistosity and of mylonitization continue in the western gneisses to the very last exposures in the west, but the further continuation of the gneisses onto the shelf to the west of Shetland is indicated by the magnetic and gravity anomaly patterns there (Flinn 1969; Bott and Watts 1970) and relic high grade minerals occur more frequently to the west in the gneisses (Pringle 1970).

The western gneisses of Shetland and their continuation to the edge of the continental shelf to the west have the same appearance of being an autochthonous basement to the Caledonides as have the Lewisian gneisses of northwest Scotland. However, the Caledonian front in Shetland differs from that in northwest Scotland in the absence of thrusts of Moine-thrust type and of Cambro–Ordovician rocks. The association of slides, blastomylonites, and widespread garnet indicates formation at a depth greater than that at which the Moine thrust was formed. The steepness of the slides does not accord with most recent predictions of the attitude of the Moine thrust at depth. The Moine Thrust may be a younger structure than the slides in Shetland and does not appear on shore in the Shetland area.

The East Mainland Succession

The East Mainland Succession which underlies most of Shetland east of the Walls Boundry fault (Fig. 2) can be divided into four lithologically distinguishable units (Flinn et al. 1972). The Yell Sound division, the westernmost unit, underlies Yell and further south forms a narrow strip along the east side of the Walls Boundary fault. The rocks are mostly steeply dipping and show a maximum outcrop width of 11 km or 16 km if the rocks beneath the waters of Yell Sound are included.

The division is largely composed of psammites with a major quartzite and intercalated hornblende schists. Garnet is a characteristic and often profuse constituent of both the psammites and the hornblende schists. The psammites vary in appearance from granulitic to schistose, more by variation of the size than the amount of the mica flakes. Occasional widely-scattered slightly pelitic psammites contain kyanite and staurolite but never sillimanite except in thermal aureoles. The division contains a number of belts of gneiss up to several hundred metres wide. Some of these have developed from psammite by recrystallization, others by porphyroblastic development of oligoclase, by development of quartz–oligoclase lenses, or by impregnation by microcline. Microcline is otherwise absent from the psammites. Other gneiss bands have been interpreted as tectonized early granite intrusions. The division has been injected by a considerable number of pegmatites and aplites of several generations. In some areas pegmatite injections make up more than 50 per cent of the rocks.

The Yell Sound division has been equated with the Moine of Scotland (Flinn et al. 1972, p. 34) and in particular with the Glenfinnan division (Dr. A. L. Harris personal communication). The other three divisions of the East Mainland Succession have been equated with the Dalradian of Scotland.

The Scatsta division, possibly the equivalent of the Lower Dalradian or Appin Group, follows the Yell Sound division to the east. The upper part of the division is largely composed of impure quartzites with interbanded semipelitic schists. Several bands of Al–Fe-rich pelitic schist containing staurolite, kyanite, and garnet also occur. These can be followed for considerable distances, whether exposed or not, as they give rise to positive magnetic anomalies of about 1000 γ at ground level. At the base of the upper part of the division is a sharply defined band of microcline augen gneiss with an appearance similar to that of many tectonized early granites. It can be followed for 25 km on the west side of the Nesting fault but is only seen locally to the east of it. The lower part of this division is composed of garnet–biotite–muscovite granulites and gneisses, a limestone and calc-silicate band which is best developed east of the Nesting fault, and some bands of the magnetic schist containing kyanite, staurolite, and garnet. The pelitic schists of the Scatsta division to the east of the Nesting fault in Lunnasting, Fetlar, and Unst developed massive amounts of fibrolite prior to the crystallization of kyanite and staurolite. To the west of the fault, sillimanite occurs only in later-formed thermal aureoles.

The position and nature of the boundary between the Yell Sound division and the Scatsta division (i.e. between the Moine and Dalradian of Shetland) presents a number of problems. Close to the boundary is a remarkable band of microcline gneiss in which 3 to 4 cm diameter single crystal microline porphyroblasts occur in matrices varying from semipelitic, granulitic to gneissic. The band can be followed 70 km across Shetland from sea to sea, and varies in width from several metres to a kilometre, and in some places appears to be repeated in several parallel bands. To the west of the Nesting fault the two divisions appear to be conformable. To the east of the fault complications appear. In places the boundary takes the form of a half kilometre thick 30°W dipping zone of blastomylonitized hornblende-banded gneiss of 'Lewisian Inlier' type associated with bands of 1 mm diameter

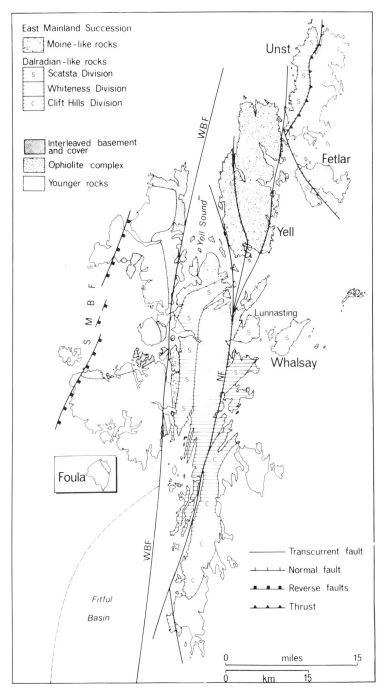

Fig. 2 Metamorphic rocks and their distribution in Shetland. SMBF = St. Magnus Bay fault; WBF = Walls Boundary
fault; NF = Nesting fault

oligoclase-porphyroblast granulite and of brown
hornblende schists not found elsewhere in Shetland.

The Whiteness division to the east of the Scatsta
division begins with the Weisdale limestone and
ends with the Laxfirth limestone, both about half a
kilometre thick. It contains two other limestones, the
half kilometre thick Whiteness limestone and the
200 m Girlsta limestone, Between the Weisdale
and Whiteness limestones the rocks are mostly bedded, laminated biotite–muscovite–quartz–oligoclase

semipelites and psammites. Interbedded are mica
schists, calc-silicate granulites with thin marbles and
hornblende schists. Between the Whiteness limestone and the Girlsta limestone, the rocks are similar
but a discontinuous zone up to 2.5 km wide has been
recrystallized to a homogeneous biotite–muscovite–quartz–oligoclase gneiss ± microcline. This
zone is the Colla Firth Permeation Belt and associated with it are veins and masses of pegmatite and
schistose aplogranite which have a Rb–Sr age of

530 Ma using $\lambda^{87}Rb = 1.39 \times 10^{-1} \, yr^{-1}$ (Flinn and Pringle 1976). Calc-silicate rocks adjacent to the gneiss belt and within it have developed diopside, microcline, and garnet while mica schists have locally developed sillimanite and garnet. The gneiss belt was evidently formed after the regional tectonizing metamorphism which in the neighbourhood of the belt produced amphibole and epidote in the calc-silicates, and muscovite and biotite in the schists. The rocks between the Girlsta and Laxfirth limestones change northward from laminated bedded semipelites containing green–brown biotite, muscovite, quartz, oligoclase, and clastic quartz grains to a calc-silicate granulite and hornblende schist dominated succession.

The Clift Hills division starts with a metavolcaniclastic band locally spilitic and including a horizon that gives a ground magnetic anomaly of about 1000 γ. Chlorite, hornblende, epidote, and carbonate are the chief minerals. These rocks are followed by 3 km of semipelitic phyllites which, in places, can be seen to be very fine-grained rocks with centi-

metre-scale graded banding of distal turbidite appearance. Interbedded with these are several impure limestones and a number of quartzite bands and quartzite beds of grain-flow turbidite appearance showing crude grading and sole markings indicating a northwards flow. To the east of these rocks is a band of Al–Fe-rich phyllite similar in composition to the kyanite–staurolite–garnet–biotite–muscovite magnetic schists of the Scatsta division, but it contains muscovite, chlorite, quartz, and chloritoid with some kyanite. The easternmost exposed unit of the succession is a greenschist composed of spilitic volcaniclastics, lavas, pillow lavas and serpentinized ultrabasic rocks which in places show spiniflex-like textures (personal communication D. Moffat). The Clift Hills division has been correlated with the Upper Dalradian or Southern Highland Group of Scotland (Miller and Flinn 1966, p. 112).

The Ophiolite Complex

In northeast Shetland an ophiolite complex occurs

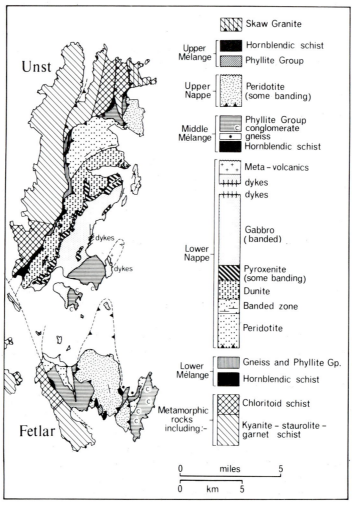

Fig. 3   The ophiolite complex of northeast Shetland

(Fig. 3) but is partly covered by the sea (Read 1934; Flinn 1958; Flinn *et al.* 1979; H. M. Prichard this volume). The main unit is a slice of rock 6 km wide and over 20 km long composed of a standard ophiolite sequence: peridotite, dunite, pyroxenite, gabbro, sheeted dyke complex, and cap rock metavolcanic rocks.

The peridotite is a structureless mass up to 2 km thick containing scattered irregular masses of dunite a metre or two in diameter. It is heavily serpentinized but in some thin sections relics indicate an excess of clinopyroxene over orthopyroxene (see also Phillips 1927, p. 628–9). However, Amin's report (Amin 1952) of less clinopyroxene than orthopyroxene and most analyses (Flinn 1970 p. 522; Thomas 1980) indicate a harzburgite nature. Between the peridotite and dunite layer is a half-kilometre thick zone of banded rock in which bands of dunite from 10 to 100 cm thick become increasingly frequent in the peridotite towards the dunite layer, and the peridotite itself shows a fine striping due to varying proportions of pyroxene and olivine. No regular grading can be detected in either the striping or the banding. The dunite layer which is up to 2 km thick is heavily serpentinized and contains sac-form chromite bodies and chromite schlieren. Between the dunite and the gabbro is an irregular pyroxenite-rich layer up to 1 km thick. It is dominantly a wehrlite showing banding several centimetres thick due to varying proportions of olivine and pyroxene, and also contains bands of clinopyroxenite and of serpentinized dunite. The pyroxenite also occurs as lenses in the dunite below and the gabbro above. The gabbro layer is several kilometres thick. The rock is rather fine grained and has been recrystallized to actinolitic amphibole and altered feldspar. In places, the gabbro is rhythmically banded on a scale of one or two centimetres due to variation in the proportions of light and dark coloured minerals but the bands are not regularly graded. On the east side of the main ophiolite unit about 3 km from the base of the gabbro, the gabbro is intruded by dykes of fine-grained basic rocks composed of hornblende and altered plagioclase which show chilled contacts against the gabbro. The dykes have a strike of about 045° where they are regularly developed, and have a vertical dip, but they cut obliquely through the easterly-dipping rhythmic banding of the gabbro. In the same area the gabbro is cut by veins of altered trondhjemite rock. In an isolated exposure a kilometre further east, poorly banded gabbro sparsely cut by dykes gives way suddenly across an irregular surface to structureless metavolcanic rocks composed of green hornblende and plagioclase with possible breccia texture and with veining on both sides of the contact.

The gabbro is cut in places by brown hornblende pegmatites. The hornblende has an Ar–Ar age of about 470 Ma (Dr. A. E. Mussett personal communication).

None of the rocks in the ophiolite succession show the tectonite fabrics reported to be characteristic of such rocks but they are heavily shattered and are often sheared. This is especially true of the gabbro layer which is in places a coherent breccia, often so strongly brecciated as to destroy the rhythmic banding. The gabbro is cut by shear zones in which the rock has been transformed to a chlorite phyllite in zones up to several hundred metres wide. The ultrabasic rocks are cut by several generations of serpentinite veins and in places are crushed into conglomerate-like masses, but shears are much less well developed than in the gabbro except near the base where they form steatite zones.

The banding in the main ophiolite unit is folded (Fig. 3). In the north of the unit the banding dips steeply to the north–northwest. In the base of the gabbro several kilometres to the southeast it is vertical and strikes south–southwest, while farther south near the top of the gabbro the banding dips 45° due east. The folding is non-cylindrical and probably preceded the emplacement of the unit. Alternatively, the arrangement of the banding could be in an original structure of the type proposed by Casey and Karson (1981) but distorted by later deformation.

A *mélange* zone underlies the main ophiolite unit to the west, separating it from the metasediments of the East Mainland Succession. Another *mélange* zone lies across the exposed surface of the main ophiolite unit. Both mélanges are composed of metasediments (the phyllite Group) containing ophiolite debris, exotic hornblende schist masses, and variably altered slices of metasediments matching those to the west in the East Mainland Succession.

The second *mélange* is overlain in the south of the complex in Fetlar by a great mass of peridotite serpentine indistinguishable from that in the main ophiolite unit. It is underlain by a thrust dipping 30–40° to the west on the east side and the same angle to the east on the west side. Banding in the east side of this peridotite mass dips 30° or more to the east. The Clibberswick peridotite, a similar mass, occurs in the north of the complex bounded by a thrust dipping steeply east to vertical and separated from the main ophiolite unit by rocks of the *mélange*.

The Phyllite Group has developed by metamorphism and deformation from very finely, intensely laminated, graphitic and micaceous mudstones, alternating with clastic sandstones, and includes a variety of polymictic conglomerates and mud flake conglomerates of probable turbidite origin. Among the constituents of the conglomerates are pebbles of typical gabbro, of quartz–albite porphyry identical with intrusions in the gabbro, and of albite granophyre not found *in situ* but of ophiolite affinity. Most of the conglomerates are little more than a few metres thick at the most, but one, the Funzie conglomerate on the east side of Fetlar, is an immense deposit with a thickness to be measured in kilometres. It contains pebbles of ophiolite affinity but is mostly composed of quartzite pebbles.

The rocks of the Phyllite Group form a series of tectonic lenses in the *mélanges* so that no succession can be established. Flinn (1958) interpreted the Phyllite Group as the product of the erosion of the ophiolite slices or nappes as they were thrust up on to the surface. Continued thrusting resulted in these erosion products being overriden, metamorphosed, and deformed. The phyllites appear to lie on the truncated upper surface of the main ophiolite unit and do not appear to have been thrust into place, although the junctions are highly schistose.

Among the lenses of phyllite are some which appear to be composed of metamorphosed volcanic material including both greenstone and felsic material. It has not yet been established whether these rocks are composed of erosional detritus from the ophiolite or are the result of later volcanic events; nor has their relationship to the hornblende schists been established.

Much of the Phyllite Group was formed during the obduction of the complex but some, like the hornblende schists, may be a part of the ocean floor sedimentary and volcanic succession forming a cap rock to the ophiolite. It is also possible that some overlie unconformably the obducted complex and suffered later deformation and metamorphism, a relationship seen in West Norway (Sturt *et al*. 1980).

The upper surfaces of the *mélange* zones are the thrusts which carried the ophiolite slices into place and are often closely associated with the hornblende schists which are usually interpreted as ocean floor volcanics or gabbro metamorphosed by the still hot ultrabasic mantle rocks as they were thrust over them at the beginning of the ophiolite emplacement. However, the peridotite seems to have already been serpentinized before thrusting and to have recrystallized to antigorite serpentine in a zone above the thrust as a result of the thrusting. The hornblende schists are generally composed of hornblende and epidote, but are locally metamorphosed to green hornblende and garnet, and even to brown hornblende and clinopyroxene and possibly are of too high a grade for this origin.

More widespread is a greenschist facies metamorphism associated with strong tectonite fabrics. The effects of this metamorphism can be seen especially well in the rocks of the Phyllite Group, and

also in the gabbro, the hornblende schist, and the metasediments of the East Mainland Succession which have been converted locally to fine-grained tectonized chloritic phyllites. Pelitic schists in the East Mainland Succession beneath the Ophiolite Complex regionally contain kyanite, staurolite, garnet, biotite, muscovite, and relic fibrolite but have locally been converted to chloritoid–chlorite-muscovite rocks.

The deformation in the *mélanges* is very variable. In places the rocks seem to be little deformed, yet in others several phases of folding can be recognized. The phyllites vary from s-tectonites to l–tectonites. The lineation and all the fold axes (except those belonging to late kink folds) trend north–northeast. The conglomerates have all been elongated parallel to this lineation so that the total resultant deformation of the rocks has been a constriction which has caused folding on north–northeast axes and an elongation in the same direction. This deformation is probably the result of the *mélange* zones being squeezed between the more competent ophiolite nappes, and consequently does not give a direct indication of the direction of transport of the ophiolites during their emplacement. The folds are often upright but some are overturned to the west–north-west, which may indicate a movement of the ophiolite nappes in that direction at a late phase in their emplacement.

These metamorphisms are associated with different stages in the emplacement of the complex. The hornblende schists were probably formed first at considerable depth and were carried up to the surface as relics attached to the base of the rising ophiolite nappes. An ophiolite nappe appears to have reached the surface, been subjected to erosion, and at least partly covered by the erosion products before being buried to considerable depth, presumably by the emplacement above it of another nappe carrying beneath it not only further slices of hornblende schist but also slices of the metamorphic basement. The deformation and burial resulting from this stacking of the nappes led to a greenschist facies metamorphism affecting the sediments between the nappes, the basement on which they rest, and the nappes themselves. The sediments and shear zones in the nappes have suffered a tectonizing metamorphism, and the basement and basal zones of the nappes a static recrystallization. The final stages of the nappe emplacement have caused a widespread cataclasis and shearing.

The relationship of the Skaw granite to the Ophiolite Complex is not clear. It occurs in the northeast corner of Unst in contact with the metasediments underlying the complex, but the enclaves and xenoliths of metasediments within the granite do not match the metasediments with which the granite is now in contact. In its eastern part the granite contains closely-spaced rectangular phenocrysts of K-feldspar up to 8 cm long, but for some hundreds of metres east of its western margin the granite and its inclusions have been deformed to an l–tectonite gneiss of mylonite aspect in which the K-feldspar phenocrysts have become rounded augen. The granite together with some relics of its aureole may have been thrust into place (Read 1934) but the age of the thrusting relative to the emplacement of the ophiolite nappes has not been determined. A K–Ar age of 354 Ma (Miller and Flinn, 1966) has been obtained for biotite in the granite.

Devonian Sandstones

The seas all round Shetland are probably underlain by Old Red Sandstone deposits. However, such rocks can be seen on shore only on the east and west sides of the islands (and in tiny outcrops in the north). The outcrops can be divided into three groups on their geographical distribution (Fig. 4) and their lithologies (Mykura 1976).

The *eastern group* extends from Lerwick to Sumburgh Head, and includes sediments deposited in the western margins of ephemeral lakes. It is otherwise composed of flagstones and sandstones, often pebbly, often cross-bedded in both fluvial and aeolian forms, which onlap onto the metamorphic rocks to the west with breccias at the unconformity and alluvial fan conglomerates extending out into the basin to the east (Allen 1981a, 1981b). The *western group* lies west of the St. Magnus Bay fault (Flinn *et al*. 1968, p. 17), on Foula, on Papa Stour, and on the Mainland. On Foula, sandstones lie unconformably on the metamorphic rocks. They show cross-bedding, are often pebbly and include siltstones. At Melby on the Mainland the sandstones pass up into volcanic rocks, well developed on Papa Stour and Esha Ness. These vary from basalt to rhyolite and include ignimbrites and agglomerates as well as lavas. The *central group*, the Walls Sandstone, forms the Walls Peninsula between the St. Magnus Bay fault and the Walls Boundary fault and forms Fair Isle lying to the east of the latter fault. The succession exceeds 10 km in thickness and is composed of sandstones, frequently cross-bedded or laminated with interbedded silts and shales, with conglomerates and breccias at and near the unconformity with the underlying rock on the north coast of the Walls Peninsula. A group of basic and intermediate lavas, tuffs and agglomerates, together with ignimbrites and felsite intrusions, occur interleaved with the sandstones within the succession. The rocks of this group, unlike those of the other two groups, have been folded and are intruded and metamorphosed by Devonian granite complexes and associated dykes (Mykura and Phemister 1976; Melvin 1976).

The sandstones of the eastern and western groups are more similar to the Old Red Sandstones of Orkney and Caithness than to those of the central group which were not recognized as Devonian until long after the others. The western group has been correlated with the Old Red Sandstone of Orkney by means of fossil fish and were found to contain Middle Devonian spores (Donovan *et al*. 1978). The eastern group is also generally considered to be of Middle to Upper Devonian age but Allen (1979) lists a series of features which tend to require its deposition in a basin separate from the Orcadian one. The lithological differences distinguishing the Walls Sandstones from the east and west groups have not been analysed and the conditions of deposition so far proposed for the group do not seem to differ from those proposed for the other groups. Nevertheless, it is generally accepted that the three groups cannot have been formed at the same time in their present close proximity. Interpretations have postulated considerable post-Devonian horizontal displacements on the St. Magnus Bay and Walls Boundary faults in order to isolate the three deposition basins from each other in Devonian times (Donovan *et al*. 1976; Mykura 1976). However, the St. Magnus Bay fault is a reverse fault and not a transcurrent fault (Mykura and Phemister 1976, p. 139; Flinn 1977, p. 242) and the Walls Boundary fault cuts through the central group. Furthermore, these reconstructions (Donovan *et al*. 1976, p. 550) require that a proposed 150 km offset across Shetland dies out north of the Scottish coast little more than 200 km away, and even so the separation of the basins of deposition thus achieved is minimal.

It is generally agreed that the central group is Middle to Lower Devonian in age. Since the central group is older than the other two and since the St. Magnus Bay fault is a reverse fault raising the central group up into contact with the western group, it is suggested that the western group, a part of the Orcadian basin, originally extended unconformably over the central group.

All these sandstones and volcanics were deposited in intramontane basins on deeply eroded Caledonian schists and gneisses and the basement gneisses to the west. In places the sandstones now bury the metamorphic rocks to the depths at which they were originally formed. The Middle Devonian unconformity of Dunrossness cuts the sillimanite–kyanite–andalusite-containing aureole of the 400 Ma old Spiggie granite. Therefore, the granite was emplaced, ero-

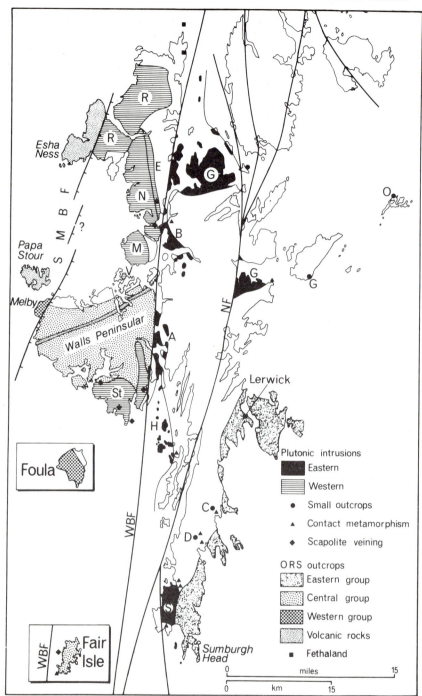

Fig. 4   Distribution of Devonian sediments and intrusives in Shetland. R = Ronas Hill granite; N = Mangaster Voe diorite; E = Eastern granite; M = Muckle Roe granite; V = Vementry granite; St = Sandsting granite; G = Graven complex; O = Out Skerries granite; B = Brae complex; A = Aith complex; H = Hildesay granite; C = Cunningsburgh granite; D = Channerwick granite; S = Spiggie granite. NF = Nesting fault; WBF = Walls Boundary fault; SMBF = St. Magnus Bay fault

sion brought the granite to the surface and then deposition buried it to a considerable depth all in the course of the Devonian period.

The Plutonic Complexes

The plutonic complexes exposed on Shetland fall into two main groups lying, respectively, east and west of the Walls Boundary fault (Fig. 4). The eastern complexes give K–Ar ages of about 400 Ma and are all cut by both the Walls Boundary fault and the Nesting fault. The Old Red sandstones of the eastern group are cut neither by these complexes nor by their associated dyke rocks and the Spiggie granite is unconformably overlain by the sandstones. The western complexes give K–Ar ages nearer to 350 Ma, and some of the dykes associated with them and the Sandsting granite itself cut the Old Red sandstones of the central group. Besides these late Caledonian complexes there are several premetamorphic, now gneissose, granites in the Yell Sound division. A partially tectonized, very coarsely porphyritic Skaw granite (Read 1934) forms a tectonic slice near the Ophiolite Complex; and a series of schistose aplogranites, referred to above, varying in size from a fraction of a metre to a kilometre across and giving a Rb–Sr age 530 Ma is associated with the permeation gneiss belt in the Whiteness Division (Flinn and Pringle 1976; Mykura 1976).

The Eastern Complexes (Mykura 1976)

The Graven complex is a hornblende granodiorite with appinitic affinities. The granodiorite is characterized by the presence of innumerable tiny xenoliths of hornblendite and in places passes into complex associations of hornblendites and hornblende diorites veined by and reacting with the granitic rocks. These associations sometimes take the form of igneous breccias which themselves can show intrusive relations to the country rocks. The complex is rich also in xenoliths and enclaves of country rock, the larger of which retain their orientations relative to one another so that bands of gneiss in the country rocks (the Yell Sound division) can be followed through the complex. Among the inclusions are masses of coarse pegmatite which could be an early phase of the complex but are more likely to be part of the pegmatite injection belt in the Yell Sound division. Contact metamorphism is almost confined to enclaves; the Yell Sound psammites within the complex are often very rich in sillimanite. The granodiorite cuts the lamprophyres and porphyrites of the area and is itself cut by thin pegmatite stringers.

The Brae Complex immediately south of the Graven Complex is cut by the dykes which are cut by the Graven granodiorite. The Brae Complex is a series of plugs formed by the multiple intrusion of rocks varying from 2-pyroxene mica diorite to ultramafic rocks on the one hand and granitic rocks on the other. Lithologically the complex is remarkably similar to the Garabal Hill complex of Scotland.

The Aith–Spiggie Complex which lies along the east side of the Walls Boundary fault is cut by many associated faults but much of it appears to be hidden beneath the sea. The main unit is a quartz-blebby epidote granodiorite which at its margins changes variously to porphyritic adamellite and hornblende syenite, or is bounded by complex hornblende-rich masses of very basic rocks including hornblendites, hornblende diorites, monzonites, and serpentinites.

These complexes are all post-metamorphic and post-tectonic (except for faulting). The large granodiorites contain in places weak fabrics probably created during their emplacement.

The lamprophyres, microdiorites, and porphyrites cut all the rocks east of the Walls Boundary fault including the ophiolite complex, but excluding the Graven granodiorite and the Old Red sandstones. They are so variable in grain-size and in degree of alteration that few are exactly alike and there is no sharp division between the main types. Some of the hornblende microdiorites and porphyrites that cut the ophiolite complex are schistose and appear to have been emplaced during the final phases of emplacement of the complex. Many of the dykes in the area east and northeast of the Aith–Spiggie Complex have schistose fabrics. They appear to have been intruded at the same time or just before the complex which has metamorphosed and in places strongly deformed the country rocks. Only a few dykes cut the complex.

The Western Complexes

The Northmaven Complex (Phemister 1979) is composed of two drusy granophyric granites, the Ronas Hill and Muckle Roe granites, connected by the Mangaster Voe diorite, a net-veined complex, and the Eastern granite which forms the eastern boundary of the net-veined complex. None of these complexes cut the Old Red sandstones, but they are cut by a diverse series of dykes, some of which cut the Walls sandstones to the south.

The Ronas Hill granite is considered to be a sheet-type intrusion emplaced in a number of stages. The main component is a red, drusy granophyre containing stilpnomelane. Other components are diorite and biotite granite. The biotite has a K–Ar age of 350 Ma (Miller and Flinn 1966). The net-veined complex has an outcrop of about 40 km². It is an apparently structureless mass of dolerite, basalt, and gabbro which has been intensely veined in a random manner by a granite similar to the non-drusy parts of the Ronas Hill granite. The veining is extraordinarily variable in pattern and in the degree of reaction between the granite and the basic rocks, which in places have been transformed into a diorite in which the plagioclase feldspars are zoned with labradorite cores and oligoclase rims. The net-veined complex is bounded to the east by the dyke-like mass of granophyre called the Eastern granite and to the south by the plug-like Muckle Roe granophyre.

The Sandsting complex to the south is cut by the Walls Boundary fault and itself cuts and metamorphoses the Walls sandstones. It does not appear to be cut by any of the dykes which cut the North-maven complex. According to Mykura (1976, p. 49–51) the complex is a sheet-like mass of granite varying to granodiorite and granophyre and contains a mass of diorite. Biotites have given K–Ar ages of 334 Ma (Miller and Flinn 1966) and 360 Ma (Mykura and Phemister 1976, p. 211), while hornblende from the diorite has given a K–Ar age of 369 Ma (Mykura and Phemister 1976, p. 211).

The Walls Boundary Fault System

The Walls Boundary fault is a major transcurrent fault which appears to have been active both in Caledonian and post-Caledonian times (Flinn 1977). In order to study the Caledonian fault it is necessary first to study the later movements.

The fault-plane is frequently exposed in Shetland as a near vertical, gouge-filled fracture which in one place can be seen to have large-scale horizontal fluting. In plan, the fault has two slight inflections but is otherwise extraordinarily straight (Fig. 4). It lies in a zone of intensely fractured rock up to a kilometre wide, in which evidence of significant offset is limited to a few north–south fractures. The cataclasis does not appear to be associated with a systematic distortion of the rocks.

The natural extrapolation of the Walls Boundary fault southwards past the west side of Fair Isle towards the Great Glen Fault in Scotland takes it along or very close to the fault-bounded east side of the Permo-Triassic filled Fitful basin. The offset on the fault is not easily determined. Restoration of a 65 km dextral movement on the fault aligns a positive magnetic anomaly over the Sandsting granite with a similar anomaly in the sea north of Fair Isle. It also brings the Devonian Walls sandstones into proximity with the lithologically similar sandstones in Fair Isle, and scapolite veining in the two areas into proximity. Such a restoration also brings the Fitful Basin, truncated on its east side, into near contact with the Moray Firth basin truncated on its west side, both containing Mesozoic sediments. However, these two basins have been held to be the result of normal faulting, down to the west on the Walls Boundary fault to create the Fitful basin (Bott and Browitt 1975), and down to the east on the Great Glen fault to create the Moray Firth basin (Bacon and Chesher 1975), although these two faults are a natural continuation of each other and neither shows signs of normal faulting on land.

If the Walls Boundary fault is indeed a major dextral transcurrent fault of Mesozoic or later age, then the nearby Nesting fault is probably of the same age. It is a dextral transcurrent fault showing an offset of 16 km, and appears to have acted as a relief fracture across a large scale deflection of the course of the Walls Boundary fault which would otherwise have blocked further offset. Restoration of this offset does not provide a perfect match across the line of the fault. In particular, the pelitic rocks on the east side of the fault are rich in sillimanite whereas none occurs on the west side of the fault. It is probable that the displacement on the Nesting fault was oblique; the eastern block rose relative to the western as the transcurrent movement took place. The St. Magnus Bay fault which occupies a position on the west side of the Walls Boundary fault similar to that of the Nesting fault on the east side is not, however, a transcurrent fault, but an east-dipping fault on which the eastern block moved up dip relative to the western block (Flinn 1977, p. 242).

All three faults require a nearly east–west compression to cause them but it is not yet clear when this faulting took place. The Walls Boundary fault appears to cut Mesozoic rocks, yet during the Mesozoic era, certainly in Jurassic times, the grabens on either side of the Shetland platform were forming as a result of a dominantly east–west extension. According to some writers (Ziegler 1981) this tension started in the Permian period and ended in the Cretaceous. According to Hay (1978) it started in the Jurassic period and ended in the Cretaceous. It seems likely that the Walls Boundary fault preceded the graben formation. However, it should be noted that Tertiary movement has been identified on the Great Glen fault (Holgate 1969) to the southwest along the line of the Walls Boundary fault.

The dextral movement on the Walls Boundary fault system was accompanied by the formation of analcite and laumontite in the movement planes, and the formation of these minerals together with the widespread occurrence of cataclasis indicates that at the time of the movement the surface at present exposed in Shetland lay at a depth of no more than several kilometres and was possibly very much shallower.

Evidence of earlier movement on the fault is provided at the present surface by mylonites formed at much greater depths, up to 15 km below the contemporary surface. Mylonites occur in places along the Walls Boundary fault plane as slices often a few centimetres across, but one major slice contains a segment of abandoned fault plane represented by a band of mylonite 10 m or more wide.

A major sinistral transcurrent movement on the Walls Boundary fault predating the dextral movement and possibly associated with the formation of these mylonites is required by the lack of matching between the crystalline rocks on either side of the fault, even after the dextral movement proposed above is restored to match the Devonian features. No matching of the Caledonoid metamorphic rocks on either side of the fault is possible but the magnitude and the sense of the offset is indicated by the geology of the two sides. On the west side of the fault the Caledonian front is exposed just south of its probable intersection with the fault. On the east side of the fault, across from the Caledonian front, are Caledonoid or at least Moine-like, garnet-rich psammites frequently gneissified and containing kyanite. Restoration of the late dextral movement brings the Caledonian front into contact with Upper Dalradian-like rocks in the south of Shetland. It is suggested that the early sinistral offset on the fault was at least 100 km plus an amount equal to the later dextral offset. Evidence in Shetland sets no upper limit to the magnitude of the offset, but any large offset must have passed along the fault into the Great Glen fault system which appears to cut the Caledonian front in southwest Scotland. Too great a sinistral offset there would result in the Caledonian front intersecting the Dalradian rocks of that area.

The sinistral movement on the Walls Boundary fault must be younger than the formation of the Caledonian front in Shetland and seems likely to have predated the deposition of the Old Red sandstones there. A substantial part, at least, of the movement seems to have taken place later than 400 Ma ago since no part of the 400 Ma eastern plutonic complexes appears on the west side of the fault in Shetland although it truncates them.

Acknowledgements

T. Robinson and R. Thomas provided information on the geology of northwest and northeast Shetland respectively and commented on an earlier version of the text.

References

Allen, P. A. 1979. *Sedimentological Aspects of the Devonian Strata of S.E. Shetland*. Ph.D. thesis, Univ. Cambridge (unpubl.).

Allen, P. A. 1981a Devonian lake margin environments and processes, S.E. Shetland, Scotland. *J. geol. Soc. London*, **138**, 1–4.

Allen, P. A. 1981b. Sediments and processes on a small stream flow dominated, Devonian alluvial fan, Shetland Islands. *Sedim. Geol.*, **29**, 31–66.

Amin, M. S. 1952. Notes on the ultrabasic body of Unst, Shetland Islands. *Geol. Mag.*, **91**, 399–406.

Bacon, M. and Chesher, J. 1975. Evidence against post Hercynian transcurrent movement on the Great Glen fault. *Scott. J. geol.*, **11**, 74–82.

Bott, M. H. P. and Browitt, C. W. A. 1975. Interpretation of the geophysical observations between the Orkney and Shetland Islands. *J. geol. Soc. London*, **131**, 35–371.

Bott, M. H. P. and Watts, A. B. 1970. Deep sedimentary basin formed on the Shetland–Hebridean Continental Shelf Margin. *Nature, London*, **225**, 265–268.

Casey, J. R. and Karson, J. A. 1981. Magma chamber profiles from the Bay of Islands ophiolite complex. *Nature, London*, **292**, 295–301.

Day, G. A., Cooper, B. A., Anderson, C., Burgers, W. F. J., Ronnivick, M. C. and Schönech, H. 1981. Regional seismic structure map of the North Sea. In: Illing, L. V. and Hobson, G. D. (Eds), *Petroleum Geology of the Continental Shelf of north-west Europe*, Heyden, London. 76–84.

Donovan, R. N., Archer, R., Turner, P. and Tarling, D. H. 1976. Devonian palaeogeography of the Orcadian basin and the Great Glen fault. *Nature, London*, **259**, 550–551.

Donovan, R. N., Collins, A., Rowlands, M. A. and Archer, R. 1978. The age of the sediments on Foula, Shetland. *Scott. J. geol.*, **14**, 87–88.

Flinn, D. 1958. The nappe structure of north-east Shetland. *J. geol. Soc. London*, **114**, 107–136.

Flinn, D. 1969. A geological interpretation of the aeromagnetic maps of the continental shelf around Orkney and Shetland. *Geol. J.*, **6**, 279–292.

Flinn, D. 1970. Some aspects of the geochemistry of the metamorphic rocks of Unst and Fetlar, Shetland. *Proc. Geol. Ass.*, **81**, 509–527.

Flinn, D. 1977. Transcurrent faults and associated cataclasis in Shetland. *J. geol. Soc. London*, **133**, 231–248.

Flinn, D., Frank, P. L., Brook, M. and Pringle, I. R. 1979. Basement–cover relations in Shetland. In Harris, A. L., Holland, C. H. and Leake, B. E. (Eds), *The Caledonides of the British Isles—reviewed*, Scottish Academic Press, Edinburgh, 109–115.

Flinn, D., May, F., Roberts, J. L. and Treagus, J. E. 1972. A revision of the stratigraphic succession of the East Mainland of Shetland. *Scott. J. geol*, **8**, 335–343.

Flinn, D., Miller, J. A., Evans, A. C. and Pringle, I. R. 1968. On the age of the sediments and contemporaneous volcanic rocks of western Shetland. *Scott. J. geol.*, **4**, 10–19.

Flinn, D. and Pringle, I. R. 1976. Age of the migmatisation in the Dalradian of Shetland. *Nature, London*, **259**, 299–300.

Frank, P. L. 1977. *The Structure and Metamorphism of the Walls Metamorphic Series, Shetland*. Ph.D. thesis, Univ. Liv. (unpubl.).

Frost, R. T. C. and Fitch, F. J. 1981. The age and nature of the crystalline basement of the North Sea basin. In Illing, L. V. and Hobson, G. D. (Eds), *Petroleum Geology of the Continental Shelf of North-west Europe*, Heyden, London, 43–57.

Hay, J. T. C. 1978. Structural development in the northern North Sea. *J. Petrol. geol.*, **1**, 65–77.

Holgate, N. 1969. Palaeozoic and Tertiary transcurrent movements on the Great Glen fault. *Scott. J. geol.*, **5**, 97–139.

Johnson, R. J. and Dingwall, R. G. 1981. The Caledonides: their influence on the stratigraphy of the northwestern European continental shelf. In Illing, L. V., and Hobson, G. D. (Eds), *Petroleum Geology of the Continental Shelf of North-west Europe*, Heyden, London, 85–97.

Melvin, J. 1976. *Sedimentological Studies in Upper*

Palaeozoic Sandstones near Bude, Cornwall and Walls, Shetland. Ph.D. thesis, Univ. Edinburgh (unpubl.).

Miller, J. A. and Flinn, D. 1966. A survey of the age relations of Shetland rocks. *Geol. J.*, **5**, 95–116.

Mykura, W. 1976. *British Regional Geology: Orkney and Shetland*. OHMS, Edinburgh, 149.

Mykura, W. and Phemister, J. 1976. The geology of Western Shetland. *Mem. Geol. Surv. G.B.*

Phemister, J. 1979. The Old Red Sandstone intrusive complex of northern Northmaven, Shetland. *I.G.S. Ref. 78/2*, OHMS.

Phillips, C. F. 1927. The serpentines and associated rocks and minerals of the Shetland Islands. *Q. J. geol. Soc. London*, **83**, 622–652.

Prichard, H. M. 1983. The Shetland Ophiolite in the Caledonide Orogen–Scandinavia and related areas. In this volume.

Pringle, I. R. 1970. The structural geology of the North Roe area, Shetland. *Geol. J.*, **7**, 147–170.

Read, H. H. 1934. The metamorphic geology of Unst in the Shetland Islands. *Q. J. geol. Soc. London*, **90**, 637–688.

Ridd, M. F. 1981. Petroleum geology west of the Shetlands. In Illing, L. V. and Hobson, G. D. (Eds), *Petroleum Geology of the Continental Shelf of North-west Europe*, Heyden, London, 414–425.

Robinson, T. 1983. *Basement/cover relations in west Shetland*. Ph.D. thesis, Univ. Liv. (unpubl.).

Sturt, B. A., Thon, A. and Furnes, H. 1980. The geology and preliminary geochemistry of the Karmøy ophiolite, S.W. Norway. In *Ophiolites. Proceedings of the International Ophiolite Symposium*, Cyprus, 1979.

Thomas, C. H. 1980. Analysis of rock samples from the Shetland ophiolite belt. *Rep. 113 Anal Chem Unit I.G.S.*

Ziegler, P. A. 1981. Evolution of sedimentary basins in north-west Europe. In Illing, L. V. and Hobson, G. D. (Eds), *Petroleum Geology of the Continental Shelf of North-west Europe*, Heyden, London, 3–39.

Ziegler, P. A. this volume. The late Caledonian framework of Western and Central Europe.

The Caledonide Orogen—Scandinavia and Related Areas
Edited by D. G. Gee and B. A. Sturt
© 1985 John Wiley & Sons Ltd

The Shetland ophiolite

H. M. Prichard

Department of Earth Sciences, The Open University, Walton Hall, Milton Keynes, MK7 6AA, England

ABSTRACT

A basic and ultrabasic igneous complex outcrops on the northeastern part of the Shetland Islands and forms the lower portion of a Caledonian ophiolite complex. As such it is an important link between the Norwegian ophiolites 350 km to the east, and those of the Scottish mainland and Newfoundland.

The complex is situated on the islands of Unst and Fetlar and the sequence is best developed in the north, on Unst, where disruption during emplacement has been less severe. The basal harzburgite tectonite is extensively serpentinized but fresh olivine, orthopyroxene, and clinopyroxene remnants are present. Orthopyroxene constitutes up to 15 per cent of the harzburgite while clinopyroxene rarely exceeds 5 per cent of the mode and the remainder is olivine with a uniform composition of $Fo_{90.5}$. In the dunite that overlies the harzburgite, olivine varies in composition from Fo_{90-88}. Podiform chromite occurs in the dunite overlying the harzburgite and in lenses of dunite within the harzburgite. The dunite grades up through wehrlites into clinopyroxenites which are overlain by massive and phase-layered gabbros. The highest exposed level of the ophiolite consists of gabbros intruded by a metadolerite dyke swarm which, in places, makes up as much as 50 per cent of the outcrop.

A metamorphic aureole is developed in the metasediments immediately beneath the ophiolite. Locally, at the contact of the aureole with the serpentinite, garnet clinopyroxene amphibolites are present which grade downwards, over a distance of 100–200 m, into greenschists.

Introduction

Since the Penrose Conference in 1972 an ophiolite complex has been defined as a group of basic and ultrabasic rocks with a specific order and composition, including harzburgite, layered cumulates, high level intrusives, a basic sheeted dyke complex, and basic volcanics (Coleman 1977; Gass 1980). The harzburgite is thought to represent residual mantle. Lenses of dunite, sometimes containing podiform chromite, occur within the harzburgite and it has been suggested that they have been produced by crystallization *in situ* from diapirs of magma rising through the harzburgite, Allen (1975) and Gass and Smewing (1979). The harzburgite is overlain by ultramafic and mafic rocks forming a magmatic sequence which crystallized in a magma chamber. Lavas at the top of the ophiolite sequence are fed from the magma chamber through a dyke complex. It is now widely accepted that ophiolite complexes were once part of the ocean crust and formed at a constructive plate margin.

The Shetland ophiolite complex was first recognized as such by Garson and Plant (1973). The complex is incomplete and has been tectonically dismembered but the characteristic lower units of tectonized harzburgite overlain by cumulate dunite, wehrlite, and gabbro are clearly recognizable. On the eastern side of Unst, the most northerly island in the Shetlands, the greatest exposed thickness and best preserved sequence is found (Fig. 1). Here the complex has been tilted on its side and emplaced on top of metasediments, with the basal thrust contact running down the centre of the island. The northern and western parts of the complex are harzburgite which forms the lowest exposed unit of the ophiolite complex. This is overlain to the south and east by dunite, wehrlite, clinopyroxenite, and gabbro. The gabbro on the eastern coast of Unst is intruded by a swarm of doleritic dykes and represents the highest level of the ophiolite. In the southwest of the complex on Unst the rocks have been more highly altered and deformed and the sequence of rock types is not so clear. The ophiolite complex extends southwards from Unst across the centre of the island of Fetlar where the rock types do not form a continuous sequence and the harzburgite in the centre of Fetlar tectonically overlies gabbros and metasediments. The structure of the complex has been described by Flinn (1958).

The following paragraphs describe the main rock units in the ophiolite complex.

The ultramafics

The ultramafic part of the ophiolite is represented by harzburgite, dunite, wehrlite, and clinopyroxenite. The harzburgite is the lowest member and is characterized by phase zoning with segregations, less than 1 m thick, distinguished by varying proportions of orthopyroxene and olivine. These segregations are approximately parallel to the outcrop of the boundary between the harzburgite and the overlying dunite. 'Dykes' composed of pyroxene (now serpentinized) cross-cut the harzburgite at all levels. 'Dykes' of dunite (Fig. 2) often 5–10 cm wide are commonly parallel to the phase zoning but also cross cut it at acute angles (the dip and strike of some of these in the harzburgite are given in Fig. 1). On Unst these 'dykes' become more abundant in the upper parts of the harzburgite. Lenses of dunite, sometimes containing massive podiform chromite, occur at all levels in the harzburgite (Fig. 3). Some dunite lenses are barren but many contain thin layers enriched in chromite and thin dunite 'dykes' often have a row of chromite crystals along their centres.

On Unst the harzburgite is overlain by the dunite. On Fetlar there are no large areas of dunite exposed but numerous lenses of dunite, arranged *en echelon*, outcrop in the centre of the harzburgite. This concentration of dunite lenses is similar to that just below the junction of the harzburgite and dunite on Unst and it seems likely therefore that the dunite once overlay the harzburgite on Fetlar but the present erosion level is now below the dunite/harzburgite junction.

On Unst the base of the magmatic layered sequences is represented by dunite which gradually gives way upwards to wehrlite and pyroxenite. The dunite comprises almost entirely massive dunite often with discontinuous layers of chromite. Such magmatic layers thicken into massive layers of chromitite over

a metre wide, these have been quarried, and two or three parallel layers often thicken at the same point laterally (unpublished mining records, A. Sandison). The chromite is similar in form and composition to podiform chromite (Thayer 1964; Neary and Brown 1979; Prichard and Neary 1981). The thickest chromitite layers occur in the north of the complex, in the centre of the outcrop of the dunite and there is a concentration of smaller pods in the dunite just above the dunite/harzburgite junction (Fig. 3). In mid-Unst, chromitite layers are situated in the upper part of the dunite and rarely as discontinuous layers within the wehrlite. Occasionally dunite veins or 'dykes' cross-cut the massive units but are only distinguishable when they cut a sequence including layers of chromite (Fig. 4).

Dunite is dominant at the base of the ultramafic magmatic sequence and clinopyroxene is virtually absent whereas in the upper parts wehrlite and clinopyroxenite layers become more abundant. Superimposed on the overall gradual increase in clinopyroxene content are rhythmic oscillations of dunite, wehrlite, and pyroxenite. Although layers consisting predominantly of dunite, wehrlite, or pyroxenite are often 10–20 m thick, all three rock types frequently occur over the distance of a metre. The orientation of the phase layering is shown on Fig. 1.

The gabbro and dykes

The transition between pyroxenite at the top of the ultramafics and the overlying gabbro is sharp, although the contact is not well exposed. The pyroxenite contains no plagioclase and the pyroxene gabbro is almost entirely olivine-free. Higher in the gabbro, lenses of layered plagioclase-free pyroxenites and wehrlites occur which also have sharp contacts with the gabbro. Within the gabbro phase layering is quite common and consists of alternations of pyroxene-rich and plagioclase-rich layers whose boundaries may be sharp or transitional. The layers often have quite different orientations (Fig. 1) and the thickness ranges from 2 cm to 2 m (Fig. 5). The grain size of the gabbro is very variable especially near the top of the gabbro (Fig. 6). Wispy discontinuous layering and net veining of felsic late stage differentiates are also developed in the upper gabbro (Figs. 7 and 8). The gabbro is cut at all levels by plagioclase-hornblende gabbro pegmatites which at the base of the gabbro, sometimes incorporate randomly-orientated blocks of angular pyroxenite (Fig. 9).

A swarm of doleritic dykes intrudes the highest levels of the gabbro (Fig. 10). In places dykes make up 50 per cent of the exposure and are occasionally

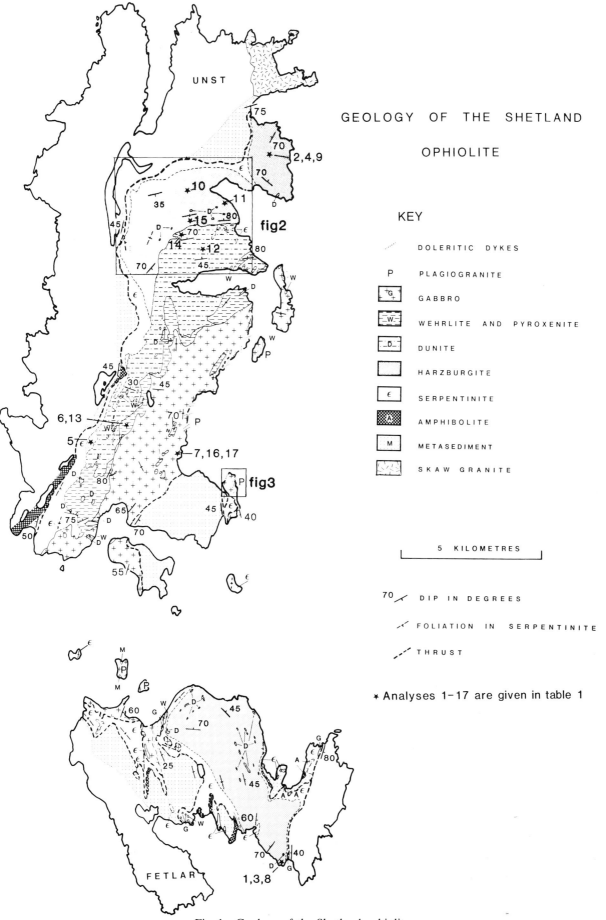

Fig. 1   Geology of the Shetland ophiolite

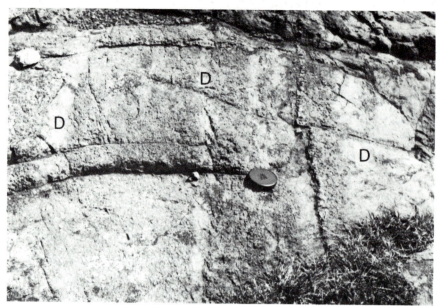

Fig. 2   Dunite 'dykes' (D) in harzburgite in the upper part of the harzburgite north of Baltasound

adjacent to each other. There appears to be several different generations present. Some dykes have no chilled margin and are intruded by apophyses of gabbro suggesting that they were intruded whilst the gabbro was still hot and plastic. Other dykes do have chilled margins and in rare instances dykes have brecciated margins. The gabbros occasionally show small scale displacement of whispy layering whereas the adjacent dykes are not displaced, indicating that at least some movement occurred soon after the crystallization of the gabbro and before dyke intrusion, (Fig. 11). Fig. 1 shows the trends of the dykes

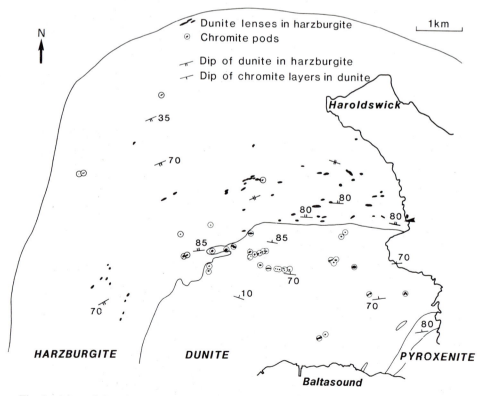

Fig. 3   Map of the ultramafic rocks north of Baltasound, Unst (for position see Fig. 1)

Fig. 4   'Dyke' of dunite (D) crosses dunite and chromite layers in the dunite, northeast of Baltasound

Fig. 5   Phase-layered gabbro, north Fetlar

Fig. 6   Pegmatitic and microgabbro in the upper gabbro on the east coast of Unst

Fig. 7   Whispy discontinuous layering in the upper gabbro on the east coast of Unst

Fig. 8   Net veining of leucocratic material in upper gabbro on the east coast of Unst

Fig. 9   Blocks of pyroxenite in gabbro at the base of the gabbro, Mid-Unst

Fig. 10   Faulted whispy layering in upper gabbro is cut by later doleritic dykes, east coast of Unst

which are often vertical. The original trends have in at least some cases been rotated by late shearing, possibly during emplacement.

One of the best exposures of the dyke swarm is on the extreme southeastern peninsula of Unst (Fig. 3). Here a slice of gabbro is overlain by a slice of serpentinized ultramafic rock and the relationship with the rest of the ophiolite complex on Unst is unclear. Within the gabbro on the northern half of this peninsula lies a dyke swarm with an approximate strike of 060° (Fig. 12). The crystal size of the gabbro in the vicinity of the dykes varies from coarse pegmatitic to fine-grained microgabbroic, in zones that do not have a dyke or pillow structure. The microgabbro is massive but is cut by shears and contains abundant veins of epidote, pyrite, and some chalcopyrite. Its origin is uncertain but as it is associated with the dyke swarm at the top of the gabbro it probably represents magmas that have chilled against the roof of the magma chamber.

Fig. 11   Dyke swarm composing 50 per cent of the rock, intrudes upper gabbro on Huney Island

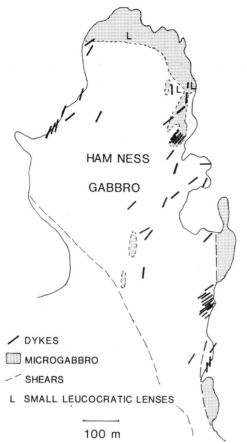

HAM NESS
GABBRO

/ DYKES
▨ MICROGABBRO
◠ SHEARS
L SMALL LEUCOCRATIC LENSES

100 m

Fig. 12 Map of dykes in gabbro on Ham Ness (for position see Fig. 1)

Petrology

Different minerals in the ophiolite sequence have suffered varying degrees and types of alteration. For example olivine may be 75 per cent fresh or completely serpentinized. The primary mineralogy and texture can usually still be deduced provided that the textures have not been totally destroyed by later deformation. Recognition of the original rock type is most difficult for the ultramafic rocks. It is often almost impossible to distinguish between a completely serpentinized wehrlite or harzburgite but in less altered samples it is evident that the wehrlites commonly lack the minor spinel component of the harzburgites.

Orthopyroxene constitutes up to 12 per cent of the harzburgites, whereas clinopyroxene is present in small quantities in most samples (rarely greater than 5 per cent of the mode) usually as clusters around the larger orthopyroxene crystals. The orthopyroxene is usually more altered than the clinopyroxene. In a very altered rock both types of pyroxene form bastites and as in the original mineralogy the bastites after orthopyroxene are

larger than those after clinopyroxene. The least depleted, clinopyroxene-rich harzburgite examined to date contains less than 5 per cent clinopyroxene and comes from the most southeastern part of the harzburgite on Fetlar, analyses 1, 3, 7, Table 1.

The olivine and pyroxene crystals are generally granular and orthopyroxene occasionally encloses both olivine and clinopyroxene. Some olivine grains show strain extinction and occasional twinning. Kink bands are virtually absent in harzburgitic pyroxenes although the cleavage planes may be slightly bent. However, some of the pyroxenite dykes which cut the harzburgite are composed of strongly-deformed crystals with abundant kink bands.

The dunite overlying the harzburgite is composed of unzoned olivine crystals and the first clinopyroxene in the upper part of this dunite forms as an intercumulus phase. The clinopyroxene in the dunite, wehrlite, and pyroxenite is generally very fresh.

Most of the gabbro was originally composed of clinopyroxene and plagioclase but pale brown hornblende pegmatites occur where the hornblende shows no sign of pseudomorphing pyroxene. The gabbros are very altered, with clinopyroxene only rarely preserved as relict centres surrounded by actinolite. Both the gabbro and the dykes usually form a mosaic of albite, actinolite, chlorite, epidote, quartz, calcite, and opaque minerals. Many of the dykes were originally doleritic but some are more felsic, containing hornblende phenocrysts; others are more basic, containing serpentine and accessory spinel. Recent major and trace element analysis of these dykes suggests that they have island arc type geochemistry (Prichard and Spray in preparation).

Mineral Chemistry

Olivine, orthopyroxene, clinopyroxene, chromite, and amphibole analyses are given in Table 1. The stratigraphic distribution of the igneous rock types and their mineralogy is given in Fig. 13.

The orthopyroxene in the harzburgite is $En_{90.5}$–$En_{91.5}$ and the olivine is approximately constant at $Fo_{90.5}$. These ratios are similar to those described in harzburgites from Troodos, Cyprus (Fo_{90}–Fo_{92} and En_{90}–En_{92}; Menzies and Allan 1974) and from the Semail ophiolite, Oman (Fo_{91} and En_{91}; Smewing 1980). In Shetland, the olivine in the overlying dunite varies from Fo_{90}–Fo_{88}. The olivine analyses show that the nickel content of the olivines decreases upwards from the harzburgite through the dunite and into the pyroxenite.

The chromites have a range of compositions with Cr_2O_3 values varying from 40 to 60 per cent. Mas-

Table 1

| | Orthopyroxene | | | | Clinopyroxene | | |
	1	2	3	4	5	6	7
SiO_2	55.42	55.60	52.98	53.93	53.44	53.24	52.01
TiO_2	0.02	—	0.06	0.02	0.02	—	0.19
Al_2O_3	3.53	1.16	3.46	1.43	1.80	1.40	1.05
FeO	5.82	5.69	2.02	1.78	1.88	2.13	7.31
MnO	0.14	0.11	0.08	0.09	0.08	0.13	0.22
MgO	32.50	33.67	16.62	17.35	17.64	17.22	14.91
CaO	1.40	1.43	23.42	24.19	23.26	24.11	22.62
Na_2O	0.01	—	0.11	0.15	0.05	—	0.29
Cr_2O_3	0.82	0.88	1.05	0.69	0.86	0.77	0.05
NiO	0.11	0.19	0.05	—	0.05	—	—
Totals	99.77	98.73	99.85	99.63	99.08	99.00	98.65
Mg/(Mg + Fe)	90.81	91.33	93.75	94.95	95.05	93.51	—
Mg:	88.42	88.85	48.66	48.70	50.00	48.16	—
Ca:	2.63	2.72	48.13	48.70	47.40	48.49	
Fe:	8.95	8.43	3.21	2.59	2.60	3.34	

| | Olivine | | | | | |
	8	9	10	11	12	13
SiO_2	41.12	41.52	41.50	41.41	40.19	40.16
TiO_2	0.02	0.02	0.02	0.02	0.02	0.02
FeO	9.25	9.25	8.81	8.72	10.71	11.80
MnO	0.12	0.17	0.12	0.14	0.11	0.19
MgO	49.37	49.16	49.11	49.08	48.08	47.64
NiO	0.41	0.41	0.42	0.36	0.33	0.18
Totals	100.29	100.53	99.98	99.73	99.44	99.99
Mg/(Mg + Fe)	90.57	90.36	90.82	90.86	88.89	88.00

| | Chromite | | | Amphibole | |
	Massive 14	Disseminated 15		Green fibrous 16	Pale brown 17
SiO_2	0.08	0.08	SiO_2	47.37	47.11
TiO_2	0.18	0.18	TiO_2	0.20	1.54
Al_2O_3	18.84	18.30	Al_2O_3	10.60	9.30
FeO	14.33	18.13	FeO	12.84	12.44
MnO	0.12	0.19	MnO	0.30	0.31
MgO	15.06	13.22	MgO	12.88	13.64
Cr_2O_3	51.56	50.03	CaO	12.24	11.56
NiO	0.14	0.05	Na_2O	1.68	1.90
			K_2O	0.37	0.07
Total	100.31	100.18	Total	98.48	97.87

1 and 2: orthopyroxene in harzburgite; 3: clinopyroxene from the most clinopyroxene rich harzburgite; 4 and 5: clinopyroxene in clinopyroxenite; 7: clinopyroxene in the upper part of the gabbro; 8: olivine in the most clinopyroxene rich harzburgite; 9, 10, and 11: olivine in harzburgite; 12: olivine in dunite; 13: olivine in clinopyroxenite; 14: massive chromite; and 15: disseminated chromite, from the dunite just above the harzburgite/dunite junction; 16: green fibrous amphibole; and 17: pale brown amphibole, from the top of the gabbro on the east coast of Unst.
Analyses were obtained using a wavelength electron microprobe, Cambridge MK9.

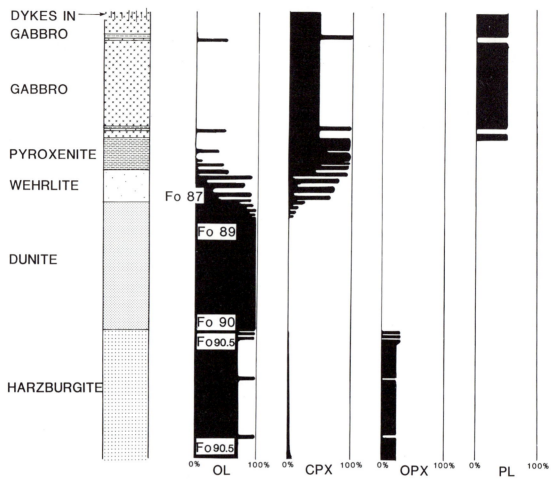

Fig. 13   Diagrammatic interpretation of the variations in phases present in the Shetland ophiolite sequence

sive chromitite usually has a higher Mg/(Mg + Fe^{++}) ratio than the disseminated chromite.

Clinopyroxene from the wehrlite is richer in CaO and MgO but poorer in FeO than from pyroxene gabbro. This would be expected if the pyroxene gabbro crystallized from a more evolved magma than the wehrlite, at a higher level in the complex.

The Basal Contact

Immediately above the basal thrust contact of the Shetland ophiolite complex is a zone of serpentinite in which it is difficult to recognize any primary unaltered mineralogy other than fresh chrome-spinel. Occasionally fresh olivine is present but most commonly serpentinization is complete. The serpentinites have also been altered to talc, carbonates, and in places have been silicified. In the north, on Unst, the serpentinite grades upwards into harzburgite, and was probably originally a harzburgite. In south Unst and on Fetlar, the igneous sequence is not clear and the original mineralogy of the serpentinite is very difficult to determine.

The presence of metamorphic rocks immediately beneath the ultramafic rocks was first recognized by Read during his 6″ mapping of Unst for HM Geological Survey in the 1920's. Williams and Smyth (1973) suggested that the amphibolites beneath the Shetland ophiolite complex form part of an aureole similar to those beneath ophiolite complexes in Newfoundland. The contact between the basal serpentinite and underlying metamorphic rocks is poorly exposed, but amphibolites are present beneath the serpentinite in the south of Unst and beneath the serpentinite at the base of the harzburgite in central Fetlar. The metamorphic grade of the amphibolites decreases away from the basal contact of the ophiolite. On Fetlar 2–3 m of high grade garnet, clinopyroxene-bearing amphibolites are developed immediately adjacent to the contact; these grade away into amphibolites and then metasediments and metavolcanics in greenschist facies.

Metamorphic aureoles with garnetiferous, pyroxene-bearing amphibolites occurring near the ultramafics and passing away from the contact with the basal thrust of the ophiolite complex into greenschist facies within 1000 m have also been described in Newfoundland from the St. Anthony Complex (Jamieson 1981) and the Bay of Islands Complex (Malpas 1979), from Tethyan ophiolite complexes (Woodcock and Robertson 1976) and from Oman (Searle and Malpas 1980). In the Shetland ophiolite complex the aureole is seldom complete and in the north, amphibolites beneath the basal thrust contact are completely absent with the serpentinite immediately overlying metasediments. It seems likely that the aureole rocks have been disrupted and in places completely removed by shearing perhaps occurring during the final stages of the emplacement of the ophiolite.

Deformation and Metamorphism

Small isoclinal folds of less than 1 m wavelength occur in the ultramafic rocks. In harzburgite the folds are picked out by the segregation of more and less pyroxene-rich zones and in the overlying dunite they are visible in the chromite layering. The spinels in both the harzburgite and in the overlying dunite have been deformed showing a strong foliation. Structural analysis of these ultramafic rocks has been studied by I. D. Bartholomew (1984). Disruption of the ophiolite complex during emplacement has not only caused deformation at the base of the complex but also within the complex are ultramafic rocks and gabbros which have been highly deformed by brittle fracturing along localized fault zones.

The whole complex has been overprinted by greenschist metamorphism. Chlorite and actinolite are abundant in the gabbros and dykes; some of these minerals may have been produced by an oceanic crust greenschist metamorphism (Cann 1970). However, the aureole rocks are often retrogressed to greenschist metamorphism with only a few remnants of amphibolite remaining and the underlying metasediments and metavolcanics are in greenschist facies metamorphism. Within the ultramafic rocks, magnesium chlorite rims are present around many disseminated chrome-spinels. Therefore, any ocean crust metamorphism is overprinted by a regional retrogressive event, causing widespread greenschist facies metamorphism, which was first recognized in this area by Read (1936). It has been suggested that this regional metamorphism can be correlated with the 425 ± 30 Ma K/Ar age of micas from the metasediments beneath the ophiolite complex (Miller and Flinn 1966). If this is so then the ophiolite complex was emplaced prior to 425 ± 30 Ma and is overprinted by this Caledonian metamorphism.

Conclusions

The Shetland basic and ultrabasic complex forms an incomplete or dismembered ophiolite complex with the lower parts of the sequence well preserved. The harzburgite represents residual mantle; undepleted mantle lherzolite being absent. The dunite, wehrlite, and pyroxenite overlying the harzburgite may be interpreted as a sequence crystallized at the base of the magma chamber and the gabbro formed above after greater fractionation of the magma. The top of the chamber is represented by a dyke swarm which in a complete ophiolite complex would normally underlie a sheeted dyke complex and associated lavas.

The isolated nature of the Shetland Islands makes correlation with other Caledonian ophiolite complexes difficult. In Norway most described ophiolites (Furnes et al. 1980) do not appear to contain a major development of ultrabasic rocks with the exception of Leka. This occurrence bears many similarities to the Shetland ophiolite having an ultrabasic residual mantle tectonite, overlain by a layered sequence, followed by gabbros (Prestwick 1980). The Shetland ophiolite complex apparently more closely corresponds to some of the ophiolite complexes in western Newfoundland, for example the lower parts of the Bay of Island ophiolite complex (Malpas 1977).

The Shetland complex is one of several ophiolite complexes in Scotland (Garson and Plant 1973) and has the best preserved Caledonian ultrabasic and basic ophiolite sequence in the British Isles.

Acknowledgements

I would like to thank Professor I. G. Gass, Drs. S. J. Lippard, C. R. Neary and J. D. Smewing for making helpful and constructive suggestions. This work was supported by an EEC raw materials research and development grant.

References

Allen, C. R. 1975. *The Petrology of a Portion of the Troodos Plutonic Complex, Cyprus*, unpublished Ph.D. thesis, Cambridge University, 315 pp.

Bartholomew, I. D. 1984. *The primary structures and fabrics of the upper mantle and lower oceanic crust from ophiolite complexes*, unpublished Ph.D. thesis, the Open University. Milton Keynes, 523 pp.

Cann, J. R. 1970. New model for the structure of the ocean crust. *Nature, London*, No. 5249, **266**, 928–930.

Coleman, R. G. 1977. Ophiolite: Oceanic Lithosphere? Springer-Verlag, New York, 220 pp.

Flinn, D. 1958. On the nappe structure of north-east Shetland, *Q.J. geol. Soc. London*, **114**, 107–136.

Furnes, H., Roberts, D., Sturt, B. A., Thon, A. and Gale G. H. 1980. Ophiolite fragments in the Scandinavian Caledonides. *Proc. Int. Oph. symp. Cyprus, Geol. Surv. Dept. Cyprus*, 582–600.

Garson, M. S. and Plant, J. A. 1973. Alpine type ultramafic rocks and episodic mountain building in the Scottish Highlands. *Nature Phys. Sci.*, **242**, 34–38.

Gass, I. G. 1980. The Troodos massif: its role in the unravelling of the ophiolite problem and its significance in the understanding of constructive plate margin processes. *Proc. Int. Oph. Symp. Cyprus, Geol. Surv. Dept. Cyprus.*, 23–35.

Gass, I. G. and Smewing, J. D. 1979. Ophiolites: obducted oceanic lithosphere. In Emiliani, C. (Ed.), *The Sea*, Vol. 7, Wiley, New York, 339–362.

Jamieson, R. A. 1981. Metamorphism during ophiolite emplacement—the petrology of the St Anthony Complex. *J. Petrol.*, **22**, 3, 397–449.

Malpas, J. G. 1977. Petrology and tectonic significance of Newfoundland ophiolites, with examples from the Bay of Islands. In Coleman, R. G. and Irwin, W. P. (eds), *North American Ophiolites. Oregon Depart. Geo. Min. Indust. Bull.*, **95**, 13–23.

Malpas, J. G. 1979. The dynamothermal aureole of the Bay of Islands Ophiolite Suite. *Can. J. Earth Sci.*, **16**, 2086–2101.

Menzies, M. A. and Allan, C. R. 1974. Plagioclase Iherzolite–residual mantle relationships within two eastern Mediterranean ophiolites. *Contrib. mineral. Petrol.*, **45**, 197–213.

Miller, J. A. and Flinn, D. 1966. A survey of the age relations of Shetland Rocks. *Geol. J.*, **5**, 95–116.

Neary, C. R. and Brown, M. A. 1979. Chromites from Al'Ays Complex Saudi Arabia and the Semail Complex, Oman. In Al-Shanti, A. M. S. (ed.), *Evolution and Mineralization of the Arabian Shield. I.A.G. Bull.*, **2**, 193–205.

Prestvik, T. 1980. The Caledonian ophiolite complex of Leka, north central Norway. *Proc. Int. Oph. Symp. Cyprus, Geol. Surv. Dept. Cyprus*, 555–566.

Prichard, H. M. and Neary, C. R. 1981. Chromite in the Shetland Islands ophiolite complex. In *An International Symposium on Metallogeny of Mafic and Ultramafic Complexes*. UNESCO. Pub. Athens. UNESCO project 169, **3**, 343–360.

Read, H. H. 1936. The metamorphic history of Unst, Shetland. *Proc. Geol. Assoc. London*, **47**, 283–293.

Sandison, A. Unpublished records of chromite quarrying in Unst, Shetland, A. Sandison and Sons, Baltasound, Unst, Shetland.

Searle, M. P. and Malpas, J. 1980. The structure and metamorphism of rock beneath the Semail ophiolite of Oman and their significance in ophiolite obduction. *Phil. Trans. R. Soc. Edin.*, **71**, 247–262.

Smewing, J. D. 1980. An Upper Cretaceous ridge-transform intersection in the Oman ophiolite. *Proc. Int. Oph. Symp. Cyprus, Geol. Surv. Dept. Cyprus*, 407–413.

Thayer, T. P. 1964. Principle features and origin of podiform chromite deposits and some observation on the Guleman Soridag district, Turkey. *Econ. Geol.*, 1497–1524.

Williams, H. and Smyth, W. R. 1973. Metamorphic aureoles beneath ophiolite suites and Alpine peridotites: tectonic implications with west Newfoundland examples. *Am. J. Sci.*, **273**, 594–621.

Woodcock, N. H. and Robertson, A. H. F. 1976. Origins of some ophiolite-related metamorphic rocks of the 'Tethyan' belt. *Geology*, **5**, 373–376.

The Caledonide Orogen—Scandinavia and Related Areas
Edited by D. G. Gee and B. A. Sturt
© 1985 John Wiley & Sons Ltd

The Ligerian Orogeny: a Proto-Variscan event, related to the Siluro-Devonian evolution of the Tethys I ocean

J. Cogné and **J. P. Lefort**

Centre Armoricain d'Etude Structurale des Socles, (L.P. CNRS) Institut de Géologie de l'Université de Rennes, 35042 Rennes Cédex (France)

ABSTRACT

Evidence for a post-Ordovician pre-Carboniferous orogenic cycle in southwest France suggests that contemporaneous events occurred in the Caledonides and in the Ligero–Moldanubian domain. This orogenic cycle, known in France as Ligerian, was previously often considered as Caledonian because it began with widespread Ordovician magmatism. It has also been attributed to the Acadian as the earliest deformations are cogenetic with this Appalachian phase, and because there is a paired metamorphic belt of Devonian age. At times it has been regarded as part of the Hercynian evolution, since it ends at that time.

However, the magmatism is related to subduction, the older deformation is related to collision, and the Hercynian evolution is related to ensialic shortening. Different geodynamic frameworks are indicated.

Geophysical data also suggests that the Caledonian and Ligerian orogenies developed on opposite sides of the Avalon Prong; thus there is no relationship between these two tectonometamorphic events, which must be regarded as being caused by the closing of two different oceans. The Ligerian orogeny is only dependent on the evolution of the Tethys I Ocean.

Introduction

Many plate tectonic models have been proposed for the Hercynian orogeny of western and central Europe but only a few of them have properly considered its most active zone in western France, which is usually passed over quickly and referred to as complex metamorphic Precambrian basement, somewhat rejuvenated during Carboniferous times. In western Europe, this zone, which is often referred to as 'Moldanubian', displays a tectonothermal pattern which developed between the Late Precambrian and the Lower Carboniferous. Geological literature is difficult to synthesize as most authors have attempted to correlate this episode either with the Cadomian orogeny (Cogné 1962) or with the Caledonian cycle (Autran 1978) while a few refer it mainly to the Hercynian.

The term 'Caledono–Variscan', which implies totally unrealistic geological configurations from a plate tectonic point of view, has also been used. The South Armorican orogeny is now regarded as having

formed in a converging plate framework which ended during Carboniferous time with the climax of Hercynian mountain building. This latest event was a result of the collision between Gondwanaland and the Cadomian (Hadrynian) basement of western and central Europe and compressed within it the 'Ligerian cordillera' (Cogné 1976; Autran and Cogné 1980; Cogné and Wright 1980). Large parts of the Variscan system are thus the result of the superimposition of younger Carboniferous Hercynian events on an older pre-Carboniferous Ligerian orogenic belt.

At this point two remarks are called for. The term 'Ligeria' was invented by Pruvost (1949) to describe the palaeogeographic area characterized in the Southern Armorican Massif (Loire region) by regressions, transgressions, and non-sequences between the Silurian and Middle Devonian. Pruvost attributed these superficial events to the development of a 'Ligerian Cordillera', which had not yet then been identified on structural or palaeo-geodynamic grounds.

Cogné (1962), having recognized that deep-seated metamorphic and structural events in this same area were pre-Hercynian, suggested that they were related to a Ligerian Cordillera. Yet there was evidence to suggest that the same events were Cadomian (Cogné 1957), only the latest vertical movements being responsible for Ligerian palaeogeography. Hence the term 'Ligerian Cordillera' came to be used for a structure considered to have formed in Cadomian times.

Recent geochronological and structural work has now shown that all these tectonic and metamorphic events immediately precede the Hercynian orogeny and are of Devonian age. Thus the term 'Ligerian Cordillera' as proposed by Pruvost (1949), seems particularly appropriate for the events which paved the way for the Hercynian tectonics (Cogné 1976; Autran and Cogné 1980).

This does not exclude the possibility of Cadomian events in the basement underlying the Ligerian Cordillera (Cogné and Wright 1980). There is no structural evidence, however, for them in the Ligerian mobile belts of the Southern Armorican Massif or the western Massif Central (France). They are detectable, in contrast, in surrounding zones, especially the basement underlying Pruvost's (1949) palaeographic Ligeria (Brioverian of the Mauges, unconformably overlain by Cambrian to Lower Ordovician, or by Upper Devonian rocks); but even here, most of the deep-seated phenomena observed in the Brioverian rocks, immediately underlying the Palaeozoic, appear to be Devonian and associated with deep-seated 'subfluence' (as in the Champtoceaux 'nappe').

The main aim of this paper is to set out the early history of the Variscan orogeny in the Atlantic province of western Europe.

The Ligerian Cordillera

Devonian age for the main tectonothermal event

Stratigraphical data

North of Vendée and Anjou, the widespread Upper Silurian and Lower Devonian regression was followed by a mid-Devonian transgression. These younger sediments often rest on what was regarded as a much older metamorphic terrain as well as the slightly older strata. More recent studies (cf. Autran and Cogné 1980) however, show that some of the metamorphic rocks, which were originally regarded mainly as being uplifted Late-Precambrian basement, are Cambrian, Ordovician, and even as young as Silurian in age. The first sediments which post-date

this orogenic event are Givetian in age in Vendée and Givetian to Frasnian in the western part of Massif Central.

Geochronological data

All the early intermediate to acid igneous intrusive rocks within these schists are now orthogneisses. They were emplaced in the sediments prior to, or during the early part of the regional metamorphism and are dated at between 590 to 430 Ma (cf. Autran and Cogné 1980). i.e. between Upper Hadrynian and Silurian time. The post-metamorphic granitic rocks range between 360 and 345 Ma in age and thus were emplaced between end Devonian and end Tournaisian times.

A major suite of synkinematic intrusive rocks, the South Armorican leucogranites, which were emplaced during the Hercynian tectogenesis, are always associated with major shear zones and range in age from 340 to 300 Ma. These are characteristic of the ensialic evolution of the Late Variscan period (Vidal 1976; Peucat et al. 1979). The metamorphic terrain, which was produced between the two first intrusive periods, is regarded as being associated with a deep orogenic evolution, and yields consistent age dates throughout the whole of the Ligerian zone. These metamorphic Ligerian rocks formed in two main metamorphic regimes depending on the structural level: either in a high pressure environment ranging from high to low temperature in which blueschists, eclogites, or granulites dated at between 420 and 375 Ma formed (Gebauer et al. 1978; Peucat et al. 1977, 1978; Pin et al. 1978); or they show an intermediate pressure Barrovian metamorphism which is locally associated with the production of anatectic granites at about 375 Ma (Vidal 1976). These dates thus range between the Upper Silurian and Middle Devonian and coincide in time with the regression described by Pruvost (1949).

Deformation

Although many shear zones, of Carboniferous age and related to the Hercynian collision event, obscure the Ligerian structures, two superimposed foliations older than the transgressive Devonian sediments are always seen. The anatectic phase is usually contemporaneous with the second deformational episode. Furthermore, large flat-lying shear zones occur throughout the area and indicate the one-time presence of late nappe or thrust tectonics. In the Haut Allier, deep-rooted shear zones bounding thick slices of crust have been recognized (J. Grolier 1971) and in central and southern Limousin, large thrusts are associated with recumbent folds facing north or east (Guillot 1981). In Brittany, the Champtoceaux nappe, which displays a high grade metamorphism, is bounded by major sinistral shear zones that are older than the Frasnian sediments (Lagarde 1978; Diot and Blaise 1978). Further westward, a large number of recumbent folds have been recognized in the metamorphic rocks and these were subsequently tightened and refolded by the Hercynian deformation (Audren pers. comm. 1982).

Structural zones of the Ligerian Cordillera

The northern margin of the Ligerian orogenic zone (Fig. 1)

This margin is marked in Brittany by the South Armorican shear zone which was active in Carboniferous times. Because it thoroughly reworked a large area of Ligerian and Cadomian rocks, it is no longer possible to recognize the older transition between the outer part of the Ligerian orogenic zone and its Cadomian foreland to the north. However, a few aligned K-rich subalkaline plutons of Ordovician or Early Silurian age have been recognized (Cogné

and Vidal 1972) and these are sometimes associated with tholeiitic gabbros of continental origin (Le Metour and Bernard-Griffiths 1980). Both are characteristic of a rifting episode, which is regarded as being related to the opening of the Tethys I ocean that lay to the south during Palaeozoic times. Some other distentional events occurred subsequently and are probably related to Ligerian subduction beneath Cadomian basement.

In the Chateaulin basin a Lower Devonian mafic volcanism of tholeiitic affinity has been recognized and in Saint-Georges-sur-Loire the volcanics and sediments are characteristic of a subsiding furrow or a backarc basin.

The calc-alkaline batholith

A discontinuous line of calc-alkaline plutons has a clear trend between Massif Central and Vendée, but is seen only in a part of Brittany, primarily because it is cut across by the South Armorican Shear Zone but also because of the deep level of erosion seen in Morbihan, suggesting greater uplift there. Most of the plutons are Upper Devonian or Lower Visean in

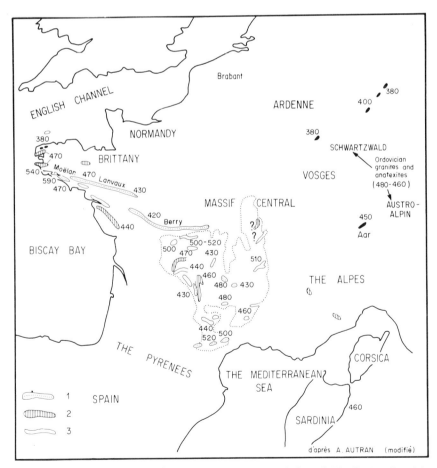

Fig. 1   Cambrian to Silurian intrusive rocks (partly after Autran and Cogné 1980). 1—Granitic rocks with K-rich alkaline affinity; 2—Trondhjemites with associated tonalites; 3—Al-rich leucocratic granitoid rocks

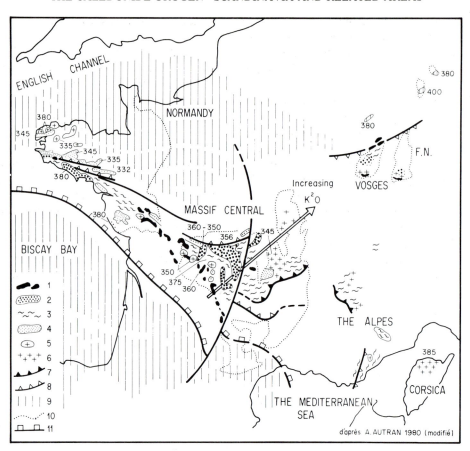

Fig. 2   Upper Devonian to Visean intrusive rocks (partly after Lameyre *et al*. 1980). Ligerian orogeny: 1—Diorites, quartz-diorites and tonalites; 2—Al-rich monzonites (deep level); 3—Devonian anatectic zones. Hercynian orogeny: 4—Al-rich leucogranites; 5—High level granites and granodiorites; 6—Very K-rich granitic rocks; 7—Devonian thrusts; 8—Tournaisian to Visean thrusts; 9—Pre-Ligerian basement; 10—Boundary between basement and cover; 11—South Ligerian suture

age (Figs. 2 and 3) (Autran and Lameyre, in Lameyre *et al*. 1980). The K_2O/SiO_2 ratio increases toward the northeast and has been used to suggest that a subduction zone, related to the generation of the magma, was dipping in that direction.

In this central zone, the tectonothermal pattern which preceded the emplacement of the Devonian intrusions appears to be consistent throughout and is commonly characterized by a rapid metamorphic evolution which passed from a high pressure regime (M_1) to a typical Barrovian metamorphism (M_2). Following M_1 there was a progressive but possibly irregular reduction of pressure, and widespread up-doming of anatectic zones occurred. Isograds associated with M_1 and M_2 are not coincident. These features of the tectonothermal history of the Ligerian central zone are typical of an Alpine metamorphic style and are in complete contrast to the characteristically ensialic Hercynian orogeny which was subsequently superimposed across the Ligerian arc.

The southern margin of the Ligerian orogenic zone

This margin is exposed south of the Massif Central in the Lot area, where large south-facing thrusts are associated with a great mass of low-grade micaschists characterized by flat-lying planar fabrics (Briand and Gay 1978; Pin 1979). To the west, this boundary passes beneath a Mesozoic and Cenozoic cover before it turns in the direction of the South Armorican shelf. In this area, between Vendée and Audierne Bay, the Ligerian structural and metamorphic history is well displayed where there are no Hercynian granites. In the Groix graben, in western Vendée, and in Belle-Ile (Audren and Lefort 1977; Lefort *et al*. 1982a, b) similar mica schists have been recognized and these are regarded as representing the western extension of the epimetamorphic Ligerian rocks which are better seen in the southern part of the Massif Central. Offshore, to the south of Brittany, these are often associated with volcano-sedimentary rocks and rhyolitic flows, which clearly show the proximity of a volcanic arc that once reached above sea level. In these the foliation is flat lying and folds face south. West of Audierne Bay, the level of erosion seen in the Ligerian rocks is higher, and the Palaeozoic cover, which is also weakly metamorphosed, thickens towards the west (Lefort and Peucat 1974) before it passes beneath a Mesozoic and Cenozoic cover. Mafic and ultramafic rocks seen onshore, and known from geophysics offshore, have undergone deep-seated metamorphism which is not yet fully understood. They could be in part considered, as in Champtoceaux, as the result of pre-Variscan rifting (Cogné and Wright 1980), while the associated effusive rocks could be regarded as either the western extension of the main volcanic arc or as an equivalent of the backarc basin seen at Saint-Georges-sur-Loire.

The South Ligerian suture

Blueschists (glaucophane–lawsonite–jadeite), which outcrop in the centre of the Groix graben, have been long considered as marking the site of the South Armorican suture. It is now known, however, that these rocks have been affected by two major deformational episodes; and the whole of the blueschist belt is now regarded as a flat-lying obduction sole or a klippe (Quinquis 1980; Quinquis and Choukroune 1981), which was emplaced northwards during Silurian or Lower Devonian times (Peucat and Cogné 1977). This new geological interpretation has been substantiated by geomagnetic investigations which suggest that the blueschists, in their present position, are not rooted (Lefort and Segoufin 1978). Core samples of high temperature metamorphic rocks which has been recovered adjacent to the blue-

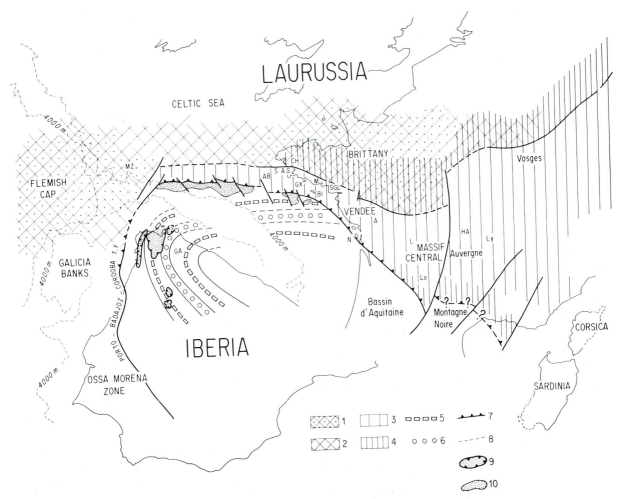

Fig. 3   The Ligerian orogenic environment (shown on a Triassic fit, Lefort 1980). 1—Pre-Ligerian basement; 2—Inferred Pre-Ligerian basement; 3—Devonian orogenic zone; 4—Dinantian orogenic zone; 5–Lower Palaeozoic basins in Spain; 6–Upper Precambrian ridge in Spain; 7—South Armorican suture; 8—Present plate boundaries; 9—Klippes of oceanic crust; 10—Remnants of oceanic crust (after magnetic data). A: Anjou; AB: Audierne Bay; BI: Belle-Île; C: Champtoceaux; CH: Chateaulin; GA: Galicia; GX: Groix; HA: Haut-Allier; L: Limousin; Lo: Lot; LY: Lyonnais; M: Morbihan; MZ: Meriadzec; N: Noirmoutier; SASZ: South Armorican Shear Zone; SGL: Saint-Georges-Sur-Loire

schist belt (Audren and Lefort 1977) are regarded as lying structurally beneath the blueschist thrust sheet. Because of the proposed northward direction of thrusting and the suggested dip direction of the probably related subduction system, the Ligerian suture must occur to the south, i.e. beneath Mesozoic and Cenozoic cover.

About 60 km southwest of Brittany a belt of magnetic, gravity, and seismic anomalies suggests the presence of a large mafic body at depth (Lefort 1979), and increasing in width to the west. This geophysical disruption appears to separate two different areas of basement. To the north, all the gravity and magnetic linears are about parallel to the trend of the orogenic belt seen onshore, whereas to the south, the geophysical trends are oblique and can be linked with the arcuate structures of the Ibero-Armorican arc (Lefort 1979).

Between Noirmoutier Island and the Southern Massif Central a pronounced linear magnetic and gravity gradient can be used to extend this suture beneath the cover in the Aquitaine basin. This boundary also may be followed southeast of the Massif Central where it is sinistrally offset where it intersects the Sillon Houiller. It then passes north of Montagne Noire and could extend as far as the Mediterranean coastline, where a similar linear magnetic body is known at depth approximately on this line. Northeast of this line Devonian deformation is severe and widespread, whereas to the southwest (Montagne Noire, the Pyrénées, and the Aquitaine basin) there are almost no structures older than Carboniferous in age (Autran and Cogné 1980).

Geodynamic evolution of the Ligerian orogeny in southwestern France

The evolution of the Ligerian Cordillera developed, during Silurian and Devonian times, over about 60 Ma. Although the younger intrusions of Ordovician age are contemporaneous with the Scandinavian Caledonian event, they did not result from crustal shortening but were a consequence of a widespread rifting which accompanied the opening of the Tethys I ocean between Cambrian (?) and Ordovician times. Even though it is often difficult to discriminate between distensions related to a previous rifting event, and those related to a subsequent subduction event, it is now clear that the well dated metamorphic and intrusive phenomena of Siluro-Devonian age are related to subduction of an oceanic plate. This plate appears to have dipped to the northeast beneath the Cadomian–Armorican craton.

The Porto–Badajoz–Cordoba Transform Fault

West of Brittany, the South Armorican suture is interrupted by two N 40° deep faults, probably related to a single structural zone, which have been delineated by seismic reflection data. These faults can be followed to the south using a reconstruction with Biscay Bay closed.

In the north they separate the Galicia Banks from the Spanish mainland and in the south link with the Porto–Cordoba shear zone; the later can be followed southwards to Ossa Morena. This zone which is 600 km long played an important part in the pre-Mesozoic history of the Iberian Peninsula, since it marks the line of separation, during Lower and Middle Palaeozoic, of two faunal realms: the centro-Iberian domain is of Armorican (Brittanian) affinities, whereas the Southwest Iberian zone had an African character (Paris and Robardet 1977).

This main fault zone had a complex evolution (Lefort and Ribeiro 1980):

1. Early dextral movement has been inferred during Cambro-Ordovician times from the distribution of the en echelon folds of the Sardic deformational episode close to the lineament.
2. During Ordovician times an abundant alkaline and peralkaline volcanism was restricted to the area of the fault zone which suggests the one-time presence of a narrow intracontinental rift. This phase is regarded as a mainly tensional relaxation prior to subsequent sinistral movement.
3. During Upper Devonian and Carboniferous time, field evidence suggests that the Porto–Badajoz–Cordoba fault zone moved sinistrally. At the same time obduction of oceanic crust from the northwest is considered to have been responsible for widespread deformation in the upper part of the deep parautochthonous Palaeozoic nappes seen in northwestern Spain; the remnant klippes contain mafic and ultramafic rocks which are most likely of Ordovician or Silurian age.
4. The different tectonic movements along the Porto–Cordoba fault zone can be used to explain why the lineament is marked by a belt of blastomylonitic rocks up to 5 km wide.
5. Gravity data show that a pronounced horizontal gradient follows the fault. It would appear that deformation with the emplacement of basic rocks along this geological line, affected the whole thickness of the crust.

Because of this lengthly and complex evolution it is considered that the Porto–Badajoz–Cordoba lineament is probably a fossil intracontinental transform fault. The timing of the evolution of the fault zone is now very well known: it had a dextral component while the South Armorican Ocean was opening, whereas it had a sinistral component of movement when the South Armorican Ocean was closing (Lefort and Ribeiro 1980).

The geotectonic framework of the Ligerian orogeny

The proposed new positioning of the South Armorican suture and Porto-Badajoz transform fault allows us to delineate more precisely two of the boundaries of the 'Iberian' plate before it collided with 'Armorica' (Fig. 4). However, the original shape of the transform fault is difficult to reconstruct because it is now considered that the Hercynian collision tightened the Porto–Badajoz fault and the Ibero–Armorican arc (Lefort 1979; Perroud 1980). The palaeogeographical relationships between the Iberian plate and the surrounding continents are themselves poorly understood; the only clear stratigraphic and palaeontologic similarities suggest that the centro-Iberian zone and mid-Brittany were not far from one another during Ordovician time. This would indicate that the South Armorican Ocean was

merely a narrow sea at that time or was not yet open (Paris and Robardet 1977; Hamman and Henry 1978).

The South Armorican suture apparently represents a segment of a longer suture which can be followed, using geophysical data, between the Mediterranean coastline and Massachusetts (Lefort and Van der Voo 1981). Between the southern end of the Porto–Cordoba transform fault in Spain and Massachusetts the Laurussian suture crops out five times at the surface: in Aracena (Spain), Beja (Portugal), Cobequid mountains (Nova Scotia), Passamaquoddy Bay (Maine), and perhaps Cap Ann (Massachusetts). Evidence suggesting an Upper Palaeozoic age for the suture has been recognized in Spain, Portugal, and North America. In these regions volcanic rocks related to a previous oceanic distensional phase are usually of Ordovician or Silurian age and remnants of Upper Silurian or Devonian volcanic arcs, or miscellaneous effects of subduction, are known (Lefort 1983). Eastward the southern limit of Laurussia is less well documented and has been rendered much more speculative by Alpine rejuvenation; this is the reason why the location of the Austro-Alpine block (Fig. 4) is only tentative in our reconstruction.

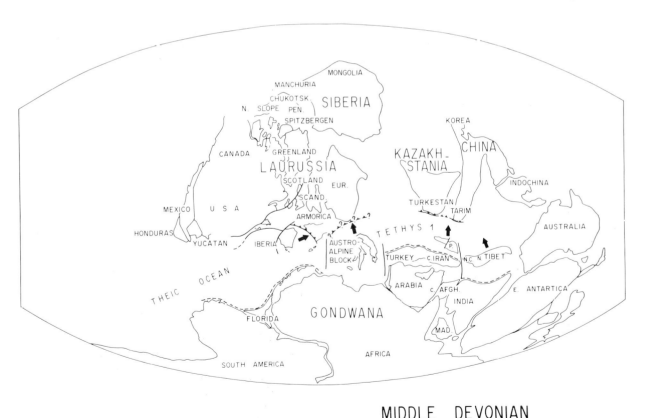

MIDDLE DEVONIAN

Fig. 4 Relationship between the Theic and Tethys I oceans, during Middle Devonian time (modified after Scotese *et al*. 1979; Mollweide projection). Dashed line: Approximate location of the northern boundary of Gondwanaland; black triangles: Devonian subduction zones; solid line: possible transform faults; large arrows: direction of microplate displacements

The framework of Scotese et al. (1979), on which our reconstruction is based, has been largely accepted and is only slightly modified for southeastern Europe and northeastern Gondwanaland. If this reconstruction is correct, during Middle Devonian time the connection between the Theic Ocean (located between Gondwana and America) and Tethys I (located between Gondwana and Asia) was situated south of Iberia, and the South Armorican ocean was possibly a gulf connected with a larger ocean located between southeastern Spain and Gondwanaland.

From the palaeomagnetic-data, 'Armorica' is considered to have been detached from Gondwanaland after the Late Cambrian and to have joined with Laurentia during Middle Palaeozoic time thus causing the Caledonian orogeny (Van der Voo et al. 1980; Lefort 1981). 'Iberia' which seems to have been a southern microplate satellite of 'Armorica', joined this composite plate later, in Late Devonian time (i.e. before the collision of Gondwana and Laurussia). Other microplates in the Tethys I ocean such as Paghman–Nurestan, Chitral, and Northern Thibet (Boulin 1981) behaved much the same. All of these microplates, because of the inferred northward dipping subduction zones, are regarded as having collided first with Laurussia, Kazakhstania, or China, causing restricted 'Caledonian' or 'Acadian' orogenies before the final closing with Gondwana.

Thus the pre-Variscan events known in southern Europe and southern Asia, from which Ligeria constitutes the westernmost extension, seem to have been completely independent of the Caledonian orogenesis and the closing of a Greenland–Scandinavian Iapetus Ocean. Although some structural and metamorphic events are contemporaneous with those recognized in the northern Caledonides, the Ligerian orogeny seems to have been related to the closing of a separate ocean, i.e. distinct from Iapetus. The term Caledonian must, therefore, be abandoned for southern Europe in any other than a chronological sense, since the Ligerian orogeny is more likely part of the beginning of the Variscan cycle and not an integral part of the Caledonian geodynamic framework.

References

Audren, C. and Le Metour, J. 1976. Mobilisation anatectique et déformation: les migmatites du golfe du Morbihan (Bretagne mériodinale). Bull. Soc. géol. Fr., (7) XVIII, 4, 1041–1049.

Audren, C. and Lefort, J. P. 1977. Géologie du plateau continental sud-armoricain entre lesîles de Glénan et de Noirmoutier. Implications géodynamiques. Bull. Soc. Géol. Fr., (7) 19, 395–404.

Autran, A. 1978. Synthèse provisoire des événements orogéniques calédoniens en France. In Caledonian Appalachian Orogen of the North Atlantic Regions, P.I.C.G. n°27, Pap. Geol. Surv. Canada, 78–13, 159–175.

Autran, A. and Cogne, J. 1980. La zone interne de l'orogène varisque dans l'Oust de la France et sa place dans le développement de la chaîne hercynienne. Congrès géologique international, Paris, Colloque C6, 90–111. Soc. géol. du Nord et B.R.G.M. édit.

Bayer, R. and Matte, Ph. 1979. Is the mafic/ultramafic massif of cabo Ortegal (Northwest Spain) a nappe emplaced during a Variscan obduction? A new gravity interpretation. Tectonophysics, 57, 9–18.

Bernard-Griffiths, J. 1976. Essai sur les âges au strontium dans une série métamorphique: le Bas-Limousin. Thèse, 243, Ann. Scient. Univ., Clermont, 55, 27.

Boulin, J. 1981. Afghanistan structure, greater India concept and Eastern Tethys evolution. Tectonophysics, 72, 261–287.

Briand, B. and Gay, M. 1978. La série inverse de Saint Genies d'Olt: évolution métamorphique et structurale. Bull. B.R.G.M., S¹, 3, 167–186.

Cogné, J. 1957. Schistes cristallins et granites en Bretagne méridionale: le domaine de l'Anticlinal de Cornouaille. Mém. expl. Carte géol. Fr., 1960, 382.

Cogné, J. 1962. Le Briovérien: esquisse des caractères stratigraphiques, métamorphiques, structuraux et paléogéographiques de l'Antécambrien récent dans le Massif armoricain. Bull. Soc. géol. Fr., (7) IV, 413–430.

Cogné, J. 1976. La chaîne hercynienne ouest-européenne correspond-elle à un orogène par collision? Propositions pour une interprétation géodynamique globale. Colloque Int. CNRS, Géologie de l'Himalaya, 268, 111–129.

Cogné, J. 1978. Vers un essai d'interprétation paléogéodynamique de l'orogène cadomien d'Europe occidentale: sa place dans l'évolution des régions médio et ouest-européennes sous-jacentes au domaine varisque. In PreCambrian in Younger Fold Belts, P.I.G.G. n. 22, Cluj Napoca 1978, Ann. Inst. Geol. Geophys. Bucarest, LVII, 119–121.

Cogné, J. and Vida, Ph. 1972. Résultats géochronologiques récents en Bretagne méridionale: signification de l'axe structural Moëlan–Lanvaux à l'Ordovicien. C.R. somm. Soc. Géol. Fr., 117.

Cogné, J. and Wright, A. E. 1980. L'orogène cadomien: vers un essai d'interprétation paléogéodynamique unitaire des phénomènes orogéniques fini-précambriens d'Europe moyenne et occidenale, et leur signification à l'origine de la croûte et du mobilisme varisque puis alpin. Congrès Géol. Int. Paris, Colloque C6, 29–55, Soc. Géol. Nord et B.R.G.M. edit.

Diot, H. and Blaise, J. 1978. Etude structurale dans le Précambrien et le Paléozoïque de la partie méridionale du domaine ligérien (S–E du Massif armoricain) (Mauges, synclinal d'Ancenis et sillon houiller de la Basse Loire). Bull. Soc. géol. min. Bret., C, X, 1, 31–50.

Gebauer, D., Bernard-Griffiths, J., Krebbs, O. and Grunenfelder, M. 1978. U–Pb systematics of zircons and monazite from a mafic complex and its country rocks (Sauviat, French Central Massif). U.S.G.S., Open File report, 78/701, 131–132.

Grolier, J. 1971. La tectonique du socle du Massif central. In Symposium J. Jung, 215–268 (Ed. Plain Air Service, Clermont-Ferrand).

Guillot, P. L. 1981. Etude géologique de la série métamorphique du Bas Limousin (Massif Central Français). Thèse Univ., Orléans.

Hamman, W. and Henry, J. L. 1978. Quelques espèces de Calymenella, Eohomalonotus et Kerfornella (Trilobita, Ptychopariida) de l'Ordovicien du Massif Armoricain et de la Péninsule Ibérique. *Senckenbergiana Lethaea*, **59**, 416, 401–429.

Lagarde, J. L. 1978. La déformation des roches dans les domaines de la schistosité subhorizontale. Applications à la nappe du Canigou-Roc de France (Pyrénées) et au complexe métamorphique de Champtoceaux (Massif armoricain). *Thèse 3ème cycle, Univ.*, Rennes.

Lameyre, J., Autran, A., Barriere, M., Didier, J., Fluck P., Giraud, P., Bonin, J. and Orsini, B. 1980. Les granitoïdes de France. In Autran, A. and Dercourt, J. (Eds), *Géologie de la France*, B.R.G.M.

Lefort, J. P. 1979. Iberian–Armorican arc and Hercynian orogeny in Western Europe. *Geology*, 7, 384–388.

Lefort, J. P. 1983. A new geophysical criterion to correlate the Acadian and Hercynian Orogenies of Western Europe and Eastern North America. In Hatcher, Williams, and Zietz (Eds), *Tectonics and Geophysics of Mountain Chains. Geol. Soc. Am. Mem.,* **158**, 3–18.

Lefort, J. P. and Peucat, J. J. 1974, Le socle antémésozoïque submergé à l'Ouest de la Baie d'Audierne, Finistère. *Comptes Rendus de l'Académie des Sciences*, Paris, 279, D, 635–637.

Lefort, J. P. and Segoufin, J. 1978. Etude géologique de quelques structures magnétiques reconnues dans le socle péri-armoricain. *Bull. Soc. géol. Fr.*, (7) XX, 2, 185–192.

Lefort, J. P. and Ribeiro, A. 1980. La faille Porto–Badajoz–Cordoue a-t-elle contrôlé l'évolution de l'océan paléozoïque sud-armoricain? *Bull. Soc. géol. Fr.*, (7) XXII, 3, 455–462.

Lefort, J. P., Audren, C., Jegouzo, P., Max, M. D., Grant, P. and Rattey, P. 1982a. Disposition of structures in the South Brittany (France). Belt of high pressure metamorphism. *The sixth int. Sc. Symp. of world, underwat., Fed.*, Edinburgh, 285–291.

Lefort, J. P., Audren, C. and Max, M. D. 1982b. The Southern Armorican orogenic belt: the result of a crustal shortening related to reactivation of a Devonian paleosuture during carboniferous time. *Tectonophysics*, 359–377.

Lefort, J. P. and Van der Voo, R. 1981. A kinematic model for the collision and complete suturing between Gondwanaland and Laurussia in the Carboniferous. *J. Geol.*, **89**, 5, 537–550.

Le Metour, J. and Bernard-Griffiths, J. 1980. Age limite Ordovicien–Silurien de la mise en place du massif hypovolcanique de Thouars (Vendée). Implications géologiques. *Bull. B.R.G.M.*, S1, 4.

Paris, F. and Robardet, M. 1977. Paléogéographie et relations ibéro-armoricaines au Paléozoïque antécarbonifère. *Bull Soc. géol. Fr.*, (7) XIV, 5, 1121–1126.

Perroud, H. 1980. Contribution à l'étude paléomagnétique de l'arc ibéro-armoricain. *Thèse 3ème cycle*, Rennes, 91.

Peucat, J. J. and Cogné, J. 1977. Geochronology of some bluechists from Ile de Groix (France). *Nature*, London, **268**, 131–132.

Peucat, J. J., Le Metour, J., and Audren, C. 1978. Arguments géochronologiques en faveur de l'existence d'une double ceinture métamorphique siluro-dévonienne en Bretagne méridionale. *Bull. Soc. géol. Fr.*, (7) XX, 2, 163–167.

Peucat, J. J., Charlot, R., Mifdal, A., Chantraine, J. and Autran, A. 1979. Définition géochronologique de la phase bretonne en Bretagne centrale: étude Rb/Sr de granites du domaine centre-armoricain. *Bull. B.R.G.M.*, S1, 4, 349–356.

Pin, C. 1979. Géochronologie U–Pb et microtectonique des séries métamorphiques de la région de Marvejols. *Thèse 3ème cycle, Univ.*, Montpellier, 210.

Pin, C. and Lancelot, J. 1978. U–Pb evidences of bimodal magmatism of early paleozoic age in the Massif Central: the leptyno-amphibolitic group of Marvejols. *U.S.G.S. Open file report*, **78–101**, 337.

Pruvost, P. 1949. Les mers et les terres de Bretagne aux temps paleozoïques. *Annales Hebert et Haug*, **VII**, 345–360.

Quinquis, J. 1980. Schistes bleus et déformation progressive: l'exemple de l'île de Groix (Massif Armoricain). *Thèse 3ème cycle, Univ.*, Rennes, 145.

Quinquis, H. and Choukroune, P. 1981. Les schistes bleus del l'Ile de Groix dans la chaine hercynienne: contraintes cinématiques. *Bull Soc. géol. Fr.*, **4**, 409–418.

Scotese, C. R., Bambach, R. K., Barton, C., Van der Voo, R. and Ziegler, A. 1979. Paleozoic base maps. *J. Geol.*, **87**, 3, 217–277.

Van der Voo, R., Briden, J. C. and Duff, B. A. 1980. Late Precambrian and Paleozoic paleomagnetism of the Atlantic-bordering continent. 26e C.G.I., *Coll. C6, Géologie de l'Europe*, 201–212.

Vidal, Ph. 1976. L'évolution polyorogénique du Massif armoricain: apport de la géochronologie et de la géochimie du strontium. *Thèse, Univ.*, Rennes, 140.

The Caledonide Orogen—Scandinavia and Related Areas
Edited by D. G. Gee and B. A. Sturt
© 1985 John Wiley & Sons Ltd

The Caledonian events in Sardinia

S. Barca*, L. Carmignani†, T. Cocozza‡, M. Franceschelli†, C. Ghezzo‡, I. Memmi‡, N. Minzoni§, P. C. Pertusati† and C. A. Ricci‡

*Dipartimento di Scienze della Terra, Università di Cagliari, Italy
†Dipartimento di Scienze della Terra, Università di Pisa, Italy
‡Dipartimento di Scienze della Terra, Università di Siena, Italy
§Istituto di Mineralogia e Petrografia, Università di Ferrara, Italy

ABSTRACT

The main features of the metamorphic basement of Sardinia have been produced during the Hercynian orogeny, which developed in an ensialic environment involving Cambrian or even older pre-existing continental crust and its Lower Ordovician–Lower Tournaisian volcano-sedimentary cover.

Evidence of Caledonian events is scarce and essentially preserved in the cover sequence:

1. The Middle Cambrian–Lower Ordovician unconformity of southern Sardinia (Sardic phase);
2. The products of post-orogenic, acidic infracrustal magmatism of Ordovician age (meta-rhyolites and meta-rhyodacites of pre-Caradocian age; orthogneiss of radiometric age 430–460 Ma by Rb/Sr whole rock method).

No metamorphic effect of clear Caledonian age has been documented up to now, but one cannot exclude a Caledonian age for the rare fragments of a pre-Hercynian metamorphic basement, previously attributed to the Assintic orogeny.

These elements, which are common features of the Hercynian massives of southern Europe, suggest that a 'Caledonian' (in age) orogenic belt existed in this area, even if it was possibly independent of the Caledonian orogenic belt of northern Europe.

Introduction

In the classical view, the Mediterranean areas are placed far from the Caledonian orogenic belt, but in southern Europe there is evidence to suggest the existence of a relationship with the Caledonide orogen (cf. Zwart and Dornsiepen 1978, 1980; Bourrouilh *et al.* 1980).

In this paper we present a review of the geological evidence that we have collected in Sardinia. This island shows the same evidence as that in the other Mediterranean areas; it was only marginally affected by the Alpine orogeny (Alvarez and Cocozza 1973) and therefore it is a favourable region to evaluate the effects of the Caledonian orogeny in southern Europe.

The Hercynian Orogeny

Before presenting the evidence for Caledonian events in Sardinia, it is necessary to outline briefly the effects of the Hercynian orogeny that have strongly overprinted the earlier events and now dominate the geology of the island (Fig. 1). In fact, the geological features of the basement must be attributed to the Hercynian orogeny which had a polyphase history. It began with large-scale crustal shortening along an ensialic shear zone, without involvement of oceanic crust, coupled with regional metamorphic recrystallization; it continued with general uplift, accompanied and followed by intrusion of widespread plutonic masses (Carmignani *et al.* 1981; Di Simplicio *et al.* 1975; Ferrara *et al.* 1978).

Reprint requests to: Prof. C. A. Ricci, Dipartimento di Scienze della Terra, Via delle Cerchia, 3, 53100 Siena (Italy)

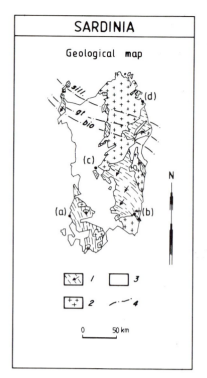

Fig. 1 The geology of Sardinia. Key: 1—Trend of fold axes and overturning direction of the major Hercynian tectonic phase. 2—Hercynian granitoids. 3—Post-Hercynian volcano-sedimentary cover. 4—Hercynian metamorphic isograds (chl = chorite; bio = biotite; gt = garnet; stau = staurolite; sill = sillimanite)

The axes of the first tectonometamorphic Hercynian phase show a variable strike from NW–SE to N–S (Carmignani *et al*. 1978). From southwest to northeast the structural style varies concomitantly with the intensity of the metamorphism. In the southwest the folds are open and tectonic transport is slight, as is the grade of the metamorphic recrystallization (Carmignani *et al*. 1980). Moving northeastwards, the folds become tighter till they are isoclinal. Concomitantly, the amount of tectonic transport increases up to several tens of kilometers and the generally syntectonic metamorphism passes from low greenschist facies up to amphibolite facies with migmatites (Carmignani *et al*. 1980; Di Simplicio *et al*. 1975).

The age of the Hercynian orogeny must fall between the Westphalian–Stefanian (age of the first post-orogenetic lacustrine basins) and the Lower Tournaisian (age of the youngest tectonized fossil-bearing strata) (Vai and Cocozza 1974). This dating agrees with the radiometric data, which place the peak of the metamorphism in the northeast at 344 ± 7 Ma and the age of the basement uplift at about 300 to 310 Ma (emplacement age of the granitoids and cooling age of the metamorphic minerals) (Del Moro *et al*. 1975; Ferrara *et al*. 1978).

Pre-Hercynian History

Basement sequences

Evidence of pre-Hercynian events is of different kind and variable certainty and concerns both the basement and the cover sequences (Cocozza 1969) as shown in Fig. 2.

In the southwestern area (Fig. 2a), there is a clear unconformity between the Middle–Upper (?) Cambrian and the Lower Ordovician rocks (Arenig?). This marks the well known tectonic movements of the 'Sardinian phase' of the Caledonian orogeny of Stille (Cocozza 1979). The Cambrian rocks were folded by this tectonic phase, but they do not show evidence of synchronous development, neither of penetrative deformations nor of metamorphism.

The character of this tectonic phase, which also occurs in Spain, is still largely unsolved (Cocozza 1969; Dunnet 1969; Poll and Zwart 1964; Poll 1966; Arthaud 1970; Tempier 1978), both because the folds, striking E–W, were probably gentle and because they were refolded by two intense E–W and N–S Hercynian tectonometamorphic phases.

An unconformity between Cambrian and Upper Ordovician in southeastern Sardinia was pointed out by Calvino (1961), and recently confirmed by Naud (1979; 1981). In the whole of central Sardinia (lower Flumendosa Valley, Mandas, Asuni, and Orani) below the Ordovician–Devonian cover (generally affected by chlorite grade metamorphism), there are sequences of slightly higher metamorphic grade (up to biotite grade) which have unclear relationships with the cover. Some authors (for instance Naud) consider these sequences as Cambrian, making up a Caledonian basement; they have, therefore, interpreted the contact (see Fig. 2b) between them as an unconformity. Others, on the basis of the lithological analogy of these lower sequences with the Siluro–Devonian series, interpret these sequences as part of a deeper Hercynian nappe and, therefore, consider this contact to be a thrust.

In a similar structural position, in the Grighini mountains (see Fig. 2c), below the Ordovician–Devonian cover (affected by a chlorite grade Hercynian metamorphism) a higher grade complex outcrops, consisting of garnet and staurolite bearing micaschists and orthogneisses of pre-Hercynian age. The age of this basement is unknown (Carmignani *et al*. 1981). Although it was considered in the past as pre-Cambrian, some of us do not exclude the possibility that it might be considered as Caledonian.

The northeastern area (see Fig. 2d) is affected by a prograde metamorphism of Hercynian age increasing northwards from biotite to sillimanite plus muscovite and/or K-feldspar grade (Ricci 1972; Di Simplicio *et al*. 1975).

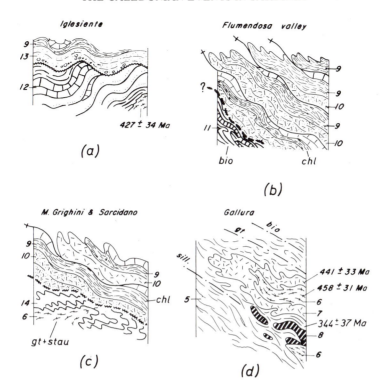

Fig. 2   Evidence of Caledonian orogenesis. Locations of (a), (b), (c) and (d) are shown in Fig. 1. Key: 5—Micaschists and migmatites. 6—Orthogneisses. 7—Augen gneisses. 8—Granulitic and/or eclogitic rocks. 9—Post-Caradocian metasediments. 10—Pre-Caradocian metasediments and metavolcanics. 11—Phyllites and marbles (low Flumendosa valley, Mandas, Asuni). 12—Cambrian and PreCambrian (?) of Iglesiente–Sulcis. 13—'Puddinga': transgressive conglomerate on the Cambrian sequence of Iglesiente. 14—Micaschists

In the migmatites of northern Sardinia, some blocks and masses of metabasic rocks occur, containing relics of a previous intermediate pressure granulite and/or eclogite facies metamorphism (Miller *et al.* 1976; Ghezzo *et al.* 1979). The age of this dry metamorphism is still an open question. In southern Europe the supposed age of the granulitic rocks involved in the Hercynian chain varies from pre-Cambrian to eo-Hercynian even if most of the radiometric dating points to a Caledonian age (Zwart and Dornsiepen 1978; Jäger 1977; Hunziker and Zingg 1980). Our evidence indicates that it pre-dates the widespread amphibolite facies metamorphism of the northeastern area, which developed during the second Hercynian tectonometamorphic phase. According to Prof. Sturt's field observations (personal communication), the eclogites of northern Sardinia appear to be intruded by orthogneisses similar to those we dated around 460 Ma, and therefore the eclogite facies metamorphism should pre-date the Ordovician acidic magmatism.

Cover sequences

With regard to the cover sequences we note that they are characterized by the abundance of meta-igneous acidic rocks. In central and southeast Sardinia (see Fig. 2b and 2c), they consist of a thick unit (up to 1000 m) of meta-rhyolitic and meta-rhyodacitic lava flows, lava domes, ignimbrites and sediments derived from these volcanic rocks (Di Simplicio *et al.* 1975). These rocks always lie just below the Caradocian and in southern Sardinia they are post-Late Cambrian in age (Barca *et al.* 1982). On chronological and petrological grounds the Capo Spartivento orthogneiss (427 ± 34 Ma) (Cocozza *et al.* 1977; Scharbert 1978), the granodioritic orthogneiss (458 ± 31 Ma) and the augen-gneiss (meta-rhyolites) (441 ± 33 Ma) of northeast Sardinia (Ferrara *et al.* 1978) must also be related to this igneous event (see Fig. 2a and 2d).

The petrologic affinity of this igneous activity is clearly calc-alkaline. We have never found rocks of possible per-alkaline affinity. The initial strontium ratios range from 0.708 to 0.712. These values are typical of melts produced in the continental crust (Cocozza *et al.* 1977; Scharbert 1978; Di Simplicio *et al.* 1975; Ferrara *et al.* 1978).

The acidic igneous activity was followed by Upper Ordovician–Lower Silurian basic magmatism. The geochemical data now available are scarce (Ricci and Sabatini 1978; Beccaluva *et al.* 1981), but they suggest that most of this volcanism was represented by within-plate alkali-basalts and was probably connected with crustal thinning and the formation of continental rifts and narrow basins.

Concluding Remarks

The most relevant data regarding the Caledonian events in Sardinia are: (1) The unconformity (without clear evidence of synchronous penetrative deformation and metamorphism) between Middle–Upper (?) Cambrian and Ordovician rocks in southern Sardinia; (2) The evidence of a pre-Hercynian basement, which may be Caledonian; (3) The widespread calc-alkaline acidic magmatism of Ordovician age.

The Upper Ordovician–Lower Silurian basic volcanism, taking into account its petrologic affinity, was clearly independent to any orogenic compressive event and can be disregarded in this context.

These three elements are all common features of most of the crystalline basement of southern Europe. Among these elements, the most relevant in suggesting a Caledonian age of orogeny in southern Europe, is the Ordovician acidic magmatism. Three things regarding this magmatism appear, in fact, well established: (1) It was widespread and voluminous; (2) It is calc-alkaline in affinity; (3) It originated in the continental crust.

All these characters point to this magmatism being of late-orogenetic type affinity. But, is a late-orogenetic affinity of a magmatism sufficient to establish an orogenic event? Probably not, on its own!

The Upper Ordovician unconformity produced by the Sardic phase could also be inferred to favour an orogenic event. But probably not on its own!

However, the two pieces of evidence could support each other to suggest an orogeny.

Evidently, it would be necessary to find more consistent evidence, for instance radiometric evidence of a Caledonian age of metamorphism. The recent discovery in Sardinia of relics of granulitic and/or eclogitic rocks and the evidence of a pre-Hercynian metamorphic basement are all elements which suggest to us that we must continue our researches on this island for the effects of Caledonian orogenesis in southern Europe, especially by radiometric dating.

In conclusion, we recognize Caledonian movements and chronologically related acidic late-orogenic magmatism, but we are unable to recognize a Caledonian orogenic belt in southern Europe and, therefore, the relationship with the Caledonian orogenic belt *sensu stricto* of northern Europe remains for us quite obscure.

Possibly there are two or more Caledonian orogenic belts in Europe, maybe of different ages.

References

Alvarez, W. and Cocozza, T. 1973. The tectonics of central eastern Sardinia and the possible continuation of the Alpine chain to the south of Corsica. In *Palaeogeografia del Terziario sardo nell'ambito del Mediterraneo occidentale. Suppl. Rend. Sem. Fac. Sci. Univ. Cagliari.,* 1–34.

Arthaud, F. 1970. Etude tectonique et microtectonique comparée de deux domaines hercyniens: les nappes de la Montagne Noire (France) et l'anticlinorium de l'Iglesiente (Sardaigne). *Publ. Ustela, Série Géol. Struct.,* **1**, 175 pp.

Barca, S. and Di Gregorio F. 1980. La successionne ordoviciano–siluriana inferiore nel Sarrabus (Sardegna sud-orientale). *Mem. Soc. Geol. It.,* **20**, 189–202.

Barca, S., Del Rio, M. and Pittau Demelia, P. in press. Acritarchs in the 'Arenarie di S. Vito' of south-east Sardinia: stratigraphic and geological implications. *Boll. Soc. Geol. It.*

Barca, S., Carmignani, L., Cocozza, T., Ghezzo, C., Minzoni, N., Pertusati, P. C. and Ricci, C. A. 1981. Geotraverse B in Sardinia: stratigraphic correlation forms. In Karamata, S. and Sassi, F. P. (Eds), *IGCP N.5 Newsletter,* **3**, 7–21.

Barca, S., Cocozza, T., Del Rio, M. and Pittau Demelia, P. in press. Discovery of Lower Ordovician Acritarchs in the 'Postgotlandiano' sequence of southwestern Sardinia (Italy): age and tectonic implications. *Boll. Soc. Geol. It.*

Beccaluva. L., Leone, F., Maccioni, L. and Macciotta, G. 1981. Petrology and tectonic setting of the Palaeozoic basic rocks from Iglesiente–Sulcis (Sardinia, Italy). *N. Jb. Miner. Abh.,* **149** (2), 184–201.

Bourrouilh, R., Cocozza, T., Demange, M., Durand-Delga, M., Gueirard, S., Guitard, G., Julivert, M., Martinez, F. J., Massa, D., Mirouse, R. and Orsini, J. B. 1980. Essai sur l'évolution paléogéographique, structurale et métamorphique du Paléozoïque du Sud de la France et de l'Ouest de le Méditerranée. *26ᵉ Congrès Géol. Intern.,* Paris 1980, 159–188.

Calvino, F. 1961. Lineamenti strutturali del Sarrabus–Gerei. *Boll. Serv. Geol. d'It.,* **81** (1959), 489–556.

Carannante, G., Cocozza, T. and D'Argenio, B. in press. Late Precambrian–Cambrian geodynamic setting and tectono-sedimentary evolution of Sardinia (Italy). *AAPG.*

Carmignani, L., Cocozza, T., Minzoni, N. and Pertusati, P. C. 1978. The Hercynian Orogenic revolution in Sardinia. *Z. dt. geol. Ges.,* **129**, 485–493.

Carmignani, L., Cocozza, T., Minzoni, N. and Pertusati, P. C. 1980. Explanatory notes of cross sections through Sardinia Hercynian range. In Sassi, F. P. (Ed.), *IGCPN. 5 Newsletter,* **2**, 93–96.

Carmignani, L., Cocozza, T., Minzoni, N. and Pertusati, P. C. 1981. Structural and palaeogeographic lineaments of the Variscan cycle in Sardinia. *Geol. en Mijnbouw,* **60**, 171–181.

Cocozza, T. 1969. Slumping e brecce intraformazionali nel Cambrico medio della Sardegna. *Boll. Soc. Geol. It.* **88**, 71–80.

Cocozza, T. 1979. The Cambrian of Sardinia. *Mem. Soc. Geol. It.*, **20**, 163–187.

Cocozza, T. and Jacobacci, A. 1975. Geological outline of Sardinia. In Squires, C. (Ed.), Geology of Italy. *The Earth Sc. Soc. Libyan Arab Republic*, 49–81.

Cocozza, T., Conti, L., Cozzupoli, D., Lombardi, G., Scharbert, S. and Traverse, G. 1977. Rb/Sr age and geo-petrologic evolution of crystalline rocks in the Southern Sulcis (Sardinia). *N. Jb. Geol. Paläont. Mh.*, 1977, **H. 2**, 95–102.

Debrenne, F. and Naud, G. 1981. Méduses et traces fossiles supposées précambriennes dans la formation de San Vito, Sarrabus, Sud-Est de la Sardaigne. *Bull. Soc. Géol. France*, **23** (1), 23–31.

Del Moro, A., Di Simplicio, P., Ghezzo, C., Guasparri, G., Rita, F. and Sabatini, G. 1975. Radiometric data and intrusive sequence in the Sardinian Batholith. *N. Jb. Miner. Abh.*, **126**, 28–44.

Di Simplicio, P., Ferrara, G., Ghezzo, C. Guasparri, G., Pellizzer, R., Ricci, C. A., Rita, F. and Sabatini, G. 1975. Il metamorfismo e il magmatismo paleozoico in Sardegna. *Rend. Soc. It. Miner. Petr.*, **30**, 979–1068.

Dunnet, D. 1969. *Deformation in the Palaeozoic rocks of Iglesiente, S.W. Sardinia*. Ph.D. Thesis, University of London n. 29163, 412 pp.

Ferrara, P., Ricci, C. A., and Rita, F. 1978. Isotopic ages and tectono-metamorphic history of the metamorphic basement of North-Eastern Sardinia. *Contrib. Miner. Petr.*, **68**, 99–106.

Ghezzo, C., Memmi, I. and Ricci, C. A. 1979. Un evento granulitico nel· basamento metamorfico della Sardegna nord-orientale. *Mem. Soc. Geol. It.*, **20**, 23–38.

Hunziker, J. C. and Zingg, A. 1980. Lower Palaeozoic amphibolite granulite facies metamorphism in the Ivrea zone (Southern Alps, Northern Italy). *Schweiz. mineral. petrogr. Mitt.*, **60**, 181–213.

Jäger, E. 1977. The evolution of central and West European continent. In La chaine varisque d'Europe moyenne et occidentale. *Coll. Inter. Rennes*, 25 Sept. 6 Oct., 1974. Ed. CNRS, 227–239.

Miller, C., Sassi, F. P. and Armari, G. 1976. On the occurrence of altered eclogitic rocks in north-western Sardinia and their implication. *N. Jb. Geol. Paläont. Mh.*, **11**, 683–689.

Naud, G. 1979. Tentative de syntèse sur l'évolution geodynamique de la Sardaigne antepermienne. *Mem. Soc. Geol. It.*, **20**, 85–96.

Naud, G. 1981. Confirmation de l'existence de la discordance angulaire anté-ordovicienne dans le Sarrabus (Sardaigne sub-orientale): conséquences géodinamiques. *C.R. Acad. Sc. Paris*, **292**, 1153–1156.

Poll, J. J. K. and Zwart, H. J. 1964. On the tectonics of the Sulcis area, S. Sardinia. *Geol. en Mijnbouw*, **43**, (4), 144–146.

Poll, J. J. K. 1966. The geology of the Rosas–Terraseo area, Sulcis, South Sardinia. *Over. U. Leidse Geol. Maded.*, **35**, 197–208.

Rasetti, F. 1972. Cambrian Trilobite faunas of Sardinia. *Mem. Acc. Naz. Lincei*, **11**, 100 pp.

Ricci, C. A. 1972. Geo-petrological features of the Sardinian crystalline basement. The metamorphic formations. *Min. Petr. Acta*, **18**, 235–244.

Ricci, C. A. and Sabatini, G. 1978. Petrogenetic affinity and geodynamic significance of metabasic rocks from Sardinia, Corsica and Provence. *N. Jb. Mineral. Mh.*, **1**, 23–38.

Scharbert, S. 1978. Supplementary remarks on 'Rb/Sr age and geopetrologic evolution of crystalline rocks in southern Sulcis (Sardinia)' by T. Cocozza *et al.* (1977). *N. Jb. Geol. Paläont. Mh.*, 1978, **1**, 59–64.

Schneider, H.-H. 1974. Revision of the Lower Palaeozoic in Sardinia especially of the Sardinian conglomerate. *N. Jb. Geol. Paläont. Abh.*, **146**, 78–103.

Tempier, C. 1978. Les événements Calédoniens dans les massifs varisques du Sud-Est de la France, Corse et Sardaigne. In *Caledonian–Appalachian orogen of the north Atlantic Region. Geol. Surv. of Canada*, **78–13**, 177–181.

Vai, G. B. and Cocozza, T. 1974. Il 'Postgotlandiano' sardo, Unità sinorogenetica ercinica. *Boll. Soc. Geol. It.*, **92**, 61–72.

Vardabasso, S. 1956. La fase sarda dell'orogenesi caledonica in Sardegna. *Geotektonisches Symp. z. Ehren von H. Stille, Deutsch. Geol. Gesell., Stuttgart*, 120–127.

Vardabasso, S. 1961. Considerazioni paleogeografiche sul Cambrico dell'Isola di Sardegna. *Atti XX Congr. Geol. Intern., Ciudad de Mexico*, 86–90.

Zwart, H. J. and Dornsiepen, V. F. 1978. The tectonic framework of Central and Western Europe. *Geol en Mijnbouw*, **57** (4), 627–654.

Zwart, H. J. and Dornsiepen, V. F. 1980. The Variscan and pre-Variscan tectonic evolution of Central and Western Europe: a tentative model. *26ᵉ Congrès Géol. Intern., Paris 1980*, 227–232.

The Caledonide Orogen—Scandinavia and Related Areas
Edited by D. G. Gee and B. A. Sturt
© 1985 John Wiley & Sons Ltd

The Caledonides of Kazakhstan, Siberia, and Mongolia: a review of structure, development history, and palaeotectonic environments

A. A. Mossakovsky and **A. B. Dergunov**

Geological Institute of the USSR Academy of Sciences, Pyzhevsky per. 7, SU-109017 Moscow, USSR

ABSTRACT

The Central Asian Caledonides were formed at the expense of the Kazakhstan–Siberian oceanic basin, which was a northern branch of the Central Asian Palaeozoic Ocean. The closure of this basin commenced at its margins which are occupied by the early Caledonides. Inwards follows the late Caledonides and a central Variscan zone subdividing the Caledonides into two separate regions. The early Caledonides comprise miogeosynclinal and eugeosynclinal parts. In the Vendian and early Cambrian, both continental margins were subject to a tensional regime which caused the formation of a peculiar mosaic of volcanic rift troughs and intervening sedimentary terrains. In the east, folding commenced at the boundary between the middle and late Cambrian, in the west in the late Ordovician. Meanwhile, abyssal conditions prevailed in the rest of the oceanic basin. The rock piles of most of the late Caledonides therefore contain lower sections of Vendian, Cambrian, and, locally, early Ordovician tholeiites and abyssal siliceous sediments deposited on an oceanic substratum. Flyschoids follow upwards. The two continental margins developed differently. In the west, a wedge of terrigenous sediments covered a passive continental slope and rise. In the east, marginal island arcs developed along an east-dipping subduction zone. Island arc volcanism is Vendian in the southern and Vendian to early Ordovician in the northern arcs. By the middle Cambrian, a west-dipping Benioff zone and an oceanic island arc were formed, along an uplifted block of ocean floor in the present Chingiz–Tarbagatai Zone. This island arc subdivided the oceanic basin into western back-arc and eastern relict oceanic subbasins. The rise of a late Cambrian to early Ordovician mountain belt along the eastern continental margin facilitated the influx of detritus into the eastern subbasin. Major folding and thrusting occurred in the late Ordovician. Subsequently, eugeosynclinal troughs on a folded substratum and relict oceanic seas arose temporarily in the former eastern subbasin of the now disappearing ocean. After the late Ordovician, only the western back-arc subbasin persisted as a major sea. A new E-dipping subduction zone was formed along its northeastern margin and turbidite sedimentation continued until the Devonian. The rest of the Caledonian parts of the Kazakhstan–Siberian Foldbelt were developed finally by intense folding and thrusting at the boundary between the early and late Silurian and at the end of the Silurian. Devonian and later molasse was deposited in depressions within the eroding folded belt. The late stages of Caledonian deformation were accompanied by renewed, Variscan, rift opening of a new oceanic basin between Kazakhstan and Siberia.

Introduction

The present paper reviews the structure and palaeotectonic evolution of the Caledonides in Kazakhstan, Siberia, and Mongolia. Naturally, it utilizes the results of numerous previous investigators.

Geological study in the Caledonides of Central Asia commenced at the end of the 19th century. However, fundamental regional work was carried out mainly during the nineteen-forties, fifties, and sixties. It employed the principles of classical geosynclinal theory and contributed to the development of the historical–geological concept for the development of heterochronous and heterogeneous fold belts. Results obtained by A. A. Bogdanov, V. A. Kuznetsov, N. A. Shtreis, V. N. Nekhoroshev, R. A. Borukaev, V. A. Unksov, N. S. Zaitsev, and others were incorporated in several issues of the Tectonic Map of the USSR, the Tectonic Map of Eurasia (1966), and maps of the various Central Asian regions.

From the end of the sixties onwards, the previous approach underwent gradual revision. This reassessment among Soviet geologists was greatly influenced by A. V. Peive and his school at the Geological Institute of the USSR Academy of Sciences in Moscow (GIN AN USSR). It was stimulated by mobilistic tectonic ideas, results obtained from the study of ophiolite complexes and new conclusions emphasizing the importance of oceanic crust in the development of continents and particularly eugeosynclinal belts.

Because of the similarity of ophiolite sequences and the lower parts of eugeosynclinal rock piles to the crust of recent oceans, island arcs, and marginal seas, investigators such as L. P. Zonenshain, A. S. Perfiliev, V. V. Volkov, A. A. Mossakovsky, A. B. Dergunov, N. N., Kheraskov, S. G. Samygin, and R. M. Antonyuk concluded that the eugeosynclinal folded zones of the Central Asian Caledonides incorporated remains of former relict ancient and newly rift-generated oceanic basins. In Kazakhstan and the Altai–Sayan region, they identified zones of spreading (Antonyuk 1974; Zonenshain 1974) and subduction (Mossakovsky 1972) as well as previous oceanic sutures marked by extensive contraction and attendant crushing during the various tectonic stages of the early and middle Palaeozoic.

Peive et al. (1972, 1976, 1980) employed the Caledonides of Siberia and Central Asia as an example to demonstrate the evolution of eugeosynclinal belts. In their model, oceanic crust is transformed into continental crust by a protracted and complex process. The three stages of this process are named the oceanic, the transitional, and the continental. The oceanic stage is characterized by oceanic crust and palaeogeographical environments similar to those of recent oceans. The transitional stage is marked by a transitional type of crust with local, insular patches of a granitic–metamorphic crustal layer in an island arc–marginal sea palaeoenvironment. Finally, during the continental stage there exists a coherent granitic–metamorphic layer within a crust of continental type. This concept was developed further by work at GIN AN USSR. It provides the basis for the recent publication 'Tectonics of Northern Eurasia' (Peive et al. 1980) and the new Tectonic Map of Northern Eurasia (1980). This map indicates the periods of formation of conti-

nental crust and the times of development of the granitic–metamorphic crustal layer in the various orogens, among them the Caledonides of Kazakhstan, Siberia, and Central Asia.

As the fundamental tectonic concepts and the tectonic development of the Caledonides were revised during the last two decades, new syntheses were made also in regard to magmatism, metamorphism, and geochronology. Among the results of this work are numerous new general and specialized geological maps of the USSR as a whole and of the various Caledonide regions. These maps cover among other subjects palaeotectonics, palaeogeography, sedimentary facies, magmatism, the configuration of metamorphic belts, and the distribution of metamorphic facies.

Configuration and Structure of the Caledonides in Kazakhstan, Siberia, and Mongolia

General aspects

Location

The Central Asian Caledonides form part of the vast Kazakhstan–Siberian Palaeozoic Foldbelt, which occupies the core of the Eurasian supercontinent (Fig. 1). In the west, this belt is delimited by the East European Craton and the adjacent Variscan Urals–Tien Shan Foldbelt. In the east, it borders on the Siberian–Baikalian Craton and its Cadomian

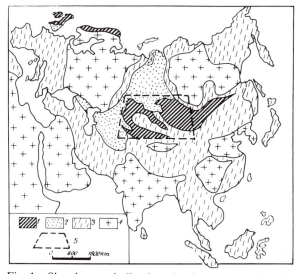

Fig. 1   Sketch-map indicating the location of the Central Asian Caledonides. Legend; 1–Caledonides; 2—Mesozoic to Quaternary cover of the Epipalaeozoic West Siberian Platform; 3—Pre- and Post-Caledonian orogens undivided (Cadomian, Variscan, Mesozoic, and Alpine); 4—Ancient cratons; 5—boundaries of the considered region. (cf. Fig. 2)

fringe. The northern part of the Kazakh-stan–Siberian Foldbelt is overlain by the Mesozoic to Quaternary cover of the West Siberian Platform. In the south, Caledonian structures in the Chinese and Mongolian PR are cut by the Variscan Foldbelt of the Southern and Gobi Tien Shans. This belt trends approximately east–west.

Within the Palaeozoic Kazakhstan–Siberian belt, the Caledonides form two large massifs separated by the NW-striking Irtysh–Zaysan Variscan Zone (Figs 1 and 2). Towards the south, this zone passes into the South Mongolian Variscides. The western Caledonide massif occupies Central Kazakhstan (Kokchetav, Ulutau, Yerementau, Bet-Pak-Dala, and the Chu-Ili Mountains) and Northern Tien Shan (Fig. 2). The larger, eastern Caledonide massif comprises the southwestern folded margin of the Siberian Platform (Eastern Sayan, Khamardaban, Prikosogol, Sangilen, Northern and Central Mongolia) and the Altai–Sayan region (Western Sayan, the Kuznetsk Alatau, Gornaya Shoria, Salair, the Mongolian Altai, and adjacent parts of the basement of the West Siberian Platform).

General characteristics

The Caledonides of Central Asia are composed of continuous upper Proterozoic (upper Riphean to Vendian) and lower Palaeozoic eu- and miogeo-synclinal sequences made up of ophiolites, volcanic, sedimentary–volcanogenic complexes, and siliceous, terrigenous, and calcareous sediments reaching thick-nesses in excess of 10 km. This rock pile is completely and unevenly folded and faulted. It exhibits zonal metamorphism ranging from greenschist to amphibolite facies and has been intruded by large granitoid massifs of various types and ages (Leontiev et al. 1981; Pavlenko et al. 1974; Lepezin 1978).

A feature distinguishing the Caledonides from the adjacent Variscan regions is the common occurrence of lower to middle Devonian, red, terrestrial molasse, and mixed units of volcanics and molasse of 'Old Red' type. Molasse fills numerous depressions and troughs, e.g. the Minusinsk Basin (signature 10, Fig. 2). It rests unconformably on a lower Palaeozoic basement and is traversed by germano-type faults. In the early Palaeozoic, geological development was remarkably similar in the Caledo-nian and some of the Variscan regions (e.g. the Lake Balkhash province) of the Kazakhstan–Siberian Foldbelt. Many scientists therefore conclude that the Central Asian Caledonides and part of the Varis-cides had a common early history in a large, late Precambrian–Early Palaeozoic, synclinal basin (Mar-kova 1964; Tikhomirov 1974; Peive et al. 1976; Kheraskova 1979). Later differences must have

arisen as a result of the protracted, multistage closing of this basin during the Palaeozoic. This evolutionary model is supported by the clear zonal arrangement of the early and late Caledonian and Variscan folded zones traditionally distinguished in Kazakhstan and Siberia.

The early Caledonian folded zones occupy the peripheral parts of the Kazakhstan–Siberian Fold-belt. In the west, they comprise Kokchetav, Ulutau, and Northern Tien Shan; in the east, they are found in the Kuznetsk Alatau, Eastern Sayan, Eastern Tuva, and Mongolia. The late Caledonides form two zones farther inside the foldbelt in Gorny Altai, the Mongolian Altai, Western Sayan, Chingiz, and Tar-bagatay (Fig. 2). The Variscides, finally, occupy the innermost, axial part of the belt.

As has been pointed out by many investigators (e.g. Kuznetsov 1954; Bogdanov 1958), the concen-tric closing of the Kazakhstan–Siberian Foldbelt is, in fact, its most striking characteristic. It identifies the Central Asian Caledonides as an object of singu-lar interest to Caledonide studies in general. In this region, both margins of the Caledonian oceanic basin and the entire axial zone of the orogen are available for direct examination. In the Caledonides of North America, Greenland, Africa, and Scan-dinavia, in contrast, the axial zone is virtually mis-sing, whilst it is extremely narrow in the British Isles.

Early Caledonides

The eastern early Caledonides

The early Caledonian folded zone in the easten part of the Altai–Sayan region and Mongolia (the so-called Salairides) features a continuous sequence of Vendian to lower or middle Cambrian deposits which was deformed at the boundary between the middle and late Cambrian. This Vendian to middle Cambrian sequence is covered discordantly by upper Cambrian and Ordovician marine lower molasse.

The eugeosynclinal nature of many of the early Caledonide foldbelts is most obvious in the Ozer-naya Zone ('The Lake Zone') of western Mongolia (signature 1, Fig. 2). The Vendian to lower Cam-brian succession of this area is almost wholly made up of the spilite–diabase ('ocean-floor tholeiitic' in western terminology) and abyssal siliceous–clayey associations, which occur together with ophiolites and serpentinitic mélanges (Fig. 3). In contrast, in the Kuznetsk Alatau, Gornaya Shoria, Eastern Sayan, and Tuva, that is farther to the northeast (Fig. 2), in the eastern part of the Altai–Sayan re-gion, only a few zones exhibit typically eugeosynclinal characteristics, whereas considerable areas were dominated by clearly miogeosynclinal petrogenetic

regimes (Mossakovsky 1963). In the latter case, most structural and lithological variation in the Vendian to lower or middle Cambrian rock sequences is due to a complex mosaic of narrow, linear, divergently trending volcanic belts, and isometric areas typified by calcareous and calcareous–terrigenous clastic sedimentation. The lithologies comprise reef and other limestones, dolomite, impure siliceous limestone, and phosphorite-bearing rocks. Sometimes, as for instance in eastern Tuva, calcareous sedimentation was accompanied by the eruption of felsic quartz porphyries. Volcanism in the linear belts intervening between the blocks of the mosaic was more mafic in character. It contributed to form a sequence containing the spilite–diabase, spilite–keratophyre, greenschist–siliceous shale, greywacke, andesite, and reef limestone associations. There also occur small bodies of ultramafics, mafics, and serpentinite mélanges, and numerous intrusions of gabbro, plagioclase granite, granodiorite, and monzonite. In the isometric ares, the sediments have undergone intense, post-folding palingenesis and granitization. Very large granitic batholiths were formed at the end of the Cambrian (Polyakov 1971).

The nature of the basement below the Caledonides of the eastern Altai–Sayan region is not known with certainty as no primary contacts appear to exist between most of the Vendian–Cambrian sequence and its substratum. References in the literature to depositional contacts are ambiguous and when checked in the field, these contacts as a rule have proved to be tectonic. However, there exist some questionable older age-determinations (Klyarovsky 1972) on the underlying rocks and indirect evidence such as differnces in metamorphic grade and sedimentary contacts at the easternmost margin of the early Caledonides in Eastern Sayan, Eastern Tuva, and Khamardaban where Vendian–Cambrian limestones rest unconformably on Precambrian metamorphic rocks. On the basis of these indications, most geologists support the opinion that the early Caledonides were formed upon a deformed, volcanic–terrigenous–calcareous upper Proterozoic substratum, which had been metamorphosed in or below the amphibolite facies prior to Vendian deposition. Such rocks are exposed in the Tomsk Precambrian Window of the Kuznetsk Alatau.

The western early Caledonides

The early Caledonian folded zones of Central Kazakhstan and Northern Tien Shan differ from the eastern early Caledonides by containing a complete Ordovician stratigraphic column. Folding occurred in the late Ordovician or at the boundary between the Ordovician and the Silurian (Markova 1964; Tectonics of Eurasia 1966). Most of the multistage granitoids of the western early Caledonides must also have been intruded during this period. However, their K–Ar ages are 450–500 Ma.

The Silurian is represented by continental lower molasse and associated andesitic–dacitic volcanics. It is overlain unconformably by a lower Devonian red upper molasse of 'Old Red' type (Mazarovich 1976). (Both are included in signature 10, Fig. 2.)

Vertical and lateral lithological variation in the Vendian to lower Palaeozoic rocks of the early Caledonides in west central Kazakhstan is very similar to that in the eastern early Caledonides (Fig. 3). Also in the west there occur divergently striking linear zones containing mixed volcanosedimentary rock sequences (the Zhalair–Nayman, Selety, Kalmykul' and other zones). Vast areas of limestone, marl, and terrigenous sediments form the Ulutau, Northern Tien Shan, Kokchetav, and Akchatau–Dzhungaria complexes, which intervene between the linear zones. In the volcanic–sedimentary belts, the rocks are quartz diabases, spilitic diabases, jasperiferous basaltoids, basalts, andesitic basalts, andesites, terrigenous cherty sequences,

Fig. 2 Tectonic sketch-map of the Caledonides in the Kazakhstan–Siberian Foldbelt. Legend: 1—relicts of Vendian and early Palaeozoic oceanic crust (mainly Cambrian); 2—alpinotype ultramafics and serpentinite mélanges; 3—volcanic island arc rocks on a substratum of piled up oceanic crust (Vendian to Ordovician, locally lower Silurian); 4—terrigenous and terrigenous–siliceous complexes on the oceanic crust of marginal seas; 5—terrigenous complexes of the continental slope and rise (Vendian and Cambrian, locally also Ordovician and Silurian); 6—marginal volcanic island arcs along the interface between oceanic and continental crust (Vendian/lower Cambrian, locally Cambrian and Ordovician); 7—pre-Vendian sialic basement complexes along the margins of the Kazakhstan–Siberian Ocean; 8—calcareous, siliceous–calcareous and terrigenous shelf-sea sediments (Vendian to middle Cambrian, locally Ordovician); 9—rifts in the continental margins with newly formed oceanic and transitional crust (Vendian, Cambrian, and Ordovician); 10—Middle and Upper Palaeozoic molasse and volcanic–molasse basins (Devonian, Carboniferous, and Permian); 11—Mesozoic and later cover of the West Siberian Platform and internal basins; 12—boundaries of the principal tectonic zones and provinces; 13—faults, including wrench faults (a) and overthrusts (b). The digits in the map are: 1—Bet-Pak-Dala; 2—Chu–Ili Mountains; 3—Sarysu–Teniz watershed; 4—Dzhungarian Alatau; 5—Greater Karatau; 6—Baikonur Trough; 7—Selety Trough; 8—Kalmykul' Trough; 9—Aktau–Mointy Massif; 10—Spassk Tectonic Belt; 11—Central Kazakhstan Deep Fault; 12—Talitsk uplift; 13—Mras Zone; 14—Tel'bes Zone; 15—Chernoinyus Zone; 16—Kizir Zone; 17—Sangilen; 18—Prikosogol; 19—Khamardaban; 20—Bateni Ridge

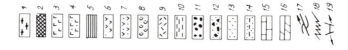

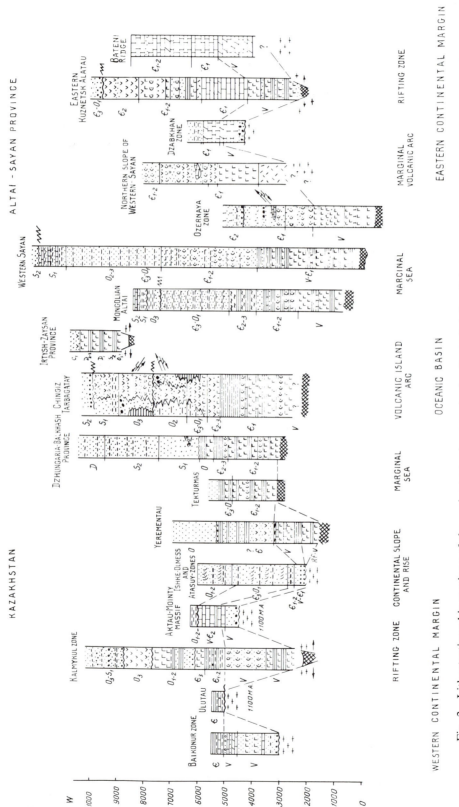

Fig. 3   Lithostratigraphic sections of the major tectonic zones in the Central Asian Caledonides. Legend: 1—pre-Vendian continental basement; 2—oceanic basement; 3 to 5—spilite–diabase (=ocean-floor tholeiitic) rock associations comprising 3—diabases and spilites, 4—mafic tuffs and 5—siliceous rocks (jasper, chert, siliceous pelites); 6 to 8—basalt–andesite/basalt (=island arc volcanic) associations comprising 6—andesites and andesitic basalts, 7—tuffs, 8—tuffites and tuffitic psammites; 9—felsic volcanics; 10—pelites; 11—conglomerates; 12—olistostromes; 13—psammites; 14—flyschoid pelitic–psammitic sequences; 15—limestone; 18—phases of folding; 19—rifting

flysch, and greywacke. Occasional alpinotype ultramafics have usually been altered into serpentinite mélanges. In the sedimentary complexes there mostly occur impure limestones, reef limestones, terrigenous jasperiferous rocks, and flyschoid greywackes. A Vendian quartz-arkose is particularly important because it has a depositional, unconformable contact with metamorphic basement in the cores of the Kokchetav and Ulutau massifs. The radiometric age of the basement is 1100 Ma. No tectonic breaks exist between this basement contact and Vendian reef limestones in the overlying sedimentary formations (e.g. Zaitsev and Filatova 1971; Zaitsev and Kheraskova 1979). These must therefore be autochthonous.

Late Caledonides

In general, the eugeosynclinal sequences of the late Caledonides are stratigraphically more complete and continuous than the sediments of the early Caledonian folded zones. They comprise the Vendian, the entire Cambrian and Ordovician, and the lower Silurian. Intense folding occurred at the boundary between the early and late Silurian and at the end of the late Silurian. In some places as, for instance, the Talitsk–Mongolian Altai Zone, there was also a prominent late Ordovician phase of folding. Granitoid magmatism is predominantly Silurian and Devonian, but some Ordovician plagioclase granite and granodiorite intrusions can be found occasionally. Lithogically, the late Caledonian folded zones of the Kazakhstan–Siberian Foldbelt can be subdivided into two principal types. One is characterized by essentially bipartite abyssal marine and flyschoid rock piles, whereas the other carries volcanics throughout the rock succession and is more complex in its tectonic evolution.

Folded zones of the first type

In the late Caledonian folded zones of Western Sayan, Gorny Altai, the Mongolian Altai, and northeastern Kazakhstan, the lower Palaeozoic deposits reach thicknesses of 12 to 15 km. They are essentially terrigenous (Fig. 3).

The Western Sayan Zone has recently been the subject of intense lithological and structural study by Kheraskov (1975, 1979). In this area, the basal Vendian–lowermost Cambrian rocks belong to a deep-sea spilite–chert–clay facies. They are associated with very large massifs of alpinotype ultramafic and gabbroic rocks forming typical ophiolite assemblages. Upwards, the ophiolitic sequence is gradually replaced by lower to middle Cambrian 'green tuffs' (= sandstones and siltstones, greywacke, polymict

tuffs, and tuffites, and andesite–porphyritic lavas). Sometimes, these rocks have been metamorphosed in the greenschist facies. There follow several kilometres of a terrigenous rhythmic succession made up of green flyschoid upper Cambrian and lower Ordovician sediments. Still higher stratigraphical levels feature multicoloured Ordovician flyschoids and calcareous flysch rocks of the lower Silurian and the lowermost parts of the upper Silurian.

During the Silurian folding, these rocks were affected by strong tangential strain, which resulted in major faulting and extensive overthrusting. Overthrusts are most frequent in the peripheral parts of the Western Sayan Synclinorium and therefore generate an overall fan-shaped structural pattern with a two-sided vergence.

In Central Kazakhstan, the late Caledonian folded zones of the first type are structurally somewhat more complex. They contain very large volumes of the lower, abyssal marine rocks, which are exposed, for example, in the Tekturmas Zone described by Kheraskova (1979). In this case, the deep-sea rocks contain jasperiferous basaltic and tuffite–jasper sequences, which are repeated twice in the stratigraphic column. The deep-sea succession ranges from the Vendian through the Cambrian into the lower Ordovician.

Whilst the Tekturmas Zone formed a deep trough as late as the middle Silurian, other late Caledonian folded zones containing the same bipartite abyssal marine and terrigenous flyschoid succession exhibit shallow-water, tectonically disturbed geosynclinal developments from the upper Ordovician onwards. The Talitsk–Mongolian Altai Zone (Volochkovich and Leontiev 1964) may serve as an example. In this zone there was intense folding already in the middle or early upper Ordovician (Nekhoroshev 1958). After this time, sedimentation occurred in shallow waters where upper Ordovician and Silurian pure and impure calcareous rocks, black shales, and multicoloured terrigenous sediments were deposited in a series of isolated, (geo-)synclinal basins superimposed upon a substratum of folded rocks older than the upper Ordovician. Numerous sedimentation breaks and unconformities developed during the early and late Silurian. Nevertheless, as is the case also in the other late Caledonian folded zones, the Talitsk–Mongolian Altai Zone attained its final structural configuration only after the main regional folding at the end of the late Silurian.

Late Caledonian folded zones of the second type

The two principal characteristics of this type of folded zones are the diversity and long duration of their volcanic activity and the complexity of the multistage tectonic development. The Chingiz and Tarbagatay Folded Zones in east central Kazakhstan, and the Salair Zone in the Altai–Sayan region belong to the second type of late Caledonian regional structures.

In the Chingiz and Tarbagatay Zones (Fig. 3), volcanism commenced in the Vendian or early Cambrian. During a first stage, the melanocratic oceanic basement was covered by the deep-sea spilitic–calcareous rock associations. Along the northeastern and southwestern fringes of the Chingiz–Tarbagatay belt, these rocks are replaced by jasper–spilite and tuff–jasper facies. In the middle Cambrian, the chemistry of volcanism changed. Typically calc-alkaline, island arc andesitic, andesitic–basaltic, and andesitic–dioritic igneous associations were formed from that time onwards. Coarse pyroclastics characterize this stage of volcanism which was dominated by eruptions from central volcanoes. A new culmination of calc-alkaline igneous activity followed in the middle Ordovician. The cumulative thickness of volcanic rocks accumulated during the three principal eruptive cycles exceeds 6 kilometres.

As has been established by S. G. Samygin (personal communication 1981), igneous activity was accompanied by the tectonic piling of rock masses and extensive overthrusting toward the northeast. A consequence of thrusting was the development of thick middle Ordovician olistostromes in the northeastern part of the Chingiz Zone.

The local, island arc character of the middle Cambrian to Ordovician volcanism in the Chingiz and Tarbagatay Zones is substantiated by its geological setting. In the southwest, in the vicinity of the Dzhungaria–Balkhash Province, and in the northeast, in the direction of the Irtysh–Zaysan Province the volcanics give way to upper Cambrian and Ordovician terrigenous polymictic greywackes, tuffitic–siliceous formations, and abyssal siliceous deposits. According to Samygin, the southwestern rocks accumulated in a back-arc basin, whereas the rocks in the northeast belong to a fore-arc oceanic environment.

At the beginning of the late Ordovician, the Chingiz and Tarbagatay Zones were folded and tectonically thoroughly reconstructed. This deformation was coincident with deformation in the Talitsk–Mongolian Altai Zone and other parts of the Altai–Sayan late Caledonides. The Chingiz–Tarbagatay area continued as a system of islands also during the latest Ordovician and the Silurian, but the direction of tectonic transport was now reversed toward the southwest. A lower Silurian andesite–dacite formation and predominantly greywackoid sediments were formed in the southwestern Chingiz Zone. They were folded during the late Silurian main phase of the late Caledonian deformation.

Late distortion of the Caledonides

The original structural relationships between the early and late Caledonian folded zones in Kazakhstan, Siberia, and Mongolia were disturbed by middle to late Palaeozoic, predominantly Variscan movements. In the east, these events were connected with the formation of a N–S trending system of large left-lateral faults deviating toward the northwest and northeast in the central part of the Altai–Sayan region. At the end of the Palaeozoic these faults were transformed into arcuate thrusts which are convex toward the west. Thrusting was due to extensive crustal shortening.

A west Caledonian equivalent is found in the NNE-striking, right-lateral Central Kazakhstan Deep Fault and the associated E–W system of thrusts in the Spassk Tectonic Belt (locality 11, Fig. 2) and the system of associated thrusts. On the east side of this fault, the Chingiz–Tarbagatai Zone was displaced toward the south. Still farther west, a set of NW-facing overthrusts at the Sarysu–Teniz watershed complicated the structure of west–central Kazakhstan.

A very important role may have been played by E–W stretching connected with the formation of the Irtysh–Zaysan Variscan geosynclinal system in the latest Ordovician and the Silurian. The Variscan belt split the previously coherent Central Asian Caledonides into two separate parts in Kazakhstan and Siberia–Mongolia.

Tectonic Evolution

Variations of stratigraphy, composition, and facies development of the Vendian and early Palaeozoic sediments, the chemistry of the associated igneous rocks, and the diversity of relationships with earlier Precambrian units demonstrate that the various Caledonian folded zones are distinctly different in character. Some of them were formed on a continental Precambrian crust, whereas others have a melanocratic oceanic substratum overlain by basal Caledonian abyssal siliceous and tholeiitic formations. Apart from these fundamental differences there exists great palaeoenvironmental variation determining the peculiarities of the individual folded zones and their subareas. During the last decade, the complexities and interrelationships of palaeotectonic environments have been the subject of intense study.

The following summary is based on partly still unpublished work and conclusions by L. P. Zonenshain, N. N. Kheraskov and V. V. Volkov in the Altai–Sayan region, A. B. Dergunov, N. N. Kheraskov and A. S. Perfiliev in Mongolia, S. G. Samygin, T. N. Kheraskova and N. G. Markova in central Kazakhstan, and S. G. Samygin and I. A. Rotarash in eastern Kazakhstan.

As suggested by Zonenshain (1974), the Central Asian oceanic basin was the factor determining the Vendian and early Cambrian palaeotectonic environments in the Kazakhstan–Siberian Foldbelt. The north-trending branch of this ocean formed a wide and deep embayment occupying the axial part of the future folded belt. It covered present Mongolia, Western Sayan, Gorny Altai, Eastern Kazakhstan, and the eastern part of Central Kazakhstan (Fig. 2). In the east and west, this ocean was fringed by complex continental margins. Presently available data indicate that a continental margin existed also in the north, in the present substratum of the West Siberian Platform. The western continental margin occupied west Central Kazakhstan (Ulutau, Kokchetav, Northern Tien Shan, and the Aktau–Mointy massif). The continental mass in the east was the Siberian–Baikalian craton which extended as far as the Dzabkhan Zone in Central Mongolia, Eastern Tuva and Sangilen, Eastern Sayan, the Kuznetsk Alatau, and Eastern Salair.

Vendian to early Palaeozoic continental margins

The basement of the one-time continental margins consists of Precambrian gneisses, quartzites, crystalline schists, and amphibolites which yield radiometric ages above 1000 Ma (signature 7 on Fig. 2). A great stratigraphic and angular disconformity separates the basement from a relatively thin, terrigenous psammitic–pelitic and calcareous cover (signature 5). In the eastern Caledonides (Gornaya Shoria, the Batenevsk Range of the Kuznetsk Alatau, Eastern Tuva, etc.), the age of the cover is Vendian to middle Cambrian. In the west, the cover ranges from Vendian to lower Ordovician (Ulutau, the Aktau–Mointy massif, Greater Karatau, etc.).

A characteristic feature of both continental margins is their extensive rifting, which was responsible for the formation of a system of divergently oriented narrow, trough-shaped eugeosynclinal zones (the Baikonur, Kalmykul, Zhalair–Nayman, and other Vendian to early Palaeozoic troughs in the west, and the Tel'bes, Mras, Chernoyus, Kizir, and other Vendian to Cambrian volcanic zones of the Kuznetsk Alatau, Eastern Sayan, and Gornaya Shoria in the east). Some of these zones still exhibit a substratum

of ocean-floor type. Rifting origin of the troughs is demonstrated by distinctly linear configurations and faulted margins, which were preferred loci of igneous and hydrothermal activity. Coarse debris is common in the lower parts of the rock piles. A characteristic feature testifying to the primary linearity and narrowness of the troughs is the persistence of sedimentary and igneous formations along the strike and their distinct thinning in cross-strike directions.

The rocks filling the eugeosynclinal troughs (signature 9) differ somewhat from classical eugeosynclinal facies (signatures 1 and 4). They comprise complexly deformed and partly metamorphosed, relatively thick sequences of greenstones, Spilite–keratophyres (altered diabase-albitophyric and diabase-orthophyric rocks) and andesites, siliceous beds, greywackes and carbonate rocks. Particularly at low stratigraphical levels the volcanics are diversified in composition and high in alkalis especially potassium. In both west (Kheraskova 1979) and east (Belousov et al. 1969) they compare well with volcanics associated with recent epicontinental rifting.

According to Belousov et al. (1969), the series of successive Vendian and Cambrian volcanic zones is marked by a regular increase of the potassium content of the basalts, which was concomitant with the gradual advance of the ocean–continent boundary toward the east, deep into the eastern continental margin.

Along with alpinotype ultramafics, alkaline ultramafic rocks are frequent in the eugeosynclinal troughs.

Some differences exist between the eastern and western riftogenic eugeosynclinal zones of the Central Asian Caledonides. As has been mentioned above, deposition in the eugeosynclinal troughs of the eastern Caledonides ceased in the middle Cambrian, whereas in the west it continued until the Ordovician. In the east, folding of the trough fillings occurred at the boundary between the middle and late Cambrian. In the west, this folding occurred at the end of the Ordovician. Other very important differences are related to fundamental characteristics of the two oceanic margins. They will be considered in the next section.

Finally, it is important to emphasize that massive development of riftogenic, eugeosynclinal troughs occurred in both the east and the west. This indicates that both fringes of the Kazakhstan–Siberian Ocean suffered equally strong Vendian and early Cambrian stretching. This extension is responsible for the abundance of faults and tectonic breaks controlling the peculiar mosaic pattern of crustal blocks along the continental margins.

The Vendian and early Palaeozoic Kazakhstan–Siberian oceanic basin

The original extent and configuration of this oceanic basin can be inferred from the distribution of abyssal tholeiites and abyssal siliceous rocks such as jaspers, chert, radiolarites, and siliceous pelites (signature 1, Fig. 2). As a rule, these rocks are found in close association with ophiolitic complexes (signature 2) which are fault-displaced fragments of oceanic crust.

The ocean floor

The Vendian and lower Palaeozoic tholeiitic basalts of the spilite–diabase association are chemically similar to the tholeiites of recent ocean floors and mid-oceanic ridges. Comparative geochemical studies have been carried out in Eastern Kazakhstan (Antonyuk 1974; Kheraskova 1979), Western Sayan (Kheraskov 1975, 1979) and Western Mongolia (Dergunov 1980; Dergunov and Kheraskova 1981). However, along with predominant tholeiitic basalts, the Vendian–lower Paleozoic volcanic successions also comprise some trachybasalts and alkali olivine basalts (Kheraskova 1979). The presence of such rocks may be due to a systematic difference between the volcanic associations of Caledonian and recent floors. Analogous differences have been found in Caledonian abyssal siliceous sediments, which are similar to but not identical with the siliceous sediments of recent oceans (Kheraskova 1979).

The metamorphic oceanic substratum of the Vendian to lower Palaeozoic volcanics is most frequently represented by serpentinite mélanges in which ultramafics and gabbros occur together with eclogite-like rocks, garnet amphibolites, and blocks and fragments of diabases and gabbro-diabases derived from sheeted-dyke complexes.

In most studied sections, tholeiitic and abyssal siliceous Caledonian rocks occupy the lowermost stratigraphic levels resting, apparently, on an oceanic basement. In the Ozernay Zone of Western Mongolia, the Mongolian and Gorny Altai, Western Sayan, and Yerementau in northwest Central Kazakhstan, the abyssal rocks are Vendian and early Cambrian in age. In the Chingiz and Tarbagatay Zones, they belong to the lower Cambrian, but older segments may also be present. In the Tekturmas Zone and the Balkhash Anticlinorium their ages span the entire Cambrian and part of the Ordovician. These ages may be taken to suggest a chronologically heterogeneous floor of the Kazakhstan–Siberian Ocean, which is Vendian in its easternmost and westernmost parts (Altai, Western Sayan, and Western Mongolia, and Yerementau respectively), early Cambrian closer to the centre,

and Cambrian to Ordovician in the central Dzhungaria–Balkhash Province (based on unpublished datings of radiolarians in the Balkhash Anticlinorium by B. Nazarov personal communication 1981). Attractive as such an inference may be, it appears premature at present since the Vendian and Cambrian rocks are not known in sufficient biostratigraphic detail. This applies particularly to the virtually barren basaltic–siliceous formations.

In any case, there can hardly be any doubt that at the end of the Vendian and in the early Cambrian, the Kazakhstan–Siberian Ocean was fully developed in a little differentiated, clearly outlined deep-water basin. This was a gulf with a diameter of at least 1600–2000 km generated, essentially, by a tensional tectonic regime. In the south, the Kazakhstan–Siberian Ocean passed into the Central Asian Ocean. In the west and east, it was delimited by wide continental margins, which were the sites of extensive rifting and island arc igneous activity. In the axial part of the present Chingiz Ridge, which was close to the centre of the oceanic basin, there probably existed a submarine uplift of the ocean floor. A lower Cambrian spilitic–calcareous association was formed at this site. According to T. N. Kheraskova and S. G. Samygin (personal communication), its basalts are geochemically identical with the basalts occurring in elevated blocks of recent ocean floors.

The western continental slopes and rises

In the west and northwest, the boundary between the oceanic basin and its continental margin had the shape of a relatively simple embayment which was convex toward the west. It extended from the Yerementau mountains in the north along the western margin of the Aktau–Mointy Massif, then swinging southeastwards along the northern edge of the Dzungaria Alatau (Fig. 2). This boundary is marked by the occurrence of a thick complex of terrigenous psammites of Vendian, Cambrian, and early to middle Ordovician ages (signature 5), unconformably overlying a basement of Precambrian quartzites and crystalline schist. According to Kheraskova (1979) and Zaitsev and Kheraskova (1979), the psammitic complex is mostly made up of inequigranular, indistinctly bedded quartz-rich sandstones, sandy–pebbly mixtites, and siltstones with rare siliceous interbeds. Facies developments characteristic of underwater debris cones and valley-lake systems can be distinguished. The direction of detritus flow was toward the east. Immature turbidites and fluxoturbidites accumulated in the west. The rocks in the east are mature products of suspension flows characteristic of the distal part of debris cones. They exhibit good sorting and graded lamination. The Caledonian terrigenous complex thus resembles the sediments of recent continental slopes and rises. The features described may suggest that the Vendian to early Palaeozoic western limit of the Kazakhstan–Siberian oceanic basin was a passive continental margin of the Atlantic type.

Eastern island arcs

In contrast to the west, the eastern delimitation of the oceanic basin had an irregularly angular, gulf-like configuration (though certainly not as irregular as would appear from the intricacies of present-day structure). Still more important, it was characterized by a subduction-type junction between oceanic and continental crust. As has been noted by many researchers (Volkov 1966, 1981; Zonenshain 1974; Kheraskov 1979), Vendian and early Cambrian volcanic island arcs existed along the continental margin extending from the eastern slopes of Salair toward eastern Gornay Altai, skirting Western Sayan along its E–W trending northern and southern margins, then turning toward the south and southeast parallel to the margin of the Dzabkhan Zone (Fig. 2). These marginal island arcs are built of volcanics of the contrasting spilite–keratophyric (metabasaltic–albitophyric according to Belousov et al. 1969) and andesitic associations (signature 6). The prominence of metadacitic and metarhyolitic rocks is a distinctive feature. Andesitic volcanism belonged predominantly to the highly explosive, pyroclast-rich, central volcano type. Geochemically, the basaltic members of these sequences are similar to the basalts and andesitic basalts of recent island arcs (Belousov et al. 1969; Kheraskova 1975, 1979). The keratophyres and albitophyres are products of remelting of the sialic substratum.

Island arc volcanism along the eastern margin of the Kazakhstan–Siberian Ocean was not all of the same age. At the boundary between the Ozernaya and Dzabkhan Zones in the extreme southeast (Fig. 2), island arc volcanism occurred in the Vendian only (Dergunov 1980). Along the edges of the Western Sayan oceanic reentrant, volcanic island arcs were active until the end of the early Cambrian (Kheraskov 1975). In the eastern part of Gorny Altai (the western segment of the Biysk–Katun Zone), the thick island arc volcanic formations belong to the Vendian and lower to middle Cambrian (Volkov 1966; Belousov et al. 1969). In central and eastern Salair, island arc volcanism did not cease until the end of the early Ordovocian (Belousov et al. 1969; Naletov and Sidorenko 1970). Northern island arcs thus tended to be more long lived than southern ones. This regularity is quite obvious, but the reasons for this are unclear. They may be related

to the successive tectonic differentiation of the eastern part of the Kazakhstan–Siberian oceanic basin. In this process the oceanic crust of relict marginal sea basins appears to have been preserved for a longer time in the north. To the west of Salair, it did not disappear before the end of the Ordovician.

Synopsis of Vendian and early Cambrian developments

In summary, the Vendian and early Cambrian palaeotectonic environment of the Kazakhstan–Siberian Ocean was characterized by the predominance of tensional regimes, which contributed to the existence of a little-differentiated oceanic basin covered by Vendian and lower Cambrian abyssal volcanic and siliceous deposits. The western continental slope and margin of this basin were buried under a thick, terrigenous sedimentary wedge, whereas subduction of oceanic crust beneath the continental margin commenced in the east. In the Vendian and early Cambrian this resulted in the formation of a chain of marginal island arcs.

Middle Cambrian to Ordovician developments

The middle Cambrian was the time when large-scale tectonic differentiation commenced in the Kazakhstan–Siberian Ocean. A large, NNW-trending volcanic island arc developed in the central part of the oceanic basin and divided it into western (Central Kazakhstan) and eastern (Altai–Sayan and western Mongolia) subbasins. The island arc formed as a result of the piling-up and thickening of oceanic crust in the present Chingiz–Tarbagatay region.

The Chingiz–Tarbagatay island arc and the central Kazakhstan oceanic subbasin

From the middle Cambrian to the early Silurian, the Chingiz–Tarbagatay arc was the site of intense volcanism forming thick, calc-alkaline, andesitic, andesitic–basaltic, and andesitic–dacitic island arc sequences (signature 3, Fig. 2). Toward the west and east, this facies passes into abyssal siliceous and terrigenous deposits formed in back-arc and fore-arc oceanic basins (Fig. 4B). The origin and subsequent evolution of the Chingiz–Tarbagatay island arc was apparently due to the formation of a west-dipping Benioff zone in the previously existing uplifted block of ocean floor.

In the late Cambrian and early to middle Ordovician, subduction of oceanic crust was accompanied by nappe thrusting in the hanging wall of the Benioff zone. As mentioned previously, these thrusts were directed toward the east and caused the formation of

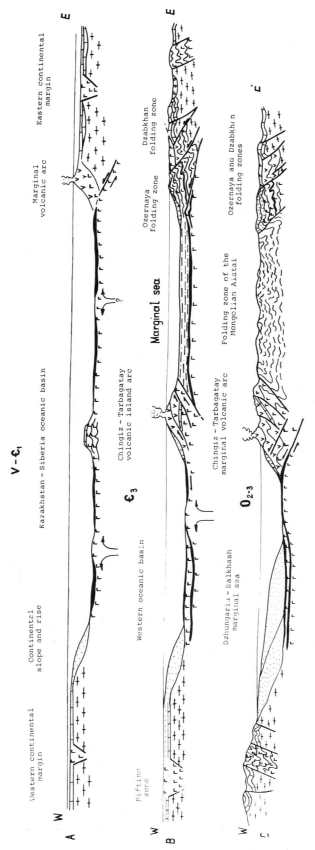

Fig. 4   Sections showing reconstructions of the palaeotectonic environments during the most important development stages of the Kazakhstan–Siberian oceanic basin. The scale is arbitrary. Section A: Vendian to lower Cambrian; Section B: Upper Cambrian; Section C: end of the Ordovician

olistostromes in the adjacent parts of the eastern fore-arc basin (S. G. Samygin personal communications 1981). From the middle Cambrian onwards, the oceanic subbasin to the west of the Chingiz–Tarbagatay island arc assumed a back-arc character which persisted throughout the late Cambrian and the early and middle Ordovician. As in the early Cambrian, crustal extension dominated the tectonic regime. Along the western margin of the basin and also in the western Chingiz area there was terrigenous turbidite sedimentation on the continental slope and the slope of the island arc. In the central parts of the basin, in the Dzhungaria–Balkhash province, the deposition of volcanics of the spilite–diabase association and siliceous abyssal sediments continued as before.

The eastern oceanic subbasin

After the early Cambrian, the large oceanic basin to the east of the Chingiz–Tarbagatay island arc developed into a relict oceanic structure in which the formation of new oceanic crust had virtually ceased. A slow, gradual closing of the basin commenced. It was effected by the overthrusting of the eastern continental margin from the east and the Chingiz–Tarbagatay island arc from the west. In the middle Cambrian, this process was accompanied by island arc volcanism in the west as well as the east, which affected the character of the sediments in the oceanic basin where greywackes, volcanomictites, and tuffaceous–siliceous deposits gained increased prominence.

In the late Cambrian and early Ordovician, an extensive mountain range developed along the eastern rim of the ocean. Its uplift was preceded by crustal shortening and initial Caledonian folding in the eastern continental margin and some adjacent parts of the ocean floor (the Ozernaya Zone). Huge masses of commonly sialic detritus (signature 4, Fig. 2) began to fill the reduced, relict oceanic basin. Thick terrigenous flyschoid turbidites accumulated in its eastern part, especially in Western Sayan. At the same time, a peculiar, very thick deep-water formation of oligomictic plagioclase- and quartz-rich, sandy, silty, and clayey sediments was developed in the central parts of the basin (the Mongolian and Gorny Altai).

Final development of the Caledonian foldbelt

The second half of the Ordovician was a new turning point in the evolution of the Kazakhstan–Siberian Foldbelt. During this period, Caledonian folding extended to the Chingiz–Tarbagatay island arc, both continental margins and almost the entire eastern

oceanic subbasin. Remaining exceptions were the central parts of the Central Kazakhstan oceanic subbasin in the west, and Western Sayan, some areas of Gorny Altai and Salair in the east. Most of the Kazakhstan–Siberian oceanic basin was replaced by a region of transitional crust. In the midst of extensive areas with a granite–metamorphic crustal layer at the base (Fig. 4C), there remained some, mostly isolated, residual depressions with an oceanic type of crust (the Dzhungaria–Balkhash province, Western Sayan, the Anuisk–Chuisk area, etc.). The newly folded regions formed the basement of superimposed (geo-)synclinal troughs and monogeosynclines (Dergunov *et al.* 1980) with epicontinental sedimentation of terrigenous–calcareous and black shale facies. In Western Sayan, Gorny Altai, and Western Salair, multicoloured terrigenous, calcareous and flyschoid complexes of late Ordovician and Silurian age accumulated in land-locked, relict oceanic seas. These complexes contain endemic assemblages of brachiopod fauna (Rozman 1977).

After the late Ordovician, only the Dzhungaria–Balkhash oceanic subbasin retained its size, the character of a back-arc marginal sea and a connection with the Central Asian Palaeozoic Ocean. The sedimentation of terrigenous turbidites (signature 4, Fig. 2) continued during the late Ordovician, the Silurian, and the Devonian. Along its northeastern margin, a volcanic island arc existed in the southwestern part of the Chingiz Zone. However, the Chingiz island arc now had a typically marginal character in the vicinity of an extensive late Ordovician foldbelt comprising Chingiz, Tarbagatay, Gorny Altai and the Mongolian Altai. According to S. G. Samygin (personal communication 1981), thrusting and nappe transport were directed toward the southwest, the arc itself owing its existence to the formation of a new, late Ordovician Benioff zone, which dipped eastwards, underneath the now folded Chingiz Zone.

The final development of the Caledonian part of the Kazakhstan–Siberian Foldbelt was accomplished during the late Silurian phase of folding. Residual oceanic seas, such as those in Western Sayan, and the island arc in the southwestern Chingiz Zone were replaced by folded zones.

In a broader geodynamic context, the late Silurian folding was connected with the opening of the Irtysh–Zaysan middle Palaeozoic ocean, where initial rifting appears to have occurred already in the early Silurian or perhaps even as early as the very end of the Ordovician.

By the beginning of the Devonian, two large Caledonian continental massifs had originated as a result of late Silurian folding. Devonian red continental molasse and associated subordinate basaltic–andesitic–rhyolitic volcanics were forming everywhere in these areas. In addition, there was the Irtysh–Zaysan oceanic basin separating the two Caledonian regions. Their tectonic history in the Variscan is, however, a subject of an independent study.

Acknowledgement

Authors sincerely thank Prof. R. Gorbachev for his efforts in improving the Englsih text of the manuscript.

References

Antonyuk, R. M. 1974. Oceanic crust of the eugeosynclinal region in east Central Kazakhstan. In *Tectonics of the Uralo-Mongolian Foldbelt*, Nauka, Moscow, 67–74.

Belousov, A. F., Kochkin, Yu. N. and Polyakova, Z. G. 1969. *Volcanic Complexes of the Mountainous Altai, Mountainous Shoria and the Salair Ridge*, Nauka, Moscow.

Bogdanov, A. A. 1958. On the tectonic zonation of the Ural–Sayan Palaeozoic folded region. *Nauchnye Doklady Vysshey Shkoly—Geologicheskiye i Geograficheskiye Nauki*, **1958: 1**, 3–6.

Bogdanov, A. A. 1959. Main features of the Palaeozoic structure of Central Kazakhstan. *Byulleten' Moskovskogo Obshchestva Ispytateley Prirody. Otdel Geologicheskiy*, **1959: 1**, 3–18.

Dergunov, A. B. 1980. The structure of the Caledonides and the development of the Earth's crust in Western Mongolia and the Altai–Sayany region. In *Tectonic Problems of the Earth's Crust*, Nauka, Moscow, 184–195.

Dergunov, A. B., Luvsandanzan, B. and Pavlenko, V. S. 1980. Geology of Western Mongolia, Nauka, Moscow.

Dergunov, A. B. and Kheraskova, T. N. 1981. On the composition of volcanites during the early development of the Central Asian Caledonides (Central Kazakhstan, the Altai–Sayany region, Western Mongolia). *Byulleten' Moskovskogo Obshchestva Ispytateley Prirody. Otdel Geologicheskiy*, **56**, 35–53.

Kheraskov, N. N. 1977. Lithological associations and the initial phases of geosynclinal development of the Western Sayan, Nauka, Moscow.

Kheraskov, N. N. 1975. Lithological associations and the stages of geosynclinal development of the Western Sayany. *Geotektonika*, **1975: 1**, 35–53.

Kheraskova, T. N. 1979. Siliceous rocks of the lower Palaeozoic in Central Kazakhstan. In *Sedimentation and Volcanism in Geosynclinal Basins*, Nauka, Moscow, 5–37.

Klyarovsky, V. M. 1972. Geochronology of the Mountainous Districts of the South-Western Margin of the Siberian Platform, Nauka, Novosibirsk.

Kuznetsov, V. A. 1954. The geotectonic zonation of the Altai–Sayany folded region. In *Problems of the Geology of Asia*, Izdatel'stvo AN USSR, Moscow, 202–227.

Leontiev, V. N., Litvinovskiy, B. A., Gavrilova, S. P. and Zakharov, A. A. 1981. *Palaeozoic Granitic Magmatism of the Central Asian Foldbelt*, Nauka, Novosibirsk.

Lepezin, G. G. 1978. The Metamorphic Complexes of the Altai–Sayany region, Nauka, Novosibirsk.

Markova, N. G. 1964. Regularities of the distribution of heterochronous folded zones as illustrated by the case of Central Kazakhstan. In *The Folded Regions of Eurasia*, Nauka, Moscow, 120–147.

Mazarovich, O. A. 1976. The Geology of Devonian Molases (Formational Analysis of the Devonian in the Central Kazakhstan Caledonides and General Genetic Problems of Molasse Formation), Nedra, Moscow.

Mossakovsky, A. A. 1963. Tectonic Development of the Minusinsk Basins and their Metamorphic margins during the Precambrian and the Palaeozoic, Gosgeoltekhizdat, Moscow.

Mossakovsky, A. A. 1972. The Palaeozoic orogenic volcanism of Eurasia (Principal rock associations and the tectonic controls of their distribution). *Geotektonika*, **1972: 1**, 6–28.

Mossakovsky, A. A. 1975. The Orogenic Structure and Volcanism of the Eurasian Palaeozoides, Nauka, Moscow.

Naletov, B. F. and Sidorenko, T. F. 1969. The lower Ordovician volcanic complex in the north-western part of the Kuznetsk Alatau. *Geologiya i Geofizika*, **1969: 11**, 55–60.

Nekhoroshev, V. P. 1958. Geology of the Altai, Gosgeoltekhizdat, Moscow.

Pavlenko, A S. Filippov. L. V. and Orlova, L. P. 1974. Granitoid Formations of the Central Asian Foldbelt: their Petrology, Geochemistry and Metal Contents, Nauka, Moscow.

Peive, A. V., Shtreis, N. A., Mossakovsky, A. A., Perfiliev, A. S., Ruzhentsev, S. V. Bogdanov, N. A., Burtman, V. S., Knipper, A. L., Makarychev, G. I., Markov, M. S. and Suvorov, A. I. 1972. The Palaeozoides of Eurasia and some problems of the evolution of the geosynclinal process. *Sovetskaya Geologiya*, **1972: 12**, 7–25.

Peive, A. V., Yanshin, A. L., Zonenshain, L. P., Knipper, A. L., Markov, M. S., Mossakovsky, A. A., Perfiliev, A. S., Pushcharovsky, Yu, M., Schlezinger, A. E. and Shtreis, N. A. 1976. The formation of continental crust in Northern Eurasia (Aspects in connection with the compilation of the new tectonic map). *Geotektonika*, **1976: 5**, 6–23.

Peive, A. V., Zonenshain, L. P., Knipper, A. L., Markov, M. S., Mossakovsky, A. A., Perfiliev, A. S., Pushcharovsky, Yu, M., Fedorovsky, V. S., Shrteis, N. A. and Yanshin, A. L. 1980. Tectonics of Northern Eurasia, Nauka, Moscow. [Explanatory text of the 'Tectonic Map of Northern Eurasia'].

Polyakov, G. V. 1971. Palaeozoic Magmatism and Iron Mineralization of the Central South Asia, Nauka, Novisibirsk.

Rozman, Kh. S. 1977. Biostratigraphy and Zoogeography of the Upper Ordovician in Northern Asia and North America. Nauka, Moscow. Trudy GIN AN USSR, vol. **305**.

Tectonics of Eurasia 1966. Nauka, Moscow.

Tectonic Map of Eurasia 1966. GUGK, Moscow.

Tectonic Map of Northern Eurasia 1980. NPO Aerogeologiya, Moscow.

Tikhomirov, V. G. 1975. Palaeozoic magmatism and Tectonics of Central Kazakhstan, Nedra, Moscow.

Volkov, V. V. 1966. Major Regularities of the Geological Development of the Mountainous Altai [Gorny Altai] (Late Precambrian and Early Palaeozoic), Nauka, Novosibirsk.

Volkov, V. V. 1981. Tectonics and volcanism of the early Palaeozoic geosynclinal belt in Central Asia. In *Problems of the Evolution of Geological Processes*, Nauka, Novosibirsk, 56–65.

Volochkovich, K. L. and Leontiev, A. N. 1964. *The Talitsk–Mongolian Altai Metallogenic Zone*, Nauka, Moscow.

Zaitsev, Yu, A. and Filatova, L. I. 1971. New data on the structure of the Ulu–Tau Precambrian (in connection with the development of a consistent stratigraphical classification of the Precambrian of Central Kazakhstan). In *Problems of the Geology of Central Kazakhstan*, Izdatel'stvo Moskovskogo Gosudarstvennogo Universiteta, Moscow, 21–91.

Zaitsev, Yu. A. and Kheraskova, T. N. 1979. *The Vendian of Central Kazakhstan*, Izdatel'stvo Moskovskogo Gosudarstvennogo Universiteta, Moscow.

Zonenshain, L. P. 1974. A model of the development of geosynclinal belts (based on the case of the Central Asian Foldbelt). In *Tectonics of the Uralo–Mongolian Foldbelt*, Nauka, Moscow, 11–35.

The Appalachian collage

J. D. Keppie

Department of Mines and Energy, P.O.B. 1087, Halifax, Nova Scotia, Canada B3J 2XI

ABSTRACT

A map of the northern Appalachians is presented which shows that it is made up of a collage of terranes originating outboard of cratonic North America. A terrane is defined as an area characterized by an internal continuity of geology, including stratigraphy, faunal provinces, structure, metamorphism, igneous petrology, metallogeny, and palaeomagnetic record, that is distinct from that of neighbouring terranes and cannot be explained by facies changes. The term composite terrane is applied to terranes that amalgamated before accretion to the continental margin. Terrane boundaries are generally faults or melanges representing sutures, now largely cryptic, with complex structural histories often involving transcurrent motions. Constraints on the time of accretion are provided by the age of the youngest strata unique to a particular terrane and the age of overstepping strata, straddling plutons, thrusts, straddling regional metamorphism, and exotic pebbles. Deformation is often synchronous with accretion as a result of collision, whereas metamorphism and plutonism are generally longer lived.

Using these criteria, it is possible to unravel the accretionary history of the northern Appalachians. During latest Precambrian times, the Avalon composite terrane was formed by the collision of several terranes resulting in the Cadomian Orogeny. In Late Cambrian–earliest Ordovician times the Boundary composite terrane was formed by collision of several terranes producing the Penobscot Disturbance. During Arenig Caradocian Taconian Orogeny, terranes in the northwestern half of the Appalachians were accreted to cratonic North America. At the same time, terranes were accreted to the northern margin of the Avalon composite terrane. During the Devonian Acadian Orogeny, terranes were accreted to both sides of the Avalon composite terrane to form the Acadia composite terrane. This was followed in Mid Devonian times by 1500 km sinistral transcurrent movement of the Acadia Composite terrane to its present location. The climatic Hercynian/Alleghanian Orogeny resulted from the collision of Gondwana with North America in the Late Carboniferous and Permian.

Introduction

Although a variety of models have been proposed for the tectonic evolution of the Appalachian Orogen most have been based upon the assumption that in the Early Palaeozoic it represented a two-sided symmetrical system with the North American craton and the Avalon microcontinent bordering an ocean, Iapetus, (Williams 1964, 1978, 1979; Williams *et al.* 1972; Wilson 1966; Dewey 1969; Bird and Dewey 1970; Hatcher 1972). It has been gener-ally assumed that the Avalon microcontinent rifted from the North American craton during the Late Precambrian times to form the Iapetus Ocean followed by closing of Iapetus in Ordovician–Devonian times. This model forms the basis for the subdivision of the Appalachian Orogen into tectonostratigraphic zones and the assumption that these zones can be traced along the entire length of the orogen (Williams 1978). Most of these zones have recently been renamed terranes by Williams and Hatcher (1982). Thus, the Humber Zone represents the ancient continental margin of eastern North America, the Dunnage Zone or Terrane includes the vestiges of Iapetus, and the Gander Zone or Terrane forms the eastern continental margin of Iapetus built upon the western side of the Avalon Zone or Terrane interpreted as a microcontinent. The Meguma Zone or Terrane was interpreted as a continental rise prism southeast of the Avalon Zone by Schenk (1970) or as an intracratonic intradeep by Keppie (1982a). Recent palaeomagnetic results suggest that large transcurrent movements have taken place along faults oriented roughly parallel to the length of the Appalachians (Morris 1976; Kent and Opdyke 1978, 1979, 1980; van der Voo *et al.* 1979; van der Voo and Scotese 1981; Brochwicz-Lewinski *et al.* 1981; Kent 1982; Spariosu *et al.* 1983; Scotese *et al.* 1983). Concurrently, it has been suggested that much of the Appalachian Orogen represents a collage of terranes (Fig. 1) that collided and were accreted to the North American craton from the Ordovician through Late Palaeozoic times (Keppie 1977, 1981, 1982a and b). It should be noted that the definition of a terrane presented here, while similar to that of Coney *et al.* (1980), differs from that used by Williams and Hatcher (1982), and should not be confused with it. Most of the terranes of Williams and Hatcher (1982) are in fact superterranes in which a number of terranes have been grouped. It follows from the terrane concept presented here that southeast of the cratonic North American Palaeozoic miogeocline the orogen cannot be subdivided into zones extending along its entire length, and that a two-sided symmetrical model is altogether too simple. This model has been modified recently by Williams and Hatcher (1982) who propose that the Dunnage Zone was caught between the Humber and Gander Zones followed by sequential accretion of the Avalon and Meguma Zones. However, this model is also based upon the zonal subdivision, which, in general, is too broad to allow a precise genetic model for the Appalachian Orogen to be generated. In contrast, this paper returns to basics and represents a preliminary attempt to outline the terranes and define their nature, followed by an inferred accretionary history.

Modern plate tectonic analogues form the basis for a tectonic collage interpretation of the Appalachian Orogen. The present oceans include a variety of tectonic elements in addition to normal oceanic lithosphere, such as microcontinents, submarine plateaux, ridges, rises, and terraces, seamounts, oceanic and cratonic volcanic arc complexes, and small ocean basins, many filled with thick sediments (Ben-Avraham *et al.* 1981). During subduction these tectonic elements migrate towards the continental margins where many resist subduction and then become accreted as the trench steps oceanward. In general, subduction is usually oblique to plate boundaries causing transcurrent movements to take place along faults that parallel the strike of the developing orogen (Fitch 1972; Monger and Price 1979; Monger *et al.* 1982). Such transcurrent faulting is both pre- and post-accretionary and may involve displacements of hundreds of kilometres and significant rotations about vertical axes (Irving 1979). Thus, in the absence of positive evidence, it may be reasonable to assume that terranes that are presently adjacent within an orogen probably originally evolved separately.

General Principles

In order to clearly understand the terrane model for the northern Appalachian Orogen, it is essential to define a terrane and a terrane boundary, and to outline the constraints that can be placed upon times of terrane amalgamation.

A terrane is defined as an area characterized by an internal continuity of geology, including stratigraphy, faunal provinces, structure, metamorphism, igneous petrology, metallogeny, and palaeomagnetic record, that is distinct from that of neighbouring terranes and cannot be explained by facies changes.

Note that the definition of a terrane is based upon the geological record and does not rely on any genetic or plate tectonic interpretation. It should be noted that the distinct geological history unique to a terrane may vary from a small to a large part of the geological record. Thus, for example, a terrane may have originated as part of a craton from which it was separated for a time and subsequently accreted back to the same craton. Alternatively, a terrane may have spent most of its geological history at a considerable distance from the continent with which it was finally accreted. Thus a terrane may be of local origin or far-travelled. In some cases two or more terranes amalgamate before accretion to the continental margin. Such terranes are here termed composite terranes and are characterized by the continuity of a younger geological record across several older disparate terranes.

Terrane boundaries are generally faults or melanges representing sutures, now largely cryptic, with complex structural histories often involving transcurrent motions.

The melange boundaries are often interpreted as trench. complexes, but other origins are possible.

In attempting to reconstruct the accretionary history of an orogen, constraints upon the time of terrane accretion are required. A variety of geological constraints have been used in the Nova Scotia Appalachians (Keppie 1981, 1982 a and b) and in the Cordillera (Coney *et al.* 1980). A lower limit to the time of accretion is provided by the age of the youngest strata unique to a particular terrane. Deformation is often synchronous with accretion as a consequence of collision. Thus, the time of deformation gives the time of accretion and the spatial distribution of the deformation indicates which terranes were involved in the collision. Metamorphism and plutonism may accompany the deformation, but are generally much longer lived. Upper limits on the time of accretion are provided by the following criteria:

1. The age of the base of *overstepping strata* deposited across two or more neighbouring terranes;
2. The age of *straddling plutons* which intrude across the boundary between adjacent terranes;
3. The age of *thrusting* which places one terrane structurally above another;
4. The age of a *regional metamorphism* superimposed on adjacent terranes;
5. The age of the strata in one terrane containing *exotic pebbles* derived from rocks unique to another terrane.

Where such geological constraints are lacking, either through circumstances or lack of data, it is necessary to invoke correlation and deduction.

Northern Appalachian Terranes

The distribution of the major terranes, overstepping strata, and dated plutons in the northern Appalachians are shown on the map (Fig. 1)*. This is accompanied by a listing of the terranes, a brief description of the nature of each terrane and its boundaries, and an upper limit on the age of the main movement on individual boundaries. Time and space diagrams for the northern Appalachians are presented in Figs. 2 and 3*. The outline and nature of some terranes within the northern Appalachians are poorly defined and many of the constraints upon times of terrane amalgamation or accretion remain to be established. Therefore Fig. 1 is a preliminary terrane map and this paper provides only a working hypothesis.

A well-defined Cambro-Ordovician miogeocline built upon the eastern margin of cratonic North America may be traced along the entire western

margin of the Appalachians (Williams 1978). All areas east of the miogeocline are classed as terranes or underlain by terranes because their palaeotectonic positions relative to North America are uncertain, and they may be proximal or exotic. Also considered as terranes are those units which originated southeast of the miogeocline and have been thrust over the autochthonous rocks of ancient cratonic North America. In the northern Appalachians, these comprise the Humber Arm, Hare Bay, and Taconic Allochthons and the Lower St. Lawerence Valley Nappes (Fig. 1). They consist largely of continental rise sediments and oceanic lithosphere and probably originated adjacent to the North American continental margin (Bird and Dewey 1970; Williams 1979). Associated wildflysch in front and beneath these allochthons suggests that the nappe emplacement began during the Arenig.

Another class of northern Appalachian terranes includes those with a Grenvillian basement with or without its autochthonous miogeoclinal cover, which appear to have originated as part of cratonic North America but were subsequently rifted or underwent significant horizontal transcurrent displacement. Examples include Cape Ray and Fleur de Lys terranes in Newfoundland.

The southeastern margin of the cratonic North American miogeocline is marked by the presence of ophiolitic melange along its entire exposed length and has been termed the Baie Verte–Brompton Line (Williams 1977; Williams and St. Julien 1978). Several large ophiolite complexes (i.e. terranes) occur along this line in the eastern townships of Quebec and at Baie Verte in Newfoundland, but they are generally too small to be shown on the map (Fig. 1). Also, remobilized Grenvillian gneisses occur within this zone, the largest cropping out in the Chester, Athens, and Rayponda Domes of New England (Robinson and Hall 1980; Williams 1978). On the North American mainland, exposed portions of the Brompton Line are followed by a gravity gradient (Bird and Dewey 1970; Haworth *et al.* 1980). This geophysical signature continues through the Gaspe Peninsula suggesting to Bird and Dewey (1970) and this author that the Brompton Line may be extrapolated through the area where it is covered by the overlying strata of the Gaspe Synclinorium. Alternatively, Williams and St. Julien (1978) and Williams (1978) extrapolate the Brompton Line south of the Macquereau Dome based upon a correlation of these rocks with the rift facies at the precarious edge of the North American craton. While this interpretation may be valid, the gravity data suggests the Macquereau was rifted away from North America to form a distinct terrane. The Baie Verte–Brompton Line and its associated ophiolites

*Figures 1, 2 and 3 are in colour and included in the slip case.

have been variously interpreted as the western margin of Iapetus Ocean deformed during nappe emplacement and ophiolite obduction (Williams 1977; Laurent 1975) or as part of a marginal basin (Dewey and Bird 1971; Kidd 1977; Kidd et al. 1978).

A wide variety of Cambro-Ordovicain terranes occur between the Baie Verte–Brompton Line and the Avalon composite terrane. These include: microcontinental blocks (ChL, RI, LP, MC), microcontinental submarine plateaux or shelves (MB, SS, N, BI), oceanic volcanic arcs (AW, ND), cratonic volcanic arcs (BH, W, RA, BL), periarc basins (BV, E), small ocean basins (EM, PM, F, D, BW, CL), and some terranes of debatable origin (MM, CC, LP).

The Avalon composite terrane is made up of several distinct Late Precambrian terranes including continental volcanic rocks (AP, B), ensialic volcanic arcs (CB, C and possibly CH & K), and interarc basins (AH, CC) (Strong et al. 1978; Giles and Ruitenberg 1977; Keppie 1982b). On the mainland, widespread Late Ordovician, Silurian, and Devonian, flyschoid and molassic, overstepping strata blanket much of the older rocks, obscuring the older terranes over large areas. In Newfoundland, on the other hand, overstepping strata are more limited.

Southeast of the Avalon composite terrane lies the Meguma terrane with Cambro-Ordovician flysch deposited in an intradeep or trough within an orogenic zone (Keppie 1982a, b).

Accretionary History

At present, the accretionary history of the northern Appalachians is poorly understood because of uncertain constraints on time limits of amalgamation and the extensive overstepping strata that obscure older terranes. Also, although the identification of terranes relies essentially upon geological record and maps, reconstructions of the origin and relative movements of terranes are dependent largely upon Palaeomagnetic data. Even though such palaeomagnetic data is becoming available, it relates mainly to signatures acquired during the Late palaeozoic by which time most terranes were already amalgamated. Thus, the accretionary model presented here (Figs 1–3) represents only an introductory step towards a comprehensive model.

Recent palaeomagnetic data (Kent and Opdyke 1978, 1979, 1980; Spariosu and Kent 1981, 1983; Kent 1982; Spariosu et al. 1983; Scotese et al. 1983) for Silurian, Devonian, and Carboniferous rocks suggests that 1500 km sinistral movement took place between the Acadia composite terrane and the North American craton during mid-Carboniferous

times. More recent palaeomagnetic data suggest a mid-Devonian age for this movement (Kent D. V. pers. comm.) The distribution of palaeomagnetic sampling sites is still rather dispersed. Nevertheless, they limit the major fault movement to the area between Passamaquoddy Bay on the coast between U.S.A. and Canada, and Traveller Mountain–Plaster Rock in central Maine–New Brunswick. Likely candidates include the Pendar Brook Fault, the Fredericton Fault, or a fault along the northern side of Miramichi terrane (Fig. 1). The Fredericton Fault appears to be the most likely candidate for the locus of this movement. The Fredericton Fault passes westwards into the Norumberga Fault in eastern Maine which cuts Devonian plutons (Loiselle and Ayuso 1980). The westward extension of this fault is uncertain, but it appears to pass along the northern boundary of the Cushing gneisses and thence into the Flint Hill, Clinton, and Honey Hill Faults (Fig. 1). An upper limit on the movement on the Fredericton–Norumbega Fault is provided by the Westphalian age of the overstepping Pictou strata. The northeasterly extension of this fault is obscured by the Gulf of St. Lawrence. In Newfoundland it is still unclear how much of the island was part of the Acadia composite terrane which moved c. 1500 km during the mid-Devonian. The Avalon Platform terrane was attached to the Little Passage and Bois Island terranes by earliest Carboniferous times because the terrane boundaries are sealed by the Ackley City batholith (Fig. 1) yielding a whole rock Rb/Sr isochron age of 357 ± 5 Ma (Bell et al. 1977). Farther west, the Silurian Botwood Group has yielded poles (LaPointe 1979) similar to those derived from Siluro-Devonian rocks in the Acadia composite terrane in coastal Maine (Kent and Opdyke 1980). The similar character of Late Ordovician and Silurian rocks on most terranes from the Davidsville–Burgeo undivided terrane to the Fleur de Lys terrane suggests they were also part of the Acadia composite terrane. In this connection the anomalous gravity gradient over the northern Long Range (Haworth et al. 1980) and the extensive Devonian and Carboniferous deformation in western Newfoundland (Keppie et al. 1982) may suggest that it was involved either peripherally or as an integral part of the emplacement of the Acadia composite terrane. Thus the Cabot Fault or a fault running up the western side of Newfoundland are the most likely correlatives of the Fredericton–Norumbega Faults. Further palaeomagnetic results may resolve this question.

An outline of the accretionary history of the three segments, north and south of the Fredericton–Norumbega Fault and Newfoundland, is presented below. Space does not allow a detailed description of individual terranes and their boundaries. These data are summarized in the legend to Fig. 1.

Terranes north of the Fredericton–Norumbega Fault

During Cambro-Ordovician times, terranes in this area are interpreted to include continental rise sediments (T, SL), microcontinental islands (ChL), and platforms (MB) (Lyons *et al*. 1982), oceanic volcanic arcs (AW), cratonic volcanic arcs (BH, W, and Boundary), small ocean basins (EM, PM), and the Miramichi terrane which may be either a cratonic volcanic arc or a cratonic volcanic arc rift. The extensive overstepping strata in this part of the northern Appalachians largely obscures terrane boundaries, making it difficult to construct a detailed accretionary history. Nevertheless, two accretionary events are evident in this area (Figs. 1 and 2). The first involved amalgamation of four terranes (Chain Lakes, Bronson Hill, Pennington Mountain, and Weeksboro) in Late Cambrian or Early Ordovician time to form the Boundary composite terrane. These terranes are overstepped by the abyssal flysch of the unfossiliferous Dead River Formation believed to be of Arenig or older age on stratigraphic and structural evidence (Moench *et al*. 1981), thereby providing an upper limit on the time of amalgamation. The collision of these four terranes appears to be recorded by the Penobscot Disturbance (Neuman 1967; Hall 1969; Osberg 1983), a deformational event restricted to these four terranes. The absence of any effects of the Penobscot Disturbance in the North American miogeocline suggests that the amalgamation of these four terranes occurred at some place removed from cratonic North America. Convergence between the four terranes is explicable in terms of postulated subduction-related phenomena: namely the inferred volcanic arc origin for the Bronson Hill and Weeksboro terranes and the interpreted trench complex (Hurricane Mountain Formation) between the Bronson Hill and Chain Lakes terranes (Robinson and Hall 1980; Boone *et al*. 1981).

The second accretionary event involves the successive addition of terranes to the southern margin of cratonic North America during the Mid–Late Ordovician Taconian Orogeny (Fig. 2). In the northwest, the time of accretion is provided by the Arenig–Caradocian age of olistostrome genetically related to the emplacement of the Taconic and St. Lawrence Valley allochthons (Bird and Dewey 1970; St. Julien and Hubert 1975), and a 465 ± 5 Ma age for associated Taconian metamorphism (Sut-

ter and Dallmeyer 1982). An upper limit for this event is provided by the Caradocian age of the overstepping Magog Group flysch (St. Julien and Hubert 1975). The time of deformation and the age of the base of overstepping strata appears to become younger towards the southeast (Keppie *et al*. 1983; Boucot *et al*. 1964). Thus the time of deformation in the Elmtree–Macquereau and Miramichi terranes is bracketed between the mid-Caradocian age of the youngest deformed rocks and the late Llandoverian (Elmtree) or Ludlovian (Miramichi) age of the oldest overstepping strata (Fyffe 1982) (Figs. 1 and 2). The extent of any transcurrent motion between cratonic North America and these terranes is unknown. Convergence between cratonic North America and these terranes is explicable in terms of postulated subduction-related phenomena: Ordovician volcanic arcs in the Boundary composite terrane (Robinson and Hall 1980), Ascot–Weedon terrane (St. Julien and Hubert 1975) and possibly also in the Miramichi terrane (Bird and Dewey 1970). One possible explanation for the limited volume of volcanic arc lavas is that convergence was slow, possibly as a result of highly oblique relative plate motions. Ordovician plutonism is mainly confined to the Boundary composite volcanic arc terrane.

Terranes south of the Fredericton–Norumbega Fault

The backbone of this area is composed of various Precambrian terranes (B, I, CB, C, CH, K, AH, CC). These Precambrian terranes are inferred to have been amalgamated into the Avalon composite terrane by Early Cambrian times because of the lithological similarity of small outliers of Cambrian–Early Ordovician strata which locally unconformably overlie the various Precambrian sequences. This continental to submarine shelf overstep sequence containing an Atlantic fauna characterizes the Avalon composite terrane. These Cambro-Ordovician rocks do not overstep terrane boundaries and other tight constraints are generally absent. Thus, it is possible that these terranes may have been distributed along strike within the same Cambro-Ordovician facies belt and were not juxtaposed until later (Fig. 2). A reasonable upper time limit on amalgamation is only provided by a 530 ± 44 Ma whole-rock Rb–Sr isochron age for the Cheticamp pluton (Keppie and Smith 1978; Cormier 1972) which intrudes across the thrust boundary between the Cheticamp (CC) and Cape Breton (CB) terranes. On the other hand, the presence of moderate to strong, Late Precambrian Cadomian deformation in many of these terranes (B, CB, C, CH, K, AH, CC) (Rast and Skehan 1981; Ruitenberg and McCutcheon 1982; Keppie 1982; Keppie *et al.* 1982, 1983) supports their collision at this time. This accretionary–deformational event appears to be bracketed isotopically between 620 and 570 Ma (Keppie *et al.* 1983). The general absence of such deformation in the Avalon Peninsula terrane of Newfoundland suggests that it may not have been involved in this collisional event and thus may have been far removed from the other mainland Precambrian terranes during Cambro-Ordovician times.

During the Silurian and Devonian, terranes were successively added to both sides of the Precambrian core of the Avalon composite terrane. On the north side of the Avalon composite terrane lie the St. Stephen and Nashoba terranes. The Wheaton Brook thrust forms the boundary between the St. Stephen terrane and the Avalon composite terrane. The thrust is of Acadian age and intruded by the Early Carboniferous Mt. Champlain pluton (McCutcheon 1981). The Bloody Bluff Fault separates the Nashoba terrane (Lyons *et al.* 1982) and the Avalon composite terrane and is sealed by the Salem gabbro of unknown age (Cameron and Naylor 1976). Lithological and faunal similarities between Silurian and Early Devonian rocks on the Avalon composite terrane and the St. Stephen and Nashoba terranes (Boucot *et al.* 1974; Watkins and Boucot 1975) suggest that they were juxtaposed during this time interval although post-accretionary movements have disturbed direct overstepping relationships. A lower limit on the time of accretion is provided by the contrasting depositional environments suggested for Cambro-Ordovician euxinic shales of the Cookson Formation in the St. Stephen terrane and the shelf sediments of the same age in the Saint John Group on the Avalon composite terrane. Thus, the time of accretion is inferred to be Mid to Late Ordovician. This inferred accretionary event coincides with the fragmentary evidence for Taconian deformation and metamorphism in Penobscot Bay, southern Maine, and in the northern Antigonish Highlands (Stewart and Wones 1974; Murphy *et al.* 1980).

The Pendar Brook Fault separates the St. Stephen and Fredericton terranes. These terranes contain different Silurian stratigraphic assemblages: terrestrial to shallow marine sediments and volcanic rocks versus deep water turbidites respectively (Ruitenberg and Ludman 1978). The Pendar Brook fault is sealed by 400–385 Ma plutons (Wones *et al.* 1981), bracketing the time of accretion as Late Silurian–Early Devonian. Accretion of the Fredericton terrane coincides with Acadian deformational

events. Thus, terranes appear to be successively accreted to the northern margin of the Avalon composite terrane.

On the southeast side of the Avalon composite terrane and separated from it by the Minas Geofracture lies the Meguma terrane (Keppie 1982b). Exotic Torbook pebbles from the Meguma terrane have been found in earliest Carboniferous rocks in the Cape Breton terrane and Late Devonian metamorphic isograds have been superimposed across the Minas Geofracture (Keppie 1982b). These provide a Late Devonian upper limit for the time of accretion of the Meguma terrane. Accretion of the Meguma terrane appears to have been accompanied by the Early–Mid Devonian Acadian Orogeny. The Cape Cod terrane is poorly exposed (Robinson and Hall 1980). It could be part of either the Avalon composite terrane, the Meguma terrane or a separate entity.

Thus by Mid-Devonian times all of the terranes south of the Frederiction Fault had been amalgamated into the Acadia composite terrane. However, palaeomagnetic data cited earlier indicate that the Acadia composite terrane lay adjacent to Georgia and Florida during the Early Devonian and subsequently moved as one coherent block into its present geographic position during the Devonian along a major sinistral fault. The present precision of the palaomagnetic data is such that some intraterrane movement could be accommodated. An upper limit for this sinistral movement is provided by the overstep of the Westphalian Pictou Group across the Fredericton Fault in New Brunswick.

Newfoundland

In Newfoundland, except for the local occurrence of Late Cambrian–Early Ordovician deformation in the Twillingate area of the Notre Dame terrane (Williams *et al.* 1976), two accretionary events Mid-Ordovician and Devonian, may be recognized Figs. 1 and 3). The Late Cambrian–Early Ordovician event may perhaps be correlated with the Penobscot Disturbance suggesting a possible genetic connection with the Boundary composite terrane.

A Mid-Ordovician accretionary event records the addition of terranes (FdL, CR, S, E, and DB, and possibly BI and LP also) to the Long Range terrane resulting in the Taconian Orogeny. In western Newfoundland, the emplacement of the previously assembled Humber Arm and Hare Bay allochthonous terranes onto the North American continental margin is dated by the Arenig-Llanvirn age of the genetically related olistostrome (Bird and Dewey

1970; Williams 1979). An upper limit on this event is provided by the Caradocian age of the overstepping Long Point Formation (Rodgers 1965). Further east, on the Burlington Peninsula, the time of accretion is given by the 460 → 20 Ma U–Pb zircon age of the Dunamagon granite (Dallmeyer 1981) which straddles the boundary between the Burlington and Mings Bight terrancs. Elsewhere overstepping sequences resting directly across terrane boundaries are generally absent due to post-accretionary movements. However, the similarity of Late Ordovician (Caradocian and later) and Silurian rocks on most terranes from the Fleur de Lys to the Davidsville–Burgeo undivided terrane suggests Middle Ordovician amalgamation. Convergence between these terranes is explicable in terms of subduction-related phenomena such as the volcanic arc origins inferred for the Burlington, Notre Dame, and Exploit terranes (Williams 1979).

Finally, a Devonian time of accretion is inferred for the Bois Island, Little Passage, and Avalon Platform terranes with the Davidsville–Burgeo undivided terrane resulting in the Acadian Orogeny. The oblique emplacement of the Avalon Platform terrane produced a ductile shear zone in the adjacent area (Hanmer 1982) and extensive overthrusting farther west (Karlstrom et al. 1982). Dating in this shear zone indicates that the mylonitic foliation formcd between 365 and 420 Ma (Dalmeyer et al. 1981). An upper limit on the time of accretion is also provided by sealing of terrane boundaries by straddling plutons ranging in age from 383 ± 15 Ma for the Middle Ridge pluton to 357 ± 5 Ma for the Ackley City batholith (Bell et al. 1977), and by the record of superposed metamorphic effects across the Dover Fault yielding 350–390 Ma $^{40}Ar/^{39}Ar$ on biotites and phyllites (Dallmeyer et al. 1981).

Discussion

Oblique subduction resulting in obduction and oblique accretion of terranes and associated transcurrent faulting provide the basis for solutions to several intractable northern Appalachian problems, such as the origin of the Chain Lakes terrane. This terrane is unique within the Appalachian Orogen and is characterized by diamictitic and gneissic basement rocks yielding an U–Pb zircon age of 1600 Ma (Naylor et al. 1973), i.e. pre-Grenvillian. A small fragment of the Chain Lakes massif has also been recorded on the Brompton Line (Keppie et al. 1982). The southern margin of the Chain Lakes massif is inferred to be welded to an ophiolitic sequence (Boone et al. 1981), a junction which could be interpreted as

an Atlantic-type continent/ocean boundary or a metamorphosed fault. The origin of the Chain Lakes terrane is uncertain, however, palaeomagnetic data suggest that the Baltic Shield lay adjacent to this part of the North American craton during the Late Precambrian and Early Cambrian (Ueno et al. 1975). The Swedish part of the Baltic Shield is composed mainly of the Svecokarelian Orogen formed between 1950 and 1750 Ma with magmatic activity continuing until c. 1300 Ma (Lundquist 1979). In southwest Sweden, the Svecokarelian basement is overprinted by Grenvillian tectonothermal events. It is possible that the basement rocks of the Chain Lakes terrane represent fragments of the Svecokarelian Orogen left near the North American craton when the Baltic Shield separated. It follows from this hypothesis that (i) the Grenvillian Orogeny may well be the consequence of the collision of the Svecokarelian basement with cratonic North America, (ii) Iapetus must have formed close to the cryptic suture between the Grenvillian and Svecokarelian basements; and (iii) the Ascot–Weedon (AW) terrane must have initially been a small ocean basin produced by subsidiary rifting on the margin of Iapetus, and subsequently closed during the Taconian Orogeny by relatively minor subduction reflected by the small amount of volcanic arc activity.

Transcurrent movements associated with oblique subduction and following accretion can dismember an originally systematic arrangement of tectonic elements resulting in a chaotic juxtaposing of terranes. Extremely oblique subduction probably could explain the paucity of Late Ordovician to Early Devonian volcanic arcs in the Appalachians. Thus, transcurrent movements with limited subduction could occur in the mainland Appalachians while predominantly subduction took place elsewhere such as in Newfoundland, Ireland, and Great Britain producing volcanic arcs transverse to the strike of the orogen (Stillman and Francis 1979; Mitchell and McKerrow 1975). Remnants of these transverse trends appear to have been preserved in the Notre Dame Bay area of Newfoundland (Fig. 1) and in the Siluro-Devonian Andean volcanic arc in northern Maine and the Gaspe Peninsula (Osberg 1978), even though syn- and post-accretionary deformation would cause the volcanic arcs to be rotated towards the strike of the orogen. Thus, the Maine–Gaspe volcanic arc recorded a palaeomagnetically determined, post-Early Devonian, 10° to 30° clockwise rotation (Spariosu and Kent 1983) consistent with the oblique dextral subduction which may be inferred from the trend of the volcanic arc. This agrees with postulated dextral oblique subduction associated with Late Ordovician and Silurian volcanic arcs in the British Isles (Phillips *et al*. 1976). The location of the northern Maine–Gaspe volcanic arc far from a contemporary continental margin is understood only when the Acadia composite terrane is removed to its palaeomagnetically determined Siluro-Devonian position c. 1500 km southward along strike (Kent and Opdyke 1978, 1979). This allows a Siluro-Devonian continental margin to be situated along the Fredericton Fault.

By analogy with the northern Maine–Gaspe volcanic arc, the transverse trend of the Ordovician volcanic arcs in the Notre Dame Bay area of Newfoundland may be interpreted in terms of a sinistral component to the oblique subduction. No palaeomagnetic data are available yet to test this hypothesis.

Another interesting tectonic problem of the northern Appalachians is the origin of the Gander Zone in Newfoundland. It has generally been interpreted as a continental rise prism on the southeastern side of Iapetus adjacent to the Avalon microcontinent (Kennedy 1975; Williams 1979). However, recent work in southern Newfoundland, where the grade of metamorphism within the Gander Zone is lower, has documented the presence of two distinct microcontinental terranes, the Little Passage and Bois Island terranes (Coleman-Sadd 1980). The presence of these terranes where a continental rise prism could have been predicted adjacent to the Avalon terrane is explicable in terms of extensive Acadian ductile shearing and transcurrent faulting along and west of the Dover Fault (Hanmer 1982).

In conclusion, the terrane map of the Appalachians provides the basic building block on which evolutionary models may be constructed. However, much geological data still remain to be collected to provide better definition of the terranes, their nature, constraints on the accretionary history, and the tectonic evolution of individual terranes.

Furthermore, the terrane map shows that there are many 'sutures' between terranes rather than one unique Iapetus suture as proposed by some authors (e.g. Brown 1973; McKerrow and Cocks 1977). Also, there is a dearth of palaeomagnetic data for the Early Palaeozoic and Precambrian which is essential to provide constraints on palinspastic reconstructions.

Acknowledgements

I would like to thank R. D. Dallmeyer, R. A. Price, and P. H. Osberg for their critical reviews of the manuscript and helpful discussions. I am also grateful to S. Saunders for typing the manuscript and to Maritime Resource Management Services for the cartography. This paper is published with the permission of the Director of Geological Surveys Division of the Nova Scotia Department of Mines and Energy.

References

Bell, K., Blenkinsop, J. and Strong, D. F. 1977. The geochronology of some granite bodies from eastern Newfoundland and its bearing on Appalachian evolution. *Can. J. Earth Sci.*, **14**, 456–476.

Ben-Avraham, Z., Nur, A., Jones, D. and Cox, A. 1981. Continental accretion: from oceanic plateau to allochthonous terranes. *Science*, **213**, 47–54.

Bird, J. M. and Dewey, J. F. 1970. Lithosphere plate–continental margin tectonics and the evolution of the Appalachian Orogen. *Bull. Geol. Soc. Am.*, **81**, 1031–1060.

Boone, G. M., Boudette, E. L. and Moench, R. H. 1981. Geologic outline map of pre-Silurian stratigraphic units, north–central Maine to northern New Hampshire. *Geol. Soc. Am. Abstract*, **13**(3), 123.

Boucot, A. J., Dewey, J. F., Dineley, D. L., Fyson, W. K., Hickox, C. F., McKerrow, W. S. and Ziegler, A. M. 1974. Geology of the Arisaig Area. *Geol. Soc. Am. Spec. Pap.*, **139**, 191 p.

Boucot, A. J., Field, M. T., Fletcher, R., Forbes, W. H., Naylor, R. S. and Pavlides, L. 1964. Reconnaissance Bedrock geology of the Presque Isle Quadrangle, Maine. *Maine Geol. Surv. Ser.*, No. 2, 123 p.

Brochwicz-Lewinski, W., Pozaryski, W. and Tomczyk, H. 1981. Wielkoskalowe Ruchy Przesuwcze Wzdluz SW Brzegu Platformy Wschodnioeuropejskiej We Wczesnym Palaeozoiku. *Przeglad Geol.*, **28**(8), 385–397.

Brown, P. A. 1973. Possible cryptic suture in southwest Newfoundland. *Nature, London, (Phys. Sci.)*, **245**, 9–10.

Cameron, B. and Naylor, R. S. 1976. General geology of southeastern New England. In Cameron, B. (Ed.), *Geology of Southeastern New England. New Engl. Int. Geol. Conf. Field Trip Guidebook.*, **68**, 13–27.

Coleman-Sadd, S. P. 1980. Geology of south–central Newfoundland and evolution of the eastern margin of Iapetus. *Am. J. Sci.*, **280A**, 991–1017.

Coney, P. J., Jones, D. L. and Monger, J. W. H. 1980. Cordilleran suspect terranes. *Nature*, London, **288**, 329–333.

Cormier, R. F. 1972. Radiometric ages of granitic rocks, Cape Breton Island, Nova Scotia. *Can. J. Earth Sci.*, **9**, 1074–1086.

Dallmeyer, R. D. 1981. Geochronology of the Newfoundland Appalachians: recent advances. *Geol. Soc. Am. Abstr.*, **13**, p. 127.

Dallmeyer, R. D., Blackwood, R. F. and Odom, A. L. 1981. Age and origin of the Dover Fault: tectonic boundary between the Gander and Avalon Zones of the northeastern Newfoundland Appalachians. *Can. J. Earth Sci.*, **18**, 1431–1442.

Dewey, J. F. 1969. The evolution of the Caledonian/Appalachian Orogen. *Nature*, London, **222**, 124–128.

Dewey, J. F. and Bird, J. M. 1971. Origin and emplacement of the ophiolite suite. Appalachian ophiolites in Newfoundland. *J. Geophys. Res.*, **76**, 3179–3206.

Fitch, T. J. 1972. Plate convergence, transcurrent faults and internal deformation adjacent to southeast Asia and the western Pacific. *J. Geophys. Res.*, 77, 4432–4460.

Fyffe, L. R. 1982. Taconian and Acadian structural trends in Central and Northern New Brunswick. *Geol. Assoc. Can. Spec. Pap.*, 24, 117–130.

Giles, P. S. and Ruitenberg, A. A. 1977. Stratigraphy, paleogeography and tectonic setting of the Coldbrook Group in the Caledonian Highlands of southern New Brunswick. *Can. J. Earth Sci.*, 14, 1263–1275.

Hall, B. A. 1969. Pre-Middle Ordovician unconformity in northern New England and Quebec. In Kay, M. (Ed.), *North Atlantic—Geology and Continental Drift. Amer. Assoc. Petr. Geol. Mem.*, 12, 467–476.

Hanmer, S. 1982. Tectonic significance of the Gander Zone, Newfoundland: an Acadian ductile shear zone. *Can. J. Earth Sci.*, 18, 121–135.

Hatcher, R. D. Jr. 1972. Developmental model for the Southern Appalachians. *Bull. Geol. Soc. Am.*, 83, 2735–2760.

Haworth, R. T., Daniels, D. L., Williams, H. and Zietz, I. 1980. Bouger gravity anomaly map of the Appalachian Orogen. *Memorial Univ. of Newfoundland, Map No. 3*.

Irving, E. 1979. Paleopoles and paleolatitudes of North America and speculations about displaced terrains. *Can. J. Earth Sci.*, 16, 669–694.

Karlstrom, K. E., van der Pluijm, A. and Williams, P. F. 1982. Structural interpretation of the eastern Notre Dame Bay area, Newfoundland: regional post-Middle Silurian, thrusting and asymmetrical folding. *Can. J. Earth Sci.*, 19, 2325–2341.

Karson, J. and Dewey, J. F. 1978. Coastal complex, western Newfoundland: an early Ordovician fracture zone. *Bull. Geol. Soc. Am.*, 89, 1037–1049.

Kennedy, M. J. 1975. Repetitive orogeny in the northeastern Appalachians—new plate models based upon Newfoundland examples. *Tectonophysics*, 28, 39–87.

Kent, D. V. 1982. Paleomagnetic evidence for post-Devonian displacement of the Avalon platform (Newfoundland). *J. Geophys. Res.*, 87, 8709–8716.

Kent, D. V. and Opdyke, N. D. 1978. Paleomagnetism of the Devonian Catskill red beds: evidence for motion of the coastal New England–Canadian Maritime region relative to North America. *J. Geophys. Res.*, 83, 4441–4450.

Kent, D. V. and Opdyke, N. D. 1979. The Early Carboniferous paleomagnetic field of North America and its bearing on the tectonics of the northern Appalachians. *Earth Planet. Sci. Lett.*, 44, 365–372.

Kent, D. V. and Opdyke, N. D. 1980. Paleomagnetism of Siluro-Devonian rocks from eastern Maine. *Can. J. Earth Sci.*, 17, 1653–1665.

Keppie, J. D. 1977. Plate tectonic interpretation of Palaeozoic world maps (with emphasis on circum-Atlantic Orogens and southern Nova Scotia). *Nova Scotia Dept. Mines Pap.*, 77–3, 1–45.

Keppie, J. D. 1981. The Appalachian Collage. *Terra cognita*, 1, 54.

Keppie, J. D. 1982a. Tectonic Map of Nova Scotia, *Nova Scotia Dept. Mines and Energy*, Scale 1:500,000.

Keppie, J. D. 1982b. The Minas Geofracture. *Geol. Assoc. Can. Spec. Pap.*, 24, 263–280.

Keppie, J. D. and Smith P. K. 1978. Compilation of isotopic age data of Nova Scotia. *Nova Scotia Dept. Mines and Energy Rept.*, 78–4.

Keppie, J. D., St. Julien, P., Hubert, C., Beland, J., Skid-more, B., Ruitenberg, A. A., Fyffe, L. R., McCutcheon, S. R., Williams, H. and Bursnall, J. 1982. Structural map of the Canadian Appalachians. *Memorial University of Newfoundland, Map No. 4*, Scale 1:1,000,000.

Keppie, J. D., St. Julien, P., Hubert, C., Beland, J., Skid-more, B., Ruitenberg, A. A., Fyffe, L. R., McCutcheon, S. R. Williams, H. and Bursnall, J. 1983. Times of deformation in the Canadian Appalachians. In *NATO–ASI/IGCP Atlantic Canada 1982 volume, C 116*, Reidel, Holland. 307–314.

Kidd, W. S. F. 1977. The Baie Verte Lineament, Newfoundland: ophiolite complex floor and mafic volcanic fill of a small Ordovician marginal basin. *American Geophysical Union, Maurice Ewing Series 1*, pp. 407–418.

Kidd, W. S. F., Dewey, J. F. and Bird, J. M. 1978. The Mings Bight ophiolite complex, Newfoundland: Appalachian ocean crust and mantle. *Can. J. Earth Sci.*, 15, 781–804.

LaPointe, P. L. 1979. Paleomagnetism and orogenic history of the Botwood Group and Mount Pleasant batholith, Central Mobile Belt, Newfoundland. *Can. J. Earth Sci.*, 16, 866–876.

Laurent, R. 1975. Occurrences and origin of the ophiolites in southern Quebec, Northern Appalachians. *Can. J. Earth Sci.*, 12, 443–455.

Loiselle, M. C. and Ayuso, R. A. 1980. Geochemical characteristics of granitoids across the Merrimack Synclinorium, eastern and central Maine. In Wones, D. (ed.), *IGCP-Caledonide Orogen Project Proc. VPISU Mem.*, 2, 117–121.

Lundquist, T. 1979. The Precambrian of Sweden. *Sver. Geol. Und.*, C 768, 87 pp.

Lyons, J. B., Boudette, E. L., and Aleinikoff, J. N. 1982. The Avalonian and Gander Zones in central eastern New England. *Geol. Assoc. Can. Spec. Pap.*, 24, 43–66.

McCutcheon, S. R. 1981. Revised stratigraphy of the Long Reach area, southern New Brunswick: evidence for major, northwestward directed Acadian thrusting. *Can. J. Earth Sci.*, 18, 646–656.

McKerrow, W. S. and Cocks, L. R. M. 1977. The location of the Iapetus Ocean suture in Newfoundland. *Can. J. Earth Sci.*, 14, 488–499.

Mitchell, A. H. G. and McKerrow, W. S. 1975. Analogous evolution of the Burma orogen and the Scottish Caledonides. *Bull. Geol. Soc. Am.*, 86, 305–315.

Moench, R. H., Boudette, E. L. and Boone, G. M. 1981. Two stratigraphically separate pre-Silurian volcanic sequences in northern New England: a summary of stratigraphy and problems. *Geol. Soc. Am. Abstr.*, 13(3), 166.

Monger, J. W. H. and Price, R. A. 1979. Geodynamic evolution of the Canadian Cordillera—progress and problems. *Can. J. Earth Sci.*, 16, 770–791.

Monger, J. W. H., Price, R. A. and Tempelman-Kluit, D. J. 1982. Tectonic accretion and the origin of the two major metamorphic and plutonic welts in the Canadian Cordillera. *Geology*, 10, 70–75.

Morris, W. A. 1976. Transcurrent motion determined paleomagnetically in the northern Appalachians and Caledonides and the Acadian Orogeny, *Can. J. Earth Sci.*, 13, 1236–1243.

Murphy, J. B., Keppie, J. D. and Hynes, A. 1980. Geology of the northern Antigonish Highlands. *Nova Scotia Dept. Mines and Energy Rept.*, 80–1, 103–108.

Naylor, R. S., Boone, G. M., Boudette, E. L., Ashenden,

D. D. and Robinson, P. 1973. Pre-Ordovician rocks in the Bronson Hill and Boundary Mountain anticlinoria, New England, U.S.A. *EOS*, **54**, 495.

Neuman, R. B. 1967. Bedrock geology of the Shin Pond and Staceyville quadrangles, Penobscot County, Maine. *U.S. Geol. Surv. Pap.*, **524–I**, 37 p.

Osberg, P. H. 1978. Synthesis of the geology of the northern Appalachians, U.S.A. *Geol. Surv. Can. Pap.*, **78–13**, 137–147.

Osberg, P. H. 1983. Timing of Orogenic events in the U.S. Appalachians. In *NATO–ASI/IGCP Atlantic Canada 1982 volume C 116*, Reidel, Holland, 315–338.

Phillips, W. E. A., Stillman, C. J. and Murphy, T. 1976. A Caledonian plate tectonic model. *J. Geol. Soc. London*, **132**, 579–609.

Rast, N. and Skehan, J. W. S. J. 1981. Possible correlation of Precambrian rocks of Newport, Rhode Island, with those of Anglesey, Wales. *Geology*, **9**, 596–601.

Robinson, P. and Hall L. M. 1980. Tectonic synthesis of southern New England. In Wones, D. (Ed.), *IGCP–Caledonide Orogen Project Proc. VPISU Mem.*, **2**, 73–82.

Rodgers, J. 1965. Long Point and Clam Bank Formations, western Newfoundland. *Geol. Assoc. Can. Proc.*, **16**, 83–94.

Ruitenberg, A. A. and Ludman, A. 1978. Stratigraphy and tectonic setting of early Paleozoic sedimentary rocks of the Wirral–Big Lake area, southwestern New Brunswick and southeastern Maine. *Can. J. Earth Sci.*, **15**, 22–32.

Ruitenberg, A. A. and McCutcheon, S. R. 1982. Acadian and Hercynian structural evolution of southern New Brunswick. *Geol. Assoc. Spec. Pap.*, **24**, 131–148.

Scotese, C. R., van der Voo, R., Johnson, R. E. and Giles, P. S. 1983. Paleomagnetic results from the Carboniferous of Nova Scotia. *Amer. Geophys. Union Geodynamics Series*, **12**, 63–81.

Schenk, P. E. 1970. Regional variation in the flysch-like Meguma Group (Lower Paleozoic) of Nova Scotia, compared to recent sedimentation off the Scotia Shelf. *Geol. Assoc. Can, Spec. Pap.*, **7**, 127–153.

Spariosu, D. J. and Kent, D. V. 1981. Paleomagnetism of Lower Carboniferous redbeds and volcanics from western New Brunswick. *EOS, Trans AGU*, **62**, 264.

Spariosu, D. J. and Kent, D. V. 1983. Paleomagnetism of the Lower Devonian Traveler Felsite and the Acadian Orogeny in the New England Appalachians. *Bull. Geol. Soc. Am.*,

Spariosu, D. J., Kent, D. V. and Keppie, J. D. 1983. Late paleozoic motions of the Meguma Zone, Nova Scotia: new paleomagnetic evidence. *Amer. Geophys. Union. Geodynamics Series*, **12**, 82–98

St. Julien, P. and Hubert, C. 1975. Evolution of the Taconian Orogen in the Quebec Appalachians. *Am. J. Sci.*, **275A**, 337–362.

Stewart, D. L. and Wones, D. R. 1974. Bedrock geology of northern Penobscot Bay area. In Osberg, P. H. (Ed.), *Geology of East-Central and North-Central Maine.*

New Engl. Int. Geol. Conf. Field Trip Guide, 223–239.

Stillman, C. J. and Francis, E. H. 1979. Caledonide volcanism in Britain and Ireland. *J. Geol. Soc. London*, **8**, 557–578.

Strong, D. F., O'Brien, S. J., Taylor, S. W., Strong, P. G. and Wilton, D. H. 1978. Aborted Proterozoic rifting in eastern Newfoundland. *Can. J. Earth Sci.*, **15**, 117–131.

Sutter, J. F. and Dallmeyer, R. D. 1982. Timing of metamorphism in the Appalachian Orogen based upon ^{40}Ar/^{39}Ar dates of hornblende and biotite. *Bull. Geol. Soc. Am.*, **14**, 87.

Ueno, H., Irving, E. and McNutt, R. H. 1975. Palaeomagnetism of the Whitestone Anorthosite and Diorite, the Grenville Polar track, and relative motions of Laurentian and Baltic Shields. *Can. J. Earth Sci.*, **12**, 209–226.

Van der Voo, R. and Scotese, C. 1981. Paleomagnetic evidence for a large ($\approx$2000 km) sinistral offset along the Great Glen fault during Carboniferous time. *Geology*, **9**, 583–589.

Van der Voo, R., French, A. N. and French, R. B. 1979. A paleomagnetic pole position from the folded Upper Devonian Catskill redbeds, and its tectonic implications. *Geology*, **7**, 345–348.

Watkins, R. and Boucot, A. J. 1975. Evolution of Silurian brachiopod communities along the southeastern coast of Acadia. *Bull. Geol. Soc. Am.*, **86**, 243–254.

Williams, H. 1964. The Appalachians in Northeastern Newfoundland—a two-sided symmetrical system. *Am. J. Sci.*, **262**, 1137–1158.

Williams, H. 1977. Ophiolitic melange and its significance in the Fleur de Lys Supergroups, northern Appalachians. *Can. J. Earth Sci.*, **14**, 987–1003.

Williams, H. 1978. *Tectonic–Lithofacies Map of the Appalachian Orogen*, Map No. 1, Memorial Univ. of Newfoundland.

Williams, H. 1979. Appalachian Orogen in Canada. *Can. J. Earth Sci.*, **16**, 792–807.

Williams, H. and Hatcher, R. D., Jr. 1982. Suspect terranes and accretionary history of the Appalachian Orogen. *Geology*, **10**, 530–536.

Williams, H. and St. Julien, P. 1978. The Baie Verte–Brompton Line in Newfoundland and regional correlations in the Canadian Appalachians. *Geol. Surv. Can. Pap.*, **78–1A**, 225–229.

Williams, H., Dallmeyer, R. D. and Wanless, R. K. 1976. Geochronology of the Twillingate Granite and Herring Neck Group, Notre Dame Bay, Newfoundland. *Can. J. Earth Sci.*, **13**, 1591–1601.

Williams, H., Kennedy, M. J. and Neale, E. R. W. 1972. The Appalachian Structure Province. In Price, R. A. and Douglas, R. J. W. (Eds), *Variations in tectonic styles in Canada. Geol. Assoc. Can. Spec. Pap.*, **11**, 181–261.

Wilson, J. T. 1966. Did the Atlantic close and then reopen? *Nature*, London, **211**, 676–681.

Wones, D. R., Loiselle, M., Ayuso, R. A., Sinha, A. K., Scambos, T. A. and Andrew, A. 1981. Source models for Caledonian plutons, eastern and central Maine. *Geol. Soc. Am. Abstr.*, **13**, 184.

The Caledonide Orogen—Scandinavia and Related Areas
Edited by D. G. Gee and B. A. Sturt
© 1985 John Wiley & Sons Ltd

Early Caledonian orogenesis in northwestern Europe

P. D. Ryan[*] and **B. A. Sturt**[†]

* Dept. of Geology, University College, Galway, Ireland
† Geological Institute, Bergen University, N–5014 Bergen, Norway

ABSTRACT

It is now obvious that major climactic orogenic events occurred in the British Isles (Grampian) and Scandinavia (Finnmarkian) during Late Cambrian to Early Ordovician time, involving polyphase deformation and regional metamorphism of mainly Barrovian type. Inherent to this time period was the widespread obduction of ophiolite complexes in Scandinavia, Shetland and Girvan. This broadly coincides with major obduction of Iapetus-derived ophiolites in the Appalachian segment of the orogen, although here the timing of obduction was somewhat later. It appears probable that subduction mélanges and/or high pressure/low temperature metamorphism associated with these climactic events are present in Spitsbergen, Girvan, Mayo, and Anglesey. The evidence for the timing of these events and their spatial relationships is reviewed and their implications for tectonic modelling are discussed.

Introduction

Early Caledonian orogenic events, which affected Eocambrian tillite or tilloid-bearing successions and other rock units of equivalent age and were completed before mid-Ordovician times, have long been recognized within the northern British Caledonides. However, evidence for the existence of widespread, broadly coeval, orogenic activity within the Scandinavian Caledonides is only now emerging. This paper reviews the evidence for the nature and time-ing of these events in Scandinavia and then compares them with the available data from the northern British Caledonides and other parts of the Caledonides of northwest Europe.

Evidence for the age of early Caledonian events has been mainly drawn from two sources: stratigraphic evidence for the youngest age of rocks affected by these events or the oldest age of rocks incorporating detritus derived from early Caledonian terrains, or resting unconformably upon them; and radiometric dates of certain specific early Caledonide events. To overcome the problem of cooling ages, this synthesis relies mainly upon ages given as the time of emplacement of igneous rocks whose position in the structural chronology of early Caledonian events is known.

Any synthesis of the nature involving the comparison of bio- and chronostratigraphic data is limited by the accuracy with which the Lower Palaeozoic time scale is known. Unfortunately, there is no clear consensus at the present time and individual estimates for the absolute ages of the Lower Palaeozoic biostratigraphic zones vary: compare Bluck *et al*. (1980), Churkin *et al*. (1977), Cowie and Cribb (1978), Faure (1977), Gale *et al*. (1979), Harland *et al*. (1982), McKerrow *et al*. (1980) and Patchett *et al*. (1980). It is not our intention to comment upon the relative merits of the recently published age scales cited above. However, in this work we used the geologic timescale of Harland *et al*. (1982) as it fully covers the stratigraphic range of early Caledonian events; the ages given for the various epoch boundaries (Table 1) overlap, within the limits of error, nearly all of those suggested by other authors.

In the text, all ages quoted have been, where necessary, recalculated using the same standardized decay constants (Steiger and Jäger 1977) employed by Harland *et al*. (1982).

Table 1 Comparison of the standard Welsh and Oslo successions with absolute ages for each epoch or series boundary with their estimated error based on Harland *et al.* (1982) and Hemmingsmoen (1960)

	Welsh	Oslo	Age Ma	'Error' Ma
System	Epochs	Series		
SIL			438	±6
ORDOVICIAN	Ashgill	Tretaspis (4c–5b)	448	±6
	Caradoc	Chasmops (4aβ–4b)	458	±8
	Llandeilo	Ogygiocaris (4aβ)	468	±8
	Llanvirn		478	±8
	Arenig	Asaphus (3b–c)	488	±10
	Tremadoc	Ceratopyge (2e–3a)	505	±16
CAMBRIAN	Merioneth	Olenid (2a–d)	525	±18
	St. David's	Paradoxides (1c–d)	540	±14
	Caerfai (Comley)	Olenellus (1a–b)	590	?
PREC.				

Early Caledonian Orogenesis in the Scandinavian Caledonides

The Finnmarkian orogenic phase

The Finnmarkian orogenic phase has its type expression in the northernmost part of Norway (Fig. 1). Here a characteristic lithofacies sequence unconformably overlies Precambrian basement rocks in autochthonous, parautochthonous, and allochthonous positions (Ramsay *et al.* this volume). The lower parts of this sequence contain late Precambrian tillite in the autochthon and the parautochthon, although tillites have not been reported from the higher-grade allochthonous nappes. In the foreland the stratigraphy is continuous through from the tillite-bearing horizons to the Tremadoc (Reading 1965), though the highest datable stratigraphic level in the allochthon is Lower and/or Middle Cambrian (Holland and Sturt 1970).

The Finnmarkian zone is characterized by a series of distinctive, generally thin-skinned nappes, though major recumbent folds approaching the dimension of the Tay Nappe, in the Dalradian, occur in different nappe units. Each nappe unit contains pre-Caledonian rocks which are covered unconformably by the Caledonian succession (Ramsay *et al.*, this volume). The pre-Caledonian units vary from Karelian supercrustals to pre-Karelian high-grade gneisses, and there are indications of Grenville metamorphism in the basement plinths of certain of the highest nappes (Sturt *et al.* 1978).

The nappes are metamorphosed in Barrovian amphibolite and upper greenschist facies, and the metamorphic maximum (750–700°C at 7–9 kbar, Sturt and Taylor 1972) was achieved between the two major regional fold producing deformations D1 and D2. In the northwest of the region the unique feature of Seiland Province magmatism is expressed in a relatively small area some 90 × 50 km with its great range of igneous rocks. These consist of layered gabbros with varied parent magmas (Robins and Gardner 1975), ultramafic rocks (Bennet 1974; Sturt *et al.* 1980), alkaline rocks (Sturt and Ramsay 1965; Robins and Gardner 1975), and dioritic and granite rocks (Sturt and Taylor 1972; Speedyman 1972). The igneous complex is synorogenic and straddles the two major deformation stages (D1 and D2), and represents an integral part of Finnmarkian tectonothermal evolution. On Sørøy, the D1 Haraldseng and Haskvik granite dykes fall within the range 540–528 Ma, whilst the D2 Baarvik syenite dyke is dated at 490 Ma (Sturt *et al* 1978). Late D2 nepheline syenites have been dated by the K–Ar method and nephelines yield ages of 491–480 Ma (Sturt *et al.* 1967) which is very close to the younger limit of the Rb–Sr dating (Fig. 2). The pattern of K–Ar and Rb–Sr mineral ages from the region are consistent with reheating during Scandian orogenesis, followed by uplift and cooling (Sturt *et al.* 1978).

In the autochthon, close to the thrust front, and in the low-grade parautochthonous, Late Precambrian slates (Stappogeide and Friarfjord Formations) and Riphean slates north of the fault (Kongsfjord Formation) bear a pronounced slaty cleavage (S1). Rb–Sr whole rock isochron studies of these (Sturt *et al.* 1978; Taylor and Pickering 1981) yield consistent results between 520–490 Ma which is in the range of the extended D2 formation within the allochthon (Fig. 2). Few post-orogenic intrusions have been identified in the type area, although the Trollvik granite of North Troms clearly cuts across a Finnmarkian deformation fabric (Zwaan personal communication 1981). This granite has been dated by the Rb–Sr whole rock isochron method to 453 ± 27 Ma (Dangla *et al.* 1978).

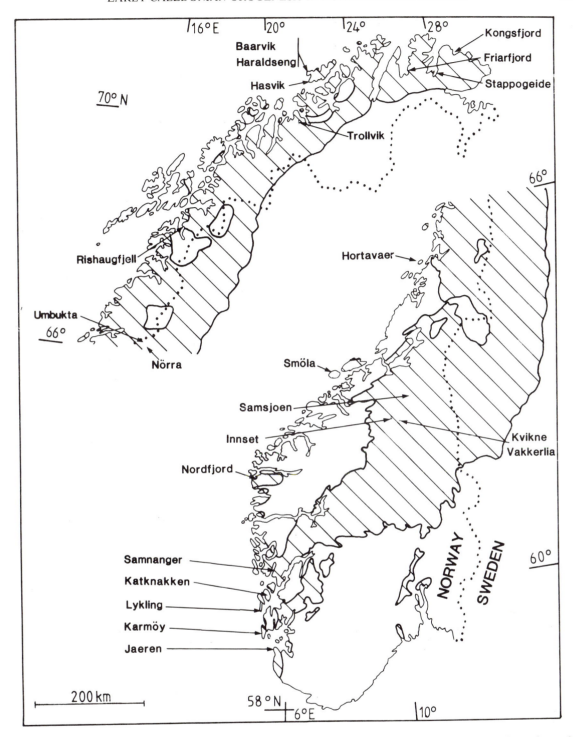

Fig. 1   Map showing the Scandinavian Caledonides (oblique ruling) and the location of the various sites where Finnmarkian or early Caledonian orogenic activity has been recorded

The ages of the syn-Finnmarkian plutons all fall within the range 540–490 Ma. D1 in the allochthon, which postdated the deposition of the Early Middle Cambrian, occurred c. 540–530 Ma whilst the earliest fabric in the parautochthon and autochthon is dated at 520–490 Ma and affects strata as young as Tremadocian. This diachroneity is attributed to the marginwards migration of the deformation front (Sturt et al. 1975; Ramsay et al., this volume). The available radiometric and stratigraphic constraints upon the timing of the Finnmarkian orogenic phase correspond well with one another and place this event in the interval Mid-Cambrian to Early Ordovician.

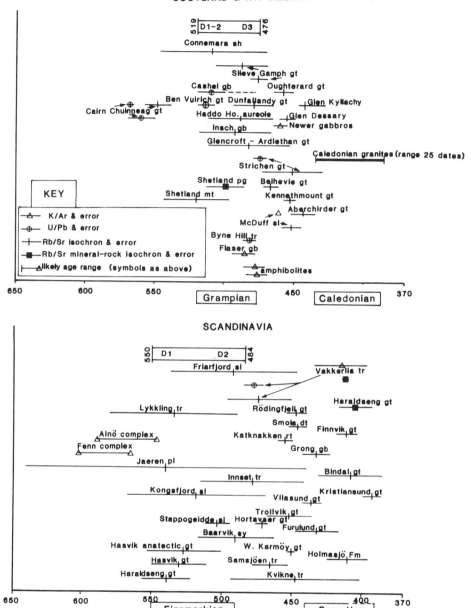

SCOTLAND & NW IRELAND

SCANDINAVIA

ENGLAND, WALES & SE IRELAND

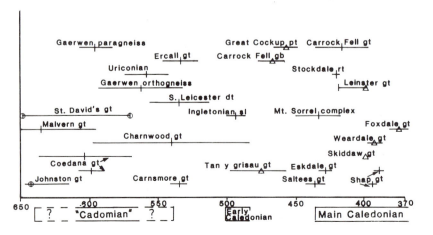

Evidence for the belt-length nature of early Caledonian orogenesis within the Scandinavian Caledonides

Evidence is now available that early Caledonian orogenesis, broadly coeval with or equivalent to the Finnmarkian orogenic phase, extends over a much larger proportion of the Scandinavian Caledonides than hitherto recognised (Figs. 1 and 2). This evidence has come from two main sources: firstly, study of the recently discovered ophiolitic assemblages; and secondly, from isotopic and petrological analyses of polymetamorphic terrains.

Ophiolitic assemblages

The discovery that the Scandinavian Caledonides contain major ophiolite complexes (Prestvik and Roaldset 1978; Sturt and Thon 1978a), and that these have an almost belt-length distribution (Furnes et al. 1980, and this volume) has given an added impetus to stratigraphic and geotectonic reassessment. It has been realized for some years that a major unconformity is present beneath the Upper Ordovician/Silurian succession of western Norway (Sturt and Thon 1978b). This unconformity, often overlain by sediments of initially continental character, may occur either above a Finnmarkian deformed and metamorphosed sequence of earlier Caledonian strata (N. Troms, eastern Trødelag), above metasedimentary rocks of unknown age (Nordland, Sunnhordaland) or truncating the majority of the so far discovered ophiolites. Such observations encouraged Durner et al. (1980) to suggest an intimate relationship between Finnmarkian orogenesis and widespread obduction of Iapetus oceanic crust.

The base of the unconformable succession overlying the ophiolites is markedly diachronous and in many areas the oldest palaeontological control is supplied by the calcareous facies of the Ashgill (Sturt et al. this volume). This may, however, be underlain by many hundreds of metres of predominantly clastic sediments filling a considerable palaeotopography (Thon, this volume; Sturt et al. this volume; Minsaas and Sturt 1981).

The period of major uplift and erosion of many of the ophiolitic complexes, prior to the deposition of the sedimentary cover, is broadly coincident with magmatism of essentially calc-alkaline aspect: e.g. the West Karmøy igneous complex (Sturt and Thon 1978c), the Kattnakken volcanics on Stord and their possible correlatives on Bømlo (Lippard 1976; Nordås et al. this volume), Smøla (Roberts 1980); and the formation of back arc basins (Furnes et al. this volume). The available data concerning the ages of these complexes is reviewed in Fig. 2.

The cap-rock sequence above the Karmøy ophiolite (Sturt et al. 1979; Solli 1981) is significant in this discussion. The Torvastad Group represents a succession, several thousand metres in thickness, of predominantly pelagic sediments with major developments of 'off-axis' greenstones. They have undergone polyphasal folding and metamorphism in the upper part of the greenschist facies (garnet grade) prior to uplift, erosion and deposition of the Upper Ordovician sequence. Recent Rb–Sr whole-rock dating of a similar metasedimentary sequence in the Stavanger area (Roddick and Jorde 1981) is consistent with an early Caledonian age for the metamorphism (Fig. 2).

Unambiguous evidence for the early Caledonian age of the Lykling ophiolite is provided by the 535 ± 46 Ma isochron obtained from the overlying quartz keratophyres (Furnes et al. 1983). These rocks are part of a bimodal sequence including extrusive greenstones and are believed to be the products of ensimatic arc volcanism (Amaliksen 1983; Nordås et al. this volume).

Fig. 2   Ages of early Caledonian igneous intrusions from northwest Europe compiled from sources cited in the text and the references therein. The following structural interpretations have been placed upon the dated plutons (see text for discussion). Scotland and northwest Ireland: Cairn Chunneag granite = pre-tectonic; Connemara schists, Cashel gabbro, Ben Vuirich, Slieve Gamph, and Dunfallandy Hill granites, Shetland migmatite = early tectonic; Haddo House aureole, Insch and Newer gabbros, Slieve Gamph pluton, Shetland pegmatite – late tectonic; Oughterard, Glencroft, Ardlethan, Strichen, Belhelvie, Kennethmount, and Aberchirder granites = post-tectonic. The Glen Dessary and Glen Kyllachy intrusions bracket the amphibolite facies event in the axial zone of the orogen. Scandinavia: Alnø and Fenn alkaline complexes developed within the Baltic Shield may be = pre-tectonic; Haraldseng and Hasvik granites, Lykling trondhjemite and Jæren phyllite = early tectonic; Friarfjord, Kongsfjord, and Stappogeide slates and Baarvik syenite = late tectonic; Vakkerlia, Innset, Samsjøen and Kvikne trondhjemites, Rødningsfjell, Trollvik and West Karmøy granites, Smøla diorite and Kattnakken rhyolites = post-tectonic. All other intrusions shown relate to main or late phases of Caledonian orogenesis. England, Wales and southeast Ireland: Malvern, Johnstone, St David's and Coedana granites and Gaerwen paragneiss = 'Cadomian'; Uriconian volcanics, Gaerwen orthogneiss, Ercall granophyre, S Leicester diorite, Charnwood and Carnsmore granite = ? 'Cadomian' or ? Early Caledonian; Ingletonian slate = Early Caledonian; Tany y Grisau granite, Carrock Fell gabbro and Great Cockup picrite = post-Early Caledonian and related to Ordovician vulcanism; all other plutons are related to main and late phases of Caledonian orogenesis. All Rb–Sr whole rock isochrons have been recalculated using 1.42×10^{-11}. yr^{-1} for λ ^{87}Rb

Polymetamorphic terrains

As the result of the Uppsala Symposium, new information is available which points directly or indirectly to relics of early Caledonian orogenesis. Hardenby *et al.* (1981) recognize two cycles of metamorphism in Trøndelag, an earlier predating Hovin sedimentation and a later of Mid-(?) Silurian age. Klingspor and Gee (1981) present a series of Rb–Sr isochron ages (478–447 Ma) and a zircon age of $477 \pm ^8_5$ from trondhjemites (Samsjoen, Innset, Vakkerlia, and Kvikne) which cross-cut folded and metamorphosed sediments (Fig. 2). They argue that the age of deformation is conveniently bracketed between deposition of the Tremadoc Nordaunevoll schist and approximately 480 Ma; which is broadly coeval with Finnmarkian D2.

Other indications of possible early Caledonian orogenesis are reported, from a number of areas between Trøndelag and Lyngen, based on a variety of evidence; i.e. unconformities, age of cross-cutting granites etc. They are presented, for example, in the contributions of Tragheim (1981) from the Rishaugfjellet–Tysfjord area, Flodberg and Stigh (1981) from Umbukta, and in Senior and Otten (1981). Further south, suggestions of a Finnmarkian age for the deformation and metamorphism of the Samnanger Complex metasediments have been made by Færseth *et al.* (1977). Nickelsen *et al.* (this volume) appeal to post-Llanvirnian deposition of Ordovician turbidites associated with nappe translation, which suggests that a classification of Caledonian orogenesis into two phases, namely Finnmarkian and Scandian, is itself an oversimplification. A possible Finnmarkian age for the cleavage in the Jæren phyllite of the Stavange region is reported by Roddick and Jorde (1982; see also Fig. 2).

The stratigraphic analysis of Thon (this volume) shows that the Dalsfjord nappe (Nordfjord) must have achieved its initial emplacement prior to ophiolite obduction and the deposition of a Middle–Upper Ordovician cover sequence. This pattern is not dissimilar to that claimed for the Finnmarkian type area.

As Nicholson (1979) pointed out, caution should be applied in assigning metamorphic assemblages bereft of faunal control and insufficient geochronological evidence to a specific orogenic event. It would appear, however, that the balance of the evidence is strongly in favour of a two-stage orogenic evolution of the Scandinavian Caledonides. The authors consider both stages, which are separated by a considerable time period, to have the status of major orogenic events. A difficulty in establishing the belt-length extent of early Caledonian orogenesis has been the non-recognition of a major hiatus between Finnmarkian and equivalent terrains and the succeeding Ordovician–Silurian sequences. This is perhaps mainly due to the fact that the metasedimentary cover sequence is often of a similar metamorphic grade to the subjacent basement sequence of Finnmarkian rocks (Ramsay *et al.* this volume).

Early Caledonian Events in the Northern British Caledonides

The Grampian orogeny

The Grampian orogeny was the set of events linked with the deformation and metamorphism of the Dalradian Supergroup during Late Cambrian to Early Ordovician times along the northwest margin of the Iapetus Ocean (Lambert and McKerrow 1976). This event has until recently been considered by most authors to have affected the whole of the orthotectonic Caledonides; however, new isotopic data that are reviewed below suggests that the Grampian orogeny might not be recorded north of the Great Glen Fault.

The earliest structures in the Dalradian of Scotland are related to the development of the Tay fold nappe (Bradbury *et al.* 1976). The large-scale inversions associated with this nappe are recorded in the highest stratigraphic levels, suggesting that the deformation of the Dalradian was not vertically diachronous (Dunning 1972). In the northeast, late D1 thrusting, that postdated the formation of the Tay Nappe, was responsible for the emplacement of the Banff Nappe, which contains >700 Ma basement gneisses and a cover sequence of Dalradian supracrustals (Ramsay and Sturt 1979).

In the Moines, north of the Great Glen Fault, the earliest fabrics of Grenville or Morarian age are cut by the pre-D2 Cairn Chuinneag granite, which has been dated at 560 Ma (Pidgeon and Johnson 1974); this is taken as providing a maximum age for the D2 event. The D2 event in the Moines is related to isoclinal folding and the development of early mylonites in the Moine Thrust Zone and is correlated with Grampian D1 in the Dalradian to the south (Mendum 1979). Van Breemen *et al.* (1979) report a 456 ± 5 Ma age for the syn-D2, D3 Glen Dessary syenite in the western Moines and do not record an isotopic event between 740 Ma and 460 Ma which could be correlated with the early Caledonian D1.

The D1 Tay Nappe is traced into northeast Ireland and then westwards into Donegal (Pitcher and Berger 1972) although Hutton (1979) argues that there are no major D1 structures in northwest

Donegal. D1 thrusting in northwest Mayo involved both pre-Caledonian basement and Moine and Dalradian supracrustals (Max and Long 1981).

In the southwest and central Ox Mountains, there is no consensus on which tectonic events should be attributed to the Grampian orogeny and consequently whether the syn-D3, D4 Slieve Gamph pluton (c. 480 Ma (Max et al. 1976; Pankhurst et al. 1976)) can be used to date D1 in the undisputed Dalradian terrains in Ireland. No major D1 structures are recognized in Connemara, where basic plutons were emplaced in the D1–D2 interval, one of which the Cashel Lough Wheelaun has been dated at 510 ± 10 Ma (Pidgeon 1969). This age, which might be up to 20 Ma too old (Pankhurst and Pidgeon 1976), provides a minimum date for the D1 event in western Ireland.

In the Dalradian of Scotland, the D2 folds are generally upright but of variable orientation. The Ben Vuirich granite (514 ± 67 (Pidgeon and Aftalion 1978)) was emplaced post-D2 but prior to the pre-D3 main peak of metamorphism. The Dunfallandy Hill granite (491 ± 15 Ma) occupies a similar position in the structural succession, the slightly younger age being attributed to metamorphic resetting (Bradbury et al. 1976). Metamorphism was of Barrovian type attaining a maximum of 550–620°C at 9–12 kbars (Wells and Richardson 1979) except in the northeast where it was synchronous with the intrusion of the Newer Gabbros and was of Buchan type attaining a maximum of 620°C at 5–6 kbars (Harte and Hudson 1979). The earliest Newer Gabbros are dated at 480–490 Ma (Pankhurst 1970).

In the Moines the metamorphic peak, which varies from greenschist facies in the west to upper amphibolite facies in the east, was attained prior to the development of open to tight D3 folds. Although this D3 in the Moines is correlated with the Grampian D2 further south (Mendum 1979; Johnson et al. 1979) the age of the Glen Dessary Syenite (Van Breemen et al. 1979) suggests these events may not be coeval. Van Breeman and Piasecki (1983) record an age for the late D3 Glen Kyllachy Granite of 443 ± 15 in the Moinian of the Central Highlands, south of the Great Glen fault. They suggest that these two plutons date an event within the axial zone of the orogen that is coeval with the closure of a Caradocian basin along the Highland Boundary Fault. An alternative model is presented by Soper and Barber (1982) in which the whole of the Moines are involved in a crustal duplex which was overthrust northwards during 50 Ma period starting c. 470 Ma.

In Donegal, the Barrovian metamorphic peak (middle or upper greenschist to lower amphibolite facies) was attained post-D2 but pre-D3. Further south, in northwest Mayo, the main peak of Barro-

vian metamorphism, which was broadly coeval with the dominant D2 folds, was predated by a low temperature, high pressure event (Gray and Yardley 1979). The metasediments in the Central Ox Mountains were subject to a simple progressive Barrovian metamorphism, like those in northwest Mayo, attaining a maximum at 6–7 kbar (Yardley et al. 1979). Major isoclines were developed during D2 in Connemara which were post-dated by the metamorphic climax of 550°–650°C at 5 kbar (Yardley 1976).

In both Scotland and northwestern Ireland, D3 folds are often associated with a crenulation cleavage. In Aberdeenshire, these folds were postdated by a phase of granite emplacement, the oldest of which is the 475 ± 5 Ma Strichen Granite (Pankhurst 1974). In Connemara, by contrast, at least three major fold-thrust nappes were formed during D3. Pressure, temperature studies indicate the possibility of a sudden rapid uplift at this time with the maximum pressure decreasing to 4 kbars (Yardley 1976). Major, upright D4 folds were postdated by the emplacement of the Oughterard Granite (c. 469 Ma, Leake 1978).

The ages of the syn-Grampian plutons (Fig. 3) are all within the range 514–476 Ma and only few of the ages of pre- or post-Grampian plutons fall within the limits of error. The structural and stratigraphic evidence is more limited. The youngest Dalradian sediments give a maximum age for the development of Grampian structures at high stratigraphic levels. The youngest macrofaunas are of probable Early to Mid Cambrian age (Harris and Pitcher 1975). It has been suggested that the microfauna may range into the Lower Ordovician (Arenig–Llanvirn, Downie et al. 1971), although Bliss (1977) states that the microfauna is not younger than Early Cambrian. Problems have also been encountered in fixing a minimum age for this event. Cobbing et al. (1965) argue that the Dalradian is overlain unconformably by Arenig strata in Tyrone, although Phillips et al. (1976) do not accept this interpretation. Dewey (1961) claimed that the Arenig–Llanvirn sediments of South Mayo contained high grade metamorphic detritus from the Connemara Dalradian and therefore proposed that the Grampian event was pre-Arenig. Recently, several authors (Stillman 1979; Ryan et al. 1980; Yardley and Senior 1982) have argued that Grampian basic magmatism in Connemara may be coeval with eruption of the Tremadoc to earliest Arenig pillow basalts at the base of the South Mayo succession. However, it has yet to be clearly established that South Mayo and Connemara were adjacent during the Early Ordovician.

In northwestern Scotland an apparently continuous Lower Cambrian to Arenig Durness limestone

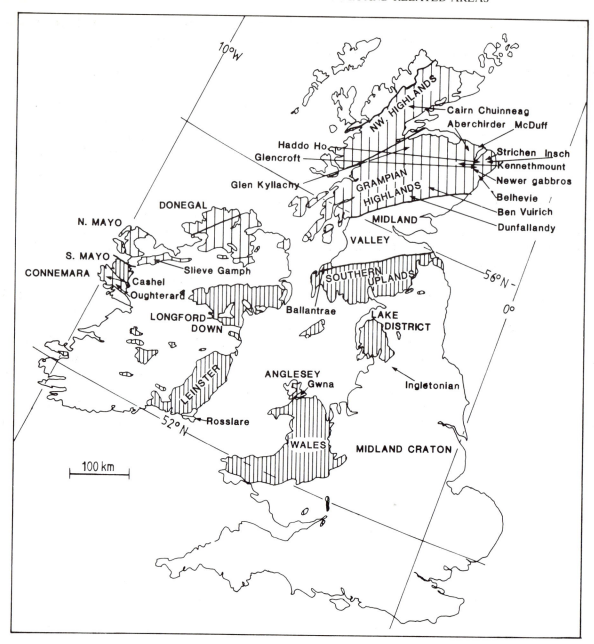

Fig. 3   Map showing the orthotectonic (spaced vertical ruling) and paratectonic (close vertical ruling) Caledonides and in addition the locations of the Early Caledonian or Grampian age dates discussed in the text

succession of the Hebridean Craton, whose lower members are involved in earliest deformation along the Moine thrust zone, has been regarded as inconsistent with a pre-Arenig Grampian event. However, Palmer *et al*. (1980) have argued that the Middle and Upper Cambrian are missing from the Durness succession and that this break may reflect Grampian tectonism further south. However, the geometric relationship of the Durness succession to the Grampian orogen is presently unknown (see Soper and Barber 1982).

Early Caledonian ophiolitic assemblages

The Grampian orogeny appears to be at least partly coeval with the development and obduction of ophiolitic assemblages (sensu Moores 1982).

In Shetland an ophiolite was obducted onto a thick metasedimentary sequence correlated with the Moine and Dalradian which underwent deformation and metamorphism at 516 Ma (Flinn 1981). This complex has a metamorphic aureole at its base and is structurally underlain by a mélange containing both metasedimentary and ophiolitic detritus which is intruded by the Skaw granite (Pritchard 1981). A possible dismembered early Caledonian ophiolite is recorded within the Highland Boundary Complex adjacent to the Highland Boundary Fault (Henderson and Robertson 1981). Curry et al. (1982) have reported Lower Arenig (c. 480 Ma) fossils from limestones within this complex which indicate that its generation and obduction post-dated the Grampian metamorphic peak. Ophiolitic mélange is recorded in a north-dipping thrust zone along the south shores of Clew Bay, western Ireland (Ryan et al. 1983). This complex is of unknown age although structural arguments indicate it was overthrust by Dalradian strata (Southern Highland Group) during the local Grampian D1 event (Phillips 1973). This complex is situated on a major magnetic linear that can be traced eastwards into the Highland Boundary Fault of Scotland (Max et al. 1983).

In the Ballantrae complex, Bluck et al. (1980) have argued that mid-Arenig ophiolite generation (Byne Hill trondhjemite and flazer gabbro, Fig. 2) and obduction (amphibolites, Fig. 2) were approximately synchronous with D3 in the central Highlands. The complex metamorphic assemblages recorded from the ophiolitic complex (beerbachites and glaucophane schists) developed prior to obduction during earliest Ordovician times.

Comparison of the Finnmarkian and Grampian Orogenic Episodes

The Finnmarkian and Grampian orogenies both involved the formation of large-scale fold-nappes with associated Barrovian metamorphism and the emplacement of large syntectonic igneous complexes. Available isotopic and stratigraphic data suggest that the Finnmarkian event began some 20 Ma before the Grampian but that both orogenies terminated around 480 Ma and were broadly coeval with the generation and obduction of ophiolitic complexes. The Grampian orogeny is restricted to the orthotectonic British Caledonides and possibly only to the zone bounded by the Great Glen and Southern Uplands faults. The geographic extent of the Finnmarkian orogenic phase is a subject of present research. Evidence for an orogenic event that predates end Silurian nappe emplacement is now recorded in most of the major nappe units. However, whilst these data are consistent with the bipartite division of orogenesis within the Scandinavian Caledonides proposed above, the possibility that this is an oversimplification cannot be ruled out.

Early Caledonian Events in England, Wales and Southeast Ireland

No early Caledonian event equivalent to the Grampian or Finnmarkian is generally recognized in the paratectonic Caledonides of southern Britain. The essential pattern is of a Late Precambrian to Silurian sequence with shelf facies in central England and basinal facies in the north and west. These rest upon a continental basement, the Midland Craton, which had been influenced by a Late Precambrian phase of crustal growth (Thorpe 1979; Patchett et al. 1980). Large breaks occurred in sedimentation, particularly during the Ordovician, associated with low grade regional metamorphism and magmatism (Johnston et al. 1979) A 490 Ma slaty cleavage, broadly coincident with the early Caledonian of the orthotectonic belt, is recorded in Ingletonian rocks (Fig. 2) which are overlain unconformably by Caradocian strata (O'Nions et al. 1973). In Wales, excluding Anglesey, a Late Precambrian magmatic event was followed by the development of the Welsh Caledonian basin with its thick Late Precambrian to Silurian succession containing local unconformities and Ordovician calc-alkaline volcanics (Coward and Siddans 1979). Southeast Ireland had a broadly similar development to Wales and the Lake District with relatively continuous flysch deposition and associated volcanism from Late Precambrian (? Cambrian) to Silurian times (Brück et al. 1979).

High grade Late Precambrian metamorphic fabrics within the paratectonic Caledonides are recorded in Anglesey and southeast Ireland at Rosslare. Most authors relate the development of these complexes to the Cadomian event of the Massif Armorican of Brittany, although the age obtained from the Gaerwen orthogneiss (Beckinsdale and Thorpe 1979) is generally younger. Early Caledonian tectonism within these complexes is perhaps recorded in the (mid-Cambrian–Arenig?) Gwna mélange of Anglesey, which contains ophiolitic blueschist detritus, and possibly in other members of the Bedded Series (see Barber and Max 1979 and discussion therein). The Rosslare complex contains a Late Precambrian to Cambrian pillow lava flysch sequence which was deformed prior to the emplacement of the Carnsmore granite at 535 Ma (Max, personal communication 1980). Nutt and Smith (1981) suggest that Anglesey is separated from the Welsh mainland by a major strike-slip zone that was active from Cambrian to Devonian times.

A review of the radiometric evidence from this region is given in Fig. 2 (see also Fig. 3 for localities). The pattern shows Cadomian metamorphism and associated plutonism followed by sporadic magmatism throughout early Palaeozoic times. Possible correlatives of the Grampian–Finnmarkian events are the deformation of the Ingletonian, the development of the Gwna mélange, the pre-535 Ma folding in Rosslare and perhaps the Gaerwen orthogneiss.

Related Areas

A parallel for the early Caledonian evolution of the British and Scandinavian orthotectonic Caledonides is found in Spitsbergen. According to Ohta (1979) a continental prism assemblage which was subjected to c. 530 Ma Barrovian metamorphism is bordered to the southwest by a blueschist ophiolitic mélange (K–Ar minimum age 470 Ma). This association has been interpreted as relating to an early Caledonian subduction cycle (Ohta 1979). Both of these terrains were uplifted and eroded prior to Silurian times (Scrutton et al. 1976). The relationship of Spitsbergen to the adjacent Caledonian massif is discussed by Harland (this volume).

In the Caledonides of east Greenland an early granite suite was emplaced c. 510–475 Ma that may have been immediately post-dated by an early phase of migmatization (Higgins and Phillips 1979). This metamorphic complex, whose development at depth apparently overlapped sedimentation at higher structural levels, cooled slowly until final nappe emplacement on to the Greenland Shield in the Late Silurian (Higgins and Phillips 1979).

Ziegler (this volume) provides radiometric evidence for an eastward arm of the Caledonides extending into north Germany and Poland, that was active during Silurian times and possibly earlier. This separated the Baltic Shield from the Precambrian London Platform.

Brochwicz–Lewinski et al. (1981) show that major Caledonian deformation and metamorphism in the Malopolska Massif and its extension into Romania, occurred in Late Cambrian to earliest Ordovician times, thus showing clear contemporaneity with early Caledonian orogenesis in the classical Caledonian section of northwest Europe. Brochwicz–Lewinsky et al. (1981, Fig. 1) further consider that these terrains were originally contiguous and were separated by large-scale sinistral strike-slip in post Grampian–Finnmarkian times.

Diachroneity

The available evidence indicates that the Grampian and Finnmarkian events took place over some 43 Ma and 66 Ma respectively. Important diachroneity must exist within these relatively long time intervals. In the type Finnmarkian terrain the time difference between the D1 event within the allochthon and parauthochthon (c. 30 Ma) appears to reflect the time of eastwards transport of the Finnmarkian nappes. Strike parallel diachroneity has been claimed within the Grampian terrains of Ireland and Scotland (see Pankhurst et al. 1976 for a review); however, the presently available data have not allowed a consensus to be reached in this matter.

Discussion

The purpose of this review has been to present the available age constraints for early Caledonian orogenesis in northwestern Europe. The Finnmarkian and Grampian tectonothermal cycles probably represent collisional events in Late Cambrian to earliest Ordovician times, involving both continental prism assemblages and obducted oceanic floor. Metamorphism of Barrovian type was the general rule within the continental prism, and this was associated with distinctive syntectonic magmatic patterns. Post-orogenic granites truncate these belts and provide a minimum age constraint at around 480 Ma. These granites were coeval with the widespread development of marginal arc volcanism and the evolution of back arc and retro arc basins. Determination of the maximum age of these orogenic events is inhibited by poor fossil control within the metamorphic assemblages. However, available data suggests that the Finnmarkian event initiated some 20 Ma prior to the Grampian.

This review argues that early Caledonian orogenesis was more widespread in Scandinavia than previously realized. This throws into even sharper contrast the present day asymmetry of early Caledonian fold belts both along strike (compare for example the early Caledonian geology of England with that of Scandinavia) and across strike (compare for example the early Caledonian geology of Scotland and northwest Ireland with that of England and southeast Ireland). If these orogenic events were the result of one or more collisional events in Late Cambrian–Early Ordovician times, the present geometry of these orogens is consistent with them having been dismembered since that time. The age of the dismemberment of these early Caledonian terrains and their docking into their present locations is presently poorly understood, although in some areas (e.g. western Ireland (Max et al. 1983)) it is believed to have started in earliest Ordovician times. Furthermore the evidence from the Polish Caledonides (Brochwicz–Lewinski et al. 1981) indicates dismemberment of the Grampian–Finnmarkian orogen by strike-slip faulting of Ordovician to Devonian age. In the authors' opinion it is apparent that the plate distributions for Early Caledonian evolution have a complexity not considered in most traditional models. Much further work is necessary to establish the geometry of late Cambrian–early Ordovician plate configurations, and it becomes more apparent that simplistic models based on a two-sided Iapetus Ocean are perhaps less relevant than Tethyan analogues.

References

Amaliksen, K. G. 1983. *The Geology of the Lykkling Ophiolitic Complex, Bømlo, Southwestern Norway.* Unpublished Cand Real thesis, Univ. of Bergen, 417 pp.

Barber, A. J. and Max, M. D. 1979. A new look at the Mona complex (Anglesey, North Wales). *J. geol. Soc. London*, **136**, 407–432.

Beckinsdale, R. D. and Thorpe, R. S. 1979. Rubidium–strontium whole-rock isochron evidence for the age of metamorphism and magmatism in the Mona Complex. *J. geol. Soc. London*, **136**, 433–439.

Bennet, M. C. 1974. The emplacement of high temperature peridotite in the Seiland Province of the Norwegian Caledonides. *J. geol. Soc. London*, **130**, 205–229.

Bliss, G. M. 1977. *The Micropalaeontology of the Dalradian.* Unpubl. PhD thesis, Univ. of London.

Bluck, B. J., Halliday, A. M., Aftalion, M. and MacIntyre, R. M. 1980. The age and origin of the Ballantrae ophiolite and its significance to the Caledonian orogeny and the Ordovician time scale. *Geology* **8**, 492–495.

Bradbury, H. J., Smith, R. A. and Harris, A. L. 1976. Older granites as time-markers in Dalradian evolution *J. geol. Soc. London*, **132**, 677–684.

Breemen, O., van and Piasecki, M. A. J. 1983. The Glen Kyllachy granite and its bearing on the nature of the Caledonian orogeny in Scotland. *J. geol. Soc. London*, **140**, 47–62 and 961–964.

Breemen, O., van, and Johnson, M. W. R. 1979. Age of the Loch Borolan complex, Assynt and the late movements along the Moine thrust zone. *J. geol. Soc. London*, **136**, 489–495.

Breeman, O., van, Aftalion, M., Pankhurst, R. J. and Richardson, S. W. 1979. Age of the Glen Dessary Syenite, Inverness-shire: diachronous Palaeozoic metamorphism across the Great Glen. *Scot. J. Geol.* **15**, 49–62.

Brochwicz-Lewinski, W., Pozaryski, W. and Tomczyk, H. 1981. Large-scale strike slip movements along the SW margin of the East European Platform in the Early Palaeozoic. *Przeglad Geologiczny*, **8**, 385–397 (in Polish with English abstract).

Brück, P. M., Colthurst, J. R. J., Feely, M., Gardiner, P. R. R., Penny, S. R., Reeves, T. J., Shannon, P. M., Smith, D. G. and Vanguestaine, M. 1979. South-east Ireland: Lower Palaeozoic stratigraphy and depositional history. In Harris, A. L., Holland, C. H. and Leake, B. E. (Eds), *The Caledonides of the British Isles—reviewed. Geol. Soc. London Spec. Publ.*, **8**, 533–544.

Churkin, M., Jr., Carter, C. and Johson, B. R. 1977. Subdivision of Ordovician time scale using accumulation rates of graptolitic shales. *Geology*, **5**, 452–456.

Cobbing, E. J., Manning, P. I. and Griffith, A. E. 1965. Ordovician–Dalradian unconformity in Tyrone. *Nature, London*, **206**, 1132–1135.

Coward, M. P. and Siddans, A. W. B. 1979. The tectonic evolution of the Welsh Caledonides. In Harris, A. L., Holland, C. H. and Leake, B. E. (Eds), *The Caledonides of the British Isles—reviewed. Geol. Soc. London Spec. Publ.*, **8**, 187–198.

Cowie, J. W. and Cribb, S. J. 1978. The Cambrian System. In Cohee, G. V., Glaessner, M. F. and Hedburg, H. D. (Eds), *The Geological Time Scale. Am. Ass. Petrol. Geol.*, 355–362.

Curry, G. B., Ingham, J. K., Bluck, B. J. and Williams, A. 1982. The significance of a reliable Ordovician age for some Highland Border rocks in Central Scotland. *J. geol. Soc. London*, **139**, 451–454.

Dangla, P., Demange, J. C., Ploquin, A., Quernaudel, J. M. and Sonet, J. 1978. Données geochronologiques sur les Caledonides Scandinaves septrionales. (Troms, Norvége du Nord). *C.R. Acad. Sci. Paris*, **286**, 1653–1656.

Dewey, J. F. 1961. A note concerning the age of metamorphism of the Dalradian rocks of western Ireland. *Geol. Mag.*, **91**, 399–405.

Downie, C., Lister, T. R., Harris, A. L. and Fettes, D. J. 1971. A palynological investigation of the Dalradian rocks of Scotland. *Inst. geol. Sci. Rep.*, **71/9**.

Dunning, F. W. 1972. Dating events in the metamorphic Caledonides. Impressions of the symposium held in Edinburgh, September 1971. *Scott. J. geol.*, **8**, 179–192.

Færseth, R. B., Thon, A., Larsen, S. G., Sivertsen, A. and Elvestad, L. 1977. Geology of the Lower Palaeozoic Rocks in the Samnanger–Osterøy area, Major Bergen Arc. Western Norway. *Norges geol. Unders.*, **334**, 19–58.

Faure, G. 1977. Principles of Isotope Geology. John Wiley, New York, 464 pp.

Flinn, D. 1981. The Caledonides of Shetland (abs.). *Terra Cognita*, **1**, 42.

Flodberg, K. and Stigh, J. 1981. The Umbukta ultramafic and mafic complex in the Rödingsfjället nappe, Northern Swedish Caledonide (abs.). *Terra Cognita*, **1**, 42.

Furnes, H., Roberts, D., Sturt, B. A., Thon, A. and Gale, G. H. 1980. Ophiolite fragments in the Scandinavian Caledonides. *Proc. Int. Ophiolite Symp. Cyprus 1979*, 582–600.

Furnes, H., Ryan, P. D., Grenne, T., Roberts, D., Sturt, B. A. and Prestvik, T. this volume. Geological and geochemical classification of the ophiolitic fragments in the Scandinavian Caledonides.

Furnes, H., Austrheim, H., Amaliksen, K. G. and Nordås, J. 1983. Evidence for an incipient early Caledonian (Cambrian) orogenic phase in southwestern Norway. *Geol. Mag.*, **120**, 607–612.

Gale, N. M., Beckinsdale, R. D. and Wadge, A. J. 1979. A Rb–Sr whole rock isochron for the Stockdale rhyolite of the English lake district and a revised mid-Palaeozoic time-scale. *J. geol. Soc. London*, **136**, 235–242.

Gray, J. R. and Yardley, B. W. D. 1979. A Caledonian blueschist from the Irish Dalradian. *Nature, London*, **278**, 736–737.

Harland, W. B. this volume. Caledonide Svalbard.

Harland, W. B., Cox, A. V., Llewellyn, P. G. Pickton, C. A. G. Smith, A. G. and Walters, R. 1982. A Geologic Time Scale. Cambridge University Press, 131 pp.

Harris, A. L. and Pitcher, W. S. 1975. The Dalradian Supergroup. In Harris, A. L. *et al.* (Eds), *A Correlation *of the Precambrian Rocks in the British Isles. Spec. Rep. Geol. Soc. London*, **6**, 52–75.

Hardenby, C., Lagerblad, B. and Andréasson, P.-G. 1981. Structural development of the northern Trondheim nappe complex, Central Scandinavian Caledonides. *Terra Cognita*, **1**, 50.

Harte, B. and Hudson, N. F. C. 1979. Pelite facies series and the temperature and pressure of Dalradian metamorphism in east Scotland. In Harris, A. L., Holland, C. H. and Leake, B. E. (Eds), *The Caledonides of the British Isles—reviewed. Geol. Soc. London, Spec. Publ.*, **8**, 323–338.

Henningsmoen, G. 1960. Cambro-Silurian deposits of the Olso region. *Norges geol. Unders.* **208**, 130–150.

Henderson, W. G. and Robertson, A. H. F. 1982. The Highland Border rocks and their relation to marginal basin development in the Scottish Caledonides. *J. geol. Soc. London*, **139**, 433–450.

Higgins, A. K. and Phillips, W. E. A. 1979. East Greenland Caledonides—an extension of the British Isles. In Harris, A. L., Holland, C. H. and Leake, B. E. (Eds), *The Caledonides of the British Isles—reviewed. Geol. Soc. London, Spec. Publ.*, **8**, 19–32.

Holland, C. H. and Sturt, B. A. 1970. On the occurrence of archaeocyathids in the Caledonian metamorphic rocks of Sørøy, and their stratigraphic significance. *Norsk geol. Tidsskr.*, **50**, 345–355.

Hutton, D. H. W. 1979. Dalradian structure in the Creeslough area, N.W. Donegal, Ireland. In Harris, A. L., Holland, C. H. and Leake, B. E. (Eds), *The Caledonides of the British Isles—reviewed. Geol. Soc. London, Spec. Publ.*, **8**, 239–242.

Johnson, M. R. W., Sanderson, D. J. and Soper, N. J. 1979. Deformation in the Caledonides of England, Ireland and Scotland. In Harris, A. L., Holland, C. H. and Leake, B. E. (Eds), *The Caledonides of the British Isles—reviewed. Geol. Soc. London, Spec. Publ.*, **8**, 165–186.

Minsaas, O. and Sturt, B. A. 1981. The Ordovician clastic sequence immediately overlying the Lyngen Gabbro complex, and its environmental significance (abs.). *Terra Cognita* **1**, 59–60.

Klingspor, I. and Gee, D. G. 1981. Isotopic age-determination studies of the Trøndelag trondhjemites (abs.). *Terra Cognita*, **1**, 55.

Lambert, R. St. J. and McKerrow, W. S. 1976. The Grampian Orogeny. *Scott. J. geol.*, **12**, 271–292.

Leake, B. E. 1978. Granite emplacement: the granites of Ireland and their origin. In Bowes, D. R. and Leake, B. E. (Eds), *Crustal Evolution in Northwestern Britain and Adjacent Regions. Geol. J. Spec. Issue*, **10**, 221–248.

Lippard, S. J. 1976. Preliminary investigations of some Ordovician volcanics from Stord, W. Norway. *Norges geol. Unders.*, **327**, 41–66.

Max, M. D., Long, C. B. and Sonet, J. 1976. The geological setting and age of the Ox Mountains Granodiorite. *Bull. geol. Surv. Ireland*, **2**, 27–35.

Max, M. D. and Long, C. B. 1981. The nature of basement and its relationships with Caledonian supercrustals in Ireland (abs.). *Terra Cognita*, **1**, 59.

Max, M. D., Ryan, P. D. and Imnandar, D. 1983. A magnetic deep structural interpretation of Ireland. *Tectonics*, **2**, 431–452.

Mendum, J. R. 1979. Caledonian thrusting in N.W. Scotland. In Harris, A. L., Holland, C. H. and Leake, B. E. (Eds), *The Caledonides of the British Isles—reviewed. Geol. Soc. London, Spec. Publ.*, **8**, 291–298.

McKerrow, W. S., Lambert, R. St. J. and Chamberlain, V.

E. 1980. The Ordovician, Silurian and Devonian time scales. *Earth Planet. Sci. Lett.*, **51**, 1–8.

Moores, E. M. 1982. Origin and emplacement of Ophiolites. *Rev. Geophys. Space Phys.*, **20**, 735–760.

Nicholson, R. 1979. Caledonian correlations: Britain and Scandinavia. In Harris, A. L., Holland, C. H. and Leake, B. E. (Eds), *Caledonides of the British Isles—reviewed. Geol. Soc. London Spec. Publ.*, **8**, 3–18.

Nordås, J., Amaliksen, K. G., Brekke, H., Suthren, R. J., Furnes, H., Sturt, B. A. and Robins, B. this volume. Lithostratigraphy and petrochemistry of Caledonian rocks on Bømlo, southwest Norway.

Nutt, M. J. C. and Smith, E. G. 1981. Transcurrent faulting and the anomalous position of pre-Carboniferous Anglesey. *Nature*, London, **290**, 492–494.

Ohta, Y. 1979. Blue schists from Motalafjella, Western Spitsbergen. *Norsk Polarinstitutt Skrifter*, **167**, 171–217.

O'Nions, R. K., Oxburgh, E. R., Hawkesworth, C. J. and Macintyre, R. M. 1973. New isotopic and structural evidence on the age of the Ingletonian: probable Cambrian of northern England. *J. geol. Soc. London*, **129**, 445–452.

Palmer, T. J., McKerrow, W. S. and Cowie, J. W. 1980. Sedimentological evidence for a stratigraphical break in the Durness Group. *Nature*, London, **287**, 720–722.

Pankhurst, R. J. 1970. The geochronology of the basic igneous complexes. *Scott. J. geol.*, **6**, 83–107.

Pankhurst, R. J. 1974. Rb–Sr whole rock chronology of Caledonian events in northeast Scotland. *Geol. Soc. Am. Bull.*, **85**, 345–350.

Pankhurst, R. J. and Pidgeon, R. T. 1976. Inherited isotope systems and the source region pre-history of early Caledonian granites in the Dalradian Series of Scotland. *Earth Planet. Sci. Lett.*, **31**, 55–68.

Pankhurst, R. J., Andrews, J. R., Phillips, W. E. A., Saunders, I. S. and Taylor, W. E. G. 1976. Age and structural setting of the Slieve Gamph Igneous Complex, Co. Mayo, Eire. *J. geol. Soc. London*, **132**, 327–334.

Patchett, P. J., Gale, N. H., Goodwin, R. and Humm, M. J. 1980. Rb–Sr whole rock isochron ages of late Precambrian to Cambrian igneous rocks from sothern Britain. *J. geol. Soc. London*, **137**, 649–656.

Phillips, W. E. A. 1973. The pre-Silurian rocks of Clare Island, Co. Mayo and the age of metamorphism of the Dalradian in Ireland. *J. geol. Soc. London*, **129**, 585–606.

Phillips, W. E. A., Stillman, C. J. and Murphy, T. 1976. A Caledonian plate tectonic model. *J. geol. Soc. London*, **132**, 579–609.

Pidgeon, R. T. 1969. Zircon U–Pb ages from the Galway granite and the Dalradian of Connemara, Ireland. *Scott. J. geol.*, **5**, 375–392.

Pidgeon, R. T. and Johnson, M. R. W. 1974. A comparison of zircon U–Pb and whole-rock Rb–Sr systems in three phases of the Cairn Chuinneag granite, Northern Scotland. *Earth Planet. Sci. Lett.*, **24**, 105–112.

Pidgeon, R. T. and Aftalion, M. 1978. Cogenetic and inherited zircon U–Pb systems in granites: Palaeozoic granites of Scotland and England. In Bowes, D. R. and Leake, B. E. (Eds), *Crustal Evolution in Northwestern Britain and Adjacent Regions. Geol. J. Spec. Issue*, **10**, 183–220.

Pitcher, W. S. and Berger, A. R. 1972. The Geology of Donegal: a study of Granite Emplacement and Unroofing. John Wiley, New York, 435 pp.

Prestvik, T. and Roaldset, E. 1978. Rare earth element abundances in Caledonian metavolcanics from the island of Leka, Norway. *Geochem. J.*, **12**, 89–100.

Pritchard, H. M. 1981. The Shetland Ophiolite (abs.). *Terra Cognita*, **1**, 65.

Ramsay, D. and Sturt, B. A. 1979. The status of the Banff nappe. In Harris, A. L., Holland, C. H. and Leake, B. E. (Eds), *The Caledonides of the British Isles—reviewed. Geol. Soc. London, Spec. Publ.*, **8**, 145–152.

Ramsay, D. M., Sturt, B. A., Jansen, Ø., Andersen, T. B. and Sinha-Roy, S. this volume. The tectonostratigraphy of western Porsangerhalvøya, Finnmark, North Norway.

Ramsay, D. M., Sturt, B. A., Zwaan, K. B. and Roberts, D. this volume. Caledonides of northern Norway.

Reading, H. G. 1965. Eocambrian and lower Palaeozoic geology of the Digermul Peninsula, Tanafjord Finnmark. *Norges geol. Unders.*, **234**, 167–191.

Roberts, D. 1980. Petrochemistry and palaeogeographic setting of Ordovician volcanic rocks of Smøla, Central Norway. *Norges geol. Unders.*, **359**, 43–60.

Robins, B. and Gardiner, P. M. 1974. Synorogenic basic intrusions in the Seiland petrographic province, northern Norway—a descriptive account. *Norges geol. Unders.*, **312**, 91–130.

Roddick, J. C. and Jorde, K. 1981. Rb–Sr whole rock isochron dates on rocks of the Jæren district, SW Norway. *Norges geol. Unders.*, **365**, 55–67.

Ross, R. J. Jr., Naeser, C. W., Izett, G. A., Whittington, H. B., Hughes, C. P., Rickards, R. B., Zalasiewicz, J., Sheldon, P. R., Jenkins, C. J., Cocks, L. R. M., Basset, M. G., Toghill, P., Dean, W. T. and Ingram, J. K. 1978. Fission track dating of Lower Palaeozoic volcanic ashes. *U.S. Geol. Surv. open file rep.*, 78–701.

Rundle, C. C. 1979. Ordovician intrusions in the English Lake District. *J. geol. Soc. London*, **136**, 29–38.

Ryan, P. D., Floyd, P. A. and Archer, J. B. 1980. The stratigraphy and petrochemistry of the Lough Nafooey Group (Tremadocian), western Ireland. *J. geol. Soc. London*, **137**, 443–458.

Ryan, P. D., Sawal, V. K. and Rowlands, A. S. 1983. Ophiolitic mélange separates ortho- and paratectonic Caledonides in western Ireland. *Nature*, London, **301**, 50–52.

Scrutton, C. T., Horsfield, W. T. and Harland, W. B. 1976. Silurian fossils from western Spitsbergen. *Geol. Mag.*, **113**, 519–523.

Senior, A. and Otten, M. T. 1981. The Artfjället gabbro and its bearing on the evolution of the Storfjället nappe, Central Swedish Caledonides (abs.). *Terra Cognita*, **1**, 72.

Solli, T. 1981. *The Geology of the Torvastad Group, the Cap Rocks to the Karmøy Ophiolite.* Unpubl. Cand Real thesis, Univ. of Bergen.

Soper, N. J. and Barber, A. J. 1982. A model for the deep structure of the Moine thrust zone. *J. geol. Soc. London*, **139**, 127–138 and **140**, 519–525.

Speedyman, D. L. 1972. Mechanism of emplacement of plutonic igneous rocks in the Husfjord area of Sørøy, west Finnmark. *Norges geol. Unders.*, **272**, 35–42.

Steiger, R. H. and Jäger, E. 1977. Subcommission on geochronology: convention on the use of decay constants in geochronology and cosmochronology. *Earth Plant. Sci. Lett.*, **36**, 359–362.

Stillman, C. J. 1979. Caledonide vulcanism in Ireland. In Wones, D. R. (Ed.), *The Caledonides in the U.S.A. Virginia Polytechnic Institute Memoir*, **2**, 279–284.

Sturt, B. A. and Ramsay, D. M. 1965. The alkaline complex of the Breivikbotn area, Sørøy, northern Norway. *Norges geol. Unders.*, **231**, 1–143.

Sturt, B. A., Miller, J. A. and Fitch, F. J. 1967. The age of the alkaline rocks from west Finnmark, northern Norway, and their bearing on the dating of the Caledonian orogeny. *Norsk geol. Tidsskr.*, **47**, 255–273.

Sturt, B. A. and Taylor, J. 1972. The timing and environment of emplacement of the Storelv gabbro, Sørøy. *Norges geol. Unders.*, **272**, 1–34.

Sturt, B. A., Pringle, I. R. and Roberts, D. 1975. Caledonian nappe sequence of Finnmark, northern Norway and the timing of orogenic deformation and metamorphism. *Bull. geol. Soc. Am.*, **86**, 710–718.

Sturt, B. A., Pringle, I. R. and Ramsay, D. M. 1978. The Finnmarkian phase of the Caledonian Orogeny. *J. geol, Soc. London*, **135**, 597–610.

Sturt, B. A. and Thon, A. 1978a. An ophiolite complex of probable early Caledonian age discovered on Karmøy. *Nature*, London, **275**, 538–539.

Sturt, B. A. and Thon, A. 1978b. A major early Caledonian igneous complex and a profound unconformity in the Lower Palaeozoic sequence of Karmøy, southwest Norway. *Norsk geol. Tidsskr.*, **58**, 221–228.

Sturt, B. A. and Thon, A. 1978c. The Caledonides of Southern Norway. *Geol. Surv. Canada Paper*, **78–13**, 39–48.

Sturt, B. A., Andersen, T. B. and Furnes, H. this volume. The Skei Group, Leka: an unconformable clastic sequence overlying the Leka Ophiolite.

Sturt, B. A., Thon, A. and Furnes, H. 1979. The Karmøy ophiolite, southwest Norway. *Geology*, **7**, 316–320.

Sturt, B. A., Thon, A. and Furnes, H. 1980. The geology and preliminary geochemistry of the Karmøy ophiolite, S.W. Norway. *Proc. Int. Ophiolite symp. Cyprus 1979*, 538–554.

Taylor, P. N. and Pickering, K. T. 1981. Rb–Sr isotopic age determination of the late Precambrian Kongsfjord Formation and the timing of deformation in the Barents Sea Group, East Finnmark. *Norges geol. Unders.*, **367**, 105–110.

Thon, A. this volume. The Gullfjellet ophiolite complex and the structural evolution of the major Bergen arc, west Norwegian Caledonides.

Thon, A. this volume. Late Ordovician and Early Silurian cover sequences to the west Norwegian ophiolite fragments: stratigraphy and structural evolution.

Thorpe, R. S. 1979. Late Precambrian igneous activity in southern Britain. In Harris, A. L., Holland, C. H. and Leake, B. E. (Eds), *The Caledonides of the British Isles—reviewed. Geol. Soc. London, Spec. Publ.*, **8**, 579–584.

Tragheim, D. 1981. Polyphase nappe-emplacement histories in Nordland, Arctic Norway (abs.). *Terra Cognita*, **1**, 80.

Wells, P. R. A. and Richardson, S. W. 1979. Thermal evolution of metamorphic rocks in the Central Highlands of Scotland. In Harris, A. L., Holland, C. H. and Leake, B. E. (Eds), *The Caledonides of the British Isles—reviewed. Geol. Soc. London, Spec. Publ.*, **8**, 339–344.

Yardley, B. W. D. 1976. Deformation and metamorphism of Dalradian rocks and the evolution of the Connemara Cordillera. *J. geol. Soc. London*, **132**, 521–542.

Yardley, B. W. D., Long, C. B. and Max, M. D. 1979. Patterns of metamorphism in the Ox Mountains and adjacent parts of western Ireland. In Harris, A. L., Holland, C. H. and Leake, B. E. (Eds), *The Caledonides of the British Isles—reviewed. Geol. Soc. London. Spec. Publ.*, **8**, 369–374.

Yardley, B. W. D. and Senior, A. 1982. Basic magmatism in Connemara, Ireland: evidence for a volcanic arc? *J. geol. Soc. London*, **139**, 67–70.

Ziegler, P. A. this volume. Late Caledonian framework of western and central Europe.

Locally this was accompanied by a mafic–felsic alkaline rift volcanism. In the course of the Late Namurian and Early Westphalian, the individual grabens gradually ceased to subside differentially and rift-related volcanic activity abated (Francis 1978). During the Late Westphalian the sedimentary fill of these tensional basins became folded and uplifted to various degrees in response to tangential stresses transmitted through the crust of the North European Craton during the terminal phases of the Variscan orogeny.

The southern continuation of the Devonian and Carboniferous wrench and rift induced basins of the British Isles is formed by the Saint Anthony and the Bay of Fundy basins of the Canadian Maritime provinces (Howie and Barss 1975) (Fig. 4). These basins, which are partly superimposed on the Acadian fold belt, began to subside during the Late Middle and Late Devonian. Their development is intimately related to the Arctic–North Atlantic wrench and rift system. They continued to subside during Late Carboniferous and Early Permian time but became partly inverted during the Alleghenian diastrophism as a result of foreland compression.

Geodynamic Considerations

The Devonian and Early Carboniferous evolution of the Variscan geosynclinal system can be explained in terms of back-arc rifting. The principal mechanism governing back-arc rifting and sea-floor spreading is thought to be secondary convection currents in the mantle wedge immediately above the subducting lithospheric slab. Tensional stresses exerted by this convective system on the crust of the overriding plate are apparently only able to induce back-arc rifting and sea-floor spreading when the convergence rate between the subducting and the overriding plates is relatively low or if they diverge. This may be accompanied by partial decoupling of the two plates at the Benioff zone. If convergence rates are relatively high, stronger coupling at the Benioff zone causes the compressive deformation of the back-arc areas and the overpowering of the back-arc convective system (Uyeda 1981; Hsui and Toksöz 1981; Zonenshain and Savostin 1981). As convergence rates between plates are not constant but appear to change through time, period of back-arc extension and compression can alternate with each other.

Following the Late Caledonian diastrophism, conditions for back-arc extension were apparently fulfilled between Laurasia and the subducting Proto-Tethys plate. Back-arc rifting set in during the Early Devonian and persisted till Early Visean time.

It affected wide areas to the north of the Ligerian–Moldanubian Cordillera but apparently did not proceed to the point of crustal separation and the opening of a back-arc oceanic basin.

Back-arc rifting was partly contemporaneous with the major Middle Devonian to earliest Carboniferous sinistral translation betwen Laurentia–Greenland and Fennoscandia–Baltica.

The Acadian–Ligerian orogeny probably reflects a temporary increase in the convergence rate between the Proto-Tethys and the Laurasian plate. Partial coupling at the respective B-subduction zone, combined with the collision of the Avalon-Meguma and the Austro-Alpine microcratons, exerted compressive stresses on the back-arc areas. Although these stresses were of sufficient magnitude to interrupt back-arc extension in the Central Armorican–Saxothuringian Basin they did not affect the Cornwall–Rhenish–East Sudetic Basin.

Following the rather short-lived Acadian–Ligerian diastrophism back-arc extension resumed to govern the subsidence of the Central Armorican–Saxothuringian Basin. This probably reflects a renewed decrease in the convergence rate between the Proto-Tethys and the Fennoscandian–Baltic subplate. This is in keeping with the postulated Middle and Late Devonian sinistral translation between Laurentia–Greenland and Fennoscandia–Baltic.

A second period of apparently temporary coupling at the Proto-Tethys B-subduction zone occurred at the transition from the Devonian to the Carboniferous and gave rise to the Bretonian orogenic pulse. Once more only the southern parts of the Central Armorican–Saxothuringian Basin were affected by compressional stresses whilst back-arc rifting persisted in the Cornwall–Rhenish–East Sudetic Basin.

During the Tournaisian and Early Visean, the development of the Variscan geosynclinal system was again governed by back-arc extension. Moreover, with the Early Dinantian termination of the Arctic–North Atlantic translation, the area of the future Norwegian–Greenland sea and the Barents Sea Shelf also became dominated by regional crustal extension.

The onset of the Late Visean Variscan main orogeny presumably reflects a renewed increase in the convergence rate between the Laurasia and the Proto-Tethys plates. This induced the collision of the Aquitaine–Pyrenean microcontinent and, somewhat later, of Gondwana with Laurasia. Resultant compressional stresses caused the termination of back-arc extension in the Variscan geosynclinal system. Whilst the Central Armorican–Saxothuringian Basin was destroyed by folding and became inactive at the transition from the Visean to the Namurian, the A-subduction zone along the northern margin of the Normannian–Mid-German High remained active till Late Carboniferous time.

Crustal shortening, accompanied by décollement thrusting, ceased in the Variscan fold belt with the Late Westphalian–Early Stephanian Asturian phase. However, in the Appalachian–Mauretanides and the Urals, crustal shortening continued into Permian time. This was accompanied by the development of a complex right-lateral wrench system that transected the Variscan fold belt (Arthaud and Matte 1977; Ziegler 1978; Fig. 5).

The Devonian and Carboniferous rifts of the central and northern British Isles occupy an intermediate position between the back-arc rifts of the Variscan geosynclinal system and the Arctic–North Atlantic wrench–rift system.

The rapid Early Devonian subsidence of the Mid-

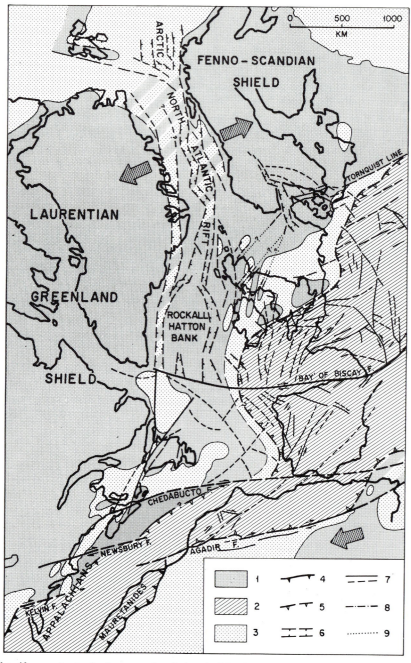

Fig. 5   Permo-Carboniferous tectonic framework of Arctic–North Atlantic realm (continental fit after Le Pichon 1977). 1. Continental cratons and intrabasinal highs; 2. Hercynian fold belt; 3. Carboniferous basins in Hercynian foreland; 4. Alleghenian deformation front; 5. Asturian deformation front; 6. Grabens, rifts; 7. Wrench faults; 8. Dyke swarms; 9. Alignments

land Valley and the Northumberland–Solway–Dublin grabens parallels the development of the Cornwall–Rhenish–East Silesian and the Central Armorican–Saxothuringian basins and may also be related to back-arc extension. During the Middle and Late Devonian the development of the rifts of the central and northern British Isles was dominated by the Arctic–North Atlantic translation, which locally gave rise to intraplate compressional deformations. Crustal extension resumed, however, at the onset of the Carboniferous and persisted into the Late Carboniferous whilst rifting in the Variscan geosynclinal system ceased with the Late Visean onset of the Variscan paroxysm. In this respect the Carboniferous rifts of Ireland and the United Kingdom together with the Saint Anthony and the Bay of Fundy basins could be regarded as having formed part of the Arctic–North Atlantic rift system.

Conclusions

The Late Caledonian paroxysm, which is variously dated as terminating at the transition from the Silurian to the Devonian, intra-Downtonian, and locally even Late Gédinnian, resulted in the consolidation of the Arctic–North Atlantic Caledonides, the North German–Polish and the Mid-European Caledonides; however, the Ligerian–Arverno–Vosgian—Moldanubian Cordillera was also strongly affected by the Late Caledonian diastrophism.

The Arctic–North Atlantic Caledonides are the result of the collision between Laurentia–Greenland and Fennoscandia–Baltica and therefore can be classified as an Alpine–Himalayan type fold belt. Their structural style indicates that major underplating occurred during their Ordovician and Silurian development.

In contrast, the evolution of the North German–Polish and Mid-European Caledonides, as well as of the Ligerian—Moldanubian Cordillera, which together embrace a number of Gondwana-derived microcratons, was associated with the Proto-Tethys subduction zone. Therefore, these fold belts could be regarded as Pacific-type orogens. The structural style of these fold belts is unknown due to limited exposure and partly also due to their over-printing by younger orogenic events.

In Europe distinct phases of back-arc extension separate the Late Caledonian, the Early Middle Devonian Acadian–Ligerian and the latest Devonian–Early Carboniferous Bretonian diastrophisms. The Acadian–Ligerian and the Bretonian orogens are essentially of the Pacific type as they result from the subduction of an oceanic plate beneath a continent. Subduction of the oceanic Proto-Tethys plate

was, however, accompanied by the northward rafting of the Avalon–Meguma and the Austro-Alpine microcontinents into the respective arc–trench systems and their accretion to Laurasia. Both the Acadian–Ligerian and the Bretonian orogenies affected only relatively narrow zones along the southern margin of the Laurasian continent. However, the contemporaneous sinistral translation between Fennoscandia–Baltica and Laurentia–Greenland induced anorogenic intraplate deformations that need to be distinguished from the subduction related deformations of the Acadian–Ligerian and Bretonic orogens.

The Late Visean to Early Stephanian Variscan orogeny, which was caused by the collision of Gondwana with Laurasia, is of the Alpine–Himalayan type. It affected much broader areas around the actual collision zone than either the Ligerian or the Bretonian diastrophism. The amount of crustal shortening that occurred during the Variscan orogeny is difficult to quantify but probably involved a significant amount of underplating.

When considering the ages, the spatial distribution and the plate tectonic settings of the different fold belts that developed in the course of the progressive consolidation of Pangea, a clear distinction can be made between Caledonian, Acadian–Lingerian, Bretonian, Variscan, and Appalachian–Mauretanian orogenic systems. The type localities of the respective orogens are generally areas either where they are best developed, preserved, and exposed, or where they were first recognized. In most of these type localities, the respective fold belts were not overprinted by later orogenic systems. However, from the above discussion, it is evident that in particular the Variscan orogeny overprinted large parts of the Caledonian and Acadian–Ligerian orogens, and probably the entire Bretonian orogen.

In the author's view, this does not constitute a reason for broadening the term 'Caledonian' to include the Acadian–Ligerian or even the Bretonian and Innunitian orogens. It is felt that a strong case can be made for retaining the established terms of Caledonian, Acadian–Ligerian, Bretonian, and Variscan orogenies, albeit with their loosely-defined time and palaeogeographic connotations.

References

Arthaud, F. and Matte, P. 1977. Late Palaeozoic strike-slip faulting in Southern Europe and Northern Africa: results of a right-lateral shear zone between the Appalachians and the Urals. *Bull. geol. Soc. Am.*, **88**, 1305–1320.

Autran, A. and Cogné, I. 1980. La zone interne de l'orogène varisque dans l'ouest de la France et sa place dans le développement de la chaîne Hercynieene. In Cogné, J. and Slansky, M. (Eds), *Géologie de l'Europe du Précambrien aux Bassins Sédimentaires Post-Hercyniens. Mém. du BRGM*, **108**, 90–111.

Bally, A. W. and Snelson, S. 1980. Realms of subsidence. *Can. Soc. Petrol. geol. Spec. Mem*, **6**, 1–94.

Floyd, P. A. 1982. Chemical variations in Hercynian basalts relative to plate tectonics. *J. geol. Soc. London*, **139**, 505–520.

Francis, E. H. 1978. Igneous activity in a fractured craton: Carboniferous volcanism in Northern Britain. In Bowes, D. R. and Leake, B. E. (Eds), *Crustal Evolution in NW Britain and Adjacent Regions. Geol. J. Spec. Issue*, **10**, 279–296.

Haller, J. 1971. Geology of the East Greenland Caledonides, Regional Geology Series, Interscience Publishers, London, 413 pp.

Harland, W. B. 1978. The Caledonides of Svalbard. In *IGCP Project 27, Caledonian–Appalachian Orogen of the North Atlantic Region. Geol. Surv. Can. Pap.*, **78–13**, 3–11.

Harland, W. B., Cutbill, J. S., Friend, P. F., Gobbett, D. J., Holliday, D. W., Maton, P. I., Parker, J. R. and Wallis, R. H. 1974. The Billefjorden Faultzone, Spitsbergen. *Skr. Norsk Polarinst.*, **161**, 72.

House, M. R., Richardson, J. B., Chaloner, W. G., Allen, J. R., Holland, C. H. and Westoll, T. S. 1977. A correlation of Devonian rocks of the British Isles. *Geol. Soc. London Spec. Report*, **8**, 110.

Howie, R. D. and Barss, M. S. 1975. Upper Palaeozoic rocks of the Atlantic Provinces, Gulf of St. Lawrence and adjacent continental shelf. *Geol. Surv. Can., Pap.*, **74–30**, 35–50.

Hsui, A. T. and Toksöz, M. N. 1981. Back-arc spreading: trench migration, continental pull or induced convention? *Tectonophysics*, **74**, 89–98.

Kent, D. V. and Opdyke, N. D. 1979. The Early Carboniferous palaeomagnetic field of North America and its bearing on tectonics of the Northern Appalachians. *Earth Planet. Sci. Lett.*, **44**, 365–372.

Le Pichon, X. 1977. The fit of the continents around the North Atlantic Ocean. *Tectonophysics*, **38**, 169–209.

Morris, W. A. 1976. Transcurrent motions determined paleomagnetically in the Northern Appalachians and Caledonides and the Acadian Orogeny. In *Canadian J. Earth Sci.*, **13(9)**, 1236–1243.

Rønnevik, H. C. 1981. Geology of the Barents Sea. In Illing. L. V. and Hobson, G. D. (Eds), *Petroleum Geology of the Continental Shelf of North-West Europe*. Institute of Petroleum, London, 395–406.

Sawkins, F. J. and Burke, K. 1980. Extensional tectonics and mid-Palaeozoic massive sulphide occurrences in Europe. *Geol. Rdsch.*, **69**, 349–360.

Uyeda, S. 1981. Subduction zones and back-arc basins, a review. *Geol. Rdsch.*, **70**, 552–569.

Vischer, A. 1943. Die Postdevonisch Tektonik von Ostgrönland Zwischen 74° und 75°N. Br. Kuhn Ø Wollaston Forland, Clavering Ø und Angrenzende Gebiete. *Meddr. Grønland*, **133(1)**, 195.

Voo, R. van der and Channel, J. E. T. 1980. Palaeomagnetism in orogenic belts. *Rev. Geophys. Space Phys.*, **18**, 455–481.

Voo, R. van der and Scotese, C. 1981. Palaeomagnetic evidence for a large (c. 2000 km) sinistral offset along the Great Glen fault during the Carboniferous. *Geology*, **9**, 583–589.

Ziegler, P. A. 1978. North-western Europe: tectonics and basin development. *Geologie en Mijnbouw*, **57(4)**, 589–626.

Ziegler, P. A. 1982. Geological Atlas of Western and Central Europe, Shell Internationale Petroleum Maatschappij B.V., Elsevier Scient. Publ. Co., Amsterdam, pp. 130.

Zonenshain, L. P. and Savostin, L. A. 1981. Movements of lithospheric plates relative to subduction zones, formation of marginal seas and active continental margins. *Tectonophysics*, **74**, 57–87.

Index

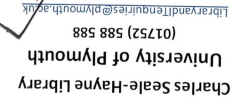